MW00744337

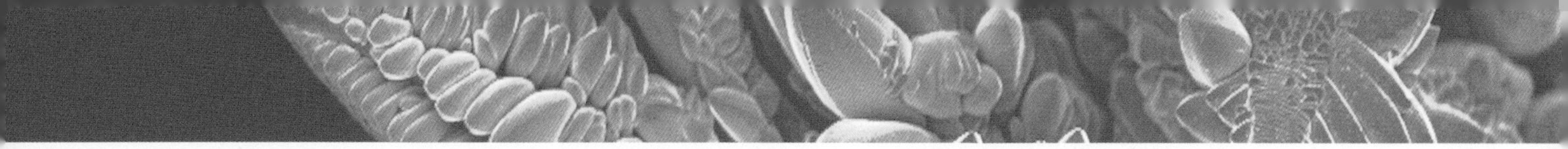

The Science and Engineering of Materials

Enhanced Seventh Edition

Donald R. Askeland
University of Missouri—Rolla, Emeritus

Wendelin J. Wright
Bucknell University

CENGAGE

Australia • Brazil • Canada • Mexico • Singapore • United Kingdom • United States

CENGAGE

***The Science and Engineering of Materials*, Enhanced Seventh Edition**

Donald R. Askeland and Wendelin J. Wright

Product Director: Mark Santee

Senior Product Manager: Timothy L. Anderson

Learning Designer: MariCarmen Constable

Associate Content Manager: Alexander Sham

Senior Product Assistant: Anna Goulart

Senior Marketing Manager: Scott Chrysler

Senior Digital Delivery Lead: Nikkita Kendrick

IP Analyst: Deanna Ettinger

IP Project Manager: Kelli Besse

Production Service: MPS Limited

Compositor: MPS Limited

Designer: Nadine Ballard

Manufacturing Planner: Ron Montgomery

Cover Designer: Nadine Ballard

Cover Image: Courtesy of the Materials Research Society Science as Art Competition and Peter Cameron Sherrell, Imperial College London

Library of Congress Control Number: 2020907981

Student Edition:
ISBN: 978-0-357-44786-4

Loose-leaf Edition:
ISBN: 978-0-357-44801-4

Cengage
200 Pier 4 Blvd
Boston, MA 02210
USA

Printed in the United States of America
Print Number: 01 Print Year: 2020

To Mary Sue and Tyler
–Donald R. Askeland

To my mom, Patricia Wright,
my first writing teacher
–Wendelin J. Wright

CONTENTS

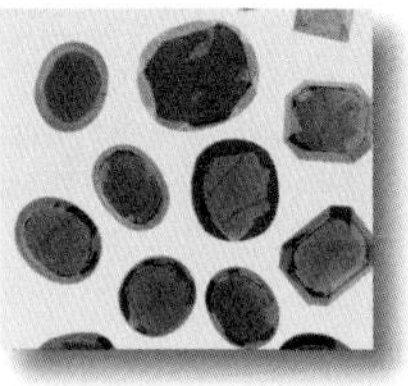

Chapter 4 Imperfections in the Atomic and Ionic Arrangements 103

Chapter 5 Atom and Ion Movements in Materials 141

Chapter 6 Mechanical Properties: Part One 181

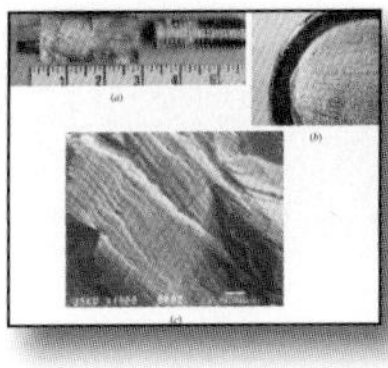

Chapter 7 Mechanical Properties: Part Two 229

Chapter 8 Strain Hardening and Annealing 269

Chapter 9 Principles of Solidification 305

Chapter 16 Polymers 571

Chapter 17 Composites: Teamwork and Synergy in Materials 617

Chapter 18 Construction Materials 659

Chapter 19 Electronic Materials 679

Chapter 20 Magnetic Materials 723

Chapter 21 Photonic Materials 753

Chapter 22 Thermal Properties of Materials 781

PREFACE

The copper age, the iron age, the silicon age . . . eras defined by the materials found in nature, but manipulated by the engineers of their day. The fundamental principles of structure, defects, kinetics, and processing are generally applicable to all materials, while over time our understanding has advanced and incorporated new ideas. As a result, the observable and macroscopic behavior of materials, spanning such varied characteristics as mechanical strength and toughness, electrical conductivity, refractive index, and corrosion resistance, is both understood more deeply and related more directly to underlying atomic-level phenomena.

Our tools for characterizing and manipulating materials have also grown vastly more sophisticated, allowing for deeper insights into materials structures and phenomena. At the edge of innovation we find the discovery—or even the creation—of entirely new materials, often made possible by new processing techniques, by circumventing equilibrium to cause materials to exist in metastable states, and by developing the tools to assemble, form, and study materials at the nanoscale. It is now routine, for instance, to examine materials at the near-atomic level for both structure and composition, using techniques such as high resolution transmission electron microscopy, grazing incidence x-ray diffraction, and electron energy loss spectroscopy. At the same time, materials processing has advanced to the point where thin films just a few atomic layers thick can in many instances be grown or deposited, while three-dimensional structures with dimensions of only a few tens of nanometers or less can also be manufactured. The entire electronics industry, for instance, is based on these types of advances. Flat screen televisions, high-speed wireless data systems, portable computation and telecommunication devices, automobiles and other transportation systems . . . these and countless other technologies are all dependent on our contemporary understanding of materials.

While not all students who study materials science will be practicing materials engineers, most engineers will in fact work with a diverse materials set, comprising metals, ceramics, plastics, composites, and semiconductors, and across lengths scales from the nanoscale to the macroscale, all within a context of myriad and diverse applications. Materials are an enabling component of what engineers imagine, design, and build. The ability to *innovate* and to incorporate materials *safely* in a design is rooted in an understanding of how to manipulate materials properties and functionality through the control of materials structure and processing techniques. The objective of this textbook, then, is to describe the foundations and applications of materials science for college-level engineering students as predicated upon the structure-processing-properties paradigm.

The challenge of any textbook is to provide the proper balance of breadth and depth for the subject at hand, to provide rigor at the appropriate level, to provide meaningful examples and up to date content, and to stimulate the intellectual excitement of the reader. Our goal here is to provide enough *science* so that the reader may understand basic materials phenomena, and enough *engineering* to prepare a wide range of students for competent professional practice.

Cover Art

The cover art for the enhanced seventh edition of the text is a field-emission scanning electron microscope image showing crystals of tungsten-doped copper sulfide grown using a technique known as chemical vapor deposition. These nanoscale crystals were fabricated as part of an effort to make atomically thin tungsten sulfide photocatalysts on porous metal foams. Photocatalysts cause the intentional degradation of materials subjected to ultraviolet radiation. When the sulfur content was stoichiometrically controlled, the desired atomically thin photocatalysts were produced, whereas excess sulfur caused the formation of these fractal-like crystals. The color is false; it has been mapped to the brightness of the otherwise black/white micrograph. The image is titled "Nanosucculents" by Peter Cameron Sherrell of Imperial College London.

Audience and Prerequisites

This text is intended for an introductory science of materials class taught at the sophomore or junior level. A first course in college level chemistry is assumed, as is some coverage of first-year college physics. A calculus course is helpful, but certainly not required. The text does not presume that students have taken other introductory engineering courses such as statics, dynamics, or mechanics of materials.

New in the Seventh Edition

The beginning of each chapter includes learning objectives to guide students in their studies. New problems have been added to the end of each chapter to increase the number of problems by 15%. The breadth of Chapter 15 on ceramic materials has been extended to include crystalline ceramics, silica and silicate compounds, and other topics of interest to provide a more comprehensive view of this important class of engineering materials. Other portions of the chapter have been revised for clarity. The cost of common engineering materials in Chapter 14 has been updated. As always, we have taken great care to provide the most error-free text possible.

The enhanced seventh edition includes new digital resources and solutions for end of chapter problems. Select problems have also been updated.

Knovel™ Problems

At the conclusion of the end of chapter problems, you will find a special section with problems that require the use of Knovel (**app.knovel.com/web/**). Knovel is an online aggregator of engineering references including handbooks, encyclopedias, dictionaries, textbooks, and databases from leading technical publishers and engineering societies such as the American Society of Mechanical Engineers (ASME) and the American Institute of Chemical Engineers (AIChE).

The Knovel problems build on material found in the textbook and require familiarity with online information retrieval. The problems are also available online at **login.cengage.com.** In addition, the solutions are accessible by registered instructors. If your institution does not have a subscription to Knovel or if you have any questions about Knovel, please visit **www.elsevier.com/solutions/knovel-engineering-information**.

The Knovel problems were created by a team of engineers led by Sasha Gurke, senior vice president and cofounder of Knovel.

Supplements for the Instructor

Supplements to the text include a Solution and Answer Guide that provides complete solutions to all problems, annotated Lecture Note Powerpoint™ slides, and an image library of all figures in the book. These can be found on the password-protected Instructor's Resources website for the book at login.cengage.com.

Acknowledgments

We thank all those who have contributed to the success of past editions and also the reviewers who provided detailed and constructive feedback on the sixth edition:

Cynthia W. Barnicki, Milwaukee School of Engineering

Deborah Chung, University of Buffalo, SUNY

Margaret Pinnell, University of Dayton

Stephen W. Stafford, University of Texas, El Paso

We also thank Jeffrey Florando of Lawrence Livermore National Laboratory for comments on the revisions to Chapter 15.

We are grateful to the team at Cengage Learning that has carefully guided this enhanced seventh edition through all stages of the publishing process. In particular, we wish to thank Timothy Anderson, Senior Product Manager; MariCarmen Constable, Learning Designer; Alexander Sham, Associate Content Manager; and Anna Goulart, Senior Product Assistant.

Wendelin Wright thanks Xiaojun Gu for his careful review of the proofs of the text as well as Briana King and Joanne Mowen for their assistance during the revision process. Elyssa Penson of Bucknell University wrote activities and answers that are available online for the "Have You Ever Wondered" questions at the beginning of each chapter. Professor Wright is grateful to John Bravman of Bucknell University for his feedback, contributed illustrations, patience, and constant support.

Donald R. Askeland
University of Missouri — Rolla, Emeritus

Wendelin J. Wright
Bucknell University

ABOUT THE AUTHORS

Donald R. Askeland is a Distinguished Teaching Professor Emeritus of Missouri University of Science and Technology. He received his degrees from the Thayer School of Engineering at Dartmouth College and the University of Michigan prior to joining the faculty at the University of Missouri–Rolla in 1970. Dr. Askeland taught a number of courses in materials and manufacturing engineering to students in a variety of engineering and science curricula. He received a number of awards for excellence in teaching and advising at UMR. He served as a Key Professor for the Foundry Educational Foundation and received several awards for his service to that organization. His teaching and research were directed primarily to metals casting and joining, in particular lost foam casting, and resulted in over 50 publications and a number of awards for service and best papers from the American Foundry Society.

Emily Paine/Bucknell University

Wendelin J. Wright is Professor at Bucknell University with a joint appointment in the departments of Mechanical Engineering and Chemical Engineering. She received her B.S., M.S., and Ph.D. in Materials Science and Engineering from Stanford University. Prior to assuming her position at Bucknell, Professor Wright was a faculty member at Santa Clara University.

Professor Wright's research interests focus on the mechanical behavior of materials, particularly of metallic glasses. She is the recipient of the 2003 Walter J. Gores Award for Excellence in Teaching, which is Stanford University's highest teaching honor, a 2005 Presidential Early Career Award for Scientists and Engineers, and a 2010 National Science Foundation CAREER Award. Professor Wright is a licensed professional engineer in metallurgy in California. She is married to John Bravman and is the mother of two young sons, Cole and Cooper.

DIGITAL RESOURCES

CENGAGE | WEBASSIGN

New Digital Solution for Your Engineering Classroom

WebAssign is a powerful digital solution designed by educators to enrich the engineering teaching and learning experience. With a robust computational engine at its core, WebAssign provides extensive content, instant assessment, and superior support.

WebAssign's powerful question editor allows engineering instructors to create their own questions or modify existing questions. Each question can use any combination of text, mathematical equations and formulas, sound, pictures, video, and interactive HTML elements. Numbers, words, phrases, graphics, and sound or video files can be randomized so that each student receives a different version of the same question.

In addition to common question types such as multiple choice, fill-in-the-blank, essay, and numerical, you can also incorporate robust answer entry palettes (mathPad, chemPad, calcPad, physPad, Graphing Tool) to input and grade symbolic expressions, equations, matrices, and chemical structures using powerful computer algebra systems.

WebAssign Offers Engineering Instructors the Following

- The ability to create and edit algorithmic and numerical exercises.
- The opportunity to generate randomized iterations of algorithmic and numerical exercises. When instructors assign numerical WebAssign homework exercises (engineering math exercises), the WebAssign program offers them the ability to generate and assign their students differing versions of the same engineering math exercise. The computational engine extends beyond and provides the luxury of solving for correct solutions/answers.
- The ability to create and customize numerical questions, allowing students to enter units, use a specific number of significant digits, use a specific number of decimal places, respond with a computed answer, or answer within a different tolerance value than the default.
  Visit **www.webassign.com/instructors/features/** to learn more. To create an account, instructors can go directly to the signup page at **www.webassign.net/signup.html**.

WebAssign Features for Students

- **Review Concepts at Point of Use**
  Within WebAssign, a "Read It" button at the bottom of each question links students to corresponding sections of the textbook, enabling access to the MindTap Reader at the precise moment of learning. A "Watch It" button allows a short video to play. These videos help students understand and review the problem they need to complete, enabling support at the precise moment of learning.

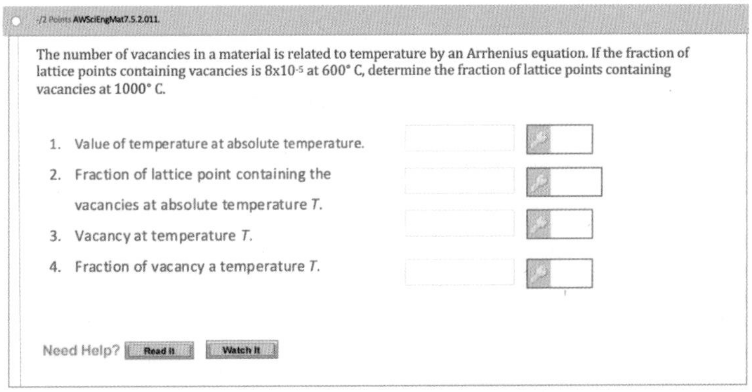

- **My Class Insights**
  WebAssign's built-in study feature shows performance across course topics so that students can quickly identify which concepts they have mastered and which areas they may need to spend more time on.
- **Ask Your Teacher**
  This powerful feature enables students to contact their instructor with questions about a specific assignment or problem they are working on.

MindTap Reader

Available via WebAssign and our digital subscription service, Cengage Unlimited, **MindTap Reader** is Cengage's next-generation eBook for engineering students.

The MindTap Reader provides more than just text learning for the student. It offers a variety of tools to help our future engineers learn chapter concepts in a way that resonates with their workflow and learning styles.

- **Personalize their experience**
  Within the MindTap Reader, students can highlight key concepts, add notes, and bookmark pages. These are collected in My Notes, ensuring they will have their own study guide when it comes time to study for exams.

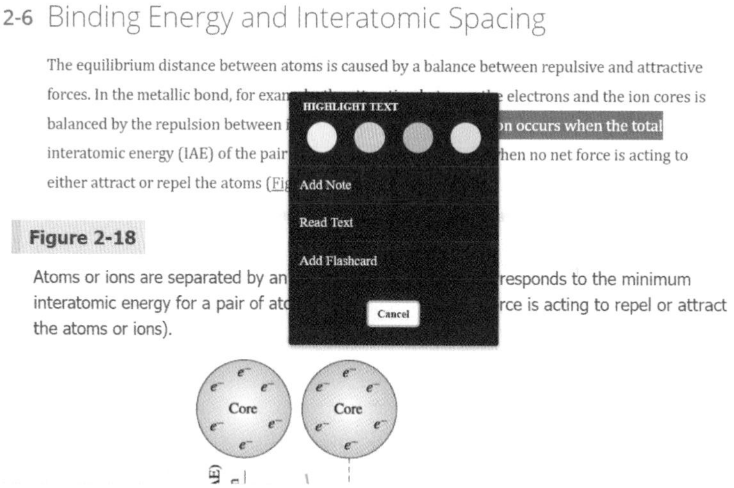

- **Flexibility at their fingertips**
  With access to the book's internal glossary, students can personalize their study experience by creating and collating their own custom flashcards. The ReadSpeaker feature reads text aloud to students, so they can learn on the go—wherever they are.

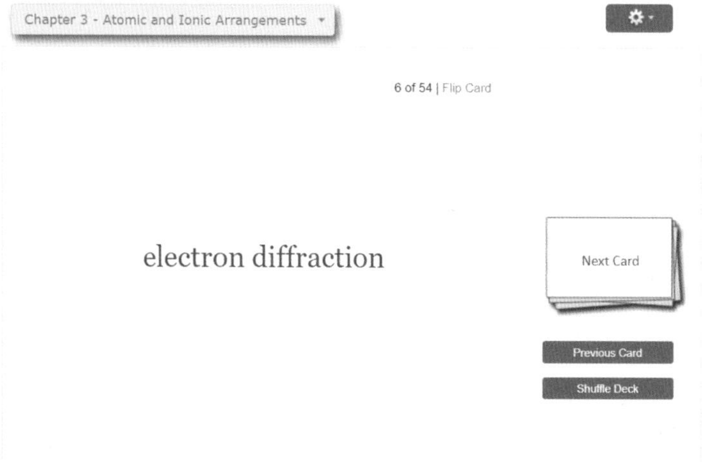

The Cengage Mobile App

Available on iOS and Android smartphones, the Cengage Mobile App provides convenience. Students can access their entire textbook anyplace and anytime. They can take notes, highlight important passages, and have their text read aloud whether they are online or off.

To learn more and download the mobile app, visit **www.cengage.com/mobile-app/.**

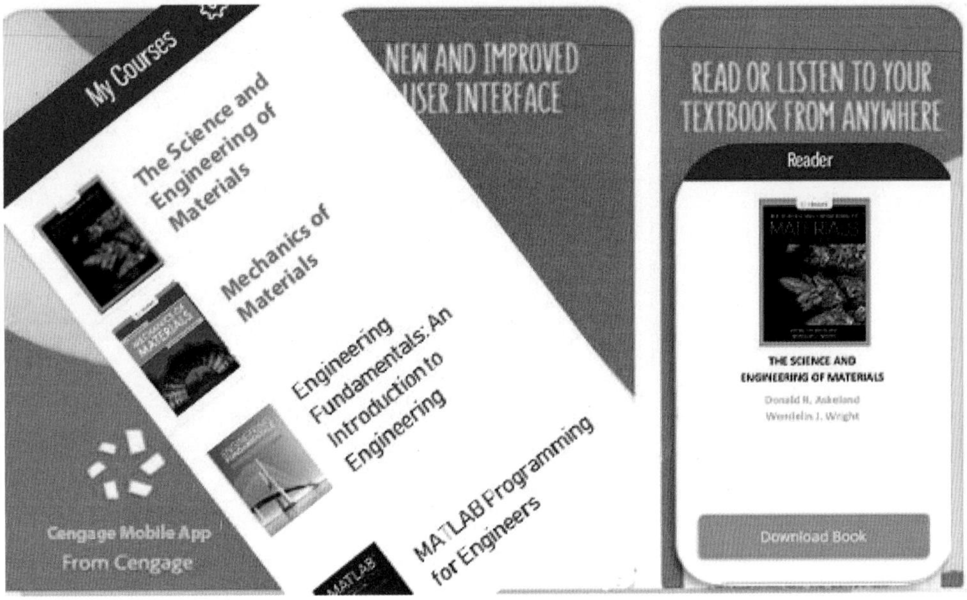

CENGAGE
UNLIMITED

All-You-Can-Learn Access with Cengage Unlimited

Cengage Unlimited is the first-of-its-kind digital subscription that gives students total and on-demand access to all the digital learning platforms, eBooks, online homework, and study tools Cengage has to offer—in one place, for one price. With Cengage Unlimited, students get access to their WebAssign courseware, as well as content in other Cengage platforms and course areas from day one. That's 70 disciplines and 675 courses worth of material, including engineering.

With Cengage Unlimited, students get **unlimited access** to a library of thousands of products. To learn more, visit **www.cengage.com/unlimited.**

The Science and Engineering of Materials

Enhanced Seventh Edition

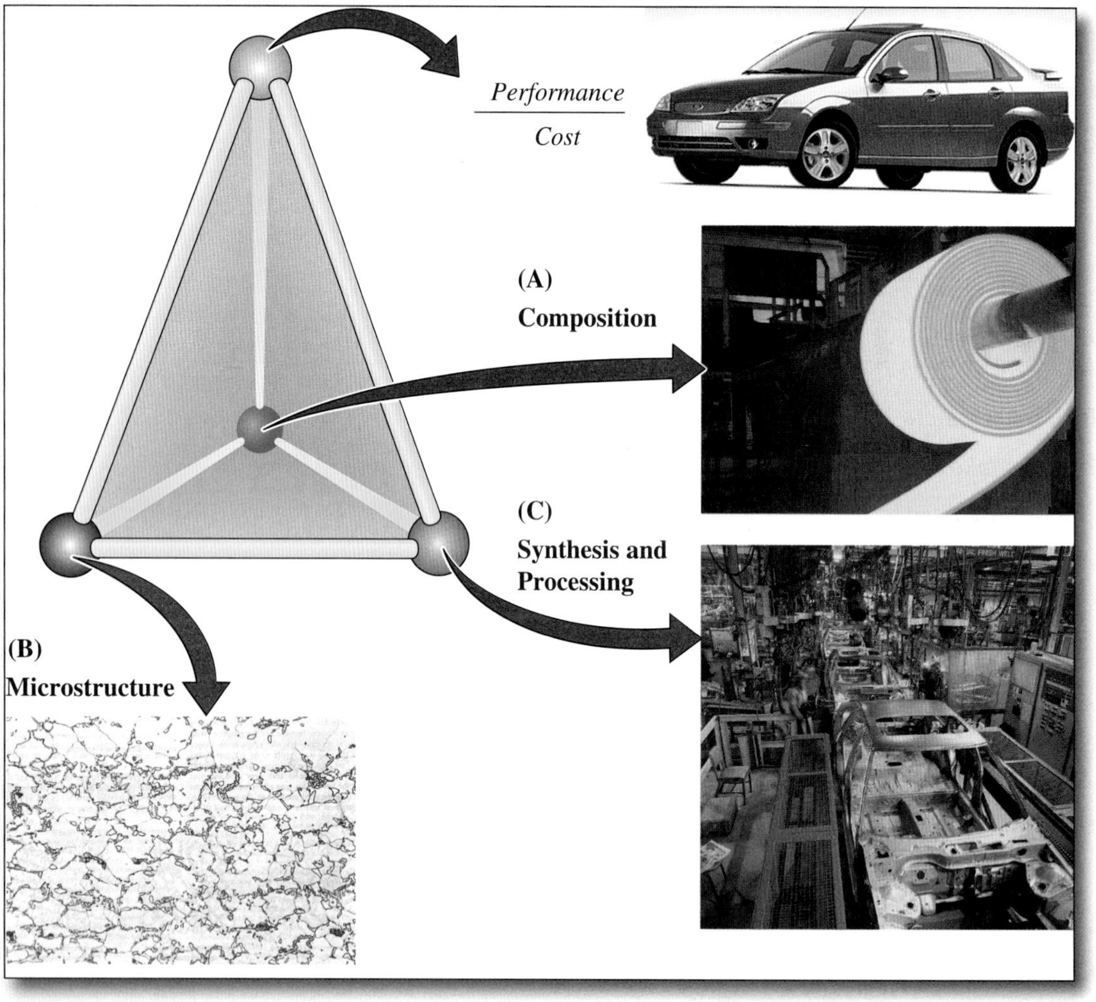

The principal goals of a materials scientist and engineer are to (1) make existing materials better and (2) invent or discover new phenomena, materials, devices, and applications. Breakthroughs in the materials science and engineering field are applied to many other fields of study such as biomedical engineering, physics, chemistry, environmental engineering, and information technology. The materials science and engineering tetrahedron shown here represents the heart and soul of this field, and its use is illustrated for the production of steel for an automotive chassis. As shown in this diagram, a materials scientist and engineer's main objective is to develop materials or devices that have the best performance for a particular application. In most cases, the performance-to-cost ratio, as opposed to the performance alone, is of utmost importance. This concept is shown as the apex of the tetrahedron and the three corners are representative of *A*—the composition, *B*—the microstructure, and *C*—the synthesis and processing of materials. These are all interconnected and ultimately affect the performance-to-cost ratio of a material or a device. The accompanying micrograph shows the microstructure of a dual phase steel. The microstructure of dual phase steels is engineered to absorb energy during automotive collisions. Hard particles of a phase called martensite (dark) are dispersed in a matrix of relatively soft, ductile ferrite (light).

For materials scientists and engineers, materials are like a palette of colors to an artist. Just as an artist can create different paintings using different colors, materials scientists create and improve upon different materials using different elements of the periodic table, and different synthesis and processing routes. (*Michael Shake/Shutterstock.com / Digital Vision/Getty Images / Digital Vision/Getty Images /* Metals Handbook, *Desk Edition (1998), ASM International, Materials Park, OH 44073-0002. Reprinted with permission of ASM International. All rights reserved. www.asminternational.org.*)

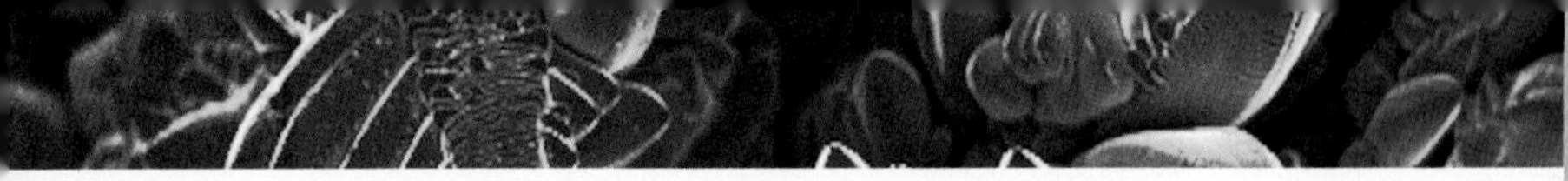

CHAPTER

1

Introduction to Materials Science and Engineering

Have You Ever Wondered?

- *What do materials scientists and engineers study?*
- *From a materials stand point, how do you significantly improve the fuel efficiency of a commercial jet airliner?*
- *Can we make flexible and lightweight electronic circuits using plastics?*
- *Why do jewelers add copper to gold?*
- *What is a "smart material?"*

Chapter Learning Objectives

The key objectives of this chapter are to

- Describe the primary concepts that define materials science and engineering.
- Explain the role of materials science in the design process.
- Classify materials by properties.
- Classify materials by function.

In this chapter, we will first introduce you to the field of materials science and engineering using different real-world examples. We will then provide an introduction to the classification of materials. Although most engineering programs require students to take a materials science course, you should approach your study of materials science as more than a mere requirement. A thorough knowledge of materials science and engineering will make you a better engineer and designer. Materials science underlies all technological advances, and an understanding of the basics of materials and their applications will not only make you a better engineer, but will help you during the design process. In order to be a good designer, you must learn what materials will be appropriate to use in different applications. You need to be capable of choosing the right material for your application based on its properties, and you must recognize how and why these properties might change over time and due to processing. Any engineer can look up materials properties in a book or search databases for a material that meets design specifications, but the *ability to innovate* and to *incorporate materials safely* in a design is rooted in an understanding of how to manipulate materials properties and functionality through the control of the material's structure and processing techniques.

The most important aspect of materials is that they are enabling; materials make things happen. For example, in the history of civilization, materials such as stone, iron, and bronze played a key role in mankind's development. In today's fast-paced world, the discovery of silicon single crystals and an understanding of their properties have enabled the information age.

In this book, we provide compelling examples of real-world applications of engineered materials. The diversity of applications and the unique uses of materials illustrate why a good engineer needs to understand and know how to apply the principles of materials science and engineering.

1-1 What is Materials Science and Engineering?

Materials science and engineering (MSE) is an interdisciplinary field that studies and manipulates the composition and structure of materials across length scales to control materials properties through synthesis and processing. The term **composition** means the chemical make-up of a material. The term **structure** means the arrangement of atoms, as seen at different levels of detail. Materials scientists and engineers not only deal with the development of materials, but also with the **synthesis** and **processing** of materials and manufacturing processes related to the production of components. The term "synthesis" refers to how materials are made from naturally occurring or man-made chemicals. The term "processing" means how materials are shaped into useful components to cause changes in the properties of different materials. One of the most important functions of materials scientists and engineers is to establish the relationships between a material or a device's properties and performance and the microstructure of that material, its composition, and the way the material or the device was synthesized and processed. In **materials science**, the emphasis is on the underlying relationships between the synthesis and processing, structure, and properties of materials. In **materials engineering**, the focus is on how to translate or transform materials into useful devices or structures.

One of the most fascinating aspects of materials science involves the investigation of a material's structure. The structure of materials has a profound influence on many properties of materials, even if the overall composition does not change! For example, if you take a pure copper wire and bend it repeatedly, the wire not only becomes harder but also becomes increasingly brittle! Eventually, the pure copper wire becomes so hard and brittle that it will break! The electrical resistivity of the wire will also increase as we bend it repeatedly. In this simple example, take note that we did not change the material's composition (i.e., its chemical make-up). The changes in the material's properties are due to a change in its internal structure. If you look at the wire after bending, it will look the same as before; however, its structure has been changed at the microscopic scale. The structure at the microscopic scale is known as the **microstructure**. If we can understand what has changed microscopically, we can begin to discover ways to control the material's properties.

Let's consider one example using the **materials science and engineering tetrahedron** shown in Figure 1-1. (Another example is shown on the chapter opening page.) For most of the history of commercial air travel, the fuselages of airplanes have been made using aluminum alloys. The fuselage material must possess sufficiently high strength, but must also be lightweight and formable into aerodynamic contours. Aluminum is one material that meets these requirements. In 2011, passengers began traveling on Boeing's 787 Dreamliner aircraft. One of the primary innovations of the Boeing 787 is the extensive use of **composites**; composite materials are formed by incorporating multiple components in a material in such a way that the properties of the resultant material are unique and not otherwise attainable. Composite materials comprise half of the Dreamliner's total weight, and in fact, the fuselage of the Boeing 787 is made from carbon fiber-reinforced plastic. Carbon fiber-reinforced plastic is a composite of carbon fiber in a polymer epoxy resin matrix.

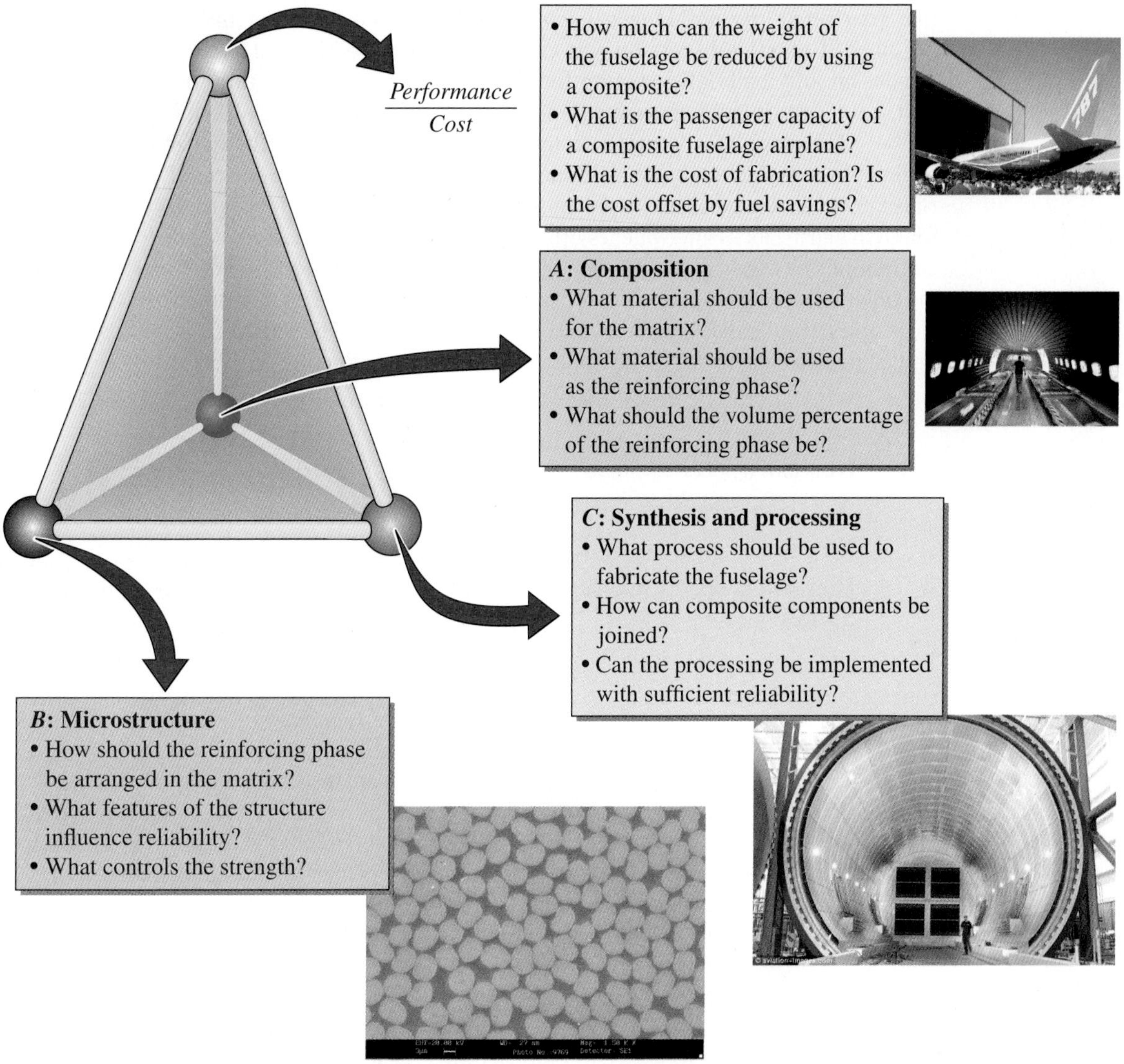

Figure 1-1 Application of the materials science and engineering tetrahedron to carbon fiber-reinforced plastic for the fabrication of aircraft fuselages. The composition, microstructure, and synthesis/processing are all interconnected and affect the performance-to-cost ratio. Clockwise from upper right: the Boeing 787; the interior of an empty Boeing 787 fuselage; a giant autoclave used to bake carbon fiber-reinforced plastic sections; carbon fiber in an epoxy matrix. (*Bloomberg via Getty Images / Srinivasa, Vinod, Shivakumar, Vinay, Nayaka, Vinay, Jagadeeshaiaih, Sunil, Seethram, Murali, Shenoy, Raghavendra, & Nafidi, Abdelhakim. (2010). Fracture morphology of carbon fiber reinforced plastic composite laminates.* Materials Research, *13(3), 417-424. Retrieved January 6, 2014, from www.scielo.br/scielo.php?script=sci_arttext&pid=S1516-14392010000300022&lng=en&tlng=en. 10.1590/S1516-14392010000300022./ AFP/Getty Images / Aviation Images*)

After decades of success with their various models of aircraft, Boeing invested billions of dollars to develop a commercial airplane based on a new class of materials. Why would Boeing do this? The driving force behind the move to carbon fiber-reinforced plastic was to reduce the weight of the fuselage, thereby increasing fuel efficiency. This significantly increases the performance to cost ratio of the aircraft.

The switch to a composite material involved numerous technical challenges. What would the composite material be? How would the composite fuselage be formed? Decades of data are available for the growth of cracks in aluminum under the cyclic loading of take-offs and landings. Would the composite fuselage be reliable? Would a carbon fiber-reinforced plastic also have the corrosion-resistance that aluminum has or would delamination between the fibers and plastic occur? Aluminum jets have structural panels that are riveted together. How can various structural components made from composites be joined? From this discussion, you can see that many issues need to be considered during the design and materials selection for any product and that the ratio of performance to cost, composition, microstructure, and synthesis and process are all critical factors.

Let's look at one more example of the application of the materials science and engineering tetrahedron by considering the use in microelectronic devices of a class of materials known as semiconducting polymers (Figure 1-2). Many types of displays such as those found in alarm clocks and watches utilize light emitting diodes (LEDs) made from inorganic compounds based on gallium arsenide (GaAs) and other materials; however, semiconducting polymers also have been used more recently. The advantages of using plastics for microelectronics include their flexibility and ease of processing. The questions materials scientists and engineers must answer with applications of semiconducting polymers are

- What are the relationships between the structure of polymers and their electrical properties?
- How can devices be made using these plastics?
- Will these devices be compatible with existing silicon chip technology?
- How robust are these devices?
- How will the performance and cost of these devices compare with traditional devices?

These are just a few of the factors that engineers and scientists must consider during the development, design, and manufacturing of semiconducting polymer devices.

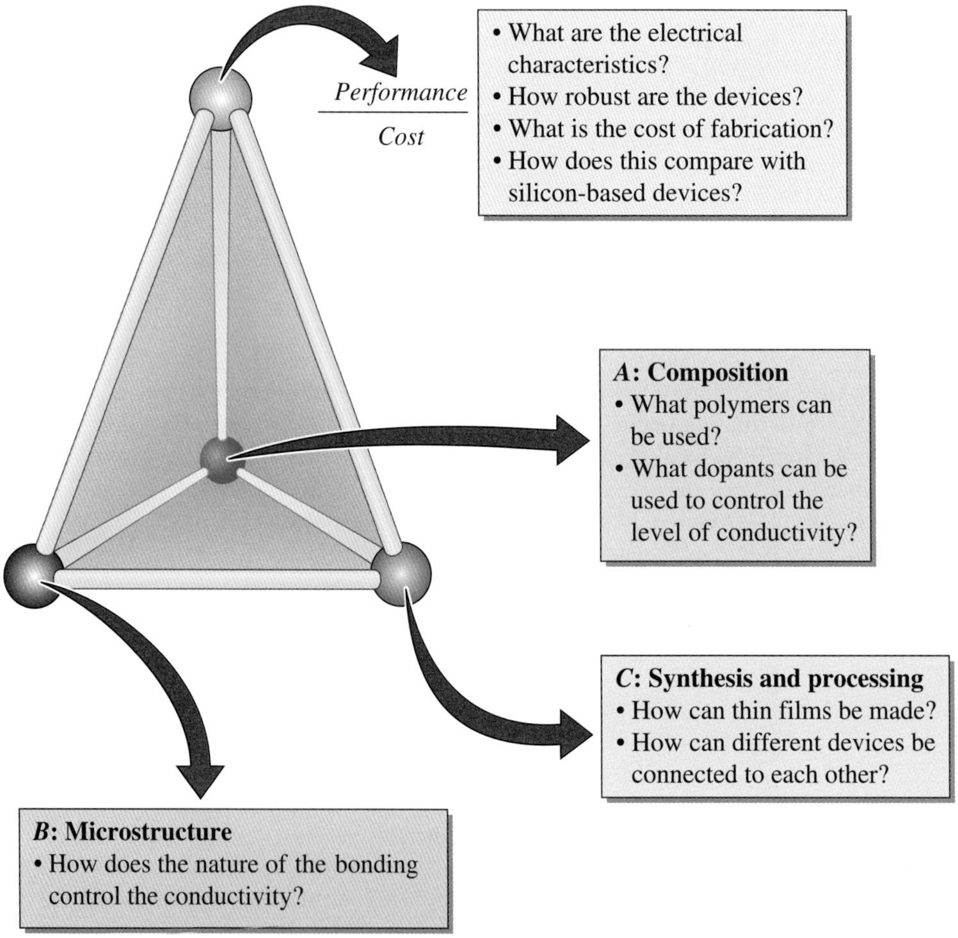

Figure 1-2 Application of the tetrahedron of materials science and engineering to semiconducting polymers for microelectronics.

1-2 Classification of Materials

There are different ways of classifying materials. One way is to describe five groups (Table 1-1):

1. **metals** and **alloys**;
2. **ceramics**, **glasses**, and **glass-ceramics**;
3. **polymers** (plastics);
4. **semiconductors**; and
5. **composite** materials.

Table 1-1 Representative examples, applications, and properties for each category of materials

	Examples of Applications	Properties
Metals and Alloys		
Copper	Electrical conductor wire	High electrical conductivity, good formability
Gray cast iron	Automobile engine blocks	Castable, machinable, vibration-damping
Alloy steels	Wrenches, automobile chassis	Significantly strengthened by heat treatment
Ceramics and Glasses		
SiO_2-Na_2O-CaO	Window glass	Optically transparent, thermally insulating
Al_2O_3, MgO, SiO_2	Refractories (i.e., heat-resistant lining of furnaces) for containing molten metal	Thermally insulating, withstand high temperatures, relatively inert to molten metal
Barium titanate	Capacitors for microelectronics	High ability to store charge
Silica	Optical fibers for information technology	Low optical losses
Polymers		
Polyethylene	Food packaging	Easily formed into thin, flexible, airtight film
Epoxy	Encapsulation of integrated circuits	Electrically insulating and moisture resistant
Phenolics	Adhesives for joining plies in plywood	Strong, moisture resistant
Semiconductors		
Silicon	Transistors and integrated circuits	Unique electrical behavior
GaAs	Optoelectronic systems	Converts electrical signals to light, used in lasers, laser diodes, etc.
Composites		
Graphite-epoxy	Aircraft components	High strength-to-weight ratio
Tungsten carbide-cobalt (WC-Co)	Carbide cutting tools for machining	High hardness, yet good shock resistance
Titanium-clad steel	Reactor vessels	Low cost and high strength of steel with the corrosion resistance of titanium

Materials in each of these groups possess different structures and properties. The differences in strength, which are compared in Figure 1-3, illustrate the wide range of properties from which engineers can select. Since metallic materials are extensively used for load-bearing applications, their mechanical properties are of great practical interest. We briefly introduce these properties here. The term "stress" refers to load or force per unit area. "Strain" refers to elongation or change in dimension divided by the original dimension. Application of "stress" causes "strain." If the strain goes away after the load or applied stress is removed, the strain is said to be "elastic." If the strain remains after the stress is removed, the strain is said to be "plastic." When the deformation is elastic, stress and strain are linearly related; the slope of the stress strain diagram is known as the elastic or Young's modulus. The level of stress needed to initiate plastic deformation is known as the "yield strength." The maximum percent deformation that can be achieved is a measure of the ductility of a metallic material. These concepts are discussed further in Chapters 6 and 7.

Metals and Alloys Metals include aluminum, magnesium, zinc, iron, titanium, copper, and nickel. An alloy is a metal that contains additions of one or more metals or non-metals, e.g., steel is an alloy of iron with carbon additions. In general, metals have good electrical and thermal conductivities. Metals and alloys have relatively high strength, high stiffness, ductility or formability, and shock resistance. They are particularly useful for structural or load-bearing applications. Although pure metals are occasionally used, alloys provide improvement in a particular desirable property or permit better combinations of properties. For example, pure gold is a soft metal; thus, jewelers add copper to gold to improve strength so that gold jewelry is not easily damaged.

Ceramics Ceramics can be defined as inorganic nonmetallic materials. Beach sand and rocks are examples of naturally occurring ceramics. Advanced ceramics are materials made by refining naturally occurring ceramics and other special processes. Advanced ceramics are used in substrates that house computer chips, sensors and

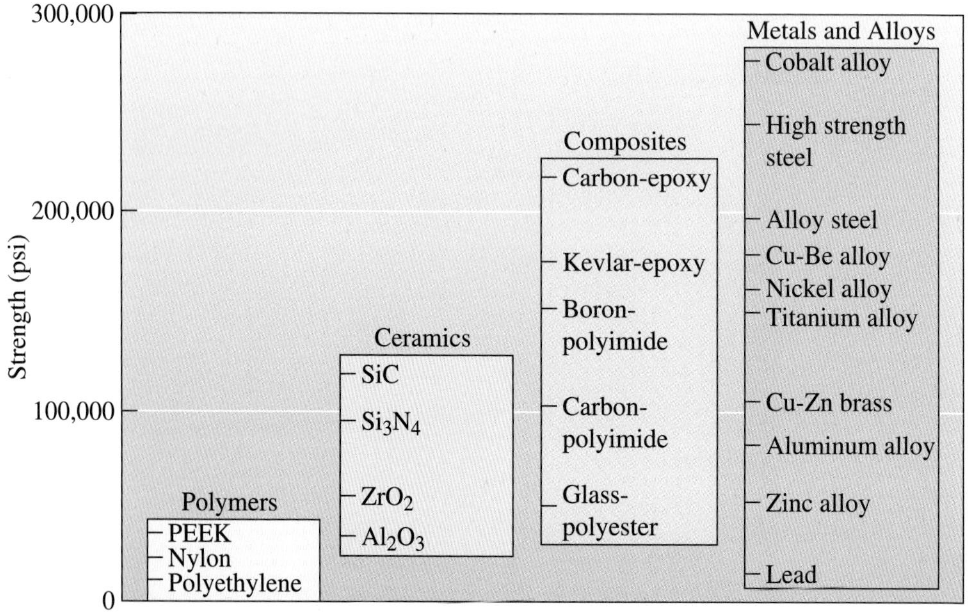

Figure 1-3 Representative strengths of various categories of materials. Compressive strengths are shown for ceramics.

actuators, capacitors, wireless communications, spark plugs, inductors, and electrical insulation. Some ceramics are used as barrier coatings to protect metallic substrates in turbine engines. Ceramics are also used in such consumer products as paints, plastics, and tires, and for industrial applications such as the oxygen sensors used in cars. Traditional ceramics are used to make bricks, tableware, bathroom fixtures, refractories (heat-resistant materials), and abrasives. In general, ceramics do not conduct heat well; they must be heated to very high temperatures before melting. Ceramics are strong and hard, but also very brittle. We normally prepare fine powders of ceramics and mold into different shapes. New processing techniques make ceramics sufficiently resistant to fracture that they can be used in load-bearing applications, such as impellers in turbine engines. Ceramics have exceptional strength under compression. Can you believe that an entire fire truck can be supported using four ceramic coffee cups?

Glasses and Glass-Ceramics Glass is an amorphous material, often, but not always, derived from a molten liquid. The term "amorphous" refers to materials that do not have a regular, periodic arrangement of atoms. Amorphous materials will be discussed in Chapter 3. The fiber optics industry is founded on optical fibers based on high-purity silica glass. Glasses are also used in houses, cars, screens for computers, televisions, and smart phones, and hundreds of other applications. Glasses can be thermally treated (tempered) to make them stronger. Forming glasses and then nucleating (forming) small crystals within them by a special thermal process creates materials that are known as glass-ceramics. Zerodur™ is an example of a glass-ceramic material that is used to make the mirror substrates for large telescopes (e.g., the Chandra and Hubble telescopes). Glasses and glass-ceramics are usually processed by melting and casting.

Polymers Polymers are typically organic materials. They are produced using a process known as **polymerization**. Polymeric materials include rubber (elastomers) and many types of adhesives. Polymers typically are good electrical and thermal insulators although there are exceptions. Although they have lower strengths than metals or ceramics, polymers have very good **strength-to-weight ratios**. They are typically not suitable for use at high temperatures. Many polymers have very good resistance to corrosive chemicals. Polymers have thousands of applications ranging from bulletproof vests, compact discs (CDs), ropes, and liquid crystal displays (LCDs) to clothes and coffee cups. **Thermoplastic** polymers, in which the long molecular chains are not rigidly connected, have good ductility and formability; **thermosetting** polymers are stronger but more brittle because the molecular chains are tightly linked (Figure 1-4). Polymers are

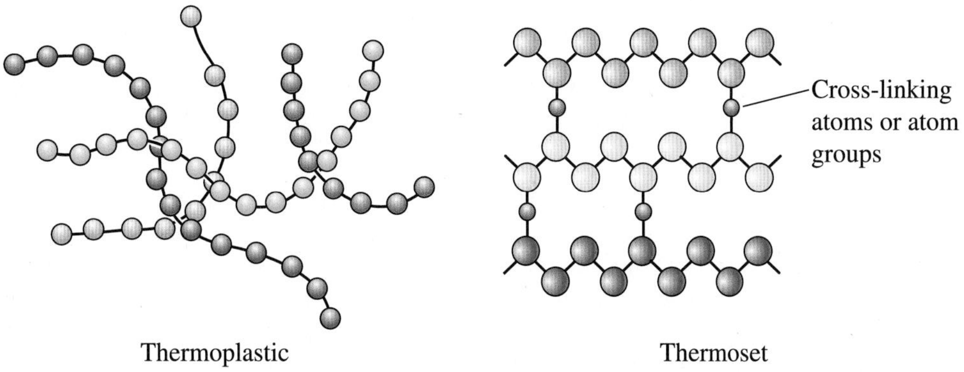

Figure 1-4 Polymerization occurs when small molecules, represented by the circles, combine to produce larger molecules, or polymers. The polymer molecules can have a structure that consists of many chains that are entangled but not connected (thermoplastics) or can form three-dimensional networks in which chains are cross-linked (thermosets).

used in many applications, including electronic devices. Thermoplastics are made by shaping their molten form. Thermosets are typically cast into molds. **Plastics** contain additives that enhance the properties of polymers.

Semiconductors

Silicon, germanium, and gallium arsenide-based semiconductors such as those used in computers and electronics are part of a broader class of materials known as electronic materials. The electrical conductivity of semiconducting materials is between that of ceramic insulators and metallic conductors. Semiconductors have enabled the information age. In some semiconductors, the level of conductivity can be controlled to produce electronic devices such as transistors and diodes that are used to build integrated circuits. In many applications, we need large single crystals of semiconductors. These are grown from molten materials. Often, thin films of semiconducting materials are also made using specialized processes.

Composite Materials

The main idea in developing composites is to blend the properties of different materials. These are formed from two or more materials, producing properties not found in any single material. Concrete, plywood, and fiberglass are examples of composite materials. Fiberglass is made by dispersing glass fibers in a polymer matrix. The glass fibers make the polymer stiffer, without significantly increasing its density. With composites, we can produce lightweight, strong, ductile, temperature-resistant materials or we can produce hard, yet shock-resistant, cutting tools that would otherwise shatter. Advanced aircraft and aerospace vehicles rely heavily on composites. As discussed earlier in this chapter, the Boeing 787 uses carbon fiber-reinforced plastic in many structural components instead of aluminum, leading to high fuel efficiency. Sports equipment such as bicycles, golf clubs, tennis rackets, and the like also make use of different kinds of composite materials that are light and stiff.

1-3 Functional Classification of Materials

We can classify materials based on whether the most important function they perform is mechanical (structural), biological, electrical, magnetic, or optical. This classification of materials is shown in Figure 1-5. Some examples of each category are shown. These categories can be broken down further into subcategories.

Aerospace

Light materials such as wood and an aluminum alloy (that accidentally strengthened the engine even more by picking up copper from the mold used for casting) were used in the Wright brothers' historic flight. NASA's space shuttle made use of aluminum powder for booster rockets and silica for the space shuttle tiles. The fuselage and wings of Boeing's 787 aircraft are largely composed of carbon-fiber-reinforced plastic.

Biomedical

Our bones and teeth are made, in part, from a naturally formed ceramic known as hydroxyapatite. A number of artificial organs, bone replacement parts, cardiovascular stents, orthodontic braces, and other components are made using different plastics, titanium alloys, and nonmagnetic stainless steels. Ultrasonic imaging systems make use of ceramics known as lead zirconium titanate (PZT).

Electronic Materials

As mentioned before, semiconductors, such as those made from silicon, are essential to making integrated circuits for computer chips. Barium titanate ($BaTiO_3$), tantalum oxide (Ta_2O_5), and other dielectric materials are used to make ceramic capacitors and other devices. Copper, aluminum, and other metals are used as conductors in power transmission and in microelectronics.

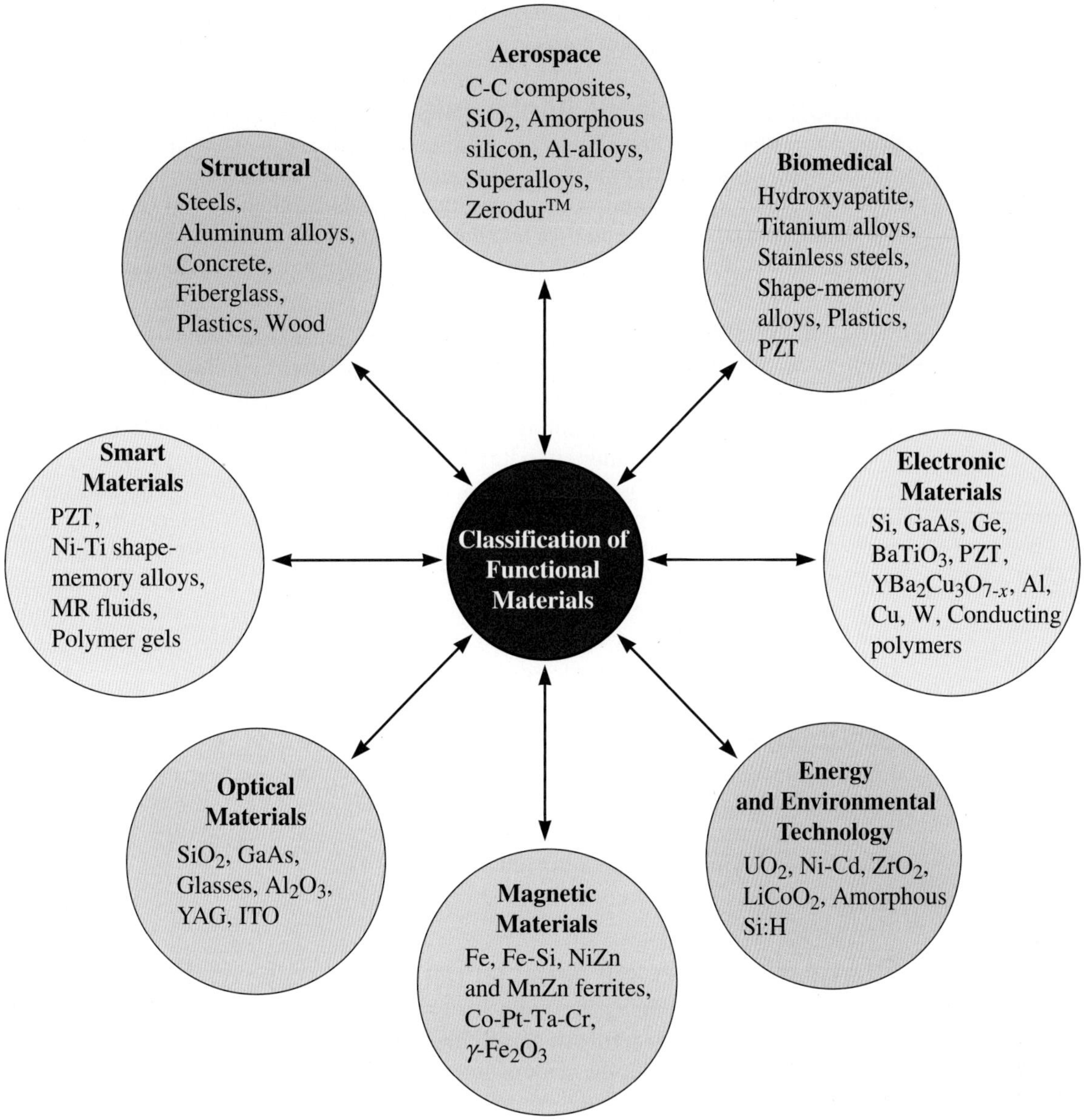

Figure 1-5 Functional classification of materials. Notice that metals, plastics, and ceramics occur in different categories. A limited number of examples in each category are provided.

Energy Technology and Environmental Technology The nuclear industry uses materials such as uranium dioxide and plutonium as fuel. Numerous other materials, such as glasses and stainless steels, are used in handling nuclear materials and managing radioactive waste. Technologies related to batteries and fuel cells make use of many ceramic materials such as zirconia (ZrO_2) and polymers. Battery technology has gained significant importance owing to the need for many electronic devices that require longer lasting and portable power. Fuel cells are used in some electric cars. The oil and petroleum industry widely uses zeolites, alumina, and other materials as catalyst substrates. They use Pt, Pt/Rh, and many other metals as catalysts. Many membrane technologies for purification of liquids and gases make use of ceramics and plastics. Solar power is generated using materials such as amorphous silicon (a:Si:H).

Magnetic Materials Computer hard disks make use of many ceramic, metallic, and polymeric materials. Computer hard disks are made using cobalt-platinum-tantalum-chromium (Co-Pt-Ta-Cr) alloys. Many magnetic ferrites are used to make inductors and components for wireless communications. Steels based on iron and silicon are used to make transformer cores.

Photonic or Optical Materials Silica is used widely for making optical fibers. More than ten million kilometers of optical fiber have been installed around the world. Optical materials are used for making semiconductor detectors and lasers used in fiber optic communications systems and other applications. Similarly, alumina (Al_2O_3) and yttrium aluminum garnets (YAG) are used for making lasers. Amorphous silicon is used to make solar cells and photovoltaic modules. Polymers are used to make liquid crystal displays (LCDs).

Smart Materials A **smart material** can sense and respond to an external stimulus such as a change in temperature, the application of a stress, or a change in humidity or chemical environment. Usually a smart material-based system consists of sensors and actuators that read changes and initiate an action. An example of a passively smart material is lead zirconium titanate (PZT) and shape-memory alloys. When properly processed, PZT can be subjected to a stress, and a voltage is generated. This effect is used to make such devices as spark generators for gas grills and sensors that can detect underwater objects such as fish and submarines. Other examples of smart materials include magnetorheological or MR fluids. These are magnetic paints that respond to magnetic fields. These materials are being used in suspension systems of automobiles, including models by General Motors, Ferrari, and Audi. Still other examples of smart materials and systems are photochromic glasses and automatic dimming mirrors.

Structural Materials These materials are designed for carrying some type of stress. Steels, concrete, and composites are used to make buildings and bridges. Steels, glasses, plastics, and composites also are used widely to make automotives. Often in these applications, combinations of strength, stiffness, and toughness are needed under different conditions of temperature and loading.

1-4 Classification of Materials Based on Structure

As mentioned before, the term "structure" means the arrangement of a material's atoms; the structure at a microscopic scale is known as "microstructure." We can view these arrangements at different scales, ranging from a few angstrom units to a millimeter. We will learn in Chapter 3 that some materials may be **crystalline** (the material's atoms are arranged in a periodic fashion) or they may be amorphous (the arrangement of the material's atoms does not have long-range order). Some crystalline materials may be in the form of one crystal and are known as **single crystals**. Others consist of many crystals or **grains** and are known as **polycrystalline**. The characteristics of crystals or grains (size, shape, etc.) and that of the regions between them, known as the **grain boundaries**, also affect the properties of materials. We will further discuss these concepts in later chapters. The microstructure of a dual phase steel is shown on the chapter opening page.

1-5 Environmental and Other Effects

The structure-property relationships in materials fabricated into components are often influenced by the surroundings to which the material is subjected during use. This can include exposure to high or low temperatures, cyclical stresses, sudden impact, corrosion, or oxidation. These effects must be accounted for in design to ensure that components do not fail unexpectedly.

Temperature Changes in temperature dramatically alter the properties of materials (Figure 1-6). Metals and alloys that have been strengthened by certain heat treatments or forming techniques may suddenly lose their strength when heated. A tragic reminder of this is the collapse of the World Trade Center towers on September 11, 2001. Although the towers sustained the initial impact of the collisions, their steel structures were weakened by elevated temperatures caused by fire, ultimately leading to the collapse.

High temperatures change the structure of ceramics and cause polymers to melt or char. Very low temperatures, at the other extreme, may cause a metal or polymer to fail in a brittle manner, even though the applied loads are low. This low-temperature embrittlement was a factor that caused the *Titanic* to fracture and sink. Similarly, the 1986 *Challenger* accident, in part, was due to embrittlement of rubber O-rings. The reasons why some polymers and metallic materials become brittle are different. We will discuss these concepts in later chapters.

Corrosion Most metals and polymers react with oxygen or other gases, particularly at elevated temperatures. Metals and ceramics may disintegrate, and polymers and non-oxide ceramics may oxidize. Materials also are attacked by corrosive liquids, leading to premature failure. The engineer faces the challenge of selecting materials or coatings that prevent these reactions and permit operation in extreme environments. In space applications, we may have to consider the effect of radiation.

Fatigue For many applications, components must be designed such that the load on the material is not enough to cause permanent deformation. When we load and unload the material thousands of times, even at low loads, small cracks may begin to develop, and materials fail as these cracks grow. This is known as **fatigue failure**. In designing load-bearing components, the possibility of fatigue must be accounted for.

Strain Rate Many of you are aware of the fact that Silly Putty®, a silicone-based (not silicon) plastic, can be stretched significantly if we pull it slowly (small rate of strain). If you pull it fast (higher rate of strain), it snaps. A similar behavior can occur with many metallic materials. Thus, in many applications, the level and rate of strain have to be considered.

In many cases, the effects of temperature, fatigue, stress, and corrosion may be interrelated, and other outside effects can affect the material's performance.

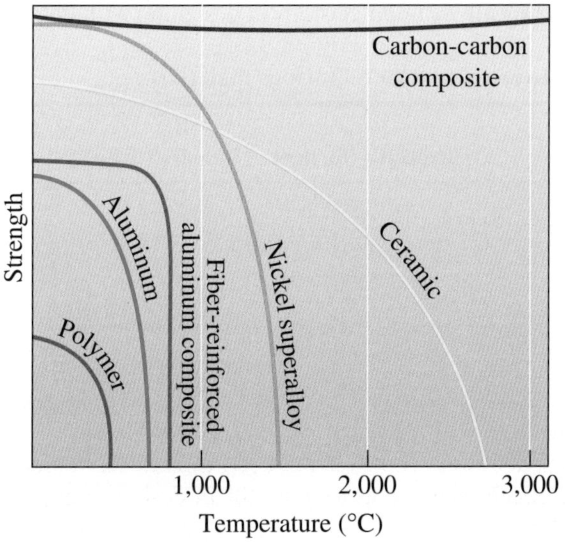

Figure 1-6 Increasing temperature normally reduces the strength of a material. Polymers are suitable only at low temperatures. Some composites, such as carbon-carbon composites, special alloys, and ceramics, have excellent properties at high temperatures.

1-6 Materials Design and Selection

When a material is designed for a given application, a number of factors must be considered. The material must acquire the desired **physical** and **mechanical properties**, must be capable of being processed or manufactured into the desired shape, and must provide an economical solution to the design problem. Satisfying these requirements in a manner that protects the environment—perhaps by encouraging recycling of the materials—is also essential. In meeting these design requirements, the engineer may have to make a number of trade-offs in order to produce a serviceable, yet marketable, product.

As an example, material cost is normally calculated on a cost-per-pound basis. We must consider the **density** of the material, or its weight-per-unit volume, in our design and selection (Table 1-2). Aluminum may cost more than steel on a weight basis, but it has only one-third the density of steel. Although parts made from aluminum may have to be thicker, the aluminum part may be less expensive than the one made from steel because of the weight difference.

In some instances, particularly in aerospace applications, weight is critical, since additional vehicle weight increases fuel consumption. By using materials that are lightweight but very strong, aerospace vehicles or automobiles can be designed to improve fuel utilization. Many advanced aerospace vehicles use composite materials instead of aluminum. These composites, such as carbon-epoxy, are more expensive than the traditional aluminum alloys; however, the fuel savings yielded by the higher strength-to-weight ratio of the composite (Table 1-2) may offset the higher initial cost of the aircraft as is also true for the Boeing 787. There are literally thousands of applications in which similar considerations apply. Usually the selection of materials involves trade-offs between many properties.

By this point of our discussion, we hope that you can appreciate that the properties of materials depend not only on composition, but also how the materials are made (synthesis and processing) and, most importantly, their internal structure. This is why it is not a good idea for an engineer to refer to a handbook and select a material for a given application. The handbooks may be a good starting point. A good engineer will consider: the effects of how the material was made, the exact composition of the candidate material for the application being considered, any processing that may have to be done for shaping the material or fabricating a component, the structure of the material after processing into a component or device, the environment in which the material will be used, and the cost-to-performance ratio.

Earlier in this chapter, we discussed the need for you to know the principles of materials science and engineering. If you are an engineer and you need to decide which materials you will choose to fabricate a component, the knowledge of principles of materials science and engineering will empower you with the fundamental concepts. These will allow you to make technically sound decisions in designing with engineered materials.

Table 1-2 Strength-to-weight ratios of various materials

Material	Strength ($lb/in.^2$)	Density ($lb/in.^3$)	Strength-to-Weight Ratio (in.)
Polyethylene	1000	0.030	0.03×10^6
Pure aluminum	6500	0.098	0.07×10^6
Al_2O_3	30,000	0.114	0.26×10^6
Epoxy	15,000	0.050	0.30×10^6
Heat-treated alloy steel	240,000	0.280	0.86×10^6
Heat-treated aluminum alloy	86,000	0.098	0.88×10^6
Carbon-carbon composite	60,000	0.065	0.92×10^6
Heat-treated titanium alloy	170,000	0.160	1.06×10^6
Kevlar-epoxy composite	65,000	0.050	1.30×10^6
Carbon-epoxy composite	80,000	0.050	1.60×10^6

Summary

- Materials science and engineering (MSE) is an interdisciplinary field concerned with inventing new materials and devices and improving existing materials by developing a deeper understanding of the microstructure-composition-synthesis-processing relationships.
- Engineered materials are materials designed and fabricated considering MSE principles.
- The properties of engineered materials depend upon their composition, structure, synthesis, and processing. An important performance index for materials or devices is their performance-to-cost ratio.
- The structure of a material refers to the arrangement of atoms or ions in the material.
- The structure at a microscopic level is known as the microstructure.
- Many properties of materials depend strongly on the structure, even if the composition of the material remains the same. This is why the structure-property or microstructure-property relationships in materials are extremely important.
- Materials are classified as metals and alloys, ceramics, glasses and glass-ceramics, composites, polymers, and semiconductors.
- Metals and alloys have good strength, good ductility, and good formability. Metals have good electrical and thermal conductivities. Metals and alloys play an indispensable role in many applications such as automotives, buildings, bridges, aerospace, and the like.
- Ceramics are inorganic crystalline materials. They are strong, serve as good electrical and thermal insulators, are often resistant to damage by high temperatures and corrosive environments, but are mechanically brittle. Modern ceramics form the underpinnings of many microelectronic and photonic technologies.
- Glasses are amorphous, inorganic solids that are typically derived from a molten liquid. Glasses can be tempered to increase strength. Glass-ceramics are formed by annealing glasses to nucleate small crystals that improve resistance to fracture and thermal shock.
- Polymers have relatively low strength; however, the strength-to-weight ratio is very favorable. Polymers are not suitable for use at high temperatures. They have very good corrosion resistance, and—like ceramics—provide good electrical and thermal insulation. Polymers may be either ductile or brittle, depending on structure, temperature, and strain rate.
- Semiconductors possess unique electrical and optical properties that make them essential for manufacturing components in electronic and communications devices.
- Composites are made from two or more different types of materials. They provide unique combinations of mechanical and physical properties that cannot be found in any single material.
- Functional classification of materials includes aerospace, biomedical, electronic, energy and environmental, magnetic, optical (photonic), and structural materials.
- Materials can also be classified as crystalline or amorphous. Crystalline materials may be single crystal or polycrystalline.
- Properties of materials can depend upon the temperature, level and type of stress applied, strain rate, oxidation and corrosion, and other environmental factors.
- Selection of a material having the needed properties and the potential to be manufactured economically and safely into a useful product is a complicated process requiring the knowledge of the structure-property-processing-composition relationships.

Glossary

Alloy A metallic material that is obtained by chemical combinations of different elements (e.g., steel is made from iron and carbon).

Ceramics A group of crystalline inorganic materials characterized by good strength, especially in compression, and high melting temperatures.

Composites A group of materials formed from mixtures of metals, ceramics, or polymers in such a manner that unusual combinations of properties are obtained (e.g., fiberglass).

Composition The chemical make-up of a material.

Crystalline material A material composed of one or many crystals. In each crystal, atoms or ions show a long-range periodic arrangement.

Density Mass per unit volume of a material, usually expressed in units of g/cm^3 or $lb/in.^3$

Fatigue failure Failure of a material due to repeated loading and unloading.

Glass An amorphous material derived from the molten state, typically, but not always, based on silica.

Glass-ceramics A special class of materials obtained by forming a glass and then heat treating it to form small crystals.

Grain boundaries Regions between grains of a polycrystalline material.

Grains Crystals in a polycrystalline material.

Materials engineering An engineering-oriented field that focuses on how to transform materials into useful devices or structures.

Materials science A field of science that emphasizes studies of relationships between the microstructure, synthesis and processing, and properties of materials.

Materials science and engineering An interdisciplinary field concerned with inventing new materials and improving existing materials by developing a deeper understanding of the structure-property-processing-composition relationships between different materials.

Materials science and engineering tetrahedron A tetrahedron diagram showing how the performance-to-cost ratio of materials depends upon their composition, microstructure, synthesis, and processing.

Mechanical properties Properties of a material, such as strength, that describe how well a material withstands applied forces, including tensile or compressive forces, impact forces, cyclical or fatigue forces, or forces at high temperatures.

Metal An element that has metallic bonding and generally good ductility, strength, and electrical conductivity.

Microstructure The structure of a material at the microscopic length scale.

Physical properties Characteristics such as color, elasticity, electrical conductivity, thermal conductivity, magnetism, and optical behavior that generally are not significantly influenced by forces acting on a material.

Plastics Polymers containing other additives.

Polycrystalline material A material composed of many crystals (as opposed to a single-crystal material that has only one crystal).

Polymerization The process by which organic molecules are joined into giant molecules, or polymers.

Polymers A group of materials normally obtained by joining organic molecules into giant molecular chains or networks.

Processing Different ways for shaping materials into useful components or changing their properties.

Semiconductors A group of materials (e.g., Si, GaAs) having electrical conductivity between metals and typical ceramics.

Single crystal A crystalline material that comprises only one crystal (there are no grain boundaries).

Smart material A material that can sense and respond to an external stimulus such as change in temperature, application of a stress, or change in humidity or chemical environment.

Strength-to-weight ratio The strength of a material divided by its density; materials with a high strength-to-weight ratio are strong but lightweight.

Structure The arrangements of atoms or ions in a material. The structure of materials has a profound influence on many properties of materials, even if the overall composition does not change.

Synthesis The process by which materials are made from naturally occurring or other chemicals.

Thermoplastics A special group of polymers in which molecular chains are entangled but not interconnected. They can be easily melted and formed into useful shapes. Normally, these polymers have a chainlike structure (e.g., polyethylene).

Thermosets A special group of polymers that decompose rather than melt upon heating. They are normally quite brittle due to a relatively rigid, three-dimensional network structure (e.g., polyurethane).

Problems

Section 1-1 What is Materials Science and Engineering?

1-1 Define materials science and engineering (MSE).

1-2 What is the importance of the engineering tetrahedron for materials engineers?

1-3 Define the following terms:
(a) composition;
(b) structure;
(c) synthesis;
(d) processing; and
(e) microstructure.

1-4 Explain the difference between the terms materials science and materials engineering.

Section 1-2 Classification of Materials

1-5 The myriad materials in the world primarily fall into four basic categories; what are they? What are materials called that have one or more different types of material fabricated into one component? Give one example.

1-6 What are some of the materials and mechanical properties of metals and alloys?

1-7 What is a ceramic, and what are some of the properties that you expect from a ceramic?

1-8 Make comparisons between thermoplastics and thermosetting polymers (a) on the basis of mechanical characteristics upon heating, and (b) according to possible molecular structures.

1-9 Give three examples of composites that can be fabricated.

1-10 For each of the following classes of materials, give two *specific* examples that are a regular part of your life:
(a) metals;
(b) ceramics;
(c) polymers; and
(d) semiconductors.
Specify the object that each material is found in and explain why the material is used in each specific application. *Hint:* One example answer for part (a) would be that aluminum is a metal used in the base of some pots and pans for even heat distribution. It is also a lightweight metal that makes it useful in kitchen cookware. Note that in this partial answer to part (a), a specific metal is described for a specific application.

1-11 Describe the enabling materials property of each of the following and why it is so:
(a) steel for I-beams in skyscrapers;
(b) a cobalt chrome molybdenum alloy for hip implants;
(c) polycarbonate for eyeglass lenses; and
(d) bronze for artistic castings.

1-12 Describe the enabling materials property of each of the following and why it is so:
(a) aluminum for airplane bodies;
(b) polyurethane for teeth aligners (invisible braces);
(c) steel for the ball bearings in a bicycle's wheel hub;
(d) polyethylene terephthalate for water bottles; and
(e) glass for wine bottles.

1-13 What properties should an engineer consider for a total knee replacement of a deteriorated knee joint with an artificial prosthesis when selecting the materials for this application?

1-14 Write one paragraph about why single-crystal silicon is currently the material of choice for microelectronics applications.

Write a second paragraph about potential alternatives to single-crystal silicon for solar cell applications. Provide a list of the references or websites that you used. You must use at least three references.

1-15 Coiled springs should be very strong and stiff. Silicon nitride (Si_3N_4) is a strong, stiff material. Would you select this material for a spring? Explain.

1-16 Temperature indicators are sometimes produced from a coiled metal strip that uncoils a specific amount when the temperature increases. How does this work; from what kind of material would the indicator be made; and what are the important properties that the material in the indicator must possess?

Section 1-3 Functional Classification of Materials

Section 1-4 Classification of Materials Based on Structure

Section 1-5 Environmental and Other Effects

1-17 What is the purpose of the classification for functional materials?

1-18 Explain the difference between crystalline and amorphous materials. Give an example of each that you use in your daily life.

1-19 If you were given a material and were asked to determine whether it is crystalline or amorphous, how would you determine it?

1-20 List six materials performance problems that may lead to failure of components.

1-21 Steel is often coated with a thin layer of zinc if it is to be used outside. What characteristics do you think the zinc provides to this coated, or galvanized, steel? What precautions should be considered in producing this product? How will the recyclability of the product be affected?

1-22 The relationship between structure and materials properties can be influenced by the service conditions (environmental conditions). Name two engineering disasters that have had tragic results and why they happened.

Section 1-6 Materials Design and Selection

1-23 What is the difference between physical and mechanical properties? List three examples for each one.

1-24 The type of jet engine used on most large commercial aircraft is called a turbofan jet engine because it has a large wheel at the front of the engine that propels air rearward. Most of this air bypasses the engine, but bypass air significantly increases thrust and efficiency of these engines. Some engine manufacturers are now using carbon fiber-epoxy composites rather than traditional aluminum blades.

(a) What materials properties do you think an engineer must consider when selecting a material for this application? Be as specific as possible.

(b) What benefits do you think carbon fiber epoxy composites have compared to aluminum alloys? What limitations or possible downsides could there be to using a carbon fiber epoxy composite?

1-25 You are an engineer working for a manufacturer of land-based gas turbines. These turbines are similar to jet engines, but they are used on land to provide power for electricity generation and gas compression pipeline applications. Suppose that you would like to apply a ceramic-based thermal barrier coating to the turbine blades in the first-stage turbine to increase the operating temperature and efficiency of the engine.

(a) What difficulties might engineers experience in trying to design a ceramic coating that will be applied to a super alloy metal blade?

(b) What properties should be taken into consideration when choosing a suitable ceramic material for a coating? Be as thorough as possible.

1-26 We would like to produce a transparent canopy for an aircraft. If we were to use a traditional window glass canopy, rocks or birds might cause it to shatter. Design a material that would minimize damage or at least keep the canopy from breaking into pieces.

1-27 You would like to design an aircraft that can be flown by human power nonstop for a distance of 30 km. What types of material properties would you recommend? What materials might be appropriate?

1-28 You would like to place a three-foot diameter microsatellite into orbit. The satellite will contain delicate electronic equipment that will send and receive radio signals from earth. Design the outer shell within which the electronic equipment is contained. What properties will be required, and what kind of materials might be considered?

1-29 What properties should the head of a carpenter's hammer possess? How would you manufacture a hammer head?

1-30 You would like to select a material for the electrical contacts in an electrical switching device that opens and closes frequently and forcefully. What properties should the contact material possess? What type of material might you recommend? Would Al_2O_3 be a good choice? Explain.

1-31 Aluminum has a density of 2.7 g/cm^3. Suppose you would like to produce a composite material based on aluminum having a density of 1.5 g/cm^3. Design a material that would have this density. Would introducing beads of polyethylene, with a density of 0.95 g/cm^3, into the aluminum be a likely possibility? Explain.

1-32 You would like to be able to identify different materials without resorting to chemical analysis or lengthy testing procedures. Describe some possible testing and sorting techniques you might be able to use based on the physical properties of materials.

1-33 You would like to be able to physically separate different materials in a scrap recycling plant. Describe some possible methods that might be used to separate materials such as polymers, aluminum alloys, and steels from one another.

1-34 Some pistons for automobile engines might be produced from a composite material containing small, hard silicon carbide particles in an aluminum alloy matrix. Explain what benefits each material in the composite may provide to the overall part. What problems might the different properties of the two materials cause in producing the part?

1-35 Investigate the origins and applications for a material that has been invented or discovered since you were born *or* investigate the development of a product or technology that has been invented since you were born that was made possible by the use of a novel material. Write one paragraph about this material or product. Provide a list of the references or websites that you used. You must use at least three references.

Knovel® Problems

All problems in the final section of each chapter require the use of the Knovel website (app.knovel.com/web). These three problems are designed to provide an introduction to Knovel, its website, and the interactive tools available on it.

K1-1
- Convert 7750 kg/m^3 to lb/ft^3 using the Unit Converter.
- Using the Periodic Table, determine the atomic weight of magnesium.
- What is the name of Section 4 in *Perry's Chemical Engineers' Handbook (Seventh Edition)*?
- Find a book title that encompasses the fundamentals of chemistry as well as contains interactive tables of chemical data.

K1-2
- Using the basic search option in Knovel, find as much physical and thermodynamic data associated with ammonium nitrate as possible. What applications does this chemical have?
- Using the Basic Search, find the formula for the volume of both a sphere and a cylinder.
- Using the Data Search, produce a list of five chemicals with a boiling point between 300 and 400 K.

K1-3
- Using the Equation Plotter, determine the enthalpy of vaporization of pure acetic acid at 360 K.
- What is the pressure (in atm) of air with a temperature of 200°F and a water content of 10^{-2} lb water/(lb air)?
- Find three grades of polymers with a melting point greater than 325°C.

Diamond and graphite both comprise pure carbon, but their materials properties vary considerably. These differences arise from differences in the arrangements of the atoms in the solids and differences in the bonding between atoms. Covalent bonding in diamond leads to high strength and stiffness, excellent thermal conductivity, and poor electrical conductivity. (*Özer Öner/Shutterstock.com*) The atoms in graphite are arranged in sheets. Within the sheets, the bonding between atoms is covalent, but between the sheets, the bonds are less strong. Thus graphite can easily be sheared off in sheets as occurs when writing with a pencil. (*Ronald van der Beek/Shutterstock.com*) Graphite's thermal conductivity is much lower than that of diamond, and its electrical conductivity is much higher.

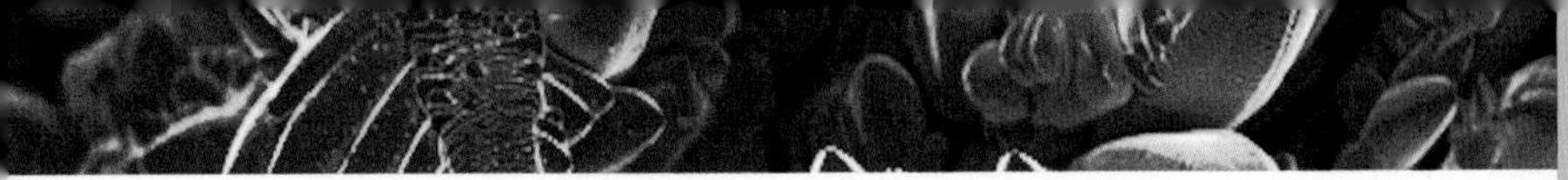

CHAPTER

2

Atomic Structure

Have You Ever Wondered?

- *What is nanotechnology?*
- *Why is carbon, in the form of diamond, one of the hardest materials known, but as graphite is very soft and can be used as a solid lubricant?*
- *How is silica, which forms the main chemical in beach sand, used in an ultrapure form to make optical fibers?*

Chapter Learning Objectives

The key objectives of this chapter are to

- Define four quantum numbers for electrons and explain the significance of each.
- Explain the arrangement of the elements in the periodic table.
- State the electronic structure of the elements.
- Explain the role of electronegativity in bonding.
- Define four different mechanisms of bonding in materials.
- Describe how interactions between atoms or between ions influence materials properties.
- Explain how allotropes can display dramatically different materials properties based on their structure even though they are composed of the same element.

Materials scientists and engineers have developed a set of instruments in order to characterize the **structure** of materials at various length scales. We can examine and describe the structure of materials at five different levels:

1. atomic structure;
2. short- and long-range atomic arrangements;
3. nanostructure;
4. microstructure; and
5. macrostructure.

The features of the structure at each of these levels may have distinct and profound influences on a material's properties and behavior.

The goal of this chapter is to examine **atomic structure** (the nucleus consisting of protons and neutrons and the electrons surrounding the nucleus) in order to lay a foundation for understanding how atomic structure affects the properties, behavior, and resulting applications of engineering materials. We will see that the structure of atoms affects the types of bonds that hold materials together. These different types of bonds directly affect the suitability of materials for real-world engineering applications. The diameter of atoms typically is measured using the angstrom unit (Å or 10^{-10} m).

It also is important to understand how atomic structure and bonding lead to different atomic or ionic arrangements in materials. A close examination of atomic arrangements allows us to distinguish between materials that are **amorphous** (those that lack a long-range ordering of atoms or ions) or **crystalline** (those that exhibit periodic geometrical arrangements of atoms or ions). Amorphous materials have only **short-range atomic arrangements**, while crystalline materials have short- and **long-range atomic arrangements**. In short-range atomic arrangements, the atoms or ions show a particular order only over relatively short distances (1 to 10 Å). For crystalline materials, the long-range atomic order is in the form of atoms or ions arranged in a three-dimensional pattern that repeats over much larger distances (from ~10 nm to cm).

Materials science and engineering is at the forefront of **nanoscience** and **nanotechnology**. Nanoscience is the study of materials at the nanometer length scale, and nanotechnology is the manipulation and development of devices at the nanometer length scale. The **nanostructure** is the structure of a material at a **length scale** of 1 to 100 nm. Controlling nanostructure is becoming increasingly important for advanced materials engineering applications.

The **microstructure** is the structure of materials at a **length scale** of 100 to 100,000 nm or 0.1 to 100 micrometers (often written as μm and pronounced as "microns"). The microstructure typically refers to features such as the grain size of a crystalline material and others related to defects in materials. (A *grain* is a single crystal in a material composed of many crystals.)

Macrostructure is the structure of a material at a macroscopic level where the length scale is >100 μm. Features that constitute macrostructure include porosity, surface coatings, and internal and external microcracks.

We will conclude the chapter by considering some of the **allotropes** of carbon. We will see that, although both diamond and graphite are made from pure carbon, they have different materials properties. The key to understanding these differences is to understand how the atoms are arranged in each allotrope.

2-1 The Structure of Materials: Technological Relevance

In today's world, information technology (IT), biotechnology, energy technology, environmental technology, and many other areas require smaller, lighter, faster, portable, more efficient, reliable, durable, and inexpensive devices. We want batteries that are smaller, lighter, and last longer. We need cars that are relatively affordable, lightweight, safe, highly fuel efficient, and "loaded" with many advanced features, ranging from global positioning systems (GPS) to sophisticated sensors for airbag deployment.

Some of these needs have generated considerable interest in nanotechnology and **micro-electro-mechanical systems** (MEMS). As a real-world example of MEMS technology, consider a small accelerometer sensor obtained by the micro-machining of silicon. This sensor is used to measure acceleration in automobiles. The information is processed by a computer and then used for controlling airbag deployment. Properties and behavior of materials at these "micro" levels can vary greatly when compared to those in their "macro" or bulk state. As a result, understanding the nanostructure and microstructure are areas that are receiving considerable attention.

The applications shown in Table 2-1 and the accompanying figures (Figures 2-1 through 2-6) illustrate how important the different levels of structure are to materials behavior. The applications are broken out by their levels of structure and their length scales (the approximate characteristic length that is important for a given application).

We now turn our attention to the details concerning the structure of atoms, the bonding between atoms, and how these form a foundation for the properties of materials. Atomic structure influences how atoms are bonded together. An understanding of this helps categorize materials as metals, semiconductors, ceramics, or polymers. It also permits us to draw some general conclusions concerning the mechanical properties and physical behaviors of these four classes of materials.

Table 2-1 Levels of structure

Level of Structure	Example of Technologies
Atomic Structure ($\sim 10^{-10}$ m or 1 Å)	Diamond is based on carbon-carbon covalent bonds. Materials with this type of bonding are expected to be relatively hard. Thin films of diamond are used for providing a wear-resistant edge in cutting tools.  Figure 2-1 Diamond-coated cutting tools. (*Courtesy of NCD Technologies*)
Atomic Arrangements: Long-Range Order (LRO) ($\sim$10 nm to cm)	When ions in lead-zirconium-titanate $[Pb(Zr_xTi_{1-x})O_3]$(PZT) are arranged such that they exhibit tetragonal and/or rhombohedral crystal structures (see Chapter 3), the material is piezoelectric (i.e., it develops a voltage when subjected to pressure or stress). PZT ceramics are used widely for many applications including gas igniters, ultrasound generation, and vibration control. 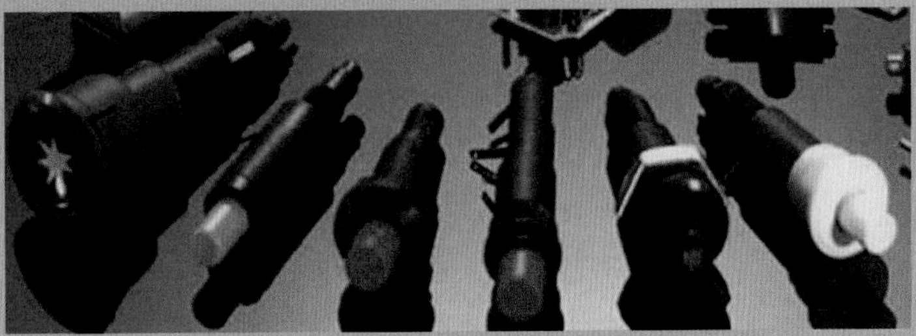 Figure 2-2 Piezoelectric PZT-based gas igniters. When the piezoelectric material is stressed (by applying a pressure), a voltage develops and a spark is created between the electrodes. (*Courtesy of Morgan Technical Ceramics Ltd UK*)

Table 2-1 Levels of structure *(Continued)*

Level of Structure	Example of Technologies
Atomic Arrangements: Short-Range Order (SRO) (1 to 10 Å)	Ions in silica (SiO_2) glass exhibit only a short-range order in which Si^{4+} and O^{2-} ions are arranged in a particular way (each Si^{4+} is bonded with four O^{2-} ions in a tetrahedral coordination, with each O^{2-} ion being shared by two tetrahedra). This order, however, is not maintained over long distances, thus making silica glass amorphous. Amorphous glasses based on silica and certain other oxides form the basis for the entire fiber-optic communications industry.  Figure 2-3 Optical fibers based on a form of silica that is amorphous. (*Nick Rowe/ Photodisc Green/GettyImages*)
Nanostructure (~10^{-9} to 10^{-7} m, 1 to 100 nm)	Nano-sized particles (~5–10 nm) of iron oxide are used in ferrofluids or liquid magnets. An application of these liquid magnets is as a cooling (heat transfer) medium for loudspeakers. 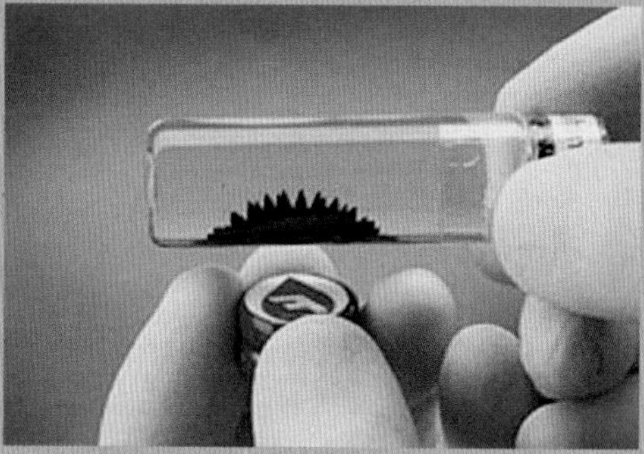 Figure 2-4 Ferrofluid. (*Courtesy of Ferrotec USA Corporation*)
Microstructure (~>10^{-7} to 10^{-4} m, 0.1 to 100 μm)	The mechanical strength of many metals and alloys depends strongly on the grain size. The grains and grain boundaries in this accompanying micrograph of steel are part of the microstructural features of this crystalline material. In general, at room temperature, a finer grain size leads to higher strength. Many important properties of materials are sensitive to the microstructure.  Figure 2-5 Micrograph of stainless steel showing grains and grain boundaries. (*Micrograph courtesy of Dr. A.J. Deardo, Dr. M. Hua and Dr. J. Garcia*)

Table 2-1 Levels of structure *(Continued)*

Level of Structure	Example of Technologies
Macrostructure (~>10^{-4} m, ~>100,000 nm or 100 μm)	Relatively thick coatings, such as paints on automobiles and other applications, are used not only for aesthetics, but also to provide corrosion resistance.

Figure 2-6 A number of organic and inorganic coatings protect the car from corrosion and provide a pleasing appearance. (*George Dolgikh/Shutterstock.com*)

2-2 The Structure of the Atom

The concepts mentioned next are covered in typical introductory chemistry courses. We are providing a brief review. An atom is composed of a nucleus surrounded by electrons. The nucleus contains neutrons and positively charged protons and carries a net positive charge. The negatively charged electrons are held to the nucleus by an electrostatic attraction. The electrical charge q carried by each electron and proton is 1.60×10^{-19} coulomb (C).

The **atomic number** of an element is equal to the number of protons in each atom. Thus, an iron atom, which contains 26 protons, has an atomic number of 26. The atom as a whole is electrically neutral because the numbers of protons and electrons are equal.

Most of the mass of the atom is contained within the nucleus. The mass of each proton and neutron is 1.67×10^{-24} g, but the mass of each electron is only 9.11×10^{-28} g. The **atomic mass unit**, or amu, is 1/12 the mass of carbon 12 (i.e., the carbon atom with twelve **nucleons**–six protons and six neutrons). The **atomic mass** M is equal to the total mass of the average number of protons and neutrons in the atom in atomic mass units and is also the mass in grams of the Avogadro number N_A of atoms. The quantity $N_A = 6.022 \times 10^{23}$ atoms/mol is the number of atoms or molecules in a mole. Therefore, the atomic mass has units of g/mol. As an example, one mole of iron contains 6.022×10^{23} atoms and has a mass of 55.847 g. Calculations including a material's atomic mass and the Avogadro number are helpful to understanding more about the structure of a material. Example 2-1 illustrates how to calculate the number of atoms for silver, a metal and a good electrical conductor.

Example 2-1 *Calculating the Number of Atoms in Silver*

Calculate the number of atoms in 100 g of silver.

SOLUTION

The number of atoms can be calculated from the atomic mass and the Avogadro number. From Appendix A, the atomic mass, or weight, of silver is 107.868 g/mol. The number of atoms is

$$\text{Number of Ag atoms} = \frac{(100\text{ g})(6.022 \times 10^{23}\text{ atoms/mol})}{107.868\text{ g/mol}}$$
$$= 5.58 \times 10^{23}$$

2-3 The Electronic Structure of the Atom

Electrons occupy discrete energy levels within the atom. Each electron possesses a particular energy with no more than two electrons in each atom having the same energy. Since each element possesses a different set of these energy levels, the differences between the energy levels also are unique. Both the energy levels and the differences between them are known with great precision for every element, forming the basis for many types of **spectroscopy**. Using a spectroscopic method, the identity of elements in a sample may be determined.

Quantum Numbers

The energy level to which each electron belongs is identified by four **quantum numbers**. The four quantum numbers are the principal quantum number n, the azimuthal or secondary quantum number l, the magnetic quantum number m_l, and the spin quantum number m_s.

The principal quantum number reflects the grouping of electrons into sets of energy levels known as shells. Azimuthal quantum numbers describe the energy levels within each shell and reflect a further grouping of similar energy levels, usually called orbitals. The magnetic quantum number specifies the orbitals associated with a particular azimuthal quantum number within each shell. Finally, the **spin quantum number** (m_s) is assigned values of $+1/2$ and $-1/2$, which reflect the two possible values of "spin" of an electron.

According to the **Pauli Exclusion Principle**, within each atom, no two electrons may have the same four quantum numbers, and thus, each electron is designated by a unique set of four quantum numbers. The number of possible energy levels is determined by the first three quantum numbers.

1. The principal quantum number n is assigned integer values 1, 2, 3, 4, 5, . . . that refer to the quantum shell to which the electron belongs. A **quantum shell** is a *set* of fixed energy levels to which electrons belong.

   Quantum shells are also assigned a letter; the shell for $n = 1$ is designated K, for $n = 2$ is L, for $n = 3$ is M, and so on. These designations were carried over from the nomenclature used in optical spectroscopy, a set of techniques that predates the understanding of quantized electronic levels.
2. The *number* of energy levels in *each* quantum shell is determined by the **azimuthal quantum number** l and the magnetic quantum number m_l. The azimuthal quantum numbers are assigned $l = 0, 1, 2, \ldots, n - 1$. For example, when $n = 2$, there are two azimuthal quantum numbers, $l = 0$ and $l = 1$. When $n = 3$, there are three azimuthal quantum numbers, $l = 0$, $l = 1$, and $l = 2$. The azimuthal quantum numbers are designated by lowercase letters; one speaks, for instance, of the d orbitals:

   s for $l = 0$ d for $l = 2$

   p for $l = 1$ f for $l = 3$
3. The number of values for the magnetic quantum number m_l gives the number of energy levels, or orbitals, for each azimuthal quantum number. The total number of magnetic quantum numbers for each l is $2l + 1$. The values for m_l are given by whole numbers between $-l$ and $+l$. For example, if $l = 2$, there are $2(2) + 1 = 5$ magnetic quantum numbers with values $-2, -1, 0, +1$, and $+2$. The combination of l and m_l specifies a particular orbital in a shell.
4. No more than two electrons with opposing electronic spins ($m_s = +1/2$ and $-1/2$) may be present in each orbital.

By carefully considering the possible numerical values for n, l, and m_l, the range of *possible* quantum numbers may be determined. For instance, in the K shell (that is, $n = 1$), there is just a single s orbital (as the only allowable value of l is 0 and m_l is 0). As a result, a K shell may contain no more than two electrons. As another example, consider an M shell. In this case, $n = 3$, so l takes values of 0, 1, and 2 (there are s, p, and d orbitals present). The values of l reflect that there is a single s orbital ($m_l = 0$, a single value), three p orbitals ($m_l = -1, 0, +1$, or three values), and five d orbitals ($m_l = -2, -1, 0, +1, +2$, or five discrete values).

The shorthand notation frequently used to denote the electronic structure of an atom combines the numerical value of the principal quantum number, the lowercase letter notation for the azimuthal quantum number, and a superscript showing the number of electrons in each type of orbital. The shorthand notation for neon, which has an atomic number of ten, is

$$1s^2 2s^2 2p^6$$

Deviations from Expected Electronic Structures The energy levels of the quantum shells do not fill in strict numerical order. The **Aufbau Principle** is a graphical device that predicts deviations from the expected ordering of the energy levels. The Aufbau principle is shown in Figure 2-7. To use the Aufbau Principle, write the possible combinations of the principal quantum number and azimuthal quantum number for each quantum shell. The combinations for each quantum shell should be written on a single line. As the principal quantum number increases by one, the number of combinations within each shell increases by one (i.e., each row is one entry longer than the prior row). Draw arrows through the rows on a diagonal from the upper right to the lower left as shown in Figure 2-7. By following the arrows, the order in which the energy levels of each quantum level are filled is predicted.

For example, according to the Aufbau Principle, the electronic structure of iron, atomic number 26, is

$$1s^2 2s^2 2p^6 3s^2 3p^6 4s^2 3d^6$$

Conventionally, the principal quantum numbers are arranged from lowest to highest when writing the electronic structure. Thus, the electronic structure of iron is written

$$1s^2 2s^2 2p^6 3s^2 3p^6 \quad \boxed{3d^6 4s^2}$$

The unfilled $3d$ level (there are five d orbitals, so in shorthand $d^1, d^2, \ldots, d^{10}$ are possible) causes the magnetic behavior of iron.

Note that not all elements follow the Aufbau principle. A few, such as copper, are exceptions. According to the Aufbau Principle, copper should have the electronic structure $1s^2 2s^2 2p^6 3s^2 3p^6 3d^9 4s^2$, but copper actually has the electronic structure

$$1s^2 2s^2 2p^6 3s^2 3p^6 3d^{10} \quad \boxed{4s^1}$$

Generally, electrons will occupy each orbital of a given energy level singly before the orbitals are doubly occupied. For example, nitrogen has the electronic structure

$$1s^2 \quad \boxed{2s^2 2p^3}$$

Each of the three p orbitals in the L shell contains one electron rather than one orbital containing two electrons, one containing one electron, and one containing zero electrons.

Valence The **valence** of an atom is the number of electrons in an atom that participate in bonding or chemical reactions. Usually, the valence is the number of electrons in the

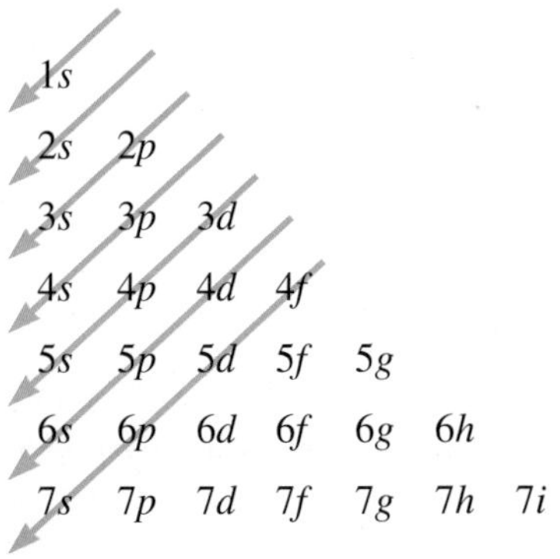

Figure 2-7 The Aufbau Principle. By following the arrows, the order in which the energy levels of each quantum level are filled is predicted: $1s$, $2s$, $2p$, $3s$, $3p$, etc. Note that the letter designations for $l = 4, 5, 6$ are g, h, and i.

outer *s* and *p* energy levels. The valence of an atom is related to the ability of the atom to enter into chemical combination with other elements. Examples of the valence are

Mg: $1s^2 2s^2 2p^6$ $\boxed{3s^2}$ valence = 2

Al: $1s^2 2s^2 2p^6$ $\boxed{3s^2 3p^1}$ valence = 3

Si: $1s^2 2s^2 2p^6$ $\boxed{3s^2 3p^2}$ valence = 4

Valence also depends on the immediate environment surrounding the atom or the neighboring atoms available for bonding. Phosphorus has a valence of five when it combines with oxygen, but the valence of phosphorus is only three—the electrons in the 3*p* level—when it reacts with hydrogen. Manganese may have a valence of 2, 3, 4, 6, or 7!

Atomic Stability and Electronegativity

If an atom has a valence of zero, the element is inert (non-reactive). An example is argon, which has the electronic structure

$$1s^2 2s^2 2p^6 \ \boxed{3s^2 3p^6}$$

Other atoms prefer to behave as if their outer *s* and *p* levels are either completely full, with eight electrons, or completely empty. Aluminum has three electrons in its outer *s* and *p* levels. An aluminum atom readily gives up its outer three electrons to empty the 3*s* and 3*p* levels. The atomic bonding and the chemical behavior of aluminum are determined by how these three electrons interact with surrounding atoms.

On the other hand, chlorine contains seven electrons in the outer 3*s* and 3*p* levels. The reactivity of chlorine is caused by its desire to fill its outer energy level by accepting an electron.

Electronegativity describes the tendency of an atom to gain an electron. Atoms with almost completely filled outer energy levels—such as chlorine—are strongly electronegative and readily accept electrons. Atoms with nearly empty outer levels—such as sodium—readily give up electrons and have low electronegativity. High atomic number elements also have low electronegativity because the outer electrons are at a greater distance from the positive nucleus, so that they are not as strongly attracted to the atom. Electronegativities for some elements are shown in Figure 2-8. Elements with low electronegativity (i.e., <2.0) are sometimes described as **electropositive**.

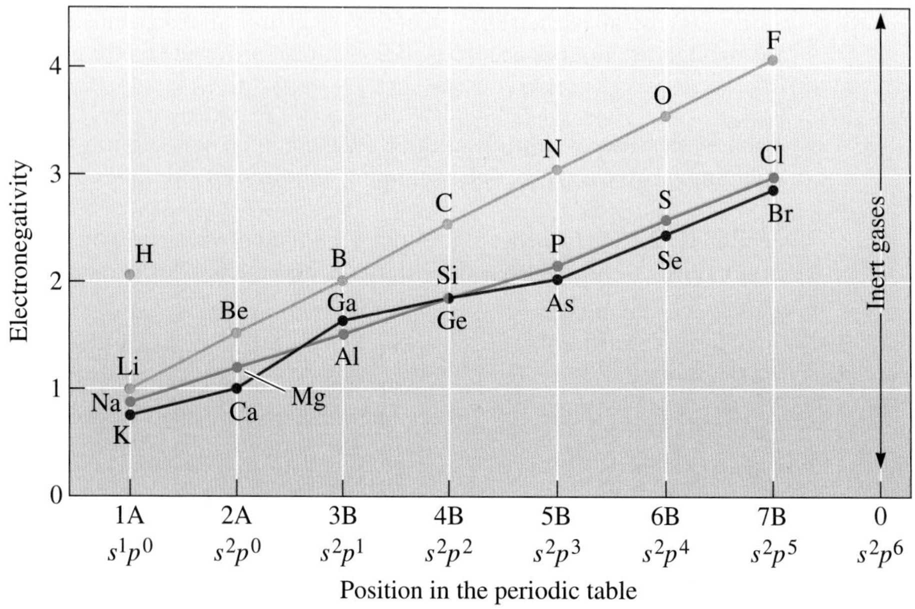

Figure 2-8 The electronegativities of selected elements relative to the position of the elements in the periodic table.

2-4 The Periodic Table

The periodic table contains valuable information about specific elements and can also help identify trends in atomic size, melting point, chemical reactivity, and other properties. The familiar periodic table (Figure 2-9) is constructed in accordance with the electronic structure of the elements. Rows in the periodic table correspond to quantum shells, or principal quantum numbers. Columns typically refer to the number of electrons in the outermost *s* and *p* energy levels and correspond to the most common valence. In engineering, we are mostly concerned with

(a) Polymers (plastics) (primarily based on carbon, which appears in Group 4B);

(b) Ceramics (typically based on combinations of many elements appearing in Groups 1 through 5B, and such elements as oxygen, carbon, and nitrogen); and

(c) Metallic materials (typically based on elements in Groups 1 or 2 or transition metal elements).

Many technologically important semiconductors appear in Group 4B (e.g., silicon, diamond, germanium). Semiconductors also can be combinations of elements from Groups 2B and 6B [e.g., cadmium selenide (CdSe), based on cadmium from Group 2 and selenium from Group 6]. These are known as **II–VI** (two-six) **semiconductors**. Similarly, gallium arsenide (GaAs) is a **III–V** (three-five) **semiconductor** based on gallium from Group 3B and arsenic from Group 5B. Many **transition elements** (e.g., titanium, vanadium, iron, nickel, cobalt, etc.) are particularly useful for magnetic and optical materials due to their electronic configurations that allow multiple valences.

The ordering of the elements in the periodic table and the origin of the Aufbau Principle become even clearer when the rows for the Lathanoid and Actinoid series are inserted into their correct positions (see Figure 2-10) rather than being placed below the periodic table to conserve space. Figure 2-10 indicates the particular orbital being filled by each additional electron as the atomic number increases. Note that exceptions are indicated for those elements that do not follow the Aufbau Principle.

Trends in Properties The periodic table contains a wealth of useful information (e.g., atomic mass, atomic number of different elements, etc.). It also points to trends in atomic size, melting points, and chemical reactivity. For example, carbon (in its diamond form) has the highest melting point (3550°C). Melting points of the elements below carbon decrease [i.e., silicon (1410°C), germanium (937°C), tin (232°C), and lead (327°C)]. Note that the melting temperature of Pb is higher than that of Sn. The periodic table indicates trends and not exact variations in properties.

We can discern trends in other properties from the periodic table. Diamond is a material with a very large bandgap (i.e., it is not a very effective conductor of electricity). This is consistent with the fact that carbon (in diamond form) has the highest melting point among Group 4B elements, which suggests the interatomic forces are strong (see Section 2-6). As we move down the column, the bandgap decreases (the bandgaps of Si and Ge are 1.11 and 0.67 eV, respectively). Moving farther down, one form of tin is a semiconductor. Another form of tin is metallic. If we look at Group 1A, we see that lithium is highly electropositive (i.e., an element with atoms that want to participate in chemical interactions by donating electrons and are therefore highly reactive). Likewise, if we move down Column 1A, we can see that the chemical reactivity of elements increases.

Thus, the periodic table gives us useful information about formulas, atomic numbers, and atomic masses of elements. It also helps us in predicting or rationalizing trends in properties of elements and compounds. This is why the periodic table is very useful to both scientists and engineers.

Figure 2-9 Periodic table of elements.

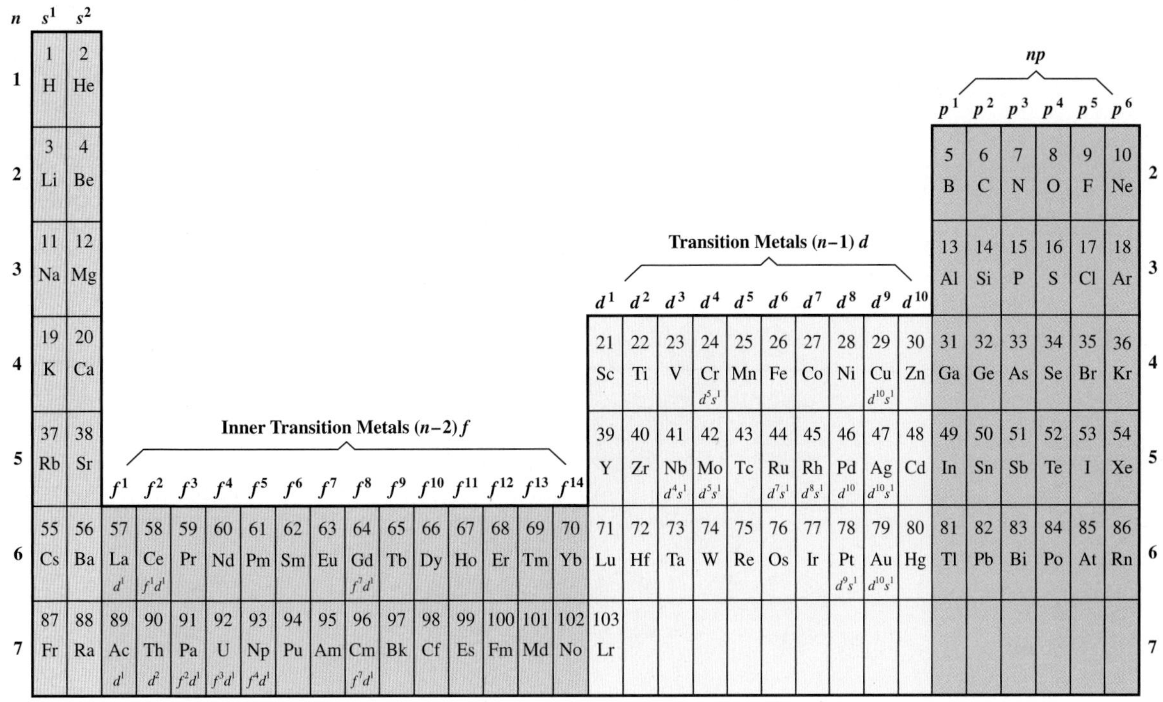

Figure 2-10 The periodic table for which the rows of the Lathanoid and Actinoid series are inserted into their correct positions. The column heading indicates the particular orbital being filled by each additional electron as the atomic number increases.

2-5 Atomic Bonding

There are four important mechanisms by which atoms are bonded in engineered materials. These are

1. **metallic bonds**;
2. **covalent bonds**;
3. **ionic bonds**; and
4. **van der Waals bonds**.

The first three types of bonds are relatively strong and are known as **primary bonds** (relatively strong bonds between adjacent atoms resulting from the transfer or sharing of outer orbital electrons). The van der Waals bonds are secondary bonds and originate from a different mechanism and are relatively weaker. Let's look at each of these types of bonds.

The Metallic Bond

The metallic elements have electropositive atoms that donate their valence electrons to form a "sea" of electrons surrounding the atoms (Figure 2-11). Aluminum, for example, gives up its three valence electrons, leaving behind a core consisting of the nucleus and inner electrons. Since three negatively charged electrons are missing from this core, it has a positive charge of three. The valence electrons move freely within the electron sea and become associated with several atom cores. The positively charged ion cores are held together by mutual attraction to the electrons, thus producing a strong metallic bond.

Because their valence electrons are not fixed in any one position, most pure metals are good electrical conductors at relatively low temperatures (less than approximately 300 K). Under the influence of an applied voltage, the valence electrons move, causing a current to flow if the circuit is complete.

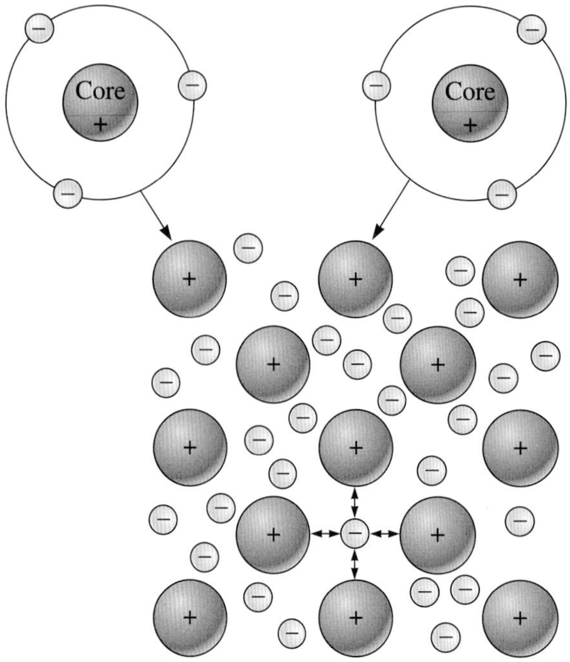

Figure 2-11 The metallic bond forms when atoms give up their valence electrons, which then form an electron sea. The positively charged atom cores are bonded by mutual attraction to the negatively charged electrons.

Metals show good ductility since the metallic bonds are non-directional. There are other important reasons related to microstructure that can explain why metals actually exhibit *lower strengths* and *higher ductility* than what we may anticipate from their bonding. **Ductility** refers to the ability of a material to be stretched or bent permanently without breaking. We will discuss these concepts in greater detail in Chapter 6. In general, the melting points of metals are relatively high. From an optical properties viewpoint, metals make good reflectors of visible radiation. Owing to their electropositive character, many metals such as iron tend to undergo corrosion or oxidation. Many pure metals are good conductors of heat and are effectively used in many heat transfer applications. We emphasize that metallic bonding is *one of the factors* in our efforts to rationalize the trends observed with respect to the properties of metallic materials. As we will see in some of the following chapters, there are other factors related to microstructure that also play a crucial role in determining the properties of metallic materials.

The Covalent Bond Materials with **covalent bonding** are characterized by bonds that are formed by sharing of valence electrons among two or more atoms. For example, a silicon atom, which has a valence of four, achieves a complete outer energy shell of eight electrons by sharing its valence electrons with four surrounding silicon atoms, as in Figure 2-12(a) and (b). Each instance of sharing represents one covalent bond; thus, each silicon atom is bonded to four neighboring atoms by four covalent bonds. In order for the covalent bonds to be formed, the silicon atoms must be arranged so the bonds have a fixed **directional relationship** with one another. A directional relationship is formed when the bonds between atoms in a covalently bonded material form specific angles, depending on the material. In the case of silicon, this arrangement produces a tetrahedron, with angles of 109.5° between the covalent bonds [Figure 2-12(c)].

Covalent bonds are very strong. As a result, covalently bonded materials are very strong and hard. For example, diamond, silicon carbide (SiC), silicon nitride (Si_3N_4), and boron nitride (BN) all have covalent bonds. These materials also exhibit very high melting points, which means they could be useful for high-temperature applications. On the other hand, the high temperatures needed for processing present a challenge. The

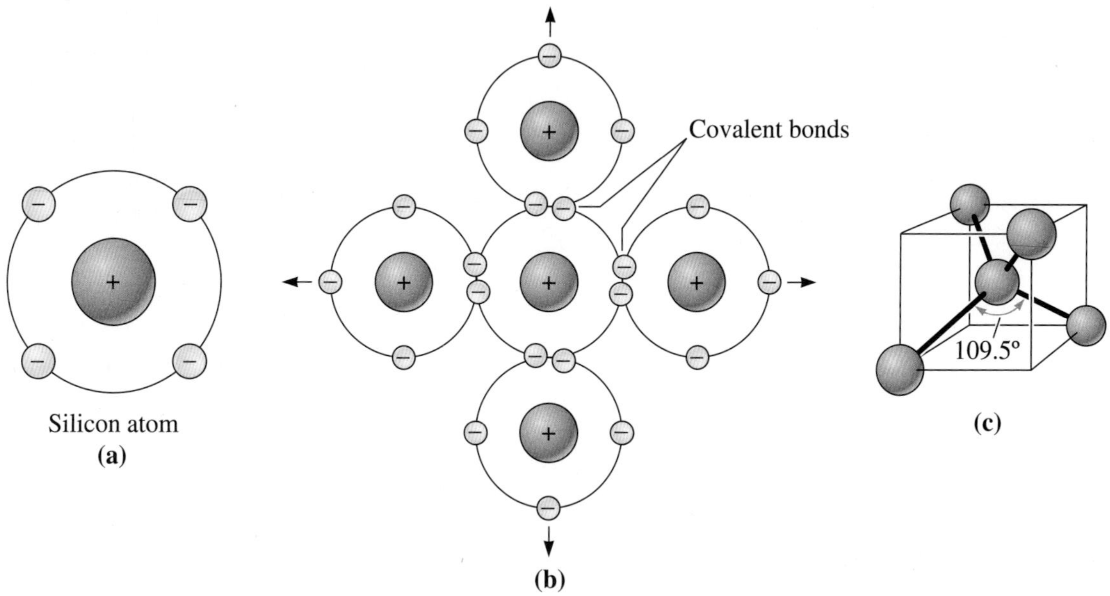

Figure 2-12 (a) Covalent bonding requires that electrons be shared between atoms in such a way that each atom has its outer *sp* orbitals filled. (b) In silicon, with a valence of four, four covalent bonds must be formed. (c) Covalent bonds are directional. In silicon, a tetrahedral structure is formed with angles of 109.5° found between any pair of covalent bonds.

materials bonded in this manner typically have limited ductilities because the bonds tend to be directional. The electrical conductivity of many covalently bonded materials (i.e., silicon, diamond, and many ceramics) is not high since the valence electrons are locked in bonds between atoms and are not readily available for conduction. With some of these materials such as Si, we can get useful and controlled levels of electrical conductivity by deliberately introducing small levels of other elements known as dopants. Conductive polymers are also a good example of covalently bonded materials that can be turned into semiconducting materials. The development of conducting polymers that are lightweight has captured the attention of many scientists and engineers for developing flexible electronic components.

We cannot simply predict whether or not a material will be high or low strength, ductile or brittle, simply based on the nature of bonding! We need additional information on the atomic structure, microstructure, and macrostructure of the material; however, the nature of bonding does point to a trend for materials with certain types of bonding and chemical compositions. Example 2-2 explores how one such bond of oxygen and silicon join to form silica.

Example 2-2 *How Do Oxygen and Silicon Atoms Join to Form Silica?*

Assuming that silica (SiO_2) has 100% covalent bonding, describe how oxygen and silicon atoms in silica are joined.

SOLUTION

Silicon has a valence of four and shares electrons with four oxygen atoms, thus giving a total of eight electrons for each silicon atom. Oxygen has a valence of six and shares electrons with two silicon atoms, giving oxygen a total of eight electrons. Figure 2-13 illustrates one of the possible structures. Similar to silicon, a tetrahedral structure is produced. We will discuss later in this chapter how to account for the ionic and covalent nature of bonding in silica.

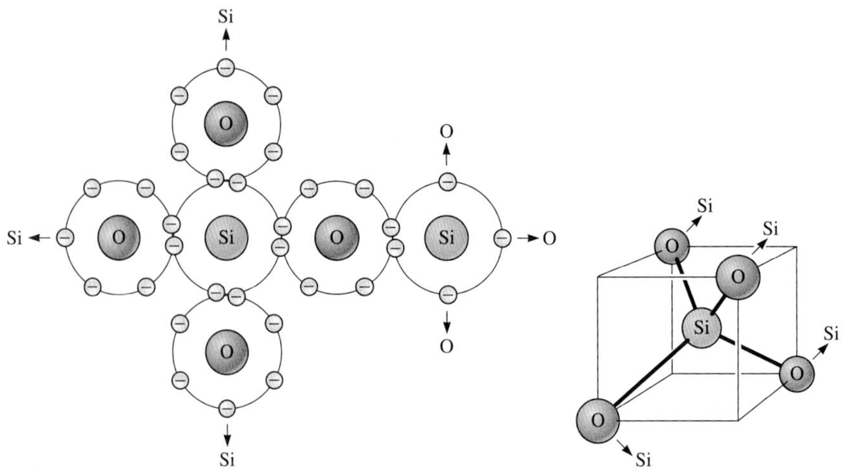

Figure 2-13 The tetrahedral structure of silica, which contains covalent bonds between silicon and oxygen atoms (for Example 2-2).

The Ionic Bond When more than one type of atom is present in a material, one atom may donate its valence electrons to a different atom, filling the outer energy shell of the second atom. Both atoms now have filled (or emptied) outer energy levels, but both have acquired an electrical charge and behave as ions. The atom that contributes the electrons is left with a net positive charge and is called a **cation**, while the atom that accepts the electrons acquires a net negative charge and is called an **anion**. The oppositely charged ions are then attracted to one another and produce the **ionic bond**. For example, the attraction between sodium and chloride ions (Figure 2-14) produces sodium chloride (NaCl), or table salt.

Van der Waals Bonding The origin of van der Waals forces between atoms and molecules is quantum mechanical in nature, and a meaningful discussion is beyond the scope of this book. We present here a simplified picture. If two electrical charges $+q$ and $-q$ are separated by a distance d, the dipole moment is defined as the product of q and d. Atoms are electrically neutral. Also, the centers of the positive charge (nucleus) and negative charge (electron cloud) coincide. Therefore, a neutral atom has no dipole moment. When a neutral atom is exposed to an internal or external electric field, the atom may become polarized (i.e., the centers of positive and negative charges separate). This creates or induces a dipole moment (Figure 2-15). In some molecules, the dipole moment does not have to be induced—it exists by virtue of the

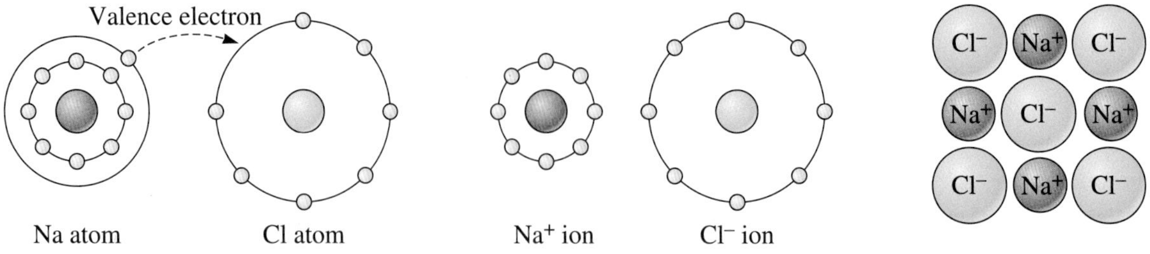

Figure 2-14 An ionic bond is created between two unlike atoms with different electronegativities. When sodium donates its valence electron to chlorine, each becomes an ion, attraction occurs, and the ionic bond is formed.

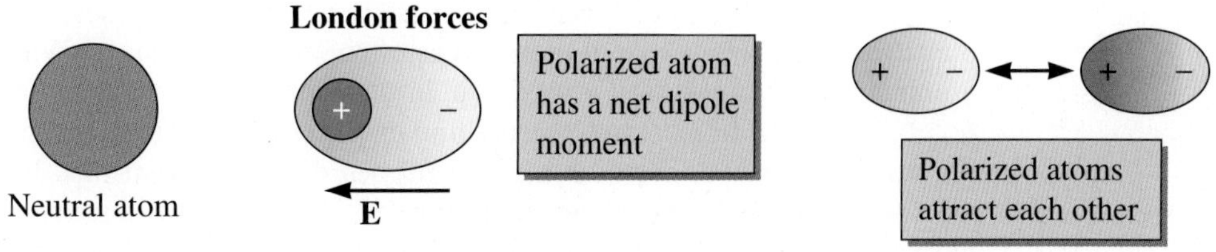

Figure 2-15 Illustration of London forces, a type of a van der Waals force, between atoms.

direction of bonds and the nature of atoms. These molecules are known as **polarized molecules**. An example of such a molecule that has a permanently built-in dipole moment is water (Figure 2-16).

Molecules or atoms in which there is either an induced or permanent dipole moment attract each other. The resulting force is known as the van der Waals force. Van der Waals forces between atoms and molecules have their origin in interactions between dipoles that are induced or in some cases interactions between permanent dipoles that are present in certain polar molecules. These forces are present in every material.

There are three types of **van der Waals** interactions, namely London forces, Keesom forces, and Debye forces. If the interactions are between two dipoles that are induced in atoms or molecules, we refer to them as **London forces** (e.g., carbon tetrachloride) (Figure 2-15). When an induced dipole (that is, a dipole that is induced in what is otherwise a non-polar atom or molecule) interacts with a molecule that has a permanent dipole moment, we refer to this interaction as a **Debye interaction**. An example of a Debye interaction would be forces between water molecules and those of carbon tetrachloride.

If the interactions are between molecules that are permanently polarized (e.g., water molecules attracting other water molecules or other polar molecules), we refer to these as **Keesom interactions**. The attraction between the positively charged regions of one water molecule and the negatively charged regions of a second water molecule provides an attractive bond between the two water molecules (Figure 2-16).

The bonding between molecules that have a permanent dipole moment, known as the Keesom force, is often referred to as a **hydrogen bond**, where hydrogen atoms represent one of the polarized regions. Thus, hydrogen bonding is essentially a Keesom force and is a type of van der Waals force. The relatively strong Keesom force between water molecules is the reason why the surface tension (72 mJ/m^2 or dyne/cm at room temperature) and the boiling point of water (100°C) are much higher than those of many organic liquids of comparable molecular weight (surface tension ~20 to 25 dyne/cm, boiling points up to 80°C).

Note that van der Waals bonds are **secondary bonds**, but the atoms within the molecule or group of atoms are joined by strong covalent or ionic bonds. Heating water to the boiling point breaks the van der Waals bonds and changes water to steam, but much higher temperatures are required to break the covalent bonds joining oxygen and hydrogen atoms.

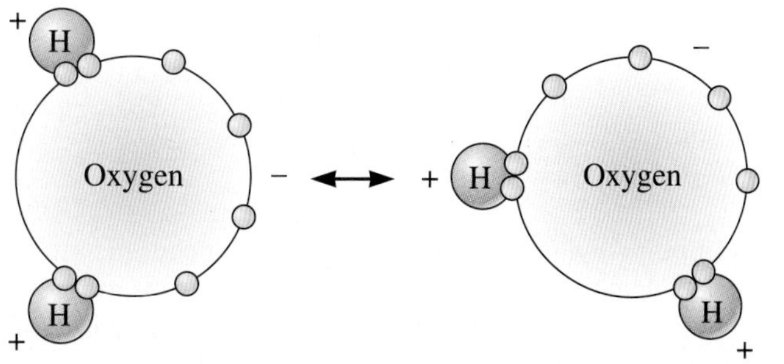

Figure 2-16 The Keesom interactions are formed as a result of polarization of molecules or groups of atoms. In water, electrons in the oxygen tend to concentrate away from the hydrogen. The resulting charge difference permits the molecule to be weakly bonded to other water molecules.

Although termed "secondary" based on the bond energies, van der Waals forces play a very important role in many areas of engineering. Van der Waals forces between atoms and molecules play a vital role in determining the surface tension and boiling points of liquids. In materials science and engineering, the surface tension of liquids and the surface energy of solids come into play in different situations. For example, when we want to process ceramic or metal powders into dense solid parts, the powders often have to be dispersed in water or organic liquids. Whether we can achieve this dispersion effectively depends upon the surface tension of the liquid and the surface energy of the solid material. Surface tension of liquids also assumes importance when we are dealing with processing of molten metals and alloys (e.g., casting) and glasses.

Van der Waals bonds can dramatically change the properties of certain materials. For example, graphite and diamond have very different mechanical properties, in part due to the presence of van der Waals bonds in graphite as will be discussed in Section 2-7. In many plastic materials, molecules contain polar parts or side groups (e.g., cotton or cellulose, PVC, Teflon). Van der Waals forces provide an extra binding force between the chains of these polymers (Figure 2-17).

Polymers in which van der Waals forces are stronger tend to be relatively stiffer and exhibit relatively higher glass-transition temperatures (T_g). The glass-transition temperature is a temperature below which some polymers tend to behave as brittle materials (i.e., they show poor ductility). As a result, polymers with van der Waals

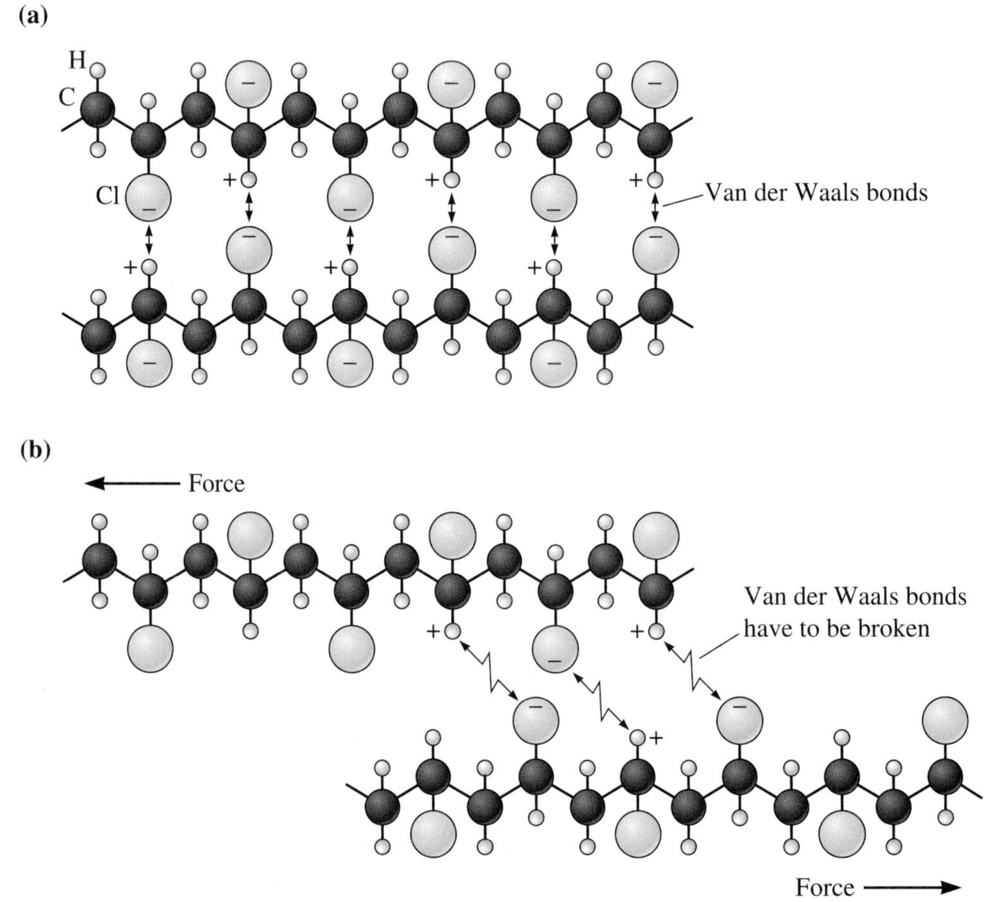

Figure 2-17 (a) In polyvinyl chloride (PVC), the chlorine atoms attached to the polymer chain have a negative charge and the hydrogen atoms are positively charged. The chains are weakly bonded by van der Waals bonds. This additional bonding makes PVC stiffer. (b) When a force is applied to the polymer, the van der Waals bonds are broken, and the chains slide past one another.

bonding (in addition to the covalent bonds in the chains and side groups) are relatively brittle at room temperature (e.g., PVC). In processing such polymers, they need to be "plasticized" by adding other smaller polar molecules that interact with the polar parts of the long polymer chains, thereby lowering the T_g and enhancing flexibility.

Mixed Bonding In most materials, bonding between atoms is a mixture of two or more types. Iron, for example, is bonded by a combination of metallic and covalent bonding that prevents atoms from packing as efficiently as we might expect.

Compounds formed from two or more metals (**intermetallic compounds**) may be bonded by a mixture of metallic and ionic bonds, particularly when there is a large difference in electronegativity between the elements. Because lithium has an electronegativity of 1.0 and aluminum has an electronegativity of 1.5, we would expect AlLi to have a combination of metallic and ionic bonding. On the other hand, because both aluminum and vanadium have electronegativities of 1.5, we would expect Al_3V to be bonded primarily by metallic bonds.

Many ceramic and semiconducting compounds, which are combinations of metallic and nonmetallic elements, have a mixture of covalent and ionic bonding. As the electronegativity difference between the atoms increases, the bonding becomes more ionic. The fraction of bonding that is covalent can be estimated from the following equation:

$$\text{Fraction covalent} = \exp(-0.25\Delta E^2) \tag{2-1}$$

where ΔE is the difference in electronegativities. Example 2-3 explores the nature of the bonds found in silica.

Example 2-3 *Determining if Silica is Ionically or Covalently Bonded*

In Example 2-2, we used silica (SiO_2) as an example of a covalently bonded material. In reality, silica exhibits ionic and covalent bonding. What fraction of the bonding is covalent? Give examples of applications in which silica is used.

SOLUTION

From Figure 2-9, the electronegativity of silicon is 1.8 and that of oxygen is 3.5. The fraction of the bonding that is covalent is

$$\text{Fraction covalent} = \exp[-0.25(3.5 - 1.8)^2] = 0.486$$

Although the covalent bonding represents only about half of the bonding, the directional nature of these bonds still plays an important role in the eventual structure of SiO_2.

Silica has many applications. Silica is used for making glasses and optical fibers. We add nanoparticles of silica to tires to enhance the stiffness of the rubber. High-purity silicon crystals are made by reducing silica to silicon.

2-6 Binding Energy and Interatomic Spacing

The equilibrium distance between atoms is caused by a balance between repulsive and attractive forces. In the metallic bond, for example, the attraction between the electrons and the ion cores is balanced by the repulsion between ion cores. Equilibrium separation occurs when the total interatomic energy (IAE) of the pair of atoms is at a minimum, or when no net force is acting to either attract or repel the atoms (Figure 2-18).

The **interatomic spacing** in a solid metal is *approximately* equal to the atomic diameter, or twice the atomic radius r. We cannot use this approach for ionically

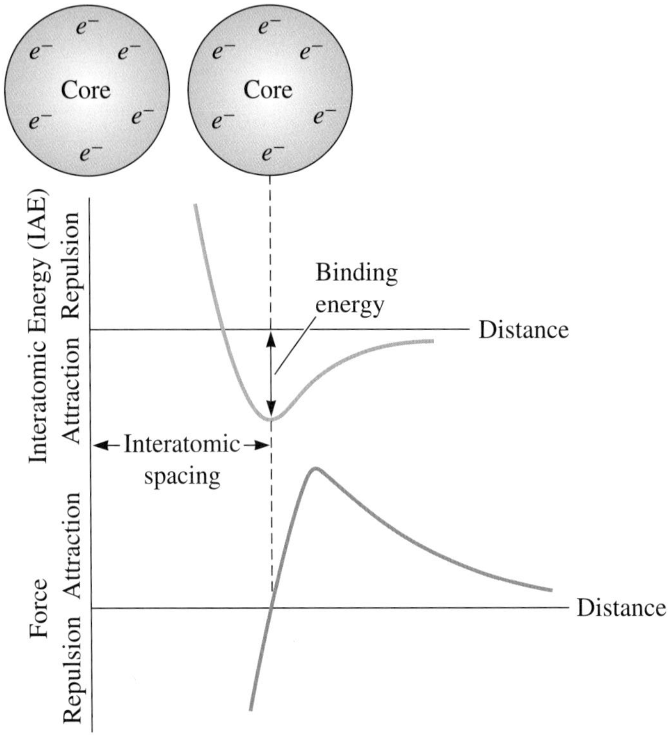

Figure 2-18 Atoms or ions are separated by an equilibrium spacing that corresponds to the minimum interatomic energy for a pair of atoms or ions (or when zero force is acting to repel or attract the atoms or ions).

bonded materials, however, since the spacing is the sum of the two different ionic radii. Atomic and ionic radii for the elements are listed in Appendix B and will be used in the next chapter.

The minimum energy in Figure 2-18 is the **binding energy**, or the energy required to create or break the bond. Consequently, materials having a high binding energy also have a high strength and a high melting temperature. Ionically bonded materials have a particularly large binding energy (Table 2-2) because of the large difference in electronegativities between the ions. Metals have lower binding energies because the electronegativities of the atoms are similar.

Other properties can be related to the force-distance and energy-distance expressions in Figure 2-18. For example, the **modulus of elasticity** of a material [the slope (E) of the stress-strain curve in the elastic region, also known as Young's modulus] is related to the slope of the force-distance curve (Figure 2-19). A steep slope, which correlates with a higher binding energy and a higher melting point, means that a greater force is required to stretch the bond; thus, the material has a high modulus of elasticity.

An interesting point that needs to be made is that not all properties of engineered materials are microstructure sensitive. Modulus of elasticity is one such property. If we have two aluminum samples that have essentially the same chemical composition but different grain size, we expect that the modulus of elasticity of these samples will be about the same; however, the **yield strengths**, the level of stress at which the material begins to permanently deform, of these samples will be quite different. The yield strength, therefore, is a microstructure sensitive property. We will learn in subsequent chapters that,

Table 2-2 Binding energies for the four bonding mechanisms

Bond	Binding Energy (kcal/mol)
Ionic	150–370
Covalent	125–300
Metallic	25–200
Van der Waals	<10

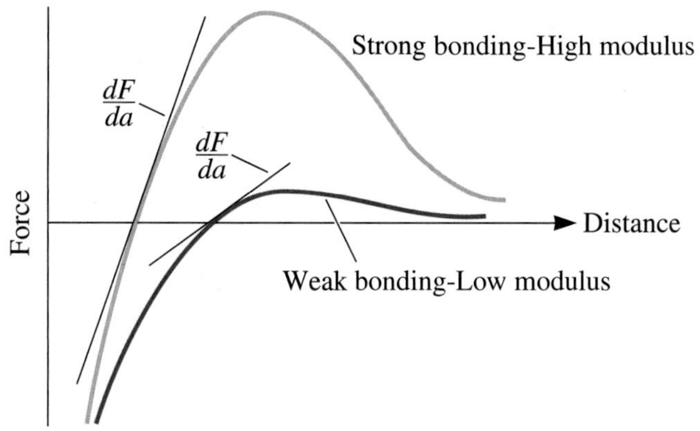

Figure 2-19 The force-distance (F–a) curve for two materials, showing the relationship between atomic bonding and the modulus of elasticity. A steep dF/da slope gives a high modulus.

compared to other mechanical properties such as yield strength and tensile strength, the modulus of elasticity does not depend strongly on the microstructure. The modulus of elasticity can be linked directly to the stiffness of bonds between atoms (Chapter 15). Thus, the modulus of elasticity depends primarily on the atoms that make up the material.

Another property that can be linked to the binding energy or interatomic force-distance curves is the **coefficient of thermal expansion** (CTE). The CTE, often denoted as α, is the fractional change in linear dimension of a material per degree of temperature. It can be written $\alpha = (1/L)(dL/dT)$, where L is length and T is temperature. The CTE is related to the asymmetry of the energy-separation curve (Figure 2-20). In order for the equilibrium separation between atoms to increase, energy must be supplied to the material. For a symmetric energy-separation curve, as energy is supplied, the average equilibrium separation between atoms remains approximately the same. Such a material has a low coefficient of thermal expansion. Materials with a low coefficient of thermal expansion maintain their dimensions more closely when the temperature changes. A material with a large thermal-expansion coefficient has an asymmetric energy-separation curve. As energy is supplied, the average equilibrium separation between the atoms increases. It is important to note that there are microstructural features (e.g., *anisotropy*, or varying properties in different crystallographic directions, in thermal expansion) that also have a significant effect on the overall thermal-expansion coefficient of an engineered material.

Materials that have very low expansion are useful in many applications where the components are expected to repeatedly undergo relatively rapid heating and cooling. For example, cordierite ceramics (used as catalyst support in catalytic

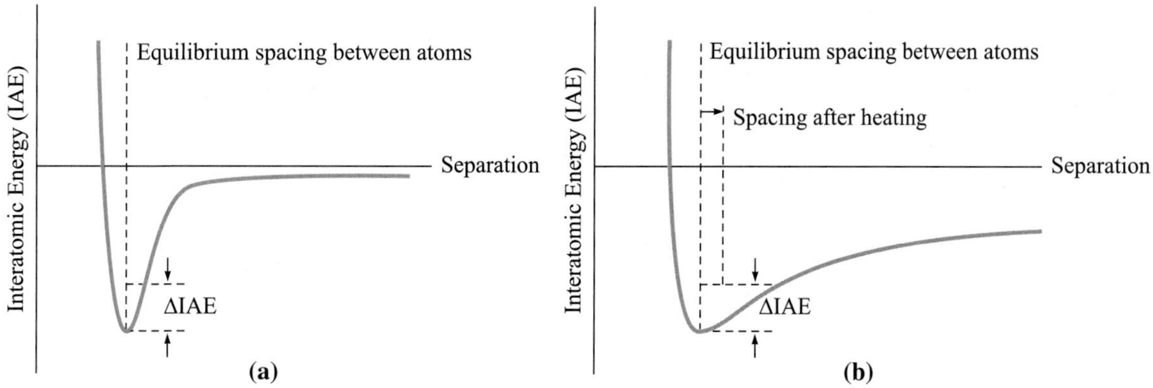

Figure 2-20 The thermal-expansion coefficient is related to the shape of the interatomic energy (IAE)—separation curve for two atoms. (a) A symmetric curve has a low (as shown, essentially zero) thermal-expansion coefficient. (b) An asymmetric curve has a relatively higher thermal-expansion coefficient. The minima of these curves are shown at the same energy to highlight the symmetry/asymmetry, but typically a material with a low CTE also has a high binding energy, i.e., a high melting temperature and deep well.

converters in cars), ultra-low expansion (ULE) glasses, Visionware™, and other glass-ceramics developed by Corning, have very low thermal-expansion coefficients. In the case of thin films or coatings on substrates, we are not only concerned about the actual values of thermal-expansion coefficients but also the difference between thermal-expansion coefficients of the substrate and the film or coating. Too much difference between these causes development of stresses that can lead to delamination or warping of the film or coating.

2-7 The Many Forms of Carbon: Relationships Between Arrangements of Atoms and Materials Properties

Carbon is one of the most abundant elements on earth. Carbon is an essential component of all living organisms, and it has enormous technological significance with a wide range of applications. For example, carbon dating is a process by which scientists measure the amount of a radioactive isotope of carbon that is present in fossils to determine their age, and now some of the most cutting-edge technologies exploit one of the world's strongest materials: carbon nanotubes. And of course, a small amount of carbon (e.g., 0.5 wt%) converts iron into steel.

Pure carbon exists as several **allotropes**, meaning that pure carbon exists in different forms (or has different arrangements of its atoms) depending on the temperature and pressure. We will learn more about allotropes in Chapter 3. Two allotropes of carbon are very familiar to us: diamond and graphite, while three other forms of carbon have been discovered much more recently: single layers of carbon atoms known as graphene, buckminsterfullerene also known as "buckyballs," and carbon nanotubes. In fact, there are other allotropes of carbon that will not be discussed here.

The allotropes of carbon all have the same composition—they are pure carbon—and yet they display dramatically different materials properties. The key to understanding these differences is to understand how the atoms are arranged in each allotrope.

In this chapter, we learned that carbon has an atomic number of six, meaning that it has six protons. Thus, a neutral carbon atom has six electrons. Two of these electrons occupy the innermost quantum shell (completely filling it), and four electrons occupy the quantum shell with the principal quantum number $n = 2$. Each carbon atom has four valence electrons and can share electrons with up to four different atoms. Thus, carbon atoms can combine with other elements as well as with other carbon atoms. This allows carbon to form many different compounds of varying structure as well as several allotropes of pure carbon.

Popular culture values one of the allotropes of carbon above all others—the diamond. Figure 2-21(a) is a diagram showing the repeat unit of the structure of diamond. Each sphere is an atom, and each thick line represents a covalent bond between carbon atoms. We will learn to make diagrams such as this in Chapter 3. Diamond is a crystal, meaning that its atoms are arranged in a regular, repeating array. Figure 2-21(a) shows that each carbon atom is bonded to four other carbon atoms. These bonds are covalent, meaning that each carbon atom shares each one of its outermost electrons with an adjacent carbon atom, so each atom has a full outermost quantum shell. Recall from Section 2-5 that covalent bonds are strong bonds. Figure 2-21(b) again shows the diamond structure, but the view has been rotated from Figure 2-21(a) by 45 degrees around a vertical axis. From this view, we see hexagonal-shaped channels in the structure of diamond.

The predominantly covalent bonding in diamond profoundly influences its macroscopic properties. Diamond is one of the highest melting-point materials known with a melting temperature of 3550°C (6420°F). This is due to the strong covalent bonding between atoms. Diamond also has one of the highest known thermal conductivities [2000 W/(m-K)]. For comparison, aluminum (which is an excellent thermal conductor) has

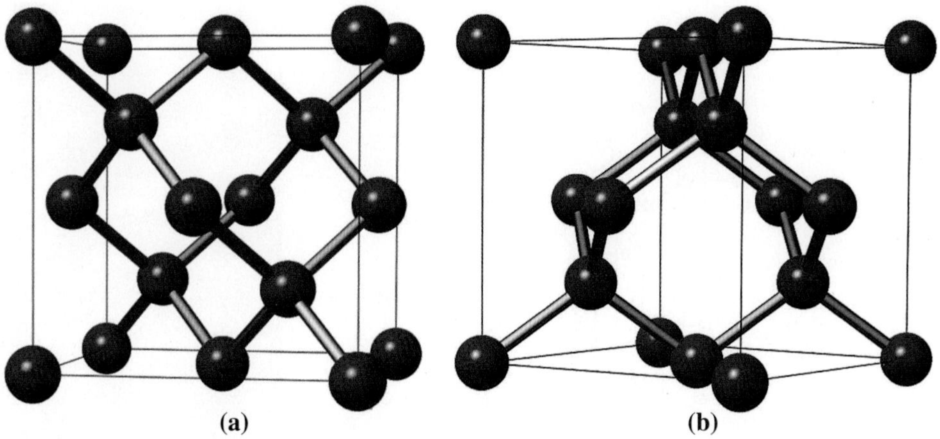

Figure 2-21 The structure of diamond. (a) The repeat unit of the diamond crystal, also known as the unit cell of diamond. (b) The same image as (a) rotated by 45° around a vertical axis.

a thermal conductivity of only 238 W/(m-K). The high thermal conductivity of diamond is due to the rigidity of its covalently bonded structure. Diamond is the stiffest material with an elastic modulus of 1100 GPa. (In Chapter 6, we will learn more about the elastic modulus; for now, we will simply say that diamond is about ten times stiffer than titanium and more than fifteen times stiffer than aluminum.) As a material is heated, the atoms vibrate with more energy. When the bonds are stiff, the vibrations are transferred efficiently between atoms, thereby conducting heat. On the other hand, diamond is an electrical insulator. All of the valence electrons of each carbon atom are shared with the neighboring atoms, leaving no free electrons to conduct electricity. (Usually an electrical insulator is also a poor conductor of heat because it lacks free electrons, but diamond is an exception due to its extraordinary stiffness.) Diamond is one of the hardest substances known, which is why it is often used in cutting tools in industrial applications (the cutting surface needs to be harder than the material being cut).

Diamond's less illustrious (and lustrous) relative is graphite. Graphite, like pure diamond, contains only carbon atoms, but we know from the experience of writing with graphite pencils that the properties of graphite are significantly different from that of diamond. In graphite, the carbon atoms are arranged in layers. In each layer, the carbon atoms are arranged in a hexagonal pattern, as shown in Figure 2-22(a). Recall that in diamond, each carbon atom is covalently bonded to four others, but in each graphite layer, each carbon atom is bonded covalently to only three others. There is a fourth bond between the layers, but this is a much weaker van der Waals bond. Also, the spacing between the graphite layers is 2.5 times larger than the spacing between the carbon atoms in the plane.

Figure 2-22(b) is an image acquired using an instrument known as an atomic force microscope (AFM). An AFM scans a sharp tip over the surface of a sample. The deflection of the cantilever tip is tracked by a laser and position-sensitive photodetector. In contact mode AFM, an actuator moves the sample with respect to the tip in order to maintain a constant deflection. In this way, the surface of the sample is mapped as a function of height. Figure 2-22(b) shows the carbon atoms in a single graphite layer. Individual carbon atoms are visible. We see that although our ball and stick models of crystals may seem somewhat crude, they are, in fact, accurate representations of the atomic arrangements in materials.

Like diamond, graphite has a high melting point. To heat a solid to the point at which it becomes a liquid, the spacing between atoms must be increased. In graphite, it is not difficult to separate the individual layers, because the bonds between the layers are weak (in fact, this is what you do when you write with a graphite pencil—layers are separated and

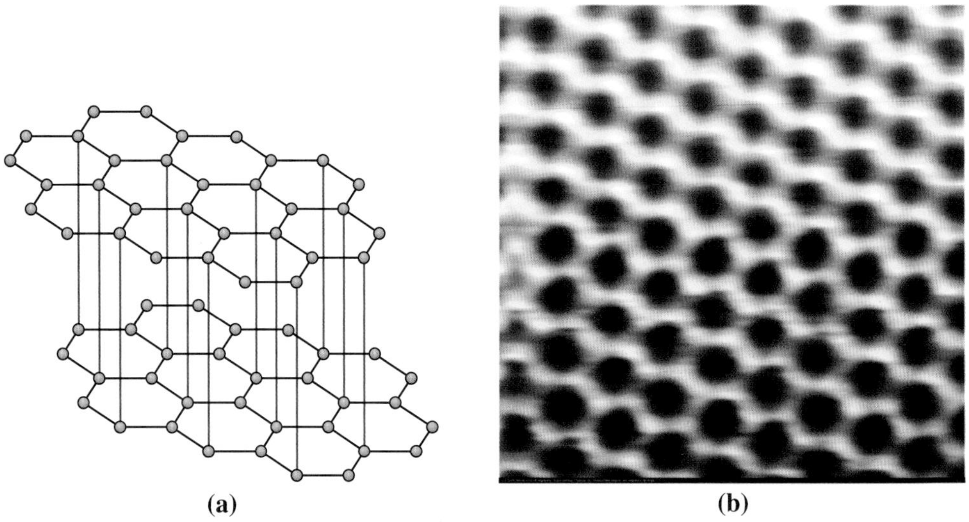

Figure 2-22 The structure of graphite. (a) The carbon atoms are arranged in layers, and in each layer, the carbon atoms are arranged in a hexagonal pattern. (b) An atomic force micrograph of graphite. (*University of Augsburg*)

left behind on your paper), but each carbon atom has three strong bonds in the layer that cause the graphite to have a high melting point. Graphite has a lower density than diamond because of its layer structure—the atoms are not packed as closely together. Unlike diamond, graphite is electrically conductive. This is because the fourth electron of each carbon atom, which is not covalently bonded in the plane, is available to conduct electricity.

A single layer of graphite is known as graphene and represents a distinct allotrope of carbon. It is an excellent thermal conductor, and like graphite, graphene is electrically conductive and strong. Notably, because it comprises a single layer, graphene is transparent. There is great excitement about graphene's potential to be used in transistors since electrons in graphene travel at speeds close to the speed of light. Graphene may also be used for liquid crystal display (LCD) screens. Since Figure 2-22(b) shows the surface of graphite, the appearance is nearly identical to graphene.

Buckminsterfullerene, an allotrope of carbon, was discovered in 1985. Each molecule of buckminsterfullerene or "buckyball" contains sixty carbon atoms and is known as C60. A buckyball can be envisioned by considering a two-dimensional pattern of twelve regular pentagons and twenty regular hexagons arranged as in Figure 2-23(a). If this pattern is folded into a three-dimensional structure by wrapping the center row into a circle and folding each end over to form end caps, then the polygons fit together perfectly—like a soccer ball! This is a highly symmetrical structure with sixty corners, and if a carbon atom is placed at each corner, then this is a model for the C60 molecule.

Buckminsterfullerene was named after the American mathematician and architect R. Buckminster Fuller who patented the design of the geodesic dome. Passing a large current of about 150 amps through a carbon rod creates buckyballs. Buckyballs are found in soot created in this fashion as well as in naturally occurring soot, like the carbon residue from a burning candle.

Figure 2-23(b) is a model of a single buckyball. Each of the 60 carbon atoms has two single bonds and one double bond. In fact, there are forms other than C60, e.g., C70, that form a class of carbon materials known generally as fullerenes. Buckyballs can enclose other atoms within them, appear to be quite strong, and have interesting magnetic and electronic properties.

Carbon nanotubes, a fourth allotrope of carbon, can be envisioned as sheets of graphite rolled into tubes with hemispherical fullerene caps on the ends. A single sheet of graphite, or graphene, can be rolled in different directions to produce nanotubes with different configurations, as shown in Figure 2-24(a). Carbon nanotubes may be

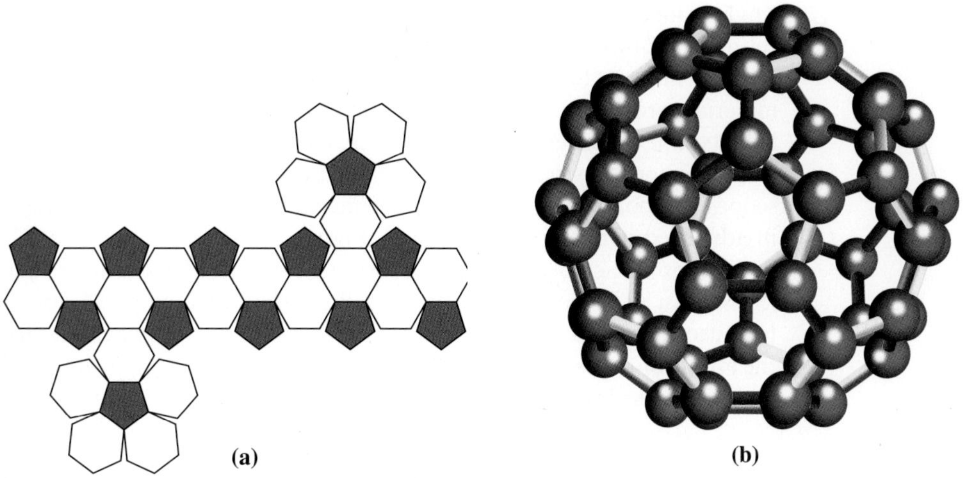

Figure 2-23 The structure of the buckyball. (a) The formation of buckminsterfullerene from twelve regular pentagons and twenty regular hexagons. (b) A "ball and stick" model of the C60 molecule. (*lculig/Shutterstock.com*)

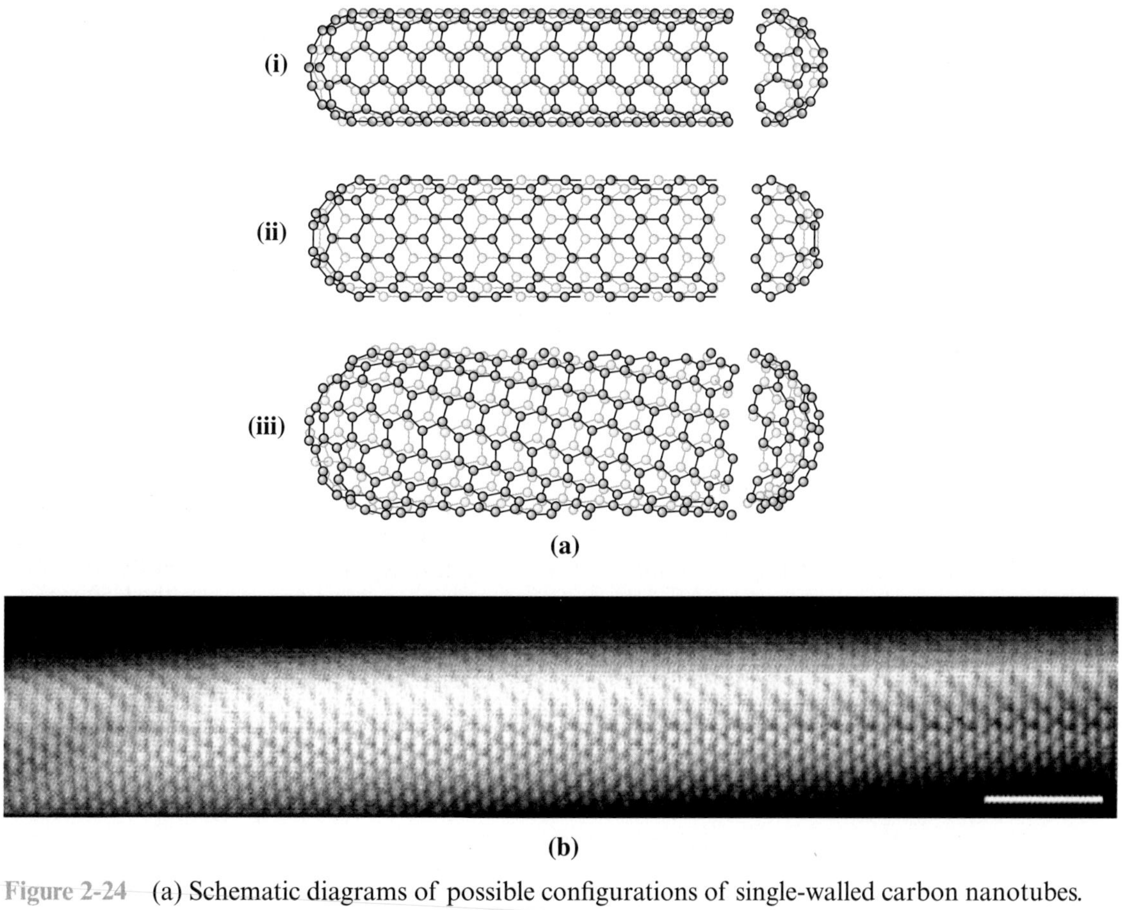

Figure 2-24 (a) Schematic diagrams of possible configurations of single-walled carbon nanotubes. (i) An armchair nanotube; (ii) a zigzag nanotube; (iii) a nanotube that is neither an armchair nanotube nor a zigzag nanotube is known as a chiral nanotube. (*Courtesy of Figure 6 from* Carbon Nanomaterials *by Andrew R. Barron.*) (b) A scanning tunneling micrograph of a carbon nanotube. [*Reprinted by permission from Macmillan Publishers Ltd:* Nature. *2004 Nov 18; 432(7015), "Electrical generation and absorption of phonons in carbon nanotubes," LeRoy et al, copyright 2004.*]

single-walled or multi-walled. Multi-walled carbon nanotubes consist of multiple concentric nanotubes. Carbon nanotubes are typically 1 to 25 nm in diameter and on the order of microns long. Carbon nanotubes with different configurations display different materials properties. For example, the electrical properties of nanotubes depend on the helicity and diameter of the nanotubes. Carbon nanotubes are currently being used as reinforcement to strengthen and stiffen polymers and as tips for atomic force microscopes. Carbon nanotubes also are being considered as possible conductors of electricity in advanced nanoelectronic devices. Figure 2-24(b) is an image of a single-walled carbon nanotube acquired using an instrument known as a scanning tunneling microscope (STM). An STM scans a sharp tip over the surface of a sample. A voltage is applied to the tip. Electrons from the tip tunnel or "leak" to the sample when the tip is in proximity to the atoms of the sample. The resulting current is a function of the tip to sample distance, and measurements of the current can be used to map the sample surface.

Summary

- Similar to composition, the structure of a material has a profound influence on the properties of a material.
- The structure of materials can be understood at various levels: atomic structure, long- and short-range atomic arrangements, nanostructure, microstructure, and macrostructure. Engineers concerned with practical applications need to understand the structure at both the micro and macro levels. Given that atoms and atomic arrangements constitute the building blocks of advanced materials, we need to understand the structure at an atomic level. There are many emerging novel devices centered on micro-electro-mechanical systems (MEMS) and nanotechnology. As a result, understanding the structure of materials at the nanoscale is also very important for some applications.
- The electronic structure of the atom, which is described by a set of four quantum numbers, helps determine the nature of atomic bonding, and, hence, the physical and mechanical properties of materials.
- Atomic bonding is determined partly by how the valence electrons associated with each atom interact. Types of bonds include metallic, covalent, ionic, and van der Waals. Most engineered materials exhibit mixed bonding.
- A metallic bond is formed as a result of atoms of low electronegativity elements donating their valence electrons to the formation of a "sea" of electrons. Metallic bonds are non-directional and relatively strong. As a result, most pure metals show a high Young's modulus and are ductile. They are good conductors of heat and electricity and reflect visible light.
- A covalent bond is formed when electrons are shared between two atoms. Covalent bonds are found in many polymeric and ceramic materials. These bonds are strong, and most inorganic materials with covalent bonds exhibit high levels of strength, hardness, and limited ductility. Most plastic materials based on carbon-carbon and carbon-hydrogen bonds show relatively lower strengths and good levels of ductility. Most covalently bonded materials tend to be relatively good electrical insulators. Some covalently bonded materials such as Si and Ge are semiconductors.
- The ionic bonding found in many ceramics is produced when an electron is "donated" from one electropositive atom to an electronegative atom, creating positively charged cations and negatively charged anions. As in covalently bonded materials, these materials tend to be mechanically strong and hard, but brittle. Melting points of ionically bonded materials are relatively high. These materials are typically electrical insulators. In some cases, though, the microstructure of these materials can be tailored so that significant ionic conductivity is obtained.

- The van der Waals bonds are formed when atoms or groups of atoms have a nonsymmetrical electrical charge, permitting bonding by an electrostatic attraction. The asymmetry in the charge is a result of dipoles that are induced or dipoles that are permanent.
- The binding energy is related to the strength of the bonds and is particularly high in ionically and covalently bonded materials. Materials with a high binding energy often have a high melting temperature, a high modulus of elasticity, and a low coefficient of thermal expansion.
- Not all properties of materials are microstructure sensitive, and modulus of elasticity is one such property.
- Carbon exists as several allotropes including diamond, graphite, graphene, buckminsterfullerene, and carbon nanotubes. All are composed of pure carbon, but their materials properties differ dramatically due to the different arrangements of atoms in their structures.

Glossary

Allotropy The characteristic of an element being able to exist in more than one crystal structure, depending on temperature and pressure.

Amorphous material A material that does not have long-range order for the arrangement of its atoms (e.g., silica glass).

Anion A negatively charged ion produced when an atom, usually of a non-metal, accepts one or more electrons.

Atomic mass The mass of the Avogadro number of atoms, g/mol. Normally, this is the average number of protons and neutrons in the atom. Also called the atomic weight.

Atomic mass unit The mass of an atom expressed as 1/12 the mass of a carbon atom with twelve nucleons.

Atomic number The number of protons in an atom.

Aufbau Principle A graphical device used to determine the order in which the energy levels of quantum shells are filled by electrons.

Avogadro number The number of atoms or molecules in a mole. The Avogadro number is 6.022×10^{23} per mole.

Azimuthal quantum number A quantum number that designates different energy levels in principal shells. Also called the secondary quantum number.

Binding energy The energy required to separate two atoms from their equilibrium spacing to an infinite distance apart. The binding energy is a measure of the strength of the bond between two atoms.

Cation A positively charged ion produced when an atom, usually of a metal, gives up its valence electrons.

Coefficient of thermal expansion The fractional change in linear dimension of a material per degree of temperature.

Composition The chemical make-up of a material.

Covalent bond The bond formed between two atoms when the atoms share their valence electrons.

Crystalline materials Materials in which atoms are arranged in a periodic fashion exhibiting a long-range order.

Debye interactions Van der Waals forces that occur between two molecules, with only one molecule having a permanent dipole moment.

Directional relationship The bonds between atoms in covalently bonded materials form specific angles, depending on the material.

Ductility The ability of materials to be permanently stretched or bent without breaking.

Electronegativity The relative tendency of an atom to accept an electron and become an anion. Strongly electronegative atoms readily accept electrons.

Electropositive The tendency for atoms to donate electrons, thus being highly reactive.

Glass-transition temperature A temperature above which many polymers and inorganic glasses no longer behave as brittle materials.

Hydrogen bond A Keesom interaction (a type of van der Waals bond) between molecules in which a hydrogen atom is involved (e.g., bonds *between* water molecules).

Interatomic spacing The equilibrium spacing between the centers of two atoms.

Intermetallic compound A compound such as Al_3V comprising two or more metallic atoms.

Ionic bond The bond formed between two different atom species when one atom (the cation) donates its valence electrons to the second atom (the anion). An electrostatic attraction binds the ions together.

Keesom interactions Van der Waals forces that occur between molecules that have permanent dipole moments.

Length scale A relative distance or range of distances used to describe materials-related structure, properties, or phenomena.

London forces Van der Waals forces that occur between molecules that do not have permanent dipole moments.

Long-range atomic arrangements Repetitive three-dimensional patterns with which atoms or ions are arranged in crystalline materials.

Macrostructure Structure of a material at a macroscopic level. The length scale is $\sim >100{,}000$ nm. Typical features include porosity and internal or external microcracks.

Magnetic quantum number A quantum number that indicates the orbitals for each azimuthal quantum number.

Metallic bond The electrostatic attraction between the valence electrons and the positively charged ion cores.

Micro-electro-mechanical systems Miniaturized devices typically prepared by micromachining.

Microstructure Structure of a material at a length scale of ~100 to 100,000 nm.

Modulus of elasticity The slope of the stress-strain curve in the elastic region (E). Also known as Young's modulus.

Nanoscale A length scale of 1–100 nm.

Nanostructure Structure of a material at the nanoscale (~ length scale 1–100 nm).

Nanotechnology A set of technologies based on nanoscale devices, phenomena, and materials.

Nucleon A proton or neutron.

Pauli exclusion principle No more than two electrons in a material can have the same energy.

Polarized molecules Molecules that have developed a dipole moment by virtue of an internal or external electric field.

Primary bonds Strong bonds between adjacent atoms resulting from the transfer or sharing of outer orbital electrons.

Quantum numbers The numbers that assign electrons in an atom to discrete energy levels. The four quantum numbers are the principal quantum number n, the azimuthal quantum number l, the magnetic quantum number m_l, and the spin quantum number m_s.

Quantum shell A set of fixed energy levels to which electrons belong.

Secondary bond Weak bonds, such as van der Waals bonds, that typically join molecules to one another.

Short-range atomic arrangements Atomic arrangements up to a distance of a few nm.

Spectroscopy The science that analyzes the emission and absorption of electromagnetic radiation.

Spin quantum number A quantum number that indicates the spin of an electron.

Structure Description of spatial arrangements of atoms or ions in a material.

Transition elements A set of elements with partially filled *d* and *f* orbitals. These elements usually exhibit multiple valences and are useful for electronic, magnetic, and optical applications.

III–V semiconductor A semiconductor that is based on Group 3B and 5B elements (e.g., GaAs).

II–VI semiconductor A semiconductor that is based on Group 2B and 6B elements (e.g., CdSe).

Valence The number of electrons in an atom that participate in bonding or chemical reactions. Usually, the valence is the number of electrons in the outer *s* and *p* energy levels.

Van der Waals bond A secondary bond developed between atoms and molecules as a result of interactions between dipoles that are induced or permanent.

Yield strength The level of stress above which a material permanently deforms.

Problems

Section 2-1 The Structure of Materials: Technological Relevance

2-1 What is meant by the term *composition* of a material?

2-2 What is meant by the term *structure* of a material?

2-3 What are the different levels of structure of a material?

2-4 Why is it important to consider the structure of a material when designing and fabricating engineering components?

2-5 What is the difference between the microstructure and the macrostructure of a material?

Section 2-2 The Structure of the Atom

2-6 (a) Aluminum foil used for storing food weighs about 0.3 g per square inch. How many atoms of aluminum are contained in one square inch of the foil?
(b) Using the densities and atomic weights given in Appendix A, calculate and compare the number of atoms per cubic centimeter in (i) lead and (ii) lithium.

2-7 (a) Using data in Appendix A, calculate the number of iron atoms in one ton (2000 pounds) of iron.
(b) Using data in Appendix A, calculate the volume in cubic centimeters occupied by one mole of boron.

2-8 In order to plate a steel part having a surface area of 200 in.^{2} with a 0.002-in.-thick layer of nickel: (a) How many atoms of nickel are required? (b) How many moles of nickel are required?

Section 2-3 The Electronic Structure of the Atom

2-9 Define *electronegativity*.

2-10 Write the electronic configuration of the following elements: (a) tungsten, (b) cobalt, (c) zirconium, (d) uranium, and (e) aluminum.

2-11 Write the electron configuration for the element Tc.

2-12 Assuming that the Aufbau Principle is followed, what is the expected electronic configuration of the element with atomic number $Z = 116$?

2-13 Using the Aufbau Principle, what is the expected electronic configuration of the hypothetical element with atomic number $Z = 123$?

2-14 Suppose an element has a valence of 2 and an atomic number of 27. Based only on the quantum numbers, how many electrons must be present in the 3*d* energy level?

Section 2-4 The Periodic Table

2-15 The periodic table of elements can help us better rationalize trends in properties of elements and compounds based on elements from different groups. Search the literature and obtain the coefficients of thermal expansion for the elements from Group 4B. Establish a trend and see if it correlates with the melting temperatures and other properties (e.g., bandgap) of these elements.

2-16 Bonding in the intermetallic compound Ni_3Al is predominantly metallic. Explain why there will be little, if any, ionic bonding component. The electronegativity of nickel is about 1.9.

2-17 Plot the melting temperatures of elements in the 4A to 8–10 columns of the periodic table versus atomic number (i.e., plot melting temperatures of Ti through Ni, Zr through Pd, and Hf through Pt). Discuss these relationships, based on atomic bonding and binding energies: (a) as the atomic number increases in each row of the periodic table and (b) as the atomic number increases in each column of the periodic table.

2-18 Plot the melting temperature of the elements in the 1A column of the periodic table versus atomic number (i.e., plot melting temperatures of Li through Cs). Discuss this relationship, based on atomic bonding and binding energy.

Section 2-5 Atomic Bonding

2-19 Compare and contrast metallic and covalent primary bonds in terms of (a) the nature of the bond, (b) the valence of the atoms involved, and (c) the ductility of the materials bonded in these ways.

2-20 Differentiate the three principle bonding mechanisms in solids. What is van der Waal's bonding? What are the relative binding energies of the different mechanisms?

2-21 What type of bonding does KCl have? Fully explain your reasoning by referring to the electronic structure and electronic properties of each element.

2-22 The compound aluminum phosphide (AlP) is a compound semiconductor having mixed ionic and covalent bonding. Calculate the fraction of the bonding that is ionic.

2-23 Calculate the fraction of bonding of MgO that is ionic.

2-24 Calculate the fraction of bonding that is covalent for silica (SiO_2).

2-25 Calculate the fraction of bonding that is ionic in nature for zirconia (ZrO_2).

2-26 What is the type of bonding in diamond? Are the properties of diamond commensurate with the nature of the bonding?

2-27 What are the bonding mechanisms in thermoplastics?

2-28 Why are covalently bonded materials generally less dense than those that are bonded ionically or metallically?

2-29 Materials such as silicon carbide (SiC) and silicon nitride (Si_3N_4) are used for grinding and polishing applications. Rationalize the choice of these materials for this use.

2-30 Explain the role of van der Waals forces in determining the properties of PVC plastic.

2-31 Calculate the fractions of ionic bonds in silicon carbide (SiC) and in silicon nitride (Si_3N_4).

2-32 One particular form of boron nitride (BN) known as cubic boron nitride is a very hard material and is used in grinding applications. Calculate the fraction of the bonding that is covalent in this material.

2-33 One form of boron nitride (BN) has properties similar to diamond. Another form of boron nitride known as hexagonal boron nitride is used as a solid lubricant. Explain how this may be possible by comparing this situation with that encountered in two forms of carbon, namely diamond and graphite.

2-34 Is there a trend in the number of electrons in the outermost energy shell of atoms or ions that have formed bonds?

Section 2-6 Binding Energy and Interatomic Spacing

Section 2-7 The Many Forms of Carbon: Relationships Between Arrangements of Atoms and Materials Properties

2-35 In order to increase the operating temperature of an engine, it is suggested that some of the aluminum components be coated with a ceramic. What kinds of problems could this pose? How could you overcome these problems?

2-36 Aluminum and silicon are side by side on the periodic table. Compare the melting temperatures of the two elements and explain the difference in terms of atomic bonding.

2-37 Titanium is stiffer than aluminum, has a lower thermal-expansion coefficient than aluminum, and has a higher melting temperature than aluminum. On the same graph, carefully and schematically draw the potential well curves for both metals. Be explicit in showing how the physical properties are manifest in these curves.

2-38 Would you expect iron or silicon nitride (SiN) to have a higher modulus of elasticity? Why?

2-39 Beryllium and magnesium, both in the 2A column of the periodic table, are lightweight metals. Which would you expect to have the higher modulus of elasticity? Explain, considering binding energy and atomic radii and using appropriate sketches of force versus interatomic spacing.

2-40 Boron has a much lower coefficient of thermal expansion than aluminum, even though both are in the 3B column of the periodic table. Schematically draw the potential wells of both, highlighting the expected differences between them.

2-41 Would you expect MgO or magnesium to have the higher modulus of elasticity? Explain.

2-42 Would you expect Al_2O_3 or aluminum to have the higher coefficient of thermal expansion? Explain.

2-43 Aluminum and silicon are side-by-side in the periodic table. Which would you expect to have the higher modulus of elasticity? Explain.

2-44 Explain why the modulus of elasticity of simple thermoplastic polymers, such as polyethylene and polystyrene, is expected to be very low compared to that of metals and ceramics.

2-45 Steel is coated with a thin layer of ceramic to help protect against corrosion. What do you expect to happen to the coating when the temperature of the steel is increased significantly? Explain.

2-46 Why is the modulus of elasticity considered a relatively structure-insensitive property?

2-47 An aluminum-alloy bar of length 2 meters at room temperature (300 K) is exposed to a temperature of 100°C ($\alpha = 25 \times 10^{-6}$ K^{-1}). What will be the length of this bar at 100°C?

2-48 Name at least four allotropes of carbon. Why is graphite electrically conductive while diamond is not if both are pure forms of carbon?

2-49 Bond hybridization in carbon leads to numerous crystalline forms. With only six electrons, how is this possible? Explain.

2-50 Describe the technological significance of each of the allotropes of carbon discussed in this section. What is the enabling materials property of each allotrope?

Design Problems

2-51 You wish to introduce ceramic fibers into a metal matrix to produce a composite material, which is subjected to high forces and large temperature changes. What design parameters might you consider to ensure that the fibers will remain intact and provide strength to the matrix? What problems might occur?

2-52 Turbine blades used in jet engines can be made from such materials as nickel-based superalloys. We can, in principle, even use ceramic materials such as zirconia or other alloys based on steels. In some cases, the blades also may have to be coated with a thermal barrier coating to minimize exposure of the blade material to high temperatures. What design parameters would you consider in selecting a material for the turbine blade and for the coating that would work successfully in a turbine engine? Note that different parts of the engine are exposed to different temperatures, and not all blades are exposed to relatively high operating temperatures. What problems might occur? Consider factors such as temperature and humidity in the environment in which the turbine blades must function.

2-53 You want to design a material for making a mirror for a telescope that will be launched in space. Given that the temperatures in space can change considerably, what material will you consider using? Remember that this material should not expand or contract at all, if possible. It also should be as strong and have as low a density as possible, and one should be able to coat it so that it can serve as a mirror.

2-54 You want to use a material that can be used for making a catalytic converter substrate. The job of this material is to be a carrier for the nanoparticles of metals (such as platinum and palladium), which are the actual catalysts. The main considerations are that this catalyst-support material must be able to withstand the constant, cyclic heating and cooling to which it will be exposed. (*Note:* The gases from automobile exhaust reach temperatures up to 500°C, and the material will get heated up to high temperatures and then cool down when the car is not being used.) What kinds of materials can be used for this application?

K Knovel® Problems

K2-1 A 2 in.-thick steel disk with an 80 in. diameter has been plated with a 0.0009 in. layer of zinc.

- What is the area of plating in $in.^2$?
- What is the weight of zinc required in lb?
- How many moles of zinc are required?
- Name a few different methods used for zinc deposition on a steel substrate.
- Which method should be selected in this case?

Quartz, also known as silica with the chemical formula SiO_2, is the mineral found in sand. Quartz is one of the most abundant materials on earth. On a weight basis, the cost of sand is very cheap; however, silica is the material that is refined to make electronics grade silicon, one of the most pure and nearly perfect materials in the world. Silicon wafers are the substrates for microchips, such as those found in the processor of your computer. While a ton of sand may cost only tens of dollars, a ton of silicon microchips is worth billions of dollars.

The image above shows a quartz crystal. A crystalline material is one in which the atoms are arranged in a regular, repeating array. The facets of the crystals reflect the long-range order of the atomic arrangements. (*Galyna Andrushko/Shutterstock.com*)

CHAPTER

3

Atomic and Ionic Arrangements

Have You Ever Wondered?

- *What is amorphous silicon, and how is it different from the silicon used to make computer chips?*
- *What are liquid crystals?*
- *If you were to pack a cubical box with uniform-sized spheres, what is the maximum packing possible?*
- *How can we calculate the density of different materials?*

Chapter Learning Objectives

The key objectives of this chapter are to

- Visualize how the atoms are arranged in the face-centered cubic, body-centered cubic, and hexagonal close-packed crystal structures and define the spatial relationships among the atoms in these structures. Given the parameters for a crystal structure other than those listed, visualize the atomic arrangements.
- Calculate the properties of crystalline materials based on the crystal structure (e.g., mass density and the relative sizes of interstitial sites).
- Identify close-packed planes and directions in a crystal structure.
- Identify and use Miller indices for directions and planes in a crystal coordinate system.

Arrangements of atoms and ions play an important role in determining the microstructure and properties of a material. The main objectives of this chapter are to

(a) learn to classify materials based on atomic/ionic arrangements; and
(b) describe the arrangements in crystalline solids according to the concepts of the **lattice, basis**, and **crystal structure**.

For crystalline solids, we will illustrate the concepts of Bravais lattices, unit cells, and crystallographic directions and planes by examining the arrangements of atoms or ions in many technologically important materials. These include metals (e.g., Cu, Al, Fe, W, Mg), semiconductors (e.g., Si, Ge, GaAs), advanced ceramics (e.g., ZrO_2, Al_2O_3, $BaTiO_3$), diamond, and other materials. We will develop the nomenclature used to characterize atomic or ionic arrangements in crystalline materials. We will examine the use of **x-ray diffraction** (XRD), **transmission electron microscopy** (TEM), and **electron diffraction**. These techniques allow us to probe the arrangements of atoms/ions in different materials. We will present an overview of different types of **amorphous materials** such as amorphous silicon, metallic glasses, polymers, and inorganic glasses.

Chapter 2 highlighted how interatomic bonding influences certain properties of materials. This chapter will underscore the influence of atomic and ionic arrangements on the properties of engineered materials. In particular, we will concentrate on "perfect" arrangements of atoms or ions in crystalline solids.

The concepts discussed in this chapter will prepare us for understanding how *deviations* from these perfect arrangements in crystalline materials create what are described as atomic level defects. The term **defect** in this context refers to a lack of perfection in atomic or ionic order of crystals and not to any flaw in an engineered material. In Chapter 4, we will describe how these atomic level defects actually enable the development of formable, strong steels used in cars and buildings, aluminum alloys for aircraft, solar cells and photovoltaic modules for satellites, and many other technologies.

3-1 Short-Range Order versus Long-Range Order

In different states of matter, we can find four types of atomic or ionic arrangements (Figure 3-1).

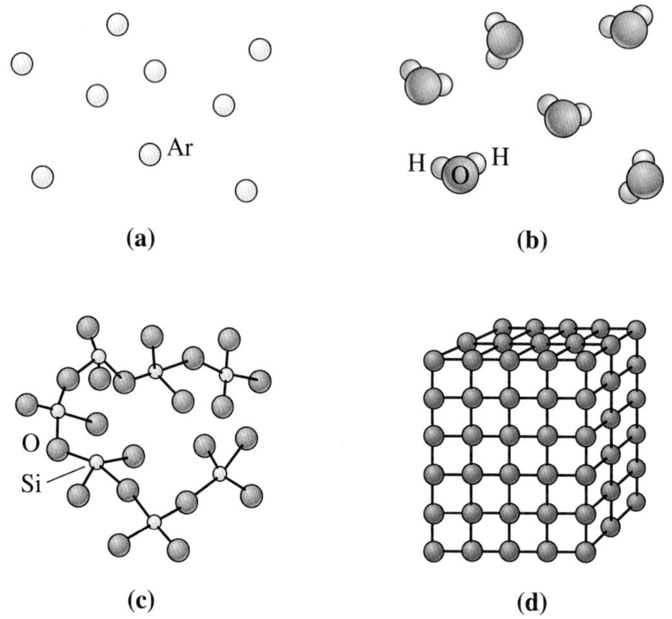

Figure 3-1 Levels of atomic arrangements in materials: (a) Inert monoatomic gases have no regular ordering of atoms. Some materials, including water vapor (b) and silicate glass (c), have short-range order. (d) Metals, alloys, many ceramics, and some polymers have regular ordering of atoms/ions that extends through the material.

No Order In monoatomic gases, such as argon or plasma created in a fluorescent tubelight, atoms or ions have no orderly arrangement.

Short-Range Order A material displays **short-range order (SRO)** if the special arrangement of the atoms extends only to the atom's nearest neighbors. Each water molecule in steam has short-range order due to the covalent bonds between the hydrogen and oxygen atoms; that is, each oxygen atom is joined to two hydrogen atoms, forming an angle of 104.5° between the bonds. There is no long-range order, however, because the water molecules in steam have no special arrangement with respect to each other's position.

A similar situation exists in materials known as inorganic glasses. In Section 2-5, we described the tetrahedral structure in silica that satisfies the requirement that four oxygen ions be bonded to each silicon ion. As will be discussed later, individual tetrahedral units are joined together in a random manner in a glass. These tetrahedra may share corners, edges, or faces. Thus, beyond the basic unit of a $(SiO_4)^{4-}$ tetrahedron, there is no periodicity in their arrangement. In contrast, in quartz or other forms of crystalline silica, the $(SiO_4)^{4-}$ tetrahedra are indeed connected in different periodic arrangements.

Many polymers also display short-range atomic arrangements that closely resemble the silicate glass structure. Polymers will be discussed in Chapter 16.

Long-Range Order Most metals and alloys, semiconductors, ceramics, and some polymers have a crystalline structure in which the atoms or ions display **long-range order (LRO)**; the special atomic arrangement extends over length scales $\sim >10$ nm. The atoms or ions in these materials form a regular repetitive, grid-like pattern, in three dimensions. We refer to these materials as **crystalline materials**. If a crystalline material consists of only one large crystal, we refer to it as a *single crystal*. Single crystals are useful in many electronic and optical applications. For example, computer chips are made from silicon in the form of large (up to 12-inch diameter) single crystals [Figure 3-2(a)]. Similarly, many useful optoelectronic devices are made from crystals of lithium niobate ($LiNbO_3$). Single crystals can also be made as thin films and used for many electronic and other applications. Certain types of turbine blades may be made from single crystals of nickel-based superalloys. **A polycrystalline material** is composed of many small crystals with varying orientations in space. These smaller crystals are known as **grains**. The borders between crystals, where the crystals are in misalignment, are known as **grain boundaries**. Figure 3-2(b) shows the microstructure of a polycrystalline stainless steel material.

(a)

(b)

Figure 3-2 (a) Photograph of a silicon single crystal. (*Petr Sobolev/iStock/Thinkstock*) (b) Micrograph of a polycrystalline stainless steel showing grains and grain boundaries. (*Micrograph courtesy of Dr. A.J. Deardo, Dr. M. Hua and Dr. J. Garcia*)

Many crystalline materials we deal with in engineering applications are polycrystalline (e.g., steels used in construction, aluminum alloys for aircrafts, etc.). We will learn in later chapters that many properties of polycrystalline materials depend upon the physical and chemical characteristics of both grains and grain boundaries. The properties of single crystal materials depend upon the chemical composition and specific directions within the crystal (known as the crystallographic directions). Long-range order in crystalline materials can be detected and measured using techniques such as **x-ray diffraction** or **electron diffraction** (see Section 3-9).

Liquid crystals (LCs) are polymeric materials that have a special type of order. Liquid crystal polymers behave as amorphous materials (liquid-like) in one state. When an external stimulus (such as an electric field or a temperature change) is provided, some polymer molecules undergo alignment and form small regions that are crystalline, hence the name "liquid crystals." These materials have many commercial applications in liquid crystal display (LCD) technology.

3-2 Amorphous Materials

Any solid that exhibits only a short-range order of atoms or ions is an **amorphous material**; that is, a noncrystalline one. In general, most materials want to form periodic arrangements since this configuration maximizes the thermodynamic stability of the material. Amorphous materials tend to form when, for one reason or other, the kinetics of the process by which the material was made did not allow for the formation of periodic arrangements. **Glasses**, which typically form in ceramic and polymer systems, are amorphous materials. Similarly, certain types of polymeric or colloidal gels, or gel-like materials, are also considered amorphous. Amorphous materials often offer a unique blend of properties since the atoms or ions are not assembled into their "regular" and periodic arrangements. Note that often many engineered materials labeled as "amorphous" may contain a fraction that is crystalline.

Similar to inorganic glasses, many plastics are amorphous. During processing, relatively large chains of polymer molecules get entangled with each other, like spaghetti. Entangled polymer molecules do not organize themselves into crystalline materials. During processing of polymeric beverage bottles, mechanical stress is applied to the preform of the bottle [e.g., the manufacturing of a standard 2-liter soft drink bottle using polyethylene terephthalate (PET plastic)]. This process is known as **blow-stretch forming**. The radial (blowing) and longitudinal (stretching) stresses during bottle formation actually untangle some of the polymer chains, causing **stress-induced crystallization**. The formation of crystals adds to the strength of the PET bottles.

Compared to plastics and inorganic glasses, metals and alloys tend to form crystalline materials rather easily. As a result, special efforts must be made to quench the metals and alloys quickly in order to prevent crystallization; for some alloys, a cooling rate of $>10^6$°C/s is required to form **metallic glasses**. This technique of cooling metals and alloys very fast is known as **rapid solidification**. Other special compositions require cooling rates of only tens of degrees per second or less. Many metallic glasses have both useful and unusual properties. The mechanical properties of metallic glasses will be discussed in Chapter 6.

3-3 Lattice, Basis, Unit Cells, and Crystal Structures

A typical solid contains on the order of 10^{23} atoms/cm^3. In order to communicate the spatial arrangements of atoms in a crystal, it is clearly not necessary or practical to specify the position of each atom. We will discuss two complementary methodologies for simply describing the three-dimensional arrangements of atoms in a crystal. We will refer to these as the **lattice and basis concept** and the **unit cell** concept. These concepts rely on the principles of **crystallography**. In Chapter 2, we discussed the structure of the atom. An atom consists of a nucleus of protons and neutrons surrounded by electrons, but for the purpose of describing the arrangements of atoms in a solid, we will envision the atoms as hard spheres, much like ping-pong balls. We will begin with the lattice and basis concept.

A lattice is a collection of points, called **lattice points**, which are arranged in a periodic pattern so that the surroundings of each point in the lattice are identical. A lattice is a purely mathematical construct and is infinite in extent. A lattice may be one-, two-, or three-dimensional. In one dimension, there is only one possible lattice: it is a line of points with the points separated from each other by an equal distance, as shown in Figure 3-3(a). A group of one or more atoms located in a particular way with respect to each other and associated with each lattice point is known as the **basis** or **motif**. The basis must contain at least one atom, but it may contain many atoms of one or more types. A basis of one atom is shown in Figure 3-3(b). We obtain a **crystal structure** by placing the atoms of the basis on every lattice point (i.e., crystal structure = lattice + basis), as shown in Figure 3-3(c). A hypothetical one-dimensional crystal that has a basis of two different atoms is shown in Figure 3-3(d). The larger atom is located on every lattice point with the smaller atom located a fixed distance above each lattice point. Note that it is not necessary that one of the basis atoms be located on each lattice point, as shown in Figure 3-3(e). Figures 3-3(d) and (e) represent the same one-dimensional crystal; the atoms are simply shifted relative to one another. Such a shift does not change the atomic arrangements in the crystal.

There is only one way to arrange points in one dimension such that each point has identical surroundings—an array of points separated by an equal distance as discussed above. There are five distinct ways to arrange points in two dimensions

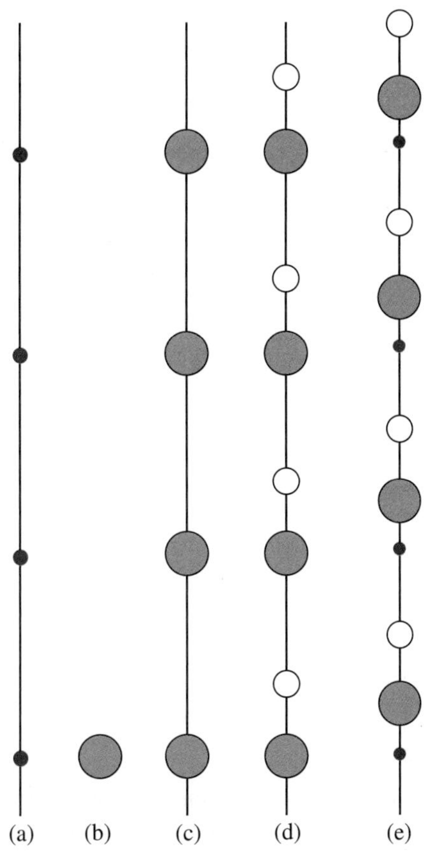

Figure 3-3 Lattice and basis. (a) A one-dimensional lattice. The lattice points are separated by an equal distance. (b) A basis of one atom. (c) A crystal structure formed by placing the basis of (b) on every lattice point in (a). (d) A crystal structure formed by placing a basis of two atoms of different types on the lattice in (a). (e) The same crystal as shown in (d); however, the basis has been shifted relative to each lattice point.

such that each point has identical surroundings; thus, there are five two-dimensional lattices. There are only fourteen unique ways to arrange points in three dimensions. These unique three-dimensional arrangements of lattice points are known as the **Bravais lattices**, named after Auguste Bravais (1811–1863), who was an early French crystallographer. The fourteen Bravais lattices are shown in Figure 3-4. As stated previously, a lattice is infinite in extent, so a single unit cell is shown for each lattice. The unit cell is a subdivision of a lattice that still retains the overall characteristics of the entire lattice. Lattice points are located at the corners of the unit cells and, in some cases, at either the faces or the center of the unit cell.

The fourteen Bravais lattices are grouped into seven **crystal systems**. The seven crystal systems are known as cubic, tetragonal, orthorhombic, rhombohedral (also known as trigonal), hexagonal, monoclinic, and triclinic. Note that for the cubic crystal system, we have simple cubic (SC), face-centered cubic (FCC), and body-centered cubic (BCC) Bravais lattices. These names describe the arrangement of lattice points in the unit cell. Similarly, for the tetragonal crystal system, we have simple tetragonal and body-centered tetragonal lattices. Again, remember that the concept of a lattice is mathematical and does not mention atoms, ions, or molecules. It is only when a basis is associated with

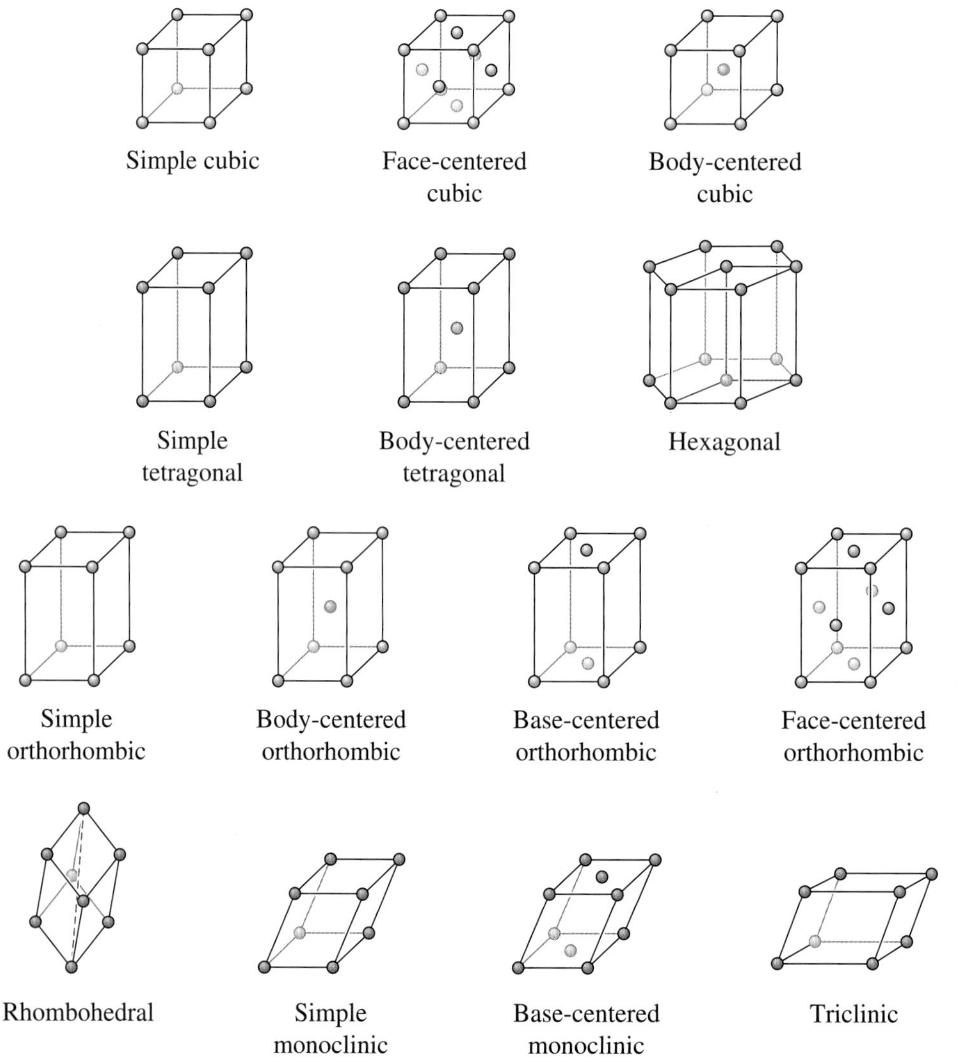

Figure 3-4 The fourteen types of Bravais lattices grouped in seven crystal systems. The actual unit cell for a hexagonal system is shown in Figures 3-6 and 3-11.

a lattice that we can describe a crystal structure. For example, if we take the face-centered cubic lattice and position a basis of one atom on every lattice point, then the face-centered cubic crystal structure is reproduced.

Note that although we have only fourteen Bravais lattices, we can have an infinite number of bases. Hundreds of different crystal structures are observed in nature or can be synthesized. Many different materials can have the same crystal structure. For example, copper and nickel have the face-centered cubic crystal structure for which only one atom is associated with each lattice point. In more complicated structures, particularly polymer, ceramic, and biological materials, several atoms may be associated with each lattice point (i.e., the basis is greater than one), forming very complex unit cells.

Unit Cell Our goal is to develop a notation to model crystalline solids that simply and completely conveys how the atoms are arranged in space. The unit cell concept complements the lattice and basis model for representing a crystal structure. Although the methodologies of the lattice and basis and unit cell concepts are somewhat different, the end result—a description of a crystal—is the same.

Our goal in choosing a unit cell for a crystal structure is to find the single repeat unit that, when duplicated and translated, reproduces the entire crystal structure. For example, imagine the crystal as a three-dimensional puzzle for which each piece of the puzzle is exactly the same. If we know what one puzzle piece looks like, we know what the entire puzzle looks like, and we don't have to put the entire puzzle together to solve it. We just need one piece! To understand the unit cell concept, we start with the crystal. Figure 3-5(a) depicts a hypothetical two-dimensional crystal that consists of atoms all of the same type.

Next, we add a grid that mimics the symmetry of the arrangements of atoms. There is an infinite number of possibilities for the grid, but by convention, we usually choose the simplest. For the square array of atoms shown in Figure 3-5(a), we choose a square grid as is shown in Figure 3-5(b). Next, we select the repeat unit of the grid, which is also known as the unit cell. This is the unit that, when duplicated and translated by integer multiples of the axial lengths of the unit cell, recreates the entire crystal. The unit cell is shown in Figure 3-5(c); note that for each unit cell, there is only one quarter of an atom at each corner in two dimensions. We will always draw full circles to represent atoms, but it is understood that only the fraction of the atom that is contained inside the unit cell contributes to the total number of atoms per unit cell. Thus, there is (1/4 atom) / corner $*$ 4 corners $=$ 1 atom per unit cell, as shown in Figure 3-5(c). It is also important to note that if there is an atom at one corner of a unit cell, there must be an atom at every corner of the unit cell in order to maintain the translational symmetry. Each unit cell has its own origin, as shown in Figure 3-5(c).

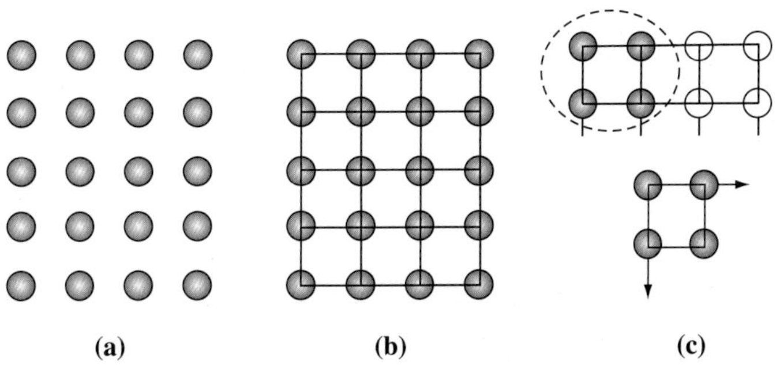

Figure 3-5 The unit cell. (a) A two-dimensional crystal. (b) The crystal with an overlay of a grid that reflects the symmetry of the crystal. (c) The repeat unit of the grid known as the unit cell. Each unit cell has its own origin.

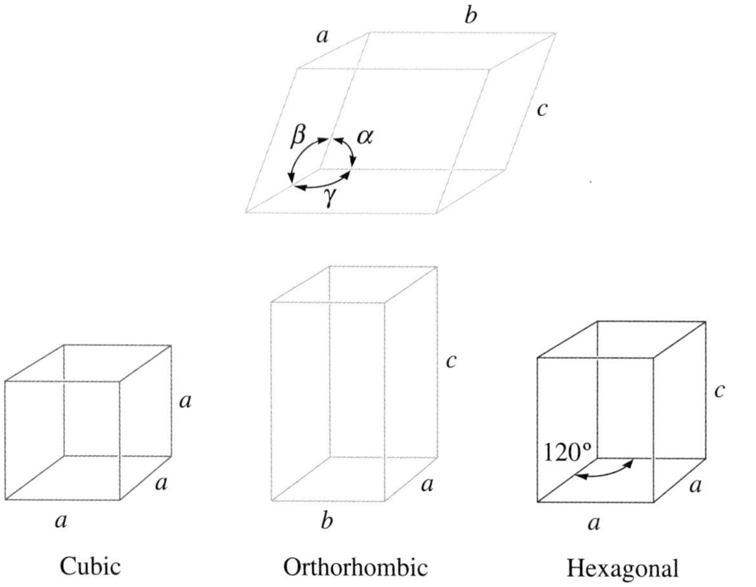

Figure 3-6 Definition of the lattice parameters and their use in cubic, orthorhombic, and hexagonal crystal systems.

Lattice Parameters and Interaxial Angles

The **lattice parameters** are the axial lengths or dimensions of the unit cell and are denoted by convention as a, b, and c. The angles between the axial lengths, known as the interaxial angles, are denoted by the Greek letters α, β, and γ. By convention, α is the angle between the lengths b and c, β is the angle between a and c, and γ is the angle between a and b, as shown in Figure 3-6. (Notice that for each combination, there is a letter a, b, and c whether it be written in Greek or Roman letters.)

In a cubic crystal system, only the length of one of the sides of the cube needs to be specified (it is sometimes designated a_0). The length is often given in nanometers (nm) or angstrom (Å) units, where

$$1 \text{ nanometer (nm)} = 10^{-9} \text{ m} = 10^{-7} \text{ cm} = 10 \text{ Å}$$
$$1 \text{ angstrom (Å)} = 0.1 \text{ nm} = 10^{-10} \text{ m} = 10^{-8} \text{ cm}$$

The lattice parameters and interaxial angles for the unit cells of the seven crystal systems are presented in Table 3-1.

Table 3-1 Characteristics of the seven crystal systems

Structure	Axes	Angles between Axes	Volume of the Unit Cell
Cubic	$a = b = c$	All angles equal 90°.	a^3
Tetragonal	$a = b \neq c$	All angles equal 90°.	a^2c
Orthorhombic	$a \neq b \neq c$	All angles equal 90°.	abc
Hexagonal	$a = b \neq c$	Two angles equal 90°. The angle between a and b equals 120°.	$0.866a^2c$
Rhombohedral or trigonal	$a = b = c$	All angles are equal and none equals 90°.	$a^3\sqrt{1 - 3\cos^2\alpha + 2\cos^3\alpha}$
Monoclinic	$a \neq b \neq c$	Two angles equal 90°. One angle (β) is not equal to 90°.	$abc \sin\beta$
Triclinic	$a \neq b \neq c$	All angles are different and none equals 90°.	$abc\sqrt{1 - \cos^2\alpha - \cos^2\beta - \cos^2\gamma + 2\cos\alpha\cos\beta\cos\gamma}$

To fully define a unit cell, the lattice parameters or ratios between the axial lengths, interaxial angles, and atomic coordinates must be specified. In specifying atomic coordinates, whole atoms are placed in the unit cell. The coordinates are specified as fractions of the axial lengths. Thus, for the two-dimensional cell represented in Figure 3-5(c), the unit cell is fully specified by the following information:

Axial lengths: $a = b$

Interaxial angle: $\gamma = 90°$

Atomic coordinate: (0, 0)

Again, only 1/4 of the atom at each origin (0, 0) contributes to the number of atoms per unit cell; however, each corner acts as an origin and contributes 1/4 atom per corner for a total of one atom per unit cell. [Do you see why with an atom at (0, 0) for each unit cell it would be repetitive to also give the coordinates of (1, 0), (0, 1), and (1, 1)?]

Similarly, a cubic unit cell with an atom at each corner is fully specified by the following information:

Axial lengths: $a = b = c$

Interaxial angles: $\alpha = \beta = \gamma = 90°$

Atomic coordinate: (0, 0, 0)

Now in three dimensions, each corner contributes 1/8 atom per each of the eight corners for a total of one atom per unit cell. Note that the number of atomic coordinates required is equal to the number of atoms per unit cell. For example, if there are two atoms per unit cell, with one atom at the corners and one atom at the body-centered position, two atomic coordinates are required: (0, 0, 0) and (1/2, 1/2, 1/2).

Number of Atoms per Unit Cell

Each unit cell contains a specific number of lattice points. When counting the number of lattice points belonging to each unit cell, we must recognize that, like atoms, lattice points may be shared by more than one unit cell. A lattice point at a corner of one unit cell is shared by eight cells; only one-eighth of each corner belongs to one particular cell. Thus, the number of lattice points from all corner positions in one unit cell is

$$\left(\frac{1/8 \text{ lattice point}}{\text{corner}}\right)\left(\frac{8 \text{ corners}}{\text{cell}}\right) = \frac{1 \text{ lattice point}}{\text{unit cell}}$$

Corners contribute 1/8 of a point, faces contribute 1/2, and body-centered positions contribute a whole point [Figure 3-7(a)].

The number of atoms per unit cell is the product of the number of atoms per lattice point and the number of lattice points per unit cell. The structures of simple cubic (SC), body-centered cubic (BCC), and face-centered cubic (FCC) unit cells are shown in Figure 3-7(b).

Atomic Radius versus Lattice Parameter

Directions in the unit cell along which atoms are in continuous contact are **close-packed directions**. In simple structures, particularly those with only one atom per lattice point, we use these directions to calculate the relationship between the apparent size of the atom and the size of the unit cell. By geometrically determining the length of the direction relative to the lattice parameters, and then adding the number of **atomic radii** along this direction, we can determine the desired relationship.

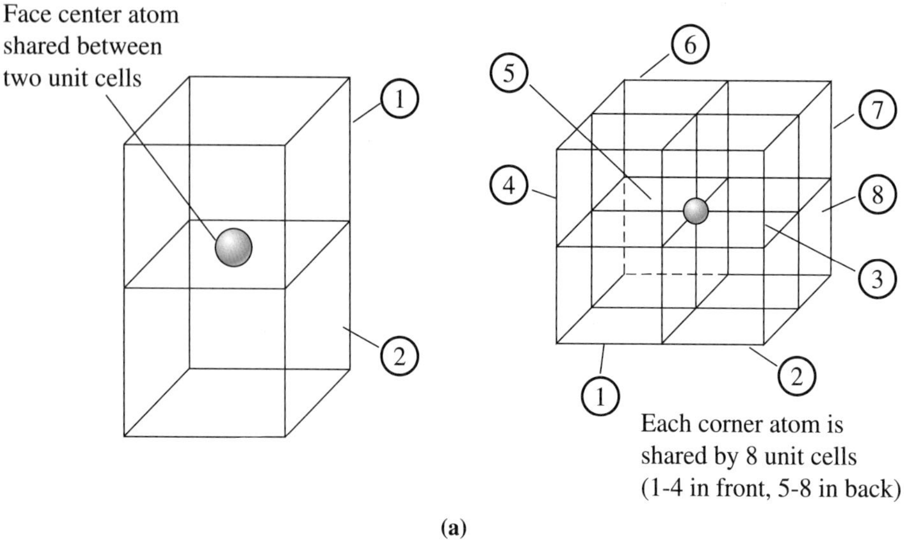

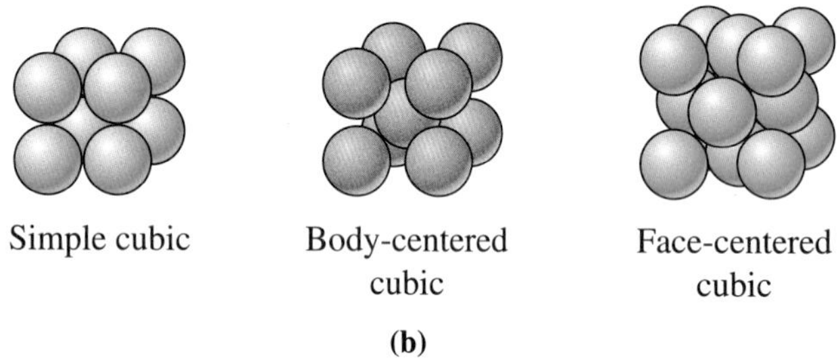

Figure 3-7 (a) Illustrations showing sharing of face and corner atoms. (b) The models for simple cubic (SC), body-centered cubic (BCC), and face-centered cubic (FCC) unit cells, assuming only one atom per lattice point.

Example 3-1 *Determining the Number of Lattice Points in Cubic Crystal Systems*

Determine the number of lattice points per unit cell in the cubic crystal systems. If there is only one atom located at each lattice point, calculate the number of atoms per unit cell.

SOLUTION

In the SC unit cell, lattice points are located only at the corners of the cube:

$$\frac{8 \text{ corners}}{\text{unit cell}} * \frac{1/8 \text{ lattice point}}{\text{corner}} = \frac{1 \text{ lattice point}}{\text{unit cell}}$$

In BCC unit cells, lattice points are located at the corners and the body-center of the cube:

$$\frac{8 \text{ corners}}{\text{unit cell}} * \frac{1/8 \text{ lattice point}}{\text{corner}} + \frac{1 \text{ body-center}}{\text{unit cell}} * \frac{1 \text{ lattice point}}{\text{body-center}} = \frac{2 \text{ lattice points}}{\text{unit cell}}$$

In FCC unit cells, lattice points are located at the corners and face-centers of the cube:

$$\frac{8 \text{ corners}}{\text{unit cell}} * \frac{1/8 \text{ lattice point}}{\text{corner}} + \frac{6 \text{ face-centers}}{\text{unit cell}} * \frac{1/2 \text{ lattice point}}{\text{face-center}} = \frac{4 \text{ lattice points}}{\text{unit cell}}$$

Since we are assuming there is only one atom located at each lattice point, the number of atoms per unit cell is one, two, and four, for the SC, BCC, and FCC unit cells, respectively.

Example 3-2 *The Cesium Chloride Structure*

Crystal structures usually are assigned names of a representative element or compound that has that structure. Cesium chloride (CsCl) is an ionic, crystalline compound. A unit cell of the CsCl crystal structure is shown in Figure 3-8. The unit cell is cubic. Chlorine anions are located at the corners of the unit cell, and a cesium cation is located at the body-centered position of each unit cell. Describe this structure as a lattice and basis and also fully define the unit cell for cesium chloride.

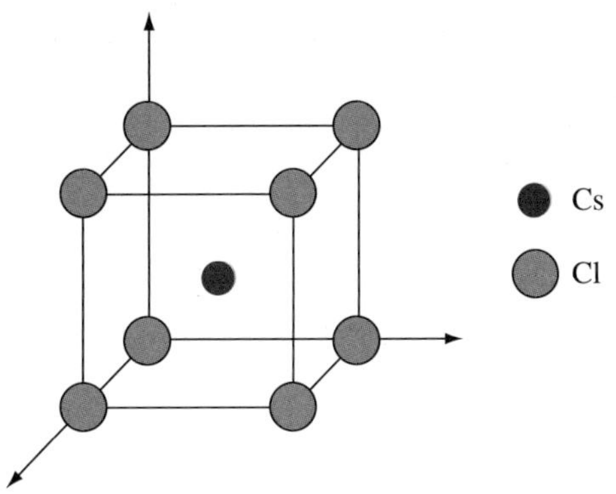

Figure 3-8 The CsCl crystal structure. *Note*: ion sizes not to scale.

SOLUTION

The unit cell is cubic; therefore, the lattice is either SC, FCC, or BCC. There are no ions located at the face-centered positions; therefore, the lattice is either SC or BCC. Each Cl anion is surrounded by eight Cs cations at the body-centered positions of the adjoining unit cells. Each Cs cation is surrounded by eight Cl anions at the corners of the unit cell. Thus, the corner and body-centered positions do not have identical surroundings; therefore, they both cannot be lattice points. The lattice must be simple cubic.

The simple cubic lattice has lattice points only at the corners of the unit cell. The cesium chloride crystal structure can be described as a simple cubic lattice with a basis of two ions, Cl (0, 0, 0) and Cs (1/2, 1/2, 1/2). Note that the coordinates are listed as fractions of the axial lengths, which for a cubic crystal structure are equal. The basis ion of Cl (0, 0, 0) placed on every lattice point (i.e., each corner of the unit cell) fully accounts for every Cl ion in the structure. The basis ion of Cs (1/2, 1/2, 1/2), located at the body-centered position with respect to each lattice point, fully accounts for every Cs ion in the structure.

Thus there are two ions per unit cell in CsCl:

$$\frac{1 \text{ lattice point}}{\text{unit cell}} * \frac{2 \text{ ions}}{\text{lattice point}} = \frac{2 \text{ ions}}{\text{unit cell}}$$

To fully define a unit cell, the lattice parameters or ratios between the axial lengths, interaxial angles, and atomic coordinates must be specified. The CsCl unit cell is cubic; therefore,

$$\text{Axial lengths: } a = b = c$$

$$\text{Interaxial angles: } \alpha = \beta = \gamma = 90°$$

The Cl anions are located at the corners of the unit cell, and the Cs cations are located at the body-centered positions. Thus,

Coordinates: Cl (0, 0, 0) and Cs (1/2, 1/2, 1/2)

Counting ions for the unit cell,

$$\frac{8 \text{ corners}}{\text{unit cell}} * \frac{1/8 \text{ Cl ion}}{\text{corner}} + \frac{1 \text{ body-center}}{\text{unit cell}} * \frac{1 \text{ Cs ion}}{\text{body-center}} = \frac{2 \text{ ions}}{\text{unit cell}}$$

As expected, the number of atoms per unit cell is the same regardless of the method used to count the atoms.

Example 3-3 Determining the Relationship between Atomic Radius and Lattice Parameters

Determine the relationship between the atomic radius and the lattice parameter in the SC, BCC, and FCC structures.

SOLUTION

If we refer to Figure 3-9, we find that atoms touch along the edge of the cube in SC structures. The corner atoms are centered on the corners of the cube, so

$$a_0 = 2r \tag{3-1}$$

In BCC structures, atoms touch along the body diagonal, which is $\sqrt{3}a_0$ in length. There are two atomic radii from the center atom and one atomic radius from each of the corner atoms on the body diagonal, so

$$a_0 = \frac{4r}{\sqrt{3}} \tag{3-2}$$

In FCC structures, atoms touch along the face diagonal of the cube, which is $\sqrt{2}a_0$ in length. There are four atomic radii along this length—two radii from the face-centered atom and one radius from each corner, so

$$a_0 = \frac{4r}{\sqrt{2}} \tag{3-3}$$

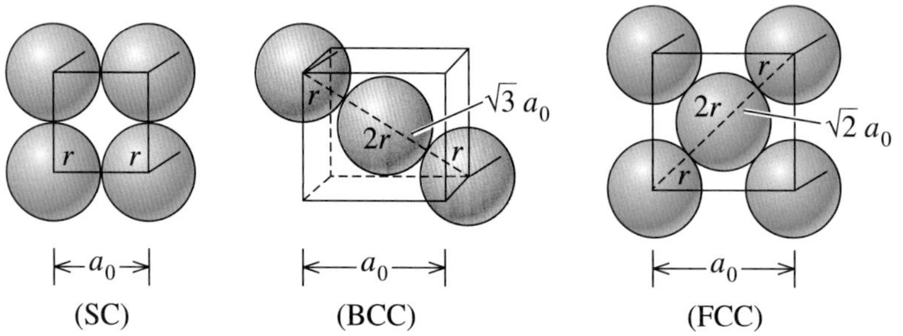

Figure 3-9 The relationships between the atomic radius and the lattice parameter in cubic systems (for Example 3-3).

The Hexagonal Lattice and Unit Cell The image of the hexagonal lattice in Figure 3-4 reflects the underlying symmetry of the lattice, but unlike the other images in Figure 3-4, it does not represent the unit cell of the lattice. The hexagonal unit cell is shown in Figure 3-6. If you study the image of the hexagonal lattice in

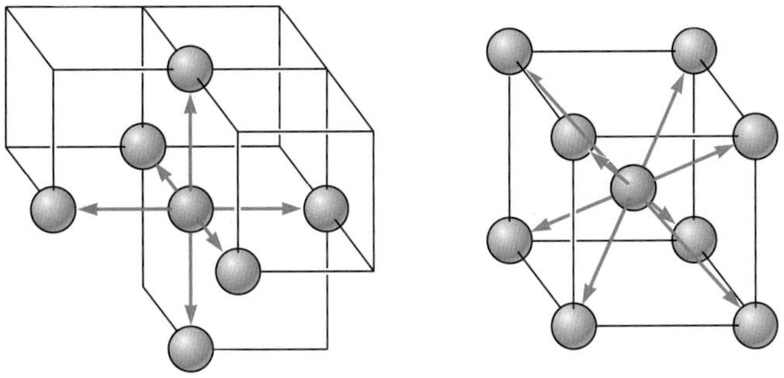

(a) (b)

Figure 3-10 Illustration of the coordination number in (a) SC and (b) BCC unit cells. Six atoms touch each atom in SC, while eight atoms touch each atom in the BCC unit cell.

Figure 3-4, you can find the hexagonal unit cell. The lattice parameters for the hexagonal unit cell are

Axial lengths: $a = b \neq c$

Interaxial angles: $\alpha = \beta = 90°$, $\gamma = 120°$

When the atoms of the unit cell are located only at the corners, the atomic coordinate is (0, 0, 0).

Coordination Number

The **coordination number** is the number of atoms touching a particular atom, or the number of nearest neighbors for that particular atom. This is one indication of how tightly and efficiently atoms are packed together. For ionic solids, the coordination number of cations is defined as the number of nearest anions. The coordination number of anions is the number of nearest cations. We will discuss the crystal structures of different ionic solids and other materials in Section 3-7.

In cubic structures containing only one atom per lattice point, atoms have a coordination number related to the lattice structure. By inspecting the unit cells in Figure 3-10, we see that each atom in the SC structure has a coordination number of six, while each atom in the BCC structure has eight nearest neighbors. In Section 3-5, we will show that each atom in the FCC structure has a coordination number of twelve, which is the maximum.

Packing Factor

The **packing factor** or **atomic packing fraction** is the fraction of space occupied by atoms, assuming that the atoms are hard spheres. The general expression for the packing factor is

$$\text{Packing factor} = \frac{(\text{number of atoms/cell})(\text{volume of each atom})}{\text{volume of unit cell}} \tag{3-4}$$

Example 3-4 illustrates how to calculate the packing factor for the FCC unit cell.

Example 3-4 *Calculating the Packing Factor*

Calculate the packing factor for the FCC unit cell.

SOLUTION

In the FCC unit cell, there are four lattice points per cell; with one atom per lattice point, there are also four atoms per cell. The volume of one atom is $4\pi r^3/3$, and the volume of the unit cell is a_0^3, where r is the radius of the atom and a_0 is the lattice parameter.

$$\text{Packing factor} = \frac{(4\ \text{atoms/cell})\left(\frac{4}{3}\pi r^3\right)}{a_0^3}$$

Since for FCC unit cells, $a_0 = 4r/\sqrt{2}$:

$$\text{Packing factor} = \frac{(4)\left(\frac{4}{3}\pi r^3\right)}{(4r/\sqrt{2})^3} = \frac{\pi}{\sqrt{18}} \cong 0.74$$

The packing factor of $\pi/\sqrt{18} \cong 0.74$ in the FCC unit cell is the most efficient packing possible. BCC cells have a packing factor of 0.68, and SC cells have a packing factor of 0.52. Notice that the packing factor is independent of the radius of atoms, as long as we assume that all atoms have a fixed radius. What this means is that it does not matter whether we are packing atoms in unit cells or packing basketballs or ping-pong balls in a cubical box. The maximum achievable packing factor is $\pi/\sqrt{18}$! This discrete geometry concept is known as **Kepler's conjecture**. Johannes Kepler proposed this conjecture in the year 1611, and it remained an unproven conjecture until 1998 when Thomas C. Hales actually proved this to be true.

The FCC arrangement represents a **close-packed structure** (CP) (i.e., the packing fraction is the highest possible with atoms of one size). The SC and BCC structures are relatively open. We will see in the next section that it is possible to have a hexagonal structure that has the same packing efficiency as the FCC structure. This structure is known as the hexagonal close-packed structure (HCP). Metals with only metallic bonding are packed as efficiently as possible. Metals with mixed bonding, such as iron, may have unit cells with less than the maximum packing factor. No commonly encountered engineering metals or alloys have the SC structure.

Density The theoretical **density** of a material can be calculated using the properties of the crystal structure. The general formula is

$$\text{Density } \rho = \frac{\text{(number of atoms/cell)(atomic mass)}}{\text{(volume of unit cell)(Avogadro number)}} \tag{3-5}$$

If a material consists of different types of atoms or ions, this formula has to be modified to reflect these differences. Example 3-5 illustrates how to determine the density of BCC iron.

Example 3-5 *Determining the Density of BCC Iron*

Determine the density of BCC iron, which has a lattice parameter of 0.2866 nm.

SOLUTION

There are two atoms per unit cell in BCC iron.

$$\text{Volume of unit cell} = (2.866 \times 10^{-8}\text{ cm})^3 = 2.354 \times 10^{-23}\text{ cm}^3$$

$$\text{Density } \rho = \frac{\text{(number of atoms/cell)(atomic mass)}}{\text{(volume of unit cell)(Avogadro number)}}$$

$$\rho = \frac{(2\text{ atoms/cell})(55.847\text{ g/mol})}{(2.354 \times 10^{-23}\text{ cm}^3\text{/cell})(6.022 \times 10^{23}\text{ atoms/mol})} = 7.879\text{ g/cm}^3$$

The measured density is 7.870 g/cm^3. The slight discrepancy between the theoretical and measured densities is a consequence of defects in the material. As mentioned before, the term "defect" in this context means imperfections with regard to the atomic arrangements.

The Hexagonal Close-Packed Structure The hexagonal close-packed structure (HCP) is shown in Figure 3-11. The lattice is hexagonal with a basis of two atoms of the same type: one located at (0, 0, 0) and one located at (2/3, 1/3, 1/2). (These coordinates are always fractions of the axial lengths *a, b*, and *c* even if the axial lengths are not equal.) The hexagonal lattice has one lattice point per unit cell located at the corners of the unit cell. In the HCP structure, two atoms are associated with every lattice point; thus, there are two atoms per unit cell.

An equally valid representation of the HCP crystal structure is a hexagonal lattice with a basis of two atoms of the same type: one located at (0, 0, 0) and one located at (1/3, 2/3, 1/2). The (2/3, 1/3, 1/2) and (1/3, 2/3, 1/2) coordinates are equivalent, meaning that they cannot be distinguished from one another.

In metals with an ideal HCP structure, the a_0 and c_0 axes are related by the ratio $c_0/a_0 = \sqrt{8/3} = 1.633$. Most HCP metals, however, have c_0/a_0 ratios that differ slightly from the ideal value because of mixed bonding. Because the HCP structure, like the FCC structure, has the most efficient packing factor of 0.74 and a coordination number of 12, a number of metals possess this structure. Table 3-2 summarizes the characteristics of crystal structures of some metals.

Structures of ionically bonded materials can be viewed as formed by the packing (cubic or hexagonal) of anions. Cations enter into the interstitial sites or holes that remain after the packing of anions. Section 3-7 discusses this in greater detail.

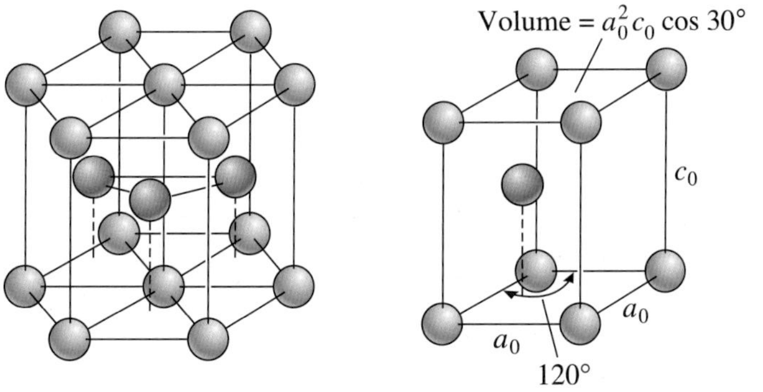

Figure 3-11 The hexagonal close-packed (HCP) structure (left) and its unit cell.

Table 3-2 Crystal structure characteristics of some metals at room temperature

Structure	a_0 versus r	Atoms per Cell	Coordination Number	Packing Factor	Examples
Simple cubic (SC)	$a_0 = 2r$	1	6	0.52	Polonium (Po), α-Mo
Body-centered cubic (BCC)	$a_0 = 4r/\sqrt{3}$	2	8	0.68	Fe, W, Mo, Nb, Ta, K, Na, V, Cr
Face-centered cubic (FCC)	$a_0 = 4r/\sqrt{2}$	4	12	0.74	Cu, Au, Pt, Ag, Pb, Ni
Hexagonal close-packed (HCP)	$a_0 = 2r$ $c_0 \approx 1.633a_0$	2	12	0.74	Ti, Mg, Zn, Be, Co, Zr, Cd

3-4 Allotropic or Polymorphic Transformations

Materials that can have more than one crystal structure are called allotropic or polymorphic. The term **allotropy** is normally reserved for this behavior in pure elements, while the term **polymorphism** is used for compounds. We discussed the allotropes of carbon in Chapter 2. Some metals, such as iron and titanium, have more than one crystal structure. At room temperature, iron has the BCC structure, but at higher temperatures, iron transforms to an FCC structure. These transformations result in changes in properties of materials and form the basis for the heat treatment of steels and many other alloys.

Many ceramic materials, such as silica (SiO_2) and zirconia (ZrO_2), also are polymorphic. A volume change may accompany the transformation during heating or cooling; if not properly controlled, this volume change causes the brittle ceramic material to crack and fail. For zirconia (ZrO_2), for instance, the stable form at room temperature (~25°C) is monoclinic. As we increase the temperature, more symmetric crystal structures become stable. At 1170°C, the monoclinic zirconia transforms into a tetragonal structure. The tetragonal form is stable up to 2370°C. At that temperature, zirconia transforms into a cubic form. The cubic form remains stable from 2370°C to the melting temperature of 2680°C. Zirconia also can have the orthorhombic form when high pressures are applied.

Ceramic components made from pure zirconia typically will fracture as the temperature is lowered and as zirconia transforms from the tetragonal to monoclinic form because of volume expansion (the cubic to tetragonal phase change does not cause much change in volume). As a result, pure monoclinic or tetragonal polymorphs of zirconia are not used. Instead, materials scientists and engineers have found that adding dopants such as yttria (Y_2O_3) make it possible to stabilize the cubic phase of zirconia, even at room temperature. This yttria stabilized zirconia (YSZ) contains up to 8 mol% Y_2O_3. Stabilized zirconia formulations are used in many applications, including thermal barrier coatings (TBCs) for turbine blades and electrolytes for oxygen sensors and solid oxide fuel cells. Virtually every car produced today uses an oxygen sensor that is made using stabilized zirconia compositions. Stabilized zirconia is similar in appearance to diamond but is an economic alternative to diamond for jewelry. Example 3-6 illustrates how to calculate volume changes in polymorphs of zirconia.

Example 3-6 *Calculating Volume Changes in Polymorphs of Zirconia*

Calculate the percent volume change as zirconia transforms from a tetragonal to monoclinic structure. The lattice constants for the monoclinic unit cell are $a = 5.156$, $b = 5.191$, and $c = 5.304$ Å, respectively. The angle β for the monoclinic unit cell is 98.9°. The lattice constants for the tetragonal unit cell are $a = 5.094$ and $c = 5.304$ Å. Does the zirconia expand or contract during this transformation? What is the implication of this transformation on the mechanical properties of zirconia ceramics?

SOLUTION

From Table 3-1, the volume of a tetragonal unit cell is given by

$$V = a^2c = (5.094)^2(5.304) = 137.63 \text{ Å}^3$$

and the volume of a monoclinic unit cell is given by

$$V = abc \sin\beta = (5.156)(5.191)(5.304)\sin(98.9°) = 140.25 \text{ Å}^3$$

Thus, there is an expansion of the unit cell as ZrO_2 transforms from a tetragonal to monoclinic form.

The percent change in volume = (final volume − initial volume)/(initial volume) $*$ 100 = (140.25 − 137.63 Å^3)/(137.63 Å^3) $*$ 100 = 1.9%

Most ceramics are very brittle and cannot withstand more than a 0.1% change in volume. (We will discuss mechanical behavior of materials in Chapters 6, 7, and 8.) The conclusion here is that ZrO_2 ceramics cannot be used in their monoclinic form since when zirconia does transform to the tetragonal form, it will most likely fracture. Therefore, ZrO_2 is often stabilized in a cubic form using different additives such as CaO, MgO, and Y_2O_3.

3-5 Points, Directions, and Planes in the Unit Cell

Coordinates of Points We can locate certain points, such as atom positions, in the lattice or unit cell by constructing the right-handed coordinate system in Figure 3-12. Distance is measured in terms of the number of lattice parameters we must move in each of the x, y, and z coordinates to get from the origin to the point in question. The coordinates are written as the three distances, with commas separating the numbers.

Directions in the Unit Cell Certain directions in the unit cell are of particular importance. **Miller indices** for directions are the shorthand notation used to describe these directions. The procedure for finding the Miller indices for directions is as follows:

1. Using a right-handed coordinate system, determine the coordinates of two points that lie on the direction.
2. Subtract the coordinates of the "tail" point from the coordinates of the "head" point to obtain the number of lattice parameters traveled in the direction of each axis of the coordinate system.
3. Clear fractions and/or reduce the results obtained from the subtraction to lowest integers.
4. Enclose the numbers in square brackets [ ]. If a negative sign is produced, represent the negative sign with a bar over the number.

Example 3-7 illustrates a way of determining the Miller indices of directions.

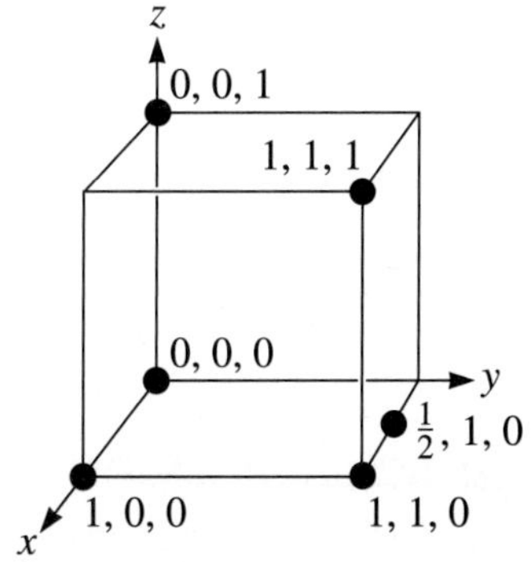

Figure 3-12 Coordinates of selected points in the unit cell. The number refers to the distance from the origin in terms of lattice parameters.

Example 3-7 Determining Miller Indices of Directions

Determine the Miller indices of directions *A, B,* and *C* in Figure 3-13.

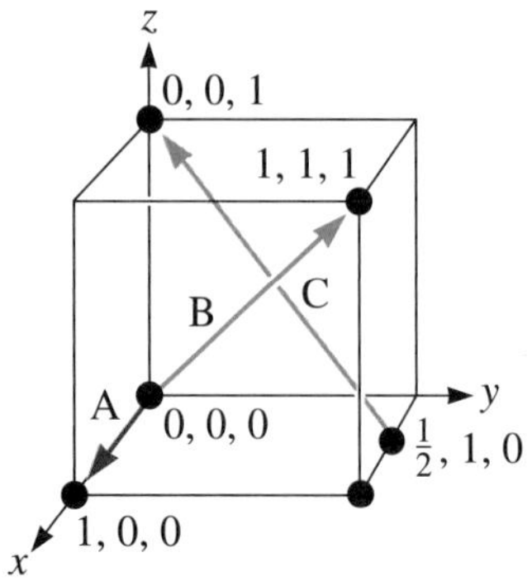

Figure 3-13 Crystallographic directions and coordinates (for Example 3-7).

SOLUTION

Direction *A*

1. Two points are 1, 0, 0, and 0, 0, 0
2. 1, 0, 0 − 0, 0, 0 = 1, 0, 0
3. No fractions to clear or integers to reduce
4. [100]

Direction *B*

1. Two points are 1, 1, 1 and 0, 0, 0
2. 1, 1, 1 − 0, 0, 0 = 1, 1, 1
3. No fractions to clear or integers to reduce
4. [111]

Direction *C*

1. Two points are 0, 0, 1 and $\frac{1}{2}$, 1, 0
2. 0, 0, 1 $-\frac{1}{2}$, 1, 0 = $-\frac{1}{2}$, −1, 1
3. $2(-\frac{1}{2}, -1, 1) = -1, -2, 2$
4. $[\bar{1}\bar{2}2]$

Several points should be noted about the use of Miller indices for directions:

1. Because directions are vectors, a direction and its negative are not identical; [100] is not equal to $[\bar{1}00]$. They represent the same line, but opposite directions.
2. A direction and its multiple are *identical*; [100] is the same direction as [200].
3. Certain groups of directions are *equivalent*; they have their particular indices because of the way we construct the coordinates. For example, in a cubic system, a [100] direction is a [010] direction if we redefine the coordinate system as shown in Figure 3-14. We may refer to groups of equivalent directions as **directions of a form** or **family**. The special brackets ⟨⟩ are used to indicate this collection of directions. All of the directions of the form ⟨110⟩ are listed in Table 3-3. We expect a cubic material to have the same properties in each of these twelve directions of the form ⟨110⟩.

Significance of Crystallographic Directions Crystallographic directions are used to indicate a particular orientation of a single crystal or of an oriented polycrystalline material. Knowing how to describe these can be useful in many

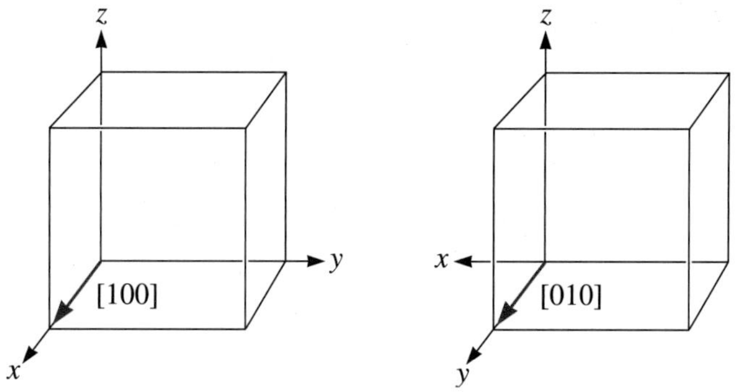

Figure 3-14 Equivalency of crystallographic directions of a form in cubic systems.

Table 3-3 Directions of the form ⟨110⟩ in cubic systems

⟨110⟩ =	[110]	$[\bar{1}\bar{1}0]$
	[101]	$[\bar{1}0\bar{1}]$
	[011]	$[0\bar{1}\bar{1}]$
	$[1\bar{1}0]$	$[\bar{1}10]$
	$[10\bar{1}]$	$[\bar{1}01]$
	$[01\bar{1}]$	$[0\bar{1}1]$

applications. Metals deform more easily, for example, in directions along which atoms are in closest contact. Another real-world example is the dependence of the magnetic properties of iron and other magnetic materials on the crystallographic directions. It is much easier to magnetize iron in the [100] direction compared to the [111] or [110] directions. This is why the grains in Fe-Si steels used in magnetic applications (e.g., transformer cores) are oriented in the [100] or equivalent directions.

Repeat Distance, Linear Density, and Packing Fraction

Another way of characterizing directions is by the **repeat distance,** or the distance between lattice points along the direction. For example, we could examine the [110] direction in an FCC unit cell (Figure 3-15); if we start at the (0, 0, 0) location, the next lattice point is at the center of a face, or a (1/2, 1/2, 0) site. The distance between lattice points is therefore one-half of the face diagonal, or $\frac{1}{2}\sqrt{2}\,a_0$. In copper, which has a lattice parameter of 0.3615 nm, the repeat distance is 0.2556 nm.

The **linear density** is the number of lattice points per unit length along the direction. In copper, there are two repeat distances along the [110] direction in each unit cell; since this distance is $\sqrt{2}a_0 = 0.5112$ nm, then

$$\text{Linear density} = \frac{2 \text{ repeat distances}}{0.5112 \text{ nm}} = 3.91 \text{ lattice points/nm}$$

Note that the linear density is the reciprocal of the repeat distance.

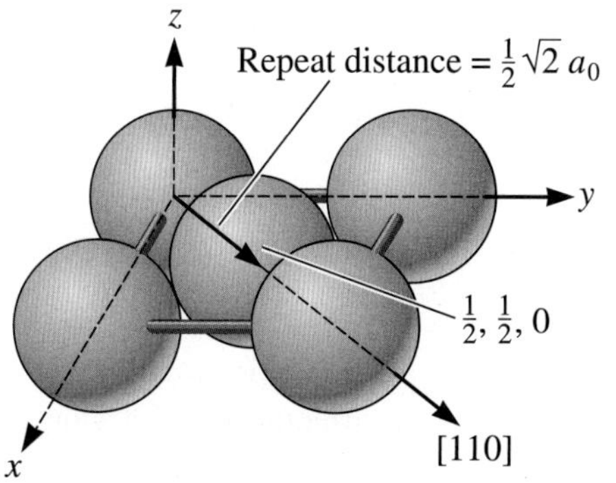

Figure 3-15 Determining the repeat distance, linear density, and packing fraction for a [110] direction in FCC copper.

Finally, we can compute the packing fraction of a particular direction, or the fraction actually covered by atoms. For copper, in which one atom is located at each lattice point, this fraction is equal to the product of the linear density and twice the atomic radius. For the [110] direction in FCC copper, the atomic radius $r = \sqrt{2}a_0/4 = 0.1278$ nm. Therefore, the packing fraction is

$$\begin{aligned}\text{Packing fraction} &= (\text{linear density})(2r) \\ &= (3.91)(2)(0.1278) \\ &= 1.0\end{aligned}$$

A value of 1.0 is consistent with the [110] direction being a close-packed direction in the FCC structure.

Planes in the Unit Cell

Certain planes of atoms in a crystal also carry particular significance. For example, metals deform along planes of atoms that are most tightly packed together. The surface energy of different faces of a crystal depends upon the particular crystallographic planes. This becomes important in crystal growth. In thin film growth of certain electronic materials (e.g., Si or GaAs), we need to be sure the substrate is oriented in such a way that the thin film can grow on a particular crystallographic plane.

Miller indices are used as a shorthand notation to identify these important planes, as described in the following procedure.

1. Identify the points at which the plane intercepts the x, y, and z axes in terms of the number of lattice parameters. If the plane passes through the origin, the origin of the coordinate system must be moved to that of an adjacent unit cell.
2. Take reciprocals of these intercepts.
3. Clear fractions but do not reduce to lowest integers.
4. Enclose the resulting numbers in parentheses (). Again, negative numbers should be written with a bar over the number.

The following example shows how Miller indices of planes can be obtained.

Example 3-8 *Determining Miller Indices of Planes*

Determine the Miller indices of planes A, B, and C in Figure 3-16.

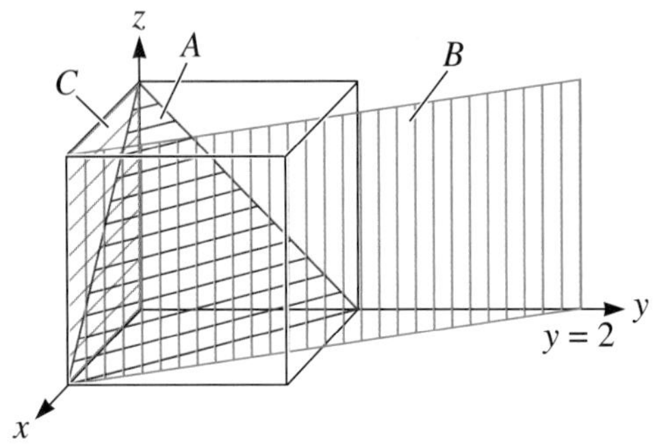

Figure 3-16 Crystallographic planes and intercepts (for Example 3-8).

SOLUTION

Plane *A*

1. $x = 1, y = 1, z = 1$
2. Take the reciprocals: 1, 1, 1
3. No fractions to clear
4. (111)

Plane *B*

1. The plane never intercepts the z axis, so $x = 1$, $y = 2$, and $z = \infty$
2. Take the reciprocals: 1, ½, 0
3. Clear fractions: 2, 1, 0
4. (210)

Plane *C*

1. We must move the origin since the plane passes through (0, 0, 0). Let's move the origin one lattice parameter in the y-direction. Then $x = \infty$, $y = -1$, and $z = \infty$.
2. Take the reciprocals: 0, −1, 0
3. No fractions to clear
4. $(0\bar{1}0)$

Several important aspects of the Miller indices for planes should be noted:

1. Planes and their negatives are identical (this was not the case for directions) because they are parallel. Therefore, $(020) = (0\bar{2}0)$.
2. Planes and their multiples are not identical (again, this is the opposite of what we found for directions). We can show this by defining planar densities and planar packing fractions. The **planar density** is the number of atoms per unit area with centers that lie on the plane; the packing fraction is the fraction of the area of that plane actually covered by these atoms. Example 3-9 shows how these can be calculated.
3. In each unit cell, **planes of a form** or **family** represent groups of equivalent planes. These planes have similar indices, and we represent these groups of similar planes with the notation { }. The planes of the form {110} in cubic systems are shown in Table 3-4.
4. In cubic systems, a direction that has the same indices as a plane is perpendicular to that plane.

Table 3-4 Planes of the form {110} in cubic systems

$$\{110\} = \begin{cases} (110) \\ (101) \\ (011) \\ (1\bar{1}0) \\ (10\bar{1}) \\ (01\bar{1}) \end{cases}$$

Note: The negatives of the planes are not unique planes.

Example 3-9 *Calculating the Planar Density and Packing Fraction*

Calculate the planar density and planar packing fraction for the (010) and (020) planes in simple cubic polonium, which has a lattice parameter of 0.334 nm.

SOLUTION

The two planes are drawn in Figure 3-17. On the (010) plane, the atoms are centered at each corner of the cube face, with 1/4 of each atom actually in the face of the unit cell.

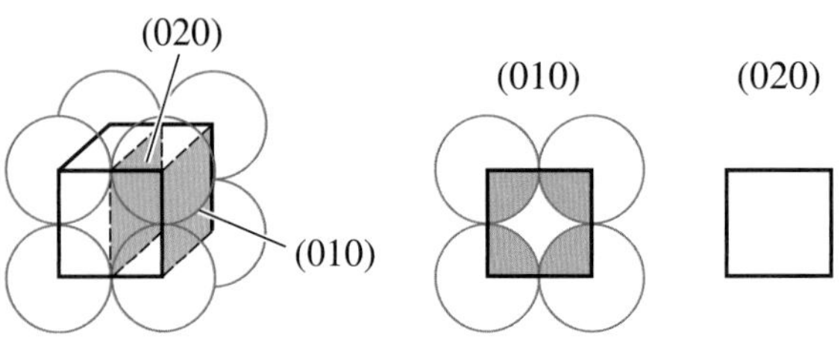

Figure 3-17 The planar densities of the (010) and (020) planes in SC unit cells are not identical (for Example 3-9).

Thus, the total number of atoms on each face is one. The planar density is

$$\text{Planar density (010)} = \frac{\text{atoms per face}}{\text{area of face}} = \frac{\text{1 atom per face}}{(0.334 \text{ nm})^2}$$
$$= 8.96 \text{ atoms/nm}^2 = 8.96 \times 10^{14} \text{ atoms/cm}^2$$

The planar packing fraction is given by

$$\text{Packing fraction (010)} = \frac{\text{area of atoms per face}}{\text{area of face}} = \frac{(1 \text{ atom})(\pi r^2)}{(a_0)^2}$$
$$= \frac{\pi r^2}{(2r)^2} = 0.79$$

No atoms are centered on the (020) planes. Therefore, the planar density and the planar packing fraction are both zero. The (010) and (020) planes are not equivalent!

Construction of Directions and Planes

To construct a direction or plane in the unit cell, we simply work backwards. Example 3-10 shows how we might do this.

Example 3-10 *Drawing a Direction and Plane*

Draw (a) the $[1\bar{2}1]$ direction and (b) the $(\bar{2}10)$ plane in a cubic unit cell.

SOLUTION

a. Because we know that we will need to move in the negative y-direction, let's locate the origin at (0, +1, 0). The "tail" of the direction will be located at this new origin. A second point on the direction can be determined by moving +1 in the x-direction, −2 in the y-direction, and +1 in the z-direction [Figure 3-18(a)].

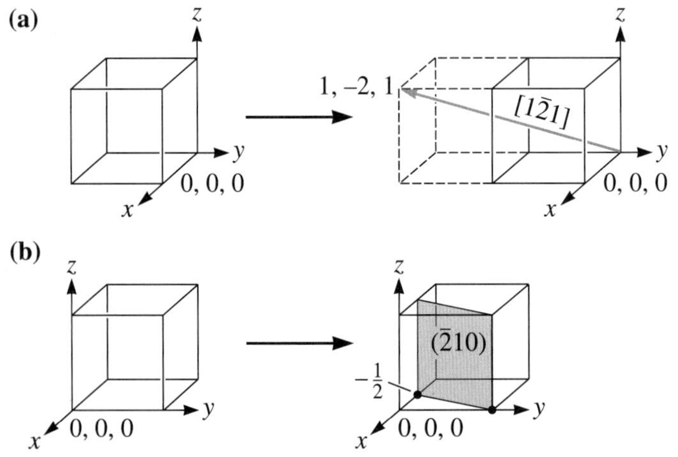

Figure 3-18 Construction of a (a) direction and (b) plane within a unit cell (for Example 3-10).

b. To draw in the $(\bar{2}10)$ plane, first take reciprocals of the indices to obtain the intercepts, that is

$$x = \frac{1}{-2} = -\frac{1}{2}; \; y = \frac{1}{1} = 1; \; z = \frac{1}{0} = \infty$$

Since the x-intercept is in a negative direction, and we wish to draw the plane within the unit cell, let's move the origin $+1$ in the x-direction to $(1, 0, 0)$. Then we can locate the x-intercept at $-1/2$ and the y-intercept at $+1$. The plane will be parallel to the z-axis [Figure 3-18(b)].

Miller Indices for Hexagonal Unit Cells

A special set of **Miller-Bravais indices** has been devised for hexagonal unit cells because of the unique symmetry of the system (Figure 3-19). The coordinate system uses four axes instead of three, with the a_3 axis being redundant. The axes a_1, a_2, and a_3 lie in a plane that is perpendicular to the fourth axis. The procedure for finding the indices of planes is exactly the same as before, but four intercepts are required, giving indices of the form $(hkil)$. Because of the redundancy of the a_3 axis and the special geometry of the system, the first three integers in the designation, corresponding to the a_1, a_2, and a_3 intercepts, are related by $h + k = -i$.

Directions in HCP cells are denoted with either the three-axis or four-axis system. With the three-axis system, the procedure is the same as for conventional Miller indices; examples of this procedure are shown in Example 3-11. A more complicated procedure, by which the direction is broken up into four vectors, is needed for the four-axis system. We determine the number of lattice parameters we must move in each direction to get from the "tail" to the "head" of the direction, while for consistency still making sure that $h + k = -i$. This is illustrated in Figure 3-20, showing that the [010] direction is the same as the $[\bar{1}2\bar{1}0]$ direction.

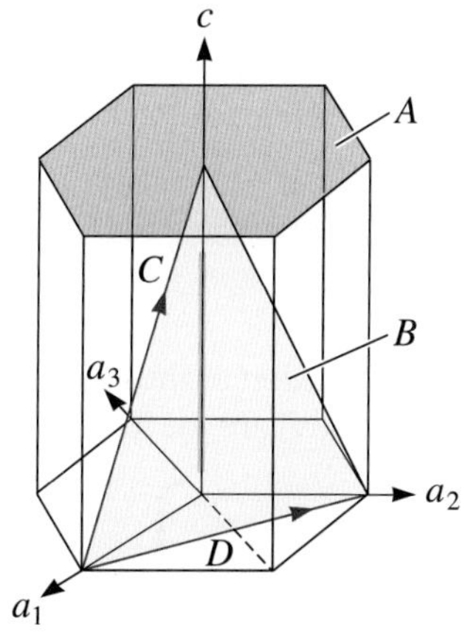

Figure 3-19 Miller-Bravais indices are obtained for crystallographic planes in HCP unit cells by using a four-axis coordinate system. The planes labeled *A* and *B* and the directions labeled *C* and *D* are those discussed in Example 3-11.

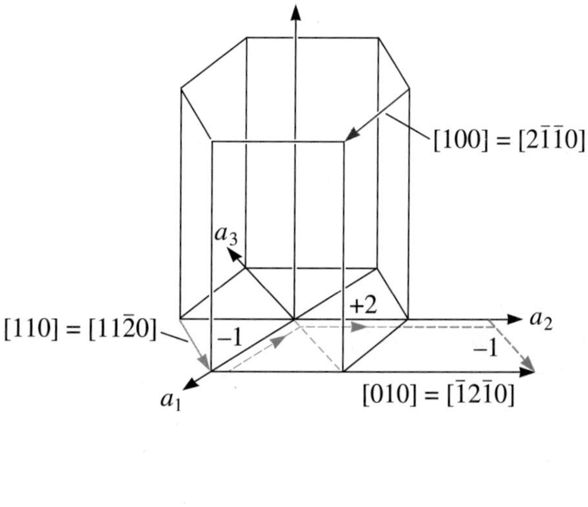

Figure 3-20 Typical directions in the HCP unit cell, using both three- and four-axis systems. The dashed lines show that the $[\bar{1}2\bar{1}0]$ direction is equivalent to a [010] direction.

Example 3-11 ***Determining the Miller-Bravais Indices for Planes and Directions***

Determine the Miller-Bravais indices for planes A and B and directions C and D in Figure 3-19.

SOLUTION

Plane A

1. $a_1 = a_2 = a_3 = \infty$, $c = 1$
2. Take the reciprocals: 0, 0, 0, 1
3. No fractions to clear
4. (0001)

Plane B

1. $a_1 = 1$, $a_2 = 1$, $a_3 = -\frac{1}{2}$, $c = 1$
2. Take the reciprocals: 1, 1, −2, 1
3. No fractions to clear.
4. $[11\bar{2}1]$

Direction C

1. Two points are 0, 0, 1 and 1, 0, 0.
2. 0, 0, 1 − 1, 0, 0 = −1, 0, 1
3. No fractions to clear or integers to reduce.
4. $[\bar{1}01]$ or $[\bar{2}113]$

Direction D

1. Two points are 0, 1, 0 and 1, 0, 0.
2. 0, 1, 0 − 1, 0, 0 = −1, 1, 0
3. No fractions to clear or integers to reduce.
4. $[\bar{1}10]$ or $[\bar{1}100]$

We can also convert the three-axis notation to the four-axis notation for directions by the following relationships, where h', k', and l' are the indices in the three-axis system:

$$\left.\begin{aligned} h &= \frac{1}{3}(2h' - k') \\ k &= \frac{1}{3}(2k' - h') \\ i &= -\frac{1}{3}(h' + k') \\ l &= l' \end{aligned}\right\} \tag{3-6}$$

After conversion, the values of h, k, i, and l may require clearing of fractions or reducing to lowest integers.

Close-Packed Planes and Directions

In examining the relationship between atomic radius and lattice parameter, we looked for close-packed directions, where

Table 3-5 Close-packed directions and planes

Structure	Directions	Planes
SC	⟨100⟩	None
BCC	⟨111⟩	None
FCC	⟨110⟩	{111}
HCP*	[100], [010], [110]	(001), (002)

*In the four-axis notation, the close-packed directions and planes are <11$\bar{2}$0> and (0001).

atoms are in continuous contact. We can now assign Miller indices to these close-packed directions, as shown in Table 3-5.

We can also examine FCC and HCP unit cells more closely and discover that there is at least one set of close-packed planes in each. Close-packed planes are shown in Figure 3-21. Notice that a hexagonal arrangement of atoms is produced in two dimensions. The close-packed planes are easy to find in the HCP unit cell; they are the (0001) and (0002) planes of the HCP structure and are given the special name **basal planes**. In fact, we can build up an HCP unit cell by stacking together close-packed planes in an . . . *ABABAB* . . . **stacking sequence** (Figure 3-21). Atoms on plane *B*, the (0002) plane, fit into the valleys between atoms on plane *A*, the bottom (0001) plane. If another plane identical in orientation to plane *A* is placed in the valleys of plane *B* directly above plane *A*, the HCP structure is created. Notice that all of the possible close-packed planes are parallel to one another. Only the basal planes—(0001) and (0002)—are close-packed.

From Figure 3-21, we find the coordination number of the atoms in the HCP structure. The center atom in a basal plane touches six other atoms in the same plane. Three atoms in a lower plane and three atoms in an upper plane also touch the same atom. The coordination number is twelve.

In the FCC structure, close-packed planes are of the form {111} (Figure 3-22). When parallel (111) planes are stacked, atoms in plane *B* fit over valleys in plane *A* and atoms in plane *C* fit over valleys in both planes *A* and *B*. The fourth plane fits directly over atoms in plane *A*. Consequently, a stacking sequence . . . *ABCABCABC* . . . is produced using the (111) plane. Again, we find that each atom has a coordination number of twelve.

Unlike the HCP unit cell, there are four sets of nonparallel close-packed planes—(111), (11$\bar{1}$), (1$\bar{1}$1), and ($\bar{1}$11)—in the FCC cell. This difference between the FCC and HCP unit cells—the presence or absence of intersecting close-packed planes—affects the mechanical behavior of metals with these structures.

Isotropic and Anisotropic Behavior

Because of differences in atomic arrangements in the planes and directions within a crystal, some properties also vary with direction. A material is crystallographically **anisotropic** if its properties depend on the crystallographic direction along which the property is measured. For example, the modulus of elasticity of aluminum is 75.9 GPa (11×10^6 psi) in ⟨111⟩ directions, but only 63.4 GPa (9.2×10^6 psi) in ⟨100⟩ directions. If the properties are identical in all directions, the material is crystallographically **isotropic**. Note that a material such as aluminum,

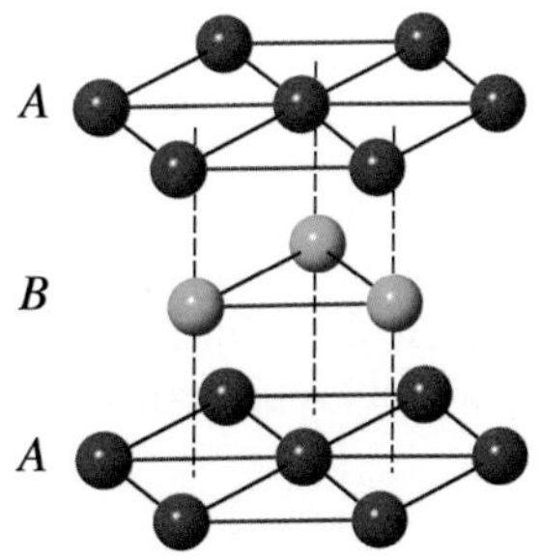

Figure 3-21 The *ABABAB* stacking sequence of close-packed planes produces the HCP structure.

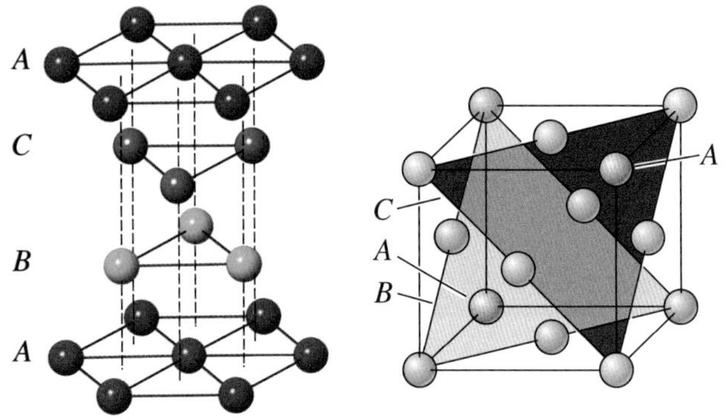

Figure 3-22 The *ABCABCABC* stacking sequence of close-packed planes produces the FCC structure.

which is crystallographically anisotropic, may behave as an isotropic material if it is in a polycrystalline form. This is because the random orientations of different crystals in a polycrystalline material will mostly cancel out any effect of the anisotropy as a result of crystal structure. In general, most polycrystalline materials will exhibit isotropic properties. Materials that are single crystals or in which many grains are oriented along certain directions (naturally or deliberately obtained by processing) will typically have anisotropic mechanical, optical, magnetic, and dielectric properties.

Interplanar Spacing The distance between two adjacent parallel planes of atoms with the same Miller indices is called the **interplanar spacing** (d_{hkl}). The interplanar spacing in *cubic* materials is given by the general equation

$$d_{hkl} = \frac{a_0}{\sqrt{h^2 + k^2 + l^2}} \tag{3-6}$$

where a_0 is the lattice parameter and h, k, and l represent the Miller indices of the adjacent planes being considered. The interplanar spacings for noncubic materials are given by more complex expressions.

3-6 Interstitial Sites

In all crystal structures, there are small holes between the usual atoms into which smaller atoms may be placed. These locations are called **interstitial sites**.

An atom, when placed into an interstitial site, touches two or more atoms in the lattice. This interstitial atom has a coordination number equal to the number of atoms it touches. Figure 3-23 shows interstitial locations in the SC, BCC, and

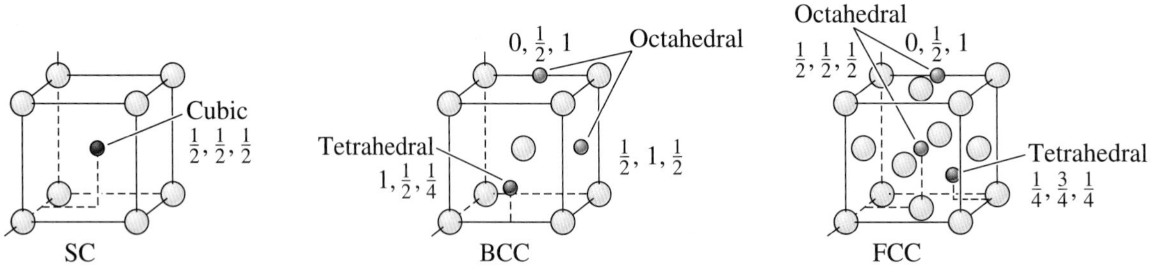

Figure 3-23 The location of the interstitial sites in cubic unit cells. Only representative sites are shown.

Example 3-12 *Calculating Octahedral Sites*

Demonstrate that the number of octahedral sites that *uniquely* belong to one FCC unit cell is four.

SOLUTION

The octahedral sites include the twelve edges of the unit cell, with the coordinates

$$\tfrac{1}{2},0,0 \quad \tfrac{1}{2},1,0 \quad \tfrac{1}{2},0,1 \quad \tfrac{1}{2},1,1$$

$$0,\tfrac{1}{2},0 \quad 1,\tfrac{1}{2},0 \quad 1,\tfrac{1}{2},1 \quad 0,\tfrac{1}{2},1$$

$$0,0,\tfrac{1}{2} \quad 1,0,\tfrac{1}{2} \quad 1,1,\tfrac{1}{2} \quad 0,1,\tfrac{1}{2}$$

plus the body-centered position, (1/2, 1/2, 1/2). Each of the sites on the edge of the unit cell is shared between four unit cells, so only 1/4 of each site belongs uniquely to each unit cell. Therefore, the number of sites belonging uniquely to each cell is

$$\frac{12 \text{ edges}}{\text{cell}} \cdot \frac{1/4 \text{ site}}{\text{edge}} + \frac{1 \text{ body-center}}{\text{cell}} \cdot \frac{1 \text{ site}}{\text{body-center}} = 4 \text{ octahedral sites/cell}$$

FCC structures. The **cubic site**, with a coordination number of eight, occurs in the SC structure at the body-centered position. **Octahedral sites** give a coordination number of six (not eight). They are known as octahedral sites because the atoms contacting the interstitial atom form an octahedron. **Tetrahedral sites** give a coordination number of four. As an example, the octahedral sites in BCC unit cells are located at the centers of the faces and the centers of the edges of the cube; a small atom placed in a face-centered octahedral site touches the four atoms at the corners of the face, the atom at the center of the unit cell, plus another atom at the center of the adjacent unit cell, giving a coordination number of six. In FCC unit cells, octahedral sites occur at the center of each edge of the cube, as well as at the body-center of the unit cell.

Counting in the same manner in which atoms are counted in a unit cell, there are six octahedral sites in the BCC structure and four octahedral sites in the FCC structure per unit cell. There are twelve tetrahedral sites in the BCC structure, and eight tetrahedral sites in the FCC structure per unit cell.

Interstitial atoms or ions with radii that are slightly larger than the radius of the interstitial site may enter that site, pushing the surrounding atoms slightly apart. Atoms with radii smaller than the radius of the hole are not allowed to fit into the interstitial site because the ion would "rattle" around in the site. If the interstitial atom becomes too large, it prefers to enter a site having a larger coordination number (Table 3-6). Therefore, an atom with a radius ratio between 0.225 and 0.414 enters a tetrahedral site; if its radius ratio is somewhat larger than 0.414, it enters an octahedral site instead.

Many ionic crystals (see Section 3-7) can be viewed as being generated by close packing of larger anions. Cations then can be viewed as smaller ions that fit into the interstitial sites of the close-packed anions. Thus, the radius ratios described in Table 3-6 also apply to the ratios of the radius of the cation to that of the anion. The packing in ionic crystals is not as tight as that in FCC or HCP metals.

Table 3-6 The coordination number and the radius ratio. The atoms/ions are drawn less than full for clarity

Coordination Number	Location of Interstitial	Radius Ratio	Representation
2	Linear	0–0.155	
3	Center of triangle	0.155–0.225	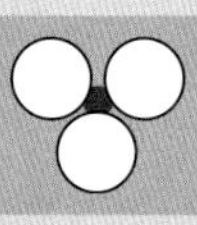
4	Center of tetrahedron	0.225–0.414	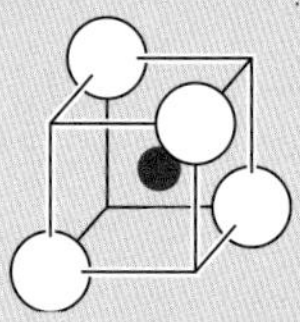
6	Center of octahedron	0.414–0.732	
8	Center of cube	0.732–1.000	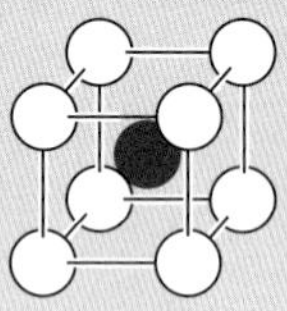

3-7 Crystal Structures of Ionic Materials

Ionic materials must have crystal structures that ensure electrical neutrality, yet permit ions of different sizes to be packed efficiently. As mentioned before, ionic crystal structures can be viewed as close-packed structures of anions. Anions form tetrahedra or octahedra, allowing the cations to fit into their appropriate interstitial sites. In some cases, it may be easier to visualize coordination polyhedra of cations with anions going to the interstitial sites. Recall from Chapter 2 that very often in real materials with engineering applications, the bonding is never 100% ionic. We still use this description of the crystal structure, though, to discuss the crystal structure of most ceramic materials. The following factors need to be considered in order to understand crystal structures of ionically bonded solids.

Ionic Radii The crystal structures of ionically bonded compounds often can be described by placing the anions at the normal lattice points of a unit cell, with the cations then located at one or more of the interstitial sites described in Section 3-6 (or vice versa). The ratio of the sizes of the ionic radii of anions and cations influences both the manner of packing and the coordination number (Table 3-6). Note that the radii of atoms and ions are different. For example, the radius of an oxygen atom is 0.6 Å; however, the radius of an oxygen anion (O^{2-}) is 1.32 Å. This is because an oxygen anion has acquired two additional electrons and has become larger. As a general rule, anions are larger than cations. Cations, having acquired a positive charge by losing electrons, are expected to be smaller. Strictly speaking, the radii of cations and anions also depend upon the coordination number. For example, the radius of an Al^{3+} ion is 0.39 Å when the coordination number is four (tetrahedral coordination); however, the radius of Al^{3+} is 0.53 Å when the coordination number is 6 (octahedral coordination). Also, note that the coordination number for cations is the number of nearest anions and vice versa. The radius of an atom also depends on the coordination number. For example, the radius of an iron atom in the FCC and BCC polymorphs is different! This tells us that atoms and ions are not "hard spheres" with fixed atomic radii. Appendix B in this book contains the atomic and ionic radii for different elements.

Electrical Neutrality The overall material has to be electrically neutral. If the charges on the anion and the cation are identical and the coordination number for each ion is identical to ensure a proper balance of charge, then the compound will have a formula AX (A: cation, X: anion). As an example, each cation may be surrounded by six anions, while each anion is, in turn, surrounded by six cations. If the valence of the cation is $+2$ and that of the anion is -1, then twice as many anions must be present, and the formula is AX_2. The structure of the AX_2 compound must ensure that the coordination number of the cation is twice the coordination number of the anion. For example, each cation may have eight anion nearest neighbors, while only four cations touch each anion.

Connection between Anion Polyhedra As a rule, the coordination polyhedra (formed by the close packing of anions) will share corners, as opposed to faces or edges. This is because in corner-sharing polyhedra, electrostatic repulsion between cations is reduced considerably leading to the formation of a more stable crystal structure. A number of common structures in ceramic materials are described in the following discussions. Compared to metals, ceramic structures are more complex. The lattice constants of ceramic materials tend to be larger than those for metallic materials because electrostatic repulsion between ions prevents close packing of both anions and cations.

Example 3-13 *Radius Ratio for CsI*

Cesium iodide (CsI) emits light when subjected to short wavelength radiation. CsI is used as optical crystals in a variety of applications. For CsI, (a) verify that the compound has the cesium chloride structure and (b) calculate the packing factor for the compound.

SOLUTION

a. From Appendix B, $r_{Cs^+} = 0.167$ nm and $r_{I^-} = 0.220$ nm, so

$$\frac{r_{Cs^+}}{r_{I^-}} = \frac{0.167\text{ nm}}{0.220\text{ nm}} = 0.759$$

Since $0.732 < 0.759 < 1.000$, the coordination number for each type of ion is eight, and the CsCl structure is likely.

b. The ions touch along the body diagonal of the unit cell, so

$$\sqrt{3}a_0 = 2r_{Cs^+} + 2r_{I^-} = 2(0.167) + 2(0.220) = 0.774$$

$$a_0 = 0.447\text{ nm}$$

$$\text{Packing Factor} = \frac{\frac{4}{3}\pi r_{Cs^+}^3\,(1\text{ Cs ion}) + \frac{4}{3}\pi r_{I^-}^3\,(1\text{ I ion})}{a_0^3}$$

$$\text{Packing Factor} = \frac{\frac{4}{3}\pi\,(0.167\text{ nm})^3\,(1\text{ Cs ion}) + \frac{4}{3}\pi\,(0.220\text{ nm})^3\,(1\text{ I ion})}{(0.447\text{ nm})^3} = 0.72$$

Sodium Chloride Structure The radius ratio for sodium and chloride ions is $r_{Na^+}/r_{Cl^-} = 0.097\text{ nm}/(0.181\text{ nm}) = 0.536$; the sodium ion has a charge of $+1$; the chloride ion has a charge of -1. Therefore, based on the charge balance and radius

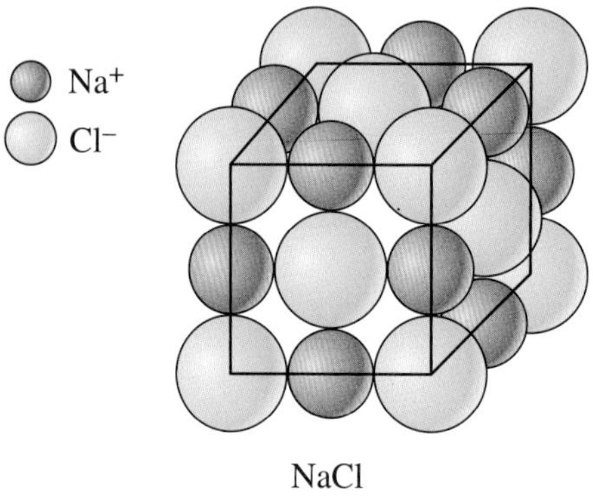

Figure 3-24 The sodium chloride structure, an FCC unit cell with two ions (Na^+ and Cl^-) per lattice point. *Note*: ion sizes not to scale.

ratio, each anion and cation must have a coordination number of six. The FCC structure, with Cl^- ions at FCC positions and Na^+ ions at the four octahedral sites, satisfies these requirements (Figure 3-24). We can also consider this structure to be FCC with two ions—one Na^+ and one Cl^-—associated with each lattice point. Many ceramics, including magnesium oxide (MgO), calcium oxide (CaO), and iron oxide (FeO) have this structure.

Zincblende Structure Zincblende is the name of the crystal structure adopted by ZnS. Although the Zn ions have a charge of +2 and S ions have a charge of −2, zincblende cannot have the sodium chloride structure because

$$\frac{r_{Zn^{+2}}}{r_{S^{-2}}} = 0.074 \text{ nm}/(0.184 \text{ nm}) = 0.402$$

This radius ratio demands a coordination number of four, which in turn means that the zinc ions enter tetrahedral sites in a unit cell (Figure 3-25). The FCC structure, with S anions at the normal lattice points and Zn cations at half of the tetrahedral sites, can accommodate the restrictions of both charge balance and coordination number. A variety of materials, including the semiconductor GaAs and many other III–V semiconductors (Chapter 2), have this structure.

Example 3-14 *Illustrating a Crystal Structure and Calculating Density*

Show that MgO has the sodium chloride crystal structure and calculate the density of MgO.

SOLUTION

From Appendix B, $r_{Mg^{+2}} = 0.066$ nm and $r_{O^{-2}} = 0.132$ nm, so

$$\frac{r_{Mg^{+2}}}{r_{O^{-2}}} = \frac{0.066 \text{ nm}}{0.132 \text{ nm}} = 0.50$$

Since $0.414 < 0.50 < 0.732$, the coordination number for each ion is six, and the sodium chloride structure is possible.

The atomic masses are 24.312 and 16.00 g/mol for magnesium and oxygen, respectively. The ions touch along the edge of the cube, so

$$a_0 = 2r_{Mg^{+2}} + 2r_{O^{-2}} = 2(0.066) + 2(0.132) = 0.396 \text{ nm} = 3.96 \times 10^{-8} \text{ cm}$$

$$\rho = \frac{(4\ Mg^{+2})\ (24.312 \text{ g/mol}) + (4\ O^{-2})\ (16.00 \text{ g/mol})}{(3.96 \times 10^{-8} \text{ cm})^3 (6.022 \times 10^{23}/\text{mol})} = 4.31 \text{ g/cm}^3$$

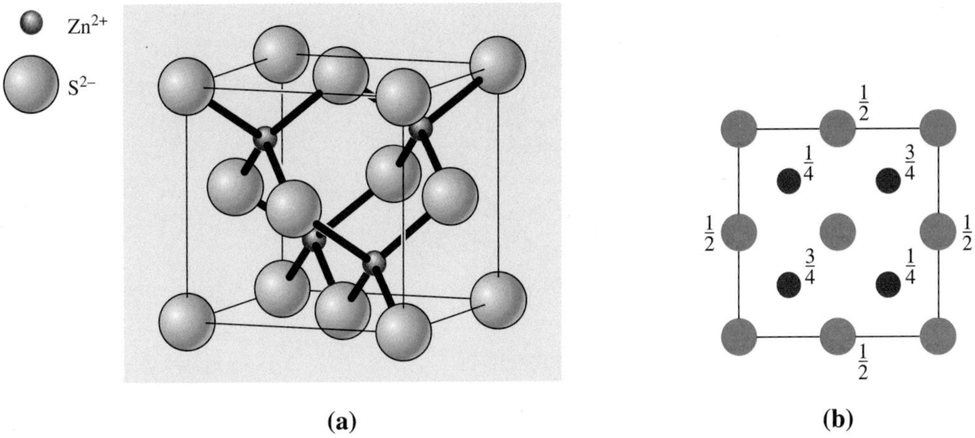

Figure 3-25 (a) The zincblende unit cell, (b) plan view. The fractions indicate the positions of the atoms out of the page relative to the height of the unit cell.

Example 3-15 *Calculating the Theoretical Density of GaAs*

The lattice constant of gallium arsenide (GaAs) is 5.65 Å. Show that the theoretical density of GaAs is 5.33 g/cm^3.

SOLUTION

For the "zincblende" GaAs unit cell, there are four Ga and four As atoms per unit cell.

Each mole of Ga has a mass of 69.72 g. Therefore, the mass of four Ga atoms is
4 atoms $*$ (69.72 g/mol) / (6.022×10^{23} atoms/mol).
Each mole of As has a mass of 74.92 g. Therefore, the mass of four As atoms is
4 atoms $*$ (74.92 g/mol) / (6.022×10^{23} atoms/mol).
These atoms occupy a volume of $(5.65 \times 10^{-8})^3$ cm^3.

$$\rho = \frac{\text{mass}}{\text{volume}} = \frac{4(69.72 + 74.92)/(6.022 \times 10^{23})}{(5.65 \times 10^{-8})^3} = 5.33 \text{ g/cm}^3$$

Fluorite Structure The fluorite structure is FCC, with anions located at all eight of the tetrahedral positions (Figure 3-26). Thus, there are four cations and eight anions per cell, and the ceramic compound must have the formula AX_2, as in calcium fluorite, or CaF_2. In the designation AX_2, A is the cation and X is the anion. The coordination number of the calcium ions is eight, but that of the fluoride ions is four, therefore ensuring a balance of charge. One of the polymorphs of ZrO_2 known as cubic zirconia exhibits this crystal structure. Other compounds that exhibit this structure include UO_2, ThO_2, and CeO_2.

Corundum Structure This is one of the crystal structures of alumina known as alpha alumina (α-Al_2O_3). In alumina, the oxygen anions pack in a hexagonal arrangement, and the aluminum cations occupy some of the available octahedral positions (Figure 3-27). Alumina is probably the most widely used ceramic material. Applications include, but are not limited to, spark plugs, refractories, electronic packaging substrates, and abrasives.

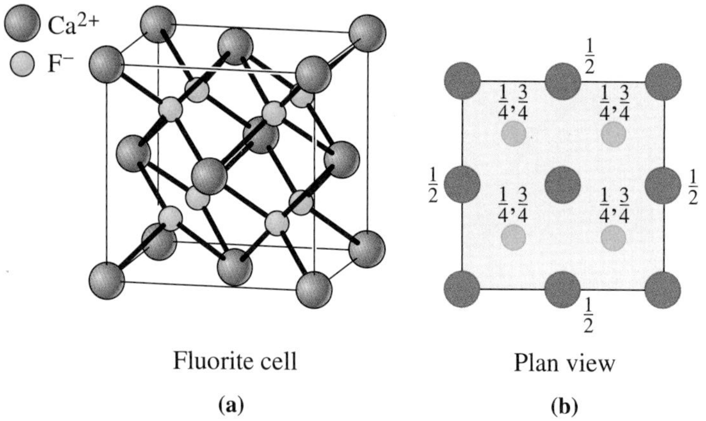

Figure 3-26 (a) Fluorite unit cell, (b) plan view. The fractions indicate the positions of the atoms out of the page relative to the height of the unit cell.

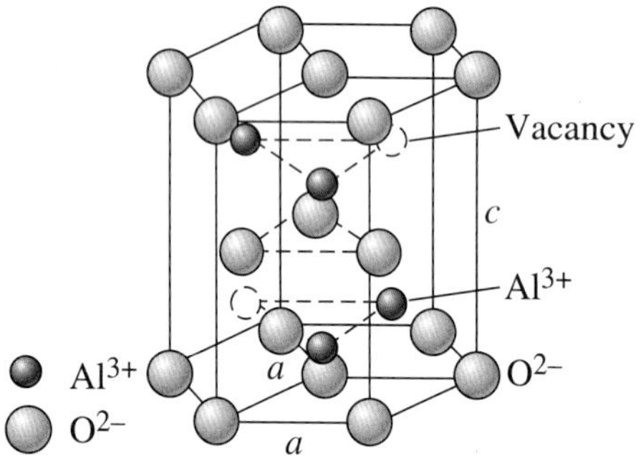

Figure 3-27 Corundum structure of alpha-alumina (α-Al_2O_3).

Example 3-16 *The Perovskite Crystal Structure*

Perovskite is a mineral containing calcium, titanium, and oxygen. The unit cell is cubic and has a calcium ion at each corner, an oxygen ion at each face center, and a titanium ion at the body-centered position. The ions contribute to the unit cell in the usual way (1/8 ion contribution for each ion at the corners, etc.).

(a) Describe this structure as a lattice and a basis. (b) How many ions of each type are there per unit cell? (c) An alternate way of drawing the unit cell of perovskite has calcium at the body-centered position of each cubic unit cell. What are the positions of the titanium and oxygen ions in this representation of the unit cell? (d) By counting the number of ions of each type per unit cell, show that the formula for perovskite is the same for both unit cell representations.

SOLUTION

(a) The lattice must belong to the cubic crystal system. Since different types of ions are located at the corner, face-centered, and body-centered positions, the lattice must be simple cubic. The structure can be described as a simple cubic lattice with a basis of Ca (0, 0, 0), Ti (1/2, 1/2, 1/2), and O (0, 1/2, 1/2), (1/2, 0, 1/2), and (1/2, 1/2, 0). The unit cell is shown in Figure 3-28.

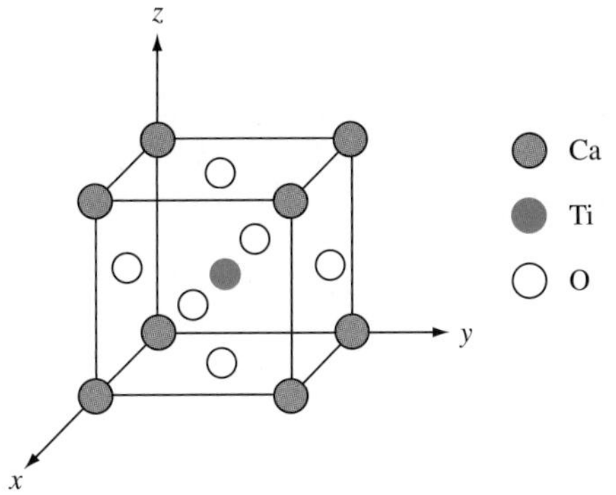

Figure 3-28 The perovskite unit cell.

(b) There are two methods for calculating the number of ions per unit cell. Using the lattice and basis concept,

$$\frac{1 \text{ lattice point}}{\text{unit cell}} * \frac{5 \text{ ion}}{\text{lattice point}} = \frac{5 \text{ ions}}{\text{unit cell}}$$

Using the unit cell concept,

$$\frac{8 \text{ corners}}{\text{unit cell}} * \frac{1/8 \text{ Ca ion}}{\text{corner}} + \frac{1 \text{ body-center}}{\text{unit cell}} * \frac{1 \text{ Ti ion}}{\text{body-center}}$$

$$+ \frac{6 \text{ face-centers}}{\text{unit cell}} * \frac{1/2 \text{ O ion}}{\text{face-center}} = \frac{5 \text{ ions}}{\text{unit cell}}$$

As expected, the number of ions per unit cell is the same regardless of which method is used. The chemical formula for perovskite is $CaTiO_3$ (calcium titanate). Compounds with the general formula ABO_3 and this structure are said to have the perovskite crystal structure. One of the polymorphs of barium titanate, which is used to make capacitors for electronic applications, and one form of lead zirconate exhibit this structure.

(c) If calcium is located at the body-centered position rather than the corners of the unit cell, then titanium must be located at the corners of the unit cell, and the oxygen ions must be located at the edge centers of the unit cell, as shown in Figure 3-29. Note that this is equivalent to shifting each ion in the basis given in part (a) by the vector [1/2 1/2 1/2]. The Ca ion is shifted from (0, 0, 0) to (1/2, 1/2, 1/2), and the Ti ion is shifted from (1/2, 1/2, 1/2) to (1, 1, 1), which is equivalent to the origin of an adjacent unit cell or (0, 0, 0). Note that the crystal has not been changed; only the coordinates of the ions in the basis are different. Another lattice and basis description of perovskite is thus a simple cubic lattice with a basis of Ca (1/2, 1/2, 1/2), Ti (0, 0, 0), and O (1/2, 0, 0), (0, 1/2, 0), and (0, 0, 1/2).

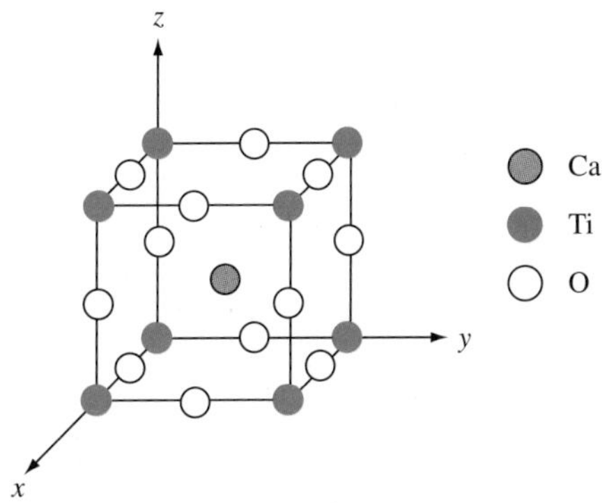

Figure 3-29 An alternate representation of the unit cell of perovskite.

Using the lattice and basis concept to count the number of atoms per unit cell,

$$\frac{1 \text{ lattice point}}{\text{unit cell}} * \frac{5 \text{ ions}}{\text{lattice point}} = \frac{5 \text{ ions}}{\text{unit cell}}$$

Using the unit cell concept,

$$\frac{1 \text{ body-center}}{\text{unit cell}} * \frac{1 \text{ Ca ion}}{\text{body-center}} + \frac{8 \text{ corners}}{\text{unit cell}} * \frac{1/8 \text{ Ti ion}}{\text{corner}}$$
$$+ \frac{12 \text{ edge-centers}}{\text{unit cell}} * \frac{1/4 \text{ O ion}}{\text{edge-center}} = \frac{5 \text{ ions}}{\text{unit cell}}$$

Again we find that the chemical formula is $CaTiO_3$.

3-8 Covalent Structures

Covalently bonded materials frequently have complex structures in order to satisfy the directional restraints imposed by the bonding.

Diamond Cubic Structure Elements such as silicon, germanium, α-Sn, and carbon (in its diamond form) are bonded by four covalent bonds and produce a **tetrahedron** [Figure 3-30(a)]. As these tetrahedral groups are combined, a large cube can be constructed [Figure 3-30(b)]. This large cube contains eight smaller cubes that are the size of the tetrahedral cube. The large cube is the **diamond cubic** (DC) unit cell. The atoms at the corners of the tetrahedral cubes provide atoms at the regular FCC lattice points. Four additional atoms are present within the DC unit cell from atoms at the centers of four of the tetrahedral cubes. We can describe the DC crystal structure as an FCC lattice with two atoms associated with each lattice point (or a basis of two). Therefore, there are eight atoms per unit cell.

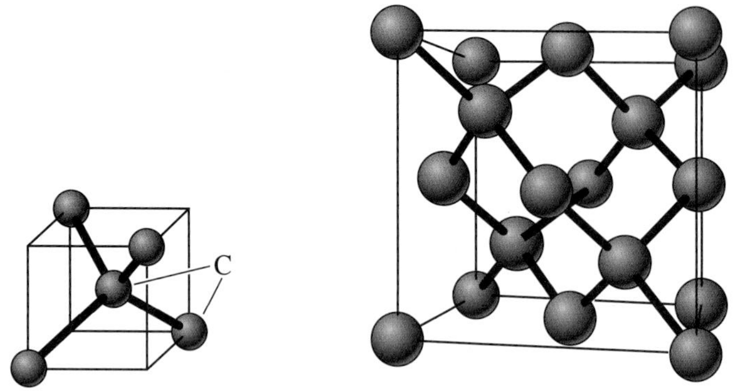

Diamond

(a) (b)

Figure 3-30 (a) Tetrahedron and (b) the diamond cubic (DC) unit cell. This open structure is produced because of the requirements of covalent bonding.

Example 3-17 *Determining the Packing Factor for the Diamond Cubic Structure*

Describe the diamond cubic structure as a lattice and a basis and determine its packing factor.

SOLUTION

The diamond cubic structure is a face-centered cubic lattice with a basis of two atoms of the same type located at (0, 0, 0) and (1/4, 1/4, 1/4). The basis atom located at (0, 0, 0) accounts for the atoms located at the FCC lattice points, which are (0, 0, 0), (0, 1/2, 1/2), (1/2, 0, 1/2), and (1/2, 1/2, 0) in terms of the coordinates of the unit cell. By adding the vector [1/4 1/4 1/4] to each of these points, the four additional atomic coordinates in the interior of the unit cell are determined to be (1/4, 1/4, 1/4), (1/4, 3/4, 3/4), (3/4, 1/4, 3/4), and (3/4, 3/4, 1/4). There are eight atoms per unit cell in the diamond cubic structure:

$$\frac{4 \text{ lattice points}}{\text{unit cell}} * \frac{2 \text{ atoms}}{\text{lattice point}} = \frac{8 \text{ atoms}}{\text{unit cell}}$$

The atoms located at the (1/4, 1/4, 1/4) type positions sit at the centers of tetrahedra formed by atoms located at the FCC lattice points. The atoms at the (1/4, 1/4, 1/4) type positions are in direct contact with the four surrounding atoms. Consider the distance between the center of the atom located at (0, 0, 0) and the center of the atom located at (1/4, 1/4, 1/4). This distance is equal to one-quarter of the body diagonal or two atomic radii, as shown in Figure 3-31. Thus,

$$\frac{a_0\sqrt{3}}{4} = 2r$$

or

$$a_0 = \frac{8r}{\sqrt{3}}$$

The packing factor is the ratio of the volume of space occupied by the atoms in the unit cell to the volume of the unit cell:

$$\text{Packing factor} = \frac{(8 \text{ atoms/cell})\left(\frac{4}{3}\pi r^3\right)}{a_0^3}$$

$$\text{Packing factor} = \frac{(8 \text{ atoms/cell})\left(\frac{4}{3}\pi r^3\right)}{\left(8r/\sqrt{3}\right)^3}$$

$$\text{Packing factor} = 0.34$$

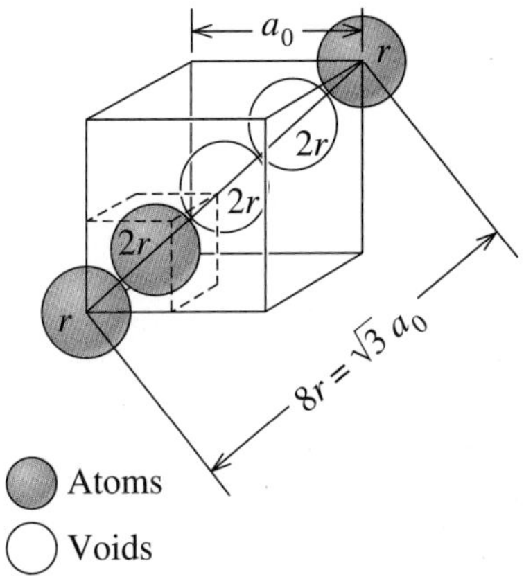

Figure 3-31 Determining the relationship between the lattice parameter and atomic radius in a diamond cubic cell (for Example 3-17).

This is a relatively open structure compared to close-packed structures. In Chapter 5, we will learn that the openness of a structure is one of the factors that affects the rate at which different atoms can diffuse in a given material.

Example 3-18 *Calculating the Radius, Density, and Atomic Mass of Silicon*

The lattice constant of Si is 5.4307 Å. Calculate the radius of a silicon atom and the theoretical density of silicon. The atomic mass of Si is 28.09 g/mol.

SOLUTION

Silicon has the diamond cubic structure. As shown in Example 3-17 for the diamond cubic structure,

$$r = \frac{a_0\sqrt{3}}{8}$$

Therefore, substituting $a_0 = 5.4307$ Å, the radius of the silicon atom = 1.176 Å. This is the same radius listed in Appendix B. For the density, we use the same approach as in Example 3-15. Recognizing that there are eight Si atoms per unit cell, then

$$\text{density} = \frac{\text{mass}}{\text{volume}} = \frac{8(28.08)/(6.022 \times 10^{23})}{(5.4307 \times 10^{-8}\ \text{cm})^3} = 2.33\ \text{g/cm}^3$$

This is the same density value listed in Appendix A.

Crystalline Silica In a number of its forms, silica (or SiO_2) has a crystalline ceramic structure that is partly covalent and partly ionic. Figure 3-32 shows the crystal structure of one of the forms of silica, β-cristobalite, which is a complicated structure with an FCC lattice. The ionic radii of silicon and oxygen are 0.042 nm and 0.132 nm, respectively, so the radius ratio is $r_{Si^{4+}}/r_{O^{2-}} = 0.318$ and the coordination number is four.

Crystalline Polymers A number of polymers may form a crystalline structure. The dashed lines in Figure 3-33 outline the unit cell for the lattice of polyethylene. Polyethylene is obtained by joining C_2H_4 molecules to produce long polymer chains that form an orthorhombic unit cell. Some polymers, including nylon, can have several polymorphic forms. Most engineered plastics are partly amorphous and may develop crystallinity during processing. It is also possible to grow single crystals of polymers.

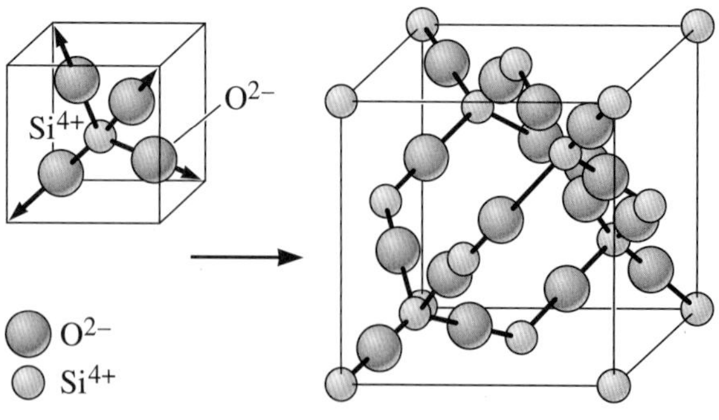

Figure 3-32 The silicon-oxygen tetrahedron and the resultant β-cristobalite form of silica.

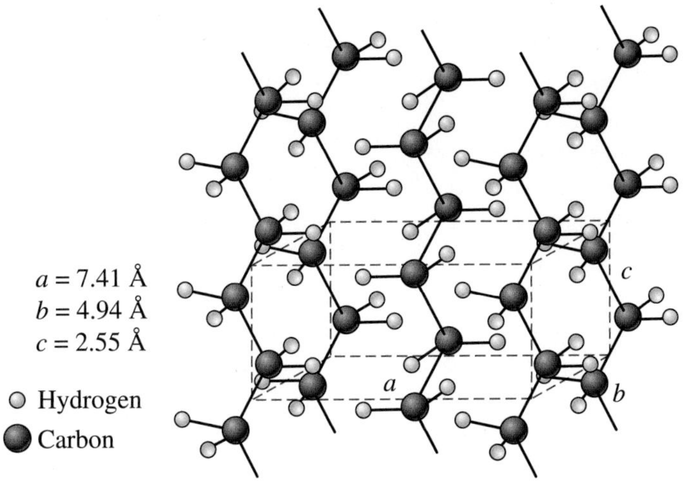

Figure 3-33 The unit cell of crystalline polyethylene (not to scale).

3-9 Diffraction Techniques for Crystal Structure Analysis

A crystal structure of a crystalline material can be analyzed using **x-ray diffraction (XRD)** or electron diffraction. Max von Laue (1879–1960) won the Nobel Prize in Physics in 1914 for his discovery related to the diffraction of x-rays by a crystal. William Henry Bragg (1862–1942) and his son William Lawrence Bragg (1890–1971) won the 1915 Nobel Prize in Physics for their contributions to XRD.

When a beam of x-rays having a single wavelength on the same order of magnitude as the atomic spacing in the material strikes that material, x-rays are scattered in all directions. Most of the radiation scattered from one atom cancels out radiation scattered from other atoms; however, x-rays that strike certain crystallographic planes at specific angles are reinforced rather than annihilated. This phenomenon is called **diffraction**. The x-rays are diffracted, or the beam is reinforced, when conditions satisfy **Bragg's law**,

$$\sin \theta = \frac{\lambda}{2d_{hkl}} \tag{3-7}$$

where the angle θ is half the angle between the diffracted beam and the original beam direction, λ is the wavelength of the x-rays, and d_{hkl} is the interplanar spacing between the planes that cause constructive reinforcement of the beam (see Figure 3-34).

When the material is prepared in the form of a fine powder, there are always at least some powder particles (crystals or aggregates of crystals) with planes (hkl) oriented at the proper θ angle to satisfy Bragg's law. Therefore, a diffracted beam, making an angle of 2θ with the incident beam, is produced. In a *diffractometer*, a moving x-ray detector records the 2θ angles at which the beam is diffracted, giving a characteristic diffraction pattern (see Figure 3-35). If we know the wavelength of the x-rays, we can determine the interplanar spacings and, eventually, the identity of the planes that cause the diffraction. In an XRD instrument, x-rays are produced by bombarding a metal target with a beam of high-energy electrons. Typically, x-rays emitted from copper have a wavelength $\lambda \cong 1.54060$ Å ($K\alpha_1$ line) and are used.

Typically, XRD analysis can be conducted relatively rapidly (~30 minutes to 1 hour per sample), on bulk or powdered samples and without extensive sample preparation. This technique can also be used to determine whether the material consists of many grains oriented in a particular crystallographic direction (texture) in bulk materials and thin films. Typically, a well-trained technician can conduct the analysis as well as interpret the powder diffraction data rather easily. As a result, XRD

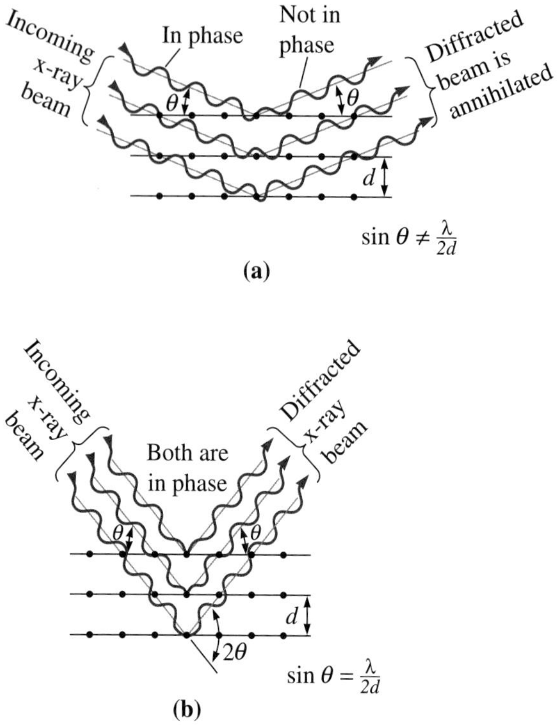

Figure 3-34 (a) Destructive and (b) reinforcing interactions between x-rays and the crystalline material. Reinforcement occurs at angles that satisfy Bragg's law.

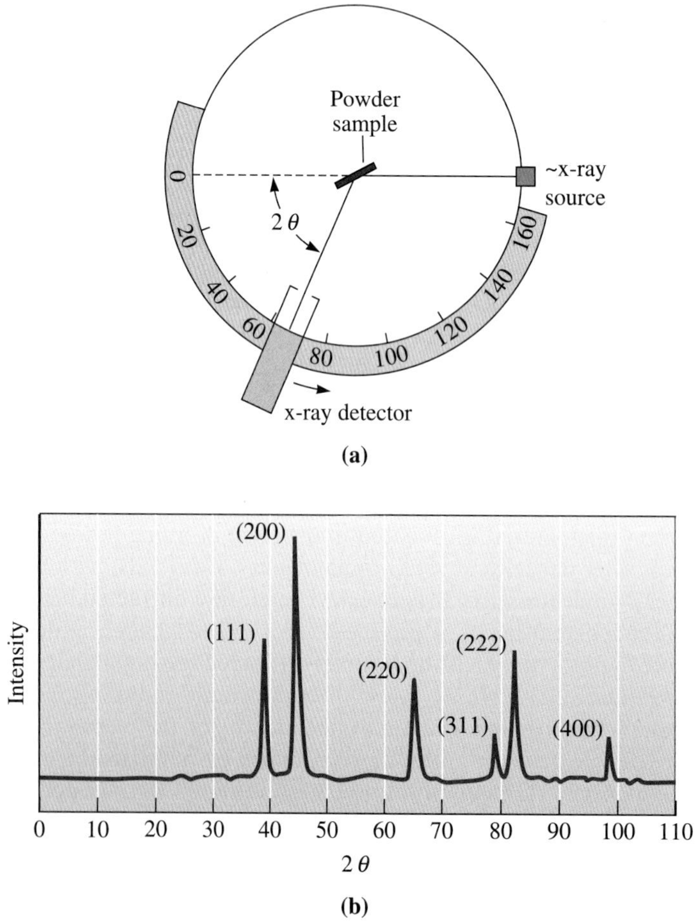

Figure 3-35 (a) Diagram of a diffractometer, showing a powder sample and the incident and diffracted beams. (b) The diffraction pattern obtained from a sample of gold powder.

is used in many industries as one tool for product quality control purposes. Analysis of single crystals and materials containing several phases can be more involved and time consuming.

To identify the crystal structure of a cubic material, we note the pattern of the diffracted lines—typically by creating a table of $\sin^2\theta$ values. By combining Equation 3-7 with Equation 3-6 for the interplanar spacing, we find that:

$$\sin^2\theta = \frac{\lambda^2}{4a_0^2}(h^2 + k^2 + l^2) \tag{3-8}$$

In simple cubic metals, all possible planes will diffract, giving an $h^2 + k^2 + l^2$ pattern of 1, 2, 3, 4, 5, 6, 8, 9 . . . . [Note that 7 is not included because there is no combination (hkl) that gives $h^2 + k^2 + l^2$ equal to 7.] In body-centered cubic metals, diffraction occurs only from planes having an even $h^2 + k^2 + l^2$ sum of 2, 4, 6, 8, 10, 12, 14, 16, ... For face-centered cubic metals, more destructive interference occurs, and planes having $h^2 + k^2 + l^2$ sums of 3, 4, 8, 11, 12, 16, . . . will diffract. By calculating the values of $\sin^2\theta$ and then finding the appropriate pattern, the crystal structure can be determined for metals having one of these simple structures, as illustrated in Example 3-19.

Example 3-19 *Examining X-ray Diffraction Data*

The results of an x-ray diffraction experiment using x-rays with $\lambda = 0.7107$ Å (radiation obtained from a molybdenum target) show that diffracted peaks occur at the following 2θ angles:

Peak	2θ (°)	Peak	2θ (°)
1	20.20	5	46.19
2	28.72	6	50.90
3	35.36	7	55.28
4	41.07	8	59.42

Determine the crystal structure, the indices of the plane producing each peak, and the lattice parameter of the material.

SOLUTION

We can first determine the $\sin^2\theta$ value for each peak, then divide through by the lowest denominator, 0.0308.

Peak	2θ (°)	$\sin^2\theta$	$\sin^2\theta/0.0308$	$h^2 + k^2 + l^2$	(hkl)
1	20.20	0.0308	1	2	(110)
2	28.72	0.0615	2	4	(200)
3	35.36	0.0922	3	6	(211)
4	41.07	0.1230	4	8	(220)
5	46.19	0.1539	5	10	(310)
6	50.90	0.1847	6	12	(222)
7	55.28	0.2152	7	14	(321)
8	59.42	0.2456	8	16	(400)

When we do this, we find a pattern of $\sin^2\theta/0.0308$ values of 1, 2, 3, 4, 5, 6, 7, and 8. If the material were simple cubic, the 7 would not be present because no planes have an $h^2 + k^2 + l^2$ value of 7. Therefore, the pattern must really be 2, 4, 6, 8, 10, 12, 14, 16, . . . and the material must be body-centered cubic. The (hkl) values listed give these required $h^2 + k^2 + l^2$ values.

We could then use 2θ values for any of the peaks to calculate the interplanar spacing and thus the lattice parameter. Picking peak 8:

$$2\theta = 59.42^\circ \quad \text{or} \quad \theta = 29.71^\circ$$

$$d_{400} = \frac{\lambda}{2\sin\theta} = \frac{0.7107}{2\sin(29.71^\circ)} = 0.71699 \text{ Å}$$

$$a_0 = d_{400}\sqrt{h^2 + k^2 + l^2} = (0.71699 \text{ Å})(4) = 2.868 \text{ Å}$$

This is the lattice parameter for body-centered cubic iron.

Electron Diffraction and Microscopy

Louis de Broglie theorized that electrons behave like waves. In electron diffraction, we make use of high-energy (~100,000 to 400,000 eV) electrons. These electrons are diffracted from electron transparent samples of materials. The electron beam that exits from the sample is also used to form an image of the sample. Thus, transmission electron microscopy and electron diffraction are used for imaging microstructural features and determining crystal structures.

A 100,000 eV electron has a wavelength of about 0.004 nm! This ultra-small wavelength of high-energy electrons allows a **transmission electron microscope** (TEM) to image the microstructure at a very fine scale. If the sample is too thick, electrons cannot be transmitted through the sample, and an image or a diffraction pattern will not be observed. Therefore, in transmission electron microscopy and electron diffraction, the sample has to be made such that portions of it are electron transparent. Figure 3-36 shows a TEM image and an electron diffraction pattern from selected areas of an aluminum alloy sample. The large bright spots correspond to the grains of the matrix. The smaller spots originate from small crystals of another phase.

Another advantage to using a TEM is the high spatial resolution. Using TEM, it is possible to determine differences between different crystalline regions and between amorphous and crystalline regions at very small length scales (~1–10 nm). This analytical technique and its variations [e.g., high-resolution electron microscopy (HREM), scanning transmission electron microscopy (STEM), etc.] are also used to determine the orientation of different grains and other microstructural features discussed in later chapters. Advanced

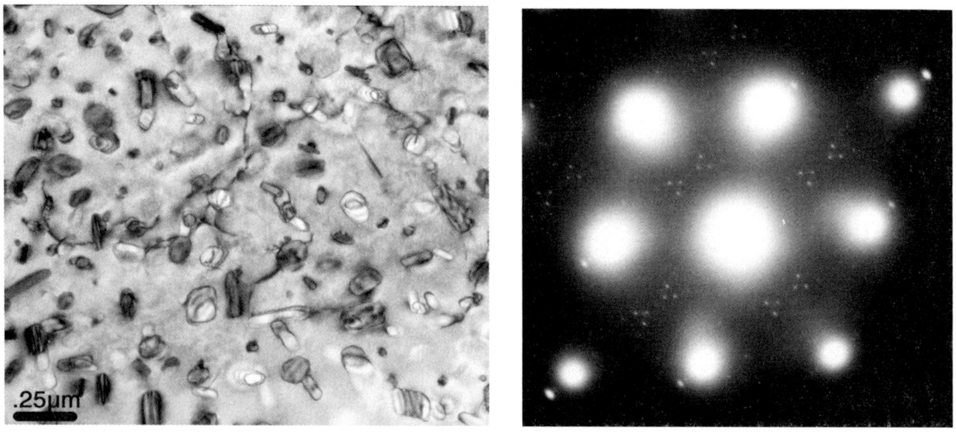

Figure 3-36 A TEM micrograph of an aluminum alloy (Al-7055). The diffraction pattern at the right shows large bright spots that represent diffraction from the aluminum matrix grains. The smaller spots originate from the nanoscale crystals of another compound that is present in the aluminum alloy. (*Courtesy of Dr. Jörg M.K. Wiezorek, University of Pittsburgh.*)

and specialized features associated with TEM also allow chemical mapping of elements in a given material. Some of the disadvantages associated with TEM include

(a) the time-consuming preparation of samples that are almost transparent to the electron beam;
(b) considerable amount of time and skill are required for analysis of the data from a thin, three-dimensional sample that is represented by a two-dimensional image and diffraction pattern;
(c) only a very small volume of the sample is examined; and
(d) the equipment is relatively expensive and requires great care in use.

In general, TEM has become a widely used and accepted research method for analysis of microstructural features at micro- and nano-length scales.

Summary

- Atoms or ions may be arranged in solid materials with either a short-range or long-range order.
- Amorphous materials, such as silicate glasses, metallic glasses, amorphous silicon, and many polymers, have only a short-range order. Amorphous materials form whenever the kinetics of a process involved in the fabrication of a material does not allow the atoms or ions to assume the equilibrium positions. These materials often offer novel properties.
- Crystalline materials, including metals and many ceramics, have both long- and short-range order. The long-range periodicity in these materials is described by the crystal structure.
- The atomic or ionic arrangements of crystalline materials are described by seven general crystal systems, which include fourteen specific Bravais lattices. Examples include simple cubic, body-centered cubic, face-centered cubic, and hexagonal lattices.
- A lattice is a collection of points organized in a unique manner. The basis or motif refers to one or more atoms associated with each lattice point. A crystal structure is defined by the combination of a lattice and a basis. Although there are only fourteen Bravais lattices, there are hundreds of crystal structures.
- A crystal structure is characterized by the lattice parameters of the unit cell, which is a subdivision of the crystal structure that still describes the lattice. By convention, the smallest possible subdivision is chosen. Other characteristics include the number of atoms or ions per unit cell, the coordination number (or number of nearest neighbors) of the atoms or ions in the unit cell, and the packing factor of the atoms or ions in the unit cell.
- Allotropic and polymorphic materials have more than one possible crystal structure. The properties of materials can depend strongly on the particular allotrope or polymorph.
- The atoms of metals having the face-centered cubic and hexagonal close-packed crystal structures are arranged in a manner that occupies the greatest fraction of space. The FCC and HCP structures achieve the closest packing by different stacking sequences of close-packed planes of atoms.
- The greatest achievable packing fraction with spheres of one size is 0.74 and is independent of the radius of the spheres (i.e., atoms and basketballs pack with the same efficiency as long as we deal with a constant radius of an atom and a fixed size basketball).

- Points, directions, and planes within the crystal structure can be identified in a formal manner by the assignment of coordinates and Miller indices.
- Various materials' properties may differ when measured along different directions or planes within a crystal; in this case, the crystal is said to be anisotropic. If the properties are identical in all directions, the material is isotropic. The effect of crystallographic anisotropy may be masked in a polycrystalline material because of the random orientation of grains.
- Interstitial sites, or holes between the atoms or ions in a crystal structure, can be filled by other atoms or ions. The crystal structures of many ceramic materials can be understood by considering how these sites are occupied. Atoms or ions located in interstitial sites play an important role in strengthening materials, influencing the physical properties of materials, and controlling the processing of materials.
- Crystal structures of many ionic materials form by the packing of anions [e.g., oxygen ions (O^{2-})]. Cations fit into coordination polyhedra formed by anions. These polyhedra typically share corners. The conditions of charge neutrality and stoichiometry must be met. Crystal structures of many ceramic materials (e.g., Al_2O_3 and ZrO_2) can be rationalized from these considerations.
- Crystal structures of covalently bonded materials tend to be open. Examples include diamond cubic (e.g., Si, Ge).
- XRD and electron diffraction are used for the determination of the crystal structures of crystalline materials. Transmission electron microscopy can also be used for imaging of microstructural features in materials at smaller length scales.

Glossary

Allotropy The characteristic of an element being able to exist in more than one crystal structure, depending on temperature and pressure.

Amorphous materials Materials, including glasses, that have no long-range order or crystal structure.

Anisotropic Having different properties in different directions.

Atomic radius The apparent radius of an atom, typically calculated from the dimensions of the unit cell, using close-packed directions (depends upon coordination number).

Basal plane The special name given to the close-packed plane in hexagonal close-packed unit cells.

Basis A group of atoms associated with a lattice point (same as motif).

Blow-stretch forming A process used to form plastic bottles.

Bragg's law The relationship describing the angle at which a beam of x-rays of a particular wavelength diffracts from crystallographic planes of a given interplanar spacing.

Bravais lattices The fourteen possible lattices that can be created in three dimensions using lattice points.

Close-packed directions Directions in a crystal along which atoms are in contact.

Close-packed structure Structures showing a packing fraction of 0.74 (FCC and HCP).

Coordination number The number of nearest neighbors to an atom in its atomic arrangement.

Crystal structure The arrangement of the atoms in a material into a regular, repeatable lattice. A crystal structure is fully described by a lattice and a basis.

Crystal systems Cubic, tetragonal, orthorhombic, hexagonal, monoclinic, rhombohedral, and triclinic arrangements of points in space that lead to fourteen Bravais lattices and hundreds of crystal structures.

Crystallography The formal study of the arrangements of atoms in solids.

Crystalline materials Materials comprising one or many small crystals or grains.

Cubic site An interstitial position that has a coordination number of eight.

Defect A microstructural feature representing a disruption in the perfect periodic arrangement of atoms/ions in a crystalline material.

Density Mass per unit volume of a material, usually in units of g/cm^3.

Diamond cubic The crystal structure of carbon, silicon, and other covalently bonded materials.

Diffraction The constructive interference, or reinforcement, of a beam of x-rays or electrons interacting with a material.

Directions of a form or **directions of a family** Crystallographic directions that all have the same characteristics. Denoted by $\langle\rangle$ brackets.

Electron diffraction A method to determine the level of crystallinity at relatively small length scales. Usually conducted in a transmission electron microscope.

Glasses Solid, non-crystalline materials (typically derived from the molten state) that have only short-range atomic order.

Grain A small crystal in a polycrystalline material.

Grain boundaries Regions between grains of a polycrystalline material.

Interplanar spacing Distance between two adjacent parallel planes with the same Miller indices.

Interstitial sites Locations between the host atoms or ions in a crystal into which another—usually different—atom or ion is placed. Typically, the size of this interstitial location is smaller than the atom or ion that is to be introduced.

Isotropic Having the same properties in all directions.

Kepler's conjecture A conjecture made by Johannes Kepler in 1611 that stated that the maximum packing fraction with spheres of uniform size could not exceed $\pi/\sqrt{18}$.

Lattice A collection of points that divide space into smaller equally sized segments.

Lattice parameters The lengths of the sides of the unit cell and the angles between those sides.

Lattice points Points that make up the lattice. The surroundings of each lattice point are identical.

Linear density The number of lattice points per unit length along a direction.

Liquid crystals Polymeric materials that are typically amorphous but can become partially crystalline when an external electric field is applied. The effect of the electric field is reversible.

Long-range order A regular, repetitive arrangement of atoms in a solid that extends over length scales of $\sim$10 nm or larger.

Metallic glass Amorphous metals or alloys obtained using rapid solidification.

Miller-Bravais indices A special shorthand notation to describe the crystallographic planes in hexagonal close-packed unit cells.

Miller indices A shorthand notation to describe certain crystallographic directions and planes in a unit cell.

Motif A group of atoms affiliated with a lattice point (same as basis).

Octahedral site An interstitial position that has a coordination number of six.

Packing factor The fraction of space in a unit cell occupied by atoms.

Planar density The number of atoms per unit area with centers that lie on the plane.

Planes of a form or **planes of a family** Crystallographic planes that all have the same characteristics, although their orientations are different. Denoted by { } braces.

Polycrystalline material A material comprising many grains.

Polymorphism Compounds exhibiting more than one type of crystal structure, depending on temperature and pressure.

Rapid solidification A technique used to cool metals and alloys very quickly.

Repeat distance The distance from one lattice point to the adjacent lattice point along a direction.

Short-range order The regular and predictable arrangement of the atoms over a short distance—usually one or two atom spacings.

Stacking sequence The sequence in which close-packed planes are stacked. If the sequence is *ABABAB*, a hexagonal close-packed unit cell is produced; if the sequence is *ABCABCABC*, a face-centered cubic structure is produced.

Stress-induced crystallization The process of forming crystals by the application of an external stress.

Tetrahedral site An interstitial position that has a coordination number of four.

Transmission electron microscopy A technique for imaging and analysis of microstructures using a high-energy electron beam.

Unit cell A subdivision of the lattice that still retains the overall characteristics of the entire lattice.

X-ray diffraction A technique for analysis of crystalline materials using a beam of x-rays.

Problems

Section 3-1 Short-Range Order versus Long-Range Order

Section 3-2 Amorphous Materials

3-1 What is a crystalline material?

3-2 What is a single crystal?

3-3 State any two applications in which single crystals are used.

3-4 What is a polycrystalline material?

3-5 What is a liquid crystal material?

3-6 What is an amorphous material?

3-7 Why do some materials assume an amorphous structure?

3-8 Explain how it is possible for a substance to exhibit short-range, but not long-range order.

3-9 Can an alloy exist in both crystalline and amorphous forms?

3-10 Approximately how many grains are shown in Figure 3-2(b)?

Section 3-3 Lattice, Basis, Unit Cells, and Crystal Structures

3-11 Define the terms lattice, unit cell, basis, and crystal structure.

3-12 List the seven different crystal systems and the types of Bravais lattices that are associated with their groups.

3-13 Explain why there is no face-centered tetragonal Bravais lattice.

3-14 Calculate the atomic radius in cm for the following:
(a) BCC metal with $a_0 = 0.3294$ nm; and
(b) FCC metal with $a_0 = 4.0862$ Å.

3-15 Calculate the lattice parameter of each of the following elements using their atomic radii: (a) iron, (b) aluminum, (c) copper, and (d) magnesium.

3-16 Determine the crystal structure for the following:
(a) a metal with $a_0 = 4.9489$ Å, $r = 1.75$ Å, and one atom per lattice point; and
(b) a metal with $a_0 = 0.42906$ nm, $r = 0.1858$ nm, and one atom per lattice point.

3-17 The density of potassium, which has the BCC structure, is 0.855 g/cm^3. The atomic weight of potassium is 39.09 g/mol. Calculate
(a) the lattice parameter; and
(b) the atomic radius of potassium.

3-18 The density of thorium, which has the FCC structure, is 11.72 g/cm^3. The atomic weight of thorium is 232 g/mol. Calculate
(a) the lattice parameter; and
(b) the atomic radius of thorium.

3-19 A metal having a cubic structure has a density of 2.6 g/cm^3, an atomic weight of 87.62 g/mol, and a lattice parameter of 6.0849 Å. One atom is associated with each lattice point. Determine the crystal structure of the metal.

3-20 A metal having a cubic structure has a density of 1.892 g/cm^3, an atomic weight of 132.91 g/mol, and a lattice parameter of 6.13 Å. One atom is associated with each lattice point. Determine the crystal structure of the metal.

3-21 Indium has a tetragonal structure, with $a_0 = 0.32517$ nm and $c_0 = 0.49459$ nm. The density is 7.286 g/cm^3, and the atomic weight is 114.82 g/mol. Does indium have the simple tetragonal or body-centered tetragonal structure?

3-22 Cobalt has a hexagonal close-packed structure, with $a_0 = 0.25071$ nm and $c_0 = 0.40686$ nm. The density is 8.832 g/cm^3, and the atomic weight is 58.93 g/mol. Determine
(a) the volume of the unit cell; and
(b) the number of atoms in each unit cell.

3-23 Gallium has an orthorhombic structure, with $a_0 = 0.45258$ nm, $b_0 = 0.45186$ nm, and $c_0 = 0.76570$ nm. The atomic radius is 0.1218 nm. The density is 5.904 g/cm^3, and the atomic weight is 69.72 g/mol. Determine
(a) the number of atoms in each unit cell; and
(b) the packing factor in the unit cell.

3-24 Beryllium has a hexagonal crystal structure, with $a_0 = 0.22858$ nm and $c_0 = 0.35842$ nm. The atomic radius is 0.1143 nm, the density is 1.848 g/cm^3, and the atomic weight is 9.01 g/mol. Determine
(a) the number of atoms in each unit cell; and
(b) the packing factor in the unit cell.

3-25 A typical paper clip weighs 0.59 g and consists of BCC iron. Calculate
(a) the number of unit cells; and
(b) the number of iron atoms in the paper clip. (See Appendix A for required data.)

3-26 Aluminum foil used to package food is approximately 0.001 inch thick. Assume that all of the unit cells of the aluminum are arranged so that a_0 is perpendicular to the foil surface. For a 4 in. × 4 in. square of the foil, determine
(a) the total number of unit cells in the foil; and
(b) the thickness of the foil in number of unit cells. (See Appendix A.)

3-27 Sodium has the body–centered cubic crystal structure and a lattice parameter (axial length) of 4.2906×10^{-8} cm.
(a) Calculate the mass density (g/cm^3) for sodium.
(b) Calculate the atomic radius of a sodium atom according to the hard sphere model of the BCC crystal structure.

3-28 Calculate the density for zinc (HCP) if the c/a ratio is 1.856 and $r = 1.332$ Å.

3-29 Thoria or thorium dioxide can be described as an FCC lattice with a basis of Th (0, 0, 0) and O (1/4, 1/4, 1/4) and O (1/4, 1/4, 3/4). Thorium and thorium dioxide are radioactive materials. Thorium dioxide is commonly used in tungsten electrodes for arc welding. Thoria improves the high temperature properties of the electrodes and promotes emission of electrons from the electrode tip.
(a) How many atoms of each type are there in the conventional unit cell of thorium dioxide?
(b) Is this consistent with the chemical formula of thorium dioxide? Explain.
(c) What is the coordination number (number of nearest neighbors) for each Th atom?
(d) What is the coordination number (number of nearest neighbors) for each O atom?
(e) What type of interstitial site do the oxygen atoms occupy?

3-30 Zinc has the hexagonal close–packed crystal structure. The lattice parameters for zinc are $a = 0.26648$ nm and $c = 0.49470$ nm, and the atomic radius is 0.1332 nm. *Note that zinc does not have the ideal packing factor.*
(a) What is the number of atoms per unit cell in the hexagonal close–packed structure?
(b) Determine the packing factor of the Zn unit cell.
(c) Is the c/a ratio for zinc greater than or less than the ideal HCP c/a ratio? Will slip be harder or easier in zinc compared to the ideal HCP structure? Explain your answer fully.

3-31 Rutile is the name given to a crystal structure commonly adopted by compounds of the form AB_2, where A represents a metal atom and B represents oxygen atoms. One form of rutile has atoms of element A at the unit cell coordinates (0, 0, 0) and (1/2, 1/2, 1/2) and atoms of element B at (1/4, 1/4, 0), (3/4, 3/4, 0), (3/4, 1/4, 1/2), and (1/4, 3/4, 1/2). The unit cell parameters are $a = b \neq c$ and $a = b = \gamma = 90°$. Note that the lattice parameter c is typically smaller than the lattice parameters a and b for the rutile structure.
(a) How many atoms of element A are there per unit cell?
(b) How many atoms of element B are there per unit cell?
(c) Is your answer to part (b) consistent with the stoichiometry of an AB_2 compound? Explain.

(d) Draw the unit cell for rutile. Use a different symbol for each type of atom. Provide a legend indicating which symbol represents which type of atom.

(e) For the simple tetragonal lattice, $a = b \neq c$ and $\alpha = \beta = \gamma = 90°$. There is one lattice point per unit cell located at the corners of the simple tetragonal lattice. Describe the rutile structure as a simple tetragonal lattice and a basis.

3-32 Consider the CuAu crystal structure. It can be described as a simple cubic lattice with a basis of Cu (0, 0, 0), Cu (1/2, 1/2, 0), Au (1/2, 0, 1/2), and Au (0, 1/2, 1/2).

(a) How many atoms of each type are there per unit cell?

(b) Draw the unit cell for CuAu. Use a different symbol for each type of atom. Provide a legend indicating which symbol represents which type of atom.

(c) Give an alternative lattice and basis representation for CuAu for which one atom of the basis is Au (0, 0, 0).

(d) A related crystal structure is that of Cu_3Au. This unit cell is similar to the face-centered cubic unit cell with Au at the corners of the unit cell and Cu at all of the face-centered positions. Describe this structure as a lattice and a basis.

(e) The Cu_3Au crystal structure is similar to the FCC crystal structure, but it does not have the face-centered cubic lattice. Explain briefly why this is the case.

3-33 Nanowires are high aspect-ratio metal or semiconducting wires with diameters on the order of 1 to 100 nanometers and typical lengths of 1 to 100 microns. Nanowires likely will be used in the future to create high-density electronic circuits.

Nanowires can be fabricated from ZnO. ZnO has the wurtzite structure. The wurtzite structure is a hexagonal lattice with four atoms per lattice point at Zn(0, 0, 0), Zn(2/3, 1/3, 1/2), O(0, 0, 3/8), and O(2/3, 1/3, 7/8).

(a) How many atoms are there in the conventional unit cell?

(b) If the atoms were located instead at Zn(0, 0, 0), Zn(1/3, 2/3, 1/2), O(0, 0, 3/8), and O(1/3, 2/3, 7/8), would the structure be different? Please explain.

(c) For ZnO, the unit cell parameters are $a = 3.24$ Å and $c = 5.19$ Å. (*Note:* This is not the ideal HCP c/a ratio.) A typical ZnO nanowire is 20 nm in diameter and 5 mm long. Assume that the nanowires are cylindrical. Approximately how many atoms are there in a single ZnO nanowire?

3-34 Calculate the atomic packing fraction for the hexagonal close-packed crystal structure for which $c = \left(\sqrt{\frac{8}{3}}\right)a$. Remember that the base of the unit cell is a parallelogram.

3-35 Magnesium has the hexagonal close–packed crystal structure. The lattice parameters for magnesium are $a = 0.32087$ nm and $c = 0.5209$ nm, and the atomic radius is 0.1604 nm. *Note that magnesium does not have the ideal atomic packing fraction.*

(a) What is the number of atoms per unit cell in the hexagonal close–packed structure?

(b) Determine the atomic packing fraction of the Mg unit cell.

(c) Is the c/a ratio for Mg greater than or less than the ideal HCP c/a ratio? Will slip be harder or easier in Mg compared to the ideal HCP structure? Explain your answer fully.

Section 3-4 Allotropic or Polymorphic Transformations

3-36 What is the difference between an allotrope and a polymorph?

3-37 What are the different polymorphs of zirconia?

3-38 A number of metals undergo an allotropic transformation from one crystal structure to another at a specific temperature. There is generally a volume change that accompanies the transformation. What is the practical significance of such data?

3-39 Above 882°C, titanium has a BCC crystal structure with $a = 0.332$ nm. Below this temperature, titanium has an HCP structure with $a = 0.29503$ nm and $c = 0.46831$ nm. Determine the percent volume change when BCC titanium transforms to HCP titanium. Is this a contraction or expansion?

3-40 α-Mn has a cubic structure with $a_0 = 0.8931$ nm and a density of 7.47 g/cm^3. β-Mn has a different cubic structure with $a_0 = 0.6326$ nm and a density of 7.26 g/cm^3. The atomic weight of manganese is 54.938 g/mol, and the atomic radius is 0.112 nm. Determine the percent volume change that would occur if α-Mn transforms to β-Mn.

3-41 Calculate the theoretical density of the three polymorphs of zirconia. The lattice constants for the monoclinic form are a = 5.156, b = 5.191, and c = 5.304 Å, respectively. The angle β for the monoclinic unit cell is 98.9°. The lattice constants for the tetragonal unit cell are a = 5.094 and c = 5.304 Å, respectively. Cubic zirconia has a lattice constant of 5.124 Å. Each unit cell contains four formula units of ZrO_2.

3-42 From the information in this chapter, calculate the volume change that will occur when the cubic form of zirconia transforms into a tetragonal form.

3-43 Monoclinic zirconia cannot be used effectively for manufacturing oxygen sensors or other devices. Explain.

3-44 What is meant by the term stabilized zirconia?

3-45 State any two applications of stabilized zirconia ceramics.

Section 3-5 Points, Directions, and Planes in the Unit Cell

3-46 Explain the significance of crystallographic directions using an example of an application.

3-47 Why are Fe-Si alloys used in magnetic applications "grain oriented"?

3-48 Draw the planes (123), (323), and $(\bar{3}\bar{2}2)$.

3-49 Determine the Miller indices of the directions for the following points: (a) from (1, 0, 2) to (2, 4, 1); (b) from (2, 1, 3) to (5, 4, 2); and (c) from (3, 1, 3) to (9, 1, 5).

3-50 Indicate the directions (a) [111], (b) [025], and (c) [414] within a unit cell.

3-51 Determine the Miller indices for the directions in the cubic unit cell shown in Figure 3-37.

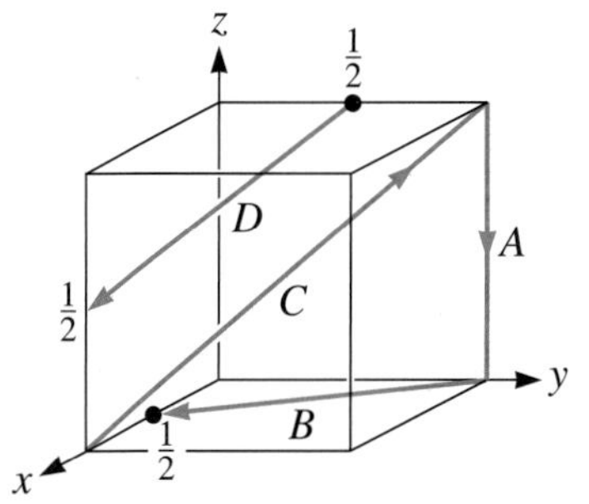

Figure 3-37 Directions in a cubic unit cell (for Problem 3-51).

3-52 Determine the indices for the directions in the cubic unit cell shown in Figure 3-38.

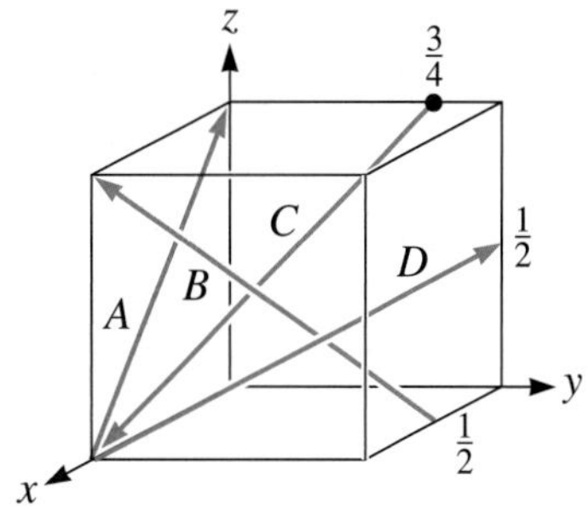

Figure 3-38 Directions in a cubic unit cell (for Problem 3-52).

3-53 Indicate the planes a) (100), b) (134), and c) (101) within a unit cell.

3-54 Determine the indices for the planes in the cubic unit cell shown in Figure 3-39.

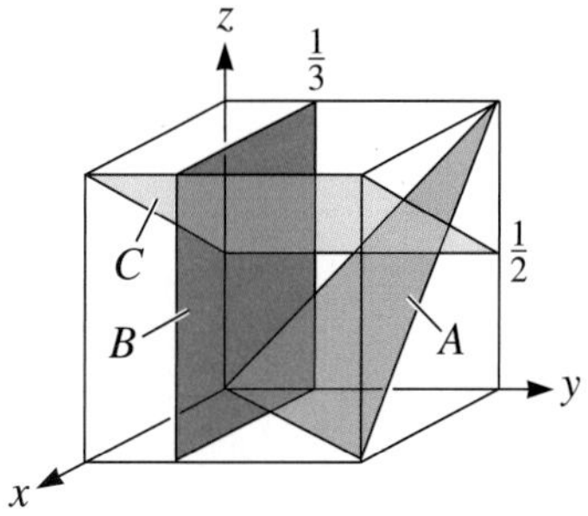

Figure 3-39 Planes in a cubic unit cell (for Problem 3-54).

3-55 Determine the indices for the planes in the cubic unit cell shown in Figure 3-40.

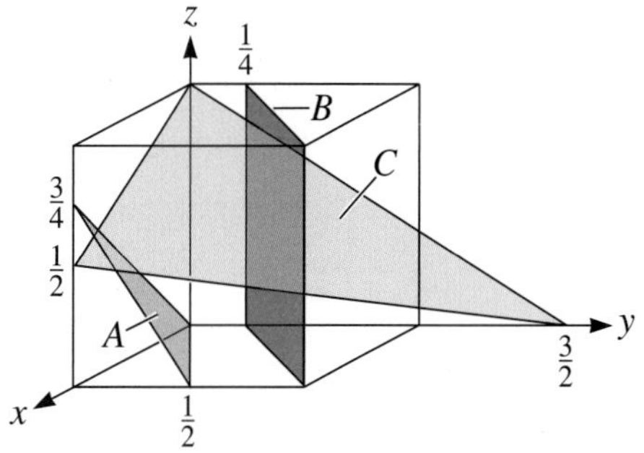

Figure 3-40 Planes in a cubic unit cell (for Problem 3-55).

3-56 Determine the indices for the directions in the hexagonal lattice shown in Figure 3-41, using both the three-digit and four-digit systems.

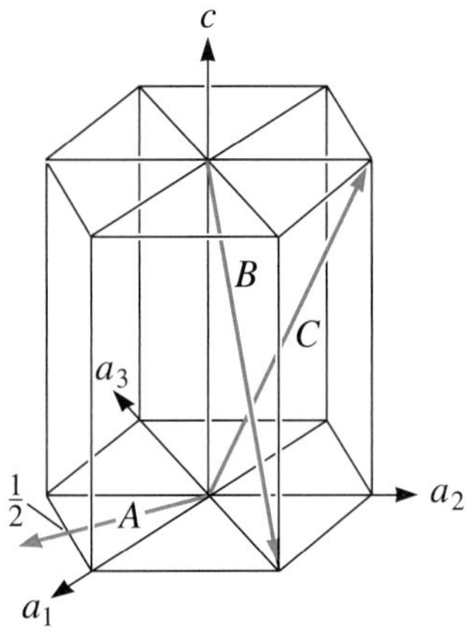

Figure 3-41 Directions in a hexagonal lattice (for Problem 3-56).

3-57 Determine the indices for the directions in the hexagonal lattice shown in Figure 3-42, using both the three-digit and four-digit systems.

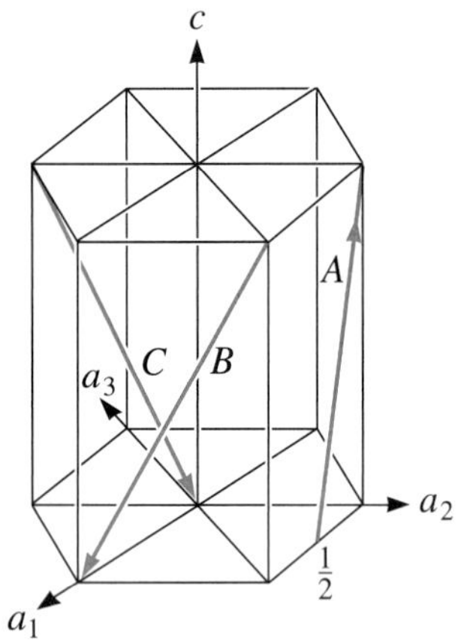

Figure 3-42 Directions in a hexagonal lattice (for Problem 3-57).

3-58 Determine the indices for the planes in the hexagonal lattice shown in Figure 3-43.

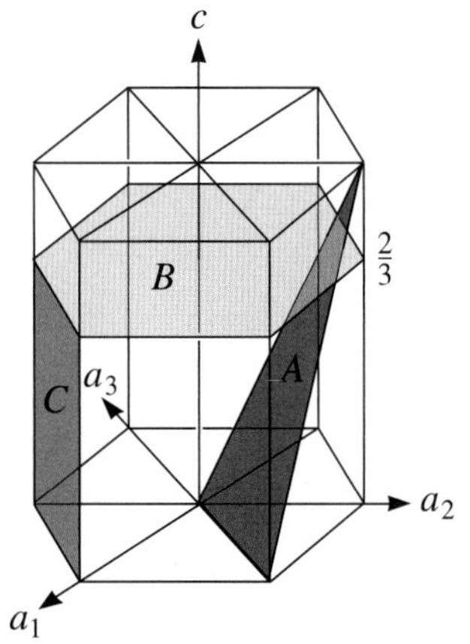

Figure 3-43 Planes in a hexagonal lattice (for Problem 3-58).

3-59 Determine the indices for the planes in the hexagonal lattice shown in Figure 3-44.

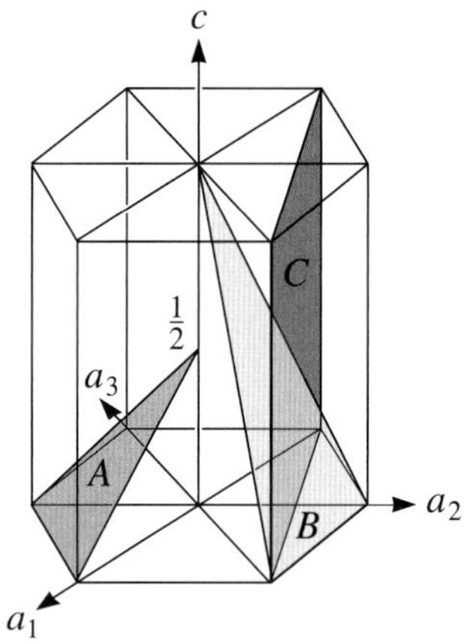

Figure 3-44 Planes in a hexagonal lattice (for Problem 3-59).

3-60 Sketch the following planes and directions within a cubic unit cell:
(a) [101] (b) $[0\bar{1}0]$ (c) $[12\bar{2}]$ (d) [301]
(e) $[\bar{2}01]$ (f) $[2\bar{1}3]$ (g) $(0\bar{1}\bar{1})$ (h) (102)
(i) (002) (j) $(1\bar{3}0)$ (k) $(\bar{2}12)$ (l) $(31\bar{2})$

3-61 Sketch the following planes and directions within a cubic unit cell:
(a) $[1\bar{1}0]$ (b) $[\bar{2}\bar{2}1]$ (c) [410] (d) $[0\bar{1}2]$
(e) $[\bar{3}\bar{2}1]$ (f) $[1\bar{1}1]$ (g) $(11\bar{1})$ (h) $(01\bar{1})$
(i) (030) (j) $(\bar{1}21)$ (k) $(11\bar{3})$ (l) $(0\bar{4}1)$

3-62 Sketch the following planes and directions within a hexagonal unit cell:
(a) $[01\bar{1}0]$ (b) $[11\bar{2}0]$ (c) $[\bar{1}011]$
(d) (0003) (e) $(\bar{1}010)$ (f) $(01\bar{1}1)$

3-63 Sketch the following planes and directions within a hexagonal unit cell:
(a) $[\bar{2}110]$ (b) $[11\bar{2}1]$ (c) $[10\bar{1}0]$
(d) $(1\bar{2}10)$ (e) $(\bar{1}\bar{1}22)$ (f) $(12\bar{3}0)$

3-64 What are the indices of the six directions of the form ⟨110⟩ that lie in the $(11\bar{1})$ plane of a cubic cell?

3-65 What are the indices of the four directions of the form ⟨111⟩ that lie in the $(\bar{1}01)$ plane of a cubic cell?

3-66 Determine the number of directions of the form ⟨110⟩ in a tetragonal unit cell and compare to the number of directions of the form ⟨110⟩ in an orthorhombic unit cell.

3-67 Determine the angle between the [110] direction and the (110) plane in a tetragonal unit cell; then determine the angle between the [011] direction and the (011) plane in a tetragonal cell. The lattice parameters are $a_0 = 4.0$ Å and $c_0 = 5.0$ Å. What is responsible for the difference?

3-68 Determine the Miller indices of the plane that passes through three points having the following coordinates:
(a) (0, 0, 1); (1, 0, 0); and (1/2, 1/2, 0)
(b) (1/2, 0, 1); (1/2, 0, 0); and (0, 1, 0)
(c) (1, 0, 0); (0, 1, 1/2); and (1, 1/2, 1/4)
(d) (1, 0, 0); (0, 0, 1/4); and (1/2, 1, 0)

3-69 Calculate and compare the linear densities for the ⟨100⟩, ⟨110⟩, and ⟨111⟩ directions in a BCC unit cell. Which direction is the most close-packed (dense)?

3-70 Calculate and compare the linear densities for the ⟨100⟩, ⟨110⟩, and ⟨111⟩ directions in an FCC unit cell. Which direction is the most close-packed (dense)?

3-71 Calculate and compare the planar densities for the {100}, {110}, and {111} planes in a BCC unit cell. Which plane is the most close-packed (dense)?

3-72 Calculate and compare the planar densities for the {100}, {110}, and {111} planes in an FCC unit cell. Which plane is the most densely packed?

3-73 Determine the repeat distance, linear density, and packing fraction for FCC nickel, which has a lattice parameter of 0.35167 nm, in the [100], [110], and [111] directions. Which of these directions is close packed?

3-74 Determine the repeat distance, linear density, and packing fraction for BCC lithium, which has a lattice parameter of 0.35089 nm, in the [100], [110], and [111] directions. Which of these directions is close packed?

3-75 Determine the repeat distance, linear density, and packing fraction for HCP magnesium in the $[\bar{2}110]$ direction and the $[11\bar{2}0]$ direction. The lattice parameters for HCP magnesium are given in Appendix A.

3-76 Determine the planar density and packing fraction for FCC nickel in the (100), (110), and (111) planes. Which, if any, of these planes are close packed?

3-77 Determine the planar density and packing fraction for BCC lithium in the (100), (110), and (111) planes. Which, if any, of these planes are close packed?

3-78 Suppose that FCC rhodium is produced as a 1 mm-thick sheet, with the (111) plane parallel to the surface of the sheet. How many (111) interplanar spacings d_{111} thick is the sheet? See Appendix A for necessary data.

3-79 In an FCC unit cell, how many d_{111} are present between the (0, 0, 0) point and the (1, 1, 1) point?

3-80 What are the stacking sequences in the FCC and HCP structures?

Section 3-6 Interstitial Sites

3-81 Determine the minimum radius of an atom that will just fit into
(a) the tetrahedral interstitial site in FCC nickel; and
(b) the octahedral interstitial site in BCC lithium.

3-82 What are the coordination numbers for octahedral and tetrahedral sites?

3-83 The atomic packing fraction for the FCC crystal structure is 0.74. This is the densest arrangement of atoms achievable in three dimensions when all atoms have the same radius R. If atoms in the hard sphere model occupy 74% of the volume of the unit cell, then the other 26% is empty space.

Imagine that the empty spaces located at the center of each edge and at the body-centered position of the FCC unit cell are filled with smaller spheres of radius r such that $r = 0.414R$. The smaller atoms fit perfectly in between the atoms of radius R.
(a) By counting in the usual way, how many atoms of radius r are there per FCC unit cell?
(b) What are the coordinates of these atoms of radius r? Do not double count atoms. Provide the same number of coordinates as the number of atoms of radius r per unit cell.
(c) What is the atomic packing fraction for this structure?

Section 3-7 Crystal Structures of Ionic Materials

3-84 What is the radius of an atom that will just fit into the octahedral site in FCC copper without disturbing the crystal structure?

3-85 Using the ionic radii given in Appendix B, determine the coordination number expected for the following compounds:
(a) Y_2O_3 (b) UO_2 (c) BaO (d) Si_3N_4
(e) GeO_2 (f) MnO (g) MgS (h) KBr

3-86 A particular unit cell is cubic with ions of type A located at the corners and face-centers of the unit cell and ions of type B located at the midpoint of each edge of the cube and at the body-centered position. The ions contribute to the unit cell in the usual way (1/8 ion contribution for each ion at the corners, etc.).

(a) How many ions of each type are there per unit cell?
(b) Describe this structure as a lattice and a basis. Check to be sure that the number of ions per unit cell given by your description of the structure as a lattice and a basis is consistent with your answer to part (a).
(c) What is the coordination number of each ion?
(d) What is the name commonly given to this crystal structure?

3-87 Would you expect NiO to have the cesium chloride, sodium chloride, or zincblende structure? Based on your answer, determine
(a) the lattice parameter;
(b) the density; and
(c) the packing factor.

3-88 Would you expect UO_2 to have the sodium chloride, zincblende, or fluorite structure? Based on your answer, determine
(a) the lattice parameter;
(b) the density; and
(c) the packing factor.

3-89 Would you expect BeO to have the sodium chloride, zincblende, or fluorite structure? Based on your answer, determine
(a) the lattice parameter;
(b) the density; and
(c) the packing factor.

3-90 Would you expect CsBr to have the sodium chloride, zincblende, fluorite, or cesium chloride structure? Based on your answer, determine
(a) the lattice parameter;
(b) the density; and
(c) the packing factor.

3-91 Sketch the ion arrangement of the (110) plane of ZnS (with the zincblende structure) and compare this arrangement to that on the (110) plane of CaF_2 (with the fluorite structure). Compare the planar packing fraction on the (110) planes for these two materials.

3-92 MgO, which has the sodium chloride structure, has a lattice parameter of 0.396 nm. Determine the planar density and the planar packing fraction for the (111) and (222) planes of MgO. What ions are present on each plane?

3-93 Consider a crystal structure that has a unit cell with the following axial lengths, interaxial angles, and atomic coordinates: $a = b = c = 2R$; $\alpha = \beta = \gamma = 90°$; A at (0, 0, 0) with radius R; and B at (1/2, 1/2, 1/2) with radius $r < R$.
(a) Draw the unit cell for this structure. Use large open circles for atom A and small filled circles for atom B.
(b) Write this structure as a lattice and a basis.
(c) How many atoms are there of each type per unit cell?
(d) Given that $a = b = c = 2R$, derive the radius ratio r/R such that an atom B with radius r will exactly fit into the largest interstitial space formed by the A atoms with radius R.
(e) Derive the atomic packing fraction (APF) for this structure. Begin by writing the definition of the APF.
(f) Calculate the mass density of this material. The atomic weight of element A is 208.98 g/mol and 127.60 g/mol for element B. The lattice parameter is 3.356×10^{-8} cm.

3-94 Consider a crystal structure that consists of a BCC lattice with a basis of a large atom of type A with radius R at (0, 0, 0) and a small atom of type B with radius r at (0, 0, 1/2).
(a) Draw the unit cell for this structure. Use large open circles for atom A and small filled circles for atom B.
(b) How many of each type of atom are there in the conventional unit cell?
(c) Describe a unit cell for this structure based on the drawing in (a) by giving the axial lengths and ratios (a, b, and c), interaxial angles (α, β, and γ), and coordinates of the atoms.
(d) Assume that for atoms of type A the relationship between R and a is $4R = a\sqrt{3}$. Also assume that atom B has a radius r such that $r/R = 0.2$. Calculate the atomic packing fraction of this unit cell.

Section 3-8 Covalent Structures

3-95 Calculate the theoretical density of α-Sn. Assume α-Sn has the diamond cubic structure and obtain the atomic radius information from Appendix B.

3-96 Calculate the theoretical density of Ge. Assume Ge has the diamond cubic structure and obtain the radius information from Appendix B.

Section 3-9 Diffraction Techniques for Crystal Structure Analysis

3-97 A diffracted x-ray beam is observed from the (220) planes of iron at a 2θ angle of 99.1° when x-rays of 0.15418 nm wavelength are used. Calculate the lattice parameter of the iron.

3-98 A diffracted x-ray beam is observed from the (311) planes of aluminum at a 2θ angle of 78.3° when x-rays of 0.15418 nm wavelength are used. Calculate the lattice parameter of the aluminum.

3-99 Figure 3-45 shows the results of an x-ray diffraction experiment in the form of the intensity of the diffracted peak versus the 2θ diffraction angle. If x-rays with a wavelength of 0.15418 nm are used, determine
(a) the crystal structure of the metal;
(b) the indices of the planes that produce each of the peaks; and
(c) the lattice parameter of the metal.

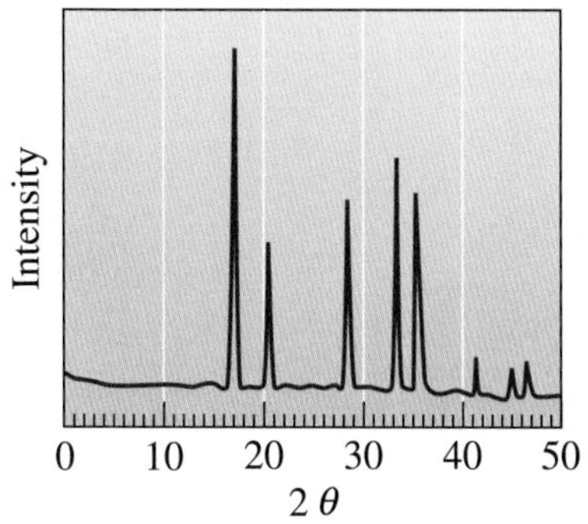

Figure 3-45 XRD pattern (for Problem 3-99).

3-100 Figure 3-46 shows the results of an x-ray diffraction experiment in the form of the intensity of the diffracted peak versus the 2θ diffraction angle. If x-rays with a wavelength of 0.07107 nm are used, determine
(a) the crystal structure of the metal;
(b) the indices of the planes that produce each of the peaks; and
(c) the lattice parameter of the metal.

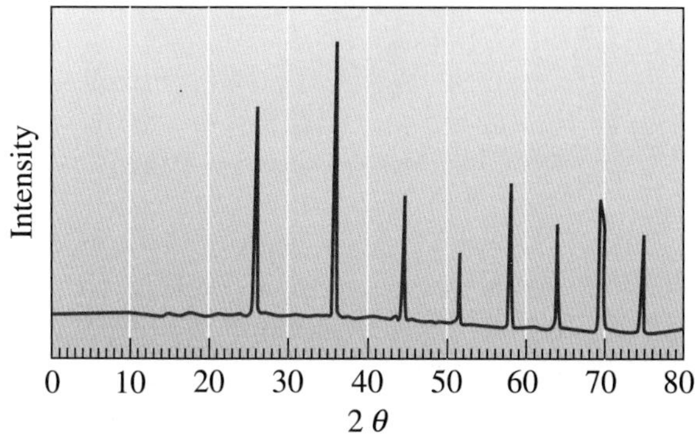

Figure 3-46 XRD pattern (for Problem 3-100).

3-101 A sample of zirconia contains cubic and monoclinic polymorphs. What will be a good analytical technique to detect the presence of these two different polymorphs?

Design Problems

3-102 You would like to sort iron specimens, some of which are FCC and others BCC. Design an x-ray diffraction method by which this can be accomplished.

3-103 You want to design a material for making kitchen utensils for cooking. The material should be transparent and withstand repeated heating and cooling. What kind of materials could be used to design such transparent and durable kitchen-ware?

Computer Problems

3-104 Table 3-1 contains formulas for the volume of different types of unit cells. Write a computer program to calculate the unit cell volume in the units of Å^3 and nm^3. Your program should prompt the user to input the (a) type of unit cell, (b) necessary lattice constants, and (c) angles. The program then should recognize the inputs made and use the appropriate formula for the calculation of unit cell volume.

3-105 Write a computer program that will ask the user to input the atomic mass, atomic radius, and cubic crystal structure for an element. The program output should be the packing fraction and the theoretical density.

K Knovel® Problems

K3-1 Determine the crystal lattice parameters and mass densities for GaN, GaP, GaAs, GaSb, InN, InP, InAs, and InSb semiconductors. Compare the data for lattice parameters from at least two different sources.

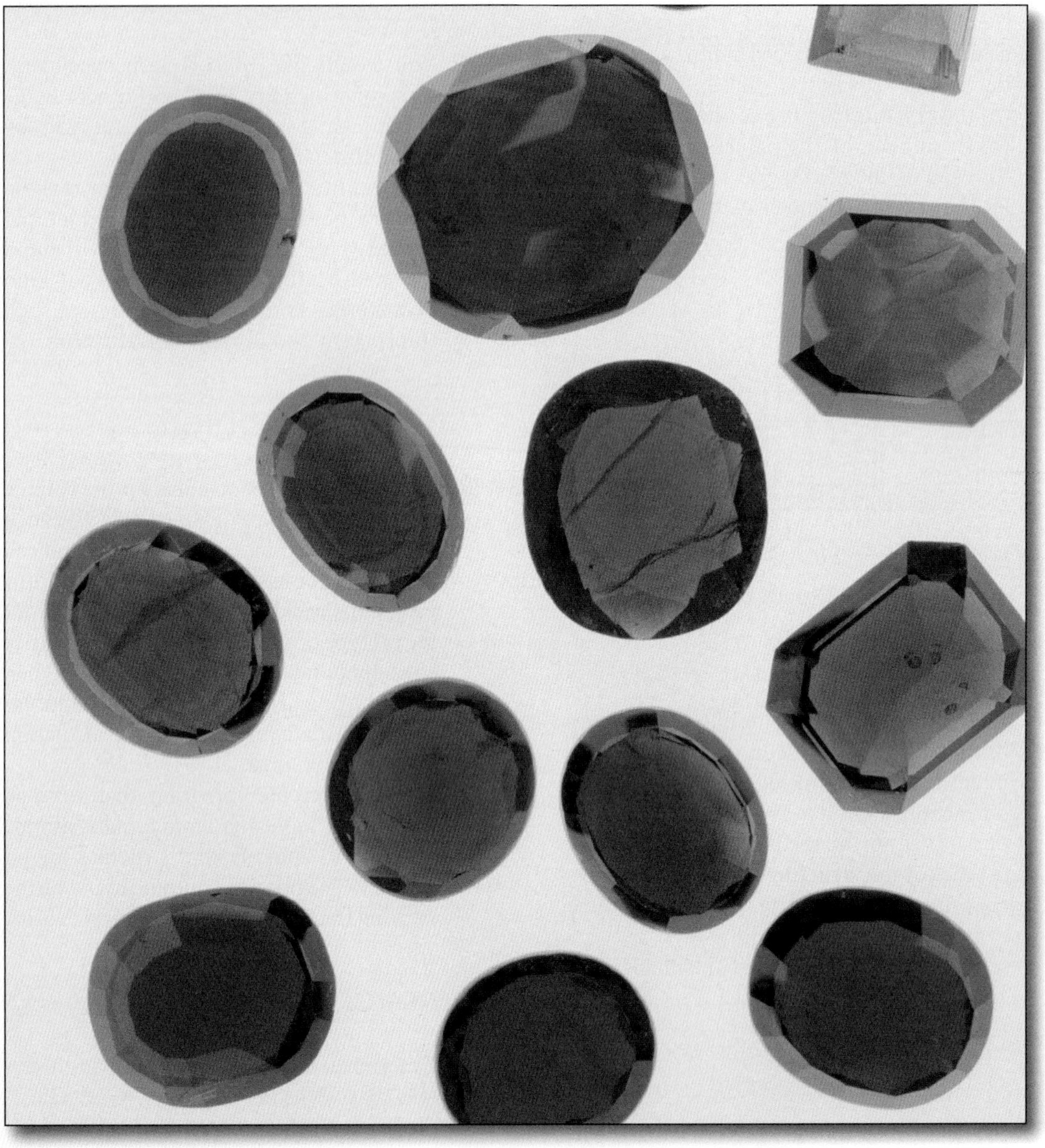

What makes a ruby red? The addition of about 1% chromium oxide in alumina creates defects. An electronic transition between defect levels produces the red ruby crystal. Similarly, incorporation of Fe^{2+} and Ti^{4+} makes the blue sapphire. (*Lawrence Lawry/Photodisc/Getty Images*)

CHAPTER

4

Imperfections in the Atomic and Ionic Arrangements

Have You Ever Wondered?

- *Why do silicon crystals used for semiconductor wafers contain trace amounts of dopants such as phosphorus or boron?*
- *What makes steel considerably harder and stronger than pure iron?*
- *Why do we use high-purity copper as a conductor in electrical applications?*
- *Why do FCC metals (such as copper and aluminum) tend to be more ductile than BCC and HCP metals?*
- *How can metals be strengthened?*

Chapter Learning Objectives

The key objectives of this chapter are to

- Describe the various point defects in a crystalline material (vacancy, substitutional defect, interstitial defect).
- Explain the concept and geometry of the edge dislocation and the screw dislocation and how dislocations strengthen crystalline materials.
- Identify the likely slip system for a given crystal structure.
- Explain how grain boundaries strengthen polycrystalline materials.
- Explain the role of grain size in strengthening polycrystalline materials.
- Explain the process of work hardening and how it affects the strength of crystalline materials.

The arrangement of the atoms or ions in engineered materials contains imperfections or defects. These defects often have a profound effect on the properties of materials. In this chapter, we introduce the three basic types of imperfections: point defects, line defects (or dislocations), and surface defects. These imperfections only represent defects in or deviations from the perfect or ideal atomic or ionic arrangements expected in a given crystal structure. The material is not considered defective from a technological viewpoint. In many applications, the presence of such defects is useful. Often the defects may be created intentionally to produce a desired set of electronic, magnetic, optical, or mechanical properties. For example, pure iron is relatively soft, yet, when we add a small amount of carbon, we create defects in the crystal structure of iron and transform it into a plain carbon steel that exhibits considerably higher strength. Similarly, a crystal of pure alumina is transparent and colorless, but, when we add a small amount of chromium, it creates a special defect, resulting in a beautiful red ruby crystal. The effect of point defects, however, is not always desirable. When we want to use copper as a conductor for microelectronics, we use the highest purity available. This is because even small levels of impurities will cause an orders of magnitude increase in the electrical resistivity of copper!

4-1 Point Defects

Point defects are localized disruptions in otherwise perfect atomic or ionic arrangements in a crystal structure. Even though we call them point defects, the disruption affects a region involving several atoms or ions. These imperfections, shown in Figure 4-1, may be introduced by movement of the atoms or ions when they gain energy by heating, during processing of the material, or by the intentional or unintentional introduction of impurities. Typically, **impurities** are elements or compounds that are present from raw materials or processing. For example, silicon crystals grown in quartz crucibles contain oxygen as an impurity. **Dopants**, on the other hand, are elements or compounds that are deliberately added, in known concentrations, at specific locations in the microstructure, with an intended beneficial effect on properties or processing. In general, the effect of impurities is deleterious, whereas the effect of dopants on the properties of materials is useful. Phosphorus and boron are examples of dopants that are added to silicon crystals to improve the electrical properties of pure silicon.

A point defect typically involves one atom or ion, or a pair of atoms or ions, and thus is different from **extended defects**, such as dislocations or grain boundaries. An important "point" about point defects is that although the defect occurs at one or two sites, their presence is "felt" over much larger distances in the crystalline material.

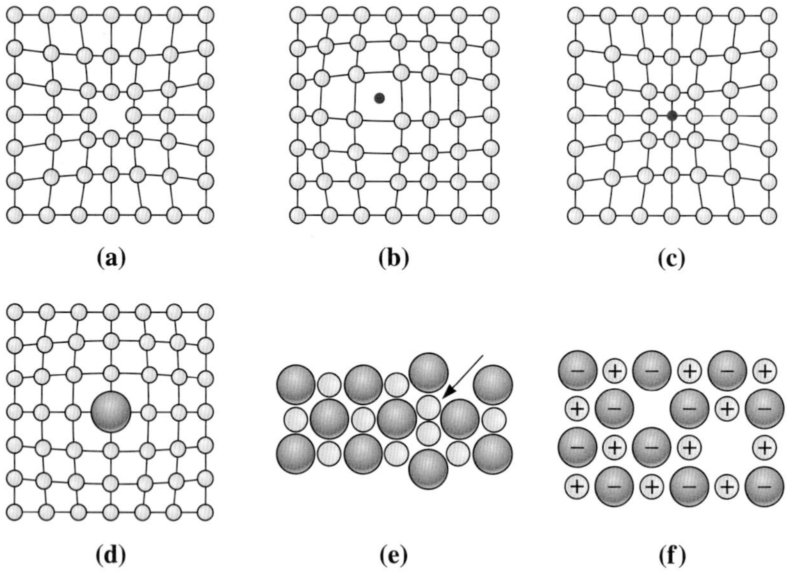

Figure 4-1 Point defects: (a) vacancy, (b) interstitial atom, (c) small substitutional atom, (d) large substitutional atom, (e) Frenkel defect, and (f) Schottky defect. All of these defects disrupt the perfect arrangement of the surrounding atoms.

Vacancies A **vacancy** is produced when an atom or an ion is missing from its normal site in the crystal structure as in Figure 4-1(a). When atoms or ions are missing (i.e., when vacancies are present), the overall randomness or entropy of the material increases, which increases the thermodynamic stability of a crystalline material. All crystalline materials have vacancy defects. Vacancies are introduced into metals and alloys during solidification, at high temperatures, or as a consequence of radiation damage. Vacancies play an important role in determining the rate at which atoms or ions move around or diffuse in a solid material, especially in pure metals. We will see this in greater detail in Chapter 5.

At room temperature (~298 K), the concentration of vacancies is small, but the concentration of vacancies increases exponentially as the temperature increases, as shown by the following Arrhenius type behavior:

$$n_v = n \exp\left(\frac{-Q_v}{RT}\right) \tag{4-1}$$

where

n_v is the number of vacancies per cm^3;

n is the number of atoms per cm^3;

Q_v is the energy required to produce one mole of vacancies, in cal/mol or joules/mol;

R is the gas constant, $1.987 \dfrac{\text{cal}}{\text{mol} \cdot \text{K}}$ or $8.314 \dfrac{\text{J}}{\text{mol} \cdot \text{K}}$; and

T is the temperature in degrees kelvin.

Due to the large thermal energy near the melting temperature, there may be as many as one vacancy per 1000 atoms. Note that this equation provides the equilibrium concentration of vacancies at a given temperature. It is also possible to retain the concentration of vacancies produced at a high temperature by quenching the material rapidly. Thus, in many situations, the concentration of vacancies observed at room temperature is not the equilibrium concentration predicted by Equation 4-1.

Example 4-1 *The Effect of Temperature on Vacancy Concentrations*

Calculate the concentration of vacancies in copper at room temperature (25°C). What temperature will be needed to heat treat copper such that the concentration of vacancies produced will be 1000 times more than the equilibrium concentration of vacancies at room temperature? Assume that 20,000 cal are required to produce a mole of vacancies in copper.

SOLUTION

The lattice parameter of FCC copper is 0.36151 nm. There are four atoms per unit cell; therefore, the number of copper atoms per cm^3 is

$$n = \frac{4 \text{ atoms/cell}}{(3.6151 \times 10^{-8} \text{ cm})^3} = 8.466 \times 10^{22} \text{ copper atoms/cm}^3$$

At room temperature, $T = 25 + 273 = 298$ K:

$$n_v = n \exp\left(\frac{-Q_v}{RT}\right)$$

$$= \left(8.466 \times 10^{22} \frac{\text{atoms}}{\text{cm}^3}\right) \exp\left[\frac{-20{,}000 \dfrac{\text{cal}}{\text{mol}}}{\left(1.987 \dfrac{\text{cal}}{\text{mol} \cdot \text{K}}\right)(298 \text{ K})}\right]$$

$$= 1.814 \times 10^{8} \text{ vacancies/cm}^3$$

We wish to find a heat treatment temperature that will lead to a concentration of vacancies that is 1000 times higher than this number, or $n_v = 1.814 \times 10^{11}$ vacancies/cm^3.

We could do this by heating the copper to a temperature at which this number of vacancies forms:

$$n_v = 1.814 \times 10^{11} = n \exp\left(\frac{-Q_v}{RT}\right)$$

$$= (8.466 \times 10^{22}) \exp[-20{,}000/(1.987T)]$$

$$\exp\left(\frac{-20{,}000}{1.987T}\right) = \frac{1.814 \times 10^{11}}{8.466 \times 10^{22}} = 2.14 \times 10^{-12}$$

$$\frac{-20{,}000}{1.987T} = \ln(2.14 \times 10^{-12}) = -26.87$$

$$T = \frac{20{,}000}{(1.987)(26.87)} = 375 \text{ K} = 102^\circ\text{C}$$

By heating the copper slightly above 100°C, waiting until equilibrium is reached, and then rapidly cooling the copper back to room temperature, the number of vacancies trapped in the structure may be one thousand times greater than the equilibrium number of vacancies at room temperature. Thus, vacancy concentrations encountered in materials are often dictated by both thermodynamic and kinetic factors.

Example 4-2 *Vacancy Concentrations in Iron*

Calculate the theoretical density of iron, and then determine the number of vacancies needed for a BCC iron crystal to have a density of 7.874 g/cm^3. The lattice parameter of iron is 2.866×10^{-8} cm.

SOLUTION

The theoretical density of iron can be calculated from the lattice parameter and the atomic mass. Since the iron is BCC, two iron atoms are present in each unit cell.

$$\rho = \frac{(2 \text{ atoms/cell})(55.847 \text{ g/mol})}{(2.866 \times 10^{-8} \text{ cm})^3 (6.022 \times 10^{23} \text{ atoms/mol})} = 7.879 \text{ g/cm}^3$$

This calculation assumes that there are no imperfections in the crystal. Let's calculate the number of iron atoms and vacancies that would be present in each unit cell for a density of 7.874 g/cm^3:

$$\rho = \frac{(X \text{ atoms/cell})(55.847 \text{ g/mol})}{(2.866 \times 10^{-8} \text{ cm})^3 (6.022 \times 10^{23} \text{ atoms/mol})} = 7.874 \text{ g/cm}^3$$

$$X \text{ atoms/cell} = \frac{(7.874 \text{ g/cm}^3)(2.866 \times 10^{-8} \text{ cm})^3 (6.022 \times 10^{23} \text{ atoms/mol})}{(55.847 \text{ g/mol})} = 1.99878$$

There should be $2.00 - 1.99878 = 0.00122$ vacancies per unit cell. The number of vacancies per cm^3 is

$$\text{Vacancies/cm}^3 = \frac{0.00122 \text{ vacancies/cell}}{(2.866 \times 10^{-8} \text{ cm})^3} = 5.18 \times 10^{19}$$

Note that other defects such as grain boundaries in a polycrystalline material also contribute to a density lower than the theoretical value.

Interstitial Defects An **interstitial defect** is formed when an extra atom or ion is inserted into the crystal structure at a normally unoccupied position, as in Figure 4-1(b). The interstitial sites were illustrated in Table 3-6. Interstitial atoms or ions, although much smaller than the atoms or ions located at the lattice points, are still larger than the interstitial sites that they occupy; consequently, the surrounding crystal region is compressed and distorted. Interstitial atoms such as hydrogen are often present as impurities, whereas carbon atoms are intentionally added to iron to produce steel. For small concentrations, carbon atoms occupy interstitial sites in the iron crystal structure, introducing a stress in the localized region of the crystal in their vicinity. As we will see, the introduction of interstitial atoms is one important way of increasing the strength of metallic materials. Unlike vacancies, once introduced, the number of interstitial atoms or ions in the structure remains nearly constant, even when the temperature is changed.

Example 4-3 *Sites for Carbon in Iron*

In FCC iron, carbon atoms are located at *octahedral* sites, which occur at the center of each edge of the unit cell at sites such as (0, 0, 1/2) and at the center of the unit cell (1/2, 1/2, 1/2). In BCC iron, carbon atoms enter *tetrahedral* sites, such as (0, 1/2, 1/4). The lattice parameter is 0.3571 nm for FCC iron and 0.2866 nm for BCC iron. Assume that carbon atoms have a radius of 0.071 nm. (a) Would we expect a greater distortion of the crystal by an interstitial carbon atom in FCC or BCC iron? (b) What would be the atomic percentage of carbon in each type of iron if all the interstitial sites were filled?

SOLUTION

(a) We can calculate the size of the interstitial site in BCC iron at the (0, 1/2, 1/4) location with the help of Figure 4-2(a). The radius R_{BCC} of the iron atom is

$$R_{BCC} = \frac{\sqrt{3}a_0}{4} = \frac{(\sqrt{3})(0.2866 \text{ nm})}{4} = 0.1241 \text{ nm}$$

From Figure 4-2(a), we find that

$$\left(\frac{1}{2}a_0\right)^2 + \left(\frac{1}{4}a_0\right)^2 = (r_{\text{interstitial}} + R_{BCC})^2$$

$$(r_{\text{interstitial}} + R_{BCC})^2 = 0.3125a_0^2 = (0.3125)(0.2866 \text{ nm})^2 = 0.02567 \text{ nm}^2$$

$$r_{\text{interstitial}} = \sqrt{0.02567 \text{ nm}^2} - 0.1241 \text{ nm} = 0.0361 \text{ nm}$$

For FCC iron, the interstitial site such as the (0, 0, 1/2) lies along ⟨001⟩ directions. Thus, the radius of the iron atom and the radius of the interstitial site are [Figure 4-2(b)]

$$R_{FCC} = \frac{\sqrt{2}a_0}{4} = \frac{(\sqrt{2})(0.3571 \text{ nm})}{4} = 0.1263 \text{ nm}$$

$$2r_{\text{interstitial}} + 2R_{FCC} = a_0$$

$$r_{\text{interstitial}} = \frac{0.3571 \text{ nm} - (2)(0.1263) \text{ nm}}{2} = 0.0523 \text{ nm}$$

The interstitial site in BCC iron is smaller than the interstitial site in FCC iron. Although both are smaller than the carbon atom, carbon distorts the BCC crystal structure more than the FCC structure. As a result, fewer carbon atoms are expected to enter interstitial positions in BCC iron than in FCC iron.

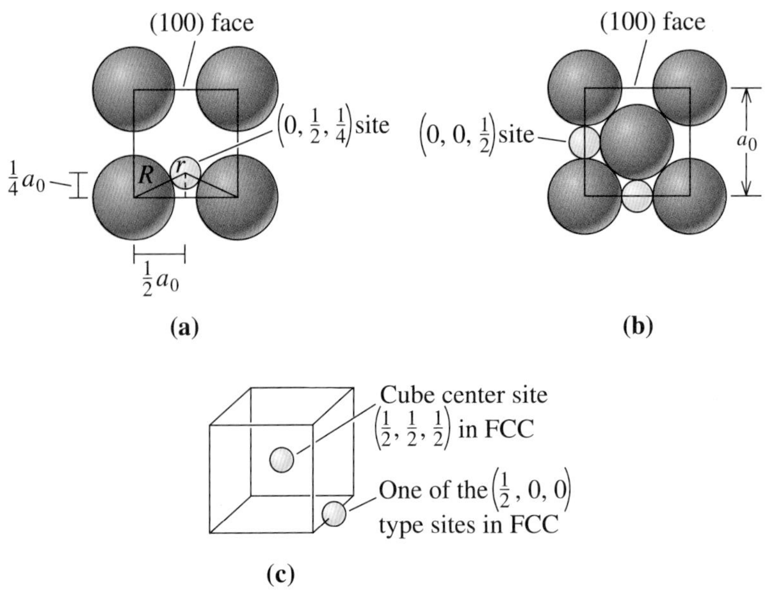

Figure 4-2 (a) The location of the $\left(0, \frac{1}{2}, \frac{1}{4}\right)$ interstitial site in BCC metals. (b) $\left(0, 0, \frac{1}{2}\right)$ site in FCC metals. (c) Edge centers and cube centers are some of the interstitial sites in the FCC structure. (For Example 4-3.)

(b) In BCC iron, two iron atoms are expected in each unit cell. We can find a total of 24 interstitial sites of the type (0, 1/2, 1/4); however, since each site is located at a face of the unit cell, only half of each site belongs uniquely to a single cell. Thus, there are

$$(24 \text{ sites})\left(\frac{1}{2}\right) = 12 \text{ interstitial sites per unit cell}$$

If all of the interstitial sites were filled, the atomic percentage of carbon contained in the iron would be

$$\text{at \% C} = \frac{12 \text{ C atoms}}{12 \text{ C atoms} + 2 \text{ Fe atoms}} \times 100 = 86\%$$

In FCC iron, four iron atoms are expected in each unit cell, and the number of octahedral interstitial sites is

$$(12 \text{ edges})\left(\frac{1}{4}\right) + 1 \text{ center} = 4 \text{ interstitial sites per unit cell [Figure 4-2(c)]}$$

Again, if all the octahedral interstitial sites were filled, the atomic percentage of carbon in the FCC iron would be

$$\text{at \% C} = \frac{4 \text{ C atoms}}{4 \text{ C atoms} + 4 \text{ Fe atoms}} \times 100 = 50\%$$

As we will see in a later chapter, the maximum atomic percentage of carbon present in the two forms of iron under equilibrium conditions is

BCC: 0.1%

FCC: 9.1%

Because of the strain imposed on the iron crystal structure by the interstitial atoms—particularly in BCC iron—the fraction of the interstitial sites that can be occupied is quite small.

Substitutional Defects

A **substitutional defect** is introduced when one atom or ion is replaced by a different type of atom or ion as in Figure 4-1(c) and (d). The substitutional atoms or ions occupy normal lattice sites. Substitutional atoms or ions may either be larger than the normal atoms or ions in the crystal structure, in which case the surrounding interatomic spacings are reduced, or smaller, causing the surrounding atoms to have larger interatomic spacings. In either case, the substitutional defects disturb the surrounding crystal. Again, the substitutional defect can be introduced either as an impurity or as a deliberate alloying addition, and, once introduced, the number of defects is relatively independent of temperature.

Examples of substitutional defects include incorporation of dopants such as phosphorus or boron into Si. Similarly, if we add copper to nickel, copper atoms will occupy crystallographic sites where nickel atoms would normally be present. The substitutional atoms will often increase the strength of the metallic material. Substitutional defects also appear in ceramic materials. For example, if we add MgO to NiO, Mg^{2+} ions occupy Ni^{2+} sites, and O^{2-} ions from MgO occupy O^{2-} sites of NiO. Whether atoms or ions go into interstitial or substitutional sites depends upon the size and valence of these guest atoms or ions compared to the size and valence of the host ions. The size of the available sites also plays a role in this as discussed in Chapter 3, Section 6.

4-2 Other Point Defects

An **interstitialcy** is created when an atom identical to those at the normal lattice points is located in an interstitial position. These defects are most likely to be found in crystal structures having a low packing factor.

A **Frenkel defect** is a vacancy-interstitial pair formed when an ion jumps from a normal lattice point to an interstitial site, as in Figure 4-1(e) leaving behind a vacancy. Although this is usually associated with ionic materials, a Frenkel defect can occur in metals and covalently bonded materials. A **Schottky defect**, Figure 4-1(f), is unique to ionic materials and is commonly found in many ceramic materials. When vacancies occur in an ionically bonded material, a stoichiometric number of anions and cations must be missing from regular atomic positions if electrical neutrality is to be preserved. For example, one Mg^{2+} vacancy and one O^{2-} vacancy in MgO constitute a Schottky pair. In ZrO_2, for one Zr^{4+} vacancy, there will be two O^{2-} vacancies.

An important substitutional point defect occurs when an ion of one charge replaces an ion of a different charge. This might be the case when an ion with a valence of +2 replaces an ion with a valence of +1 (Figure 4-3). In this case, an extra positive charge is introduced into the structure. To maintain a charge balance, a vacancy might be created where a +1 cation normally would be located. Again, this imperfection is observed in materials that have pronounced ionic bonding.

Thus, in ionic solids, when point defects are introduced, the following rules have to be observed:

(a) a charge balance must be maintained so that the crystalline material as a whole is electrically neutral;

(b) a mass balance must be maintained; and

(c) the number of crystallographic sites must be conserved.

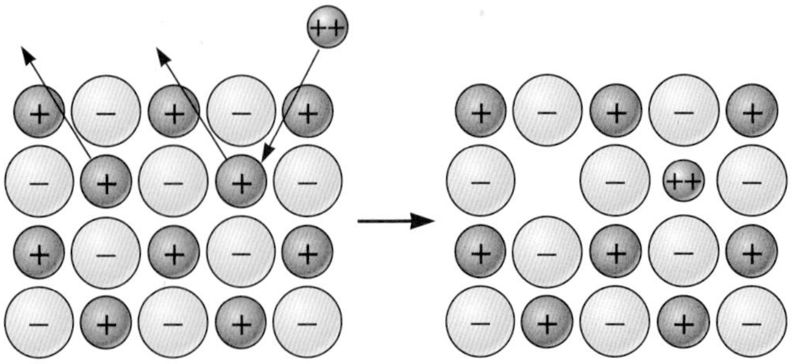

Figure 4-3 When a divalent cation replaces a monovalent cation, a second monovalent cation must also be removed, creating a vacancy.

For example, in nickel oxide (NiO), if one oxygen ion is missing, it creates an oxygen ion vacancy (designated as $V_O^{\bullet\bullet}$). Each dot (·) in the superscript position indicates an *effective* positive charge of one. To maintain stoichiometry, mass balance, and charge balance, we must also create a nickel ion vacancy (designated as V_{Ni}''). Each accent (′) in the superscript indicates an *effective* charge of −1.

We use the **Kröger-Vink notation** to write the defect chemistry equations. The main letter in this notation describes a vacancy or the name of the element. The superscript indicates the effective charge on the defect, and the subscript describes the location of the defect. A dot (·) indicates an effective charge of +1, and an accent (′) represents an effective charge of −1. Sometimes x is used to indicate no net charge. Any free electrons or holes are indicated by e and h, respectively. (Holes will be discussed in Chapter 19.) Clusters of defects or defects that have association are shown in parentheses. Associated defects, which can affect mass transport in materials, are sometimes neutral and hard to detect experimentally. Concentrations of defects are shown in square brackets.

The following examples illustrate the use of the Kröger-Vink notation for writing **defect chemical reactions**. Sometimes, it is possible to write multiple valid defect chemistry reactions to describe the possible defect chemistry. In such cases, it is necessary to take into account the energy that is needed to create different defects, and an experimental verification is necessary. This notation is useful in describing defect chemistry in semiconductors and many ceramic materials used as sensors, dielectrics, and in other applications.

Example 4-4 *Application of the Kröger-Vink Notation*

Write the appropriate defect reactions for (a) incorporation of magnesium oxide (MgO) in nickel oxide (NiO) and (b) formation of a Schottky defect in alumina (Al_2O_3).

SOLUTION

(a) MgO is the guest, and NiO is the host material. We will assume that Mg^{2+} ions will occupy Ni^{2+} sites, and oxygen anions from MgO will occupy O^{2-} sites of NiO.

$$MgO \xrightarrow{NiO} Mg_{Ni}^{x} + O_O^{x}$$

We need to ensure that the equation has charge, mass, and site balance. On the left-hand side, we have one Mg, one oxygen, and no net charge. The same is true on the right-hand side. The site balance can be a little tricky—one Mg^{2+} occupies one Ni^{2+} site. Since we are introducing MgO in NiO, we use one Ni^{2+} site, and, therefore, we must use one O^{2-} site. We can see that this is true by examining the right-hand side of this equation.

(b) A Schottky defect in alumina will involve two aluminum ions and three oxygen ions. When there is a vacancy at an aluminum site, a +3 charge is missing, and the site has an effective charge of −3. Thus V_{Al}''' describes one vacancy of Al^{3+}. Similarly, $V_O^{\bullet\bullet}$ represents an oxygen ion vacancy. For site balance in alumina, we need to ensure that for every two aluminum ion sites used, we use three oxygen ion sites. Since we have vacancies, the mass on the right-hand side is zero, and so we write the left-hand side as null. Therefore, the defect reaction will be

$$\text{null} \xrightarrow{\text{Al}_2\text{O}_3} 2V_{\text{Al}}''' + 3V_{\text{O}}^{\bullet\bullet}$$

Example 4-5 *Point Defects in Stabilized Zirconia for Solid Electrolytes*

Write the appropriate defect reactions for the incorporation of calcium oxide (CaO) in zirconia (ZrO_2) using the Kröger-Vink notation.

SOLUTION

We will assume that Ca^{2+} will occupy Zr^{4+} sites. If we send one Ca^{2+} to Zr^{4+}, the site will have an effective charge of -2 (instead of having a charge of $+4$, we have a charge of $+2$). We have used one Zr^{4+} site, and site balance requires *two oxygen sites*. We can send one O^{2-} from CaO to one of the O^{2-} sites in ZrO_2. The other oxygen site must be used and since mass balance must also be maintained, we will have to keep this site vacant (i.e., an oxygen ion vacancy will have to be created).

$$\text{CaO} \xrightarrow{\text{ZrO}_2} \text{Ca}_{\text{Zr}}'' + \text{O}_{\text{O}}^{x} + V_{\text{O}}^{\bullet\bullet}$$

The concentration of oxygen vacancies in ZrO_2 (i.e., $[V_{O}^{\bullet\bullet}]$) will increase with increasing CaO concentration. These oxygen ion vacancies make CaO-stabilized ZrO_2 an ionic conductor. This allows the use of this type of ZrO_2 in oxygen sensors used in automotives and solid oxide fuel cells.

4-3 Dislocations

Dislocations are line imperfections in an otherwise perfect crystal. They typically are introduced into a crystal during solidification of the material or when the material is deformed permanently. Although dislocations are present in all materials, including ceramics and polymers, *they are particularly useful in explaining deformation and strengthening in metallic materials*. We can identify three types of dislocations: the screw dislocation, the edge dislocation, and the mixed dislocation.

Screw Dislocations

The **screw dislocation** (Figure 4-4) can be illustrated by cutting partway through a perfect crystal and then skewing the crystal by one atom spacing. If we follow a crystallographic plane one revolution around the axis along which the crystal was skewed, starting at point x and traveling equal atom spacings in each direction in a clockwise fashion, we finish at point y one atom spacing below our starting point. If a screw dislocation were not present, the loop would close. The vector required to complete the loop is the **Burgers vector b**. If we continued our rotation, we would trace out a spiral path. The axis, or line around which we trace out this path, is the screw dislocation. The Burgers vector is parallel to the screw dislocation.

Edge Dislocations

An **edge dislocation** (Figure 4-5) can be illustrated by slicing partway through a perfect crystal, spreading the crystal apart, and partly filling the cut with an extra half plane of atoms. The bottom edge of this inserted plane represents the edge dislocation. If we describe a clockwise loop around the edge dislocation, starting at point x and traveling an equal number of atom spacings in each direction, we finish at point y one atom spacing from the starting point. If an edge dislocation were not present, the loop

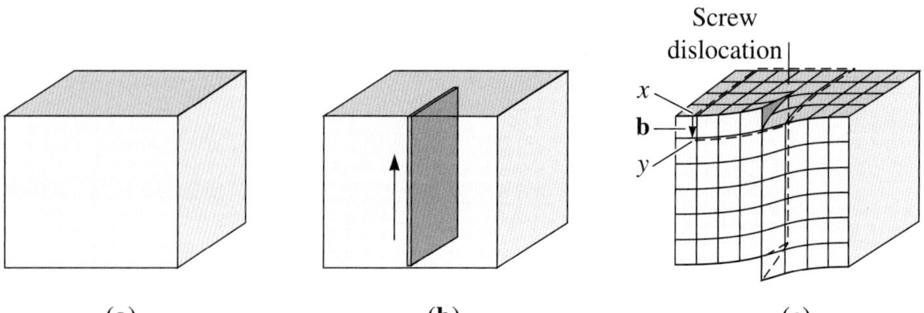

Figure 4-4 The perfect crystal (a) is cut and sheared one atom spacing, (b) and (c). The line along which shearing occurs is a screw dislocation. A Burgers vector **b** is required to close a loop of equal atom spacings around the screw dislocation.

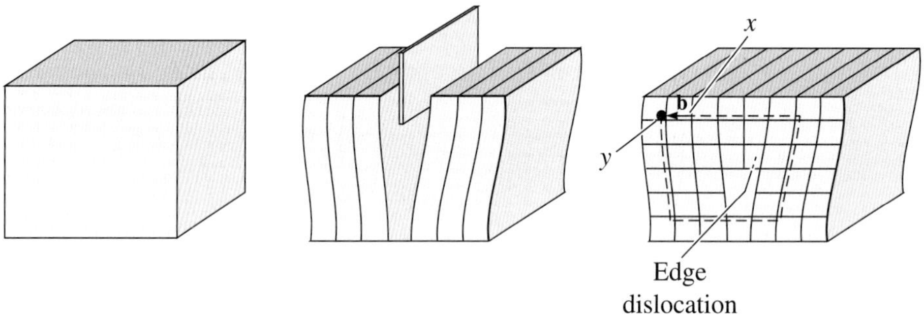

Figure 4-5 The perfect crystal in (a) is cut and an extra half plane of atoms is inserted (b). The bottom edge of the extra half plane is an edge dislocation (c). A Burgers vector **b** is required to close a loop of equal atom spacings around the edge dislocation. (*Based on* Fundamentals of Physcial Metallurgy, *J. D. Verhoeven, John Wiley & Sons, Inc., 1975.*)

would close. The vector required to complete the loop is, again, the Burgers vector. In this case, the Burgers vector is perpendicular to the dislocation. By introducing the dislocation, the atoms above the dislocation line are squeezed too closely together, while the atoms below the dislocation are stretched too far apart. The surrounding region of the crystal has been disturbed by the presence of the dislocation. A " $\perp$ " symbol is often used to denote an edge dislocation. The long axis of the " $\perp$ " points toward the extra half plane. Unlike an edge dislocation, a screw dislocation cannot be visualized as an extra half plane of atoms.

Mixed Dislocations As shown in Figure 4-6, **mixed dislocations** have both edge and screw components, with a transition region between them. The Burgers vector, however, remains the same for all portions of the mixed dislocation.

Stresses When discussing the motion of dislocations, we need to refer to the concept of stress, which will be covered in detail in Chapter 6. For now, it suffices to say that stress is force per unit area. Stress has units of lb/in.^{2} known as psi (pounds per square inch) or N/m^2 known as the pascal (Pa). A normal stress arises when the applied force acts perpendicular to the area of interest. A shear stress τ arises when the force acts in a direction parallel to the area of interest.

Dislocation Motion Consider the edge dislocation shown in Figure 4-7(a). A plane that contains both the dislocation line and the Burgers vector is known as a **slip plane**. When a sufficiently large shear stress acting parallel to the Burgers vector is

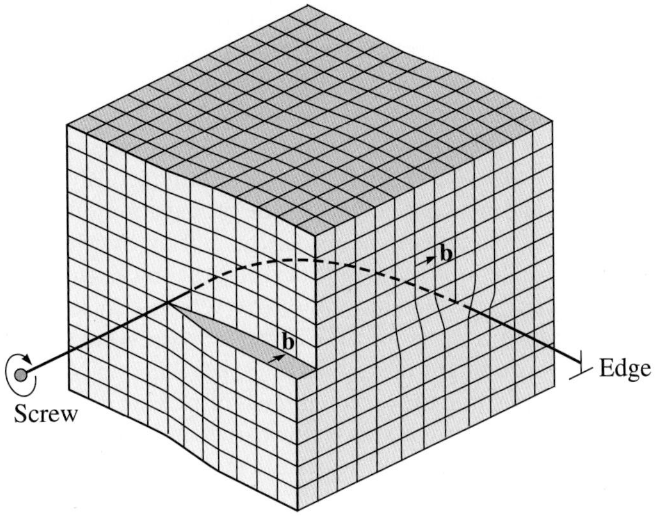

Figure 4-6 A mixed dislocation. The screw dislocation at the front face of the crystal gradually changes to an edge dislocation at the side of the crystal. (*Based on W. T. Read,* Dislocations in Crystals, *McGraw-Hill, 1953.*)

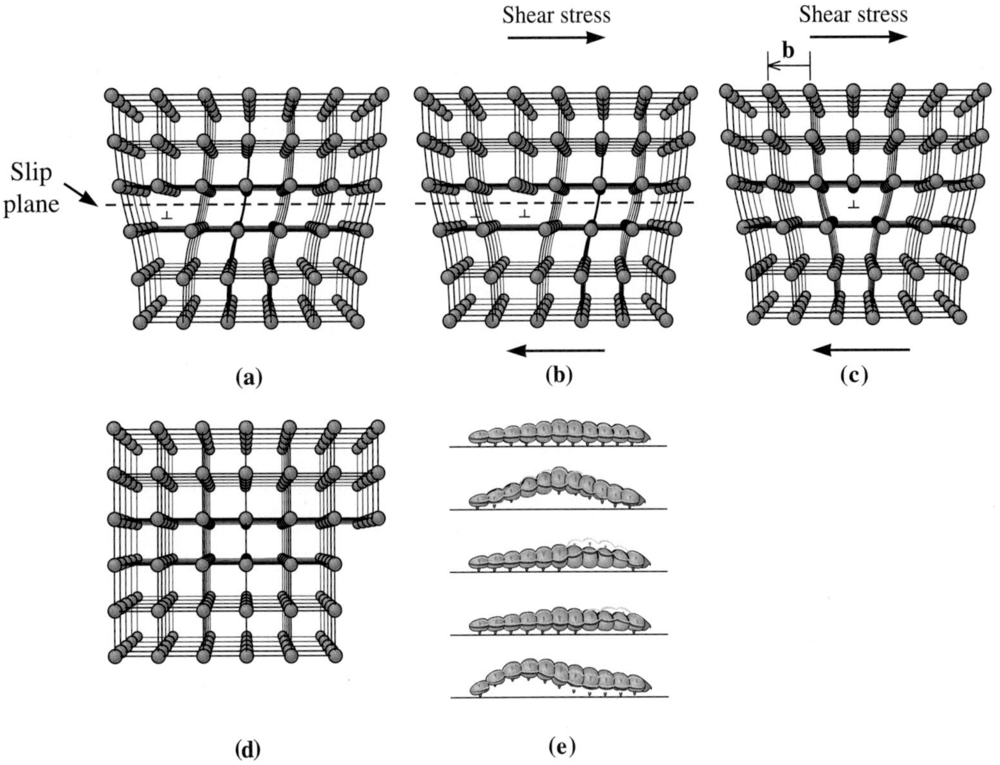

Figure 4-7 (a) When a shear stress is applied to the dislocation in (a), the atoms are displaced, (b) causing the dislocation to move one Burgers vector in the slip direction. (c) Continued movement of the dislocation eventually creates a step (d), and the crystal is deformed. (e) The motion of a caterpillar is analogous to the motion of a dislocation. (*Based on A. G. Guy,* Essentials of Materials Science, *McGraw-Hill, 1976.*)

applied to a crystal containing a dislocation, the dislocation can move through a process known as **slip**. The bonds across the slip plane between the atoms in the column to the right of the dislocation shown are broken. The atoms in the column to the right of the dislocation below the slip plane are shifted slightly so that they establish bonds with the atoms of the edge dislocation. In this way, the dislocation has shifted to the right [Figure 4-7(b)]. If this process continues, the dislocation moves through the crystal [Figure 4-7(c)] until it produces a step on the exterior of the crystal [Figure 4-7(d)] in the **slip direction** (which is parallel to the Burgers vector). (Note that the combination of a slip plane and a slip direction comprises a **slip system**.) The top half of the crystal has been displaced by one Burgers vector relative to the bottom half; the crystal has been plastically (or permanently) deformed. This is the fundamental process that occurs many, many times as you bend a paper clip with your fingers. The plastic deformation of metals is primarily the result of the propagation of dislocations.

This process of progressively breaking and reforming bonds requires far less energy than the energy that would be required to instantaneously break all of the bonds across the slip plane. The crystal deforms via the propagation of dislocations because it is an energetically favorable process. Consider the motion by which a caterpillar moves [Figure 4-7(e)]. A caterpillar only lifts some of its legs at any given time rather than lifting all of its legs at one time in order to move forward. Why? Because lifting only some of its legs requires less energy; it is easier for the caterpillar to do. Another way to visualize this is to think about how you might move a large carpet that is positioned incorrectly in a room. If you want to reposition the carpet, instead of picking it up off the floor and moving it all at once, you might form a kink in the carpet and push the kink in the direction in which you want to move the carpet. The width of the kink is analogous to the Burgers vector. Again, you would move the carpet in this way because it requires less energy—it is easier to do.

Slip Figure 4-8(a) is a schematic diagram of an edge dislocation that is subjected to a shear stress τ that acts parallel to the Burgers vector and perpendicular to the dislocation line. In this drawing, the edge dislocation is propagating in a direction opposite to the direction of propagation shown in Figure 4-7. A component of the shear stress must act parallel to the Burgers vector and perpendicular to the dislocation line in order for the edge dislocation to move. The dislocation line moves in a direction parallel to the Burgers vector. Figure 4-8(b) shows a screw dislocation. For a screw dislocation, a component of the shear stress must act parallel to the Burgers vector (and thus the dislocation line) in order for the dislocation to move. The dislocation moves in a direction perpendicular to the Burgers vector, and the slip step that is produced is parallel to the Burgers vector. Since the Burgers vector of a screw dislocation is parallel to the dislocation line, specification of the Burgers vector and dislocation line does not define a slip plane for a screw dislocation.

During slip, a dislocation moves from one set of surroundings to an identical set of surroundings. The **Peierls-Nabarro stress** (Equation 4-2) is required to move the dislocation from one equilibrium location to another,

$$\tau = c \exp\left(-\frac{kd}{b}\right) \tag{4-2}$$

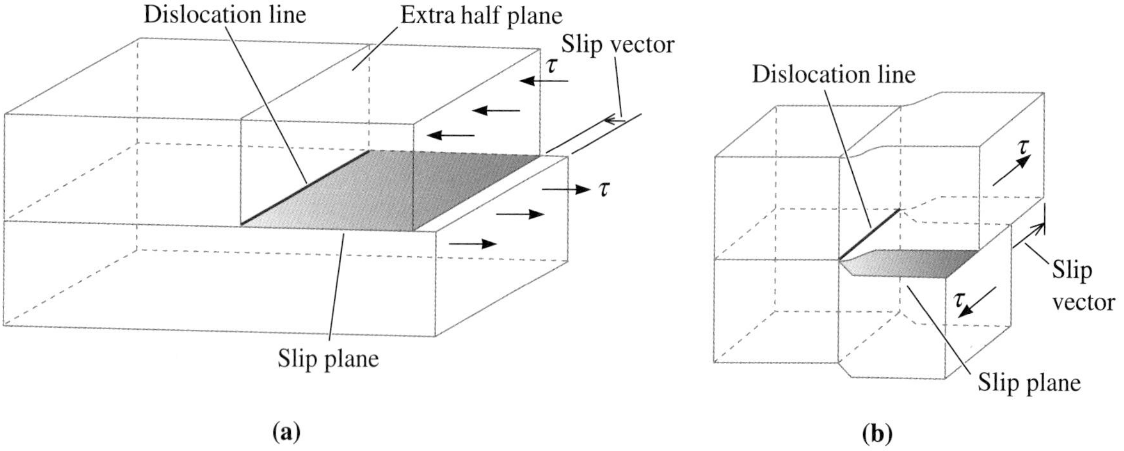

Figure 4-8 Schematic of the dislocation line, slip plane, and slip (Burgers) vector for (a) an edge dislocation and (b) a screw dislocation. (*Based on* Fundamentals of Physical Metallurgy, *J. D. Verhoeven, John Wiley & Sons, Inc., 1975.*)

where τ is the shear stress required to move the dislocation, d is the interplanar spacing between adjacent slip planes, b is the magnitude of the Burgers vector, and both c and k are constants for the material. The dislocation moves in a slip system that requires the least expenditure of energy. Several important factors determine the most likely slip systems that will be active:

1. The stress required to cause the dislocation to move increases exponentially with the length of the Burgers vector. Thus, the slip direction should have a small repeat distance or high linear density. The close-packed directions in metals and alloys satisfy this criterion and are the usual slip directions.
2. The stress required to cause the dislocation to move decreases exponentially with the interplanar spacing of the slip planes. Slip occurs most easily between planes of atoms that are smooth (so there are smaller "hills and valleys" on the surface) and between planes that are far apart (or have a relatively large interplanar spacing). Planes with a high planar density fulfill this requirement. Therefore, the slip planes are typically close-packed planes or those as closely packed as possible. Common slip systems in several materials are summarized in Table 4-1.
3. Dislocations do not move easily in materials such as silicon, which have covalent bonds. Because of the strength and directionality of the bonds, the materials typically fail in a brittle manner before the force becomes high enough to cause appreciable slip. Dislocations also play a relatively minor role in the deformation of polymers. Most polymers contain a substantial volume fraction of material that is amorphous and, therefore, does not contain dislocations. Permanent deformation in polymers primarily involves the stretching, rotation, and disentanglement of long chain molecules.
4. Materials with ionic bonding, including many ceramics such as MgO, also are resistant to slip. Movement of a dislocation disrupts the charge balance around the anions and cations, requiring that bonds between anions and cations be broken. During slip, ions with a like charge must also pass close together, causing repulsion. Finally, the repeat distance along the slip direction, or the Burgers vector, is larger than that in metals and alloys. Again, brittle failure of ceramic materials typically occurs due to the presence of flaws such as small pores before the applied level of stress is sufficient to cause dislocations to move. Ductility in ceramic materials can be obtained by

   (a) phase transformations (known as transformation plasticity, an example is fully stabilized zirconia);

   (b) mechanical twinning;

Table 4-1 Slip planes and directions in metallic structures

Crystal Structure	Slip Plane	Slip Direction
BCC metals	{110} {112} {123}	⟨111⟩
FCC metals	{111}	⟨110⟩
HCP metals	(001), (002)* $\{11\bar{2}0\}$, $\{10\bar{1}0\}$, $\{10\bar{1}1\}$ See Note	[100], [010], [110] ⟨110⟩ or $\langle 10\bar{2}0\rangle$
MgO, NaCl (ionic)	{110}	⟨110⟩
Silicon (covalent)	{111}	⟨110⟩

Note: These planes are active in some metals and alloys or at elevated temperatures.

**In the four-axis notation, the slip planes and directions are $<11\bar{2}0>$ and (0001).*

(c) dislocation motion; and

(d) grain boundary sliding.

We will discuss some of these concepts later in this chapter. Typically, higher temperatures and compressive stresses lead to higher ductility. Recently, it has been shown that certain ceramics such as strontium titanate ($SrTiO_3$) can exhibit considerable ductility. Under certain conditions, ceramics can exhibit very large deformations. This is known as superplastic behavior.

Example 4-6 *Burgers Vector Calculation*

Calculate the length of the Burgers vector in copper.

SOLUTION

Copper has an FCC crystal structure. The lattice parameter of copper is 0.36151 nm. The close-packed directions, or the directions of the Burgers vector, are of the form $\langle 110 \rangle$. The repeat distance along the $\langle 110 \rangle$ directions is one-half the face diagonal, since lattice points are located at corners and centers of faces.

$$\text{Face diagonal} = \sqrt{2}a_0 = (\sqrt{2})(0.36151) = 0.51125 \text{ nm}$$

The length of the Burgers vector, or the repeat distance, is

$$b = \frac{1}{2}(0.51125) \text{ nm} = 0.25563 \text{ nm}$$

Example 4-7 *Identification of Preferred Slip Planes*

The planar density of the (112) plane in BCC iron is 9.94×10^{14} atoms/cm^2. Calculate (a) the planar density of the (110) plane and (b) the interplanar spacings for both the (112) and (110) planes. On which plane would slip normally occur?

SOLUTION

The lattice parameter of BCC iron is 0.2866 nm or 2.866×10^{-8} cm. For the (110) plane in the BCC structure, one-fourth of the four corner atoms plus the center atom lie within an area of a_0 times $\sqrt{2}a_0$.

(a) The planar density is

$$\text{Planar density (110)} = \frac{\text{atoms}}{\text{area}} = \frac{2}{(\sqrt{2})(2.866 \times 10^{-8} \text{ cm})^2}$$
$$= 1.72 \times 10^{15} \text{ atoms/cm}^2$$

$$\text{Planar density (112)} = 0.994 \times 10^{15} \text{ atoms/cm}^2 \text{ (from problem statement)}$$

(b) The interplanar spacings are

$$d_{110} = \frac{2.866 \times 10^{-8}}{\sqrt{1^2 + 1^2 + 0}} = 2.027 \times 10^{-8} \text{ cm}$$

$$d_{112} = \frac{2.866 \times 10^{-8}}{\sqrt{1^2 + 1^2 + 2^2}} = 1.170 \times 10^{-8} \text{ cm}$$

The planar density is higher and the interplanar spacing is larger for the (110) plane than for the (112) plane; therefore, the (110) plane is the preferred slip plane.

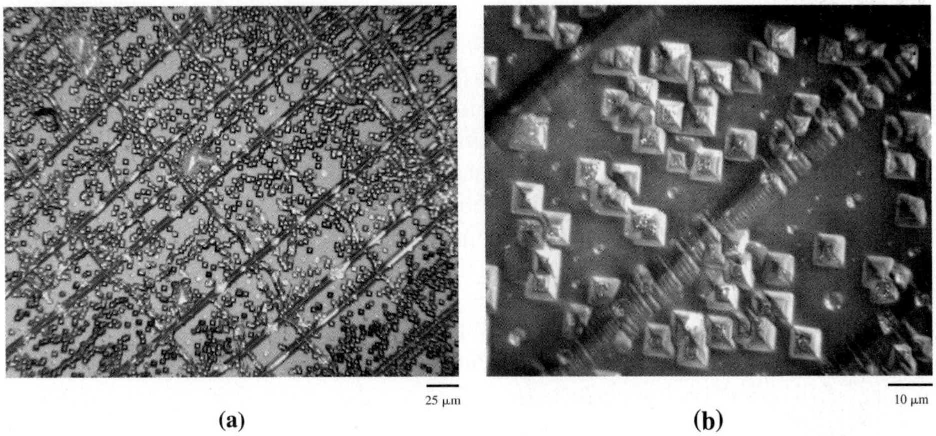

Figure 4-9 Optical images of etch pits in sodium chloride (NaCl) that was deformed and annealed for two hours at 450°C. The etch pits correspond to intersection points of dislocations with the {001} surface. The straight lines of dislocations are in the ⟨110⟩ directions. (a) Low magnification. (b) High magnification. (*Courtesy of Professor Greg Rohrer, Carnegie Mellon University*)

When a metallic material is "etched" (a chemical reaction treatment that involves exposure to an acid or a base), the areas where dislocations intersect the surface of the crystal react more readily than the surrounding material. These regions appear in the microstructure as **etch pits**. Figure 4-9 shows the etch pit distribution on a surface of a sodium chloride crystal.

4-4 Significance of Dislocations

Dislocations are most significant in metals and alloys since they provide a mechanism for plastic deformation. **Plastic deformation** refers to irreversible deformation or change in shape that occurs when the force or stress that caused it is removed. The applied stress causes dislocation motion, and the cumulative effect of slip of numerous dislocations is plastic deformation. There are, however, other mechanisms that cause permanent deformation. We will examine these in later chapters. Plastic deformation is to be distinguished from **elastic deformation**, which is a temporary change in shape that occurs while a force or stress remains applied to a material. In elastic deformation, the shape change is a result of stretching of interatomic or interionic bonds, and no dislocation motion occurs. Slip can occur in some ceramics and polymers; however, other factors (e.g., porosity in ceramics, entanglement of chains in polymers, etc.) dominate the near-room temperature mechanical behavior of polymers and ceramics. Amorphous materials such as silicate glasses do not have a periodic arrangement of ions and hence do not contain dislocations.

The slip process is particularly important in understanding the mechanical behavior of metals. First, slip explains why the strength of metals is much lower than the value predicted from the metallic bond. If slip occurs, only a tiny fraction of all of the metallic bonds across the interface need to be broken at any one time, and the force required to deform the metal is small. It can be shown that the actual strength of metals is 10^3 to 10^4 times *lower* than that expected from the strength of metallic bonds.

Second, slip provides ductility in metals. If no dislocations were present, an iron bar would be brittle and could not be shaped by metalworking processes, such as forging, into useful shapes.

Third, we control the mechanical properties of a metal or alloy by interfering with the movement of dislocations. An obstacle introduced into the crystal prevents a dislocation from slipping unless we apply higher forces. Thus, the presence of dislocations helps strengthen metallic materials.

Enormous numbers of dislocations are found in materials. The **dislocation density**, or total length of dislocations per unit volume, is usually used to represent the amount of dislocations present. Dislocation densities of 10^6 cm/cm^3 are typical of the softest metals, while densities up to 10^{12} cm/cm^3 can be achieved by deforming the material.

Dislocations also influence electronic and optical properties of materials. For example, the resistance of pure copper increases with increasing dislocation density. We mentioned previously that the resistivity of pure copper also depends strongly on small levels of impurities. Similarly, we prefer to use silicon crystals that are essentially dislocation free since this allows the charge carriers such as electrons to move more freely through the solid. Normally, the presence of dislocations has a deleterious effect on the performance of photo detectors, light emitting diodes, lasers, and solar cells. These devices are often made from compound semiconductors such as gallium arsenide-aluminum arsenide (GaAs-AlAs), and dislocations in these materials can originate from concentration inequalities in the melt from which crystals are grown or stresses induced because of thermal gradients that the crystals are exposed to during cooling from the growth temperature.

4-5 Schmid's Law

We can understand the differences in behavior of metals that have different crystal structures by examining the force required to initiate the slip process. Suppose we apply a unidirectional force F to a cylinder of metal that is a single crystal (Figure 4-10). We can orient the slip plane and slip direction to the applied force by defining the angles λ and ϕ. The angle between the slip direction and the applied force is λ, and ϕ is the angle between the normal to the slip plane and the applied force. Note that the sum of angles ϕ and λ can be, but does not have to be, 90°.

In order for the dislocation to move in its slip system, a shear force acting in the slip direction must be produced by the applied force. This resolved shear force F_r is given by

$$F_r = F \cos \lambda$$

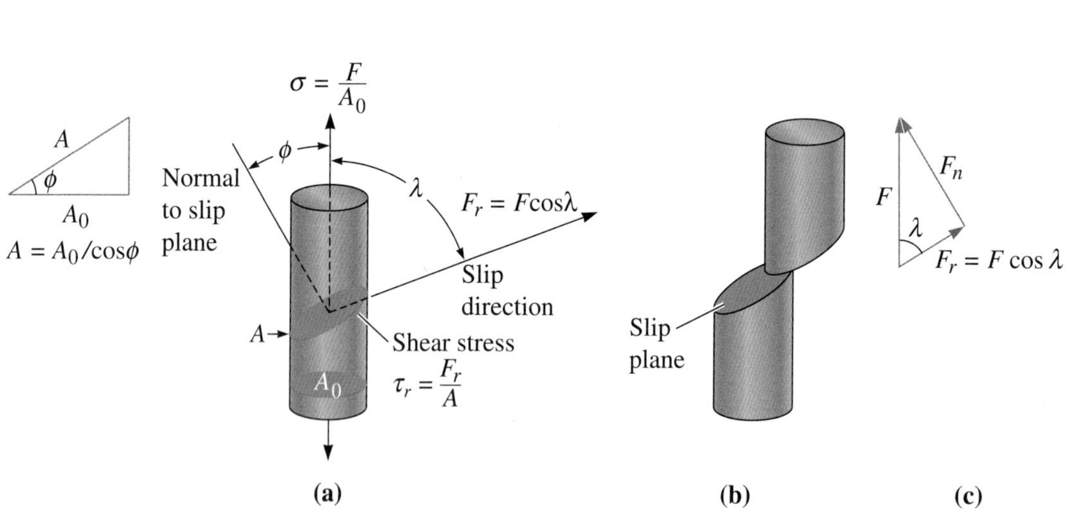

Figure 4-10 (a) A resolved shear stress τ is produced on a slip system. [Note: $(\phi + \lambda)$ does not have to equal 90°.] (b) Movement of dislocations on the slip system deforms the material. (c) Resolving the force.

If we divide the equation by the area of the slip plane, $A = A_0/\cos\phi$, we obtain the following equation known as **Schmid's law**:

$$\tau_r = \sigma \cos \phi \cos \lambda \qquad (4\text{-}3)$$

where

$$\tau_r = \frac{F_r}{A} = \text{resolved shear stress in the slip direction}$$

and

$$\sigma = \frac{F}{A_0} = \text{normal stress applied to the cylinder}$$

Example 4-8 *Calculation of Resolved Shear Stress*

Apply Schmid's law for a situation in which the single crystal is oriented so that the slip plane is perpendicular to the applied tensile stress.

SOLUTION

Suppose the slip plane is perpendicular to the applied stress σ, as in Figure 4-11. Then, $\phi = 0$, $\lambda = 90°$, $\cos \lambda = 0$, and therefore, $\tau_r = 0$. As noted before, the angles ϕ and λ can but do not always sum to 90°. Even if the applied stress σ is enormous, no resolved shear stress develops along the slip direction and the dislocation cannot move. (You could perform a simple experiment to demonstrate this with a deck of cards. If you push on the deck at an angle, the cards slide over one another, as in the slip process. If you push perpendicular to the deck, however, the cards do not slide.) Slip cannot occur if the slip system is oriented so that either λ or ϕ is 90°.

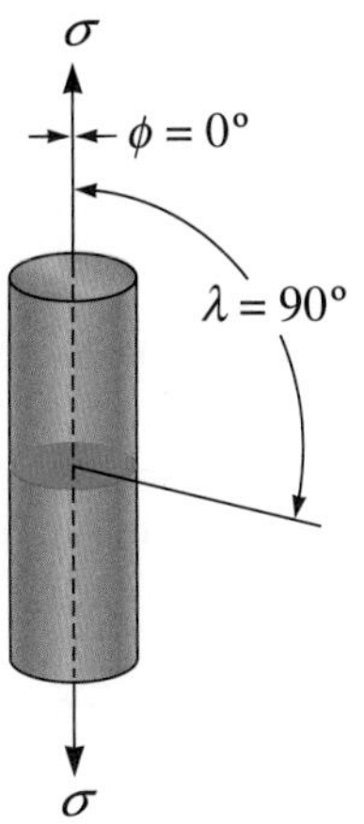

Figure 4-11 When the slip plane is perpendicular to the applied stress σ, the angle λ is 90°, and no shear stress is resolved.

The **critical resolved shear stress** τ_{crss} is the shear stress required for slip to occur. Thus slip occurs, causing the metal to plastically deform, when the *applied* stress (σ) produces a *resolved* shear stress (τ_r) that equals the critical resolved shear stress:

$$\tau_r = \tau_{crss} \qquad (4\text{-}4)$$

Example 4-9 *Design of a Single Crystal Casting Process*

We wish to produce a rod composed of a single crystal of pure aluminum, which has a critical resolved shear stress of 148 psi. We would like to orient the rod in such a manner that, when an axial stress of 500 psi is applied, the rod deforms by slip in a 45° direction to the axis of the rod and actuates a sensor that detects the overload. Design the rod.

SOLUTION

Dislocations begin to move when the resolved shear stress τ_r equals the critical resolved shear stress, 148 psi. From Schmid's law:

$$\tau_r = \sigma \cos \lambda \cos \phi$$

$$148 \text{ psi} = (500 \text{ psi}) \cos \lambda \cos \phi$$

Because we wish slip to occur at a 45° angle to the axis of the rod, $\lambda = 45°$, and

$$\cos \phi = \frac{148}{500 \cos 45°} = 0.4186$$

$$\phi = 65.3°$$

Therefore, we must produce a rod that is oriented such that $\lambda = 45°$ and $\phi = 65.3°$. Note that ϕ and λ do not add to 90°.

We might do this by a solidification process. We could orient a seed crystal of solid aluminum at the bottom of a mold. Liquid aluminum could be introduced into the mold. The liquid solidifies at the seed crystal, and a single crystal rod of the proper orientation is produced.

4-6 Influence of Crystal Structure

We can use Schmid's law to compare the properties of metals having the BCC, FCC, and HCP crystal structures. Table 4-2 lists three important factors that we can examine. We must be careful to note, however, that this discussion describes the behavior of nearly perfect single crystals. Real engineering materials are seldom single crystals and always contain large numbers of defects. Also they tend to be polycrystalline. Since different crystals or grains are oriented in different random directions, we cannot apply Schmid's law to predict the mechanical behavior of polycrystalline materials.

Critical Resolved Shear Stress If the critical resolved shear stress in a metal is very high, the applied stress σ must also be high in order for τ_r to equal τ_{crss}. A higher τ_{crss} implies a higher stress is necessary to plastically deform a metal, which in turn indicates the metal has a high strength! In FCC metals, which have close-packed {111} planes, the critical resolved shear stress is low—about 50 to 100 psi in a nearly perfect crystal. On the other hand, BCC crystal structures contain no close-packed planes, and we must exceed a higher critical resolved shear stress—on the order of 10,000 psi in nearly

Table 4-2 Summary of factors affecting slip in metallic structures

Factor	FCC	BCC	HCP $\left(\frac{c}{a} \geq 1.633\right)$
Critical resolved shear stress (psi)	50–100	5,000–10,000	50–100[a]
Number of slip systems	12	48	3[b]
Cross-slip	Can occur	Can occur	Cannot occur[b]
Summary of properties	Ductile	Strong	Relatively brittle

[a]*For slip on basal planes.*

[b]*By alloying or heating to elevated temperatures, additional slip systems are active in HCP metals, permitting cross-slip to occur and thereby improving ductility.*

perfect crystals—before slip occurs. Thus, BCC metals tend to have high strengths and lower ductilities compared to FCC metals.

We would expect the HCP metals, because they contain close-packed basal planes, to have low critical resolved shear stresses. In fact, in HCP metals such as zinc that have a *c/a* ratio greater than or equal to the theoretical ratio of 1.633, the critical resolved shear stress is less than 100 psi, just as in FCC metals. In HCP titanium, however, the *c/a* ratio is less than 1.633; the close-packed planes are spaced too closely together. Slip now occurs on the "prism" planes or faces of the hexagon, and the critical resolved shear stress is then as great as or greater than in BCC metals.

Number of Slip Systems

If at least one slip system is oriented to give the angles λ and ϕ near 45°, then τ_r equals τ_{crss} at low applied stresses. Ideal HCP metals possess only one set of parallel close-packed planes, the (001) planes, which are parallel to the (002) planes [together these are known as the (0001) planes in the four-axis notation]. One set of close-packed planes containing three close-packed directions comprises three slip systems. Consequently, the probability of the close-packed planes and directions being oriented with λ and ϕ near 45° is very low. The HCP crystal may fail in a brittle manner without a significant amount of slip; however, in HCP metals with a low *c/a* ratio, or when HCP metals are properly alloyed, or when the temperature is increased, other slip systems become active, making these metals less brittle than expected.

On the other hand, FCC metals contain four nonparallel close-packed planes of the form $\{111\}$ and three close-packed directions of the form $\langle 110 \rangle$ within each plane, giving a total of twelve slip systems. At least one slip system is always favorably oriented for slip to occur at low applied stresses, permitting FCC metals to have high ductilities.

Finally, BCC metals have as many as 48 slip systems that are nearly close-packed. Several slip systems are always properly oriented for slip to occur, allowing BCC metals to have ductility.

Cross-Slip

Consider a screw dislocation moving on one slip plane that encounters an obstacle and is blocked from further movement. This dislocation can shift to a second intersecting slip system, also properly oriented, and continue to move. This is called **cross-slip**. In many HCP metals, no cross-slip can occur because the slip planes are parallel (i.e., not intersecting); therefore, polycrystalline HCP metals tend to be brittle.

Fortunately, additional slip systems become active when HCP metals are alloyed or heated, thus improving ductility. Cross-slip is possible in both FCC and BCC metals because a number of intersecting slip systems are present. Consequently, cross-slip helps maintain ductility in these metals.

Example 4-10 *Ductility of HCP Metal Single Crystals and Polycrystalline Materials*

A single crystal of magnesium, which has the HCP crystal structure, can be stretched into a ribbon-like shape four to six times its original length; however, *polycrystalline* Mg and other metals with the HCP structure show limited ductilities. Use the values of critical resolved shear stress for metals with different crystal structures and the nature of deformation in polycrystalline materials to explain this observation.

SOLUTION

From Table 4-2, we note that for HCP metals such as Mg, the critical resolved shear stress is low (50-100 psi). We also note that slip in HCP metals will occur readily on the basal plane—the primary slip plane. When a single crystal is deformed, assuming the basal plane is suitably oriented with respect to the applied stress, a large deformation can occur. This explains why single crystal Mg can be stretched into a ribbon four to six times the original length. When we have polycrystalline Mg, the deformation is not as simple. Each crystal must deform such that the strain developed in any one crystal is accommodated by its neighbors.

In HCP metals, there are no intersecting slip systems; thus, dislocations cannot glide from one slip plane in one crystal (grain) onto another slip plane in a neighboring crystal. As a result, polycrystalline HCP metals such as Mg show limited ductility.

4-7 Surface Defects

Surface defects are the boundaries, or planes, that separate a material into regions. For example, each region may have the same crystal structure but different orientations.

Material Surface The exterior dimensions of the material represent surfaces at which the crystal abruptly ends. Each atom at the surface no longer has the proper coordination number, and atomic bonding is disrupted. The exterior surface may also be very rough, may contain tiny notches, and may be much more reactive than the bulk of the material.

In nanostructured materials, the ratio of the number of atoms or ions at the surface to that in the bulk is very high. As a result, these materials have a large surface area per unit mass. Similar to nanoscale materials, the surface area-to-volume ratio is high for porous materials, gels, and ultra-fine powders. In petroleum refining and many other areas of technology, we make use of high surface area catalysts for enhancing the kinetics of chemical reactions. The reduction in surface area is the thermodynamic driving force for **sintering** of ceramics and metal powders.

Grain Boundaries The microstructure of many engineered ceramic and metallic materials consists of many grains. A **grain** is a portion of the material within which the arrangement of the atoms is nearly identical; however, the orientation of the atom arrangement, or crystal structure, is different for each adjoining grain. Three grains are shown schematically in Figure 4-12(a); the arrangement of atoms in each grain is identical but the grains are oriented differently. A **grain boundary**, the surface that separates the individual grains, is a narrow zone in which the atoms are not properly spaced. That is to say, the atoms are so close together at some locations in the grain boundary that they cause a region of compression, and in other areas they are so far apart that they cause a region of tension. Figure 4-12(b) shows a micrograph of grains in a stainless steel sample. Figure 4-12(c) is a high-resolution scanning tunneling electron micrograph of a grain boundary in gold. Figure 4-12(c) demonstrates that the illustration in Figure 4-12(a) greatly exaggerates the separation between atoms at grain boundaries.

One method of controlling the properties of a material is by controlling the grain size. By reducing the grain size, we increase the number of grains and, hence, increase the amount of grain boundary area. Any dislocation moves only a short distance before encountering a grain boundary, and the strength of the metallic material is increased. The **Hall-Petch equation** relates the grain size to the **yield strength**,

$$\sigma_y = \sigma_0 + Kd^{-1/2} \tag{4-5}$$

where σ_y is the yield strength, d is the average diameter of the grains, and σ_0 and K are constants for the metal. Recall from Chapter 1 that the yield strength of a metallic

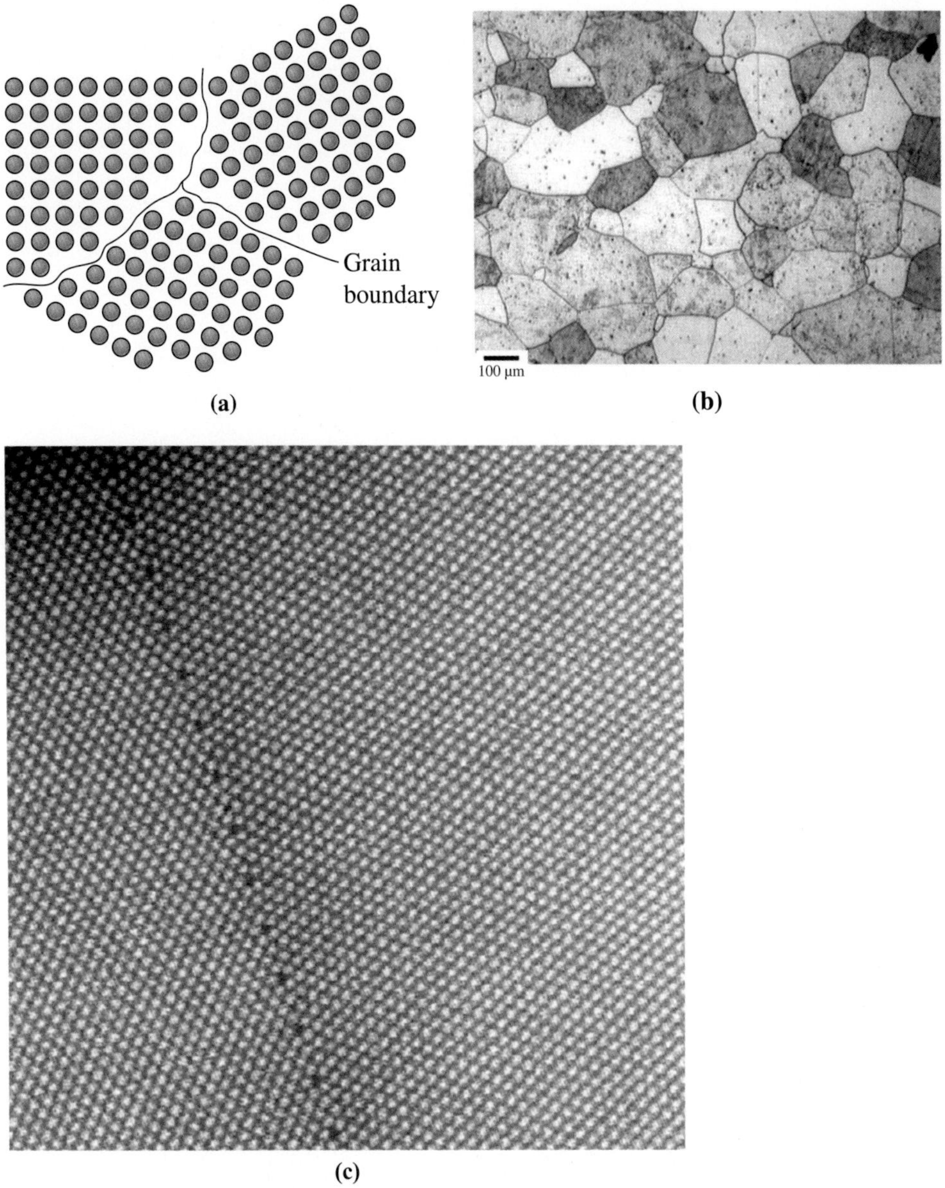

Figure 4-12 (a) The atoms near the boundaries of the three grains do not have an equilibrium spacing or arrangement as shown in this schematic. (b) Grains and grain boundaries in a stainless steel sample. (*Micrograph courtesy of Dr. A. J. Deardo, Dr. M. Hua, and Dr. J. Garcia*) (c) A high-resolution scanning transmission electron micrograph of a grain boundary in gold. (*Image courtesy of FEI taken on an FEI Titan STEM 80-300*)

material is the minimum level of stress that is needed to initiate plastic (permanent) deformation. Figure 4-13 shows this relationship in steel. The Hall-Petch equation is not valid for materials with unusually large or ultra-fine grains. In the chapters that follow, we will describe how the grain size of metals and alloys can be controlled through solidification, alloying, and heat treatment.

Optical microscopy is one technique that is used to reveal microstructural features such as grain boundaries that require less than about 2000 magnification. The process of preparing a metallic sample and observing or recording its microstructure is called **metallography**. A sample of the material is sanded and polished to a mirror-like finish. The surface is then exposed to chemical attack, or *etching*, with grain boundaries being attacked more aggressively than the remainder of the grain. Light

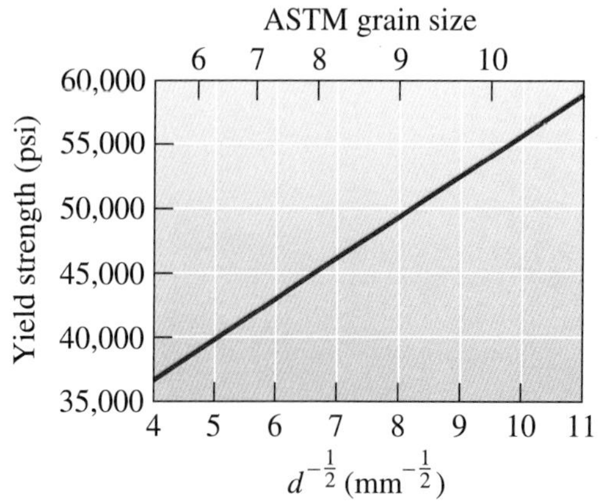

Figure 4-13 The effect of grain size on the yield strength of steel at room temperature.

Example 4-11 *Design of a Mild Steel*

The yield strength of mild steel with an average grain size of 0.05 mm is 20,000 psi. The yield stress of the same steel with a grain size of 0.007 mm is 40,000 psi. What will be the average grain size of the same steel with a yield stress of 30,000 psi? Assume the Hall-Petch equation is valid and that changes in the observed yield stress are due to changes in grain size.

SOLUTION

$$\sigma_y = \sigma_0 + Kd^{-1/2}$$

Thus, for a grain size of 0.05 mm, the yield stress is

$$20{,}000 \text{ psi } (6.895 \text{ MPa})/(1000 \text{ psi}) = 137.9 \text{ MPa}$$

(Note: 1000 psi = 6.895 MPa). Using the Hall-Petch equation

$$137.9 = \sigma_0 + \frac{K}{\sqrt{0.05}}$$

For the grain size of 0.007 mm, the yield stress is 40,000 psi (6.895 MPa)/(1000 psi) = 275.8 MPa. Therefore, again using the Hall-Petch equation:

$$275.8 = \sigma_0 + \frac{K}{\sqrt{0.007}}$$

Solving these two equations, $K = 18.44 \text{ MPa} \cdot \text{mm}^{1/2}$, and $\sigma_0 = 55.5$ MPa. Now we have the Hall-Petch equation as

$$\sigma_y = 55.5 + 18.44\, d^{-1/2}$$

If we want a yield stress of 30,000 psi or 206.9 MPa, the grain size should be 0.0148 mm.

from an optical microscope is reflected or scattered from the sample surface, depending on how the surface is etched. When more light is scattered from deeply etched features such as the grain boundaries, these features appear dark (Figure 4-14). In ceramic samples, a technique known as **thermal grooving** is often used to observe grain boundaries. It involves polishing and heating a ceramic sample to temperatures below the sintering temperature (1300°C) for a short time. Sintering is a process for forming a dense mass by heating compacted powdered material.

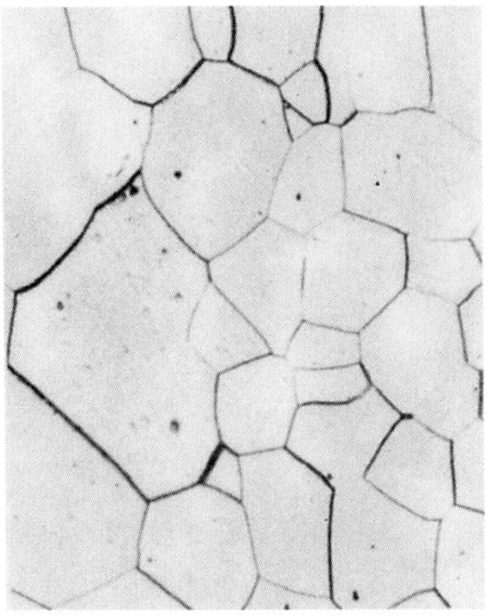

Figure 4-14 Microstructure of palladium (× 100). [*From* ASM Handbook, *Vol. 9, Metallography and Microstructure (1985), ASM International, Materials Park, OH 44073. Reprinted with permission of ASM International. All rights reserved. www.asminternational.org.*]

One manner by which grain size is specified is the **ASTM grain size number** (**ASTM** is the American Society for Testing and Materials). The number of grains per square inch is determined from a photograph of the metal taken at a magnification of 100. The ASTM grain size number n is calculated as

$$N = 2^{n-1} \tag{4-6}$$

where N is the number of grains per square inch.

A large ASTM number indicates many grains, or a fine grain size, and correlates with high strengths for metals.

When describing a microstructure, whenever possible, it is preferable to use a micrometer marker or some other scale on the micrograph, instead of stating the magnification as ×, as shown in Figure 4-14. That way, if the micrograph is enlarged or reduced, the micrometer marker scales with it, and we do not have to worry about changes in the magnification of the original micrograph. A number of sophisticated **image analysis** programs are available. Using such programs, it is possible to obtain quantitative information on average grain size, grain size distribution, porosity, second phases (Chapter 11), etc. A number of optical and scanning electron microscopes can be purchased with image analysis capabilities. The following example illustrates the calculation of the ASTM grain size number.

Example 4-12 *Calculation of ASTM Grain Size Number*

Suppose we count 16 grains per square inch in a photomicrograph taken at a magnification of 250. What is the ASTM grain size number?

SOLUTION

Consider one square inch from the photomicrograph taken at a magnification of 250. At a magnification of 100, the one square inch region from the 250 magnification image would appear as

$$1 \text{ in.}^2 \left(\frac{100}{250}\right)^2 = 0.16 \text{ in.}^2$$

and we would see

$$\frac{16 \text{ grains}}{0.16 \text{ in.}^2} = 100 \text{ grains/in.}^2$$

Substituting in Equation 4-6,

$$N = 100 \text{ grains/in.}^2 = 2^{n-1}$$
$$\log 100 = (n - 1) \log 2$$
$$2 = (n - 1)(0.301)$$
$$n = 7.64$$

Small Angle Grain Boundaries A **small angle grain boundary** is an array of dislocations that produces a small misorientation between the adjoining crystals (Figure 4-15). Because the energy of the surface is less than that of a regular grain boundary, the small angle grain boundaries are not as effective in blocking slip. Small angle boundaries formed by edge dislocations are called **tilt boundaries**, and those caused by screw dislocations are called **twist boundaries**.

Stacking Faults **Stacking faults**, which occur in FCC metals, represent an error in the stacking sequence of close-packed planes. Normally, a stacking sequence of *ABC ABC ABC* is produced in a perfect FCC crystal. Suppose instead the following sequence is produced:

ABC *<u>ABAB</u>* *CABC*

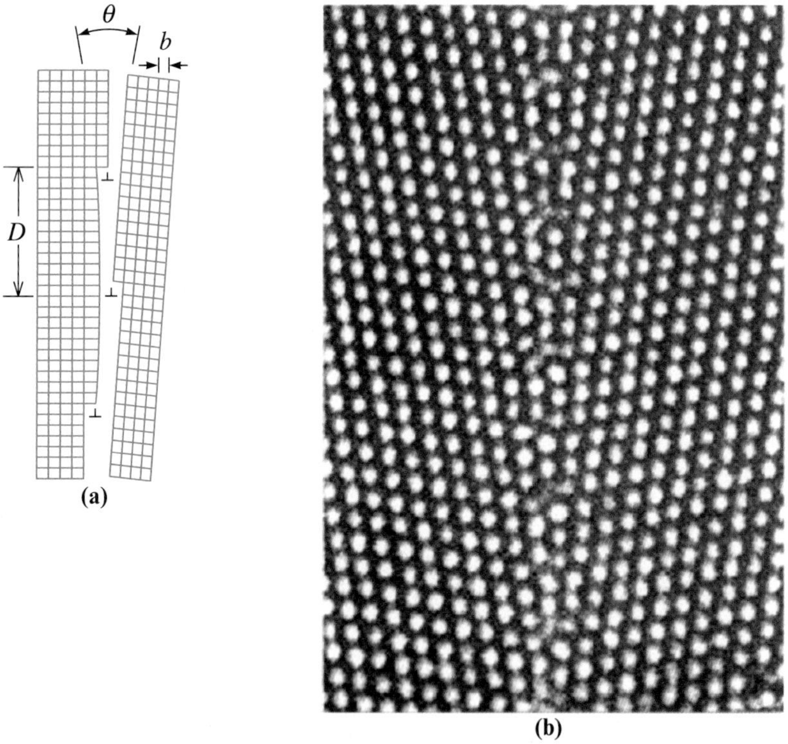

Figure 4-15 (a) The small angle grain boundary is produced by an array of dislocations, causing an angular mismatch θ between the lattices on either side of the boundary. (b) A transmission electron micrograph of a small angle grain boundary in aluminum. (*Courtesy of Ron Gronsky*)

In the portion of the sequence indicated, a type A plane replaces a type C plane. This small region, which has the HCP stacking sequence instead of the FCC stacking sequence, represents a stacking fault. Stacking faults interfere with the slip process.

Twin Boundaries

A **twin boundary** is a plane across which there is a special mirror image misorientation of the crystal structure (Figure 4-16). Twins can be produced when a shear force, acting along the twin boundary, causes the atoms to shift out of position. Twinning occurs during deformation or heat treatment of certain metals. The twin boundaries interfere with the slip process and increase the strength of the metal. Twinning also occurs in some ceramic materials such as monoclinic zirconia and dicalcium silicate.

The effectiveness of the surface defects in interfering with the slip process can be judged from the surface energies (Table 4-3). The high-energy grain boundaries are much more effective in blocking dislocations than either stacking faults or twin boundaries.

Domain Boundaries

Ferroelectrics are materials that develop spontaneous and reversible dielectric polarization (e.g., PZT or $BaTiO_3$) (see Chapter 19). Magnetic materials also develop a magnetization in a similar fashion (e.g., Fe, Co, Ni, iron oxide, etc.) (see Chapter 20). These electronic and magnetic materials contain domains. A **domain**

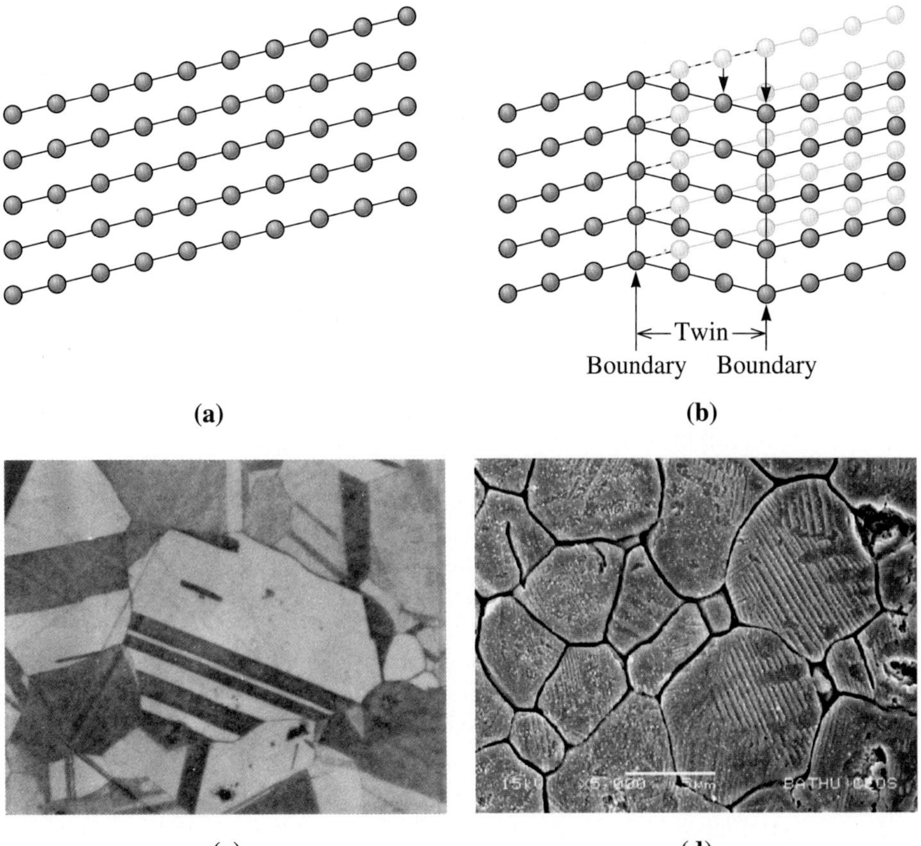

Figure 4-16 Application of stress to the (a) perfect crystal may cause a displacement of the atoms, (b) resulting in the formation of a twin. Note that the crystal has deformed as a result of twinning. (c) A micrograph of twins within a grain of brass (×250). (*Reprinted Courtesy of Don Askeland*) (d) Domains in ferroelectric barium titanate. (*Courtesy of Professor Chris Bowen*) Similar domain structures occur in magnetic materials.

Table 4-3 Energies of surface imperfections in selected metals

Surface Imperfection (ergs/cm^2)	Al	Cu	Pt	Fe
Stacking fault	200	75	95	—
Twin boundary	120	45	195	190
Grain boundary	625	645	1000	780

is a small region of the material in which the direction of magnetization or dielectric polarization is the same. In these materials, many small domains form so as to minimize the total free energy of the material. Figure 4-16(d) shows an example of domains in tetragonal ferroelectric barium titanate. The presence of domains influences the dielectric and magnetic properties of many electronic and magnetic materials. We will discuss these materials in later chapters.

4-8 Importance of Defects

Extended and point defects play a major role in influencing mechanical, electrical, optical, and magnetic properties of engineered materials. In this section, we recapitulate the importance of defects on properties of materials. We emphasize that the effect of dislocations is most important in metallic materials.

Effect on Mechanical Properties via Control of the Slip Process Any imperfection in the crystal raises the internal energy at the location of the imperfection. The local energy is increased because, near the imperfection, the atoms either are squeezed too closely together (compression) or are forced too far apart (tension).

A dislocation in an otherwise perfect metallic crystal can move easily through the crystal if the resolved shear stress equals the critical resolved shear stress. If the dislocation encounters a region where the atoms are displaced from their usual positions, however, a higher stress is required to force the dislocation past the region of high local energy; thus, the material is stronger. *Defects in materials, such as dislocations, point defects, and grain boundaries, serve as "stop signs" for dislocations.* They provide resistance to dislocation motion, and any mechanism that impedes dislocation motion makes a metal stronger. Thus, we can control the strength of a metallic material by controlling the number and type of imperfections. Three common strengthening mechanisms are based on the three categories of defects in crystals. Since dislocation motion is relatively easier in metals and alloys, these mechanisms typically work best for metallic materials. We need to keep in mind that very often the strength of ceramics in tension and at low temperatures is dictated by the level of porosity (presence of small holes). Polymers are often amorphous and hence dislocations play very little role in their mechanical behavior, as discussed in a later chapter. The strength of inorganic glasses (e.g., silicate float glass) depends on the distribution of flaws on the surface.

Strain Hardening Dislocations disrupt the perfection of the crystal structure. In Figure 4-17, the atoms below the dislocation line at point *B* are compressed, while the atoms above dislocation *B* are too far apart. If dislocation *A* moves to the right and passes near dislocation *B*, dislocation *A* encounters a region where the atoms are not properly arranged. Higher stresses are required to keep the second dislocation moving; consequently, the metal must be stronger. Increasing the number of dislocations further increases the strength of the material since increasing the dislocation density causes more stop signs for dislocation motion. The dislocation density can be shown to increase markedly as we strain or deform a material. This mechanism of increasing the strength

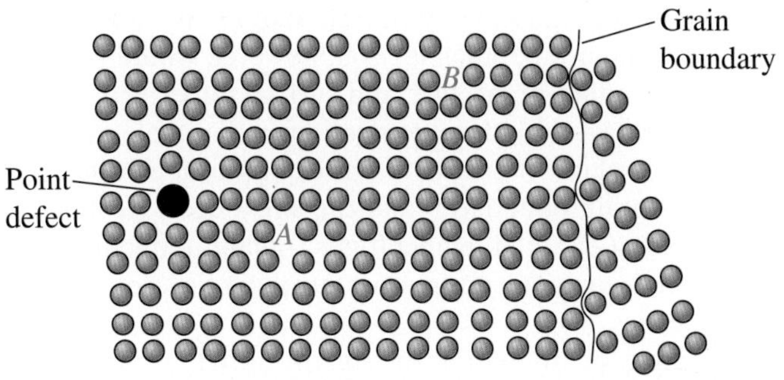

Figure 4-17 If the dislocation at point A moves to the left, its motion is hindered by the point defect. If the dislocation moves to the right, it interacts with the disturbed lattice near the second dislocation at point B. If the dislocation moves farther to the right, its motion is obstructed by a grain boundary.

of a material by deformation is known as **strain hardening**, which is discussed in Chapter 8. We can also show that dislocation densities can be reduced substantially by heating a metallic material to a relatively high temperature (below the melting temperature) and holding it there for a long period of time. This heat treatment is known as **annealing** and is used to impart ductility to metallic materials. Thus, controlling the dislocation density is an important way of controlling the strength and ductility of metals and alloys.

Solid-Solution Strengthening Any of the point defects also disrupt the perfection of the crystal structure. A solid solution is formed when atoms or ions of a guest element or compound are assimilated completely into the crystal structure of the host material. This is similar to the way salt or sugar in small concentrations dissolve in water. If dislocation A moves to the left (Figure 4-17), it encounters a disturbed crystal caused by the point defect; higher stresses are needed to continue slip of the dislocation. By intentionally introducing substitutional or interstitial atoms, we cause *solid-solution strengthening*, which is discussed in Chapter 10. This mechanism explains why plain carbon steel is stronger than pure Fe and why alloys of copper containing small concentrations of Be are much stronger than pure Cu. Pure gold or silver, both FCC metals with many active slip systems, are mechanically too soft.

Grain-Size Strengthening Surface imperfections such as grain boundaries disturb the arrangement of atoms in crystalline materials. If dislocation B moves to the right (Figure 4-17), it encounters a grain boundary and is blocked. By increasing the number of grains or reducing the grain size, **grain-size strengthening** is achieved in metallic materials. Control of grain size will be discussed in a number of later chapters.

There are two more mechanisms for strengthening of metals and alloys. These are known as **second-phase strengthening** and **precipitation strengthening**. We will discuss these in later chapters.

Example 4-13 *Design/Materials Selection for a Stable Structure*

We would like to produce a bracket to hold ceramic bricks in place in a heat-treating furnace. The bracket should be strong, should possess some ductility so that it bends rather than fractures if overloaded, and should maintain most of its strength up to 600°C. Design the material for this bracket, considering the various crystal imperfections as the strengthening mechanism.

SOLUTION

In order to serve up to 600°C, the bracket should *not* be produced from a polymer. Instead, a metal or ceramic should be considered.

In order to have some ductility, dislocations must move and cause slip. Because slip in ceramics is difficult, the bracket should be produced from a metallic material.

The metal should have a melting point well above 600°C; aluminum, with a melting point of 660°C, would not be suitable; iron, however, would be a reasonable choice.

We can introduce point, line, and surface imperfections into the iron to help produce strength, but we wish the imperfections to be stable as the service temperature increases. As discussed in Chapter 8, grains can grow at elevated temperatures, reducing the number of grain boundaries and causing a decrease in strength. Dislocations may be annihilated at elevated temperatures—again, reducing strength. The number of vacancies depends on temperature, so controlling these crystal defects may not produce stable properties.

The number of interstitial or substitutional atoms in the crystal does not, however, change with temperature. We might add carbon to the iron as interstitial atoms or substitute vanadium atoms for iron atoms at normal lattice points. These point defects continue to interfere with dislocation movement and help to keep the strength stable.

Of course, other design requirements may be important as well. For example, the steel bracket may deteriorate by oxidation or may react with the ceramic brick.

Effects on Electrical, Optical, and Magnetic Properties

In previous sections, we stated that the effect of point defects on the electrical properties of semiconductors is dramatic. The entire microelectronics industry critically depends upon the successful incorporation of substitutional dopants such as P, As, B, and Al in Si and other semiconductors. These dopant atoms allow us to have significant control of the electrical properties of semiconductors. Devices made from Si, GaAs, amorphous silicon (a:Si:H), etc., critically depend on the presence of dopant atoms. We can make *n*-type Si by introducing P atoms in Si. We can make *p*-type Si using B atoms. Similarly, a number of otherwise unsatisfied bonds in amorphous silicon are completed by incorporating H atoms.

The effect of defects such as dislocations on the properties of semiconductors is usually deleterious. Dislocations and other defects (including other point defects) can interfere with motion of charge carriers in semiconductors. This is why we make sure that the dislocation densities in single crystal silicon and other materials used in optical and electrical applications are very small. Point defects also cause increased resistivity in metals.

In some cases, defects can enhance certain properties. For example, incorporation of CaO in ZrO_2 causes an increase in the concentration of oxygen ion vacancies. This has a beneficial effect on the conductivity of zirconia and allows us to use such compositions for oxygen gas sensors and solid oxide fuel cells. Defects can convert many otherwise insulating dielectric materials into useful semiconductors! These are then used for many sensor applications (e.g., temperature, humidity, and gas sensors, etc.).

Addition of about 1% chromium oxide in alumina creates defects that make alumina ruby red. Similarly, incorporation of Fe^{2+} and Ti^{4+} makes the blue sapphire. Nanocrystals of materials such as cadmium sulfide (CdS) in inorganic glasses produce glasses that have a brilliant color.

Nanocrystals of silver halide and other crystals also allow formation of photochromic and photosensitive glasses.

Many magnetic materials can be processed such that grain boundaries and other defects make it harder to reverse the magnetization in these materials. The magnetic properties of many commercial ferrites, used in magnets for loudspeakers and devices in wireless communication networks, depend critically on the distribution of different ions on different crystallographic sites in the crystal structure. As mentioned before, the presence of domains affects the properties of ferroelectric, ferromagnetic, and ferrimagnetic materials (Chapters 19 and 20).

Summary

- Imperfections, or defects, in a crystalline material are of three general types: point defects, line defects or dislocations, and surface defects.
- The number of vacancies depends on the temperature of the material; interstitial atoms (located at interstitial sites between the normal atoms) and substitutional atoms (which replace the host atoms at lattice points) are often deliberately introduced and are typically unaffected by changes in temperature.
- Dislocations are line defects that, when a force is applied to a metallic material, move and cause a metallic material to deform.
- The critical resolved shear stress is the stress required to move the dislocation.
- The dislocation moves in a slip system, composed of a slip plane and a slip direction. The slip direction is typically a close-packed direction. The slip plane is also normally close-packed or nearly close-packed.
- In metallic crystals, the number and type of slip directions and slip planes influence the properties of the metal. In FCC metals, the critical resolved shear stress is low and an optimum number of slip planes is available; consequently, FCC metals tend to be ductile. In BCC metals, no close-packed planes are available, and the critical resolved shear stress is high; thus, BCC metals tend to be strong. The number of slip systems in HCP metals is limited, causing these metals typically to behave in a brittle manner.
- Point defects, which include vacancies, interstitial atoms, and substitutional atoms, introduce compressive or tensile strain fields that disturb the atomic arrangements in the surrounding crystal. As a result, dislocations cannot easily slip in the vicinity of point defects, and the strength of the metallic material is increased.
- Surface defects include grain boundaries. Producing a very small grain size increases the amount of grain boundary area; because dislocations cannot easily pass through a grain boundary, the material is strengthened (Hall-Petch equation).
- The number and type of crystal defects control the ease of movement of dislocations and, therefore, directly influence the mechanical properties of the material.
- Defects in materials have a significant influence on their electrical, optical, and magnetic propeties.

Glossary

Annealing A heat treatment that typically involves heating a metallic material to a high temperature for an extended period of time in order to lower the dislocation density and hence impart ductility.

ASTM American Society for Testing and Materials.

ASTM grain size number (*n*) A measure of the size of the grains in a crystalline material obtained by counting the number of grains per square inch at a magnification of 100.

Burgers vector The direction and distance that a dislocation moves in each step, also known as the slip vector.

Critical resolved shear stress The shear stress required to cause a dislocation to move and cause slip.

Cross-slip A change in the slip system of a dislocation.

Defect chemical reactions Reactions written using the Kröger-Vink notation to describe defect chemistry. The reactions must be written in such a way that mass and electrical charges are balanced and stoichiometry of sites is maintained.

Dislocation A line imperfection in a crystalline material.

Dislocation density The total length of dislocation line per cubic centimeter in a material.

Domain A small region of a ferroelectric, ferromagnetic, or ferrimagnetic material in which the direction of dielectric polarization (for ferroelectric) or magnetization (for ferromagnetic or ferrimagnetic) is the same.

Dopants Elements or compounds typically added, in known concentrations and appearing at specific places within the microstructure, to enhance the properties or processing of a material.

Edge dislocation A line defect that can be envisioned as an "extra half plane of atoms" inserted into the crystal.

Elastic deformation Deformation that is fully recovered when the stress causing it is removed.

Etch pits Holes created at locations where dislocations meet the surface.

Extended defects Defects that involve several atoms/ions and thus occur over a finite volume of the crystalline material (e.g., dislocations, stacking faults, etc.).

Ferroelectric A dielectric material that develops a spontaneous and reversible electric polarization (e.g., PZT, $BaTiO_3$).

Frenkel defect A pair of point defects produced when an ion moves to create an interstitial site, leaving behind a vacancy.

Grain One of the crystals in a polycrystalline material.

Grain boundary A surface defect representing the boundary between two grains. The crystal has a different orientation on either side of the grain boundary.

Grain-size strengthening Strengthening of a material by decreasing the grain size and therefore increasing the grain boundary area.

Hall-Petch equation The relationship between yield strength and grain size in a metallic material—i.e., $\sigma_y = \sigma_0 + Kd^{-1/2}$.

Image analysis A technique that is used to analyze images of microstructures to obtain quantitative information on grain size, shape, grain size distribution, etc.

Impurities Elements or compounds that find their way into a material, often originating from processing or raw materials and typically having a deleterious effect on the properties or processing of a material.

Interstitial defect A point defect produced when an atom is placed into the crystal at a site that is normally not a lattice point.

Interstitialcy A point defect caused when a "normal" atom occupies an interstitial site in the crystal.

Kröger-Vink notation A system used to indicate point defects in materials. The main body of the notation indicates the type of defect or the element involved. The subscript indicates the location of the point defect, and the superscript indicates the effective positive (•) or negative (′) charge.

Metallography Preparation of a metallic sample of a material by polishing and etching so that the structure can be examined using a microscope.

Mixed dislocation A dislocation that contains partly edge components and partly screw components.

Peierls-Nabarro stress The shear stress, which depends on the Burgers vector and the interplanar spacing, required to cause a dislocation to move—i.e., $\tau = c \exp(-kd/b)$

Plastic deformation Permanent deformation of a material when a load is applied, then removed.

Point defects Imperfections, such as vacancies, that are located typically at one (in some cases a few) sites in the crystal.

Precipitation strengthening Strengthening of metals and alloys by formation of precipitates inside the grains. The small precipitates resist dislocation motion.

Schmid's law The relationship between shear stress, the applied stress, and the orientation of the slip system—i.e., $\tau = \sigma \cos\lambda \cos\phi$.

Schottky defect A point defect in ionically bonded materials. In order to maintain a neutral charge, a stoichiometric number of cation and anion vacancies must form.

Screw dislocation A line defect that can be envisioned as a spiral ramp in the crystal that was produced by skewing the crystal by one atomic spacing.

Second-phase strengthening A mechanism by which grains of an additional compound or phase are introduced in a polycrystalline material. These second phase crystals resist dislocation motion, thereby causing an increase in the strength of a metallic material.

Sintering A process for forming a dense mass by heating compacted powders.

Slip Deformation of a metallic material by the movement of dislocations through the crystal.

Slip direction The direction in the crystal in which the dislocation moves. The slip direction is the same as the direction of the Burgers vector.

Slip plane The plane swept out by the dislocation line during slip. Normally, the slip plane is a close-packed plane, if one exists in the crystal structure.

Slip system The combination of the slip plane and the slip direction.

Small angle grain boundary An array of dislocations causing a small misorientation of the crystal across the surface of the imperfection.

Stacking fault A surface defect in metals caused by the improper stacking sequence of close-packed planes.

Strain hardening Strengthening of a material by increasing the number of dislocations by deformation, or cold working. Also known as "work hardening."

Substitutional defect A point defect produced when an atom is removed from a regular lattice point and replaced with a different atom, usually of a different size.

Surface defects Imperfections, such as grain boundaries, that form a two-dimensional plane within the crystal.

Thermal grooving A technique used for observing microstructures in ceramic materials that involves heating a polished sample to a temperature slightly below the sintering temperature for a short time.

Tilt boundary A small angle grain boundary formed by edge dislocations.

Twin boundary A surface defect across which there is a mirror image misorientation of the crystal structure.

Twist boundary A small angle grain boundary formed by screw dislocations

Vacancy An atom or an ion missing from its regular crystallographic site.

Yield strength The level of stress above which a material begins to show permanent deformation.

Problems

Section 4-1 Point Defects

4-1 Gold has 5.82×10^8 vacancies/cm^3 at equilibrium at 300 K. What fraction of the atomic sites is vacant at 600 K?

4-2 Calculate the number of vacancies per m^3 for gold at 900°C. The energy for vacancy formation is 0.93 eV/atom.

4-3 Calculate the number of vacancies per cm^3 expected in copper at 1080°C (just below the melting temperature). The energy for vacancy formation is 20,000 cal/mol.

4-4 The fraction of lattice points occupied by vacancies in solid aluminum at 660°C is 10^{-3}. What is the energy required to create vacancies in aluminum?

4-5 The density of a sample of FCC palladium is 11.98 g/cm^3, and its lattice parameter is 3.8902 Å. Calculate
(a) the fraction of the lattice points that contain vacancies; and
(b) the total number of vacancies in a cubic centimeter of Pd.

4-6 The density of a sample of HCP beryllium is 1.844 g/cm^3, and the lattice parameters are $a_0 = 0.22858$ nm and $c_0 = 0.35842$ nm. Calculate
(a) the fraction of the lattice points that contain vacancies; and
(b) the total number of vacancies in a cubic centimeter of Be.

4-7 BCC lithium has a lattice parameter of 3.5089×10^{-8} cm and contains one vacancy per 200 unit cells. Calculate
(a) the number of vacancies per cubic centimeter; and
(b) the density of Li.

4-8 FCC lead has a lattice parameter of 0.4949 nm and contains one vacancy per 500 Pb atoms. Calculate
(a) the density; and
(b) the number of vacancies per gram of Pb.

4-9 Cu and Ni form a substitutional solid solution. This means that the crystal structure of a Cu-Ni alloy consists of Ni atoms substituting for Cu atoms in the regular atomic positions of the FCC structure. For a Cu-30% wt% Ni alloy, what fraction of the atomic sites does Ni occupy?

4-10 Au and Ag form a substitutional solid solution. This means that the crystal structure of a Au-Ag alloy consists of Ag atoms substituting for Au atoms in the regular atomic positions of the FCC structure. For a Au-50 at% Ag alloy, what is the wt% Ag in the alloy?

4-11 A niobium alloy is produced by introducing tungsten substitutional atoms into the BCC structure; eventually an alloy is produced that has a lattice parameter of 0.32554 nm and a density of 11.95 g/cm^3. Calculate the fraction of the atoms in the alloy that are tungsten.

4-12 Tin atoms are introduced into an FCC copper crystal, producing an alloy with a lattice parameter of 3.7589×10^{-8} cm and a density of 8.772 g/cm^3. Calculate the atomic percentage of tin present in the alloy.

4-13 We replace 7.5 atomic percent of the chromium atoms in its BCC crystal with tantalum. X-ray diffraction shows that the lattice parameter is 0.29158 nm. Calculate the density of the alloy.

4-14 Suppose we introduce one carbon atom for every 100 iron atoms in an interstitial position in BCC iron, giving a lattice parameter of 0.2867 nm. For this steel, find the density and the packing factor.

4-15 The density of BCC iron is 7.882 g/cm^3, and the lattice parameter is 0.2866 nm when hydrogen atoms are introduced at interstitial positions. Calculate
(a) the atomic fraction of hydrogen atoms; and
(b) the number of unit cells on average that contain hydrogen atoms.

Section 4-2 Other Point Defects

4-16 Suppose one Schottky defect is present in every tenth unit cell of MgO. MgO has the sodium chloride crystal structure and a lattice parameter of 0.396 nm. Calculate
(a) the number of anion vacancies per cm^3; and
(b) the density of the ceramic.

4-17 ZnS has the zincblende structure. If the density is 3.02 g/cm^3 and the lattice parameter is 0.59583 nm, determine the number of Schottky defects
(a) per unit cell; and
(b) per cubic centimeter.

4-18 Suppose we introduce the following point defects.
(a) Mg^{2+} ions substitute for yttrium ions in Y_2O_3;
(b) Fe^{3+} ions substitute for magnesium ions in MgO;
(c) Li^{1+} ions substitute for magnesium ions in MgO; and
(d) Fe^{2+} ions replace sodium ions in NaCl.
What other changes in each structure might be necessary to maintain a charge balance? Explain.

4-19 Write down the defect chemistry equation for introduction of $SrTiO_3$ in $BaTiO_3$ using the Kröger-Vink notation.

Section 4-3 Dislocations

4-20 Do amorphous and crystalline materials plastically deform by the same mechanisms? Explain.

4-21 What is the Burger's vector orientation relationship with the dislocation axis for both edge and screw dislocations?

4-22 What is a slip system and what role does it play in plastic deformation?

4-23 Draw a Burgers circuit around the dislocation shown in Figure 4-18. Clearly indicate the Burgers vector that you find. What type of dislocation is this? In what direction will the dislocation move due to the applied shear stress τ? Reference your answers to the coordinate axes shown.

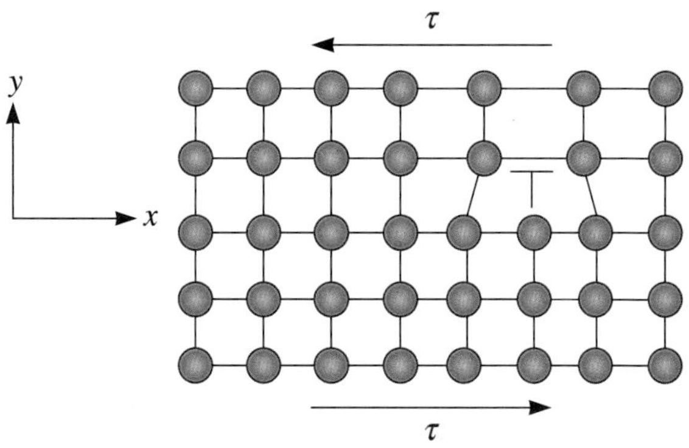

Figure 4-18 A schematic diagram of a dislocation (for Problem 4-23).

4-24 What are the Miller indices of the slip directions:
(a) on the (111) plane in an FCC unit cell?
(b) on the (011) plane in a BCC unit cell?

4-25 What are the Miller indices of the slip planes in FCC unit cells that include the [101] slip direction?

4-26 What are the Miller indices of the {110} slip planes in BCC unit cells that include the [111] slip direction?

4-27 Calculate the length of the Burgers vector in the following materials:
(a) BCC niobium;
(b) FCC silver; and
(c) diamond cubic silicon.

4-28 Determine the interplanar spacing and the length of the Burgers vector for slip on the expected slip systems in FCC aluminum. Repeat, assuming that the slip system is a (110) plane and a $[1\bar{1}1]$ direction. What is the ratio between the shear stresses required for slip for the two systems? Assume that $k = 2$ in Equation 4-2.

4-29 Determine the interplanar spacing and the length of the Burgers vector for slip on the $(110)[1\bar{1}1]$ slip system in BCC tantalum. Repeat, assuming that the slip system is a $(111)/[1\bar{1}0]$ system. What is the ratio between the shear stresses required for slip for the two systems? Assume that $k = 2$ in Equation 4-2.

4-30 The crystal shown in Figure 4-19 contains two dislocations *A* and *B*. If a shear stress is applied to the crystal as shown, what will happen to dislocations *A* and *B*?

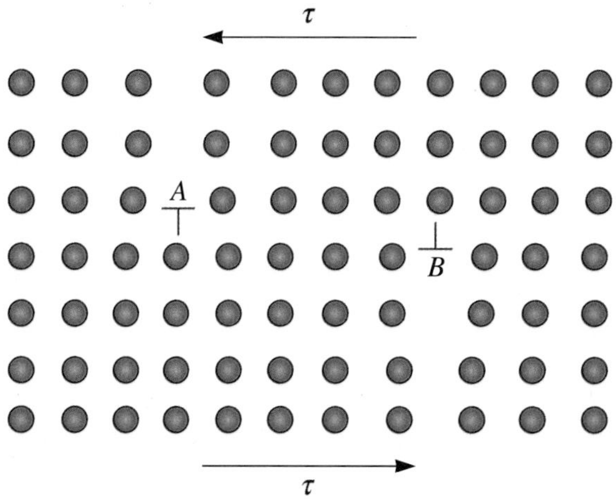

Figure 4-19 A schematic diagram of two dislocations (for Problem 4-30).

4-31 Can ceramic and polymeric materials contain dislocations?

4-32 Why is it that ceramic materials are brittle?

Section 4-4 Significance of Dislocations

4-33 What is meant by the terms plastic and elastic deformation?

4-34 Why is the theoretical strength of metals much higher than that observed experimentally?

4-35 How many grams of aluminum, with a dislocation density of 10^{10} cm/cm^3, are required to give a total dislocation length that would stretch from New York City to Los Angeles (3000 miles)?

4-36 The distance from earth to the moon is 240,000 miles. If this were the total length of dislocation in a cubic centimeter of material, what would be the dislocation density? Compare your answer to typical dislocation densities for metals.

4-37 Why would metals behave as brittle materials without dislocations?

4-38 Why is it that dislocations play an important role in controlling the mechanical properties of metallic materials, however, they do not play a role in determining the mechanical properties of glasses?

4-39 Suppose you would like to introduce an interstitial or large substitutional atom into the crystal near a dislocation. Would the atom fit more easily above or below the dislocation line shown in Figure 4-7(c)? Explain.

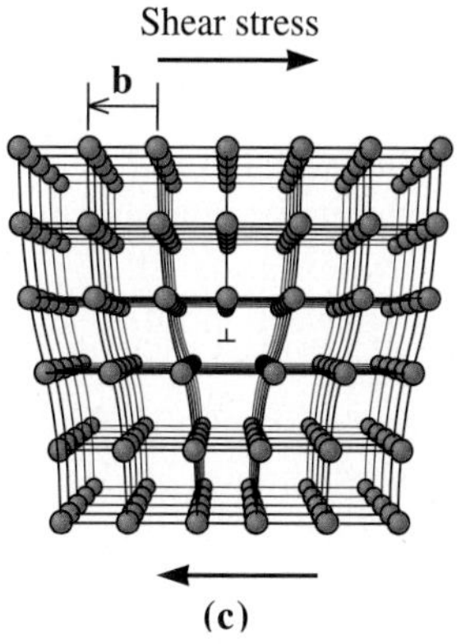

Figure 4-7(c) (Repeated for Problem 4-39.)

4-40 Compare the *c/a* ratios for the following HCP metals, determine the likely slip processes in each, and estimate the approximate critical resolved shear stress. Explain. (See data in Appendix A.)
(a) zinc;
(b) magnesium;
(c) titanium;

(d) zirconium;
(e) rhenium; and
(f) beryllium.

Section 4-5 Schmid's Law

4-41 Using Schmid's law, the resolved shear stress operating on a slip direction/plane is given by $\tau_r = \sigma \cos \lambda \cos \varphi$. If either of these angles equal 90°, what happens to the dislocations involved?

4-42 A single crystal of an FCC metal is oriented so that the [001] direction is parallel to an applied stress of 5000 psi. Calculate the resolved shear stress acting on the (111) slip plane in the $[\bar{1}10]$ and $[10\bar{1}]$ slip directions. Which slip system(s) will become active first?

4-43 A single crystal of a BCC metal is oriented so that the [001] direction is parallel to the applied stress. If the critical resolved shear stress required for slip is 12,000 psi, calculate the magnitude of the applied stress required to cause slip to begin in the $[1\bar{1}1]$ direction on the (110), (011), and $(10\bar{1})$ slip planes.

4-44 A single crystal of silver is oriented so that the (111) slip plane is perpendicular to an applied stress of 50 MPa. List the slip systems composed of close-packed planes and directions that may be activated due to this applied stress.

Section 4-6 Influence of Crystal Structure

4-45 Why is it that single crystal and polycrystalline copper are both ductile; however, only single crystal, but not polycrystalline, zinc can exhibit considerable ductility?

4-46 Explain why hexagonal close-packed metals tend to have a limited ability to be strain hardened.

4-47 Why is it that cross slip in BCC and FCC metals is easier than in HCP metals? How does this influence the ductility of BCC, FCC, and HCP metals?

4-48 Arrange the following metals in the expected order of increasing ductility: Cu, Ti, and Fe. Explain.

Section 4-7 Surface Defects

4-49 What are the imperfections in the atomic arrangements that have a significant effect on the material behavior? Give an example of each.

4-50 The strength of titanium is found to be 65,000 psi when the grain size is 6.7×10^{-4} in. and 82,000 psi when the grain size is 3.15×10^{-5} in. Determine
(a) the constants in the Hall-Petch equation; and
(b) the strength of the titanium when the grain size is reduced to 8.00×10^{-6} in.

4-51 A copper-zinc alloy has the properties shown in the table below.

Grain Diameter (mm)	Strength (MPa)
0.015	170 MPa
0.025	158 MPa
0.035	151 MPa
0.050	145 MPa

Determine
(a) the constants in the Hall-Petch equation; and
(b) the grain size required to obtain a strength of 200 MPa.

4-52 If there were 50 grains per in.^{2} in a photograph of a metal taken at 100× magnification, calculate the ASTM grain size number (n).

4-53 If the area of a photograph measured 7.812 in.^{2} and 23 grains were documented at 250 × magnification, what would the ASTM grain size number (n) be?

4-54 You have the choice to either purchase a copper alloy that has an ASTM grain size of 5 or ASTM grain size of 8. You are unable to decide if there is much of a difference between these two. Determine how many grains/in.^{2} would appear on a photograph taken at 100× for a metal given these two ASTM grain size choices. Does this seem like a significant difference?

4-55 The mill test report (MTR) on a structural steel you are considering using on an elevated pedestrian bridge over a busy roadway shows an ASTM grain size of 1.5. Would you accept this steel for this particular project? Explain.

4-56 For an ASTM grain size number of 8, calculate the number of grains per square inch
(a) at a magnification of 100 and
(b) with no magnification.

4-57 Determine the ASTM grain size number if 20 grains/square inch are observed at a magnification of 400.

4-58 Determine the ASTM grain size number if 25 grains/square inch are observed at a magnification of 50.

4-59 Determine the ASTM grain size number for the materials in Figure 4-14 and Figure 4-20.

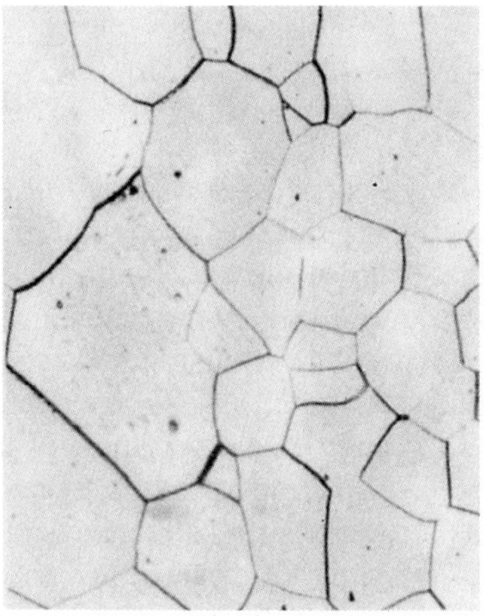

Figure 4-14 Microstructure of palladium (× 100). (*From* ASM Handbook, *Vol. 9, Metallography and Microstructure (1985), ASM International, Materials Park, OH 44073. Reprinted with permission of ASM International. All rights reserved. www.asminternational.org.*) (Repeated for Problem 4-59.)

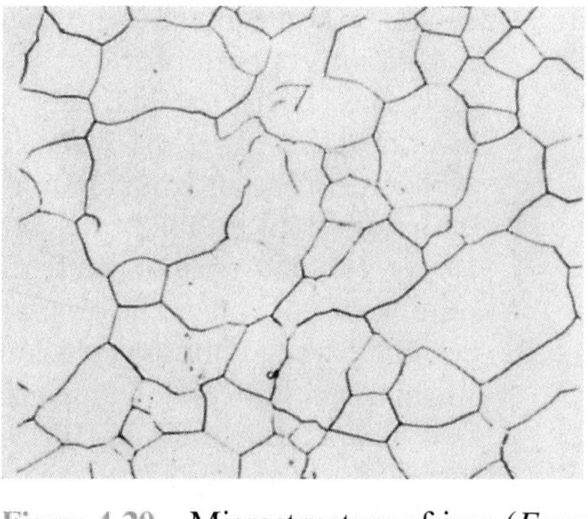

Figure 4-20 Microstructure of iron (*From* ASM Handbook, *Vol. 9, Metallography and Microstructure (1985), ASM International, Materials Park, OH 44073. Reprinted with permission of ASM International. All rights reserved. www.asminternational.org.*) (for Problem 4-59).

4-60 The drawing in Figure 4-21 shows a material with the simple cubic crystal structure containing two edge dislocations. Identify their locations by marking each with a ⊥ (oriented appropriately).

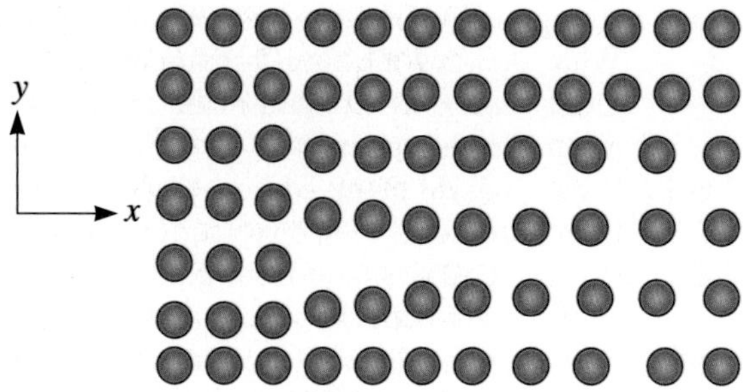

Figure 4-21 A schematic diagram of two dislocations (for Problem 4-60).

4-61 The angle θ of a tilt boundary is given by $\sin(\theta/2) = b/(2D)$ [See Figure 4-15(a).] Verify the correctness of this equation.

4-62 Calculate the angle θ of a small-angle grain boundary in FCC aluminum when the dislocations are 5000 Å apart. [See Figure 4-15(a) and the equation in Problem 4-61.]

4-63 For BCC iron, calculate the average distance between dislocations in a small-angle grain boundary tilted 0.50°. [See Figure 4-15(a).]

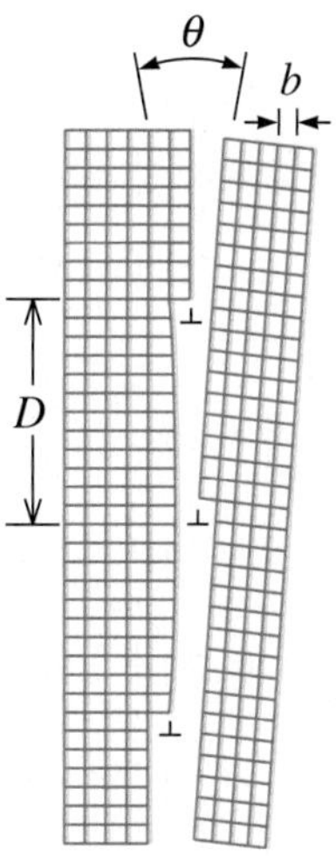

Figure 4-15(a) The small angle grain boundary is produced by an array of dislocations, causing an angular mismatch θ between the lattices on either side of the boundary. (Repeated for Problems 4-61, 4-62 and 4-63.)

Section 4-8 Importance of Defects

4-64 Every time we alloy a metal, it gets stronger. Is this true or false? Explain your answer fully.

4-65 How do the strengthening mechanisms of strain hardening, alloying, and grain size strengthening increase materials strength? What do they all have in common? Can more than one mechanism be used in a given alloy? Explain.

4-66 What makes plain carbon steel harder than pure iron?

4-67 Why is jewelry made from gold or silver alloyed with copper?

4-68 Why do we prefer to use semiconductor crystals that contain as small a number of dislocations as possible?

4-69 In structural applications (e.g., steel for bridges and buildings or aluminum alloys for aircraft), why do we use alloys rather than pure metals?

4-70 Assuming that we could obtain a 1 wt% solid solution of each of the following elements in aluminum, which of the elements would be expected to produce the greatest strengthening effect: Mg, Mn, Cu, and Zn? Why?

4-71 Do dislocations control the strength of a silicate glass? Explain.

4-72 What is meant by the term strain hardening?

4-73 To which mechanism of strengthening is the Hall-Petch equation related?

4-74 Pure copper is strengthened by the addition of a small concentration of Be. To which mechanism of strengthening is this related?

Design Problems

4-75 The density of pure aluminum calculated from crystallographic data is expected to be 2.69955 g/cm^3.

(a) Design an aluminum alloy that has a density of 2.6450 g/cm^3.

(b) Design an aluminum alloy that has a density of 2.7450 g/cm^3.

4-76 You would like a metal plate with good weldability. During the welding process, the metal next to the weld is heated almost to the melting temperature and, depending on the welding parameters, may remain hot for some period of time. Design an alloy that will minimize the loss of strength in this "heat-affected zone" during the welding process.

4-77 We need a material that is optically transparent but electrically conductive. Such materials are used for touch screen displays. What kind of materials can be used for this application? (*Hint*: Think about coatings of materials that can provide electronic or ionic conductivity; the substrate has to be transparent for this application.)

Computer Problems

4-78 *Temperature dependence of vacancy concentrations.* Write a computer program that will provide a user with the equilibrium concentration of vacancies in a metallic element as a function of temperature. The user should specify a meaningful and valid range of temperatures (e.g., 100 to 1200 K for copper). Assume that the crystal structure originally specified is valid for this range of temperature. Ask the user to input the activation energy for the formation of one mole of vacancies (Q_v). The program then should ask the user to input the density of the element and crystal structure (FCC, BCC, etc.). You can use character variables to detect the type of crystal structures (e.g., "F" or "f" for FCC, "B" or "b" for BCC, etc.). Be sure to pay attention to the correct units for temperature, density, etc. The program should ask the user if the temperature range that has been provided is in °C, °F, or K and convert the temperatures properly into K before any calculations are performed. The program should use this information to establish the number of atoms per unit volume and provide an output for this value. The program should calculate the equilibrium concentration of vacancies at different temperatures. The first temperature will be the minimum temperature specified and then temperatures should be increased by 100 K or another convenient increment. You can make use of any graphical software to plot the data showing the equilibrium concentration of vacancies as a function of temperature. Think about what scales will be used to best display the results.

4-79 *Hall-Petch equation.* Write a computer program that will ask the user to enter two sets of values of σ_y and grain size (d) for a metallic material. The program should then utilize the data to calculate the Hall-Petch equation. The program then should prompt the user to input another value of grain size and calculate the yield stress or vice versa.

4-80 *ASTM grain size number calculator.* Write a computer program that will ask the user to input the magnification of a micrograph of the sample for which the ASTM number is being calculated. The program should then ask the user for the number of grains counted and the area (in square inches) from which these grains were counted. The program then should calculate the ASTM number, taking into consideration the fact that the micrograph magnification is not 100 and the area may not have been one square inch.

Knovel® Problems

K4-1 Describe the problems associated with metal impurities in silicon devices.

K4-2 What are the processes involved in the removal of metal impurities from silicon devices by gettering?

Sintered barium magnesium tantalate (BMT) ceramic microstructure. This ceramic material is useful in making electronic components used in wireless communications. The process of sintering is driven by the diffusion of atoms or ions. (*Courtesy of H. Shirey*)

CHAPTER 5

Atom and Ion Movements in Materials

Have You Ever Wondered?

- *Aluminum oxidizes more easily than iron, so why do we say aluminum normally does not "rust?"*
- *How are the surfaces of certain steels hardened?*
- *What are two factors that influence the diffusion of food coloring in a liquid?*
- *How does a tungsten filament in a light bulb fail?*

Chapter Learning Objectives

The key objectives of this chapter are to

- Describe the driving forces for diffusion in solids.
- Calculate the number of vacancies in a solid and the diffusion coefficient using an Arrhenius relationship.
- Explain the mechanisms of diffusion in solids.
- Calculate the net flux of atoms due to a concentration gradient using Fick's first law.
- Explain the factors influencing diffusion including temperature, diffusivity, time, concentration, and crystal structure.
- Design diffusion profiles and heat treatments using Fick's second law.
- Explain how sintering and grain growth depend on diffusion.

In Chapter 4, we learned that the atomic and ionic arrangements in materials are never perfect. We also saw that most materials are not pure elements; they are alloys or blends of different elements or compounds. Different types of atoms or ions typically "diffuse," or move within the material so that the differences in their concentration are minimized. **Diffusion** refers to an observable net flux of atoms or other species. It depends upon the concentration gradient and temperature. Just as water flows from a mountain toward the sea to minimize its gravitational potential energy, atoms and ions have a tendency to move in a predictable fashion to eliminate concentration differences and produce homogeneous compositions that make the material thermodynamically more stable.

In this chapter, we will learn that temperature influences the kinetics of diffusion and that a concentration difference contributes to the overall net flux of diffusing species. The goal of this chapter is to examine the principles and applications of diffusion in materials. We will illustrate the concept of diffusion through examples of several real-world technologies dependent on the diffusion of atoms, ions, or molecules.

We will present an overview of Fick's laws that describe the diffusion process quantitatively. We will also see how the relative openness of different crystal structures and the size of atoms or ions, temperature, and concentration of diffusing species affect the rate at which diffusion occurs. We will discuss specific examples of how diffusion is used in the synthesis and processing of advanced materials as well as manufacturing of components using advanced materials.

5-1 Applications of Diffusion

Diffusion Diffusion refers to the net flux of any species, such as ions, atoms, electrons, holes (Chapter 19), and molecules. The magnitude of this flux depends upon the concentration gradient and temperature. The process of diffusion is central to a large number of today's important technologies. In materials processing technologies, control over the diffusion of atoms, ions, molecules, or other species is key. There are hundreds of applications and technologies that depend on either enhancing or limiting diffusion. The following are just a few examples.

Carburization for Surface Hardening of Steels Let's say we want a surface, such as the teeth of a gear, to be hard; however, we do not want the entire gear to be hard. The carburization process can be used to increase surface hardness. In carburization, a source of carbon, such as a graphite powder or gaseous phase containing carbon, is diffused into steel components such as gears (Figure 5-1). The increased carbon concentration on the surface of the steel increases the steel's hardness because the carbon atoms in interstitial sites hinder dislocation motion. Similar to the introduction of carbon, we can also use a process known as **nitriding**, in which nitrogen is introduced into the surface of a metallic material. Diffusion also plays a central role in the control of the phase transformations needed for the heat treatment of metals and alloys, the processing of ceramics, and the solidification and joining of materials (see Section 5-9).

Dopant Diffusion for Semiconductor Devices The entire microelectronics industry, as we know it today, would not exist if we did not have a very good understanding of the diffusion of different atoms into silicon or other semiconductors. The creation of the *p-n junction* (Chapter 19) involves diffusing dopant atoms, such as phosphorus, arsenic, antimony, boron, aluminum, etc., into precisely defined regions of silicon wafers. Some of these regions are so small that they are best measured in nanometers. A *p-n* junction is a region of the semiconductor, one side of which is doped with *n*-type dopants (e.g., As in Si) and the other side is doped with *p*-type dopants (e.g., B in Si).

Figure 5-1 Furnace for heat treating steel using the carburization process. (*Courtesy of Arrow Gear Company.*)

Conductive Ceramics In general, polycrystalline ceramics tend to be good insulators of electricity. Diffusion of ions, electrons, or holes also plays an important role in the electrical conductivity of many **conductive ceramics**, such as partially or fully-stabilized zirconia (ZrO_2) or indium tin oxide (also commonly known as ITO). Lithium cobalt oxide ($LiCoO_2$) is an example of an ionically conductive material that is used in lithium ion batteries. These ionically conductive materials are used for such products as oxygen sensors in cars, touch-screen displays, fuel cells, and batteries. The ability of ions to diffuse and provide a pathway for electrical conduction plays an important role in enabling these applications.

Creation of Plastic Beverage Bottles The occurrence of diffusion may not always be beneficial. In some applications, we may want to limit the occurrence of diffusion for certain species. For example, the diffusion of carbon dioxide (CO_2) in plastic soda bottles must be minimized. This is one of the major reasons why we use polyethylene terephthalate (PET) to make bottles that ensure the carbonated beverages they contain will not lose their fizz for a reasonable period of time!

Oxidation of Aluminum You may have heard or know that aluminum does not "rust." In reality, aluminum oxidizes (rusts) more easily than iron; however, the aluminum oxide (Al_2O_3) forms a very protective but thin coating on the aluminum's surface, preventing any further diffusion of oxygen and hindering further oxidation of the underlying aluminum. The oxide coating does not have a color and is thin and, hence, invisible. This is why we think aluminum does not rust.

Thermal Barrier Coatings for Turbine Blades In an aircraft engine, some of the nickel superalloy-based turbine blades are coated with ceramic oxides such as yttria stabilized zirconia (YSZ). These ceramic coatings protect the underlying alloy from high temperatures; hence, the name **thermal barrier coatings** (TBCs) (Figure 5-2). The diffusion of oxygen through these ceramic coatings and the subsequent oxidation of the underlying alloy play a major role in determining the lifetime and

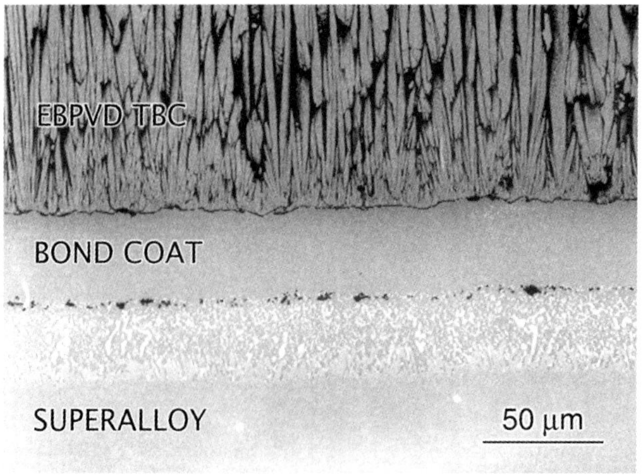

Figure 5-2 A thermal barrier coating on a nickel-based superalloy. (*Courtesy of Dr. F.S. Pettit and Dr. G.H. Meier, University of Pittsburgh.*)

durability of the turbine blades. In Figure 5-2, EBPVD means electron beam physical vapor deposition. The bond coat is either a platinum or molybdenum-based alloy. It provides adhesion between the TBC and the substrate.

Optical Fibers and Microelectronic Components Optical fibers made from silica (SiO_2) are coated with polymeric materials to prevent diffusion of water molecules. This, in turn, improves the optical and mechanical properties of the fibers.

Example 5-1 *Diffusion of Ar/He and Cu/Ni*

Consider a box containing an impermeable partition that divides the box into equal volumes (Figure 5-3). On one side, we have pure argon gas; on the other side, we have pure helium gas. Explain what will happen when the partition is opened. What will happen if we replace the Ar side with a Cu single crystal and the He side with a Ni single crystal?

SOLUTION

Before the partition is opened, one compartment has no argon and the other has no helium (i.e., there is a concentration gradient of Ar and He). When the partition is opened, Ar atoms will diffuse toward the He side, and vice versa. This diffusion will continue until the entire box has a uniform concentration of both gases. There may be some density gradient driven convective flows as well. If we took random samples of different regions in this box after a few hours, we would find a statistically uniform concentration of Ar and He. Owing to their thermal energy, the Ar and He atoms will continue to move around in the box; however, there will be no concentration gradients.

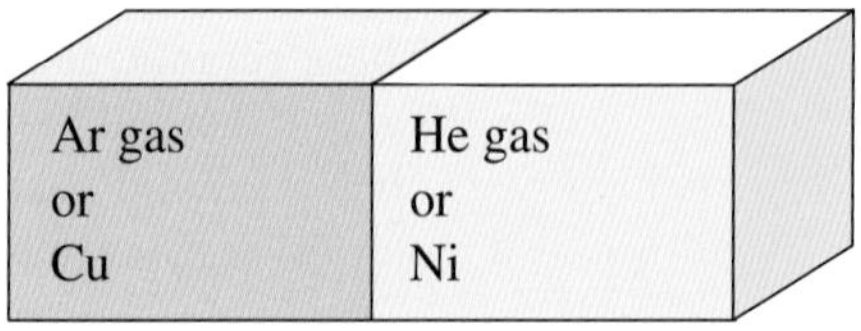

Figure 5-3 Illustration for diffusion of Ar/He and Cu/Ni (for Example 5-1).

If we open the hypothetical partition between the Ni and Cu single crystals at room temperature, we would find that, similar to the Ar/He situation, the concentration gradient exists, but the temperature is too low to see any significant diffusion of Cu atoms into the Ni single crystal and vice versa. This is an example of a situation in which a concentration gradient exists; however, because of the lower temperature, the kinetics for diffusion are not favorable. Certainly, if we increase the temperature (say to 600°C) and wait for a longer period of time (e.g., ~24 hours), we would see diffusion of Cu atoms into the Ni single crystal and vice versa. After a very long time, the entire solid will have a uniform concentration of Ni and Cu atoms. The new solid that forms consists of Cu and Ni atoms completely dissolved in each other, and the resultant material is termed a "solid solution," a concept we will study in greater detail in Chapter 10.

This example also illustrates something many of you may know by intuition. The diffusion of atoms and molecules occurs faster in gases and liquids than in solids. As we will see in Chapter 9 and other chapters, diffusion has a significant effect on the evolution of microstructure during the solidification of alloys, the heat treatment of metals and alloys, and the processing of ceramic materials.

5-2 Stability of Atoms and Ions

In Chapter 4, we showed that imperfections are present and also can be deliberately introduced into a material; however, these imperfections and, indeed, even atoms or ions in their normal positions in the crystal structures are not stable or at rest. Instead, the atoms or ions possess thermal energy, and they will move. For instance, an atom may move from a normal crystal structure location to occupy a nearby vacancy. An atom may also move from one interstitial site to another. Atoms or ions may jump across a grain boundary, causing the grain boundary to move.

The ability of atoms and ions to diffuse increases as the temperature, or thermal energy possessed by the atoms and ions, increases. The rate of atom or ion movement is related to temperature or thermal energy by the *Arrhenius equation*:

$$\text{Rate} = c_0 \exp\left(\frac{-Q}{RT}\right) \tag{5-1}$$

where c_0 is a constant, R is the gas constant $\left(1.987 \frac{\text{cal}}{\text{mol} \cdot \text{K}} \text{ or } 8.314 \frac{\text{J}}{\text{mol} \cdot \text{K}}\right)$, T is the absolute temperature (K), and Q is the **activation energy** (cal/mol or J/mol) required to cause one mole of atoms or ions to move. This equation is derived from a statistical analysis of the probability that the atoms will have the extra energy Q needed to cause movement. The rate is related to the number of atoms that move.

We can rewrite the equation by taking natural logarithms of both sides:

$$\ln(\text{Rate}) = \ln(c_0) - \frac{Q}{RT} \tag{5-2}$$

If we plot ln(Rate) of some reaction versus $1/T$ (Figure 5-4), the slope of the line will be $-Q/R$ and, consequently, Q can be calculated. The constant c_0 corresponds to the intercept at $\ln(c_0)$ when $1/T$ is zero.

Svante August Arrhenius (1859–1927), a Swedish chemist who won the Nobel Prize in Chemistry in 1903 for his research on the electrolytic theory of dissociation applied this idea to the rates of chemical reactions in aqueous solutions. His basic idea of activation energy and rates of chemical reactions as functions of temperature has since been applied to diffusion and other rate processes.

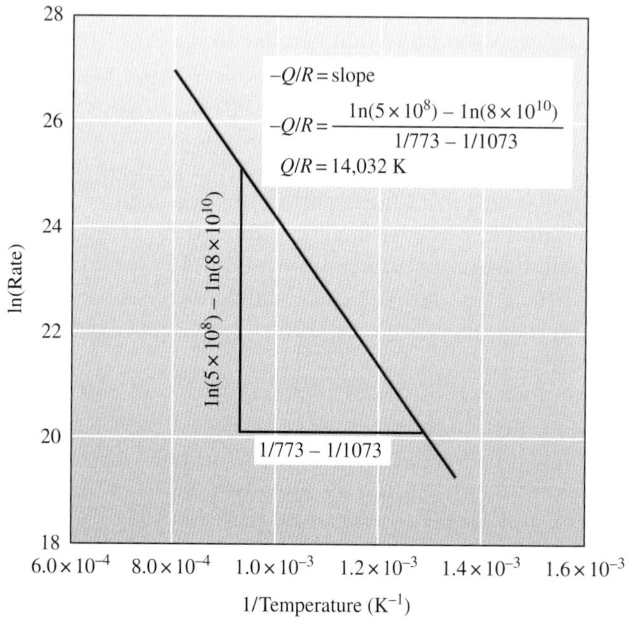

Figure 5-4 The Arrhenius plot of ln(Rate) versus $1/T$ can be used to determine the activation energy for a reaction. The data from this figure is used in Example 5-2.

Example 5-2 *Activation Energy For Interstitial Atoms*

Suppose that interstitial atoms are found to move from one site to another at the rates of 5×10^8 jumps/s at 500°C and 8×10^{10} jumps/s at 800°C. Calculate the activation energy Q for the process.

SOLUTION

Figure 5-4 represents the data on a ln(Rate) versus $1/T$ plot; the slope of this line, as calculated in the figure, gives $Q/R = 14{,}032$ K, or $Q = 27{,}880$ cal/mol. Alternately, we could write two simultaneous equations:

$$\text{Rate} = c_0 \exp\left(\frac{-Q}{RT}\right)$$

$$5 \times 10^8 \frac{\text{jumps}}{\text{s}} = c_0 \exp\left[\frac{-Q}{\left(1.987 \frac{\text{cal}}{\text{mol} \cdot \text{K}}\right)(500\text{ K} + 273\text{ K})}\right]$$
$$= c_0 \exp(-0.000651\ Q)$$

$$8 \times 10^{10} \frac{\text{jumps}}{\text{s}} = c_0 \exp\left[\frac{-Q}{\left(1.987 \frac{\text{cal}}{\text{mol} \cdot \text{K}}\right)(800\text{ K} + 273\text{ K})}\right]$$
$$= c_0 \exp(-0.000469\ Q)$$

Note the temperatures were converted into K. Since

$$c_0 = \frac{5 \times 10^8}{\exp(-0.000651\ Q)}\left(\frac{\text{jumps}}{\text{s}}\right)$$

then

$$8 \times 10^{10} = \frac{(5 \times 10^8)\exp(-0.000469\ Q)}{\exp(-0.000651\ Q)}$$

$$160 = \exp[(0.000651 - 0.000469)Q] = \exp(0.000182\ Q)$$

$$\ln(160) = 5.075 = 0.000182\ Q$$

$$Q = \frac{5.075}{0.000182} = 27{,}880\text{ cal/mol}$$

5-3 Mechanisms for Diffusion

As we saw in Chapter 4, defects known as vacancies exist in materials. The disorder these vacancies create (i.e., increased entropy) helps minimize the free energy and, therefore, increases the thermodynamic stability of a crystalline material. In materials containing vacancies, atoms move or "jump" from one lattice position to another. This process, known as **self-diffusion**, can be detected by using radioactive tracers. As an example, suppose we were to introduce a radioactive isotope of gold (Au^{198}) onto the surface of standard gold (Au^{197}). After a period of time, the radioactive atoms would move into the standard gold. Eventually, the radioactive atoms would be uniformly distributed throughout the entire standard gold sample. Although self-diffusion occurs continually in all materials, its effect on the material's behavior is generally not significant.

Interdiffusion

Diffusion of unlike atoms in materials also occurs (Figure 5-5). Consider a nickel sheet bonded to a copper sheet. At high temperatures, nickel atoms gradually diffuse into the copper, and copper atoms migrate into the nickel. Again, the nickel and copper atoms eventually are uniformly distributed. Diffusion of different atoms in different directions is known as **interdiffusion**. There are two important mechanisms by which atoms or ions can diffuse (Figure 5-6): vacancy diffusion and interstitial diffusion.

Vacancy Diffusion

In self-diffusion and diffusion involving substitutional atoms, an atom leaves its lattice site to fill a nearby vacancy (thus creating a new vacancy at the original lattice site). As diffusion continues, we have counterflows of atoms and vacancies, called **vacancy diffusion**. The number of vacancies, which increases as the temperature increases, influences the extent of both self-diffusion and diffusion of substitutional atoms.

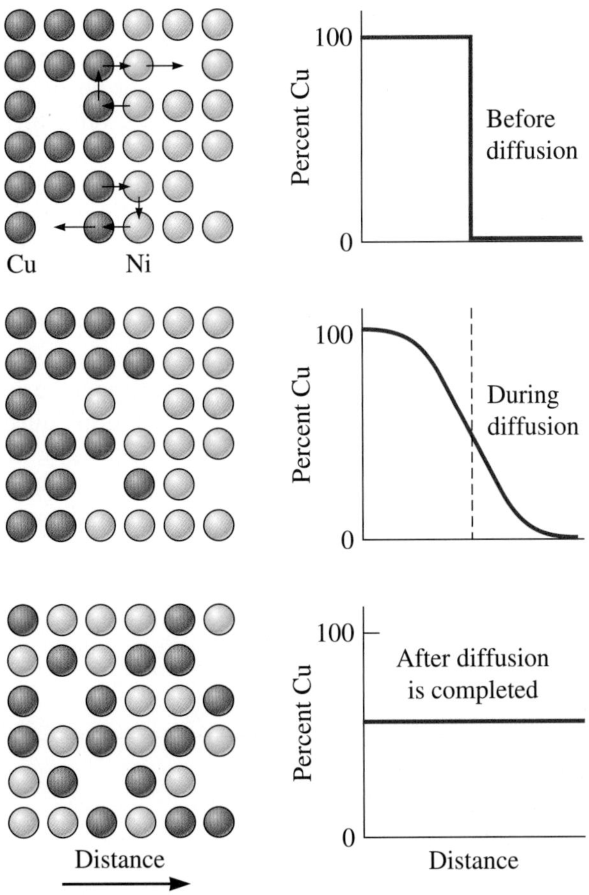

Figure 5-5 Diffusion of copper atoms into nickel. Eventually, the copper atoms are randomly distributed throughout the nickel.

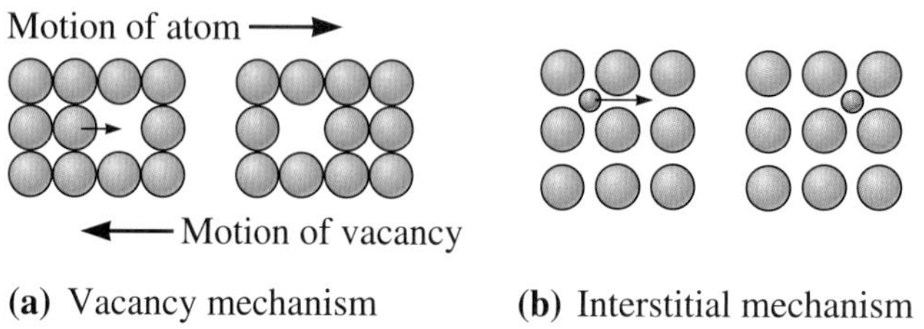

Figure 5-6 Diffusion mechanisms in materials: (a) vacancy or substitutional atom diffusion and (b) interstitial diffusion.

Interstitial Diffusion When a small interstitial atom or ion is present in the crystal structure, the atom or ion moves from one interstitial site to another. No vacancies are required for this mechanism. Partly because there are many more interstitial sites than vacancies, **interstitial diffusion** occurs more easily than vacancy diffusion. Interstitial atoms that are relatively smaller can diffuse faster. In Chapter 3, we saw that the structure of many ceramics with ionic bonding can be considered as a close packing of anions with cations in the interstitial sites. In these materials, smaller cations often diffuse faster than larger anions.

5-4 Activation Energy for Diffusion

A diffusing atom must squeeze past the surrounding atoms to reach its new site. In order for this to happen, energy must be supplied to allow the atom to move to its new position, as shown schematically for vacancy and interstitial diffusion in Figure 5-7. The atom is originally in a low-energy, relatively stable location. In order to move to a new location, the atom must overcome an energy barrier. The energy barrier is the activation energy Q. The thermal energy supplies atoms or ions with the energy needed to exceed this barrier. Note that the symbol Q is often used for activation energies for different processes (rate at which atoms jump, a chemical reaction, energy needed to produce vacancies, etc.), and we should be careful in understanding the specific process or phenomenon to which the general term for activation energy Q is being applied, as the value of Q depends on the particular phenomenon.

Normally, less energy is required to squeeze an interstitial atom past the surrounding atoms; consequently, activation energies are lower for interstitial diffusion than for vacancy diffusion. Typical values for activation energies for diffusion of different

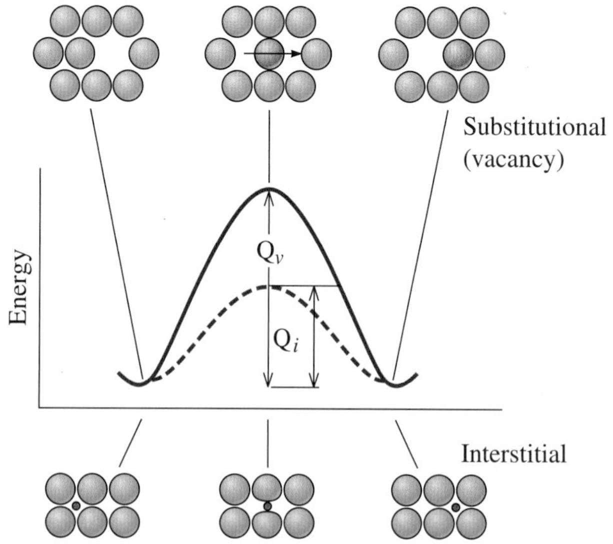

Figure 5-7 The activation energy Q is required to squeeze atoms past one another during diffusion. Generally, more energy is required for a substitutional atom than for an interstitial atom.

Table 5-1 Diffusion data for selected materials

Diffusion Couple	Q (cal/mol)	D_0 (cm^2/s)
Interstitial diffusion:		
C in FCC iron	32,900	0.23
C in BCC iron	20,900	0.011
N in FCC iron	34,600	0.0034
N in BCC iron	18,300	0.0047
H in FCC iron	10,300	0.0063
H in BCC iron	3,600	0.0012
Self-diffusion (vacancy diffusion):		
Pb in FCC Pb	25,900	1.27
Al in FCC Al	32,200	0.10
Cu in FCC Cu	49,300	0.36
Fe in FCC Fe	66,700	0.65
Zn in HCP Zn	21,800	0.1
Mg in HCP Mg	32,200	1.0
Fe in BCC Fe	58,900	4.1
W in BCC W	143,300	1.88
Si in Si (covalent)	110,000	1800.0
C in C (covalent)	163,000	5.0
Heterogeneous diffusion (vacancy diffusion):		
Ni in Cu	57,900	2.3
Cu in Ni	61,500	0.65
Zn in Cu	43,900	0.78
Ni in FCC iron	64,000	4.1
Au in Ag	45,500	0.26
Ag in Au	40,200	0.072
Al in Cu	39,500	0.045
Al in Al_2O_3	114,000	28.0
O in Al_2O_3	152,000	1900.0
Mg in MgO	79,000	0.249
O in MgO	82,100	0.000043

Based on several sources, including Adda, Y. and Philibert, J., La Diffusion dans les Solides, *Vol. 2, 1966.*

atoms in different host materials are shown in Table 5-1. We use the term **diffusion couple** to indicate a combination of an atom of a given element (e.g., carbon) diffusing in a host material (e.g., BCC Fe). A low-activation energy indicates easy diffusion. In self-diffusion, the activation energy is equal to the energy needed to create a vacancy and to cause the movement of the atom. Table 5-1 also shows values of D_0, which is the pre-exponential term and the constant c_0 from Equation 5-1, when the rate process is diffusion. We will see later that D_0 is the diffusion coefficient when $1/T = 0$.

5-5 Rate of Diffusion [Fick's First Law]

Adolf Eugen Fick (1829–1901) was the first scientist to provide a quantitative description of the diffusion process.

The rate at which atoms, ions, particles or other species diffuse in a material can be measured by the **flux** J. Here we are mainly concerned with diffusion of ions or atoms. The flux J is defined as the number of atoms passing through a plane of unit area per unit time (Figure 5-8). **Fick's first law** explains the net flux of atoms:

$$J = -D\frac{dc}{dx} \tag{5-3}$$

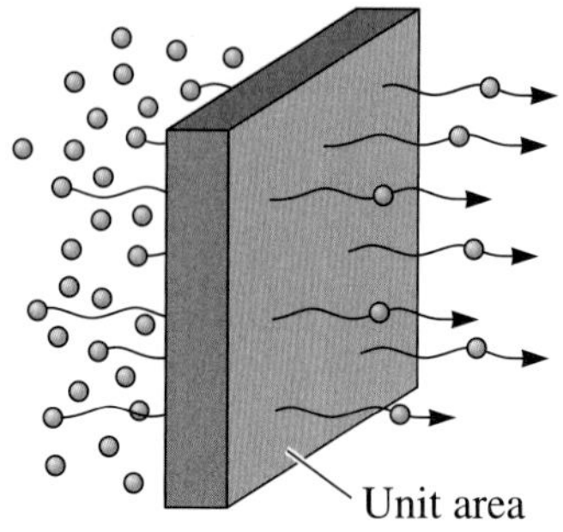

Figure 5-8 The flux during diffusion is defined as the number of atoms passing through a plane of unit area per unit time.

where J is the flux, D is the **diffusivity** or **diffusion coefficient** $\left(\frac{\text{cm}^2}{\text{s}}\right)$, and dc/dx is the **concentration gradient** $\left(\frac{\text{atoms}}{\text{cm}^3 \cdot \text{cm}}\right)$. Depending upon the situation, concentration may be expressed as atom percent (at%), weight percent (wt%), mole percent (mol%), atom fraction, or mole fraction. The units of concentration gradient and flux will change accordingly.

Several factors affect the flux of atoms during diffusion. If we are dealing with diffusion of ions, electrons, holes, etc., the units of J, D, and $\frac{dc}{dx}$ will reflect the appropriate species that are being considered. The negative sign in Equation 5-3 tells us that the flux of diffusing species is from higher to lower concentrations, so that if the $\frac{dc}{dx}$ term is negative, J will be positive. Thermal energy associated with atoms, ions, etc., causes the random movement of atoms. At a microscopic scale, the thermodynamic driving force for diffusion is the concentration gradient. A net or an observable flux is created depending upon temperature and the concentration gradient.

Concentration Gradient The concentration gradient shows how the composition of the material varies with distance: Δc is the difference in concentration over the distance Δx (Figure 5-9). A concentration gradient may be created when two materials of different composition are placed in contact, when a gas or liquid is in contact with a solid material, when nonequilibrium structures are produced in a material due to processing, and from a host of other sources.

The flux at a particular temperature is constant only if the concentration gradient is also constant—that is, the compositions on each side of the plane in Figure 5-8 remain unchanged. In many practical cases, however, these compositions vary as atoms are redistributed, and thus the flux also changes. Often, we find that the flux is initially high and then gradually decreases as the concentration gradient is reduced by diffusion. The examples that follow illustrate calculations of flux and concentration gradients for

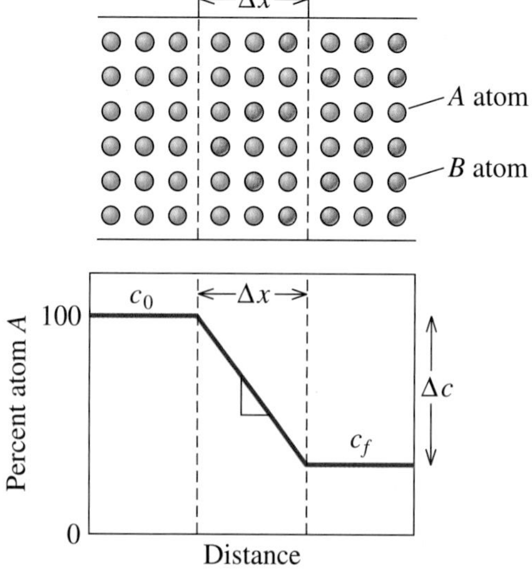

Figure 5-9 Illustration of the concentration gradient.

diffusion of dopants in semiconductors and in ceramics, but only for the case of constant concentration gradient. Later in this chapter, we will consider non-steady state diffusion with the aid of Fick's second law.

Example 5-3 *Semiconductor Doping*

One step in manufacturing transistors, which function as electronic switches in integrated circuits, involves diffusing impurity atoms into a semiconductor material such as silicon. Suppose a silicon wafer 0.1 cm thick, which originally contains one phosphorus atom for every 10 million Si atoms, is treated so that there are 400 phosphorus atoms for every 10 million Si atoms at the surface (Figure 5-10). Calculate the concentration gradient (a) in atomic percent/cm and (b) in $\frac{\text{atoms}}{\text{cm}^3 \cdot \text{cm}}$. The lattice parameter of silicon is 5.4307 Å.

SOLUTION

a. First, let's calculate the initial and surface compositions in atomic percent:

$$c_i = \frac{1 \text{ P atom}}{10^7 \text{ atoms}} \times 100 = 0.00001 \text{ at\% P}$$

$$c_s = \frac{400 \text{ P atoms}}{10^7 \text{ atoms}} \times 100 = 0.004 \text{ at\% P}$$

$$\frac{\Delta c}{\Delta x} = \frac{0.00001 - 0.004 \text{ at\% P}}{0.1 \text{ cm}} = -0.0399 \frac{\text{at\% P}}{\text{cm}}$$

b. To find the gradient in terms of $\frac{\text{atoms}}{\text{cm}^3 \cdot \text{cm}}$, we must find the volume of the unit cell. The crystal structure of Si is diamond cubic (DC). The lattice parameter is 5.4307×10^{-8} cm. Thus,

$$V_{\text{cell}} = (5.4307 \times 10^{-8} \text{ cm})^3 = 1.6 \times 10^{-22} \frac{\text{cm}^3}{\text{cell}}$$

The volume occupied by 10^7 Si atoms, which are arranged in a DC structure with 8 atoms/cell, is

$$V = \left[\frac{10^7 \text{ atoms}}{8 \frac{\text{atoms}}{\text{cell}}}\right]\left[1.6 \times 10^{-22}\left(\frac{\text{cm}^3}{\text{cell}}\right)\right] = 2 \times 10^{-16} \text{ cm}^3$$

The compositions in atoms/cm^3 are

$$c_i = \frac{1 \text{ P atom}}{2 \times 10^{-16} \text{ cm}^3} = 0.005 \times 10^{18} \frac{\text{P atoms}}{\text{cm}^3}$$

$$c_s = \frac{400 \text{ P atoms}}{2 \times 10^{-16} \text{ cm}^3} = 2 \times 10^{18} \frac{\text{P atoms}}{\text{cm}^3}$$

Thus, the composition gradient is

$$\frac{\Delta c}{\Delta x} = \frac{0.005 \times 10^{18} - 2 \times 10^{18} \frac{\text{P atoms}}{\text{cm}^3}}{0.1 \text{ cm}}$$

$$= -1.995 \times 10^{19} \frac{\text{P atoms}}{\text{cm}^3 \cdot \text{cm}}$$

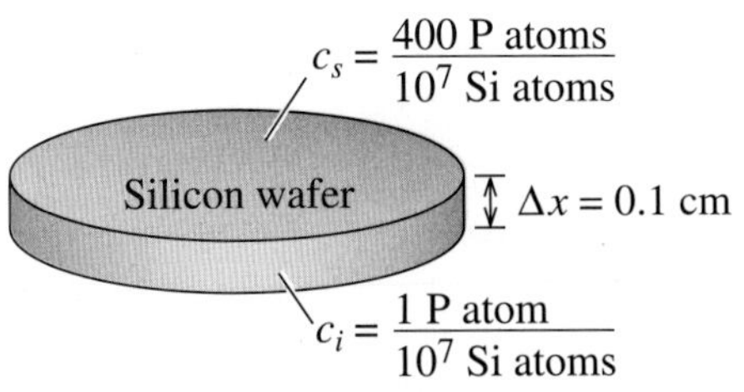

Figure 5-10 Silicon wafer showing a variation in concentration of P atoms (for Example 5-3).

Example 5-4 Diffusion of Nickel in Magnesium Oxide

A 0.05 cm layer of magnesium oxide (MgO) is deposited between layers of nickel and tantalum to provide a diffusion barrier that prevents reactions between the two metals (Figure 5-11). At 1400°C, nickel ions diffuse through the MgO ceramic to the tantalum. Determine the number of nickel ions that pass through the MgO per second. At 1400°C, the diffusion coefficient of nickel ions in MgO is 9×10^{-12} cm²/s, and the lattice parameter of nickel at 1400°C is 3.6×10^{-8} cm.

SOLUTION

The composition of nickel at the Ni/MgO interface is 100% Ni, or

$$c_{\text{Ni/MgO}} = \frac{4 \text{ Ni atoms}}{(3.6 \times 10^{-8} \text{ cm})^3} = 8.573 \times 10^{22} \frac{\text{Ni atoms}}{\text{cm}^3}$$

The composition of nickel at the Ta/MgO interface is 0% Ni. Thus, the concentration gradient is

$$\frac{\Delta c}{\Delta x} = \frac{0 - 8.573 \times 10^{22} \frac{\text{Ni atoms}}{\text{cm}^3}}{0.05 \text{ cm}} = -1.715 \times 10^{24} \frac{\text{Ni atoms}}{\text{cm}^3 \cdot \text{cm}}$$

The flux of nickel atoms through the MgO layer is

$$J = -D\frac{\Delta c}{\Delta x} = -(9 \times 10^{-12} \text{ cm}^2/\text{s})\left(-1.715 \times 10^{24} \frac{\text{Ni atoms}}{\text{cm}^3 \cdot \text{cm}}\right)$$

$$J = 1.543 \times 10^{13} \frac{\text{Ni atoms}}{\text{cm}^2 \cdot \text{s}}$$

The total number of nickel atoms crossing the 2 cm × 2 cm interface per second is

$$\text{Total Ni atoms per second} = (J)(\text{Area}) = \left(1.543 \times 10^{13} \frac{\text{Ni atoms}}{\text{cm}^2 \cdot \text{s}}\right)(2 \text{ cm})(2 \text{ cm})$$
$$= 6.17 \times 10^{13} \text{ Ni atoms/s}$$

Although this may appear to be very rapid, in one second, the volume of nickel atoms removed from the Ni/MgO interface is

$$\frac{6.17 \times 10^{13} \frac{\text{Ni atoms}}{\text{s}}}{8.573 \times 10^{22} \frac{\text{Ni atoms}}{\text{cm}^3}} = 7.2 \times 10^{-10} \frac{\text{cm}^3}{\text{s}}$$

The thickness by which the nickel layer is reduced each second is

$$\frac{7.2 \times 10^{-10} \frac{\text{cm}^3}{\text{s}}}{4 \text{ cm}^2} = 1.8 \times 10^{-10} \frac{\text{cm}}{\text{s}}$$

For one micrometer (10^{-4} cm) of nickel to be removed, the treatment requires

$$\frac{10^{-4} \text{ cm}}{1.8 \times 10^{-10} \frac{\text{cm}}{\text{s}}} = 556{,}000 \text{ s} = 154 \text{ h}$$

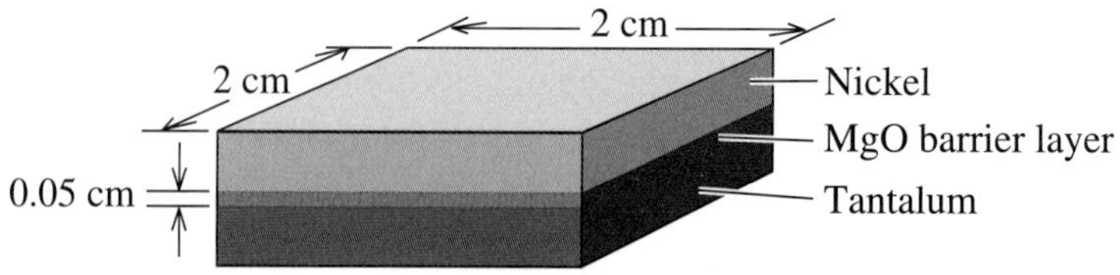

Figure 5-11 Diffusion couple (for Example 5-4).

5-6 Factors Affecting Diffusion

Temperature and the Diffusion Coefficient The kinetics of diffusion are strongly dependent on temperature. The diffusion coefficient D is related to temperature by an Arrhenius-type equation,

$$D = D_0 \exp\left(\frac{-Q}{RT}\right) \tag{5-4}$$

where Q is the activation energy (in units of cal/mol or J/mol) for diffusion of the species under consideration (e.g., Al in Si), R is the gas constant $\left(1.987 \frac{\text{cal}}{\text{mol}\cdot\text{K}} \text{ or } 8.314 \frac{\text{J}}{\text{mol}\cdot\text{K}}\right)$ and T is the absolute temperature (K). D_0 is the pre-exponential term, similar to c_0 in Equation 5-1.

D_0 is a constant for a given diffusion system and is equal to the value of the diffusion coefficient at $1/T = 0$ or $T = \infty$. Typical values for D_0 are given in Table 5-1, while the temperature dependence of D is shown in Figure 5-12 for some metals and ceramics. Covalently bonded materials, such as carbon and silicon (Table 5-1), have unusually high activation energies, consistent with the high strength of their atomic bonds. Figure 5-13 shows the diffusion coefficients for different dopants in silicon.

In ionic materials, such as some of the oxide ceramics, a diffusing ion only enters a site having the same charge. In order to reach that site, the ion must physically squeeze past adjoining ions, pass by a region of opposite charge, and move a relatively long distance

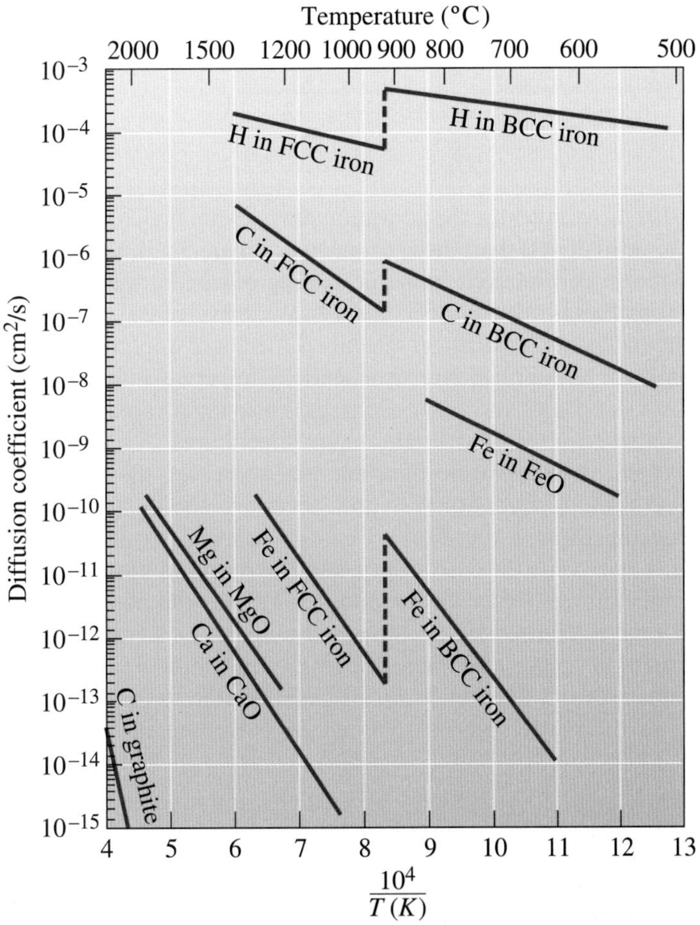

Figure 5-12 The diffusion coefficient D as a function of reciprocal temperature for some metals and ceramics. In this Arrhenius plot, D represents the rate of the diffusion process. A steep slope denotes a high activation energy.

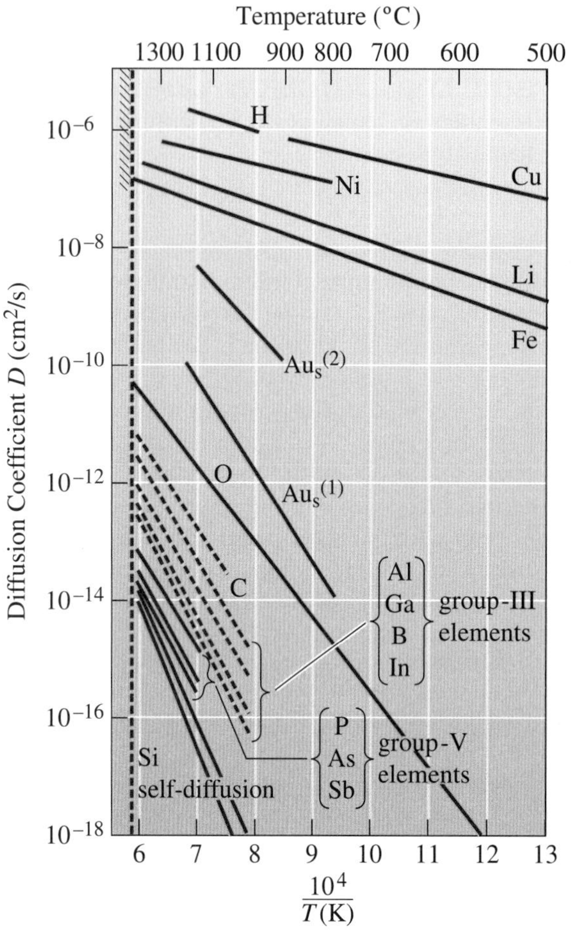

Figure 5-13 Diffusion coefficients for different dopants in silicon. (*Based on "Diffusion and Diffusion Induced Defects in Silicon," by U. Gosele. In R. Bloor, M. Flemings, and S. Mahajan (Eds.),* Encyclopedia of Advanced Materials, *Vol. 1, 1994, Pergamon Press.*)

(Figure 5-14). Consequently, the activation energies are high and the rates of diffusion are lower for ionic materials than those for metals (Figure 5-15). We take advantage of this in many situations. For example, in the processing of silicon, we create a thin layer of silica (SiO_2) on top of a silicon wafer (Chapter 19). We then create a window by removing part of the silica layer. This window allows selective diffusion of dopant atoms such as phosphorus and boron because the silica layer is essentially impervious to the dopant atoms. Slower diffusion in most oxides and other ceramics is also an advantage in applications in which components are required to withstand high temperatures.

When the temperature of a material increases, the diffusion coefficient D increases (according to Equation 5-4) and, therefore, the flux of atoms increases as well. At higher temperatures, the thermal energy supplied to the diffusing atoms permits the atoms to overcome the activation energy barrier and more easily move to new sites. At low temperatures—often below about 0.4 times the absolute melting temperature of the material—diffusion is very slow and may not be significant. For this reason, the heat treatment of metals and the processing of ceramics are done at high temperatures, where atoms move rapidly to complete reactions or to reach equilibrium conditions. Because less thermal energy is required to overcome the smaller activation energy

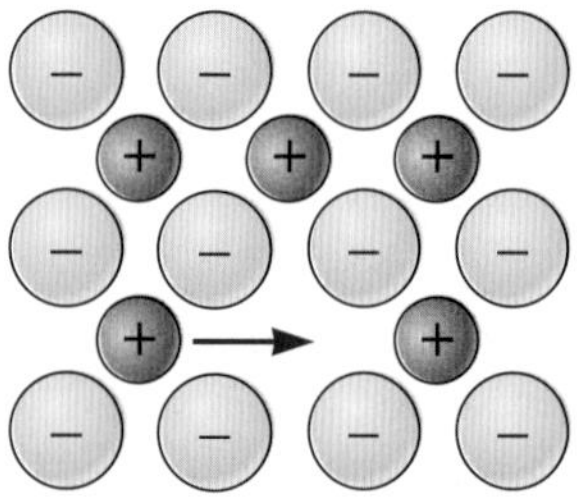

Figure 5-14 Diffusion in ionic compounds. Anions can only enter other anion sites. Smaller cations tend to diffuse faster.

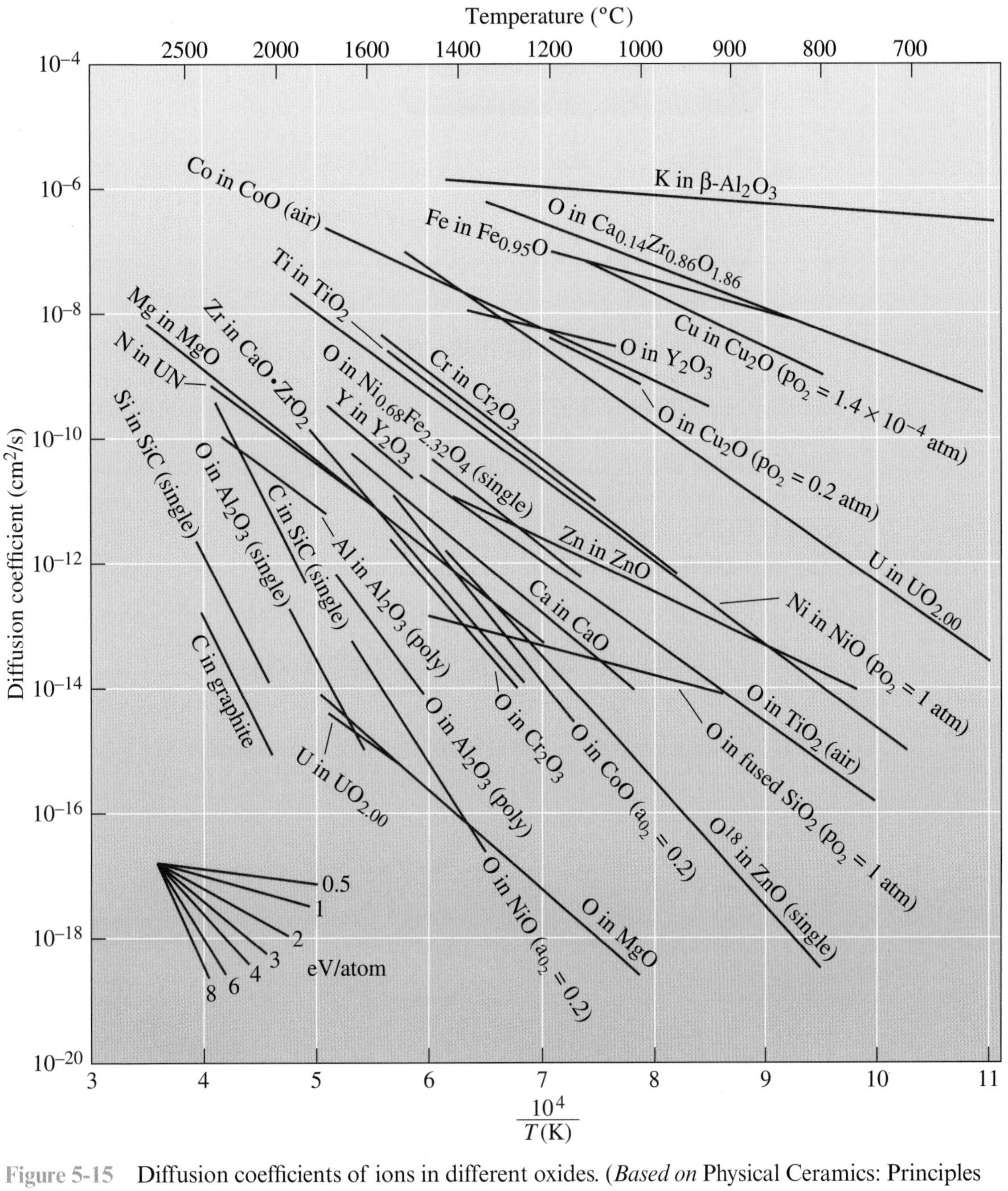

Figure 5-15 Diffusion coefficients of ions in different oxides. (*Based on* Physical Ceramics: Principles for Ceramic Science and Engineering, *Y. M. Chiang, D. Birnie, and W. D. Kingery, John Wiley & Sons, Inc., 1997.*)

barrier, a small activation energy Q increases the diffusion coefficient and flux. The following example illustrates how Fick's first law and concepts related to the temperature dependence of D can be applied to design an iron membrane.

Example 5-5 *Design of an Iron Membrane*

An impermeable cylinder 3 cm in diameter and 10 cm long contains a gas that includes 5×10^{19} N atoms per cm^3 and 5×10^{19} H atoms per cm^3 on one side of an iron membrane (Figure 5-16). Gas is continuously introduced to the pipe to ensure a constant

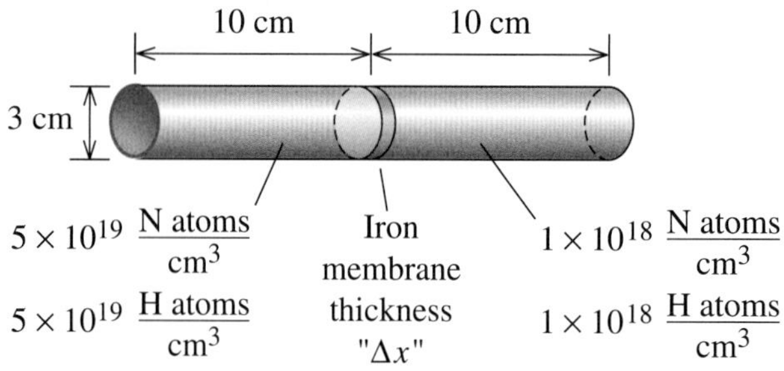

Figure 5-16 Design of an iron membrane (for Example 5-5).

concentration of nitrogen and hydrogen. The gas on the other side of the membrane contains a constant 1×10^{18} N atoms per cm^3 and 1×10^{18} H atoms per cm^3. The entire system is to operate at 700°C, at which iron has the BCC structure. Design an iron membrane that will allow no more than 1% of the nitrogen to be lost through the membrane each hour, while allowing 90% of the hydrogen to pass through the membrane per hour.

SOLUTION

The total number of nitrogen atoms in the container is

$$(5 \times 10^{19} \text{ N atoms/cm}^3)(\pi/4)(3 \text{ cm})^2(10 \text{ cm}) = 3.534 \times 10^{21} \text{ N atoms}$$

The maximum number of atoms to be lost is 1% of this total, or

$$\text{N atom loss per h} = (0.01)(3.534 \times 10^{21}) = 3.534 \times 10^{19} \text{ N atoms/h}$$

$$\text{N atom loss per s} = (3.534 \times 10^{19} \text{ N atoms/h})/(3600 \text{ s/h}) = 9.817 \times 10^{15} \text{ N atoms/s}$$

The flux is then

$$J = \frac{9.817 \times 10^{15} \text{ N atoms/s}}{\left(\frac{\pi}{4}\right)(3 \text{ cm})^2} = 1.389 \times 10^{15} \frac{\text{N atoms}}{\text{cm}^2 \cdot \text{s}}$$

Using Equation 5-4 and values from Table 5-1, the diffusion coefficient of nitrogen in BCC iron at 700°C = 973 K is

$$D = D_0 \exp\left(\frac{-Q}{RT}\right)$$

$$D_N = 0.0047 \frac{\text{cm}^2}{\text{s}} \exp\left[\frac{-18{,}300 \frac{\text{cal}}{\text{mol}}}{\left(1.987 \frac{\text{cal}}{\text{mol} \cdot \text{K}}\right)(973 \text{ K})}\right]$$

$$= \left(0.0047 \tfrac{\text{cm}^2}{\text{s}}\right)(7.748 \times 10^{-5}) = 3.642 \times 10^{-7} \tfrac{\text{cm}^2}{\text{s}}$$

From Equation 5-3:

$$J = -D\left(\frac{\Delta c}{\Delta x}\right) = 1.389 \times 10^{15} \frac{\text{N atoms}}{\text{cm}^2 \cdot \text{s}}$$

$$\Delta x = -\frac{D\Delta c}{J} = \frac{(-3.642 \times 10^{-7} \text{ cm}^2/\text{s})\left(1 \times 10^{18} - 5 \times 10^{19} \frac{\text{N atoms}}{\text{cm}^3}\right)}{1.389 \times 10^{15} \frac{\text{N atoms}}{\text{cm}^2 \cdot \text{s}}}$$

$$\Delta x = 0.013 \text{ cm} = \text{minimum thickness of the membrane}$$

In a similar manner, the maximum thickness of the membrane that will permit 90% of the hydrogen to pass can be calculated as

$$\text{H atom loss per h} = (0.90)(3.534 \times 10^{21}) = 3.180 \times 10^{21} \text{ H atoms/h}$$

$$\text{H atom loss per s} = 8.836 \times 10^{17} \text{ H atoms/s}$$

$$J = 1.250 \times 10^{17} \frac{\text{H atoms}}{\text{cm}^2 \cdot \text{s}}$$

From Equation 5-4 and Table 5-1,

$$D_H = 0.0012 \frac{\text{cm}^2}{\text{s}} \exp\left[\frac{-3{,}600 \frac{\text{cal}}{\text{mol}}}{\left(1.987 \frac{\text{cal}}{\text{mol} \cdot \text{K}}\right)(973\ K)}\right] = 1.864 \times 10^{-4} \text{ cm}^2/\text{s}$$

Since

$$\Delta x = -\frac{D\Delta c}{J}$$

$$\Delta x = \frac{\left(-1.864 \times 10^{-4} \frac{\text{cm}^2}{\text{s}}\right)\left(1 \times 10^{18} - 5 \times 10^{19} \frac{\text{H atoms}}{\text{cm}^3}\right)}{1.250 \times 10^{17} \frac{\text{H atoms}}{\text{cm}^2 \cdot \text{s}}}$$

$$= 0.073 \text{ cm} = \text{maximum thickness}$$

An iron membrane with a thickness between 0.013 and 0.073 cm will be satisfactory.

Types of Diffusion In **volume diffusion**, the atoms move through the crystal from one regular or interstitial site to another. Because of the surrounding atoms, the activation energy is large and the rate of diffusion is relatively slow.

Atoms can also diffuse along boundaries, interfaces, and surfaces in the material. Atoms diffuse easily by **grain boundary diffusion** because the atom packing is disordered and less dense in the grain boundaries. Because atoms can more easily squeeze their way through the grain boundary, the activation energy is low (Table 5-2). **Surface diffusion** is easier still because there is even less constraint on the diffusing atoms at the surface.

Time Diffusion requires time. The units for flux are $\frac{\text{atoms}}{\text{cm}^2 \cdot \text{s}}$. If a large number of atoms must diffuse to produce a uniform structure, long times may be required, even at high temperatures. Times for heat treatments may be reduced by using higher temperatures or by making the **diffusion distances** (related to Δx) as small as possible.

We find that some rather remarkable structures and properties are obtained if we prevent diffusion. Steels quenched rapidly from high temperatures to prevent diffusion form nonequilibrium structures and provide the basis for sophisticated heat treatments. Similarly, in forming metallic glasses, we have to quench liquid metals at a very high cooling rate. This is to avoid diffusion of atoms by decreasing their thermal energy

Table 5-2 The effect of the type of diffusion for thorium in tungsten and for self-diffusion in silver

Diffusion Type	Diffusion Coefficient (*D*) Thorium in Tungsten		Silver in Silver	
	D_0 (cm²/s)	*Q* (cal/mol)	D_0 (cm²/s)	*Q* (cal/mol)
Surface	0.47	66,400	0.068	8,900
Grain boundary	0.74	90,000	0.24	22,750
Volume	1.00	120,000	0.99	45,700

and encouraging them to assemble into nonequilibrium amorphous arrangements. Melts of silicate glasses, on the other hand, are viscous and diffusion of ions through these is slow. As a result, we do not have to cool these melts very rapidly to attain an amorphous structure. There is a myth that many old buildings contain windowpanes that are thicker at the bottom than at the top because the glass has flowed down over the years. Based on kinetics of diffusion, it can be shown that even several hundred or thousand years will not be sufficient to cause such flow of glasses at near-room temperature. In the following example, we examine different mechanisms for diffusion.

Example 5-6 *Tungsten Thorium Diffusion Couple*

Consider a diffusion couple between pure tungsten and a tungsten alloy containing 1 at% thorium. After several minutes of exposure at 2000°C, a transition zone of 0.01 cm thickness is established. What is the flux of thorium atoms at this time if diffusion is due to (a) volume diffusion, (b) grain boundary diffusion, and (c) surface diffusion? (See Table 5-2.)

SOLUTION

The lattice parameter of BCC tungsten is 3.165 Å. Thus, the number of tungsten atoms/cm^3 is

$$\frac{\text{W atoms}}{\text{cm}^3} = \frac{2 \text{ atoms/cell}}{(3.165 \times 10^{-8} \text{ cm})^3\text{/cell}} = 6.3 \times 10^{22}$$

In the tungsten-1 at% thorium alloy, the number of thorium atoms is

$$c_{\text{Th}} = (0.01)(6.3 \times 10^{22}) = 6.3 \times 10^{20} \frac{\text{Th atoms}}{\text{cm}^3}$$

In the pure tungsten, the number of thorium atoms is zero. Thus, the concentration gradient is

$$\frac{\Delta c}{\Delta x} = \frac{0 - 6.3 \times 10^{20} \frac{\text{Th atoms}}{\text{cm}^3}}{0.01 \text{ cm}} = -6.3 \times 10^{22} \frac{\text{Th atoms}}{\text{cm}^3 \cdot \text{cm}}$$

a. Volume diffusion

$$D = 1.0 \frac{\text{cm}^2}{\text{s}} \exp\left[\frac{-120{,}000 \frac{\text{cal}}{\text{mol}}}{\left(1.987 \frac{\text{cal}}{\text{mol} \cdot \text{K}}\right)(2273 \text{ K})}\right] = 2.89 \times 10^{-12} \text{ cm}^2\text{/s}$$

$$J = -D\frac{\Delta c}{\Delta x} = -\left(2.89 \times 10^{-12} \frac{\text{cm}^2}{\text{s}}\right)\left(-6.3 \times 10^{22} \frac{\text{Th atoms}}{\text{cm}^3 \cdot \text{cm}}\right)$$

$$= 1.82 \times 10^{11} \frac{\text{Th atoms}}{\text{cm}^2 \cdot \text{s}}$$

b. Grain boundary diffusion

$$D = 0.74 \frac{\text{cm}^2}{\text{s}} \exp\left[\frac{-90{,}000 \frac{\text{cal}}{\text{mol}}}{\left(1.987 \frac{\text{cal}}{\text{mol} \cdot \text{K}}\right)(2273 \text{ K})}\right] = 1.64 \times 10^{-9} \text{ cm}^2\text{/s}$$

$$J = -\left(1.64 \times 10^{-9} \frac{\text{cm}^2}{\text{s}}\right)\left(-6.3 \times 10^{22} \frac{\text{Th atoms}}{\text{cm}^3 \cdot \text{cm}}\right) = 1.03 \times 10^{14} \frac{\text{Th atoms}}{\text{cm}^2 \cdot \text{s}}$$

c. Surface diffusion

$$D = 0.47 \frac{\text{cm}^2}{\text{s}} \exp\left[\frac{-66{,}400 \frac{\text{cal}}{\text{mol}}}{\left(1.987 \frac{\text{cal}}{\text{mol} \cdot \text{K}}\right)(2273 \text{ K})}\right] = 1.94 \times 10^{-7} \text{ cm}^2\text{/s}$$

$$J = -\left(1.94 \times 10^{-7} \frac{\text{cm}^2}{\text{s}}\right)\left(-6.3 \times 10^{22} \frac{\text{Th atoms}}{\text{cm}^3 \cdot \text{cm}}\right) = 1.22 \times 10^{16} \frac{\text{Th atoms}}{\text{cm}^2 \cdot \text{s}}$$

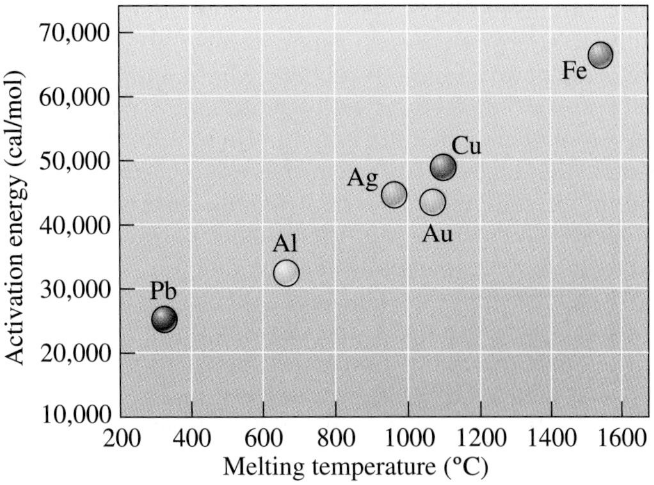

Figure 5-17 The activation energy for self-diffusion increases as the melting point of the metal increases.

Dependence on Bonding and Crystal Structure A number of factors influence the activation energy for diffusion and, hence, the rate of diffusion. Interstitial diffusion, with a low-activation energy, usually occurs much faster than vacancy or substitutional diffusion. Activation energies are usually lower for atoms diffusing through open crystal structures than for close-packed crystal structures. Because the activation energy depends on the strength of atomic bonding, it is higher for diffusion of atoms in materials with a high melting temperature (Figure 5-17).

We also find that, due to their smaller size, cations (with a positive charge) often have higher diffusion coefficients than those for anions (with a negative charge). In sodium chloride, for instance, the activation energy for diffusion of chloride ions (Cl^-) is about twice that for diffusion of sodium ions (Na^+).

Diffusion of ions also provides a transfer of electrical charge; in fact, the electrical conductivity of ionically bonded ceramic materials is related to temperature by an Arrhenius equation. As the temperature increases, the ions diffuse more rapidly, electrical charge is transferred more quickly, and the electrical conductivity is increased. As mentioned before, some ceramic materials are good conductors of electricity.

Dependence on Concentration of Diffusing Species and Composition of Matrix The diffusion coefficient (D) depends not only on temperature, as given by Equation 5-4, but also on the concentration of diffusing species and composition of the matrix. The reader should consult more advanced textbooks for more information.

5-7 Permeability of Polymers

In polymers, we are most concerned with the diffusion of atoms or small molecules between the long polymer chains. As engineers, we often cite the **permeability** of polymers and other materials, instead of the diffusion coefficients. Polymers that have a polar group (e.g., ethylene vinyl alcohol) have higher permeability for water vapor than for oxygen gas. Polyethylene, on the other hand, has much higher permeability for oxygen than for water vapor. In general, the more compact the structure of polymers, the lesser the permeability. For example, low-density polyethylene has a higher permeability than high-density polyethylene. Polymers used for food and other applications need to have the appropriate barrier properties. For example, polymer films are typically used as packaging to store food. If air diffuses through the film, the food may spoil. Similarly, care has to be exercised in the storage of ceramic or metal powders that are sensitive to atmospheric water vapor, nitrogen, oxygen, or carbon dioxide. For example, zinc oxide powders used in rubbers, paints, and ceramics must be stored in polyethylene bags to avoid reactions with atmospheric water vapor.

Diffusion of some molecules into a polymer can cause swelling problems. For example, in automotive applications, polymers used to make O-rings can absorb considerable amounts of oil, causing them to swell. On the other hand, diffusion is required to enable dyes to uniformly enter many of the synthetic polymer fabrics. Selective diffusion through polymer membranes is used for desalinization of water. Water molecules pass through the polymer membrane, and the ions in the salt are trapped.

In each of these examples, the diffusing atoms, ions, or molecules penetrate between the polymer chains rather than moving from one location to another within the chain structure. Diffusion will be more rapid through this structure when the diffusing species is smaller or when larger voids are present between the chains. Also diffusion tends to be faster through an amorphous polymer than through a crystalline polymer because an amorphous polymer has no long-range order and, therefore, a lower density.

5-8 Composition Profile [Fick's Second Law]

Fick's second law, which describes the dynamic, or non-steady state, diffusion of atoms, is the differential equation

$$\frac{\partial c}{\partial t} = \frac{\partial}{\partial x}\left(D\frac{\partial c}{\partial x}\right) \tag{5-5}$$

If we assume that the diffusion coefficient D is not a function of location x and the concentration (c) of diffusing species, we can write a simplified version of Fick's second law as follows

$$\frac{\partial c}{\partial t} = D\left(\frac{\partial^2 c}{\partial x^2}\right) \tag{5-6}$$

The solution to this equation depends on the boundary conditions for a particular situation. One solution is

$$\frac{c_s - c_x}{c_s - c_0} = \text{erf}\left(\frac{x}{2\sqrt{Dt}}\right) \tag{5-7}$$

where c_s is a constant concentration of the diffusing atoms at the surface of the material, c_0 is the initial uniform concentration of the diffusing atoms in the material, and c_x is the concentration of the diffusing atom at location x below the surface after time t. These concentrations are illustrated in Figure 5-18. In these equations, we have assumed basically a one-dimensional model (i.e., we assume that atoms or other diffusing species are moving only in the direction x). The function "erf" is the error function and can be evaluated from Table 5-3 or Figure 5-19. Note that most standard

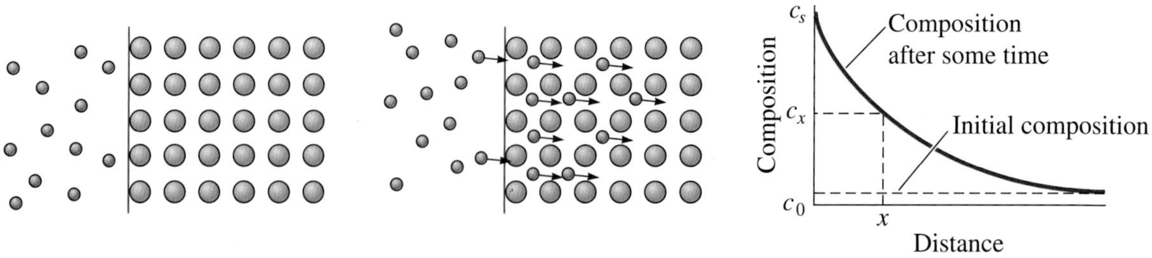

Figure 5-18 Diffusion of atoms into the surface of a material illustrating the use of Fick's second law.

Table 5-3 Error function values for Fick's second law

Argument of the Error Function $\frac{x}{2\sqrt{Dt}}$	Value of the Error Function erf $\frac{x}{2\sqrt{Dt}}$
0	0
0.10	0.1125
0.20	0.2227
0.30	0.3286
0.40	0.4284
0.50	0.5205
0.60	0.6039
0.70	0.6778
0.80	0.7421
0.90	0.7969
1.00	0.8427
1.50	0.9661
2.00	0.9953

Note that error function values are available in many software packages.

spreadsheet and other software programs (e.g., Excel™ and MATLAB®) also provide error function values.

The mathematical definition of the error function is as follows:

$$\text{erf}\,(x) = \frac{2}{\sqrt{\pi}}\int_0^x \exp(-y^2)dy \tag{5-8}$$

In Equation 5-8, y is known as the argument of the error function. We also define a complementary error function as follows:

$$\text{erfc}(x) = 1 - \text{erf}(x) \tag{5-9}$$

This function is used in certain solution forms of Fick's second law.

As mentioned previously, depending upon the boundary conditions, different solutions (i.e., different equations) describe the solutions to Fick's second law. These solutions to Fick's second law permit us to calculate the concentration of one diffusing species as a function of time (t) and location (x). Equation 5-7 is *a possible solution* to Fick's law that describes the variation in concentration of different species near the surface of

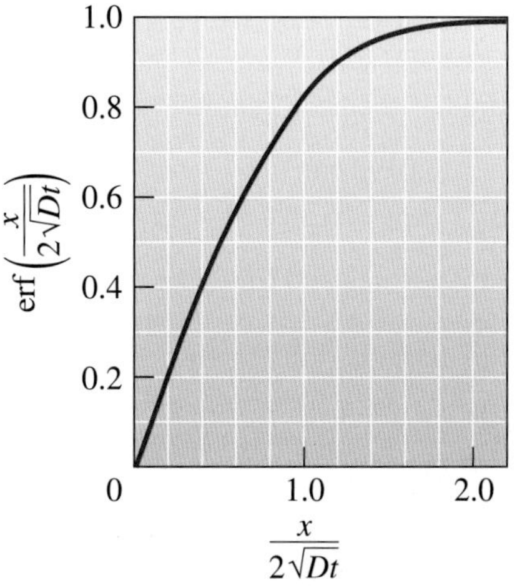

Figure 5-19 Graph showing the argument and value of the error function encountered in Fick's second law.

the material as a function of time and distance, provided that the diffusion coefficient D remains constant and the concentrations of the diffusing atoms at the surface (c_s) and at large distance (x) within the material (c_0) remain unchanged. Fick's second law can also assist us in designing a variety of materials processing techniques, including **carburization** as described in the following examples.

Example 5-7 *Design of a Carburizing Treatment*

The surface of a 0.1% C steel gear is to be hardened by carburizing. In gas carburizing, the steel gears are placed in an atmosphere that provides 1.2% C at the surface of the steel at a high temperature (Figure 5-1). Carbon then diffuses from the surface into the steel. For optimum properties, the steel must contain 0.45% C at a depth of 0.2 cm below the surface. Design a carburizing heat treatment that will produce these optimum properties. Assume that the temperature is high enough (at least 900°C) so that the iron has the FCC structure.

SOLUTION

Since the boundary conditions for which Equation 5-7 was derived are assumed to be valid, we can use this equation:

$$\frac{c_s - c_x}{c_s - c_0} = \text{erf}\left(\frac{x}{2\sqrt{Dt}}\right)$$

We know that $c_s = 1.2\%$ C, $c_0 = 0.1\%$ C, $c_x = 0.45\%$ C, and $x = 0.2$ cm. From Fick's second law:

$$\frac{c_s - c_x}{c_s - c_0} = \frac{1.2\%\ \text{C} - 0.45\%\ \text{C}}{1.2\%\ \text{C} - 0.1\%\ \text{C}} = 0.68 = \text{erf}\left(\frac{0.2\ \text{cm}}{2\sqrt{Dt}}\right) = \text{erf}\left(\frac{0.1\ \text{cm}}{\sqrt{Dt}}\right)$$

From Table 5-3, we find that

$$\frac{0.1\ \text{cm}}{\sqrt{Dt}} = 0.71 \text{ or } Dt = \left(\frac{0.1}{0.71}\right)^2 = 0.0198\ \text{cm}^2$$

Any combination of D and t with a product of 0.0198 cm^2 will work. For carbon diffusing in FCC iron, the diffusion coefficient is related to temperature by Equation 5-4:

$$D = D_0 \exp\left(\frac{-Q}{RT}\right)$$

From Table 5-1:

$$D = 0.23 \exp\left[\frac{-32{,}900\ \text{cal/mol}}{\left(1.987 \frac{\text{cal}}{\text{mol} \cdot \text{K}}\right)T}\right] = 0.23 \exp\left(\frac{-16{,}558}{T}\right)$$

Therefore, the temperature and time of the heat treatment are related by

$$t = \frac{0.0198\ \text{cm}^2}{D} = \frac{0.0198\ \text{cm}^2}{0.23 \exp(-16{,}558/T)} = \frac{0.0861}{\exp(-16{,}558/T)}$$

Some typical combinations of temperatures and times are

If $T = 900°\text{C} = 1173$ K, then $t = 116{,}273$ s $= 32.3$ h
If $T = 1000°\text{C} = 1273$ K, then $t = 38{,}362$ s $= 10.7$ h
If $T = 1100°\text{C} = 1373$ K, then $t = 14{,}876$ s $= 4.13$ h
If $T = 1200°\text{C} = 1473$ K, then $t = 6{,}560$ s $= 1.82$ h

The exact combination of temperature and time will depend on the maximum temperature that the heat treating furnace can reach, the rate at which parts must be produced, and the economics of the tradeoffs between higher temperatures versus longer times. Another factor to consider is changes in microstructure that occur in the rest of the material. For example, while carbon is diffusing into the surface, the rest of the microstructure can begin to experience grain growth or other changes.

Example 5-8 shows that one of the consequences of Fick's second law is that the same concentration profile can be obtained for different processing conditions, so long as the term Dt is constant. This permits us to determine the effect of temperature on the time required for a particular heat treatment to be accomplished.

Example 5-8 *Design of a More Economical Heat Treatment*

We find that 10 h are required to successfully carburize a batch of 500 steel gears at 900°C, where the iron has the FCC structure. We find that it costs \$1000 per hour to operate the carburizing furnace at 900°C and \$1500 per hour to operate the furnace at 1000°C. Is it economical to increase the carburizing temperature to 1000°C? What other factors must be considered?

SOLUTION

We again assume that we can use the solution to Fick's second law given by Equation 5-7:

$$\frac{c_s - c_x}{c_s - c_0} = \operatorname{erf}\left(\frac{x}{2\sqrt{Dt}}\right)$$

The temperatures of interest are 900°C = 1173 K and 1000°C = 1273 K. To achieve the same carburizing treatment at 1000°C as at 900°C:

$$D_{1273}t_{1273} = D_{1173}t_{1173}$$

For carbon diffusing in FCC iron, the activation energy is 32,900 cal/mol. Since we are dealing with the ratios of times, it does not matter whether we substitute for the time in hours or seconds. It is, however, always a good idea to use units that balance out. Therefore, we will show time in seconds. Note that temperatures must be converted into kelvin.

$$D_{1273}t_{1273} = D_{1173}t_{1173}$$

$$D = D_0 \exp(-Q/RT)$$

$$t_{1273} = \frac{D_{1173}t_{1173}}{D_{1273}}$$

$$= \frac{D_0 \exp\left[-\dfrac{32{,}900 \frac{\text{cal}}{\text{mol}}}{\left(1.987 \frac{\text{cal}}{\text{mol}\cdot\text{K}}\right)(1173\ \text{K})}\right](10\ \text{hr})(3600\ \text{s/hr})}{D_0 \exp\left[-\dfrac{32{,}900 \frac{\text{cal}}{\text{mol}}}{\left(1.987 \frac{\text{cal}}{\text{mol}\cdot\text{K}}\right)(1273\ \text{K})}\right]}$$

$$t_{1273} = \frac{\exp(-14.1156)(10)(3600)}{\exp(-13.0068)}$$

$$= 11{,}878\ \text{s}$$

$$t_{1273} = 3.299\ \text{h} = 3\ \text{h and } 18\ \text{min}$$

Notice, we did not need the value of the pre-exponential term D_0 since it canceled out.

At 900°C, the cost per part is (\$1000/h) (10 h)/500 parts = \$20/part. At 1000°C, the cost per part is (\$1500/h) (3.299 h)/500 parts = \$9.90/part.

Considering only the cost of operating the furnace, increasing the temperature reduces the heat-treating cost of the gears and increases the production rate. Another factor to consider is if the heat treatment at 1000°C could cause some other microstructural or other changes. For example, would increased temperature cause grains to grow significantly? If this is the case, we would be weakening the bulk of the material. How does the increased temperature affect the life of the other equipment such as the furnace itself and any accessories? How long would the cooling take? Will cooling from a higher temperature cause residual stresses? Would the product still meet all other specifications? These and other questions should be considered. The point is, as engineers, we need to ensure that the solution we propose is not only technically sound and economically sensible, it should recognize and make sense for the system as a whole. A good solution is often simple, solves problems for the system, and does not create new problems.

Example 5-9 *Silicon Device Fabrication*

Devices such as transistors are made by doping semiconductors. The diffusion coefficient of phosphorus in Si is $D = 6.5 \times 10^{-13}$ cm^2/s at a temperature of 1100°C. Assume the source provides a surface concentration of 10^{20} atoms/cm^3 and the diffusion time is one hour. Assume that the silicon wafer initially contains no P.

Calculate the depth at which the concentration of P will be 10^{18} atoms/cm^3. State any assumptions you have made while solving this problem.

SOLUTION

We assume that we can use one of the solutions to Fick's second law (i.e., Equation 5-7):

$$\frac{c_s - c_x}{c_s - c_0} = \text{erf}\left(\frac{x}{2\sqrt{Dt}}\right)$$

We will use concentrations in atoms/cm^3, time in seconds, and D in $\frac{\text{cm}^2}{\text{s}}$. Notice that the left-hand side is dimensionless. Therefore, as long as we use concentrations in the same units for c_s, c_x, and c_0, it does not matter what those units are.

$$\frac{c_s - c_x}{c_s - c_0} = \frac{10^{20}\,\frac{\text{atoms}}{\text{cm}^3} - 10^{18}\,\frac{\text{atoms}}{\text{cm}^3}}{10^{20}\,\frac{\text{atoms}}{\text{cm}^3} - 0\,\frac{\text{atoms}}{\text{cm}^3}} = 0.99$$

$$= \text{erf}\left[\frac{x}{2\sqrt{\left(6.5 \times 10^{-13}\,\frac{\text{cm}^2}{\text{s}}\right)(3600\text{ s})}}\right]$$

$$= \text{erf}\left(\frac{x}{9.67 \times 10^{-5}}\right)$$

From the error function values in Table 5-3 (or from your calculator/computer), if $\text{erf}(z) = 0.99$, $z = 1.82$; therefore,

$$1.82 = \frac{x}{9.67 \times 10^{-5}}$$

or

$$x = 1.76 \times 10^{-4}\text{ cm}$$

or

$$x = (1.76 \times 10^{-4}\text{ cm})\left(\frac{10^4\,\mu\text{m}}{\text{cm}}\right)$$

$$x = 1.76\,\mu\text{m}$$

Note that we have expressed the final answer in micrometers since this is the length scale that is appropriate for this application. The main assumptions we made are (1) the D value does not change while phosphorus gets incorporated in the silicon wafer and (2) the diffusion of P is only in one dimension (i.e., we ignore any lateral diffusion).

Limitations to Applying the Error-Function Solution Given by Equation 5-7 Note that in the equation describing Fick's second law (Equation 5-7):

(a) It is assumed that D is independent of the concentration of the diffusing species;

(b) the surface concentration of the diffusing species (c_s) is always constant.

There are situations under which these conditions may not be met and hence the concentration profile evolution will not be predicted by the error-function solution given by Equation 5-7. If the boundary conditions are different from the ones we assumed, different solutions to Fick's second law must be used.

5-9 Diffusion and Materials Processing

We briefly discussed applications of diffusion in processing materials in Section 5-1. Many important examples related to solidification, phase transformations, heat treatments, etc., will be discussed in later chapters. In this section, we provide more information to highlight the importance of diffusion in the processing of engineered materials. Diffusional processes become very important when materials are used or processed at elevated temperatures.

Melting and Casting One of the most widely used methods to process metals, alloys, many plastics, and glasses involves melting and casting of materials into a desired shape. Diffusion plays a particularly important role in solidification of metals and alloys. During the growth of single crystals of semiconductors, for example, we must ensure that the differences in the diffusion of dopants in both the molten and solid forms are accounted for. This also applies for the diffusion of elements during the casting of alloys. Similarly, diffusion plays a critical role in the processing of glasses. In inorganic glasses, for instance, we rely on the fact that diffusion is slow and inorganic glasses do not crystallize easily. We will examine this topic further in Chapter 9.

Sintering Although casting and melting methods are very popular for many manufactured materials, the melting points of many ceramic and some metallic materials are too high for processing by melting and casting. These relatively refractory materials are manufactured into useful shapes by a process that requires the consolidation of small particles of a powder into a solid mass (Chapter 15). **Sintering** is the high-temperature treatment that causes particles to join, gradually reducing the volume of pore space between them. Sintering is a frequent step in manufacturing of ceramic components (e.g., alumina, barium titanate, etc.) as well as in the production of metallic parts by **powder metallurgy**—a processing route by which metal powders are pressed and sintered into dense, monolithic components. A variety of composite materials such as tungsten carbide-cobalt based cutting tools are produced using this technique. With finer particles, many atoms or ions are at the surface for which the atomic or ionic bonds are not satisfied. As a result, a collection of fine particles of a certain mass has higher energy than that for a solid cohesive material of the same mass. Therefore, the driving force for solid state sintering of powdered metals and ceramics is the *reduction in the total surface area* of powder particles. When a powdered material is compacted into a shape, the powder particles are in contact with one another at numerous sites, with a significant amount of pore space between them. In order to reduce the total energy of the material, atoms diffuse to the points of contact, bonding the particles together and eventually causing the pores to shrink.

Lattice diffusion from the bulk of the particles into the neck region causes densification. Surface diffusion, gas or vapor phase diffusion, and lattice diffusion from curved surfaces into the neck area between particles do not lead to densification (Chapter 15). If sintering is carried out over a long period of time, the pores may be eliminated and the material becomes dense (Figure 5-20). In Figure 5-21, particles of a powder of a ceramic material known as barium magnesium tantalate [$Ba(Mg_{1/3}Ta_{2/3})O_3$ or BMT] are shown. This ceramic material is useful in making electronic components known as *dielectric resonators* used in wireless communication systems. The microstructure of

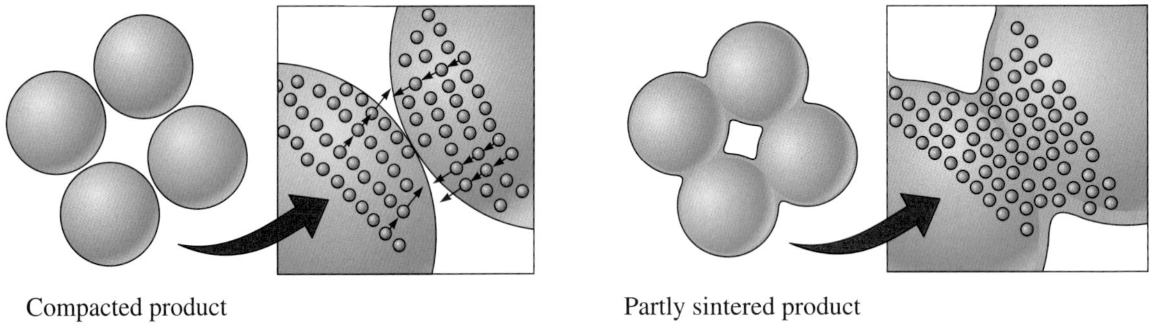

Figure 5-20 Diffusion processes during sintering and powder metallurgy. Atoms diffuse to points of contact, creating bridges and reducing the pore size.

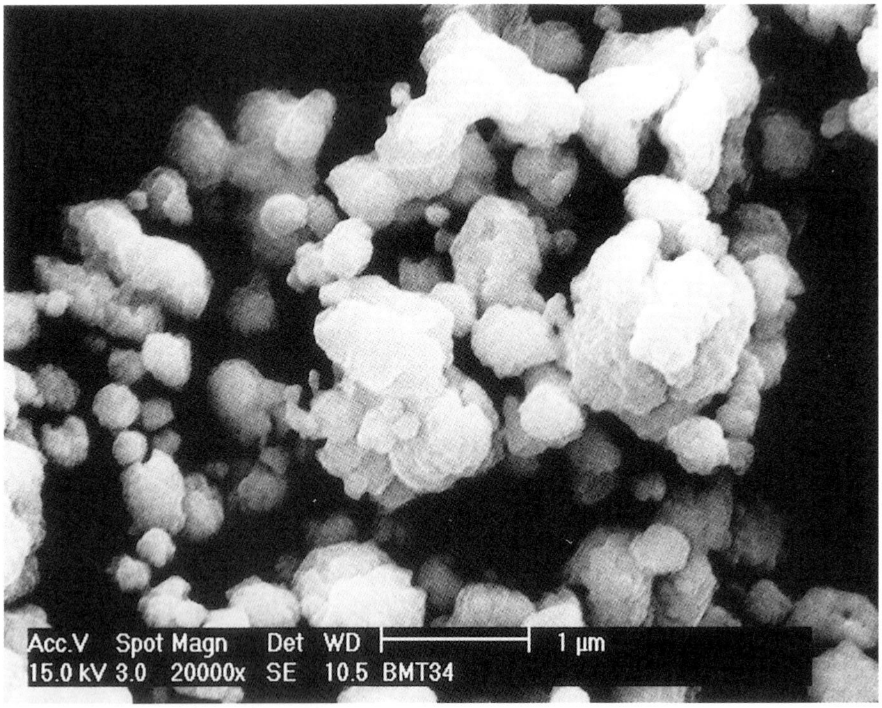

Figure 5-21 Particles of barium magnesium tantalate (BMT) [$Ba(Mg_{1/3}Ta_{2/3})O_3$] powder are shown. This ceramic material is useful in making electronic components known as dielectric resonators that are used for wireless communications. (*Courtesy of H. Shirey*)

Figure 5-22 The microstructure of BMT ceramics obtained by compaction and sintering of BMT powders. (*Courtesy of H. Shirey*)

BMT ceramics is shown in Figure 5-22, which is also the opening image of this chapter. These ceramics were produced by compacting the powders in a press and sintering the compact at a high temperature (~1500°C).

The extent and rate of sintering depends on (a) the initial density of the compacts, (b) temperature, (c) time, (d) the mechanism of sintering, (e) the average particle size, and (f) the size distribution of the powder particles. In some situations, a liquid phase forms in localized regions of the material while sintering is in process. Since diffusion of species, such as atoms and ions, is faster in liquids than in the solid state, the presence of a liquid phase can provide a convenient way for accelerating the sintering of many refractory metal and ceramic formulations. The process in which a small amount of liquid forms and assists densification is known as **liquid phase sintering**. For the liquid phase to be effective in enhancing sintering, it is important to have a liquid that can "wet" the grains, similar to how water wets a glass surface. If the liquid is non-wetting, similar to how mercury does not wet glass, then the liquid phase will not be helpful for enhancing sintering. In some cases, compounds are added to materials to cause the liquid phase to form at sintering temperatures. In other situations, impurities can react with the material and cause formation of a liquid phase. In most applications, it is desirable if the liquid phase is transient or converted into a crystalline material during cooling. This way a glassy and brittle amorphous phase does not remain at the grain boundaries.

When exceptionally high densities are needed, pressure (either uniaxial or isostatic) is applied while the material is being sintered. These techniques are known as **hot pressing**, when the pressure is unidirectional, or **hot isostatic pressing** (HIP), when the pressure is isostatic (i.e., applied in all directions). Many superalloys and ceramics such as lead lanthanum zirconium titanate (PLZT) are processed using these techniques. Hot isostatic pressing leads to high density materials with isotropic properties (Chapter 15).

Grain Growth A polycrystalline material contains a large number of grain boundaries, which represent high-energy areas because of the inefficient packing of the atoms. A lower overall energy is obtained in the material if the amount of grain boundary area is reduced by grain growth. **Grain growth** involves the movement of grain boundaries, permitting larger grains to grow at the expense of smaller grains (Figure 5-23). If you have watched froth, you have probably seen the principle of grain growth! Grain growth is similar to the way smaller bubbles in the froth disappear as they form bigger bubbles. For grain growth in materials, diffusion of atoms across the grain boundary is required, and, consequently, the growth of the grains is related to the activation energy needed for an atom to jump across the boundary. The increase in grain size can be seen from the sequence of micrographs for alumina ceramics shown in Figure 5-23. Another example for which grain growth plays a role is in the tungsten filament in a lightbulb. As the tungsten filament gets hotter, the grains grow causing it to get weaker. This grain growth, vaporization of tungsten, and oxidation via reaction with remnant oxygen contribute to the failure of tungsten filaments in a lightbulb.

The **driving force** for grain growth is reduction in grain boundary area. Grain boundaries are defects, and their presence causes the free energy of the material to increase. Thus, the thermodynamic tendency of polycrystalline materials is to transform into materials that have a larger average grain size. High temperatures or low-activation energies increase the size of the grains. Many heat treatments of metals, which include holding the metal at high temperatures, must be carefully controlled to *avoid* excessive grain growth. This is because, as the average grain size grows, the grain-boundary area decreases, and there is consequently less resistance to motion of dislocations. As a result, the strength of a metallic material will decrease with increasing grain size. We have seen this concept before in the form of the Hall-Petch equation (Chapter 4). In **normal grain**

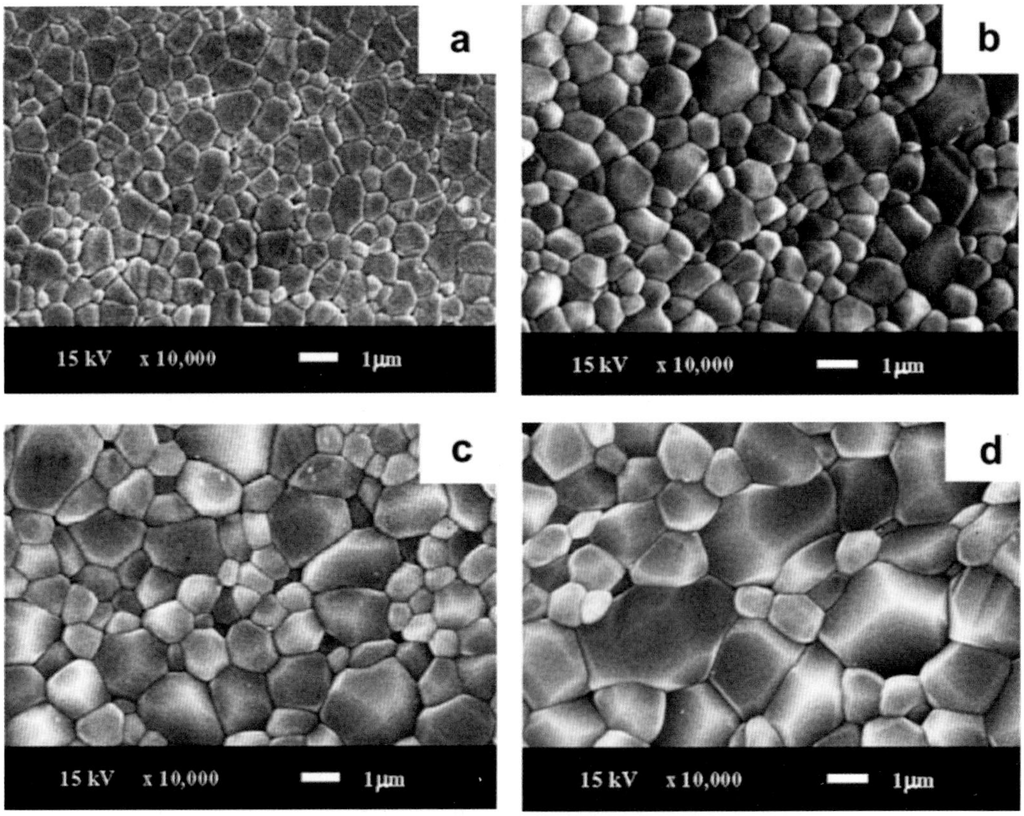

Figure 5-23 Scanning electron micrographs of alumina compacts sintered in an atmospheric furnace at 1350°C with an optimal heating rate of 10°C/min and varied soaking times of (a) 0.5 h; (b) 1.0 h; (c) 2.0 h and (d) 5.0 h. (*Reprinted from* Current Applied Physics, *Volume 9, Issue 5, Chutinan Promdej, Varong Pavarajarn, Shigetaka Wada, Thanakorn Wasanapiarnpong, Tawatchai Charinpanitkul, "Effect of hot isostatically pressed sintering on microstructure of translucent alumina compact," Pages 960–966, Copyright 2009, with permission from Elsevier.*)

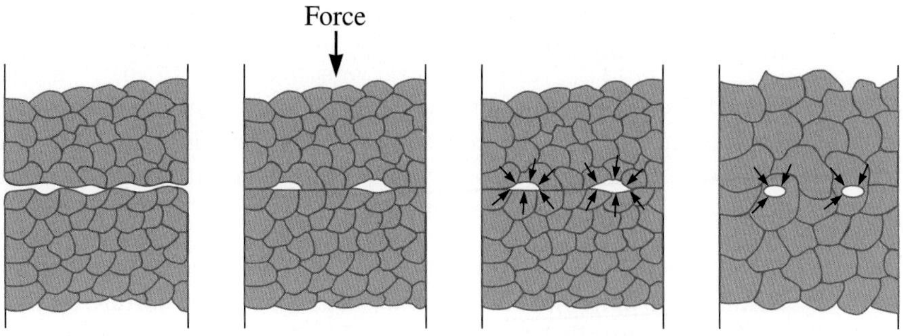

Figure 5-24 The steps in diffusion bonding: (a) Initially the contact area is small; (b) application of pressure deforms the surface, increasing the bonded area; (c) grain boundary diffusion permits voids to shrink; and (d) final elimination of the voids requires volume diffusion.

growth, the average grain size increases steadily and the width of the grain size distribution is not affected severely. In **abnormal grain growth**, the grain size distribution tends to become bi-modal (i.e., we get a few grains that are very large and then we have a few grains that remain relatively small). Certain electrical, magnetic, and optical properties of materials also depend upon the grain size of materials. As a result, in the processing of these materials, attention has to be paid to factors that affect diffusion rates and grain growth.

Diffusion Bonding A method used to join materials, called **diffusion bonding**, occurs in three steps (Figure 5-24). The first step forces the two surfaces together at a high temperature and pressure, flattening the surface, fragmenting impurities, and producing a high atom-to-atom contact area. As the surfaces remain pressed together at high temperatures, atoms diffuse along grain boundaries to the remaining voids; the atoms condense and reduce the size of any voids at the interface. Because grain boundary diffusion is rapid, this second step may occur very quickly. Eventually, however, grain growth isolates the remaining voids from the grain boundaries. For the third step—final elimination of the voids—volume diffusion, which is comparatively slow, must occur. The diffusion bonding process is often used for joining reactive metals such as titanium, for joining dissimilar metals and materials, and for joining ceramics.

Summary

- The net flux of atoms, ions, etc., resulting from diffusion depends upon the initial concentration gradient.
- The kinetics of diffusion depend strongly on temperature. In general, diffusion is a thermally activated process and the dependence of the diffusion coefficient on temperature is given by the Arrhenius equation.
- The extent of diffusion depends on temperature, time, the nature and concentration of diffusing species, crystal structure, composition of the matrix, stoichiometry, and point defects.
- Encouraging or limiting the diffusion process forms the underpinning of many important technologies. Examples include the processing of semiconductors, heat treatments of metallic materials, sintering of ceramics and powdered metals, formation

of amorphous materials, solidification of molten materials during a casting process, diffusion bonding, and barrier plastics, films, and coatings.

- Fick's laws describe the diffusion process quantitatively. Fick's first law defines the relationship between the concentration gradient and the flux of diffusing species. Fick's second law describes the variation of concentration of diffusing species under nonsteady state diffusion conditions.
- For a particular system, the amount of diffusion is related to the term Dt. This term permits us to determine the effect of a change in temperature on the time required for a diffusion-controlled process.
- The two important mechanisms for atomic movement in crystalline materials are vacancy diffusion and interstitial diffusion.
- The rate of diffusion is governed by the Arrhenius relationship—that is, the rate increases exponentially with temperature. Diffusion is particularly significant at temperatures above about 0.4 times the melting temperature (in kelvin) of the material.
- The activation energy Q describes the ease with which atoms diffuse, with rapid diffusion occurring for a low activation energy. A low-activation energy and rapid diffusion rate are obtained for (1) interstitial diffusion compared to vacancy diffusion, (2) crystal structures with a smaller packing factor, (3) materials with a low melting temperature or weak atomic bonding, and (4) diffusion along grain boundaries or surfaces.
- The total movement of atoms, or flux, increases when the concentration gradient and temperature increase.
- Diffusion of ions in ceramics is usually slower than that of atoms in metallic materials. Diffusion in ceramics is also affected significantly by non-stoichiometry, dopants, and the possible presence of liquid phases during sintering.
- Atom diffusion is of paramount importance because many materials processing techniques, such as sintering, powder metallurgy, and diffusion bonding, require diffusion. Furthermore, many of the heat treatments and strengthening mechanisms used to control structures and properties in materials are diffusion-controlled processes. The stability of the structure and the properties of materials during use at high temperatures depend on diffusion.

Glossary

Abnormal grain growth A type of grain growth observed in metals and ceramics. In this mode of grain growth, a bimodal grain size distribution usually emerges as some grains become very large at the expense of smaller grains. See "Grain growth" and "Normal grain growth."

Activation energy The energy required to cause a particular reaction to occur. In diffusion, the activation energy is related to the energy required to move an atom from one lattice site to another.

Carburization A heat treatment for steels to harden the surface using a gaseous or solid source of carbon. The carbon diffusing into the surface makes the surface harder and more abrasion resistant.

Concentration gradient The rate of change of composition with distance in a nonuniform material, typically expressed as $\frac{\text{atoms}}{\text{cm}^3 \cdot \text{cm}}$ or at% / cm.

Conductive ceramics Ceramic materials that are good conductors of electricity as a result of their ionic and electronic charge carriers (electrons, holes, or ions). Examples of such materials are stabilized zirconia and indium tin oxide.

Diffusion The net flux of atoms, ions, or other species within a material caused by temperature and a concentration gradient.

Diffusion bonding A joining technique in which two surfaces are pressed together at high pressures and temperatures. Diffusion of atoms to the interface fills in voids and produces a strong bond between the surfaces.

Diffusion coefficient A temperature-dependent coefficient related to the rate at which atoms, ions, or other species diffuse. The diffusion coefficient depends on temperature, the composition and microstructure of the host material, and the concentration of the diffusing species.

Diffusion couple A combination of elements involved in diffusion studies (e.g., if we are considering diffusion of Al in Si, then Al-Si is a diffusion couple).

Diffusion distance The maximum or desired distance that atoms must diffuse; often, the distance between the locations of the maximum and minimum concentrations of the diffusing atom.

Diffusivity Another term for the diffusion coefficient.

Driving force A cause that induces an effect. For example, an increased gradient in composition enhances diffusion; similarly reduction in surface area of powder particles is the driving force for sintering.

Fick's first law The equation relating the flux of atoms by diffusion to the diffusion coefficient and the concentration gradient.

Fick's second law The partial differential equation that describes the rate at which atoms are redistributed in a material by diffusion. Many solutions exist to Fick's second law; Equation 5-7 is one possible solution.

Flux The number of atoms or other diffusing species passing through a plane of unit area per unit time. This is related to the rate at which mass is transported by diffusion in a solid.

Grain boundary diffusion Diffusion of atoms along grain boundaries. This is faster than volume diffusion because the atoms are less closely packed in grain boundaries.

Grain growth Movement of grain boundaries by diffusion in order to reduce the amount of grain boundary area. As a result, small grains shrink and disappear and other grains become larger, similar to how some bubbles in soap froth become larger at the expense of smaller bubbles. In many situations, grain growth is not desirable.

Hot isostatic pressing A sintering process in which a uniform pressure is applied in all directions during sintering. This process is used for obtaining very high densities and isotropic properties.

Hot pressing A sintering process conducted under uniaxial pressure, used for achieving higher densities.

Interdiffusion Diffusion of different atoms in opposite directions. Interdiffusion may eventually produce an equilibrium concentration of atoms within the material.

Interstitial diffusion Diffusion of small atoms from one interstitial position to another in the crystal structure.

Liquid phase sintering A sintering process in which a liquid phase forms. Since diffusion is faster in liquids, if the liquid can wet the grains, it can accelerate the sintering process.

Nitriding A process in which nitrogen is diffused into the surface of a material, such as a steel, leading to increased hardness and wear resistance.

Normal grain growth Grain growth that occurs in an effort to reduce grain boundary area. This type of grain growth is to be distinguished from abnormal grain growth in that the grain size distribution remains unimodal but the average grain size increases steadily.

Permeability A relative measure of the diffusion rate in materials, often applied to plastics and coatings. It is often used as an engineering design parameter that describes the effectiveness of a particular material to serve as a barrier against diffusion.

Powder metallurgy A method for producing monolithic metallic parts; metal powders are compacted into a desired shape, which is then heated to allow diffusion and sintering to join the powders into a solid mass.

Self-diffusion The random movement of atoms within an essentially pure material. No net change in composition results.

Sintering A high-temperature treatment used to join small particles. Diffusion of atoms to points of contact causes bridges to form between the particles. Further diffusion eventually fills in any remaining voids. The driving force for sintering is a reduction in total surface area of the powder particles.

Surface diffusion Diffusion of atoms along surfaces, such as cracks or particle surfaces.

Thermal barrier coatings (TBC) Coatings used to protect a component from heat.

Vacancy diffusion Diffusion of atoms when an atom leaves a regular lattice position to fill a vacancy in the crystal. This process creates a new vacancy, and the process continues.

Volume diffusion Diffusion of atoms through the interior of grains.

Problems

Section 5-1 Applications of Diffusion

5-1 What is the driving force for diffusion?

5-2 Give three examples of materials processes that rely on diffusion in solids and explain how diffusion plays a critical role for *one* of those processes.

5-3 In the carburization treatment of steels, what are the diffusing species?

5-4 Why do we use PET plastic to make carbonated beverage bottles?

5-5 Why is it that aluminum metal oxidizes more readily than iron but aluminum is considered to be a metal that usually does not "rust"?

5-6 What is a thermal barrier coating? Where are such coatings used?

5-7 What is a nitriding heat treatment?

5-8 A certain mechanical component is heat treated using carburization. A commonly encountered engineering problem is that we need to machine a certain part of the component and this part of the surface should not be hardened. Explain how we can achieve this objective.

Section 5-2 Stability of Atoms and Ions

5-9 Write down the Arrhenius equation and explain the different terms.

5-10 Atoms are found to move from one lattice position to another at the rate of $5 \times 10^5 \frac{\text{jumps}}{\text{s}}$ at 400°C when the activation energy for their movement is 30,000 cal/mol. Calculate the jump rate at 750°C.

5-11 The number of vacancies in a material is related to temperature by an Arrhenius equation. If the fraction of lattice points containing vacancies is 8×10^{-5} at 600°C, determine the fraction of lattice points containing vacancies at 1000°C.

5-12 The Arrhenius equation was originally developed for comparing rates of chemical reactions. Compare the rates of a chemical reaction at 20 and 100°C by calculating the ratio of the chemical reaction rates. Assume that the activation energy for liquids in which the chemical reaction is conducted is 10 kJ/mol and that the reaction is limited by diffusion.

Section 5-3 Mechanisms for Diffusion

5-13 What are the different mechanisms for diffusion?

5-14 (a) Compare interstitial and vacancy atomic mechanisms for diffusion and (b) cite two reasons why interstitial diffusion is normally more rapid than vacancy diffusion.

5-15 How is self-diffusion of atoms in metals verified experimentally?

Section 5-4 Activation Energy for Diffusion

5-16 Why is it that the activation energy for diffusion via the interstitial mechanism is less than those for other mechanisms?

5-17 Compare the diffusion coefficients of carbon in BCC and FCC iron at the allotropic transformation temperature of 912°C and explain the reason for the difference in their values.

5-18 Compare the diffusion coefficients for hydrogen and nitrogen in FCC iron at 1000°C and explain the reason for the difference in their values.

5-19 Activation energy is sometimes expressed as (eV/atom), (e.g., as shown in Figure 5-15, which illustrates the diffusion coefficients of ions in different oxides). Convert eV/atom to J/mole.

5-20 In order to produce a certain semiconductor, a process called doping is performed in which phosphorus is diffused into germanium. If $D_0 = 2.0$ cm²/s and the activation energy is 57,500 cal/mol, determine the diffusivity of P in Ge at 800°C.

5-21 The activation energy for the diffusion of copper into silver is 193,000 J/mol. Calculate the diffusion coefficient at 1200 K given that D at 1000 K is 1.0×10^{-14} m²/s.

5-22 Determine the diffusion coefficient D for the diffusion of hydrogen into FCC iron at 800°C.

5-23 The diffusion coefficient for Cr^{3+} in Cr_2O_3 is 6×10^{-15} cm²/s at 727°C and 1×10^{-9} cm²/s at 1400°C. Calculate
(a) the activation energy; and
(b) the constant D_0.

5-24 The diffusion coefficient for O^{2-} in Cr_2O_3 is 4×10^{-15} cm²/s at 1150°C and 6×10^{-11} cm²/s at 1715°C. Calculate
(a) the activation energy; and
(b) the constant D_0.

5-25 Without referring to the actual data, can you predict whether the activation energy for diffusion of carbon in FCC iron will be higher or lower than that in BCC iron? Explain.

Section 5-5 Rate of Diffusion (Fick's First Law)

5-26 Write down Fick's first law of diffusion. Clearly explain what each term means.

5-27 What is the difference between diffusivity and the diffusion coefficient?

5-28 Determine the maximum allowable temperature that will produce a flux of less than 2000 $\frac{\text{N atoms}}{\text{cm}^2 \cdot \text{s}}$ through a BCC iron foil when the concentration gradient is $-5 \times 10^{16} \frac{\text{atoms}}{\text{cm}^3 \cdot \text{cm}}$.

5-29 Calculate the diffusion coefficients for the diffusion of hydrogen through BCC iron and FCC aluminum at room temperature (25°C). For the diffusion of H into Al, $D_0 = 0.11$ cm²/s and $Q = 9780$ cal/mol. Based on your calculations, which material would be better suited as the material for a high-pressure hydrogen storage tank?

5-30 How would the internal pressure of hydrogen in a storage tank affect the rate of diffusion? Explain.

5-31 A 1-mm-thick BCC iron foil is used to separate a region of high nitrogen gas concentration of 0.1 at % from a region of low nitrogen gas concentration at 650°C. If the flux of nitrogen through the foil is $10^{12} \frac{\text{atoms}}{\text{cm}^2 \cdot \text{s}}$, what is the nitrogen concentration in the low concentration region?

5-32 A 0.2-mm-thick wafer of silicon is treated so that a uniform concentration gradient of antimony is produced. One surface contains 1 Sb atom per 10^8 Si atoms and the other surface contains 500 Sb atoms per 10^8 Si atoms. The lattice parameter for Si is 5.4307 Å (Appendix A). Calculate the concentration gradient in
(a) at% Sb per cm; and
(b) $\frac{\text{Sb atoms}}{\text{cm} \cdot \text{cm}^3}$.

5-33 When a Cu-Zn alloy solidifies, one portion of the structure contains 25 at% zinc and another portion 0.025 mm away contains 20 at% zinc. The lattice parameter for the FCC alloy is about 3.63×10^{-8} cm. Determine the concentration gradient in
(a) at% Zn per cm;
(b) wt% Zn per cm; and
(c) $\frac{\text{Zn atoms}}{\text{cm} \cdot \text{cm}^3}$.

5-34 A 0.001 in. BCC iron foil is used to separate a high hydrogen content gas from a low hydrogen content gas at 650°C. 5×10^8 H atoms/cm³ are in equilibrium on one side of the foil, and 2×10^3 H atoms/cm³ are in equilibrium on the other side. Determine
(a) the concentration gradient of hydrogen; and
(b) the flux of hydrogen through the foil.

5-35 Compute the number of kilograms of hydrogen that pass per hour through a 5-mm-thick sheet of palladium having an area of 0.20 m^2 at 500°C. Assume a diffusion coefficient of 1.0×10^{-8} m^2/s and that the concentrations at the high and low pressure sides of the plate are 2.4 and 0.6 kg of hydrogen per m^3 of palladium. Assume steady state conditions.

5-36 A 1-mm-thick sheet of FCC iron is used to contain nitrogen in a heat exchanger at 1200°C. The concentration of N at one surface is 0.04 at%, and the concentration at the second surface is 0.005 at%. Determine the flux of nitrogen through the foil in N atoms/($cm^2 \cdot$ s).

5-37 A 4-cm-diameter, 0.5-mm-thick spherical container made of BCC iron holds nitrogen at 700°C. The concentration at the inner surface is 0.05 at% and at the outer surface is 0.002 at%. Calculate the number of grams of nitrogen that are lost from the container per hour.

5-38 A BCC iron structure is to be manufactured that will allow no more than 50 g of hydrogen to be lost per year through each square centimeter of the iron at 400°C. If the concentration of hydrogen at one surface is 0.05 H atom per unit cell and 0.001 H atom per unit cell at the second surface, determine the minimum thickness of the iron.

5-39 Determine the maximum allowable temperature that will produce a flux of less than 2000 $\frac{\text{H atoms}}{\text{cm}^2 \cdot \text{s}}$ through a BCC iron foil when the concentration gradient is $-5 \times 10^{16} \frac{\text{atoms}}{\text{cm} \cdot \text{cm}^3}$. (Note the negative sign for the flux.)

Section 5-6 Factors Affecting Diffusion

5-40 Write down the equation that describes the dependence of D on temperature.

5-41 In solids, the process of diffusion of atoms and ions takes time. Explain how this is used to our advantage while forming metallic glasses.

5-42 Why is it that inorganic glasses form upon relatively slow cooling of melts, while rapid solidification is necessary to form metallic glasses?

5-43 In ionic materials, is diffusion expected to occur at a faster rate for cations or anions? Explain.

5-44 Use the diffusion data in the table below for atoms in iron to answer the questions that follow. Assume metastable equilibrium conditions and trace amounts of C in Fe. The gas constant in SI units is 8.314 J/(mol · K).

Diffusion Couple	Diffusion Mechanism	Q (J/mol)	D_0 (m^2/s)
C in FCC iron	Interstitial	1.38×10^5	2.3×10^{-5}
C in BCC iron	Interstitial	8.74×10^4	1.1×10^{-6}
Fe in FCC iron	Vacancy	2.79×10^5	6.5×10^{-5}
Fe in BCC iron	Vacancy	2.46×10^5	4.1×10^{-4}

(a) Plot the diffusion coefficient as a function of inverse temperature ($1/T$) showing all four diffusion couples in the table.

(b) Recall the temperatures for phase transitions in iron, and for each case, indicate on the graph the temperature range over which the diffusion data is valid.

(c) Why is the activation energy for Fe diffusion higher than that for C diffusion in iron?

(d) Why is the activation energy for diffusion higher in FCC iron when compared to BCC iron?

(e) Does C diffuse faster in FCC Fe than in BCC Fe? Support your answer with a numerical calculation and state any assumptions made.

5-45 The plot below has three lines representing grain boundary, surface, and volume self-diffusion in a metal. Match the lines labeled A, B, and C with the type of diffusion. Justify your answer by calculating the activation energy for diffusion for each case.

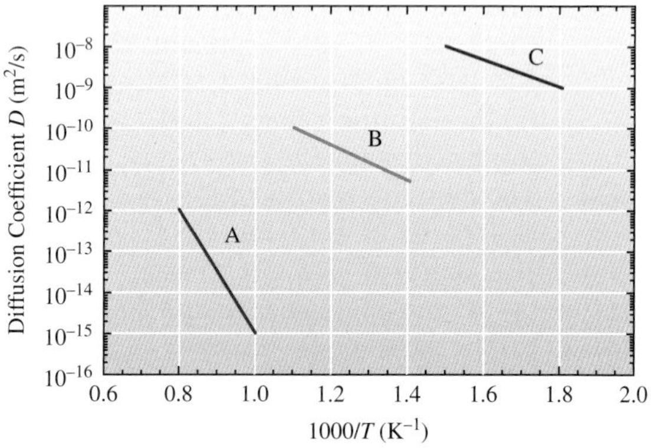

Figure 5-25 Plot for Problem 5-45.

Section 5-7 Permeability of Polymers

5-46 What are barrier polymers?

5-47 What factors, other than permeability, are important in selecting a polymer for making plastic bottles?

5-48 Amorphous PET is more permeable to CO_2 than PET that contains microcrystallites. Explain why.

Table 5-4 Error function (erf) values

z	erf (z)	z	erf (z)	z	erf (z)
0	0	0.55	0.5633	1.3	0.9340
0.025	0.0282	0.60	0.6039	1.4	0.9523
0.05	0.0564	0.65	0.6420	1.5	0.9661
0.10	0.1125	0.70	0.6778	1.6	0.9763
0.15	0.1680	0.75	0.7112	1.7	0.9838
0.20	0.2227	0.80	0.7421	1.8	0.9891
0.25	0.2763	0.85	0.7707	1.9	0.9928
0.30	0.3286	0.90	0.7969	2.0	0.9953
0.35	0.3794	0.95	0.8209	2.2	0.9981
0.40	0.4284	1.0	0.8427	2.4	0.9993
0.45	0.4755	1.1	0.8802	2.6	0.9998
0.50	0.5205	1.2	0.9103	2.8	0.9999

Section 5-8 Composition Profile (Fick's Second Law)

5-49 Pure zinc is to be diffused into copper by dipping copper into molten zinc at 450°C. Calculate how long it would take to obtain 10 wt% zinc at a depth of 0.5 mm beneath the copper surface. Is this commercially feasible? What practical problems might arise if we raise the temperature to 1000°C? Use Table 5-4 for error function values as needed. At 450°C, the diffusion coefficient $D = 6.3 \times 10^{-13}$ cm²/s.

5-50 Nitriding is a process in which nitrogen is allowed to diffuse into the surface of steel for the purpose of increasing the surface hardness of a component. It has been determined that a satisfactory nitrogen case depth is produced in BCC iron after 1 hour at 700°C. How much time is required to produce a satisfactory case depth if nitriding is carried out at 600°C?

5-51 Determine the carburizing time necessary to achieve a carbon concentration of 0.30 wt% at a position 4-mm into a steel alloy that initially contains 0.10 wt% carbon. The surface concentration is to be maintained at 0.90 wt% carbon, and the treatment is to be conducted at 1100°C. Use the diffusion data for FCC iron and the error function values in Table 5-4 as needed.

5-52 One method of fabricating complex aerospace components is diffusion bonding. We wish to join a nickel sheet, intended to rapidly transfer lightning strike data away from critical locations, to an FCC iron (stainless steel) plate. We find that we can accomplish a good bond in 8 hours when the two metals are joined at 800°C. How long will it take to obtain an equally good bond if we decide to perform the bonding at 600°C? (Assume that the quality of the bond is determined by the rate of diffusion into the FCC iron).

5-53 Determine the carburizing time necessary to achieve a carbon concentration of 0.45 wt% at a position 2-mm into a steel alloy that initially contains 0.20 wt% carbon. The surface concentration is to be maintained at 1.30 wt% at 1000°C. Use Table 5-4 for error function values as needed.

5-54 Nitrogen from a gaseous phase is to be diffused into pure iron (BCC) at 700°C. If the surface concentration is maintained at 0.10 wt% N, what will the concentration 1 mm from the surface be after 10 hrs? Use Table 5-4 for error function values as needed.

5-55 The surface of 0.10 wt% carbon steel is to be strengthened by carburizing. The steel is placed in an atmosphere that provides a maximum of 1.2 wt% carbon at the surface of the steel at a high temperature. Carbon then diffuses 0.2 cm into the surface of the steel giving a carbon concentration of 0.45 wt% carbon at that depth. How long will the process of carburizing take if the diffusion coefficient is 2×10^{-7} cm²/s? Use Table 5-4 for error function values as needed.

5-56 Transistors are made by doping single crystal silicon with different types of impurities to generate n- and p-type regions. Phosphorus and boron are typical n- and p-type dopant species, respectively. Assuming that a thermal treatment at 1100°C for 1 h is used to cause diffusion of the dopants, calculate the constant surface concentration of P and B needed to achieve a concentration of 10^{18} atoms/cm^3 at a depth of 0.1 mm from the surface for both n- and p-type regions. The diffusion coefficients of P and B in single crystal silicon at 1100°C are 6.5×10^{-13} cm^2/s and 6.1×10^{-13} cm^2/s, respectively.

5-57 Consider a 2-mm-thick silicon wafer to be doped using antimony. Assume that the dopant source (gas mixture of antimony chloride and other gases) provides a constant concentration of 10^{22} atoms/m^3. We need a dopant profile such that the concentration of Sb at a depth of 1 micrometer is 5×10^{21} atoms/m^3. What is the required time for the diffusion heat treatment? Assume that the silicon wafer initially contains no impurities or dopants. Assume that the activation energy for diffusion of Sb into silicon is 380 kJ/mole and D_0 for Sb diffusion into Si is 1.3×10^{-3} m^2/s. Assume $T = 1250$°C.

5-58 Consider doping of silicon with gallium. Assume that the diffusion coefficient of gallium in Si at 1100°C is 7×10^{-13} cm^2/s. Calculate the concentration of Ga at a depth of 2.0 micrometer if the surface concentration of Ga is 10^{23} atoms/cm^3. The diffusion times are 1, 2, and 3 hours.

5-59 Compare the rate at which oxygen ions diffuse in alumina (Al_2O_3) with the rate at which aluminum ions diffuse in Al_2O_3 at 1500°C. Explain the difference.

5-60 A carburizing process is carried out on a 0.10% C steel by introducing 1.0% C at the surface at 980°C, where the iron is FCC. Calculate the carbon content at 0.01 cm, 0.05 cm, and 0.10 cm beneath the surface after 1 h.

5-61 Iron containing 0.05% C is heated to 912°C in an atmosphere that produces 1.20% C at the surface and is held for 24 h. Calculate the carbon content at 0.05 cm beneath the surface if
(a) the iron is BCC; and
(b) the iron is FCC.
Explain the difference.

5-62 What temperature is required to obtain 0.50% C at a distance of 0.5 mm beneath the surface of a 0.20% C steel in 2 h, when 1.10% C is present at the surface? Assume that the iron is FCC.

5-63 A 0.15% C steel is to be carburized at 1100°C, giving 0.35% C at a distance of 1 mm beneath the surface. If the surface composition is maintained at 0.90% C, what time is required?

5-64 A 0.02% C steel is to be carburized at 1200°C in 4 h, with the carbon content 0.6 mm beneath the surface reaching 0.45% C. Calculate the carbon content required at the surface of the steel.

5-65 A 1.2% C tool steel held at 1150°C is exposed to oxygen for 48 h. The carbon content at the steel surface is zero. To what depth will the steel be decarburized to less than 0.20% C?

5-66 A 0.80% C steel must operate at 950°C in an oxidizing environment for which the carbon content at the steel surface is zero. Only the outermost 0.02 cm of the steel part can fall below 0.75% C. What is the maximum time that the steel part can operate?

5-67 A steel with the BCC crystal structure containing 0.001% N is nitrided at 550°C for 5 h. If the nitrogen content at the steel surface is 0.08%, determine the nitrogen content at 0.25 mm from the surface.

5-68 What time is required to nitride a 0.002% N steel to obtain 0.12% N at a distance of 0.002 in. beneath the surface at 625°C? The nitrogen content at the surface is 0.15%.

5-69 We can successfully perform a carburizing heat treatment at 1200°C in 1 h. In an effort to reduce the cost of the brick lining in our furnace, we propose to reduce the carburizing temperature to 950°C. What time will be required to give us a similar carburizing treatment?

5-70 During freezing of a Cu-Zn alloy, we find that the composition is nonuniform. By heating the alloy to 600°C for 3 h, diffusion of zinc helps to make the composition more uniform. What temperature would be required if we wished to perform this homogenization treatment in 30 minutes?

5-71 To control junction depth in transistors, precise quantities of impurities are introduced at relatively shallow depths by ion implantation and diffused into the silicon substrate in a subsequent thermal treatment. This can be approximated as a finite source diffusion problem. Applying the appropriate boundary conditions, the solution to Fick's second law under these conditions is

$$c(x, t) = \frac{Q}{\sqrt{\pi Dt}} \exp\left(\frac{-x^2}{4Dt}\right)$$

where Q is the initial surface concentration with units of atoms/cm^2. Assume that we implant 10^{14} atoms/cm^2 of phosphorus at the surface of a silicon wafer with a background boron concentration of 10^{16} atoms/cm^3 and this wafer is subsequently annealed at 1100°C. The diffusion coefficient of phosphorus in silicon at 1100°C is 6.5×10^{-13} cm^2/s.

(a) Plot a graph of the concentration c (atoms/cm^3) versus x (cm) for anneal times of 5 minutes, 10 minutes, and 15 minutes.

(b) What is the anneal time required for the phosphorus concentration to equal the boron concentration at a depth of 1 μm?

Section 5-9 Diffusion and Materials Processing

5-72 Arrange the following materials in increasing order of self-diffusion coefficient: Ar gas, water, single crystal aluminum, and liquid aluminum at 700°C.

5-73 Most metals and alloys can be processed using the melting and casting route, but we typically do not choose to apply this method for the processing of specific metals (e.g., W) and most ceramics. Explain.

5-74 What is sintering? What is the driving force for sintering?

5-75 Why does grain growth occur? What is meant by the terms normal and abnormal grain growth?

5-76 Why is the strength of many metallic materials expected to decrease with increasing grain size?

5-77 A ceramic part made from MgO is sintered successfully at 1700°C in 90 minutes. To minimize thermal stresses during the process, we plan to reduce the temperature to 1500°C. Which of the following will limit the rate at which sintering can be performed: diffusion of magnesium ions or diffusion of oxygen ions? What time will be required at the lower temperature?

5-78 A Cu-Zn alloy has an initial grain diameter of 0.01 mm. The alloy is then heated to various temperatures, permitting grain growth to occur. The times required for the grains to grow to a diameter of 0.30 mm are shown below.

Temperature (°C)	Time (minutes)
500	80,000
600	3,000
700	120
800	10
850	3

Determine the activation energy for grain growth. Does this correlate with the diffusion of zinc in copper? (*Hint*: Note that rate is the reciprocal of time.)

5-79 What are the advantages of using hot pressing and hot isostatic pressing compared to using normal sintering?

5-80 A sheet of gold is diffusion-bonded to a sheet of silver in 1 h at 700°C. At 500°C, 440 h are required to obtain the same degree of bonding, and at 300°C, bonding requires 1530 years. What is the activation energy for the diffusion bonding process? Does it appear that diffusion of gold or diffusion of silver controls the bonding rate? (*Hint*: Note that rate is the reciprocal of time.)

Design Problems

5-81 The wear resistance of a steel shaft is to be improved by hardening its surface. This is to be accomplished by increasing the nitrogen content within an outer surface layer as a result of nitrogen diffusion into the steel; the nitrogen is to be supplied from an external nitrogen-rich gas at an elevated and constant temperature. The initial nitrogen content of the steel is 0.002 wt%, whereas the surface concentration will be maintained at 0.45 wt%. For this treatment to be effective, a nitrogen content of 0.12 wt% must be established at a position 0.45 mm below the surface. Design an appropriate heat treatment in terms of temperature between 475°C and 625°C and time. Use $D_0 = 3.0 \times 10^{-7}$ m^2/s and $Q = 76{,}150$ J/mol.

5-82 Design a spherical tank, with a wall thickness of 2 cm that will ensure that no more than 50 kg of hydrogen will be lost per year. The tank, which will operate at 500°C, can be made from nickel, aluminum, copper, or iron. The diffusion coefficient of hydrogen and the cost per pound for each available material is listed below.

Diffusion Data for Problem 5-82

Material	D_0 (cm^2/s)	Q (cal/mol)	Cost ($/lb)
Nickel	0.0055	8,900	4.10
Aluminum	0.16	10,340	0.60
Copper	0.011	9,380	1.10
Iron (BCC)	0.0012	3,600	0.15

5-83 A steel gear initially containing 0.10% C is to be carburized so that the carbon content at a depth of 0.05 in. is 0.50% C. We can generate a carburizing gas at the surface that contains anywhere from 0.95% C to 1.15% C. Design an appropriate carburizing heat treatment.

5-84 When a valve casting containing copper and nickel solidifies under nonequilibrium conditions, we find that the composition of the alloy varies substantially over a distance of 0.005 cm. Usually we are able to eliminate this concentration difference by heating the alloy for 8 h at 1200°C; however, sometimes this treatment causes the alloy to begin to melt, destroying the part. Design a heat treatment that will eliminate the non-uniformity without melting. Assume that the cost of operating the furnace per hour doubles for each 100°C increase in temperature.

5-85 Assume that the surface concentration of phosphorus being diffused in silicon is 10^{21} atoms/cm^3. We need to design a process, known as the pre-deposition step, such that the concentration of P (c_1 for step 1) at a depth of 0.25 μm is 10^{13} atoms/cm^3. Assume that this is conducted at a temperature of 1000°C and the diffusion coefficient of P in Si at this temperature is 2.5×10^{-14} cm^2/s. Assume that the process is carried out for a total time of 8 minutes. Calculate the concentration profile (i.e., c_1 as a function of depth, which in this case is given by the following equation). Notice the use of the complementary error function.

$$c_1(x, t_1) = c_s\left[1 - \text{erf}\left(\frac{x}{4Dt}\right)\right]$$

Use different values of x to generate and plot this profile of P during the pre-deposition step.

Computer Problems

5-86 *Calculation of Diffusion Coefficients.* Write a computer program that will ask the user to provide the data for the activation energy Q and the value of D_0. The program should then ask the user to input a valid range of temperatures. The program, when executed, provides the values of D as a function of temperature in increments chosen by the user. The program should convert Q, D_0, and temperature to the proper units. For example, if the user provides temperatures in °F, the program should recognize that and convert the temperature into K. The program should also carry a cautionary statement about the standard assumptions made. For example, the program should caution the user that effects of any possible changes in crystal structure in the temperature range specified are not accounted for. Check the validity of your programs using examples in the book and also other problems that you may have solved using a calculator.

5-87 *Comparison of Reaction Rates.* Write a computer program that will ask the user to input the activation energy for a chemical reaction. The program should then ask the user to provide two temperatures for which the reaction rates need to be compared. Using the value of the gas constant and activation energy, the program should then provide a ratio of the reaction rates. The program should take into account different units for activation energy.

5-88 *Carburization Heat Treatment.* Write a computer program for calculating carburization heat treatments. The program should ask the user to provide an input for the carbon concentration at the surface (c_s), and the concentration of carbon in the bulk (c_0). The program should also ask the user to provide the temperature and a value for D (or values for D_0 and Q, that will allow for D to be calculated). The program should then provide an output of the concentration profile in tabular form. The distances at which concentrations of carbon are to be determined can be provided by the user or defined by the person writing the program. The program should also be able to handle calculation of heat treatment times if the user provides a level of concentration that is needed at a specified depth. This program will require calculation of the error function. The programming language or spreadsheet you use may have a built-in function that calculates the error function. If that is the case, use that function. You can also calculate the error function as the expansion of the following series:

$$\text{erf}(z) = 1 - \frac{e^{-z^2}}{\sqrt{\pi}z}\left(1 - \frac{1}{2z^2} - \frac{1 \times 3}{(2z^2)^2} + \frac{1 \times 3 \times 5}{(2z^2)^3} + \cdots\right)$$

or use an approximation

$$\text{erf}(z) = 1 - \left[\left(\frac{1}{\sqrt{\pi}}\right)\frac{e^{-z^2}}{z}\right]$$

In these equations, z is the argument of the error function. Also, in certain situations, you will know the value of the error function (if all concentrations are provided) and you will have to figure out the argument of the error function. This can be handled by having part of the program compare different values for the argument of the error function and by minimizing the difference between the value of the error function you require and the value of the error function that was approximated.

Knovel® Problems

K5-1 Compare the carbon dioxide permeabilities of low-density polyethylene (LDPE), polypropylene, and polyethylene terephthalate (PET) films at room temperature.

Some materials can become brittle when temperatures are low and/or strain rates are high. The special chemistry of the steel used on the *Titanic* and the stresses associated with the fabrication and embrittlement of this steel when subjected to lower temperatures have been identified as factors contributing to the failure of the ship's hull. (*John Parrot/Stocktrek Images/Stocktrek Images/Getty Images.*)

CHAPTER

6

Mechanical Properties: Part One

Have You Ever Wondered?

- *Why can Silly Putty® be stretched a considerable amount when pulled slowly, but snaps when pulled fast?*
- *Why can we load the weight of a fire truck on four ceramic coffee cups, yet ceramic cups tend to break easily when we drop them on the floor?*
- *What materials-related factors played an important role in the sinking of the Titanic?*
- *What factors played a major role in the 1986 Challenger space shuttle accident?*

Chapter Learning Objectives

The key objectives of this chapter are to

- Identify the critical features of a uniaxial stress strain curve (elastic modulus, yield stress, ultimate tensile strength, necking, failure stress, strain to failure, elastic region, plastic region, toughness).
- Construct a stress strain curve given load and displacement data and calculate the ductility of the sample.
- Calculate engineering and true stress and strain and explain why two systems of stress and strain exist.
- Predict how the mechanical properties of a material will change when subjected to different temperatures and strain rates.
- Calculate the flexural strength and flexural modulus for a material subjected to a bend test.
- Define the materials property of hardness and explain how it is measured at various length scales.
- Explain why metallic glasses are extraordinarily strong but brittle.

The mechanical properties of materials depend on their composition and microstructure. In Chapters 2, 3, and 4, we learned that a material's composition, nature of bonding, crystal structure, and defects have a profound influence on the strength and ductility of metallic materials. In this chapter, we will begin to evaluate other factors that affect the mechanical properties of materials, such as how lower temperatures can cause many metals and plastics to become brittle. Lower temperatures contributed to the brittleness of the plastic used for O-rings in the solid rocket boosters, causing the 1986 *Challenger* accident. In 2003, the space shuttle *Columbia* was lost because of the impact of debris on the ceramic tiles and failure of carbon-carbon composites. Similarly, the special chemistry of the steel used on the *Titanic* and the stresses associated with the fabrication and embrittlement of this steel when subjected to lower temperatures have been identified as factors contributing to the failure of the ship's hull. Some researchers have shown that weak rivets and design flaws also contributed to the failure.

The main goal of this chapter is to introduce the basic concepts associated with mechanical properties. We will learn terms such as hardness, stress, strain, elastic and plastic deformation, viscoelasticity, and strain rate. We will also review some of the testing procedures that engineers use to evaluate many of these properties. These concepts will be discussed using illustrations from real-world applications.

6-1 Technological Significance

In many of today's emerging technologies, the primary emphasis is on the mechanical properties of the materials used. For example, in aircraft manufacturing, aluminum alloys or carbon-reinforced composites used for aircraft components must be lightweight, strong, and able to withstand cyclic mechanical loading for a long and predictable period of time. Steels used in the construction of structures such as buildings and bridges must have adequate strengths so that these structures can be built without compromising safety. The plastics used for manufacturing pipes, valves, flooring, and the like also must have adequate mechanical strength. Materials such as pyrolytic graphite or cobalt chromium alloys, used for prosthetic heart valves, must not fail. Similarly, the performance of baseballs, cricket bats, tennis rackets, golf clubs, skis, and other sports equipment depends not only on the strength and weight of the materials used, but also on their ability to perform under "impact" loading. The importance of mechanical properties is easy to appreciate in many of these "load-bearing" applications.

In many applications, the mechanical properties of the material play an important role, even though the primary function is electrical, magnetic, optical, or biological. For example, an optical fiber must have a certain level of strength to withstand the stresses encountered in its application. A biocompatible titanium alloy used for a bone implant must have enough strength and toughness to survive in the human body for many years without failure. A scratch-resistant coating on optical lenses must resist mechanical abrasion. An aluminum alloy or a glass-ceramic substrate used as a base for building magnetic hard drives must have sufficient mechanical strength so that it will not break or crack during operation that requires rotation at high speeds. Similarly, electronic packages used to house semiconductor chips and the thin-film structures created on the semiconductor chip must be able to withstand stresses encountered in various applications, as well as those encountered during the heating and cooling of electronic devices. The mechanical robustness of small devices fabricated using nanotechnology is also important. Float glass used in automotive and building applications must have sufficient strength and shatter resistance. Many components designed from plastics, metals, and ceramics must not only have adequate toughness and strength at room temperature but also at relatively high and low temperatures.

For load-bearing applications, engineered materials are selected by matching their mechanical properties to the design specifications and service conditions required of the component. The first step in the selection process requires an analysis of the material's application to determine its most important characteristics. Should it be strong, stiff, or ductile? Will it be subjected to an application involving high stress or sudden intense force, high stress at elevated temperature, cyclic stresses, and/or corrosive or abrasive conditions? Once we know the required properties, we can make a preliminary selection of the appropriate material using various databases. We must, however, know how the properties listed in the handbook are obtained, know what the properties mean, and realize that the properties listed are obtained from idealized tests that may not apply exactly to real-life engineering applications. Materials with the same nominal chemical composition and other properties can show significantly different mechanical properties as dictated by microstructure. Furthermore, changes in temperature; the cyclical nature of stresses applied; the chemical changes due to oxidation, corrosion, or erosion; microstructural changes due to temperature; the effect of possible defects introduced during machining operations (e.g., grinding, welding, cutting, etc.); or other factors can also have a major effect on the mechanical behavior of materials.

The mechanical properties of materials must also be understood so that we can process materials into useful shapes using deformation processing techniques. **Deformation processing** requires a detailed understanding of the mechanical properties of materials at different temperatures and conditions of loading. We must also understand how materials processing may change materials properties, e.g., by making a metal stronger or weaker than it was prior to processing.

In the sections that follow, we define terms that are used to describe the mechanical properties of engineered materials. Different tests used to determine mechanical properties of materials are discussed.

6-2 Terminology for Mechanical Properties

There are different types of forces or "stresses" that are encountered in dealing with mechanical properties of materials. In general, we define stress as the force acting per unit area over which the force is applied. Tensile, compressive, and shear stresses are illustrated in Figure 6-1(a). **Strain** is defined as the change in dimension per unit length. Stress is typically expressed in psi (pounds per square inch) or Pa (pascals, newtons per square meter). Strain has no dimensions and is often expressed as in./in. or cm/cm.

Tensile and compressive stresses are normal stresses. A normal stress arises when the applied force acts perpendicular to the area of interest. Tension causes elongation in the direction of the applied force, whereas compression causes shortening. A shear stress arises when the applied force acts in a direction parallel to the area of interest. Many load-bearing applications involve tensile or compressive stresses. Shear stresses are often encountered in the processing of materials using such techniques as polymer extrusion. Shear stresses are also found in structural applications. Note that even a simple tensile stress applied along one direction will cause a shear stress in other directions (e.g., see Schmid's law, Chapter 4).

Elastic strain is defined as fully recoverable strain resulting from an applied stress. The strain is "elastic" if it develops instantaneously (i.e., the strain occurs as soon as the force is applied), remains as long as the stress is applied, and is recovered when the force is withdrawn. A material subjected to an elastic strain does not show any permanent deformation (i.e., it returns to its original shape after the force or stress is removed). Consider stretching a stiff metal spring by a small amount and letting go. If the spring immediately returns to its original dimensions, the strain developed in the spring was elastic.

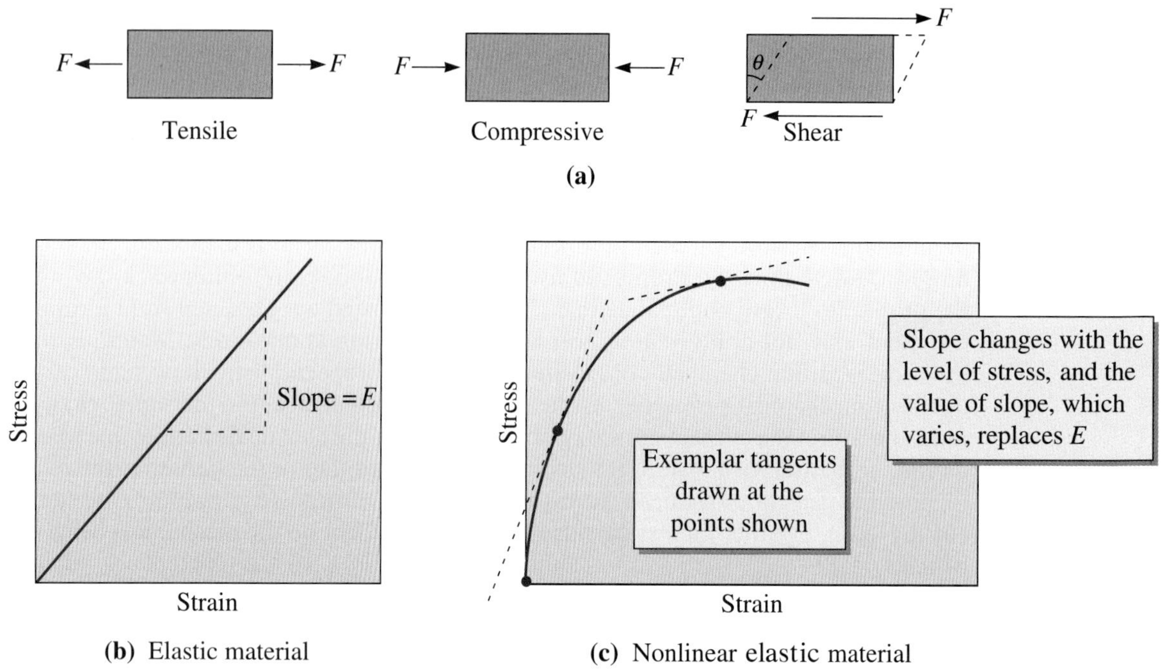

Figure 6-1 (a) Tensile, compressive, and shear stresses. F is the applied force. (b) Illustration showing how Young's modulus is defined for an elastic material. (c) For nonlinear elastic materials, we use the slope of a tangent as a varying quantity that replaces the Young's modulus.

In many materials, elastic stress and elastic strain are linearly related. The slope of a tensile stress strain curve in the linear regime defines the **Young's modulus** or **modulus of elasticity** (E) of a material [Figure 6-1(b)]. The units of E are measured in pounds per square inch (psi) or pascals (Pa) (same as those of stress). Large elastic deformations are observed in **elastomers** (e.g., natural rubber, silicones), for which the relationship between elastic strain and stress is nonlinear. In elastomers, the large elastic strain is related to the coiling and uncoiling of spring-like molecules (see Chapter 16). In dealing with such materials, we use the slope of the tangent at any given value of stress or strain as a varying quantity that replaces the Young's modulus [Figure 6-1(c)]. We define the **shear modulus** (G) as the slope of the linear part of the shear stress-shear strain curve.

Permanent or **plastic deformation** in a material is known as **plastic strain**. In this case, when the stress is removed, the material does *not* go back to its original shape. A dent in a car is plastic deformation! Note that the word "plastic" here does not refer to strain in a plastic (polymeric) material, but rather to permanent strain in any material.

The rate at which strain develops in a material is defined as the **strain rate**. Units of strain rate are s^{-1}. You will learn later in this chapter that the rate at which a material is deformed is important from a mechanical properties perspective. Many materials considered to be ductile behave as brittle solids when the strain rates are high. Silly Putty® (a silicone polymer) is an example of such a material. When the strain rates are low, Silly Putty® can show significant ductility. When stretched rapidly (at high strain rates), we do not allow the untangling and extension of the large polymer molecules and, hence, the material snaps. When materials are subjected to high strain rates, we refer to this type of loading as **impact loading**.

A **viscous material** is one in which the strain develops over a period of time and the material does not return to its original shape after the stress is removed. The development of strain takes time and is not in phase with the applied stress. Also, the material will remain deformed when the applied stress is removed (i.e., the strain will be plastic). A **viscoelastic** (or **anelastic**) material can be thought of as a material with a response between that of a viscous material and an elastic material. The term "anelastic" is typically used for metals, while the term "viscoelastic" is usually associated with polymeric materials. Many plastics (solids and molten) are viscoelastic. A common example of a viscoelastic material is Silly Putty®.

In a viscoelastic material, the development of a permanent strain is similar to that in a viscous material. Unlike a viscous material, when the applied stress is removed, part of the strain in a viscoelastic material will recover over a period of time. Recovery of strain refers to a change in shape of a material after the stress causing deformation is removed. A qualitative description of development of strain as a function of time in relation to an applied force in elastic, viscous, and viscoelastic materials is shown in Figure 6-2. In viscoelastic materials held under constant strain, if we wait, the level of stress decreases over a period of time. This is known as **stress relaxation**. Recovery of strain and stress relaxation are different terms and should not be confused. A common example of stress relaxation is provided by the nylon strings in a tennis racket. We know that the level of stress, or the "tension," as the tennis players call it, decreases with time.

In the sections that follow, we will discuss different mechanical properties of solid materials and some of the testing methods to evaluate these properties.

6-3 The Tensile Test: Use of the Stress Strain Diagram

The tensile test is popular since the properties obtained can be applied to design different components. The tensile test measures the resistance of a material to a static or slowly applied force. The strain rates in a tensile test are typically small (10^{-4} to 10^{-2} s^{-1}). A test setup is shown in Figure 6-3; a typical specimen has a diameter of 0.505 in. and a gage length of 2 in. The specimen is placed in the testing machine and a force F, called the **load**, is applied. A universal testing machine on which tensile and compressive tests can be performed often is used. A **strain gage** or **extensometer** is used to measure the amount that the specimen stretches between the gage marks when the force is applied. Thus, the change in length of the specimen (Δl) is measured with respect to the original length (l_0). Information concerning the strength, Young's modulus, and ductility of a material can be obtained from such a tensile test. Typically, a tensile test is conducted on metals, alloys, and plastics. Tensile tests can be used for ceramics; however, these are not very popular because the sample may fracture while it is being aligned. The following discussion mainly applies to tensile testing of metals and alloys. We will briefly discuss the stress strain behavior of polymers as well.

Figure 6-4 shows *qualitatively* the stress strain curves for a typical (a) metal, (b) thermoplastic material, (c) elastomer, and (d) ceramic (or glass) under relatively small strain rates. The scales in this figure are qualitative and different for each material. In practice, the actual magnitude of stresses and strains will be very different. The temperature of the plastic material is assumed to be above its **glass-transition temperature** (T_g). The temperature of the metal is assumed to be room temperature. Metallic and thermoplastic materials show an initial elastic region followed by a nonlinear plastic region. A separate curve for elastomers (e.g., rubber or silicones) is also included since the behavior of these materials is different from other polymeric materials. For elastomers, a large portion of the deformation is elastic and nonlinear. On the other hand, ceramics and glasses show only a linear elastic region and almost no plastic deformation at room temperature.

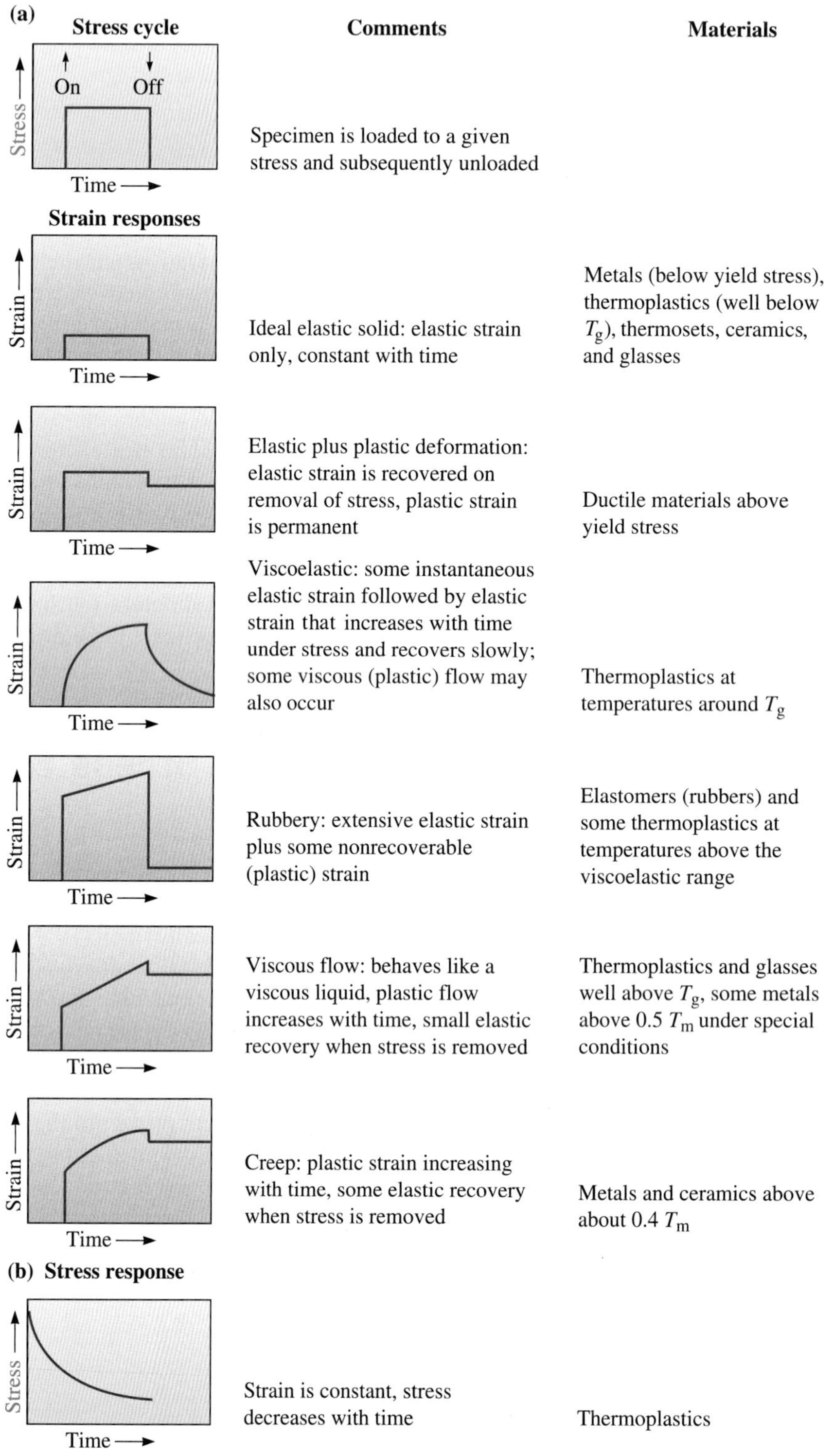

Figure 6-2 (a) Various types of strain response to an imposed stress where T_g = glass-transition temperature and T_m = melting point. [*Based on* Materials Principles and Practice, *by C. Newey and G. Weaver (Eds.), Elsevier, 1991.*] (b) Stress relaxation in a viscoelastic material. *Note the vertical axis is stress.* Strain is constant.

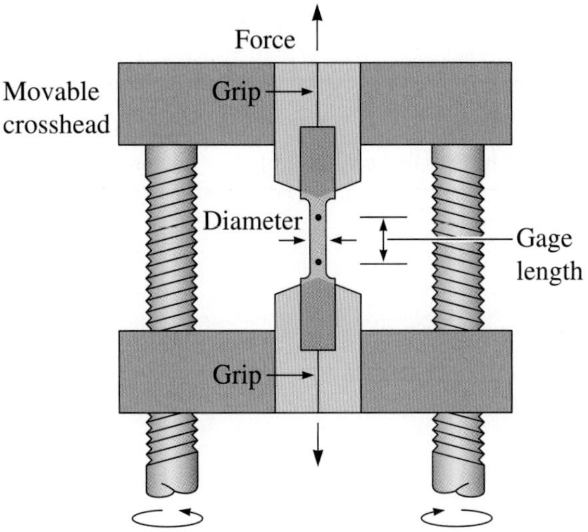

Figure 6-3 A unidirectional force is applied to a specimen in the tensile test by means of the moveable crosshead. The crosshead movement can be performed using screws or a hydraulic mechanism.

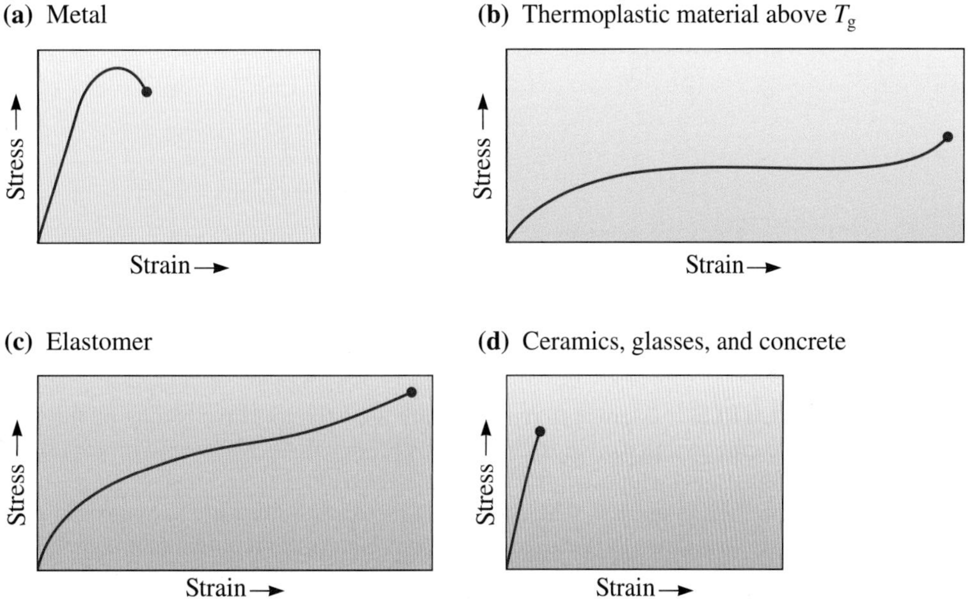

Figure 6-4 Tensile stress strain curves for different materials. Note that these are *qualitative*. The magnitudes of the stresses and strains should not be compared. Failure occurs at the points marked by circles.

When a tensile test is conducted, the data recorded includes load or force as a function of change in length (Δl) as shown in Table 6-1 for an aluminum alloy test bar. These data are then subsequently converted into stress and strain. The stress strain curve is analyzed further to extract properties of materials (e.g., Young's modulus, yield strength, etc.).

Engineering Stress and Strain The results of a single test apply to all sizes and cross-sections of specimens for a given material if we convert the force to stress and the distance between gage marks to strain. If force and displacement are used, the results depend on the sample geometry (i.e., larger specimens require higher forces and experience greater extensions than smaller specimens of the same material). The results for

Table 6-1 The results of a tensile test of a 0.505-in. diameter aluminum alloy test bar, initial length (l_0) = 2 in.

		Calculated	
Load (lb)	Change in Length (in.)	Stress (psi)	Strain (in./in.)
0	0.000	0	0
1,000	0.001	4,993	0.0005
3,000	0.003	14,978	0.0015
5,000	0.005	24,963	0.0025
7,000	0.007	34,948	0.0035
7,500	0.030	37,445	0.0150
7,900	0.080	39,442	0.0400
8,000 (maximum load)	0.120	39,941	0.0600
7,950	0.160	39,691	0.0800
7,600 (fracture)	0.205	37,944	0.1025

stress and strain are the same for the same material regardless of specimen size. **Engineering stress** and **engineering strain** are defined by the following equations:

$$\text{Engineering stress} = S = \frac{F}{A_0} \tag{6-1}$$

$$\text{Engineering strain} = e = \frac{\Delta l}{l_0} \tag{6-2}$$

where A_0 is the *original* cross-sectional area of the specimen before the test begins, l_0 is the *original* distance between the gage marks, and Δl is the change in length after force F is applied. The conversions from load and sample length to stress and strain are included in Table 6-1. The stress strain curve (Figure 6-5) is used to record the results of a tensile test.

Units Many different units are used to report the results of the tensile test. The most common units for stress are pounds per square inch (psi) and megapascals (MPa). The units for strain include inch/inch, centimeter/centimeter, and meter/meter, and thus, strain is often written as unitless. The conversion factors for stress are summarized in Table 6-2. Because strain is dimensionless, no conversion factors are required to change the system of units.

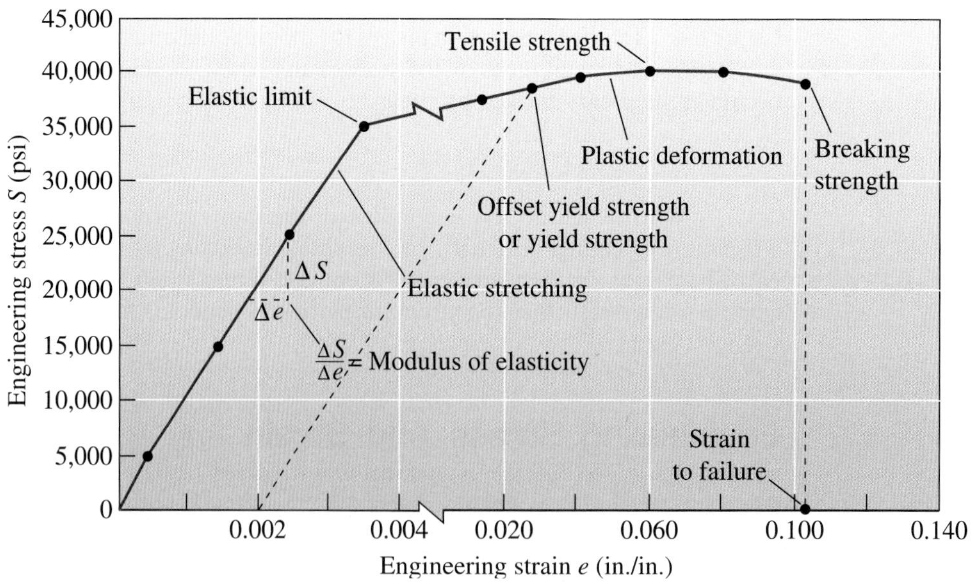

Figure 6-5 The engineering stress strain curve for an aluminum alloy from Table 6-1.

Table 6-2 Units and conversion factors

1 pound (lb) = 4.448 Newtons (N)
1 psi = pounds per square inch
1 MPa = megapascal = meganewtons per square meter (MN/m^2)
= newtons per square millimeter (N/mm^2) = 10^6 Pa
1 GPa = 1000 MPa = gigapascal
1 ksi = 1000 psi = 6.895 MPa
1 psi = 0.006895 MPa
1 MPa = 0.145 ksi = 145 psi

Example 6-1 *Tensile Testing of Aluminum Alloy*

Convert the load and change in length data in Table 6-1 to engineering stress and strain and plot a stress strain curve.

SOLUTION

For the 1000-lb load:

$$S = \frac{F}{A_0} = \frac{1000 \text{ lb}}{(\pi/4)(0.505 \text{ in.})^2} = 4{,}993 \text{ psi}$$

$$e = \frac{\Delta l}{l_0} = \frac{0.001 \text{ in.}}{2.000 \text{ in.}} = 0.0005 \text{ in./in.}$$

The results of similar calculations for each of the remaining loads are given in Table 6-1 and are plotted in Figure 6-5.

Example 6-2 *Design of a Suspension Rod*

An aluminum rod is to withstand an applied force of 45,000 pounds. The engineering stress strain curve for the aluminum alloy to be used is shown in Figure 6-5. To ensure safety, the maximum allowable stress on the rod is limited to 25,000 psi, which is below the yield strength of the aluminum. The rod must be at least 150 in. long but must deform elastically no more than 0.25 in. when the force is applied. Design an appropriate rod.

SOLUTION

From the definition of engineering strain,

$$e = \frac{\Delta l}{l_0}$$

For a rod that is 150 in. long, the strain that corresponds to an extension of 0.25 in. is

$$e = \frac{0.25 \text{ in.}}{150 \text{ in.}} = 0.00167$$

According to Figure 6-5, this strain is purely elastic, and the corresponding stress value is approximately 17,000 psi, which is below the 25,000 psi limit. We use the definition of engineering stress to calculate the required cross-sectional area of the rod:

$$S = \frac{F}{A_0}$$

Note that the stress must not exceed 17,000 psi, or consequently, the deflection will be greater than 0.25 in. Rearranging,

$$A_0 = \frac{F}{S} = \frac{45{,}000 \text{ lb}}{17{,}000 \text{ psi}} = 2.65 \text{ in.}^2$$

The rod can be produced in various shapes, provided that the cross-sectional area is 2.65 in.^{2} For a round cross section, the minimum diameter to ensure that the stress is not too high is

$$A_0 = \frac{\pi d^2}{4} = 2.65 \text{ in.}^2 \quad \text{or} \quad d = 1.84 \text{ in.}$$

Thus, one possible design that meets all of the specified criteria is a suspension rod that is 150 in. long with a diameter of 1.84 in.

6-4 Properties Obtained from the Tensile Test

Yield Strength As we apply stress to a material, the material initially exhibits elastic deformation. The strain that develops is completely recovered when the applied stress is removed. As we continue to increase the applied stress, the material eventually "yields" to the applied stress and exhibits both elastic and plastic deformation. The critical stress value needed to initiate plastic deformation is defined as the **elastic limit** of the material. In metallic materials, this is usually the stress required for dislocation motion, or slip, to be initiated. In polymeric materials, this stress will correspond to disentanglement of polymer molecule chains or sliding of chains past each other. The **proportional limit** is defined as the level of stress above which the relationship between stress and strain is not linear.

In most materials, the elastic limit and proportional limit are quite close; however, neither the elastic limit nor the proportional limit values can be determined precisely. Measured values depend on the sensitivity of the equipment used. We, therefore, define them at an **offset strain value** (typically, but not always, 0.002 or 0.2%). We then draw a line parallel to the linear portion of the engineering stress strain curve starting at this offset value of strain. The stress value corresponding to the intersection of this line and the engineering stress strain curve is defined as the **offset yield strength**, also often stated as the **yield strength**. The 0.2% offset yield strength for gray cast iron is 40,000 psi as shown in Figure 6-6(a). Engineers normally prefer to use the offset yield strength for design purposes because it can be reliably determined.

For some materials, the transition from elastic deformation to plastic flow is rather abrupt. This transition is known as the **yield point phenomenon**. In these materials,

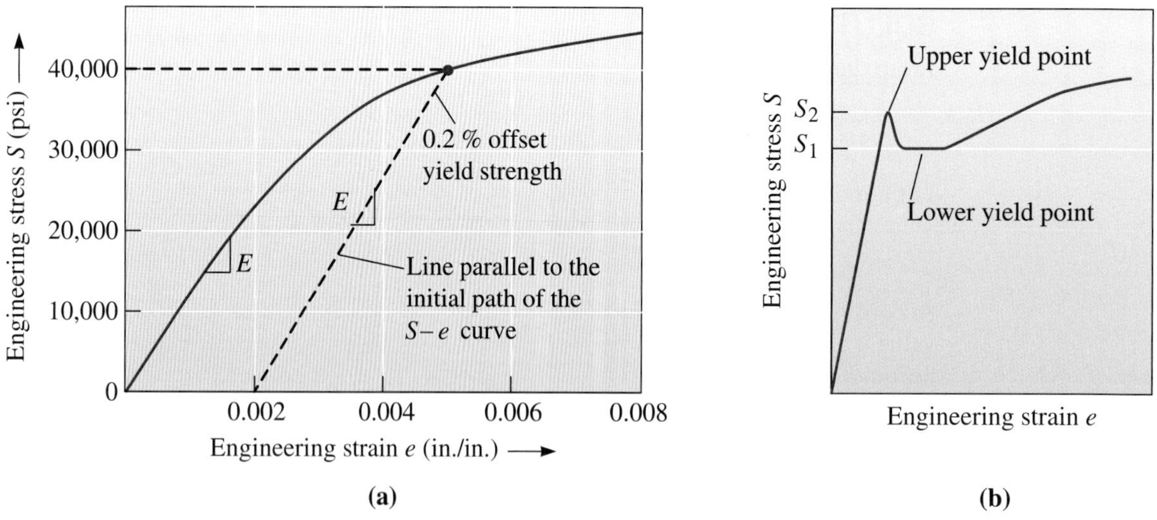

Figure 6-6 (a) Determining the 0.2% offset yield strength and (b) upper and lower yield point behavior in a low-carbon steel.

as plastic deformation begins, the stress value drops first from the *upper yield point* (S_2) [Figure 6-6(b)]. The stress value then oscillates around an average value defined as the *lower yield point* (S_1). For these materials, the yield strength is usually defined from the 0.2% strain offset.

The stress strain curve for certain low-carbon steels displays the yield point phenomenon [Figure 6-6(b)]. The material is expected to plastically deform at stress S_1; however, small interstitial atoms clustered around the dislocations interfere with slip and raise the yield point to S_2. Only after we apply the higher stress S_2 do the dislocations slip. After slip begins at S_2, the dislocations move away from the clusters of small atoms and continue to move very rapidly at the lower stress S_1.

When we design parts for load-bearing applications, we prefer little or no plastic deformation. As a result, we must select a material such that the design stress is considerably lower than the yield strength at the temperature at which the material will be used. We can also make the component cross-section larger so that the applied force produces a stress that is well below the yield strength. On the other hand, when we want to shape materials into components (e.g., take a sheet of steel and form a car chassis), we need to apply stresses that are well above the yield strength.

Tensile Strength The stress obtained at the highest applied force is the **tensile strength** (S_{UTS}), which is the maximum stress on the engineering stress strain curve. This value is also commonly known as the **ultimate tensile strength**. In many ductile materials, deformation does not remain uniform. At some point, one region deforms more than others and a large local decrease in the cross-sectional area occurs (Figure 6-7). This locally deformed region is called a "neck." This phenomenon is known as **necking**. Because the cross-sectional area becomes smaller at this point, a lower force is required to continue its deformation, and the engineering stress, calculated from the *original* area A_0, decreases. The tensile strength is the stress at which necking begins in ductile metals. In compression testing, the materials will bulge; thus necking is seen only in a tensile test.

Figure 6-8 shows typical yield strength values for various engineering materials. Ultra-pure metals have a yield strengths of $\sim 1 - 10$ MPa. On the other hand, the yield strength of alloys is higher. Strengthening in alloys is achieved using various mechanisms (e.g., grain size, solid-solution formation, strain hardening, etc.). The yield strength of a particular metal or alloy is usually the same for tension and compression. The yield strengths of plastics and elastomers are generally lower than those for metals and alloys, ranging up to about $10 - 100$ MPa. The values for ceramics (Figure 6-8) are for compressive strength

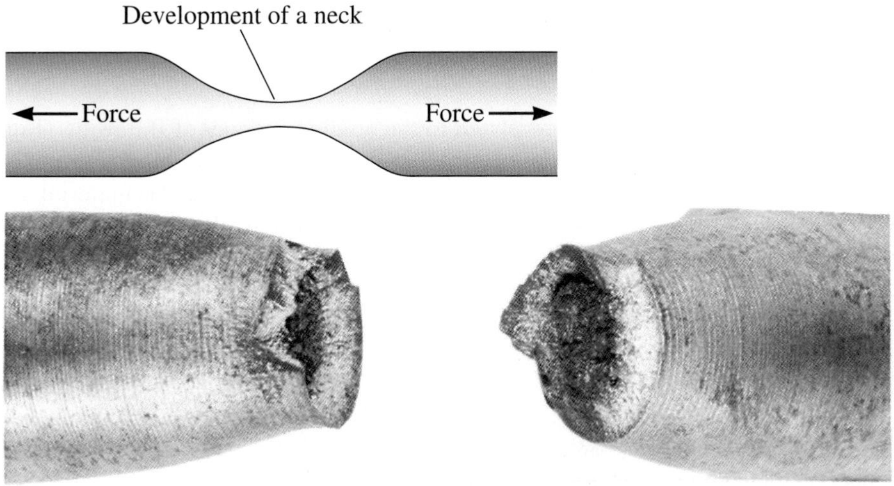

Figure 6-7 Localized deformation of a ductile material during a tensile test produces a necked region. The micrograph shows a necked region in a fractured sample. (*Reprinted Courtesy of Don Askeland.*)

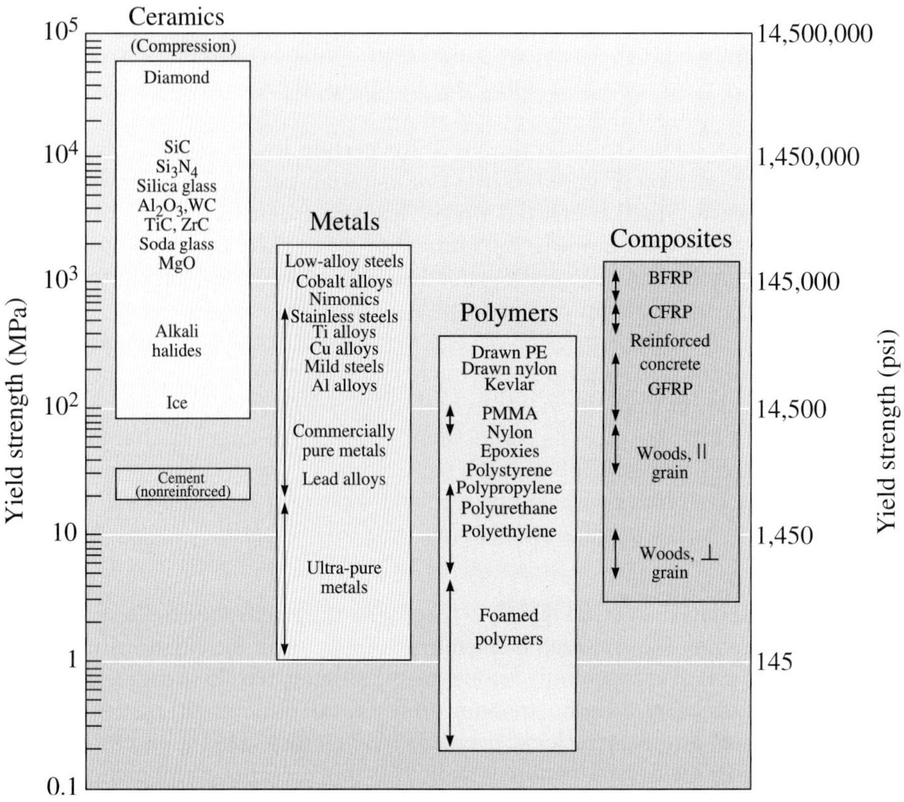

Figure 6-8 Typical yield strength values for different engineering materials. Note that values shown are in MPa and psi. (*Based on* Engineering Materials I, *Second Edition, M. F. Ashby and D. R. H. Jones, Elsevier, 1996.*)

(obtained using a hardness test). The tensile strengths of most ceramics are much lower (~100–200 MPa). The tensile strength of glasses is about ~70 MPa and depends on surface flaws.

Elastic Properties

The modulus of elasticity, or Young's modulus (E), is the slope of the stress strain curve in the elastic region. This relationship between stress and strain in the elastic region is known as **Hooke's Law**:

$$E = \frac{S}{e} \tag{6-3}$$

The modulus is closely related to the binding energies of the atoms (Figure 2-18). A steep slope in the force-distance graph at the equilibrium spacing (Figure 2-19) indicates that high forces are required to separate the atoms and cause the material to stretch elastically. Thus, the material has a high modulus of elasticity. Binding forces, and thus the modulus of elasticity, are typically higher for high-melting-point materials (Table 6-3). In metallic materials, the modulus of elasticity is considered a microstructure *insensitive* property since the value is dominated by the stiffness of atomic bonds. Grain size or other microstructural features do not have a large effect on the Young's modulus. Note that Young's modulus does depend on such factors as orientation of a single crystal material (i.e., it depends upon crystallographic direction). For ceramics, the Young's modulus depends on the level of porosity. The Young's modulus of a composite depends upon the stiffness and amounts of the individual components.

The **stiffness** of a component is proportional to its Young's modulus. (The stiffness also depends on the component dimensions.) A component with a high modulus of elasticity will show much smaller changes in dimensions if the applied stress causes only elastic deformation when compared to a component with a lower elastic modulus. Figure 6-9

Table 6-3 Elastic properties and melting temperature (T_m) of selected materials

Material	T_m (°C)	E (psi)	Poisson's ratio (ν)
Pb	327	2.0×10^6	0.45
Mg	650	6.5×10^6	0.29
Al	660	10.0×10^6	0.33
Cu	1085	18.1×10^6	0.36
Fe	1538	30.0×10^6	0.27
W	3410	59.2×10^6	0.28
Al_2O_3	2020	55.0×10^6	0.26
Si_3N_4		44.0×10^6	0.24

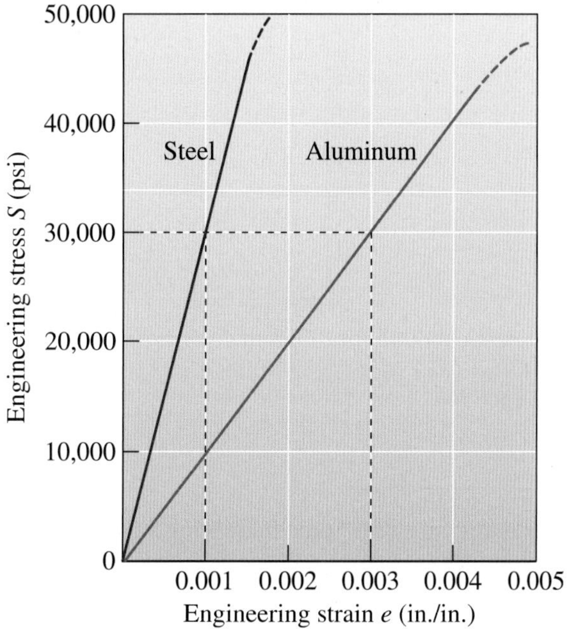

Figure 6-9 Comparison of the elastic behavior of steel and aluminum. For a given stress, aluminum deforms elastically three times as much as does steel (i.e., the elastic modulus of aluminum is about three times lower than that of steel).

compares the elastic behavior of steel and aluminum. If a stress of 30,000 psi is applied to each material, the steel deforms elastically 0.001 in./in.; at the same stress, aluminum deforms 0.003 in./in. The elastic modulus of steel is about three times higher than that of aluminum.

Figure 6-10 shows the ranges of elastic moduli for various engineering materials. The modulus of elasticity of plastics is much smaller than that for metals or ceramics and glasses. For example, the modulus of elasticity of nylon is 2.7 GPa ($\sim 0.4 \times 10^6$ psi); the modulus of glass fibers is 72 GPa ($\sim 10.5 \times 10^6$ psi). The Young's modulus of composites such as glass fiber-reinforced composites (GFRC) or carbon fiber-reinforced composites (CFRC) lies between the values for the matrix polymer and the fiber phase (carbon or glass fibers) and depends upon their relative volume fractions. The Young's modulus of many alloys and ceramics is higher, generally ranging up to 400 GPa (~60,000 ksi). Ceramics, because of the strength of ionic and covalent bonds, typically have the highest elastic moduli.

Poisson's ratio, ν, relates the longitudinal elastic deformation produced by a simple tensile or compressive stress to the lateral deformation that occurs simultaneously:

$$\nu = -\frac{e_{\text{lateral}}}{e_{\text{longitudinal}}} \tag{6-4}$$

For many metals in the elastic region, the Poisson's ratio is typically about 0.3 (Table 6-3). During a tensile test, the ratio increases beyond yielding to about 0.5, since during plastic

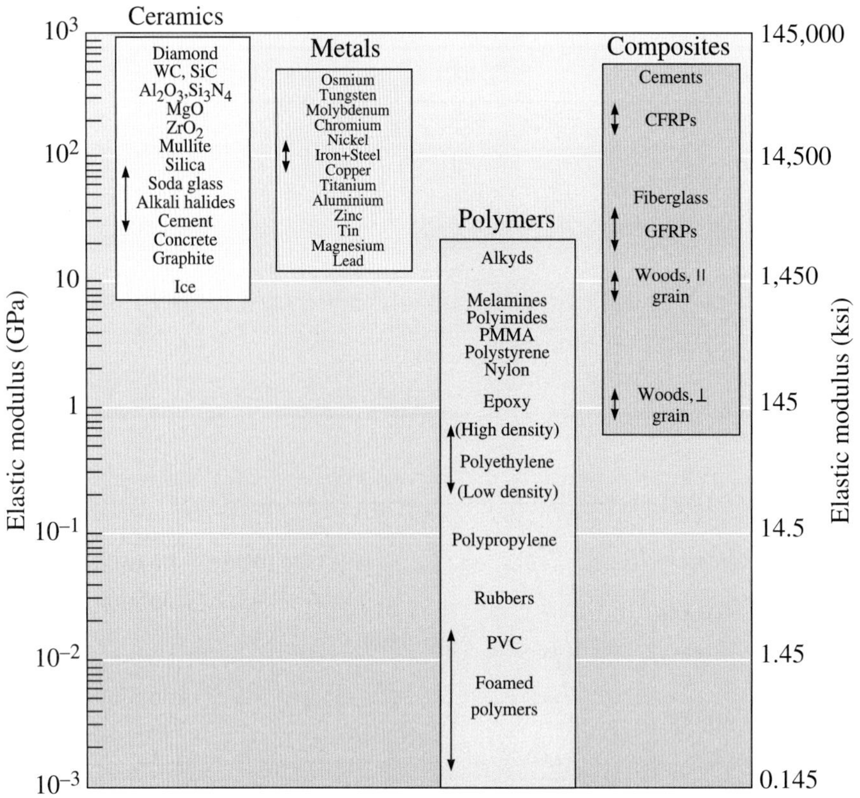

Figure 6-10 Range of elastic moduli for different engineering materials. Note: values are shown in GPa and ksi. (*Based on* Engineering Materials I, *Second Edition, M. F. Ashby and D. R. H. Jones, Elsevier, 1996.*)

deformation, volume remains constant. Some interesting structures, such as some honeycomb structures and foams, exhibit negative Poisson's ratios.

The **modulus of resilience** (E_r), the area contained under the elastic portion of a stress strain curve, is the elastic energy that a material absorbs prior to yielding and subsequently releases when the load is removed. For linear elastic behavior:

$$E_r = \left(\frac{1}{2}\right)(\text{yield strength})(\text{strain at yielding}) \tag{6-5}$$

The ability of a spring or a golf ball to perform satisfactorily depends on a high modulus of resilience.

Tensile Toughness

The energy absorbed by a material prior to fracture is known as **tensile toughness** and is sometimes measured as the area under the true stress strain curve (also known as the **work of fracture**). We will define true stress and true strain in Section 6-5. Since it is simpler to measure engineering stress strain, engineers often equate tensile toughness to the area under the engineering stress strain curve.

Example 6-3 *Young's Modulus of an Aluminum Alloy*

Calculate the modulus of elasticity of the aluminum alloy for which the engineering stress strain curve is shown in Figure 6-5. Calculate the length of a bar of initial length 50 in. when a tensile stress of 30,000 psi is applied.

SOLUTION

When a stress of 34,948 psi is applied, a strain of 0.0035 in./in. is produced. Thus,

$$\text{Modulus of elasticity} = E = \frac{S}{e} = \frac{34{,}948 \text{ psi}}{0.0035} = 10 \times 10^6 \text{ psi}$$

Note that any combination of stress and strain in the elastic region will produce this result. From Hooke's Law,

$$e = \frac{S}{E} = \frac{30{,}000 \text{ psi}}{10 \times 10^6 \text{ psi}} = 0.003 \text{ in./in.}$$

From the definition of engineering strain,

$$e = \frac{\Delta l}{l_0}$$

Thus,

$$\Delta l = e(l_0) = (0.003 \text{ in./in.})(50 \text{ in.}) = 0.15 \text{ in.}$$

When the bar is subjected to a stress of 30,000 psi, the total length is given by

$$l = \Delta l + l_0 = 0.15 \text{ in.} + 50 \text{ in.} = 50.15 \text{ in.}$$

Ductility Ductility is the ability of a material to be permanently deformed without breaking when a force is applied. There are two common measures of ductility. The **percent elongation** quantifies the permanent plastic deformation at failure (i.e., the elastic deformation recovered after fracture is not included) by measuring the distance between gage marks on the specimen before and after the test. Note that the strain after failure is smaller than the strain at the breaking point, because the elastic strain is recovered when the load is removed. The percent elongation can be written as

$$\% \text{ Elongation} = \frac{l_f - l_0}{l_0} \times 100 \qquad (6\text{-}6)$$

where l_f is the distance between gage marks after the specimen breaks.

A second approach is to measure the percent change in the cross-sectional area before and after the test. The **percent reduction** in area describes the amount of thinning undergone by the specimen during the test:

$$\% \text{ Reduction in area} = \frac{A_0 - A_f}{A_0} \times 100 \qquad (6\text{-}7)$$

where A_f is the final cross-sectional area at the fracture surface after the specimen has failed.

Ductility is important to both designers of load-bearing components and manufacturers of components (bars, rods, wires, plates, I-beams, fibers, etc.) utilizing deformation processing.

Example 6-4 *Ductility of an Aluminum Alloy*

The aluminum alloy in Example 6-1 has a final length after failure of 2.195 in. and a final diameter of 0.398 in. at the fracture surface. Calculate the ductility of this alloy.

SOLUTION

$$\% \text{ Elongation} = \frac{l_f - l_0}{l_0} \times 100 = \frac{2.195 - 2.000}{2.000} \times 100 = 9.75\%$$

$$
\begin{aligned}
\% \text{ Reduction in area} &= \frac{A_0 - A_f}{A_0} \times 100 \\
&= \frac{(\pi/4)(0.505)^2 - (\pi/4)(0.398)^2}{(\pi/4)(0.505)^2} \times 100 \\
&= 37.9\%
\end{aligned}
$$

The final length is less than 2.205 in. (see Table 6-1) because, after fracture, the elastic strain is recovered.

Effect of Temperature

Mechanical properties of materials depend on temperature (Figure 6-11). Yield strength, tensile strength, and modulus of elasticity decrease at higher temperatures, whereas ductility commonly increases. A materials fabricator may wish to deform a material at a high temperature (known as *hot working*) to take advantage of the higher ductility and lower required stress.

We use the term "high temperature" here with a note of caution. A high temperature is defined relative to the melting temperature. Thus, 500°C is a high temperature for aluminum alloys; however, it is a relatively low temperature for the processing of steels. In metals, the yield strength decreases rapidly at higher temperatures due to a decreased dislocation density and an increase in grain size via grain growth or a related process known as recrystallization (both processes are described later in Chapter 8). Similarly, any strengthening that may have occurred due to the formation of ultra-fine precipitates may also decrease as the precipitates begin to either grow in size or dissolve into the matrix. We will discuss these effects in greater detail in later chapters. When temperatures are reduced, many, but not all, metals and alloys become brittle.

Increased temperatures also play an important role in forming polymeric materials and inorganic glasses. In many polymer-processing operations, such as extrusion or the blow-stretch process (Chapter 16), the increased ductility of polymers at higher temperatures is advantageous. Again, a word of caution concerning the use of the term "high temperature." For polymers, the term "high temperature" generally means a temperature higher than the glass-transition temperature (T_g). For our purpose, the glass-transition temperature is a temperature below which materials behave

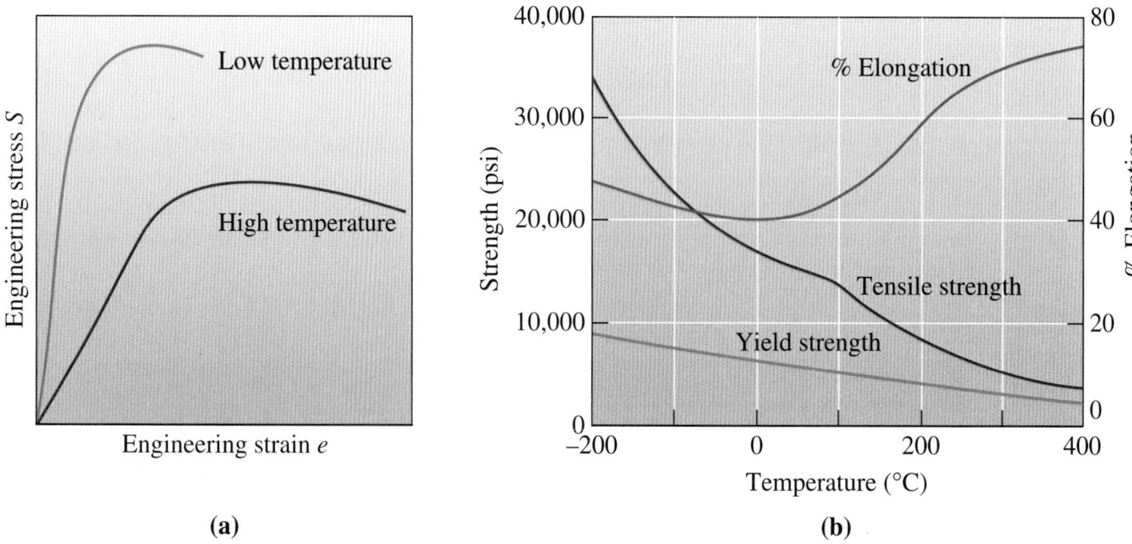

Figure 6-11 The effect of temperature (a) on the stress strain curve and (b) on the tensile properties of an aluminum alloy.

as brittle materials. Above the glass-transition temperature, plastics become ductile. The glass-transition temperature is not a fixed temperature, but depends on the rate of cooling as well as the polymer molecular weight distribution. Many plastics are ductile at room temperature because their glass-transition temperatures are *below* room temperature. To summarize, many polymeric materials will become harder and more brittle as they are exposed to temperatures that are below their glass-transition temperatures. The reasons for loss of ductility at lower temperatures in polymers and metallic materials are different; however, this is a factor that played a role in the failures of the *Titanic* in 1912 (metal failure) and the *Challenger* in 1986 (polymer failure).

Ceramic and glassy materials are generally considered brittle at room temperature. As the temperature increases, glasses can become more ductile. As a result, glass processing (e.g., fiber drawing or bottle manufacturing) is performed at high temperatures.

6-5 True Stress and True Strain

The decrease in engineering stress beyond the tensile strength on an engineering stress strain curve is related to the definition of engineering stress. We used the original area A_0 in our calculations, but this is not precise because the area continually changes. We define true stress and true strain by the following equations:

$$\text{True stress} = \sigma = \frac{F}{A} \tag{6-8}$$

$$\text{True strain} = \varepsilon = \int_{l_0}^{l} \frac{dl}{l} = \ln\left(\frac{l}{l_0}\right) \tag{6-9}$$

where A is the instantaneous area over which the force F is applied, l is the instantaneous sample length, and l_0 is the initial length. In the case of metals, plastic deformation is essentially a constant-volume process (i.e., the creation and propagation of dislocations results in a negligible volume change in the material). When the constant volume assumption holds, we can write

$$A_0 l_0 = Al \text{ or } A = \frac{A_0 l_0}{l} \tag{6-10}$$

and using the definitions of engineering stress S and engineering strain e, Equation 6-8 can be written as

$$\sigma = \frac{F}{A} = \frac{F}{A_0}\left(\frac{l}{l_0}\right) = S\left(\frac{l_0 + \Delta l}{l_0}\right) = S(1 + e) \tag{6-11}$$

It can also be shown that

$$\varepsilon = \ln(1 + e) \tag{6-12}$$

Thus, it is a simple matter to convert between the engineering stress strain and true stress strain systems. Note that the expressions in Equations 6-11 and 6-12 are not valid after the onset of necking, because after necking begins, the distribution of strain along the gage length is not uniform. After necking begins, Equation 6-8 must be used to calculate the true stress and the expression

$$\varepsilon = \ln\left(\frac{A_0}{A}\right) \tag{6-13}$$

must be used to calculate the true strain. Equation 6-13 follows from Equations 6-9 and 6-10. After necking, the instantaneous area A is the cross-sectional area of the neck A_{neck}.

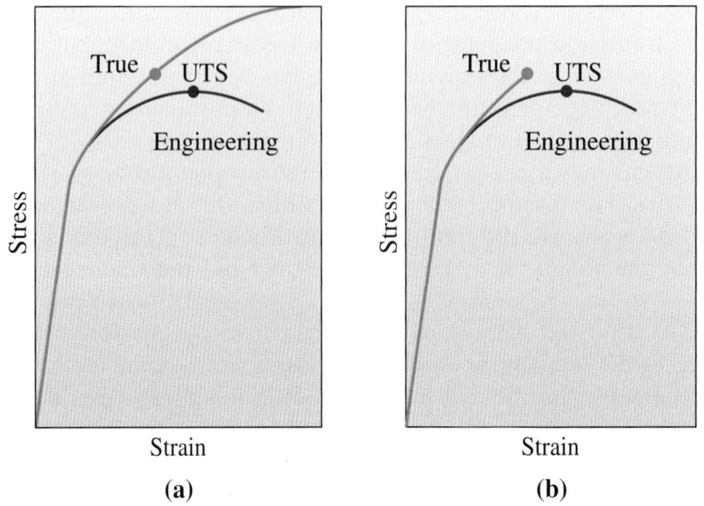

Figure 6-12 (a) The relation between the true stress–true strain diagram and engineering stress–engineering strain diagram. The curves are nominally identical to the yield point. The true stress corresponding to the ultimate tensile strength (UTS) is indicated by circles. (b) Typically true stress strain curves must be truncated at the true stress corresponding to the ultimate tensile strength, since the cross-sectional area at the neck is unknown.

The true stress strain curve is compared to the engineering stress strain curve in Figure 6-12(a). There is no maximum in the true stress–true strain curve.

Note that it is difficult to measure the instantaneous cross-sectional area of the neck. Thus, true stress strain curves are typically truncated at the true stress that corresponds to the ultimate tensile strength, as shown in Figure 6-12(b).

Example 6-5 *True Stress and True Strain*

Compare engineering stress and strain with true stress and strain for the aluminum alloy in Example 6-1 at (a) the maximum load and (b) fracture. The diameter at maximum load is 0.4905 in. and at fracture is 0.398 in.

SOLUTION

(a) At the maximum load,

$$\text{Engineering stress } S = \frac{F}{A_0} = \frac{8000 \text{ lb}}{(\pi/4)(0.505 \text{ in.})^2} = 39{,}941 \text{ psi}$$

$$\text{Engineering strain } e = \frac{\Delta l}{l_0} = \frac{2.120 - 2.000}{2.000} = 0.060 \text{ in./in.}$$

$$\text{True stress} = \sigma = S(1 + e) = 39{,}941 \text{ psi } (1 + 0.060) = 42{,}337 \text{ psi}$$

$$\text{True strain} = \ln(1 + e) = \ln(1 + 0.060) = 0.058 \text{ in./in.}$$

The maximum load is the last point at which the expressions used here for true stress and true strain apply. Note that the same answers are obtained for true stress and strain if the instantaneous dimensions are used:

$$\sigma = \frac{F}{A} = \frac{8000 \text{ lb}}{(\pi/4)(0.4905 \text{ in})^2} = 42{,}337 \text{ psi}$$

$$\varepsilon = \ln\left(\frac{A_0}{A}\right) = \ln\left[\frac{(\pi/4)(0.505 \text{ in.})^2}{(\pi/4)(0.4905 \text{ in.})^2}\right] = 0.058 \text{ in./in.}$$

Up until the point of necking in a tensile test, the engineering stress is less than the corresponding true stress, and the engineering strain is greater than the corresponding true strain.

(b) At fracture,

$$S = \frac{F}{A_0} = \frac{7600 \text{ lb}}{(\pi/4)(0.505 \text{ in.})^2} = 37{,}944 \text{ psi}$$

$$e = \frac{\Delta l}{l_0} = \frac{2.205 - 2.000}{2.000} = 0.103 \text{ in./in.}$$

$$\sigma = \frac{F}{A_f} = \frac{7600 \text{ lb}}{(\pi/4)(0.398 \text{ in.})^2} = 61{,}088 \text{ psi}$$

$$\varepsilon = \ln\left(\frac{A_0}{A_f}\right) = \ln\left[\frac{(\pi/4)(0.505 \text{ in.})^2}{(\pi/4)(0.398 \text{ in.})^2}\right] = 0.476 \text{ in./in.}$$

It was necessary to use the instantaneous dimensions to calculate the true stress and strain, since failure occurs past the point of necking. After necking, the true strain is greater than the corresponding engineering strain.

6-6 The Bend Test for Brittle Materials

In ductile metallic materials, the engineering stress strain curve typically goes through a maximum; this maximum stress is the tensile strength of the material. Failure occurs at a lower engineering stress after necking has reduced the cross-sectional area supporting the load. In more brittle materials, failure occurs at the maximum load, where the tensile strength and breaking strength are the same (Figure 6-13).

In many brittle materials, the normal tensile test cannot easily be performed because of the presence of flaws at the surface. Often, simply placing a brittle material in the grips of the tensile testing machine causes cracking. These materials may be tested using the **bend test** (Figure 6-14). By applying the load at three points and causing bending, a tensile force acts on the material opposite the midpoint. Fracture begins at this location. The **flexural strength**, or **modulus of rupture**, describes the material's strength:

$$\text{Flexural strength for three-point bend test } \sigma_{\text{bend}} = \frac{3FL}{2wh^2} \qquad \text{(6-14)}$$

where F is the fracture load, L is the distance between the two outer points, w is the width of the specimen, and h is the height of the specimen. The flexural strength has units of

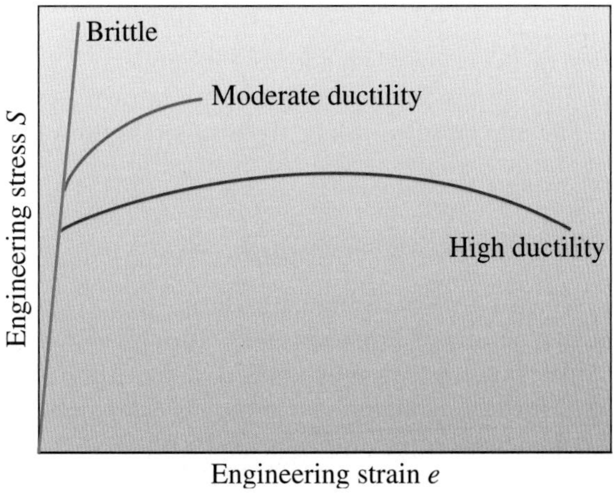

Figure 6-13 The engineering stress strain behavior of brittle materials compared with that of more ductile materials.

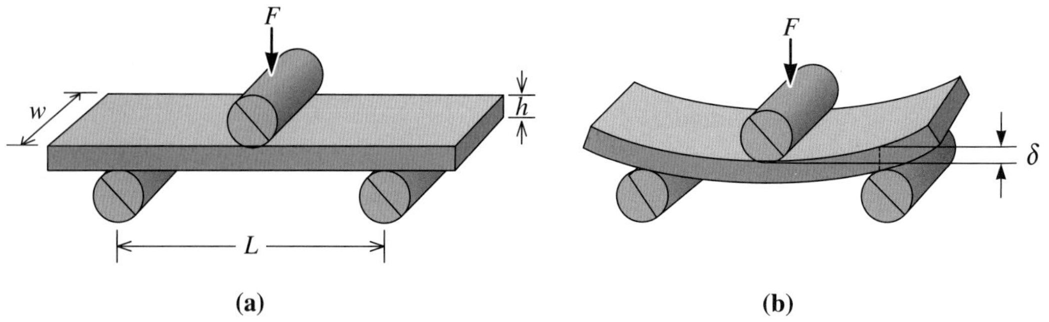

Figure 6-14 (a) The bend test often used for measuring the strength of brittle materials, and (b) the deflection δ obtained by bending.

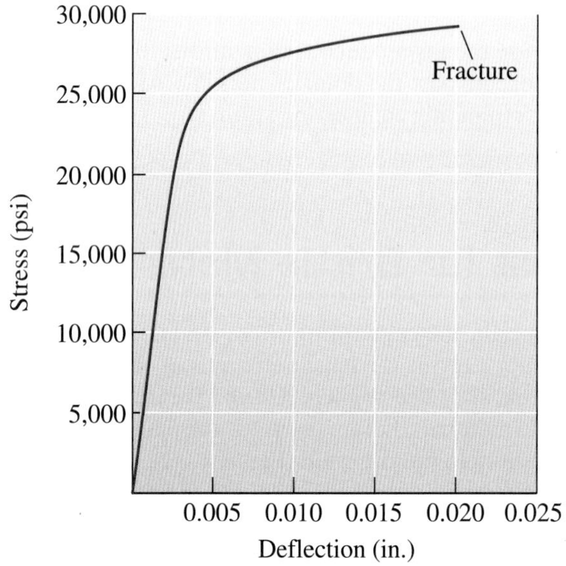

Figure 6-15 Stress-deflection curve for an MgO ceramic obtained from a bend test.

stress. The results of the bend test are similar to stress strain curves; however, the stress is plotted versus deflection rather than versus strain (Figure 6-15). The corresponding bending moment diagram is shown in Figure 6-16(a).

The modulus of elasticity in bending, or the **flexural modulus** (E_{bend}), is calculated as

$$\text{Flexural modulus } E_{bend} = \frac{L^3F}{4wh^3\delta} \tag{6-15}$$

where δ is the deflection of the beam when a force F is applied.

This test can also be conducted using a setup known as the four-point bend test [Figure 6-16(b)]. The maximum stress or flexural stress for a four-point bend test is given by

$$\sigma_{bend} = \frac{3FL}{4wh^2} \tag{6-16}$$

for the specific case in which $L_1 = L/4$ in Figure 6-16(b).

Note that the derivations of Equations 6-14 through 6-16 assume a linear stress strain response (and thus cannot be correctly applied to many polymers). The four-point bend test is better suited for testing materials containing flaws because the bending moment between the inner platens is constant [Figure 6-16(b)].

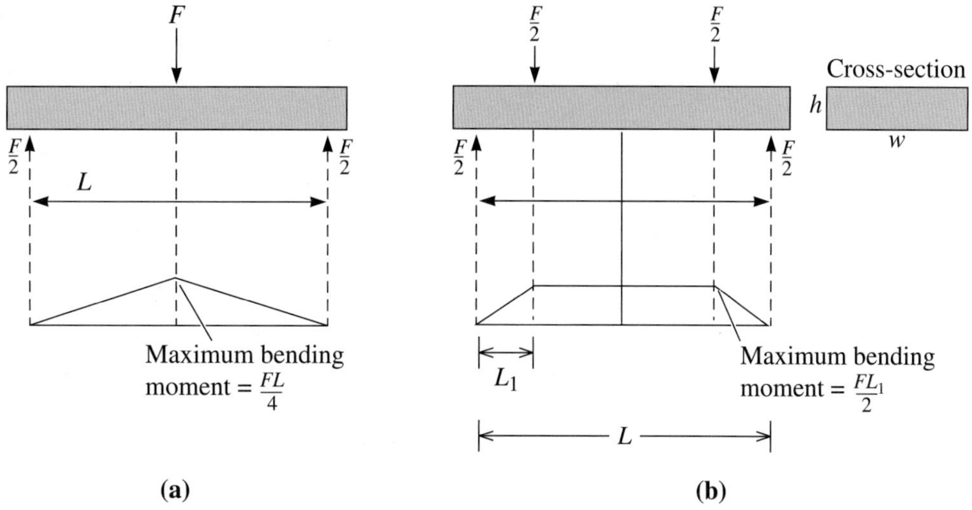

Figure 6-16 (a) Three-point and (b) four-point bend test setup.

Table 6-4 Comparison of the tensile, compressive, and flexural strengths of selected ceramic and composite materials

Material	Tensile Strength (psi)	Compressive Strength (psi)	Flexural Strength (psi)
Polyester-50% glass fibers	23,000	32,000	45,000
Polyester-50% glass fiber fabric	37,000	27,000[a]	46,000
Al_2O_3 (99% pure)	30,000	375,000	50,000
SiC (pressureless-sintered)	25,000	560,000	80,000

[a]*A number of composite materials are quite poor in compression.*

Since cracks and flaws tend to remain closed in compression, brittle materials such as concrete are often incorporated into designs so that only compressive stresses act on the part. Often, we find that brittle materials fail at much higher compressive stresses than tensile stresses (Table 6-4). This is why it is possible to support a fire truck on four coffee cups; however, ceramics have very limited mechanical toughness. Hence, when we drop a ceramic coffee cup, it can break easily.

Example 6-6 *Flexural Strength of Composite Materials*

The flexural strength of a composite material reinforced with glass fibers is 45,000 psi, and the flexural modulus is 18×10^6 psi as determined using a three-point bend test. A sample, which is 0.5 in. wide, 0.375 in. high, and 8 in. long, is supported between two rods 5 in. apart. Determine the force required to fracture the material and the deflection of the sample at fracture, assuming that no plastic deformation occurs.

SOLUTION

Based on the description of the sample, $w = 0.5$ in., $h = 0.375$ in., and $L = 5$ in. From Equation 6-14:

$$45{,}000 \text{ psi} = \frac{3FL}{2wh^2} = \frac{(3)(F)(5 \text{ in.})}{(2)(0.5 \text{ in.})(0.375 \text{ in.})^2} = 106.7F$$

$$F = \frac{45{,}000}{106.7} = 422 \text{ lb}$$

Therefore, the deflection, from Equation 6-15, is

$$18 \times 16^6 \text{ psi} = \frac{L^3F}{4wh^3\delta} = \frac{(5 \text{ in.})^3(422 \text{ lb})}{(4)(0.5 \text{ in.})(0.375 \text{ in.})^3\delta}$$

$$\delta = 0.0278 \text{ in.}$$

In this calculation, we assumed a linear relationship between stress and strain and also that there is no viscoelastic behavior.

6-7 Hardness of Materials

The **hardness test** measures the resistance to penetration of the surface of a material by a hard object. Hardness as a term is not defined precisely. Hardness, depending upon the context, can represent resistance to scratching or indentation and a qualitative measure of the strength of the material. In general, in **macrohardness** measurements, the load applied is ~2 N. A variety of hardness tests have been devised, but the most commonly used are the Rockwell test and the Brinell test. Different indenters used in these tests are shown in Figure 6-17.

In the *Brinell hardness test*, a hard steel sphere (usually 10 mm in diameter) is forced into the surface of the material. The diameter of the impression, typically 2 to 6 mm, is measured and the Brinell hardness number (abbreviated as HB or BHN) is calculated from the following equation:

$$\text{HB} = \frac{2F}{\pi D\left[D - \sqrt{D^2 - D_i^2}\right]} \tag{6-17}$$

where F is the applied load in kilograms, D is the diameter of the indenter in millimeters, and D_i is the diameter of the impression in millimeters. The Brinell hardness has units of kg/mm^2.

The *Rockwell hardness test* uses a small-diameter steel ball for soft materials and a diamond cone, or Brale, for harder materials. The depth of penetration of the indenter is automatically measured by the testing machine and converted to a Rockwell hardness number (HR). Since an optical measurement of the indentation dimensions is not needed, the Rockwell test tends to be more popular than the Brinell test. Several variations of the Rockwell test are used, including those described in Table 6-5. A Rockwell *C* (HRC) test is used for hard steels, whereas a Rockwell *F* (HRF) test might be selected for aluminum. Rockwell tests provide a hardness number that has no units.

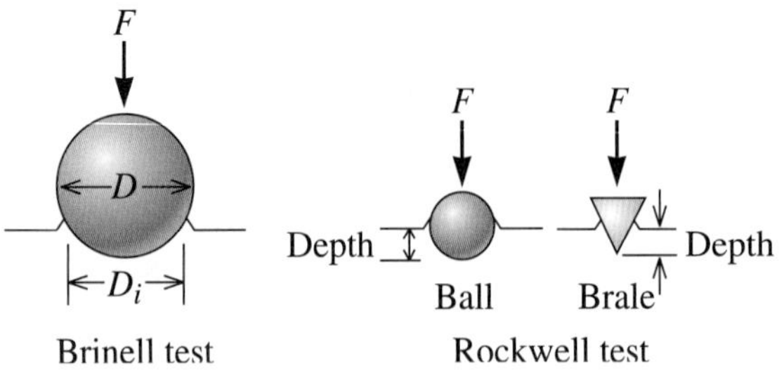

Figure 6-17 Indenters for the Brinell and Rockwell hardness tests.

Table 6-5 Comparison of typical hardness tests

Test	Indenter	Load	Application
Brinell	10-mm ball	3000 kg	Cast iron and steel
Brinell	10-mm ball	500 kg	Nonferrous alloys
Rockwell *A*	Brale	60 kg	Very hard materials
Rockwell *B*	1/16-in. ball	100 kg	Brass, low-strength steel
Rockwell *C*	Brale	150 kg	High-strength steel
Rockwell *D*	Brale	100 kg	High-strength steel
Rockwell *E*	1/8-in. ball	100 kg	Very soft materials
Rockwell *F*	1/16-in. ball	60 kg	Aluminum, soft materials
Vickers	Diamond square pyramid	10 kg	All materials
Knoop	Diamond elongated pyramid	500 g	All materials

Hardness numbers are used primarily as a *qualitative* basis for comparison of materials, specifications for manufacturing and heat treatment, quality control, and correlation with other properties of materials. For example, Brinell hardness is related to the tensile strength of steel by the approximation:

$$\text{Tensile strength (psi)} = 500 \text{ HB} \tag{6-18}$$

where HB has units of kg/mm^2.

Hardness correlates well with wear resistance. A separate test is available for measuring the wear resistance. A material used in crushing or grinding of ores should be very hard to ensure that the material is not eroded or abraded by the hard feed materials. Similarly, gear teeth in the transmission or the drive system of a vehicle should be hard enough that the teeth do not wear out. Typically we find that polymer materials are exceptionally soft, metals and alloys have intermediate hardnesses, and ceramics are exceptionally hard. We use materials such as tungsten carbide-cobalt composite (WC-Co), known as "carbide," for cutting tool applications. We also use microcrystalline diamond or diamond-like carbon (DLC) materials for cutting tools and other applications.

The Knoop hardness (HK) test is a **microhardness test**, forming such small indentations that a microscope is required to obtain the measurement. In these tests, the load applied is less than 2 N. The Vickers test, which uses a diamond pyramid indenter, can be conducted either as a macro or microhardness test. Microhardness tests are suitable for materials that may have a surface that has a higher hardness than the bulk, materials in which different areas show different levels of hardness, or samples that are not macroscopically flat.

6-8 Nanoindentation

The hardness tests described in the previous section are known as macro or microhardness tests because the indentations have dimensions on the order of millimeters or microns. The advantages of such tests are that they are relatively quick, easy, and inexpensive. Some of the disadvantages are that they can only be used on macroscale samples and hardness is the only materials property that can be directly measured. Nanoindentation is hardness testing performed at the nanometer length scale. A small diamond tip is used to indent the material of interest. The imposed load and displacement are continuously measured with micronewton and sub-nanometer resolution, respectively. Both

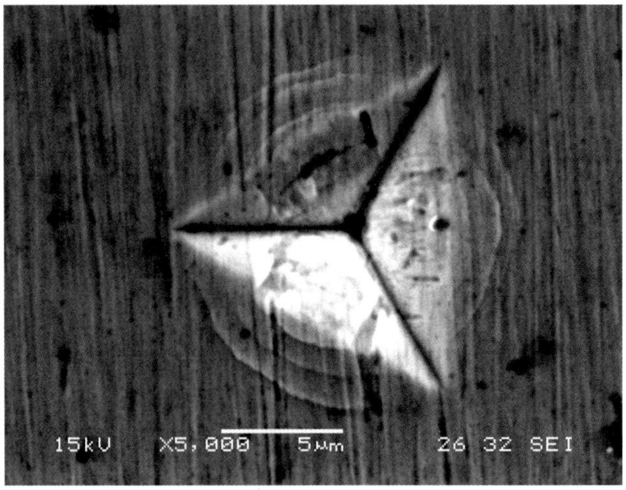

Figure 6-18 An indentation in a $Zr_{41.2}Ti_{13.8}Cu_{12.5}Ni_{10.0}Be_{22.5}$ bulk metallic glass made using a Berkovich tip in a nanoindenter. (*Courtesy of Gang Feng, Villanova University*)

load and displacement are measured throughout the indentation process. Nanoindentation techniques are important for measuring the mechanical properties of thin films on substrates (such as for microelectronics applications) and nanophase materials and for deforming free-standing micro and nanoscale structures. Nanoindentation can be performed with high positioning accuracy, permitting indentations within selected grains of a material. Nanoindenters incorporate optical microscopes and sometimes a scanning probe microscope capability. Both hardness and elastic modulus are measured using nanoindentation.

Nanoindenter tips come in a variety of shapes. A common shape is known as the Berkovich indenter, which is a three-sided pyramid. An indentation made by a Berkovich indenter is shown in Figure 6-18. The indentation in the figure measures 12.5 μm on each side and about 1.6 μm deep.

The first step of a nanoindentation test involves performing indentations on a calibration standard. Fused silica is a common calibration standard because it has homogeneous and well-characterized mechanical properties (elastic modulus $E = 72$ GPa and Poisson's ratio $v = 0.17$). The purpose of performing indentations on the calibration standard is to determine the projected contact area of the indenter tip A_c as a function of indentation depth. For a perfect Berkovich tip,

$$A_c = 24.5\, h_c^2 \tag{6-19}$$

This function relates the cross-sectional area of the indenter to the distance from the tip h_c that is in contact with the material being indented. No tip is perfectly sharp, and the tip wears and changes shape with each use. Thus, a calibration must be performed each time the tip is used as will be discussed below.

The total indentation depth h (as measured by the displacement of the tip) is the sum of the contact depth h_c and the depth h_s at the periphery of the indentation where the indenter does not make contact with the material surface, i.e.,

$$h = h_c + h_s \tag{6-20}$$

as shown in Figure 6-19. The surface displacement term h_s is calculated according to

$$h_s = \varepsilon \frac{P_{\max}}{S} \tag{6-21}$$

where $P_{\max}$ is the maximum load and ε is a geometric constant equal to 0.75 for a Berkovich indenter. S is the unloading stiffness.

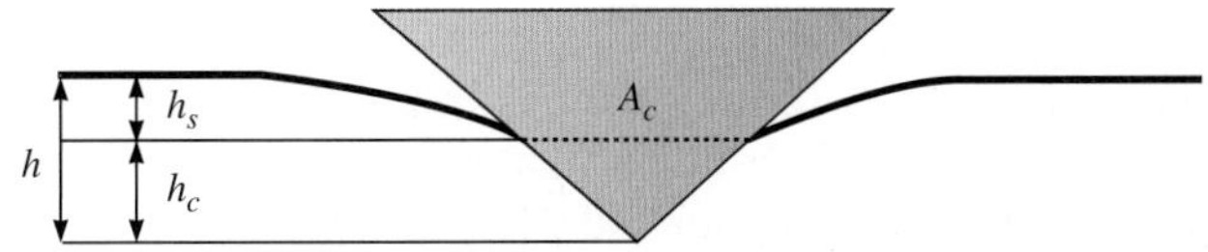

Figure 6-19 The relationship between the total indentation depth h, the contact depth h_c, and the displacement of the surface at the periphery of the indent h_s. The contact area A_c at a depth h_c is seen edge-on in this view. (Based on W.C. Oliver and G.M. Pharr, "An improved technique for determining hardness and elastic modulus using load and displacement," *Journal of Materials Research,* Volume 07, Issue 06, pp. 1564-1583, Materials Research Society, 1992.)

In nanoindentation, the imposed load is measured as a function of indentation depth h, as shown in Figure 6-20. On loading, the deformation is both elastic and plastic. As the indenter is removed from the material, the recovery is elastic. The unloading stiffness is measured as the slope of a power law curve fit to the unloading curve at the maximum indentation depth. The reduced elastic modulus E_r is related to the unloading stiffness S according to

$$E_r = \frac{\sqrt{\pi}}{2\beta} \frac{S}{\sqrt{A_c}} \tag{6-22}$$

where β is a constant for the shape of the indenter being used ($\beta = 1.034$ for a Berkovich indenter). The reduced modulus E_r is given by

$$\frac{1}{E_r} = \frac{1 - \nu^2}{E} + \frac{1 - \nu_i^2}{E_i} \tag{6-23}$$

where E and ν are the elastic modulus and Poisson's ratio of the material being indented, respectively, and E_i and ν_i are the elastic modulus and Poisson's ratio of the indenter, respectively (for diamond, $E_i = 1.141$ TPa and $\nu_i = 0.07$). Since the elastic properties of the standard are known, the only unknown in Equation 6-22 for a calibration indent is A_c.

Using Equation 6-22, the projected contact area is calculated for a particular contact depth. When the experiment is subsequently performed on the material of interest, the same tip shape function is used to calculate the projected contact area for the same contact depth. Equation 6-22 is again used with the elastic modulus of the material

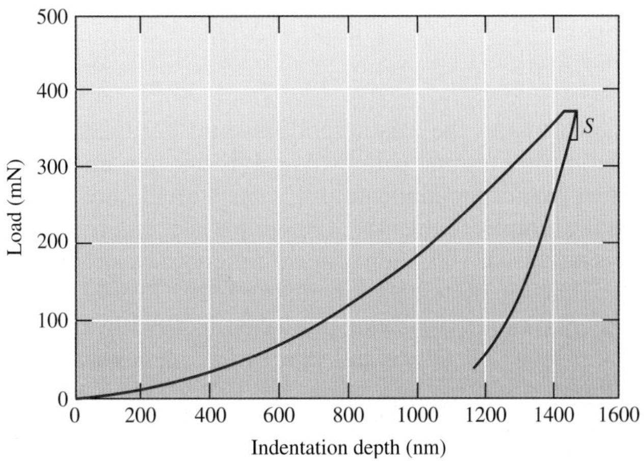

Figure 6-20 Load as a function of indentation depth for nanoindentation of MgO. The unloading stiffness S at maximum load is indicated.

being the unknown quantity of interest. (A Poisson's ratio must be assumed for the material being indented. As we saw in Table 6-3, typical values range from 0.2 to 0.4 with most metals having a Poisson's ratio of about 0.3. Errors in this assumption result in relatively little error in the elastic modulus measurement.)

The hardness of a material as determined by nanoindentation is calculated as

$$H = \frac{P_{\max}}{A_c} \tag{6-24}$$

Hardnesses (as determined by nanoindentation) are typically reported with units of GPa, and the results of multiple indentations are usually averaged to improve accuracy.

Example 6-7 *Nanoindentation of MgO*

Figure 6-20 shows the results of an indentation into single crystal (001) MgO using a diamond Berkovich indenter. The unloading stiffness at a maximum indentation depth of 1.45 μm is 1.8×10^6 N/m. A calibration on fused silica indicates that the projected contact area at the corresponding contact depth is 41 μm^2. The Poisson's ratio of MgO is 0.17. Calculate the elastic modulus and hardness of MgO.

SOLUTION

The projected contact area $A_c = 41\ \mu m^2 \times \dfrac{(1\text{ m})^2}{(10^6\ \mu m)^2} = 4.1 \times 10^{-11}\text{ m}^2$

The reduced modulus is given by

$$E_r = \frac{\sqrt{\pi}}{2\beta}\frac{S}{\sqrt{A_c}} = \frac{\sqrt{\pi}(1.8 \times 10^6\text{ N/m})}{2(1.034)\sqrt{(4.1 \times 10^{-11}\text{ m}^2)}} = 241 \times 10^9\text{ N/m}^2 = 241\text{ GPa}$$

Substituting for the Poisson's ratio of MgO and the elastic constants of diamond and solving for E,

$$\frac{1}{E_r} = \frac{1 - 0.17^2}{E} + \frac{1 - 0.07^2}{1.141 \times 10^{12}\text{ Pa}} = \frac{1}{241 \times 10^9\text{ Pa}}$$

$$\frac{0.9711}{E} = \frac{1}{241 \times 10^9\text{ Pa}} - \frac{1 - 0.07^2}{1.141 \times 10^{12}\text{ Pa}}$$

$$E = 296\text{ GPa}$$

From Figure 6-20, the load at the indentation depth of 1.45 μm is 380 mN (380×10^{-3} N). Thus, the hardness is

$$H = \frac{P_{\max}}{A_c} = \frac{380 \times 10^{-3}\text{ N}}{4.1 \times 10^{-11}\text{ m}^2} = 9.3 \times 10^9\text{ Pa} = 9.3\text{ GPa}$$

This analysis calculates the elastic modulus and hardness at the maximum load; however, an experimental technique known as dynamic nanoindentation is now usually employed. During dynamic nanoindentation, a small oscillating load is superimposed on the total load on the sample. In this way, the sample is continuously unloaded elastically as the total load is increased. This allows for continuous measurements of elastic modulus and hardness as a function of indentation depth.

This nanoindentation analysis was published in 1992 in the *Journal of Materials Research* and is known as the Oliver and Pharr method, named after Warren C. Oliver and George M. Pharr.

6-9 Strain Rate Effects and Impact Behavior

When a material is subjected to a sudden, intense blow, in which the strain rate is extremely rapid, it may behave in much more brittle a manner than is observed in the tensile test. This, for example, can be seen with many plastics and materials such as Silly Putty®. If you stretch a plastic such as polyethylene or Silly Putty® very slowly, the polymer molecules have time to disentangle or the chains to slide past each other and cause large plastic deformations. If, however, we apply an impact loading, there is insufficient time for these mechanisms to play a role and the materials break in a brittle manner. An **impact test** is often used to evaluate the brittleness of a material under these conditions. In contrast to the tensile test, in this test, the strain rates are much higher ($\dot{\varepsilon} \sim 10^3\ \mathrm{s}^{-1}$).

Many test procedures have been devised, including the *Charpy* test and the *Izod* test (Figure 6-21). The Izod test is often used for plastic materials. The test specimen may be either notched or unnotched; V-notched specimens better measure the resistance of the material to crack propagation.

In the test, a heavy pendulum, starting at an elevation h_0, swings through its arc, strikes and breaks the specimen, and reaches a lower final elevation h_f. If we know the initial and final elevations of the pendulum, we can calculate the difference in potential energy. This difference is the **impact energy** absorbed by the specimen during failure. For the Charpy test, the energy is usually expressed in foot-pounds (ft · lb) or joules (J), where 1 ft · lb = 1.356 J. The results of the Izod test are expressed in units of ft · lb/in. or J/m. The ability of a material to withstand an impact blow is often referred to as the **impact toughness** of the material. As we mentioned before, in some situations,

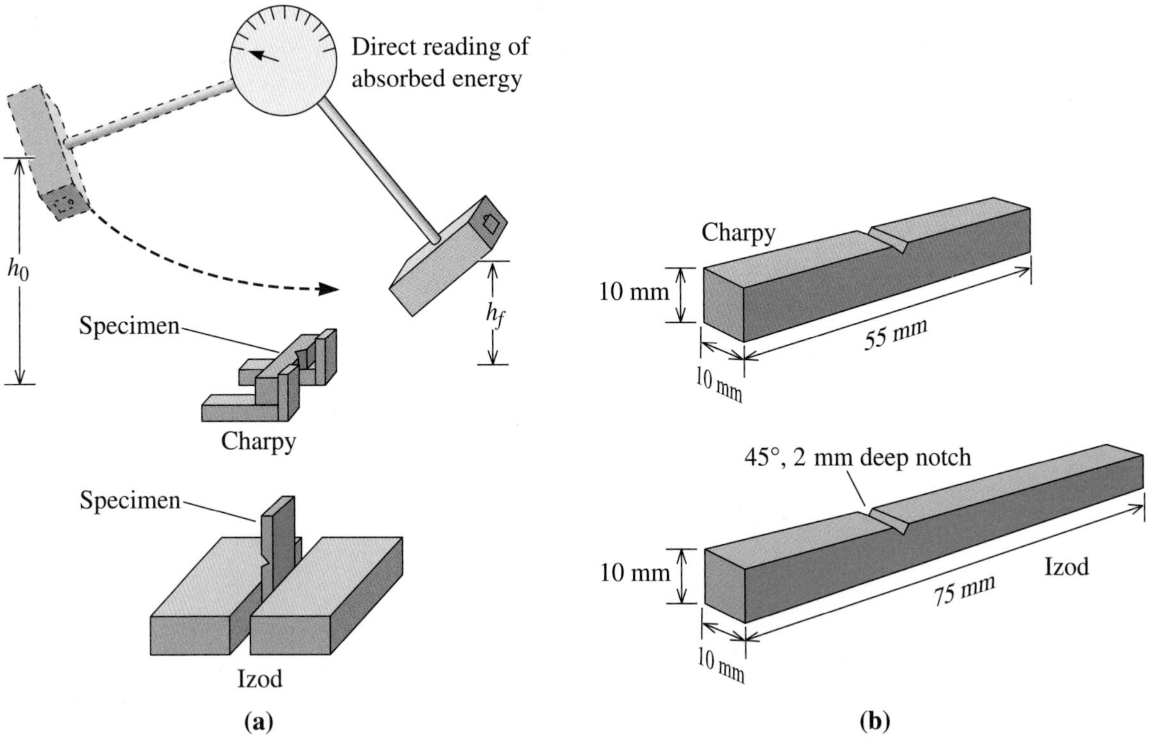

Figure 6-21 The impact test: (a) the Charpy and Izod tests, and (b) dimensions of typical specimens.

we consider the area under the true or engineering stress strain curve as a measure of **tensile toughness**. In both cases, we are measuring the energy needed to fracture a material. The difference is that, in tensile tests, the strain rates are much smaller compared to those used in an impact test. Another difference is that in an impact test, we usually deal with materials that have a notch. **Fracture toughness** of a material is defined as the ability of a material containing flaws to withstand an applied load. We will discuss fracture toughness in Chapter 7.

6-10 Properties Obtained from the Impact Test

A curve showing the trends in the results of a series of impact tests performed on nylon at various temperatures is shown in Figure 6-22. In practice, the tests will be conducted at a limited number of temperatures.

Ductile to Brittle Transition Temperature (DBTT)

The **ductile to brittle transition temperature** is the temperature at which the failure mode of a material changes from ductile to brittle fracture. This temperature may be defined by the average energy between the ductile and brittle regions, at some specific absorbed energy, or by some characteristic fracture appearance. A material subjected to an impact blow during service should have a transition temperature *below* the temperature of the material's surroundings.

Not all materials have a distinct transition temperature (Figure 6-23). BCC metals have transition temperatures, but most FCC metals do not. FCC metals have high absorbed energies, with the energy decreasing gradually and, sometimes, even increasing as the temperature decreases. As mentioned before, the effect of this transition in steel may have contributed to the failure of the *Titanic*.

In polymeric materials, the ductile to brittle transition temperature is related closely to the glass-transition temperature and for practical purposes is treated as the same. As mentioned before, the transition temperature of the polymers used in booster rocket O-rings and other factors led to the *Challenger* disaster.

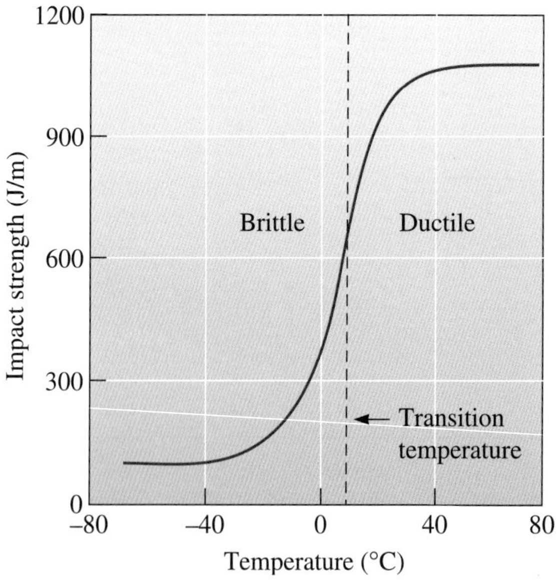

Figure 6-22 Results from a series of Izod impact tests for a tough nylon thermoplastic polymer.

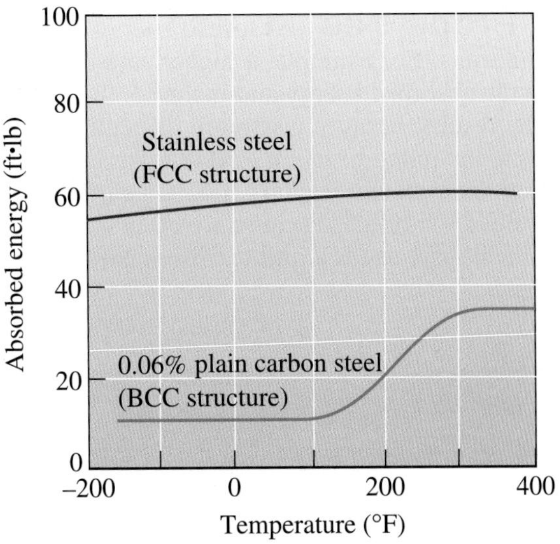

Figure 6-23 The Charpy V-notch properties for a BCC carbon steel and an FCC stainless steel. The FCC crystal structure typically leads to higher absorbed energies and no transition temperature.

Notch Sensitivity Notches caused by poor machining, fabrication, or design concentrate stresses and reduce the toughness of materials. The **notch sensitivity** of a material can be evaluated by comparing the absorbed energies of notched versus unnotched specimens. The absorbed energies are much lower in notched specimens if the material is notch-sensitive. We will discuss in Section 7-7 how the presence of notches affects the behavior of materials subjected to cyclical stress.

Relationship to the Stress Strain Diagram The energy required to break a material during impact testing (i.e., the impact toughness) is not always related to the tensile toughness [i.e., the area contained under the true stress-true strain curve (Figure 6-24)]. As noted before, engineers often consider the area under the engineering stress strain curve as tensile toughness. In general, metals with both high strength and high ductility have good tensile toughness; however, this is not always the case when the strain rates are high. For example, metals that show excellent tensile toughness may show brittle behavior under high strain rates (i.e., they may show poor impact toughness). Thus, the imposed strain rate can shift the ductile to brittle transition. Ceramics and many composites normally have poor toughness, even though they have high strength, because they display virtually no ductility. These materials show both poor tensile toughness and poor impact toughness.

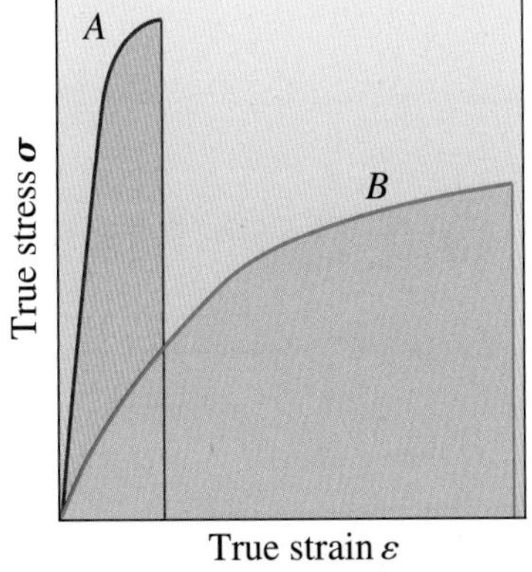

Figure 6-24 The area contained under the true stress–true strain curve is related to the tensile toughness. Although material *B* has a lower yield strength, it absorbs more energy than material *A*. The energies from these curves may not be the same as those obtained from impact test data.

Use of Impact Properties Absorbed energy and the DBTT are very sensitive to loading conditions. For example, a higher rate of application of energy to the specimen reduces the absorbed energy and increases the DBTT. The size of the specimen also affects the results; because it is more difficult for a thick material to deform, smaller energies are required to break thicker materials. Finally, the configuration of the notch affects the behavior; a sharp, pointed surface crack permits lower absorbed energies than does a V-notch. Because we often cannot predict or control all of these conditions, the impact test is a quick, convenient, and inexpensive way to compare different materials.

Example 6-8 *Design of a Sledgehammer*

Design an eight pound sledgehammer for driving steel fence posts into the ground.

SOLUTION

First, we must consider the design requirements to be met by the sledgehammer. A partial list would include

1. The handle should be light in weight, yet tough enough that it will not catastrophically break.
2. The head must not break or chip during use, even in subzero temperatures.
3. The head must not deform during continued use.
4. The head must be large enough to ensure that the user does not miss the fence post, and it should not include sharp notches that might cause chipping.
5. The sledgehammer should be inexpensive.

Although the handle could be a lightweight, tough composite material [such as a polymer reinforced with Kevlar (a special polymer) fibers], a wood handle about 30 in. long would be much less expensive and would still provide sufficient toughness. As shown in Chapter 17, wood can be categorized as a natural fiber-reinforced composite.

To produce the head, we prefer a material that has a low transition temperature, can absorb relatively high energy during impact, and yet also has enough hardness to avoid deformation. The toughness requirement would rule out most ceramics. A face-centered cubic metal, such as FCC stainless steel or copper, might provide superior toughness even at low temperatures; however, these metals are relatively soft and expensive. An appropriate choice might be a BCC steel. Ordinary steels are inexpensive, have good hardness and strength, and some have sufficient toughness at low temperatures.

In Appendix A, we find that the density of iron is 7.87 g/cm^3, or 0.28 lb/in.^{3} We assume that the density of steel is about the same. The volume of steel required is $V = 8$ lbs/(0.28 lb/in.^{3}) = 28.6 in.^{3} To ensure that we will hit our target, the head might have a cylindrical shape with a diameter of 2.5 in. The length of the head would then be 5.8 in.

6-11 Bulk Metallic Glasses and Their Mechanical Behavior

Metals, as they are found in nature, are crystalline; however, when particular multicomponent alloys are cooled rapidly, amorphous metals may form. Some alloys require cooling rates as high as 10^6 K/s in order to form an amorphous (or "glassy") structure, but recently, new compositions have been found that require cooling rates on the order of only a few degrees per second. This has enabled the production of so-called "bulk metallic glasses"—metallic glasses with thicknesses or diameters as large as 5 cm (2 inches).

Before the development of bulk metallic glasses, amorphous metals were produced by a variety of rapid solidification techniques, including a process known as melt spinning. In melt spinning, liquid metal is poured onto chilled rolls that rotate and "spin off" thin ribbons on the order of 10 μm thick. It is difficult to perform mechanical testing on ribbons; thus, the development of bulk metallic glasses enabled mechanical tests that were not previously possible. Bulk metallic glasses can be produced by several methods. One method involves using an electric arc to melt elements of high purity and then suction casting or pour casting into cooled copper molds.

As shown in Figure 6-25, metallic glasses exhibit fundamentally different stress strain behavior from other classes of materials. Most notably, metallic glasses are exceptionally high-strength materials. They typically have yield strengths on the order of 2 GPa (290 ksi), comparable to those of the highest strength steels, and fracture strengths higher than 5 GPa (725 ksi) have been reported for iron-based amorphous metal alloys. Since metallic glasses are not crystalline, they do not contain dislocations. As we learned in Chapter 4, dislocations lead to yield strengths that are lower than those that are theoretically predicted for perfect crystalline materials. The high strengths of metallic glasses are due to the lack of dislocations in the amorphous structure.

Despite their high strengths, typical bulk metallic glasses are brittle. Most metallic glasses exhibit nearly zero plastic strain in tension and only a few percent plastic strain in compression (compared to tens of percent plastic strain for crystalline metals). This lack of ductility is related to a lack of dislocations. Since metallic glasses do not contain dislocations, they do not work harden (i.e., the stress does not increase with strain after plastic deformation has commenced, as shown in Figure 6-25. Work hardening, also known as strain hardening, will be discussed in detail in Chapter 8). Thus, when deformation becomes localized, it intensifies and quickly leads to failure.

At room temperature, metallic glasses are permanently deformed through intense shearing in narrow bands of material about 10 to 100 nm thick. This creates shear offsets at the edges of the material, as shown in Figure 6-26. For the case of compression, this

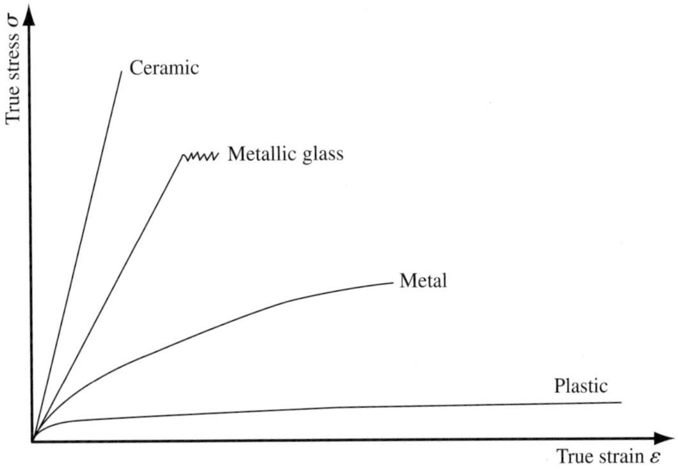

Figure 6-25 A schematic diagram of the compressive stress strain behavior of various engineering materials including metallic glasses. Serrated flow is observed in the plastic region.

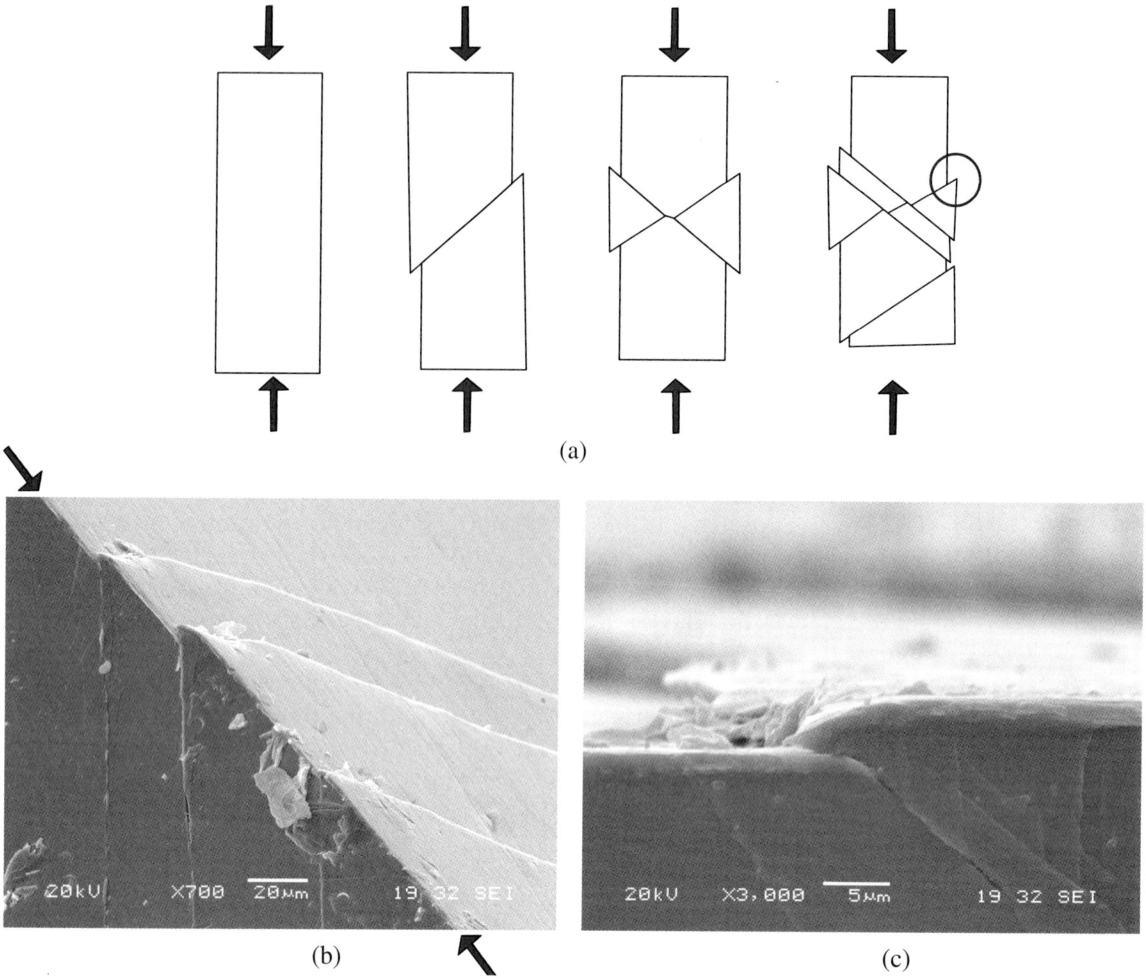

Figure 6-26 (a) A schematic diagram of shear band formation in a metallic glass showing successive stages of a compression test. (b) A scanning electron micrograph showing three shear bands in a $Zr_{41.2}Ti_{13.8}Cu_{12.5}Ni_{10.0}Be_{22.5}$ bulk metallic glass. The arrows indicate the loading direction. (*Courtesy of Wendelin Wright*) (c) A scanning electron micrograph of a shear band offset in a $Zr_{41.2}Ti_{13.8}Cu_{12.5}Ni_{10.0}Be_{22.5}$ bulk metallic glass. The location of such an offset is circled in (a). (*Courtesy of Wendelin Wright*)

results in a decrease in length in the direction of the loading axis. As plasticity proceeds, more shear bands form and propagate. More shear bands form to accommodate the increasing plastic strain until finally one of these shear bands fails the sample. The effects of shear banding can be observed in Figure 6-25. When a metallic-glass compression sample is deformed at a constant displacement rate, the load (or stress) drops as a shear band propagates. This leads to the "serrated" stress strain behavior shown in the figure. Although some crystalline materials show serrated behavior, the origin of this phenomenon is completely different in amorphous metals. Current research efforts are aimed at understanding the shear banding process in metallic glasses and preventing shear bands from causing failure.

Metallic glasses have applications in sporting goods equipment, electronic casings, defense components, and as structural materials. Metallic glasses are also suitable as industrial coatings due to their high hardness and good corrosion resistance (both again due to a lack of dislocations in the structure). All potential applications must utilize metallic glasses well below their glass-transition temperatures since they will crystallize and lose their unique mechanical behavior at elevated temperatures.

6-12 Mechanical Behavior at Small Length Scales

With the development of thin films on substrates for microelectronics applications and the synthesis of structures with nanometer scale dimensions, materials scientists have observed that materials may display different mechanical properties depending on the length scale at which the materials are tested. Several experimental techniques have been developed in order to measure mechanical behavior at small length scales. Nanoindentation, which was discussed in Section 8, is used for this purpose. Another technique is known as wafer curvature analysis.

In the wafer curvature technique, the temperature of a thin film on a substrate is cycled, typically from room temperature to several hundred degrees Celsius. Since the thin film and the substrate have different coefficients of thermal expansion, they expand (or contract) at different rates. This induces stresses in the thin film. These stresses in the film are directly proportional to the change in curvature of the film-substrate system as the temperature is cycled. An example of a wafer curvature experiment for a 0.5 μm polycrystalline aluminum thin film on an oxidized silicon substrate is shown in Figure 6-27(a). Aluminum has a larger thermal-expansion coefficient than silicon.

Figure 6-27(b) shows two cycles of the experiment. The experiment begins with the film-substrate system at room temperature. The aluminum has a residual stress of 30 MPa at room temperature due to cooling from the processing temperature. During the first cycle, the grains in the film grow as the temperature increases (Chapter 8). As the

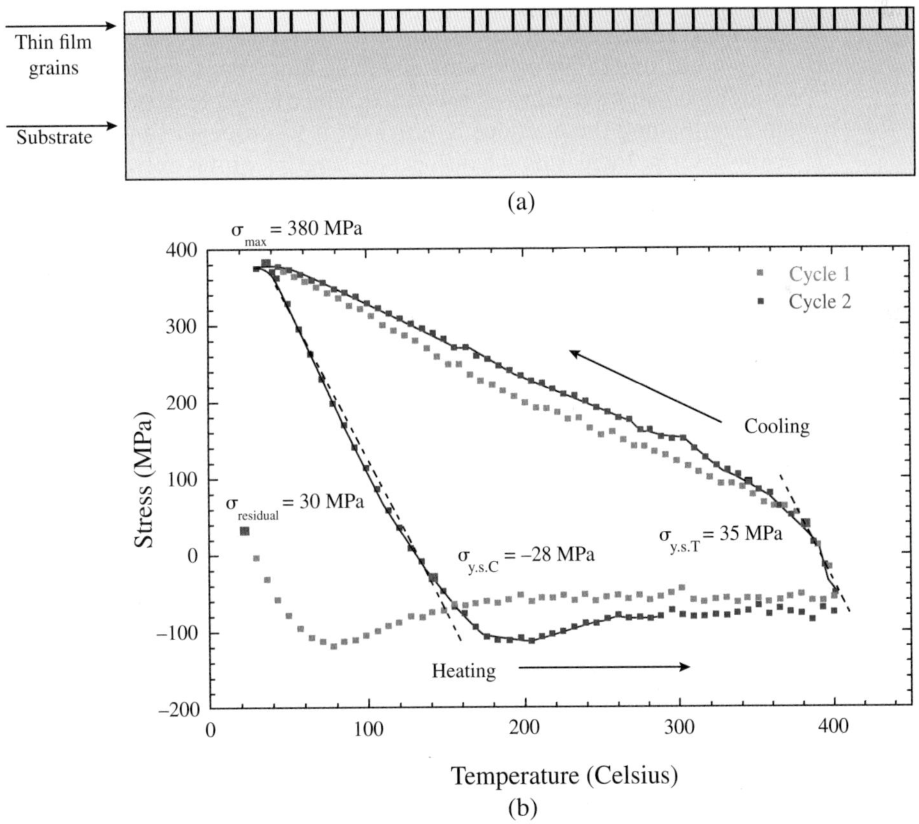

Figure 6-27 (a) A schematic diagram of a thin film on a substrate. The grains have diameters on the order of the film thickness. Note that thin films typically have thicknesses about 1/500th of the substrate thickness (some are far thinner), and so, this diagram is not drawn to scale. (b) Stress versus temperature for thermal cycling of a 0.5 μm polycrystalline aluminum thin film on an oxidized silicon substrate during a wafer curvature experiment.

grains grow, the grain boundary area decreases, and the film densifies. At the same time, the aluminum expands more rapidly than the silicon, and it is constrained from expanding by the silicon substrate to which it is bonded. Thus, the aluminum is subjected to a state of compression. As the temperature increases, the aluminum plastically deforms. The stress does not increase markedly because the temperature rise causes a decrease in strength.

As the system cools, the aluminum contracts more with a decrease in temperature than does the silicon. As the aluminum contracts, it first unloads elastically and then deforms plastically. As the temperature decreases, the stress continues to increase until the cycle is complete when room temperature is reached. A second thermal cycle is shown. Subsequent temperature cycles will be similar in shape, since the microstructure of the film does not change unless the highest temperature exceeds that of the previous cycle.

Notice that the stress sustained by the aluminum at room temperature after cycling is 380 MPa! Pure bulk aluminum (aluminum with macroscopic dimensions) is able to sustain only a fraction of this stress without failing. Thus, we have observed a general trend in materials science—for crystalline metals, smaller is stronger!

In general, the key to the strength of a crystalline metal is dislocations. As the resistance to dislocation motion increases, the strength of the metal also increases. One mechanism for strengthening in micro and nanoscale crystalline metals is the grain size. The grain size of thin films tends to be on the order of the film thickness, as shown in Figure 6-27(a). As grain size decreases, yield strength increases (see Chapter 4, Section 7 for a discussion of the Hall-Petch equation).

Dislocations distort the surrounding crystal, increasing the strain energy in the atomic bonds. Thus, dislocations have energies associated with them. For all metals, as the dislocation density (or the amount of dislocation length per unit volume in the crystal) increases, this energy increases.

In thin films, dislocations may be pinned at the interface between the thin film and the substrate to which it is bonded. Thus, it is necessary that the dislocation increase in length along the interface in order for it to propagate and for the thin film to plastically deform. Increasing the dislocation length requires energy, and the stress required to cause the dislocation to propagate is higher than it would be for a dislocation that is not constrained by an interface. This stress is inversely proportional to the film thickness; thus, as film thickness decreases, the strength increases. When two surfaces constrain the dislocation, such as when a passivating layer (i.e., one that protects the thin film surface from oxidation or corrosion) is deposited on a thin film, the effect is even more pronounced. This inverse relationship between film strength and thickness is independent from the grain-size effect discussed previously. Remember: any mechanism that interferes with the motion of dislocations makes a metal stronger.

In order to induce a nonuniform shape change in a material, such as bending a bar or indenting a material, dislocations must be introduced to the crystal structure. Such dislocations are called *geometrically necessary dislocations.* These dislocations exist in addition to the dislocations (known as *statistically stored dislocations*) that are produced by homogeneous strain; thus, the dislocation density is increased. At small length scales (such as for small indentations made using a nanoindenter), the density of the geometrically necessary dislocations is significant, but at larger length scales, the effect is diminished. Thus, the hardness of shallow indents is greater than the hardness of deep indents made in the same material. As you will learn in Chapter 8, as dislocation density increases, the strength of a metal increases. Dislocations act as obstacles to the propagation of other dislocations, and again, any mechanism that interferes with the motion of dislocations makes a metal stronger.

An increasingly common mechanical testing experiment involves fabricating compression specimens with diameters on the order of 1 micron using a tool known as the focused ion beam. Essentially, a beam of gallium ions is used to remove atoms from the surface of a material, thereby performing a machining process at the micron and sub-micron length scale. These specimens are then deformed under compression in a nanoindenter using a flat punch tip. The volume of such specimens is on the order of 2.5 μm^3. Extraordinary strengths have been observed in single-crystal pillars made from metals. This topic is an area of active research in the materials community.

6-13 Rheology of Liquids

The descriptions of stress and strain that we have developed thus far in this chapter are typically applied to solids. **Rheology** is the study of the deformation and flow of materials, particularly of liquids. If the relationship between the applied shear stress τ and the **shear strain rate** ($\dot{\gamma}$) is linear, we refer to that material as **Newtonian**. The slope of the shear stress versus the steady-state shear strain rate curve is defined as the **viscosity** (η) of the material. Water is an example of a Newtonian material. The following relationship defines viscosity:

$$\tau = \eta\dot{\gamma} \quad (6\text{-}25)$$

The units of η are Pa·s (in the SI system) or Poise (P) or g/(cm · s) in the cgs system. Sometimes the term centipoise (cP) is used, 1 cP = 10^{-2} P. Conversion between these units is given by 1 Pa·s = 10 P = 1000 cP.

The **kinematic viscosity** (ν) is defined as

$$\nu = \eta/\rho \quad (6\text{-}26)$$

where viscosity (η) has units of Poise and density (ρ) has units of g/cm^3. The kinematic viscosity unit is Stokes (St) or equivalently cm^2/s. Sometimes the unit of centiStokes (cSt) is used; 1 cSt = 10^{-2} St. Note that in this chapter ν has been used as the symbol for both Poisson's ratio and the kinematic viscosity per convention. The use of the same symbol to represent different quantities is common in scientific notation albeit confusing.

For many materials, the relationship between shear stress and shear strain rate is nonlinear. These materials are **non-Newtonian**. The stress versus steady state shear strain rate relationship in these materials can be described as

$$\tau = \eta\dot{\gamma}^m \quad (6\text{-}27)$$

where the exponent m is not equal to 1.

Non-Newtonian materials are classified as **shear thinning** (or pseudoplastic) or **shear thickening** (or dilatant). The relationships between the shear stress and shear strain rate for different types of materials are shown in Figure 6-28. If we take the slope of the line obtained by joining the origin to any point on the curve, we determine the **apparent viscosity** (η_{app}). Apparent viscosity as a function of steady-state shear strain rate is shown in Figure 6-29(a). The apparent viscosity of a Newtonian material will remain constant with changing shear strain rate. In shear thinning materials, the apparent viscosity decreases with increasing shear strain rate. In shear thickening materials, the apparent viscosity increases with increasing shear strain rate. If you have a can of paint sitting in storage, for example, the shear strain rate that the paint is subjected to is very small, and the paint behaves as if it is very viscous. When you take a brush and paint, the paint is subjected to a high shear strain rate. The paint now behaves as if it is quite thin or less viscous (i.e., it exhibits a small apparent viscosity). This is the shear thinning behavior. Some materials have "ideal plastic" behavior. For an ideal plastic material, the shear stress does not change with shear strain rate.

Many useful materials can be modeled as **Bingham plastics** and are defined by the following equations:

$$\tau = G \cdot \gamma \text{ (when } \tau \text{ is less than } \tau_{ys}) \quad (6\text{-}28)$$

$$\tau = \tau_{ys} + \eta\dot{\gamma} \text{ (when } \tau \geq \tau_{ys}) \quad (6\text{-}29)$$

This is illustrated in Figure 6-29(b) and 6-29(c). In these equations, τ_{ys} is the apparent **yield strength** obtained by interpolating the shear stress shear strain rate data to zero

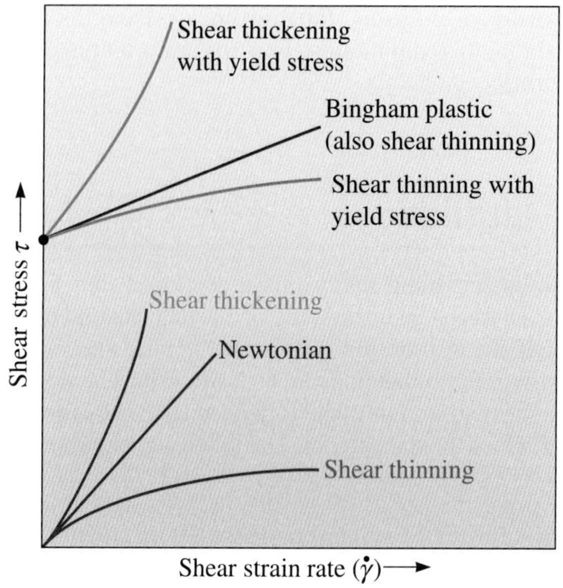

Figure 6-28 Shear stress–shear strain rate relationships for Newtonian and non-Newtonian materials.

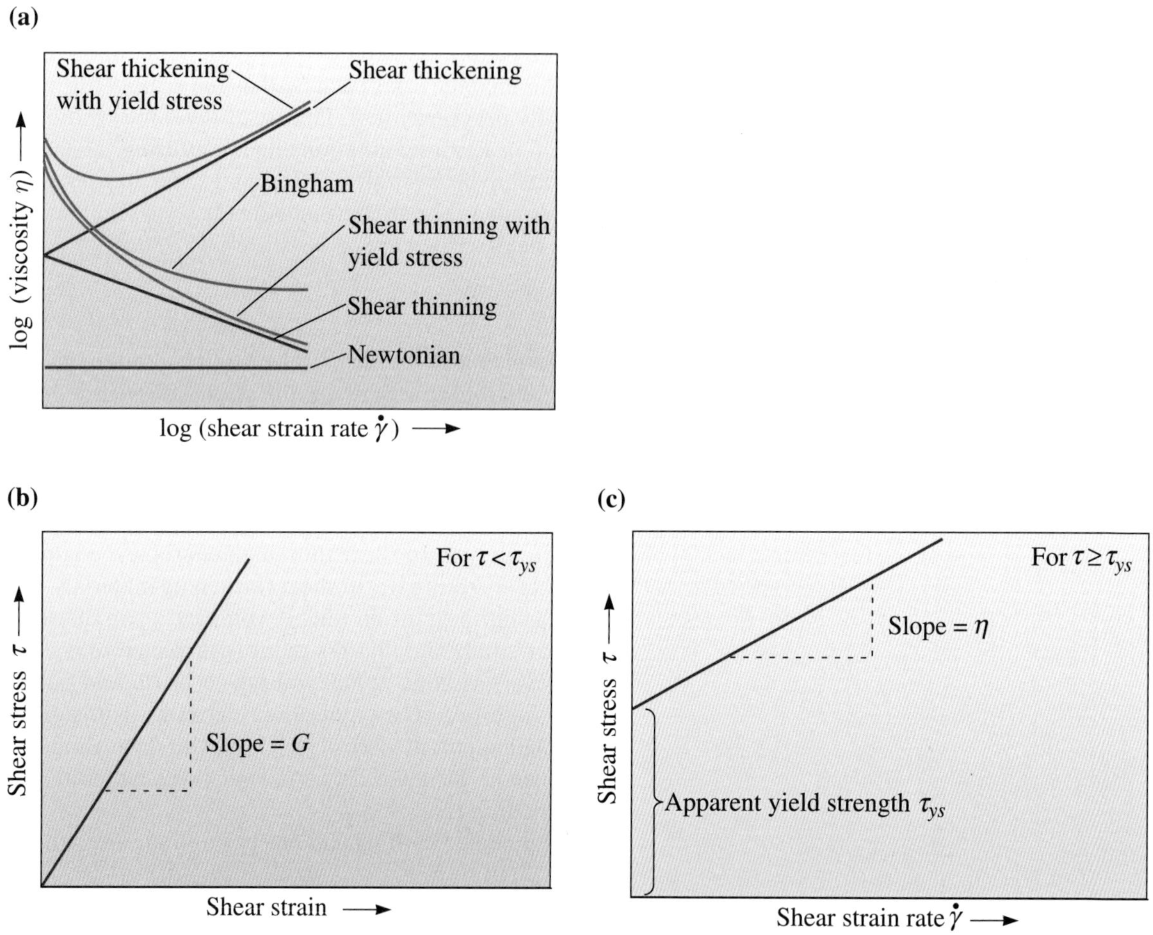

Figure 6-29 (a) Apparent viscosity as a function of shear strain rate ($\dot{\gamma}$). (b) and (c) Illustration of a Bingham plastic (Equations 6-28 and 6-29). Note the horizontal axis in (b) is shear strain.

shear strain rate. We define yield strength as the stress level that has to be exceeded so that the material deforms plastically. The existence of a true yield strength (sometimes also known as yield stress) has not been proven unambiguously for many plastics and dispersions such as paints. To prove the existence of a yield stress, separate measurements of stress versus strain are needed. For these materials, a critical yielding strain may be a better way to describe the mechanical behavior. Many ceramic slurries (dispersions such as those used in ceramic processing), polymer melts (used in polymer processing), paints and gels, and food products (yogurt, mayonnaise, ketchups, etc.) exhibit Bingham plastic-like behavior. Note that Bingham plastics exhibit shear thinning behavior (i.e., the apparent viscosity decreases with increasing shear rate).

Shear thinning materials also exhibit a **thixotropic behavior** (e.g., paints, ceramic slurries, polymer melts, gels, etc.). Thixotropic materials usually contain some type of network of particles or molecules. When a sufficiently large shear strain (i.e., greater than the critical yield strain) is applied, the thixotropic network or structure breaks and the material begins to flow. As the shearing stops, the network begins to form again, and the resistance to the flow increases. The particle or molecular arrangements in the newly formed network are different from those in the original network. Thus, the behavior of thixotropic materials is said to be time and deformation history dependent. Some materials show an increase in the apparent viscosity as a function of time and at a constant shearing rate. These materials are known as **rheopectic**.

The rheological properties of materials are determined using instruments known as a viscometer or a *rheometer*. In these instruments, a constant stress or constant strain rate is applied to the material being evaluated. Different geometric arrangements (e.g., cone and plate, parallel plate, Couette, etc.) are used.

Summary

- The mechanical behavior of materials is described by their mechanical properties, which are measured with idealized, simple tests. These tests are designed to represent different types of loading conditions. The properties of a material reported in various handbooks are the results of these tests. Consequently, we should always remember that handbook values are average results obtained from idealized tests and, therefore, must be used with some care.
- The tensile test describes the resistance of a material to a slowly applied tensile stress. Important properties include yield strength (the stress at which the material begins to permanently deform), tensile strength (the stress corresponding to the maximum applied load), modulus of elasticity (the slope of the elastic portion of the stress strain curve), and % elongation and % reduction in area (both measures of the ductility of the material).
- The bend test is used to determine the tensile properties of brittle materials. A modulus of elasticity and a flexural strength (similar to a tensile strength) can be obtained.
- The hardness test measures the resistance of a material to penetration and provides a measure of the wear and abrasion resistance of the material. A number of hardness tests, including the Rockwell and Brinell tests, are commonly used. Often the hardness can be correlated to other mechanical properties, particularly tensile strength.
- Nanoindentation is a hardness testing technique that continuously measures the imposed load and displacement with micronewton and sub-nanometer resolution, respectively. Nanoindentation techniques are important for measuring the mechanical properties of thin films on substrates and nanophase materials and for deforming free-standing micro and nanoscale structures. Both hardness and elastic modulus are measured using nanoindentation.
- The impact test describes the response of a material to a rapidly applied load. The Charpy and Izod tests are typical. The energy required to fracture the specimen is

measured and can be used as the basis for comparison of various materials tested under the same conditions. In addition, a transition temperature above which the material fails in a ductile, rather than a brittle, manner can be determined.

- Metallic glasses are amorphous metals. As such, they do not contain dislocations. A lack of dislocations leads to high strengths and low ductilities for these materials.
- Crystalline metals exhibit higher strengths when their dimensions are confined to the micro and nanoscale. Size-dependent mechanical behavior has critical implications for design and materials reliability in nanotechnology applications.
- Rheology is the study of the deformation and flow of materials, particularly of liquids. Liquid flow can be classified as Newtonian (when the shear stress is linearly related to the shear strain rate) or non-Newtonian.

Glossary

Anelastic (viscoelastic) material A material in which the total strain developed has elastic and viscous components. Part of the total strain recovers instantaneously when the applied load is removed. Some part, though, recovers over a period of time. Examples of viscoelastic materials include polymer melts and many polymers including Silly Putty®. Typically, the term anelastic is used for metallic materials.

Apparent viscosity Viscosity obtained by dividing shear stress by the corresponding value of the shear strain rate for that stress.

Bend test Application of a force to a bar that is supported on each end to determine the resistance of the material to a static or slowly applied load. Typically used for brittle materials.

Bingham plastic A material with a mechanical response given by $\tau = G\gamma$ when $\tau < \tau_{ys}$ and $\tau = \tau_{ys} + \eta\dot{\gamma}$ when $\tau \geq \tau_{ys}$.

Deformation processing Manufacturing or fabrication methods used for shaping of materials (e.g., extrusion, forging).

Dilatant (shear thickening) Materials in which the apparent viscosity increases with increasing rate of shear.

Ductile to brittle transition temperature The temperature below which a material behaves in a brittle manner in an impact test; it also depends on the strain rate.

Ductility The ability of a material to be permanently deformed without breaking when a force is applied.

Elastic deformation Deformation of the material that is recovered instantaneously when the applied load is removed.

Elastic limit The magnitude of stress at which plastic deformation commences.

Elastic strain Fully and instantaneously recoverable strain in a material.

Elastomers Natural or synthetic polymers that are composed of molecules with spring-like coils that lead to large elastic deformations (e.g., natural rubber, silicones).

Engineering strain Elongation per unit length calculated using the original dimensions.

Engineering stress The applied load, or force, divided by the original area over which the load acts.

Extensometer An instrument to measure change in length of a tensile specimen, thus allowing calculation of strain. An extensometer is often a clip that attaches to a sample and elastically deforms to measure the length change.

Flexural modulus The modulus of elasticity calculated from the results of a bend test; it is proportional to the slope of the stress-deflection curve.

Flexural strength (modulus of rupture) The stress required to fracture a specimen in a bend test.

Fracture toughness The resistance of a material to failure in the presence of a flaw.

Glass-transition temperature A temperature below which an otherwise ductile material behaves as if it is brittle. Usually, this temperature is not fixed and is affected by processing of the material.

Hardness test Measures the resistance of a material to penetration by a sharp object. Common hardness tests include the Brinell test, Rockwell test, Knoop test, and Vickers test.

Hooke's law The linear relationship between stress and strain in the elastic portion of the stress strain curve.

Impact energy The energy required to fracture a standard specimen when the load is applied suddenly.

Impact loading Application of stress at a high strain rate ($\sim > 100\ s^{-1}$).

Impact test Measures the ability of a material to absorb the sudden application of a load without breaking. The Charpy and Izod tests are commonly used impact tests.

Impact toughness Energy absorbed by a material, usually notched, during fracture, under the conditions of the impact test.

Kinematic viscosity Ratio of viscosity and density, often expressed in centiStokes.

Load The force applied to a material during testing.

Macrohardness Bulk hardness of materials measured using loads > 2 N.

Microhardness Hardness of materials typically measured using loads less than 2 N with a test such as the Knoop.

Modulus of elasticity Young's modulus, or the slope of the linear part of the stress strain curve in the elastic region. It is a measure of the stiffness of the bonds of a material and is not strongly dependent upon microstructure.

Modulus of resilience The maximum elastic energy absorbed by a material when a load is applied.

Nanoindentation Hardness testing performed at the nanometer length scale. The imposed load and displacement are measured with micronewton and sub-nanometer resolution, respectively.

Necking Local deformation causing a reduction in the cross-sectional area of a tensile specimen. Many ductile materials show this behavior. The engineering stress begins to decrease at the onset of necking.

Newtonian Materials in which the shear stress and shear strain rate are linearly related (e.g., light oil or water).

Non-Newtonian Materials in which the shear stress and shear strain rate are not linearly related; these materials are shear thinning or shear thickening (e.g., polymer melts, slurries, paints, etc.).

Notch sensitivity Measures the effect of a notch, scratch, or other imperfection on a material's properties such as toughness or fatigue life.

Offset strain value A value of strain (e.g., 0.002) used to obtain the offset yield stress.

Offset yield strength A stress value obtained graphically that describes the stress that gives no more than a specified amount of plastic deformation. Often simply stated as the yield strength.

Percent elongation The total percentage permanent increase in the length of a specimen due to a tensile test.

Percent reduction in area The total percentage permanent decrease in the cross-sectional area of a specimen due to a tensile test.

Plastic deformation or strain Permanent deformation of a material when a load is applied, then removed.

Poisson's ratio The negative of the ratio between the lateral and longitudinal strains in the elastic region.

Proportional limit A level of stress above which the relationship between stress and strain is not linear.

Pseudoplastics (shear thinning) Materials in which the apparent viscosity decreases with increasing rate of shear.

Rheopectic behavior Materials that show shear thickening and also an apparent viscosity that at a constant rate of shear increases with time.

Shear modulus The slope of the linear part of the shear stress shear strain curve.

Shear strain rate Time derivative of shear strain. See Strain rate.

Shear thickening (dilatant) Materials in which the apparent viscosity increases with increasing rate of shear.

Shear thinning (pseudoplastics) Materials in which the apparent viscosity decreases with increasing rate of shear.

Stiffness A measure of a material's resistance to elastic deformation. Stiffness is the slope of a load-displacement curve and is proportional to the elastic modulus. Stiffness depends on the geometry of the component under consideration, whereas the elastic or Young's modulus is a materials property.

Strain Elongation per unit length.

Strain gage A device used for measuring strain. A strain gage typically consists of a fine wire embedded in a polymer matrix. The strain gage is bonded to the test specimen and deforms as the specimen deforms. As the wire in the strain gage deforms, its resistance changes. The resistance change is directly proportional to the strain.

Strain rate The rate at which strain develops in or is applied to a material; it is represented by $\dot{\varepsilon}$ or $\dot{\gamma}$ for normal and shear-strain rates, respectively. Strain rate can have an effect on whether a material behaves in a ductile or brittle fashion.

Stress Force per unit area over which the force is acting.

Stress relaxation Decrease in stress for a material held under constant strain as a function of time, which is observed in viscoelastic materials. Stress relaxation is different from time-dependent recovery of strain.

Tensile strength The stress that corresponds to the maximum load in a tensile test.

Tensile test Measures the response of a material to a slowly applied uniaxial force. The yield strength, tensile strength, modulus of elasticity, and ductility are obtained.

Tensile toughness The area under the true stress-true strain tensile test curve. It is a measure of the energy required to cause fracture under tensile test conditions.

Thixotropic behavior Materials that show shear thinning and also an apparent viscosity that at a constant rate of shear decreases with time.

True strain Elongation per unit length calculated using the instantaneous dimensions.

True stress The load divided by the instantaneous area over which the load acts.

Ultimate tensile strength See Tensile strength.

Viscoelastic (or anelastic) material See Anelastic material.

Viscosity Measure of the resistance to flow, defined as the ratio of shear stress to shear strain rate (units Poise or Pa · s).

Viscous material A viscous material is one in which the strain develops over a period of time and the material does not return to its original shape after the stress is removed.

Work of fracture Area under the stress strain curve, considered a measure of tensile toughness.

Yield point phenomenon An abrupt transition, seen in some materials, from elastic deformation to plastic flow.

Yield strength A stress value obtained graphically that describes no more than a specified amount of deformation (usually 0.002). Also known as the offset yield strength.

Young's modulus The slope of the linear part of the stress strain curve in the elastic region, same as modulus of elasticity.

Problems

Section 6-1 Technological Significance

6-1 Explain the role of mechanical properties in load-bearing applications using real-world examples.

6-2 Explain the importance of mechanical properties in functional applications (e.g., optical, magnetic, electronic, etc.) using real-world examples.

6-3 Explain the importance of understanding mechanical properties in the processing of materials.

Section 6-2 Terminology for Mechanical Properties

6-4 Define "engineering stress" and "engineering strain."

6-5 Define "modulus of elasticity."

6-6 Define "plastic deformation" and compare it to "elastic deformation."

6-7 What is strain rate? How does it affect the mechanical behavior of polymeric and metallic materials?

6-8 Why does Silly Putty® break when you stretch it very quickly?

6-9 What is a viscoelastic material? Give an example.

6-10 What is an elastomer? Give an example.

6-11 What is meant by the term "stress relaxation"?

6-12 Draw a schematic diagram showing the development of strain in an elastic and viscoelastic material. Assume that the load is applied at some time $t = 0$ and taken off at some time t.

Section 6-3 The Tensile Test: Use of the Stress Strain Diagram

6-13 Draw qualitative engineering stress-engineering strain curves for a ductile polymer, a ductile metal, a ceramic, and an elastomer. Label the diagrams carefully. Rationalize your sketch for each material.

6-14 Calculate the maximum force that a 0.2-in. diameter rod of Al_2O_3, having a yield strength of 35,000 psi, can withstand with no plastic deformation. Express your answer in both lbs and N.

6-15 A cylindrical specimen of a titanium alloy having an elastic modulus of 107 GPa and an original diameter of 3.8 mm should experience only elastic deformation when a tensile load of 2000 N is applied. Compute the maximum length of the specimen for a maximum allowable elongation of 0.42 mm.

6-16 A material with a diameter of 8 mm is pulled with a force of 2500 N. Calculate the stress. If the original length is 50 mm, what is the strain under load if it is elongated to 50.15 mm?

6-17 An aluminum plate 0.5 cm thick is to withstand a force of 50,000 N with no permanent deformation. If the aluminum has a yield strength of 125 MPa, what is the minimum width of the plate?

Section 6-4 Properties Obtained from the Tensile Test and Section 6-5 True Stress and True Strain

6-18 An 850-lb force is applied to a 0.15-in. diameter nickel wire having a yield strength of 45,000 psi and a tensile strength of 55,000 psi. Determine

(a) whether the wire will plastically deform; and

(b) whether the wire will experience necking.

6-19 A force of 100,000 N is applied to an iron bar with a cross-sectional area of 10 mm $\times$ 20 mm and having a yield strength of 400 MPa and a tensile strength of 480 MPa. Determine

(a) whether the bar will plastically deform; and

(b) whether the bar will experience necking.

6-20 Define "true stress" and "true strain." Compare with engineering stress and engineering strain.

6-21 At what point does necking typically initiate during a tensile test?

6-22 Write down the formulas for calculating the stress and strain for a sample subjected to a tensile test. Assume the sample shows necking.

6-23 Derive the expression $\varepsilon = \ln(1 + e)$, where ε is the true strain and e is the engineering strain. Note that this expression is not valid after the onset of necking.

6-24 What is necking? How does it lead to reduction in engineering stress as true stress increases?

6-25 Two rods in the form of single crystals for pure tin show wildly different yield strengths at the 0.2% offset yield strength. What would be the cause of this phenomenon?

6-26 Develop an expression for the engineering strain along the loading axis of a bar subjected to tension prior to any necking that may occur. Your expression should be a function only of the bar's initial and instantaneous diameters. Assume that the volume is constant.

6-27 The cable of a hoist has a cross-sectional area of 80 mm^2. The hoist is used to lift a crate weighing 500 kg. The free length of the cable is 30 m. Assume all deformation is elastic.

(a) What is the stress on the cable?

(b) How much will the cable extend if it is made from steel ($E = 200$ GPa)?

(c) How much will the cable extend if it is made from polypropylene ($E = 1.2$ GPa)?

6-28 A force of 20,000 N will cause a 1 cm $\times$ 1 cm bar of magnesium to stretch from 10 cm to 10.045 cm. Calculate the modulus of elasticity, both in GPa and psi.

6-29 A polymer bar's dimensions are 1 in. $\times$ 2 in. $\times$ 15 in. The polymer has a modulus of elasticity of 600,000 psi. What force is required to stretch the bar elastically from 15 in. to 15.25 in.?

6-30 A steel cable 1.25 in. in diameter and 50 ft long is to lift a 20-ton load without permanently deforming. What is the length of the cable during lifting? The modulus of elasticity of the steel is 30×10^6 psi.

6-31 (a) Carbon nanotubes are one of the stiffest and strongest materials known to scientists and engineers. Carbon nanotubes have an elastic modulus of 1.1 TPa (1 TPa $= 10^{12}$ Pa). If a carbon nanotube has a diameter of 15 nm, determine the engineering stress sustained by the nanotube when subjected to a tensile load of 4 μN (1 μN $= 10^{-6}$ N) along the length of the tube. Assume that the entire cross-sectional area of the nanotube is load bearing.

(b) Assume that the carbon nanotube is only deformed elastically (not plastically) under the load of 4 μN. The carbon nanotube has a length of 10 μm (1 μm = 10^{-6} m). What is the tensile elongation (displacement) of the carbon nanotube in nanometers (1 nm = 10^{-9} m)?

6-32 A force of 4000 lbs is applied to a cylindrical metal bar that is 0.505 in. in diameter. This load produces a reduction in diameter of 0.0003 in. The deformation is purely elastic. What is the Poisson's ratio of this material if the elastic modulus is 15×10^6 psi?

6-33 A cylindrical bar of steel, 10 mm in diameter, is to be deformed elastically by application of a force along the bar axis (axial loading). Determine the force that will produce an elastic reduction of 0.003 mm in the diameter. The Poisson's ratio is 0.30, and the modulus of elasticity is 207 GPa for carbon steel.

6-34 (a) A 0.4-in.-diameter, 12-in.-long titanium bar has a yield strength of 50,000 psi, a modulus of elasticity of 16×10^6 psi, and a Poisson's ratio of 0.30. Determine the length and diameter of the bar when a 500-lb load is applied.

(b) When a tensile load is applied to a 1.5-cm-diameter copper bar, the diameter is reduced to a 1.498-cm diameter. Determine the applied load, using the data in Table 6-3.

6-35 A standard 0.505-in.-diameter tensile bar was machined to a 2.00-in.-gage length from a copper-nickel alloy and the following data were collected:

Load (lbs)	Gage Length (in.)
0	2.00000
1,000	2.00045
2,000	2.00091
3,000	2.00136
4,000	2.0020
6,000	2.020
8,000	2.052
10,000	2.112
11,000	2.280 (maximum load)
9,000	2.750 (fracture)

After fracture, the gage length was 2.75 in. and the diameter was 0.365 in. Plot the engineering stress strain curve and calculate

(a) the 0.2% offset yield strength;
(b) the tensile strength;
(c) the modulus of elasticity;
(d) the % elongation;
(e) the % reduction in area;
(f) the engineering stress at fracture;
(g) the true stress at necking; and
(h) the modulus of resilience.

6-36 A specimen of an AISI-SAE type 416 stainless steel with a 0.505-in. diameter was machined to a 2.00-in.-gage length and the following data were collected:

Load (lbs)	Gage Length (in.)
0	2.000
2,850	2.001
5,710	2.002
8,560	2.003
11,400	2.004
17,100	2.006
20,000	2.008
20,800	2.01
23,000	2.015
24,200	2.02
26,800	2.03
28,800	2.04
33,650	2.08
35,750	2.12
36,000	2.14 (maximum load)
35,850	2.16
34,050	2.20
28,000	2.23 (fracture)

After fracture, the gage length was 2.20 in. and the diameter was 0.325 in. Plot the engineering stress strain curve and calculate

(a) the 0.2% offset yield strength;
(b) the tensile strength;
(c) the modulus of elasticity;
(d) the % elongation;
(e) the % reduction in area;
(f) the engineering stress at fracture;
(g) the true stress at necking;
(h) the modulus of resilience; and
(i) the elastic and plastic strain to fracture.
(j) When the sample was loaded to 11,400 lbs, the diameter was measured

to be 0.504 in. Calculate the tranverse and axial strains at this load. Compute the Poisson's ratio.

(k) Obtain the tensile properties for type 416 stainless steel that has been quenched and tempered and compare them to your answers. Discuss the similarities and differences.

6-37 The following data were collected from a test specimen of cold-rolled and annealed brass. The specimen had an initial gage length l_0 of 35 mm and an initial cross-sectional area A_0 of 10.5 mm^2.

Load (N)	Δl (mm)
0	0.0000
66	0.0112
177	0.0157
327	0.0199
462	0.0240
797	1.72
1350	5.55
1720	8.15
2220	13.07
2690	22.77 (maximum load)
2410	25.25 (fracture)

(a) Plot the engineering stress strain curve and the true stress strain curve. Since the instantaneous cross-sectional area of the specimen is unknown past the point of necking, truncate the true stress true strain data at the point that corresponds to the ultimate tensile strength. Use of a software graphing package is recommended.

(b) Comment on the relative values of true stress strain and engineering stress strain during the elastic loading and prior to necking.

(c) If the true stress strain data were known past the point of necking, what might the curve look like?

(d) Calculate the 0.2% offset yield strength.

(e) Calculate the tensile strength.

(f) Calculate the elastic modulus using a linear fit to the appropriate data.

6-38 The following data were collected from a standard 0.505-in.-diameter test specimen of a copper alloy (l_0 = 2.0 in.):

Load (lb)	Δl (in.)
0	0.00000
3,000	0.00167
6,000	0.00333
7,500	0.00417
9,000	0.0090
10,500	0.040
12,000	0.26
12,400	0.50 (maximum load)
11,400	1.02 (fracture)

After fracture, the total length was 3.014 in. and the diameter was 0.374 in. Plot the engineering stress strain curve and calculate

(a) the 0.2% offset yield strength;
(b) the tensile strength;
(c) the modulus of elasticity;
(d) the % elongation;
(e) the % reduction in area;
(f) the engineering stress at fracture;
(g) the true stress at necking; and
(h) the modulus of resilience.

6-39 The following data were collected from a 0.4-in.-diameter test specimen of polyvinyl chloride (l_0 = 2.0 in.):

Load (lb)	Δl (in.)
0	0.00000
300	0.00746
600	0.01496
900	0.02374
1200	0.032
1500	0.046
1660	0.070 (maximum load)
1600	0.094
1420	0.12 (fracture)

After fracture, the total length was 2.09 in. and the diameter was 0.393 in. Plot the engineering stress strain curve and calculate

(a) the 0.2% offset yield strength;
(b) the tensile strength;
(c) the modulus of elasticity;
(d) the % elongation;
(e) the % reduction in area;
(f) the engineering stress at fracture; and
(g) the modulus of resilience.

6-40 The following data were collected from a 12-mm-diameter test specimen of magnesium (l_0 = 30.00 mm):

Load (N)	Δl (mm)
0	0.0000
5,000	0.0296
10,000	0.0592
15,000	0.0888
20,000	0.15
25,000	0.51
26,500	0.90
27,000	1.50 (maximum load)
26,500	2.10
25,000	2.79 (fracture)

After fracture, the total length was 32.61 mm and the diameter was 11.74 mm. Plot the engineering stress strain curve and calculate
(a) the 0.2% offset yield strength;
(b) the tensile strength;
(c) the modulus of elasticity;
(d) the % elongation;
(e) the % reduction in area;
(f) the engineering stress at fracture;
(g) the true stress at necking; and
(h) the modulus of resilience.

6-41 The following data were collected from a 20-mm-diameter test specimen of a ductile cast iron (l_0 = 40.00 mm):

Load (N)	Δl (mm)
0	0.0000
25,000	0.0185
50,000	0.0370
75,000	0.0555
90,000	0.20
105,000	0.60
120,000	1.56
131,000	4.00 (maximum load)
125,000	7.52 (fracture)

After fracture, the total length was 47.42 mm and the diameter was 18.35 mm. Plot the engineering stress strain curve and calculate
(a) the 0.2% offset yield strength;
(b) the tensile strength;
(c) the modulus of elasticity, using a linear fit to the appropriate data;
(d) the % elongation;
(e) the % reduction in area;
(f) the engineering stress at fracture;
(g) the true stress at necking; and
(h) the modulus of resilience.

6-42 Consider the tensile stress strain diagrams in Figure 6-30 labeled 1 and 2. These diagrams are typical of metals. Answer the following questions, and consider each part as a separate question that has no relationship to previous parts of the question.
(a) Samples 1 and 2 are identical except for the grain size. Which sample has the smaller grains? How do you know?
(b) Samples 1 and 2 are identical except that they were tested at different temperatures. Which was tested at the lower temperature? How do you know?
(c) Samples 1 and 2 are different materials. Which sample is tougher? Explain.
(d) Samples 1 and 2 are identical except that one of them is a pure metal and the other has a small percentage alloying addition. Which sample has been alloyed? How do you know?
(e) Given the stress strain curves for materials 1 and 2, which material has the lower hardness value on the Brinell hardness scale? How do you know?
(f) Are the stress strain curves shown true stress strain or engineering stress strain curves? How do you know?
(g) Which of the two materials represented by samples 1 and 2 would exhibit a higher shear yield strength? How do you know?

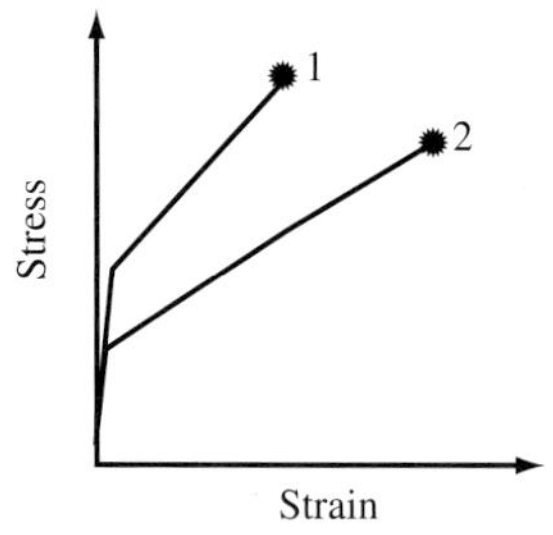

Figure 6-30 Stress strain curves (for Problem 6-42).

Section 6-6 The Bend Test for Brittle Materials

6-43 Define the terms "flexural strength" and "flexural modulus."

6-44 Why is it that we often conduct a bend test on brittle materials?

6-45 A square specimen of MgO is loaded using a three-point bend test. Compute the minimum possible thickness of the specimen without fracture, given that the applied load is 95 lbs, the flexural strength is 15 ksi, and the separation between load points is 2 in.

6-46 A bar of Al_2O_3 that is 0.25 in. thick, 0.5 in. wide, and 9 in. long is tested in a three-point bending apparatus with the supports located 6 in. apart. The deflection of the center of the bar is measured as a function of the applied load. The data are shown below. Determine the flexural strength and the flexural modulus.

Force (lb)	Deflection (in.)
14.5	0.0025
28.9	0.0050
43.4	0.0075
57.9	0.0100
86.0	0.0149 (fracture)

6-47 A three-point bend test is performed on a block of ZrO_2 that is 8 in. long, 0.50 in. wide, and 0.25 in. thick and is resting on two supports 4 in. apart. When a force of 400 lb is applied, the specimen deflects 0.037 in. and breaks. Calculate
(a) the flexural strength; and
(b) the flexural modulus, assuming that no plastic deformation occurs.

6-48 A three-point bend test is performed on a block of silicon carbide that is 10 cm long, 1.5 cm wide, and 0.6 cm thick and is resting on two supports 7.5 cm apart. The sample breaks when a deflection of 0.09 mm is recorded. The flexural modulus for silicon carbide is 480 GPa. Assume that no plastic deformation occurs. Calculate
(a) the force that caused the fracture; and
(b) the flexural strength.

6-49 (a) A thermosetting polymer containing glass beads is required to deflect 0.5 mm when a force of 500 N is applied. The polymer part is 2 cm wide, 0.5 cm thick, and 10 cm long. If the flexural modulus is 6.9 GPa, determine the minimum distance between the supports. Will the polymer fracture if its flexural strength is 85 MPa? Assume that no plastic deformation occurs.
(b) The flexural modulus of alumina is 45×10^6 psi, and its flexural strength is 46,000 psi. A bar of alumina 0.3 in. thick, 1.0 in. wide, and 10 in. long is placed on supports 7 in. apart. Determine the amount of deflection at the moment the bar breaks, assuming that no plastic deformation occurs.

6-50 Ceramics are much stronger in compression than in tension. Explain why.

6-51 Dislocations have a major effect on the plastic deformation of metals, but do not play a major role in the mechanical behavior of ceramics. Why?

6-52 What controls the strength of ceramics and glasses?

Section 6-7 Hardness of Materials and Section 6-8 Nanoindentation

6-53 What does the term "hardness of a material" mean?

6-54 Why is hardness data difficult to correlate to mechanical properties of materials in a quantitative fashion?

6-55 What is the hardest material (natural or synthetic)? Is it diamond?

6-56 Explain the terms "macrohardness" and "microhardness."

6-57 A Brinell hardness measurement, using a 10-mm-diameter indenter and a 500-kg load, produces an indentation of 2.5 mm on a steel plate.
(a) Determine the Brinell hardness number (HB) of the steel plate; and
(b) approximate the tensile strength of the steel.

6-58 A Brinell hardness measurement, using a 10-mm-diameter indenter and a 500 kg load, produces an indentation of 4.5 mm on an aluminum plate. Determine the Brinell hardness number (HB) of the metal.

6-59 When a 3000 kg load is applied to a 10-mm-diameter ball in a Brinell test of a steel, an indentation of 3.1 mm diameter is produced. Estimate the tensile strength of the steel.

6-60 Why is it necessary to perform calibrations on a standard prior to performing a nanoindentation experiment?

6-61 The elastic modulus of a metallic glass is determined to be 95 GPa using nanoindentation testing with a diamond Berkovich tip. The Poisson's ratio of the metallic glass is 0.36. The unloading stiffness as determined from the load-displacement data is 5.4×10^5 N/m. The maximum load is 120 mN. What is the hardness of the metallic glass at this indentation depth?

6-62 The elastic modulus and hardness of MgO are determined to be 306 GPa and 9.2 GPa, respectively, using nanoindentation testing with a diamond tip. The Poisson's ratio of the MgO is 0.17. The maximum load is 150 mN at this particular indentation depth. Determine the contact stiffness S. Take $\beta = 1$.

Section 6-9 Strain Rate Effects and Impact Behavior and Section 6-10 Properties Obtained from the Impact Test

6-63 Plot the Charpy V-notch data below for commercial steel. Label the upper and lower energy shelves on your plot. Would you recommend this steel for service conditions below 10°C ? Explain.

Temperature (°C)	Impact Energy (J)
50	75
40	75
30	70
20	58
10	37
0	24
−10	15
−20	10
−30	4
−40	1

6-64 The following data were obtained from a series of Charpy impact tests performed on four steels, each having a different manganese content. Plot the data and determine

(a) the transition temperature of each (defined by the mean of the absorbed energies in the ductile and brittle regions); and

(b) the transition temperature of each (defined as the temperature that provides 50 J of absorbed energy).

Test Temperature (°C)	Impact Energy (J)			
	0.30% Mn	0.39% Mn	1.01% Mn	1.55% Mn
−100	2	5	5	15
−75	2	5	7	25
−50	2	12	20	45
−25	10	25	40	70
0	30	55	75	110
25	60	100	110	135
50	105	125	130	140
75	130	135	135	140
100	130	135	135	140

6-65 Plot the transition temperature versus manganese content using the data in Problem 6-64 and discuss the effect of manganese on the toughness of steel. What is the minimum manganese allowed in the steel if a part is to be used at 0°C?

6-66 The following data were obtained from a series of Charpy impact tests performed on four ductile cast irons, each having a different silicon content. Plot the data and determine

(a) the transition temperature of each (defined by the mean of the absorbed energies in the ductile and brittle regions); and

(b) the transition temperature of each (defined as the temperature that provides 10 J of absorbed energy).

(c) Plot the transition temperature versus silicon content and discuss the effect of silicon on the toughness of the cast iron. What is the maximum silicon allowed in the cast iron if a part is to be used at 25°C?

Test Temperature (°C)	Impact Energy (J)			
	2.55% Si	2.85% Si	3.25% Si	3.63% Si
−50	2.5	2.5	2	2
−5	3	2.5	2	2
0	6	5	3	2.5
25	13	10	7	4
50	17	14	12	8
75	19	16	16	13
100	19	16	16	16
125	19	16	16	16

6-67 FCC metals are often recommended for use at low temperatures, particularly when any sudden loading of the part is expected. Explain.

6-68 A steel part can be made by powder metallurgy (compacting iron powder particles and sintering to produce a solid) or by machining from a solid steel block. Which part is expected to have the higher toughness? Explain.

6-69 What is meant by the term "notch sensitivity?"

6-70 What is the difference between a tensile test and an impact test? Explain why the toughness values measured using impact tests may not always correlate with tensile toughness measured using tensile tests.

6-71 A number of aluminum-silicon alloys have a structure that includes sharp-edged plates of brittle silicon in the softer, more ductile aluminum matrix. Would you expect these alloys to be notch-sensitive in an impact test? Would you expect these alloys to have good toughness? Explain your answers.

6-72 What is the ductile to brittle transition temperature (DBTT)?

6-73 How is tensile toughness defined in relation to the true stress strain diagram? How is tensile toughness related to impact toughness?

6-74 What caused NASA's *Challenger* 1986 accident?

6-75 What factors contributed to NASA's *Columbia* 2003 accident?

6-11 Bulk Metallic Glasses and Their Mechanical Behavior

6-76 A load versus displacement diagram is shown in Figure 6-31 for a metallic glass. A metallic glass is a non-crystalline (amorphous) metal. The sample was tested in compression. *Therefore, even though the load and displacement values are plotted as positive, the sample length was shortened during the test.* The sample had a length in the direction of loading of 6 mm and a cross-sectional area of 4 mm^2. Numerical values for the load and displacement are given at the points marked with a circle and an X. The first data point is (0, 0). Sample failure is indicated with an X. Answer the following questions.

(a) Calculate the elastic modulus.

(b) How does the elastic modulus compare to the modulus of steel?

(c) Calculate the engineering stress at the proportional limit.

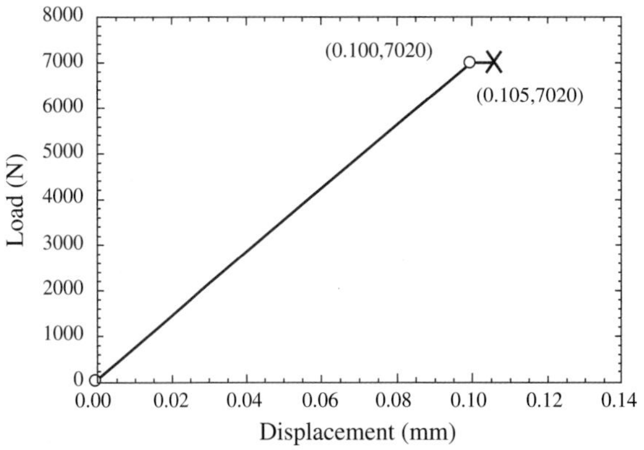

Figure 6-31 Load versus displacement for a metallic glass tested in compression (for Problem 6-76).

(d) Consider your answer to (c) to be the yield strength of the material. Is this a high yield stress or a low yield stress? Support your answer with an order of magnitude comparison for a typical polycrystalline metal.

(e) Calculate the true strain at the proportional limit. Remember that the length of the sample is decreasing in compression.

(f) Calculate the total true strain at failure.

(g) Calculate the work of fracture for this metallic glass based on engineering stress and strain.

6-12 Mechanical Behavior at Small Length Scales

6-77 Name a specific application for which understanding size-dependent mechanical behavior may be important to the design process.

6-13 Rheology of Liquids

6-78 Define the terms "viscosity," "apparent viscosity," and "kinematic viscosity."

6-79 What two equations are used to describe Bingham plastic-like behavior?

6-80 What is a Newtonian material? Give an example.

6-81 What is meant by the terms "shear thinning" and "shear thickening" materials?

6-82 Many paints and other dispersions are not only shear thinning, but also thixotropic. What does the term "thixotropy" mean?

Knovel® Problem

K6-1 A 120 in. annealed rod with a cross-sectional area of 0.86 in.^{2} was extruded from a 5083-O aluminum alloy and axially loaded. Under load, the length of the rod increased to 120.15 in. No plastic deformation occurred.

(a) Find the modulus of elasticity of the material and calculate its allowable tensile stress, assuming it to be 50% of the tensile yield stress.

(b) Calculate the tensile stress and the axial load applied to the rod.

(c) Compare the calculated tensile stress with the allowable tensile stress, and find the absolute value of elongation of the rod for the allowable stress.

The 316 stainless steel bolt failures shown here were due to mechanical fatigue. In this case, the bolts broke at the head-to-shank radius [see Figure (a)]. An optical fractograph of one of the fracture surfaces [see Figure (b)] shows the fracture initiates at one location and propagates across the bolt until final failure occurred. Beach marks and striations [see Figure (c)], typical of fatigue fractures, are present on all of the fracture surfaces. Fatigue failure of threaded fasteners is most often associated with insufficient tightening of the fastener, resulting in flexing and subsequent fracture. (*Corrosion Testing Laboratories, Bradley Krantz, Richard Corbett, Albert Olszewski, and Robert R. Odle*)

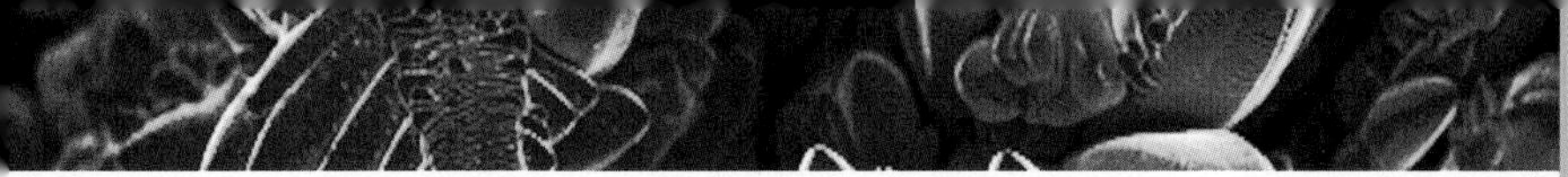

CHAPTER 7

Mechanical Properties: Part Two

Have You Ever Wondered?

- *Why is it that glass fibers of different lengths have different strengths?*
- *Can a material or component ultimately fracture even if the overall stress does not exceed the yield strength?*
- *Why do aircraft have a finite service life?*
- *Why do materials ultimately fail?*

Chapter Learning Objectives

The key objectives of this chapter are to

- Explain why flaws in a material may lead to premature fracture using energy arguments.
- Define fracture toughness and calculate it and the critical flaw size for a material.
- Define the fracture mechanics relationship between the material properties, the stress applied, and the flaw size.
- Identify the morphology of failure surfaces that indicate whether ductile or brittle fracture occurred.
- Describe the process of fatigue.
- Explain why fatigue occurs and quantify fatigue behavior.
- Describe the process of creep.

One goal of this chapter is to introduce the basic concepts associated with the fracture toughness of materials. In this regard, we will examine what factors affect the strength of glasses and ceramics and how the Weibull distribution quantitatively describes the variability in their strength. Another goal is to learn about time-dependent phenomena such as fatigue, creep, and stress corrosion. This chapter will review some of the basic testing procedures that engineers use to evaluate many of these properties and the failure of materials.

7-1 Fracture Mechanics

Fracture mechanics is the discipline concerned with the behavior of materials containing cracks or other small flaws. The term "flaw" refers to such features as small pores (holes), inclusions, or microcracks. The term "flaw" does *not* refer to atomic level defects such as vacancies or dislocations. What we wish to know is the maximum stress that a material can withstand if it contains flaws of a certain size and geometry. **Fracture toughness** measures the ability of a material containing a flaw to withstand an applied load. Note that this does *not* require a high strain rate (impact).

A typical fracture toughness test may be performed by applying a tensile stress to a specimen prepared with a flaw of known size and geometry (Figure 7-1). The stress applied to the material is intensified at the flaw, which acts as a *stress raiser.* For a simple case, the *stress intensity factor* K is

$$K = f\sigma\sqrt{\pi a} \quad (7\text{-}1)$$

where f is a geometry factor for the specimen and flaw, σ is the applied stress, and a is the flaw size as defined in Figure 7-1. For a small single edge notch [Figure 7-1(a)], $f = 1.12$. If the specimen is assumed to have an "infinite" width, $f \cong 1.0$.

By performing a test on a specimen with a known flaw size, we can determine the value of K that causes the flaw to grow and cause failure. This critical stress intensity factor is defined as the *fracture toughness* K_c:

$$K_c = K \text{ required for a crack to propagate} \quad (7\text{-}2)$$

Fracture toughness depends on the thickness of the sample: as thickness increases, fracture toughness K_c decreases to a constant value (Figure 7-2). This constant is called the *plane strain fracture toughness* K_{Ic}. It is K_{Ic} that is normally reported as the property of a material. The value of K_{Ic} does not depend upon the thickness of the sample. Table 7-1 reports the values of K_{Ic} and the yield strengths of several materials. Units for fracture toughness are ksi $\sqrt{\text{in.}}$ = 1.0989 MPa $\sqrt{\text{m}}$.

The ability of a material to resist the growth of a crack depends on a large number of factors:

1. Larger flaws reduce the permitted stress. Special manufacturing techniques, such as filtering impurities from liquid metals and hot pressing or hot isostatic pressing of powder particles to produce ceramic or superalloy components reduce flaw size and improve fracture toughness.

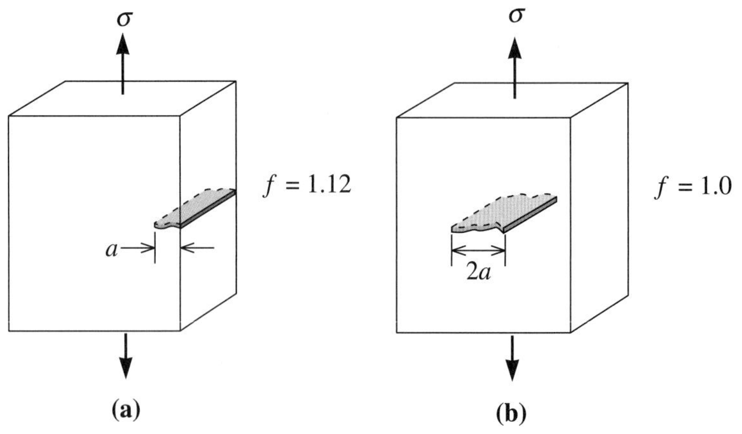

Figure 7-1 Schematic drawing of fracture toughness specimens with (a) edge and (b) internal flaws. The flaw size is defined differently for the two cases.

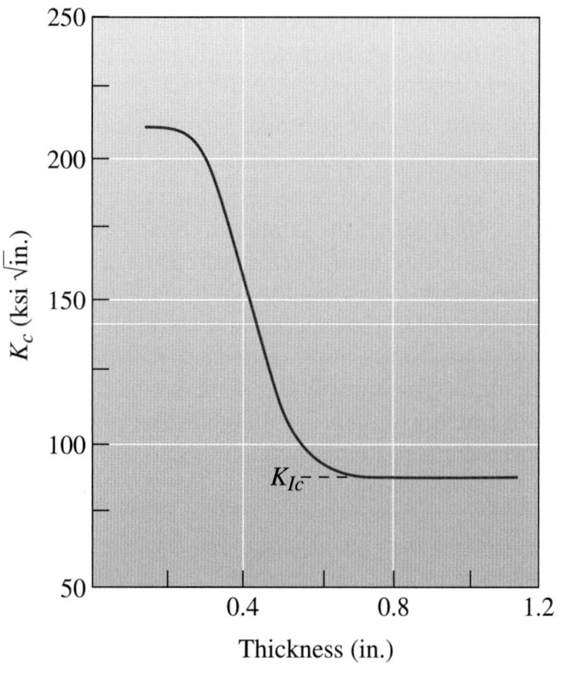

Figure 7-2 The fracture toughness K_c of a 300,000 psi yield strength steel decreases with increasing thickness, eventually leveling off at the plane strain fracture toughness K_{Ic}.

Table 7-1 The plane strain fracture toughness K_{Ic} of selected materials

Material	Fracture Toughness K_{Ic} (psi $\sqrt{\text{in.}}$)	Yield Strength or Ultimate Strength (for Brittle Solids) (psi)
Al-Cu alloy	22,000	66,000
	33,000	47,000
Ti - 6% Al - 4% V	50,000	130,000
	90,000	125,000
Ni-Cr steel	45,800	238,000
	80,000	206,000
Al_2O_3	1,600	30,000
Si_3N_4	4,500	80,000
Transformation toughened ZrO_2	10,000	60,000
Si_3N_4-SiC composite	51,000	120,000
Polymethyl methacrylate polymer	900	4,000
Polycarbonate polymer	3,000	8,400

2. The ability of a material to deform is critical. In ductile metals, the material near the tip of the flaw can deform, causing the tip of any crack to become blunt, reducing the stress intensity factor and preventing growth of the crack. Increasing the strength of a given metal usually decreases ductility and gives a lower fracture toughness. (See Table 7-1.) Brittle materials such as ceramics and many polymers have much lower fracture toughnesses than metals.
3. Thicker, more rigid pieces of a given material have a lower fracture toughness than thin materials.
4. Increasing the rate of application of the load, such as in an impact test, typically reduces the fracture toughness of the material.
5. Increasing the temperature normally increases the fracture toughness, just as in the impact test.

6. A small grain size normally improves fracture toughness, whereas more point defects and dislocations reduce fracture toughness. Thus, a fine-grained ceramic material may provide improved resistance to crack growth.
7. In certain ceramic materials, we can take advantage of stress-induced transformations that lead to compressive stresses that cause increased fracture toughness.

Fracture testing of ceramics cannot be performed easily using a sharp notch, since formation of such a notch often causes the samples to break. We can use hardness testing to measure the fracture toughness of many ceramics. When a ceramic material is indented, tensile stresses generate secondary cracks that form at the indentation and the length of secondary cracks provides a measure of the toughness of the ceramic material. In some cases, an indentation created using a hardness tester is used as a starter crack for the bend test. In general, this direct-crack measurement method is better suited for comparison, rather than absolute measurements of fracture toughness values. The fracture toughnesses and fracture strengths of many engineered materials are shown in Figure 7-3.

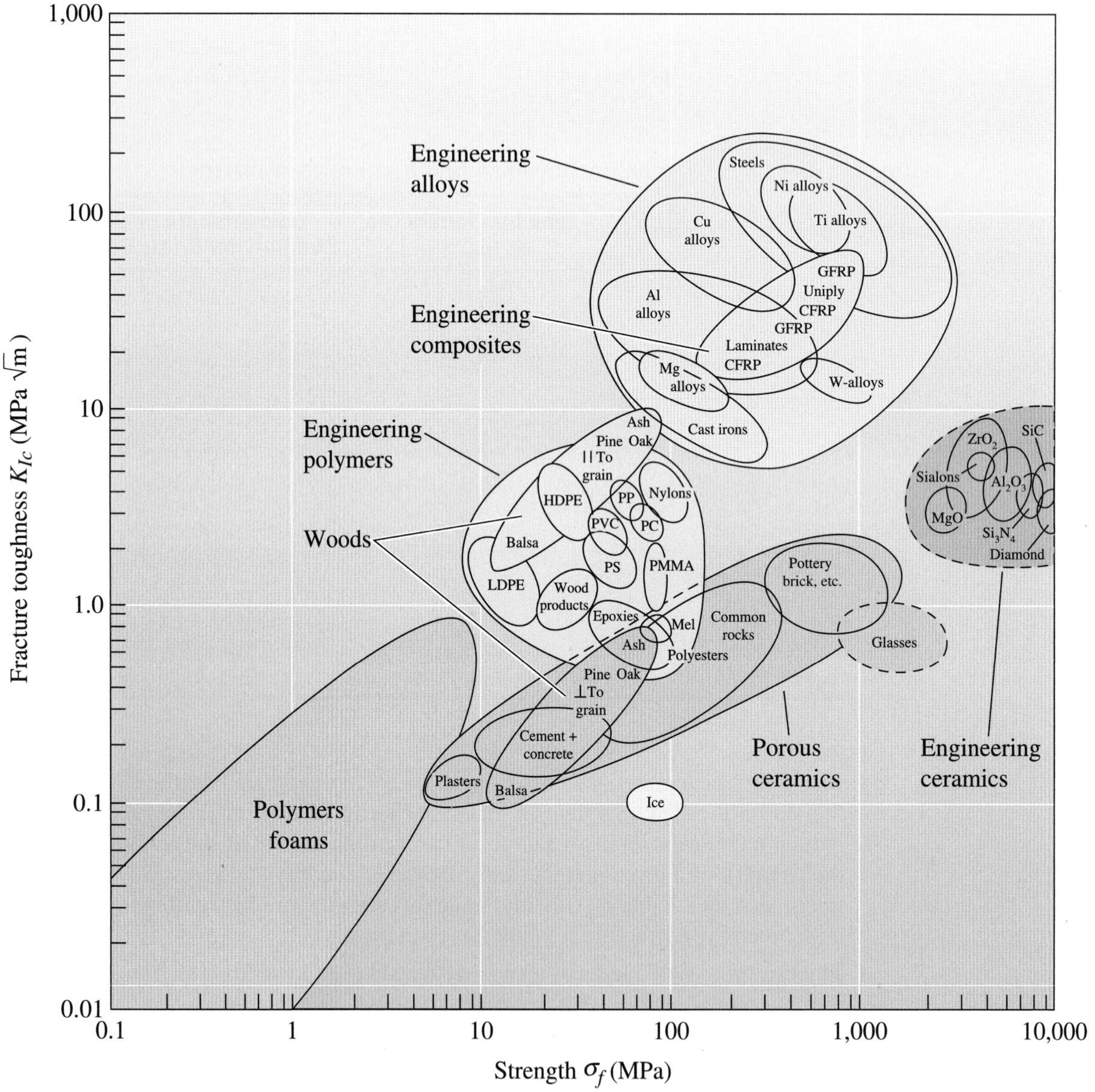

Figure 7-3 Fracture toughness versus strength of different engineered materials. [*Based on Courtney, T. H.,* Mechanical Behavior of Materials, *2/e, Long Grove, IL; Waveland Press, Inc., 2000.*]

7-2 The Importance of Fracture Mechanics

The fracture mechanics approach allows us to design and select materials while taking into account the inevitable presence of flaws. There are three variables to consider: the property of the material (K_c or K_{Ic}), the stress σ that the material must withstand, and the size of the flaw a. If we know two of these variables, the third can be determined.

Selection of a Material If we know the maximum size a of flaws in the material and the magnitude of the applied stress, we can select a material that has a fracture toughness K_c or K_{Ic} large enough to prevent the flaw from growing.

Design of a Component If we know the maximum size of any flaw and the material (and therefore its K_c or K_{Ic} has already been selected), we can calculate the maximum stress that the component can withstand. Then we can size the part appropriately to ensure that the maximum stress is not exceeded.

Design of a Manufacturing or Testing Method If the material has been selected, the applied stress is known, and the size of the component is fixed, we can calculate the maximum size of a flaw that can be tolerated. A nondestructive testing technique that detects any flaw greater than this critical size can help ensure that the part will function safely. In addition, we find that, by selecting the correct manufacturing process, we will produce flaws that are all smaller than this critical size.

Brittle Fracture Any crack or imperfection limits the ability of a ceramic to withstand a tensile stress. This is because a crack (sometimes called a **Griffith flaw**) concentrates and magnifies the applied stress. Figure 7-4 shows a crack of length a at the surface of a brittle material. The radius r of the crack is also shown. When a tensile stress σ is applied, the actual stress at the crack tip is

$$\sigma_{\text{actual}} \cong 2\sigma\sqrt{a/r} \tag{7-3}$$

For very thin cracks (r) or long cracks (a), the ratio $\sigma_{\text{actual}}/\sigma$ becomes large, or the stress is intensified. If the stress (σ_{actual}) exceeds the yield strength, the crack grows and eventually causes failure, even though the nominal applied stress σ is small.

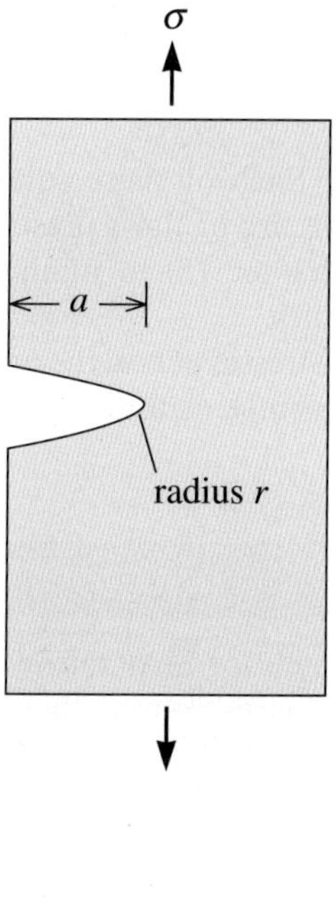

Figure 7-4 Schematic diagram of a Griffith flaw in a ceramic.

Example 7-1 *Design of a Nondestructive Test*

A large steel plate used in a nuclear reactor has a plane strain fracture toughness of 80,000 psi $\sqrt{\text{in.}}$ and is exposed to a stress of 45,000 psi during service. Design a testing or inspection procedure capable of detecting a crack at the edge of the plate before the crack is likely to grow at a catastrophic rate.

SOLUTION

We need to determine the minimum size of a crack that will propagate in the steel under these conditions. From Equation 7-1 assuming that $f = 1.12$ for a single-edge notch crack:

$$K_{Ic} = f\sigma\sqrt{\pi a}$$

$$80{,}000 \text{ psi}\sqrt{\text{in.}} = (1.12)(45{,}000 \text{ psi})\sqrt{\pi a}$$

$$a = 0.8 \text{ in.}$$

A 0.8 in. deep crack on the edge should be relatively easy to detect. Often, cracks of this size can be observed visually. A variety of other tests, such as dye penetrant inspection, magnetic particle inspection, and eddy current inspection, also detect cracks much smaller than this. If the growth rate of a crack is slow and inspection is performed on a regular basis, a crack should be discovered long before reaching this critical size.

In a different approach, we recognize that an applied stress causes an elastic strain, related to the modulus of elasticity E of the material. When a crack propagates, this strain energy is released, reducing the overall energy. At the same time, however, two new surfaces are created by the extension of the crack; this increases the energy associated with the surface. By balancing the strain energy and the surface energy, we find that the critical stress required to propagate the crack is given by the Griffith equation:

$$\sigma_{\text{critical}} \cong \sqrt{\frac{2E\gamma}{\pi a}} \tag{7-4}$$

where a is the length of a surface crack (or one-half the length of an internal crack) and γ is the surface energy per unit area. Again, this equation shows that even small flaws severely limit the strength of the ceramic.

We also note that if we rearrange Equation 7-1, which described the stress intensity factor K, we obtain

$$\sigma = \frac{K}{f\sqrt{\pi a}} \tag{7-5}$$

This equation is similar in form to Equation 7-4. Each of these equations points out the dependence of the mechanical properties on the size of flaws present in the ceramic. Development of manufacturing processes (see Chapter 15) to minimize the flaw size becomes crucial in improving the strength of ceramics.

The flaws are most important when tensile stresses act on the material. Compressive stresses close rather than open a crack; consequently, ceramics often have very good compressive strengths.

Example 7-2 *Properties of SiAlON Ceramics*

Assume that an advanced ceramic sialon (acronym for SiAlON or silicon aluminum oxynitride) has a tensile strength of 60,000 psi. Let us assume that this value is for a flaw-free ceramic. (In practice, it is almost impossible to produce flaw-free ceramics.)

A thin crack 0.01 in. deep is observed before a sialon part is tested. The part unexpectedly fails at a stress of 500 psi by propagation of the crack. Estimate the radius of the crack tip.

SOLUTION

The failure occurred because the 500 psi applied stress, magnified by the stress concentration at the tip of the crack, produced an actual stress equal to the ultimate tensile strength. From Equation 7-3,

$$\sigma_{\text{actual}} = 2\sigma\sqrt{a/r}$$

$$60{,}000 \text{ psi} = (2)(500 \text{ psi})\sqrt{0.01 \text{ in.}/r}$$

$$\sqrt{0.01/r} = 60 \qquad \text{or} \qquad 0.01/r = 3600$$

$$r = 2.8 \times 10^{-6} \text{ in.} = 7.1 \times 10^{-6} \text{ cm} = 710 \text{ Å}$$

The likelihood of being able to measure a radius of curvature of this size by any method of nondestructive testing is virtually zero. Therefore, although Equation 7-3 may help illustrate the factors that influence how a crack propagates in a brittle material, it does not help predict the strength of actual ceramic parts.

Example 7-3 *Design of a Ceramic Support*

Determine the minimum allowable thickness for a 3-in.-wide plate made from sialon that has a fracture toughness of 9000 psi $\sqrt{\text{in.}}$. The plate must withstand a tensile load of 40,000 lb. The part will be nondestructively tested to ensure that no flaws are present that might cause failure. The minimum allowable thickness of the part will depend on the minimum flaw size that can be determined by the available testing technique. Assume that three nondestructive testing techniques are available. X-ray radiography can detect flaws larger than 0.02 in.; gamma-ray radiography can detect flaws larger than 0.008 in.; and ultrasonic inspection can detect flaws larger than 0.005 in. Assume that the geometry factor $f = 1.0$ for all flaws.

SOLUTION

For the given flaw sizes, we must calculate the minimum thickness of the plate that will ensure that these flaw sizes will not propagate. From Equation 7-5, with $f = 1.0$,

$$\sigma_{\max} = \frac{K_{Ic}}{\sqrt{\pi a}} = \frac{F}{A_{\min}}$$

$$A_{\min} = \frac{F\sqrt{\pi a}}{K_{Ic}} = \frac{(40{,}000 \text{ lb})(\sqrt{\pi})(\sqrt{a})}{9{,}000 \text{ psi}\sqrt{\text{in.}}}$$

$$A_{\min} = 7.88\sqrt{a} \text{ in.}^2 \text{ and thickness} = [7.88 \text{ in.}^2/(3 \text{ in.})]\sqrt{a} = 2.63\sqrt{a}$$

Nondestructive Testing Method	Smallest Detectable Crack (in.)	Minimum Area (in.^{2})	Minimum Thickness (in.)	Maximum Stress (ksi)
X-ray radiography	0.020	1.11	0.37	36
γ-ray radiography	0.008	0.70	0.23	57
Ultrasonic inspection	0.005	0.56	0.19	72

Our ability to detect flaws, coupled with our ability to produce a ceramic with flaws smaller than our detection limit, significantly affects the maximum stress than can be tolerated and, hence, the size of the part. In this example, the part can be smaller if ultrasonic inspection is available.

The fracture toughness is also important. Had we used Si_3N_4, with a fracture toughness of 3000 psi $\sqrt{\text{in.}}$ instead of the sialon, we could repeat the calculations and show that, for ultrasonic testing, the minimum thickness is 0.56 in. and the maximum stress is only 24 ksi.

7-3 Microstructural Features of Fracture in Metallic Materials

Ductile Fracture Ductile fracture normally occurs in a **transgranular** manner (through the grains) in metals that have good ductility and toughness. Often, a considerable amount of deformation—including necking—is observed in the failed component. The deformation occurs before the final fracture. Ductile fractures are usually caused by simple overloads, or by applying too high a stress to the material.

In a simple tensile test, ductile fracture begins with the nucleation, growth, and coalescence of microvoids near the center of the test bar (Figure 7-5). **Microvoids** form when a high stress causes separation of the metal at grain boundaries or interfaces between the metal and small impurity particles (inclusions). As the local stress increases, the microvoids grow and coalesce into larger cavities. Eventually, the metal-to-metal contact area is too small to support the load and fracture occurs.

Deformation by slip also contributes to the ductile fracture of a metal. We know that slip occurs when the resolved shear stress reaches the critical resolved shear stress and that the resolved shear stresses are highest at a 45° angle to the applied tensile stress (Chapter 4, Schmid's law).

These two aspects of ductile fracture give the failed surface characteristic features. In thick metal sections, we expect to find evidence of necking, with a significant portion of the fracture surface having a flat face where microvoids first nucleated and coalesced, and a small shear lip, where the fracture surface is at a 45° angle to the applied stress. The shear lip, indicating that slip occurred, gives the fracture a cup and cone appearance (Figure 6-7 and Figure 7-6). Simple macroscopic observation of this fracture may be sufficient to identify the ductile fracture mode.

Examination of the fracture surface at a high magnification—perhaps using a scanning electron microscope—reveals a dimpled surface (Figure 7-7). The dimples are traces of

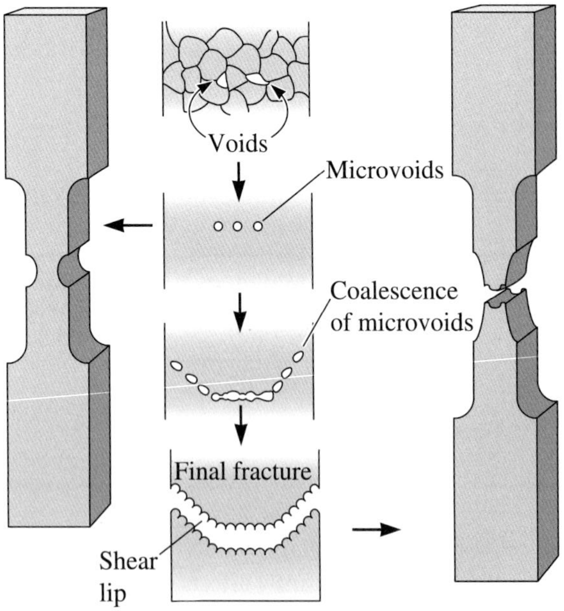

Figure 7-5 When a ductile material is pulled in a tensile test, necking begins and voids form—starting near the center of the bar—by nucleation at grain boundaries or inclusions. As deformation continues, a 45° shear lip may form, producing a final cup and cone fracture.

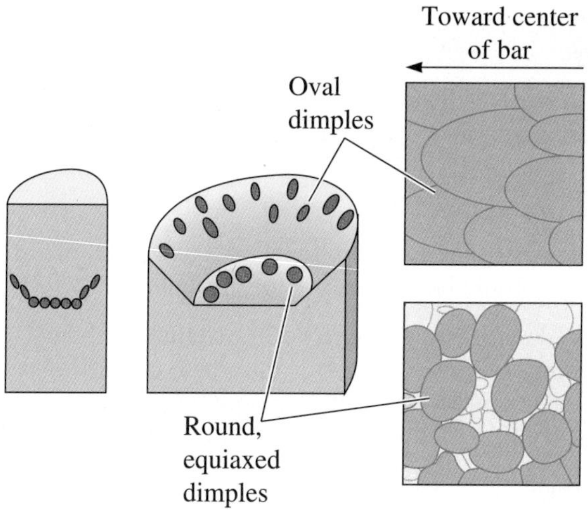

Figure 7-6 Dimples form during ductile fracture. Equiaxed dimples form in the center, where microvoids grow. Elongated dimples, pointing toward the origin of failure, form on the shear lip.

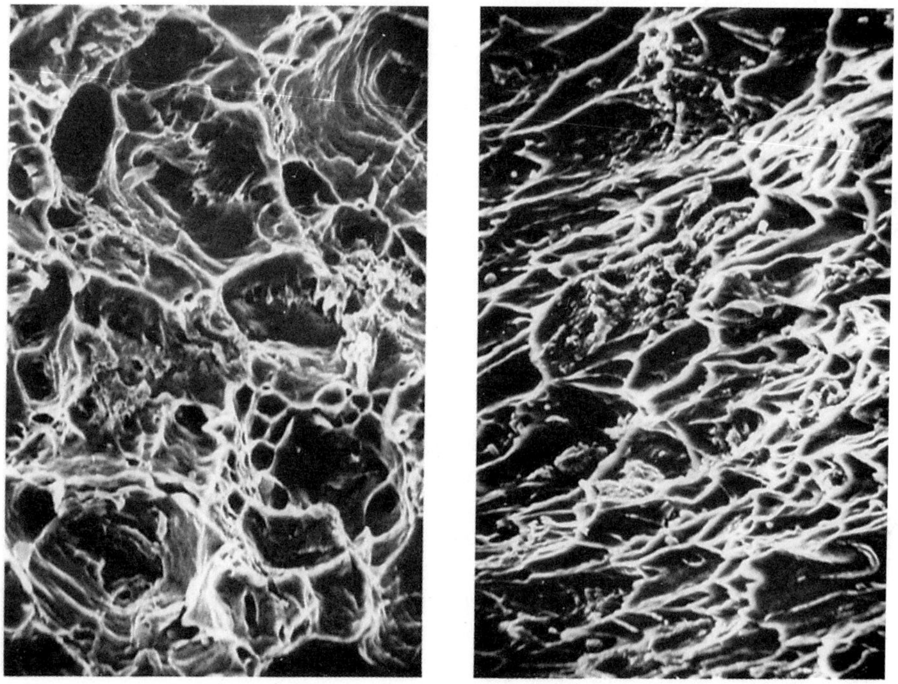

Figure 7-7 Scanning electron micrographs of an annealed 1018 steel exhibiting ductile fracture in a tensile test. (a) Equiaxed dimples at the flat center of the cup and cone, (*Reprinted Courtesy of Don Askeland*) and (b) elongated dimples at the shear lip (× 1250). (*Reprinted Courtesy of Don Askeland*)

the microvoids produced during fracture. Normally, these microvoids are round, or equiaxed, when a normal tensile stress produces the failure [Figure 7-7(a)]; however, on the shear lip, the dimples are oval-shaped, or elongated, with the ovals pointing toward the origin of the fracture [Figure 7-7(b)].

In a thin plate, less necking is observed and the entire fracture surface may be a shear face. Microscopic examination of the fracture surface shows elongated dimples rather than equiaxed dimples, indicating a greater proportion of 45° slip than in thicker metals.

Brittle Fracture Brittle fracture occurs in high-strength metals and alloys or metals and alloys with poor ductility and toughness. Furthermore, even metals that are normally ductile may fail in a brittle manner at low temperatures, in thick sections, at

Example 7-4 Hoist Chain Failure Analysis

A chain used to hoist heavy loads fails. Examination of the failed link indicates considerable deformation and necking prior to failure. List some of the possible reasons for failure.

SOLUTION

This description suggests that the chain failed in a ductile manner by a simple tensile overload. Two factors could be responsible for this failure:

1. The load exceeded the hoisting capacity of the chain. Thus, the stress due to the load exceeded the ultimate tensile strength of the chain, permitting failure. Comparison of the load to the manufacturer's specifications will indicate that the chain was not intended for such a heavy load. This is the fault of the user!
2. The chain was of the wrong composition or was improperly heat-treated. Consequently, the yield strength was lower than intended by the manufacturer and could not support the load. This may be the fault of the manufacturer!

high strain rates (such as impact), or when flaws play an important role. Brittle fractures are frequently observed when impact, rather than overload, causes failure.

In brittle fracture, little or no plastic deformation is required. Initiation of the crack normally occurs at small flaws, which cause a concentration of stress. The crack may move at a rate approaching the velocity of sound in the metal. Normally, the crack propagates most easily along specific crystallographic planes, often the {100} planes, by cleavage. In some cases, however, the crack may take an **intergranular** (along the grain boundaries) path, particularly when segregation (preferential separation of different elements) or inclusions weaken the grain boundaries.

Brittle fracture can be identified by observing the features on the failed surface. Normally, the fracture surface is flat and perpendicular to the applied stress in a tensile test. If failure occurs by cleavage, each fractured grain is flat and differently oriented, giving a crystalline or "rock candy" appearance to the fracture surface (Figure 7-8). Often, the layman claims that the metal failed because it crystallized. Of course, we know that the metal was crystalline to begin with and the surface appearance is due to the cleavage faces.

Another common fracture feature is the **Chevron pattern** (Figure 7-9), produced by separate crack fronts propagating at different levels in the material. A radiating pattern of surface markings, or ridges, fans away from the origin of the crack (Figure 7-10). The Chevron pattern is visible with the naked eye or a magnifying glass and helps us identify both the brittle nature of the failure process as well as the origin of the failure.

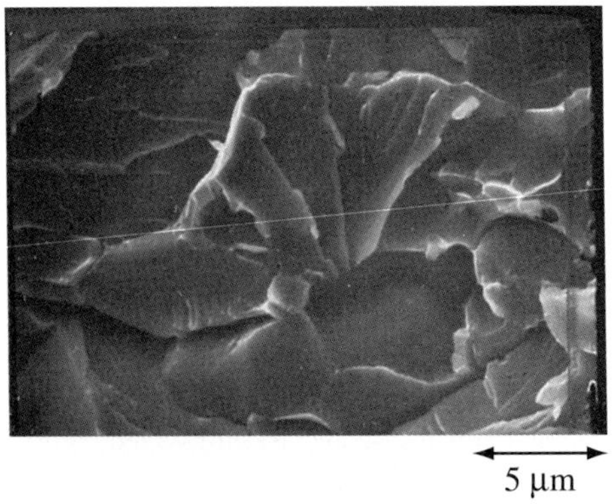

Figure 7-8 Scanning electron micrograph of a brittle fracture surface of a quenched 1010 steel. (*Courtesy of C. W. Ramsay*)

Figure 7-9 The Chevron pattern in a 0.5-in.-diameter quenched 4340 steel. The steel failed in a brittle manner by an impact blow. (*Reprinted Courtesy of Don Askeland*)

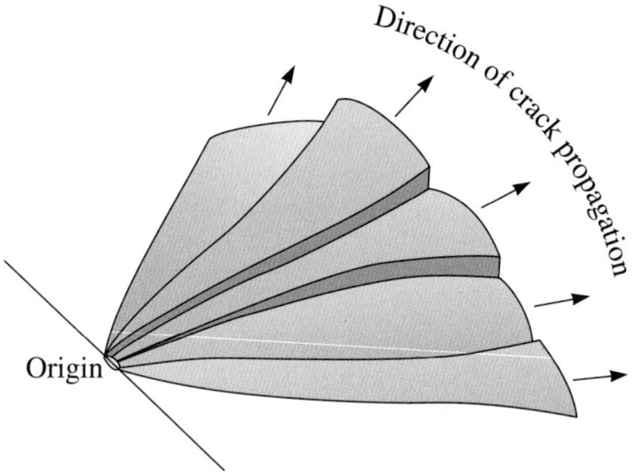

Figure 7-10 The Chevron pattern forms as the crack propagates from the origin at different levels. The pattern points back to the origin.

Example 7-5 *Automobile Axle Failure Analysis*

An engineer investigating the cause of an automobile accident finds that the right rear wheel has broken off at the axle. The axle is bent. The fracture surface reveals a Chevron pattern pointing toward the surface of the axle. Suggest a possible cause for the fracture.

SOLUTION

The evidence suggests that the axle did not break prior to the accident. The deformed axle means that the wheel was still attached when the load was applied. The Chevron pattern indicates that the wheel was subjected to an intense impact blow, which was transmitted to the axle. The preliminary evidence suggests that the driver lost control and crashed, and the force of the crash caused the axle to break. Further examination of the fracture surface, microstructure, composition, and properties may verify that the axle was manufactured properly.

7-4 Microstructural Features of Fracture in Ceramics, Glasses, and Composites

In ceramic materials, the ionic or covalent bonds permit little or no slip. Consequently, failure is a result of brittle fracture. Most crystalline ceramics fail by cleavage along widely spaced, closely packed planes. The fracture surface typically is smooth, and frequently no characteristic surface features point to the origin of the fracture [Figure 7-11(a)].

Glasses also fracture in a brittle manner. Frequently, a **conchoidal fracture** surface is observed. This surface contains a smooth mirror zone near the origin of the fracture, with tear lines comprising the remainder of the surface [Figure 7-11(b)].

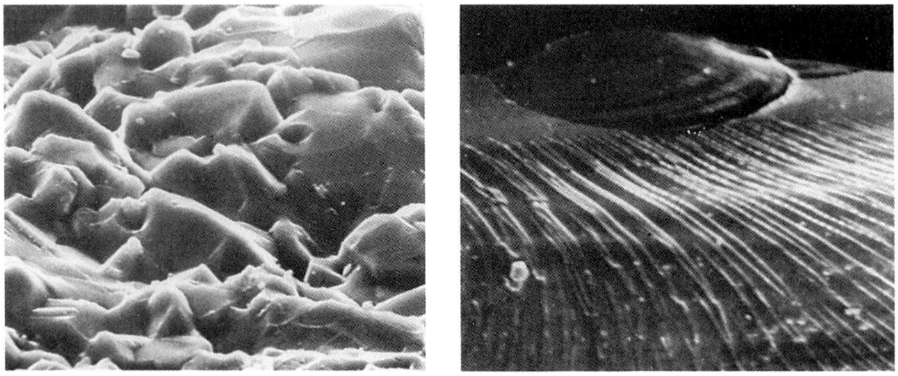

(a) (b)

Figure 7-11 Scanning electron micrographs of fracture surfaces in ceramics. (a) The fracture surface of Al_2O_3, showing the cleavage faces (× 1250) (*Reprinted Courtesy of Don Askeland*) and (b) the fracture surface of glass, showing the mirror zone (top) and tear lines characteristic of conchoidal fracture (× 300). (*Reprinted Courtesy of Don Askeland*)

The tear lines point back to the mirror zone and the origin of the crack, much like the Chevron pattern in metals.

Polymers can fail by either a ductile or a brittle mechanism. Below the glass-transition temperature (T_g), thermoplastic polymers fail in a brittle manner—much like a glass. Likewise, the hard thermoset polymers, which have a rigid, three-dimensional cross-linked structure (see Chapter 16), fail by a brittle mechanism. Some plastics with structures consisting of tangled but not chemically cross-linked chains fail in a ductile manner above the glass-transition temperature, giving evidence of extensive deformation and even necking prior to failure. The ductile behavior is a result of sliding of the polymer chains, which is not possible in thermosetting polymers.

Fracture in fiber-reinforced composite materials is more complex. Typically, these composites contain strong, brittle fibers surrounded by a soft, ductile matrix, as in boron-reinforced aluminum. When a tensile stress is applied along the fibers, the soft aluminum deforms in a ductile manner, with void formation and coalescence eventually producing a dimpled fracture surface. As the aluminum deforms, the load is no longer transmitted effectively between the fibers and matrix; the fibers break in a brittle manner until there are too few of them left intact to support the final load.

Fracture is more common if the bonding between the fibers and matrix is poor. Voids can then form between the fibers and the matrix, causing pull-out. Voids can also form between layers of the matrix if composite tapes or sheets are not properly bonded, causing **delamination** (Figure 7-12). Delamination, in this context, means the layers of different materials in a composite begin to come apart.

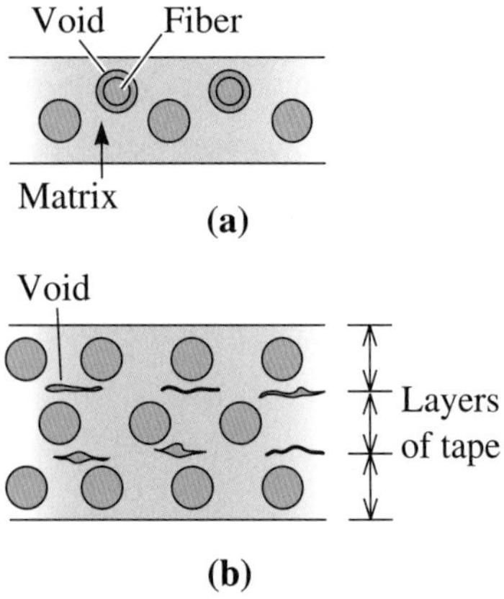

Figure 7-12 Fiber-reinforced composites can fail by several mechanisms. (a) Due to weak bonding between the matrix and fibers, voids can form, which then lead to fiber pull-out. (b) If the individual layers of the matrix are poorly bonded, the matrix may delaminate, creating voids.

Example 7-6 *Fracture in Composites*

Describe the difference in fracture mechanism between a boron-reinforced aluminum composite and a glass fiber-reinforced epoxy composite.

SOLUTION

In the boron-aluminum composite, the aluminum matrix is soft and ductile; thus, we expect the matrix to fail in a ductile manner. Boron fibers, in contrast, fail in a brittle manner. Both glass fibers and epoxy are brittle; thus the composite as a whole should display little evidence of ductile fracture.

7-5 Weibull Statistics for Failure Strength Analysis

We need a statistical approach when evaluating the strength of ceramic materials. The strength of ceramics and glasses depends upon the size and distribution of sizes of flaws. In these materials, flaws originate during processing as the ceramics are manufactured. The flaws can form, for example, during machining or grinding. Glasses can also develop microcracks as a result of interaction with water vapor in air. If we test alumina or other ceramic components of different sizes and geometries, we often find a large scatter in the measured values—even if their nominal composition is the same. Similarly, if we are testing the strength of glass fibers of a given composition, we find that, on average, shorter fibers are stronger than longer fibers. The strength of ceramics and glasses depends upon the probability of finding a flaw that exceeds a certain critical size. This probability increases as the component size or fiber length increases. As a result, the strength of larger components or fibers is likely to be lower than that of smaller components or shorter fibers. In metallic or polymeric materials, which can exhibit relatively large plastic deformations, the effect of flaws and flaw size distribution is not felt to the extent it is in ceramics and glasses. In these materials, cracks initiating from flaws get blunted by plastic deformation. Thus, for ductile materials, the distribution of strength is narrow and close to a Gaussian distribution. The strength of ceramics and glasses, however, varies considerably (i.e., if we test a large number of identical samples of silica glass or alumina ceramic, the data will show a wide scatter owing to changes in distribution of flaw sizes). The strength of brittle materials, such as ceramics and glasses, is not Gaussian; it is given by the **Weibull distribution**. The Weibull distribution is an indicator of the variability of strength of materials resulting from a distribution of flaw sizes. This behavior results from critical-sized flaws in materials with a distribution of flaw sizes (i.e., failure due to the weakest link of a chain).

The Weibull distribution shown in Figure 7-13 describes the fraction of samples that fail at different applied stresses. At low stresses, a small fraction of samples contains flaws large enough to cause fracture; most fail at an intermediate applied stress, and a few contain only small flaws and do not fail until large stresses are applied. To provide predictability, we prefer a very narrow distribution.

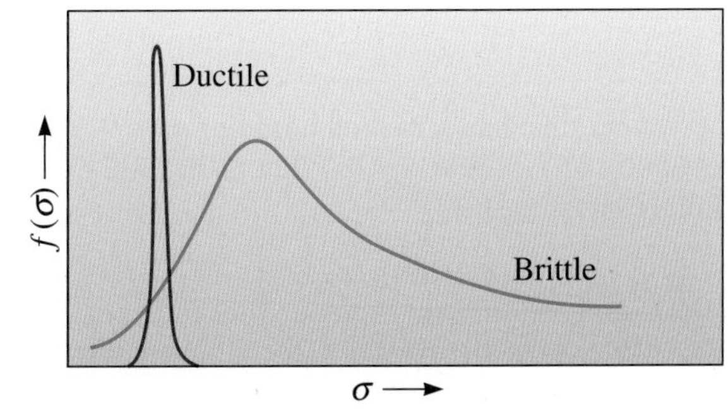

Figure 7-13 The Weibull distribution describes the fraction of the samples that fail at any given applied stress.

Consider a body of volume V with a distribution of flaws and subjected to a stress σ. If we assume that the volume, V, is made up of n elements with volume V_0 and each element has the same flaw-size distribution, it can be shown that the survival probability, $P(V_0)$, (i.e., the probability that a brittle material will not fracture under the applied stress σ) is given by

$$P(V_0) = \exp\left[-\left(\frac{\sigma - \sigma_{\min}}{\sigma_0}\right)^m\right] \tag{7-6}$$

The probability of failure, $F(V_0)$, can be written as

$$F(V_0) = 1 - P(V_0) = 1 - \exp\left[-\left(\frac{\sigma - \sigma_{\min}}{\sigma_0}\right)^m\right] \tag{7-7}$$

In Equations 7-6 and 7-7, σ is the applied stress, σ_0 is a scaling parameter dependent on specimen size and shape, and $\sigma_{\min}$ is the stress level below which the probability of failure is zero (i.e., the probability of survival is 1.0). In these equations, m is the **Weibull modulus**. In theory, Weibull modulus values can range from 0 to ∞. The Weibull modulus is a measure of the variability of the strength of the material.

The Weibull modulus m indicates the strength variability. For metals and alloys, the Weibull modulus is ~100. For traditional ceramics (e.g., bricks, pottery, etc.), the Weibull modulus is less than 3. Engineered ceramics, in which the processing is better controlled and hence the number of flaws is expected to be fewer, have a Weibull modulus of 5 to 10.

Note that for ceramics and other brittle solids, we can assume $\sigma_{\min} = 0$. This is because there is no nonzero stress level for which we can claim a brittle material will not fail. For *brittle materials*, Equations 7-6 and 7-7 can be rewritten as follows:

$$P(V_0) = \exp\left[-\left(\frac{\sigma}{\sigma_0}\right)^m\right] \tag{7-8}$$

and

$$F(V_0) = 1 - P(V_0) = 1 - \exp\left[-\left(\frac{\sigma}{\sigma_0}\right)^m\right] \tag{7-9}$$

From Equation 7-8, for an applied stress σ of zero, the probability of survival is 1. As the applied stress σ increases, $P(V_0)$ decreases, approaching zero at very high values of applied stresses. We can also describe another meaning of the parameter σ_0. In Equation 7-8, when $\sigma = \sigma_0$, the probability of survival becomes $1/e \cong 0.37$. Therefore, σ_0 is the stress level for which the survival probability is $\cong 0.37$ or 37%. We can also state that σ_0 is the stress level for which the failure probability is $\cong 0.63$ or 63%.

Equations 7-8 and 7-9 can be modified to account for samples with different volumes. It can be shown that for an equal probability of survival, samples with larger volumes will have lower strengths. This is what we mentioned before (e.g., longer glass fibers will be weaker than shorter glass fibers).

The following examples illustrate how the Weibull plots can be used for analysis of mechanical properties of materials and design of components.

Example 7-7 *Strength of Ceramics and Probability of Failure*

An advanced engineered ceramic has a Weibull modulus $m = 9$. The flexural strength is 250 MPa at a probability of failure $F = 0.4$. What is the level of flexural strength if the probability of failure has to be 0.1?

SOLUTION

We assume all samples tested had the same volume; thus the size of the sample will not be a factor in this case. We can use the symbol V for sample volume instead of V_0. We are dealing with a brittle material so we begin with Equation 7-9:

$$F(V) = 1 - P(V) = 1 - \exp\left[-\left(\frac{\sigma}{\sigma_0}\right)^m\right]$$

or

$$1 - F(V) = \exp\left[-\left(\frac{\sigma}{\sigma_0}\right)^m\right]$$

Taking the natural logarithm of both sides,

$$\ln[1 - F(V)] = -\left(\frac{\sigma}{\sigma_0}\right)^m$$

Taking natural logarithms again,

$$\ln\{-\ln[1 - F(V)]\} = m(\ln \sigma - \ln \sigma_0) \quad (7\text{-}10)$$

We can eliminate the minus sign on the left-hand side of Equation 7-10 by rewriting it as

$$\ln\left\{\ln\left[\frac{1}{1 - F(V)}\right]\right\} = m(\ln \sigma - \ln \sigma_0) \quad (7\text{-}11)$$

For $F = 0.4$, $\sigma = 250$ MPa, and $m = 9$ in Equation 7-11, we have

$$\ln\left[\ln\left(\frac{1}{1 - 0.4}\right)\right] = 9(\ln 250 - \ln \sigma_0) \quad (7\text{-}12)$$

Therefore,

$$\ln[\ln(1/0.6)] = -0.67173$$
$$= 9(5.52146 - \ln \sigma_0)$$

Therefore, $\ln \sigma_0 = 5.52146 + 0.07464 = 5.5961$. This gives us a value of $\sigma_0 = 269.4$ MPa. This is the characteristic strength of the ceramic. For a stress level of 269.4 MPa, the probability of survival is 0.37 (or the probability of failure is 0.63).

Now, we want to determine the value of σ for $F = 0.1$. As the required probability of failure (F) decreases, the stress level to which the ceramic can be subjected (σ) also decreases. We know that $m = 9$ and $\sigma_0 = 269.4$ MPa. We substitute these values into Equation 7-11:

$$\ln\left[\ln\left(\frac{1}{1 - 0.1}\right)\right] = 9(\ln \sigma - \ln 269.4)$$
$$-2.25037 = 9(\ln \sigma - 5.5961)$$
$$\therefore -0.25004 = \ln \sigma - 5.5961$$
$$\ln \sigma = 5.3461$$

or $\sigma = 209.8$ MPa. As expected, as we lowered the probability of failure to 0.1, we also decreased the level of stress that can be supported.

Example 7-8 *Weibull Modulus Parameter Determination*

Seven silicon carbide specimens were tested and the following fracture strengths were obtained: 23, 49, 34, 30, 55, 43, and 40 MPa. Estimate the Weibull modulus for the data by fitting the data to Equation 7-11. Discuss the reliability of the ceramic.

SOLUTION

First, we point out that for any type of statistical analysis, we need a large number of samples. Seven samples are not enough. The purpose of this example is to illustrate the calculation.

One simple though not completely accurate method for determining the behavior of the ceramic is to assign a numerical rank (1 to 7) to the specimens, with the specimen having the lowest fracture strength assigned the value 1. The total number of specimens is n (in our case, 7). The probability of failure F is then the numerical rank divided by $n + 1$ (in our case, 8). We can then plot $\ln\{\ln 1/[1 - F(V_0)]\}$ versus $\ln \sigma$. The following table and Figure 7-14 show the results of these calculations. Note that σ is plotted on a log scale.

*i*th Specimen	σ (MPa)	$F(V_0)$	$\ln\{\ln 1/[1 - F(V_0)]\}$
1	23	1/8 = 0.125	−2.013
2	30	2/8 = 0.250	−1.246
3	34	3/8 = 0.375	−0.755
4	40	4/8 = 0.500	−0.367
5	43	5/8 = 0.625	−0.019
6	49	6/8 = 0.750	+0.327
7	55	7/8 = 0.875	+0.732

The slope of the fitted line, or the Weibull modulus m, is (using the two points indicated on the curve)

$$m = \frac{0.5 - (-2.0)}{\ln(52) - \ln(23.5)} = \frac{2.5}{3.951 - 3.157} = 3.15$$

This low Weibull modulus of 3.15 suggests that the ceramic has a highly variable fracture strength, making it difficult to use reliably in load-bearing applications.

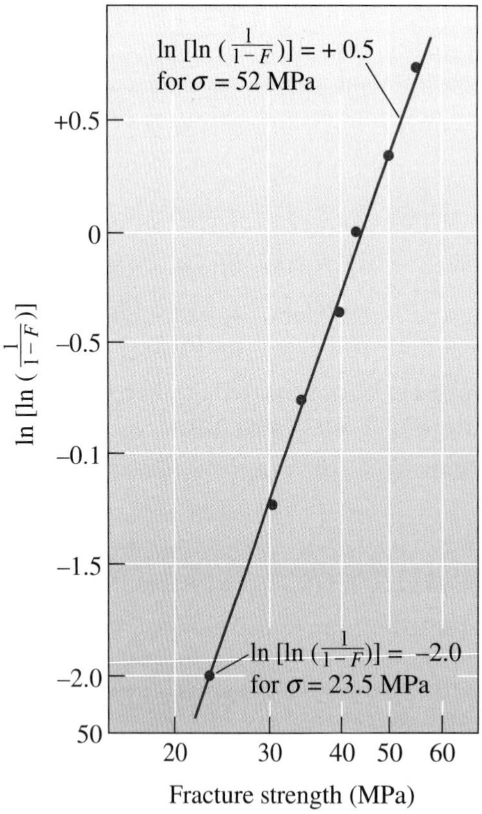

Figure 7-14 Plot of cumulative probability of failure versus fracture stress. Note the fracture strength is plotted on a log scale.

7-6 Fatigue

Fatigue is the lowering of strength or failure of a material due to repetitive stress that may be above or below the yield strength. It is a common phenomenon in load-bearing components in cars and airplanes, turbine blades, springs, crankshafts and other machinery, biomedical implants, and consumer products, such as shoes, that are subjected constantly to repetitive stresses in the form of tension, compression, bending, vibration, thermal expansion and contraction, or other stresses. These stresses are often *below* the yield strength of the material; however, when the stress occurs a sufficient number of times, it causes failure by fatigue! Quite a large fraction of components found in an automobile junkyard belongs to those that failed by fatigue. The possibility of a fatigue failure is the main reason why aircraft components have a finite life.

Fatigue failures typically occur in three stages. First, a tiny crack initiates or nucleates often at a time well after loading begins. Normally, nucleation sites are located at or near the surface, where the stress is at a maximum, and include surface defects such as scratches or pits, sharp corners due to poor design or manufacturing, inclusions, grain boundaries, or regions of high dislocation density. Next, the crack gradually propagates as the load continues to cycle. Finally, a sudden fracture of the material occurs when the remaining cross-section of the material is too small to support the applied load. Thus, components fail by fatigue because even though the overall applied stress may remain below the yield stress, at a local length scale, the stress intensity exceeds the tensile strength. For fatigue to occur, there must be a component of tensile stress. We typically are concerned with fatigue of metallic and polymeric materials.

In ceramics, we usually do not consider fatigue since ceramics typically fail because of their low fracture toughness. Any fatigue cracks that may form will lower the useful life of the ceramic since they will lower the fracture toughness. In general, we design with ceramics for static (and not cyclic) loading, and we factor in the Weibull modulus.

Polymeric materials also show fatigue failure. The mechanism of fatigue in polymers is different than that in metallic materials. In polymers, as the materials are subjected to repetitive stresses, considerable heating can occur near the crack tips and the interrelationships between fatigue and another mechanism, known as *creep* (discussed in Section 7-9), affect the overall behavior.

Fatigue failures are often easy to identify. The fracture surface—particularly near the origin—is typically smooth. The surface becomes rougher as the original crack increases in size and may be fibrous during final crack propagation. Microscopic and macroscopic examinations reveal a fracture surface including a beach mark pattern and striations (Figure 7-15). **Beach** or **clamshell marks** (Figure 7-16) are normally

Example 7-9 *Fatigue Failure Analysis of a Crankshaft*

A crankshaft in a diesel engine fails. Examination of the crankshaft reveals no plastic deformation. The fracture surface is smooth. In addition, several other cracks appear at other locations in the crankshaft. What type of failure mechanism occurred?

SOLUTION

Since the crankshaft is a rotating part, the surface experiences cyclical loading. We should immediately suspect fatigue. The absence of plastic deformation supports our suspicion. Furthermore, the presence of other cracks is consistent with fatigue; the other cracks did not have time to grow to the size that produced catastrophic failure. Examination of the fracture surface will probably reveal beach marks or fatigue striations.

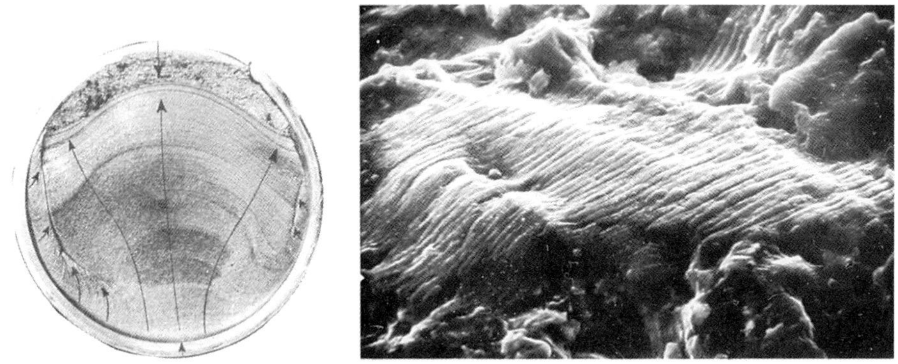

Figure 7-15 Fatigue fracture surface. (a) At low magnifications, the beach mark pattern indicates fatigue as the fracture mechanism. The arrows show the direction of growth of the crack front with the origin at the bottom of the photograph. (*C.C. Cottell, "Fatigue Failures, with Special Reference to Fracture Characteristics,"* Failure Analysis: The British Engine Technical Reports, *American Society of Metals, 1981, p. 318. Reprinted with permission of ASM International. All rights reserved. www.asminternational.org.*) (b) At very high magnifications, closely spaced striations formed during fatigue are observed (× 1000). (*Reprinted Courtesy of Don Askeland*)

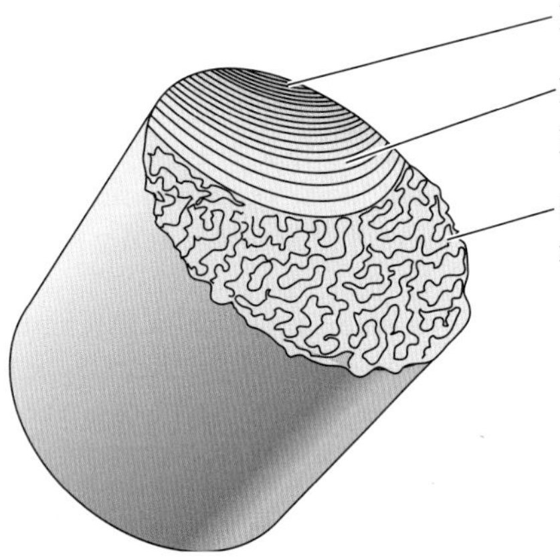

Figure 7-16 Schematic representation of a fatigue fracture surface in a steel shaft, showing the initiation region, the propagation of the fatigue crack (with beach markings), and catastrophic rupture when the crack length exceeds a critical value at the applied stress.

formed when the load is changed during service or when the loading is intermittent, perhaps permitting time for oxidation inside the crack. **Striations**, which are on a much finer scale, show the position of the crack tip after each cycle. Beach marks always suggest a fatigue failure, but—unfortunately—the absence of beach marks does not rule out fatigue failure.

A conventional and older method used to measure a material's resistance to fatigue is the **rotating cantilever beam test** (Figure 7-17). One end of a machined, cylindrical specimen is mounted in a motor-driven chuck. A weight is suspended from the opposite end. The specimen initially has a tensile force acting on the top surface, while the bottom surface is compressed. After the specimen turns 90°, the locations that were originally in tension and compression have no stress acting on them. After a half revolution of 180°, the material that was originally in tension is now in compression. Thus, the stress at any one point goes through a complete sinusoidal cycle from maximum tensile stress to maximum compressive stress. The maximum stress acting on this type of specimen is given by

$$\pm\sigma = \frac{32\,FL}{\pi d^3} \tag{7-13}$$

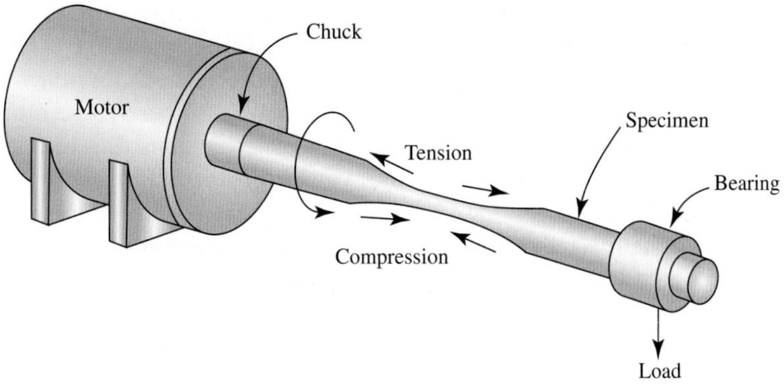

Figure 7-17 The rotating cantilever beam fatigue test.

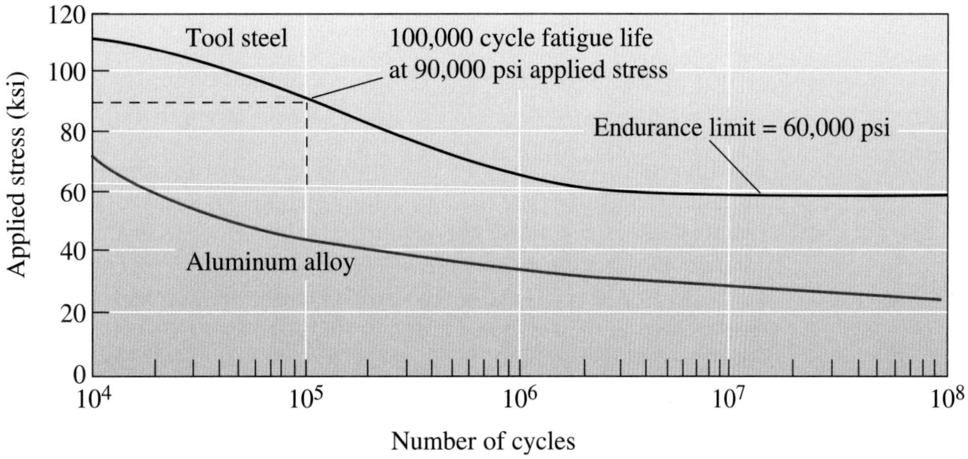

Figure 7-18 The stress-number of cycles to failure (S-N) curves for a tool steel and an aluminum alloy.

where L is the distance between the bending force location and the support, F is the load, and d is the diameter. Note that Equation 7-13 assumes that the specimen has a uniform cross-section.

Newer machines used for fatigue testing are known as direct-loading machines. In these machines, a servo-hydraulic system, an actuator, and a control system, driven by computers, applies a desired force, deflection, displacement, or strain. In some of these machines, temperature and atmosphere (e.g., humidity level) also can be controlled.

After a sufficient number of cycles in a fatigue test, the specimen may fail. Generally, a series of specimens is tested at different applied stresses. The results are presented as an **S-N curve** (also known as the **Wöhler curve**), with the stress (S) plotted versus the number of cycles (N) to failure (Figure 7-18).

7-7 Results of the Fatigue Test

The **fatigue test** can tell us how long a part may survive or the maximum allowable loads that can be applied without causing failure. The **endurance limit**, which is the stress below which there is a 50% probability that failure by fatigue will never occur, is our preferred design criterion. To prevent a tool steel part from failing (Figure 7-18), we must be sure that the applied stress is below 60,000 psi. The assumption of the existence of an endurance limit is a somewhat outdated concept. Recent research on many metals has shown that probably an endurance limit does not exist. We also need to account for the presence of corrosion, occasional overloads, and other mechanisms that may cause the material to fail below the endurance limit. Thus, values for an endurance limit should be treated with caution.

Fatigue life tells us how long a component survives at a particular stress. For example, if the tool steel (Figure 7-18) is cyclically subjected to an applied stress of 90,000 psi, the fatigue life will be 100,000 cycles. Knowing the time associated with each cycle, we can calculate a fatigue life value in years. **Fatigue strength** is the maximum stress for which fatigue will not occur within a particular number of cycles, such as 500,000,000. The fatigue strength is necessary for designing with aluminum and polymers, which do not have endurance limits.

In some materials, including steels, the endurance limit is approximately half the tensile strength. The ratio between the endurance limit and the tensile strength is known as the **endurance ratio**:

$$\text{Endurance ratio} = \frac{\text{endurance limit}}{\text{tensile strength}} \approx 0.5 \tag{7-14}$$

The endurance ratio allows us to estimate fatigue properties from the tensile test. The endurance ratio values are ~0.3 to 0.4 for metallic materials other than low and medium strength steels. *Again, recall the cautionary note that research has shown that an endurance limit does not exist for many materials.*

Most materials are **notch sensitive**, with the fatigue properties particularly sensitive to flaws at the surface. Design or manufacturing defects concentrate stresses and reduce the endurance limit, fatigue strength, or fatigue life. Sometimes highly polished surfaces are prepared in order to minimize the likelihood of a fatigue failure. **Shot peening** is a process that is used very effectively to enhance fatigue life of materials. Small metal spheres are shot at the component. This leads to a residual compressive stress at the surface similar to **tempering** of inorganic glasses.

Example 7-10 *Design of a Rotating Shaft*

A solid shaft for a cement kiln produced from the tool steel in Figure 7-18 must be 96 in. long and must survive continuous operation for one year with an applied load of 12,500 lb. The shaft makes one revolution per minute during operation. Design a shaft that will satisfy these requirements.

SOLUTION

The fatigue life required for our design is the total number of cycles N that the shaft will experience in one year:

$$N = (1 \text{ cycle/min})(60 \text{ min/h})(24 \text{ h/day})(365 \text{ days/yr})$$

$$N = 5.256 \times 10^5 \text{ cycles/yr}$$

From Figure 7-18, the applied stress therefore, must be less than about 72,000 psi. Using Equation 7-13, the diameter of the shaft is given by

$$\pm\sigma = \frac{32FL}{\pi d^3}$$

$$72{,}000 \text{ psi} = \frac{(32)(12{,}500 \text{ lb})(96 \text{ in.})}{\pi d^3}$$

$$d = 5.54 \text{ in.}$$

A shaft with a diameter of 5.54 in. should operate for one year under these conditions; however, a significant margin of safety probably should be incorporated in the design. In addition, we might consider producing a shaft that would never fail.

Let us assume the factor of safety to be 2 (i.e., we will assume that the maximum allowed stress level will be 72,000/2 = 36,000 psi). The minimum diameter required to prevent failure would now be

$$36{,}000 \text{ psi} = \frac{32(12{,}500 \text{ lb})(96 \text{ in.})}{\pi d^3}$$

$$d = 6.98 \text{ in.}$$

Selection of a larger shaft reduces the stress level and makes fatigue less likely to occur or delays the failure. Other considerations might, of course, be important. High temperatures and corrosive conditions are inherent in producing cement. If the shaft is heated or attacked by the corrosive environment, fatigue is accelerated. Thus, for applications involving fatigue of components, regular inspections of the components go a long way toward avoiding a catastrophic failure.

7-8 Application of Fatigue Testing

Components are often subjected to loading conditions that do not give equal stresses in tension and compression (Figure 7-19). For example, the maximum stress during compression may be less than the maximum tensile stress. In other cases, the loading may be between a maximum and a minimum tensile stress; here the S-N curve is presented as the stress amplitude versus the number of cycles to failure. *Stress amplitude* (σ_a) is defined as half of the difference between the maximum and minimum stresses, and *mean stress* (σ_m) is defined as the average between the maximum and minimum stresses:

$$\sigma_a = \frac{\sigma_{\max} - \sigma_{\min}}{2} \tag{7-15}$$

$$\sigma_m = \frac{\sigma_{\max} + \sigma_{\min}}{2} \tag{7-16}$$

A compressive stress is a "negative" stress. Thus, if the maximum tensile stress is 50,000 psi and the minimum stress is a 10,000 psi compressive stress, using Equations 7-15 and 7-16, the stress amplitude is 30,000 psi, and the mean stress is 20,000 psi.

As the mean stress increases, the stress amplitude must decrease in order for the material to withstand the applied stresses. The condition can be summarized by the Goodman relationship:

$$\sigma_a = \sigma_{fs}\left[1 - \left(\frac{\sigma_m}{\sigma_{UTS}}\right)\right] \tag{7-17}$$

where σ_{fs} is the desired fatigue strength for zero mean stress and σ_{UTS} is the tensile strength of the material. Therefore, in a typical rotating cantilever beam fatigue test, where the mean stress is zero, a relatively large stress amplitude can be tolerated without fatigue. If, however, an airplane wing is loaded near its yield strength, vibrations of even a small amplitude may cause a fatigue crack to initiate and grow.

Crack Growth Rate In many cases, a component may not be in danger of failure even when a crack is present. To estimate when failure might occur, the rate of propagation of

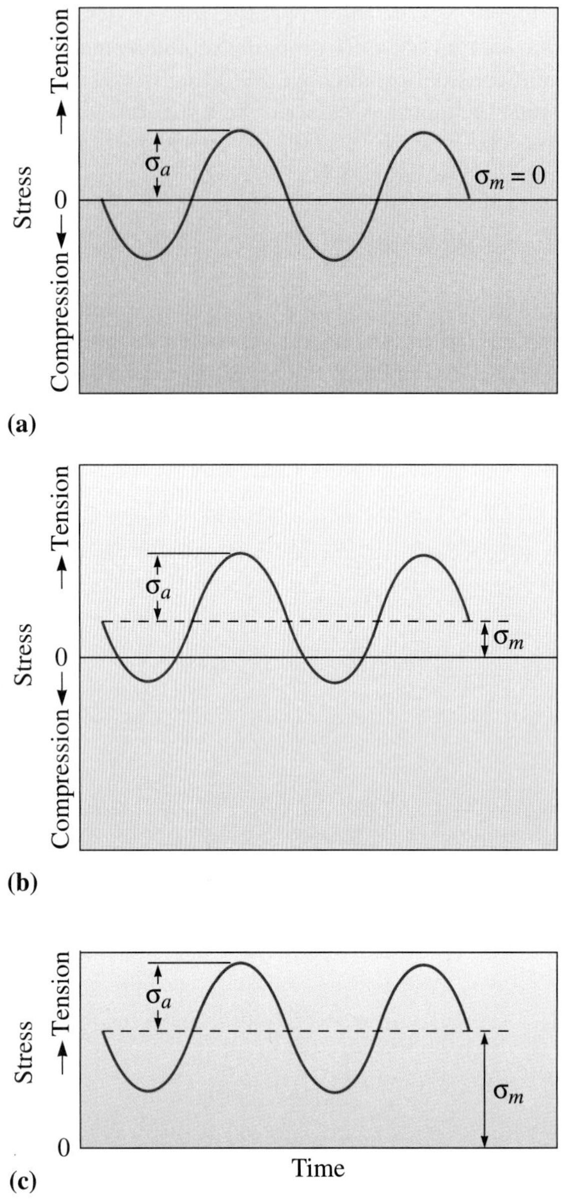

Figure 7-19 Examples of stress cycles. (a) Equal stress in tension and compression, (b) greater tensile stress than compressive stress, and (c) all of the stress is tensile.

a crack becomes important. Figure 7-20 shows the crack growth rate versus the range of the stress intensity factor ΔK, which characterizes the crack geometry and the stress amplitude. Below a threshold ΔK, a crack does not grow; for somewhat higher stress intensities, cracks grow slowly; and at still higher stress intensities, a crack grows at a rate given by

$$\frac{da}{dN} = C(\Delta K)^n \tag{7-18}$$

In this equation, C and n are empirical constants that depend upon the material. Finally, when ΔK is still higher, cracks grow in a rapid and unstable manner until fracture occurs.

The rate of crack growth increases as a crack increases in size, as predicted from the stress intensity factor (Equation 7-1):

$$\Delta K = K_{\max} - K_{\min} = f\sigma_{\max}\sqrt{\pi a} - f\sigma_{\min}\sqrt{\pi a} = f\Delta\sigma\sqrt{\pi a} \tag{7-19}$$

If the cyclical stress $\Delta\sigma$, $(\sigma_{\max} - \sigma_{\min})$, is not changed, then as crack length a increases, ΔK and the crack growth rate da/dN increase. In using this expression, one should note

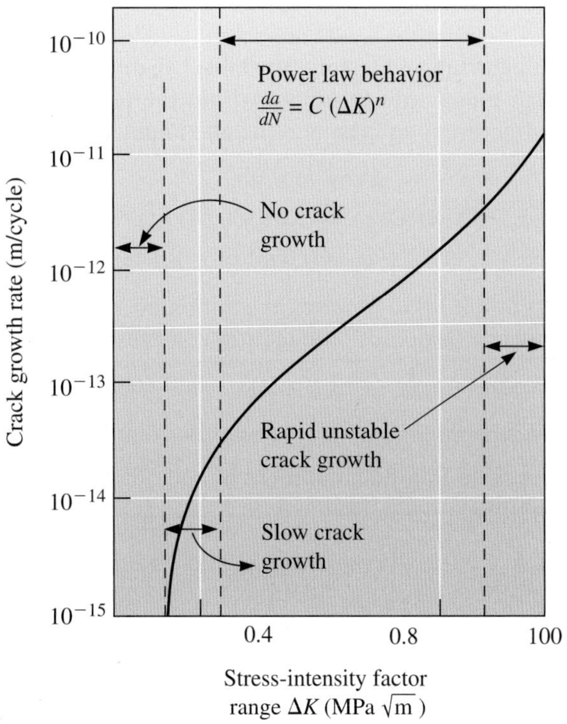

Figure 7-20 Crack growth rate versus stress intensity factor range for a high-strength steel. For this steel, $C = 1.62 \times 10^{-12}$ and $n = 3.2$ for the units shown.

that a crack will not propagate during compression. Therefore, if σ_{min} is compressive, or less than zero, σ_{min} should be set equal to zero.

Knowledge of crack growth rate is of assistance in designing components and in nondestructive evaluation to determine if a crack poses imminent danger to the structure. One approach to this problem is to estimate the number of cycles required before failure occurs. By rearranging Equation 7-18 and substituting for ΔK:

$$dN = \frac{da}{Cf^n \Delta\sigma^n \pi^{n/2} a^{n/2}}$$

If we integrate this expression between the initial size of a crack and the crack size required for fracture to occur, we find that

$$N = \frac{2[(a_c)^{(2-n)/2} - (a_i)^{(2-n)/2}]}{(2-n)Cf^n \Delta\sigma^n \pi^{n/2}} \tag{7-20}$$

where a_i is the initial flaw size and a_c is the flaw size required for fracture. If we know the material constants n and C in Equation 7-18, we can estimate the number of cycles required for failure for a given cyclical stress.

Example 7-11 *Design of a Fatigue Resistant Plate*

A high-strength steel plate (Figure 7-20), which has a plane strain fracture toughness of 80 MPa$\sqrt{m}$ is alternately loaded in tension to 500 MPa and in compression to 60 MPa. The plate is to survive for ten years with the stress being applied at a frequency of once every five minutes. Design a manufacturing and testing procedure that ensures that the component will serve as intended. Assume a geometry factor $f = 1.0$ for all flaws.

SOLUTION

To design our manufacturing and testing capability, we must determine the maximum size of any flaws that might lead to failure within the ten-year period. The critical crack size using the fracture toughness and the maximum stress is

$$K_{Ic} = f\sigma\sqrt{\pi a_c}$$
$$80 \text{ MPa}\sqrt{\text{m}} = (1.0)(500 \text{ MPa})\sqrt{\pi a_c}$$
$$a_c = 0.0081 \text{ m} = 8.1 \text{ mm}$$

The maximum stress is 500 MPa; however, the minimum stress is zero, not 60 MPa in compression, because cracks do not propagate in compression. Thus, $\Delta\sigma$ is

$$\Delta\sigma = \sigma_{\max} - \sigma_{\min} = 500 - 0 = 500 \text{ MPa}$$

We need to determine the minimum number of cycles that the plate must withstand:

$$N = [1 \text{ cycle}/(5 \text{ min})](60 \text{ min/h})(24 \text{ h/day})(365 \text{ days/yr})(10 \text{ yr})$$
$$N = 1{,}051{,}200 \text{ cycles}$$

If we assume that $f = 1.0$ for all crack lengths and note that $C = 1.62 \times 10^{-12}$ and $n = 3.2$ from Figure 7-20 in Equation 7-20, then

$$1{,}051{,}200 = \frac{2[(0.0081)^{(2-3.2)/2} - (a_i)^{(2-3.2)/2}]}{(2 - 3.2)(1.62 \times 10^{-12})(1)^{3.2}(500)^{3.2}\pi^{3.2/2}}$$

$$1{,}051{,}200 = \frac{2[18 - a_i^{-0.6}]}{(-1.2)(1.62 \times 10^{-12})(1)(4.332 \times 10^{8})(6.244)}$$

$$a_i = 1.82 \times 10^{-6} \text{ m} = 0.00182 \text{ mm for surface flaws}$$
$$2a_i = 0.00364 \text{ mm for internal flaws}$$

The manufacturing process must produce surface flaws smaller than 0.00182 mm in length. In addition, nondestructive tests must be available to ensure that cracks approaching this length are not present.

Effects of Temperature As the material's temperature increases, both fatigue life and endurance limit decrease. Furthermore, a cyclical temperature change encourages failure by thermal fatigue; when the material heats in a nonuniform manner, some parts of the structure expand more than others. This nonuniform expansion introduces a stress within the material, and when the structure later cools and contracts, stresses of the opposite sign are imposed. As a consequence of the thermally induced stresses and strains, fatigue may eventually occur. The frequency with which the stress is applied also influences fatigue behavior. In particular, high-frequency stresses may cause polymer materials to heat; at increased temperatures, polymers fail more quickly. Chemical effects of temperature (e.g., oxidation) must also be considered.

7-9 Creep, Stress Rupture, and Stress Corrosion

If we apply stress to a material at an elevated temperature, the material may stretch and eventually fail, even though the applied stress is *less* than the yield strength at that temperature. Time dependent permanent deformation under a constant load or constant stress and at high temperatures is known as **creep**. A large number of failures occurring in components used at high temperatures can be attributed to creep or a combination of creep and fatigue. Diffusion, dislocation glide or climb, or grain boundary sliding can contribute to the creep of metallic materials. In ductile metals and alloys subjected to creep, fracture is accompanied by necking, void nucleation and coalescence, or grain boundary sliding. Polymeric materials also show creep.

A material is considered failed by creep even if it has *not* actually fractured. When a material does creep and then ultimately breaks, the fracture is defined as *stress rupture.* Normally, ductile stress-rupture fractures include necking and the presence of many cracks that did not have an opportunity to produce final fracture. Furthermore, grains near the fracture surface tend to be elongated. Ductile stress-rupture failures generally occur at high creep rates and relatively low exposure temperatures and have short rupture times. Brittle stress-rupture failures usually show little necking and occur more often at low creep rates and high temperatures. Equiaxed grains are observed near the fracture surface. Brittle failure typically occurs by formation of voids at the intersection of three grain boundaries and precipitation of additional voids along grain boundaries by diffusion processes (Figure 7-21).

Stress Corrosion Stress corrosion is a phenomenon in which materials react with corrosive chemicals in the environment. This leads to the formation of cracks and a lowering of strength. Stress corrosion can occur at stresses well below the yield strength of the metallic, ceramic, or glassy material due to attack by a corrosive medium. In metallic materials, deep, fine corrosion cracks are produced, even though the metal as a whole shows little uniform attack. The stresses can be either externally applied or stored residual stresses. Stress corrosion failures are often identified by microstructural examination of the surrounding metal. Ordinarily, extensive branching of the cracks along grain boundaries is observed (Figure 7-22). The location at which cracks initiated may be identified by the presence of a corrosion product.

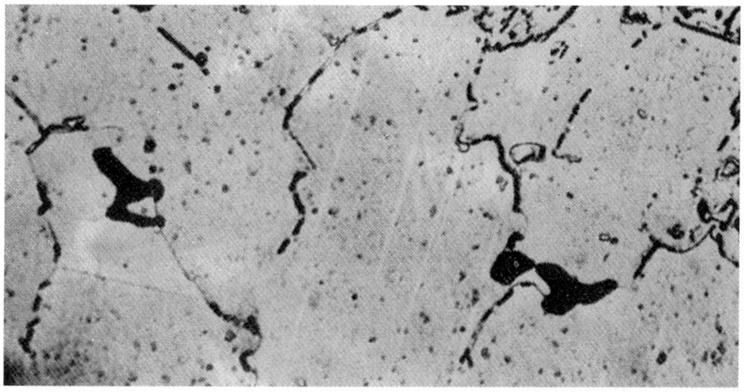

Figure 7-21 Creep cavities formed at grain boundaries in an austentic stainless steel (× 500). [*From* ASM Handbook, *Vol. 7, Metallography and Microstructure (1972), ASM International, Materials Park, OH 44073. Reprinted with permission of ASM International. All rights reserved. www.asminternational.org.*]

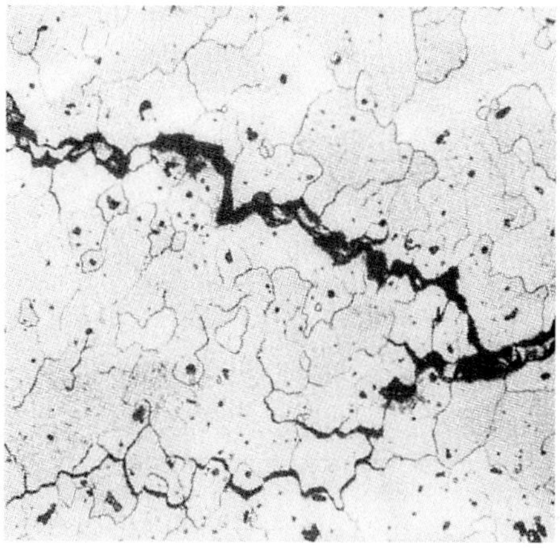

Figure 7-22 Micrograph of a metal near a stress corrosion fracture, showing the many intergranular cracks formed as a result of the corrosion process (× 200). [*From* ASM Handbook, *Vol. 7, Metallography and Microstructure (1972), ASM International, Materials Park, OH 44073. Reprinted with permission of ASM International. All rights reserved. www.asminternational.org.*]

Example 7-12 *Failure Analysis of a Pipe*

A titanium pipe used to transport a corrosive material at 400°C is found to fail after several months. How would you determine the cause for the failure?

SOLUTION

Since a period of time at a high temperature was required before failure occurred, we might first suspect a creep or stress corrosion mechanism for failure. Microscopic examination of the material near the fracture surface would be advisable. If many tiny, branched cracks leading away from the surface are noted, stress corrosion is a strong possibility. If the grains near the fracture surface are elongated, with many voids between the grains, creep is a more likely culprit.

Inorganic silicate glasses are especially prone to failure by reaction with water vapor. It is well known that the strength of silica fibers or silica glass products is very high when these materials are protected from water vapor. As the fibers or silica glass components get exposed to water vapor, corrosion reactions begin, leading to the formation of surface flaws that ultimately cause the cracks to grow when stress is applied. For bulk glasses, special heat treatments such as tempering are used to prevent them from reacting with water vapor. Tempering produces an overall compressive stress on the surface of glass. Thus, even if the glass surface reacts with water vapor, the cracks do not grow since the overall stress at the surface is compressive. If we create a flaw that will penetrate the compressive stress region on the surface, tempered glass will shatter. Tempered glass is used widely in building and automotive applications.

7-10 Evaluation of Creep Behavior

To determine the creep characteristics of a material, a constant stress is applied to a heated specimen in a **creep test**. As soon as the stress is applied, the specimen stretches elastically a small amount ε_0 (Figure 7-23), depending on the applied stress and the modulus of elasticity of the material at the high temperature. Creep testing can also

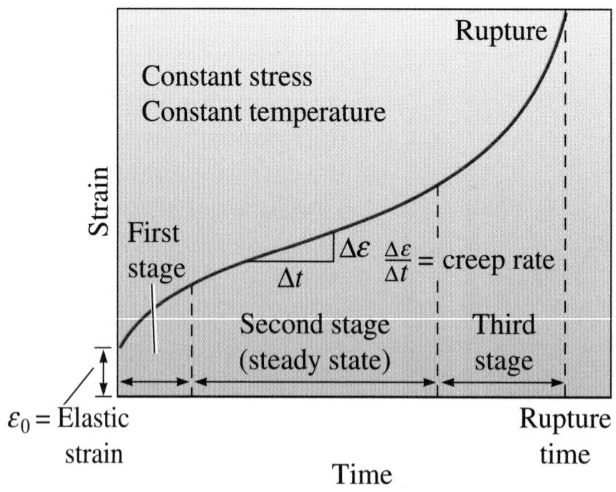

Figure 7-23 A typical creep curve showing the strain produced as a function of time for a constant stress and temperature.

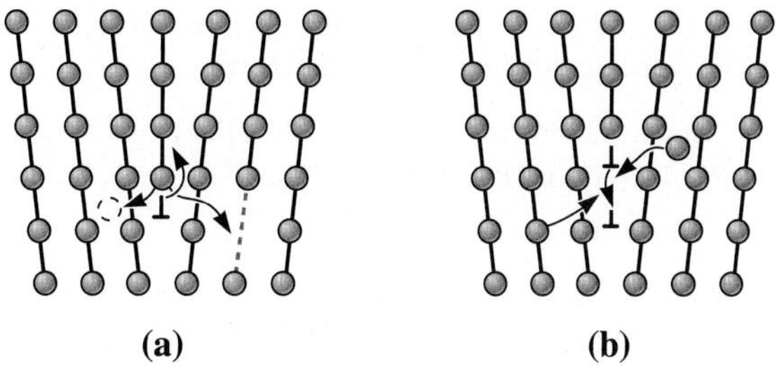

Figure 7-24 Dislocations can climb (a) when atoms leave the dislocation line to create interstitials or to fill vacancies or (b) when atoms are attached to the dislocation line by creating vacancies or eliminating interstitials.

be conducted under a constant load and is important from an engineering design viewpoint.

Dislocation Climb High temperatures permit dislocations in a metal to climb. In climb, atoms move either to or from the dislocation line by diffusion, causing the dislocation to move in a direction that is perpendicular, not parallel, to the slip plane (Figure 7-24). The dislocation escapes from lattice imperfections, continues to slip, and causes additional deformation of the specimen even at low applied stresses.

Creep Rate and Rupture Times During the creep test, strain or elongation is measured as a function of time and plotted to give the creep curve (Figure 7-23). In the first stage of creep of metals, many dislocations climb away from obstacles, slip, and contribute to deformation. Eventually, the rate at which dislocations climb away from obstacles equals the rate at which dislocations are blocked by other imperfections. This leads to the second stage, or steady-state, creep. The slope of the steady-state portion of the creep curve is the **creep rate**:

$$\text{Creep rate} = \frac{\Delta \text{ strain}}{\Delta \text{ time}} \tag{7-21}$$

Eventually, during third stage creep, necking begins and the specimen deforms at an accelerated rate until failure occurs. The time required for failure to occur is the **rupture time**. Either a higher stress or a higher temperature reduces the rupture time and increases the creep rate (Figure 7-25).

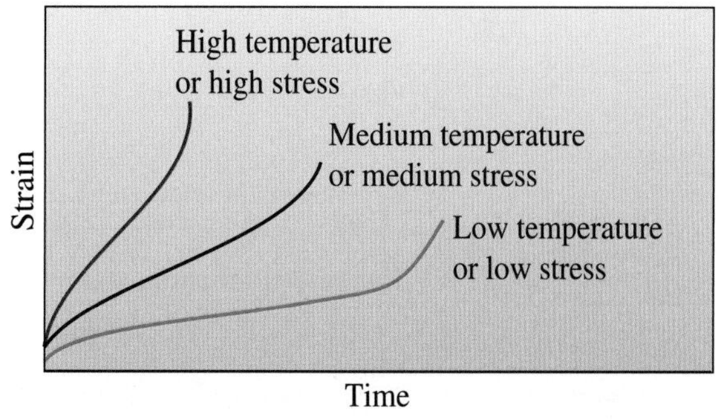

Figure 7-25 The effect of temperature or applied stress on the creep curve.

The creep rate and rupture time (t_r) follow relationships that account for the combined influence of the applied stress and temperature:

$$\text{Creep rate} = \text{C}\sigma^{\text{n}} \exp\left(-\frac{Q_c}{RT}\right) \tag{7-22}$$

$$t_r = \text{K}\sigma^{\text{m}} \exp\left(\frac{Q_r}{RT}\right), \tag{7-23}$$

where R is the gas constant, T is the temperature in kelvin and C, K, n, and m are constants for the material. Q_c is the activation energy for creep, and Q_r is the activation energy for rupture. In particular, Q_c is related to the activation energy for self-diffusion when dislocation climb is important.

In crystalline ceramics, other factors—including grain boundary sliding and nucleation of microcracks—are particularly important. Often, a noncrystalline or glassy material is present at the grain boundaries; the activation energy required for the glass to deform is low, leading to high creep rates compared with completely crystalline ceramics. For the same reason, creep occurs at a rapid rate in ceramic glasses and amorphous polymers.

7-11 Use of Creep Data

The **stress-rupture curves**, shown in Figure 7-26(a), estimate the expected lifetime of a component for a particular combination of stress and temperature. The **Larson-Miller parameter**, illustrated in Figure 7-26(b), is used to consolidate the stress-temperature-rupture time relationship into a single curve. The Larson-Miller parameter ($L.M.$) is

$$L.M. = \left(\frac{T}{1000}\right)(\text{A} + \text{B} \ln t) \tag{7-24}$$

where T is in kelvin, t is the time in hours, and A and B are constants for the material.

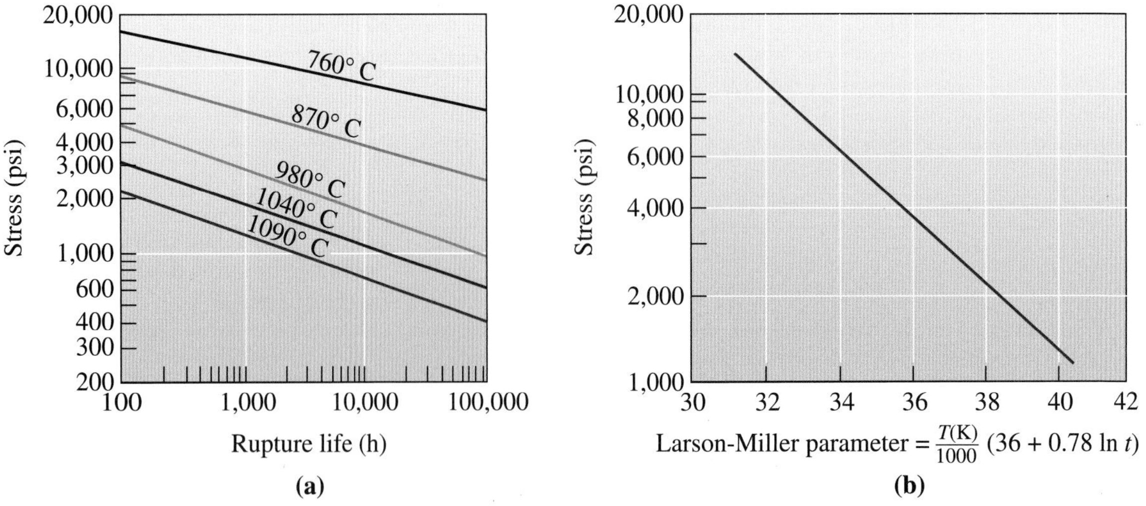

Figure 7-26 Results from a series of creep tests. (a) Stress-rupture curves for an iron-chromium-nickel alloy and (b) the Larson-Miller parameter for ductile cast iron.

Example 7-13 *Design of Links for a Chain*

Design a ductile cast iron chain (Figure 7-27) to operate in a furnace used to fire ceramic bricks. The chain will be used for five years at 600°C with an applied load of 5000 lbs.

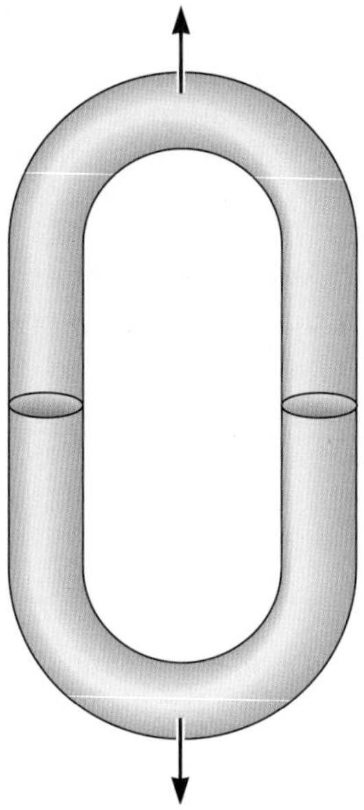

Figure 7-27 Sketch of chain link (for Example 7-13).

SOLUTION

From Figure 7-26(b), the Larson-Miller parameter for ductile cast iron is

$$L.M. = \frac{T(36 + 0.78 \ln t)}{1000}$$

The chain is to survive five years, or

$$t = (24 \text{ h/day})(365 \text{ days/yr})(5 \text{ yr}) = 43{,}800 \text{ h}$$

$$L.M. = \frac{(600 + 273)[36 + 0.78 \ln(43{,}800)]}{1000} = 38.7$$

From Figure 7-26(b), the applied stress must be no more than 1800 psi.

Let us assume a **factor of safety** of 2, this will mean the applied stress should not be more than 1800/2 = 900 psi.

The total cross-sectional area of the chain required to support the 5000 lb load is

$$A = F/S = \frac{5000 \text{ lb}}{900 \text{ psi}} = 5.56 \text{ in.}^2$$

The cross-sectional area of each "half" of the iron link is then 2.78 in.^{2} and, assuming a round cross section

$$d^2 = (4/\pi)A = (4/\pi)(2.78 \text{ in.}^2) = 3.54 \text{ in.}^2$$
$$d = 1.88 \text{ in.}$$

Summary

- Toughness refers to the ability of materials to absorb energy before they fracture. Tensile toughness is equal to the area under the true stress–true strain curve. The impact toughness is measured using the impact test. Fracture toughness describes how easily a crack or flaw in a material propagates.

- Weibull statistics are used to describe and characterize the variability in the strength of brittle materials. The Weibull modulus is a measure of the variability of the strength of a material.
- The fatigue test permits us to understand how a material performs when a cyclical stress is applied. Knowledge of the rate of crack growth can help determine fatigue life.
- Microstructural analysis of fracture surfaces can lead to better insights into the origin and cause of fracture. Different microstructural features are associated with ductile and brittle fracture as well as fatigue failure.
- The creep test provides information on the load-carrying ability of a material at high temperatures. Creep rate and rupture time are important properties obtained from these tests.

Glossary

Beach or clamshell marks Patterns often seen on a component subjected to fatigue. Normally formed when the load is changed during service or when the loading is intermittent, perhaps permitting time for oxidation inside the crack.

Chevron pattern A common fracture feature produced by separate crack fronts propagating at different levels in the material.

Climb Movement of a dislocation perpendicular to its slip plane by the diffusion of atoms to or from the dislocation line.

Conchoidal fracture Fracture surface containing a smooth mirror zone near the origin of the fracture with tear lines comprising the remainder of the surface. This is typical of amorphous materials.

Creep A time-dependent, permanent deformation at high temperatures, occurring at constant load or constant stress.

Creep rate The rate at which a material deforms when a stress is applied at a high temperature.

Creep test Measures the resistance of a material to deformation and failure when subjected to a static load below the yield strength at an elevated temperature.

Delamination The process by which different layers in a composite debond.

Endurance limit An older concept that defined a stress below which a material will not fail in a fatigue test. Factors such as corrosion or occasional overloading can cause materials to fail at stresses below the assumed endurance limit.

Endurance ratio The endurance limit divided by the tensile strength of the material. The ratio is about 0.5 for many ferrous metals. See the cautionary note on endurance limit.

Factor of safety The ratio of the stress level for which a component is designed to the actual stress level experienced.

Fatigue life The number of cycles permitted at a particular stress before a material fails by fatigue.

Fatigue strength The stress required to cause failure by fatigue in a given number of cycles, such as 500 million cycles.

Fatigue test Measures the resistance of a material to failure when a stress below the yield strength is repeatedly applied.

Fracture mechanics The study of a material's ability to withstand stress in the presence of a flaw.

Fracture toughness The resistance of a material to failure in the presence of a flaw.

Griffith flaw A crack or flaw in a material that concentrates and magnifies the applied stress.

Intergranular In between grains or along the grain boundaries.

Larson-Miller parameter A parameter used to relate the stress, temperature, and rupture time in creep.

Microvoids Development of small cavities in a material. These form when a high stress causes separation of the metal at grain boundaries or interfaces between the metal and inclusions.

Notch sensitivity Measures the effect of a notch, scratch, or other imperfection on a material's properties, such as toughness or fatigue life.

Rotating cantilever beam test A method for fatigue testing that permits fully-reversed cyclical loading.

Rupture time The time required for a specimen to fail by creep at a particular temperature and stress.

S-N curve (also known as the Wöhler curve) A graph showing the relationship between the applied stress and the number of cycles to failure in fatigue.

Shot peening A process in which metal spheres are shot at a component. This leads to a residual compressive stress at the surface of a component and enhances fatigue life.

Stress corrosion A phenomenon in which materials react with corrosive chemicals in the environment, leading to the formation of cracks and lowering of strength.

Stress-rupture curve A method of reporting the results of a series of creep tests by plotting the applied stress versus the rupture time.

Striations Patterns seen on a fracture surface of a fatigued sample. These are visible on a much finer scale than beach marks and show the position of the crack tip after each cycle.

Tempering A glass heat treatment that creates a compressive stress layer at the surface, thereby making the glass safer by hindering crack growth.

Toughness A qualitative measure of the energy required to cause fracture of a material.

Transgranular Meaning through the grains (e.g., a transgranular fracture would be fracture in which cracks go through the grains).

Weibull distribution A mathematical distribution showing the probability of failure or survival of a material as a function of the stress.

Weibull modulus A parameter related to the Weibull distribution. It is an indicator of the variability of the strength of materials resulting from a distribution of flaw sizes.

Wöhler curve See S-N curve.

Problems

Section 7-1 Fracture Mechanics

Section 7-2 The Importance of Fracture Mechanics

7-1 Alumina (Al_2O_3) is a brittle ceramic with low toughness. Suppose that fibers of silicon carbide SiC, another brittle ceramic with low toughness, could be embedded within the alumina. Would doing this affect the toughness of the ceramic matrix composite? Explain.

7-2 What are the differences between the stress intensity factor (K_I) and the plane strain fracture toughness (K_{Ic})? Explain.

7-3 A wing component on an aircraft is fabricated from an aluminum alloy that has a plane strain fracture toughness of 26 MPa$\sqrt{m}$. It has been determined that fracture results at a stress of 112 MPa when the maximum internal crack length is 8.6 mm. For this same component and alloy, compute the stress level at which fracture will occur for a critical internal crack length of 6.0 mm.

7-4 Calculate the maximum internal crack length allowable for a titanium alloy component that is loaded to a stress that is half of its yield strength. Assume that the geometry factor is 1.5. The yield strength of this alloy is 132 ksi, and the fracture toughness is 50 ksi$\sqrt{in}$.

7-5 A steel alloy with a plane strain fracture toughness of 55 MPa$\sqrt{m}$ is exposed to a stress of 1020 MPa. Will this component fracture if it is known that the largest surface crack is 0.5 mm long? Assume the geometry factor to be 1.0.

7-6 A ceramic matrix composite contains internal flaws as large as 0.001 cm in length. The plane strain fracture toughness of the composite is 45 $MPa\sqrt{m}$ and the tensile strength is 550 MPa. Will the stress cause the composite to fail before the tensile strength is reached? Assume that $f = 1$.

7-7 An aluminum alloy that has a plane strain fracture toughness of 25,000 $psi\sqrt{in.}$ fails when a stress of 42,000 psi is applied. Observation of the fracture surface indicates that fracture began at the surface of the part. Estimate the size of the flaw that initiated fracture. Assume that $f = 1.1$.

7-8 A polymer that contains internal flaws 1 mm in length fails at a stress of 25 MPa. Determine the plane strain fracture toughness of the polymer. Assume that $f = 1$.

7-9 A ceramic part for a jet engine has a yield strength of 75,000 psi and a plane strain fracture toughness of 5,000 $psi\sqrt{in.}$ To be sure that the part does not fail, we plan to ensure that the maximum applied stress is only one-third of the yield strength. We use a nondestructive test that will detect any internal flaws greater than 0.05 in. long. Assuming that $f = 1.4$, does our nondestructive test have the required sensitivity? Explain.

7-10 A manufacturing process that unintentionally introduces cracks to the surface of a part was used to produce load-bearing components. The design requires that the component be able to withstand a stress of 450 MPa. The component failed catastrophically in service.

You are a failure analysis engineer who must determine whether the component failed due to an overload in service or flaws from the manufacturing process. The manufacturer claims that the components were polished to remove the cracks and inspected to ensure that no surface cracks were larger than 0.5 mm. The manufacturer believes the component failed due to operator error.

It has been independently verified that the 5-cm diameter part was subjected to a tensile load of 1 MN (10^6 N).

The material from which the component is made has a fracture toughness of 75 $MPa\sqrt{m}$ and an ultimate tensile strength of 600 MPa. Assume external cracks for which $f = 1.12$.

(a) Who is at fault for the component failure, the manufacturer or the operator? Show your work for both cases.

(b) In addition to the analysis that you presented in (a), what features might you look for on the fracture surfaces to support your conclusion?

7-11 Explain how the fracture toughness of ceramics can be obtained using hardness testing.

Section 7-3 Microstructural Features of Fracture in Metallic Materials

Section 7-4 Microstructural Features of Fracture in Ceramics, Glasses, and Composites

7-12 Explain the terms intergranular and transgranular fractures. Use a schematic to show grains, grain boundaries, and a crack path that is typical of intergranular and transgranular fracture in materials.

7-13 What are the characteristic microstructural features associated with ductile fracture?

7-14 What are the characteristic microstructural features associated with a brittle fracture in a metallic material?

7-15 What materials typically show a conchoidal fracture?

7-16 Briefly describe how fiber-reinforced composite materials can fail.

7-17 Concrete has exceptional strength in compression, but it fails rather easily in tension. Explain why.

7-18 What controls the strength of glasses? What can be done to enhance the strength of silicate glasses?

Section 7-5 Weibull Statistics for Failure Strength Analysis

7-19 Sketch a schematic of the probability of failure of ceramics and that of metals and alloys as a function of strength. Explain the differences that you anticipate.

7-20 Why does the strength of ceramics vary considerably with the size of ceramic components?

7-21 What parameter tells us about the variability of the strength of ceramics and glasses?

7-22 Why do glass fibers of different lengths have different strengths?

7-23 Explain the significance of the Weibull distribution.

7-24* Turbochargers Are Us, a new start-up company, hires you to design their new turbocharger. They explain that they want to replace their metallic superalloy turbocharger with a high-tech ceramic that is much lighter for the same configuration. Silicon nitride may be a good choice, and you ask Ceramic Turbochargers, Inc. to supply you with bars made from a certain grade of Si_3N_4. They send you 25 bars that you break in three-point bending, using a 40-mm outer support span and a 20-mm inner loading span, obtaining the data in the following table.

(a) Calculate the bend strength using $\sigma_f = (1.5\ FS)/(t^2w)$, where F = load, S = support span (a = 40 mm loading span and b = 20 mm in this case), t = thickness, and w = width, and give the mean strength (50% probability of failure) and the standard deviation.

(b) Make a Weibull plot by ranking the strength data from lowest strength to highest strength. The lowest strength becomes $n = 1$, next $n = 2$, etc., until $n = N = 25$. Use $F = (n - 0.5)/N$ where n = rank of strength going from lowest strength $n = 1$ to highest strength $n = 25$ and $N = 25$. Note that F is the probability of failure. Plot $\ln\{\ln[1/(1 - F)]\}$ as a function of $\ln \sigma_f$ and use a linear regression to find the slope of the line, which is m: the Weibull modulus. Find the characteristic strength σ_0 (63.2% probability of failure). [*Hint*: The characteristic strength is calculated easily once you know the equation of the line. A spreadsheet program (such as Excel) greatly facilitates the calculations.]

Bar	Width (mm)	Thickness (mm)	Load (N)
1	6.02	3.99	2510
2	6.00	4.00	2615
3	5.98	3.99	2375
4	5.99	4.04	2865
5	6.00	4.05	2575
6	6.01	4.00	2605
7	6.01	4.01	2810
8	5.95	4.02	2595
9	5.99	3.97	2490
10	5.98	3.96	2650
11	6.05	3.97	2705
12	6.00	4.05	2765
13	6.02	4.00	2680
14	5.98	4.01	2725
15	6.03	3.99	2830
16	5.95	3.98	2730
17	6.04	4.03	2565
18	5.96	4.01	2650
19	5.97	4.05	2650
20	6.02	4.00	2745
21	6.01	4.00	2895
22	6.00	3.99	2525
23	6.00	3.98	2660
24	6.04	3.95	2680
25	6.02	4.05	2640

* This problem was contributed by Dr. Raymond Cutler of Ceramatek, Inc.

7.25** Your boss asks you to calculate the design stress for a brittle nickel-aluminide rod she wants to use in a high-temperature design where the cylindrical component is stressed in tension. You decide to test rods of the same diameter and length as her design in tension to avoid the correction for the effective area (or effective volume in this case). You measure an average stress of 673 MPa and a Weibull modulus (m) of 14.7 for the nickel-aluminide rods. What is the design stress in MPa if 99.999% of the parts you build must be able to handle this stress without fracturing? (*Note*: The design stress is the stress you choose as the engineer so that the component functions as you want.)

Section 7-6 Fatigue

Section 7-7 Results of the Fatigue Test

Section 7-8 Application of Fatigue Testing

7-26 Research the Aloha Airlines Flight 243 accident in 1988 using online resources and explain the factors that led to the breach and rupture of the main fuselage.

7-27 List four measures that may be taken to increase the resistance to fatigue of a metal alloy.

7-28 A fatigue test was performed in which the mean stress was 70 MPa and the stress amplitude was 210 MPa. Compute the maximum and minimum stress.

7-29 A cylindrical tool steel specimen that is 6 in. long and 0.25 in. in diameter rotates as a cantilever beam and is to be designed so that failure never occurs. Assuming that the maximum tensile and compressive stresses are equal, determine the maximum load that can be applied to the end of the beam. (See Figure 7-18.)

7-30 A 2-cm-diameter, 20-cm-long bar of an acetal polymer (Figure 7-28) is loaded on one end and is expected to survive one million cycles of loading, with equal maximum tensile and compressive stresses, during its lifetime. What is the maximum permissible load that can be applied?

7-31 A nickel-based alloy (INCONEL® alloy 801) beam (Figure 7-29) is annealed and subjected to a cyclical load of 100 N with equal maximum and minimum stresses. The beam diameter is 5 mm, and the beam length is 50 mm. What is the expected lifetime of the beam in number of cycles to failure? If the cycle frequency is 10 Hz, what is the lifetime in hours?

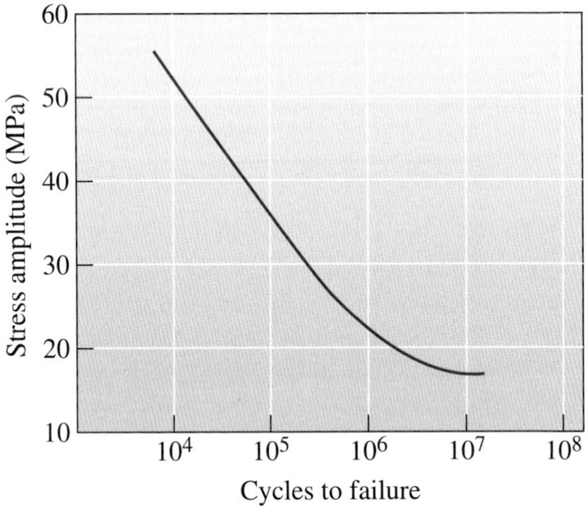

Figure 7-28 The S-N fatigue curve for an acetal polymer (for Problems 7-30, 7-33, and 7-34).

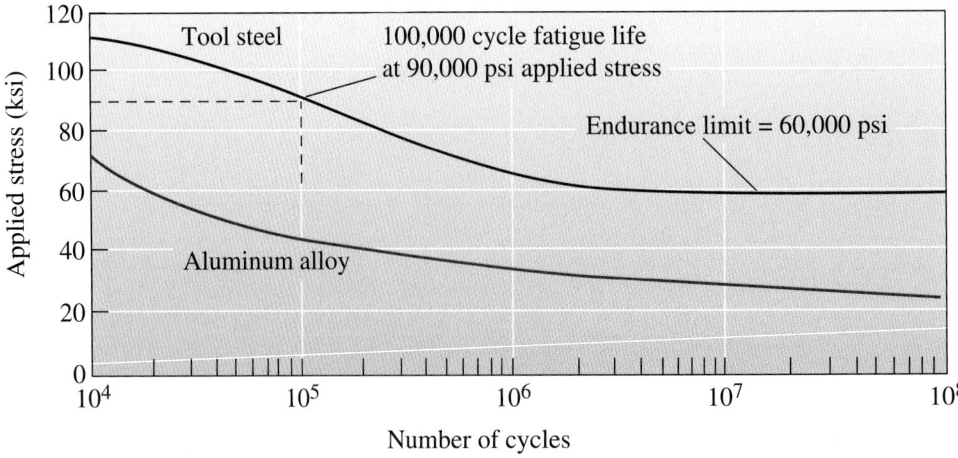

Figure 7-18 The stress-number of cycles to failure (S-N) curves for a tool steel and an aluminum alloy. (Repeated for Problems 7-29 and 7-32.)

** This problem was contributed by Dr. Raymond Cutler of Ceramatek, Inc.

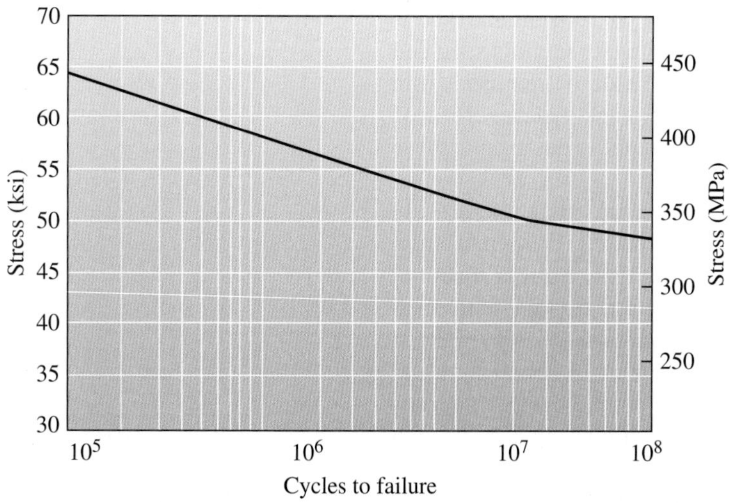

Figure 7-29 The stress-number of cycles to failure (S-N) curve for annealed INCONEL alloy 801 for Problem 7-31.

7-32 A cyclical load of 1500 lb is to be exerted at the end of a 10-in-long aluminium beam (Figure 7-18). The bar must survive for at least 10^6 cycles. What is the minimum diameter of the bar?

7-33 A cylindrical acetal polymer bar 20 cm long and 1.5 cm in diameter is subjected to a vibrational load at one end of the bar at a frequency of 500 vibrations per minute, with a load of 50 N. How many hours will the part survive before breaking? (See Figure 7-28.)

7-34 Suppose that we would like a part produced from the acetal polymer shown in Figure 7-28 to survive for one million cycles under conditions that provide for equal compressive and tensile stresses. What is the fatigue strength, or maximum stress amplitude, required? What are the maximum stress, the minimum stress, and the mean stress on the part during its use? What effect would the frequency of the stress application have on your answers? Explain.

7-35 The high-strength steel in Figure 7-20 is subjected to a stress alternating at 200 revolutions per minute between 600 MPa and 200 MPa (both tension). Calculate the growth rate of a surface crack when it reaches a length of 0.2 mm in both m/cycle and m/s. Assume that $f = 1.0$.

7-36 The high-strength steel in Figure 7-20, which has a critical fracture toughness of 80 $\text{MPa}\sqrt{\text{m}}$, is subjected to an alternating stress varying from −900 MPa (compression) to +900 MPa (tension). It is to survive for 10^5 cycles before failure occurs. Assume that $f = 1$. Calculate

(a) the size of a surface crack required for failure to occur; and

(b) the largest initial surface crack size that will permit this to happen.

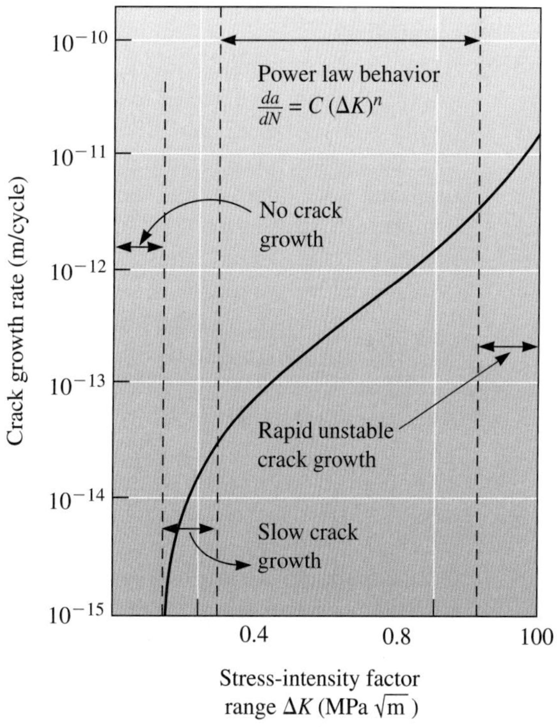

Figure 7-20 Crack growth rate versus stress intensity factor range for a high-strength steel. For this steel, $C = 1.62 \times 10^{-12}$ and $n = 3.2$ for the units shown. (Repeated for Problems 7-35 and 7-36.)

7-37 The manufacturer of a product that is subjected to repetitive cycles has specified that the product should be removed from service when any crack reaches 15% of the critical crack length required to cause fracture.

Consider a crack that is initially 0.02 mm long in a material with a fracture toughness of 55 $MPa\sqrt{m}$. The product is continuously cycled between compressive and tensile stresses of 300 MPa at a constant frequency. Assume external cracks for which $f = 1.12$. The materials constants for these units are $n = 3.4$ and $C = 2 \times 10^{-11}$.

(a) What is the critical crack length required to cause fracture?

(b) How many cycles will cause product failure?

(c) If the product is removed from service as specified by the manufacturer, how much of the useful life of the product remains?

7-38 A material containing cracks of initial length 0.010 mm is subjected to alternating tensile stresses of 25 and 125 MPa for 350,000 cycles. The material is then subjected to alternating tensile and compressive stresses of 250 MPa. How many of the larger stress amplitude cycles can be sustained before failure? The material has a fracture toughness of 25 $MPa\sqrt{m}$ and materials constants of $n = 3.1$ and $C = 1.8 \times 10^{-10}$ for these units. Assume $f = 1.0$ for all cracks.

7-39 The acrylic polymer from which Figure 7-30 was obtained has a critical fracture toughness of 2 $MPa\sqrt{m}$. It is subjected to a stress alternating between −10 and +10 MPa. Calculate the growth rate of a surface crack when it reaches a length of 5×10^{-6} m if $f = 1.0$.

7-40 Calculate the constants C and n in Equation 7-18 for the crack growth rate of an acrylic polymer. (See Figure 7-30.)

7-41 The acrylic polymer from which Figure 7-30 was obtained is subjected to an alternating stress between 15 MPa and 0 MPa. The largest surface cracks initially detected by nondestructive testing are 0.001 mm in length. If the critical fracture toughness of the polymer is 2 $MPa\sqrt{m}$, calculate the number of cycles required before failure occurs. Let $f = 1.0$. (*Hint*: Use the results of Problem 7-40.)

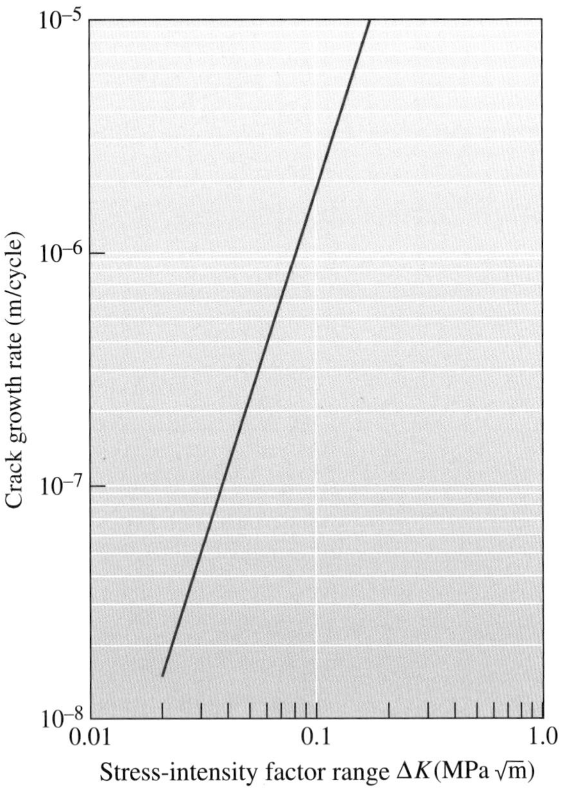

Figure 7-30 The crack growth rate for an acrylic polymer (for Problems 7-39, 7-40, and 7-41).

7-42 Explain how fatigue failure occurs even if the material does not see overall stress levels higher than the yield strength.

7-43 Verify that integration of $da/dN = C(\Delta K)^n$ will give Equation 7-20.

7-44 What is shot peening? What is the purpose of using this process?

Section 7-9 Creep, Stress Rupture, and Stress Corrosion

Section 7-10 Evaluation of Creep Behavior

Section 7-11 Use of Creep Data

7-45 Why is creep accelerated by heat?

7-46 A child's toy was left at the bottom of a swimming pool for several weeks. When the toy was removed from the water, it failed after only a few hundred cycles of loading and unloading, even though it should have been able to withstand thousands of cycles. Speculate as to why the toy failed earlier than expected.

7-47 Define the term "creep" and differentiate creep from stress relaxation.

7-48 What is meant by the terms "stress rupture" and "stress corrosion"?

7-49 What is the difference between failure of a material by creep and that by stress rupture?

7-50 Approximate the temperature at which creep deformation becomes an important consideration for each of the following metals: tin, molybdenum, iron, gold, zinc and chromium.

7-51 The activation energy for self-diffusion in copper is 49,300 cal/mol. A copper specimen creeps at 0.002 $\frac{\text{in./in.}}{\text{h}}$ when a stress of 15,000 psi is applied at 600°C. If the creep rate of copper is dependent on self-diffusion, determine the creep rate if the temperature is 800°C.

7-52 When a stress of 20,000 psi is applied to a material heated to 900°C, rupture occurs in 25,000 h. If the activation energy for rupture is 35,000 cal/mol, determine the rupture time if the temperature is reduced to 800°C.

7-53 The following data were obtained from a creep test for a specimen having a gage length of 2.0 in. and an initial diameter of 0.6 in. The initial stress applied to the material is 10,000 psi. The diameter of the specimen at fracture is 0.52 in.

Length Between Gage Marks (in.)	Time (h)
2.004	0
2.010	100
2.020	200
2.030	400
2.045	1000
2.075	2000
2.135	4000
2.193	6000
2.230	7000
2.300	8000 (fracture)

Determine
(a) the load applied to the specimen during the test;
(b) the approximate length of time during which linear creep occurs;
(c) the creep rate in $\frac{\text{in./in.}}{\text{h}}$ and in %/h; and
(d) the true stress acting on the specimen at the time of rupture.

7-54 A stainless steel is held at 705°C under different loads. The following data are obtained:

Applied Stress (MPa)	Rupture Time (h)	Creep Rate (%/h)
106.9	1200	0.022
128.2	710	0.068
147.5	300	0.201
160.0	110	0.332

Determine the exponents n and m in Equations 7-22 and 7-23 that describe the dependence of creep rate and rupture time on applied stress.

7-55 Using the data in Figure 7-26 for an iron-chromium-nickel alloy, determine the activation energy Q_r and the constant m for rupture in the temperature range 980 to 1090°C.

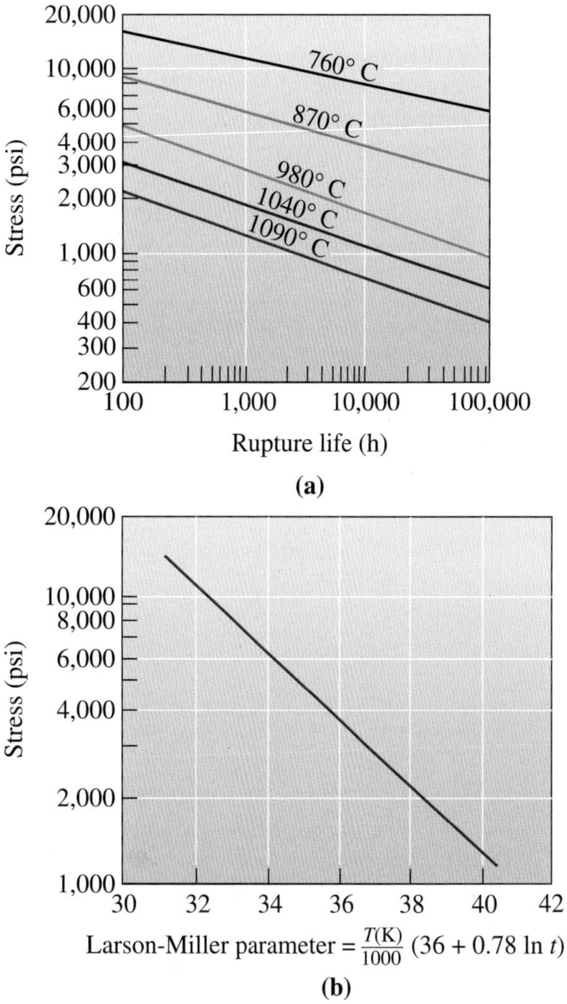

Figure 7-26 Results from a series of creep tests. (a) Stress-rupture curves for an iron-chromium-nickel alloy and (b) the Larson-Miller parameter for ductile cast iron. (Repeated for Problems 7-55 through 7-59, 7-62, and 7-63.)

7-56 A 1-in-diameter bar of an iron-chromium-nickel alloy is subjected to a load of 2500 lb. How many days will the bar survive without rupturing at 980°C? [See Figure 7-26(a).]

7-57 A 5 mm × 20 mm bar of an iron-chrominum-nickel alloy is to operate at 1040°C for 10 years without rupturing. What is the maximum load that can be applied? [See Figure 7-26(a).]

7-58 An iron-chromium-nickel alloy is to withstand a load of 1500 lb at 760°C for 6 years. Calculate the minimum diameter of the bar. [See Figure 7-26(a).]

7-59 A 1.2-in-diameter bar of an iron-chromium-nickel alloy is to operate for 5 years under a load of 4000 lb. What is the maximum operating temperature? [See Figure 7-26(a).]

7-60 A nickel-based alloy component (INCONEL alloy 601) is designed to withstand a service stress of 4000 psi at a temperature of 925°C. Determine the temperature at which the component should be operated in order to double the expected life of the component if the applied stress is to remain the same. Refer to Figure 7-31.

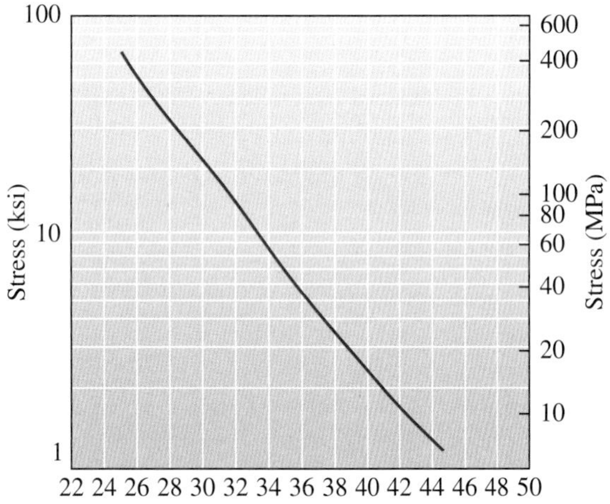

Figure 7-31 The Larsen-Miller parameter for INCONEL alloy 601 (for Problems 7-60 and 7-61). The temperature term T has units of °C, and the time t has units of hours.

7-61 For a temperature of 300°C and a lifetime of three years, what is the maximum stress that the nickel-based component of Figure 7-31 can withstand?

7-62 A 1 in. × 2 in. ductile cast-iron bar must operate for 9 years at 650°C. What is the maximum load that can be applied? [See Figure 7-26(b).]

7-63 A ductile cast-iron bar is to operate at a stress of 6000 psi for 1 year. What is the maximum allowable temperature? [See Figure 7-26(b).]

Design Problems

7-64 A hook (Figure 7-32) for hoisting containers of ore in a mine is to be designed using a nonferrous (not based on iron) material. (A nonferrous material is used because iron and steel could cause a spark that would ignite explosive gases in the mine.) The hook must support a load of 25,000 pounds, and a factor of safety of 2 should be used. We have determined that the cross-section labeled "?" is the most critical area; the rest of the device is already well overdesigned. Determine the design requirements for this device and, based on the mechanical property data given in Chapters 14 and 15 and the metal/alloy prices obtained from such sources as your local newspapers, the internet website of London Metal Exchange or *The Wall Street Journal*, design the hook and select an economical material for the hook.

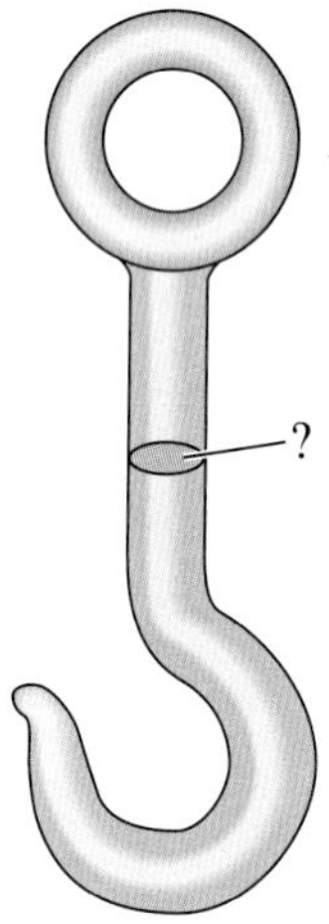

Figure 7-32 Schematic of a hook (for Problem 7-64).

7-65 A support rod for the landing gear of a private airplane is subjected to a tensile load during landing. The loads are predicted to be as high as 40,000 pounds. Because the rod is crucial and failure could lead to a loss of life, the rod is to be designed with a factor of safety of 4 (that is, designed so that the rod is capable of supporting loads four times as great as expected). Operation of the system also produces loads that may induce cracks in the rod. Our nondestructive testing equipment can detect any crack greater than 0.02 in. deep. Based on the materials given in Section 7-1, design the support rod and the material, and justify your answer.

7-66 A lightweight rotating shaft for a pump on a plane is to be designed to support a cyclical load of 15,000 pounds during service. The maximum stress is the same in both tension and compression. The endurance limits or fatigue strengths for several candidate materials are shown below. Design the shaft, including an appropriate material, and justify your solution.

Material	Endurance Limit/ Fatigue Strength (MPa)
Al-Mn alloy	110
Al-Mg-Zn alloy	225
Cu-Be alloy	295
Mg-Mn alloy	80
Be alloy	180
Tungsten alloy	320

7-67 A ductile cast-iron bar is to support a load of 40,000 lb in a heat-treating furnace used to make malleable cast iron. The bar is located in a spot that is continuously exposed to 500°C. Design the bar so that it can operate for at least 10 years without failing.

Knovel® Problems

K7-1 A hollow shaft made from AISI 4340 steel has an outer diameter D_o of 4 in. and an inner diameter D_i of 2.5 in. The shaft rotates at 46 rpm for one hour during each day. It is supported by two bearings and loaded in the middle with a load W of 5500 lbf. The distance between the bearings L is 78 in. The maximum tensile stress due to bending for this type of cyclic loading is calculated using the following equation:

$$\sigma_m = \frac{8WLD_o}{\pi(D_o^4 - D_i^4)}$$

What is the stress ratio for this type of cyclic loading? Would this shaft last for one year assuming a safety factor of 2?

In applications such as the chassis formation of automobiles, metals and alloys are deformed. The mechanical properties of materials change during this process due to strain hardening. The strain hardening behavior of steels used in the fabrication of chassis influences the steel's ability to form aerodynamic shapes. The strain hardening behavior is also important in improving the crashworthiness of vehicles. (*Digital Vision/Getty Images*)

CHAPTER

8

Strain Hardening and Annealing

Have You Ever Wondered?

- *Why does bending a copper wire make it stronger?*
- *How are aluminum beverage cans made?*
- *Why do thermoplastics get stronger when strained?*
- *What is the difference between an annealed, tempered, and laminated safety glass?*
- *How is it that the strength of the metallic material around a weld can be lower than that of the surrounding material?*

Chapter Learning Objectives

The key objectives of this chapter are to

- Quantify the strain hardening of a metal and explain why strain hardening may significantly increase the force required to deform a metal.
- Quantify the strain rate sensitivity of a metal and explain why strain rate sensitivity may influence the force required to deform a material.
- Predict how materials properties of a cold-worked polycrystalline material will change due to annealing (recovery, recrystallization, grain growth) and deformation and how this affects mechanical behavior.
- Design processes to produce metal sheet with the desired dimensions and mechanical properties using combinations of working and annealing treatments.

In this chapter, we will learn how the strength of metals and alloys is influenced by mechanical processing and heat treatments. In Chapter 4, we learned about the different techniques that can strengthen metals and alloys (e.g., increasing dislocation density, decreasing grain size, alloying, etc.). In this chapter, we will learn how to enhance the strength of metals and alloys using cold working, a process by which a metallic material is simultaneously deformed and strengthened. We will also see how hot working can be used to shape metals and alloys by deformation at high temperatures without strengthening. We will learn how the annealing heat treatment can be used to enhance ductility and counter the increase in hardness caused by cold working. The topics discussed in this chapter pertain particularly to metals and alloys.

What about polymers, glasses, and ceramics? Do they also exhibit strain hardening? We will show that the deformation of thermoplastic polymers often produces a strengthening effect, but the mechanism of deformation strengthening is completely different in polymers than in metallic materials. The strength of most brittle materials such as ceramics and glasses depends upon the flaws and flaw size distribution (Chapter 7). Therefore, inorganic glasses and ceramics do not respond well to strain hardening. We should consider different strategies to strengthen these materials. In this context, we will learn the principles of tempering and annealing of glasses.

We begin by discussing strain hardening in metallic materials in the context of stress strain curves.

8-1 Relationship of Cold Working to the Stress Strain Curve

A stress strain curve for a ductile metallic material is shown in Figure 8-1(a). If we apply a stress S_1 that is greater than the yield strength S_y, it causes permanent deformation or strain. When the stress is removed, a strain of e_1 remains. If we make a tensile test sample from the metallic material that had been previously stressed to S_1 and retest that material, we obtain the stress strain curve shown in Figure 8-1(b). Our new test specimen would begin to deform plastically or flow at stress level S_1. We define the *flow stress* as the stress that is needed to initiate plastic flow in previously deformed material. Thus, S_1 is now the flow stress of the material. If we continue to apply a stress until we reach S_2 then release the stress and again retest the metallic material, the new flow stress is S_2. Each time we apply a higher stress, the flow stress and tensile strength increase, and the ductility decreases. We eventually strengthen the metallic material until the flow stress, tensile, and breaking strengths are equal, and there is no ductility [Figure 8-1(c)]. At this point, the metallic material cannot be plastically deformed any further. Figures 8-1(d) and (e) are related to springback, a concept that is discussed later in this section.

By applying a stress that exceeds the original yield strength of the metallic material, we have **strain hardened** or **cold worked** the metallic material, while simultaneously deforming it. This is the basis for many manufacturing techniques, such as wire drawing. Figure 8-2 illustrates several manufacturing processes that make use of both cold-working and hot-working processes. We will discuss the difference between **hot working** and **cold working** later in this chapter. Many techniques for **deformation processing** are used to simultaneously shape and strengthen a material by cold working (Figure 8-2). For example, **rolling** is used to produce metal plate, sheet, or foil. **Forging** deforms the metal into a die cavity, producing relatively complex shapes such as automotive crankshafts or connecting rods. In **drawing**, a metallic rod is pulled through a die to produce a wire or fiber. In **extrusion**, a material is pushed through a die to form products of uniform cross-sections, including rods, tubes, or aluminum trim for doors or windows. *Deep drawing* is used to form the body of aluminum beverage cans. *Stretch forming* and *bending* are used to shape sheet material. Thus, cold working is an effective way of shaping metallic materials while simultaneously increasing their strength. The down side of this process is the loss of ductility. If you take a metal wire and bend it repeatedly, it will harden and eventually break because of strain hardening. Strain hardening is used in many products, especially those that are not going to be exposed to very high temperatures. For example, an aluminum beverage can derives

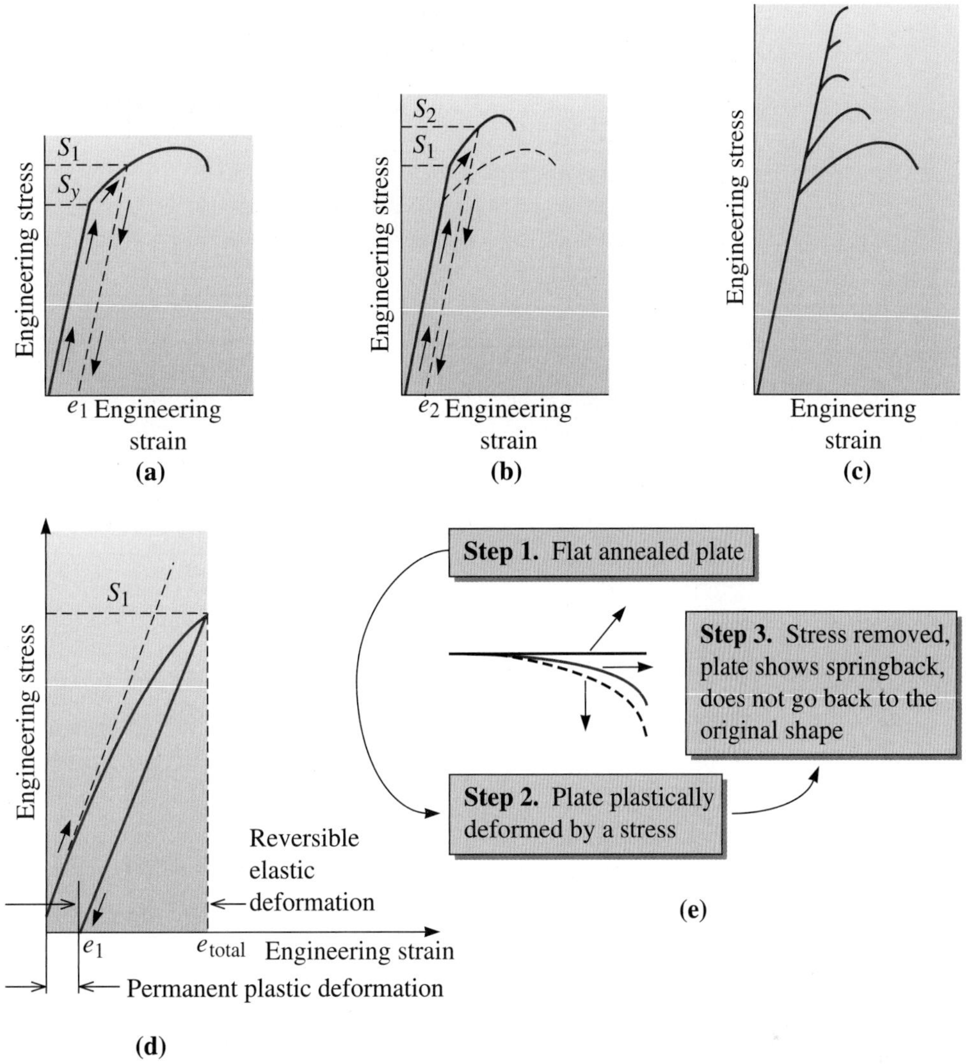

Figure 8-1 Development of strain hardening from the engineering stress strain diagram. (a) A specimen is stressed beyond the yield strength S_y before the stress is removed. (b) Now the specimen has a higher yield strength and tensile strength, but lower ductility. (c) By repeating the procedure, the strength continues to increase and the ductility continues to decrease until the alloy becomes very brittle. (d) The total strain is the sum of the elastic and plastic components. When the stress is removed, the elastic strain is recovered, but the plastic strain is not. (e) Illustration of springback. (*Based on* Engineering Materials 1, *Second Edition, M. F. Ashby and D. R. H. Jones, Elsevier, 1996.*)

almost 70% of its strength from strain hardening that occurs during its fabrication. Some of the strength of aluminum cans also comes from the alloying elements (e.g., Mg) added. Note that many of the processes, such as rolling, can be conducted using both cold and hot working. The pros and cons of using each will be discussed later in this chapter.

Strain-Hardening Exponent (*n*) The response of a metal to cold working can be quantified by the **strain-hardening** exponent *n*. The relationship between

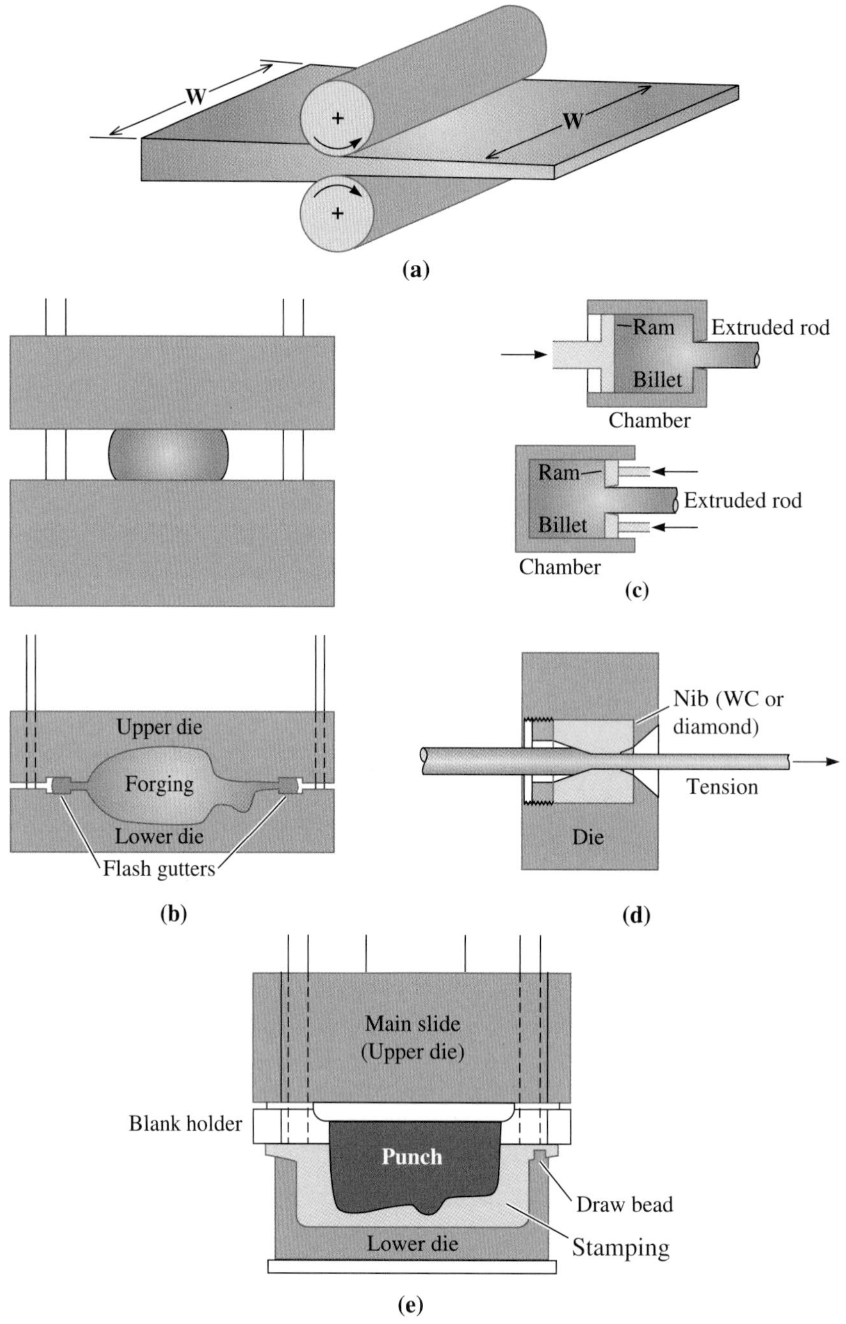

Figure 8-2 Manufacturing processes that make use of cold working as well as hot working. Common metalworking methods. (a) Rolling. (b) Forging (open and closed die). (c) Extrusion (direct and indirect). (d) Wire drawing. (e) Stamping. (*Based on* Mechanical Behavior of Materials, *by M. A. Meyers and K. K. Chawla, Cambridge University Press.*)

true stress σ, true strain ε, and the strain-hardening exponent n is governed by so-called power law behavior according to

$$\sigma = K\varepsilon^n \tag{8-1}$$

The constant K (known as the strength coefficient) is equal to the stress when $\varepsilon = 1$. Taking the natural logarithm of both sides of Equation 8-1 yields

$$\ln \sigma = \ln K + n \ln \varepsilon \tag{8-2}$$

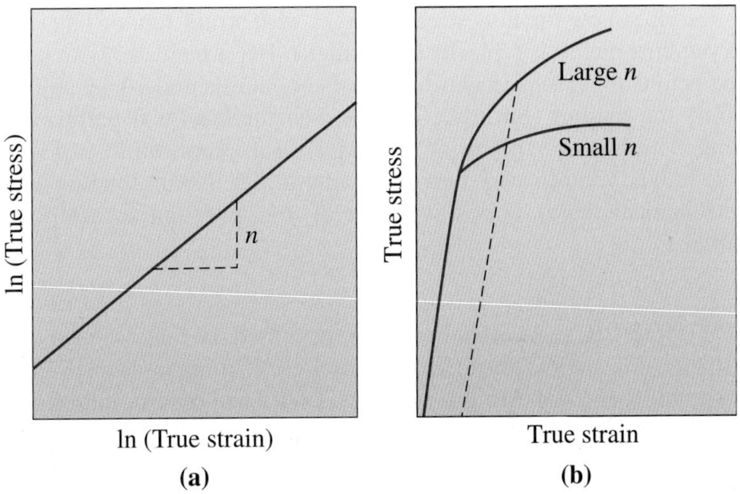

Figure 8-3 (a) The plastic portion of a true stress-true strain curve for a metal. The natural logarithms of stress and strain are plotted. The strain-hardening exponent n is the slope of this curve. (b) The true stress-true strain curves for metals with large and small strain-hardening exponents. Larger degrees of strengthening are obtained for a given strain for the metal with larger n.

The natural logarithms of true stress and true strain are plotted in Figure 8-3(a), and the strain-hardening exponent is the slope of the plastic portion. Larger degrees of strengthening are obtained for a given strain as n increases as shown in Figure 8-3(b). Additionally, necking during a tension test begins approximately when the true strain ε is equal to the strain-hardening exponent n.

For metals, strain hardening is the result of dislocation interaction and multiplication. The strain-hardening exponent is relatively low for HCP metals, but is higher for BCC and, particularly, for FCC metals (Table 8-1). Metals with a low strain-hardening exponent respond poorly to cold working. If we take a copper wire and bend it, the bent wire is stronger as a result of strain hardening.

Strain-Rate Sensitivity (m)

The **strain-rate sensitivity** (m) of stress σ is defined as

$$m = \left[\frac{\partial(\ln \sigma)}{\partial(\ln \dot{\varepsilon})}\right] \tag{8-3}$$

Table 8-1 Strain-hardening exponents and strength coefficients of typical metals and alloys

Metal	Crystal Structure	n	K (psi)
Titanium	HCP	0.05	175,000
Annealed alloy steel	BCC	0.15	93,000
Quenched and tempered medium-carbon steel	BCC	0.10	228,000
Molybdenum	BCC	0.13	105,000
Copper	FCC	0.54	46,000
Cu-30% Zn	FCC	0.50	130,000
Austenitic stainless steel	FCC	0.52	220,000

Based on G. Dieter, Mechanical Metallurgy, *McGraw-Hill, 1961, and other sources.*

This describes how the flow stress changes with strain rate $\dot{\varepsilon}$. The strain-rate sensitivity for crystalline metals is typically less than 0.1, but it increases with temperature. As mentioned before, the mechanical behavior of sheet steels under high strain rates is important not only for shaping, but also for how well the steel will perform under high-impact loading. The crashworthiness of sheet steels is an important consideration for the automotive industry. Steels that harden rapidly under impact loading are useful in absorbing mechanical energy. A positive value of m implies that a material will resist necking (Chapter 6).

Springback Another point to be noted is that when a metallic material is deformed using a stress above its yield strength to a higher level [S_1 in Figure 8-1(d)], the corresponding strain existing at stress S_1 is obtained by dropping a perpendicular line to the horizontal axis (point e_{total}). A strain equal to $(e_{total} - e_1)$ is recovered when the stress is removed since it is elastic in nature. The *elastic strain* that is recovered after a material has been *plastically* deformed is known as *springback* [Figure 8-1(e)]. The occurrence of springback is extremely important for the formation of automotive body panels from sheet steels along with many other applications. This effect is also seen in polymeric materials processed, for example, by extrusion.

Bauschinger Effect Consider a material that has been subjected to tensile plastic deformation. Then, consider two separate samples (A and B) of this material. Test sample A in tension, and sample B under compression. We notice that for the deformed material the flow stress in tension ($\sigma_{\text{flow, tension}}$) for sample A is greater than the compressive yield strength ($\sigma_{\text{flow, compression}}$) for sample B. This effect, in which a material subjected to tension shows a reduction in compressive strength, is known as the **Bauschinger effect**. Note that we are comparing the yield strength of a material under compression and tension after the material has been subjected to plastic deformation under a tensile stress. The Bauschinger effect is also seen on stress reversal. Consider a sample deformed under compression. We can then evaluate two separate samples C and D. The sample subjected to *compressive stress* (C) shows a higher flow stress than that for the sample D subjected to tensile stress. The Bauschinger effect plays an important role in mechanical processing of steels and other alloys.

8-2 Strain-Hardening Mechanisms

We obtain strengthening during deformation of a metallic material by increasing the number of dislocations. Before deformation, the dislocation density is about 10^6 cm of dislocation line per cubic centimeter of metal—a relatively small concentration of dislocations.

When we apply a stress greater than the yield strength, dislocations begin to slip (Schmid's law, Chapter 4). Eventually, a dislocation moving on its slip plane encounters obstacles that pin the dislocation line. As we continue to apply the stress, the dislocation attempts to move by bowing in the center. The dislocation may move so far that a loop is produced (Figure 8-4). When the dislocation loop finally touches itself, a new dislocation is created. The original dislocation is still pinned and can create additional dislocation loops. This mechanism for generating dislocations is called a **Frank-Read source**; Figure 8-4(e) shows an electron micrograph of a Frank-Read source.

The dislocation density may increase to about 10^{12} cm of dislocation line per cubic centimeter of metal during strain hardening. As discussed in Chapter 4, dislocation motion is the mechanism for the plastic flow that occurs in metallic materials; however,

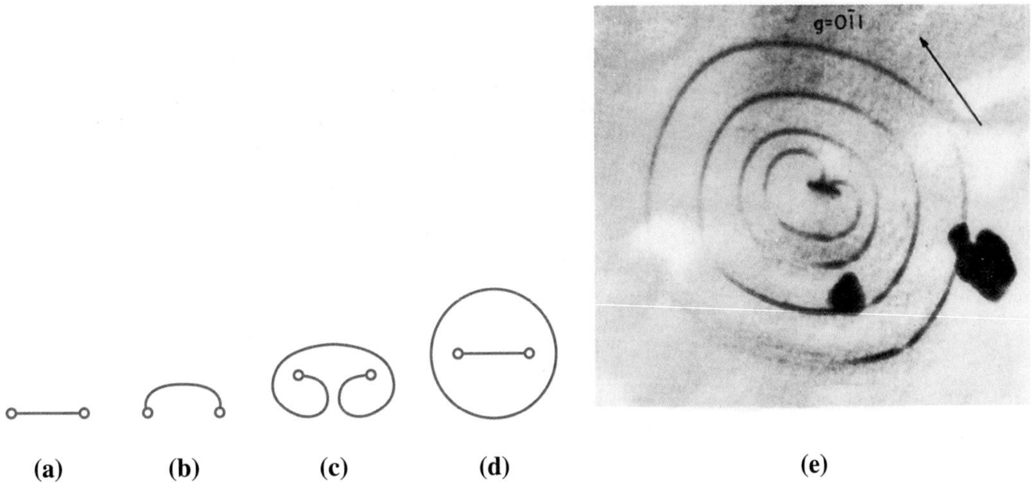

Figure 8-4 The Frank-Read source can generate dislocations. (a) A dislocation is pinned at its ends by lattice defects. (b) As the dislocation continues to move, the dislocation bows, (c) eventually bending back on itself. (d) Finally the dislocation loop forms, and a new dislocation is created. (e) Electron micrograph of a Frank-Read source (× 330,000). (*Springer,* Metallurgical Transactions A, *April 1975, Volume 6, Issue 4, pp. 921-926, "Climb sources in β'-NiAl," G. W. Marshall, J. O. Brittain, figure 3, Copyright © 1975, American Society for Metals, The Metallurgical Society of AIME. With kind permission from Springer Science and Business Media.*)

when we have too many dislocations, they interfere with their own motion. An analogy for this is when we have too many people in a room, it is difficult to move around. The result is increased strength, but reduced ductility, for metallic materials that have undergone work hardening.

Ceramics contain dislocations and even can be strain hardened to a small degree; however, dislocations in ceramics are normally not very mobile. Polycrystalline ceramics also contain porosity. As a result, ceramics behave as brittle materials and significant deformation and strengthening by cold working are not possible. Likewise, covalently bonded materials such as silicon are too brittle to work harden appreciably. Glasses are amorphous and do not contain dislocations and therefore cannot be strain hardened.

Thermoplastics are polymers such as polyethylene, polystyrene, and nylon. These materials consist of molecules that are long spaghetti-like chains. Thermoplastics will strengthen when they are deformed. This is ***not*** strain hardening due to dislocation multiplication but, instead, strengthening of these materials involves alignment and possibly localized crystallization of the long, chainlike molecules. When a stress greater than the yield strength is applied to thermoplastic polymers such as polyethylene, the van der Waals bonds (Chapter 2) between the molecules in different chains are broken. The chains straighten and become aligned in the direction of the applied stress (Figure 8-5). The strength of the polymer, particularly in the direction of the applied stress, increases as a result of the alignment of polymeric chains in the direction of the applied stress. This strength increase is seen in the stress strain curve of typical thermoplastics. Many techniques used for polymer processing are similar to those used for the fabrication of metallic materials. Extrusion, for example, is the most widely used polymer processing technique. Although many of these techniques share conceptual similarities, there are important differences between the mechanisms by which polymers become strengthened during their processing.

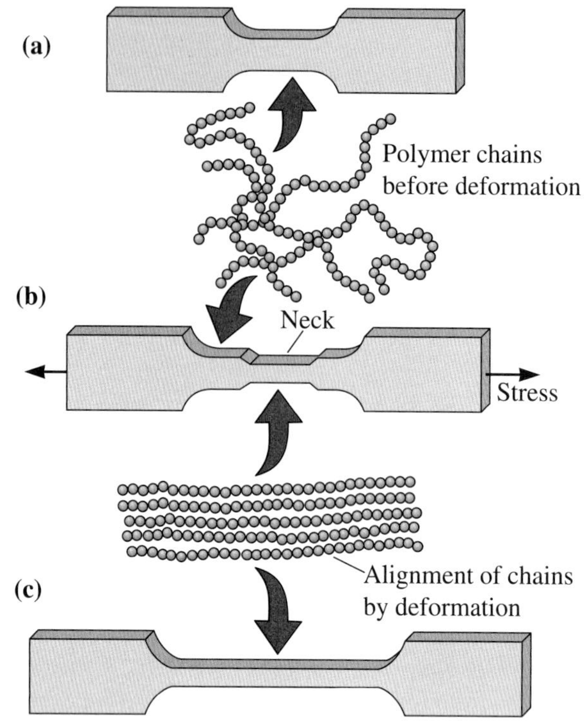

Figure 8-5 In an undeformed thermoplastic polymer tensile bar, (a) the polymer chains are randomly oriented. (b) When a stress is applied, a neck develops as chains become aligned locally. (c) The neck continues to grow until the chains in the entire gage length have aligned. The strength of the polymer is increased.

8-3 Properties versus Percent Cold Work

By controlling the amount of plastic deformation, we control strain hardening. We normally measure the amount of deformation by defining the percent cold work:

$$\text{Percent cold work} = \left[\frac{A_0 - A_f}{A_0}\right] \times 100 \tag{8-4}$$

where A_0 is the original cross-sectional area of the metal and A_f is the final cross-sectional area after deformation. For the case of cold rolling, the percent reduction in thickness is used as the measure of cold work according to

$$\text{Percent reduction in thickness} = \left[\frac{t_0 - t_f}{t_0}\right] \times 100 \tag{8-5}$$

where t_0 is the initial sheet thickness and t_f is the final thickness.

The effect of cold work on the mechanical properties of commercially pure copper is shown in Figure 8-6. As the cold work increases, both the yield and the

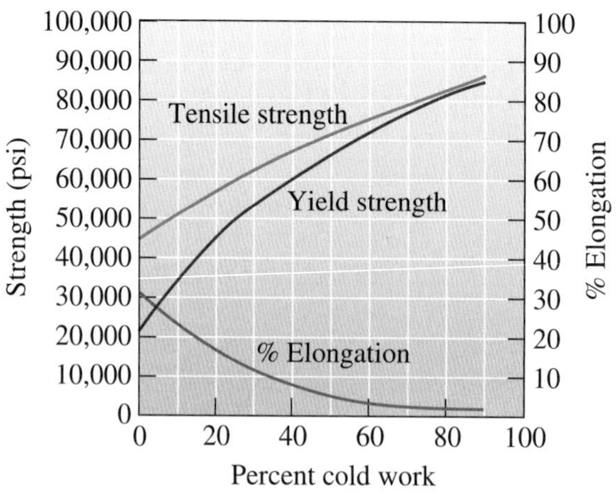

Figure 8-6 The effect of cold work on the mechanical properties of copper.

tensile strengths increase; however, the ductility decreases and approaches zero. The metal breaks if more cold work is attempted; therefore, there is a maximum amount of cold work or deformation that we can perform on a metallic material before it becomes too brittle and breaks.

Example 8-1 *Cold Working a Copper Plate*

A 1-cm-thick copper plate is cold-reduced to 0.50 cm and later further reduced to 0.16 cm. Determine the total percent cold work and the tensile strength of the 0.16 cm plate. (See Figures 8-6 and 8-7.)

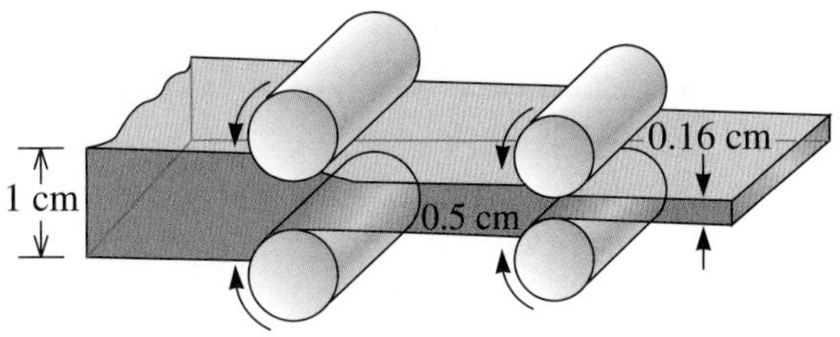

Figure 8-7 Diagram showing the rolling of a 1 cm plate to a 0.16 cm plate (for Example 8-1).

SOLUTION

Note that because the width of the plate does not change during rolling, the cold work can be expressed as the percent reduction in the thickness t.

Our definition of cold work is the percentage change between the original and final cross-sectional areas; it makes no difference how many intermediate steps are involved. Thus, the total cold work is

$$\% \text{ CW} = \left[\frac{t_0 - t_f}{t_0}\right] \times 100 = \left[\frac{1 \text{ cm} - 0.16 \text{ cm}}{1 \text{ cm}}\right] \times 100 = 84\%$$

and, from Figure 8-6, the tensile strength is about 85,000 psi.

We can predict the properties of a metal or an alloy if we know the amount of cold work during processing. We can then decide whether the component has adequate strength at critical locations.

When we wish to select a material for a component that requires certain minimum mechanical properties, we can design the deformation process. We first determine the necessary percent cold work and then, using the final dimensions we desire, calculate the original dimensions from the cold work equation.

Example 8-2 *Design of a Cold Working Process*

Design a manufacturing process to produce a 0.1-cm-thick copper plate having at least 65,000 psi tensile strength, 60,000 psi yield strength, and 5% elongation.

SOLUTION

From Figure 8-6, we need at least 35% cold work to produce a tensile strength of 65,000 psi and 40% cold work to produce a yield strength of 60,000 psi, but we need less than 45% cold work to meet the 5% elongation requirement. Therefore, any cold work between 40% and 45% gives the required mechanical properties.

To produce the plate, a cold-rolling process would be appropriate. The original thickness of the copper plate prior to rolling can be calculated from Equation 8-5, assuming that the width of the plate does not change. Because there is a range of allowable cold work—between 40% and 45%—there is a range of initial plate thicknesses:

$$\% \text{ CW}_{\text{min}} = 40 = \left[\frac{t_{\text{min}} - 0.1 \text{ cm}}{t_{\text{min}}}\right] \times 100, \quad \therefore t_{\text{min}} = 0.167 \text{ cm}$$

$$\% \text{ CW}_{\text{max}} = 45 = \left[\frac{t_{\text{max}} - 0.1 \text{ cm}}{t_{\text{max}}}\right] \times 100, \quad \therefore t_{\text{max}} = 0.182 \text{ cm}$$

To produce the 0.1-cm copper plate, we begin with a 0.167 to 0.182-cm copper plate in the softest possible condition, then cold roll the plate to a 40% to 45% reduction in thickness to achieve the 0.1 cm thickness.

8-4 Microstructure, Texture Strengthening, and Residual Stresses

During plastic deformation using cold or hot working, a microstructure consisting of grains that are elongated in the direction of the applied stress is often produced (Figure 8-8).

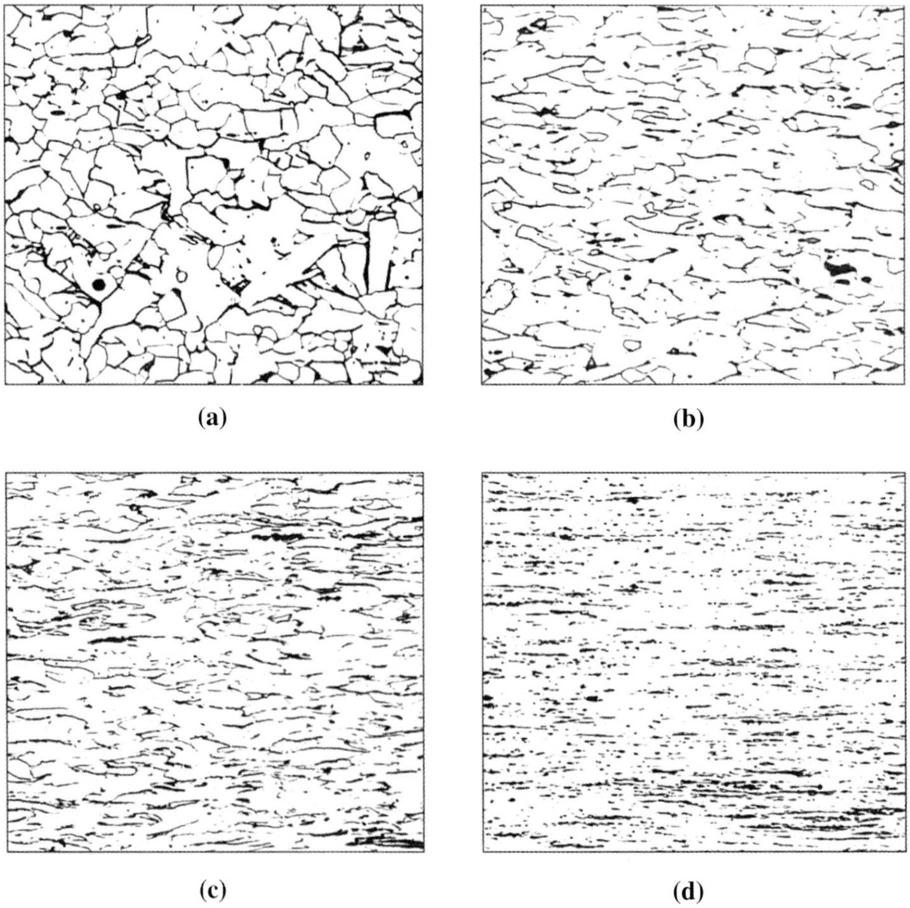

Figure 8-8 The fibrous grain structure of a low-carbon steel produced by cold working: (a) 10% cold work, (b) 30% cold work, (c) 60% cold work, and (d) 90% cold work (× 250). [*From* ASM Handbook, Vol. 9, Metallography and Microstructure, *(1985), ASM International, Materials Park, OH 44073-0002. Reprinted with permission of ASM International. All rights reserved. www.asminternational.org.*]

Anisotropic Behavior During deformation, grains rotate as well as elongate, causing certain crystallographic directions and planes to become aligned with the direction in which stress is applied. Consequently, preferred orientations, or textures, develop and cause anisotropic behavior.

In processes such as wire drawing and extrusion, a **fiber texture** is produced. The term "fibers" refers to the grains in the metallic material, which become elongated in a direction parallel to the axis of the wire or an extruded product. In BCC metals, <110> directions line up with the axis of the wire. In FCC metals, <111> or <100> directions are aligned. This gives the highest strength along the axis of the wire or the extrudate (product being extruded such as a tube), which is what we desire.

As mentioned previously, a somewhat similar effect is seen in thermoplastic materials when they are drawn into fibers or other shapes. The cause, as discussed before, is that polymer chains line up side-by-side along the length of fiber. The strength is greatest along the axis of the polymer fiber. This type of strengthening is seen in polyethylene terephthalate (PET) plastic bottles made using the *blow-stretch process.* This process causes alignment of polymer chains along the radial and length directions, leading to increased strength of PET bottles along those directions.

In processes such as rolling, grains become oriented in a preferred crystallographic direction and plane, giving a **sheet texture**. The properties of a rolled sheet or plate depend on the direction in which the property is measured. Figure 8-9 summarizes the tensile properties of a cold-worked aluminum-lithium alloy. For this alloy, strength is highest parallel to the rolling direction, whereas ductility is highest at a 45° angle to the rolling direction. The strengthening that occurs by the development of anisotropy or of a texture is known as **texture strengthening**. As pointed out in Chapter 6, the Young's modulus of materials also depends upon crystallographic direction in single crystals. For example, the Young's modulus of iron along [111] and [100] directions is

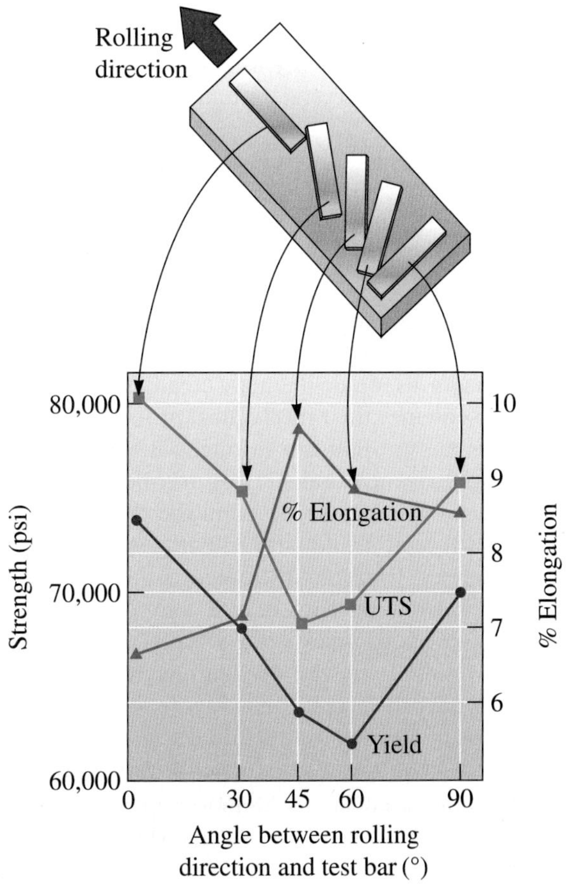

Figure 8-9 Anisotropic behavior in a rolled aluminum-lithium sheet material used in aerospace applications. The sketch relates the position of tensile bars to the mechanical properties that are obtained.

Table 8-2 Common wire drawing and extrusion and sheet textures in materials

Crystal Structure	Wire Drawing and Extrusion Fiber Texture (Direction Parallel to Wire Axis)	Sheet or Rolling Texture
FCC	<111> and <100>	{110} planes parallel to rolling plane <112> directions parallel to rolling direction
BCC	<110>	{001} planes parallel to rolling plane <110> directions parallel to rolling direction
HCP	<10$\bar{1}$0>	{0001} planes parallel to rolling plane <11$\bar{2}$0> directions parallel to rolling direction

~260 and 140 GPa, respectively. The dependence of yield strength on texture is even stronger. Development of texture not only has an effect on mechanical properties but also on magnetic and other properties of materials. For example, grain-oriented magnetic steels made from about 3% Si and 97% Fe used in transformer cores are textured via thermo-mechanical processing so as to optimize their electrical and magnetic properties. Some common fiber (wire drawing) and sheet (rolling) textures with different crystal structures are shown in Table 8-2.

Texture Development in Thin Films Orientation or crystallographic texture development also occurs in thin films. In the case of thin films, the texture is often a result of the mechanisms of the growth process and not a result of externally applied stresses. Sometimes, internally generated thermal stresses can play a role in the determination of thin-film crystallographic texture. Oriented thin films can offer better electrical, optical, or magnetic properties. **Pole figure analysis**, a technique based on x-ray diffraction (XRD) (Chapter 3), or a specialized scanning electron microscopy technique known as **orientation microscopy** are used to identify textures in different engineered materials (films, sheets, single crystals, etc.).

Residual Stresses A small portion of the applied stress is stored in the form of **residual stresses** within the structure as a tangled network of dislocations. The presence of dislocations increases the total internal energy of the structure. As the extent of cold working increases, the level of total internal energy of the material increases. Residual stresses generated by cold working may not always be desirable and can be relieved by a heat treatment known as a **stress-relief anneal** (Section 8-6). As will be discussed shortly, in some instances, we deliberately create residual compressive stresses at the surface of materials to enhance their mechanical properties.

The residual stresses are not uniform throughout the deformed metallic material. For example, high compressive residual stresses may be present at the surface of a rolled plate and high tensile stresses may be stored in the center. If we machine a small amount of metal from one surface of a cold-worked part, we remove metal that contains only compressive residual stresses. To restore the balance, the plate must distort. If there is a net compressive residual stress at the surface of a component, this may be beneficial to the mechanical properties since any crack or flaw on the surface will not likely grow. These are reasons why any residual stresses, originating from cold work or any other source, affect the ability of a part to carry a load (Figure 8-10). If a tensile stress is applied to a material that already contains tensile residual stresses, the total stress acting on the part is the sum of the applied and residual stresses. If, however, compressive stresses are stored at the surface of a metal part, an applied tensile stress must first balance the compressive residual stresses. Now the part may be capable of withstanding a larger than normal load. In Chapter 7, we learned that fatigue is a common mechanism of failure

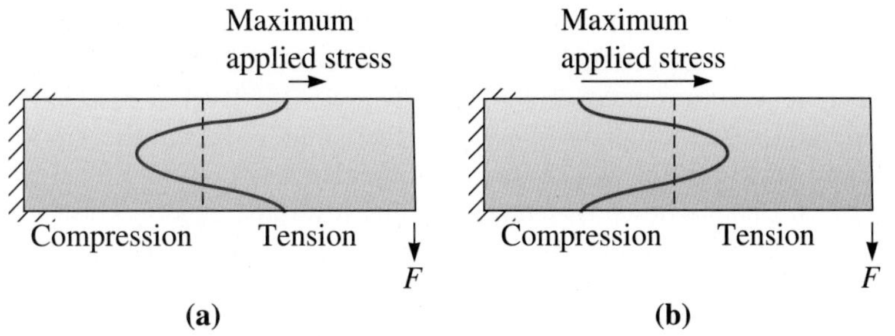

Figure 8-10 Residual stresses can be harmful or beneficial. (a) A bending force applies a tensile stress on the top of the beam. Since there are already tensile residual stresses at the top, the load-carrying characteristics are poor. (b) The top contains compressive residual stresses. Now the load-carrying characteristics are very good.

for load-bearing components. Sometimes, components that are subject to fatigue failure can be strengthened by **shot peening**. Bombarding the surface with steel shot propelled at a high velocity introduces compressive residual stresses at the surface that increase the resistance of the metal surface to fatigue failure (Chapter 7). Example 8-3 explains the use of shot peening.

Example 8-3 *Design of a Fatigue-Resistant Shaft*

Your company has produced several thousand shafts that have a fatigue strength of 20,000 psi. The shafts are subjected to high bending loads during rotation. Your sales engineers report that the first few shafts placed into service failed in a short period of time by fatigue. Design a process by which the remaining shafts can be salvaged by improving their fatigue properties.

SOLUTION

Fatigue failures typically begin at the surface of a rotating part; thus, increasing the strength at the surface improves the fatigue life of the shaft. A variety of methods might be used to accomplish this.

If the shaft is made from steel, we could carburize the surface of the part (Chapter 5). In carburizing, carbon is diffused into the surface of the shaft. After an appropriate heat treatment, the higher carbon content at the surface increases the strength of the surface and, perhaps more importantly, introduces *compressive* residual stresses at the surface.

We might consider cold working the shaft; cold working increases the yield strength of the metal and, if done properly, introduces compressive residual stresses. The cold work also reduces the diameter of the shaft and, because of the dimensional change, the shaft may not be able to perform its function.

Another alternative is to shot peen the shaft. Shot peening introduces local compressive residual stresses at the surface without changing the dimensions of the part. This process, which is also inexpensive, might be sufficient to salvage the remaining shafts.

Tempering and Annealing of Glasses Residual stresses originating during the cooling of glasses are of considerable interest. We can deal with residual stresses in glasses in two ways. First, we can reheat the glass to a high temperature known as the *annealing point* (~450°C for silicate glasses with a viscosity of $\sim 10^{13}$ poise) and let it cool slowly so that the outside and inside cool at about the same rate. The resultant glass will have little or no residual stress. This process is known as **annealing**, and the resultant glass that is nearly stress-free is known as **annealed glass**. The purpose of annealing glasses and the process known as stress-relief annealing in metallic materials is the same (i.e., to remove or significantly lower the level of residual stress). The origin of residual stress, though, is different for these materials. Another option we have in glass processing is to conduct a heat treatment that leads to compressive stresses on the surface of a glass; this is known as **tempering**. The resultant glass is known as **tempered glass**. Tempered glass is obtained by heating glass to a temperature just below the annealing point, then, deliberately letting the surface cool more rapidly than the center. This leads to a uniform compressive stress at the surface of the glass. The center region remains under a tensile stress. It is also possible to exchange ions in the glass structure and to introduce a compressive stress. This is known as *chemical tempering.*

In Chapter 7, we saw that the strength of glass depends on flaws on the surface. If we have a compressive stress at the surface of the glass and a tensile stress in the center (so as to have overall zero stress), the strength of glass is improved significantly. Any microcracks present will not grow readily, owing to the presence of a net compressive stress on the surface of the glass. If we, however, create a large impact, then the crack does penetrate through the region where the stresses are compressive, and the glass shatters.

Tempered glass has many uses. For example, side window panes and rear windshields of cars are made using tempered glass. Applications such as fireplace screens, ovens, shelving, furniture, and refrigerators also make use of tempered glass. For automobile windshields in the front, we make use of **laminated safety glass**. Front windshield glass is made from two annealed glass pieces laminated using a plastic known as polyvinyl butyral (PVB). If the windshield glass breaks, the laminated glass pieces are held together by PVB plastic. This helps minimize injuries to the driver and passengers. Also, the use of laminated safety glass reduces the chances of glass pieces cutting into the fabric of airbags that are probably being deployed simultaneously.

8-5 Characteristics of Cold Working

There are a number of advantages and limitations to strengthening a metallic material by cold working or strain hardening.

- We can simultaneously strengthen the metallic material and produce the desired final shape.
- We can obtain excellent dimensional tolerances and surface finishes by the cold working process.
- The cold-working process can be an inexpensive method for producing large numbers of small parts.
- Some metals, such as HCP magnesium, have a limited number of slip systems and are rather brittle at room temperature; thus, only a small degree of cold working can be accomplished.
- Ductility, electrical conductivity, and corrosion resistance are impaired by cold working. Since the extent to which electrical conductivity is reduced by cold working is less than that for other strengthening processes, such as introducing alloying elements (Figure 8-11), cold working is a satisfactory way to strengthen conductor materials, such as the copper wires used for transmission of electrical power.
- Properly controlled residual stresses and anisotropic behavior may be beneficial; however, if residual stresses are not properly controlled, the materials properties are greatly impaired.
- As will be seen in Section 8-6, since the effect of cold working is decreased or eliminated at higher temperatures, we cannot use cold working as a

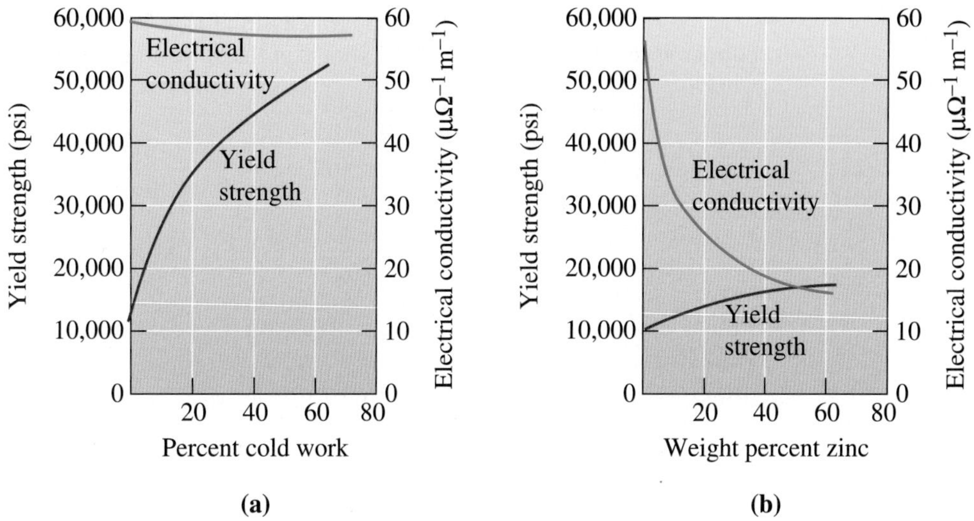

Figure 8-11 A comparison of strengthening copper by (a) cold working and (b) alloying with zinc. Note that cold working produces greater strengthening, yet has little effect on electrical conductivity.

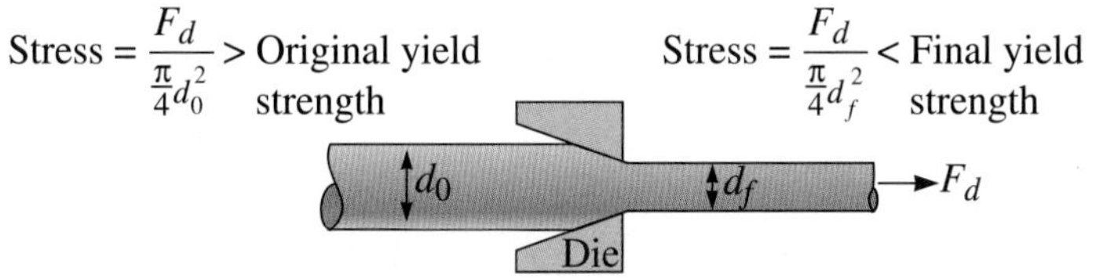

Figure 8-12 The wire-drawing process. The force F_d acts on both the original and final diameters. Thus, the stress produced in the final wire is greater than that in the original. If the wire did not strain harden during drawing, the final wire would break before the original wire was drawn through the die.

strengthening mechanism for components that will be subjected to high temperatures during service.

- Some deformation processing techniques can be accomplished only if cold working occurs. For example, wire drawing requires that a rod be pulled through a die to produce a smaller cross-sectional area (Figure 8-12). For a given draw force F_d, a different stress is produced in the original and final wire. The stress on the initial wire must exceed the yield strength of the metal to cause deformation. The stress on the final wire must be less than its yield strength to prevent failure. This is accomplished only if the wire strain hardens during drawing.

Example 8-4 *Design of a Wire-Drawing Process*

Design a process to produce 0.20-in.-diameter copper wire. The mechanical properties of the copper to be considered are shown in Figure 8-6.

SOLUTION

Wire drawing is the obvious manufacturing technique for this application. To produce the copper wire as efficiently as possible, we make the largest reduction in the diameter possible. Our design must ensure that the wire strain hardens sufficiently during drawing to prevent the drawn wire from breaking.

As an example calculation, let's assume that the starting diameter of the copper wire is 0.40 in. and that the wire is in the softest possible condition. The cold work is

$$\% \text{ CW} = \left[\frac{A_0 - A_f}{A_0}\right] \times 100 = \left[\frac{(\pi/4)d_0^2 - (\pi/4)d_f^2}{(\pi/4)d_0^2}\right] \times 100$$

$$= \left[\frac{(0.40 \text{ in.})^2 - (0.20 \text{ in.})^2}{(0.40 \text{ in.})^2}\right] \times 100 = 75\%$$

From Figure 8-6, the initial yield strength with 0% cold work is 22,000 psi. The final yield strength with 75% cold work is about 80,000 psi (with very little ductility). The draw force required to deform the initial wire is

$$F = S_y A_0 = (22{,}000 \text{ psi})(\pi/4)(0.40 \text{ in.})^2 = 2{,}765 \text{ lb}$$

The stress acting on the wire after passing through the die is

$$S = \frac{F_d}{A_f} = \frac{2765 \text{ lb}}{(\pi/4)(0.20 \text{ in.})^2} = 88{,}000 \text{ psi}$$

The applied stress of 88,000 psi is greater than the 80,000 psi yield strength of the drawn wire. Therefore, the wire breaks because the % elongation is almost zero.

We can perform the same set of calculations for other initial diameters, with the results shown in Table 8-3 and Figure 8-13.

The graph shows that the draw stress exceeds the yield strength of the drawn wire when the original diameter is about 0.37 in. To produce the wire as efficiently as possible, the original diameter should be just under 0.37 in.

Table 8-3 Mechanical properties of copper wire (see Example 8-4)

d_0 (in.)	% CW	Yield Strength of Drawn Wire (psi)	Force (lb)	Draw Stress on Drawn Wire (psi)
0.25	36	58,000	1,080	34,380
0.30	56	70,000	1,555	49,500
0.35	67	74,000	2,117	67,380
0.40	75	80,000	2,765	88,000

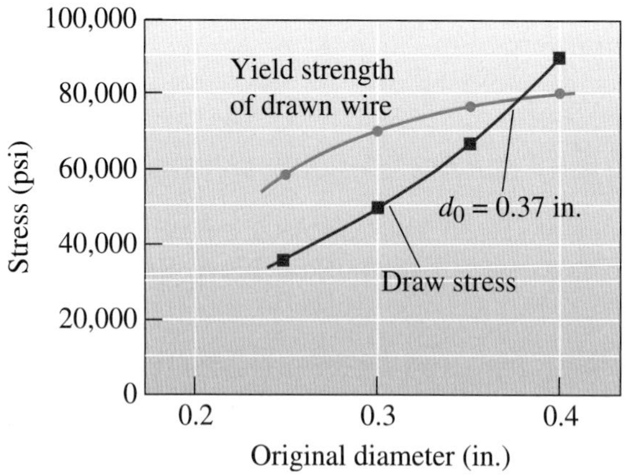

Figure 8-13 Yield strength and draw stress of wire (for Example 8-4).

8-6 The Three Stages of Annealing

Cold working is a useful strengthening mechanism, and it is an effective tool for shaping materials using wire drawing, rolling, extrusion, etc. Sometimes, cold working leads to effects that are undesirable. For example, the loss of ductility or development of residual stresses may not be desirable for certain applications. Since cold working or strain hardening results from increased dislocation density, we can assume that any treatment to rearrange or annihilate dislocations reverses the effects of cold working.

Annealing is a heat treatment used to eliminate some or all of the effects of cold working. Annealing at a low temperature may be used to eliminate the residual stresses produced during cold working without affecting the mechanical properties of the finished part, or annealing may be used to completely eliminate the strain hardening achieved during cold working. In this case, the final part is soft and ductile but still has a good surface finish and dimensional accuracy. After annealing, additional cold work can be done since the ductility is restored; by combining repeated cycles of cold working and annealing, large total deformations may be achieved. There are three possible stages in the annealing process; their effects on the properties of brass are shown in Figure 8-14.

Note that the term "annealing" is also used to describe other thermal treatments. For example, glasses may be annealed, or heat treated, to eliminate residual stresses. Cast irons and steels may be annealed to produce the maximum ductility, even though no prior cold work was done to the material. These annealing heat treatments will be discussed in later chapters.

Recovery The original cold-worked microstructure is composed of deformed grains containing a large number of tangled dislocations. When we first heat the metal, the additional thermal energy permits the dislocations to move and form the boundaries

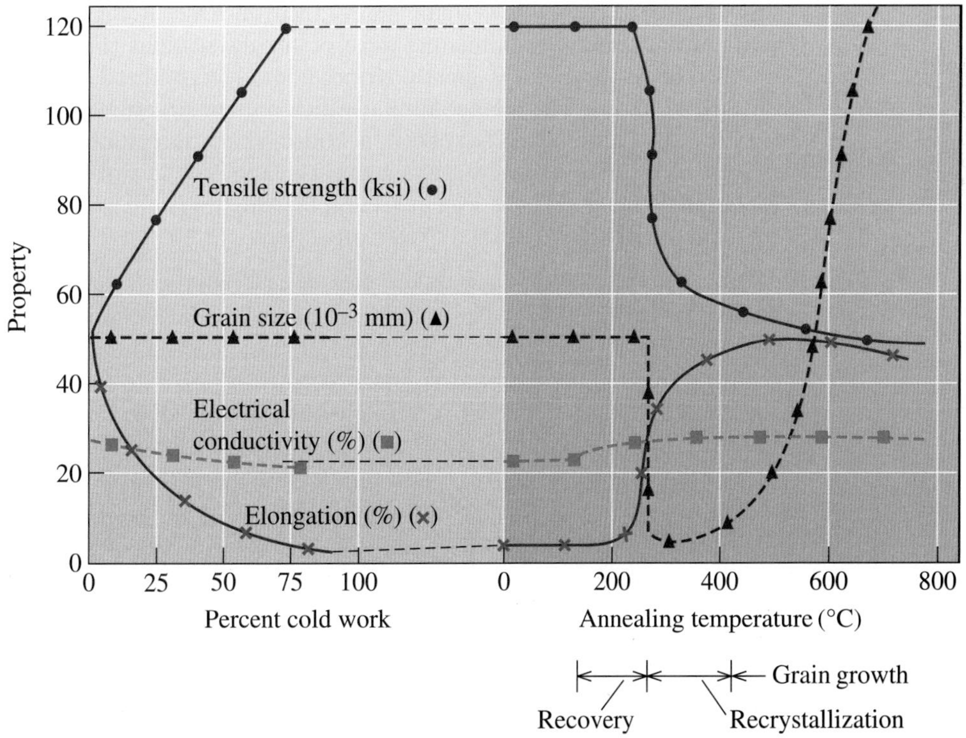

Figure 8-14 The effect of cold work on the properties of a Cu-35% Zn alloy and the effect of annealing temperature on the properties of a Cu-35% Zn alloy that is cold worked 75%. The electrical conductivity is defined relative to the International Annealed Copper Standard for which a value of 100% is equal to 5.80×10^7 siemens per meter (5.80×10^5 $ohm^{-1} \cdot cm^{-1}$).

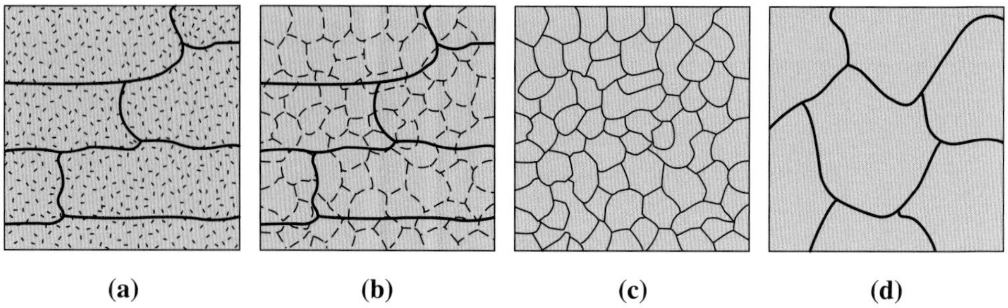

Figure 8-15 The effect of annealing temperature on the microstructure of cold-worked metals. (a) Cold worked, (b) after recovery, (c) after recrystallization, and (d) after grain growth.

of a **polygonized subgrain structure** (Figure 8-15). The dislocation density, however, is virtually unchanged. This low temperature treatment removes the residual stresses due to cold working without causing a change in dislocation density and is called **recovery**.

The mechanical properties of the metal are relatively unchanged because the number of dislocations is not reduced during recovery. Since residual stresses are reduced or even eliminated when the dislocations are rearranged, recovery is often called a stress relief anneal. In addition, recovery restores high electrical conductivity to the metal, permitting us to manufacture copper or aluminum wire for transmission of electrical power that is strong yet still has high conductivity. Finally, recovery often improves the corrosion resistance of the material.

Recrystallization When a cold-worked metal is heated above a certain temperature, rapid recovery eliminates residual stresses and produces the polygonized dislocation structure. New small grains then nucleate at the cell boundaries of the polygonized structure, eliminating most of the dislocations (Figure 8-15). Because the number of dislocations is greatly reduced, the recrystallized metal has low strength but high ductility. The temperature at which a microstructure of new grains that have very low dislocation density appears is known as the **recrystallization temperature**. The process of formation of new grains by heat treating a cold-worked material is known as **recrystallization**. As will be seen in Section 8-7, the recrystallization temperature depends on many variables and is not a fixed temperature.

Grain Growth At still higher annealing temperatures, both recovery and recrystallization occur rapidly, producing a fine recrystallized grain structure. If the temperature is high enough, the grains begin to grow, with favored grains consuming the smaller grains (Figure 8-16). This phenomenon, called *grain growth*, is driven by the reduction in grain boundary area. Illustrated for a copper-zinc alloy in Figure 8-16, grain growth is almost always undesirable. Grain growth will occur in most materials if they are subjected to a high enough temperature and, as such, is not related to cold working. Thus, recrystallization or recovery are not needed for grain growth to occur.

You may be aware that incandescent light bulbs contain filaments that are made from tungsten. The high operating temperature causes grain growth and is one of the factors leading to filament failure.

Ceramic materials, which normally do not show any significant strain hardening, show a considerable amount of grain growth. Also, abnormal grain growth can occur in some materials as a result of formation of a liquid phase during sintering. Sometimes grain growth is desirable, as is the case for alumina ceramics for making optical materials used in lighting. In this application, we want very large grains since the scattering of light from grain boundaries has to be minimized. Some researchers have also developed methods for growing single crystals of ceramic materials using grain growth.

Figure 8-16 Photomicrographs showing the effect of annealing temperature on grain size in brass. Twin boundaries can also be observed in the structures. The brass was cold-rolled to the same percent reduction in thickness and annealed at the following temperatures for one hour: (a) 500°C, (b) 700°C, and (c) 800°C. (*Courtesy of Wendelin Wright*)

8-7 Control of Annealing

In many metallic materials applications, we need a combination of strength and toughness. Therefore, we need to design processes that involve shaping via cold working. We then need to control the annealing process to obtain the desired ductility. To design an appropriate annealing heat treatment, we need to know the recrystallization temperature and the size of the recrystallized grains.

Recrystallization Temperature This is the temperature at which grains in the cold-worked microstructure begin to transform into new, equiaxed, dislocation-free grains. The driving force for recrystallization is the difference between the internal energy of a cold-worked material and that of a recrystallized material. It is important for us to emphasize that the recrystallization temperature is *not* a fixed temperature, like the melting temperature of a pure element, and is influenced by a variety of processing variables.

- The recrystallization temperature decreases when the amount of cold work increases. Greater amounts of cold work make the metal less stable and encourage nucleation of recrystallized grains. There is a minimum amount of cold work, about 30 to 40%, below which recrystallization will not occur.
- A smaller initial cold-worked grain size reduces the recrystallization temperature by providing more sites—the former grain boundaries—at which new grains can nucleate.
- Pure metals recrystallize at lower temperatures than alloys.
- Increasing the annealing time reduces the recrystallization temperature (Figure 8-17), since more time is available for nucleation and growth of the new recrystallized grains.
- Higher melting-point alloys have a higher recrystallization temperature. Since recrystallization is a diffusion-controlled process, the recrystallization temperature is roughly proportional to $0.4T_m$ (kelvin) where T_m is the melting temperature. Typical recrystallization temperatures for selected metals are shown in Table 8-4.

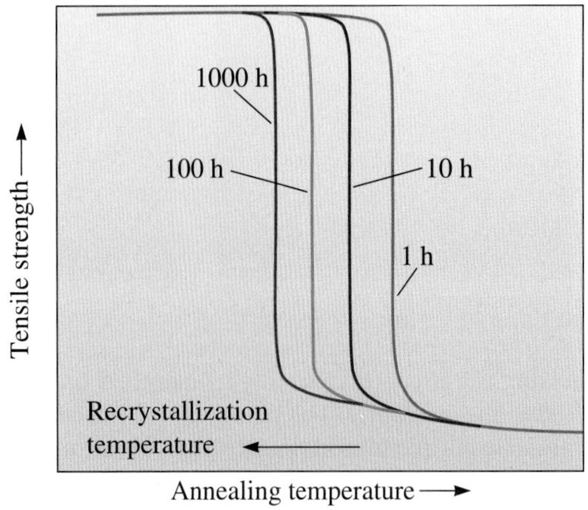

Figure 8-17 Longer annealing times reduce the recrystallization temperature. Note that the recrystallization temperature is not a fixed temperature.

Table 8-4 Typical recrystallization temperatures for selected metals

Metal	Melting Temperature (°C)	Recrystallization Temperature (°C)
Sn	232	−4
Pb	327	−4
Zn	420	10
Al	660	150
Mg	650	200
Ag	962	200
Cu	1085	200
Fe	1538	450
Ni	1453	600
Mo	2610	900
W	3410	1200

(Based on Structure and Properties of Engineering Materials *by Brick, R. M.; et al, McGraw-Hill, 1977.)*

The concept of recrystallization temperature is very important since it also defines the boundary between cold working and hot working of a metallic material. If we conduct deformation (shaping) of a material above the recrystallization temperature, we refer to it as hot working. If we conduct the shaping or deformation at a temperature below the recrystallization temperature, we refer to it as cold working. As can be seen from Table 8-4, for lead or tin deformed at 25°C, we are conducting hot working! This is why iron can be cold worked at room temperature but not lead. For tungsten being deformed at 1000°C, we are conducting cold working! In some cases, processes conducted above 0.6 times the melting temperature of a metal (in K) are considered as hot working. Processes conducted below 0.3 times the melting temperature are considered cold working, and processes conducted between 0.3 and 0.6 times T_m are considered **warm working**. These descriptions of ranges that define hot, cold, and warm working, however, are approximate and should be used with caution.

Recrystallized Grain Size A number of factors influence the size of the recrystallized grains. Reducing the annealing temperature, the time required to heat to the annealing temperature, or the annealing time reduces grain size by minimizing the opportunity for grain growth. Increasing the initial cold work also reduces final grain size by providing a greater number of nucleation sites for new grains. Finally, the presence of a second phase in the microstructure may foster or hinder recrystallization and grain growth depending on its arrangement and size.

8-8 Annealing and Materials Processing

The effects of recovery, recrystallization, and grain growth are important in the processing and eventual use of a metal or an alloy.

Deformation Processing By taking advantage of the annealing heat treatment, we can increase the total amount of deformation we can accomplish. If we are required to reduce a 5-in.-thick plate to a 0.05-in.-thick sheet, we can do the maximum permissible cold work, anneal to restore the metal to its soft, ductile condition, and then cold work again. We can repeat the cold work-anneal cycle until we approach the proper thickness. The final cold-working step can be designed to produce the final dimensions and properties required, as in the following example.

Example 8-5 *Design of a Process to Produce Copper Strip*

We wish to produce a 0.1-cm-thick, 6-cm-wide copper strip having at least 60,000 psi yield strength and at least 5% elongation. We are able to purchase 6-cm-wide strip only in a thickness of 5 cm. Design a process to produce the product we need. Refer to Figure 8-6 as needed.

SOLUTION

In Example 8-2, we found that the required properties can be obtained with a cold work of 40 to 45%. Therefore, the starting thickness must be between 0.167 cm and 0.182 cm, and this starting material must be as soft as possible—that is, in the annealed condition. Since we are able to purchase only 5-cm-thick stock, we must reduce the thickness of the 5 cm strip to between 0.167 and 0.182 cm, then anneal the strip prior to final cold working. But can we successfully cold work from 5 cm to 0.182 cm?

$$\%\,\mathrm{CW} = \left[\frac{t_0 - t_f}{t_0}\right] \times 100 = \left[\frac{5\ \mathrm{cm} - 0.182\ \mathrm{cm}}{5\ \mathrm{cm}}\right] \times 100 = 96.4\%$$

Based on Figure 8-6, a maximum of about 90% cold work is permitted. Therefore, we must do a series of cold work and anneal cycles. Although there are many possible combinations, one is as follows:

1. Cold work the 5 cm strip 80% to 1 cm:

$$80 = \left[\frac{t_0 - t_f}{t_0}\right] \times 100 = \left[\frac{5\ \mathrm{cm} - t_f}{5\ \mathrm{cm}}\right] \times 100 \quad \text{or} \quad t_f = 1\ \mathrm{cm}$$

2. Anneal the 1 cm strip to restore the ductility. If we don't know the recrystallization temperature, we can use the 0.4 T_m relationship to provide an estimate. The melting point of copper is 1085°C:

$$T_r \cong (0.4)(1085 + 273) = 543\ \mathrm{K} = 270^\circ\mathrm{C}$$

3. Cold work the 1-cm-thick strip to 0.182 cm:

$$\% \text{ CW} = \left[\frac{1 \text{ cm} - 0.182 \text{ cm}}{1 \text{ cm}}\right] \times 100 = 81.8\%$$

4. Again anneal the copper at 270°C to restore ductility.
5. Finally, cold work 45% from 0.182 cm to the final dimension of 0.1 cm. This process gives the correct final dimensions and properties.

High Temperature Service

As mentioned previously, neither strain hardening nor grain-size strengthening (Hall-Petch equation, Chapter 4) are appropriate for an alloy that is to be used at elevated temperatures, as in creep-resistant applications. When the cold-worked metal is placed into service at a high temperature, recrystallization immediately causes a catastrophic decrease in strength. In addition, if the temperature is high enough, the strength continues to decrease because of growth of the newly recrystallized grains.

Joining Processes

Metallic materials can be joined using processes such as welding. When we join a cold-worked metal using a welding process, the metal adjacent to the weld heats above the recrystallization and grain growth temperatures and subsequently cools slowly. This region is called the **heat-affected zone (HAZ)**. The structure and properties in the heat-affected zone of a weld are shown in Figure 8-18. The mechanical properties are reduced potentially catastrophically by the heat of the welding process.

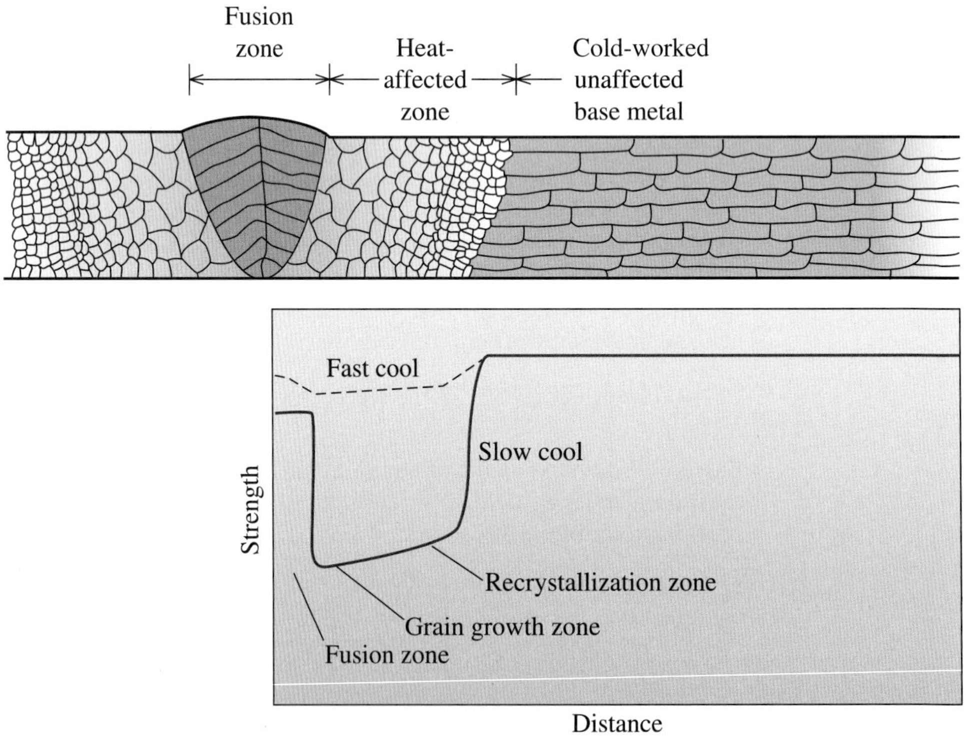

Figure 8-18 The structure and properties surrounding a fusion weld in a cold-worked metal. Only the right-hand side of the heat-affected zone is marked on the diagram. Note the loss in strength caused by recrystallization and grain growth in the heat-affected zone.

Welding processes, such as electron beam welding or laser welding, which provide high rates of heat input for brief times, and, thus, subsequent fast cooling, minimize the exposure of the metallic materials to temperatures above recrystallization and minimize this type of damage. Similarly a process known as friction stir welding provides almost no HAZ and is being commercially used for welding aluminum alloys.

8-9 Hot Working

We can deform a metal into a useful shape by hot working rather than cold working. As described previously, hot working is defined as plastically deforming the metallic material at a temperature above the recrystallization temperature. During hot working, the metallic material is continually recrystallized (Figure 8-19). As mentioned before, at room temperature, lead is well above its recrystallization temperature of −4°C, and therefore, Pb does not strain harden and remains soft and ductile at room temperature.

Lack of Strengthening No strengthening occurs during deformation by hot working; consequently, the amount of plastic deformation is almost unlimited. A very thick plate can be reduced to a thin sheet in a continuous series of operations. The first steps in the process are carried out well above the recrystallization temperature to take advantage of the lower strength of the metal. The last step is performed just above the recrystallization temperature, using a large percent deformation in order to produce the finest possible grain size.

Hot working is well suited for forming large parts, since the metal has a low yield strength and high ductility at elevated temperatures. In addition, HCP metals such as magnesium have more active slip systems at hot-working temperatures; the higher ductility permits larger deformations than are possible by cold working. Example 8-6 illustrates a design of a hot-working process.

Elimination of Imperfections Some imperfections in the original metallic material may be eliminated or their effects minimized during hot working. Gas pores can be closed and welded shut during hot working—the internal lap formed when the pore is closed is eliminated by diffusion during the forming and cooling process. Composition differences in the metal can be reduced as hot working brings the surface and center of the plate closer together, thereby reducing diffusion distances.

Anisotropic Behavior The final properties in hot-worked parts are *not* isotropic. The forming rolls or dies, which are normally at a lower temperature than the

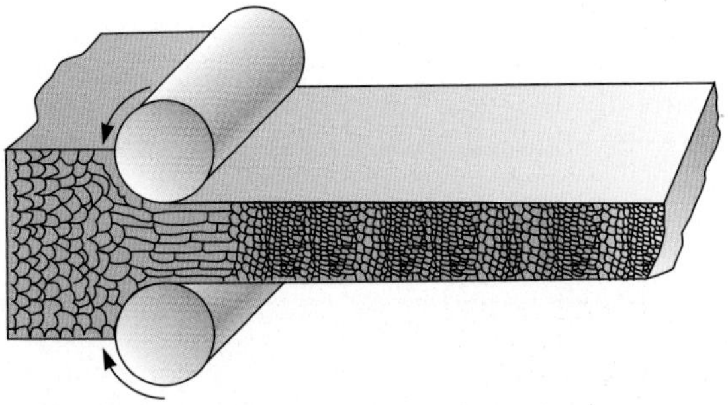

Figure 8-19 During hot working, the elongated, anisotropic grains immediately recrystallize. If the hot-working temperature is properly controlled, the final hot-worked grain size can be very fine.

metal, cool the surface more rapidly than the center of the part. The surface then has a finer grain size than the center. In addition, a fibrous structure is produced because inclusions and second-phase particles are elongated in the working direction.

Example 8-6 *Design of a Process to Produce a Copper Strip*

We want to produce a 0.1-cm-thick, 6-cm-wide copper strip having at least 60,000 psi yield strength and at least 5% elongation. We are able to purchase 6-cm-wide strip only in thicknesses of 5 cm. Design a process to produce the product we need, but in fewer steps than were required in Example 8-5.

SOLUTION

In Example 8-5, we relied on a series of cold work-anneal cycles to obtain the required thickness. We can reduce the steps by hot rolling to the required intermediate thickness:

$$\%\,\text{HW} = \left[\frac{t_0 - t_f}{t_0}\right] \times 100 = \left[\frac{5\text{ cm} - 0.182\text{ cm}}{5\text{ cm}}\right] \times 100 = 96.4\%$$

$$\%\,\text{HW} = \left[\frac{t_0 - t_f}{t_0}\right] \times 100 = \left[\frac{5\text{ cm} - 0.167\text{ cm}}{5\text{ cm}}\right] \times 100 = 96.7\%$$

Note that the formulas for hot and cold work are the same.

Because recrystallization occurs simultaneously with hot working, we can obtain these large deformations and a separate annealing treatment is not required. Thus our design might be

1. Hot work the 5 cm strip 96.4% to the intermediate thickness of 0.182 cm.
2. Cold work 45% from 0.182 cm to the final dimension of 0.1 cm. This design gives the correct dimensions and properties.

Surface Finish and Dimensional Accuracy The surface finish formed during hot working is usually poorer than that obtained by cold working. Oxygen often reacts with the metal at the surface to form oxides, which are forced into the surface during forming. Hot-worked steels and other metals are often subjected to a "pickling" treatment in which acids are used to dissolve the oxide scale. In some metals, such as tungsten and beryllium, hot working must be done in a protective atmosphere to prevent oxidation. Note that forming processes performed on Be-containing materials require protective measures, since the inhalation of Be-containing materials is hazardous.

The dimensional accuracy is also more difficult to control during hot working. A greater elastic strain must be considered, since the modulus of elasticity is lower at hot-working temperatures than at cold-working temperatures. In addition, the metal contracts as it cools from the hot-working temperature. The combination of elastic strain and thermal contraction requires that the part be made oversized during deformation. Forming dies must be carefully designed, and precise temperature control is necessary if accurate dimensions are to be obtained.

Summary

- The properties of metallic materials can be controlled by combining plastic deformation and heat treatments.
- When a metallic material is deformed by cold working, strain hardening occurs as additional dislocations are introduced into the structure. Very large increases in

strength may be obtained in this manner. The ductility of the strain hardened metallic material is reduced.

- Strain hardening, in addition to increasing strength and hardness, increases residual stresses, produces anisotropic behavior, and reduces ductility, electrical conductivity, and corrosion resistance.
- The amount of strain hardening is limited because of the simultaneous decrease in ductility; FCC metals typically have the best response to strengthening by cold working.
- Wire drawing, stamping, rolling, and extrusion are some examples of manufacturing methods for shaping metallic materials. Some of the underlying principles for these processes also can be used for the manufacturing of polymeric materials.
- The strain-hardening mechanism is not effective at elevated temperatures because the effects of cold work are eliminated by recrystallization.
- Annealing of metallic materials is a heat treatment intended to eliminate all, or a portion of, the effects of strain hardening. The annealing process may involve as many as three steps.
- Recovery occurs at low temperatures, eliminating residual stresses and restoring electrical conductivity without reducing strength. A "stress relief anneal" refers to recovery.
- Recrystallization occurs at higher temperatures and eliminates almost all of the effects of strain hardening. The dislocation density decreases dramatically during recrystallization as new grains nucleate and grow.
- Grain growth, which typically should be avoided, occurs at still higher temperatures. In cold-worked metallic materials, grain growth follows recovery and recrystallization. In ceramic materials, grain growth can occur due to high temperatures or the presence of a liquid phase during sintering.
- Hot working combines plastic deformation and annealing in a single step, permitting large amounts of plastic deformation without embrittling the material.
- Annealing of glasses leads to the removal of stresses developed during cooling. Thermal tempering of glasses is a heat treatment in which deliberate rapid cooling of the glass surface leads to a compressive stress at the surface. We use tempered or laminated glass in applications where safety is important.
- In metallic materials, compressive residual stresses can be introduced using shot peening. This treatment will lead to an increase in the fatigue life.

Glossary

Annealed glass Glass that has been treated by heating above the annealing point temperature (where the viscosity of glass becomes 10^{13} poise) and then cooled slowly to minimize or eliminate residual stresses.

Annealing In the context of metals, annealing is a heat treatment used to eliminate part or all of the effects of cold working. For glasses, annealing is a heat treatment that removes thermally induced stresses.

Bauschinger effect A material previously plastically deformed under tension shows decreased flow stress under compression.

Cold working Deformation of a metal below the recrystallization temperature. During cold working, the number of dislocations increases, causing the metal to be strengthened as its shape is changed.

Deformation processing Techniques for the manufacturing of metallic and other materials using such processes as rolling, extrusion, drawing, etc.

Drawing A deformation processing technique in which a material is pulled through an opening in a die (e.g., wire drawing).

Extrusion A deformation processing technique in which a material is pushed through an opening in a die. Used for metallic and polymeric materials.

Fiber texture A preferred orientation of grains obtained during the wire drawing process. Certain crystallographic directions in each elongated grain line up with the drawing direction, causing anisotropic behavior.

Frank-Read source A pinned dislocation that, under an applied stress, produces additional dislocations. This mechanism is at least partly responsible for strain hardening.

Heat-affected zone The volume of material adjacent to a weld that is heated during the welding process above some critical temperature at which a change in the structure, such as grain growth or recrystallization, occurs.

Hot working Deformation of a metal above the recrystallization temperature. During hot working, only the shape of the metal changes; the strength remains relatively unchanged because no strain hardening occurs.

Laminated safety glass Two pieces of annealed glass held together by a plastic such as polyvinyl butyral (PVB). This type of glass can be used in car windshields.

Orientation microscopy A specialized technique, often based on scanning electron microscopy, used to determine the crystallographic orientation of different grains in a polycrystalline sample.

Pole figure analysis A specialized technique based on x-ray diffraction, used for the determination of preferred orientation of thin films, sheets, or single crystals.

Polygonized subgrain structure The arrangement of dislocations that forms during the recovery stage of annealing in preparation for forming new grain boundaries during recrystallization.

Recovery A low-temperature annealing heat treatment designed to eliminate residual stresses introduced during deformation without reducing the strength of the cold-worked material. This is the same as a stress-relief anneal.

Recrystallization A medium-temperature annealing heat treatment designed to eliminate all of the effects of strain hardening produced during cold working. Dislocation-free and new grains form.

Recrystallization temperature A temperature above which essentially dislocation-free and new grains emerge from a material that was previously cold worked. This depends upon the extent of cold work, time of heat treatment, etc., and is not a fixed temperature.

Residual stresses Stresses introduced in a material during processing. These can originate as a result of cold working or differential thermal expansion and contraction.

Sheet texture A preferred orientation of grains obtained during the rolling process. Certain crystallographic directions line up with the rolling direction, and certain preferred crystallographic planes become parallel to the sheet surface.

Shot peening Introducing compressive residual stresses at the surface of a part by bombarding that surface with steel shot to improve the overall performance of the material.

Strain hardening Strengthening of a material by increasing the number of dislocations by deformation. Also known as "work hardening."

Strain-hardening exponent A parameter that describes the effect that strain has on the resulting strength of the material. A material with a high strain-hardening coefficient obtains high strength with only small amounts of deformation or strain.

Strain rate The rate at which a material is deformed.

Strain-rate sensitivity The rate at which stress changes as a function of strain rate. A material may behave much differently if it is slowly pressed into a shape rather than smashed rapidly into a shape by an impact blow.

Stress-relief anneal The recovery stage of the annealing heat treatment during which residual stresses are relieved without altering the strength and ductility of the material.

Tempered glass A glass, mainly for applications where safety is particularly important, obtained by either heat treatment and quenching or by the chemical exchange of ions. Tempering results in a net compressive stress at the surface of the glass.

Tempering In the context of glass making, tempering refers to a heat treatment that leads to a compressive stress on the surface of a glass. This compressive stress layer makes tempered glass safer. In the context of processing of metallic materials, tempering refers to a heat treatment used to soften the material and to increase its toughness.

Texture strengthening Increase in the yield strength of a material as a result of preferred crystallographic texture.

Thermomechanical processing Processes involved in the manufacturing of metallic components using mechanical deformation and various heat treatments.

Thermoplastics A class of polymers that consist of large, long spaghetti-like molecules that are intertwined (e.g., polyethylene, nylon, PET, etc.).

Warm working A term used to indicate the processing of metallic materials in a temperature range that is between those that define cold and hot working (usually a temperature between 0.3 to 0.6 of the melting temperature in K).

Problems

Section 8-1 Relationship of Cold Working to the Stress Strain Curve

8-1 Using a stress strain diagram, explain what the term "strain hardening" means.

8-2 What is meant by the term "springback"? What is the significance of this term from a manufacturing viewpoint?

8-3 What does the term "Bauschinger effect" mean?

8-4 What manufacturing techniques make use of the cold-working process?

8-5 What is the difference between extrusion and drawing?

8-6 Consider the tensile stress strain curves in Figure 8-20 labeled 1 and 2 and answer the following questions. These curves are typical of metals. Consider each part as a separate question that has no relationship to previous parts of the question.

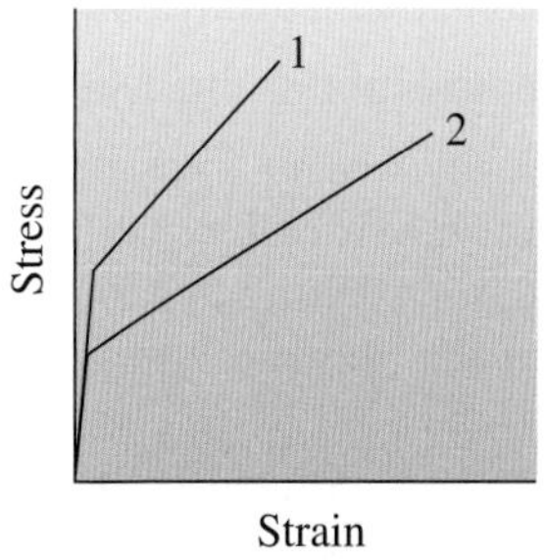

Figure 8-20 Stress strain curves (for Problem 8-6).

(a) Which material has the larger work-hardening exponent? How do you know?

(b) Samples 1 and 2 are identical except that they were tested at different strain rates. Which sample was tested at the higher strain rate? How do you know?

(c) Assume that the two stress strain curves represent successive tests of the same sample. The sample was loaded, then unloaded before necking began, and then the sample was reloaded. Which sample represents the first test: 1 or 2? How do you know?

8-7 Figure 8-21 is a plot of true stress versus true strain for a metal. For total imposed strains of $\varepsilon = 0.1$, 0.2, 0.3 and 0.4, determine the elastic and plastic components of the strain. The modulus of elasticity of the metal is 100 GPa.

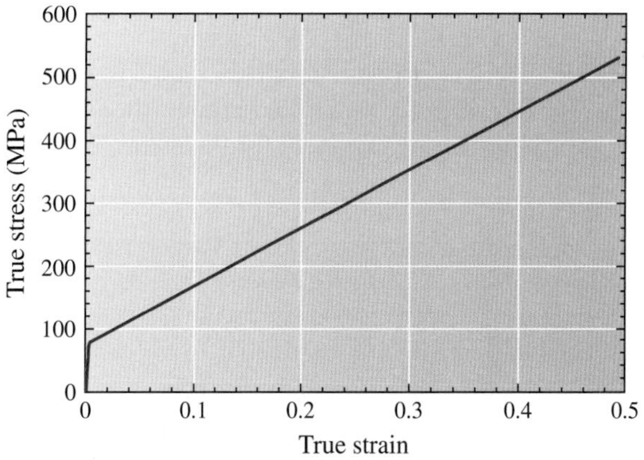

Figure 8-21 A true stress versus true strain curve for a metal (for Problem 8-7).

8-8 Define the terms: strain-hardening exponent (n) and strain-rate sensitivity (m). Use appropriate equations.

8-9 An annealed copper sample experiences a true strain of 0.07 caused by a true stress of 164 MPa. If this is in the plastic deformation region and the strain-hardening exponent is 0.44, what is the strength coefficient?

8-10 A 0.505-in.-diameter metal bar with a 2-in. gage length l_0 is subjected to a tensile test. The following measurements are made in the plastic region:

Force (lb)	Change in Gage length (in.) (Δl)	Diameter (in.)
27,500	0.2103	0.4800
27,000	0.4428	0.4566
25,700	0.6997	0.4343

Determine the strain-hardening exponent for the metal. Is the metal most likely to be FCC, BCC, or HCP? Explain.

8-11 A 1.33-cm-diameter metal bar with a 3-cm gage length (l_0) is subjected to a tensile test. The following measurements are made in the plastic region:

Force (N)	Change in Gage length (cm)	Diameter (cm)
16,240	0.6642	1.2028
19,066	1.4754	1.0884
19,273	2.4663	0.9848

Determine the strain-hardening coefficient for the metal. Is the metal most likely to be FCC, BCC, or HCP? Explain.

8-12 A true stress-true strain curve is shown in Figure 8-22. Determine the strain-hardening exponent for the metal.

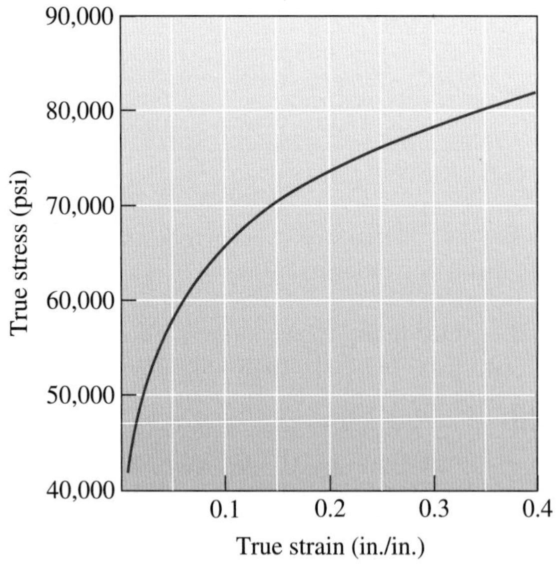

Figure 8-22 True stress-true strain curve (for Problem 8-12).

8-13 Figure 8-23 shows a plot of the natural logarithm of the true stress versus the natural logarithm of the true strain for a Cu-30% Zn sample tested in tension. Only the plastic portion of the stress strain curve is shown. Determine the strength coefficient K and the work-hardening exponent n.

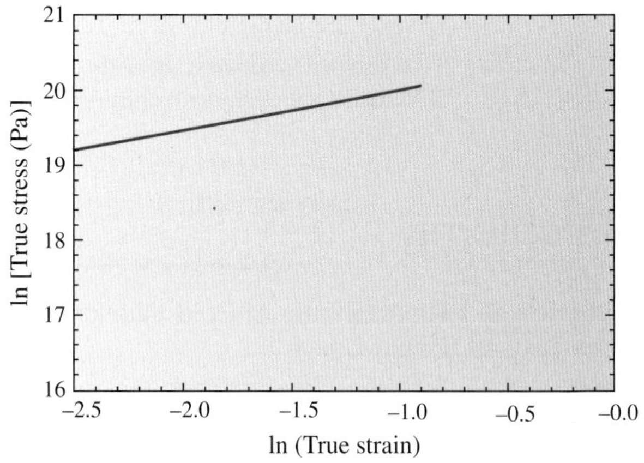

Figure 8-23 The natural logarithm of the true stress versus the natural logarithm of the true strain for a Cu-30% Zn sample tested in tension. Only the plastic portion of the stress strain curve is shown.

8-14 A Cu-30% Zn alloy tensile bar has a strain hardening coefficient of 0.50. The bar, which has an initial diameter of 1 cm and an initial gage length of 3 cm, fails at an engineering stress of 120 MPa. At the moment of fracture, the gage length is 3.5 cm and the diameter is 0.926 cm. No necking occurred. Calculate the true stress when the true strain is 0.05 cm/cm.

8-15 What would a strain hardening exponent of zero mean mathematically and practically?

Section 8-2 Strain-Hardening Mechanisms

8-16 Explain why many metallic materials exhibit strain hardening.

8-17 Does strain hardening depend upon grain size? Does it depend upon dislocation density?

8-18 Compare and contrast strain hardening with grain-size strengthening. What causes resistance to dislocation motion in each of these mechanisms?

8-19 Strain hardening is normally not a consideration in ceramic materials. Explain why.

8-20 Thermoplastic polymers such as polyethylene show an increase in strength when subjected to stress. Explain how this strengthening occurs.

8-21 Bottles of carbonated beverages are made using PET plastic. Explain how stress-induced crystallization increases the strength of PET bottles made by the blow-stretch process (see Chapters 3 and 5).

Section 8-3 Properties versus Percent Cold Work

8-22 Write down the equation that defines percent cold work. Explain the meaning of each term.

8-23 A 0.25-in.-thick copper plate is to be cold worked 63%. Find the final thickness.

8-24 A 0.25-in.-diameter copper bar is to be cold worked 63% in tension. Find the final diameter.

8-25 A 2-in.-diameter copper rod is reduced to a 1.5 in. diameter, then reduced again to a final diameter of 1 in. In a second case, the 2-in.-diameter rod is reduced in one step from a 2 in. to a 1 in. diameter. Calculate the % CW for both cases.

8-26 If a 25-mm diameter steel bar is cold extruded first to 22 mm, then from 22 mm to a final diameter of 20 mm, determine (a) the amount of cold work for each step, and (b) compare the sum of the amounts of cold work for each step to the cold work process from 25 mm directly to 20 mm. (See the next problem to fully understand your solution.)

8-27 If the amounts of extrusion in the previous problem were expressed as true strains, (a) determine the true strain for each of the extrusion steps, and (b) compare the sum of the true strain for each extrusion to the true strain from 25 mm to 20 mm.

8-28 A 2-mm-diameter, pure copper wire is reduced to a 1.75-mm diameter by cold working. What is its tensile strength? (See Figure 8-6.)

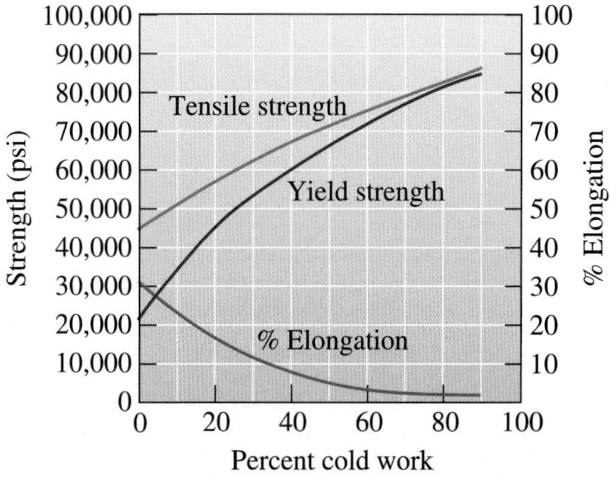

Figure 8-6 The effect of cold work on the mechanical properties of copper. (Repeated for Problem 8-28.)

8-29 A 3105 aluminum plate is reduced from 1.75 in. to 1.15 in. Determine the final properties of the plate. Note 3105 designates a particular aluminum alloy. (See Figure 8-24.)

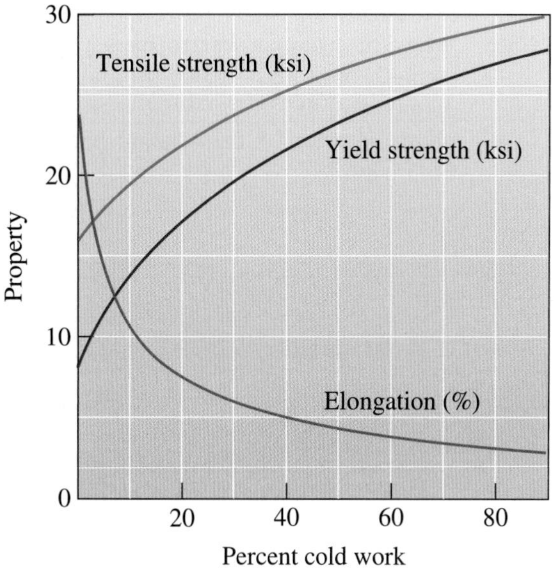

Figure 8-24 The effect of percent cold work on the properties of a 3105 aluminum alloy (for Problems 8-29, 8-31, 8-37 and 8-54).

8-30 A Cu-30% Zn brass bar is reduced from a 1 in. diameter to a 0.45 in. diameter. Determine the final properties of the bar. (See Figure 8-25.)

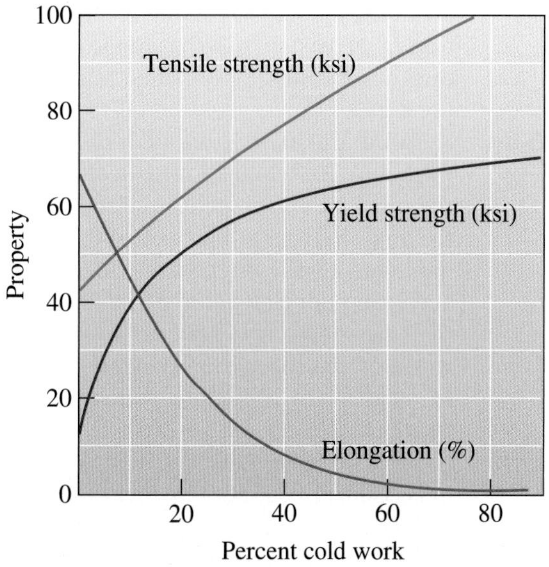

Figure 8-25 The effect of percent cold work on the properties of a Cu-30% Zn brass (for Problems 8-30 and 8-33 to 8-35).

8-31 A 3105 aluminum bar is reduced from a 1 in. diameter, to a 0.8 in. diameter, to a 0.6-in. diameter, to a final 0.4 in. diameter. Determine the % CW and the properties after each step of the process. Calculate the total percent cold work. Note 3105 designates a particular aluminum alloy. (See Figure 8-24.)

8-32 We want a copper bar to have a tensile strength of at least 70,000 psi and a final diameter of 0.375 in. What is the minimum diameter of the original bar? (See Figure 8-6.)

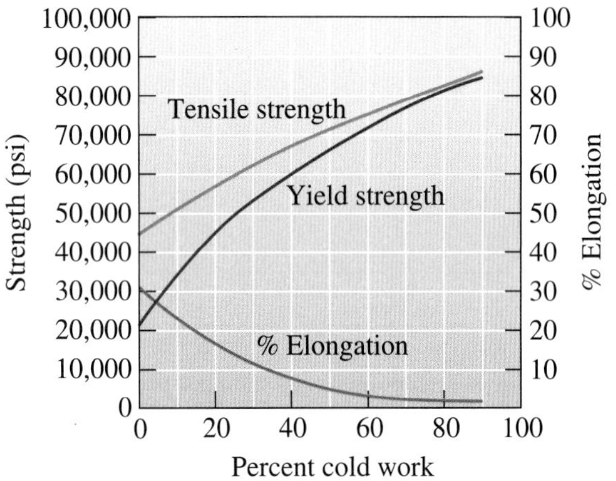

Figure 8-6 The effect of cold work on the mechanical properties of copper. (Repeated for Problems 8-32, 8-36, 8-53, and 8-82.)

8-33 We want a Cu-30% Zn brass plate originally 1.2 in. thick to have a yield strength greater than 50,000 psi and a % elongation of at least 10%. What range of final thicknesses must be obtained? (See Figure 8-25.)

8-34 A rolled Cu-30 wt% Zn plate 0.500-in.-thick has 2% elongation as-received by the supplier. The desired specifications for the final sheet are a thickness of 0.125 in., minimum tensile strength of 70,000 psi, and minimum elongation of 7%. Assume that the rolling is conducted so that the width of the sheet is unchanged. Specify all steps in the procedure, including heat treatments. (See Figure 8-25.)

8-35 A piece of cold-worked Cu-30 wt% Zn brass is identified, but the extent to which it has been cold-worked is unknown. Tensile testing indicates that the tensile strength is 60 ksi with an elongation of 30%. Determine the amount of cold-work and tensile strength of the brass in the annealed condition. (See Figure 8-25.)

8-36 We want a copper sheet to have at least 50,000 psi yield strength and at least 10% elongation, with a final thickness of 0.12 in. What range of original thickness must be used? (See Figure 8-6.)

8-37 A 3105 aluminum plate previously cold worked 20% is 2 in. thick. It is then cold worked further to 1.3 in. Calculate the total percent cold work and determine the final properties of the plate. (*Note*: 3105 designates a particular aluminum alloy.) (See Figure 8-24.)

8-38 An aluminum-lithium (Al-Li) strap 0.2 in. thick and 2 in. wide is to be cut from a rolled sheet as described in Figure 8-9. The strap must be able to support a 35,000 lb load without plastic deformation. Determine the range of orientations from which the strap can be cut from the rolled sheet.

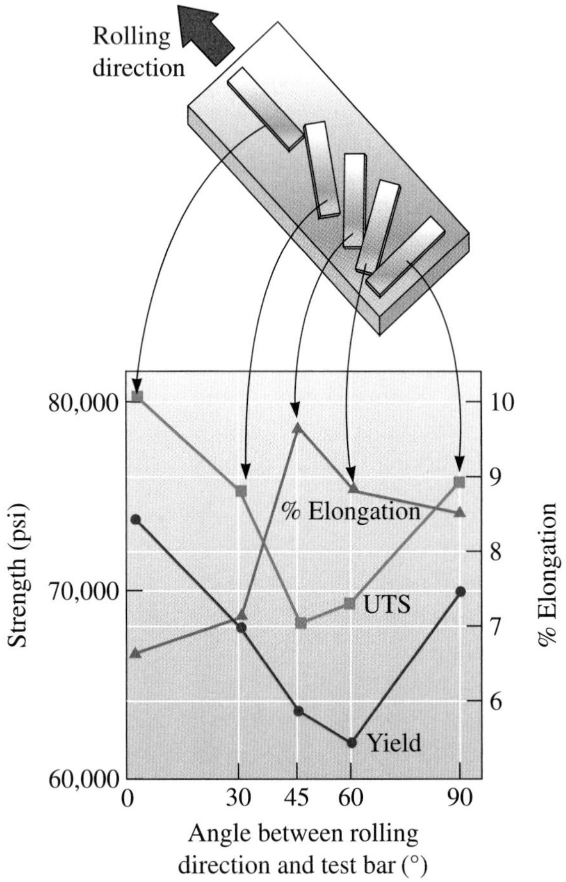

Figure 8-9 Anisotropic behavior in a rolled aluminum-lithium sheet material used in aerospace applications. The sketch relates the position of tensile bars to the mechanical properties that are obtained. (Repeated for Problem 8-38.)

Section 8-4 Microstructure, Texture Strengthening, and Residual Stresses

8-39 Does the yield strength of metallic materials depend upon the crystallographic texture materials develop during cold working? Explain.

8-40 Does the Young's modulus of a material depend upon crystallographic direction in a crystalline material? Explain.

8-41 What do the terms "fiber texture" and "sheet texture" mean?

8-42 One of the disadvantages of the cold-rolling process is the generation of residual stresses. Explain how we can eliminate residual stresses in cold-worked metallic materials.

8-43 What is shot peening?

8-44 What is the difference between tempering and annealing of glasses?

8-45 How is laminated safety glass different from tempered glass? Where is laminated glass used?

8-46 What is thermal tempering? How is it different from chemical tempering? State applications of tempered glasses.

8-47 Explain factors that affect the strength of glass most. Explain how thermal tempering helps increase the strength of glass.

8-48 Residual stresses are not always undesirable. True or false? Justify your answer.

Section 8-5 Characteristics of Cold Working

8-49 What is cold working? What features does a material have after it has been cold-worked?

8-50 Cold working cannot be used as a strengthening mechanism for materials that are going to be subjected to high temperatures during their use. Explain why.

8-51 Aluminum cans made by deep drawing derive considerable strength during their fabrication. Explain why.

8-52 Such metals as zinc cannot be effectively strengthened using cold working. Explain why.

8-53 We want to draw a 0.3-in.-diameter copper wire having a yield strength of 20,000 psi into a 0.25-in. diameter wire.
(a) Find the draw force, assuming no friction;
(b) Will the drawn wire break during the drawing process? Show why. (See Figure 8-6.)

8-54 A 3105 aluminum wire is to be drawn to give a 1-mm-diameter wire having a yield strength of 20,000 psi. Note 3105 designates a particular aluminum alloy.
(a) Find the original diameter of the wire;
(b) calculate the draw force required; and
(c) determine whether the as-drawn wire will break during the process. (See Figure 8-24.)

8-55 The electrical conductance of a Cu-40 wt% Zn wire is measured. How much less conductive is it than a 60% cold-worked pure Cu wire? (See Figure 8-11.)

Section 8-6 The Three Stages of Annealing

8-56 Explain the three stages of the annealing of metallic materials.

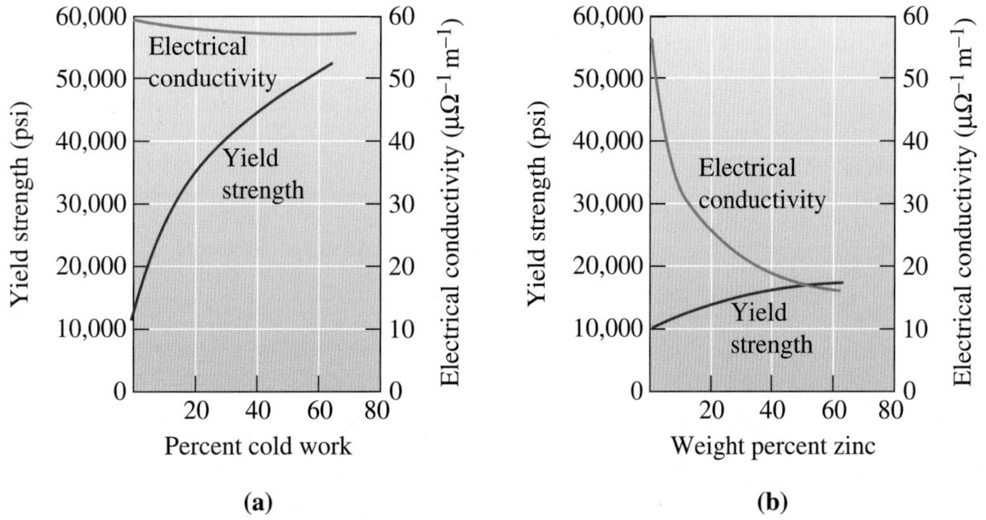

Figure 8-11 A comparison of strengthening copper by (a) cold working and (b) alloying with zinc. Note that cold working produces greater strengthening, yet has little effect on electrical conductivity. (Repeated for Problem 8-55.)

8-57 What is the driving force for recrystallization?

8-58 In the recovery state, the residual stresses are reduced; however, the strength of the metallic material remains unchanged. Explain why.

8-59 What is the driving force for grain growth?

8-60 Treating grain growth as the third stage of annealing, explain its effect on the strength of metallic materials.

8-61 Why is it that grain growth is usually undesirable? Cite an example where grain growth is actually useful.

8-62 Find a piece or photo of galvanized steel and explain the origin of the pattern on the surface.

8-63 Are annealing and recovery always prerequisites to grain growth? Explain.

8-64 A titanium alloy contains a very fine dispersion of Er_2O_3 particles. What is the effect of these particles on the grain growth temperature and the size of the grains at any particular annealing temperature? Explain.

8-65 Explain why a tungsten filament used in an incandescent light bulb ultimately fails.

8-66 Is it possible to obtain a single crystal from an initially polycrystalline material through a process of cold work, recrystallization, and grain growth? Yes or no? Fully explain your reasoning.

8-67 For a Cu 35% Zn alloy, a tensile strength of not less than 70 ksi and an elongation of 25% are desired. If the metal is cold worked to 75% in forming the desired shape, how high should it be heated when annealing to reach these properties? (See Figure 8-14.)

8-68 Samples of cartridge brass (Cu-30% Zn) were cold rolled and then annealed for one hour. The data shown in the table below were obtained.

Annealing Temperature (°C)	Grain Size (μm)	Yield Strength (MPa)
400	15	159
500	23	138
600	53	124
700	140	62
800	505	48

(a) Plot the yield strength and grain size as a function of annealing temperature on the same graph. Use two vertical axes, one for yield strength and one for grain size.

(b) For each temperature, state which stages of the annealing process occurred. Justify your answers by referring to features of the plot.

8-69 The following data were obtained when a cold-worked metal was annealed.

(a) Estimate the recovery, recrystallization, and grain growth temperatures;

(b) recommend a suitable temperature for a stress-relief heat treatment;

(c) recommend a suitable temperature for a hot-working process; and

(d) estimate the melting temperature of the alloy.

Annealing Temperature (°C)	Electrical Conductivity ($ohm^{-1} \cdot cm^{-1}$)	Yield Strength (MPa)	Grain Size (mm)
400	3.04×10^5	86	0.10
500	3.05×10^5	85	0.10
600	3.36×10^5	84	0.10
700	3.45×10^5	83	0.098
800	3.46×10^5	52	0.030
900	3.46×10^5	47	0.031
1000	3.47×10^5	44	0.070
1100	3.47×10^5	42	0.120

8-70 The following data were obtained when a cold-worked metallic material was annealed:

(a) Estimate the recovery, recrystallization, and grain growth temperatures;

(b) recommend a suitable temperature for obtaining a high-strength, high-electrical conductivity wire;

(c) recommend a suitable temperature for a hot-working process; and

(d) estimate the melting temperature of the alloy.

Annealing Temperature Size (°C)	Residual Stresses (psi)	Tensile Strength (psi)	Grain Size (in.)
250	21,000	52,000	0.0030
275	21,000	52,000	0.0030
300	5000	52,000	0.0030
325	0	52,000	0.0030
350	0	34,000	0.0010
375	0	30,000	0.0010
400	0	27,000	0.0035
425	0	25,000	0.0072

8-71 What is meant by the term "recrystallization?" Explain why the yield strength of a metallic material goes down during this stage of annealing.

Section 8-7 Control of Annealing

8-72 How do we distinguish between the hot working and cold working of metallic materials?

8-73 Why is it that the recrystallization temperature is not a fixed temperature for a given material?

8-74 Why does increasing the time for a heat treatment mean recrystallization will occur at a lower temperature?

8-75 Two sheets of steel were cold worked 20% and 80%, respectively. Which one would likely have a lower recrystallization temperature? Why?

8-76 Give examples of two metallic materials for which deformation at room temperature will mean "hot working."

8-77 Give examples of two metallic materials for which mechanical deformation at 900°C will mean "cold working."

8-78 Plot the recrystallization temperatures in Table 8-4 as a function of the melting temperatures. Are they proportional and, if they are, what is the proportionality constant? Make sure to convert the temperatures to kelvin first.

8-79 Add another column to Table 8-4 containing the range (°C) of warm working temperatures. Remember to convert to and from degrees kelvin in performing your calculations.

Table 8-4 Typical recrystallization temperatures for selected metals (Repeated for Problem 8-78.)

Metal	Melting Temperature (°C)	Recrystallization Temperature (°C)
SN	232	−4
Pb	327	−4
Zn	420	10
Al	660	150
Mg	650	200
Ag	962	200
Cu	1085	200
Fe	1538	450
Ni	1453	600
Mo	2610	900
W	3410	1200

(Based on Structure and Properties of Engineering Materials *by Brick, R. M.; et al, McGraw-Hill, 1977.)*

8-80 Consider the tensile stress strain curves in Figure 8-20 labeled 1 and 2 and answer the following questions. These diagrams are typical of metals. Consider each part as a separate question that has no relationship to previous parts of the question.

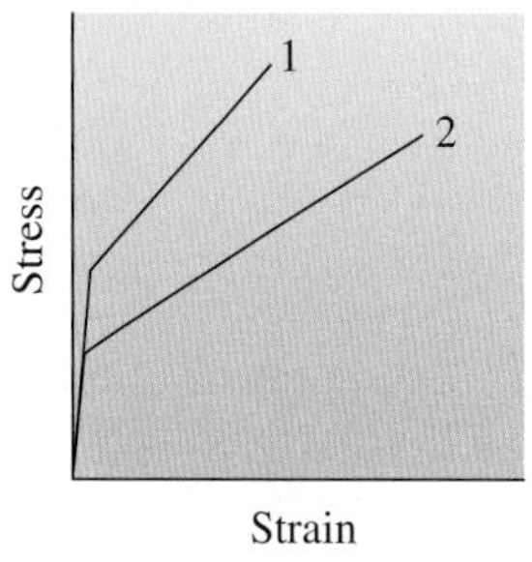

Figure 8-20 Stress strain curves (for Problem 8-80).

(a) Which of the two materials represented by samples 1 and 2 can be cold rolled to a greater extent? How do you know?

(b) Samples 1 and 2 have the same composition and were processed identically, except that one of them was cold worked more than the other. The stress strain curves were obtained after the samples were cold worked. Which sample has the lower recrystallization temperature: 1 or 2? How do you know?

(c) Samples 1 and 2 are identical except that they were annealed at different temperatures for the same period of time. Which sample was annealed at the higher temperature: 1 or 2? How do you know?

(d) Samples 1 and 2 are identical except that they were annealed at the same temperature for different periods of time. Which sample was annealed for the shorter period of time: 1 or 2? How do you know?

Section 8-8 Annealing and Materials Processing

8-81 We wish to produce a 0.3-in. thick plate of 3105 aluminum having a tensile strength of at least 25,000 psi and a % elongation of at least 5%. The original thickness of the plate is 3 in. The maximum cold work in each step is 80%. Describe the cold working and annealing steps required to make this product. Compare this process with what you would recommend if you could do the initial deformation by hot working. (See Figure 8-24.)

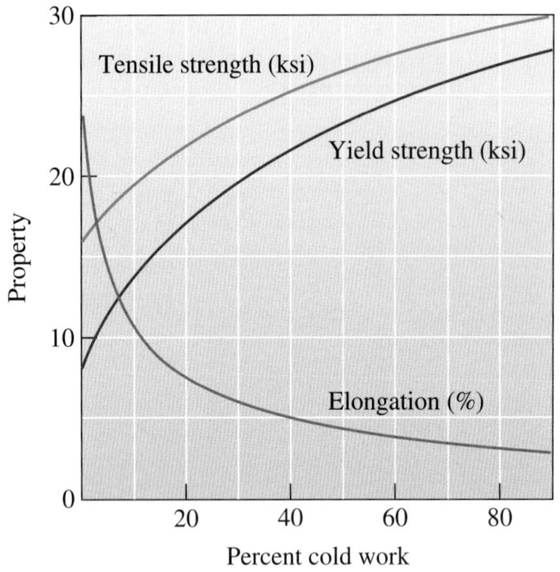

Figure 8-24 The effect of percent cold work on the properties of a 3105 aluminum alloy (for Problem 8-81).

8-82 We wish to produce a 0.2-in.-diameter wire of copper having a minimum yield strength of 60,000 psi and a minimum % elongation of 5%. The original diameter of the rod is 2 in. and the maximum cold work in each step is 80%. Describe the cold working and annealing steps required to make this product. Compare this process with what you would recommend if you could do the initial deformation by hot working. (See Figure 8-6.)

8-83 What is a heat-affected zone? Why do some welding processes result in a joint where the material in the heat-affected zone is weaker than the base metal?

8-84 What welding techniques can be used to avoid loss of strength in the material in the heat-affected zone? Explain why these techniques are effective.

Section 8-9 Hot Working

8-85 The amount of plastic deformation that can be performed during hot working is almost unlimited. Justify this statement.

8-86 Compare and contrast hot working and cold working.

Design Problems

8-87 Design, using one of the processes discussed in this chapter, a method to produce each of the following products. Should the process include hot working, cold working, annealing, or some combination of these? Explain your decisions.

(a) paper clips;
(b) I-beams that will be welded to produce a portion of a bridge;
(c) copper tubing that will connect a water faucet to the main copper plumbing;
(d) the steel tape in a tape measure; and
(e) a head for a carpenter's hammer formed from a round rod.

8-88 We plan to join two sheets of cold-worked copper by soldering. Soldering involves heating the metal to a high enough temperature that a filler material melts and is drawn into the joint (Chapter 9). Design a soldering process that will not soften the copper. Explain. Could we use higher soldering temperatures if the sheet material were a Cu-30% Zn alloy? Explain.

8-89 We wish to produce a 1-mm-diameter copper wire having a minimum yield strength of 60,000 psi and a minimum % elongation of 5%. We start with a 20-mm-diameter rod. Design the process by which the wire can be drawn. Include important details and explain.

Computer Problems

8-90 *Design of a Wire Drawing Process.* Write a program that will effectively computerize the solution to solving Example 8-4. The program should ask the user to provide a value of the final diameter for the wire (e.g., 0.20 cm). The program should assume a reasonable value for the initial diameter (d_0) (e.g., 0.40 cm), and calculate the extent of cold work using the proper formula. Assume that the user has access to the yield strength versus % cold work curve. The user is then asked to enter the value of the yield strength for 0% cold work. Use this value to calculate the forces needed for drawing and the stress acting on the wire as it comes out of the die. The program should then ask the user to provide the value of the yield strength of the wire for the amount of cold work calculated for the assumed initial diameter and the final diameter needed. As in Example 8-4, the program should repeat these calculations until obtaining a value of d_0 that will be acceptable.

Knovel® Problems

K8-1 An axially loaded compression member is cold formed from an AISI 1025 steel plate. A cross-section of the member is shown in Figure K8-1. The metal at the corners of the member was strengthened by cold work. Determine the ultimate and yield strengths of the material prior to forming; the yield strength of the material at the corners of the section after cold forming; and the average yield strength of the section, accounting for the flat portions and the corners strengthened by cold work. Assume a compact section and a reduction factor $\rho = 1.0$ (Figure K8-1).

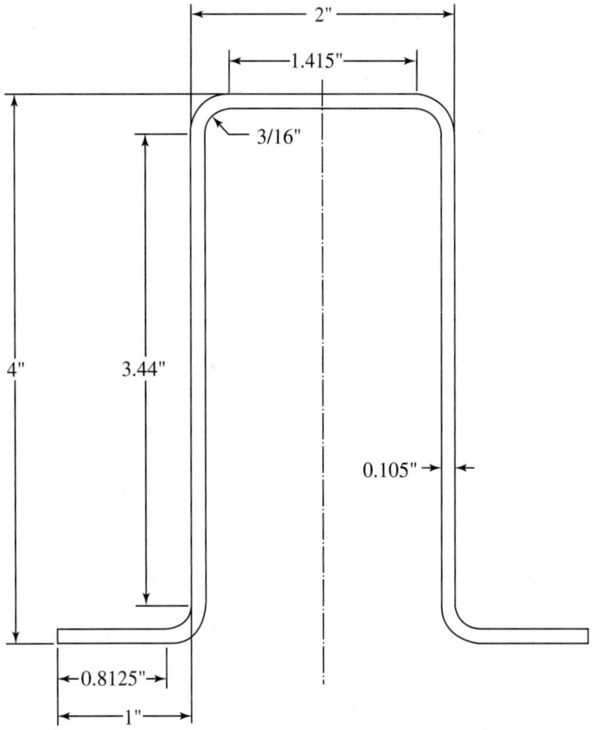

Figure K8-1 A cross-section of the compression member (for Problem K8-1).

The photo on the left shows a casting process known as green sand molding (© *Peter Bowater/Alamy*). Clay-bonded sand is packed around a pattern in a two-part mold. The pattern is removed, leaving behind a cavity, and molten metal is poured into the cavity. Sand cores can produce internal cavities in the casting. After the metal solidifies, the part is shaken out from the sand. The photo on the right shows a process known as investment casting (© *Jim Powell/Alamy*). In investment casting, a pattern is made from a material that is easily melted such as wax. The pattern is then "invested" in a slurry, and a structure is built up around the pattern using particulate. The pattern is removed using heat, and molten metal is poured into the cavity. Unlike sand casting, the pattern is not reusable. Sand casting and investment casting are only two of a wide variety of casting processes.

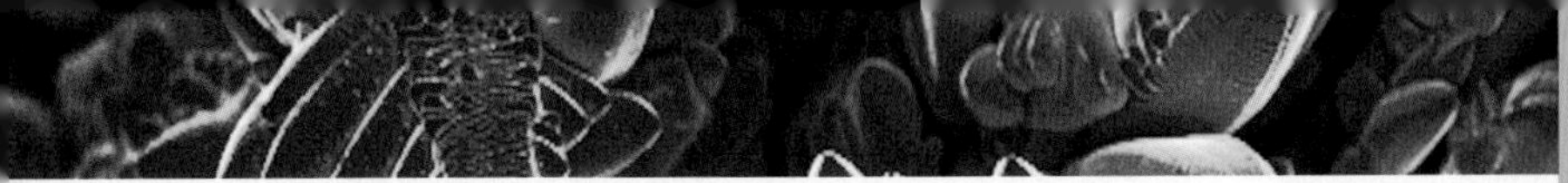

CHAPTER

9

Principles of Solidification

Have You Ever Wondered?

- *Whether water really does "freeze" at 0°C and "boil" at 100°C?*
- *Is there a specific melting temperature for an alloy or a thermoplastic material?*
- *What factors determine the strength of a cast product?*
- *Why are most glasses and glass-ceramics processed by melting and casting?*

Chapter Learning Objectives

The key objectives of this chapter are to

- Develop the expression that describes the total free energy change during solidification, and calculate the size of the critical radius for homogeneous nucleation of a solid from a liquid.
- Explain why liquids typically solidify via heterogeneous nucleation.
- Describe the factors that influence planar and dendritic modes of solid growth from a liquid.
- Calculate the solidification time for a casting according to Chvorinov's rule.
- Draw cooling curves for pure metals that are and are not well inoculated.
- Describe the structures and defects that occur in castings.
- Describe the most common casting processes used in solidification.

Of all the processing techniques used in the manufacturing of materials, solidification is probably the most important. All metallic materials, as well as many ceramics, inorganic glasses, and thermoplastic polymers, are liquid or molten at some point during processing. Like water freezes to ice, molten materials solidify as they cool below their freezing temperature. In Chapter 3, we learned how materials are classified based on their atomic, ionic, or molecular order. During the solidification of materials that crystallize, the atomic arrangement changes from a short-range order (SRO) to a long-range order (LRO). The solidification of crystalline materials requires two steps. In the first step, ultra-fine crystallites, known as the nuclei of a solid phase, form from the liquid. In the second step, which can overlap with the first, the ultra-fine solid crystallites begin to grow as atoms from the liquid are attached to the nuclei until no liquid remains. Some materials, such as inorganic silicate glasses, will become solid without developing a long-range order (i.e., they remain amorphous). Many polymeric materials may develop partial crystallinity during solidification or processing.

The solidification of metallic, polymeric, and ceramic materials is an important process to study because of its effect on the properties of the materials involved. In this chapter, we will study the principles of solidification as they apply to pure metals. We will discuss solidification of alloys and more complex materials in subsequent chapters. We will first discuss the technological significance of solidification and then examine the mechanisms by which solidification occurs. This will be followed by an examination of the microstructure of cast metallic materials and its effect on the material's mechanical properties. We will also examine the role of casting as a materials shaping process. We will examine how techniques such as welding, brazing, and soldering are used for joining metals. Applications of the solidification process in single crystal growth and the solidification of glasses and polymers also will be discussed.

9-1 Technological Significance

The ability to use heat to produce, melt, and cast metals such as copper, bronze, and steel is regarded as an important hallmark in the development of mankind. The use of fire for reducing naturally occurring ores into metals and alloys led to the production of useful tools and other products. Today, thousands of years later, **solidification** is still considered one of the most important manufacturing processes. Several million pounds of steel, aluminum alloys, copper, and zinc are produced through the casting process each month in the United States. The solidification process is also used to manufacture specific components (e.g., aluminum alloys for automotive wheels). Industry also uses the solidification process as a **primary processing** step to produce metallic slabs or ingots (a simple, and often large casting that later is processed into useful shapes). The ingots or slabs are then hot and cold worked through **secondary processing** steps into more useful shapes (i.e., sheets, wires, rods, plates, etc.). Solidification also is applied when joining metallic materials using techniques such as welding, brazing, and soldering.

We also use solidification for processing inorganic glasses; silicate glass, for example, is processed using the float-glass process. High-quality optical fibers and other materials, such as fiberglass, also are produced from the solidification of molten glasses. During the solidification of inorganic glasses, amorphous rather than crystalline materials are produced. In the manufacturing of glass-ceramics, we first shape the materials by casting amorphous glasses and then crystallize them using a heat treatment to enhance their strength. Many thermoplastic materials such as polyethylene, polyvinyl chloride (PVC), polypropylene, and the like are processed into useful shapes (i.e., fibers, tubes, bottles, toys, utensils, etc.) using a process that involves melting and solidification. Therefore, solidification is an extremely important technology used to control the properties of many melt-derived products as well as a tool for the manufacturing of modern engineered materials. In the sections that follow, we first discuss the nucleation and growth processes.

9-2 Nucleation

In the context of solidification, the term **nucleation** refers to the formation of the first nanocrystallites from molten material. For example, as water begins to freeze, nanocrystals, known as **nuclei**, form first. In a broader sense, the term nucleation refers to the initial stage of formation of one phase from another phase. When a vapor condenses into liquid, the nanoscale-sized drops of liquid that appear when the condensation begins are referred to as nuclei. Later, we will also see that there are many systems in which the nuclei of a solid (β) will form from a second solid material (α) (i.e., an α- to β-phase transformation). What is interesting about these transformations is that, in most engineered materials, many of them occur while the material is in the solid state (i.e., there is no melting involved). Therefore, although we discuss nucleation from a solidification perspective, it is important to note that the phenomenon of nucleation is general and is associated with phase transformations.

We expect a material to solidify when the liquid cools to just below its freezing (or melting) temperature because the energy associated with the crystalline structure of the solid is then less than the energy of the liquid. This energy difference between the liquid and the solid is the free energy per unit volume ΔG_v and is the driving force for solidification.

When the solid forms, however, a solid-liquid interface is created [Figure 9-1(a)]. A surface free energy σ_{sl} is associated with this interface. Thus, the total change in energy ΔG, shown in Figure 9-1(b), is

$$\Delta G = \frac{4}{3}\pi r^3 \Delta G_v + 4\pi r^2 \sigma_{sl} \tag{9-1}$$

where $\frac{4}{3}\pi r^3$ is the volume of a spherical solid of radius r, $4\pi r^2$ is the surface area of a spherical solid, σ_{sl} is the surface free energy of the solid-liquid interface (in this case), and ΔG_v is the free energy change per unit volume, which is negative since the phase transformation is assumed to be thermodynamically feasible. Note that σ_{sl} is not a strong function of r and is assumed constant. It has units of energy per unit area. ΔG_v also does not depend on r.

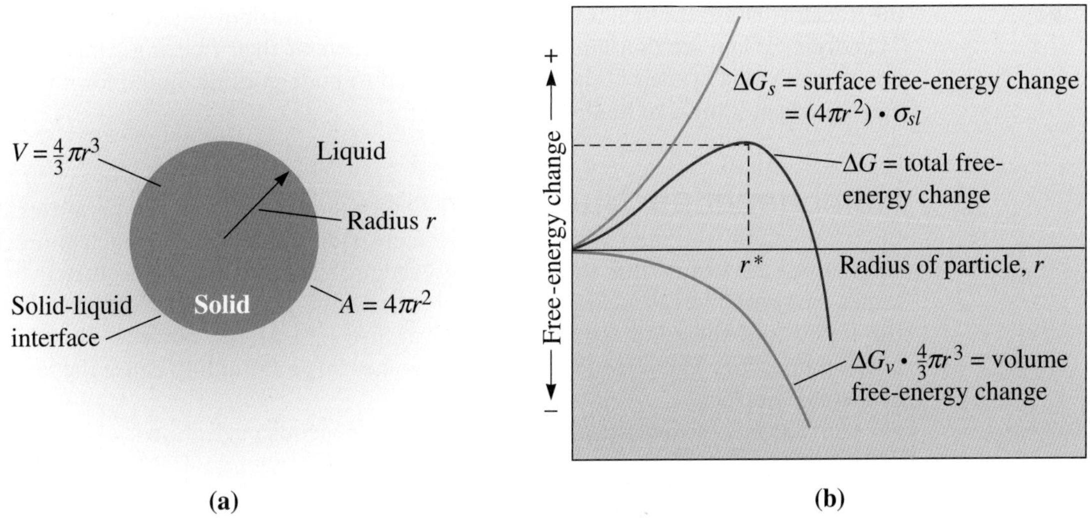

Figure 9-1 (a) An interface is created when a solid forms from the liquid. (b) The total free energy of the solid-liquid system changes with the size of the solid. The solid is an embryo if its radius is less than the critical radius and is a nucleus if its radius is greater than the critical radius.

An **embryo** is a tiny particle of solid that forms from the liquid as atoms cluster together. The embryo is unstable and may either grow into a stable nucleus or redissolve.

In Figure 9-1(b), the top curve shows the parabolic variation of the total surface energy ($4\pi r^2 \cdot \sigma_{sl}$). The bottom most curve shows the total volume free energy change term $\left(\frac{4}{3}\pi r^3 \cdot \Delta G_v\right)$. The curve in the middle shows the variation of ΔG. It represents the sum of the other two curves as given by Equation 9-1. At the temperature at which the solid and liquid phases are predicted to be in thermodynamic equilibrium (i.e., at the freezing temperature), the free energy of the solid phase and that of the liquid phase are equal ($\Delta G_v = 0$), so the total free energy change (ΔG) will be positive. When the solid is very small with a radius less than the **critical radius** for nucleation (r^*) [Figure 9-1(b)], further **growth** causes the total free energy to increase. The critical radius (r^*) is the minimum size of a crystal that must be formed by atoms clustering together in the liquid before the solid particle is stable and begins to grow.

The formation of embryos is a statistical process. Many embryos form and redissolve. If by chance, an embryo forms with a radius that is larger than r^*, further growth causes the total free energy to decrease. The new solid is then stable and sustainable since nucleation has occurred, and growth of the solid particle—which is now called a nucleus—begins. At the thermodynamic melting or freezing temperatures, the probability of forming stable, sustainable nuclei is extremely small. Therefore, solidification does not begin at the thermodynamic melting or freezing temperature. If the temperature continues to decrease below the equilibrium freezing temperature, the liquid phase that should have transformed into a solid becomes thermodynamically increasingly unstable. Because the temperature of the liquid is below the equilibrium freezing temperature, the liquid is considered undercooled. The **undercooling** (ΔT) is the difference between the equilibrium freezing temperature and the actual temperature of the liquid. As the extent of undercooling increases, the thermodynamic driving force for the formation of a solid phase from the liquid overtakes the resistance to create a solid-liquid interface.

This phenomenon can be seen in many other phase transformations. When one solid phase (α) transforms into another solid phase (β), the system has to be cooled to a temperature that is below the thermodynamic phase transformation temperature (at which the energies of the α and β phases are equal). When a liquid is transformed into a vapor (i.e., boiling water), a bubble of vapor is created in the liquid. In order to create the transformation though, we need to **superheat** the liquid above its boiling temperature! Therefore, we can see that liquids do not really freeze at their freezing temperature and do not really boil at their boiling point! We need to undercool the liquid for it to solidify and superheat it for it to boil!

Homogeneous Nucleation

As a liquid cools to temperatures below the equilibrium freezing temperature, two factors combine to favor nucleation. First, since atoms are losing their thermal energy, the probability of forming clusters to form larger embryos increases. Second, the larger volume free energy difference between the liquid and the solid reduces the critical size (r^*) of the nucleus. **Homogeneous nucleation** occurs when the undercooling becomes large enough to cause the formation of a stable nucleus.

The size of the critical radius r^* for homogeneous nucleation is given by

$$r^* = \frac{2\sigma_{sl}T_m}{\Delta H_f \, \Delta T} \tag{9-2}$$

where ΔH_f is the latent heat of fusion per unit volume, T_m is the equilibrium solidification temperature in kelvin, and $\Delta T = (T_m - T)$ is the undercooling when the liquid temperature is T. The latent heat of fusion represents the heat given up during the liquid-to-solid transformation. As the undercooling increases, the critical radius required for nucleation

Table 9-1 Values for freezing temperature, latent heat of fusion, surface energy, and maximum undercooling for selected materials

Material	Freezing Temperature (T_m) (°C)	Heat of Fusion (ΔH_f) (J/cm³)	Solid-Liquid Interfacial Energy (σ_{sl}) (J/cm²)	Typical Undercooling for Homogeneous Nucleation (ΔT) (°C)
Ga	30	488	56×10^{-7}	76
Bi	271	543	54×10^{-7}	90
Pb	327	237	33×10^{-7}	80
Ag	962	965	126×10^{-7}	250
Cu	1085	1628	177×10^{-7}	236
Ni	1453	2756	255×10^{-7}	480
Fe	1538	1737	204×10^{-7}	420
NaCl	801			169
CsCl	645			152
H_2O	0			40

decreases. Table 9-1 presents values for σ_{sl}, ΔH_f, and typical undercoolings observed experimentally for homogeneous nucleation.

The following example shows how we can calculate the critical radius of the nucleus for the solidification of copper.

Example 9-1 *Calculation of Critical Radius for the Solidification of Copper*

Calculate the size of the critical radius and the number of atoms in the critical nucleus when solid copper forms by homogeneous nucleation. Comment on the size of the nucleus and assumptions we made while deriving the equation for the radius of the nucleus.

SOLUTION

From Table 9-1 for Cu:

$$\Delta T = 236\text{°C} \qquad T_m = 1085 + 273 = 1358 \text{ K}$$

$$\Delta H_f = 1628 \text{ J/cm}^3$$

$$\sigma_{sl} = 177 \times 10^{-7} \text{ J/cm}^2$$

Thus, r^* is given by

$$r^* = \frac{2\sigma_{sl}T_m}{\Delta H_f \Delta T} = \frac{(2)(177 \times 10^{-7} \text{ J/cm}^2)(1358 \text{ K})}{(1628 \text{ J/cm}^3)(236 \text{ K})} = 1.2512 \times 10^{-7} \text{ cm} = 1.25 \text{ nm}$$

Note that a temperature difference of 1°C is equal to a temperature change of 1 K, or $\Delta T = 236\text{°C} = 236 \text{ K}$.

The lattice parameter for FCC copper is $a_0 = 0.3615 \text{ nm} = 3.615 \times 10^{-8}$ cm. Thus, the unit cell volume is given by

$$V_{\text{unit cell}} = (a_0)^3 = (3.615 \times 10^{-8} \text{ cm})^3 = 4.7242 \times 10^{-23} \text{ cm}^3$$

The volume of a nucleus with the critical radius is given by

$$V_{r^*} = \frac{4}{3}\pi(r^*)^3 = \left(\frac{4}{3}\pi\right)(1.2512 \times 10^{-7} \text{ cm})^3 = 8.2054 \times 10^{-21} \text{ cm}^3$$

The number of unit cells in the critical nucleus is

$$\frac{V_{r^*}}{V_{\text{unit cell}}} = \frac{8.2054 \times 10^{-21}\ \text{cm}^3}{4.7242 \times 10^{-23}\ \text{cm}^3} = 174 \text{ unit cells}$$

Since there are four atoms in each FCC unit cell, the number of atoms in the critical nucleus must be

$$(4 \text{ atoms/cell})(174 \text{ cells/nucleus}) = 696 \text{ atoms/nucleus}$$

In these types of calculations, we assume that a nucleus that is made from only a few hundred atoms still exhibits properties similar to those of bulk materials. This is not strictly correct and, as such, is a weakness of the classical theory of nucleation.

Heterogeneous Nucleation

From Table 9-1, we can see that water will not solidify into ice via homogeneous nucleation until we reach a temperature of −40°C (undercooling of 40°C)! Except in controlled laboratory experiments, homogeneous nucleation never occurs in liquids. Instead, impurities in contact with the liquid, either suspended in the liquid or on the walls of the container that holds the liquid, provide a surface on which the solid can form (Figure 9-2). Now, a radius of curvature greater than the critical radius is achieved with very little total surface between the solid and liquid. Relatively few atoms must cluster together to produce a solid particle that has the required radius of curvature. Much less undercooling is required to achieve the critical size, so nucleation occurs more readily. Nucleation on preexisting surfaces is known as **heterogeneous nucleation**. This process is dependent on the contact angle (θ) for the nucleating phase and the surface on which nucleation occurs. The same type of phenomenon occurs in solid-state transformations.

Rate of Nucleation

The *rate of nucleation* (the number of nuclei formed per unit time) is a function of temperature. Prior to solidification, of course, there is no nucleation and, at temperatures above the freezing point, the rate of nucleation is zero. As the temperature drops, the driving force for nucleation increases; however, as the temperature decreases, atomic diffusion becomes slower, hence slowing the nucleation process. Thus, a typical rate of nucleation reaches a maximum at some temperature below the equilibrium transformation (freezing) temperature. In heterogeneous nucleation, the rate of nucleation is dictated by the concentration of the nucleating agents. By considering the rates of nucleation and growth, we can predict the overall rate of a phase transformation.

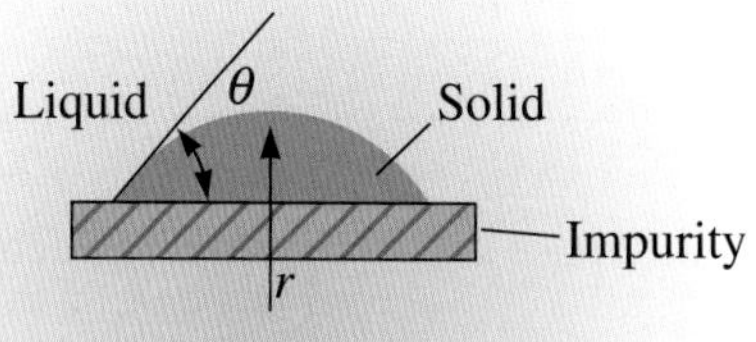

Figure 9-2 A solid forming on an impurity can assume the critical radius with a smaller increase in the surface energy. Thus, heterogeneous nucleation can occur with relatively low undercoolings.

9-3 Applications of Controlled Nucleation

Grain-Size Strengthening When a metal casting freezes, impurities in the melt and the walls of the mold in which solidification occurs serve as heterogeneous nucleation sites. Sometimes we intentionally introduce nucleating particles into the liquid. Such practices are called **grain refinement** or **inoculation**. Chemicals added to molten metals to promote nucleation and, hence, a finer grain size, are known as grain refiners or **inoculants**. For example, a combination of 0.03% titanium and 0.01% boron is added to many liquid-aluminum alloys. Tiny particles of an aluminum titanium compound (Al_3Ti) or titanium diboride (TiB_2) form and serve as sites for heterogeneous nucleation. Grain refinement or inoculation produces a large number of grains. The greater grain boundary area provides grain-size strengthening in metallic materials. This was discussed using the Hall-Petch equation in Chapter 4.

Second-Phase Strengthening Strengthening materials using ultra-fine precipitates is known as **dispersion strengthening** or **second-phase strengthening**; it is used extensively in enhancing the mechanical properties of many alloys. This process involves **solid-state phase transformations** (i.e., one solid transforming to another). The grain boundaries as well as atomic level defects within the grains of the parent phase (α) often serve as nucleation sites for heterogeneous nucleation of the new phase (β). This nucleation phenomenon plays a critical role in strengthening mechanisms. This will be discussed in Chapters 10 and 11.

Glasses For rapid cooling rates and/or high viscosity melts, there may be insufficient time for nuclei to form and grow. When this happens, the liquid structure is locked into place and an amorphous—or glassy—solid forms. The complex crystal structure of many ceramic and polymeric materials prevents nucleation of a solid crystalline structure even at slow cooling rates. Some alloys with special compositions have sufficiently complex crystal structures, such that they may form amorphous materials if cooled rapidly from the melt. These materials are known as metallic glasses (Chapter 6). Good metallic glass formers are multi-component alloys, often with large differences in the atomic sizes of the elemental constituents. This complexity limits the solid solubilities of the elements in the crystalline phases, thus requiring large chemical fluctuations to form the critical-sized crystalline nuclei. Metallic glasses were initially produced via **rapid solidification processing** in which cooling rates of 10^{6}°C/s were attained by forming continuous, thin metallic ribbons about 0.0015 in. thick. (Heat can be extracted quickly from ribbons with a large surface area to volume ratio.)

Bulk metallic glasses with diameters on the order of 0.5 in. are now produced using a variety of processing techniques for compositions that require cooling rates on the order of only tens of degrees per second. Many bulk metallic glass compositions have been discovered, including $Pd_{40}Ni_{40}P_{20}$ and $Zr_{41.2}Ti_{13.8}Cu_{12.5}Ni_{10.0}Be_{22.5}$. Many metallic glasses have strengths in excess of 250,000 psi while retaining fracture toughnesses of more than 10,000 psi$\sqrt{\text{in.}}$ Excellent corrosion resistance, magnetic properties, and other physical properties make these materials attractive for a wide variety of applications.

Other examples of materials that make use of controlled nucleation are colored glass and **photochromic glass** (glass that can change color or tint upon exposure to sunlight). In these otherwise amorphous materials, nanocrystallites of different materials are deliberately nucleated. The crystals are small and, hence, do not make the glass opaque. They do have special optical properties that make the glass brightly colored or photochromic.

Many materials formed from a vapor phase can be cooled quickly so that they do not crystallize and, therefore, are amorphous (i.e., amorphous silicon), illustrating that amorphous or non-crystalline materials do *not* always have to be formed from melts.

Glass-ceramics The term **glass-ceramics** refers to engineered materials that begin as amorphous glasses and end up as crystalline ceramics with an ultra-fine grain size. These materials are then nearly free from porosity, mechanically stronger, and often much more resistant to thermal shock. Nucleation does not occur easily in silicate glasses; however, we can help by introducing nucleating agents such as titania (TiO_2) and zirconia (ZrO_2). Engineered glass-ceramics take advantage of the ease with which glasses can be melted and formed. Once a glass is formed, we can heat it to deliberately form ultra-fine crystals, obtaining a material that has considerable mechanical toughness and thermal shock resistance. The crystallization of glass-ceramics continues until all of the material crystallizes (up to 99.9% crystallinity can be obtained). If the grain size is kept small (~ 50–100 nm), glass-ceramics can often be made transparent. All glasses eventually will crystallize as a result of exposure to high temperatures for long lengths of time. In order to produce a glass-ceramic, however, the crystallization must be carefully controlled.

9-4 Growth Mechanisms

Once the solid nuclei of a phase form (in a liquid or another solid phase), growth begins to occur as more atoms become attached to the solid surface. In this discussion, we will concentrate on the nucleation and growth of crystals from a liquid. The nature of the growth of the solid nuclei depends on how heat is removed from the molten material. Let's consider casting a molten metal in a mold, for example. We assume we have a nearly pure metal and not an alloy (as solidification of alloys is different in that in most cases, it occurs over a range of temperatures). During the solidification process, two types of heat must be removed: the specific heat of the liquid and the latent heat of fusion. The **specific heat** is the heat required to change the temperature of a unit mass of the material by one degree. The specific heat must be removed first, either by radiation into the surrounding atmosphere or by conduction into the surrounding mold, until the liquid cools to its freezing temperature. This is simply a cooling of the liquid from one temperature to a temperature at which nucleation begins.

We know that to melt a solid we need to supply heat. Therefore, when solid crystals form from a liquid, heat is generated! This type of heat is called the **latent heat of fusion** (ΔH_f). The latent heat of fusion must be removed from the solid-liquid interface before solidification is completed. The manner in which we remove the latent heat of fusion determines the material's growth mechanism and final structure of a casting.

Planar Growth When a well-inoculated liquid (i.e., a liquid containing nucleating agents) cools under equilibrium conditions, there is no need for undercooling since heterogeneous nucleation can occur. Therefore, the temperature of the liquid ahead of the **solidification front** (i.e., solid-liquid interface) is greater than the freezing temperature. The temperature of the solid is at or below the freezing temperature. During solidification, the latent heat of fusion is removed by conduction from the solid-liquid interface. Any small protuberance that begins to grow on the interface is surrounded by liquid above the freezing temperature (Figure 9-3). The growth of the protuberance then stops until the remainder of the interface catches up. This growth mechanism, known as **planar growth**, occurs by the movement of a smooth solid-liquid interface into the liquid.

Dendritic Growth When the liquid is not inoculated and the nucleation is poor, the liquid has to be undercooled before the solid forms (Figure 9-4). Under these conditions, a small solid protuberance called a **dendrite**, which forms at the interface,

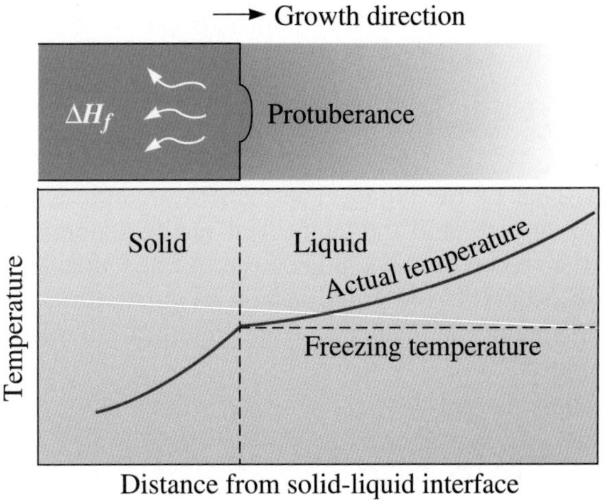

Figure 9-3 When the temperature of the liquid is above the freezing temperature, a protuberance on the solid-liquid interface will not grow, leading to maintenance of a planar interface. Latent heat is removed from the interface through the solid.

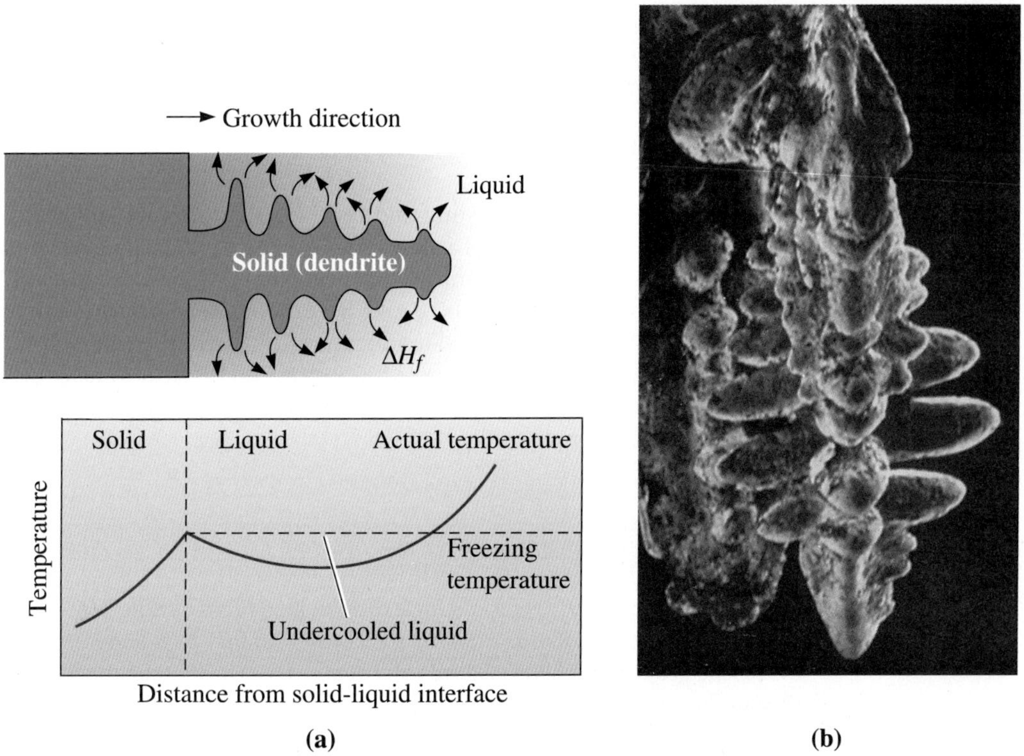

Figure 9-4 (a) If the liquid is undercooled, a protuberance on the solid-liquid interface can grow rapidly as a dendrite. The latent heat of fusion is removed by raising the temperature of the liquid back to the freezing temperature. (b) Scanning electron micrograph of dendrites in steel (× 15). (*Reprinted Courtesy of Don Askeland*)

is encouraged to grow since the liquid ahead of the solidification front is undercooled. The word dendrite comes from the Greek word *dendron* that means tree. As the solid dendrite grows, the latent heat of fusion is conducted into the undercooled liquid, raising the temperature of the liquid toward the freezing temperature. Secondary and tertiary dendrite arms can also form on the primary stalks to speed the evolution of the latent heat. Dendritic growth continues until the undercooled liquid warms to the freezing temperature. Any remaining liquid then solidifies by planar growth. The difference between planar and dendritic growth arises because of the different sinks for the latent heat of fusion. The container or mold must absorb the heat in planar growth, but the undercooled liquid absorbs the heat in dendritic growth.

In pure metals, dendritic growth normally represents only a small fraction of the total growth and is given by

$$\text{Dendritic fraction} = f = \frac{c\Delta T}{\Delta H_f} \tag{9-3}$$

where c is the specific heat of the liquid. The numerator represents the heat that the undercooled liquid can absorb, and the latent heat in the denominator represents the total heat that must be given up during solidification. As the undercooling ΔT increases, more dendritic growth occurs. If the liquid is well-inoculated, undercooling is almost zero and growth occurs mainly via the planar front solidification mechanism.

9-5 Solidification Time and Dendrite Size

The rate at which growth of the solid occurs depends on the cooling rate, or the rate of heat extraction. A higher cooling rate produces rapid solidification, or short solidification times. The time t_s required for a simple casting to solidify completely can be calculated using ***Chvorinov's rule***:

$$t_s = B\left(\frac{V}{A}\right)^n \tag{9-4}$$

where V is the volume of the casting and represents the amount of heat that must be removed before freezing occurs, A is the surface area of the casting in contact with the mold and represents the surface from which heat can be transferred away from the casting, n is a constant (usually about 2), and B is the **mold constant**. The mold constant depends on the properties and initial temperatures of both the metal and the mold. This rule basically accounts for the geometry of a casting and the heat transfer conditions. The rule states that, for the same conditions, a casting with a small volume and relatively large surface area will cool more rapidly.

Example 9-2 *Redesign of a Casting for Improved Strength*

Your company currently is producing a disk-shaped brass casting 2 in. thick and 18 in. in diameter. You believe that by making the casting solidify 25% faster the improvement in the tensile properties of the casting will permit the casting to be made lighter in weight. Design the casting to permit this. Assume that the mold constant is 22 min/in.^{2} for this process and $n = 2$.

SOLUTION

One approach would be to use the same casting process, but reduce the thickness of the casting. The thinner casting would solidify more quickly and, because of the faster cooling, should have improved mechanical properties. Chvorinov's rule helps us calculate the required thickness. If d is the diameter and x is the thickness of the casting, then the volume, surface area, and solidification time of the 2-in. thick casting are

$$V = (\pi/4)d^2x = (\pi/4)(18\text{ in.})^2(2\text{ in.}) = 508.9\text{ in.}^3$$

$$A = 2(\pi/4)d^2 + \pi dx = 2(\pi/4)(18\text{ in.})^2 + \pi(18\text{ in.})(2\text{ in.}) = 622\text{ in.}^2$$

$$t = B\left(\frac{V}{A}\right)^2 = (22\text{ min/in.}^2)\left(\frac{508.9\text{ in.}^3}{622\text{ in.}^2}\right)^2 = 14.73\text{ min}$$

The solidification time t_r of the redesigned casting should be 25% shorter than the current time:

$$t_r = 0.75t = (0.75)(14.73\text{ min}) = 11.05\text{ min}$$

Since the casting conditions have not changed, the mold constant B is unchanged. The V/A ratio of the new casting is

$$t_r = B\left(\frac{V_r}{A_r}\right)^2 = (22 \text{ min/in.}^2)\left(\frac{V_r}{A_r}\right)^2 = 11.05 \text{ min}$$

$$\left(\frac{V_r}{A_r}\right)^2 = 0.502 \text{ in.}^2 \quad \text{or} \quad \frac{V_r}{A_r} = 0.709 \text{ in.}$$

If x is the required thickness for our redesigned casting, then

$$\frac{V_r}{A_r} = \frac{(\pi/4)d^2x}{2(\pi/4)d^2 + \pi dx} = \frac{(\pi/4)(18 \text{ in.})^2(x)}{2(\pi/4)(18 \text{ in.})^2 + \pi(18 \text{ in.})(x)} = 0.709 \text{ in.}$$

Therefore, $x = 1.68$ in.

This thickness provides the required solidification time, while reducing the overall weight of the casting by more than 15%.

Solidification begins at the surface, where heat is dissipated into the surrounding mold material. The rate of solidification of a casting can be described by how rapidly the thickness d of the solidified skin grows:

$$d = k_{\text{solidification}}\sqrt{t} - c_1 \tag{9-5}$$

where t is the time after pouring, $k_{\text{solidification}}$ is a constant for a given casting material and mold, and c_1 is a constant related to the pouring temperature.

Effect on Structure and Properties

The solidification time affects the size of the dendrites. Normally, dendrite size is characterized by measuring the distance between the secondary dendrite arms (Figure 9-5). The **secondary dendrite arm spacing** (SDAS) is reduced when the casting freezes more rapidly. The finer, more extensive dendritic network serves as a more efficient conductor of the latent heat to the undercooled liquid. The SDAS is related to the solidification time by

$$\text{SDAS} = k t_s^m \tag{9-6}$$

where m and k are constants depending on the composition of the metal. This relationship is shown in Figure 9-6 for several alloys. Small secondary dendrite arm spacings are associated with higher strengths and improved ductility (Figure 9-7).

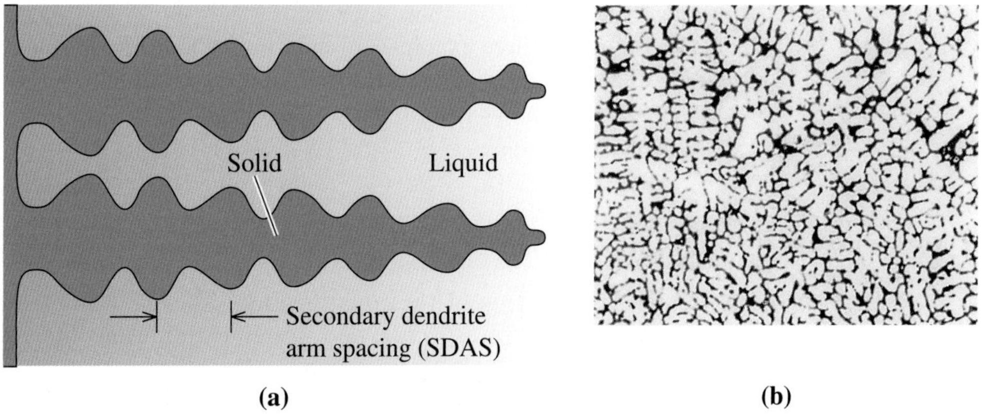

Figure 9-5 (a) The secondary dendrite arm spacing (SDAS). (b) Dendrites in an aluminum alloy (× 50). [*From* ASM Handbook, *Vol. 9, Metallography and Microstructure (1985), ASM International, Materials Park, OH 44073-0002. Reprinted with permission of ASM International. All rights reserved. www.asminternational.org.*]

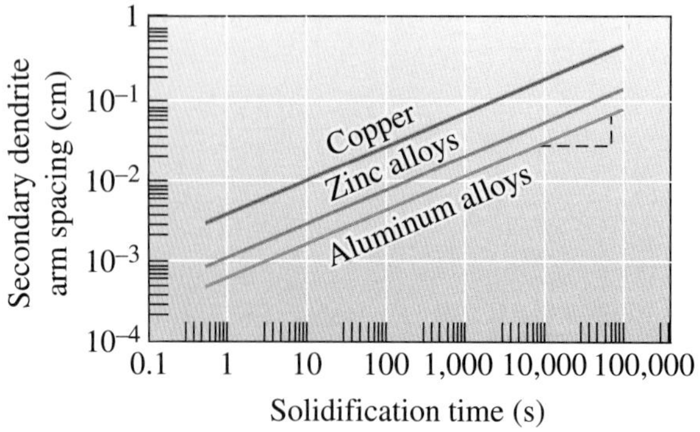

Figure 9-6 The effect of solidification time on the secondary dendrite arm spacings of copper, zinc, and aluminum.

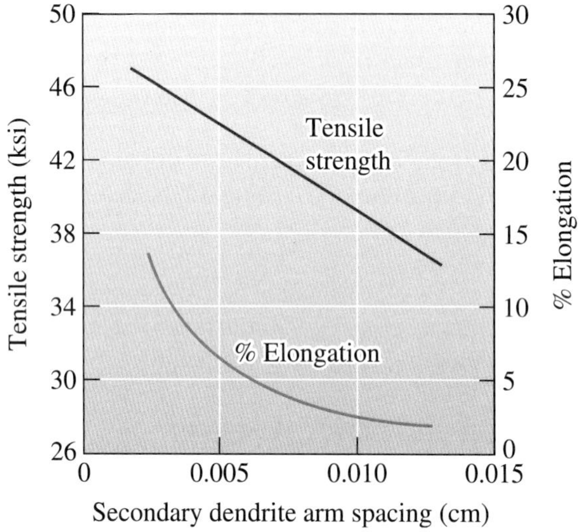

Figure 9-7 The effect of the secondary dendrite arm spacing on the mechanical properties of an aluminum alloy casting.

Rapid solidification processing is used to produce exceptionally fine secondary dendrite arm spacings; a common method is to produce very fine liquid droplets that freeze into solid particles. This process is known as spray atomization. The tiny droplets freeze at a rate of about 10^4°C/s, producing powder particles that range from ~5–100 μm in diameter. This cooling rate is not rapid enough to form a metallic glass, but does produce a fine dendritic structure. By carefully consolidating the solid droplets by powder metallurgy processes, improved properties in the material can be obtained. Since the particles are derived from a melt, many complex alloy compositions can be produced in the form of chemically homogenous powders.

The following three examples discuss how Chvorinov's rule, the relationship between SDAS and the time of solidification, and the relationship between SDAS and mechanical properties can be used to design casting processes.

Example 9-3 *Secondary Dendrite Arm Spacing for Aluminum Alloys*

Determine the constants in the equation that describe the relationship between secondary dendrite arm spacing and solidification time for aluminum alloys (Figure 9-6).

SOLUTION

We could obtain the value of the SDAS at two times from the graph and calculate k and m using simultaneous equations; however, if the scales on the ordinate and abscissa are equal for powers of ten (as in Figure 9-6), we can obtain the slope m from the log-log plot by directly measuring the slope of the graph. In Figure 9-6, we can mark five equal units on the vertical scale and 12 equal units on the horizontal scale. The slope is

$$m = \frac{5}{12} = 0.42$$

The constant k is the value of SDAS when $t_s = 1$ s, since

$$\log \text{SDAS} = \log k + m \log t_s$$

If $t_s = 1$ s, $(m \log t_s) = 0$, and SDAS = k. From Figure 9-6:

$$k = 7 \times 10^{-4} \frac{\text{cm}}{\text{s}}$$

Example 9-4 *Time of Solidification*

A 4-in.-diameter aluminum bar solidifies to a depth of 0.5 in. beneath the surface in 5 minutes. After 20 minutes, the bar has solidified to a depth of 1.5 in. How much time is required for the bar to solidify completely?

SOLUTION

From our measurements, we can determine the constants $k_{\text{solidification}}$ and c_1 in Equation 9-5:

$$0.5 \text{ in.} = k_{\text{solidification}}\sqrt{5 \text{ min}} - c_1 \quad \text{or} \quad c_1 = k\sqrt{5} - 0.5$$

$$1.5 \text{ in.} = k_{\text{solidification}}\sqrt{20 \text{ min}} - c_1 = k\sqrt{20} - (k\sqrt{5} - 0.5)$$

$$1.5 = k_{\text{solidification}}(\sqrt{20} - \sqrt{5}) + 0.5$$

$$k_{\text{soldification}} = \frac{1.5 - 0.5}{4.472 - 2.236} = 0.447 \frac{\text{in.}}{\sqrt{\text{min}}}$$

$$c_1 = (0.447)\sqrt{5} - 0.5 = 0.5 \text{ in.}$$

Solidification is complete when $d = 2$ in. (half the diameter, since freezing is occurring from all surfaces):

$$2 \text{ in.} = 0.447\sqrt{t} - 0.5 \text{ in}$$

$$\sqrt{t} = \frac{2 + 0.5}{0.447} = 5.59 \sqrt{\text{min}}$$

$$t = 31.25 \text{ min}$$

In actual practice, we would find that the total solidification time is somewhat longer than 31.25 min. As solidification continues, the mold becomes hotter and is less effective in removing heat from the casting.

Example 9-5 *Design of an Aluminum Alloy Casting*

Design the thickness of an aluminum alloy casting with a length of 12 in., a width of 8 in., and a tensile strength of 42,000 psi. The mold constant in Chvorinov's rule for aluminum alloys cast in a sand mold is 45 min/in.^{2}. Assume that data shown in Figures 9-6 and 9-7 can be used.

SOLUTION

In order to obtain a tensile strength of 42,000 psi, a secondary dendrite arm spacing of about 0.007 cm is required (see Figure 9-7). From Figure 9-6 we can determine that the solidification time required to obtain this spacing is about 300 s or 5 min.

From Chvorinov's rule

$$t_s = B\left(\frac{V}{A}\right)^2$$

where B = 45 min/in.^{2} Let x be the thickness of the casting. Since the length is 12 in. and the width is 8 in.,

$$V = (8)(12)(x) = 96x$$

$$A = (2)(8)(12) + (2)(x)(8) + (2)(x)(12) = 40x + 192$$

$$5 \text{ min} = (45 \text{ min/in.}^2)\left(\frac{96x}{40x + 192}\right)^2$$

$$\frac{96x}{40x + 192} = \sqrt{(5/45)} = 0.333$$

$$96x = 13.33x + 64$$

$$x = 0.77 \text{ in.}$$

9-6 Cooling Curves

We can summarize our discussion at this point by examining cooling curves. A cooling curve shows how the temperature of a material (in this case, a pure metal) decreases with time [Figure 9-8 (a) and (b)]. The liquid is poured into a mold at the pouring temperature, point A. The difference between the pouring temperature and the freezing temperature is the superheat. The specific heat is extracted by the mold until the liquid reaches the freezing temperature (point B). If the liquid is not well-inoculated, it must be undercooled (point B to C). The slope of the cooling curve before solidification begins is the cooling rate $\frac{\Delta T}{\Delta t}$. As nucleation begins (point C), the latent heat of fusion is given off,

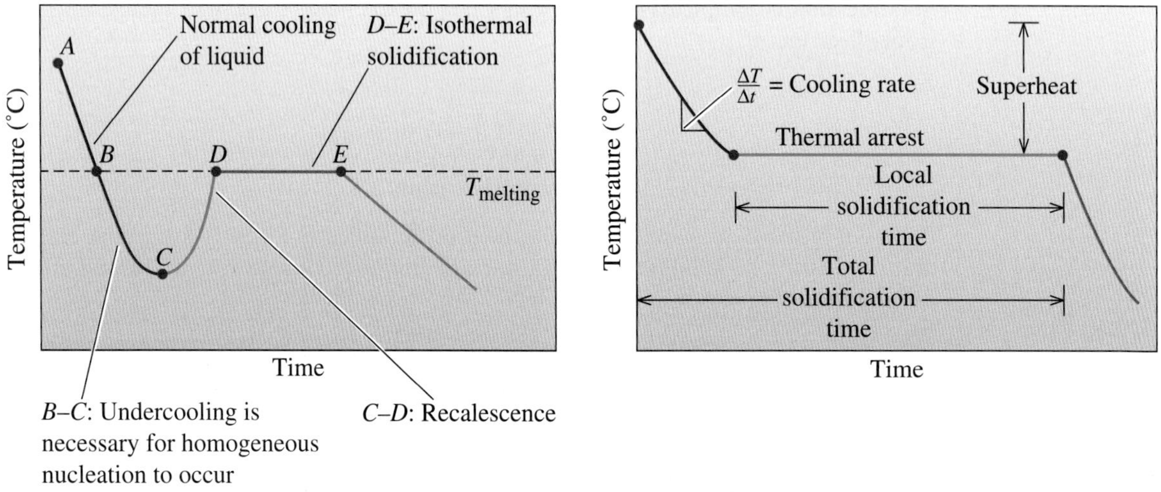

(a) (b)

Figure 9-8 (a) Cooling curve for a pure metal that has not been well-inoculated. The liquid cools as specific heat is removed (between points A and B). Undercooling is necessary (between points B and C). As nucleation begins (point C), latent heat of fusion is released causing an increase in the temperature of the liquid. This process is known as recalescence (point C to point D). The metal continues to solidify at a constant temperature (T_{melting}). At point E, solidification is complete. The solid casting continues to cool from this point. (b) Cooling curve for a well-inoculated, but otherwise pure, metal. No undercooling is needed. Recalescence is not observed. Solidification begins at the melting temperature.

and the temperature rises. This increase in temperature of the undercooled liquid as a result of nucleation is known as **recalescence** (point C to D). Solidification proceeds isothermally at the melting temperature (point D to E) as the latent heat given off from continued solidification is balanced by the heat lost by cooling. This region between points D and E, where the temperature is constant, is known as the **thermal arrest**. A thermal arrest, or plateau, is produced because the evolution of the latent heat of fusion balances the heat being lost because of cooling. At point E, solidification is complete, and the solid casting cools from point E to room temperature.

If the liquid is well-inoculated, the extent of undercooling and recalescence is usually very small and can be observed in cooling curves only by very careful measurements. If effective heterogeneous nuclei are present in the liquid, solidification begins at the freezing temperature [Figure 9-8 (b)]. The latent heat keeps the remaining liquid at the freezing temperature until all of the liquid has solidified and no more heat can be evolved. Growth under these conditions is planar. The **total solidification time** of the casting is the time required to remove both the specific heat of the liquid and the latent heat of fusion. Measured from the time of pouring until solidification is complete, this time is given by Chvorinov's rule. The **local solidification time** is the time required to remove only the latent heat of fusion at a particular location in the casting; it is measured from when solidification begins until solidification is completed. The local solidification times (and the total solidification times) for liquids solidified via undercooled and inoculated liquids will be slightly different.

We often use the terms "melting temperature" and "freezing temperature" while discussing solidification. It would be more accurate to use the term "melting temperature" to describe when a solid turns completely into a liquid. For pure metals and compounds, this happens at a fixed temperature (assuming fixed pressure) and without superheating. "Freezing temperature" or "freezing point" can be defined as the temperature at which solidification of a material is complete.

9-7 Cast Structure

In manufacturing components by casting, molten metals are often poured into molds and permitted to solidify. The mold produces a finished shape, known as a *casting*. In other cases, the mold produces a simple shape called an **ingot**. An ingot usually requires extensive plastic deformation before a finished product is created. A *macrostructure*, sometimes referred to as the **ingot structure**, consists of as many as three regions (Figure 9-9). (Recall that in Chapter 2 we used the term "macrostructure" to describe the structure of a material at a macroscopic scale. Hence, the term "ingot structure" may be more appropriate.)

Chill Zone

The **chill zone** is a narrow band of randomly oriented grains at the surface of the casting. The metal at the mold wall is the first to cool to the freezing temperature. The mold wall also provides many surfaces at which heterogeneous nucleation takes place.

Columnar Zone

The **columnar zone** contains elongated grains oriented in a particular crystallographic direction. As heat is removed from the casting by the mold material, the grains in the chill zone grow in the direction opposite to that of the heat flow, or from the coldest toward the hottest areas of the casting. This tendency usually means that the grains grow perpendicular to the mold wall.

Grains grow fastest in certain crystallographic directions. In metals with a cubic crystal structure, grains in the chill zone that have a <100> direction perpendicular to the

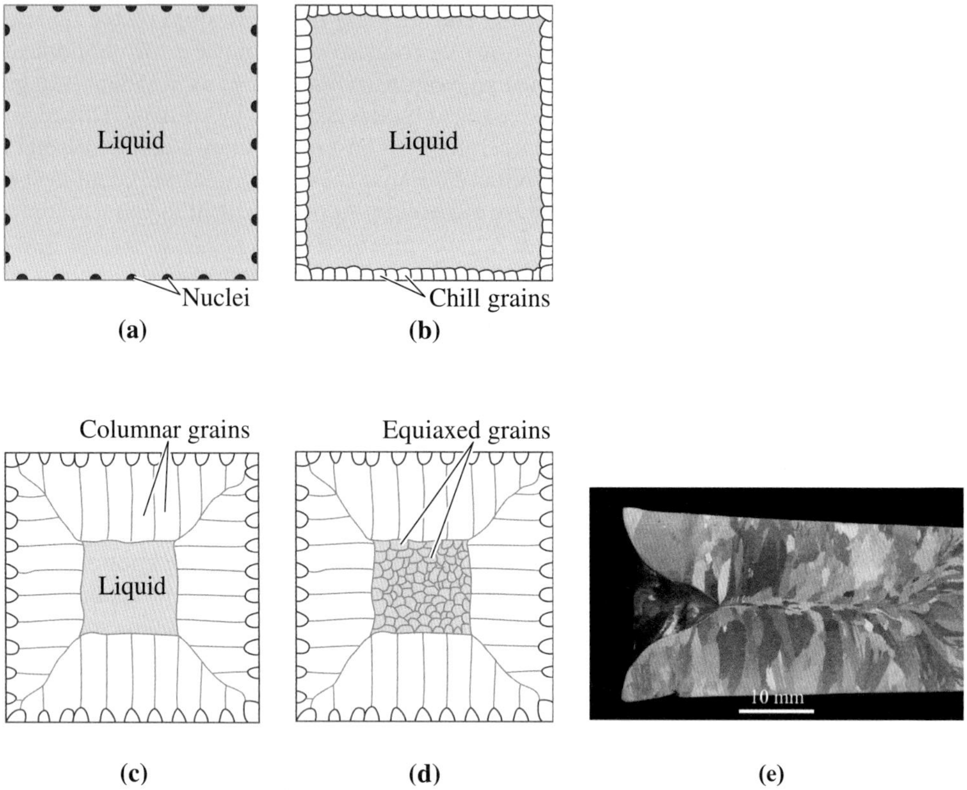

Figure 9-9 Development of the ingot structure of a casting during solidification: (a) nucleation begins, (b) the chill zone forms, (c) preferred growth produces the columnar zone, (d) additional nucleation creates the equiaxed zone, and (e) the ingot structure of a commercially pure aluminum casting. (*Courtesy of Bill Clyne*)

mold wall grow faster than other less favorably oriented grains (Figure 9-10). Eventually, the grains in the columnar zone have <100> directions that are parallel to one another, giving the columnar zone anisotropic properties. This formation of the columnar zone is influenced primarily by growth—rather than nucleation—phenomena. The grains may be composed of many dendrites if the liquid is originally undercooled. The solidification may proceed by planar growth of the columnar grains if no undercooling occurs.

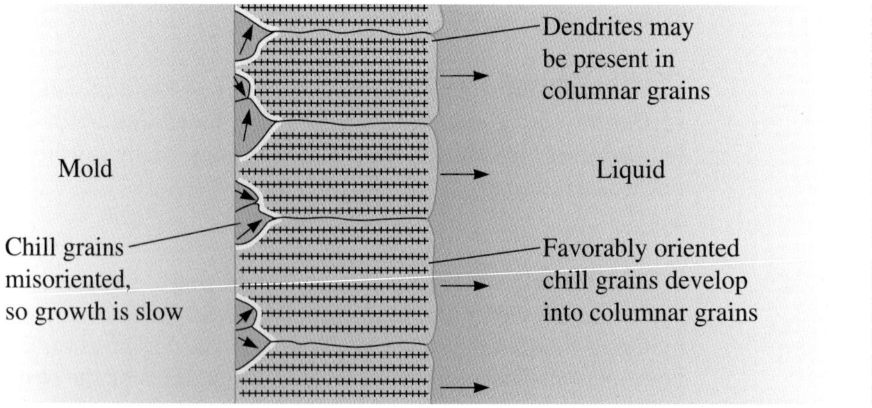

Figure 9-10 Competitive growth of the grains in the chill zone results in only those grains with favorable orientations developing into columnar grains.

Equiaxed Zone Although the solid may continue to grow in a columnar manner until all of the liquid has solidified, an equiaxed zone frequently forms in the center of the casting or ingot. The **equiaxed zone** contains new, randomly oriented grains, often caused by a low pouring temperature, alloying elements, or grain refining or inoculating agents. Small grains or dendrites in the chill zone may also be torn off by strong convection currents that are set up as the casting begins to freeze. These also provide heterogeneous nucleation sites for what ultimately become equiaxed grains. These grains grow as relatively round, or equiaxed, grains with a random orientation, and they stop the growth of the columnar grains. The formation of the equiaxed zone is a nucleation-controlled process and causes that portion of the casting to display isotropic behavior.

By understanding the factors that influence solidification in different regions, it is possible to produce castings that first form a "skin" of a chill zone and then dendrites. Metals and alloys that show this macrostructure are known as **skin-forming alloys**. We also can control the solidification such that no skin or advancing dendritic arrays of grains are seen; columnar to equiaxed switchover occurs near to the mold walls. The result is a casting with a macrostructure consisting predominantly of equiaxed grains. Metals and alloys that solidify in this fashion are known as **mushy-forming alloys** since the cast material seems like a mush of solid grains floating in a liquid melt. Many aluminum and magnesium alloys show this type of solidification. Often, we encourage an all-equiaxed structure and thus create a casting with isotropic properties by effective grain refinement or inoculation. In a later section, we will examine one case (turbine blades) where we control solidification to encourage all columnar grains and hence anisotropic behavior.

Cast ingot structure and microstructure are important particularly for components that are directly cast into a final shape. In many situations though, as discussed in Section 9-1, metals and alloys are first cast into ingots, and the ingots are subsequently subjected to thermomechanical processing (e.g., rolling, forging, etc.). During these steps, the cast macrostructure is broken down and a new microstructure will emerge, depending upon the thermomechanical process used (Chapter 8).

9-8 Solidification Defects

Although there are many defects that potentially can be introduced during solidification, shrinkage and porosity deserve special mention. If a casting contains pores (small holes), the cast component can fail catastrophically when used for load-bearing applications (e.g., turbine blades).

Shrinkage Almost all materials are more dense in the solid state than in the liquid state. During solidification, the material contracts, or shrinks, as much as 7% (Table 9-2).

Often, the bulk of the **shrinkage** occurs as **cavities** if solidification begins at all surfaces of the casting, or *pipes*, if one surface solidifies more slowly than the others (Figure 9-11). The presence of such pipes can pose problems. For example, if in the production of zinc ingots a shrinkage pipe remains, water vapor can condense in it. This water can lead to an explosion if the ingot gets introduced in a furnace in which zinc is being remelted for such applications as hot-dip galvanizing.

A common technique for controlling **cavity** and **pipe shrinkage** is to place a **riser**, or an extra reservoir of metal, adjacent and connected to the casting. As the casting solidifies and shrinks, liquid metal flows from the riser into the casting to fill the shrinkage void. We need only to ensure that the riser solidifies after the casting and that there is an internal liquid channel that connects the liquid in the riser to the last liquid to solidify in the casting. Chvorinov's rule can be used to help design the size of the riser. The following example illustrates how risers can be designed to compensate for shrinkage.

Table 9-2 Shrinkage during solidification for selected materials

Material	Shrinkage (%)
Al	7.0
Cu	5.1
Mg	4.0
Zn	3.7
Fe	3.4
Pb	2.7
Ga	+3.2 (expansion)
H_2O	+8.3 (expansion)
Low-carbon steel	2.5–3.0
High-carbon steel	4.0
White Cast Iron	4.0–5.5
Gray Cast Iron	+1.9 (expansion)

In part, based on DeGarmo, E. P., Black, J. T., and Koshe, R. A. Materials and Processes in Manufacturing, *Prentice Hall, 1997.*

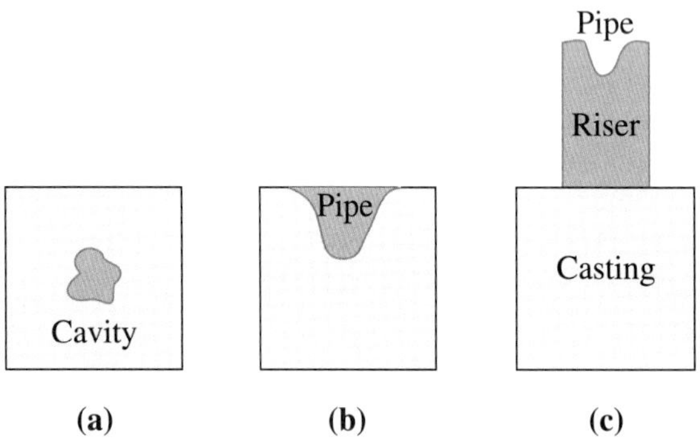

Figure 9-11 Several types of macroshrinkage can occur including (a) cavities and (b) pipes. (c) Risers can be used to help compensate for shrinkage.

Example 9-6 *Design of a Riser for a Casting*

Design a cylindrical riser, with a height equal to twice its diameter, that will compensate for shrinkage in a 2 cm × 8 cm × 16 cm casting (Figure 9-12).

SOLUTION

We know that the riser must freeze after the casting. To be conservative, we typically require that the riser take 25% longer to solidify than the casting. Therefore,

$$t_r = 1.25t_c \quad \text{or} \quad B\left(\frac{V}{A}\right)_r^2 = 1.25\,B\left(\frac{V}{A}\right)_c^2$$

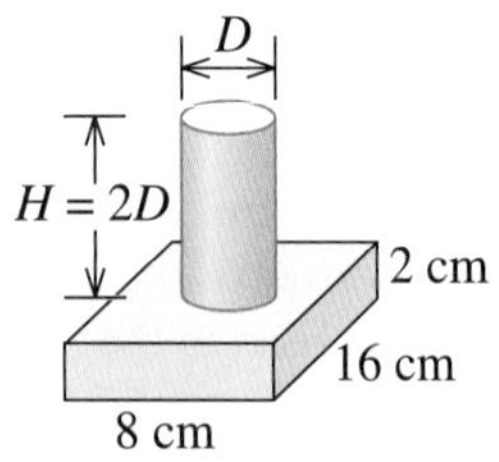

Figure 9-12 The geometry of the casting and riser (for Example 9-6).

The subscripts r and c stand for riser and casting, respectively. The mold constant B is the same for both casting and riser, so

$$\left(\frac{V}{A}\right)_r = \sqrt{1.25}\left(\frac{V}{A}\right)_c$$

The volume of the casting is

$$V_c = (2 \text{ cm})(8 \text{ cm})(16 \text{ cm}) = 256 \text{ cm}^3$$

The area of the riser adjoined to the casting must be subtracted from the total surface area of the casting in order to calculate the surface area of the casting in contact with the mold:

$$A_c = (2)(2 \text{ cm})(8 \text{ cm}) + (2)(2 \text{ cm})(16 \text{ cm}) + (2)(8 \text{ cm})(16 \text{ cm}) - \frac{\pi D^2}{4} = 352 \text{ cm}^2 - \frac{\pi D^2}{4}$$

where D is the diameter of the cylindrical riser. We can write equations for the volume and area of the cylindrical riser, noting that the cylinder height $H = 2D$:

$$V_r = \frac{\pi D^2}{4}H = \frac{\pi D^2}{4}(2D) = \frac{\pi D^3}{2}$$

$$A_r = \frac{\pi D^2}{4} + \pi DH = \frac{\pi D^2}{4} + \pi D(2D) = \frac{9}{4}\pi D^2$$

where again we have not included the area of the riser adjoined to the casting in the area calculation. The volume to area ratio of the riser is given by

$$\left(\frac{V}{A}\right)_r = \frac{(\pi D^3/2)}{(9\pi D^2/4)} = \frac{2}{9}D$$

and must be greater than that of the casting according to

$$\left(\frac{V}{A}\right)_r = \frac{2}{9}D > \sqrt{1.25}\left(\frac{V}{A}\right)_c$$

Substituting,

$$\frac{2}{9}D > \sqrt{1.25}\left(\frac{256 \text{ cm}^3}{352 \text{ cm}^2 - \pi D^2/4}\right)$$

Solving for the smallest diameter for the riser:

$$D = 3.78 \text{ cm}$$

Although the volume of the riser is less than that of the casting, the riser solidifies more slowly because of its compact shape.

Interdendritic Shrinkage This consists of small shrinkage pores between dendrites (Figure 9-13). This defect, also called **microshrinkage** or **shrinkage porosity**, is difficult to prevent by the use of risers. Fast cooling rates may reduce problems with **interdendritic shrinkage**; the dendrites may be shorter, permitting liquid to flow through the dendritic network to the solidifying interface. In addition, any shrinkage that remains may be finer and more uniformly distributed.

Gas Porosity Many metals dissolve a large quantity of gas when they are molten. Aluminum, for example, dissolves hydrogen. When the aluminum solidifies, however, the solid metal retains in its crystal structure only a small fraction of the hydrogen since the solubility of hydrogen in the solid is remarkably lower than that in the liquid (Figure 9-14). The excess hydrogen that cannot be incorporated in the solid

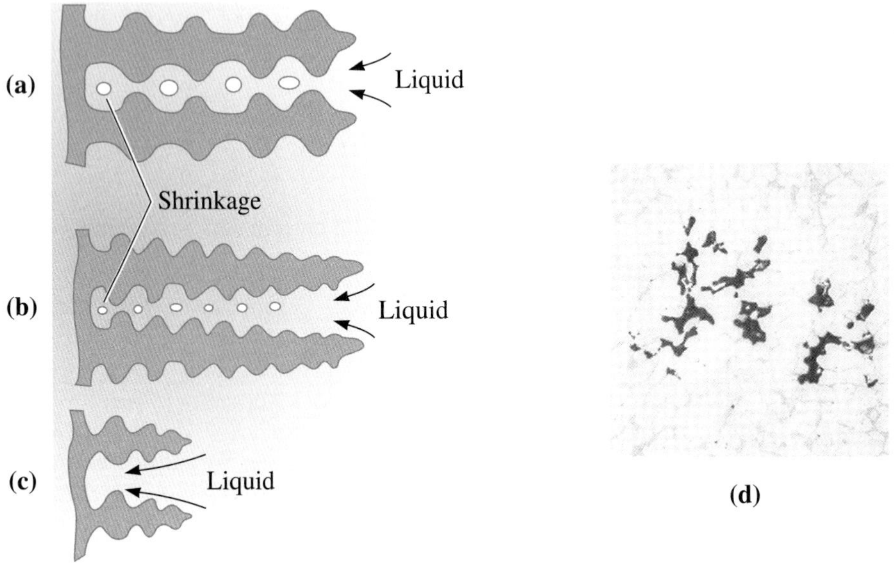

Figure 9-13 (a) Shrinkage can occur between the dendrite arms. (b) Small secondary dendrite arm spacings result in smaller, more evenly distributed shrinkage porosity. (c) Short primary arms can help avoid shrinkage. (d) Interdendritic shrinkage in an aluminum alloy (× 80). (*Reprinted Courtesy of Don Askeland*)

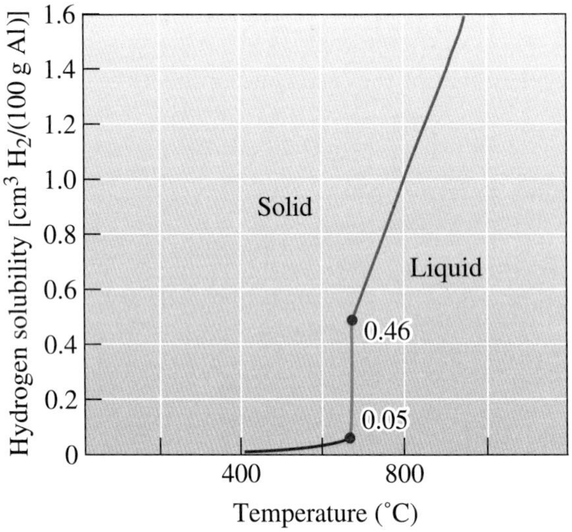

Figure 9-14 The solubility of hydrogen gas in aluminum when the partial pressure of H_2 = 1 atm.

metal or alloy crystal structure forms bubbles that may be trapped in the solid metal, producing **gas porosity**. The amount of gas that can be dissolved in molten metal is given by **Sievert's law**:

$$\text{Percent of gas} = K\sqrt{p_{gas}} \tag{9-7}$$

where p_{gas} is the partial pressure of the gas in contact with the metal and K is a constant which, for a particular metal-gas system, increases with increasing temperature. We can minimize gas porosity in castings by keeping the liquid temperature low, by adding materials to the liquid to combine with the gas and form a solid, or by ensuring that the partial pressure of the gas remains low. The latter may be achieved by placing the molten metal in a vacuum chamber or bubbling an inert gas through the metal. Because p_{gas} is low in the vacuum, the gas leaves the metal, enters the vacuum, and is carried away. **Gas flushing** is a process in which bubbles of a gas, inert or reactive, are injected into a molten metal to remove undesirable elements from molten metals and alloys. For example, hydrogen in aluminum can be removed using nitrogen or chlorine. The following example illustrates how a degassing process can be designed.

In the manufacturing of **stainless steel**, a process known as **argon oxygen decarburization** (AOD) is used to lower the carbon content of the melt without oxidizing chromium or nickel. In this process, a mixture of argon (or nitrogen) and oxygen gases is forced into molten stainless steel. The carbon dissolved in the molten steel is oxidized by the oxygen gas via the formation of carbon monoxide (CO) gas; the CO is carried away by the inert argon (or nitrogen) gas bubbles. These processes need very careful control since some reactions (e.g., oxidation of carbon to CO) are exothermic (generate heat).

Example 9-7 *Design of a Degassing Process for Copper*

After melting at atmospheric pressure, molten copper contains 0.01 weight percent oxygen. To ensure that your castings will not be subject to gas porosity, you want to reduce the weight percent to less than 0.00001% prior to pouring. Design a degassing process for the copper.

SOLUTION

We can solve this problem in several ways. In one approach, the liquid copper is placed in a vacuum chamber; the oxygen is then drawn from the liquid and carried away into the vacuum. The vacuum required can be estimated from Sievert's law:

$$\frac{\%\,O_{\text{initial}}}{\%\,O_{\text{vacuum}}} = \frac{K\sqrt{p_{\text{initial}}}}{K\sqrt{p_{\text{vacuum}}}} = \sqrt{\left(\frac{1\text{ atm}}{p_{\text{vacuum}}}\right)}$$

$$\frac{0.01\%}{0.00001\%} = \sqrt{\left(\frac{1\text{ atm}}{p_{\text{vacuum}}}\right)}$$

$$\frac{1\text{ atm}}{p_{\text{vacuum}}} = (1000)^2 \text{ or } p_{\text{vacuum}} = 10^{-6}\text{ atm}$$

Another approach would be to introduce a copper-15% phosphorus alloy. The phosphorus reacts with oxygen to produce P_2O_5, which floats out of the liquid, by the reaction:

$$5O + 2P \rightarrow P_2O_5$$

Typically, about 0.01 to 0.02% P must be added to remove the oxygen.

9-9 Casting Processes for Manufacturing Components

Figure 9-15 summarizes four of the dozens of commercial casting processes. In some processes, the molds can be reused; in others, the mold is expendable. **Sand casting** processes include green sand molding, for which silica (SiO_2) sand grains bonded with wet clay are packed around a removable pattern. Ceramic casting processes use a fine-grained ceramic material as the mold, as slurry containing the ceramic may be poured around a reusable pattern, which is removed after the ceramic hardens. In **investment casting**, the ceramic slurry of a material such as colloidal silica (consisting of ceramic nanoparticles) coats a wax pattern. After the ceramic hardens (i.e., the colloidal silica dispersion gels), the wax is melted and drained from the ceramic shell, leaving behind a cavity that is then filled with molten metal. After the metal solidifies, the mold

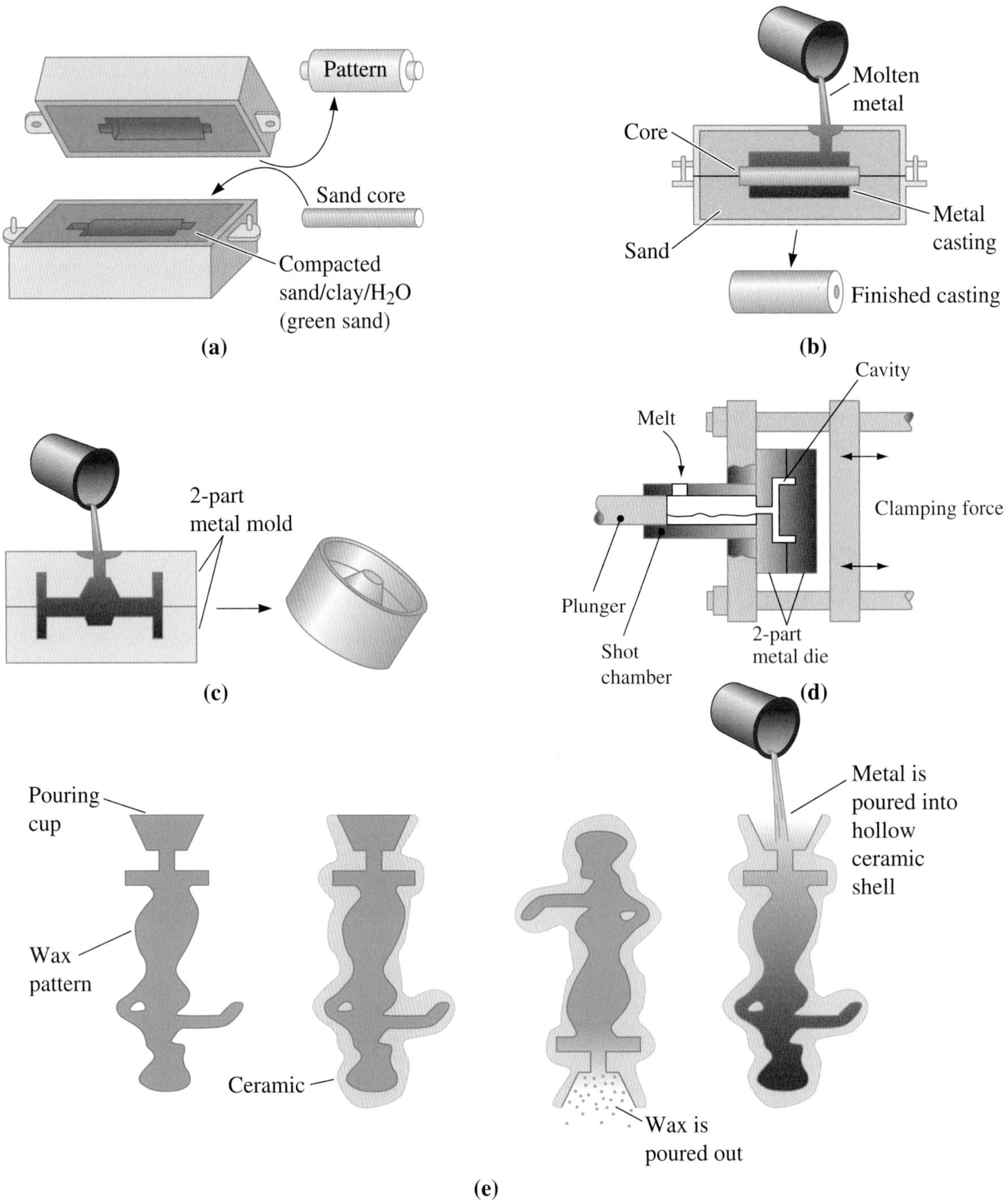

Figure 9-15 Four typical casting processes. (a) and (b) Green sand molding, in which clay-bonded sand is packed around a pattern. Sand cores can produce internal cavities in the casting. (c) The permanent mold process, in which metal is poured into an iron or steel mold. (d) Die casting, in which metal is injected at high pressure into a steel die. (e) Investment casting, in which a wax pattern is surrounded by a ceramic; after the wax is melted and drained, metal is poured into the mold.

is broken to remove the part. The investment casting process, also known as the **lost wax process**, is suited for generating complex shapes. Dentists and jewelers originally used the precision investment casting process. Currently, this process is used to produce such components as turbine blades, titanium heads of golf clubs, and parts for knee and hip prostheses. In another process known as the **lost foam process**, polystyrene beads, similar to those used to make coffee cups or packaging materials, are used to produce a foam pattern. When molten metal is poured into the mold, the polymer foam pattern melts and decomposes, with the metal taking the place of the pattern.

In the permanent mold and pressure die casting processes, a cavity is machined from metallic material. After the liquid poured into the cavity solidifies, the mold is opened, the casting is removed, and the mold is reused. The processes using metallic molds tend to give the highest strength castings because of the rapid solidification. Ceramic molds, including those used in investment casting, are good insulators and give the slowest-cooling and lowest-strength castings. Millions of truck and car pistons are made in foundries using permanent mold casting. Good surface finish and dimensional accuracy are the advantages of **permanent mold casting** techniques. High mold costs and limited complexity in shape are the disadvantages.

In **pressure die casting**, molten metallic material is forced into the mold under high pressures and is held under pressure during solidification. Many zinc, aluminum, and magnesium-based alloys are processed using pressure die casting. Extremely smooth surface finishes, very good dimensional accuracy, the ability to cast intricate shapes, and high production rates are the advantages of the pressure die casting process. Since the mold is metallic and must withstand high pressures, the dies used are expensive and the technique is limited to smaller-sized components.

9-10 Continuous Casting and Ingot Casting

As discussed in the prior section, casting is a tool used for the manufacturing of components. It is also a process for producing ingots or slabs that can be further processed into different shapes (e.g., rods, bars, wires, etc.). In the steel industry, millions of pounds of steels are produced using blast furnaces, electric arc furnaces and other processes. Figure 9-16 shows a

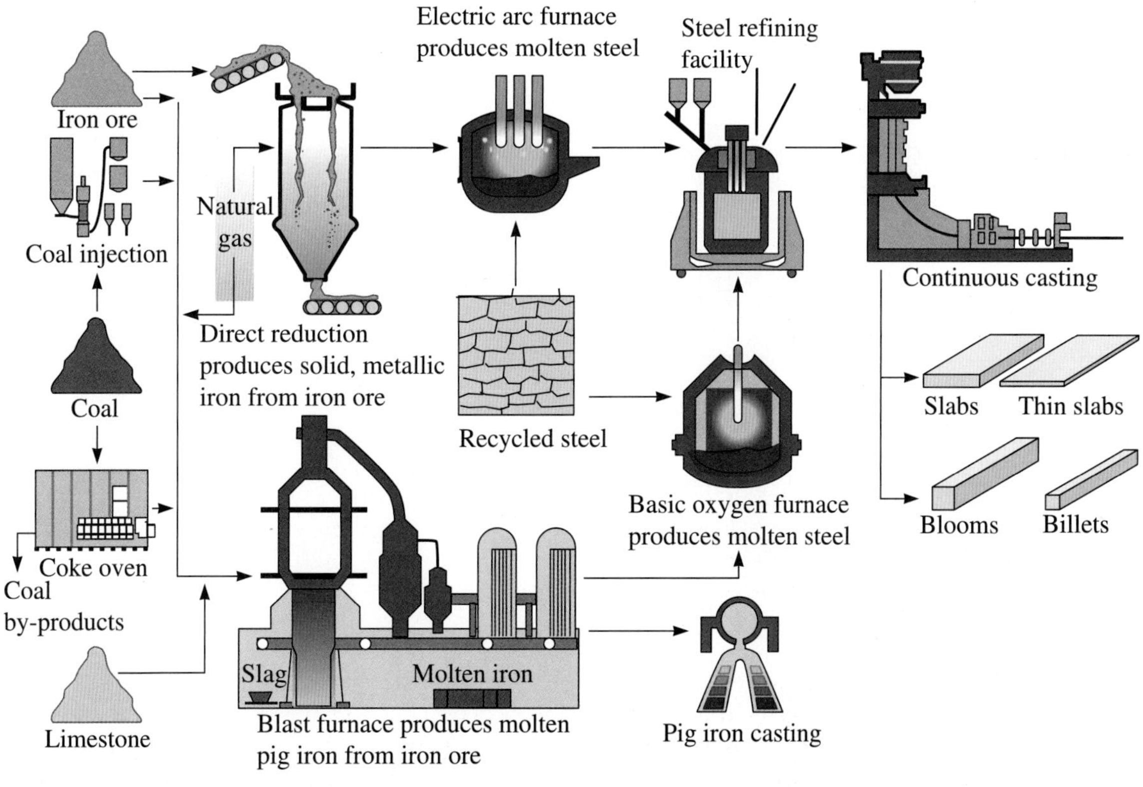

Figure 9-16 Summary of steps in the extraction of steels using iron ores, coke, and limestone. (*Source: www.steel.org. Used with permission of the American Iron and Steel Institute.*)

summary of steps to extract steels using iron ores, coke, and limestone. Although the details change, most metals and alloys (e.g., copper and zinc) are extracted from their ores using similar processes. Certain metals, such as aluminum, are produced using an electrolytic process since aluminum oxide is too stable and cannot be readily reduced to aluminum metal using coke or other reducing agents.

In many cases, we begin with scrap metals and recyclable alloys. In this case, the scrap metal is melted and processed, removing the impurities and adjusting the composition. Considerable amounts of steel, aluminum, zinc, stainless steel, titanium, and many other materials are recycled every year.

In **ingot casting**, molten steels or alloys obtained from a furnace are cast into large molds. The resultant castings, called ingots, are then processed for conversion into useful shapes via thermomechanical processing, often at another location. In the **continuous casting** process, the idea is to go from molten metallic material to some more useful "semi-finished" shape such as a plate, slab, etc. Figure 9-17 illustrates a common method for producing steel plate and bars. The liquid metal is fed from a holding vessel (a tundish) into a water-cooled oscillating copper mold, which rapidly cools the surface of the steel. The partially solidified steel is withdrawn from the mold at the same rate that additional liquid steel is introduced. The center of the steel casting finally solidifies well after the casting exits the mold. The continuously cast material is then cut into appropriate lengths by special cutting machines.

Continuous casting is cost effective for processing many steels, stainless steels, and aluminum alloys. Ingot casting is also cost effective and used for many steels where a continuous caster is not available or capacity is limited and for alloys of non-ferrous metals (e.g., zinc, copper) where the volumes are relatively small and the capital expenditure needed for a continuous caster may not be justified. Also, not all alloys can be cast using the continuous casting process.

Ingot casting and continuous casting are primary processing steps. Some of the secondary processing steps used to form ingots or slabs into useful shapes are shown in Figure 9-18.

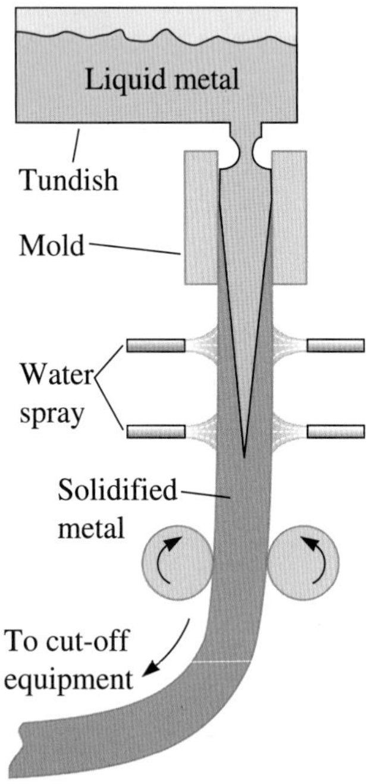

Figure 9-17 Vertical continuous casting, used in producing many steel products. Liquid metal contained in the tundish partially solidifies in a mold.

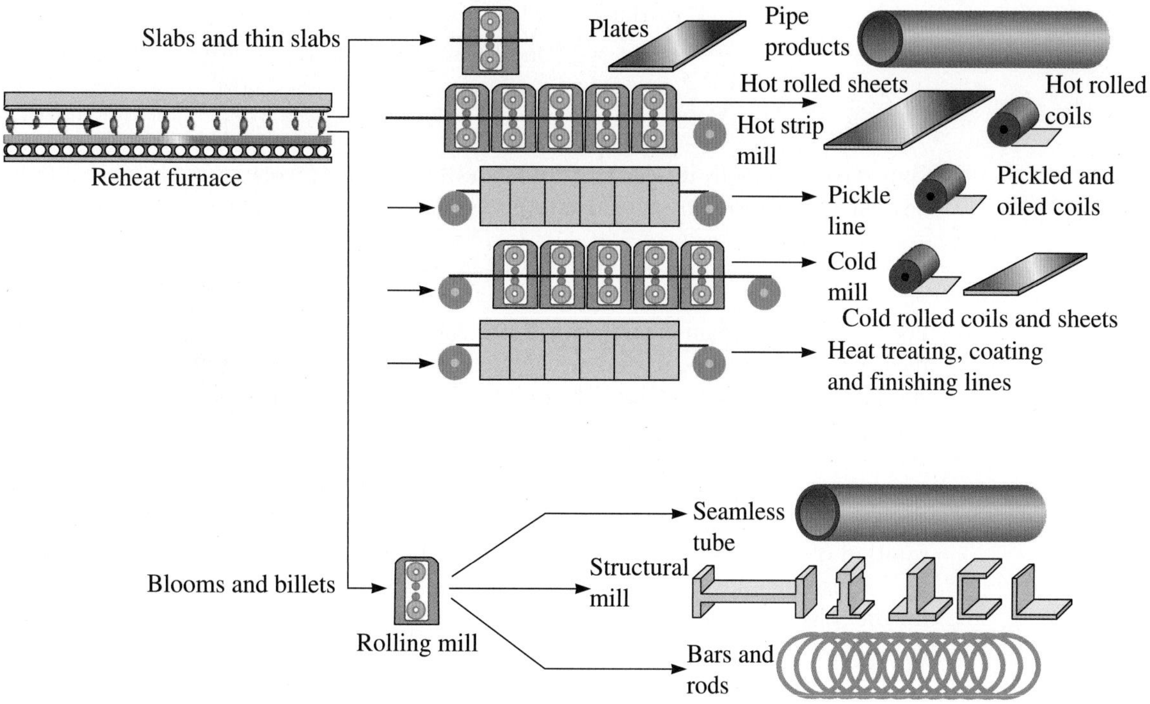

Figure 9-18 Secondary processing steps in processing of steel and alloys. (*Source: www.steel.org. Used with permission of the American Iron and Steel Institute.*)

Example 9-8 *Design of a Continuous Casting Machine*

Figure 9-19 shows a method for continuous casting of 0.25-in.-thick, 48-in.-wide aluminum plate that is subsequently rolled into aluminum foil. The liquid aluminum is introduced between two large steel rolls that slowly turn. We want the aluminum to be completely solidified by the rolls just as the plate emerges from the machine. The rolls act as a permanent mold with a mold constant B of about 5 min/in.^{2} when the aluminum is poured at the proper superheat. Design the rolls required for this process.

SOLUTION

It would be helpful to simplify the geometry so that we can determine a solidification time for the casting. Let's assume that the shaded area shown in Figure 9-19(b) represents the casting and can be approximated by the average thickness times a length and width. The average thickness is (0.50 in. + 0.25 in.)/2 = 0.375 in. Then

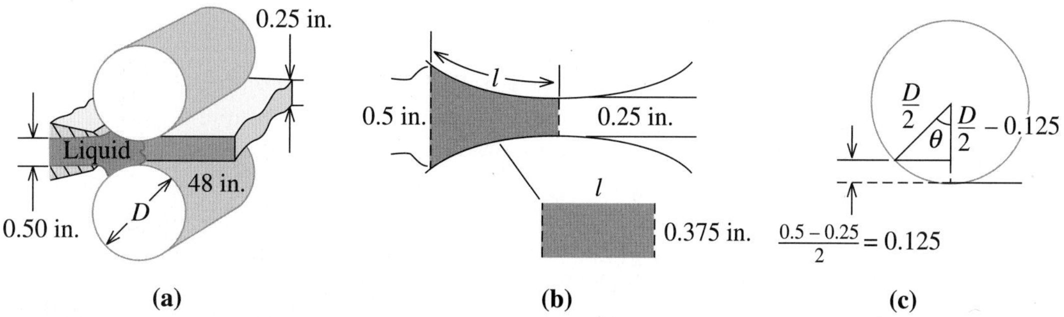

Figure 9-19 Horizontal continuous casting of aluminum (for Example 9-8).

$$V = (\text{thickness})(\text{length})(\text{width}) = 0.375lw$$

$$A = 2\,(\text{length})(\text{width}) = 2lw$$

$$\frac{V}{A} = \frac{0.375lw}{2lw} = 0.1875 \text{ in.}$$

Only the area directly in contact with the rolls is used in Chvorinov's rule, since little or no heat is transferred from other surfaces. The solidification time should be

$$t_s = B\left(\frac{V}{A}\right)^2 = (5)(0.1875)^2 = 0.1758 \text{ min}$$

For the plate to remain in contact with the rolls for this period of time, the diameter of the rolls and the rate of rotation of the rolls must be properly specified. Figure 9-19(c) shows that the angle θ between the points where the liquid enters and exits the rolls is

$$\cos\theta = \frac{(D/2) - 0.125}{(D/2)} = \frac{D - 0.25}{D}$$

The surface velocity of the rolls is the product of the circumference and the rate of rotation of the rolls, $v = \pi DR$, where R has units of revolutions/minute. The velocity v is also the rate at which we can produce the aluminum plate. The time required for the rolls to travel the distance l must equal the required solidification time:

$$t = \frac{l}{v} = 0.1758 \text{ min}$$

The length l is the fraction of the roll diameter that is in contact with the aluminum during freezing and can be given by

$$l = \frac{\pi D\theta}{360}$$

Note that θ has units of degrees. Then, by substituting for l and v in the equation for the time:

$$t = \frac{l}{v} = \frac{\pi D\theta}{360\pi DR} = \frac{\theta}{360\,R} = 0.1758 \text{ min}$$

$$R = \frac{\theta}{(360)(0.1758)} = .0158\,\theta \text{ rev/min}$$

A number of combinations of D and R provide the required solidification rate. Let's calculate θ for several diameters and then find the required R.

D (in.)	θ (°)	l (in.)	$R = 0.0158\theta$ (rev/min)	$v = \pi DR$ (in./min)
24	8.28	1.73	0.131	9.86
36	6.76	2.12	0.107	12.1
48	5.85	2.45	0.092	13.9
60	5.23	2.74	0.083	15.6

As the diameter of the rolls increases, the contact area (l) between the rolls and the metal increases. This, in turn, permits a more rapid surface velocity (v) of the rolls and increases the rate of production of the plate. Note that the larger diameter rolls do not need to rotate as rapidly to achieve these higher velocities.

In selecting our final design, we prefer to use the largest practical roll diameter to ensure high production rates. As the rolls become more massive, however, they and their supporting equipment become more expensive.

In actual operation of such a continuous caster, faster speeds could be used, since the plate does not have to be completely solidified at the point where it emerges from the rolls.

9-11 Directional Solidification [DS], Single Crystal Growth, and Epitaxial Growth

There are some applications for which a small equiaxed grain structure in the casting is not desired. Castings used for blades and vanes in turbine engines are an example (Figure 9-20). These castings are often made from titanium, cobalt, or nickel-based superalloys using precision investment casting.

In conventionally cast parts, an equiaxed grain structure is often produced; however, blades and vanes for turbine and jet engines fail along transverse grain boundaries. Better creep and fracture resistance are obtained using the **directional solidification** (DS) growth technique. In the DS process, the mold is heated from one end and cooled from the other, producing a columnar microstructure with all of the grain boundaries running in the longitudinal direction of the part. No grain boundaries are present in the transverse direction [Figure 9-20(b)].

Still better properties are obtained by using a *single crystal* (SC) technique. Solidification of columnar grains again begins at a cold surface; however, due to a helical cavity in the mold between the heat sink and the main mold cavity, only one columnar grain is able to grow to the main body of the casting [Figure 9-20(c)]. The single-crystal casting has no grain boundaries, so its crystallographic planes and directions can be directed in an optimum orientation.

Single Crystal Growth

One of the most important applications of solidification is the growth of single crystals. Polycrystalline materials cannot be used effectively in many electronic and optical applications. Grain boundaries and other defects interfere with the mechanisms that provide useful electrical or optical functions. For example, in order to utilize the semiconducting behavior of doped silicon, high-purity single crystals must be used. The current technology for silicon makes use of large (up to 12 in. diameter) single crystals. Typically, a large crystal of the material is grown [Figure 9-21(a)]. The large crystal is then cut into silicon wafers that are only a few millimeters thick [Figure 9-21(b)]. The *Bridgman* and *Czochralski processes* are some of the popular methods used for growing single crystals of silicon, GaAs, lithium niobate ($LiNbO_3$), and many other materials.

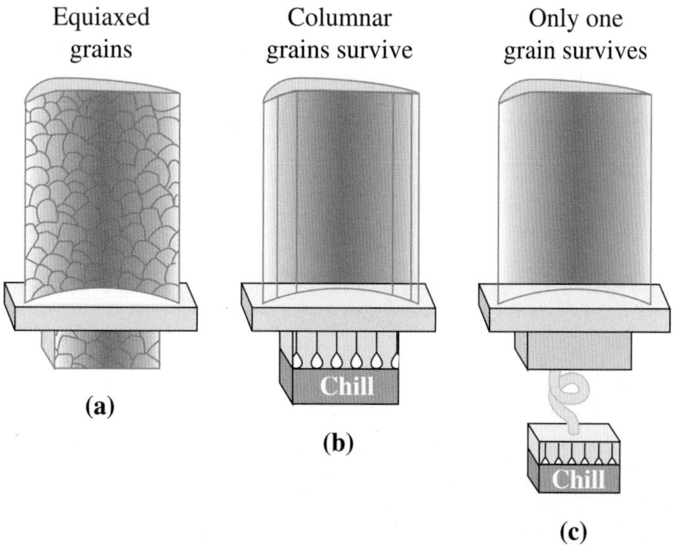

Figure 9-20 Controlling grain structure in turbine blades: (a) conventional equiaxed grains, (b) directionally solidified columnar grains, and (c) a single crystal.

(a) (b)

Figure 9-21 (a) Silicon single crystal (*Peter Sobolev/Shutterstock.com*) and (b) silicon wafer. (*Steve McAlister/Stockbyte/Getty Images*)

Crystal growth furnaces containing molten materials must be maintained at a precise and stable temperature. Often, a small crystal of a predetermined crystallographic orientation is used as a "seed." Heat transfer is controlled so that the entire melt crystallizes into a single crystal. Typically, single crystals offer considerably improved, controllable, and predictable properties at a higher cost than polycrystalline materials. With a large demand, however, the cost of single crystals may not be a significant factor compared to the rest of the processing costs involved in making novel and useful devices.

Epitaxial Growth There are probably over a hundred processes for the deposition of thin film materials. In some of these processes, there is a need to control the texture or crystallographic orientation of the polycrystalline material being deposited; others require a single crystal film oriented in a particular direction. If this is the case, we can make use of a substrate of a known orientation. Epitaxy is the process by which one material is made to grow in an oriented fashion using a substrate that is crystallographically matched with the material being grown. If the lattice matching between the substrate and the film is good (within a few %), it is possible to grow highly oriented or single crystal thin films. This is known as **epitaxial growth**.

9-12 Solidification of Polymers and Inorganic Glasses

Many polymers do not crystallize, but solidify, when cooled. In these materials, the thermodynamic driving force for crystallization may exist; however, the rate of nucleation of the solid may be too slow or the complexity of the polymer chains may be so great that a crystalline solid does not form. Crystallization in polymers is almost never complete and is significantly different from that in metals, requiring long polymer chains to become closely aligned over relatively large distances. By doing so, the polymer grows as lamellar, or plate-like, crystals (Figure 9-22). The region between each lamella contains polymer chains arranged in an amorphous manner. In addition, bundles of lamellae grow from a common nucleus, but the crystallographic orientation of the lamellae within any one

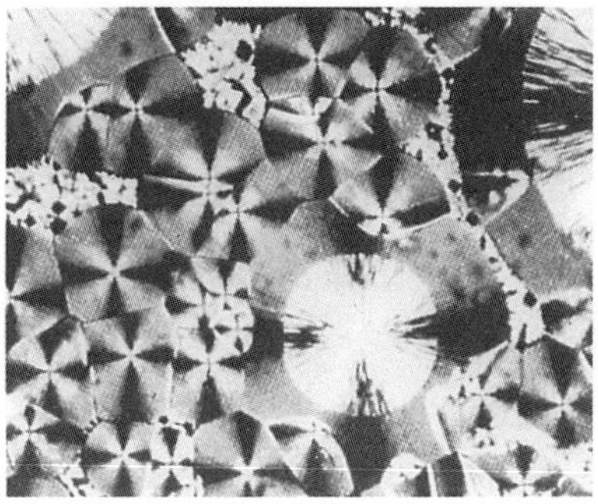

Figure 9-22 Spherulitic crystals in an amorphous matrix of nylon (× 200). (Structure and properties of engineering materials *by BRICK, R. M.; et al. Copyright 1977. Reproduced with permission of MCGRAW-HILL COMPANIES, INC.*)

bundle is different from that in another. As the bundles grow, they may produce a spheroidal shape called a **spherulite**. The spherulite is composed of many individual bundles of differently oriented lamellae. Amorphous regions are present between the individual lamellae, bundles of lamellae, and individual spherulites.

Many polymers of commercial interest develop crystallinity during their processing. Crystallinity can originate from cooling as discussed previously, or from the application of stress. For example, we have learned how PET plastic bottles are prepared using the blow-stretch process (Chapter 3). When the stress is sufficiently high, the chains untangle and straighten and then readily form crystals as they become aligned. In general, polymers such as nylon and polyethylene crystallize more easily than most other thermoplastics.

Inorganic glasses, such as silicate glasses, also do not crystallize easily for kinetic reasons. While the thermodynamic driving force exists, similar to the solidification of metals and alloys, the melts are often too viscous and the diffusion is too slow for crystallization to proceed during solidification. The float-glass process is used to melt and cast large flat pieces of glass. In this process, molten glass is made to float on molten tin. As discussed in Chapter 7, since the strength of inorganic glasses depends critically on surface flaws produced by the manufacturing process or the reaction with atmospheric moisture, most glasses are strengthened using tempering. When safety is not a primary concern, annealing is used to reduce stresses. As mentioned earlier, careful control of nucleation in glasses can lead to glass-ceramics, colored glasses, and photochromic glasses (glasses that can change their color or tint upon exposure to sunlight).

9-13 Joining of Metallic Materials

In **brazing**, an alloy, known as a filler, is used to join one metal to itself or to another metal. The brazing filler metal has a melting temperature above about 450°C. **Soldering** is a brazing process in which the filler has a melting temperature below 450°C. Lead-tin and antimony-tin alloys are the most commonly used materials for soldering. Currently, lead-free soldering materials such as Sn-Cu-Ag are being developed due to the toxicity of lead. In brazing and soldering, the metallic materials being joined do not melt; only the filler material melts. For both brazing and soldering, the composition of the filler material is different from that of the base material being joined. Various aluminum-silicon, copper, magnesium, and precious metals are used for brazing.

Solidification is also important in the joining of metals through **fusion welding**. In a fusion-welding process, a portion of the metals to be joined is melted and, in many instances, additional molten filler metal is added. The pool of liquid metal is called the **fusion zone** (Figures 9-23 and 9-24). When the fusion zone subsequently solidifies, the original pieces of metal are joined together. During solidification of the fusion zone,

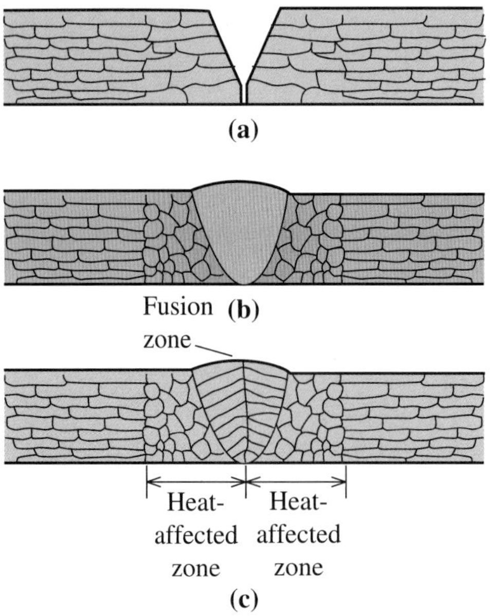

Figure 9-23 A schematic diagram of the fusion zone and solidification of the weld during fusion welding: (a) initial prepared joint, (b) weld at the maximum temperature, with joint filled with filler metal, and (c) weld after solidification.

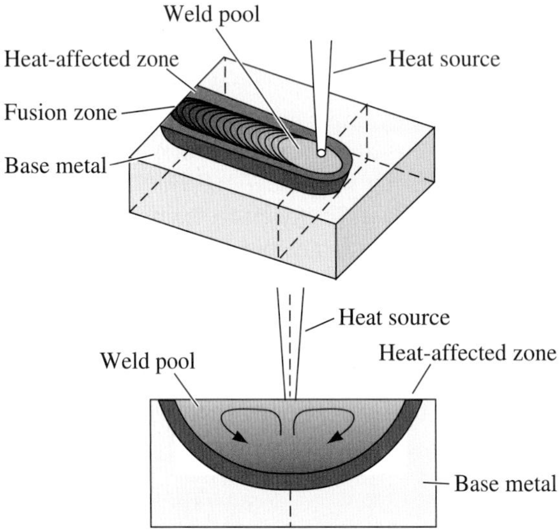

Figure 9-24 Schematic diagram showing interaction between the heat source and the base metal. Three distinct regions in the weldment are the fusion zone, the heat-affected zone, and the base metal. (*Based on "Current Issues and Problems in Welding Science," by S.A. David and T. DebRoy,* Science *24 July 1992, Vol. 257 no. 5069 pp. 497-502.*)

nucleation is not required. The solid simply begins to grow from existing grains, frequently in a columnar manner.

The structure and properties of the fusion zone depend on many of the same variables as in a metal casting. Addition of inoculating agents to the fusion zone reduces the grain size. Fast cooling rates or short solidification times promote a finer microstructure and improved properties. Factors that increase the cooling rate include increased thickness of the metal, smaller fusion zones, low original metal temperatures, and certain types of welding processes. Oxyacetylene welding, for example, uses a relatively low-intensity heat source; consequently, welding times are long and the surrounding solid metal, which becomes very hot, is not an effective heat sink. Arc-welding processes provide a more intense heat source, thus reducing heating of the surrounding metal and providing faster cooling. Laser welding and electron-beam welding utilize exceptionally intense heat sources and produce very rapid cooling rates and potentially strong welds. The friction stir welding process has been developed for Al and Al-Li alloys for aerospace applications.

Summary

- Transformation of a liquid to a solid is probably the most important phase transformation in applications of materials science and engineering.
- Solidification plays a critical role in the processing of metals, alloys, thermoplastics, and inorganic glasses. Solidification is also important in techniques used for the joining of metallic materials.
- Nucleation produces a critical-size solid particle from the liquid melt. Formation of nuclei is determined by the thermodynamic driving force for solidification and is opposed by the need to create the solid-liquid interface. As a result, solidification may not occur at the freezing temperature.
- Homogeneous nucleation requires large undercoolings of the liquid and is not observed in normal solidification processing. By introducing foreign particles into the liquid, nuclei are provided for heterogeneous nucleation. This is done in practice by inoculation or grain refining. This process permits the grain size of the casting to be controlled.
- Rapid cooling of the liquid can prevent nucleation and growth, producing amorphous solids, or glasses, with unusual mechanical and physical properties. Polymeric, metallic, and inorganic materials can be made in the form of glasses.
- In solidification from melts, the nuclei grow into the liquid melt. Either planar or dendritic modes of growth may be observed. In planar growth, a smooth solid-liquid interface grows with little or no undercooling of the liquid. Special directional solidification processes take advantage of planar growth. Dendritic growth occurs when the liquid is undercooled. Rapid cooling, or a short solidification time, produces a finer dendritic structure and often leads to improved mechanical properties of a metallic casting.
- Chvorinov's rule, $t_s = \mathrm{B}(V/A)^n$, can be used to estimate the solidification time of a casting.
- Cooling curves indicate the pouring temperature, any undercooling and recalescence, and time for solidification.
- By controlling nucleation and growth, a casting may be given a columnar grain structure, an equiaxed grain structure, or a mixture of the two. Isotropic behavior is typical of equiaxed grains, whereas anisotropic behavior is found in columnar grains.
- Porosity and cavity shrinkage are major defects that can be present in cast products. If present, they can cause cast products to fail catastrophically.
- In commercial solidification processing methods, defects in a casting (such as solidification shrinkage or gas porosity) can be controlled by proper design of the casting and riser system or by appropriate treatment of the liquid metal prior to casting.
- Sand casting, investment casting, and pressure die casting are some of the processes for casting components. Ingot casting and continuous casting are employed in the production and recycling of metals and alloys.
- The solidification process can be carefully controlled to produce directionally solidified materials as well as single crystals. Epitaxial processes make use of crystal structure matching between the substrate and the material being grown and are useful for making electronic and other devices.

Glossary

Argon oxygen decarburization A process to refine stainless steel. The carbon dissolved in molten stainless steel is reduced by blowing argon gas mixed with oxygen.

Brazing An alloy, known as a filler, is used to join two materials to one another. The composition of the filler, which has a melting temperature above 450°C, is quite different from the metal being joined.

Cavities Small holes present in a casting.

Cavity shrinkage A large void within a casting caused by the volume contraction that occurs during solidification.

Chill zone A region of small, randomly oriented grains that forms at the surface of a casting as a result of heterogeneous nucleation.

Chvorinov's rule The solidification time of a casting is directly proportional to the square of the volume-to-surface area ratio of the casting.

Columnar zone A region of elongated grains having a preferred orientation that forms as a result of competitive growth during the solidification of a casting.

Continuous casting A process to convert molten metal or an alloy into a semi-finished product such as slab.

Critical radius The minimum size that must be formed by atoms clustering together in the liquid before the solid particle is stable and begins to grow.

Dendrite The tree-like structure of the solid that grows when an undercooled liquid solidifies.

Directional solidification A solidification technique in which cooling in a given direction leads to preferential growth of grains in the opposite direction, thereby producing an anisotropic and oriented microstructure.

Dispersion strengthening Increase in strength of a metallic material by generating resistance to dislocation motion by the introduction of small clusters of a second material. (Also called second-phase strengthening.)

Embryo A particle of solid that forms from the liquid as atoms cluster together. The embryo may grow into a stable nucleus or redissolve.

Epitaxial growth Growth of a single-crystal thin film on a crystallographically matched single-crystal substrate.

Equiaxed zone A region of randomly oriented grains in the center of a casting produced as a result of widespread nucleation.

Fusion welding Joining process in which a portion of the materials must melt in order to achieve good bonding.

Fusion zone The portion of a weld heated to produce all liquid during the welding process. Solidification of the fusion zone provides joining.

Gas flushing A process in which a stream of gas is injected into a molten metal in order to eliminate a dissolved gas that might produce porosity.

Gas porosity Bubbles of gas trapped within a casting during solidification, caused by the lower solubility of the gas in the solid compared with that in the liquid.

Glass-ceramics Polycrystalline, ultra-fine grained ceramic materials obtained by controlled crystallization of amorphous glasses.

Grain refinement The addition of heterogeneous nuclei in a controlled manner to increase the number of grains in a casting.

Growth The physical process by which a new phase increases in size. In the case of solidification, this refers to the formation of a stable solid as the liquid freezes.

Heterogeneous nucleation Formation of critically-sized solid from the liquid on an impurity surface.

Homogeneous nucleation Formation of critically-sized solid from the liquid by the clustering together of a large number of atoms at a high undercooling (without an external interface).

Ingot A simple casting that is usually remelted or reprocessed by another user to produce a more useful shape.

Ingot casting Solidification of molten metal in a mold of simple shape. The metal then requires extensive plastic deformation to create a finished product.

Ingot structure The macrostructure of a casting, including the chill zone, columnar zone, and equiaxed zone.

Inoculants Materials that provide heterogeneous nucleation sites during the solidification of a material.

Inoculation The addition of heterogeneous nuclei in a controlled manner to increase the number of grains in a casting.

Interdendritic shrinkage Small pores between the dendrite arms formed by the shrinkage that accompanies solidification. Also known as microshrinkage or shrinkage porosity.

Investment casting A casting process that is used for making complex shapes such as turbine blades, also known as the lost wax process.

Lamellar A plate-like arrangement of crystals within a material.

Latent heat of fusion The heat evolved when a liquid solidifies. The latent heat of fusion is related to the energy difference between the solid and the liquid.

Local solidification time The time required for a particular location in a casting to solidify once nucleation has begun.

Lost foam process A process in which a polymer foam is used as a pattern to produce a casting.

Lost wax process A process in which a wax pattern is used to cast a metal.

Microshrinkage Small, frequently isolated pores between the dendrite arms formed by the shrinkage that accompanies solidification. Also known as interdendritic shrinkage or shrinkage porosity.

Mold constant A characteristic constant in Chvorinov's rule.

Mushy-forming alloys Alloys with a cast macrostructure consisting predominantly of equiaxed grains. They are known as such since the cast material seems like a mush of solid grains floating in a liquid melt.

Nucleation The physical process by which a new phase is produced in a material. In the case of solidification, this refers to the formation of small, stable solid particles in the liquid.

Nuclei Small particles of solid that form from the liquid as atoms cluster together. Because these particles are large enough to be stable, growth of the solid can begin.

Permanent mold casting A casting process in which a mold can be used many times.

Photochromic glass Glass that changes color or tint upon exposure to sunlight.

Pipe shrinkage A large conical-shaped void at the surface of a casting caused by the volume contraction that occurs during solidification.

Planar growth The growth of a smooth solid-liquid interface during solidification when no undercooling of the liquid is present.

Pressure die casting A casting process in which molten metal is forced into a die under pressure.

Primary processing Process involving casting of molten metals into ingots or semi-finished useful shapes such as slabs.

Rapid solidification processing Producing unique material structures by promoting unusually high cooling rates during solidification.

Recalescence The increase in temperature of an undercooled liquid metal as a result of the liberation of heat during nucleation.

Riser An extra reservoir of liquid metal connected to a casting. If the riser freezes after the casting, the riser can provide liquid metal to compensate for shrinkage.

Sand casting A casting process using sand molds.

Secondary dendrite arm spacing The distance between the centers of two adjacent secondary dendrite arms.

Second-phase strengthening See dispersion strengthening.

Secondary processing Processes such as rolling, extrusion, etc., used to process ingots or slabs and other semi-finished shapes.

Shrinkage Contraction of a casting during solidification.

Shrinkage porosity Small pores between the dendrite arms formed by the shrinkage that accompanies solidification. Also known as microshrinkage or interdendritic porosity.

Sievert's law The amount of a gas that dissolves in a metal is proportional to the square root of the partial pressure of the gas in the surroundings.

Skin-forming alloys Alloys with a microstructure that shows an outer skin of small grains in the chill zone followed by dendrites.

Soldering Soldering is a joining process in which the filler has a melting temperature below 450°C; no melting of the base materials occurs.

Solidification front Interface between a solid and liquid.

Solidification process Processing of materials involving solidification (e.g., single crystal growth, continuous casting, etc.).

Solid-state phase transformation A change in phase that occurs in the solid state.

Specific heat The heat required to change the temperature of a unit mass of the material one degree.

Spherulites Spherical crystals in a polymer produced when bundles of crystalline lamellae grow from a common nucleus.

Stainless steel A group of ferrous alloys that contain at least 11% Cr, providing extraordinary corrosion resistance.

Superheat The difference between the pouring temperature and the freezing temperature.

Thermal arrest A plateau on the cooling curve during the solidification of a material caused by the evolution of the latent heat of fusion during solidification. This heat generation balances the heat being lost as a result of cooling.

Total solidification time The time required for the casting to solidify completely after the casting has been poured.

Undercooling The temperature to which the liquid metal must cool below the equilibrium freezing temperature before nucleation occurs.

Problems

Section 9-1 Technological Significance

9-1 Give examples of materials based on inorganic glasses that are made by solidification.

9-2 What do the terms "primary" and "secondary" processing mean?

9-3 Why are ceramic materials not prepared by melting and casting?

Section 9-2 Nucleation

9-4 Define the following terms: nucleation, embryo, heterogeneous nucleation, and homogeneous nucleation.

9-5 What is the difference between homogenous nucleation and heterogeneous nucleation?

9-6 From the thermodynamic point of view, what two things must occur for solidification from liquid to solid to proceed?

9-7 Does water freeze at 0°C and boil at 100°C? Explain.

9-8 Does ice melt at 0°C? Explain.

9-9 Derive the formula for the rate of change in total energy change (i.e., the derivative of ΔG with respect to r) as a function of the radius r of the solid nucleus. Begin with Equation 9-1. Numerically, how is the critical point at which solidification proceeds defined?

9-10 Using Equation 9-2, demonstrate that interfacial energy (σ_{sl}) has units of J/m^2 in SI.

9-11 Calculate the total interfacial surface energy for 10^{16} spheres of copper, each with the critical radius r^*.

9-12 Of the ferrous elements, which has the lowest undercooling required for homogeneous nucleation? Does this have any practical significance?

9-13 If the total change in free energy of a molten metal is 5.34×10^{-17} J upon formation of the first stable solid and the free energy per unit volume is -17.7 J/cm^3, approximate the radius of the first stable spherical solid. The surface free energy of the solid-liquid interface is 100×10^{-7} J/cm^2.

9-14 Using the densities in Appendix A, convert the heats of fusion in Table 9-1 from units of J/cm^3 to kJ/kg.

9-15 Assume that instead of a spherical nucleus, we have a nucleus in the form of a cube of length x. Calculate the critical dimension x^* of the cube necessary for nucleation. Write an equation similar to Equation 9-1 for a cubical nucleus, and derive an expression for x^* similar to Equation 9-2.

9-16 Why is undercooling required for solidification? Derive an equation showing the total free energy change as a function of undercooling when the nucleating solid has the critical nucleus radius r^*.

9-17 Why is it that nuclei seen experimentally are often sphere-like but faceted? Why are they sphere-like and not like cubes or other shapes?

9-18 Explain the meaning of each term in Equation 9-2.

9-19 Suppose that liquid nickel is undercooled until homogeneous nucleation occurs. Calculate

(a) the critical radius of the nucleus required and

(b) the number of nickel atoms in the nucleus.

Assume that the lattice parameter of the solid FCC nickel is 0.356 nm.

9-20 Suppose that liquid iron is undercooled until homogeneous nucleation occurs. Calculate

(a) the critical radius of the nucleus required and

(b) the number of iron atoms in the nucleus.

Assume that the lattice parameter of the solid BCC iron is 2.92 Å.

9-21 Suppose that solid nickel was able to nucleate homogeneously with an undercooling of only 22°C. How many atoms would have to group together spontaneously for this occur? Assume that the lattice parameter of the solid FCC nickel is 0.356 nm.

9-22 Suppose that solid iron was able to nucleate homogeneously with an undercooling of only 15°C. How many atoms would have to group together spontaneously for this to occur? Assume that the lattice parameter of the solid BCC iron is 2.92 Å.

Section 9-3 Applications of Controlled Nucleation

9-23 Explain the term inoculation.

9-24 Explain how aluminum alloys can be strengthened using small amounts of titanium and boron.

9-25 Compare and contrast grain-size strengthening and strain hardening mechanisms.

9-26 What is second-phase strengthening?

9-27 Why is it that many inorganic melts solidify into amorphous materials more easily than metallic materials?

9-28 What is a glass-ceramic? How are glass-ceramics made?

9-29 What is photochromic glass?

9-30 What is a metallic glass?

9-31 How is heterogeneous nucleation a factor in making snow at ski resorts?

Section 9-4 Growth Mechanisms

9-32 What are the two steps encountered in the solidification of molten metals? As a function of time, can they overlap with one another?

9-33 During solidification, the specific heat of the material and the latent heat of fusion need to be removed. Define each of these terms.

9-34 Describe under what conditions we expect molten metals to undergo dendritic solidification.

9-35 What is a dendrite and why do dendrites form during solidification?

9-36 Use the data in Table 9-1 and the specific heat data given below to calculate the undercooling required to keep the dendritic fraction at 0.5 for each metal.

Metal	Specific Heat [J/(cm^3 · K)]
Bi	1.27
Pb	1.47
Cu	3.48
Ni	4.75

9-37 Calculate the fraction of solidification that occurs dendritically when silver nucleates
(a) at 10°C undercooling;
(b) at 100°C undercooling; and
(c) homogeneously.
The specific heat of silver is 3.25 J/(cm^3 · °C).

9-38 Calculate the fraction of solidification that occurs dendritically when iron nucleates
(a) at 10°C undercooling;
(b) at 100°C undercooling; and
(c) homogeneously.
The specific heat of iron is 5.78 J/(cm^3 · °C).

9-39 Analysis of a nickel casting suggests that 28% of the solidification process occurred in a dendritic manner. Calculate the temperature at which nucleation occurred. The specific heat of nickel is 4.1 J/(cm^3 · °C).

9-40 It is desired to increase the dendritic fraction from 0.012 to 0.025. Assuming that the specific heat of the metal being solidified is constant, determine the increase in the amount of undercooling required.

Section 9-5 Solidification Time and Dendrite Size

9-41 Write down Chvorinov's rule and explain the meaning of each term.

9-42 Find the mold constant B and exponent n in Chvorinov's rule using the following data and a log–log plot:

Shape	Dimensions (cm)	Solidification Time (s)
Cylinder	Radius = 10, Length = 30	5000
Sphere	Radius = 9	1800
Cube	Length = 6	200
Plate	Length = 30, Width = 20, Height = 1	40

9-43 A 2-in. cube solidifies in 4.6 min. Assume $n = 2$. Calculate
(a) the mold constant in Chvorinov's rule; and
(b) the solidification time for a 0.5 in. × 0.5 in. × 6 in. bar cast under the same conditions.

9-44 A 5-cm-diameter sphere solidifies in 1050 s. Calculate the solidification time for a 0.3 cm × 10 cm × 20 cm plate cast under the same conditions. Assume that $n = 2$.

9-45 Find the constants B and n in Chvorinov's rule by plotting the following data on a log–log plot:

Casting Dimensions (in.)	Solidification Time (min)
0.5 × 8 × 12	3.48
2 × 3 × 10	15.78
2.5 cube	10.17
1 × 4 × 9	8.13

9-46 Find the constants B and n in Chvorinov's rule by plotting the following data on a log–log plot:

Casting Dimensions (cm)	Solidification Time (s)
1 × 1 × 6	28.58
2 × 4 × 4	98.30
4 × 4 × 4	155.89
8 × 6 × 5	306.15

9-47 A 3-in.-diameter casting was produced. The times required for the solid-liquid interface to reach different distances beneath the casting surface were measured and are shown in the following table:

Distance from Surface (in.)	Time (s)
0.1	32.6
0.3	73.5
0.5	130.6
0.75	225.0
1.0	334.9

Determine
(a) the time at which solidification begins at the surface and
(b) the time at which the entire casting is expected to be solid.
(c) Suppose the center of the casting actually solidified in 720 s. Explain why this time might differ from the time calculated in part (b).

9-48 What solidification time is required for a casting of 9.5 cm^3 and a mold contact area of 10.0 cm^2? The mold constant is 838 s/cm^2.

9-49 An aluminum alloy plate with dimensions 20 cm × 10 cm × 2 cm needs to be cast with a secondary dendrite arm spacing of 10^{-2} cm (refer to Figure 9-6). What mold constant B is required (assume $n = 2$)?

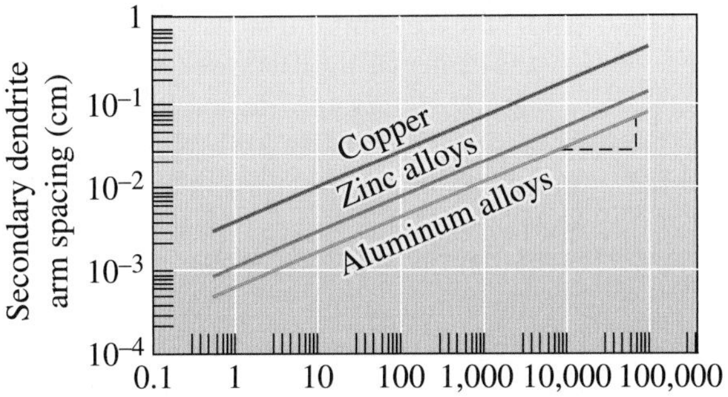

Figure 9-6 The effect of solidification time on the secondary dendrite arm spacings of copper, zinc, and aluminum. (Repeated for Problem 9-49.)

9-50 Figure 9-5(b) shows a micrograph of an aluminum alloy. Estimate
(a) the secondary dendrite arm spacing and
(b) the local solidification time for that area of the casting.

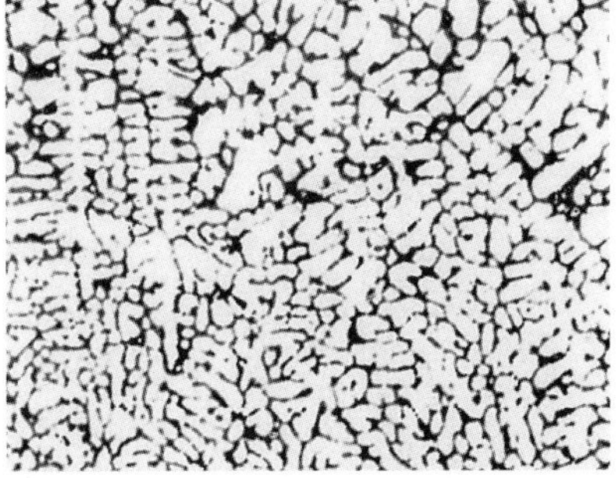

Figure 9-5 (b) Dendrites in an aluminum alloy (× 50). (*From* ASM Handbook, *Vol. 9, Metallography and Microstructure (1985), ASM International, Materials Park, OH 44073-0002. Reprinted with permission of ASM International. All rights reserved. www.asminternational.org.*)

9-51 Find the constants k and m relating the secondary dendrite arm spacing to the local solidification time by plotting the following data on a log–log plot:

Solidification Time (s)	SDAS (cm)
156	0.0176
282	0.0216
606	0.0282
1356	0.0374

9-52 Figure 9-25 shows dendrites in a titanium powder particle that has been rapidly solidified. Assuming that the size of the titanium dendrites is related to solidification time by the same relationship as in aluminum, estimate the solidification time of the powder particle.

9-53 The secondary dendrite arm spacing in an electron-beam weld of copper is 9.5×10^{-4} cm. Estimate the solidification time of the weld.

9-54 A zinc alloy has a SDAS of 0.01 cm and $m = 0.40$. What is the solidification time?

9-55 In Figure 9-7, what is the equation (slope intercept form) of the tensile strength

Figure 9-25 Dendrites in a titanium powder particle produced by rapid solidification processing (× 2200) (for Problem 9-52). (*Springer and The Metallurgical of Society of AIME,* Metallurgical Transactions, *Vol. 15A, June 1984, pp 1117-1127, "Formation of Metal Carbide Powder by Spark Machining of Reactive Metals," J.D. Ayers and K. Moore, figure number(s), Copyright © 1984, The Metallurgical of Society of AIME. With kind permission from Springer Science and Business Media.*)

line, and what SDAS would appear to give a tensile strength of zero if the line is extrapolated? Does extrapolating the trend like this make sense?

Section 9-6 Cooling Curves

9-56 Sketch a cooling curve for a poorly inoculated, pure metal and label the different regions carefully.

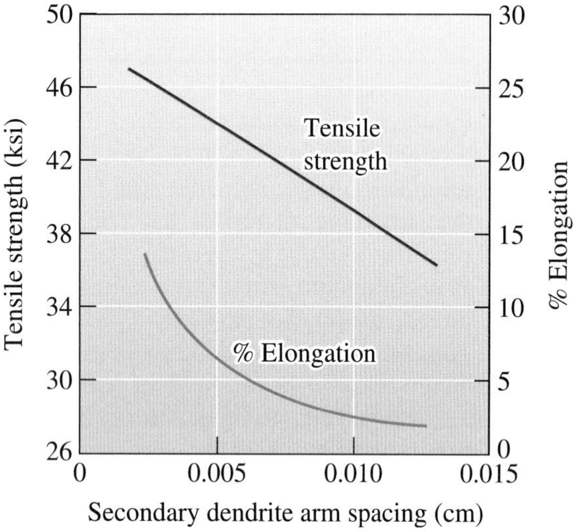

Figure 9-7 The effect of the secondary dendrite arm spacing on the properties of an aluminum casting alloy. (Repeated for Problem 9-55.)

9-57 What is meant by the term recalescence?

9-58 What is thermal arrest?

9-59 What is meant by the terms "local" and "total" solidification times?

9-60 In Figure 9-8, what do the slopes of the lines *A*-*B* and *E*-onward represent? Do these quantities represent physical properties?

9-61 A cooling curve is shown in Figure 9-26. Determine

(a) the pouring temperature;
(b) the solidification temperature;
(c) the superheat;
(d) the cooling rate, just before solidification begins;
(e) the total solidification time;
(f) the local solidification time; and
(g) the probable identity of the metal.
(h) If the cooling curve was obtained at the center of the casting sketched in the figure, determine the mold constant, assuming that $n = 2$.

9-62 A cooling curve is shown in Figure 9-27. Determine

(a) the pouring temperature;
(b) the solidification temperature;
(c) the superheat;
(d) the cooling rate, just before solidification begins;

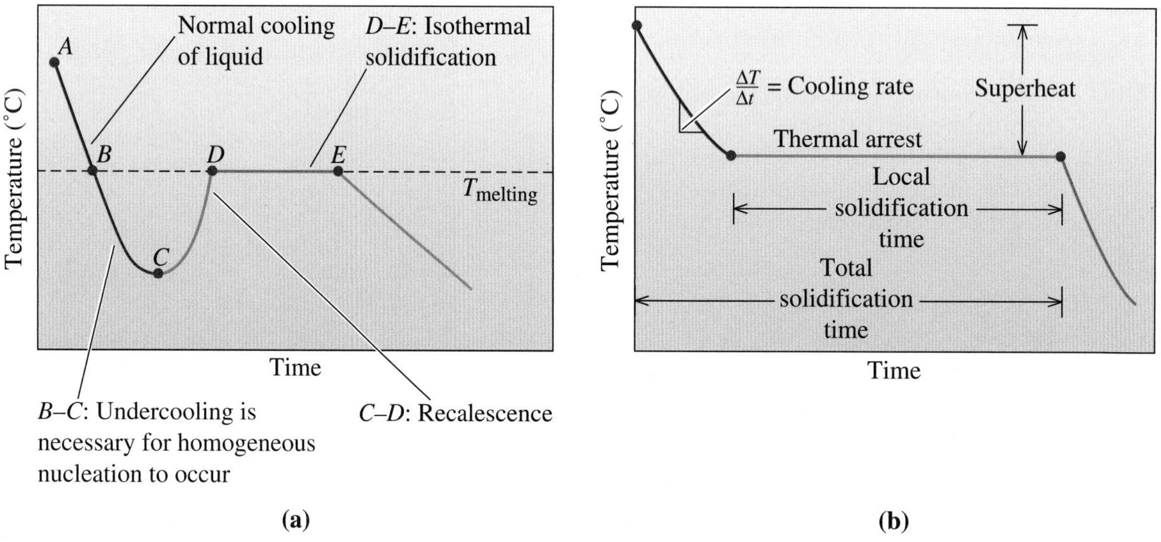

Figure 9-8 (a) Cooling curve for a pure metal that has not been well-inoculated. The liquid cools as specific heat is removed (between points *A* and *B*). Undercooling is necessary (between points *B* and *C*). As nucleation begins (point *C*), latent heat of fusion is released causing an increase in the temperature of the liquid. This process is known as recalescence (point *C* to point *D*). The metal continues to solidify at a constant temperature ($T_{melting}$). At point *E*, solidification is complete. The solid casting continues to cool from this point. (b) Cooling curve for a well-inoculated, but otherwise pure, metal. No undercooling is needed. Recalescence is not observed. Solidification begins at the melting temperature. (Repeated for Problem 9-60.)

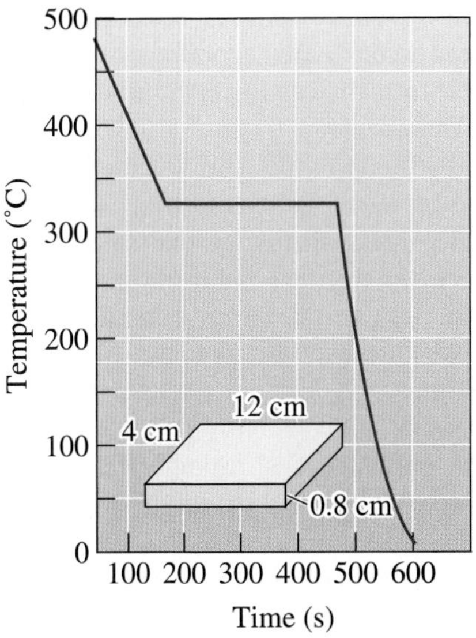

Figure 9-26 Cooling curve (for Problem 9-61).

(e) the total solidification time;
(f) the local solidification time;
(g) the undercooling; and
(h) the probable identity of the metal.
(i) If the cooling curve was obtained at the center of the casting sketched in the figure, determine the mold constant, assuming that $n = 2$.

9-63 Figure 9-28 shows the cooling curves obtained from several locations within a cylindrical aluminum casting. Determine the local solidification times and the SDAS at each location, then plot the tensile strength versus distance from the casting surface. Would you recommend that the casting be designed so that a large or small amount of material must be machined from the surface during finishing? Explain.

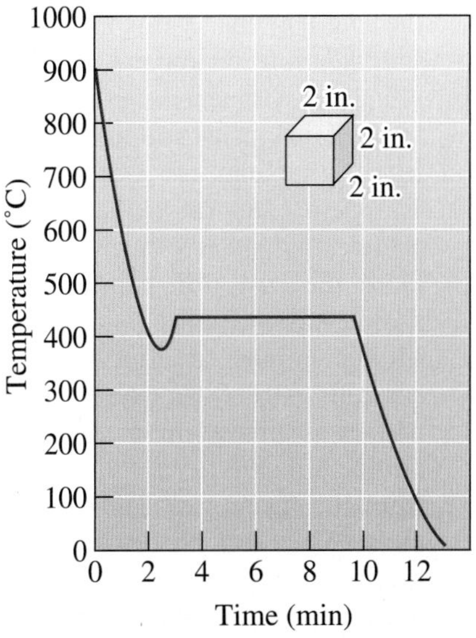

Figure 9-27 Cooling curve (for Problem 9-62).

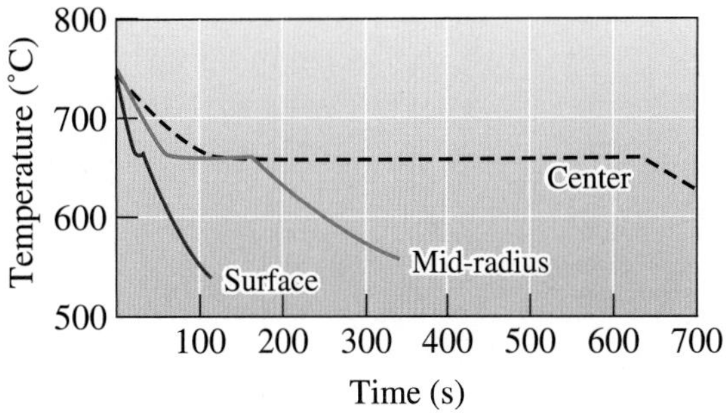

Figure 9-28 Cooling curves (for Problem 9-63).

Section 9-7 Cast Structure

9-64 What are the features expected in the macrostructure of a cast component? Explain using a sketch.

9-65 In cast materials, why does solidification almost always begin at the mold walls?

9-66 Why is it that forged components do not show a cast ingot structure?

Section 9-8 Solidification Defects

9-67 What type of defect in a casting can cause catastrophic failure of cast components such as turbine blades? What precautions are taken to prevent porosity in castings?

9-68 In general, compared to components prepared using forging, rolling, extrusion, etc., cast products tend to have lower fracture toughness. Explain why this may be the case.

9-69 What is a riser? Why should it be designed so as to freeze after the casting?

9-70 Calculate the volume, diameter, and height of the cylindrical riser required to prevent shrinkage in a 1 in. × 6 in. × 6 in. casting if the H/D of the riser is 1.0.

9-71 Calculate the volume, diameter, and height of the cylindrical riser required to prevent shrinkage in a 4 in. × 10 in. × 20 in. casting if the H/D of the riser is 1.5.

9-72 Figure 9-29 shows a cylindrical riser attached to a casting. Compare the solidification times for each casting section and the riser and determine whether the riser will be effective.

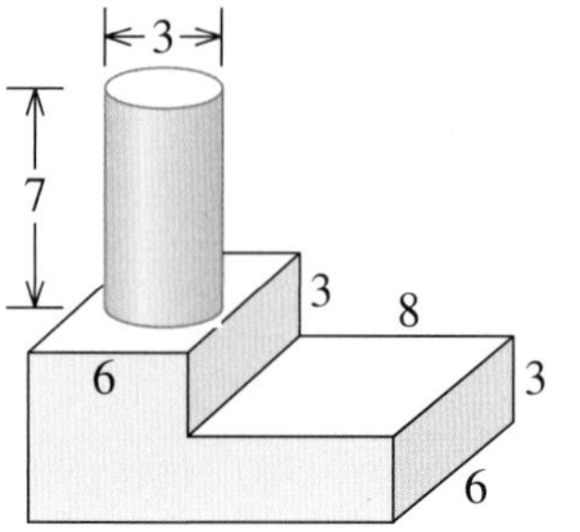

Figure 9-29 Step-block casting (for Problem 9-72).

9-73 Figure 9-30 shows a cylindrical riser attached to a casting. Compare the solidification times for each casting section and the riser and determine whether the riser will be effective.

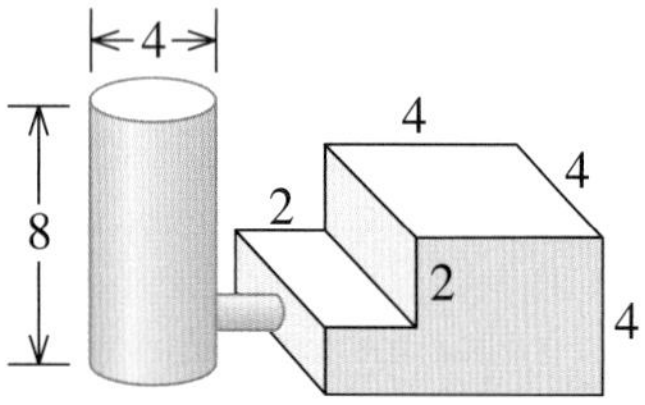

Figure 9-30 Step-block casting (for Problem 9-73).

9-74 A hollow cylindrical mold for casting aluminum ingots has a 300 mm inside diameter and is 2 m high. If the mold is filled with liquid aluminum at 935 K, what is the largest spherical cavity that may form in the ingot?

9-75 A 4-in.-diameter sphere of liquid copper is allowed to solidify, producing a spherical shrinkage cavity at the center of the casting. Compare the volume and diameter of the shrinkage cavity in the copper casting to that obtained when a 4-in. sphere of liquid iron is allowed to solidify.

9-76 A 4-in. cube of a liquid metal is allowed to solidify. A spherical shrinkage cavity with a diameter of 1.49 in. is observed in the solid casting. Determine the percent volume change that must have occurred during solidification.

9-77 A 2 cm $\times$ 4 cm $\times$ 6 cm magnesium casting is produced. After cooling to room temperature, the casting is found to weigh 80 g. Determine
(a) the volume of the shrinkage cavity at the center of the casting and
(b) the percent shrinkage that must have occurred during solidification.

9-78 A 2 in. $\times$ 8 in. $\times$ 10 in. iron casting is produced and, after cooling to room temperature, is found to weigh 43.9 lb. Determine
(a) the percent of shrinkage that must have occurred during solidification and
(b) the number of shrinkage pores in the casting if all of the shrinkage occurs as pores with a diameter of 0.05 in.

9-79 If you cool an open vessel of liquid gallium until a thin solid "cake" forms, will the cake form on the top or the bottom? Assume that the air above the liquid gallium is stagnant and does not cool the surface at all.

9-80 Give examples of materials that expand upon solidification.

9-81 What is Sievert's law? How can gas porosity in molten alloys be removed or minimized?

9-82 In the context of stainless steel making, what is argon oxygen decarburization?

9-83 Liquid magnesium is poured into a 2 cm $\times$ 2 cm $\times$ 24 cm mold and, as a result of directional solidification, all of the solidification shrinkage occurs along the 24-cm length of the casting. Determine the length of the casting immediately after solidification is complete.

9-84 A liquid cast iron has a density of 7.65 g/cm^3. Immediately after solidification, the density of the solid cast iron is found to be 7.71 g/cm^3. Determine the percent volume change that occurs during solidification. Does the cast iron expand or contract during solidification?

9-85 Molten copper at atmospheric pressure contains 0.01 wt% oxygen. The molten copper is placed in a chamber that is pumped down to 1 Pa to remove gas from the melt prior to pouring into the mold. Calculate the oxygen content of the copper melt after it is subjected to this degassing treatment.

9-86 From Figure 9-14, find the solubility of hydrogen in liquid aluminum just

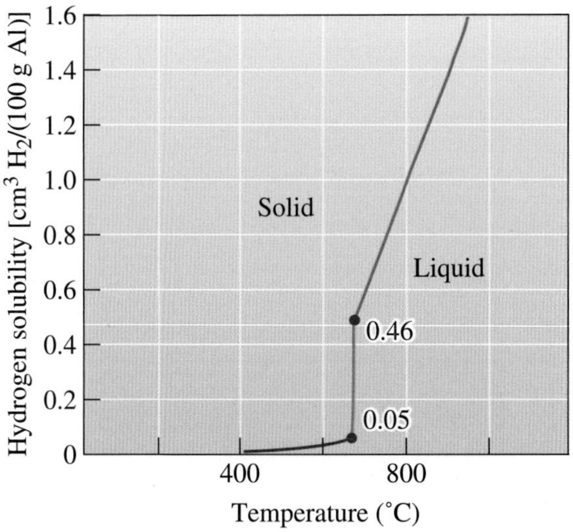

Figure 9-14 The solubility of hydrogen gas in aluminum when the partial pressure of H_2 = 1 atm. (Repeated for Problem 9-86.)

before solidification begins when the partial pressure of hydrogen is 1 atm. Determine the solubility of hydrogen (in cm^3/100 g Al) at the same temperature if the partial pressure were reduced to 0.01 atm.

9-87 The solubility of hydrogen in liquid aluminum at 715°C is found to be 1 cm^3/(100 g Al). If all of this hydrogen precipitated as gas bubbles during solidification and remained trapped in the casting, calculate the volume percent gas in the solid aluminum.

9-88 Review Example 9-7. The usual method of measuring a vacuum in the United States vacuum pump industry is in inches of mercury. For example, atmospheric pressure is 29.92 inches of mercury. Convert the pressure found in the example to in. Hg.

Section 9-9 Casting Processes for Manufacturing Components

9-89 Explain the green sand molding process.

9-90 Why is it that castings made from pressure die casting are likely to be stronger than those made using the sand casting process?

9-91 An alloy is cast into a shape using a sand mold and a metallic mold. Which casting is expected to be stronger and why?

9-92 What is investment casting? What are the advantages of investment casting? Explain why this process is often used to cast turbine blades.

9-93 Why is pressure a key ingredient in the pressure die casting process?

Section 9-10 Continuous Casting and Ingot Casting

9-94 What is an ore?

9-95 Briefly explain the steps in ingot casting and the continuous casting of steels.

9-96 Explain how metal scrap is used for making alloys.

9-97 What is an ingot?

9-98 Why has continuous casting of steels and other alloys assumed increased importance?

9-99 What are some of the secondary steps that follow the continuous casting process?

Section 9-11 Directional Solidification (DS), Single-Crystal Growth, and Epitaxial Growth

9-100 Define the term directional solidification.

9-101 Explain the role of nucleation and growth in growing single crystals.

Section 9-12 Solidification of Polymers and Inorganic Glasses

9-102 Why do most plastics contain amorphous and crystalline regions?

9-103 What is a spherulite?

9-104 How can processing influence crystallinity of polymers?

9-105 Explain why silicate glasses do not crystallize upon slow cooling whereas metals crystallize readily.

Section 9-13 Joining of Metallic Materials

9-106 Define the terms brazing and soldering.

9-107 What is the difference between fusion welding and brazing and soldering?

9-108 What is a heat-affected zone?

9-109 Explain why, while using low intensity heat sources, the strength of the material in a weld region can be reduced.

9-110 Why do laser and electron-beam welding processes lead to stronger welds?

Design Problems

9-111 Aluminum is melted under conditions that give 0.06 cm^3 H_2 per 100 g of aluminum. We have found that we must have no more than 0.002 cm^3 H_2 per 100 g of aluminum in order to prevent the formation of hydrogen gas bubbles during solidification. Design a treatment process for the liquid aluminum that will ensure that hydrogen porosity does not form.

9-112 When two 0.5-in.-thick copper plates are joined using an arc-welding process, the fusion zone contains dendrites having a SDAS of 0.006 cm; however, this process produces large residual stresses in the weld. We have found that residual stresses are low when the welding conditions produce a SDAS of more than 0.02 cm. Design a process by which we can accomplish low residual stresses. Justify your design.

9-113 Design an efficient riser system for the casting shown in Figure 9-31. Be sure to include a sketch of the system, along with appropriate dimensions.

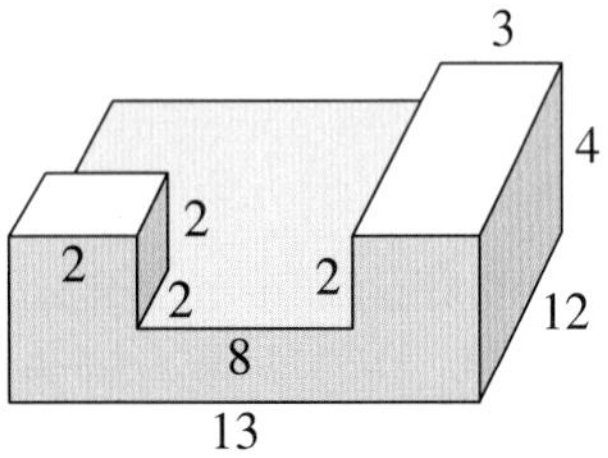

Figure 9-31 Casting to be risered (for Problem 9-113).

9-114 Design a process that will produce a steel casting having uniform properties and high strength. Be sure to include the microstructure features you wish to control and explain how you would do so.

9-115 Molten aluminum is to be injected into a steel mold under pressure (die casting). The casting is essentially a 12-in.-long, 2-in.-diameter cylinder with a uniform wall thickness, and it must have a minimum tensile strength of 40,000 psi. Based on the properties given in Figure 9-7, design the casting and process.

Computer Problems

9-116 *Critical Radius for Homogeneous Nucleation.* Write a computer program that will allow calculation of the critical radius for nucleation (r^*). The program should ask the user to provide inputs for values for σ_{sl}, T_m, the undercooling (ΔT), and the

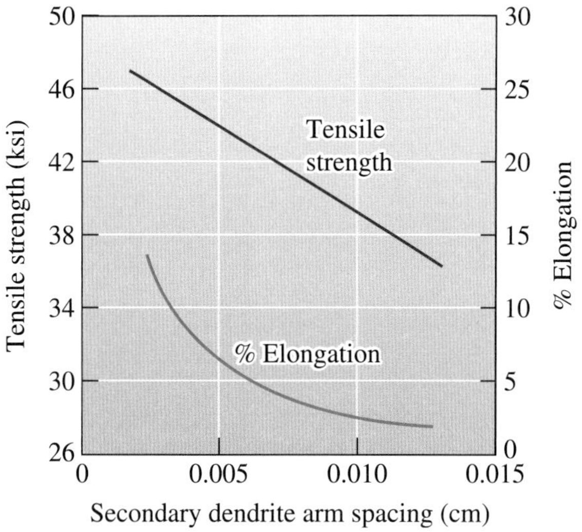

Figure 9-7 The effect of the secondary dendrite arm spacing on the properties of an aluminum casting alloy. (Repeated for Problem 9-115.)

enthalpy of fusion ΔH_f. Be sure to have the correct prompts in the program to have the values entered in the correct units.

9-117 *Free Energy for Formation of Nucleus of Critical Size via Heterogeneous Nucleation.* When nucleation occurs heterogeneously, the free energy for a nucleus of a critical size ($\Delta G^{*\text{hetero}}$) is given by

$$\Delta G^{*\text{hetero}} = \Delta G^{*\text{homo}} f(\theta), \text{ where}$$

$$f(\theta) = \frac{(2 + \cos\theta)(1 - \cos\theta)^2}{4}$$

and $\Delta G^{*\text{homo}}$ is given by $\dfrac{16\pi\sigma_{sl}^3}{3\Delta G_v^2}$,

which is the free energy for homogeneous nucleation of a nucleus of a critical size. If the contact angle (θ) of the phase that is nucleating on the pre-existing surface is 180°, there is no wetting, and the value of the function $f(\theta)$ is 1. The free energy of forming a nucleus of a critical radius is the same as that for homogeneous nucleation. If the nucleating phase wets the solid completely (i.e., $\theta = 0$), then $f(\theta) = 0$, and there is no barrier for nucleation. Write a computer program that will ask the user to provide the values of parameters needed to calculate the free energy for formation of a nucleus via homogeneous nucleation. The program should then calculate the value of $\Delta G^{*\text{hetero}}$ as a function of the contact angle (θ) ranging from 0 to 180°. Examine the variation of the free energy values as a function of contact angle.

9-118 *Chvorinov's Rule.* Write a computer program that will calculate the time of solidification for a casting. The program should ask the user to enter the volume of the casting, the surface area from which heat transfer will occur, and the mold constant. The program should then use Chvorinov's rule to calculate the time of solidification.

Knovel® Problems

K9-1 What is chilled white iron and what is it used for?

K9-2 What kinds of defects may exist in chilled white iron?

Josiah Willard Gibbs (1839–1903) was a brilliant American physicist and mathematician who conducted some of the most important pioneering work related to thermodynamic equilibrium. (*Frontispiece of* The Scientific Papers of J. Willard Gibbs, *in two volumes, eds. H. A. Bumstead and R. G. Van Name, (London and New York: Longmans, Green, and Co., 1906)*)

CHAPTER

10

Solid Solutions and Phase Equilibrium

Have You Ever Wondered?

- *Is it possible for the solid, liquid, and gaseous forms of a material to coexist?*
- *What material is commonly used to make red light-emitting diodes (LEDs)?*
- *When an alloy such as brass solidifies, which element solidifies first—copper or zinc?*

Chapter Learning Objectives

The key objectives of this chapter are to

- State and apply the Gibbs phase rule.
- State the conditions under which solid solutions can form.
- Explain the effects of solid-solution formation on the mechanical properties of metallic materials.
- Draw an isomorphous phase diagram indicating the melting temperature of each pure element and the phases present in each region.
- Calculate the amounts and composition of each phase present for any point on an isomorphous phase diagram.

We have seen that the strength of metallic materials can be enhanced using

(a) grain-size strengthening (Hall-Petch equation);
(b) cold working or strain hardening;
(c) formation of small particles of second phases; and
(d) additions of small amounts of elements.

When small amounts of elements are added, a solid material known as a solid solution may form. A **solid solution** contains two or more types of atoms or ions that are dispersed uniformly throughout the material. The impurity or **solute** atoms may occupy regular lattice sites in the crystal or interstitial sites. By controlling the amount of these point defects via the composition, the mechanical and other properties of solid solutions can be manipulated. For example, in metallic materials, the point defects created by the impurity or solute atoms disturb the atomic arrangement in the crystalline material and interfere with the movement of dislocations. The point defects cause the material to be solid-solution strengthened.

The introduction of alloying elements or impurities during processing changes the composition of the material and influences its solidification behavior. In this chapter, we will examine this effect by introducing the concept of an equilibrium phase diagram. For now, we consider a "phase" as a unique form in which a material exists. We will define the term "phase" more precisely later in this chapter. A phase diagram depicts the stability of different phases for a set of elements (e.g., Al and Si). From the phase diagram, we can predict how a material will solidify under equilibrium conditions. We can also predict what phases will be expected to be thermodynamically stable and in what concentrations such phases should be present.

10-1 Phases and the Phase Diagram

Pure metallic elements have engineering applications; for example, ultra-high purity copper or aluminum is used to make microelectronic circuitry. In most applications, however, we use **alloys**. We define an "alloy" as a material that exhibits properties of a metallic material and is made from multiple elements. A *plain carbon steel* is an alloy of iron and carbon. Corrosion-resistant **stainless steels** are alloys that usually contain iron, carbon, chromium, nickel, and some other elements. There are two types of alloys: **single-phase alloys** and **multiple-phase alloys**. In this chapter, we will examine the behavior of single-phase alloys. As a first step, let's define a "phase" and determine how the **phase rule** helps us to determine the state—solid, liquid, or gas—in which a pure material exists.

A **phase** can be defined as any portion, including the whole, of a system that is physically homogeneous within itself and bounded by a surface that separates it from any other portions. For example, water has three phases—liquid water, solid ice, and steam. A phase has the following characteristics:

1. the same structure or atomic arrangement throughout;
2. roughly the same composition and properties throughout; and
3. a definite interface between the phase and any surrounding or adjoining phases.

For example, if we enclose a block of ice in a vacuum chamber [Figure 10-1(a)], the ice begins to melt, and some of the water vaporizes. Under these conditions, we have three phases coexisting: solid H_2O, liquid H_2O, and gaseous H_2O. Each of these forms of H_2O is a distinct phase; each has a unique atomic arrangement, unique properties, and a definite boundary between each form. In this case, the phases have identical compositions.

Phase Rule

Josiah Willard Gibbs (1839–1903) was a brilliant American physicist and mathematician who conducted some of the most important pioneering work related to thermodynamic equilibrium.

Gibbs developed the **phase rule** in 1875–1876. It describes the relationship between the number of components and the number of phases for a given system and the conditions that may be allowed to change (e.g., temperature, pressure, etc.). It has the general form:

$$2 + C = F + P \text{ (when temperature and pressure both can vary)} \qquad (10\text{-}1)$$

A useful mnemonic for the Gibbs phase rule is to start with a numeric and follow with the rest of the terms alphabetically (i.e., C, F, and P) using all positive signs. In the

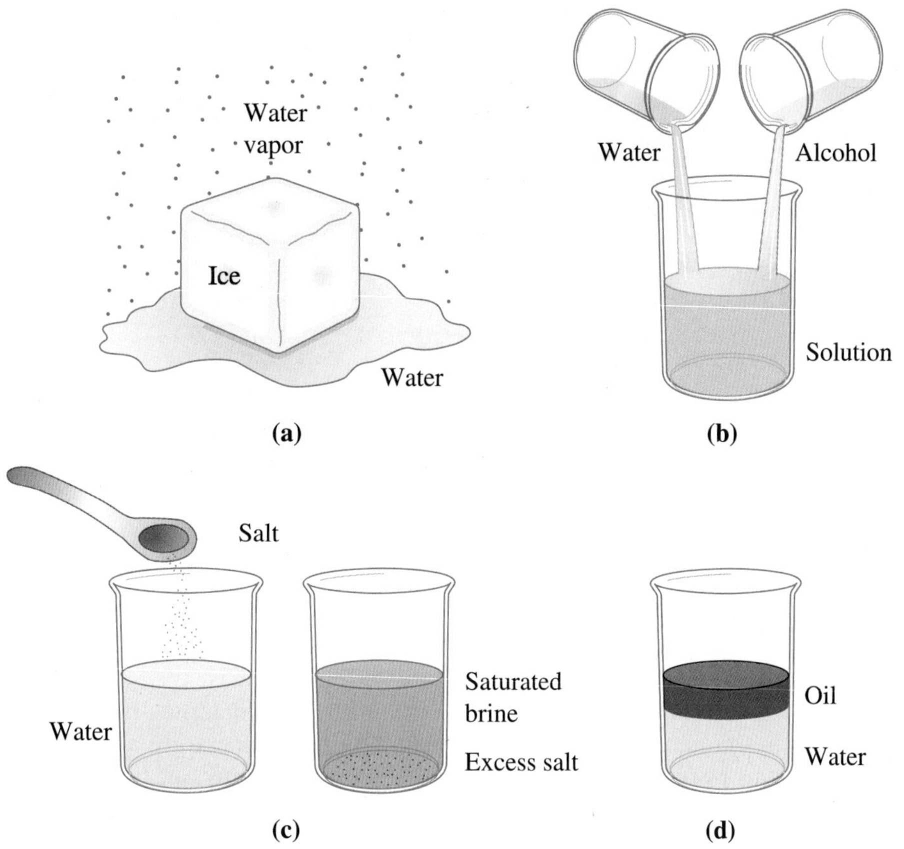

Figure 10-1 Illustration of phases and solubility: (a) The three forms of water—gas, liquid, and solid—are each a phase. (b) Water and alcohol have unlimited solubility. (c) Salt and water have limited solubility. (d) Oil and water have virtually no solubility.

phase rule, C is the number of chemically independent components, usually elements or compounds, in the system; F is the number of degrees of freedom, or the number of variables (such as temperature, pressure, or composition), that are allowed to change independently without changing the number of phases in equilibrium; and P is the number of phases present (please do not confuse P with "pressure"). The constant "2" in Equation 10-1 implies that both the temperature and pressure are allowed to change. The term "chemically independent" refers to the number of different elements or compounds needed to specify a system. For example, water (H_2O) is considered as a one component system, since the concentrations of H and O in H_2O cannot be independently varied.

It is important to note that the Gibbs phase rule assumes thermodynamic equilibrium and, more often than not in materials processing, we encounter conditions in which equilibrium is *not* maintained. Therefore, you should not be surprised to see that the number and compositions of phases seen in practice are dramatically different from those predicted by the Gibbs phase rule.

Another point to note is that phases do not always have to be solid, liquid, and gaseous forms of a material. An element, such as iron, can exist in FCC and BCC crystal structures. These two solid forms of iron are two different phases of iron that will be stable at different temperatures and pressure conditions. Similarly, ice, itself, can exist in several crystal structures. Carbon can exist in many forms, e.g., graphite or diamond. These are only two of the many possible phases of carbon as we saw in Chapter 2.

As an example of the use of the phase rule, let's consider the case of pure magnesium. Figure 10-2 shows a **unary** ($C = 1$) **phase diagram** in which the lines divide

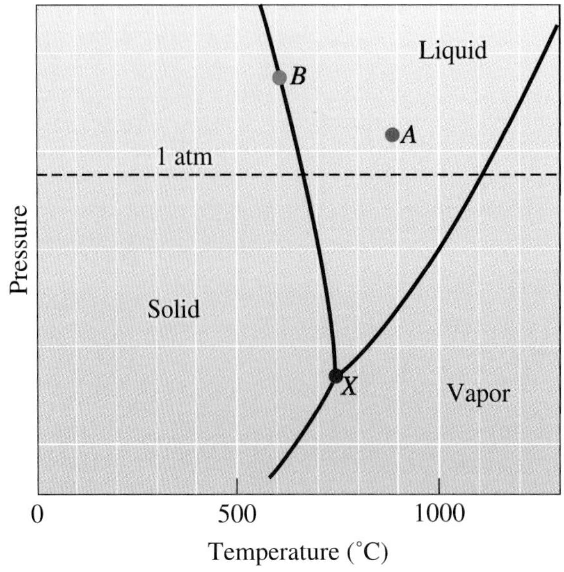

Figure 10-2 Schematic unary phase diagram for magnesium, showing the melting and boiling temperatures at one atmosphere pressure. On this diagram, point X is the triple point.

the liquid, solid, and vapor phases. This unary phase diagram is also called a pressure-temperature or **P-T diagram**. In the unary phase diagram, there is only one component; in this case, magnesium. Depending on the temperature and pressure, however, there may be one, two, or even three *phases* present at any one time: solid magnesium, liquid magnesium, and magnesium vapor. Note that at atmospheric pressure (one atmosphere, given by the dashed line), the intersection of the lines in the phase diagram give the usual melting and boiling temperatures for magnesium. At very low pressures, a solid such as magnesium can *sublime*, or go directly to a vapor form without melting, when it is heated.

Suppose we have a pressure and temperature that put us at point A in the phase diagram (Figure 10-2). At this point, magnesium is all liquid. The number of phases is one (liquid). The phase rule tells us that there are two degrees of freedom. From Equation 10-1:

$$2 + C = F + P, \quad \text{therefore, } 2 + 1 = F + 1 \text{ (i.e., } F = 2)$$

What does this mean? Within limits, as seen in Figure 10-2, we can change the pressure, the temperature, or both, and still be in an all-liquid portion of the diagram. Put another way, we must fix both the temperature and the pressure to know precisely where we are in the liquid portion of the diagram.

Consider point B, the boundary between the solid and liquid portions of the diagram. The number of components, C, is still one, but at point B, the solid and liquid coexist, or the number of phases P is two. From the phase rule Equation 10-1,

$$2 + C = F + P, \quad \text{therefore, } 2 + 1 = F + 2 \text{ (i.e., } F = 1)$$

or there is only one degree of freedom. For example, if we change the temperature, the pressure must also be adjusted if we are to stay on the boundary where the liquid and solid coexist. On the other hand, if we fix the pressure, the phase diagram tells us the temperature that we must have if solid and liquid are to coexist.

Finally, at point X, solid, liquid, and vapor coexist. While the number of components is still one, there are three phases. The number of degrees of freedom is zero:

$$2 + C = F + P, \quad \text{therefore, } 2 + 1 = F + 3 \text{ (i.e., } F = 0)$$

Now we have no degrees of freedom; all three phases coexist only if both the temperature and the pressure are fixed. A point on the phase diagram at which the solid, liquid, and gaseous phases coexist under equilibrium conditions is the **triple point** (Figure 10-2).

Example 10-1 ***Design of an Aerospace Component***

Because magnesium is a low-density material ($\rho_{Mg} = 1.738\ g/cm^3$), it has been suggested for use in an aerospace vehicle intended to enter outer space. Is this a good design?

SOLUTION

The pressure is very low in space. Even at relatively low temperatures, solid magnesium can begin to transform to a vapor, causing metal loss that could damage a space vehicle. In addition, solar radiation could cause the vehicle to heat, increasing the rate of magnesium loss.

A low-density material with a higher boiling point (and, therefore, lower vapor pressure at any given temperature) might be a better choice. At atmospheric pressure, aluminum boils at 2494°C and beryllium boils at 2770°C, compared with the boiling temperature of 1107°C for magnesium. Although aluminum and beryllium are somewhat denser than magnesium, either might be a better choice. Given the toxic effects of Be and many of its compounds when in powder form, we may want to consider aluminum first.

There are other factors to consider. In load-bearing applications, we should not only look for density but also for relative strength. Therefore, the ratio of Young's modulus to density or yield strength to density could be a better parameter to compare different materials. In this comparison, we will have to be aware that yield strength, for example, depends strongly on microstructure and that the strength of aluminum can be enhanced using aluminum alloys, while keeping the density about the same. Other factors such as oxidation during reentry into Earth's atmosphere may be applicable and will also have to be considered.

10-2 Solubility and Solid Solutions

Often, it is beneficial to know how much of each material or component we can combine without producing an additional phase. When we begin to combine different components or materials, as when we add alloying elements to a metal, solid or liquid solutions can form. For example, when we add sugar to water, we form a sugar solution. In other words, we are interested in the **solubility** of one material in another (e.g., sugar in water, copper in nickel, phosphorus in silicon, etc.).

Unlimited Solubility

Suppose we begin with a glass of water and a glass of alcohol. The water is one phase, and the alcohol is a second phase. If we pour the water into the alcohol and stir, only one phase is produced [Figure 10-1(b)]. The glass contains a solution of water and alcohol that has unique properties and composition. Water and alcohol are soluble in each other. Furthermore, they display **unlimited solubility**. Regardless of the ratio of water and alcohol, only one phase is produced when they are mixed together.

Similarly, if we were to mix any amounts of liquid copper and liquid nickel, only one liquid phase would be produced. This liquid alloy has the same composition and properties everywhere [Figure 10-3(a)] because nickel and copper have unlimited liquid solubility.

If the liquid copper-nickel alloy solidifies and cools to room temperature while maintaining thermal equilibrium, only one solid phase is produced. After solidification, the copper and nickel atoms do not separate but, instead, are randomly located within the FCC crystal structure. Within the solid phase, the structure, properties, and composition are uniform and no interface exists between the copper and nickel atoms. Therefore, copper and nickel also have unlimited solid solubility. The solid phase is a solid solution of copper and nickel [Figure 10-3(b)].

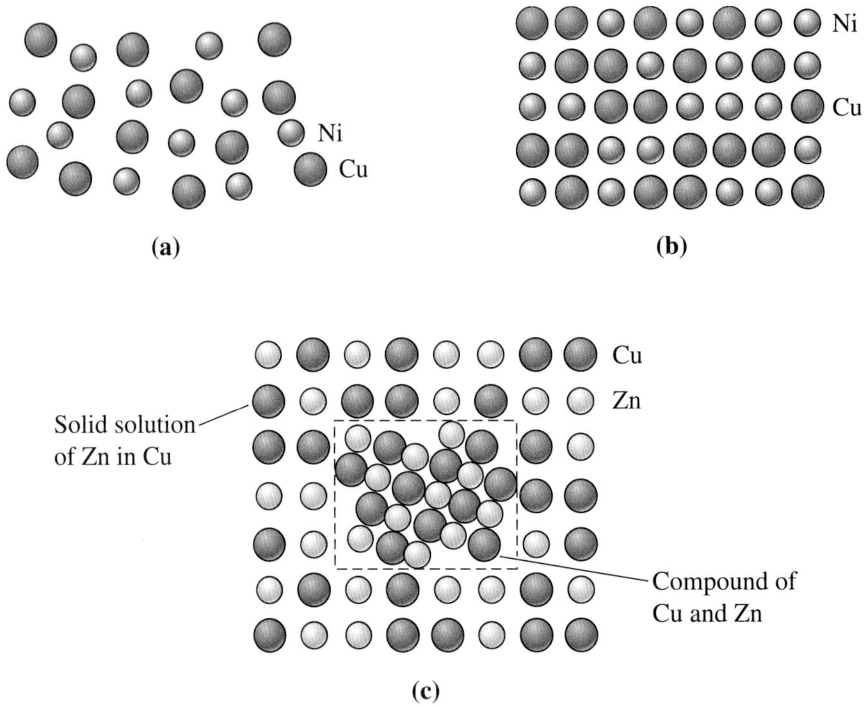

Figure 10-3 (a) Liquid copper and liquid nickel are completely soluble in each other. (b) Solid copper-nickel alloys display complete solid solubility with copper and nickel atoms occupying random lattice sites. (c) In copper-zinc alloys containing more than 30% Zn, a second phase forms because of the limited solubility of zinc in copper.

A solid solution is *not* a mixture. A mixture contains more than one type of phase, and the characteristics of each phase are retained when the mixture is formed. In contrast to this, the components of a solid solution completely dissolve in one another and do not retain their individual characteristics.

Another example of a system forming a solid solution is that of barium titanate ($BaTiO_3$) and strontium titanate ($SrTiO_3$), which are compounds found in the BaO–TiO_2–SrO ternary system. We use solid solutions of $BaTiO_3$ with $SrTiO_3$ and other oxides to make electronic components such as capacitors. Millions of multilayer capacitors are made each year using such materials (Chapter 19).

Many compound semiconductors that share the same crystal structure readily form solid solutions with 100% solubility. For example, we can form solid solutions of gallium arsenide (GaAs) and aluminum arsenide (AlAs). The most commonly used red LEDs in displays are made using solid solutions based on the GaAs-GaP system. Solid solutions can be formed using more than two compounds or elements.

Limited Solubility When we add a small quantity of salt (one phase) to a glass of water (a second phase) and stir, the salt dissolves completely in the water. Only one phase—salty water or brine—is found. If we add too much salt to the water, the excess salt sinks to the bottom of the glass [Figure 10-1(c)]. Now we have two phases—water that is saturated with salt plus excess solid salt. We find that salt has a **limited solubility** in water.

If we add a small amount of liquid zinc to liquid copper, a single liquid solution is produced. When that copper-zinc solution cools and solidifies, a single solid solution having an FCC structure results, with copper and zinc atoms randomly located at the normal lattice points. If the liquid solution contains more than about 30% Zn, some of the excess zinc atoms combine with some of the copper atoms to form a CuZn compound [Figure 10-3(c)]. Two solid phases now coexist: a solid solution of copper saturated with about 30% Zn plus a CuZn compound. The solubility of zinc in copper is limited. Figure 10-4 shows a portion of the Cu-Zn phase diagram illustrating the solubility of zinc in copper at low temperatures. The solubility increases with increasing temperature. This is similar to how we can dissolve more sugar or salt in water by increasing the temperature. Solid solutions may form either by substitutional or interstitial mechanisms. The guest atoms or ions may enter the host crystal structure at regular crystallographic positions or the interstices.

In Chapter 5, we examined how silicon can be doped with phosphorus, boron, or arsenic. All of these dopant elements exhibit limited solubility in Si (i.e., at small concentrations, they form a solid solution with Si). In fact, in order for these dopants to be effective at altering the conductivity of Si, they must form substitutional solid solutions with Si as we will learn in Chapter 19. Note that solid solutions are produced even if there is limited solubility. There does not have to be a solid solution at every possible composition for solid solutions to form within limited composition ranges (Chapter 11).

In the extreme case, there may be almost no solubility of one material in another. This is true for oil and water [Figure 10-1(d)] or for copper-lead (Cu-Pb) alloys. Note that even though materials do not dissolve into one another, they can be dispersed into one another. For example, oil-like phases and aqueous liquids can be mixed, often using surfactants (soap-like molecules), to form emulsions. Immiscibility, or lack of solubility, is seen in many molten and solid ceramic and metallic materials.

Polymeric Systems We can process polymeric materials to enhance their usefulness by employing a concept similar to the formation of solid solutions in metallic and ceramic systems. We can form materials that are known as **copolymers** that consist of different monomers. For example, acrylonitrile (A), butadiene (B), and styrene (S) monomers can be made to react to form a copolymer known as ABS. This resultant copolymer is similar to a solid solution in that it has the functionalities of the three monomers from which it is derived, blending their properties. Similar to the Cu-Ni or

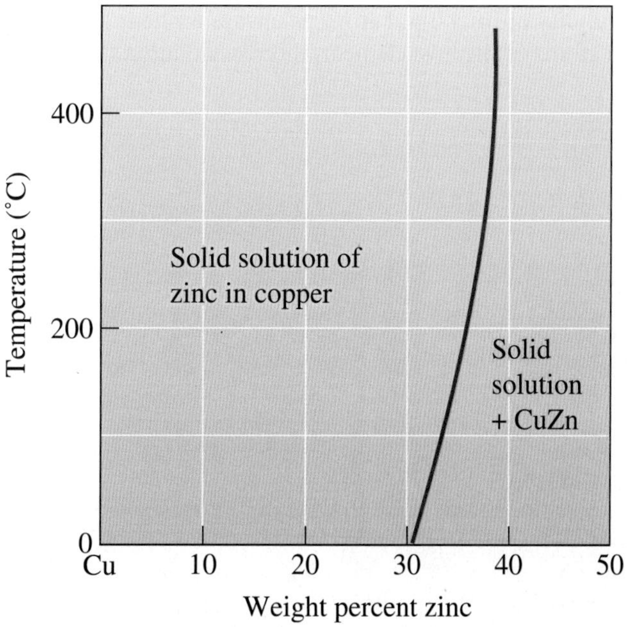

Figure 10-4 The solubility of zinc in copper. The solid line represents the solubility limit; when excess zinc is added, the solubility limit is exceeded and two phases coexist.

$BaTiO_3$-$SrTiO_3$ solid solutions, we will not be able to separate out the acrylonitrile, butadiene, or styrene from an ABS plastic. Injection molding is used to convert ABS into telephones, helmets, steering wheels, and small appliance cases. Dylark™ is another example of a copolymer. It is formed using maleic anhydride and a styrene monomer. The Dylark™ copolymer, with carbon black for UV protection, reinforced with fiberglass, and toughened with rubber, has been used for instrument panels in many automobiles (Chapter 16).

10-3 Conditions for Unlimited Solid Solubility

In order for an alloy system, such as copper-nickel to have unlimited solid solubility, certain conditions must be satisfied. These conditions, the **Hume-Rothery** rules, are as follows:

1. *Size factor*: The atoms or ions must be of similar size, with no more than a 15% difference in atomic radius, in order to minimize the lattice strain (i.e., to minimize, at an atomic level, the deviations caused in interatomic spacing).
2. *Crystal structure*: The materials must have the same crystal structure; otherwise, there is some point at which a transition occurs from one phase to a second phase with a different structure.
3. *Valence*: The ions must have the same valence; otherwise, the valence electron difference encourages the formation of compounds rather than solutions.
4. *Electronegativity*: The atoms must have approximately the same electronegativity. Electronegativity is the affinity for electrons (Chapter 2). If the electronegativities differ significantly, compounds form—as when sodium and chloride ions combine to form sodium chloride.

Hume-Rothery's conditions must be met, but they are not necessarily sufficient, for two metals or compounds to have unlimited solid solubility.

Figure 10-5 shows schematically the two-dimensional structures of MgO and NiO. The Mg^{2+} and Ni^{2+} ions are similar in size and valence and, consequently, can replace one another in a sodium chloride (NaCl) crystal structure (Chapter 3), forming a complete series of solid solutions of the form $(Mg^{2+}_x Ni^{2+}_{1-x})O$, where x = the mole fraction of Mg^{2+} or MgO.

The solubility of interstitial atoms is always limited. Interstitial atoms are much smaller than the atoms of the host element, thereby violating the first of Hume-Rothery's conditions.

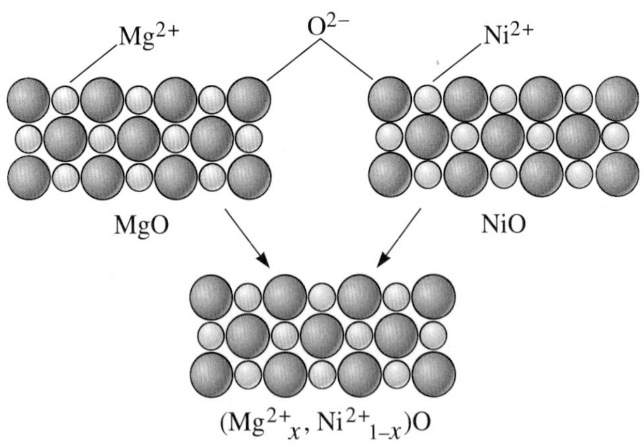

Figure 10-5 MgO and NiO have similar crystal structures, ionic radii, and valences; thus, the two ceramic materials can form solid solutions.

Example 10-2 *Ceramic Solid Solutions of MgO*

NiO can be added to MgO to produce a solid solution. What other ceramic systems are likely to exhibit 100% solid solubility with MgO?

SOLUTION

In this case, we must consider oxide additives that have metal cations with the same valence and ionic radius as the magnesium cations. The valence of the magnesium ion is +2, and its ionic radius is 0.66 Å. From Appendix B, some other possibilities in which the cation has a valence of +2 include the following:

	r (Å)	$\left[\frac{r_{ion} - r_{Mg^{2+}}}{r_{Mg^{2+}}}\right] \times 100\%$	Crystal Structure
Cd^{2+} in CdO	$r_{Cd^{2+}} = 0.97$	47	NaCl
Ca^{2+} in CaO	$r_{Ca^{2+}} = 0.99$	50	NaCl
Co^{2+} in CoO	$r_{Co^{2+}} = 0.72$	9	NaCl
Fe^{2+} in FeO	$r_{Fe^{2+}} = 0.74$	12	NaCl
Sr^{2+} in SrO	$r_{Sr^{2+}} = 1.12$	70	NaCl
Zn^{2+} in ZnO	$r_{Zn^{2+}} = 0.74$	12	NaCl

The percent difference in ionic radii and the crystal structures are also shown and suggest that the FeO-MgO system will probably display unlimited solid solubility. The CoO and ZnO systems also have appropriate radius ratios and crystal structures.

10-4 Solid-Solution Strengthening

In metallic materials, one of the important effects of solid-solution formation is the resultant **solid-solution strengthening** (Figure 10-6). This strengthening is caused by increased resistance to dislocation motion. This is one of the important reasons why brass (Cu-Zn alloy) is stronger than pure copper. Jewelry could be made from pure gold or silver; however, pure gold and pure silver are extremely soft and malleable. Jewelers add copper to gold and silver so that the jewelry will retain its shape.

In the copper-nickel system, we intentionally introduce a solid substitutional atom (nickel) into the original crystal structure (copper). The copper-nickel alloy is stronger than pure copper. Similarly, if less than 30% Zn is added to copper, the zinc behaves as a substitutional atom that strengthens the copper-zinc alloy, as compared with pure copper.

Recall from Chapter 7 that the strength of ceramics is mainly dictated by the distribution of flaws and not the propagation and interaction of dislocations; thus, solid-solution formation does not have a strong effect on the mechanical properties of ceramics. This is similar to why strain hardening was not much of a factor in enhancing the strength of ceramics (Chapter 8). The following discussion related to mechanical properties, therefore, applies mainly to metals.

Degree of Solid-Solution Strengthening The degree of solid-solution strengthening depends on two factors. First, a large difference in atomic size between the original (host or solvent) atom and the added (guest or solute) atom increases the strengthening effect. A larger size difference produces a greater disruption of the initial crystal structure, making slip more difficult (Figure 10-6).

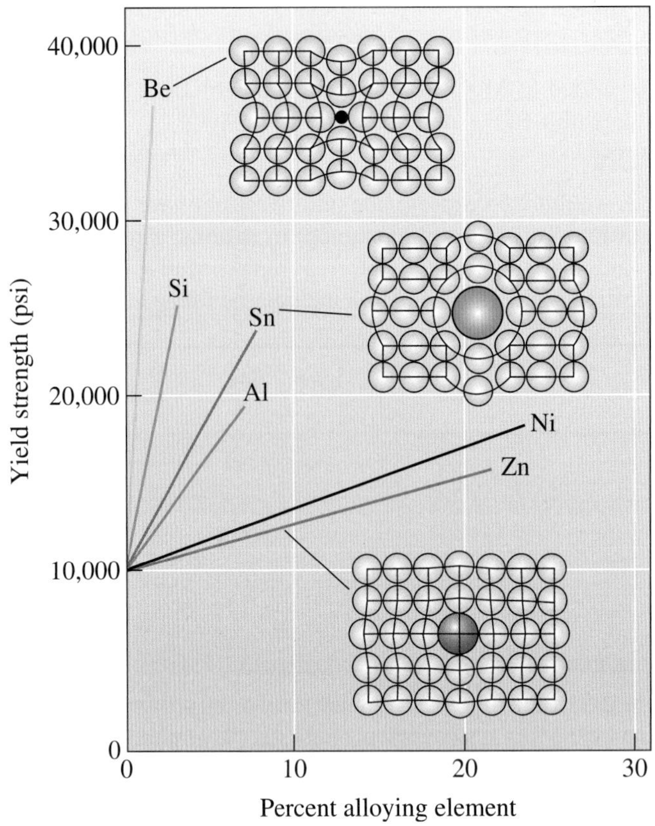

Figure 10-6 The effects of several alloying elements on the yield strength of copper. Nickel and zinc atoms are about the same size as copper atoms, but beryllium and tin atoms are much different in size from copper atoms. Increasing both the atomic size difference and the amount of alloying element increases solid-solution strengthening.

Second, the greater the amount of alloying element added, the greater the strengthening effect (Figure 10-6). A Cu-20% Ni alloy is stronger than a Cu-10% Ni alloy. Of course, if too much of a large or small atom is added, the solubility limit may be exceeded and a different strengthening mechanism, **dispersion strengthening**, is produced. In dispersion strengthening, the interface between the host phase and guest phase resists dislocation motion and contributes to strengthening. This mechanism is discussed further in Chapter 11.

Effect of Solid-Solution Strengthening on Properties

The effects of solid-solution strengthening on the properties of a metal include the following (Figure 10-7):

1. The yield strength, tensile strength, and hardness of the alloy are greater than those of the pure metals. This is one reason why we most often use alloys rather than pure metals. For example, small concentrations of Mg are added to aluminum to provide higher strength to the aluminum alloys used in making aluminum beverage cans.
2. Almost always, the ductility of the alloy is less than that of the pure metal. Only rarely, as in copper-zinc alloys, does solid-solution strengthening increase both strength and ductility.
3. Electrical conductivity of the alloy is much lower than that of the pure metal. This is because electrons are scattered by the atoms of the alloying elements more often than they are scattered by the host atoms. Solid-solution strengthening of copper or aluminum wires used for transmission of electrical power is not recommended because of this pronounced effect. Electrical conductivity of many alloys, although lower than pure metals, is often more stable as a function of temperature.
4. The resistance to creep and strength at elevated temperatures is improved by solid-solution strengthening. Many high-temperature alloys, such as those used for jet engines, rely partly on extensive solid-solution strengthening.

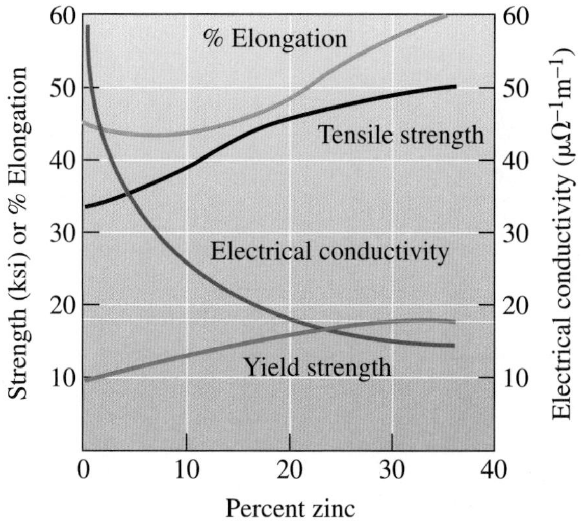

Figure 10-7 The effect of additions of zinc to copper on the properties of the solid-solution-strengthened alloy. The increase in % elongation with increasing zinc content is *not* typical of solid-solution strengthening.

Example 10-3 *Solid-Solution Strengthening*

From the atomic radii, show whether the size difference between copper atoms and alloying atoms accurately predicts the amount of strengthening found in Figure 10-6.

SOLUTION

The atomic radii and percent size difference are shown below.

Metal	Atomic Radius (Å)	$\left[\frac{r_{atom} - r_{Cu}}{r_{Cu}}\right] \times 100\%$
Cu	1.278	0
Zn	1.332	+4.2
Sn	1.405	+9.9
Al	1.432	+12.1
Ni	1.243	−2.7
Si	1.176	−8.0
Be	1.143	−10.6

For atoms larger than copper—namely, zinc, tin, and aluminum—increasing the size difference generally increases the strengthening effect. Likewise for smaller atoms, increasing the size difference increases strengthening.

10-5 Isomorphous Phase Diagrams

A **phase diagram** shows the phases and their compositions at any combination of temperature and alloy composition. When only two elements or two compounds are present in a material, a **binary phase diagram** can be constructed. **Isomorphous phase diagrams** are found in a number of metallic and ceramic systems. In the isomorphous systems, which include the copper-nickel and NiO-MgO systems [Figure 10-8(a) and (b)], only one solid phase forms; the two components in the system display complete solid solubility. As shown in the phase diagrams for the $CaO \cdot SiO_2 \cdot SrO$ and thallium-lead (Tl-Pb) systems, it is possible to have phase diagrams show a minimum or maximum point, respectively [Figure 10-8(c) and (d)]. Notice the horizontal scale can represent either mole% or weight%

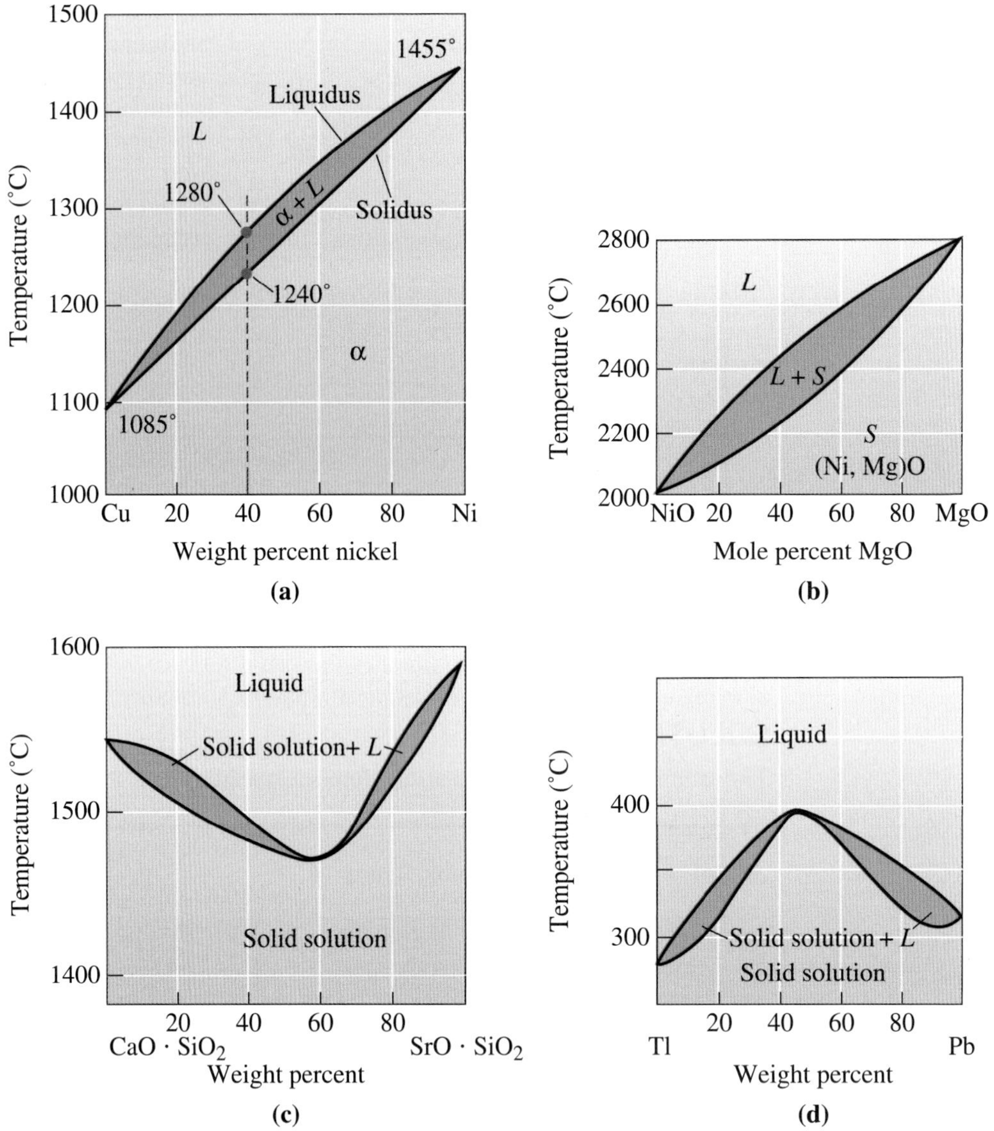

Figure 10-8 (a) and (b) The equilibrium phase diagrams for the Cu-Ni and NiO-MgO systems. The liquidus and solidus temperatures are shown for a Cu-40% Ni alloy. (c) and (d) Systems with solid-solution maxima and minima. (*Based on* Introduction to Phase Equilibria, *by C. G. Bergeron and S. H. Risbud, American Ceramic Society, 1984.*)

of one of the components. We can also plot atomic% or mole fraction of one of the components. Also, notice that the $CaO \cdot SiO_2$ and $SrO \cdot SiO_2$ diagram could be plotted as a *ternary phase diagram*. A ternary phase diagram is a phase diagram for systems consisting of three components. Here, we represent it as a *pseudo-binary diagram* (i.e., we assume that this is a diagram that represents phase equilibria between $CaO \cdot SiO_2$ and $SrO \cdot SiO_2$). In a pseudo-binary diagram, we represent equilibria between three or more components using two compounds. Ternary phase diagrams are often encountered in ceramic and metallic systems.

More recently, considerable developments have been made in phase diagrams by using computer databases containing thermodynamic properties of different elements and compounds. There are several valuable pieces of information to be obtained from phase diagrams, as follows.

Liquidus and Solidus Temperatures We define the **liquidus temperature** as the temperature above which a material is completely liquid. The upper curve in Figure 10-8(a), known as the **liquidus**, represents the liquidus temperatures for copper-nickel alloys of different compositions. We must heat a copper-nickel alloy above the liquidus temperature to produce a completely liquid alloy that can then be cast into a useful shape. The liquid alloy begins to solidify when the temperature cools to the liquidus temperature. For the Cu-40% Ni alloy in Figure 10-8(a), the liquidus temperature is 1280°C.

The **solidus temperature** is the temperature below which the alloy is 100% solid. The lower curve in Figure 10-8(a), known as the **solidus**, represents the solidus temperatures for Cu-Ni alloys of different compositions. A copper-nickel alloy is not completely solid until the material cools below the solidus temperature. If we use a copper-nickel alloy at high temperatures, we must be sure that the service temperature is below the solidus so that no melting occurs. For the Cu-40% Ni alloy in Figure 10-8(a), the solidus temperature is 1240°C.

Copper-nickel alloys melt and freeze over a range of temperatures between the liquidus and the solidus. The temperature difference between the liquidus and the solidus is the **freezing range** of the alloy. Within the freezing range, two phases coexist: a liquid and a solid. The solid is a solution of copper and nickel atoms and is designated as the α phase. For the Cu-40% Ni alloy in Figure 10-8(a), the freezing range is $1280 - 1240 = 40$°C. Note that pure metals solidify at a fixed temperature (i.e., the freezing range is zero degrees).

Phases Present Often we are interested in which phases are present in an alloy at a particular temperature. If we plan to make a casting, we must be sure that the metal is initially all liquid; if we plan to heat treat an alloy component, we must be sure that no liquid forms during the process. Different solid phases have different properties. For example, BCC Fe (indicated as the α phase on the iron-carbon phase diagram) is ferromagnetic; however, FCC iron (indicated as the γ phase on the Fe-C diagram) is not.

The phase diagram can be treated as a road map; if we know the coordinates—temperature and alloy composition—we can determine the phases present, assuming we know that thermodynamic equilibrium exists. There are many examples of technologically important situations in which we do not want equilibrium phases to form. For example, in the formation of silicate glass, we want an amorphous glass and not crystalline SiO_2 to form. When we harden steels by quenching them from a high temperature, the hardening occurs because of the formation of nonequilibrium phases. In such cases, phase diagrams will not provide all of the information we need. In these cases, we need to use special diagrams that take into account the effect of time (i.e., kinetics) on phase transformations. We will examine the use of such diagrams in later chapters.

The following two examples illustrate the applications of some of these concepts.

Example 10-4 *NiO-MgO Isomorphous System*

From the phase diagram for the NiO-MgO binary system [Figure 10-8(b)], describe a composition that can melt at 2600°C but will not melt when placed into service at 2300°C.

SOLUTION

The material must have a liquidus temperature below 2600°C, but a solidus temperature above 2300°C. The NiO-MgO phase diagram [Figure 10-8(b)] permits us to choose an appropriate composition.

To identify a composition with a liquidus temperature below 2600°C, there must be less than 60 mol% MgO in the refractory. To identify a composition with a solidus temperature above 2300°C, there must be at least 50 mol% MgO present. Consequently, we can use any composition between 50 mol% MgO and 60 mol% MgO.

Example 10-5 *Design of a Composite Material*

One method to improve the fracture toughness of a ceramic material (Chapter 7) is to reinforce the ceramic matrix with ceramic fibers. A materials designer has suggested that Al_2O_3 could be reinforced with 25% Cr_2O_3 fibers, which would interfere with the propagation of any cracks in the alumina. The resulting composite is expected to operate under load at 2000°C for several months.

Use Figure 10-9 to criticize the appropriateness of this design.

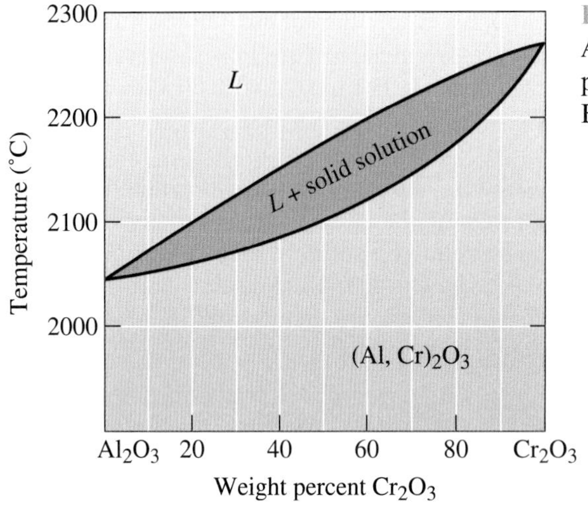

Figure 10-9 The Al_2O_3-Cr_2O_3 phase diagram (for Example 10-5).

SOLUTION

Since the composite will operate at high temperatures for a substantial period of time, the two phases—the Cr_2O_3 fibers and the Al_2O_3 matrix—must not react with one another. In addition, the composite must remain solid to at least 2000°C. The phase diagram in Figure 10-9 permits us to consider this choice for a composite.

Pure Cr_2O_3, pure Al_2O_3, and Al_2O_3-25% Cr_2O_3 have solidus temperatures above 2000°C; consequently, there is no danger of melting any of the constituents; however, Cr_2O_3 and Al_2O_3 display unlimited solid solubility. At the high service temperature, 2000°C, Al^{3+} ions will diffuse from the matrix into the fibers, replacing Cr^{3+} ions in the fibers. Simultaneously, Cr^{3+} ions will replace Al^{3+} ions in the matrix. Long before several months have elapsed, these diffusion processes will cause the fibers to completely dissolve into the matrix. With no fibers remaining, the fracture toughness will again be poor.

Composition of Each Phase

For each phase, we can specify a composition, expressed as the percentage of each element in the phase. Usually the composition is expressed in weight percent (wt%). When only one phase is present in the alloy or a ceramic solid solution, the composition of the phase equals the overall composition of the material. If the original composition of a single phase alloy or ceramic material changes, then the composition of the phase must also change.

When two phases, such as liquid and solid, coexist, their compositions differ from one another and also differ from the original overall composition. In this case, if the original composition changes slightly, the composition of the two phases is unaffected, provided that the temperature remains constant.

This difference is explained by the Gibbs phase rule. In this case, unlike the example of pure magnesium described earlier, we keep the pressure fixed at one atmosphere, which is normal for binary phase diagrams. The phase rule given by Equation 10-1 can be rewritten as

$$1 + C = F + P \text{ (for constant pressure)} \quad (10\text{-}2)$$

where, again, C is the number of independent chemical components, P is the number of phases (*not pressure*), and F is the number of degrees of freedom. We now use the number 1 instead of the number 2 because we are holding the pressure constant. This reduces the number of degrees of freedom by one. The pressure is typically, although not necessarily, one atmosphere. In a binary system, the number of components C is two; the degrees of freedom that we have include changing the temperature and changing the composition of the phases present. We can apply this form of the phase rule to the Cu-Ni system, as shown in Example 10-6.

Example 10-6 *Gibbs Rule for an Isomorphous Phase Diagram*

Determine the degrees of freedom in a Cu-40% Ni alloy at (a) 1300°C, (b) 1250°C, and (c) 1200°C. Use Figure 10-8(a).

SOLUTION

This is a binary system ($C = 2$). The two components are Cu and Ni. We will assume constant pressure. Therefore, Equation 10-2 ($1 + C = F + P$) can be used as follows.

(a) At 1300°C, $P = 1$, since only one phase (liquid) is present; $C = 2$, since both copper and nickel atoms are present. Thus,

$$1 + C = F + P \quad \therefore 1 + 2 = F + 1 \text{ or } F = 2$$

We must fix both the temperature and the composition of the liquid phase to completely describe the state of the copper-nickel alloy in the liquid region.

(b) At 1250°C, $P = 2$, since both liquid and solid are present; $C = 2$, since copper and nickel atoms are present. Now,

$$1 + C = F + P \quad \therefore 1 + 2 = F + 2 \text{ or } F = 1$$

If we fix the temperature in the two-phase region, the compositions of the two phases are also fixed. Alternately, if the composition of one phase is fixed, the temperature and composition of the second phase are automatically fixed.

(c) At 1200°C, $P = 1$, since only one phase (solid) is present; $C = 2$, since both copper and nickel atoms are present. Again,

$$1 + C = F + P \quad \therefore 1 + 2 = F + 1 \text{ or } F = 2$$

and we must fix both temperature and composition to completely describe the state of the solid.

Because there is only one degree of freedom in a two-phase region of a binary phase diagram, the compositions of the two phases are always fixed when we specify the temperature. This is true even if the overall composition of the alloy changes. Therefore, we can use a tie line to determine the compositions of the two phases. A **tie line** is a horizontal line within a two-phase region drawn at the temperature of interest (Figure 10-10). In an isomorphous system, the tie line connects the liquidus and solidus points at the specified temperature. The ends of the tie line represent the compositions of the two phases in equilibrium. Tie lines are not used in single-phase regions because we do not have two phases to "tie" in.

For any alloy with an overall or bulk composition lying between c_L and c_S, the composition of the liquid is c_L and the composition of the solid α is c_S.

The following example illustrates how the concept of a tie line is used to determine the composition of different phases in equilibrium.

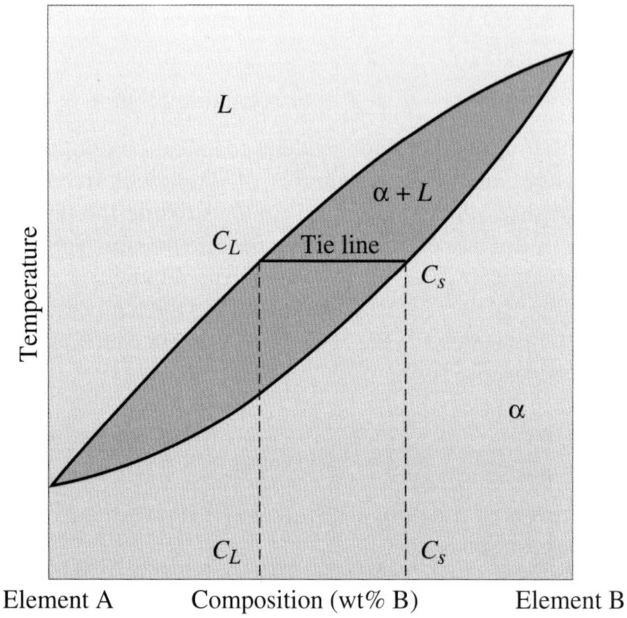

Figure 10-10
A hypothetical binary phase diagram between two elements A and B. When an alloy is present in a two-phase region, a tie line at the temperature of interest fixes the composition of the two phases. This is a consequence of the Gibbs phase rule, which provides only one degree of freedom in the two-phase region.

In Example 10-7, we find that, in the two-phase region, solid α contains more nickel and the liquid L contains more copper than the overall composition of the alloy. Generally, the higher melting point element (in this case, nickel) is concentrated in the first solid that forms.

Example 10-7 *Compositions of Phases in the Cu-Ni Phase Diagram*

Determine the composition of each phase in a Cu-40% Ni alloy at 1300°C, 1270°C, 1250°C, and 1200°C. (See Figure 10-11.)

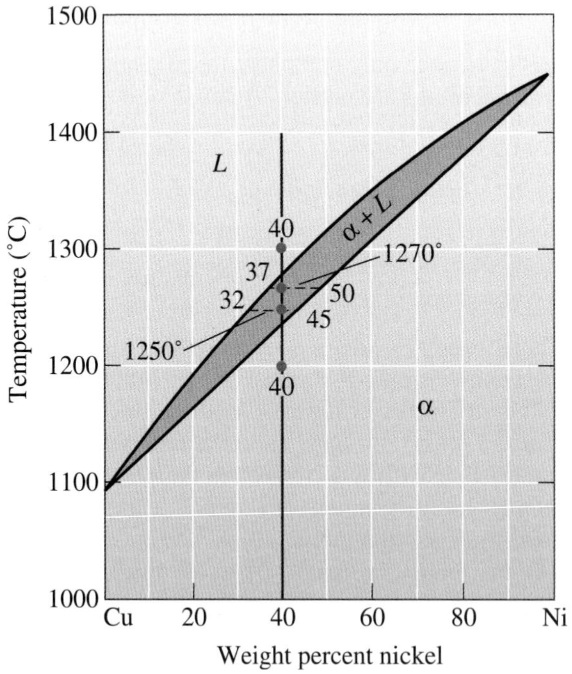

Figure 10-11
Tie lines and phase compositions for a Cu-40% Ni alloy at several temperatures (for Example 10-7).

SOLUTION

The vertical line at 40% Ni represents the overall composition of the alloy.

- 1300°C: Only liquid is present. The liquid must contain 40% Ni, the overall composition of the alloy.
- 1270°C: Two phases are present. A horizontal line within the $\alpha + L$ field is drawn. The endpoint at the liquidus, which is in contact with the liquid region, is at 37% Ni. The endpoint at the solidus, which is in contact with the α region, is at 50% Ni. Therefore, the liquid contains 37% Ni, and the solid contains 50% Ni.
- 1250°C: Again two phases are present. The tie line drawn at this temperature shows that the liquid contains 32% Ni, and the solid contains 45% Ni.
- 1200°C: Only solid α is present, so the solid must contain 40% Ni.

Amount of Each Phase (the Lever Rule)

Lastly, we are interested in the relative amounts of each phase present in the alloy. These amounts are normally expressed as weight percent (wt%). We express absolute amounts of different phases in units of mass or weight (grams, kilograms, pounds, etc.). The following example illustrates the rationale for the **lever rule**.

To calculate the amounts of liquid and solid, we construct a lever on our tie line, with the fulcrum of our lever being the original composition of the alloy. The leg of the lever *opposite* to the composition of the phase, the amount of which we are calculating, is divided by the total length of the lever to give the amount of that phase. In Example 10-8, note that the denominator represents the total length of the tie line and the numerator is the portion of the lever that is *opposite* the composition of the phase we are trying to calculate.

The lever rule in general can be written as

$$\text{Phase percent} = \frac{\text{opposite arm of lever}}{\text{total length of tie line}} \times 100 \tag{10-3}$$

We can work the lever rule in any two-phase region of a binary phase diagram. The lever rule calculation is not used in single-phase regions because the answer is trivial (there is 100% of that phase present). The lever rule is used to calculate the relative fraction or % of a phase in a two-phase mixture. The end points of the tie line we use give us the composition (i.e., the chemical concentration of different components) of each phase.

The following example reinforces the application of the lever rule for calculating the amounts of phases for an alloy at different temperatures. This is one way to track the solidification behavior of alloys, something we did not see in Chapter 9.

Sometimes we wish to express composition as atomic percent (at%) rather than weight percent (wt%). For a Cu-Ni alloy, where M_{Cu} and M_{Ni} are the molecular weights, the following equations provide examples for making these conversions:

$$\text{at\% Ni} = \left(\frac{\dfrac{\text{wt\% Ni}}{M_{Ni}}}{\dfrac{\text{wt\% Ni}}{M_{Ni}} + \dfrac{\text{wt\% Cu}}{M_{Cu}}} \right) \times 100 \tag{10-4}$$

$$\text{wt\% Ni} = \left(\frac{\text{at\% Ni} \times M_{Ni}}{\text{at\% Ni} \times M_{Ni} + \text{at\% Cu} \times M_{Cu}} \right) \times 100 \tag{10-5}$$

Example 10-8 *Application of the Lever Rule*

Calculate the amounts of α and L at 1250°C in the Cu-40% Ni alloy shown in Figure 10-12.

SOLUTION

Let's say that x = mass fraction of the alloy that is solid α. Since we have only two phases, the balance of the alloy must be in the liquid phase (L). Thus, the mass fraction of liquid will be $1 - x$. Consider 100 grams of the alloy. This alloy will consist of 40 grams of nickel at all temperatures. At 1250°C, let us write an equation that will represent the mass balance for nickel. At 1250°C, we have $100x$ grams of the α phase. We have $100(1 - x)$ grams of liquid.

$$\text{Total mass of nickel in 100 grams of the alloy} = \text{mass of nickel in liquid} + \text{mass of nickel in } \alpha$$

$$\therefore (100)(\%\text{ Ni in alloy}) = [(100)(1 - x)](\%\text{ Ni in } L) + (100)(x)(\%\text{ Ni in } \alpha)$$

$$\therefore (\%\text{ Ni in alloy}) = (\%\text{ Ni in } L)(1 - x) + (\%\text{ Ni in } \alpha)(x)$$

By multiplying and rearranging,

$$x = \frac{(\%\text{ Ni in alloy}) - (\%\text{ Ni in } L)}{(\%\text{ Ni in } \alpha) - (\%\text{ Ni in } L)}$$

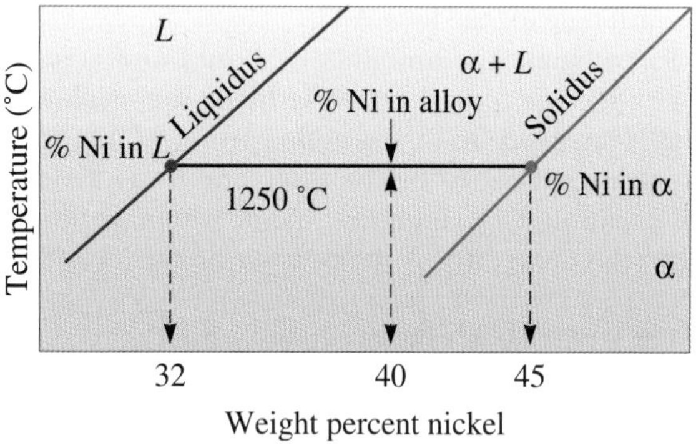

Figure 10-12 A tie line at 1250°C in the copper-nickel system that is used in Example 10-8 to find the amount of each phase.

From the phase diagram at 1250°C:

$$x = \frac{40 - 32}{45 - 32} = \frac{8}{13} = 0.62$$

If we convert from mass fraction to mass percent, the alloy at 1250°C contains 62% α and 38% L. Note that the concentration of nickel in the α phase (at 1250°C) is 45%, and the concentration of nickel in the liquid phase (at 1250°C) is 32%.

Example 10-9 *Solidification of a Cu-40% Ni Alloy*

Determine the amount of each phase in the Cu-40% Ni alloy shown in Figure 10-11 at 1300°C, 1270°C, 1250°C, and 1200°C.

SOLUTION

- **1300°C**: There is only one phase, so 100% L.
- **1270°C**: $\%\,L = \dfrac{50 - 40}{50 - 37} \times 100 = 77\%$

$$\%\,\alpha = \frac{40 - 37}{50 - 37} \times 100 = 23\%$$

- **1250°C:** $\% L = \dfrac{45 - 40}{45 - 32} \times 100 = 38\%$

  $\% \alpha = \dfrac{40 - 32}{45 - 32} \times 100 = 62\%$
- **1200°C:** There is only one phase, so 100% α.

Note that at each temperature, we can determine the composition of the phases in equilibrium from the ends of the tie line drawn at that temperature.

This may seem a little odd at first. How does the α phase change its composition? The liquid phase also changes its composition, and the amounts of each phase change with temperature as the alloy cools from the liquidus to the solidus.

10-6 Relationship Between Properties and the Phase Diagram

We have previously mentioned that a copper-nickel alloy will be stronger than either pure copper or pure nickel because of solid-solution strengthening. The mechanical properties of a series of copper-nickel alloys can be related to the phase diagram as shown in Figure 10-13.

The strength of copper increases by solid-solution strengthening until about 67% Ni is added. Pure nickel is solid-solution strengthened by the addition of copper until 33% Cu is added. The maximum strength is obtained for a Cu 67% Ni alloy, known as *Monel.* The maximum is closer to the pure nickel side of the phase diagram because pure nickel is stronger than pure copper.

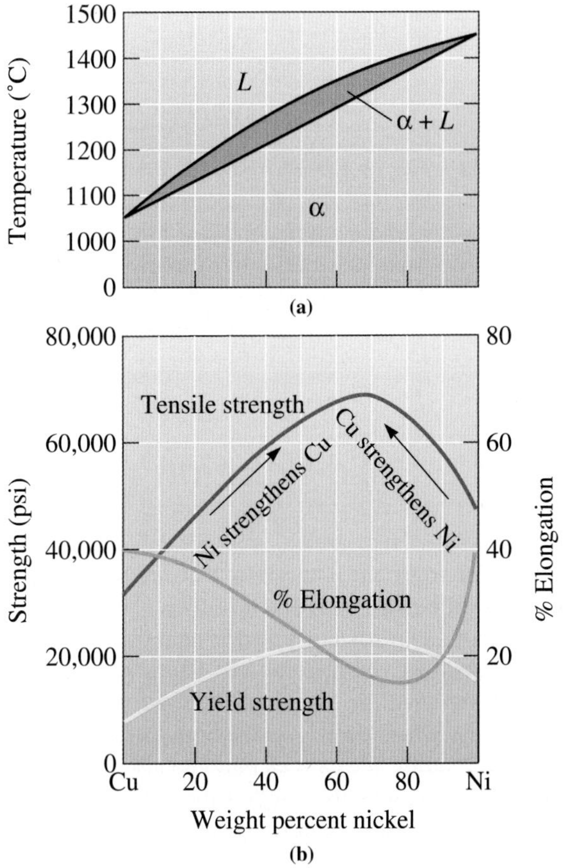

Figure 10-13 (a) The equilibrium phase diagram for the Cu-Ni system. The scale for the horizontal axis is the same as in (b). (b) The mechanical properties of copper-nickel alloys. Copper is strengthened by up to 67% Ni, and nickel is strengthened by up to 33% Cu.

Example 10-10 *Design of a Melting Procedure for a Casting*

You need to produce a Cu-Ni alloy having a minimum yield strength of 20,000 psi, a minimum tensile strength of 60,000 psi, and a minimum % elongation of 20%. You have in your inventory a Cu-20% Ni alloy and pure nickel. Design a method for producing castings having the required properties.

SOLUTION

From Figure 10-13, we determine the required composition of the alloy. To meet the required yield strength, the alloy must contain between 40 and 90% Ni; for the tensile strength, 40 to 88% Ni is required. The required % elongation can be obtained for alloys containing less than 60% Ni or more than 90% Ni. To satisfy all of these conditions, we could use Cu-40% to 60% Ni.

We prefer to select a low nickel content, since nickel is more expensive than copper. In addition, the lower nickel alloys have a lower liquidus, permitting castings to be made with less energy. Therefore, a reasonable alloy is Cu-40% Ni.

To produce this composition from the available melting stock, we must blend some of the pure nickel with the Cu-20% Ni ingot. Assume we wish to produce 10 kg of the alloy. Let x be the mass of Cu-20% Ni alloy we will need. The mass of pure nickel needed will be $10 - x$. Since the final alloy consists of 40% Ni, the total mass of nickel needed will be

$$(10 \text{ kg})\left(\frac{40\% \text{ Ni}}{100\%}\right) = 4 \text{ kg Ni}$$

Now let's write a mass balance for nickel. The sum of the nickel from the Cu-20% Ni alloy and the pure nickel must be equal to the total nickel in the Cu-40% Ni alloy being produced:

$$(x)\left(\frac{20\% \text{ Ni}}{100\%}\right) + (10 - x)\left(\frac{100\% \text{ Ni}}{100\%}\right) = 4 \text{ kg Ni}$$

$$0.2x + 10 - x = 4$$

$$6 = 0.8x$$

$$x = 7.5 \text{ kg}$$

Therefore, we need to melt 7.5 kg of Cu-20% Ni with 2.5 kg of pure nickel to produce the required alloy. We would then heat the alloy above the liquidus temperature, which is 1280°C for the Cu-40% Ni alloy, before pouring the liquid metal into the appropriate mold.

We need to conduct such calculations for many practical situations dealing with the processing of alloys because when we make them, we typically use new and recycled materials.

10-7 Solidification of a Solid-Solution Alloy

When an alloy such as Cu-40% Ni is melted and cooled, solidification requires both nucleation and growth. Heterogeneous nucleation (Chapter 9) permits little or no undercooling, so solidification begins when the liquid reaches the liquidus temperature. The phase diagram (Figure 10-14), with a tie line drawn at the liquidus temperature, indicates that the *first solid to form* has a composition of Cu-52% Ni.

Two conditions are required for growth of the solid α. First, growth requires that the latent heat of fusion (ΔH_f), which evolves as the liquid solidifies, be removed from the solid–liquid interface. Second, unlike the case of pure metals, diffusion must occur so that the compositions of the solid and liquid phases follow the solidus and liquidus curves during cooling. The latent heat of fusion (ΔH_f) is removed over a range of temperatures so that the cooling curve shows a change in slope, rather than a flat plateau (Figure 10-15). Thus, as we mentioned in Chapter 9, the solidification of alloys is different from that of pure metals.

At the start of freezing, the liquid contains Cu-40% Ni, and the first solid contains Cu-52% Ni. Nickel atoms must have diffused to and concentrated at the first solid

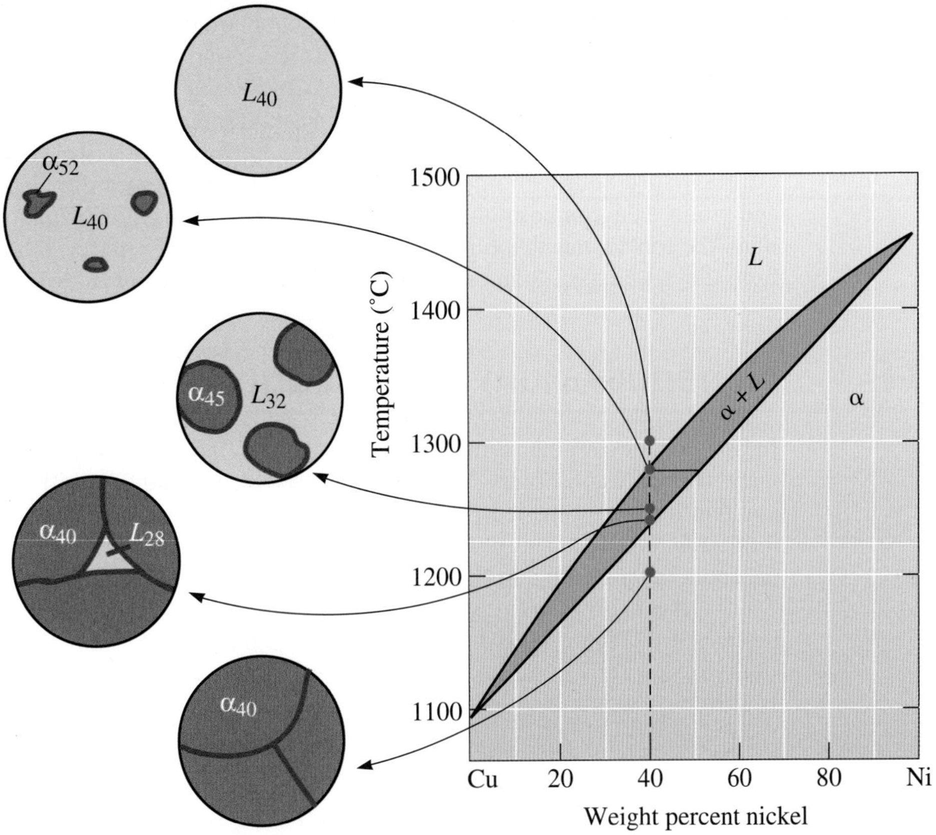

Figure 10-14 The change in structure of a Cu-40% Ni alloy during equilibrium solidification. The nickel and copper atoms must diffuse during cooling in order to satisfy the phase diagram and produce a uniform equilibrium structure.

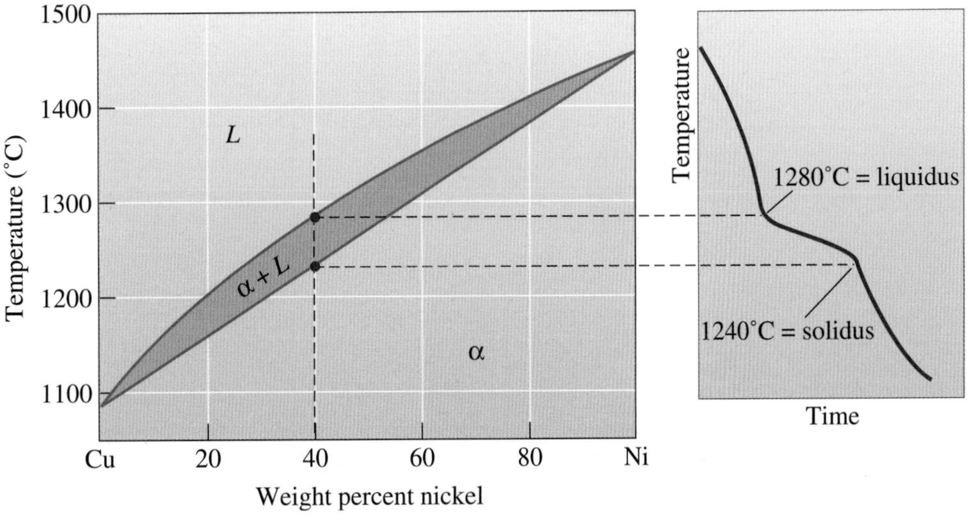

Figure 10-15 The cooling curve for an isomorphous alloy during solidification. We assume that cooling rates are low so that thermal equilibrium is maintained at each temperature. The changes in slope of the cooling curve indicate the liquidus and solidus temperatures, in this case, for a Cu-40% Ni alloy.

to form. After cooling to 1250°C, solidification has advanced, and the phase diagram tells us that now all of the liquid must contain 32% Ni and all of the solid must contain 45% Ni. On cooling from the liquidus to 1250°C, some nickel atoms must diffuse from the first solid to the new solid, reducing the nickel in the first solid. Additional nickel atoms diffuse from the solidifying liquid to the new solid. Meanwhile, copper atoms have concentrated—by diffusion—into the remaining liquid. This process must continue until we reach the solidus temperature, where the last liquid to freeze, which contains Cu-28% Ni, solidifies and forms a solid containing Cu-40% Ni. Just below the solidus, all of the solid must contain a uniform concentration of 40% Ni throughout.

In order to achieve this equilibrium final structure, the cooling rate must be extremely slow. Sufficient time must be permitted for the copper and nickel atoms to diffuse and produce the compositions given by the phase diagram. In many practical casting situations, the cooling rate is too rapid to permit equilibrium. Therefore, in most castings made from alloys, we expect composition differences to exist, a phenomenon known as **segregation**.

10-8 Nonequilibrium Solidification and Segregation

In Chapter 5, we examined the thermodynamic and kinetic driving forces for diffusion. We know that diffusion occurs fastest in gases, followed by liquids, and then solids. We also saw that increasing the temperature enhances diffusion rates. When cooling is too rapid for atoms to diffuse and produce equilibrium conditions, nonequilibrium structures are produced in the casting. Let's see what happens to our Cu-40% Ni alloy on rapid cooling.

Again, the first solid, containing 52% Ni, forms on reaching the liquidus temperature (Figure 10-16). On cooling to 1260°C, the tie line tells us that the liquid

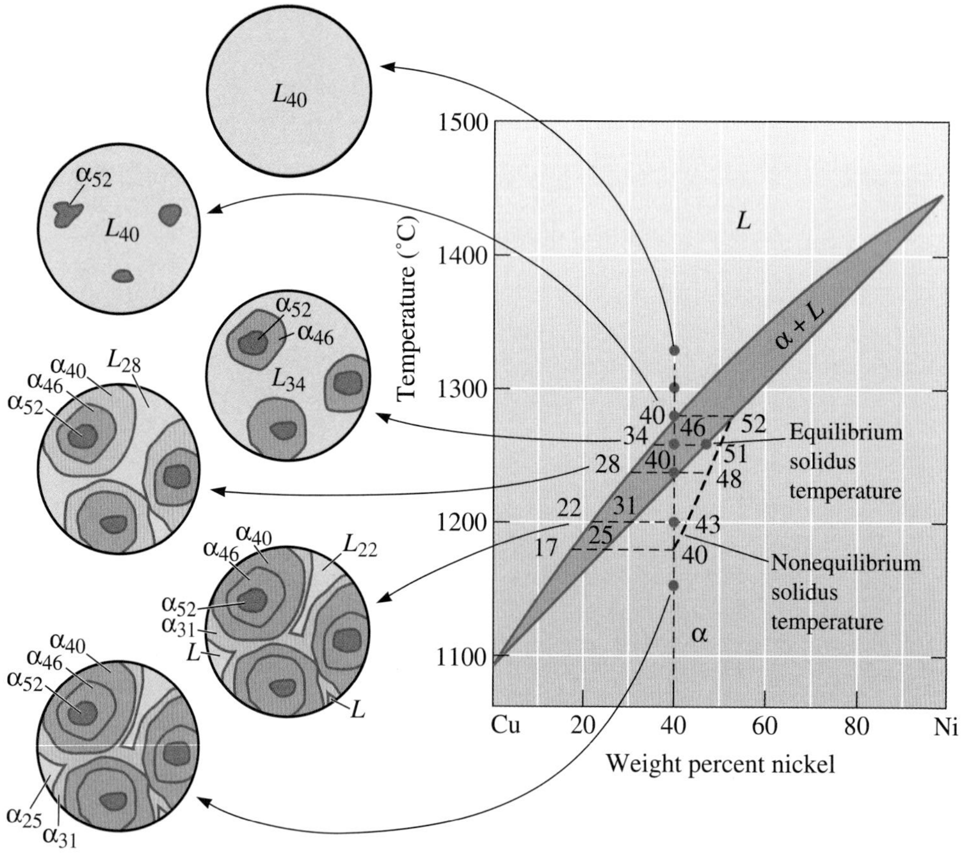

Figure 10-16 The change in structure of a Cu-40% Ni alloy during nonequilibrium solidification. Insufficient time for diffusion in the solid produces a segregated structure. Notice the nonequilibrium solidus curve.

contains 34% Ni and the solid that forms at that temperature contains 46% Ni. Since diffusion occurs rapidly in liquids, we expect the tie line to predict the liquid composition accurately; however, diffusion in solids is comparatively slow. The first solid that forms still has about 52% Ni, but the new solid contains only 46% Ni. We might find that the average composition of the solid is 51% Ni. This gives a different nonequilibrium solidus than that given by the phase diagram. As solidification continues, the nonequilibrium solidus line continues to separate from the equilibrium solidus.

When the temperature reaches 1240°C (the equilibrium solidus), a significant amount of liquid remains. The liquid will not completely solidify until we cool to 1180°C, where the nonequilibrium solidus intersects the original composition of 40% Ni. At that temperature, liquid containing 17% Ni solidifies, giving solid containing 25% Ni. The last liquid to freeze therefore contains 17% Ni, and the last solid to form contains 25% Ni. The average composition of the solid is 40% Ni, but the composition is not uniform.

The actual location of the nonequilibrium solidus line and the final nonequilibrium solidus temperature depend on the cooling rate. Faster cooling rates cause greater departures from equilibrium. The following example illustrates how we can account for the changes in composition under nonequilibrium conditions.

Example 10-11 *Nonequilibrium Solidification of Cu-Ni Alloys*

Calculate the composition and amount of each phase in a Cu-40% Ni alloy that is present under the nonequilibrium conditions shown in Figure 10-16 at 1300°C, 1280°C, 1260°C, 1240°C, 1200°C, and 1150°C. Compare with the equilibrium compositions and amounts of each phase.

SOLUTION

We use the tie line to the equilibrium solidus temperature curve to calculate compositions and percentages of phases as per the lever rule. Similarly, the nonequilibrium solidus temperature curve is used to calculate percentages and concentrations of different phases formed under nonequilibrium conditions.

Temperature	Equilibrium	Nonequilibrium
1300°C	L: 40% Ni 100% L	L: 40% Ni 100% L
1280°C	L: 40% Ni 100% L	L: 40% Ni 100% L
1260°C	L: 34% Ni $\frac{46-40}{46-34} = 50\%\ L$	L: 34% Ni $\frac{51-40}{51-34} = 65\%\ L$
	α: 46% Ni $\frac{40-34}{46-34} = 50\%\ \alpha$	α: 51% Ni $\frac{40-34}{51-34} = 35\%\ \alpha$
1240°C	L: 28% Ni ~ 0% L	L: 28% Ni $\frac{48-40}{48-28} = 40\%\ L$
	α: 40% Ni 100% α	α: 48% Ni $\frac{40-28}{48-28} = 60\%\ \alpha$
1200°C	α: 40% Ni 100% α	L: 22% Ni $\frac{43-40}{43-22} = 14\%\ L$
		α: 43% Ni $\frac{40-22}{43-22} = 86\%\ \alpha$
1150°C	α: 40% Ni 100% α	α: 40% Ni 100% α

Microsegregation The nonuniform composition produced by nonequilibrium solidification is known as segregation. **Microsegregation**, also known as **interdendritic segregation** and **coring**, occurs over short distances, often between small dendrite arms. The centers of the dendrites, which represent the first solid to freeze, are rich in the higher melting point element in the alloy. The regions between the dendrites are rich in the lower melting point element, since these regions represent the last liquid to freeze. The composition and properties of the α phase (in the case of Cu-Ni alloys) differ from one region to the next, and we expect the casting to have poorer properties as a result.

Microsegregation can cause **hot shortness**, or melting of the lower melting point interdendritic material at temperatures below the equilibrium solidus. When we heat the Cu-40% Ni alloy to 1225°C, below the equilibrium solidus but above the nonequilibrium solidus, the low nickel regions between the dendrites melt.

Homogenization We can reduce the interdendritic segregation and problems with hot shortness by means of a **homogenization heat treatment**. If we heat the casting to a temperature below the nonequilibrium solidus, the nickel atoms in the centers of the dendrites diffuse to the interdendritic regions; copper atoms diffuse in the opposite direction [Figure 10-17(a)]. Since the diffusion distances are relatively short, only a few hours are required to eliminate most of the composition differences. The homogenization time is related to

$$t = \mathrm{c}\frac{(\mathrm{SDAS})^2}{D_s} \tag{10-6}$$

where SDAS is the secondary dendrite arm spacing, D_s is the rate of diffusion of the solute in the matrix, and c is a constant. A small SDAS reduces the diffusion distance and permits short homogenization times.

Macrosegregation There exists another type of segregation, known as **macrosegregation**, which occurs over a large distance, between the surface and the center of the casting, with the surface (which freezes first) containing slightly more than the average amount of the higher melting point metal. We cannot eliminate macrosegregation by a homogenization treatment because the diffusion distances are too great. Macrosegregation can be reduced by hot working, which was discussed in Chapter 8. This is because in hot working, we are basically breaking down the cast macrostructure.

Rapidly Solidified Powders In applications in which porosity, microsegregation, and macrosegregation must be minimized, powders of complex alloys are prepared using **spray atomization** [Figure 10-17(b)]. In spray atomization, homogeneous melts of complex compositions are prepared and sprayed through a ceramic nozzle. The melt stream is broken into finer droplets and quenched using argon or nitrogen gases (gas atomization) or water (water atomization). The molten droplets solidify rapidly, generating powder particles ranging from ~10–100 μm in size. Since the solidification of droplets occurs very quickly, there is very little time for diffusion, and therefore, chemical segregation does not occur. Many complex nickel- and cobalt-based superalloys and stainless steel powders are examples of materials prepared using this technique. The spray atomized powders are blended and formed into desired shapes. The techniques used in processing such powders include sintering (Chapter 5), **hot pressing** (HP) and **hot isostatic pressing** (HIP). In HIP, sintering is conducted under an isostatic pressure (~ 25,000 psi) using, for example, argon gas. Very large (~ up to 2 foot diameter, several feet long) and smaller components can be processed using HIP. Smaller components such as disks that hold turbine blades can be machined from these. The progression of densification in a low carbon Astroalloy sample processed using spray atomized powders is shown in Figure 10-17(c).

Hot pressing is sintering under a uniaxial pressure and is used in the production of smaller components of materials that are difficult to sinter otherwise (Chapter 15). The HIP and hot pressing techniques are used for both metallic and ceramic powder materials.

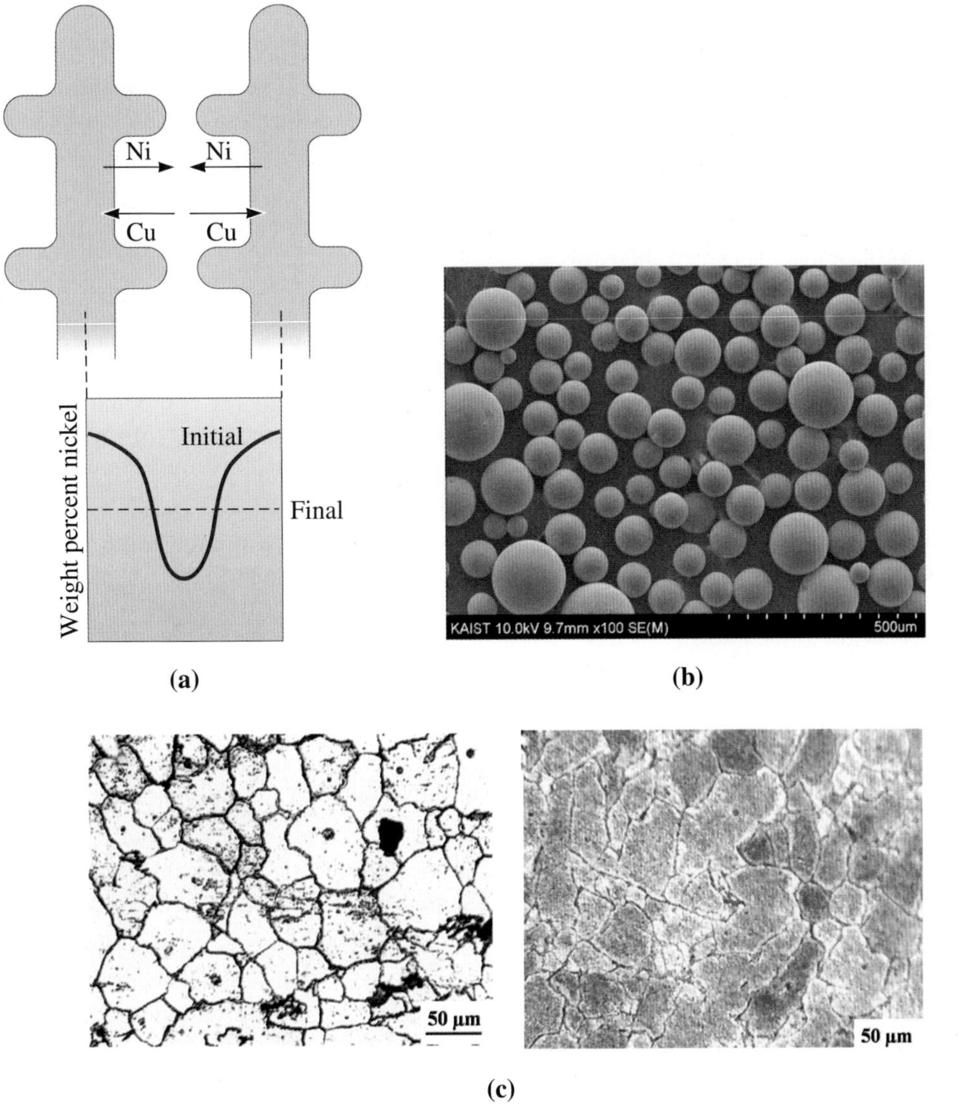

Figure 10-17 (a) Microsegregation between dendrites can be reduced by a homogenization heat treatment. Counterdiffusion of nickel and copper atoms may eventually eliminate the composition gradients and produce a homogeneous composition. (b) Spray atomized powders of superalloys. (*Reprinted from* Journal of Alloys and Compounds, *Volume 603, Youngmoo Kim, Eun-Pyo Kim, Young-Beom Song, Sung Ho Lee, Young-Sam Kwon, "Microstructure and mechanical properties of hot isostatically pressed Ti–6Al–4V alloy," Pages 207–212, Copyright 2014, with permission from Elsevier.*) (c) Micrographs of spray-formed Al - 4.5 wt.% Cu before and after HIP. Pores are clearly visible before HIP. Before HIP, the porosity was 8%, and after HIP, the porosity was less than 1%. (*S. Devaraj, S. Sankaran, R. Kumar, G. Appa Rao, "Influence of Hot Isostatic Pressing on the Microstructure and Mechanical Properties,"* Journal of Materials Engineering and Performance, *April 2014, Volume 23, Issue 4, pp 1440–1450, © 2014, ASM International.*)

Summary

- A phase is any portion, including the whole, of a system that is physically homogeneous within it and bounded by a surface that separates it from any other portions.
- A phase diagram typically shows phases that are expected to be present in a system under thermodynamic equilibrium conditions. Sometimes metastable phases may also be shown.

- Solid solutions in metallic or ceramic materials exist when elements or compounds with similar crystal structures form a single phase that is chemically homogeneous.
- Solid-solution strengthening is accomplished in metallic materials by the formation of solid solutions. The point defects created restrict dislocation motion and cause strengthening.
- The degree of solid-solution strengthening increases when (1) the amount of the alloying element increases and (2) the atomic size difference between the host material and the alloying element increases.
- The amount of alloying element (or compound) that we can add to produce solid-solution strengthening is limited by the solubility of the alloying element or compound in the host material. The solubility is limited when (1) the atomic size difference is more than about 15%, (2) the alloying element (or compound) has a different crystal structure than the host element (or compound), and (3) the valence and electronegativity of the alloying element or constituent ions are different from those of the host element (or compound).
- In addition to increasing strength and hardness, solid-solution strengthening typically decreases ductility and electrical conductivity of metallic materials. An important function of solid-solution strengthening is to provide good high-temperature properties to the alloy.
- A phase diagram in which constituents exhibit complete solid solubility is known as an isomorphous phase diagram.
- As a result of solid-solution formation, solidification begins at the liquidus temperature and is completed at the solidus temperature; the temperature difference over which solidification occurs is the freezing range.
- In two-phase regions of the phase diagram, the ends of a tie line fix the composition of each phase, and the lever rule permits the amount of each phase to be calculated.
- Microsegregation and macrosegregation occur during solidification. Microsegregation, or coring, occurs over small distances, often between dendrites. The centers of the dendrites are rich in the higher melting point element, whereas interdendritic regions, which solidify last, are rich in the lower melting point element.
- Homogenization can reduce microsegregation.
- Macrosegregation describes differences in composition over long distances, such as between the surface and center of a casting. Hot working may reduce macrosegregation.

Glossary

Alloy A material consisting of multiple elements that exhibits properties of a metallic material.

Binary phase diagram A phase diagram for a system with two components.

Copolymer A polymer that is formed by combining two or more different types of monomers, usually with the idea of blending the properties affiliated with individual polymers.

Coring Chemical segregation in cast products, also known as microsegregation or interdendritic segregation. The centers of the dendrites are rich in the higher melting point element, whereas interdendritic regions, which solidify last, are rich in the lower melting point element.

Dispersion strengthening Strengthening, typically used in metallic materials, by the formation of interfaces between phases. The interfaces provide additional resistance to dislocation motion, thereby causing strengthening of metallic materials (Chapter 11).

Freezing range The temperature difference between the liquidus and solidus temperatures.

Gibbs phase rule Describes the number of degrees of freedom, or the number of variables that must be fixed, to specify the temperature and composition of a phase ($2 + C = F + P$, where pressure and temperature can change, $1 + C = F + P$, where pressure or temperature is constant).

Homogenization heat treatment The heat treatment used to reduce the microsegregation caused by nonequilibrium solidification. This heat treatment cannot eliminate macrosegregation.

Hot isostatic pressing (HIP) Sintering of metal or ceramic powders, conducted under an isostatic pressure.

Hot pressing (HP) Sintering of metal or ceramic powders under a uniaxial pressure; used for production of smaller components of materials that are difficult to sinter otherwise.

Hot shortness Melting of the lower melting point nonequilibrium material that forms due to segregation, even though the temperature is below the equilibrium solidus temperature.

Hume-Rothery rules The conditions that an alloy or ceramic system must meet if the system is to display unlimited solid solubility. The Hume-Rothery rules are necessary but are not sufficient for materials to show unlimited solid solubility.

Interdendritic segregation See Coring.

Isomorphous phase diagram A phase diagram in which the components display unlimited solid solubility.

Lever rule A technique for determining the amount of each phase in a two-phase system.

Limited solubility When only a certain amount of a solute material can be dissolved in a solvent material.

Liquidus Curves on phase diagrams that describe the liquidus temperatures of all possible alloys.

Liquidus temperature The temperature at which the first solid begins to form during solidification.

Macrosegregation The presence of composition differences in a material over large distances caused by nonequilibrium solidification. The only way to remove this type of segregation is to break down the cast structure by hot working.

Microsegregation See Coring.

Multiple-phase alloy An alloy that consists of two or more phases.

Phase Any portion, including the whole of a system, that is physically homogeneous within it and bounded by a surface so that it is separate from any other portions.

Phase diagrams Diagrams showing phases present under equilibrium conditions and the phase compositions at each combination of temperature and overall composition. Sometimes phase diagrams also indicate metastable phases.

Phase rule See Gibbs phase rule.

P-T diagram A diagram describing thermodynamic stability of phases under different temperature and pressure conditions (same as a unary phase diagram).

Segregation The presence of composition differences in a material, often caused by insufficient time for diffusion during solidification.

Single-phase alloy An alloy consisting of one phase.

Solid solution A solid phase formed by combining multiple elements or compounds such that the overall phase has a uniform composition and properties that are different from those of the elements or compounds forming it.

Solid-solution strengthening Increasing the strength of a metallic material via the formation of a solid solution.

Solidus Curves on phase diagrams that describe the solidus temperature of all possible alloys.

Solidus temperature The temperature below which all liquid has completely solidified.

Solubility The amount of one material that will completely dissolve in a second material without creating a second phase.

Tie line A horizontal line drawn in a two-phase region of a phase diagram to assist in determining the compositions of the two phases.

Triple point A pressure and temperature at which three phases of a single material are in equilibrium.

Unary phase diagram A phase diagram in which there is only one component.

Unlimited solubility When the amount of one material that will dissolve in a second material without creating a second phase is unlimited.

Problems

10-1 Explain the principle of grain-size strengthening. Does this mechanism work at high temperatures? Explain.

10-2 Explain the principle of strain hardening. Does this mechanism work at high temperatures? Explain.

10-3 What is the principle of solid-solution strengthening? Does this mechanism work at high temperatures? Explain.

10-4 What is the principle of dispersion strengthening?

Section 10-1 Phases and the Phase Diagram

10-5 What does the term "phase" mean?

10-6 What are the different phases of water?

10-7 Ice has been known to exist in different polymorphs. Are these different phases of water?

10-8 Write down the Gibbs phase rule, assuming temperature and pressure are allowed to change. Clearly explain the meaning of each term.

10-9 What is a phase diagram?

10-10 How much of a temperature change is required to bring liquid magnesium from solid-liquid equilibrium to vapor-liquid equilibrium at 1 atmosphere? (See Figure 10-2.)

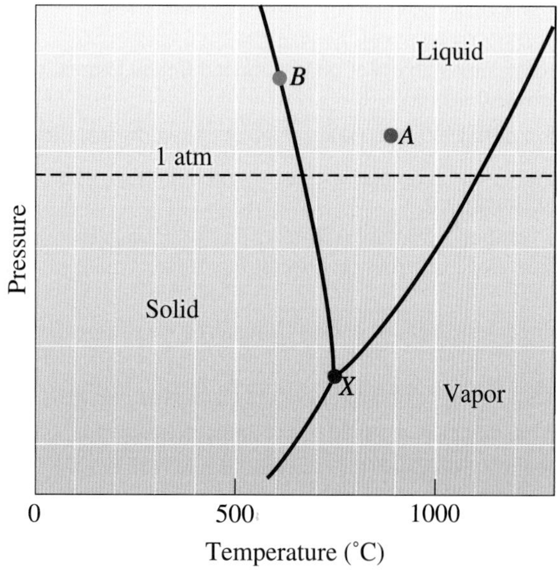

Figure 10-2 Schematic unary phase diagram for magnesium, showing the melting and boiling temperatures at one atmosphere pressure. On this diagram, point X is the triple point. (Repeated for Problem 10-10.)

10-11 The unary phase diagram for SiO_2 is shown in Figure 10-18. Locate the triple point where solid, liquid, and vapor coexist and give the temperature and the type of solid present. What do the other "triple" points indicate?

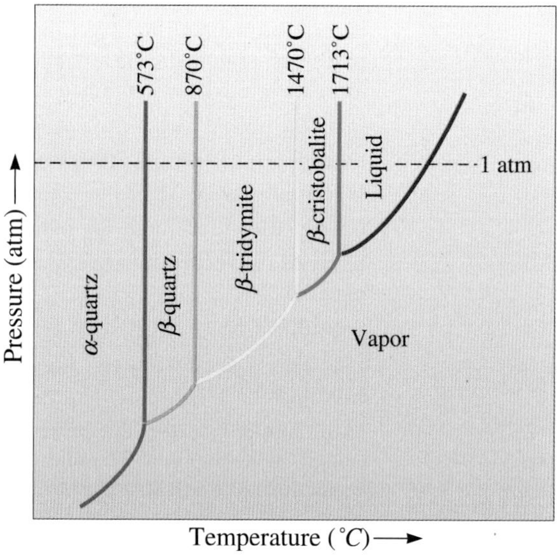

Figure 10-18 Pressure-temperature diagram for SiO_2. The dotted line shows one atmosphere pressure. (For Problem 10-11.)

10-12 Figure 10-19 shows the unary phase diagram for carbon. Based on this diagram, under what conditions can carbon in the form of graphite be converted into diamond?

10-13 Natural diamond is formed approximately 120 to 200 km below the earth's surface under high pressure and high temperature conditions. Assuming that the average density of the earth is 5500 kg/m^3, use this information and the unary phase diagram for C (Figure 10–19) to calculate the range of the earth's geothermal gradient (rate of increase of temperature with depth). Estimate the pressure below the earth's surface as ρgh where ρ is density, g is gravity, and h is depth. Note that 10 kbar = 10^9 Pa.

Section 10-2 Solubility and Solid Solutions

10-14 What is a solid solution?

10-15 How can solid solutions form in ceramic systems?

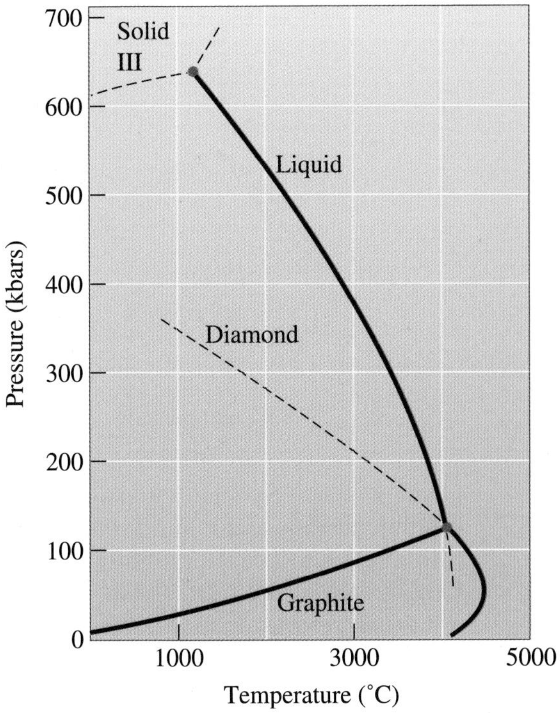

Figure 10-19 Unary phase diagram for carbon. Region for diamond formation is shown with a dotted line. (*Based on* Introduction to Phase Equilibria, *by C. G. Bergeron and S. H. Risbud, American Ceramic Society, 1984.*) (For Problems 10-12 and 10-13.)

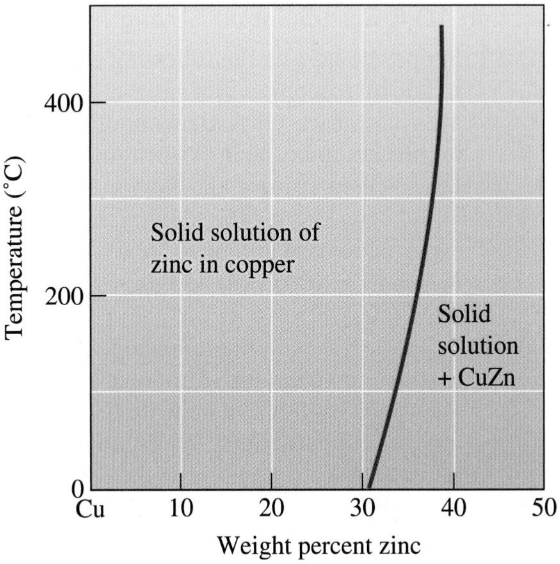

Figure 10-4 The solubility of zinc in copper. The solid line represents the solubility limit; when excess zinc is added, the solubility limit is exceeded and two phases coexist. (Repeated for Problems 10-22 and 10-23.)

10-16 Do we need 100% solid solubility to form a solid solution of one material in another?

10-17 Small concentrations of lead zirconate ($PbZrO_3$) are added to lead titanate ($PbTiO_3$). Draw a schematic of the resultant solid-solution crystal structure that is expected to form. This material, known as lead zirconium titanate (better known as PZT), has many applications ranging from spark igniters to ultrasound imaging. (See Section 3-7 for information on the perovskite crystal structure.)

10-18 Can solid solutions be formed between three elements or three compounds?

10-19 What is a copolymer? What is the advantage to forming copolymers?

10-20 Is copolymer formation similar to solid-solution formation?

10-21 What is the ABS copolymer? State some of the applications of this material.

10-22 A 15 kg block of copper is being alloyed with zinc and brought to the solubility limit at 200°C. How much zinc is required? (See Figure 10-4.)

10-23 A 115 kg total of copper and zinc (4 parts Zn to 6 parts Cu) is mixed at 400°C. How much of the zinc is in excess of solubility? (See Figure 10-4.)

Section 10-3 Conditions for Unlimited Solid Solubility

10-24 Briefly state the Hume–Rothery rules and explain the rationale.

10-25 Can the Hume–Rothery rules apply to ceramic systems? Explain.

10-26 What is the difference between limited and unlimited solid solubility?

10-27 Copper and nickel have unlimited solid solubility in each other whereas lead and iron are completely insoluble. Using the Hume–Rothery rules, justify these conditions. Show your work.

10-28 Which of the following elements would be expected to have unlimited solid solubility in gold? (a) Ag; (b) Al; and (c) Fe.

10-29 Based on Hume–Rothery's conditions, which of the following systems would be expected to display unlimited solid solubility? Explain. (a) Au-Ag; (b) Al-Cu; (c) Al-Au; (d) U-W; (e) Mo-Ta; (f) Nb-W; (g) Mg-Zn; and (h) Mg-Cd.

10-30 Identify which of the following oxides when added to $BaTiO_3$ are likely to exhibit 100% solid solubility: (a) $SrTiO_3$; (b) $CaTiO_3$; (c) $ZnTiO_3$; and (d) $BaZrO_3$. All of these oxides have a perovskite crystal structure.

Section 10-4 Solid-Solution Strengthening

10-31 Suppose 1 at% of the following elements is added to copper (forming a separate alloy with each element) without exceeding the solubility limit. Which one would be expected to give the higher strength alloy? Are any of the alloying elements expected to have unlimited solid solubility in copper? (a) Au; (b) Mn; (c) Sr; (d) Si; and (e) Co.

10-32 Suppose 1 at% of the following elements is added to aluminum (forming a separate alloy with each element) without exceeding the solubility limit. Which one would be expected to give the higher strength alloy? Why? (a) Mn; (b) Mg; (c) Cu; and (d) Zn.

10-33 Suppose 1 at% of the following elements is added to aluminum (forming a separate alloy with each element) without exceeding the solubility limit. Which one would be expected to give the smallest reduction in electrical conductivity? Are any of the alloy elements expected to have unlimited solid solubility in aluminum? (a) Li; (b) Ba; (c) Be; (d) Cd; and (e) Ga.

10-34 Which of the following oxides is expected to have the largest solid solubility in Al_2O_3? (a) Y_2O_3; (b) Cr_2O_3; and (c) Fe_2O_3.

10-35 What is the role of small concentrations of Mg in aluminum alloys used to make beverage cans?

10-36 Why do jewelers add small amounts of copper to gold and silver?

10-37 Why is it not a good idea to use solid-solution strengthening as a mechanism to increase the strength of copper for electrical applications?

Section 10-5 Isomorphous Phase Diagrams

10-38 What is a binary phase diagram and what information can be learned from it? In terms of thermodynamics, what is meant by the term "equilibrium phase diagram"?

10-39 What is the significance of the liquidus curve? Solidus curve?

10-40 What is an alloy? In general, what is the difference in the melting behavior of an alloy from that of a pure component?

10-41 Determine the degrees of freedom under the following conditions:
(a) Tl-20 wt% Pb at 325°C and 400°C;
(b) Tl-40 wt% Pb at 325°C and 400°C;
(c) Tl-90 wt% Pb at 325°C and 400°C.
Refer to the phase diagram in Figure 10-8(d).

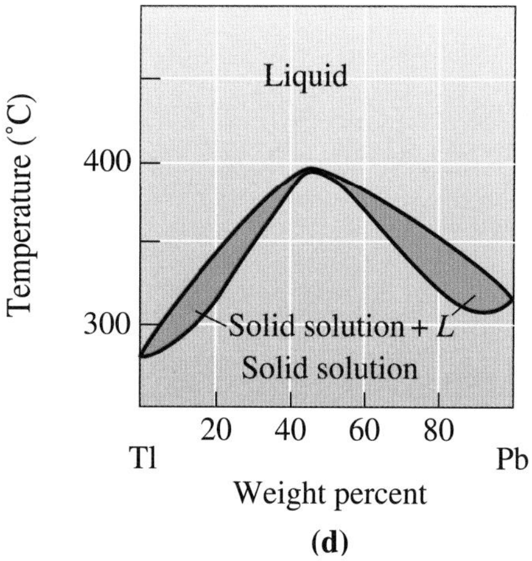

Figure 10-8(d) Systems with solid-solution maxima and minima. (*Based on Introduction to Phase Equilibria, by C. G. Bergeron and S. H. Risbud, American Ceramic Society, 1984.*)

10-42 Locate the following points in the Bi-Sb phase diagram and indicate the phases present and their relative amounts: (a) 60 at% Bi at 250°C; (b) 30 at% Bi at 500°C; and (c) 50 at% Bi at 600°C. (See Figure 10-20.)

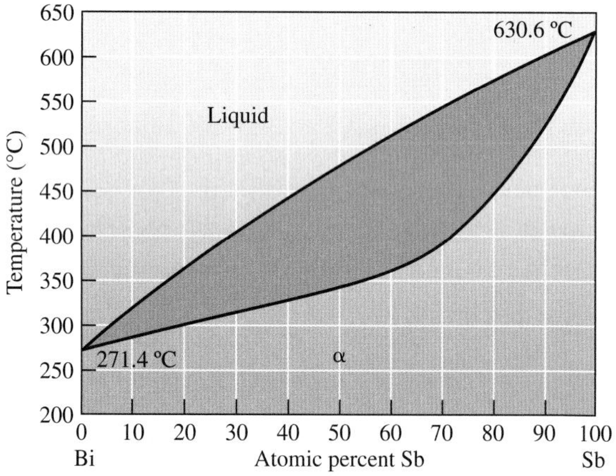

Figure 10-20 The Bi-Sb phase diagram. (*Based on Alberto Torres, Luis Hernández, Octavio Domínguez, "Effect of Antimony Additions on Corrosion and Mechanical Properties of Sn-Bi Eutectic Lead-Free Solder Alloy,"* Materials Sciences and Applications, *Vol.3 No.6(2012), http://file.scirp.org/Html/3-7700734_19815.htm*) (For Problems 10-42 and 10-68.)

10-43 Determine the composition range in which the Tl-Pb alloy at 350°C is (a) fully liquid; (b) fully solid; and (c) partly liquid and partly solid.

Refer to Figure 10-8(d) for the Tl-Pb phase diagram. Further, determine the amount of liquid and solid solution for Tl-25 wt% Pb and Tl-75 wt% Pb at 350°C and also the wt% Pb in the liquid and solid solution for both of the alloy compositions.

10-44 Determine the liquidus temperature, solidus temperature, and freezing range for the following NiO-MgO ceramic compositions: (a) NiO-30 mol% MgO; (b) NiO-45 mol% MgO; (c) NiO-60 mol% MgO; and (d) NiO-85 mol% MgO. [See Figure 10-8(b).]

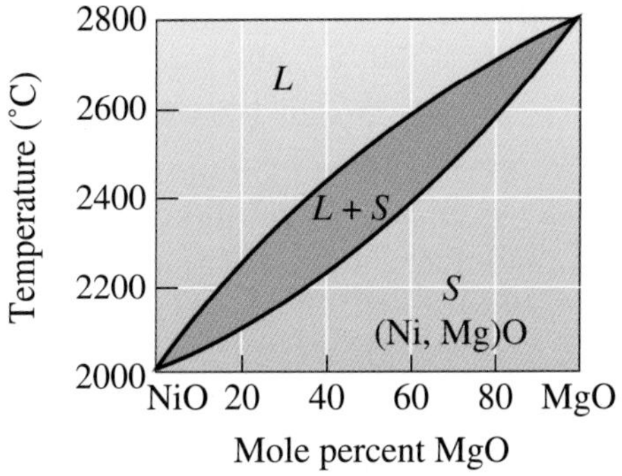

(b)

Figure 10-8(b) The equilibrium phase diagram for the NiO-MgO system. (Repeated for Problems 10-44, 10-50, 10-53, 10-57, 10-58, and 10-71.)

10-45 Determine the liquidus temperature, solidus temperature, and freezing range for the following Al_2O_3-Cr_2O_3 ceramic compositions: (a) Al_2O_3-30 wt% Cr_2O_3; (b) Al_2O_3-50 wt% Cr_2O_3; and (c) Al_2O_3-75 wt% Cr_2O_3. (See Figure 10-9.)

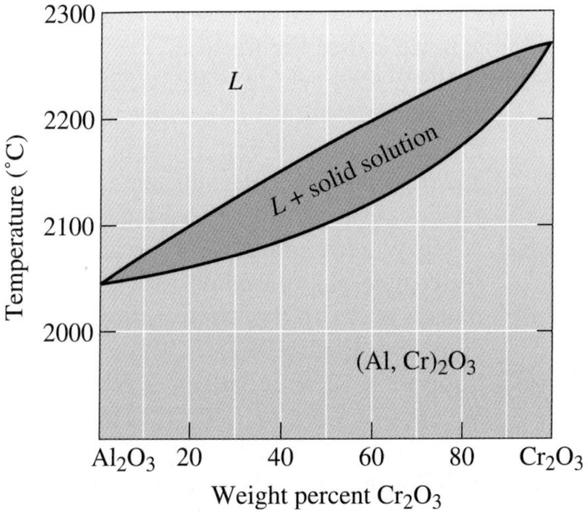

Figure 10-9 The Al_2O_3-Cr_2O_3 phase diagram. (Repeated for Problems 10-45 and 10-46.)

10-46 Determine the phases present, the compositions of each phase, and the amount of each phase in wt% for the following Al_2O_3-Cr_2O_3 ceramics at 2150°C: (a) Al_2O_3-30 wt% Cr_2O_3; (b) Al_2O_3-50 wt% Cr_2O_3; and (c) Al_2O_3-75 wt% Cr_2O_3. (See Figure 10-9.)

10-47 For a Ag-60 wt% Pd alloy determine the (a) liquidus temperature; (b) solidus temperature; and (c) freezing range. (See Figure 10-21.)

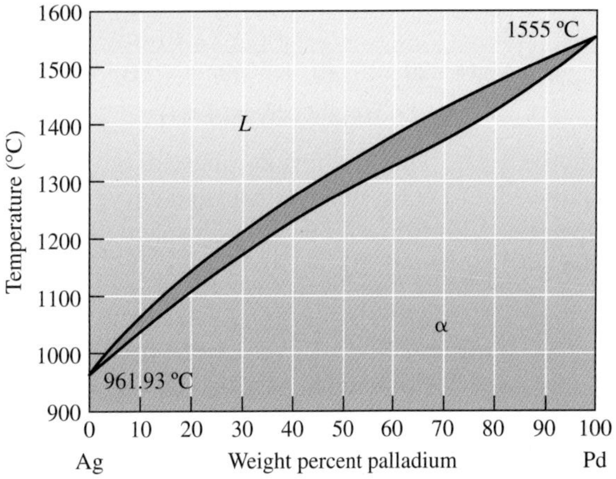

Figure 10-21 The Ag-Pd phase diagram for Problems 10-47 and 10-48. (*Based on Karakaya and Thompson,* Bulletin of Alloy Phase Diagrams, *June 1988, Volume 9, Issue 3, pp 237–243, Springer.*)

10-48 Locate the following points in the Ag-Pd phase diagram and indicate the phases present and their relative amounts: (a) 50 wt% Pd at 1300°C; (b) 80 wt% Pd at 1425°C; and (c) 90 wt% Ag at 1100°C. (See Figure 10-21.)

10-49 Determine the liquidus temperature, solidus temperature, and freezing range for the following MgO-FeO ceramic compositions: (a) MgO-25 wt% FeO; (b) MgO-45 wt% FeO; (c) MgO-65 wt% FeO; and (d) MgO-80 wt% FeO. (See Figure 10-22.)

10-50 Determine the phases present, the compositions of each phase, and the amount of each phase in mol% for the following NiO-MgO ceramics at 2400°C: (a) NiO-30 mol% MgO; (b) NiO-45 mol% MgO; (c) NiO-60 mol% MgO; and (d) NiO-85 mol% MgO. [See Figure 10-8(b).]

10-51 Determine the phases present, the compositions of each phase, and the amount of each phase in wt% for the following MgO-FeO ceramics at 2000°C: (a) MgO-25 wt%

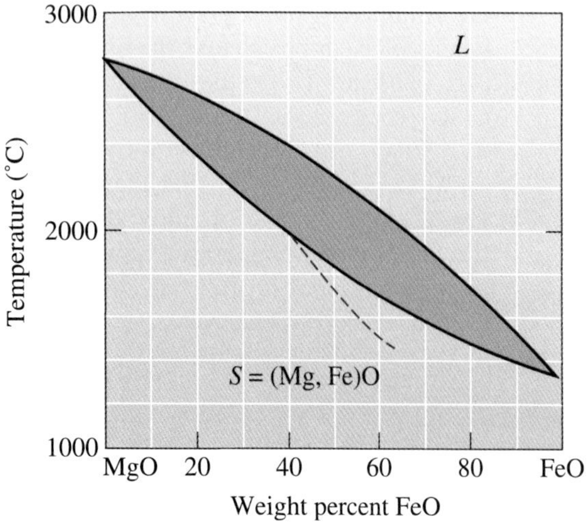

Figure 10-22 The equilibrium phase diagram for the MgO-FeO system. The dotted curve shows the nonequilibrium solidus. (For Problems 10-49, 10-51, 10-59, 10-60, 10-69, 10-73, and 10-79.)

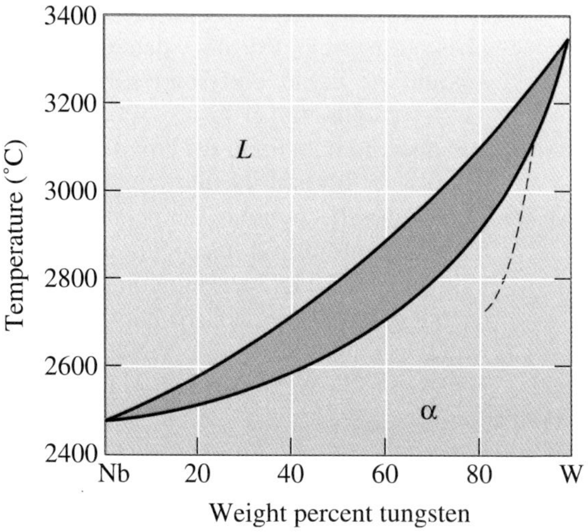

Figure 10-23 The equilibrium phase diagram for the Nb-W system. The dotted curve shows nonequilibrium solidus. (Repeated for Problems 10-54, 10-61, 10-62, 10-63, 10-64, 10-70, 10-72, 10-75, and 10-80.)

FeO; (b) MgO-45 wt% FeO; (c) MgO-60 wt% FeO; and (d) MgO-80 wt% FeO. (See Figure 10-22.)

10-52 Consider a ceramic composed of 30 mol% MgO and 70 mol% FeO. Calculate the composition of the ceramic in wt%.

10-53 A NiO-20 mol% MgO ceramic is heated to 2200°C. Determine (a) the composition of the solid and liquid phases in both mol% and wt%; (b) the amount of each phase in mol% and wt%; and (c) assuming that the density of the solid is 6.32 g/cm^3 and that of the liquid is 7.14 g/cm^3, determine the amount of each phase in vol%. [See Figure 10-8(b).]

10-54 A Nb-60 wt% W alloy is heated to 2800°C. Determine (a) the composition of the solid and liquid phases in both wt% and at%; (b) the amount of each phase in both wt% and at%; and (c) assuming that the density of the solid is 16.05 g/cm^3 and that of the liquid is 13.91 g/cm^3, determine the amount of each phase in vol%. (See Figure 10-23.)

10-55 How many grams of nickel must be added to 500 grams of copper to produce an alloy that has a liquidus temperature of 1350°C? What is the ratio of the number of nickel atoms to copper atoms in this alloy? [See Figure 10-8(a).]

10-56 How many grams of nickel must be added to 500 grams of copper to produce an alloy that contains 50 wt% α at 1300°C? [See Figure 10-8(a).]

10-57 How many grams of MgO must be added to 1 kg of NiO to produce a ceramic that has a solidus temperature of 2200°C? [See Figure 10-8(b).]

10-58 How many grams of MgO must be added to 1 kg of NiO to produce a ceramic that contains 25 mol% solid at 2400°C? [See Figure 10-8(b).]

10-59 We would like to produce a solid MgO-FeO ceramic that contains equal mol percentages of MgO and FeO at 1200°C. Determine the wt% FeO in the ceramic. (See Figure 10-22.)

10-60 We would like to produce a MgO-FeO ceramic that is 30 wt% solid at 2000°C. Determine the composition of the ceramic in wt%. (See Figure 10-22.)

10-61 A Nb-W alloy held at 2800°C is partly liquid and partly solid. (a) If possible, determine the composition of each phase in the alloy, and (b) if possible, determine the amount of each phase in the alloy. (See Figure 10-23.)

10-62 A Nb-W alloy contains 55% α at 2600°C. Determine (a) the composition of each phase, and (b) the composition of the alloy. (See Figure 10-23.)

10-63 Suppose a 1200 lb bath of a Nb-40 wt% W alloy is held at 2800°C. How many pounds of tungsten can be added to the bath before any solid forms? How many pounds of tungsten must be added to cause the entire bath to be solid? (See Figure 10-23.)

10-64 A fiber-reinforced composite material is produced, in which tungsten fibers are embedded in a Nb matrix. The composite is composed of 70 vol% tungsten. (a) Calculate the wt% of tungsten fibers in the composite, and (b) suppose the composite is heated to 2600°C and held for several years. What happens to the fibers? Explain. (See Figure 10-23.)

10-65 Suppose a crucible made from pure nickel is used to contain 500 g of liquid copper at 1150°C. Describe what happens to the system as it is held at this temperature for several hours. Explain [See Figure 10-8(a).]

Section 10-6 Relationship Between Properties and the Phase Diagram

10-66 What is brass? Explain which element strengthens the matrix for this alloy.

10-67 What is the composition of the Monel alloy?

Section 10-7 Solidification of a Solid-Solution Alloy

10-68 A Bi-60 at% Sb alloy is cooled under equilibrium conditions from the liquid state. Determine (a) the temperature at which the first solid nucleus forms and its composition; (b) the compositions and relative amounts of the phases at 450°C; and (c) the temperature at which the last liquid is present and its composition. (See Figure 10-20.)

10-69 Equal moles of MgO and FeO are combined and melted. Determine (a) the liquidus temperature, the solidus temperature, and the freezing range of the ceramic, and (b) determine the phase(s) present, their composition(s), and their amount(s) at 1800°C. (See Figure 10-22.)

10-70 Suppose 75 cm^3 of Nb and 45 cm^3 of W are combined and melted. Determine (a) the liquidus temperature, the solidus temperature, and the freezing range of the alloy, and (b) the phase(s) present, their composition(s), and their amount(s) at 2800°C. (See Figure 10-23.)

10-71 A NiO-60 mol% MgO ceramic is allowed to solidify. Determine (a) the composition of the first solid to form, and (b) the composition of the last liquid to solidify under equilibrium conditions. [See Figure 10-8(b).]

10-72 A Nb-35% W alloy is allowed to solidify. Determine (a) the composition of the first solid to form, and (b) the composition of the last liquid to solidify under equilibrium conditions. (See Figure 10-23.)

10-73 For equilibrium conditions and a MgO-65 wt% FeO ceramic, determine (a) the liquidus temperature; (b) the solidus temperature; (c) the freezing range; (d) the composition of the first solid to form during solidification; (e) the composition of the last liquid to solidify; (f) the phase(s) present, the composition of the phase(s), and the amount of the phase(s) at 1800°C; and (g) the phase(s) present, the composition of the phase(s), and the amount of the phase(s) at 1600°C. (See Figure 10-22.)

10-74 Figure 10-24 shows the cooling curve for a NiO-MgO ceramic. Determine (a) the liquidus temperature; (b) the solidus temperature; (c) the freezing range; (d) the pouring temperature; (e) the superheat; (f) the local solidification time; (g) the total solidification time; and (h) the composition of the ceramic.

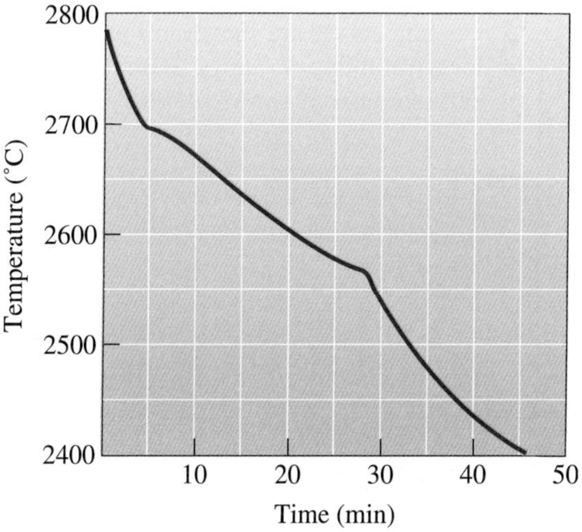

Figure 10-24 Cooling curve for a NiO-MgO ceramic (for Problem 10-74).

10-75 For equilibrium conditions and a Nb-80 wt% W alloy, determine (a) the liquidus temperature; (b) the solidus temperature; (c) the freezing range; (d) the composition of the first solid to form during solidification; (e) the composition of the last liquid to solidify; (f) the phase(s) present, the composition of the phase(s), and the amount of the phase(s) at 3000°C; and (g) the phase(s) present, the composition of the phase(s), and the amount of the phase(s) at 2800°C. (See Figure 10-23.)

10-76 Figure 10-25 shows the cooling curve for a Nb-W alloy. Determine (a) the liquidus temperature; (b) the solidus temperature;

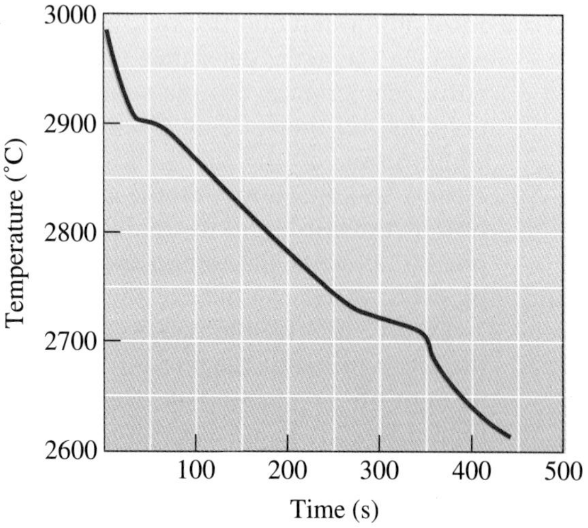

Figure 10-25 Cooling curve for a Nb-W alloy (for Problem 10-76).

(c) the freezing range; (d) the pouring temperature; (e) the superheat; (f) the local solidification time; (g) the total solidification time; and (h) the composition of the alloy.

10-77 Cooling curves are shown in Figure 10-26 for several Mo-V alloys. Based on these curves, construct the Mo-V phase diagram.

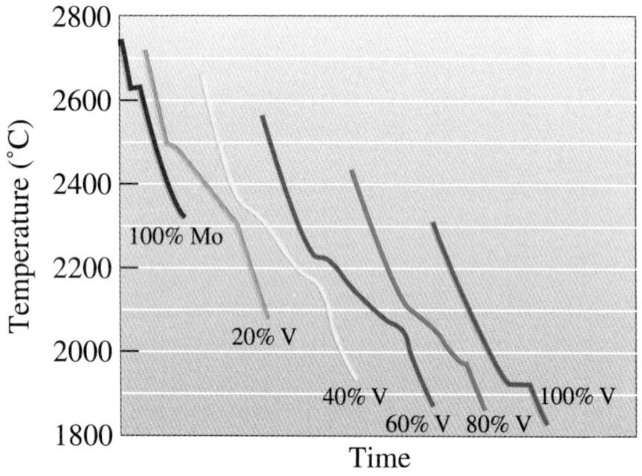

Figure 10-26 Cooling curves for a series of Mo-V alloys (for Problem 10-77).

Section 10-8 Nonequilibrium Solidification and Segregation

10-78 What are the origins of chemical segregation in cast products?

10-79 For the nonequilibrium conditions shown for the MgO-65 wt% FeO ceramic, determine (a) the liquidus temperature; (b) the nonequilibrium solidus temperature; (c) the freezing range; (d) the composition of the first solid to form during solidification; (e) the composition of the last liquid to solidify; (f) the phase(s) present, the composition of the phase(s), and the amount of the phase(s) at 1800°C; and (g) the phase(s) present, the composition of the phase(s), and the amount of the phase(s) at 1600°C. (See Figure 10-22.)

10-80 For the nonequilibrium conditions shown for the Nb-80 wt% W alloy, determine (a) the liquidus temperature; (b) the nonequilibrium solidus temperature; (c) the freezing range; (d) the composition of the first solid to form during solidification; (e) the composition of the last liquid to solidify; (f) the phase(s) present, the composition of the phase(s), and the amount of the phase(s) at 3000°C; and (g) the phase(s) present, the composition of the phase(s), and the amount of the phase(s) at 2800°C. (See Figure 10-23.)

10-81 How can microsegregation be removed?

10-82 What is macrosegregation? Is there a way to remove it without breaking up the cast structure?

10-83 What is homogenization? What type of segregation can it remove?

10-84 A copper-nickel alloy that solidifies with a secondary dendrite arm spacing (SDAS) of 0.001 cm requires 15 hours of homogenization heat treatment at 1100°C. What is the homogenization time required for the same alloy with a SDAS of 0.01 cm and 0.0001 cm? If the diffusion coefficient of Ni in Cu at 1100°C is 3×10^{-10} cm^2/s, calculate the constant c in the homogenization time equation. What assumption is made in this calculation?

10-85 What is spray atomization? Can it be used for making ceramic powders?

10-86 Suppose you are asked to manufacture a critical component based on a nickel-based superalloy. The component must not contain any porosity and it must be chemically homogeneous. What manufacturing process would you use for this application? Why?

10-87 What is hot pressing? How is it different from hot isostatic pressing?

Design Problems

10-88 Homogenization of a slowly cooled Cu-Ni alloy having a secondary dendrite arm spacing of 0.025 cm requires 8 hours at 1000°C. Design a process to produce a homogeneous structure in a more rapidly cooled Cu-Ni alloy having a SDAS of 0.005 cm.

10-89 Design a process to produce a NiO-60% MgO refractory with a structure that is 40% glassy phase at room temperature. Include all relevant temperatures.

10-90 Design a method by which glass beads (having a density of 2.3 g/cm^3) can be uniformly mixed and distributed in a Cu-20% Ni alloy (density of 8.91 g/cm^3).

10-91 Suppose that MgO contains 5 mol% NiO. Design a solidification purification method that will reduce the NiO to less than 1 mol% in the MgO.

Computer Problems

10-92 *Gibbs Phase Rule.* Write a computer program that will automate the Gibbs phase rule calculation. The program should ask the user for information on whether the pressure and temperature or only the pressure is to be held constant. The program then should use the correct equation to calculate the appropriate variable the user wants to know. The user will provide inputs for the number of components. Then, if the user wishes to provide the number of phases present, the program should calculate the degrees of freedom and vice-versa.

10-93 *Conversion of wt% to at% for a Binary System.* Write a computer program that will allow conversion of wt% into at%. The program should ask the user to provide appropriate formula weights of the elements/compounds. (See Equations 10-4 and 10-5.)

10-94 *Hume–Rothery Rules.* Write a computer program that will predict whether or not there will likely be 100% solid solubility between two elements. The program should ask the user to provide information on crystal structures of the elements or compounds, radii of different atoms or ions involved, and valence and electronegativity values. You will have to make assumptions as to how much difference in values of electronegativity might be acceptable. The program should then use the Hume–Rothery rules and provide the user with guidance on the possibility of forming a system that shows 100% solid solubility.

Knovel® Problem

K10-1 What is the solidus temperature for a silicon-germanium system containing 30 wt% Si?

K10-2 What is the composition of liquid Ag found in a 50 wt% mixture of Ni and Ag at 1273 K (1000°C)?

Soldering plays a key role in processing of printed circuit boards and other devices for microelectronics. This process often makes use of alloys based on lead and tin. Specific compositions of these alloys known as the eutectic compositions, melt at constant temperature (like pure elements). These compositions are also mechanically strong since their microstructure comprises an intimate mixture of two distinct phases. The enhancement of mechanical properties by dispersing one phase in another phase via the formation of eutectics is the central theme for this chapter. (*Photolink/Photodisc Green/Getty Images*)

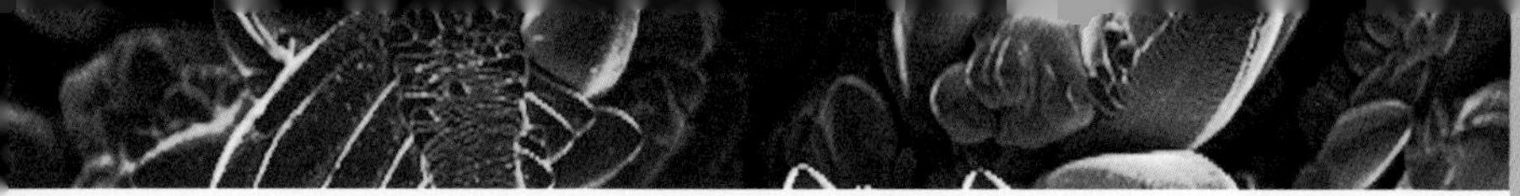

CHAPTER

11

Dispersion Strengthening and Eutectic Phase Diagrams

Have You Ever Wondered?

- *What alloys are most commonly used for soldering?*
- *What is fiberglass?*
- *Is there an alloy that freezes at a constant temperature?*
- *What is Pyrex® glass?*

Chapter Learning Objectives

The key objectives of this chapter are to

- Explain the effects of dispersion strengthening on the mechanical properties of metallic materials.
- Draw a eutectic phase diagram indicating the melting temperature of each pure element, the eutectic temperature, and the phases present in each region.
- Calculate the amounts and composition of each phase present for any point on a eutectic phase diagram.
- Sketch the likely microstructure of a eutectic system given the composition and cooling treatment history.
- Determine the composition of an alloy based on its cooling curve.
- Describe the mechanism of vapor-liquid-solid (VLS) nanowire growth.

When the solubility of a material is exceeded by adding too much of an alloying element or compound, a second phase forms and a two-phase material is produced. The boundary between the two phases, known as the **interphase interface**, is a surface where the arrangement of atoms is not perfect. In metallic materials, this boundary interferes with the slip or movement of dislocations, causing strengthening. The general term for such strengthening by the introduction of a second phase is dispersion strengthening. In this chapter, we first discuss the fundamentals of dispersion strengthening to determine the microstructure we should aim to produce. Next, we examine the types of reactions that produce multiple phase alloys. Finally, we examine methods to achieve dispersion strengthening through control of the solidification process. In this context, we will examine phase diagrams that involve the formation of multiple phases. We will concentrate on eutectic phase diagrams.

We will conclude the chapter by examining the vapor-liquid-solid method of nanowire growth, which can be understood by considering eutectic phase diagrams.

11-1 Principles and Examples of Dispersion Strengthening

Most engineered materials are composed of more than one phase, and many of these materials are designed to provide improved strength. In simple **dispersion-strengthened** alloys, small particles of one phase, usually very strong and hard, are introduced into a second phase, which is weaker but more ductile. The soft phase, usually continuous and present in larger amounts, is called the **matrix**. The hard, strengthening phase may be called the **dispersed phase** or the **precipitate**, depending on how the alloy is formed. In some cases, a phase or a mixture of phases may have a very characteristic appearance—in these cases, this phase or phase mixture may be called a **microconstituent**. For dispersion strengthening to occur, the dispersed phase or precipitate must be small enough to provide effective obstacles to dislocation movement, thus providing the strengthening mechanism.

In most alloys, dispersion strengthening is produced by phase transformations. In this chapter, we will concentrate on a solidification transformation by which a liquid freezes to simultaneously form two solid phases. This will be called a **eutectic** reaction and is of particular importance in cast irons and many aluminum alloys. In the next chapter, we will discuss the **eutectoid** reaction, by which one solid phase reacts to simultaneously form two different solid phases; this reaction is key in the control of properties of steels. In Chapter 12, we will also discuss **precipitation** (or **age) hardening**, which produces precipitates by a sophisticated heat treatment.

When increased strength and toughness are the goals of incorporating a dispersed phase, the guidelines below should be followed (Figure 11-1).

1. The matrix should be soft and ductile, while the dispersed phase should be hard and strong. The dispersed phase particles interfere with slip, while the matrix provides at least some ductility to the overall alloy.
2. The hard, dispersed phase should be discontinuous, while the soft, ductile matrix should be continuous. If the hard and brittle dispersed phase were continuous, cracks could propagate through the entire structure.
3. The dispersed phase particles should be small and numerous, increasing the likelihood that they interfere with the slip process since the area of the interphase interface is increased significantly.
4. The dispersed phase particles should be round, rather than needle-like or sharp-edged, because the rounded shape is less likely to initiate a crack or to act as a notch.
5. Higher concentrations of the dispersed phase increase the strength of the alloy.

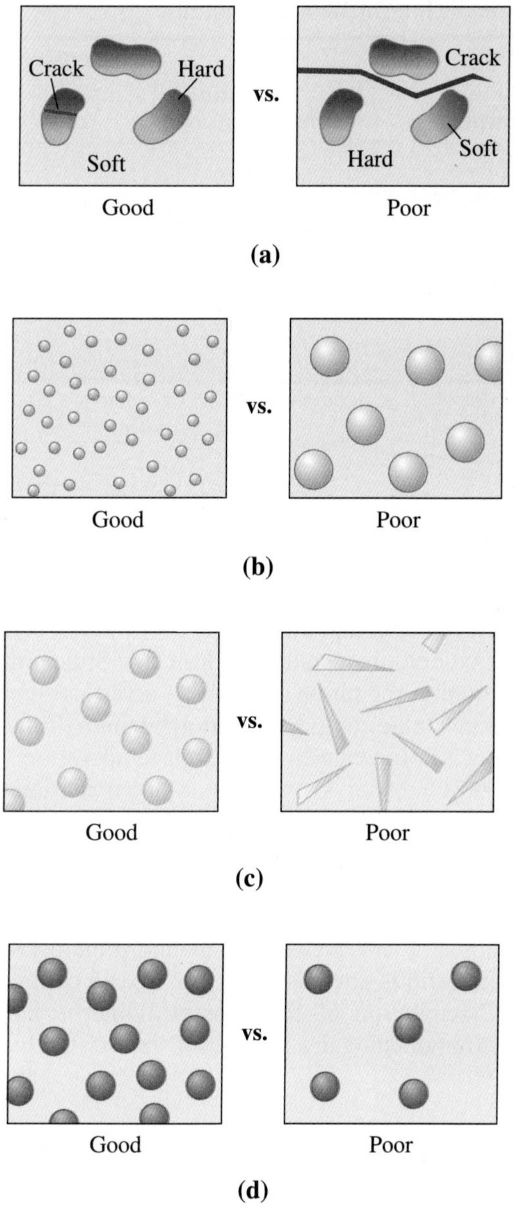

Figure 11-1 Considerations for effective dispersion strengthening: (a) The precipitate phase should be hard and discontinuous, while the matrix should be continuous and soft, (b) the dispersed phase particles should be small and numerous, (c) the dispersed phase particles should be round rather than needle-like, and (d) larger amounts of the dispersed phase increase strengthening.

11-2 Intermetallic Compounds

An **intermetallic compound** contains two or more metallic elements, producing a new phase with its own composition, crystal structure, and properties. Intermetallic compounds are almost always very hard and brittle. Intermetallics or intermetallic compounds are similar to ceramic materials in terms of their mechanical properties.

Our interest in intermetallics is twofold. First, often dispersion-strengthened alloys contain an intermetallic compound as the dispersed phase. Second, many intermetallic compounds, on their own (and not as a second phase), are being investigated and developed for high-temperature applications. In this section, we will discuss properties of intermetallics as stand-alone materials. In the sections that follow, we will discuss how intermetallic phases help strengthen metallic materials. Table 11-1 summarizes the properties of some intermetallic compounds.

Table 11-1 Properties of some intermetallic compounds

Intermetallic Compound	Crystal Structure	Melting Temperature (°C)	Density $\left(\frac{g}{cm^3}\right)$	Young's Modulus (GPa)
FeAl	Ordered BCC	1250–1400	5.6	263
NiAl	Ordered FCC (*B2)	1640	5.9	206
Ni_3Al	Ordered FCC (*$L1_2$)	1390	7.5	337
TiAl	Ordered tetragonal (*$L1_0$)	1460	3.8	94
Ti_3Al	Ordered HCP	1600	4.2	210
$MoSi_2$	Tetragonal	2020	6.31	430

*Based on Meyers, M. A., and Chawla, K. K., Mechanical Behavior of Materials, 2nd Edition. Cambridge University Press, 2009.

Stoichiometric intermetallic compounds have a fixed composition. Steels are often strengthened by a stoichiometric compound, iron carbide (Fe_3C), which has a fixed ratio of three iron atoms to one carbon atom. Note that Fe_3C does not meet the strict definition of an intermetallic compound because it does not contain two or more metallic elements, but it is commonly included in this class. Stoichiometric intermetallic compounds are represented in the phase diagram by a vertical line [Figure 11-2(a)]. An example of a useful intermetallic compound is molybdenum disilicide ($MoSi_2$). This material is used for making heating elements for high-temperature furnaces. At high temperatures (~1000 to 1600°C), $MoSi_2$ shows outstanding oxidation resistance. At low temperatures (~500°C and below), $MoSi_2$ is brittle and shows catastrophic oxidation known as pesting.

Nonstoichiometric intermetallic compounds have a range of compositions and are sometimes called **intermediate solid solutions**. In the molybdenum-rhodium system, the γ phase is a nonstoichiometric intermetallic compound [Figure 11-2(b)]. Because the molybdenum-rhodium atom ratio is not fixed, the γ phase can contain from 45 wt% to 83 wt% Rh at 1600°C. Precipitation of the nonstoichiometric intermetallic copper aluminide ($CuAl_2$) causes strengthening in a number of important aluminum alloys.

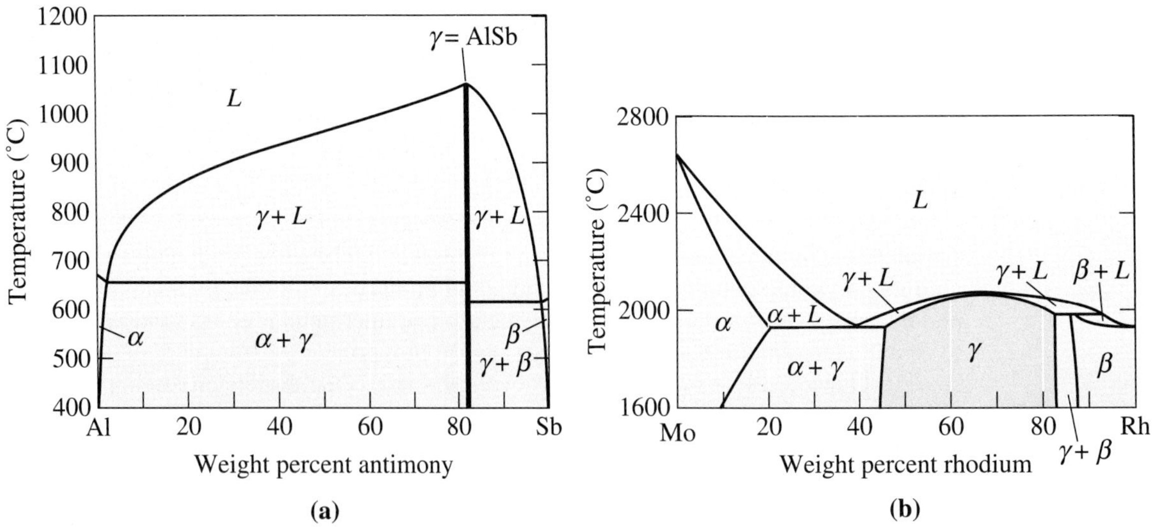

Figure 11-2 (a) The aluminum-antimony phase diagram includes a stoichiometric intermetallic compound γ. (b) The molybdenum-rhodium phase diagram includes a nonstoichiometric intermetallic compound γ.

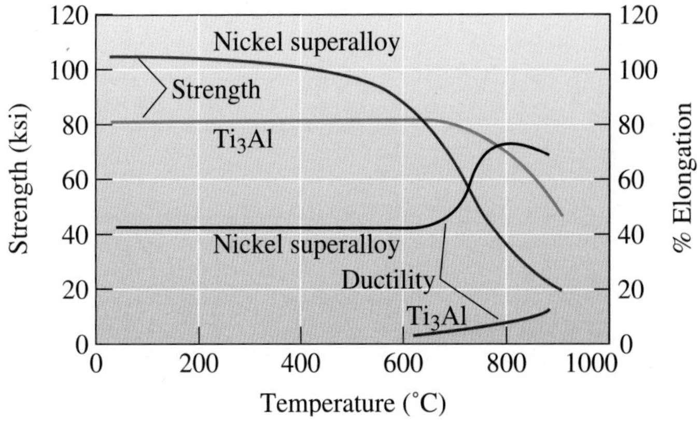

Figure 11-3 The strength and ductility of the intermetallic compound Ti_3Al compared to that of a conventional nickel superalloy. The Ti_3Al maintains its strength to higher temperatures than does the nickel superalloy.

Properties and Applications of Intermetallics Intermetallics such as Ti_3Al and Ni_3Al maintain their strength and even develop usable ductility at elevated temperatures (Figure 11-3). Low ductility, though, has impeded further development of these materials. It has been shown that the addition of small levels of boron (up to 0.2%) can enhance the ductility of polycrystalline Ni_3Al. Enhanced ductility levels could make it possible for intermetallics to be used in many high-temperature and load-bearing applications. Ordered compounds of NiAl and Ni_3Al are also candidates for supersonic aircraft, jet engines, and high-speed commercial aircraft. Not all applications of intermetallics are structural. Intermetallics based on silicon (e.g., platinum silicide) play a useful role in microelectronics and certain intermetallics such as Nb_3Sn are useful as superconductors.

11-3 Phase Diagrams Containing Three-Phase Reactions

Many binary systems produce phase diagrams more complicated than the isomorphous phase diagrams discussed in Chapter 10. The systems we will discuss here contain reactions that involve three separate phases. Five such reactions are defined in Figure 11-4. Each of these reactions can be identified in a phase diagram by the following procedure.

1. Locate a horizontal line on the phase diagram. The horizontal line, which indicates the presence of a three-phase reaction, represents the temperature at which the reaction occurs under equilibrium conditions.
2. Locate three distinct points on the horizontal line: the two endpoints plus a third point, in between the two endpoints of the horizontal line. This third point represents the composition at which the three-phase reaction occurs. In Figure 11-4, the point in between has been shown at the center; however, on a real phase diagram, this point is not necessarily at the center.
3. Look immediately above the in-between point and identify the phase or phases present; look immediately below the in-between point and identify the phase or phases present. Then write the reaction from the phase(s) above the point that are transforming to the phase(s) below the point. Compare this reaction with those in Figure 11-4 to identify the reaction.

Eutectic	$L \rightarrow \alpha + \beta$	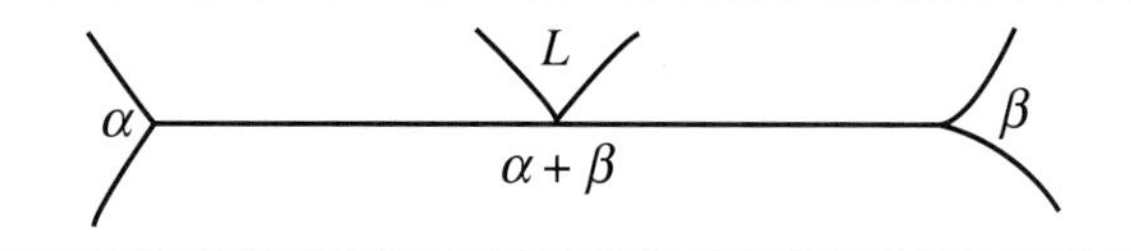
Peritectic	$\alpha + L \rightarrow \beta$	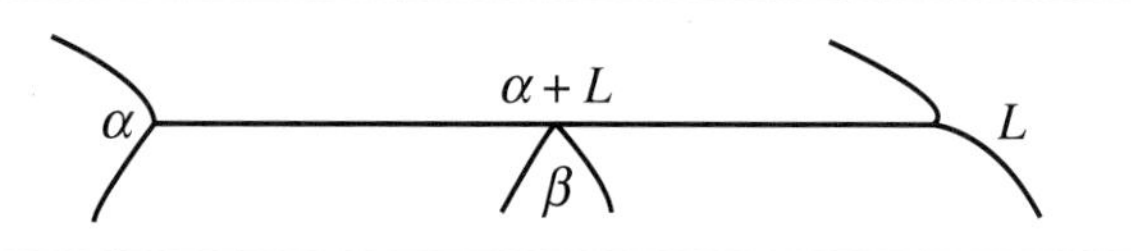
Monotectic	$L_1 \rightarrow L_2 + \alpha$	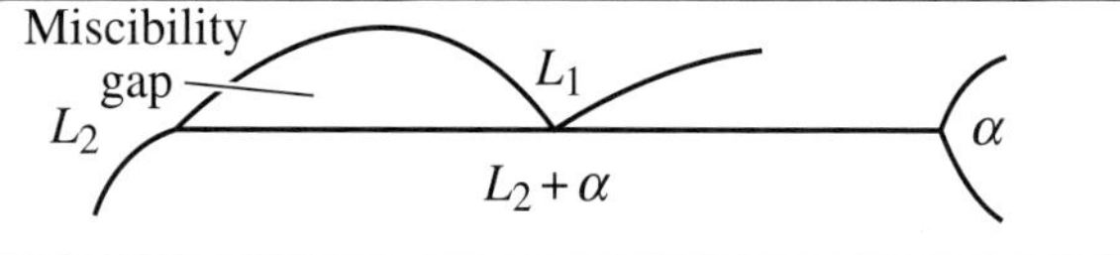
Eutectoid	$\gamma \rightarrow \alpha + \beta$	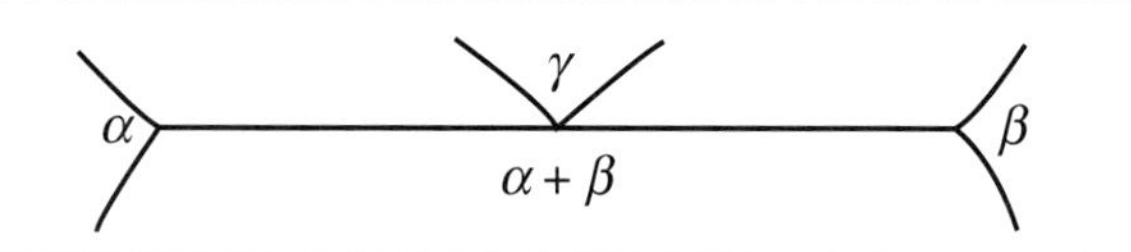
Peritectoid	$\alpha + \beta \rightarrow \gamma$	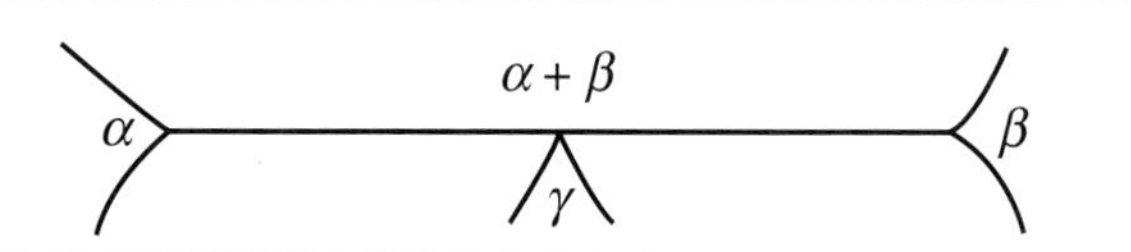

Figure 11-4 The five most important three-phase reactions in binary phase diagrams.

Example 11-1 *Identifying Three-phase Reactions*

Consider the binary phase diagram in Figure 11-5. Identify the three-phase reactions that occur.

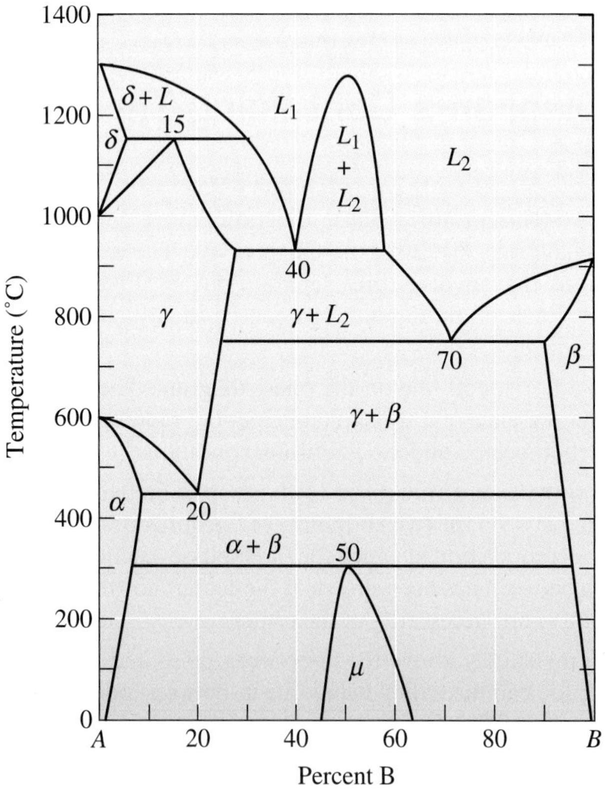

Figure 11-5
A hypothetical phase diagram (for Example 11-1).

SOLUTION

We find horizontal lines at 1150°C, 920°C, 750°C, 450°C, and 300°C. For 1150°C: This reaction occurs at 15% B (i.e., the in-between point is at 15% B). $\delta + L$ are present above the point, and γ is present below. The reaction is

$$\delta + L \rightarrow \gamma, \quad \text{a } \textit{peritectic}$$

920°C: This reaction occurs at 40% B:

$$L_1 \rightarrow \gamma + L_2, \quad \text{a } \textit{monotectic}$$

750°C: This reaction occurs at 70% B:

$$L \rightarrow \gamma + \beta, \quad \text{a } \textit{eutectic}$$

450°C: This reaction occurs at 20% B:

$$\gamma \rightarrow \alpha + \beta, \quad \text{a } \textit{eutectoid}$$

300°C: This reaction occurs at 50% B:

$$\alpha + \beta \rightarrow \mu, \quad \text{a } \textit{peritectoid}$$

The eutectic, **peritectic**, and **monotectic** reactions are part of the solidification process. Alloys used for casting or soldering often take advantage of the low melting point of the eutectic reaction. The phase diagram of monotectic alloys contains a dome, or a **miscibility gap**, in which two liquid phases coexist. In the copper-lead system, the monotectic reaction produces small globules of dispersed lead, which improve the machinability of the copper alloy. Peritectic reactions lead to nonequilibrium solidification and segregation.

In many systems, there is a **metastable miscibility gap**. In this case, the immiscibility dome extends into the sub-liquidus region. In some cases, the entire miscibility gap is metastable (i.e., the immiscibility dome is completely under the liquidus). These systems form such materials as Vycor™ and Pyrex® glasses, also known as phase-separated glasses.

The eutectoid and **peritectoid** reactions are completely solid-state reactions. The eutectoid reaction forms the basis for the heat treatment of several alloy systems, including steel (Chapter 12). The peritectoid reaction is extremely slow, often producing undesirable, nonequilibrium structures in alloys. As noted in Chapter 5, the rate of diffusion of atoms in solids is much smaller than in liquids.

Each of these three-phase reactions occurs at a fixed temperature and composition. The Gibbs phase rule for a three-phase reaction is (at a constant pressure),

$$1 + C = F + P$$
$$F = 1 + C - P = 1 + 2 - 3 = 0 \tag{11-1}$$

since there are two components C in a binary phase diagram and three phases P are involved in the reaction. When the three phases are in equilibrium during the reaction, there are no degrees of freedom. As a result, these reactions are called invariant. The temperature and the composition of each phase involved in the three-phase reaction are fixed. Note that of the five reactions discussed here, only eutectic and eutectoid reactions can lead to dispersion strengthening.

11-4 The Eutectic Phase Diagram

The lead-tin system contains only a simple eutectic reaction (Figure 11-6). This alloy system is the basis for the most common alloys used for soldering. As mentioned in Chapter 9, there is an intense effort underway to replace lead in Pb-Sn solders due to lead's toxicity. We will continue to use a Pb-Sn system, though, to discuss the eutectic phase diagram. Let's examine four classes of alloys in this system.

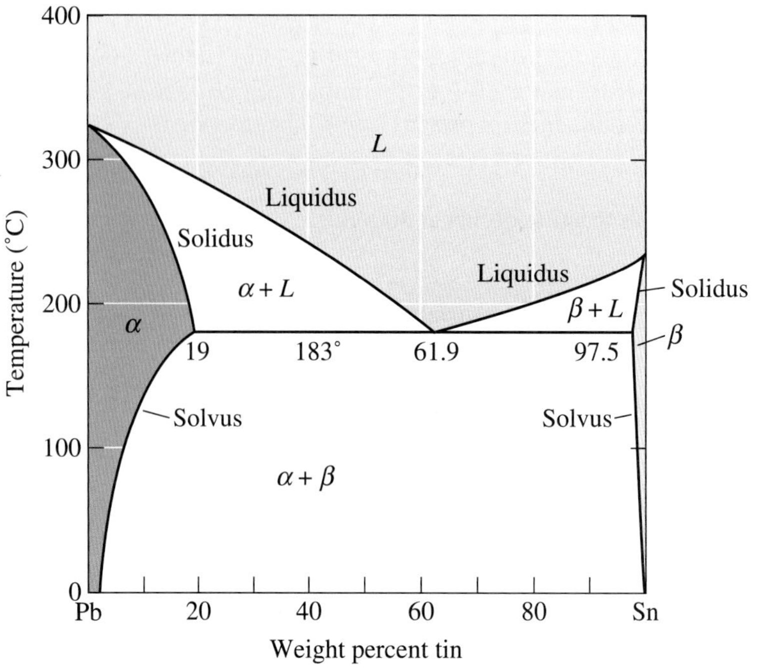

Figure 11-6 The lead-tin equilibrium phase diagram.

It is critical to note that binary phase diagrams typically do not incorporate different colors to indicate different phases. We have chosen to do so, however, for the Pb-Sn phase diagrams in this chapter to help you visualize the various microstructures that form during solidification. The single-phase regions of the Pb-Sn diagrams are colored, while the two-phase regions are white. All other phase diagrams are presented as a single color so that you are also comfortable with the more conventional presentation.

Solid-Solution Alloys Alloys that contain 0 to 2% Sn behave exactly like the copper-nickel alloys; a single-phase solid solution α forms during solidification (Figure 11-7). These alloys are strengthened by solid-solution strengthening, strain hardening, and controlling the solidification process to refine the grain structure.

Alloys That Exceed the Solubility Limit Alloys containing between 2% and 19% Sn also solidify to produce a single solid solution α; however, as the alloy continues to cool, a solid-state reaction occurs, permitting a second solid phase (β) to precipitate from the original α phase (Figure 11-8).

On this phase diagram, the α is a solid solution of tin in lead; however, the solubility of tin in the α solid solution is limited. At 0°C, only 2% Sn can dissolve in α. As the temperature increases, more tin dissolves into the lead until, at 183°C, the solubility of tin in lead has increased to 19% Sn. This is the maximum solubility of tin in lead. The solubility of tin in solid lead at any temperature is given by the **solvus** curve (Figure 11-6). As any alloy containing between 2% and 19% Sn cools below the solvus, the solubility limit is exceeded, and a small amount of β forms.

We control the properties of this type of alloy by several techniques, including solid-solution strengthening of the α portion of the structure, controlling the microstructure produced during solidification, and controlling the amount and characteristics of the β phase. These types of compositions, which form a single solid phase at high temperatures and two solid phases at lower temperatures, are suitable for age or precipitate hardening. In Chapter 12, we will learn how nonequilibrium processes are needed to make precipitation hardened alloys. A vertical line on a phase diagram (e.g., Figure 11-8) that shows a specific composition is known as an **isopleth**. Determination of reactions that occur upon the cooling of a particular composition is known as an **isoplethal study**. The following example illustrates how certain calculations related to the composition of phases and their relative concentrations can be performed.

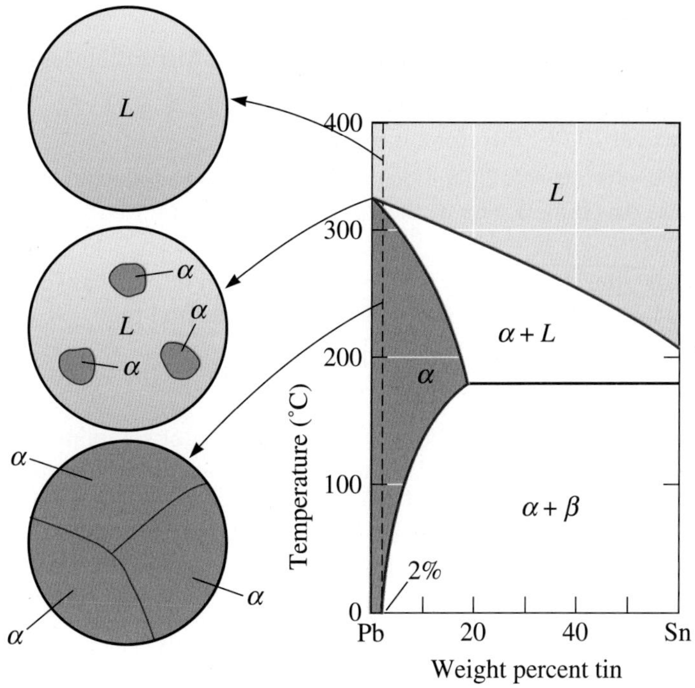

Figure 11-7 Solidification and microstructure of a Pb-2% Sn alloy. The alloy is a single-phase solid solution.

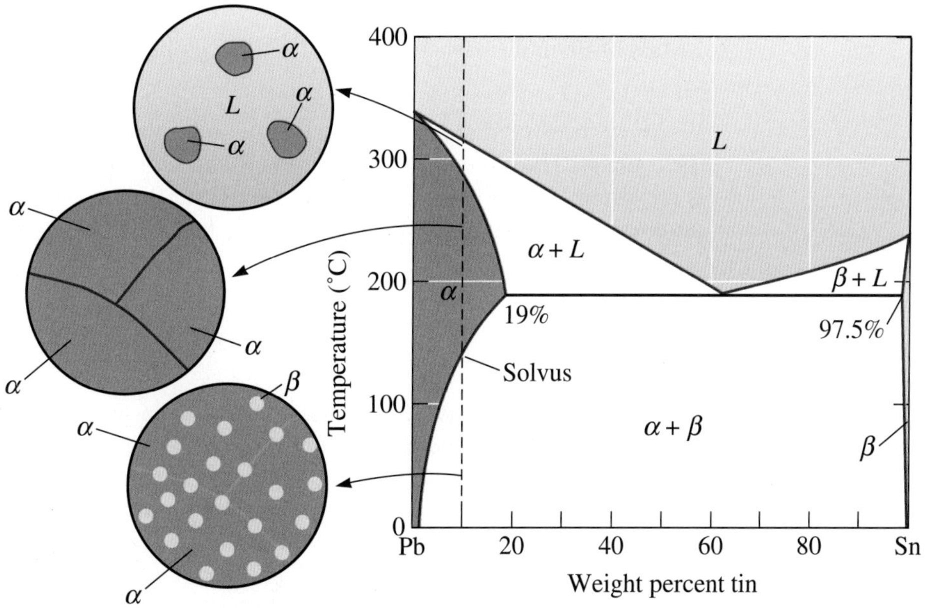

Figure 11-8 Solidification, precipitation, and microstructure of a Pb-10% Sn alloy. Some dispersion strengthening occurs as the β solid precipitates.

Example 11-2 *Phases in the Lead-Tin Phase Diagram*

Determine (a) the solubility of tin in solid lead at 100°C and (b) the maximum solubility of lead in solid tin. Also if a Pb-10% Sn alloy is cooled to 0°C, determine (c) the amount of β that forms, (d) the masses of tin contained in the α and β phases, and (e) the mass of lead contained in the α and β phases. Assume that the total mass of the Pb-10% Sn alloy is 100 grams.

SOLUTION

The phase diagram we need is shown in Figure 11-8. All percentages shown are weight %.

(a) The 100°C temperature intersects the solvus curve at 6% Sn. The solubility of tin in lead at 100°C, therefore, is 6%.

(b) The maximum solubility of lead in tin, which is found from the tin-rich side of the phase diagram, occurs at the eutectic temperature of 183°C and is 97.5% Sn or 2.5% Pb.

(c) At 0°C, the 10% Sn alloy is in the $\alpha + \beta$ region of the phase diagram. By drawing a tie line at 0°C and applying the lever rule, we find that

$$\%\beta = \frac{10 - 2}{100 - 2} \times 100 = 8.2\%$$

Note that the tie line intersects the solvus curve for solubility of Pb in Sn at a non-zero concentration of Sn. We cannot read this accurately from the diagram; thus, we assume that the right-hand point for the tie line is 100% Sn. The percent of α would be $(100 - \%\beta) = 91.8\%$. This means if we have 100 g of the 10% Sn alloy, it will consist of 8.2 g of the β phase and 91.8 g of the α phase.

(d) Note that 100 g of the alloy will consist of 10 g of Sn and 90 g of Pb. The Pb and Sn are distributed in two phases (i.e., α and β). The mass of Sn in the α phase = (2% Sn)(91.8 g of α phase) = (0.02)(91.8 g) = 1.836 g. Since tin appears in both the α and β phases, the mass of Sn in the β phase will be = (10 − 1.836) g = 8.164 g. Note that in this case, the β phase at 0°C is nearly pure Sn.

(e) Let's now calculate the mass of lead in the two phases. The mass of Pb in the α phase will be equal to the mass of the α phase minus the mass of Sn in the α phase = 91.8 g − 1.836 g = 89.964 g. We could have also calculated this as

$$\text{Mass of Pb in the } \alpha \text{ phase} = (98\%\ \text{Pb})(91.8\ \text{g of } \alpha \text{ phase}) = (0.98)(91.8\ \text{g}) = 89.964\ \text{g}$$

We know the total mass of the lead (90 g), and we also know the mass of lead in the α phase. Thus, the mass of Pb in the β phase = 90 − 89.964 = 0.036 g. This is consistent with what we said earlier (i.e., the β phase, in this case, is almost pure tin).

Eutectic Alloys The alloy containing 61.9% Sn has the eutectic composition (Figure 11-9). The word eutectic comes from the Greek word *eutectos* that means easily fused. Indeed, in a binary system showing one eutectic reaction, an alloy with the eutectic composition has the lowest melting temperature. This is the composition for which there is no freezing range (i.e., solidification of this alloy occurs at one temperature, 183°C in the Pb-Sn system). Above 183°C, the alloy is all liquid and, therefore, must contain 61.9% Sn. After the liquid cools to 183°C, the eutectic reaction begins:

$$L_{61.9\%\ \text{Sn}} \rightarrow \alpha_{19\%\ \text{Sn}} + \beta_{97.5\%\ \text{Sn}} \quad (11\text{-}2)$$

Two solid solutions—α and β—are formed during the eutectic reaction. The compositions of the two solid solutions are given by the ends of the eutectic line.

During solidification, growth of the eutectic requires both removal of the latent heat of fusion and redistribution of the two different atom species by diffusion. Since solidification occurs completely at 183°C, the cooling curve (Figure 11-10) is similar to that of a pure metal; that is, a thermal arrest or plateau occurs at the eutectic temperature. In Chapter 9, we stated that alloys solidify over a range of temperatures (between the liquidus and solidus) known as the freezing range. Eutectic compositions

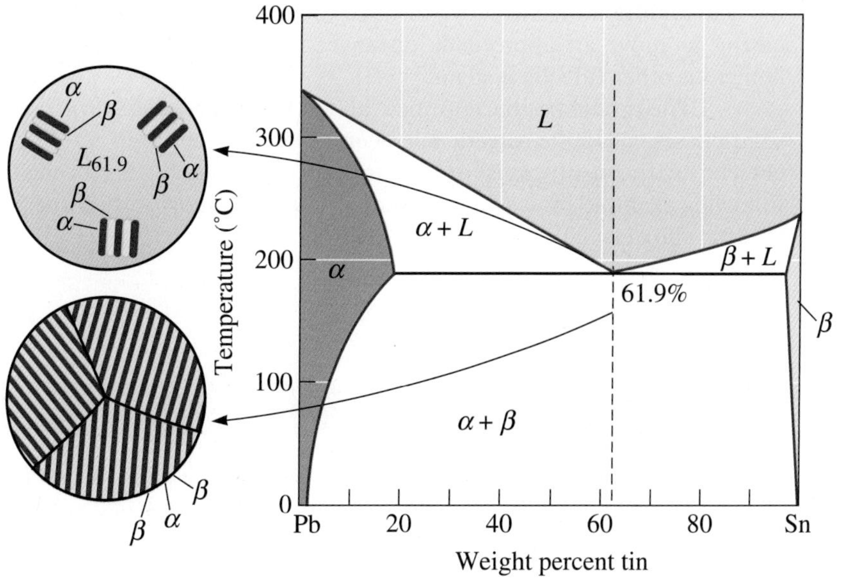

Figure 11-9 Solidification and microstructure of the eutectic alloy Pb-61.9% Sn.

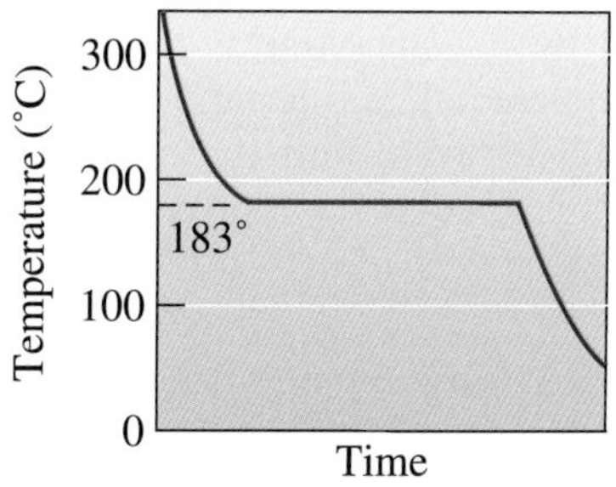

Figure 11-10 The cooling curve for a eutectic alloy is a simple thermal arrest, since eutectics freeze or melt at a single temperature.

are an exception to this rule since they transform from a liquid to a solid at a constant temperature (i.e., the eutectic temperature).

As atoms are redistributed during eutectic solidification, a characteristic microstructure develops. In the lead-tin system, the solid α and β phases grow from the liquid in a **lamellar**, or plate-like, arrangement (Figure 11-11). The lamellar structure permits

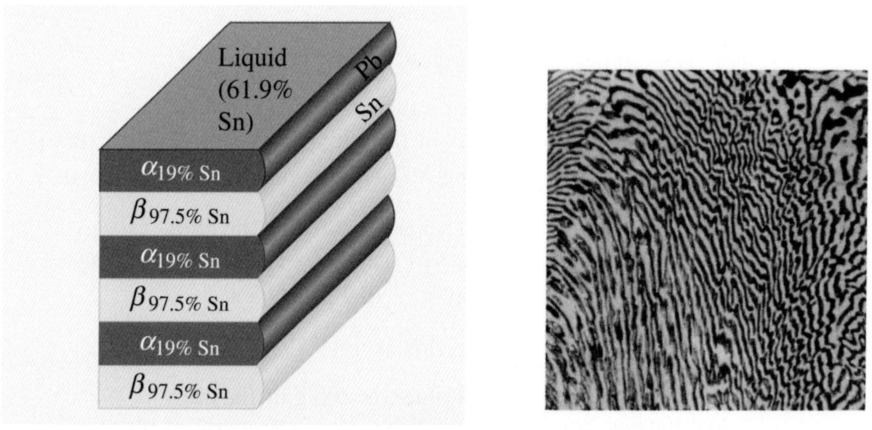

(a) **(b)**

Figure 11-11 (a) Atom redistribution during lamellar growth of a lead-tin eutectic. Tin atoms from the liquid preferentially diffuse to the β plates, and lead atoms diffuse to the α plates. (b) Photomicrograph of the lead-tin eutectic microconstituent ($\times$ 400). (*Reprinted Courtesy of Don Askeland*) the lead and tin atoms to move through the liquid, in which diffusion is rapid, without having to move an appreciable distance. This lamellar structure is characteristic of numerous other eutectic systems.

The product of the eutectic reaction has a characteristic arrangement of the two solid phases called the **eutectic microconstituent**. In the Pb-61.9% Sn alloy, 100% of the eutectic microconstituent is formed, since all of the liquid goes through the reaction. The following example shows how the amounts and compositions of the phases present in a eutectic alloy can be calculated.

Example 11-3 *Amount of Phases in the Eutectic Alloy*

(a) Determine the amount and composition of each phase in 200 g of a lead-tin alloy of eutectic composition immediately after the eutectic reaction has been completed. (b) Calculate the mass of phases present. (c) Calculate the masses of lead and tin in each phase.

SOLUTION

(a) The eutectic alloy contains 61.9% Sn. We work the lever law at a temperature just below the eutectic—say, at 182°C, since that is the temperature at which the eutectic reaction is just completed. The fulcrum of our lever is 61.9% Sn. The composition of the α is Pb-19% Sn, and the composition of the β is Pb-97.5% Sn. The ends of the tie line coincide approximately with the ends of the eutectic line.

$$\alpha\text{: (Pb} - 19\%\text{ Sn) } \%\,\alpha = \frac{97.5 - 61.9}{97.5 - 19.0} \times 100 = 45.35\%$$

$$\beta\text{: (Pb} - 97.5\%\text{ Sn) } \%\,\beta = \frac{61.9 - 19.0}{97.5 - 19.0} \times 100 = 54.65\%$$

Alternately, we could state that the weight fraction of the α phase is 0.4535 and that of the β phase is 0.5465.

A 200 g sample of the alloy would contain a total of 200 × 0.6190 = 123.8 g Sn and a balance of 76.2 g lead. The total mass of lead and tin cannot change as a consequence of conservation of mass. What changes is the mass of lead and tin in the different phases.

(b) At a temperature just below the eutectic:

The mass of the α phase in 200 g of the alloy
= mass of the alloy × fraction of the α phase
= 200 g × 0.4535 = 90.7 g

The mass of the β phase in 200 g of the alloy
= (mass of the alloy − mass of the α phase)
= 200.0 g − 90.7 = 109.3 g

We could have also written this as

The mass of the β phase in 200 g of the alloy
= mass of the alloy × fraction of the β phase
= 200 g × 0.5465 = 109.3 g

Thus, at a temperature just below the eutectic (i.e., at 182°C), the alloy contains 109.3 g of the β phase and 90.7 g of the α phase.

(c) Now let's calculate the masses of lead and tin in the α and β phases:

$$\text{Mass of Pb in the } \alpha \text{ phase} = \text{mass of the } \alpha \text{ phase in 200 g} \times (\text{weight fraction Pb in } \alpha)$$

$$\text{Mass of Pb in the } \alpha \text{ phase} = 90.7 \text{ g} \times (1 - 0.19) = 73.5 \text{ g}$$

$$\text{Mass of Sn in the } \alpha \text{ phase} = \text{mass of the } \alpha \text{ phase} - \text{mass of Pb in the } \alpha \text{ phase}$$

$$\text{Mass of Sn in the } \alpha \text{ phase} = (90.7 - 73.5 \text{ g}) = 17.2 \text{ g}$$

$$\text{Mass of Pb in the } \beta \text{ phase} = \text{mass of the } \beta \text{ phase in 200 g} \times (\text{weight fraction Pb in } \beta)$$

$$\text{Mass of Pb in the } \beta \text{ phase} = (109.3 \text{ g}) \times (1 - 0.975) = 2.7 \text{ g}$$

$$\text{Mass of Sn in the } \beta \text{ phase} = \text{total mass of Sn} - \text{mass of Sn in the } \alpha \text{ phase}$$

$$\text{Mass of Sn in the } \beta \text{ phase} = 123.8 \text{ g} - 17.2 \text{ g} = 106.6 \text{ g}$$

Notice that we could have obtained the same result by considering the total lead mass balance as follows:

$$\text{Total mass of lead in the alloy} = \text{mass of lead in the } \alpha \text{ phase} + \text{mass of lead in the } \beta \text{ phase}$$

$$76.2 \text{ g} = 73.5 \text{ g} + \text{mass of lead in the } \beta \text{ phase}$$

$$\text{Mass of lead in the } \beta \text{ phase} = 76.2 - 73.5 \text{ g} = 2.7 \text{ g}$$

Figure 11-12 summarizes the various concentrations and masses. This analysis confirms that most of the lead in the eutectic alloy gets concentrated in the α phase. Most of the tin gets concentrated in the β phase.

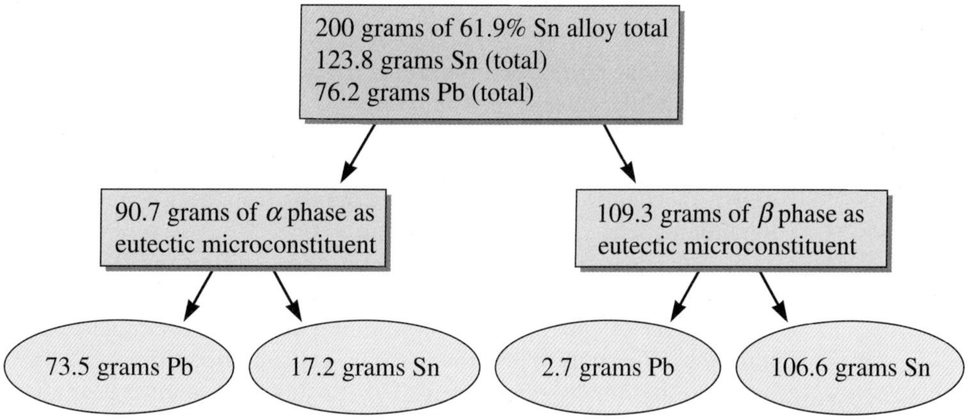

Figure 11-12 Summary of calculations (for Example 11-3).

Hypoeutectic and Hypereutectic Alloys A **hypoeutectic alloy** is an alloy with a composition between that of the left-hand end of the tie line defining the eutectic reaction and the eutectic composition. As a hypoeutectic alloy containing between 19% and 61.9% Sn cools, the liquid begins to solidify at the liquidus temperature, producing solid α; however, solidification is completed by going through the eutectic reaction (Figure 11-13). This solidification sequence occurs for compositions in which

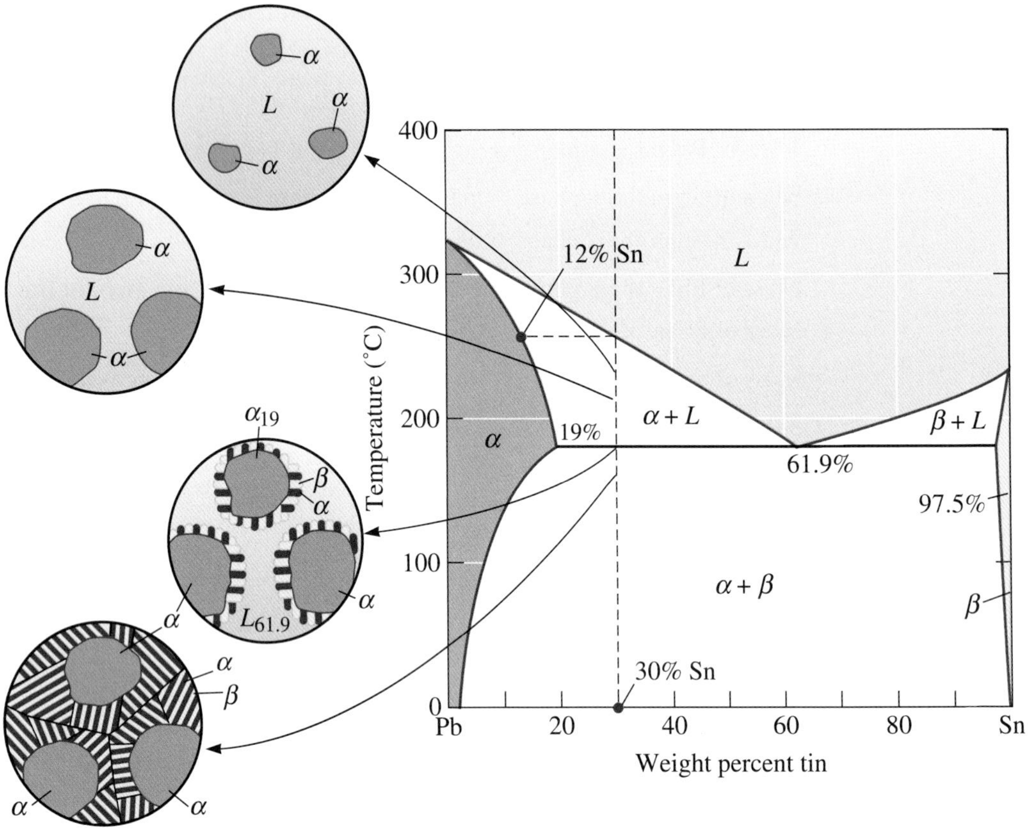

Figure 11-13 The solidification and microstructure of a hypoeutectic alloy (Pb-30% Sn).

the vertical line corresponding to the original composition of the alloy crosses both the liquidus and the eutectic.

An alloy composition between that of the right-hand end of the tie line defining the eutectic reaction and the eutectic composition is known as a **hypereutectic alloy**. In the Pb-Sn system, any composition between 61.9% and 97.5% Sn is hypereutectic.

Let's consider a hypoeutectic alloy containing Pb-30% Sn and follow the changes in structure during solidification (Figure 11-13). On reaching the liquidus temperature of 260°C, solid α containing about 12% Sn nucleates. The solid α grows until the alloy cools to just above the eutectic temperature. At 184°C, we draw a tie line and find that the solid α contains 19% Sn and the remaining liquid contains 61.9% Sn. We note that at 184°C, the liquid contains the eutectic composition! When the alloy is cooled below 183°C, all of the remaining liquid goes through the eutectic reaction and transforms to a lamellar mixture of α and β. The microstructure shown in Figure 11-14(a) results. Notice that the eutectic microconstituent surrounds the solid α that formed between the liquidus and eutectic temperatures. The eutectic microconstituent is continuous and the primary phase is dispersed between the colonies of the eutectic microconstituent.

We call the solid α phase that forms when the liquid cools from the liquidus to the eutectic the **primary** or **proeutectic microconstituent**. This solid α does not take part in the eutectic reaction. Thus, the morphology and appearance of this α phase is distinct from that of the α phase that appears in the eutectic microconstituent. Often we find that the amounts and compositions of the microconstituents are of more use to us than the amounts and compositions of the phases.

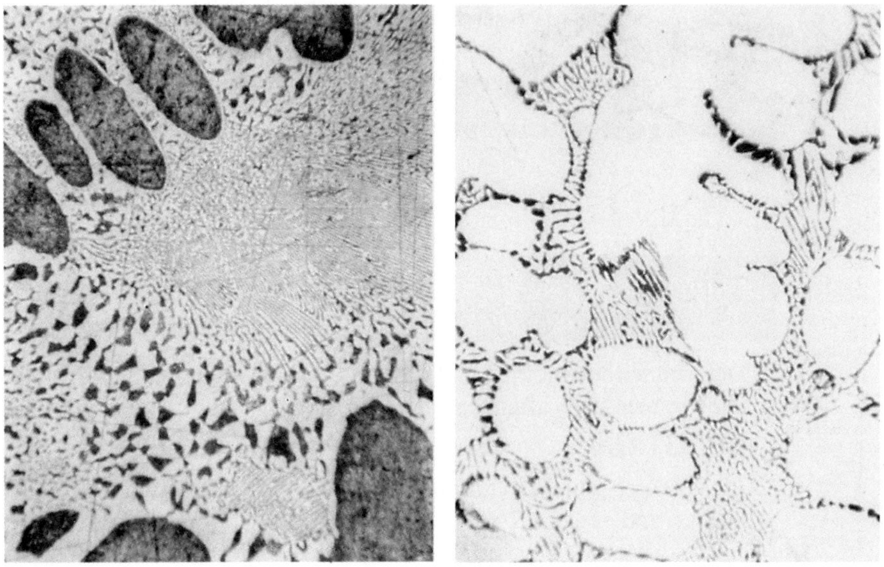

(a) (b)

Figure 11-14 (a) A hypoeutectic lead-tin alloy. (b) A hypereutectic lead-tin alloy. The dark constituent is the lead-rich solid α; the light constituent is the tin-rich solid β; and the fine plate structure is the eutectic (× 400). (*Reprinted Courtesy of Don Askeland*)

Example 11-4 *Determination of Phases and Amounts in a Pb-30% Sn Hypoeutectic Alloy*

For a Pb-30% Sn alloy, determine the phases present, their amounts, and their compositions at 300°C, 200°C, 184°C, 182°C, and 0°C.

SOLUTION

Temperature (°C)	Phases	Compositions	Amounts
300	L	L: 30% Sn	$L = 100\%$
200	$\alpha + L$	L: 55% Sn	$L = \frac{30 - 18}{55 - 18} \times 100 = 32\%$
		α: 18% Sn	$\alpha = \frac{55 - 30}{55 - 18} \times 100 = 68\%$
184	$\alpha + L$	L: 61.9% Sn	$L = \frac{30 - 19}{61.9 - 19} \times 100 = 26\%$
		α: 19% Sn	$\alpha = \frac{61.9 - 30}{61.9 - 19} \times 100 = 74\%$
182	$\alpha + \beta$	α: 19% Sn	$\alpha = \frac{97.5 - 30}{97.5 - 19} \times 100 = 86\%$
		β: 97.5% Sn	$\beta = \frac{30 - 19}{97.5 - 19} \times 100 = 14\%$
0	$\alpha + \beta$	α: 2% Sn	$\alpha = \frac{100 - 30}{100 - 2} \times 100 = 71\%$
		β: 100% Sn	$\beta = \frac{30 - 2}{100 - 2} \times 100 = 29\%$

Note that in these calculations, the fractions have been rounded off to the nearest %. This can pose problems if we were to calculate masses of different phases, in that you may not be able to preserve mass balance. It is usually a good idea not to round these percentages if you are going to perform calculations concerning amounts of different phases or masses of elements in different phases.

Example 11-5 Microconstituent Amount and Composition for a Hypoeutectic Alloy

Determine the amounts and compositions of each microconstituent in a Pb-30% Sn alloy immediately after the eutectic reaction has been completed.

SOLUTION

This is a hypoeutectic composition. Therefore, the microconstituents expected are primary α and eutectic. Note that we still have only two phases (α and β).

We can determine the amounts and compositions of the microconstituents if we look at how they form. The *primary* α microconstituent is all of the solid α that forms before the alloy cools to the eutectic temperature; the eutectic microconstituent is all of the liquid that goes through the eutectic reaction. At a temperature just above the eutectic—say, 184°C—the amounts and compositions of the two phases are

$$\alpha\text{: 19\% Sn, } \%\,\alpha = \frac{61.9 - 30}{61.9 - 19} \times 100 = 74\% = \%\text{ primary } \alpha$$

$$L\text{: 61.9\% Sn, } \%\,L = \frac{30 - 19}{61.9 - 19} \times 100 = 26\% = \%\text{ eutectic at 182°C}$$

Thus, the primary α microconstituent is obtained by determining the amount of α present at the temperature just above the eutectic. The amount of eutectic microconstituent at a temperature *just below* the eutectic (e.g., 182°C) is determined by calculating the amount of liquid *just above* the eutectic temperature (e.g., at 184°C), since all of this liquid of eutectic composition is transformed into the eutectic microconstituent. Note that at the eutectic temperature (183°C), the eutectic reaction is in progress (formation of the proeutectic α is complete); hence, the amount of the eutectic microconstituent at 183°C will change with time (starting at 0% and ending at 26% eutectic, in this case). Please be certain that you understand this example since many students tend to miss how the calculation is performed.

When the alloy cools below the eutectic to 182°C, all of the liquid at 184°C transforms to eutectic and the composition of the eutectic microconstituent is 61.9% Sn. The solid α present at 184°C remains unchanged after cooling to 182°C and is the primary microconstituent.

The cooling curve for a hypoeutectic alloy is a composite of those for solid-solution alloys and eutectic alloys (Figure 11-15). A change in slope occurs at the liquidus as primary α begins to form. Evolution of the latent heat of fusion slows the cooling rate as the solid α grows. When the alloy cools to the eutectic temperature, a thermal arrest is produced as the eutectic reaction proceeds at 183°C. The solidification sequence is similar in a hypereutectic alloy, giving the microstructure shown in Figure 11-14(b).

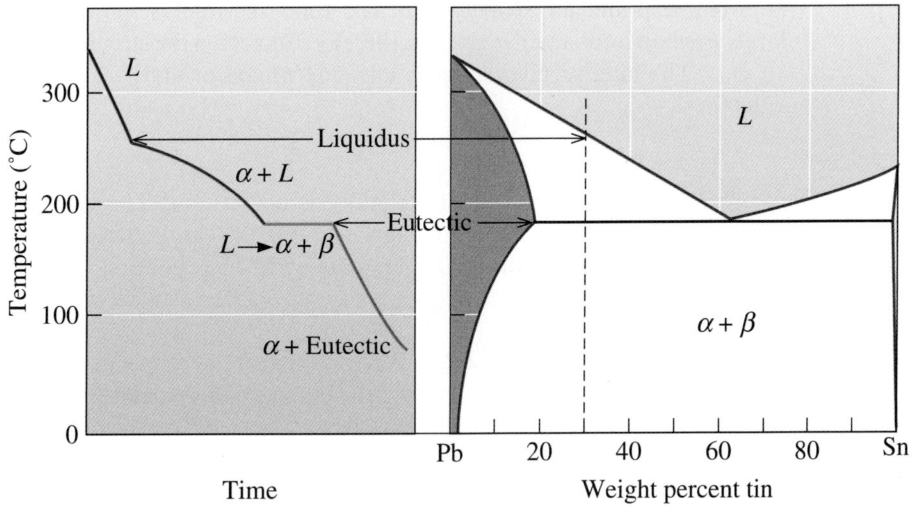

Figure 11-15 The cooling curve for a hypoeutectic Pb-30% Sn alloy.

11-5 Strength of Eutectic Alloys

Each phase in the eutectic alloy is, to some degree, solid-solution strengthened. In the lead-tin system, α, which is a solid solution of tin in lead, is stronger than pure lead (Chapter 10). Some eutectic alloys can be strengthened by cold working. We also control grain size by adding appropriate inoculants or grain refiners during solidification. Finally, we can influence the properties by controlling the amount and microstructure of the eutectic.

Eutectic Colony Size Eutectic colonies each nucleate and grow independently. Within each colony, the orientation of the lamellae in the eutectic microconstituent is identical. The orientation changes on crossing a colony boundary [Figure 11-16(a)]. We can refine the eutectic colonies and improve the strength of the eutectic alloy by inoculation (Chapter 9).

Interlamellar Spacing The **interlamellar spacing** of a eutectic is the distance from the center of one α lamella to the center of the next α lamella [Figure 11-16(b)].

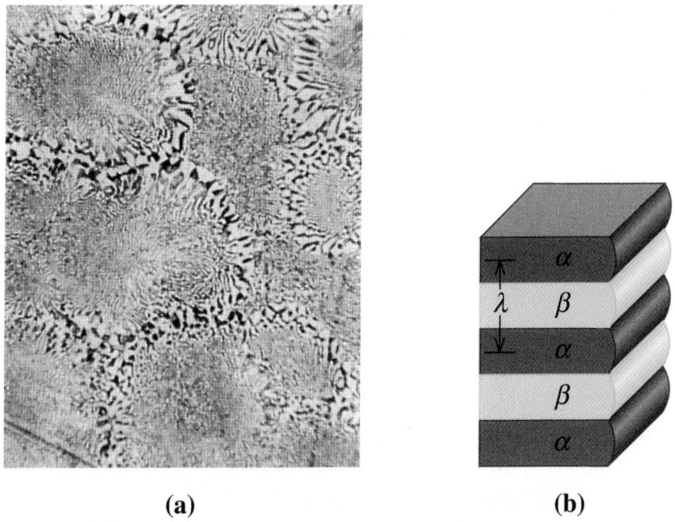

Figure 11-16 (a) Colonies in the lead-tin eutectic (×300). (*Reprinted Courtesy of Don Askeland*) (b) The interlamellar spacing in a eutectic microstructure.

A small interlamellar spacing indicates that the amount of α to β interface area is large. A small interlamellar spacing therefore increases the strength of the eutectic.

The interlamellar spacing λ is determined primarily by the growth rate of the eutectic,

$$\lambda = cR^{-1/2} \tag{11-3}$$

where R is the growth rate (cm/s) and c is a constant. The interlamellar spacing for the lead-tin eutectic is shown in Figure 11-17. We can increase the growth rate R and consequently reduce the interlamellar spacing by increasing the cooling rate or reducing the solidification time.

Amount of Eutectic We also control the properties by the relative amounts of the primary microconstituent and the eutectic. In the lead-tin system, the amount of the eutectic microconstituent changes from 0% to 100% when the tin content increases from 19% to 61.9%. With increasing amounts of the stronger eutectic microconstituent, the strength of the alloy increases (Figure 11-18). Similarly, when we increase the lead added to tin from 2.5% to 38.1% Pb, the amount of primary β in the hypereutectic alloy decreases, the amount of the strong eutectic increases, and the strength increases. When both individual phases have about the same strength, the eutectic alloy is expected to have the highest strength due to effective dispersion strengthening.

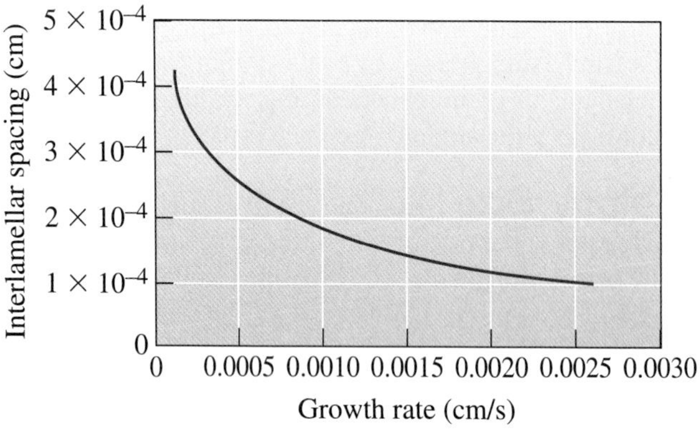

Figure 11-17 The effect of growth rate on the interlamellar spacing in the lead-tin eutectic.

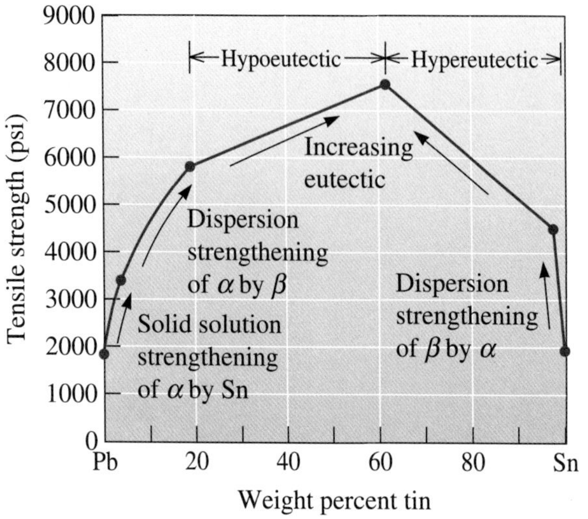

Figure 11-18 The effect of the composition and strengthening mechanism on the tensile strength of lead-tin alloys.

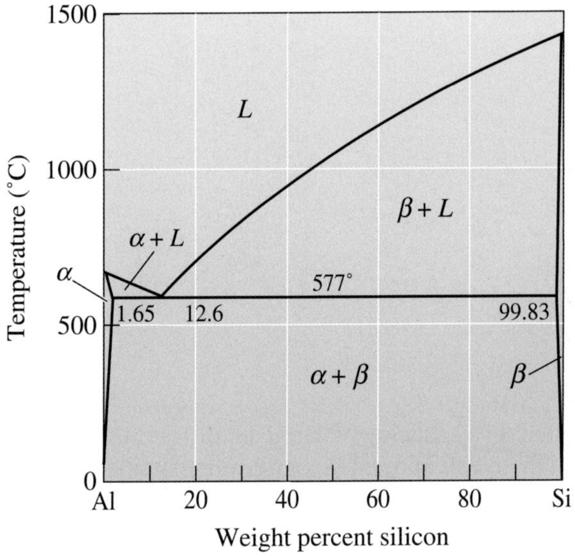

Figure 11-19 The aluminum-silicon phase diagram.

Microstructure of the Eutectic Not all eutectics give a lamellar structure. The shapes of the two phases in the microconstituent are influenced by the cooling rate, the presence of impurity elements, and the nature of the alloy.

The aluminum-silicon eutectic phase diagram (Figure 11-19) forms the basis for a number of important commercial alloys. The silicon portion of the eutectic grows as thin, flat plates that appear needle-like in a photomicrograph [Figure 11-20(a)]. The brittle silicon platelets concentrate stresses and reduce ductility and toughness.

The eutectic microstructure in aluminum-silicon alloys is altered by the addition of alloying elements in a process known as modification. **Modification** causes the silicon phase to grow as thin, interconnected rods between aluminum dendrites [Figure 11-20(b)], improving both tensile strength and percent elongation. In two dimensions, the modified silicon appears to be composed of small, round particles. Rapidly cooled alloys, such as those used for die casting, are modified naturally during solidification. At slower cooling rates, however, about 0.02% Na or 0.01% Sr must be added to cause modification.

The shape of the primary phase is also important. Often the primary phase grows in a dendritic manner; decreasing the secondary dendrite arm spacing

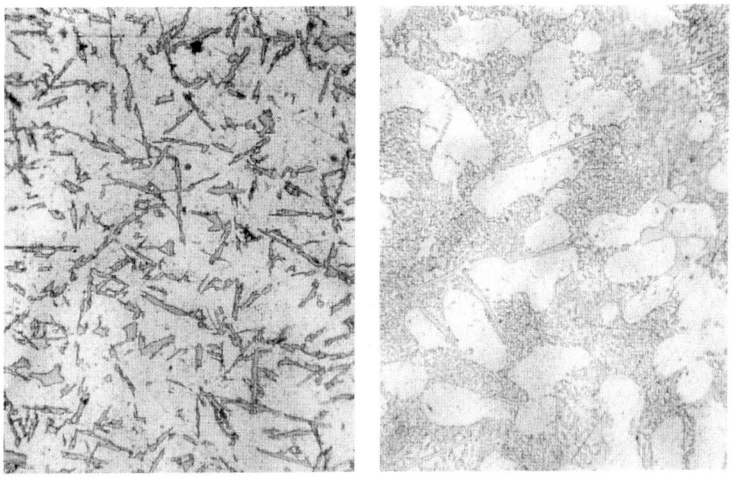

(a) (b)

Figure 11-20 Typical eutectic microstructures: (a) needle-like silicon plates in the aluminum-silicon eutectic (×100) and (b) rounded silicon rods in the modified aluminum-silicon eutectic (×100). (*Reprinted Courtesy of Don Askeland*)

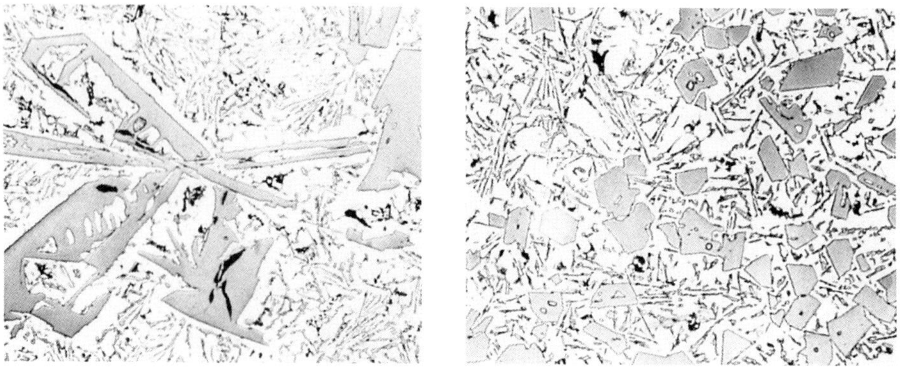

Figure 11-21 The effect of hardening with phosphorus on the microstructure of hypereutectic aluminum-silicon alloys: (a) coarse primary silicon and (b) fine primary silicon, as refined by phosphorus addition (×75). [*From* ASM Handbook, *Vol. 7, (1972), ASM International, Materials Park, OH 44073. Reprinted with permission of ASM International. All rights reserved. www.asminternational.org.*]

of the primary phase may improve the properties of the alloy. In hypereutectic aluminum-silicon alloys, coarse β is the primary phase [Figure 11-21(a)]. Because β is hard, the hypereutectic alloys are wear resistant and are used to produce automotive engine parts, but the coarse β causes poor machinability and gravity segregation (where the primary β floats to the surface of the casting during freezing). Addition of 0.05% phosphorus encourages nucleation of primary silicon, refines its size, and minimizes its deleterious qualities [Figure 11-21(b)]. The two examples that follow show how eutectic compositions can be designed to achieve certain levels of mechanical properties.

Example 11-6 *Design of Materials for a Wiping Solder*

One way to repair dents in a metal is to wipe a partly liquid-partly solid material into the dent, then allow this filler material to solidify. For our application, the wiping material should have the following specifications: (1) a melting temperature below 230°C, (2) a tensile strength in excess of 6000 psi, (3) be 60% to 70% liquid during application, and (4) the lowest possible cost. Design an alloy and repair procedure that will meet these specifications. You may wish to consider a Pb-Sn alloy.

SOLUTION

Let's see if one of the Pb-Sn alloys will satisfy these conditions. First, the alloy must contain more than 40% Sn in order to have a melting temperature below 230°C (Figure 11-6). This low temperature will make it easier for the person doing the repairs to apply the filler.

Second, Figure 11-18 indicates that the tin content must be between 23% and 80% to achieve the required 6000 psi tensile strength. In combination with the first requirement, any alloy containing between 40 and 80% Sn will be satisfactory.

Third, the cost of tin is about \$5500/ton whereas that of lead is \$550/ton. Thus, an alloy of Pb-40% Sn might be the most economical choice. There are other considerations, as well, such as what is the geometry? Can the alloy flow well under that geometry (i.e., the viscosity of the molten metal)?

Finally, the filler material must be at the correct temperature in order to be 60% to 70% liquid. As the calculations below show, the temperature must be between 200°C and 210°C:

$$\% L_{200} = \frac{40 - 18}{55 - 18} \times 100 = 60\%$$

$$\% L_{210} = \frac{40 - 17}{50 - 17} \times 100 = 70\%$$

Our recommendation, therefore, is to use a Pb-40% Sn alloy applied at 205°C, a temperature at which there will be 65% liquid and 35% primary α. As mentioned before, we should also pay attention to the toxicity of lead and any legal liabilities the use of such materials may cause. A number of new lead-free solders have been developed.

Example 11-7 *Design of a Wear-Resistant Part*

Design a lightweight, cylindrical component that will provide excellent wear resistance at the inner wall, yet still have reasonable ductility and toughness overall. Such a product might be used as a cylinder liner in an automotive engine.

SOLUTION

Many wear-resistant parts are produced from steels, which have a relatively high density, but the hypereutectic Al-Si alloys containing primary β may provide the wear resistance that we wish at one-third the weight of the steel.

Since the part to be produced is cylindrical in shape, centrifugal casting (Figure 11-22) might be an ideal method for producing it. In centrifugal casting, liquid metal is poured into a rotating mold and the centrifugal force produces a hollow shape. In addition, material that has a high density is spun to the outside wall of the casting, while material that has a lower density than the liquid migrates to the inner wall.

When we centrifugally cast a hypereutectic Al-Si alloy, primary β nucleates and grows. The density of the β phase (if we assume it to be same as that of pure Si) is, according to Appendix A, 2.33 g/cm^3, compared with a density near 2.7 g/cm^3 for aluminum. As the primary β particles precipitate from the liquid, they are spun to the

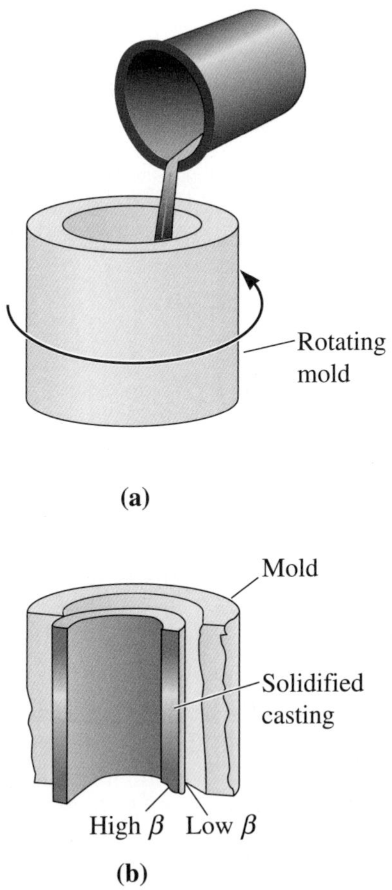

Figure 11-22 Centrifugal casting of a hypereutectic Al-Si alloy: (a) Liquid alloy is poured into a rotating mold, and (b) the solidified casting is hypereutectic at the inner diameter and eutectic at the outer diameter (for Example 11-7).

inner surface. The result is a casting that is composed of eutectic microconstituent (with reasonable ductility) at the outer wall and a hypereutectic composition, containing large amounts of primary β, at the inner wall.

A typical alloy used to produce aluminum engine components is Al-17% Si. From Figure 11-19, the total amount of primary β that can form is calculated at 578°C, just above the eutectic temperature:

$$\% \text{ Primary } \beta = \frac{17 - 12.6}{99.83 - 12.6} \times 100 = 5.0\%$$

Although only 5.0% primary β is expected to form, the centrifugal action can double or triple the amount of β at the inner wall of the casting.

11-6 Eutectics and Materials Processing

Manufacturing processes take advantage of the low melting temperature associated with the eutectic reaction. The Pb-Sn alloys are the basis for a series of alloys used to produce filler materials for soldering. If, for example, we wish to join copper pipe, individual segments can be joined by introducing the low melting point eutectic Pb-Sn alloy into the joint [Figure 11-23(a)]. The copper is heated to just above the eutectic temperature. The heated copper melts the Pb-Sn alloy, which is then drawn into the thin gap by capillary action. When the Pb-Sn alloy cools and solidifies, the copper is joined. The possibility of corrosion of such pipes and the introduction of lead into water must also be considered.

Many casting alloys are also based on eutectic alloys. Liquid can be melted and poured into a mold at low temperatures, reducing energy costs involved in

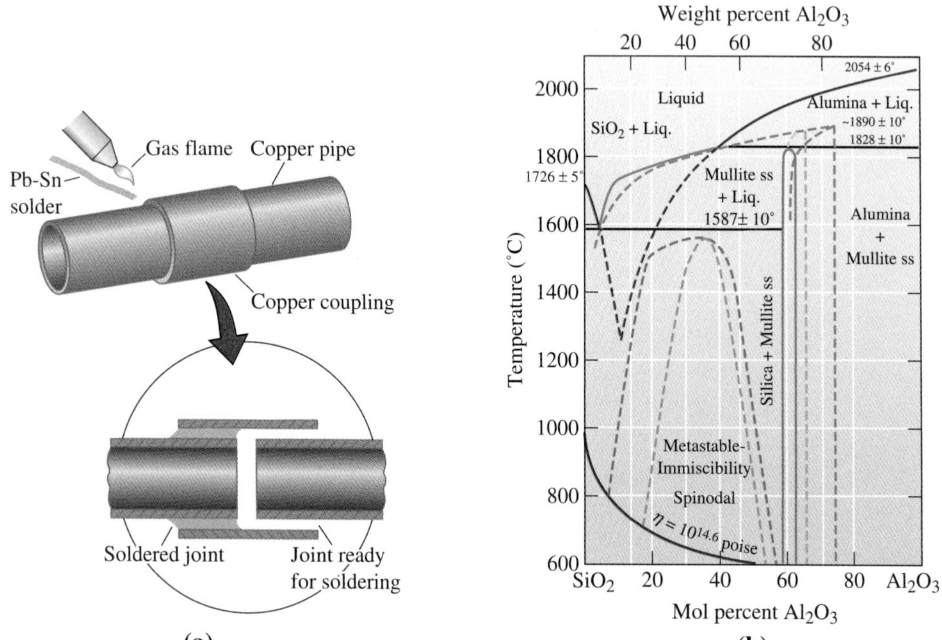

Figure 11-23 (a) A Pb-Sn eutectic alloy is often used during soldering to assemble parts. A heat source, such as a gas flame, heats both the parts and the filler material. The filler is drawn into the joint and solidifies. (b) A phase diagram for SiO_2-Al_2O_3. (*Based on* Introduction to Phase Equilibria, *by C. G. Bergeron and S. H. Risbud, American Ceramic Society, 1984.*)

melting, minimizing casting defects such as gas porosity, and preventing liquid metal-mold reactions. Cast iron (Chapter 13) and many aluminum alloys (Chapter 14) are eutectic alloys.

Although most of this discussion has been centered around metallic materials, it is important to recognize that eutectics are very important in many ceramic systems (Chapter 15). Silica and alumina are the most widely used ceramic materials. Figure 11-23(b) shows a phase diagram for the Al_2O_3-SiO_2 system. Notice the eutectic at ~1587°C. The dashed lines on this diagram show metastable extensions of the liquidus and metastable miscibility gaps. As mentioned before, the existence of these gaps makes it possible to make technologically useful products such as Vycor™ and Pyrex® glasses. A Vycor™ glass is made by first melting (approximately at 1500°C) silica (63%), boron oxide (27%), sodium oxide (7%), and alumina (3%). The glass is then formed into the desired shapes. During glass formation, the glass phase separates (because of the metastable miscibility gap) into boron oxide-rich and silica-rich regions. The boron oxide rich regions are dissolved using an acid. The porous object is sintered to form Vycor™ glass that contains 95 wt% silica, 4% boron oxide, and 1% sodium oxide. It would be very difficult to achieve a high silica glass such as this without resorting to the technique described above. Pyrex® glasses contain about 80 wt% silica, 13% boron oxide, 4% sodium oxide, and 2% alumina. These are used widely in making laboratory ware such as beakers and household products.

11-7 Nonequilibrium Freezing in the Eutectic System

Suppose we have an alloy, such as Pb-15% Sn, that ordinarily solidifies as a solid-solution alloy. The last liquid should freeze near 230°C, well above the eutectic; however, if the alloy cools too quickly, a nonequilibrium solidus curve is produced (Figure 11-24).

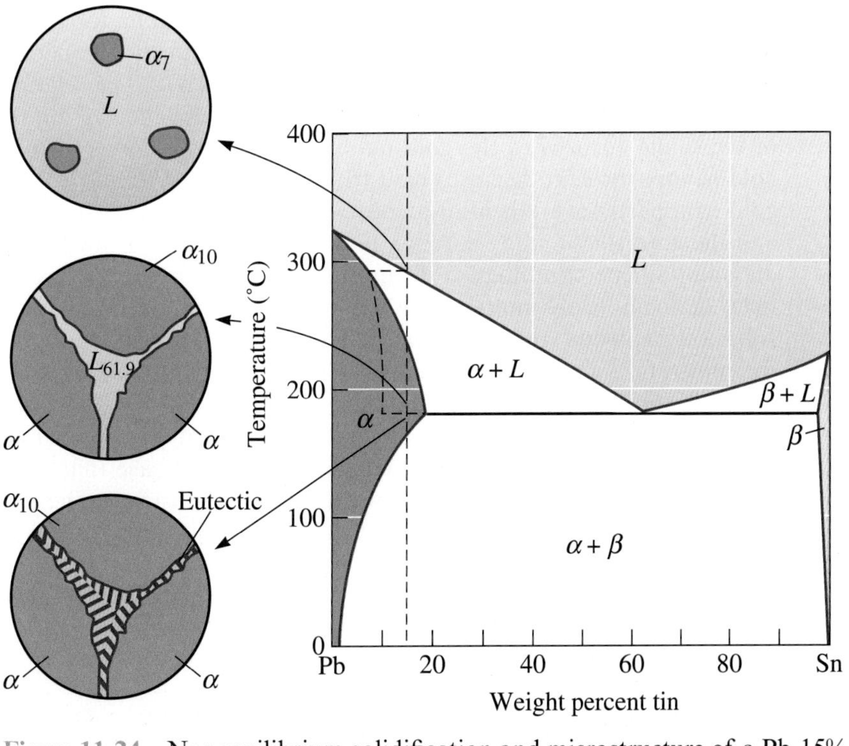

Figure 11-24 Nonequilibrium solidification and microstructure of a Pb-15% Sn alloy. A nonequilibrium eutectic microconstituent can form if the solidification is too rapid.

The primary α continues to grow until, just above 183°C, the remaining nonequilibrium liquid contains 61.9% Sn. This liquid then transforms to the eutectic microconstituent, surrounding the primary α. For the conditions shown in Figure 11-24, the amount of nonequilibrium eutectic is

$$\% \text{ eutectic} = \frac{15 - 10}{61.9 - 10} \times 100 = 9.6\%$$

When heat treating an alloy such as Pb-15% Sn, we must keep the maximum temperature below the eutectic temperature of 183°C to prevent **hot shortness** or partial melting. This concept is very important in the precipitation, or age, hardening of metallic alloys such as those in the Al-Cu system.

11-8 Nanowires and the Eutectic Phase Diagram

Nanowires are cylinders or "wires" of material with diameters on the order of 10 to 100 nm. Nanowires have generated great technological interest due to their electrical, mechanical, chemical, and optical properties. Potential applications include biological and chemical sensors, electrical conductors in nanoelectronic devices, and electrodes in batteries. A common method of fabricating nanowire materials is through the technique known as vapor-liquid-solid (VLS) growth. The growth of silicon nanowires can be understood by considering the gold-silicon binary phase diagram in Figure 11-25(a). Notice that there is no mutual solubility between gold and silicon.

In one example of VLS silicon nanowire growth, the first step is to deposit a thin layer of pure gold on a substrate. When the substrate is heated, the gold dewets from the substrate or "balls up," forming a series of gold nanoparticles on the substrate surface, as shown in Figure 11-25(b). These gold nanoparticles, also known as the nanowire catalysts, act as the template for silicon nanowire growth. Silane gas (SiH_4), the "vapor" in the VLS process for silicon nanowire growth, then flows through a chamber holding the substrate with the gold nanoparticles. The substrate is heated to a temperature above the gold-silicon eutectic temperature of 363°C but below the melting temperature of the gold (1064°C). The silane gas decomposes according to $SiH_4 \rightarrow Si + 2H_2$. The silicon that is produced from the decomposition of the silane adsorbs to the gold nanoparticles and diffuses into them, causing the gold nanoparticles to begin to melt (thus forming the "liquid" of VLS growth). The gold nanoparticles are held at a constant temperature, the silicon content increases, and the gold-silicon alloy enters a two-phase region (α-L) in the gold-silicon phase diagram. Silicon continues to diffuse into the gold nanoparticles until the nanoparticles are completely molten. The nanoparticles retain their shape as roughly hemispherical balls due to surface-energy considerations between the nanoparticles and the substrate (think of oil droplets on a Teflon® pan). As the silicon content increases due to the diffusion of silicon into the nanoparticles, a solid phase begins to form. The solid phase that forms is the silicon nanowire. It is pure silicon. The nanowire has the same diameter as the gold nanoparticle. The liquid phase that is present rides atop the silicon nanowire as it grows upward from the substrate. The nanowire continues to grow in length while silane gas is supplied to the substrate. When the nanowire growth is complete, the temperature is reduced, and the liquid phase atop the silicon nanowire solidifies on cooling through the eutectic temperature. Silicon nanowires are shown in Figure 11-25(c), and gold catalysts atop silicon nanowires are shown in Figure 11-25(d).

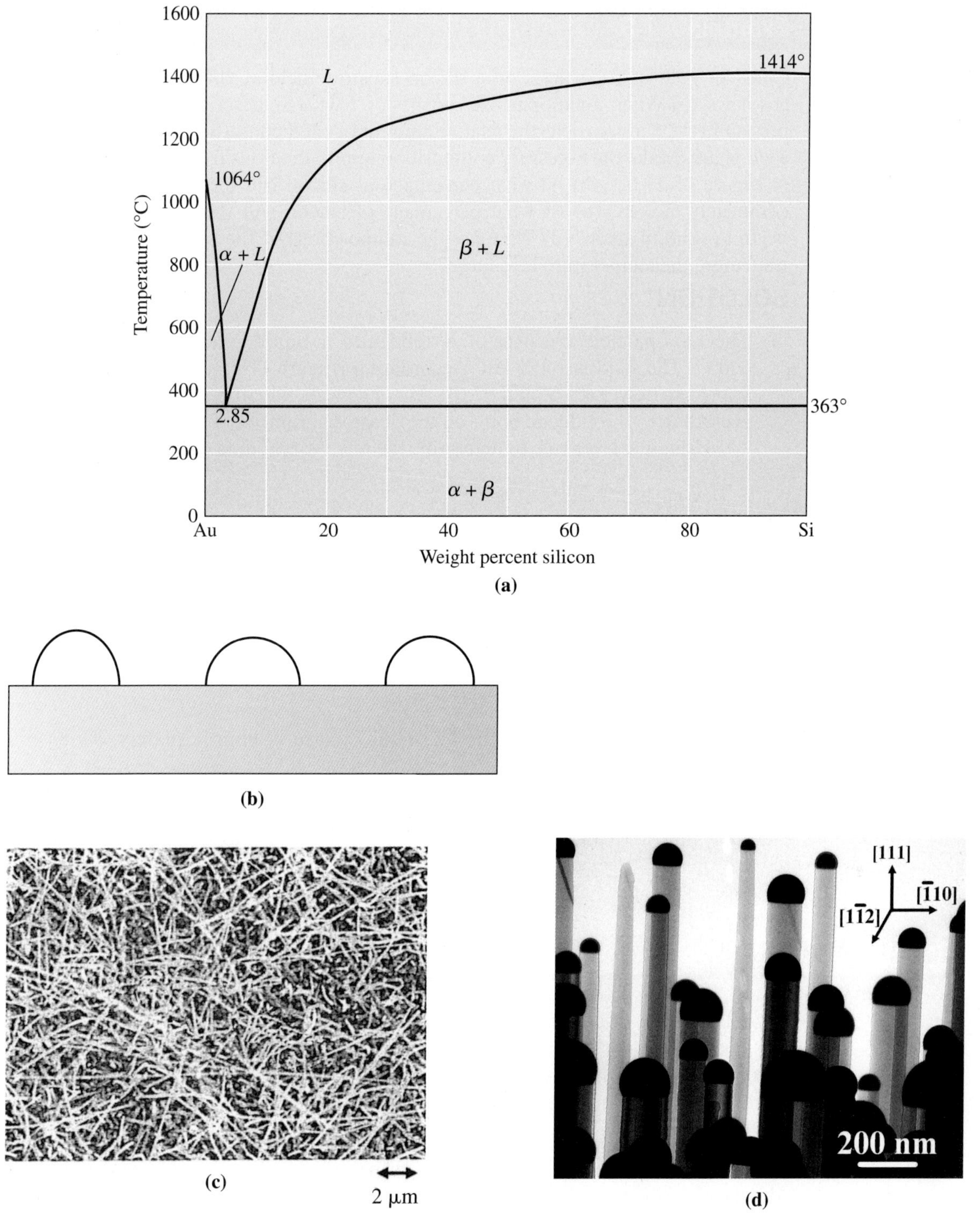

Figure 11-25 (a) The Au-Si phase diagram. (*Based on* ASM Metals Handbook, *Volume 8 Metallography, Structures and Phase Diagrams (1973), ASM International, Materials Park, OH 44073-0002.*) (b) A schematic diagram of Au nanoparticles on a substrate. (c) A scanning electron micrograph of silicon nanowires. (*Courtesy of Wendelin Wright*) (d) Gold catalysts atop silicon nanowires are visible in this high magnification image. (*Reprinted figure with permission from S. Kodambaka, J. Tersoff, M. C. Reuter, and F. M. Ross,* Physical Review Letters *96, p. 096105-2 (2006). Copyright 2006 by the American Physical Society.*)

Example 11-8 Growth of Silicon Nanowires

Consider gold nanoparticles on a silicon substrate held at 600°C over which silane gas flows. (a) When a nanoparticle consists of 1 wt% Si, describe the phases that are present in the nanoparticle, the relative amount of each phase, and the composition of each phase. Make the necessary composition approximations from the phase diagram in Figure 11-25 (a). (b) At what percentage of silicon will the nanoparticle become completely molten? (c) At what percentage of silicon will the molten nanoparticle begin to solidify again? (d) What are the compositions of the nanowire and the catalyst before the temperature is reduced?

SOLUTION

(a) The nanoparticle consists of a solid and a liquid phase at 1 wt% Si and 600°C. The solid is 100% Au. According to Figure 11-25(a), the liquid contains approximately 2.4% Si and 97.6% Au. (The wt% silicon must be less than 2.85% according to the eutectic point on the phase diagram.)

Using the lever rule to determine the relative amounts of each phase:

$$\% \text{ Solid} = \frac{2.4 - 1.0}{2.4 - 0} = 58\% \qquad \% \text{ Liquid} = \frac{1.0 - 0}{2.4 - 0} = 42\%$$

Thus, the nanoparticle consists of 58% solid (100% Au) and 42% liquid (2.4% Si and 97.6% Au).

(b) The island will completely melt when the silicon content reaches approximately 2.4%.

(c) The island will begin to solidify again when the silicon content reaches approximately 7%.

(d) The solid phase is 100% Si. The liquid phase is approximately 7% Si and 93% Au.

Summary

- By producing a material containing two or more phases, dispersion strengthening is obtained. In metallic materials, the boundary between the phases impedes the movement of dislocations and improves strength. Introduction of multiple phases may provide other benefits, including improvement of the fracture toughness of ceramics and polymers.
- For optimum dispersion strengthening, particularly in metallic materials, a large number of small, hard, discontinuous dispersed particles should form in a soft, ductile matrix to provide the most effective obstacles to dislocations. Round dispersed phase particles minimize stress concentrations, and the final properties of the alloy can be controlled by the relative amounts of these and the matrix.
- Intermetallic compounds, which normally are strong but brittle, are frequently introduced as dispersed phases.
- Phase diagrams for materials containing multiple phases normally contain one or more three-phase reactions.
- The eutectic reaction permits liquid to solidify as a mixture of two solid phases. By controlling the solidification process, we can achieve a wide range of properties. Some of the factors that can be controlled include the grain size or secondary

dendrite arm spacings of primary microconstituents, the colony size of the eutectic microconstituent, the interlamellar spacing within the eutectic microconstituent, the microstructure, or shape, of the phases within the eutectic microconstituent, and the amount of the eutectic microconstituent that forms.

- The eutectoid reaction causes a solid to transform to a mixture of two other solids. As shown in the next chapter, heat treatments to control the eutectoid reaction provide an excellent basis for dispersion strengthening.
- Nanowires can be grown from eutectic systems through a process known as vapor-liquid-solid (VLS) growth. In one example of VLS silicon nanowire growth, silane gas is passed over gold catalysts. Silicon is deposited on the gold, forming a binary system. As the silicon concentration increases, the catalysts become molten, and as the silicon concentration increases further, a solid phase, which is the silicon nanowire, forms. The diameter of the nanowires is controlled by the catalyst diameter, and the nanowire length is controlled by the growth time.

Glossary

Age hardening A strengthening mechanism that relies on a sequence of solid-state phase transformations in generating a dispersion of ultra-fine particles of a second phase. Age hardening is a form of dispersion strengthening. Also called precipitation hardening (Chapter 12).

Dispersed phase A solid phase that forms from the original matrix phase when the solubility limit is exceeded.

Dispersion strengthening Increasing the strength of a material by forming more than one phase. By proper control of the size, shape, amount, and individual properties of the phases, excellent combinations of properties can be obtained.

Eutectic A three-phase invariant reaction in which one liquid phase solidifies to produce two solid phases.

Eutectic microconstituent A characteristic mixture of two phases formed as a result of the eutectic reaction.

Eutectoid A three-phase invariant reaction in which one solid phase transforms to two different solid phases.

Hot shortness Melting of the lower melting point nonequilibrium material that forms due to segregation, even though the temperature is below the equilibrium solidus temperature.

Hyper- A prefix indicating that the composition of an alloy is between the composition defining the right-hand end of the tie line and the composition at which the relevant three-phase reaction occurs.

Hypereutectic alloy An alloy composition between that of the right-hand end of the tie line defining the eutectic reaction and the eutectic composition.

Hypo- A prefix indicating that the composition of an alloy is between the composition defining the left-hand end of the tie line and the composition at which the relevant three-phase reaction occurs.

Hypoeutectic alloy An alloy composition between that of the left-hand end of the tie line defining the eutectic reaction and the eutectic composition.

Interlamellar spacing The distance between the center of a lamella or plate of one phase and the center of the adjacent lamella or plate of the same phase.

Intermediate solid solution A nonstoichiometric intermetallic compound displaying a range of compositions.

Intermetallic compound A compound formed from two or more metals that has its own unique composition, structure, and properties.

Interphase interface The boundary between two phases in a microstructure.

Isopleth A line on a phase diagram that shows constant chemical composition.

Isoplethal study Determination of reactions and microstructural changes that are expected while studying a particular chemical composition in a system.

Lamella A thin plate of a phase that forms during certain three-phase reactions, such as the eutectic or eutectoid.

Matrix The continuous solid phase in a complex microstructure.

Metastable miscibility gap A miscibility gap that extends below the liquidus or exists completely below the liquidus. Two liquids that are immiscible continue to exist as liquids and remain unmixed. Systems such as this form the basis for Vycor™ and Pyrex® glasses.

Microconstituent A phase or mixture of phases in an alloy that has a distinct appearance. Frequently, we describe a microstructure in terms of the microconstituents rather than the actual phases.

Miscibility gap A region in a phase diagram in which two phases, with essentially the same structure, do not mix or have no solubility in one another.

Modification Addition of alloying elements that change the microstructure of the eutectic microconstituent in eutectic alloys.

Monotectic A three-phase invariant reaction in which one liquid transforms to a solid and a second liquid on cooling.

Nanowires Cylinders or "wires" of material with diameters on the order of 10 to 100 nm.

Nonstoichiometric intermetallic compound A phase formed by the combination of two components having a structure and properties different from either component. The nonstoichiometric compound has a variable ratio of the components present in the compound (see also intermediate solid solution).

Peritectic A three-phase invariant reaction in which a solid and a liquid combine to produce a second solid on cooling.

Peritectoid A three-phase invariant reaction in which two solids combine to form a third solid on cooling.

Precipitate A solid phase that forms from the original matrix phase when the solubility limit is exceeded. We often use the term precipitate, as opposed to dispersed phase particles, for alloys formed by precipitation or age hardening. In most cases, we try to control the formation of the precipitate second-phase particles to produce the optimum dispersion strengthening or age hardening. (Also called the dispersed phase.)

Precipitation hardening A strengthening mechanism that relies on a sequence of solid-state phase transformations in generating a dispersion of ultra-fine precipitates of a second phase (Chapter 12). It is a form of dispersion strengthening. Also called age hardening.

Primary microconstituent The microconstituent that forms before the start of a three-phase reaction. Also called the proeutectic microconstituent.

Solvus A solubility curve that separates a single solid-phase region from a two solid-phase region in the phase diagram.

Stoichiometric intermetallic compound A phase formed by the combination of two components into a compound having a structure and properties different from either component. The stoichiometric intermetallic compound has a fixed ratio of the components present in the compound.

Problems

Section 11-1 Principles and Examples of Dispersion Strengthening

11-1 What are the requirements of a matrix and precipitate for dispersion strengthening to be effective?

Section 11-2 Intermetallic Compounds

11-2 What is an intermetallic compound? How is it different from other compounds? For example, other than the obvious difference in composition, how is TiAl different from, for example, Al_2O_3?

11-3 Are intermetallics only used as a second phase for a material, or is it possible that they could be used as the only material for a component? Explain.

11-4 What are some of the major problems in the utilization of intermetallics for high temperature applications?

11-5 In Figure 11-2, what weight percent of rhodium permits γ to exist at the highest temperature?

Section 11-3 Phase Diagrams Containing Three-Phase Reactions

11-6 Define the terms eutectic, eutectoid, peritectic, peritectoid, and monotectic reactions.

11-7 What is an invariant reaction? Show that for a two-component system the number of degrees of freedom for an invariant reaction is zero.

11-8 Many binary phase diagrams have two scales on the abscissa (horizontal axis): one for weight percentage and the other for atomic percentage of one of the elements. Why is there interest in both scales?

11-9 Under what circumstances is the total composition the same as the phase composition or compositions? Under what circumstances is it different?

11-10 Why is it common practice to only label one-phase regions in a binary phase diagram?

11-11 A hypothetical phase diagram is shown in Figure 11-26.

(a) Are any intermetallic compounds present? If so, identify them and determine whether they are stoichiometric or nonstoichiometric.

(b) Identify the solid solutions present in the system. Is either material *A* or *B* allotropic? Explain.

(c) Identify the three-phase reactions by writing down the temperature, the reaction in equation form, the composition of each phase in the reaction, and the name of the reaction.

11-12 Consider Figure 11-5. For an equal component alloy of A and B starting at 1400°C, write each reaction as the crucible is cooled to room temperature (25°C).

11-13 The Cu-Zn phase diagram is shown in Figure 11-27.

(a) Are any intermetallic compounds present? If so, identify them and determine whether they are stoichiometric or nonstoichiometric.

(b) Identify the solid solutions present in the system.

(c) Identify the three-phase reactions by writing down the temperature, the reaction in equation form, and the name of the reaction.

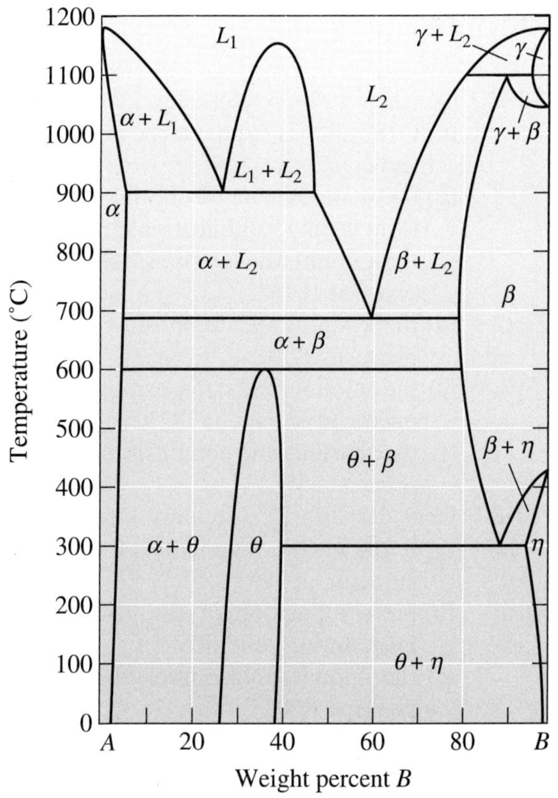

Figure 11-26 Hypothetical phase diagram (for Problem 11-11).

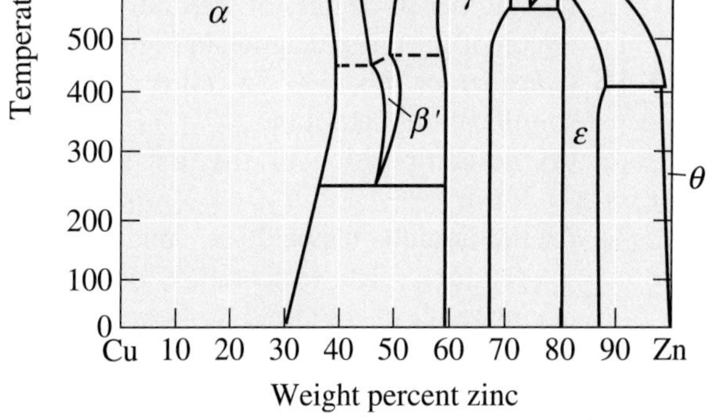

Figure 11-27 The copper-zinc phase diagram (for Problem 11-13).

11-14 The Al-Li phase diagram is shown in Figure 11-28.

(a) Are any intermetallic compounds present? If so, identify them and determine whether they are stoichiometric or nonstoichiometric. Determine the formula for each compound.

(b) Identify the three-phase reactions by writing down the temperature, the reaction in equation form, the composition of each phase in the reaction, and the name of the reaction.

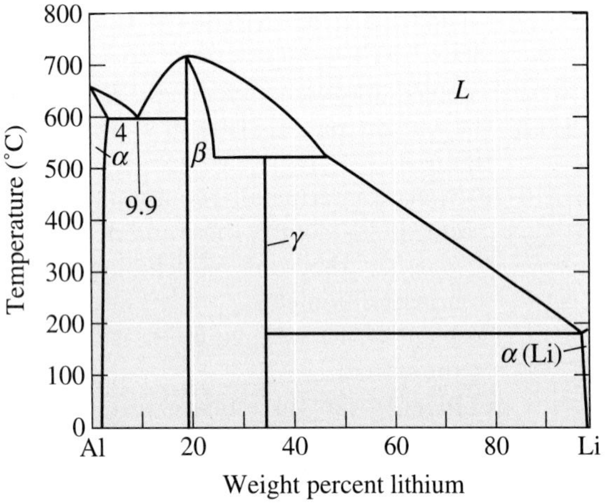

Figure 11-28 The aluminum-lithium phase diagram (for Problems 11-14 and 11-36).

11-15 An intermetallic compound is found for 38 wt% Sn in the Cu-Sn phase diagram. Determine the formula for the compound.

11-16 An intermetallic compound is found for 10 wt% Si in the Cu-Si phase diagram. Determine the formula for the compound.

Section 11-4 The Eutectic Phase Diagram

11-17 Use the Gibbs phase rule to determine the number of degrees of freedom in each region of the phase diagram in Figure 11-6.

11-18 Consider a Pb-15% Sn alloy. During solidification, determine

(a) the composition of the first solid to form;

(b) the liquidus temperature, solidus temperature, solvus temperature, and freezing range of the alloy;

(c) the amounts and compositions of each phase at 260°C;

(d) the amounts and compositions of each phase at 183°C; and

(e) the amounts and compositions of each phase at 25°C.

11-19 Consider an Al-12% Mg alloy (Figure 11-29). During solidification, determine

(a) the composition of the first solid to form;

(b) the liquidus temperature, solidus temperature, solvus temperature, and freezing range of the alloy;

(c) the amounts and compositions of each phase at 525°C;

(d) the amounts and compositions of each phase at 450°C; and

(e) the amounts and compositions of each phase at 25°C.

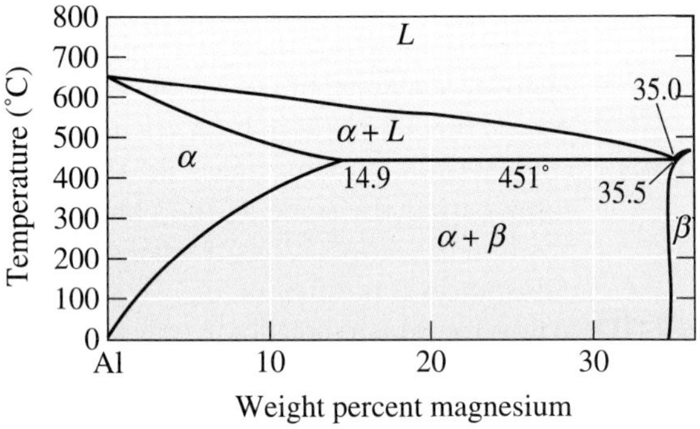

Figure 11-29 Portion of the aluminum-magnesium phase diagram (for Problems 11-19 and 11-37).

11-20 Consider a Pb-35% Sn alloy. Determine

(a) if the alloy is hypoeutectic or hypereutectic;

(b) the composition of the first solid to form during solidification;

(c) the amounts and compositions of each phase at 184°C;

(d) the amounts and compositions of each phase at 182°C;

(e) the amounts and compositions of each microconstituent at 182°C; and

(f) the amounts and compositions of each phase at 25°C.

11-21 Consider a Pb-70% Sn alloy. Determine

(a) if the alloy is hypoeutectic or hypereutectic;

(b) the composition of the first solid to form during solidification;

(c) the amounts and compositions of each phase at 184°C;

(d) the amounts and compositions of each phase at 182°C;

(e) the amounts and compositions of each microconstituent at 182°C; and

(f) the amounts and compositions of each phase at 25°C.

11-22 (a) Sketch a typical eutectic phase diagram with components A and B having similar melting points. B is much more soluble in A (maximum = 15%) than A is in B (maximum = 5%), and the eutectic composition occurs near 40% B. The eutectic temperature is 2/3 of the melting point. Label the axes of the diagram. Label all the phases. Use α and β to denote the solid phases. (b) For an overall composition of 60% B, list the sequence of phases found as the liquid is slowly cooled to room temperature.

11-23 The copper-silver phase diagram is shown in Figure 11-30. Copper has a higher melting point than silver. Refer to the silver-rich solid phase as gamma (γ) and the copper-rich solid phase as delta (δ). Denote the liquid as L.

(a) For an overall composition of 60% B (40% A) at a temperature of 800°C, what are the compositions and amounts of the phases present?

(b) For an overall composition of 30% B (70% A) at a temperature of 1000°C, what are the compositions and amounts of the phases present?

(c) Draw a schematic diagram illustrating the final microstructure of a material with a composition of 50% B (50% A) cooled to 200°C from the liquid state. Label each phase present.

11-24 Determine the phases that are present and the compositions for each phase in Cu-55 wt% Ag at 600°C. (See Problem 11-23 and Figure 11-30.)

11-25 Determine the phases that are present and the compositions for each phase in Cu-85 wt% Ag at 800°C. (See Problem 11-23 and Figure 11-30.)

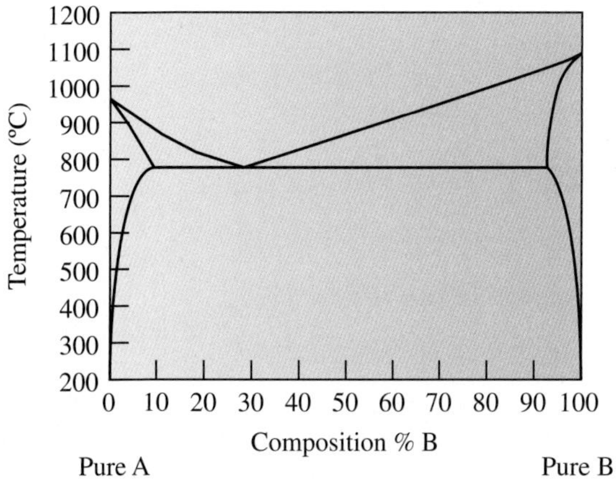

Figure 11-30 A phase diagram for elements A and B (for Problems 11-23, 11-24, 11-25, and 11-47).

11-26 Calculate the total % β and the % eutectic microconstituent at room temperature for the following lead-tin alloys: (a) 10% Sn; (b) 20% Sn; (c) 50% Sn; (d) 60% Sn; (e) 80% Sn; and (f) 95% Sn. Using Figure 11-18, plot the strength of the alloys versus the % β and the % eutectic and explain your graphs.

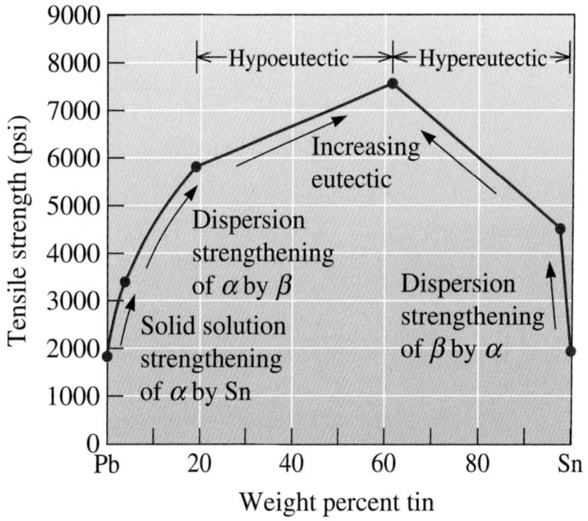

Figure 11-18 The effect of the composition and strengthening mechanism on the tensile strength of lead-tin alloys. (Repeated for Problem 11-26.)

11-27 Consider the Al-Si phase diagram. What are the percentages of the α-phase and liquid for an Al-5 wt% Si alloy at 620, 600, and 578°C? What are the percentages of α and β phases in this alloy for an Al-5 wt% Si alloy at 576 and 550°C? (See Figure 11-19.)

11-28 At the eutectic in the Al-Si phase diagram, what phase(s) is (are) present? Give a chemical analysis of the phase(s). (See Figure 11-19.)

11-29 We discussed the primary phase or primary constituent. Why would we be interested in the percentage of the primary phase in the Al-Si alloy system?

11-30 Consider an Al-4% Si alloy (Figure 11-19). Determine

(a) if the alloy is hypoeutectic or hypereutectic;

(b) the composition of the first solid to form during solidification;

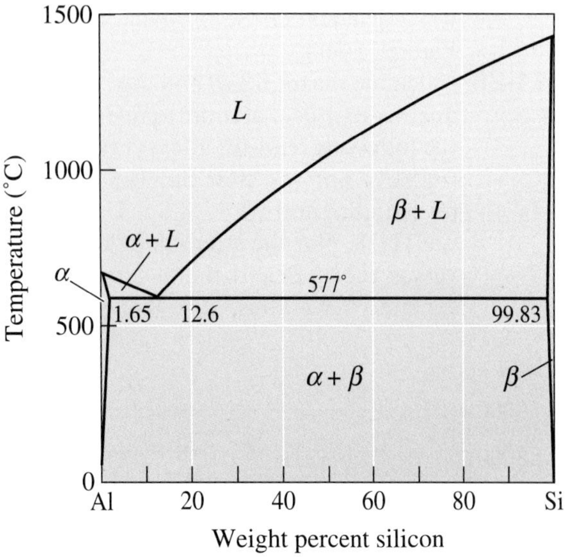

Figure 11-19 The aluminum-silicon phase diagram. (Repeated for Problems 11-27, 11-28, 11-30, 11-31, 11-33, 11-35, and 11-40.)

(c) the amounts and compositions of each phase at 578°C;
(d) the amounts and compositions of each phase at 576°C, the amounts and compositions of each microconstituent at 576°C; and
(e) the amounts and compositions of each phase at 25°C.

11-31 Consider an Al-25% Si alloy (see Figure 11-19). Determine
(a) if the alloy is hypoeutectic or hypereutectic;
(b) the composition of the first solid to form during solidification;
(c) the amounts and compositions of each phase at 578°C;
(d) the amounts and compositions of each phase at 576°C;
(e) the amounts and compositions of each microconstituent at 576°C; and
(f) the amounts and compositions of each phase at 25°C.

11-32 A Pb-Sn alloy contains 45% α and 55% β at 100°C. Determine the composition of the alloy. Is the alloy hypoeutectic or hypereutectic?

11-33 An Al-Si alloy contains 85% α and 15% β at 500°C. Determine the composition of the alloy. Is the alloy hypoeutectic or hypereutectic? (See Figure 11-19.)

11-34 A Pb-Sn alloy contains 23% primary α and 77% eutectic microconstituent immediately after the eutectic reaction has been completed. Determine the composition of the alloy.

11-35 An Al-Si alloy contains 15% primary β and 85% eutectic microconstituent immediately after the eutectic reaction has been completed. Determine the composition of the alloy. (See Figure 11-19.)

11-36 Observation of a microstructure shows that there is 28% eutectic and 72% primary β in an Al-Li alloy (Figure 11-28). Determine the composition of the alloy and whether it is hypoeutectic or hypereutectic.

11-37 Write the eutectic reaction that occurs, including the compositions of the three phases in equilibrium, and calculate the amount of α and β in the eutectic microconstituent in the Mg-Al system (Figure 11-29).

11-38 Calculate the total amount of α and β and the amount of each microconstituent in a Pb-50% Sn alloy at 182°C. What fraction of the total α in the alloy is contained in the eutectic microconstituent?

11-39 Figure 11-31 shows a cooling curve for a Pb-Sn alloy. Determine
(a) the pouring temperature;
(b) the superheat;
(c) the liquidus temperature;
(d) the eutectic temperature;
(e) the freezing range;
(f) the local solidification time;
(g) the total solidification time; and
(h) the composition of the alloy.

11-40 Figure 11-32 shows a cooling curve for an Al-Si alloy and Figure 11-19 shows

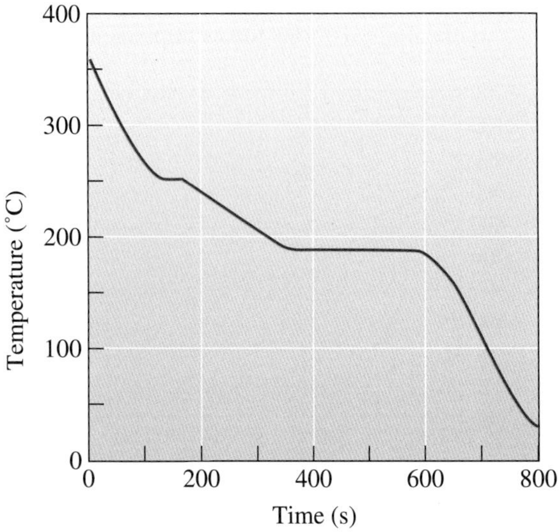

Figure 11-31 Cooling curve for a Pb-Sn alloy (for Problem 11-39).

the binary phase diagram for this system. Determine
(a) the pouring temperature;
(b) the superheat;
(c) the liquidus temperature;
(d) the eutectic temperature;
(e) the freezing range;
(f) the local solidification time;
(g) the total solidification time; and
(h) the composition of the alloy.

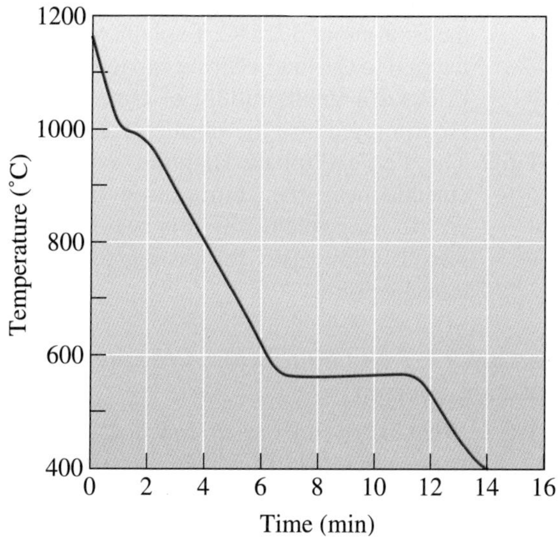

Figure 11-32 Cooling curve for an Al-Si alloy (for Problem 11-40).

11-41 Draw the cooling curves, including appropriate temperatures, expected for the following Al-Si alloys:
(a) Al-4% Si;
(b) Al-12.6% Si;
(c) Al-25% Si; and
(d) Al-65% Si.

11-42 Cooling curves are obtained for a series of Cu-Ag alloys (Figure 11-33). Use this data to produce the Cu-Ag phase diagram. The maximum solubility of Ag in Cu is 7.9%, and the maximum solubility of Cu in Ag is 8.8%. The solubilities at room temperature are near zero.

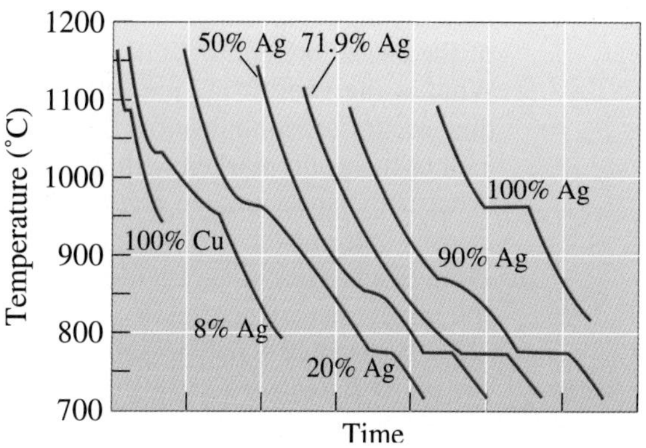

Figure 11-33 Cooling curves for a series of Cu-Ag alloys (for Problem 11-42).

Section 11-5 Strength of Eutectic Alloys

11-43 In regards to eutectic alloys, what does the term "modification" mean? How does it help properties of the alloy?

11-44 For the Pb-Sn system, explain why the tensile strength is a maximum at the eutectic composition.

11-45 Does the shape of the proeutectic phase have an effect on the strength of eutectic alloys? Explain.

11-46 The binary phase diagram for the silver and germanium system is shown in Figure 11-34.

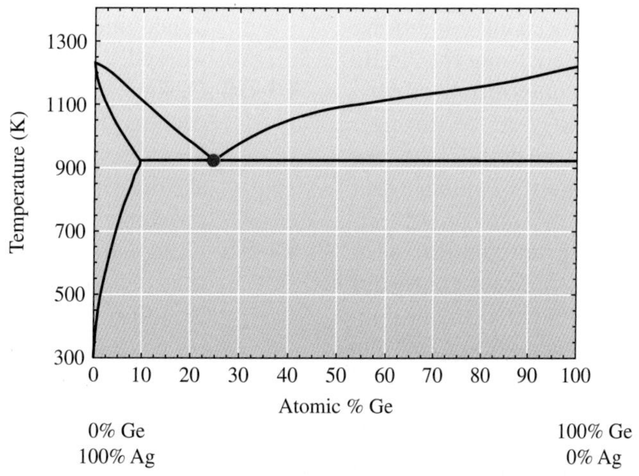

Figure 11-34 The silver–germanium phase diagram (for Problem 11-46).

(a) Schematically draw the phase diagram and label the phases present in each region of the diagram. Denote α as the Ag-rich solid phase and β as the Ge-rich solid phase. Use L to denote the liquid phase.
(b) For an overall composition of 80% Ge (20% Ag) at a temperature of 700 K, what are the compositions and amounts of the phases present?
(c) What is the transformation in phases that occurs on solidification from the melt at the point marked with a circle? What is the special name given to this transformation?
(d) Draw a schematic diagram illustrating the final microstructure of 15% Ge (85% Ag) cooled slowly to 300 K from the liquid state.
(e) Consider two tensile samples at room temperature. One is pure Ag and one is Ag with 2% Ge. Which sample would you expect to be stronger?

11-47 The copper-silver phase diagram is shown in Figure 11-30. Copper has a higher melting point than silver.
(a) Is copper element A or element B as labeled in the phase diagram?
(b) Schematically draw the phase diagram and label all phases present in each region (single phase and two phase) of the phase diagram by writing directly on your sketch. Denote the silver-rich solid phase as gamma (γ) and the copper-rich solid phase as delta (δ). Denote the liquid as L.
(c) At 600°C, the solid solution of element A in element B is stronger than the solid solution of element B in element A. Assume similar processing conditions. Is a material cooled from the liquid to 600°C with a composition of 90% A and 10% B likely to be stronger or weaker than a material with the eutectic composition? Explain your answer fully.
(d) Upon performing mechanical testing, your results indicate that your assumption of similar processing conditions in part (c) was wrong and that the material that you had assumed to be stronger is in fact weaker. Give an example of a processing condition and a description of the associated microstructure that could have led to this discrepancy.

Section 11-6 Eutectics and Materials Processing

11-48 Explain why Pb-Sn alloys are used for soldering.

11-49 Refractories used in steel making include silica brick that contain very small levels of alumina (Al_2O_3). The eutectic temperature in this system is about 1587°C. Silica melts at about 1725°C. Explain what will happen to the load bearing capacity of the bricks if a small amount of alumina gets incorporated into the silica bricks.

11-50 The Fe-Fe_3C phase diagram exhibits a eutectic near the composition used for cast irons. Explain why it is beneficial for ferrous alloys used for casting to have a eutectic point.

Section 11-7 Nonequilibrium Freezing in the Eutectic System

11-51 What is hot shortness? How does it affect the temperature at which eutectic alloys can be used?

Section 11-8 Nanowires and the Eutectic Phase Diagram

11-52 Explain the vapor-liquid-solid mechanism of nanowire growth.

Design Problems

11-53 Design a processing method that permits a Pb-15% Sn alloy solidified under nonequilibrium conditions to be hot worked.

11-54 Design a eutectic diffusion bonding process to join aluminum to silicon. Describe the changes in microstructure at the interface during the bonding process.

11-55 Design an Al-Si brazing alloy and process that will be successful in joining an Al-Mn alloy that has a liquidus temperature of 659°C and a solidus temperature of 656°C. Brazing, like soldering, involves introducing a liquid filler metal into a joint without melting the metals that are to be joined.

Computer Problems

11-56 Write a computer program to assist you in solving problems such as those illustrated in Examples 11-2 or 11-3. The input will be, for example, the bulk composition of the alloy, the mass of the alloy, and the atomic masses of the elements (or compounds) forming the binary system. The program should then prompt the user to provide temperature, number of phases, and the composition of the phases. The program should provide the user with the outputs for the fractions of phases present and the total masses of the phases, as well as the mass of each element in different phases. Start with a program that will provide a solution for Example 11-3 and then extend it to Example 11-4.

Knovel® Problems

K11-1 Find an iron-titanium phase diagram and identify the temperatures and binary alloy compositions for the three-phase points for all eutectic reactions.

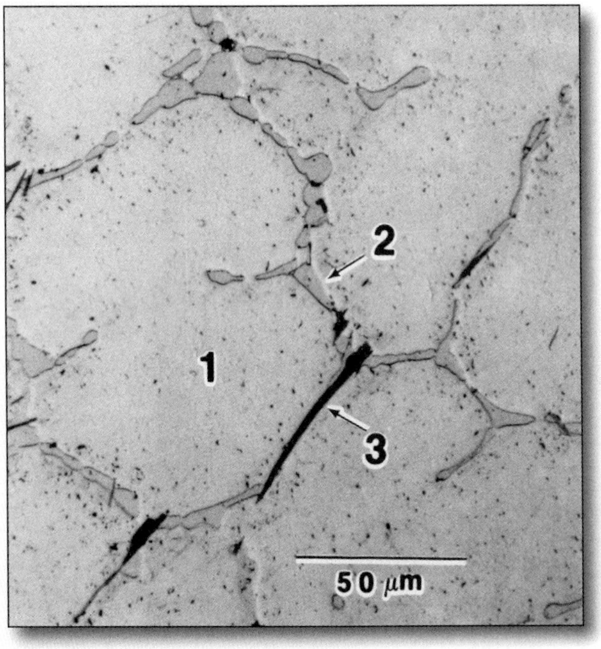

On December 17, 1903, the Wright brothers flew the first controllable airplane. This historic first flight lasted only 12 seconds and covered 120 feet, but changed the world forever. (*Topical Press Agency/Hulton Archive/Getty Images.*) The micrograph of the alloy used in the Wright brothers airplane is also shown. (*Courtesy of Dr. Frank Gayle, NIST.*) What was not known at the time was that the aluminum alloy engine that they used was inadvertently strengthened by precipitation hardening!

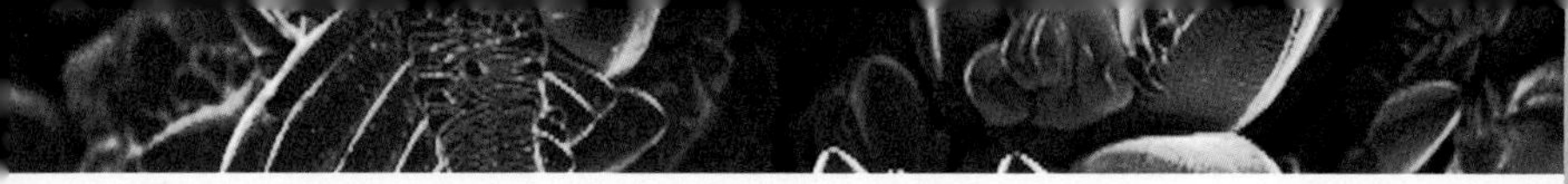

CHAPTER 12

Dispersion Strengthening by Phase Transformations and Heat Treatment

Have You Ever Wondered?

- *Who invented and flew the first controllable airplane?*
- *How do engineers strengthen aluminum alloys used in aircraft?*
- *Can alloys remember their shape?*
- *Why do some steels become very hard upon quenching from high temperatures?*
- *What alloys are used to make orthodontic braces?*
- *Would it be possible to further enhance the strength and, hence, the dent resistance of sheet steels after the car chassis is made?*
- *What are smart materials?*

Chapter Learning Objectives

The key objectives of this chapter are to

- Develop the expression that describes the total free energy change during a solid-state phase transformation and describe the kinetics of nucleation and growth during such a process.
- Design an age-hardening treatment for an appropriate alloy.
- Draw the phase diagram of a low-carbon steel between pure iron and cementite (Fe_3C).
- Predict the mechanical properties of hypoeutectoid, eutectoid, and hypereutectoid steels.
- Explain how martensite is produced and describe its structure and properties.
- Explain the relationships between processing, composition, quench rates, and annealing to control the microstructure of a material.

In Chapter 11, we examined in detail how second-phase particles can increase the strength of metallic materials. We also saw how dispersion-strengthened materials are prepared and what pathways are possible for the formation of second phases during the solidification of alloys, especially eutectic alloys. In this chapter, we will further discuss dispersion strengthening as we describe a variety of solid-state transformation processes, including precipitation or age hardening and the eutectoid reaction. We also examine how nonequilibrium phase transformations—in particular, the martensitic reaction—can provide mechanisms for strengthening.

As we discuss these strengthening mechanisms, keep in mind the characteristics that produce the most desirable dispersion strengthening, as discussed in Chapter 11:

- The matrix should be relatively soft and ductile and the precipitate, or second phase, should be strong;
- the precipitate should be round and discontinuous;
- the second-phase particles should be small and numerous; and
- in general, the more precipitate we have, the stronger the alloy will be.

As in Chapter 11, we will concentrate on how these reactions influence the strength of the materials and how heat treatments can influence other properties. We will begin with the nucleation and growth of second-phase particles in solid-state phase transformations.

12-1 Nucleation and Growth in Solid-State Reactions

In Chapter 9, we discussed nucleation of a solid from a melt. We also discussed the concepts of undercooling and homogeneous and heterogeneous nucleation. Let's now see how these concepts apply to solid-state phase transformations such as the eutectoid reaction. In order for a precipitate of phase β to form from a solid matrix of phase α, both nucleation and growth must occur. The total change in free energy required for nucleation of a spherical solid precipitate of radius r from the matrix is

$$\Delta G = \tfrac{4}{3}\pi r^3 \Delta G_{v(\alpha \rightarrow \beta)} + 4\pi r^2 \sigma_{\alpha\beta} + \tfrac{4}{3}\pi r^3 \varepsilon \qquad (12\text{-}1)$$

The first two terms include the free energy change per unit volume (ΔG_v) and the energy change needed to create the unit area of the $\alpha - \beta$ interface ($\sigma_{\alpha\beta}$), just as in solidification (Equation 9-1). The third term takes into account the **strain energy** per unit volume (ε), the energy required to permit a precipitate to fit into the surrounding matrix during the nucleation and growth of the precipitate. The precipitate does not occupy the same volume that is displaced, so additional energy is required to accommodate the precipitate in the matrix.

Nucleation

As in solidification, nucleation occurs most easily on surfaces already present in the structure, thereby minimizing the surface energy term. Thus, the precipitates heterogeneously nucleate most easily at grain boundaries and other defects.

Growth

Growth of the precipitates normally occurs by long-range diffusion and redistribution of atoms. Diffusing atoms must be detached from their original locations (perhaps at lattice points in a solid solution), move through the surrounding material to the nucleus, and be incorporated into the crystal structure of the precipitate. In some cases, the diffusing atoms might be so tightly bonded within an existing phase that the detachment process limits the rate of growth. In other cases, attaching the diffusing atoms to the precipitate—perhaps because of the lattice strain—limits growth. This result sometimes leads to the formation of precipitates that have a special relationship to the matrix structure that minimizes the strain. In most cases, however, the controlling factor is the diffusion step.

Kinetics The overall rate, or *kinetics,* of the transformation process depends on both nucleation and growth. If more nuclei are present at a particular temperature, growth occurs from a larger number of sites and the phase transformation is completed in a shorter period of time. At higher temperatures, the diffusion coefficient is higher, growth rates are more rapid, and again we expect the transformation to be completed in a shorter time, assuming an equal number of nuclei.

The rate of transformation is given by the Avrami equation (Equation 12-2), with the fraction transformed f related to time t by

$$f = 1 - \exp(-ct^{n}) \qquad (12\text{-}2)$$

where c and n are constants for a particular temperature. This **Avrami relationship**, shown in Figure 12-1, produces a sigmoidal, or S-shaped, curve. This equation can describe most solid-state phase transformations. An incubation time t_0, during which no observable transformation occurs, is the time required for nucleation. Initially, the transformation occurs slowly as nuclei form.

Incubation is followed by rapid growth as atoms diffuse to the growing precipitate. Near the end of the transformation, the rate again slows as the source of atoms available to diffuse to the growing precipitate is depleted. The transformation is 50% complete in time τ. The rate of transformation is often given by the reciprocal of τ

$$\text{Rate} = 1/\tau \qquad (12\text{-}3)$$

Effect of Temperature In many phase transformations, the material undercools below the temperature at which the phase transformation occurs under equilibrium conditions. Because both nucleation and growth are temperature-dependent, the rate of phase transformation depends on the undercooling (ΔT). The rate of nucleation is low for small undercoolings (since the thermodynamic driving force is low) and increases for larger undercoolings as the thermodynamic driving force increases at least up to a certain point (since diffusion becomes slower as temperature decreases). At the same time, the growth rate of the new phase decreases continuously (because of slower diffusion), as the undercooling increases. The growth rate follows an *Arrhenius relationship* (recall, Equation 5-1):

$$\text{Growth rate} = A \exp\left(\frac{-Q}{RT}\right) \qquad (12\text{-}4)$$

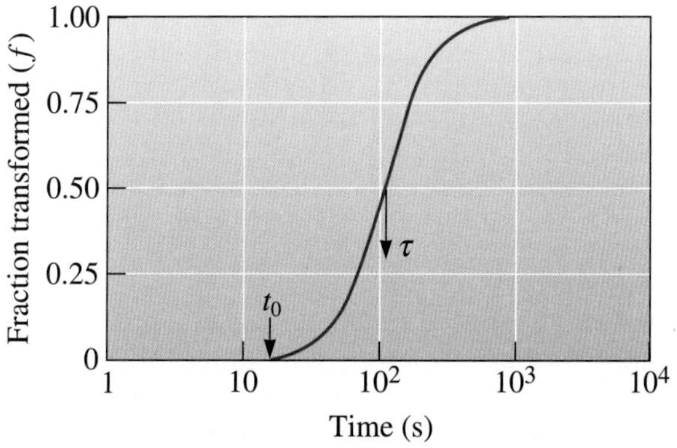

Figure 12-1 Sigmoidal curve showing the rate of transformation of FCC iron at a constant temperature. The incubation time t_0 and the time τ for 50% transformation are also shown.

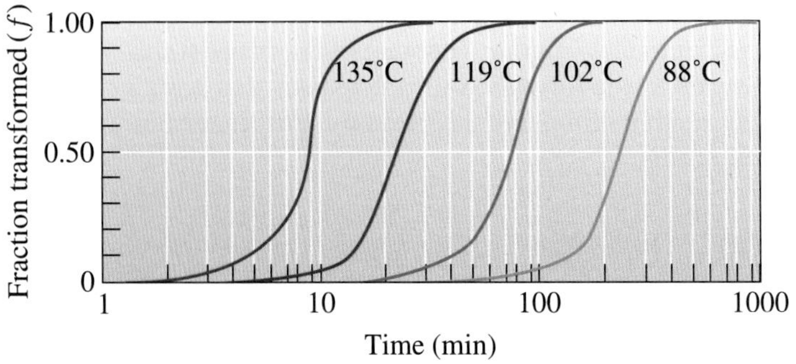

Figure 12-2 The effect of temperature on the recrystallization of cold-worked copper.

where Q is the activation energy (in this case, for the phase transformation), R is the gas constant, T is the temperature, and A is a constant.

Figure 12-2 shows sigmoidal curves at different temperatures for the recrystallization of copper; as the temperature increases, the rate of recrystallization of copper *increases*.

At any particular temperature, the overall rate of transformation is the product of the nucleation and growth rates. In Figure 12-3(a), the combined effect of the nucleation and growth rates is shown. A maximum transformation rate may be observed at a critical undercooling. The time required for transformation is inversely related to the rate of transformation; Figure 12-3(b) describes the time (on a log scale) required for the transformation. This C-shaped curve is common for many transformations in metals, ceramics, glasses, and polymers. Notice that the time required at a temperature corresponding to the equilibrium phase transformation would be ∞ (i.e., the phase transformation will not occur). This is because there is no undercooling and, hence, the rate of homogeneous nucleation is zero.

In some processes, such as the recrystallization of a cold-worked metal, we find that the transformation rate continually decreases with decreasing temperature. In this case, nucleation occurs easily, and diffusion—or growth—predominates (i.e., the growth is the rate limiting step for the transformation). The following example illustrates how the activation energy for a solid-state phase transformation such as recrystallization can be obtained from data related to the kinetics of the process.

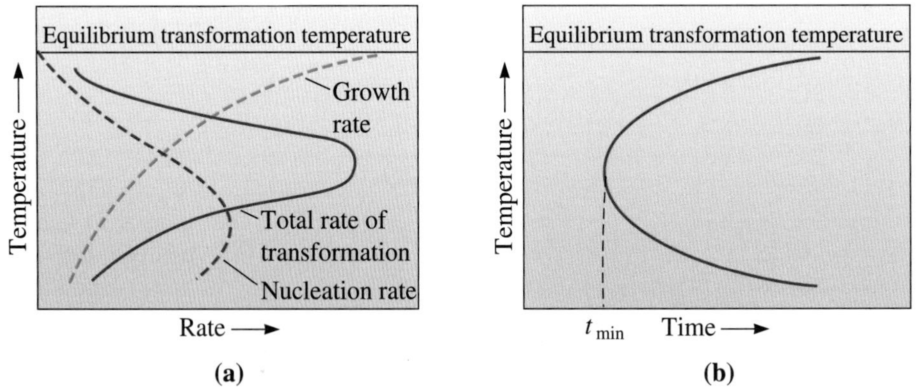

Figure 12-3 (a) The effect of temperature on the rate of a phase transformation is the product of the growth rate and nucleation rate contributions, giving a maximum transformation rate at a critical temperature. (b) Consequently, there is a minimum time (t_{min}) required for the transformation, given by the "C-curve."

Example 12-1 *Activation Energy for the Recrystallization of Copper*

Determine the activation energy for the recrystallization of copper from the sigmoidal curves in Figure 12-2.

SOLUTION

The rate of transformation is the reciprocal of the time τ required for half of the transformation to occur. From Figure 12-2, the times required for 50% transformation at several different temperatures can be calculated:

T (°C)	T (K)	τ (min)	Rate (min^{-1})
135	408	9	0.111
119	392	22	0.045
102	375	80	0.0125
88	361	250	0.0040

The rate of transformation follows an Arrhenius equation, so a plot of ln(Growth rate) versus $1/T$ (Figure 12-4 and Equation 12-4) allows us to calculate the constants in the equation. Taking the natural logarithm of both sides of Equation 12-4:

$$\ln(\text{Growth rate}) = \ln A - \frac{Q}{RT}$$

Thus, if we plot ln(Growth rate) as a function of $1/T$, we expect a straight line with a slope of $-Q/R$. A linear regression to the data is shown in Figure 12-4, which indicates that the slope $-Q/R = -10{,}494$ K. Therefore,

$$Q = (10{,}494 \text{ K}) [1.987 \text{ cal/(mol} \cdot \text{K)}] = 20{,}852 \text{ cal/mol}$$

and the constant A is calculated as

$$\ln(A) = 23.59$$

$$A = \exp(23.59) = 1.76 \times 10^{10} \text{ min}^{-1}$$

In this particular example, the rate at which the reaction occurs *increases* as the temperature *increases,* indicating that the reaction may be dominated by diffusion.

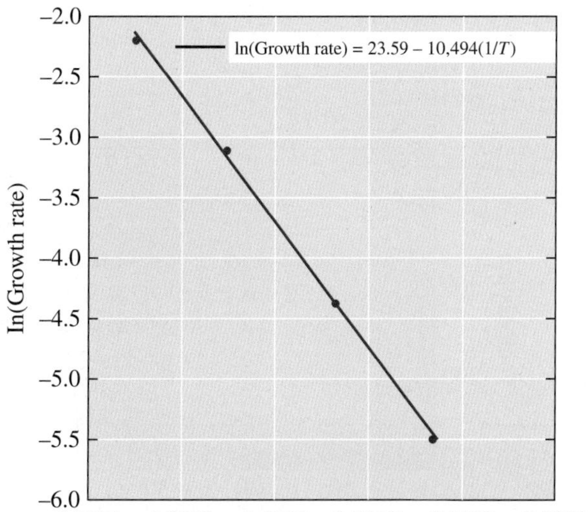

Figure 12-4 Arrhenius plot of transformation rate versus reciprocal temperature for recrystallization of copper (for Example 12-1).

12-2 Alloys Strengthened by Exceeding the Solubility Limit

In Chapter 11, we learned that lead-tin alloys containing about 2 to 19% Sn can be dispersion strengthened because the solubility of tin in lead is exceeded.

A similar situation occurs in aluminum-copper alloys. For example, the Al-4% Cu alloy (shown in Figure 12-5) is 100% α above 500°C. The α phase is a solid solution of aluminum containing copper up to 5.65 wt%. On cooling below the solvus temperature, a second phase, θ, precipitates. The θ phase, which is the hard, brittle intermetallic compound $CuAl_2$, provides dispersion strengthening. Applying the lever rule to the phase diagram shown in Figure 12-5, we can show that at 200°C and below, in a 4% Cu alloy, only about 7.5% of the final structure is θ. We must control the precipitation of the second phase to satisfy the requirements of good dispersion strengthening.

Widmanstätten Structure

The second phase may grow so that certain planes and directions in the precipitate are parallel to preferred planes and directions in the matrix, creating a basket-weave pattern known as the **Widmanstätten structure**. This growth mechanism minimizes strain and surface energies and permits faster growth rates. Widmanstätten growth produces a characteristic appearance for the precipitate. When a needle-like shape is produced [Figure 12-6(a)], the Widmanstätten precipitate may encourage the nucleation of cracks, thus reducing the ductility of the material. Conversely, some of these structures make it more difficult for cracks, once formed, to propagate, therefore providing good fracture toughness. Certain titanium alloys and ceramics obtain toughness in this way.

Interfacial Energy Relationships

We expect the precipitate to have a spherical shape in order to minimize surface energy; however, when the precipitate forms at an interface, the precipitate shape is also influenced by the **interfacial energy** of the boundary between the matrix grains and the precipitate. Assuming that the second phase is nucleating at the grain boundaries, the interfacial surface energies of the matrix-precipitate boundary (γ_{mp}) and the grain boundary energy of the matrix ($\gamma_{m,gb}$) fix a

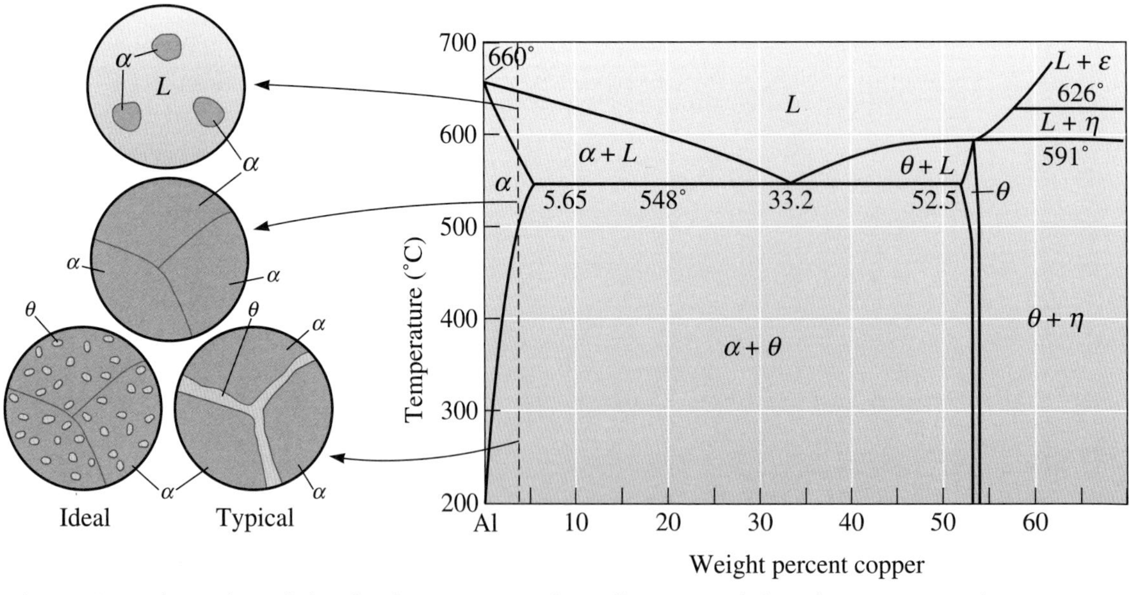

Figure 12-5 A portion of the aluminum-copper phase diagram and the microstructures that may develop during cooling of an Al-4% Cu alloy.

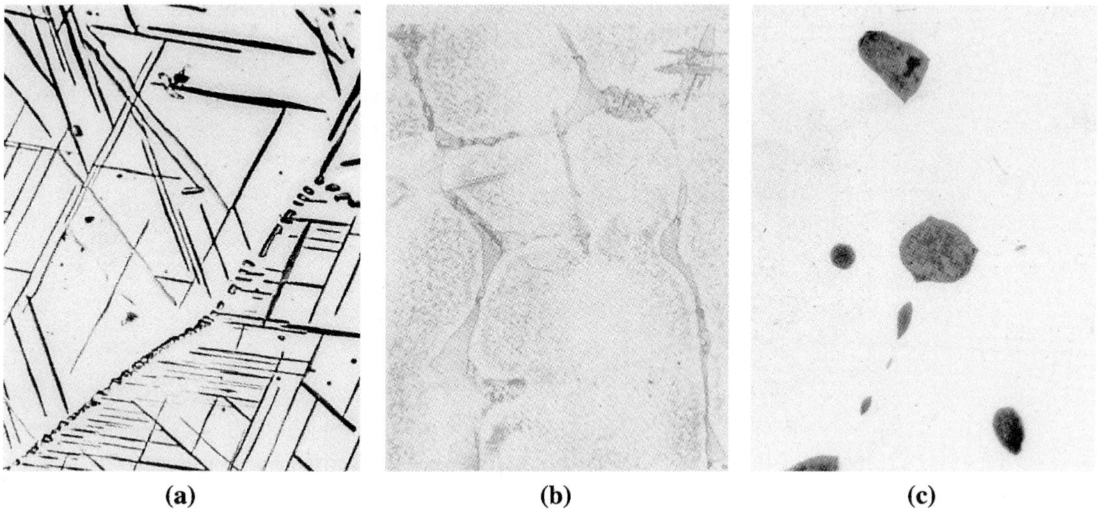

Figure 12-6 (a) Widmanstätten needles in a Cu-Ti alloy (× 420). [*From* ASM Handbook, *Vol. 9, Metallography and Microstructure (1985), ASM International, Materials Park, OH 44073-0002. Reprinted with permission of ASM International. All rights reserved. www.asminternational.org.*] (b) Continuous θ precipitate in an Al-4% Cu alloy, caused by slow cooling (× 500). (c) Precipitates of lead at grain boundaries in copper (× 500). [*Micrographs (b) and (c) (Reprinted Courtesy of Don Askeland.)*]

dihedral angle θ between the matrix-precipitate interface that, in turn, determines the shape of the precipitate (Figure 12-7). The relationship is

$$\gamma_{m,gb} = 2\gamma_{mp}\cos\frac{\theta}{2} \tag{12-5}$$

Note that this equation cannot be used when the dihedral angle is 0° or 180°.

If the precipitate phase completely wets the grain (similar to how water wets glass), then the dihedral angle is zero, and the second phase grows as a continuous layer along the grain boundaries of the matrix phase. If the dihedral angle is small, the precipitate may be continuous. If the precipitate is also hard and brittle, the thin film that surrounds the matrix grains causes the alloy to be very brittle [Figure 12-6(b)].

On the other hand, discontinuous and even spherical precipitates form when the dihedral angle is large [Figure 12-6(c)]. This occurs if the precipitate phase does not wet the matrix.

Coherent Precipitate Even if we produce a uniform distribution of discontinuous precipitate, the precipitate may not significantly disrupt the surrounding matrix structure [Figure 12-8(a)]. Consequently, the precipitate blocks slip only if it lies directly in the path of the dislocation.

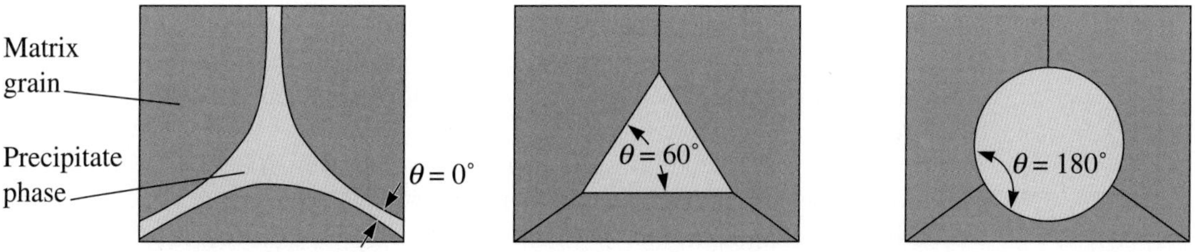

Figure 12-7 The effect of surface energy and the dihedral angle on the shape of a precipitate.

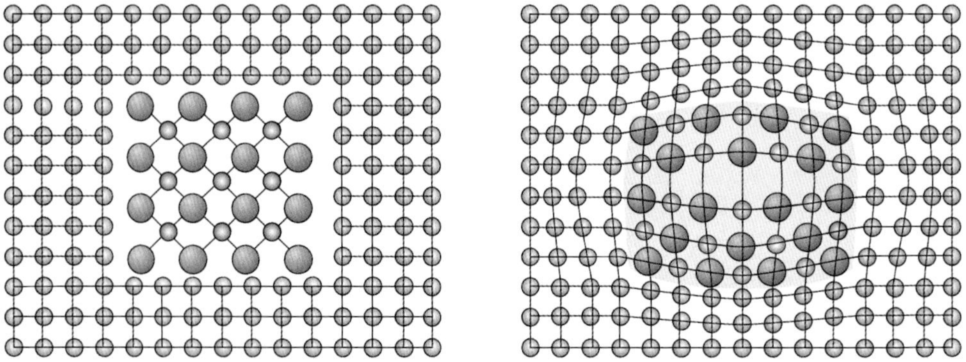

Figure 12-8 (a) A noncoherent precipitate has no relationship with the crystal structure of the surrounding matrix. (b) A coherent precipitate forms so that there is a definite relationship between the precipitate's and the matrix's crystal structures.

When a **coherent precipitate** forms, the planes of atoms in the crystal structure of the precipitate are related to—or even continuous with—the planes in the crystal structure of the matrix [Figure 12-8(b)]. Now a widespread disruption of the matrix crystal structure is created, and the movement of a dislocation is impeded even if the dislocation merely passes near the coherent precipitate. A special heat treatment, such as age hardening, may produce the coherent precipitate.

12-3 Age or Precipitation Hardening and Its Applications

Age hardening, or **precipitation hardening**, is produced by a sequence of phase transformations that leads to a uniform dispersion of nanoscale, coherent precipitates in a softer, more ductile matrix. The inadvertent occurrence of this process may have helped the Wright brothers, who, on December 17, 1903, made the first controllable flight that changed the world forever. Gayle and co-workers showed that the aluminum alloy used by the Wright brothers for making the engine of the first airplane ever flown picked up copper from the casting mold. The age hardening occurred inadvertently as the mold remained hot during the casting process. The application of age hardening started with the Wright brothers' historic flight and, even today, aluminum alloys used for aircrafts are strengthened using this technique. Age or precipitation hardening is probably one of the earliest examples of nanostructured materials that have found widespread applications.

Before we examine the details of the mechanisms of phase transformations that are needed for age hardening to occur, let's examine some of the applications of this technique. A major advantage of precipitation hardening is that it can be used to increase the yield strength of many metallic materials via relatively simple heat treatments and without creating significant changes in density. Thus, the strength-to-density ratio of an alloy can be improved substantially using age hardening. For example, the yield strength of an aluminum alloy can be increased from about 20,000 psi to 60,000 psi as a result of age hardening.

Nickel-based superalloys (alloys based on Ni, Cr, Al, Ti, Mo, and C) are precipitation hardened by precipitation of a Ni_3Al-like γ' phase that is rich in Al and Ti. Similarly, titanium alloys (e.g., Ti $-$ 6% Al $-$ 4% V), stainless steels, Be-Cu, and many steels are precipitation hardened and used for a variety of applications.

New sheet-steel formulations are designed such that precipitation hardening occurs in the material while the paint on the chassis is being "baked" or cured (~100°C). These **bake-hardenable steels** are just one example of steels that take advantage of the strengthening effect provided by age-hardening mechanisms.

A weakness associated with this mechanism is that age-hardened alloys can be used over a limited range of temperatures. At higher temperatures, the precipitates formed initially begin to grow and eventually dissolve if the temperatures are high enough (Section 12-7). This is where alloys in which dispersion strengthening is achieved by using a second phase that is insoluble are more effective than age-hardened alloys.

12-4 Microstructural Evolution in Age or Precipitation Hardening

How do precipitates form in precipitation hardening? How do they grow or age? Can the precipitates grow too much, or overage, so that they cannot provide maximum dispersion strengthening? Answers to these questions can be found by following the microstructural evolution in the sequence of phase transformations that are necessary for age hardening.

Let's use Al-Cu as an archetypal system to illustrate these ideas. The Al-4% Cu alloy is a classic example of an age-hardenable alloy. There are three steps in the age-hardening heat treatment (Figure 12-9).

Step 1: Solution Treatment In the **solution treatment**, the alloy is first heated above the solvus temperature and held until a homogeneous solid solution α is produced. This step dissolves the θ phase precipitate and reduces any microchemical segregation present in the original alloy.

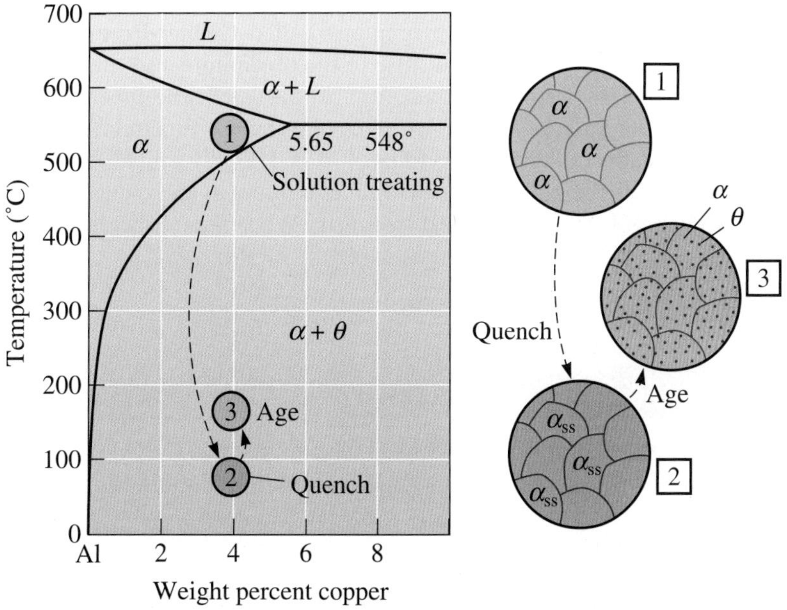

Figure 12-9 The aluminum-rich end of the aluminum-copper phase diagram showing the three steps in the age-hardening heat treatment and the microstructures that are produced.

We could heat the alloy to just below the solidus temperature and increase the rate of homogenization; however, the presence of a nonequilibrium eutectic microconstituent may cause melting (hot shortness, Chapter 11). Thus, the Al-4% Cu alloy is solution treated between 500°C and 548°C, that is, between the solvus and the eutectic temperatures.

Step 2: Quench After solution treatment, the alloy, which contains only α in its structure, is rapidly cooled, or quenched. The atoms do not have time to diffuse to potential nucleation sites, so the θ does not form. After the quench, the structure is a **supersaturated solid solution** α_{SS} containing excess copper, and it is not an equilibrium structure. It is a metastable structure. This situation is effectively the same as undercooling of water, molten metals, and silicate glasses (Chapter 9). The only difference is we are dealing with materials in their solid state.

Step 3: Age Finally, the supersaturated α is heated at a temperature below the solvus temperature. At this *aging* temperature, atoms diffuse only short distances. Because the supersaturated α is metastable, the extra copper atoms diffuse to numerous nucleation sites and precipitates grow. Eventually, if we hold the alloy for a sufficient time at the aging temperature, the equilibrium $\alpha + \theta$ structure is produced. Note that even though the structure that is formed has two equilibrium phases (i.e., $\alpha + \theta$), the morphology of the phases is different from the structure that would have been obtained by the slow cooling of this alloy (Figure 12-5). When we go through the three steps described previously, we produce the θ phase in the form of ultra-fine uniformly dispersed second-phase precipitate particles. This is what we need for effective precipitation strengthening.

The following two examples illustrate the effect of quenching on the composition of phases and a design for an age-hardening treatment.

Nonequilibrium Precipitates during Aging During aging of aluminum-copper alloys, a continuous series of other precursor precipitate phases forms prior to the formation of the equilibrium θ phase. This is fairly common in precipitation-hardened alloys. The simplified diagram in Figure 12-9 does not show these intermediate phases. At the start of aging, the copper atoms concentrate on $\{100\}$ planes in the α matrix and produce very thin precipitates called **Guinier-Preston (GP) zones**. As aging continues, more copper atoms diffuse to the precipitate and the GP-I zones thicken into thin disks, or GP-II zones. With continued diffusion, the precipitates develop a greater degree of order and are called θ'. Finally, the stable θ precipitate is produced.

The nonequilibrium precipitates—GP-I, GP-II, and θ'—are coherent precipitates. The strength of the alloy increases with aging time as these coherent phases grow in size during the initial stages of the heat treatment. When these coherent precipitates are present,

Example 12-2 *Composition of Al-4% Cu Alloy Phases*

Compare the composition of the α solid solution in the Al-4% Cu alloy at room temperature when the alloy cools under equilibrium conditions with that when the alloy is quenched.

SOLUTION

From Figure 12-9, a tie line can be drawn at room temperature. The composition of the α determined from the tie line is about 0.02% Cu; however, the composition of the α after quenching is still 4% Cu. Since α contains more than the equilibrium copper content, the α is supersaturated with copper.

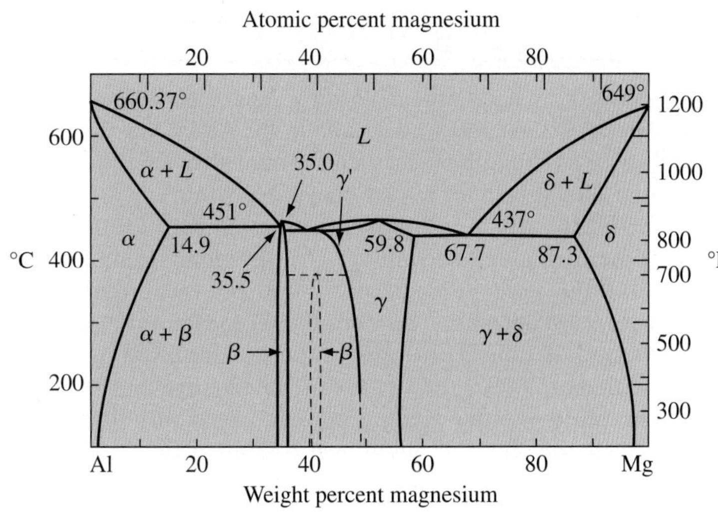

Figure 12-10 The aluminum-magnesium phase diagram.

Example 12-3 *Design of an Age-Hardening Treatment*

The magnesium-aluminum phase diagram is shown in Figure 12-10. Suppose a Mg-8% Al alloy is responsive to an age-hardening heat treatment. Design a heat treatment for the alloy.

SOLUTION

Step 1: Solution-treat at a temperature between the solvus and the eutectic to avoid hot shortness. Thus, heat between 340°C and 451°C.

Step 2: Quench to room temperature fast enough to prevent the precipitate β phase from forming.

Step 3: Age at a temperature below the solvus, that is, below 340°C, to form a fine dispersion of the β phase.

the alloy is in the aged condition. Figure 12-11 shows the structure of an aged Al-Ag alloy. This important development in the microstructure evolution of precipitation-hardened alloys is the reason the time for heat treatment during aging is very important.

When the stable noncoherent θ phase precipitates, the strength of the alloy begins to decrease. Now the alloy is in the overaged condition. The θ still provides some dispersion strengthening, but with increasing time, the θ grows larger and even the simple dispersion-strengthening effect diminishes.

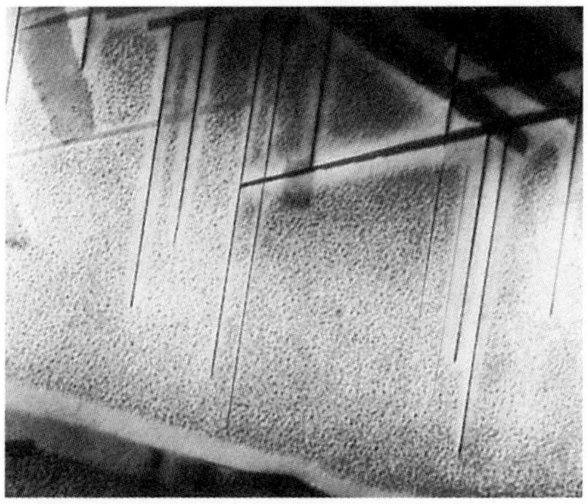

Figure 12-11 An electron micrograph of aged Al-15% Ag showing coherent γ' plates and round GP zones (× 40,000). [From ASM Handbook, *Volume 8 (1973), ASM International, Materials Park, OH 44073-0002. Reprinted with permission of ASM International. All rights reserved. www.asminternational.org.*]

12-5 Effects of Aging Temperature and Time

The properties of an age-hardenable alloy depend on both aging temperature and aging time (Figure 12-12). At 260°C, diffusion in the Al-4% Cu alloy is rapid, and precipitates quickly form. The strength reaches a maximum after less than 0.1 h exposure. Overaging occurs if the alloy is held for longer than 0.1 h (6 min).

At 190°C, which is a typical aging temperature for many aluminum alloys, a longer time is required to produce the optimum strength; however, there are benefits to using the lower temperature. First, the maximum strength increases as the aging temperature decreases. Second, the alloy maintains its maximum strength over a longer period of time. Third, the properties are more uniform. If the alloy is aged for only 10 min at 260°C, the surface of the part reaches the proper temperature and strengthens, but the center remains cool and ages only slightly. The example that follows illustrates the effect of aging heat treatment time on the strength of aluminum alloys.

Aging at either 190°C or 260°C is called **artificial aging** because the alloy is heated to produce precipitation. Some solution-treated and quenched alloys age at room temperature; this is called **natural aging**. Natural aging requires long times—often several days—to reach maximum strength; however, the peak strength is higher than that obtained in artificial aging, and no overaging occurs.

Example 12-4 *Effect of Aging Heat-Treatment Time on the Strength of Aluminum Alloys*

The operator of a furnace left for his hour lunch break without removing the Al-4% Cu alloy from the furnace used for the aging treatment. Compare the effect on the yield strength of the extra hour of aging for the aging temperatures of 190°C and 260°C.

SOLUTION

At 190°C, the peak strength of 400 MPa (60,000 psi) occurs at 2 h (Figure 12-12). After 3 h, the strength is essentially the same.

At 260°C, the peak strength of 340 MPa (50,000 psi) occurs at 0.06 h; however, after 1 h, the strength decreases to 250 MPa (35,000 psi).

Thus, the higher aging temperature gives a lower peak strength and makes the strength more sensitive to aging time.

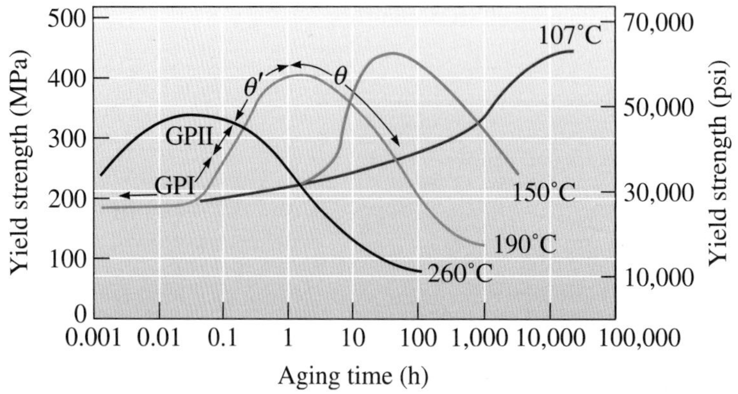

Figure 12-12 The effect of aging temperature and time on the yield strength of an Al-4% Cu alloy.

An interesting observation made by Dr. Gayle and his coworkers at NIST is a striking example of the difference between natural aging and artificial aging. Dr. Gayle and coworkers analyzed the aluminum alloy of the engine used in the Wright brothers' airplane. They found two interesting things. First, they found that the original alloy had undergone precipitation hardening as a result of being held in the mold for a period of time and at a temperature that was sufficient to cause precipitation hardening. Second, between the times when the alloy was cast in 1903 and about 1993 when the research was done (almost ninety years), the alloy had continued to age naturally! This could be seen from two different size distributions for the precipitate particles using transmission electron microscopy. In some aluminum alloys (designated as T4) used to make tapered poles or fasteners, it may be necessary to refrigerate the alloy prior to forming to avoid natural aging at room temperature. If not, the alloy would age at room temperature, become harder, and not be workable!

12-6 Requirements for Age Hardening

Not all alloys are age hardenable. Four conditions must be satisfied for an alloy to have an age-hardening response during heat treatment:

1. The alloy system must display decreasing solid solubility with decreasing temperature. In other words, the alloy must form a single phase on heating above the solvus line, then enter a two-phase region on cooling.
2. The matrix should be relatively soft and ductile, and the precipitate should be hard and brittle. In most age-hardenable alloys, the precipitate is a hard, brittle intermetallic compound.
3. The alloy must be quenchable. Some alloys cannot be cooled rapidly enough to suppress the formation of the precipitate. Quenching may, however, introduce residual stresses that cause distortion of the part (Chapter 8). To minimize residual stresses, aluminum alloys are quenched in hot water at about 80°C.
4. A coherent precipitate must form.

As mentioned before in Section 12-3, a number of important alloys, including certain stainless steels and alloys based on aluminum, magnesium, titanium, nickel, chromium, iron, and copper, meet these conditions and are age hardenable.

12-7 Use of Age-Hardenable Alloys at High Temperatures

Based on our previous discussion, we would not select an age-hardened Al-4% Cu alloy for use at high temperatures. At service temperatures ranging from 100°C to 500°C, the alloy overages and loses its strength. Above 500°C, the second phase redissolves in the matrix, and we do not even obtain dispersion strengthening. In general, the aluminum age-hardenable alloys are best suited for service near room temperature; however, some magnesium alloys may maintain their strength to about 250°C and certain nickel superalloys resist overaging at 1000°C.

We may also have problems when welding age-hardenable alloys (Figure 12-13). During welding, the metal adjacent to the weld is heated. The *heat-affected zone* (HAZ) contains two principal zones. The lower temperature zone near the unaffected base metal is exposed to temperatures just below the solvus and may overage. The higher temperature zone is solution treated, eliminating the effects of age hardening. If the solution-treated zone cools slowly, stable θ may form at the grain boundaries, embrittling the

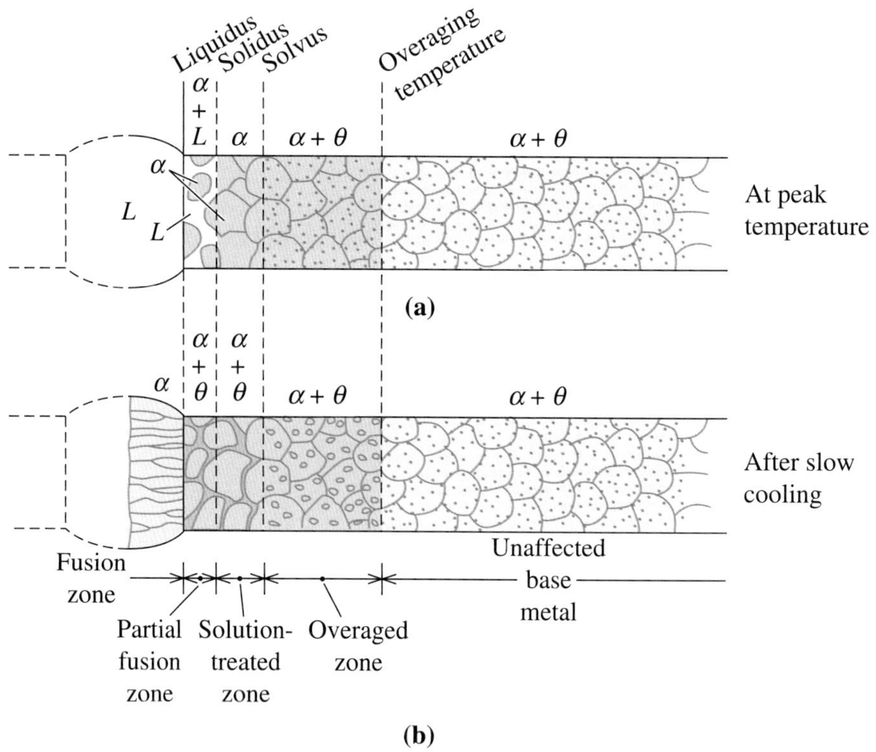

Figure 12-13 Microstructural changes that occur in age-hardened alloys during fusion welding: (a) microstructure in the weld at the peak temperature, and (b) microstructure in the weld after slowly cooling to room temperature.

weld area. Very fast welding processes such as electron-beam welding, complete reheat treatment of the area after welding, or welding the alloy in the solution-treated condition improve the quality of the weld (Chapter 9). Welding of nickel-based superalloys strengthened by precipitation hardening does not pose such problems since the precipitation process is sluggish and the welding process simply acts as a solution and quenching treatment. The process of friction stir welding has also been recently applied to welding of Al and Al-Li alloys for aerospace and aircraft applications.

12-8 The Eutectoid Reaction

In Chapter 11, we defined the eutectoid as a solid-state reaction in which one solid phase transforms to two other solid phases:

$$S_1 \rightarrow S_2 + S_3 \quad (12\text{-}6)$$

As an example of how we can use the eutectoid reaction to control the microstructure and properties of an alloy, let's examine the technologically important portion of the iron-iron carbide (Fe-Fe_3C) phase diagram (Figure 12-14), which is the basis for steels and cast irons. The formation of the two solid phases (α and Fe_3C) permits us to obtain dispersion strengthening. The ability to control the occurrence of the eutectoid reaction (this includes either making it happen, slowing it down, or avoiding it all together) is probably the most important step in the thermomechanical processing of steels. On the Fe-Fe_3C diagram, the eutectoid temperature is known as the A_1 temperature. The boundary between austenite (γ) and the two-phase field consisting of ferrite (α) and austenite is known as the A_3. The boundary between austenite (γ) and the two-phase field consisting of cementite (Fe_3C) and austenite is known as the A_{cm}.

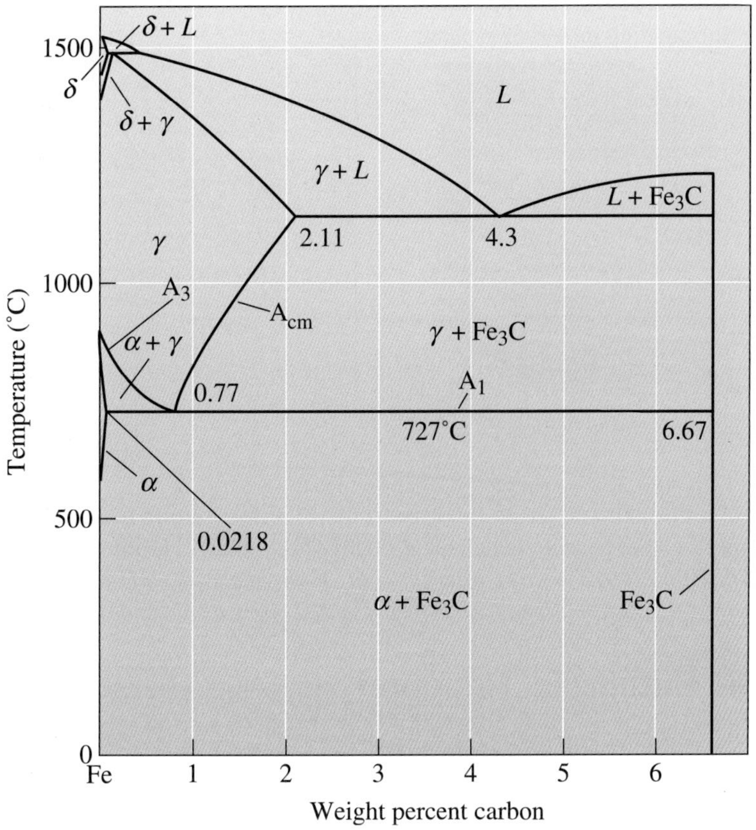

Figure 12-14 The Fe-Fe_3C phase diagram (a portion of the Fe-C diagram). The vertical line at 6.67% C is the stoichiometric compound Fe_3C.

We normally are not interested in the carbon-rich end of the Fe-C phase diagram and this is why we examine the Fe-Fe_3C diagram as part of the Fe-C binary phase diagram.

Solid Solutions Iron goes through two allotropic transformations (Chapter 3) during heating or cooling. Immediately after solidification, iron forms a BCC structure called δ-ferrite. On further cooling, the iron transforms to an FCC structure called γ, or **austenite**. Finally, iron transforms back to the BCC structure at lower temperatures; this structure is called α, or **ferrite**. Both of the ferrites (α and δ) and the austenite are solid solutions of interstitial carbon atoms in iron. Normally, when no specific reference is made, the term ferrite refers to the α ferrite, since this is the phase we encounter more often during the heat treatment of steels. Certain ceramic materials used in magnetic applications are also known as ferrites but are not related to the ferrite phase in the Fe-Fe_3C system.

Because interstitial holes in the FCC crystal structure are somewhat larger than the holes in the BCC crystal structure, a greater number of carbon atoms can be accommodated in FCC iron. Thus, the maximum solubility of carbon in austenite is 2.11% C, whereas the maximum solubility of carbon in BCC iron is much lower (i.e., ~0.0218% C in α and 0.09% C in δ). The solid solutions of carbon in iron are relatively soft and ductile, but are stronger than pure iron due to solid-solution strengthening by the carbon.

Compounds A stoichiometric compound Fe_3C, or **cementite**, forms when the solubility of carbon in solid iron is exceeded. The Fe_3C contains 6.67% C, is extremely hard and brittle (like a ceramic material), and is present in all commercial steels. By properly controlling the amount, size, and shape of Fe_3C, we control the degree of dispersion strengthening and the properties of the steel.

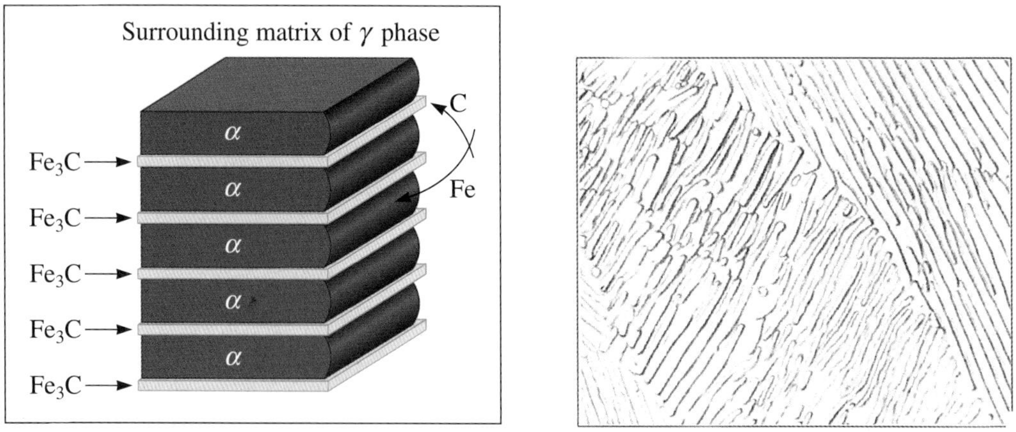

(a) (b)

Figure 12-15 Growth and structure of pearlite: (a) redistribution of carbon and iron, and (b) micrograph of the pearlite lamellae (× 2000). [*From* ASM Handbook, *Vol. 7, Metallography and Microstructure (1972), ASM International, Materials Park, OH 44073-0002. Reprinted with permission of ASM International. All rights reserved. www.asminternational.org.*]

The Eutectoid Reaction If we heat an alloy containing the eutectoid composition of 0.77% C above 727°C, we produce a structure containing only austenite grains. When austenite cools to 727°C, the eutectoid reaction begins:

$$\gamma(0.77\%\ \text{C}) \rightarrow \alpha(0.0218\%\ \text{C}) + \text{Fe}_3\text{C}(6.67\%\ \text{C}) \tag{12-7}$$

As in the eutectic reaction, the two phases that form have different compositions, so atoms must diffuse during the reaction (Figure 12-15). Most of the carbon in the austenite diffuses to the Fe_3C, and iron atoms diffuse to the α. This redistribution of atoms is easiest if the diffusion distances are short, which is the case when the α and Fe_3C grow as thin lamellae, or plates.

Pearlite The lamellar structure of α and Fe_3C that develops in the iron-carbon system is called **pearlite**, which is a microconstituent in steel. This was so named because a polished and etched pearlite shows the colorfulness of mother-of-pearl. The lamellae in pearlite are much finer than the lamellae in the lead—tin eutectic because the iron and carbon atoms must diffuse through solid austenite rather than through liquid. One way to think about pearlite is to consider it as a metal-ceramic nanocomposite. The following example shows the calculation of the amounts of the phases in the pearlite microconstituent.

Example 12-5 *Phases and Composition of Pearlite*

Calculate the amounts of ferrite and cementite present in pearlite.

SOLUTION

Since pearlite must contain 0.77% C, using the lever rule:

$$\%\,\alpha = \left(\frac{6.67 - 0.77}{6.67 - 0.0218}\right) \times 100 = 88.7\%$$

$$\%\,\text{Fe}_3\text{C} = \left(\frac{0.77 - 0.0218}{6.67 - 0.0218}\right) \times 100 = 11.3\%$$

In Example 12-5, we saw that most of the pearlite is composed of ferrite. In fact, if we examine the pearlite closely, we find that the Fe_3C lamellae are surrounded by α. The pearlite structure, therefore, provides dispersion strengthening—the continuous ferrite phase is relatively soft and ductile and the hard, brittle cementite is dispersed.

Primary Microconstituents Hypoeutectoid steels contain less than 0.77% C, and hypereutectoid steels contain more than 0.77% C. Ferrite is the primary or proeutectoid microconstituent in hypoeutectoid alloys, and cementite is the primary or proeutectoid microconstituent in hypereutectoid alloys. If we heat a hypoeutectoid alloy containing 0.60% C above 750°C, only austenite remains in the microstructure. Figure 12-16 shows what happens when the austenite cools. Just below 750°C, ferrite nucleates and grows, usually at the austenite grain boundaries. Primary ferrite continues to grow until the temperature falls to 727°C. The remaining austenite at that temperature is now surrounded by ferrite and has changed in composition from 0.60% C to 0.77% C. Subsequent cooling to below 727°C causes all of the remaining austenite to transform to pearlite by the eutectoid reaction. The structure contains two phases—ferrite and cementite—arranged as two microconstituents—primary ferrite and pearlite. The final microstructure contains islands of pearlite surrounded by the primary ferrite [Figure 12-17(a)]. This structure permits the alloy to be strong, due to the dispersion-strengthened pearlite, yet ductile, due to the continuous primary ferrite.

In hypereutectoid alloys, the primary phase is Fe_3C, which forms at the austenite grain boundaries. After the austenite cools through the eutectoid reaction, the steel contains hard, brittle cementite surrounding islands of pearlite [Figure 12-17(b)].

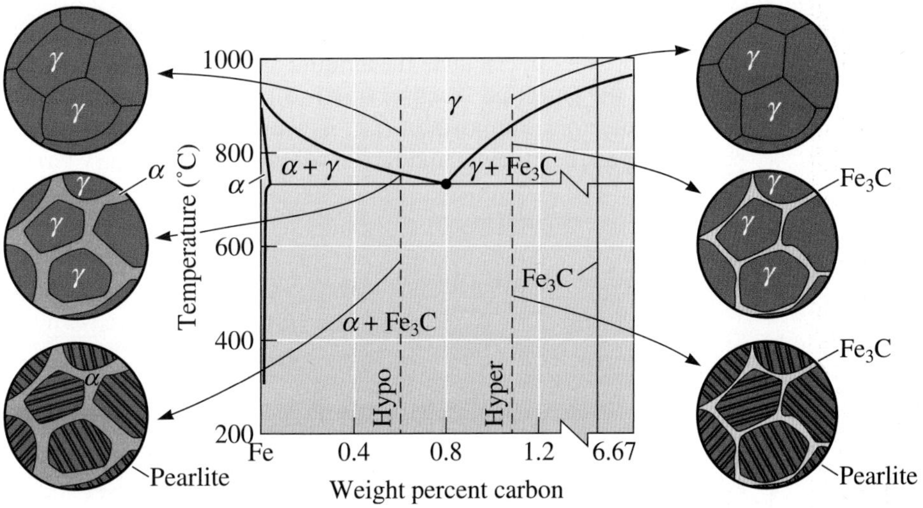

Figure 12-16 The evolution of the microstructure of hypoeutectoid and hypereutectoid steels during cooling in relationship to the Fe-Fe_3C phase diagram.

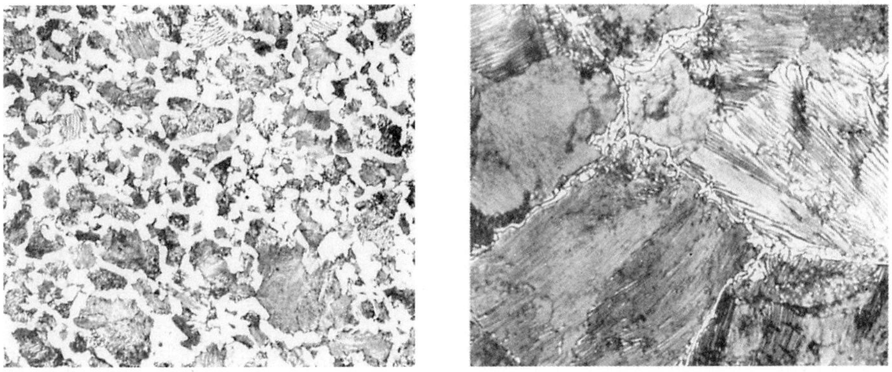

(a) (b)

Figure 12-17 (a) A hypoeutectoid steel showing primary α (white) and pearlite ($\times$ 400). (b) A hypereutectoid steel showing primary Fe_3C surrounding pearlite ($\times$ 800). [*From* ASM Handbook, *Vol. 7, (1972), ASM International, Materials Park, OH 44073-0002. Reprinted with permission of ASM International. All rights reserved. www.asminternational.org.*]

Now, because the hard, brittle microconstituent is continuous, the steel is also brittle. Fortunately, we can improve the microstructure and properties of the hypereutectoid steels by heat treatment. The following example shows the calculation for the amounts and compositions of phases and microconstituents in a plain carbon steel.

Example 12-6 *Phases in Hypoeutectoid Plain Carbon Steel*

Calculate the amounts and compositions of phases and microconstituents in a Fe-0.60% C alloy at 726°C.

SOLUTION

The phases are ferrite and cementite. Using a tie line and working the lever law at 726°C, we find

$$\alpha(0.0218\%\ \text{C})\ \%\,\alpha = \left(\frac{6.67 - 0.60}{6.67 - 0.0218}\right) \times 100 = 91.3\%$$

$$\text{Fe}_3\text{C}(6.67\%\ \text{C})\ \%\,\text{Fe}_3\text{C} = \left(\frac{0.60 - 0.0218}{6.67 - 0.0218}\right) \times 100 = 8.7\%$$

The microconstituents are primary ferrite and pearlite. If we construct a tie line just above 727°C, we can calculate the amounts and compositions of ferrite and austenite just before the eutectoid reaction starts. All of the austenite at that temperature will have the eutectoid composition (i.e., it will contain 0.77% C) and will transform to pearlite; all of the proeutectoid ferrite will remain as primary ferrite.

$$\text{Primary } \alpha(0.0218\%\ \text{C})\ \%\ \text{Primary } \alpha = \left(\frac{0.77 - 0.60}{0.77 - 0.0218}\right) \times 100 = 22.7\%$$

$$\text{Austentite just above } 727°\text{C} = \text{Pearlite: } 0.77\%\ \text{C}$$

$$\%\ \text{Pearlite} = \left(\frac{0.60 - 0.0218}{0.77 - 0.0218}\right) \times 100 = 77.3\%$$

12-9 Controlling the Eutectoid Reaction

We control dispersion strengthening in the eutectoid alloys in much the same way that we did in eutectic alloys (Chapter 11).

Controlling the Amount of the Eutectoid By changing the composition of the alloy, we change the amount of the hard second phase. As the carbon content of steel increases towards the eutectoid composition of 0.77% C, the amounts of Fe_3C and pearlite increase, thus increasing the strength. This strengthening effect eventually peaks, and the properties level out or even decrease when the carbon content is too high (Table 12-1).

Table 12-1 The effect of carbon on the strength of steels

	Slow Cooling (Coarse Pearlite)			Fast Cooling (Fine Pearlite)		
Carbon %	Yield Strength (psi)	Tensile Strength (psi)	% Elongation	Yield Strength (psi)	Tensile Strength (psi)	% Elongation
0.20	42,750	57,200	36.5	50,250	64,000	36.0
0.40	51,250	75,250	30.0	54,250	85,500	28.0
0.60	54,000	90,750	23.0	61,000	112,500	18.0
0.80	54,500	89,250	25.0	76,000	146,500	11.0
0.95	55,000	95,250	13.0	72,500	147,000	9.5

(Based on Metals Progress Materials and Processing Databook, 1981.)

Controlling the Austenite Grain Size Pearlite grows as grains or *colonies*. Within each colony, the orientation of the lamellae is identical. The colonies nucleate most easily at the grain boundaries of the original austenite. Thus, we can increase the number and decrease the size of pearlite colonies by reducing the prior austenite grain size, usually by using a low temperature to produce the austenite. Increasing the number and decreasing the size of the pearlite colonies create more obstacles to dislocation motion by creating more boundaries between colonies, thereby increasing the strength of the alloy.

Controlling the Cooling Rate By increasing the cooling rate during the eutectoid reaction, we reduce the distance that the atoms are able to diffuse. Consequently, the lamellae produced during the reaction are finer or more closely spaced. By producing fine pearlite, we increase the strength of the alloy (Table 12-1 and Figure 12-18).

Controlling the Transformation Temperature The solid-state eutectoid reaction is rather slow, and the steel may cool below the equilibrium eutectoid temperature before the transformation begins (i.e., the austenite phase can be undercooled). Lower transformation temperatures give a finer, stronger structure (Figure 12-19), influence the time required for transformation, and even alter the

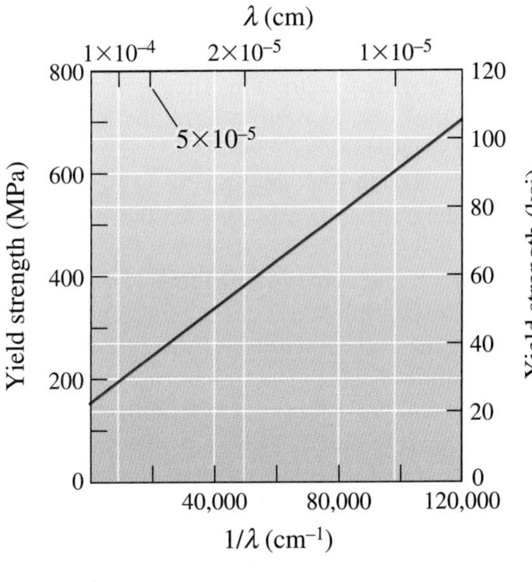

Figure 12-18 The effect of interlamellar spacing (λ) on the yield strength of pearlite.

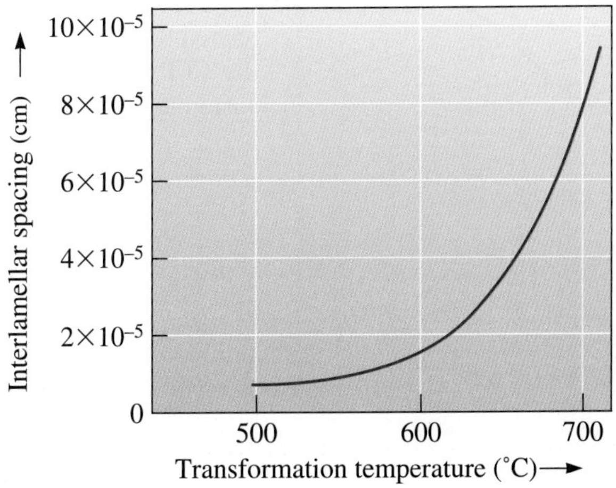

Figure 12-19 The effect of the austenite transformation temperature on the interlamellar spacing in pearlite.

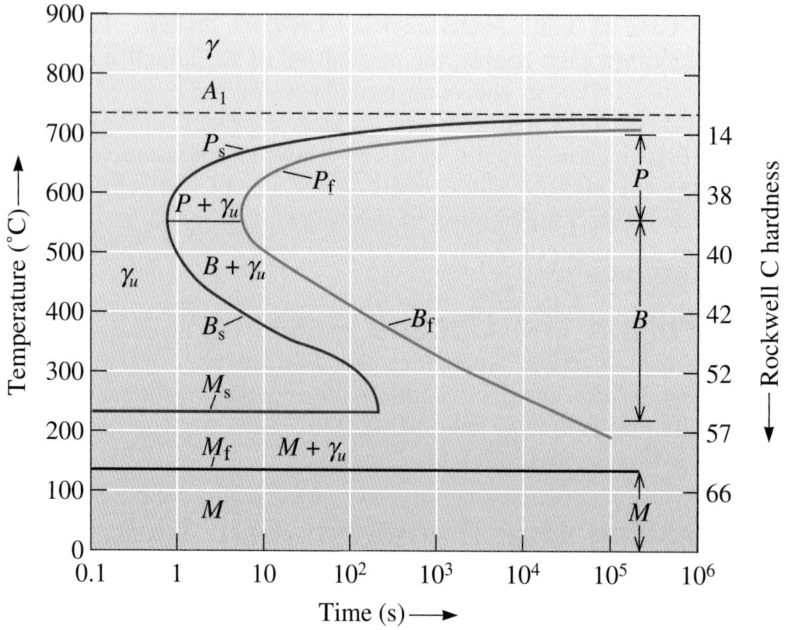

Figure 12-20 The time-temperature-transformation (TTT) diagram for a eutectoid steel, where P = Pearlite, B = Bainite, and M = Martensite. The subscripts "s" and "f" indicate the start and finish of a transformation. γ_u is unstable austenite.

arrangement of the two phases. This information is contained in the **time-temperature-transformation (TTT) diagram** (Figure 12-20). This diagram, also called the **isothermal transformation (IT)** diagram or the C-curve, permits us to predict the structure, properties, and heat treatment required in steels.

The shape of the TTT diagram is a consequence of the kinetics of the eutectoid reaction and is similar to the diagram shown by the Avrami relationship (Figure 12-3). At any particular temperature, a sigmoidal curve describes the rate at which the austenite transforms to a mixture of ferrite and cementite (Figure 12-21). An incubation time is required for nucleation. The P_s (pearlite start) curve represents the time at which austenite starts to transform to ferrite and cementite via the eutectoid transformation. The sigmoidal curve also gives the time at which the transformation is complete; this time is given by the P_f (pearlite finish) curve. When the temperature decreases from 727°C, the rate of nucleation increases, while the rate of growth of the microconstituent decreases. As in Figure 12-3, a maximum transformation rate, or minimum transformation time, is found; the maximum rate of transformation occurs near 550°C for a eutectoid steel (Figure 12-20).

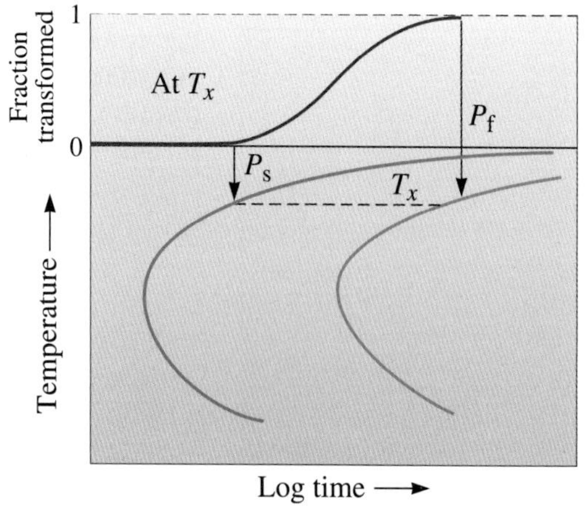

Figure 12-21 The sigmoidal curve is related to the start and finish times on the TTT diagram for steel. In this case, austenite is transforming to pearlite.

Two types of microconstituents are produced as a result of the transformation. Pearlite (P) forms above 550°C, and bainite (B) forms at lower temperatures.

Nucleation and growth of phases in pearlite: If we quench to just below the eutectoid temperature, the austenite is only slightly undercooled. Long times are required before stable nuclei for ferrite and cementite form. After the phases that form pearlite nucleate, atoms diffuse rapidly and *coarse* pearlite is produced; the transformation is complete at the pearlite finish (P_f) time. Austenite quenched to a lower temperature is more highly undercooled. Consequently, nucleation occurs more rapidly and the P_s is shorter. Diffusion is also slower, however, so atoms diffuse only short distances and *fine* pearlite is produced. Even though growth rates are slower, the overall time required for the transformation is reduced because of the shorter incubation time. Finer pearlite forms in shorter times as we reduce the isothermal transformation temperature to about 550°C, which is the *nose,* or *knee,* of the TTT curve (Figure 12-20).

Nucleation and growth of phases in bainite: At a temperature just below the nose of the TTT diagram, diffusion is very slow and total transformation times increase. In addition, we find a different microstructure! At low transformation temperatures, the lamellae in pearlite would have to be extremely thin and, consequently, the boundary area between the ferrite and Fe_3C lamellae would be very large. Because of the energy associated with the ferrite-cementite interface, the total energy of the steel would have to be very high. The steel can reduce its internal energy by permitting the cementite to precipitate as discrete, rounded particles in a ferrite matrix. This new microconstituent, or arrangement of ferrite and cementite, is called **bainite**. Transformation begins at a bainite start (B_s) time and ends at a bainite finish (B_f) time.

The times required for austenite to begin and finish its transformation to bainite increase, and the bainite becomes finer as the transformation temperature continues to decrease. The bainite that forms just below the nose of the curve is called coarse bainite, upper bainite, or feathery bainite. The bainite that forms at lower temperatures is called fine bainite, lower bainite, or acicular bainite. Figure 12-22 shows typical microstructures of bainite. Note that the morphology of bainite depends on the heat treatment used.

Figure 12-23 shows the effect of transformation temperature on the properties of eutectoid (0.77% C) steel. As the temperature decreases, there is a general trend toward higher strength and lower ductility due to the finer microstructure that is produced. The following two examples illustrate how we can design heat treatments of steels to produce desired microstructures and properties.

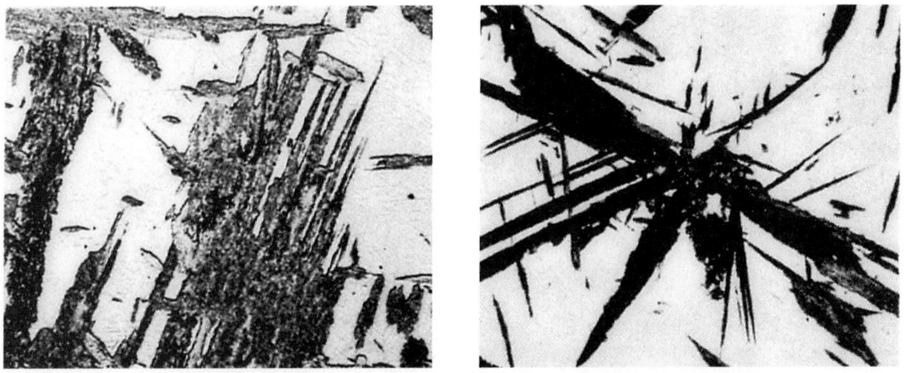

(a) (b)

Figure 12-22 (a) Upper bainite (gray, feathery plates) (× 600). (b) Lower bainite (dark needles) (× 400). [*From* ASM Handbook, *Vol. 8, (1973), ASM International, Materials Park, OH 44073-0002. Reprinted with permission of ASM International. All rights reserved. www.asminternational.org.*]

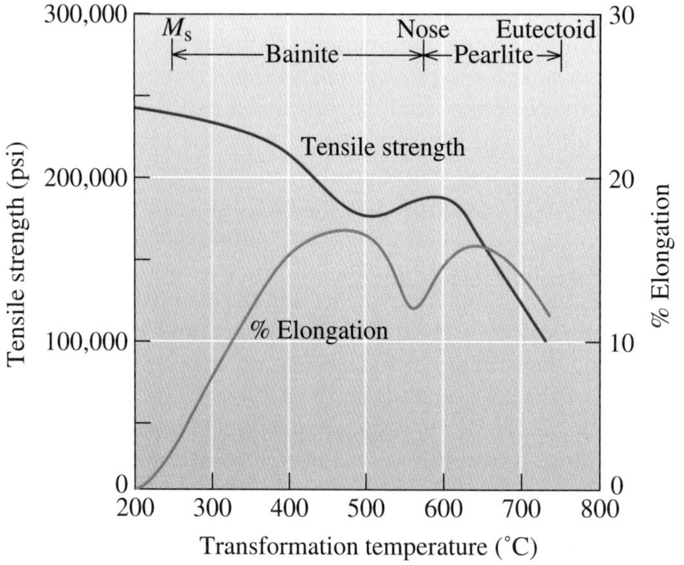

Figure 12-23 The effect of transformation temperature on the properties of a eutectoid steel.

Example 12-7 *Design of a Heat Treatment to Generate the Pearlite Microstructure*

Design a heat treatment to produce the pearlite structure shown in Figure 12-15(b).

SOLUTION

First, we need to determine the interlamellar spacing of the pearlite. If we count the number of lamellar spacings in the upper right of Figure 12-15(b), remembering that the interlamellar spacing is measured from one α plate to the next α plate, we find 14 spacings over a 2 cm distance. Due to the factor of 2000 magnification, this 2 cm distance is actually 0.001 cm. Thus,

$$\lambda = \frac{0.001 \text{ cm}}{14 \text{ spacings}} = 7.14 \times 10^{-5} \text{ cm}$$

If we assume that the pearlite is formed by an isothermal transformation, we find from Figure 12-19 that the transformation temperature must have been approximately 700°C. From the TTT diagram (Figure 12-20), our heat treatment should be

1. Heat the steel to about 750°C and hold—perhaps for 1 h—to produce all austenite. A higher temperature may cause excessive growth of austenite grains.
2. Quench to 700°C and hold for at least 10^5 s (the P_f time). We assume here that the steel cools instantly to 700°C. In practice, this does not happen, and thus, the transformation does not occur at one temperature. We may need to use the continuous cooling transformation diagrams to be more precise (see Chapter 13).
3. Cool to room temperature.

The steel should have a hardness of HRC 14 (Figure 12-20) and a yield strength of about 200 MPa (30,000 psi) from Figure 12-18.

Example 12-8 *Heat Treatment to Generate the Bainite Microstructure*

Excellent combinations of hardness, strength, and toughness are obtained from bainite. One heat treatment facility austenitized a eutectoid steel at 750°C, quenched and held the steel at 250°C for 15 min, and finally permitted the steel to cool to room temperature. Was the required bainitic structure produced?

SOLUTION

Let's examine the heat treatment using Figure 12-20. After heating at 750°C, the microstructure is 100% γ. After quenching to 250°C, unstable austenite remains for slightly more than 100 s, when fine bainite begins to grow. After 15 min, or 900 s, about 50% fine bainite has formed, and the remainder of the steel still contains unstable austenite. As we will see later, the unstable austenite transforms to martensite when the steel is cooled to room temperature, and the final structure is a mixture of bainite and hard, brittle martensite. The heat treatment was not successful! The heat treatment facility should have held the steel at 250°C for at least 10^4 s, or about 3 h.

12-10 The Martensitic Reaction and Tempering

Martensite is a phase that forms as the result of a diffusionless solid-state transformation. In this transformation, there is no diffusion and, hence, it does not follow the Avrami transformation kinetics. The growth rate in **martensitic transformations** (also known as **displacive** or **athermal transformations**) is so high that nucleation becomes the controlling step.

Cobalt, for example, transforms from an FCC to an HCP crystal structure by a slight shift in the atom locations that alters the stacking sequence of close-packed planes. Because the reaction does not depend on diffusion, the martensite reaction is an athermal transformation—that is, the reaction depends only on the temperature, not on the time. The martensite reaction often proceeds rapidly, at speeds approaching the velocity of sound in the material.

Many other alloys (such as Cu-Zn-Al, Cu-Al-Ni, and Ni-Ti) and ceramic materials show martensitic phase transformations. These transformations can also be driven by the application of mechanical stress. Other than the martensite that forms in certain types of steels, the Ni-Ti alloy, known as **nitinol**, is perhaps the best-known example of alloys that make use of martensitic phase transformations. These materials can remember their shape and are known as shape-memory alloys (SMAs). (See Section 12-11.)

Martensite in Steels

In steels with less than about 0.2% C, upon quenching, the FCC austenite can transform to a nonequilibrium supersaturated BCC martensite structure. In higher carbon steels, the martensite reaction occurs as FCC austenite transforms to BCT (body-centered tetragonal) martensite. The relationship between the FCC austenite and the BCT martensite [Figure 12-24(a)] shows that carbon atoms in the (1/2, 0, 0) type of interstitial sites in the FCC cell can be trapped during the transformation to the body-centered structure, causing the tetragonal structure to be produced. As the carbon content of the steel increases, a greater number of carbon atoms are trapped in these sites, thereby increasing the difference in length between the a- and c-axes of the martensite [Figure 12-24(b)].

The steel must be quenched, or rapidly cooled, from the stable austenite region to prevent the formation of pearlite, bainite, or primary microconstituents. The martensite reaction begins in an eutectoid steel when austenite cools below 220°C, the martensite start (M_s) temperature (Figure 12-20). The amount of martensite increases as the temperature decreases. When the temperature passes below the martensite finish temperature (M_f), the steel should contain 100% martensite. At any intermediate temperature, the amount of martensite does not change as the time at that temperature increases.

Owing to the conservation of mass, the composition of martensite must be the same as that of the austenite from which it forms. There is no long-range diffusion during the transformation that can change the composition. Thus, in iron-carbon alloys, the initial austenite composition and the final martensite composition are the same. The following example illustrates how heat treatment is used to produce a dual-phase steel.

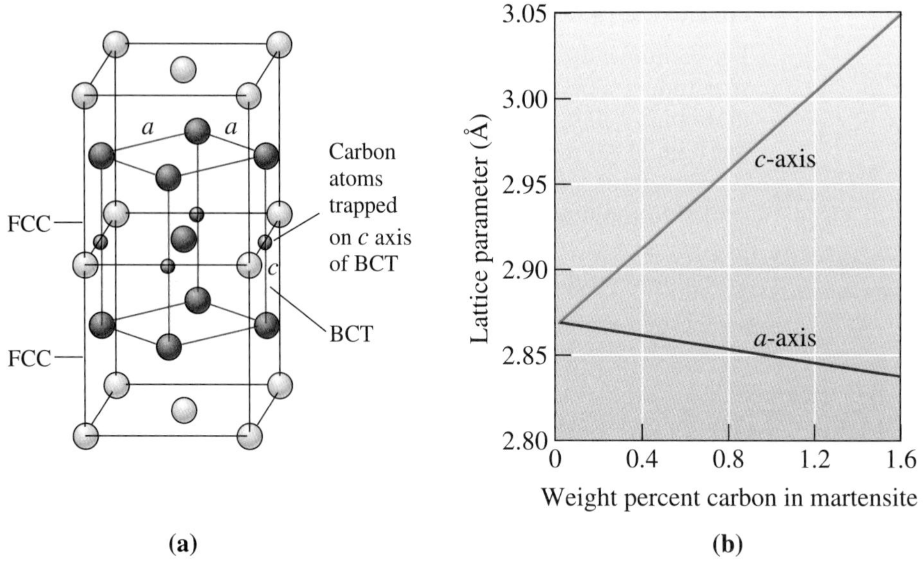

Figure 12-24 (a) The unit cell of BCT martensite is related to the FCC austenite unit cell. (b) As the percentage of carbon increases, more interstitial sites are filled by the carbon atoms, and the tetragonal structure of the martensite becomes more pronounced.

Example 12-9 *Design of a Heat Treatment for a Dual-Phase Steel*

Unusual combinations of properties can be obtained by producing a steel with a microstructure containing 50% ferrite and 50% martensite. The martensite provides strength, and the ferrite provides ductility and toughness. Design a heat treatment to produce a dual phase steel in which the composition of the martensite is 0.60% C.

SOLUTION

To obtain a mixture of ferrite and martensite, we need to heat treat a hypoeutectoid steel into the $\alpha + \gamma$ region of the phase diagram. The steel is then quenched, permitting the γ portion of the structure to transform to martensite.

The heat treatment temperature is fixed by the requirement that the martensite contain 0.60% C. From the solubility line between the γ and the $\alpha + \gamma$ regions, we find that 0.60% C is obtained in austenite when the temperature is about 750°C. To produce 50% martensite, we need to select a steel that gives 50% austenite when the steel is held at 750°C. If the carbon content of the steel is x, then

$$\%\,\gamma = \left(\frac{x - 0.02}{0.60 - 0.02}\right) \times 100 = 50 \quad \text{or} \quad x = 0.31\%\ \text{C}$$

Our final design is

1. Select a hypoeutectoid steel containing 0.31% C.
2. Heat the steel to 750°C and hold (perhaps for 1 h, depending on the thickness of the part) to produce a structure containing 50% ferrite and 50% austenite, with 0.60% C in the austenite.
3. Quench the steel to room temperature. The austenite transforms to martensite, also containing 0.60% C.

Properties of Steel Martensite

Martensite in steels is very hard and brittle, just like ceramics. The BCT crystal structure has no close-packed slip planes on which dislocations can easily move. The martensite is highly supersaturated with carbon,

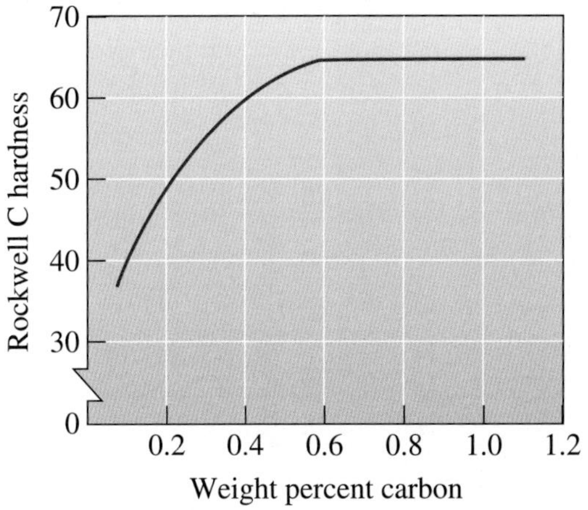

Figure 12-25 The effect of carbon content on the hardness of martensite in steels.

since iron normally contains less than 0.0218% C at room temperature, and martensite contains the amount of carbon present in the steel. Finally, martensite has a fine grain size and an even finer substructure within the grains.

The structure and properties of steel martensites depend on the carbon content of the alloy (Figure 12-25). When the carbon content is low, the martensite grows in a "lath" shape, composed of bundles of flat, narrow plates that grow side by side [Figure 12-26(a)]. This martensite is not very hard. At a higher carbon content, plate martensite grows, in which flat, narrow plates grow individually rather than as bundles [Figure 12-26(b)]. The hardness is much greater in the higher carbon, plate martensite structure, partly due to the greater distortion, or large *c*/*a* ratio, of the crystal structure.

Tempering of Steel Martensite Martensite is not an equilibrium phase. This is why it does not appear on the Fe-Fe_3C phase diagram (Figure 12-14). When martensite in a steel is heated below the eutectoid temperature, the thermodynamically stable α and Fe_3C phases precipitate. This process is called **tempering**. The decomposition of martensite in steels causes the strength and hardness of the steel to decrease while the ductility and impact properties are improved (Figure 12-27). Note that the term tempering here is different from the term we used for tempering of silicate glasses. In both tempering of glasses and tempering of steels, however, the key result is an increase in the toughness of the material.

At low tempering temperatures, the martensite may form two transition phases—a lower carbon martensite and a very fine nonequilibrium ε-carbide, or $Fe_{2.4}C$.

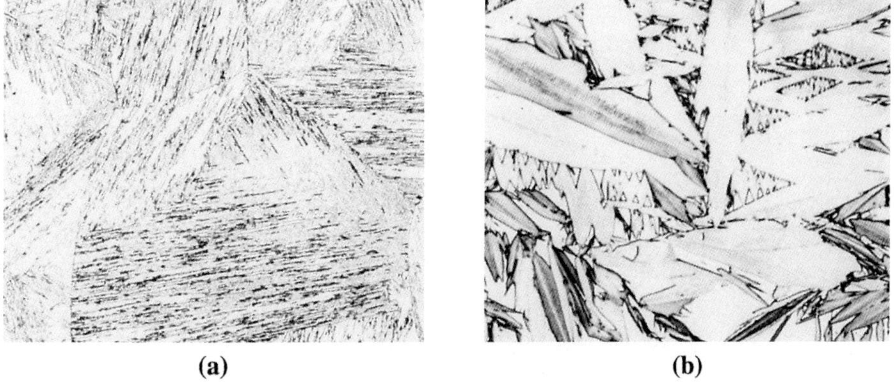

Figure 12-26 (a) Lath martensite in low-carbon steel (× 80). (b) Plate martensite in high-carbon steel (× 400). [*From* ASM Handbook, *Vol. 8, (1973), ASM International, Materials Park, OH 44073-0002. Reprinted with permission of ASM International. All rights reserved. www.asminternational.org.*]

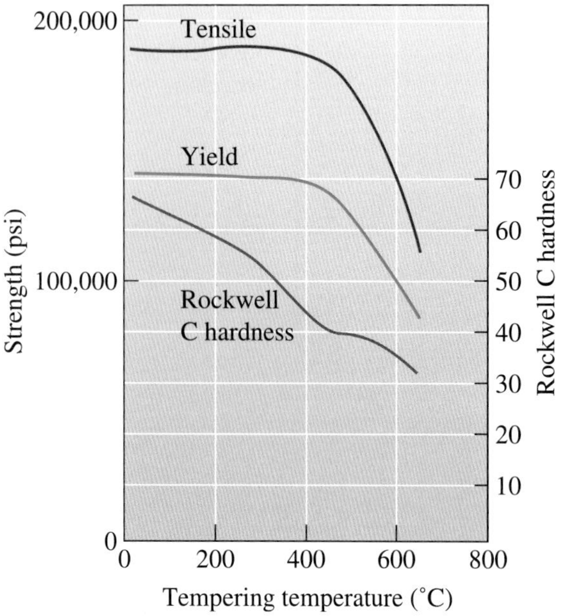

Figure 12-27 Effect of tempering temperature on the properties of a eutectoid steel.

The steel is still strong, brittle, and perhaps even harder than before tempering. At higher temperatures, the stable α and Fe_3C form, and the steel becomes softer and more ductile. If the steel is tempered just below the eutectoid temperature, the Fe_3C becomes very coarse, and the dispersion-strengthening effect is greatly reduced. By selecting the appropriate tempering temperature, a wide range of properties can be obtained. The product of the tempering process is a microconstituent called tempered martensite (Figure 12-28).

Martensite in Other Systems

The characteristics of the martensitic reaction are different in other alloy systems. For example, martensite can form in iron-based alloys that contain little or no carbon by a transformation of the FCC crystal structure to a BCC crystal structure. In certain high-manganese steels and stainless steels, the FCC structure changes to an HCP crystal structure during the martensitic transformation. In addition, the martensitic reaction occurs during the transformation of many polymorphic ceramic materials, including ZrO_2, and even in some crystalline polymers. Thus, the terms martensitic reaction and martensite are rather generic. In the context of steel properties, microstructure, and heat treatment, the term "martensite" refers to the hard and brittle BCT phase obtained upon the quenching of steels.

The properties of martensite in other alloys are also different from the properties of steel martensite. In titanium alloys, BCC titanium transforms to an HCP martensite

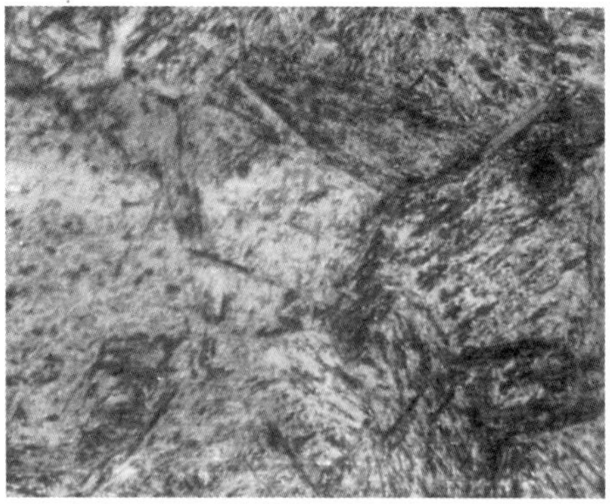

Figure 12-28 Tempered martensite in steel (× 500). [*From* ASM Handbook, *Vol. 9, Metallography and Microstructure (1985), ASM International, Materials Park, OH 44073-0002. Reprinted with permission of ASM International. All rights reserved. www.asminternational.org.*]

structure during quenching; however, the titanium martensite is softer and weaker than the original structure. The martensite that forms in other alloys can also be tempered. The martensite produced in titanium alloys can be reheated to permit the precipitation of a second phase. Unlike the case of steel, however, the tempering process *increases,* rather than decreases, the strength of the titanium alloy.

12-11 The Shape-Memory Alloys [SMAs]

The **shape-memory effect** is a unique property possessed by some alloys that undergo the martensitic reaction. These alloys can be processed using a sophisticated thermomechanical treatment to produce a martensitic structure. At the end of the treatment process, the material is deformed to a predetermined shape. The metal can then be deformed into a second shape, but when the temperature is increased, the metal changes back to its original shape! Orthodontic braces, blood-clot filters, engines, frames for eyeglasses, and actuators for smart systems have been developed using these materials. Flaps that change direction of airflow depending upon temperature have been developed and used for air conditioners.

More recently, a special class of materials known as ferromagnetic shape-memory alloys also has been developed. Examples of ferromagnetic shape-memory alloys include Ni_2MnGa, Fe-Pd, and Fe_3Pt. Unlike Ni-Ti, these materials show a shape-memory effect in response to a magnetic field. Most commercial shape-memory alloys including Ni-Ti are not ferromagnetic. We discussed in Chapter 6 that many polymers are viscoelastic, and the viscous component is recovered over time. Thus, many polymers do have a memory of their shape! Recently, researchers have developed new shape-memory plastics.

Shape-memory alloys exhibit a memory that can be triggered by stress or temperature change. **Smart materials** are materials that can sense an external stimulus (such as stress, temperature change, magnetic field, etc.) and undergo some type of change. Actively smart materials can even initiate a response (i.e., they function as a sensor and an actuator). Shape-memory alloys are a family of passively smart materials in that they merely sense a change in stress or temperature.

Shape-memory alloys also show a **superelastic** behavior. Recoverable strains up to 10% are possible. This is why shape-memory alloys have been used so successfully in such applications as orthodontic wires and eyeglass frames. In these applications, we make use of the superelastic (and not the shape memory) effect.

Example 12-10 *Design of a Coupling for Tubing*

Design a method for quickly joining titanium tubing in the field.

SOLUTION

Titanium is quite reactive and, unless special welding processes are used, may be contaminated. In the field, we may not have access to these processes. Therefore, we wish to make the joint without resorting to high-temperature processes.

We can take advantage of the shape-memory effect for this application (Figure 12-29). Ahead of time, we can set a Ni-Ti coupling into a small diameter, then deform it into a larger diameter in the martensitic state. In the field, the coupling, which is in the martensitic state, is slipped over the tubing and heated (at a low enough temperature so that the titanium tubing is not contaminated). The coupling contracts back to its predetermined shape as a result of the shape-memory effect, producing a strong mechanical bond to join the tubes.

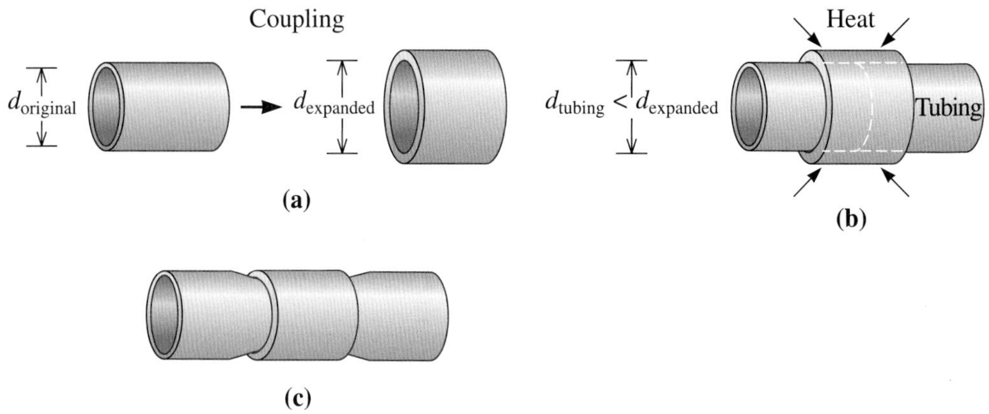

Figure 12-29 Use of shape-memory alloys for coupling tubing: A memory alloy coupling is expanded (a) so it fits over the tubing. (b) When the coupling is reheated, it shrinks back to its original diameter, (c) squeezing the tubing for a tight fit (for Example 12-10).

Summary

- Solid-state phase transformations, which have a profound effect on the structure and properties of a material, can be controlled by proper heat treatments. These heat treatments are designed to provide an optimum distribution of two or more phases in the microstructure. Dispersion strengthening permits a wide variety of structures and properties to be obtained.

- These transformations typically require both nucleation and growth of new phases from the original structure. The kinetics of the phase transformation helps us understand the mechanisms that control the reaction and the rate at which the reaction occurs, enabling us to design the heat treatment to produce the desired microstructure. Reference to appropriate phase diagrams also helps us select the necessary compositions and temperatures.

- Age hardening, or precipitation hardening, is one powerful method for controlling the optimum dispersion strengthening in many metallic alloys. In age hardening, a very fine widely dispersed coherent precipitate is allowed to form via a heat treatment that includes (a) solution treating to produce a single-phase solid solution, (b) quenching to retain that single phase, and (c) aging to permit a precipitate to form. In order for age hardening to occur, the phase diagram must show decreasing solubility of the solute in the solvent as the temperature decreases.

- The eutectoid reaction can be controlled to permit one type of solid to transform to two different types of solid. The kinetics of the reaction depends on the nucleation of the new solid phases and the diffusion of the different atoms in the material to permit the growth of the new phases.

- The most widely used eutectoid reaction occurs in producing steels from iron-carbon alloys. Either pearlite or bainite can be produced as a result of the eutectoid reaction in steel. In addition, primary ferrite or primary cementite may be present, depending on the carbon content of the alloy.

- Factors that influence the mechanical properties of the material produced by the eutectoid reaction include (a) the composition of the alloy (amount of eutectoid microconstituent), (b) the grain size of the original solid, the eutectoid microconstituent, and any primary microconstituents, (c) the fineness of the structure within the eutectoid microconstituent (interlamellar spacing), (d) the cooling rate during the phase transformation, and (e) the temperature at which the transformation occurs (the amount of undercooling).

- A martensitic reaction occurs with no long-range diffusion. Again, the best known transformation occurs in steels:
    - The amount of martensite that forms depends on the temperature of the transformation (an athermal reaction).
    - Martensite is very hard and brittle, with the hardness determined primarily by the carbon content.
    - The amount and composition of the martensite are the same as the austenite from which it forms.
- Martensite can be tempered. During tempering, a dispersion-strengthened structure is produced. In steels, tempering reduces the strength and hardness but improves the ductility and toughness.
- Since optimum properties are obtained through heat treatment, we must remember that the structure and properties may change when the material is used at or exposed to elevated temperatures. Overaging or overtempering occur as a natural extension of the phenomena governing these transformations when the material is placed into service.
- Shape-memory alloys (e.g., Ni-Ti) are a class of smart materials that can remember their shape. They also exhibit superelastic behavior.

Glossary

Age hardening A special dispersion-strengthening heat treatment. By solution treatment, quenching, and aging, a coherent precipitate forms that provides a substantial strengthening effect. (Also known as precipitation hardening.)

Artificial aging Reheating a solution-treated and quenched alloy to a temperature below the solvus in order to provide the thermal energy required for a precipitate to form.

Athermal transformation When the amount of the transformation depends only on the temperature, not on the time (same as martensitic transformation or displacive transformation).

Austenite The name given to the FCC crystal structure of iron and iron-carbon alloys.

Avrami relationship Describes the fraction of a transformation that occurs as a function of time. This describes most solid-state transformations that involve diffusion; thus martensitic transformations are not described.

Bainite A two-phase microconstituent, containing ferrite and cementite, that forms in steels that are isothermally transformed at relatively low temperatures.

Bake-hardenable steels These are steels that can show an increase in their yield stress as a result of precipitation hardening that can occur at fairly low temperatures (~100°C), conditions that simulate baking of paints on cars. This additional increase leads to better dent resistance.

Cementite The hard, brittle ceramic-like compound Fe_3C that, when properly dispersed, provides the strengthening in steels.

Coherent precipitate A precipitate with a crystal structure and atomic arrangement that have a continuous relationship with the matrix from which the precipitate is formed. The coherent precipitate provides disruption of the atomic arrangement in the matrix and provides excellent strengthening.

Dihedral angle The angle that defines the shape of a precipitate particle in the matrix. The dihedral angle is determined by the relative surface energies of the grain boundary energy of the matrix and the matrix-precipitate interfacial energy.

Displacive transformation A phase transformation that occurs via small displacements of atoms or ions and without diffusion. Same as athermal or martensitic transformation.

Ferrite The name given to the BCC crystal structure of iron that can occur as α or δ. This is not to be confused with ceramic ferrites, which are magnetic materials.

Guinier-Preston zones Clusters of atoms that precipitate from the matrix in the early stages of the age-hardening process. Although the GP zones are coherent with the matrix, they are too small to provide optimum strengthening.

Interfacial energy The energy associated with the boundary between two phases.

Isothermal transformation When the amount of a transformation at a particular temperature depends on the time permitted for the transformation.

Martensite A metastable phase formed in steel and other materials by a diffusionless, athermal transformation.

Martensitic transformation A phase transformation that occurs without diffusion. Same as athermal or displacive transformation.

Natural aging When a coherent precipitate forms from a solution-treated and quenched age-hardenable alloy at room temperature, providing optimum strengthening.

Nitinol A nickel-titanium shape-memory alloy.

Pearlite A two-phase lamellar microconstituent, containing ferrite and cementite, that forms in steels cooled in a normal fashion or isothermally transformed at relatively high temperatures.

Precipitation hardening See age hardening.

Shape-memory effect The ability of certain materials to develop microstructures that, after being deformed, can return the material to its initial shape when heated (e.g., Ni-Ti alloys).

Smart materials Materials that can sense an external stimulus (e.g., stress, pressure, temperature change, magnetic field, etc.) and initiate a response. Passively smart materials can sense external stimuli; actively smart materials have sensing and actuation capabilities.

Solution treatment The first step in the age-hardening heat treatment. The alloy is heated above the solvus temperature to dissolve any second phase and to produce a homogeneous single-phase structure.

Strain energy The energy required to permit a precipitate to fit into the surrounding matrix during nucleation and growth of the precipitate.

Superelastic behavior A large reversible elastic deformation that occurs as a result of a stress-induced martensitic transformation.

Supersaturated solid solution The solid solution formed when a material is rapidly cooled from a high-temperature single-phase region to a low-temperature two-phase region without the second phase precipitating. Because the quenched phase contains more alloying element than the solubility limit, it is supersaturated in that element.

Tempering A heat treatment used to reduce the hardness of martensite by permitting the martensite to begin to decompose to the equilibrium phases. This leads to increased toughness.

Time-temperature-transformation diagram The diagram describes the time required at any temperature for a phase transformation to begin and end. The TTT diagram assumes that the temperature is constant during the transformation.

Widmanstätten structure The precipitation of a second phase from the matrix when there is a fixed crystallographic relationship between the precipitate and matrix crystal structures. Often needle-like or plate-like structures form in the Widmanstätten structure.

Problems

Section 12-1 Nucleation and Growth in Solid-State Reactions

12-1 (a) Determine the critical nucleus size r^* for homogeneous nucleation for precipitation of phase β in a matrix of phase α. *Hint:* The critical nucleus size occurs at the maximum in the expression for $\Delta G(r)$ in Equation 12-1. (b) Plot the total free energy change ΔG as a function of the radius of the precipitate. (c) Comment on the value of r^* for homogeneous nucleation for solid-state precipitation when compared to the liquid to solid transformation.

12-2 How is the equation for nucleation of a phase in the solid state different from that for a liquid-to-solid transformation?

12-3 Determine the constants c and n in Equation 12-2 that describe the rate of crystallization of polypropylene at 140°C. (See Figure 12-30.)

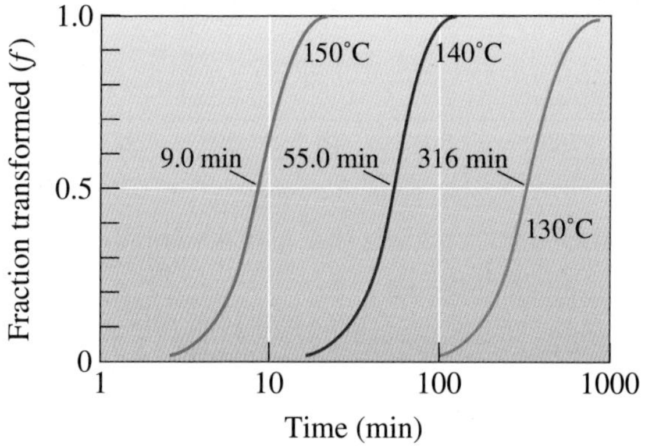

Figure 12-30 The effect of temperature on the crystallization of polypropylene (for Problems 12-3 and 12-5).

12-4 Determine the constants c and n in Equation 12-2 that describe the rate of recrystallization of copper at 135°C. (See Figure 12-2.)

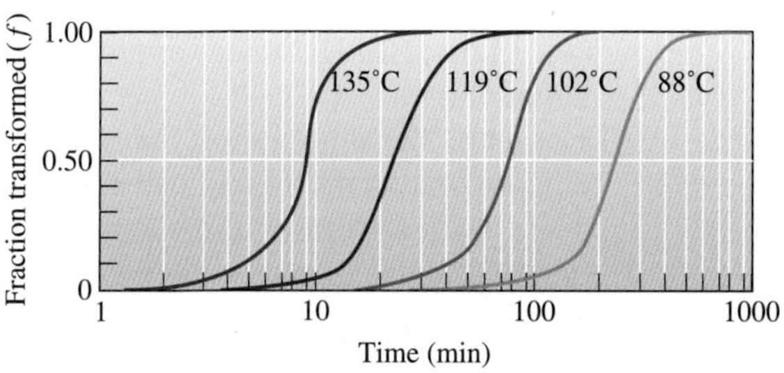

Figure 12-2 The effect of temperature on the recrystallization of cold-worked copper. (Repeated for Problems 12-4 and 12-7.)

12-5 Determine the constants c and n in Equation 12-2 that describe the rate of crystallization of polypropylene at 150°C. (See Figure 12-30.)

12-6 If an FCC sample has an observed incubation time of 10 seconds and is fully transformed at 3.5×10^4 s, what are its c and n values in Equation 12-2?

12-7 For cold-worked copper, construct a plot for the time to 50% recrystallization as a function of annealing temperature. (See Figure 12-2 for data.)

12-8 Most solid-state phase transformations follow the Avrami equation. True or false? Discuss briefly.

12-9 What step controls the rate of recrystallization of a cold-worked metal?

Section 12-2 Alloys Strengthened By Exceeding the Solubility Limit

12-10 What are some ways by which a two-phase microstructure can be made to precipitate from a single-phase solid?

12-11 Explain why the second phase in Al-4% Cu alloys nucleates and grows along the grain boundaries when cooled slowly. Is this usually desirable?

12-12 What do the terms "coherent" and "non-coherent" precipitates mean?

12-13 What properties of the precipitate phase are needed for precipitation hardening? Why?

12-14 Electromigration (diffusion of atoms/ions due to momentum transfer from high energy electrons) leads to voids in aluminum interconnects used in many semiconductor metallization processes and thus is a leading cause of device reliability issues in the industry. Propose an additive to Al that can help mitigate this issue. Refer to the appropriate phase diagram to justify your answer.

12-15 For what dihedral angle will the interfacial energy of the matrix-precipitate boundary and the grain boundary energy of the matrix be equal?

12-16 What is the relationship between the interfacial surface energies of the matrix-precipitate boundary and grain boundary when the precipitate forms a tetrahedron?

12-17 45.7 kg of aluminum comprise 35 wt% of a Cu-Al alloy. The temperature is halfway between the upper and lower bounds of the $L + \eta$ phase region at this composition. How much more aluminum must be added and how much must the resulting liquid mixture be cooled to reach the eutectic point? (See Figure 12-5.)

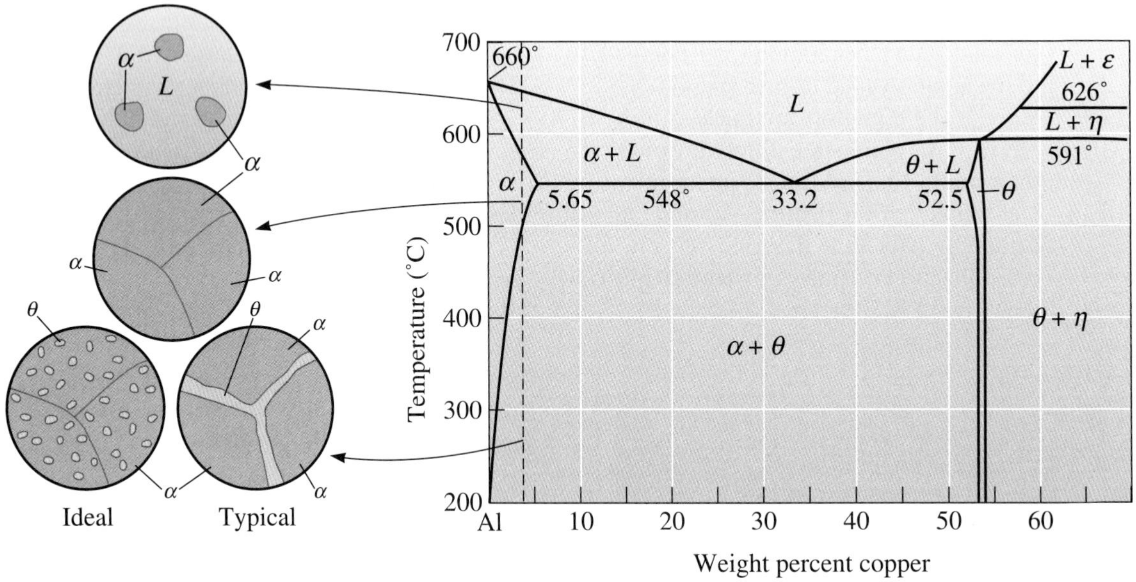

Figure 12-5 The aluminum-copper phase diagram and the microstructures that may develop during cooling of an Al-4% Cu alloy. (Repeated for Problems 12-17 and 12-31.)

Section 12-3 Age or Precipitation Hardening and Its Applications

12-18 What is precipitation hardening?

12-19 What is the difference between precipitation hardening and dispersion strengthening?

12-20 What is a supersaturated solution? How do we obtain supersaturated solutions during precipitation hardening? Why is the formation of a supersaturated solution necessary?

12-21 Why do the precipitates formed during precipitation hardening form throughout the microstructure and not just at grain boundaries?

12-22 On aging for longer times, why do the second-phase precipitates grow? What is the driving force? Compare this with driving forces for grain growth and solid-state sintering.

12-23 Why is precipitation hardening an attractive mechanism of strengthening for aircraft materials?

12-24 Why are most precipitation-hardened alloys suitable only for low-temperature applications?

12-25 Why are some alloy systems that exhibit decreasing solid solubility and are age-hardening candidates, not used commercially?

Section 12-4 Microstructural Evolution in Age or Precipitation Hardening

12-26 Explain the three basic steps encountered during precipitation hardening.

12-27 Explain how hot shortness can occur in precipitation-hardened alloys.

12-28 In precipitation hardening, does the phase that provides strengthening form directly from the supersaturated matrix phase? Explain.

12-29 (a) Recommend an artificial age-hardening heat treatment for a Cu-1.2% Be alloy. (See Figure 12-31.) Include appropriate temperatures.

(b) Compare the amount of the γ_2 precipitate that forms by artificial aging at 400°C with the amount of the precipitate that forms by natural aging.

12-30 Suppose that age hardening is possible in the Al-Mg system. (See Figure 12-10.)

(a) Recommend an artificial age-hardening heat treatment for each of the following alloys; and

(b) Compare the amount of the β precipitate that forms from your treatment of each alloy: (i) Al-4% Mg; (ii) Al-6% Mg; and (iii) Al-12% Mg.

(c) Testing of the alloys after the heat treatment reveals that little strengthening occurs as a result of the heat treatment. Which of the requirements for age hardening is likely not satisfied?

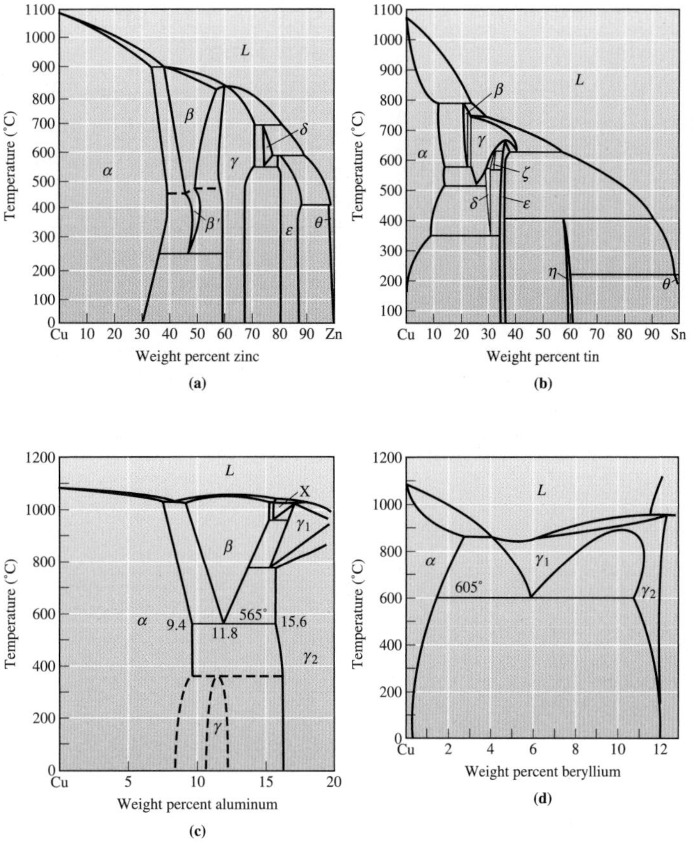

Figure 12-31 Binary phase diagrams for the (a) copper-zinc, (b) copper-tin, (c) copper-aluminum, and (d) copper-beryllium systems (for Problems 12-29 and 12-74).

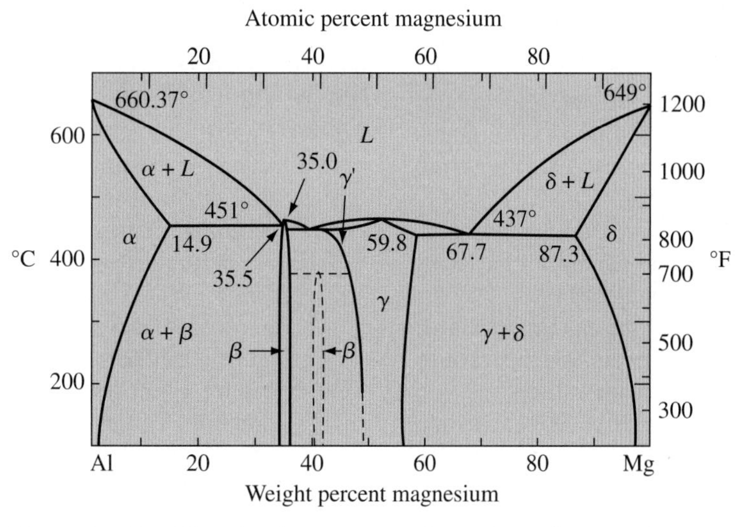

Figure 12-10 The aluminum-magnesium phase diagram. (Repeated for Problems 12-30 and 12-32.)

12-31 An Al-2.5% Cu alloy is solution-treated, quenched, and overaged at 230°C to produce a stable microstructure. If the θ precipitates as spheres with a diameter of 9000 Å and a density of 4.26 g/cm^3, determine the number of precipitate particles per cm^3. (See Figure 12-5.)

12-32 Examine Figure 12-10 and state the phases that will be present in each stage if this process path is followed: (500°C,

20 wt% Mg) to (500°C, 80 wt% Mg) to (200°C, 80 wt% Mg).

12-33 Examine the Mg-Pb phase diagram. Is an Mg-20 wt% Pb alloy suitable for age hardening? Explain. (See Figure 12-32.)

Section 12-5 Effects of Aging Temperature and Time

12-34 What is aging? Why is this step needed in precipitation hardening?

12-35 What is overaging?

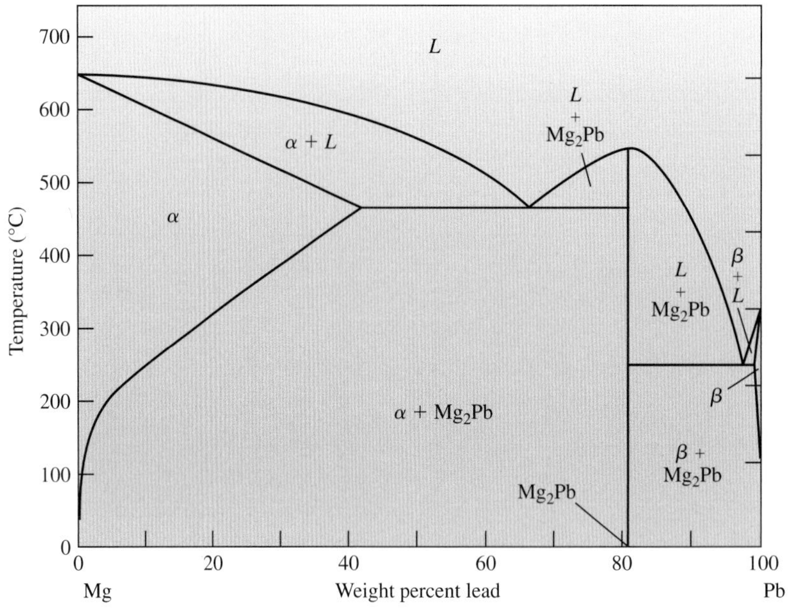

Figure 12-32 The Mg-Pb phase diagram (for Problem 12-33).

12-36 What do the terms natural aging and artificial aging mean?

12-37 In the plane flown by the Wright brothers, how was the alloy precipitation strengthened?

12-38 Why did the work of Dr. Gayle and coworkers reveal two sets of precipitates in the alloy that was used to make the Wright brothers' plane?

12-39 Speculate as to why a longer time is required to reach the maximum hardness at lower aging temperatures.

12-40 Speculate as to why the maximum hardness is higher for lower aging temperatures compared to higher temperatures.

12-41 Based on the principles of age hardening of Al-Cu alloys, rank the following Al-Cu alloys from highest to lowest for maximum yield strength achievable by age hardening and longest to shortest time required at 190°C to achieve the maximum yield strength: Al-2 wt% Cu, Al-3 wt% Cu, and Al-4 wt% Cu. Refer to the Al-Cu phase diagram. (See Figure 12-9.)

12-42 What analytical techniques would you use to characterize the nanoscale precipitates in an alloy?

12-43 Why do we have to keep some aluminum alloys at low temperatures until they are ready for forming steps?

Section 12-6 Requirements for Age Hardening

12-44 Can all alloy compositions be strengthened using precipitation hardening? Can we use this mechanism for the strengthening of ceramics, glasses, or polymers?

12-45 A conductive copper wire is to be made. Would you choose precipitation hardening as a way of strengthening this wire? Explain.

12-46 Figure 12-33 shows a hypothetical phase diagram. Determine whether each of the following alloys might be good candidates for age hardening, and explain your answer. For those alloys that might be good candidates, describe the heat treatment required, including recommended temperatures.

(a) *A*-10% *B*;
(b) *A*-20% *B*;
(c) *A*-55% *B*;
(d) *A*-87% *B*; and
(e) *A*-95% *B*.

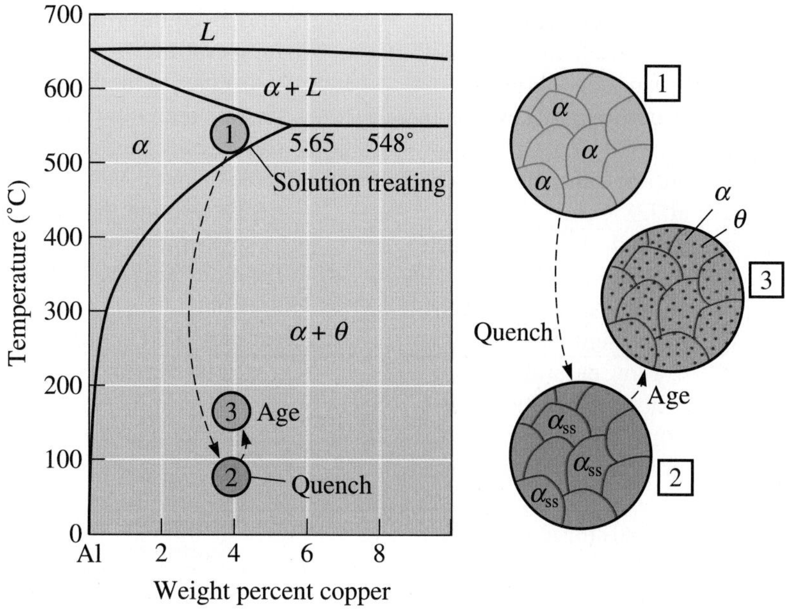

Figure 12-9 The aluminum-rich end of the aluminum-copper phase diagram showing the three steps in the age-hardening heat treatment and the microstructures that are produced. (Repeated for Problem 12-41.)

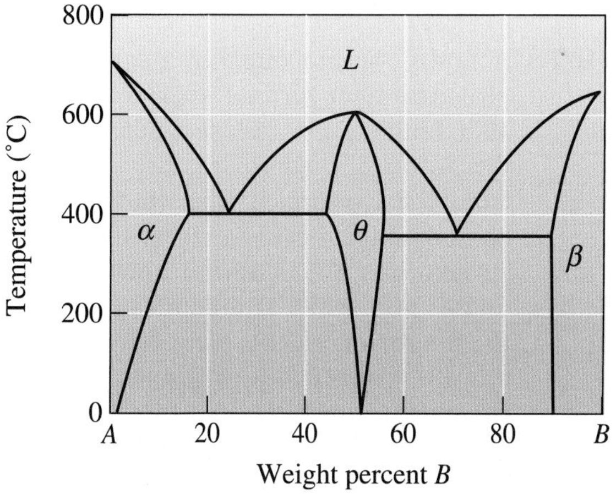

Figure 12-33 Hypothetical phase diagram (for Problem 12-46).

Section 12-7 Use of Age-Hardenable Alloys at High Temperatures

12-47 What is the major limitation of the use for precipitation-hardened alloys?

12-48 Why is it that certain aluminum (not nickel-based) alloys strengthened using age hardening can lose their strength on welding?

12-49 Would you choose a precipitation-hardened alloy to make an aluminum alloy baseball bat?

12-50 What type of dispersion-strengthened alloys can retain their strengths up to 1000°C? These alloys can be welded effectively. Explain.

Section 12-8 The Eutectoid Reaction

12-51 Write down the eutectoid reaction in the Fe-Fe_3C system.

12-52 Sketch the microstructure of pearlite formed by the slow cooling of a steel with the eutectoid composition.

12-53 Compare and contrast eutectic and eutectoid reactions.

12-54 Draw the eutectoid portion of the Fe-Fe_3C phase diagram. Be sure to indicate all of the compositions and temperatures and write the relevant reaction.

12-55 Draw the peritectic portion of the Fe-Fe_3C phase diagram. Be sure to indicate all of the compositions and temperatures and write the relevant reaction.

12-56 What are the solubilities of carbon in the α, δ, and γ forms of iron?

12-57 Define the following terms: ferrite, austenite, pearlite, and cementite.

12-58 The pearlite microstructure is similar to a ceramic-metal nanocomposite. True or false. Comment.

12-59 What do the terms hypoeutectoid and hypereutectoid steels mean?

12-60 What is the difference between a microconstituent and a phase?

12-61 Austenite and Fe_3C are in equilibrium at 1000°C and 3 wt% C. What is the weight percent of iron carbide in the two-phase alloy?

12-62 For an Fe-0.35% C alloy, determine
(a) the temperature at which austenite first begins to transform on cooling;
(b) the primary microconstituent that forms;
(c) the composition and amount of each phase present at 728°C;
(d) the composition and amount of each phase present at 726°C; and
(e) the composition and amount of each microconstituent present at 726°C.

12-63 For an Fe-1.15% C alloy, determine
(a) the temperature at which austenite first begins to transform on cooling;
(b) the primary microconstitutent that forms;
(c) the composition and amount of each phase present at 728°C;
(d) the composition and amount of each phase present at 726°C; and
(e) the composition and amount of each microconstituent present at 726°C.

12-64 A steel contains 8% cementite and 92% ferrite at room temperature. Estimate the carbon content of the steel. Is the steel hypoeutectoid or hypereutectoid?

12-65 A steel contains 18% cementite and 82% ferrite at room temperature. Estimate the carbon content of the steel. Is the steel hypoeutectoid or hypereutectoid?

12-66 A steel contains 18% pearlite and 82% primary ferrite at room temperature. Estimate the carbon content of the steel. Is the steel hypoeutectoid or hypereutectoid?

12-67 A steel contains 94% pearlite and 6% primary cementite at room temperature. Estimate the carbon content of the steel. Is the steel hypoeutectoid or hypereutectoid?

12-68 A steel contains 55% α and 45% γ at 750°C. Estimate the carbon content of the steel.

12-69 A steel contains 96% γ and 4% Fe_3C at 800°C. Estimate the carbon content of the steel.

12-70 A steel is heated until 40% austenite, with a carbon content of 0.5%, forms. Estimate the temperature and the overall carbon content of the steel.

12-71 A steel is heated until 85% austenite, with a carbon content of 1.05%, forms. Estimate the temperature and the overall carbon content of the steel.

12-72 The carbon steels listed in the table below were soaked at 1000°C for 1 hour to form austenite and were cooled slowly, under equilibrium conditions to room temperature.

Refer to the Fe-Fe_3C phase diagram to answer the following questions for each of the carbon steel compositions listed in the table below.
(a) Determine the amounts of the phases present;
(b) determine the C content of each phase; and
(c) plot the C content of pearlite, α, and Fe_3C versus yield strength. Based on the graph, discuss the factors influencing yield strength in steels with <1% C.

Carbon %	Yield Strength (MPa)
0.2	295
0.4	353
0.6	372
0.8	376
0.95	379

12-73 A faulty thermocouple in a carburizing heat-treatment furnace leads to unrealiable temperature measurement during the process. Microstructure analysis from the carburized steel that initially contained 0.2% carbon revealed that the surface had 93% pearlite and 7% primary Fe_3C, while at a depth of 0.5 cm, the microconstituents were 99% pearlite and 1% primary Fe_3C. Estimate the temperature at which the heat treatment was carried out if the carburizing heat treatment was carried out for four hours. For interstitial carbon diffusion in FCC iron, the diffusion coefficient $D_0 = 2.3 \times 10^{-5}$ m^2/s and the activation energy $Q = 1.38 \times 10^{-5}$ J/mol. It may be helpful to refer to Sections 5-6 and 5-8.

12-74 Determine the eutectoid temperature, the composition of each phase in the eutectoid reaction, and the amount of each phase present in the eutectoid microconstituent for the following systems. For the metallic systems, comment on whether you expect the eutectoid microconstituent to be ductile or brittle.
(a) ZrO_2-CaO (See Figure 12-34.)
(b) Cu-Al at 11.8% Al [See Figure 12-31(c).]
(c) Cu-Zn at 47% Zn [See Figure 12-31(a).]
(d) Cu-Be [See Figure 12-31(d).]

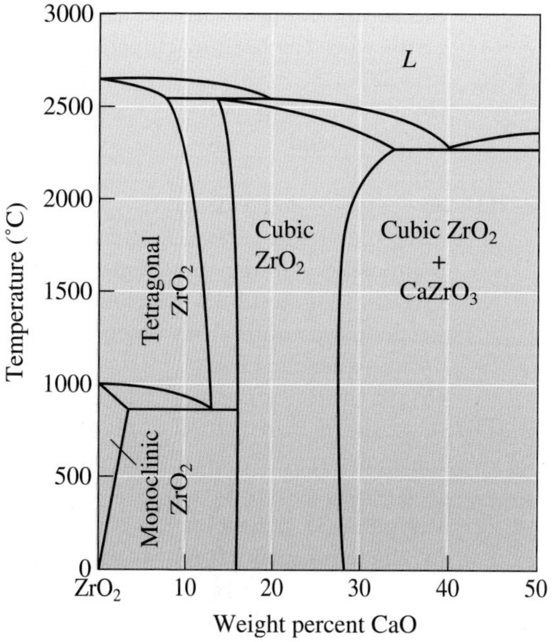

Figure 12-34 A portion of the ZrO_2-CaO phase diagram. A polymorphic phase transformation occurs for pure ZrO_2. Adding 16 to 26% CaO produces a single cubic zirconia phase at all temperatures (for Problem 12-74).

Section 12-9 Controlling the Eutectoid Reaction

12-75 Why are the distances between lamellae formed in a eutectoid reaction typically separated by distances smaller than those formed in eutectic reactions?

12-76 Compare the interlamellar spacing and the yield strength when a eutectoid steel is isothermally transformed to pearlite at
(a) 700°C; and
(b) 600°C.

12-77 Why is it that a eutectoid steel exhibits different yield strengths and % elongations, depending upon if it was cooled slowly or relatively fast?

12-78 Plot the yield and tensile strengths of the ratio of coarse to fine pearlite as a function of carbon composition according to Table 12-1.

12-79 In hypoeutectoid steel, both eutectoid and proeutectoid ferrite exist. Explain the difference between them. What will be the carbon concentration in each?

12-80 What is a TTT diagram?

12-81 Sketch and clearly label the different parts of a TTT diagram for a plain carbon steel with 0.77% carbon.

12-82 On the TTT diagram, what is the difference between the γ and γ_u phases?

12-83 How is it that bainite and pearlite do not appear in the Fe-Fe_3C phase diagram? Are these phases or microconstituents?

12-84 Why is it that we cannot make use of TTT diagrams for describing heat treatment profiles in which samples are cooled over a period of time (i.e., why are TTT diagrams suitable only for isothermal transformations)?

12-85 What is bainite? Why do steels containing bainite exhibit higher levels of toughness?

12-86 An isothermally transformed eutectoid steel is found to have a yield strength of 410 MPa. Estimate
(a) the transformation temperature and
(b) the interlamellar spacing in the pearlite.

12-87 Determine the required transformation temperature and microconstituent if a eutectoid steel is to have the following hardness values:
(a) HRC 38;
(b) HRC 42;
(c) HRC 48; and
(d) HRC 52.

12-88 Describe the hardness and microstructure in a eutectoid steel that has been heated to 800°C for 1 h, quenched to 350°C and held for 750 s, and finally quenched to room temperature.

12-89 Describe the hardness and microstructure in a eutectoid steel that has been heated to 800°C, quenched to 650°C, held for 500 s, and finally quenched to room temperature.

12-90 Describe the hardness and microstructure in a eutectoid steel that has been heated to 800°C, quenched to 300°C and held for 10 s, and finally quenched to room temperature.

12-91 Describe the hardness and microstructure in a eutectoid steel that has been heated to 800°C, quenched to 300°C and held for 10 s, quenched to room temperature, and then reheated to 400°C before finally cooling to room temperature again.

12-92 A 1080 steel was austenitized at 775°C for one hour and quenched to 550°C until it isothermally transformed at 550°C. How long must the austenite be held at 550°C to complete the transformation? What is the expected microstructure and hardness?

Section 12-10 The Martensitic Reaction and Tempering

12-93 What is the primary source of hardening (strengthening) during heat-treating for steels?

12-94 What are the similarities and differences between the heat-treating processes for age-hardened alloys and steel?

12-95 What is the difference between solid-state phase transformations such as the eutectoid reaction and the martensitic phase transformation?

12-96 What is the difference between isothermal and athermal transformations?

12-97 What step controls the rate of martensitic phase transformations?

12-98 Why does martensite not appear on the Fe-Fe_3C phase diagram?

12-99 Does martensite in steels have a fixed composition? What do the properties of martensite depend upon?

12-100 Can martensitic phase transformations occur in other alloys and ceramics?

12-101 Compare the mechanical properties of martensite, pearlite, and bainite formed from the eutectoid steel composition.

12-102 A steel containing 0.3% C is heated to various temperatures above the eutectoid temperature, held for 1 h, and then quenched to room temperature. Using Figure 12-35, determine the amount, composition, and hardness of any martensite that forms when the heating temperature is
(a) 728°C;
(b) 750°C;
(c) 790°C; and
(d) 850°C.

12-103 A steel containing 0.95% C is heated to various temperatures above the eutectoid temperature, held for 1 h, and then quenched to room temperature. Using Figure 12-35, determine the amount and the composition of any martensite that forms when the heating temperature is
(a) 728°C;
(b) 750°C;
(c) 780°C; and
(d) 850°C.

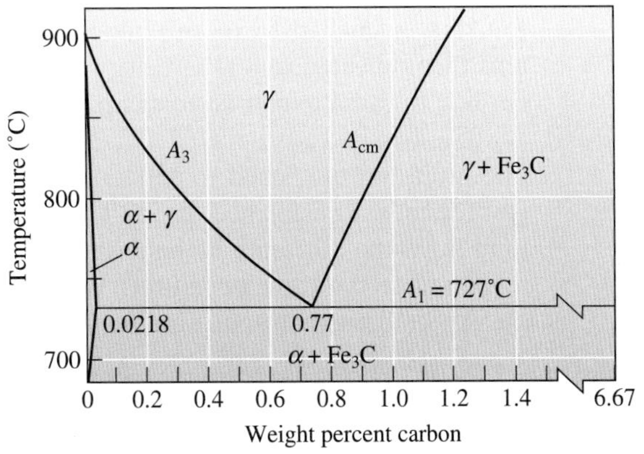

Figure 12-35 The eutectoid portion of the Fe-Fe_3C phase diagram (for Problems 12-102, 12-103, 12-104, and 12-105).

12-104 A steel microstructure contains 75% martensite and 25% ferrite; the composition of the martensite is 0.6% C. Using Figure 12-35, determine
(a) the temperature from which the steel was quenched and
(b) the carbon content of the steel.

12-105 A steel microstructure contains 92% martensite and 8% Fe_3C; the composition of the martensite is 1.10% C. Using Figure 12-35, determine
(a) the temperature from which the steel was quenched and
(b) the carbon content of the steel.

12-106 A steel containing 0.8% C is quenched to produce all martensite. Estimate the volume change that occurs, assuming that the lattice parameter of the austenite is 3.6 Å. Does the steel expand or contract during quenching?

12-107 Describe the complete heat treatment required to produce a quenched and tempered eutectoid steel having a tensile strength of at least 125,000 psi. Include appropriate temperatures.

12-108 Describe the complete heat treatment required to produce a quenched and tempered eutectoid steel having an HRC hardness of less than 50. Include appropriate temperatures.

12-109 In eutectic alloys, the eutectic microconstituent is generally the continuous one, but in the eutectoid structures, the primary microconstituent is normally continuous. By describing the changes that occur with decreasing temperature in each reaction, explain why this difference is expected.

12-110 What is the tempering of steels? Why is tempering necessary?

12-111 What phases are formed by the decomposition of martensite?

12-112 What is tempered martensite?

12-113 If tempering results in the decomposition of martensite, why should we form martensite in the first place?

12-114 Describe the changes in properties that occur when a eutectoid steel is quenched to form martensite and then tempered. Consider low, intermediate, and high tempering temperatures.

Section 12-11 The Shape-Memory Alloys (SMAs)

12-115 What is the principle by which shape-memory alloys display a memory effect?

12-116 Give examples of materials that display a shape-memory effect.

12-117 What are some of the applications of shape-memory alloys?

Design Problems

12-118 You wish to attach aluminum sheets to the frame of the twenty-fourth floor of a skyscraper. You plan to use rivets made from an age-hardenable aluminum, but the rivets must be soft and ductile in order to close. After the sheets are attached, the rivets must be very strong. Design a method for producing, using, and strengthening the rivets.

12-119 Design a process to produce a polypropylene polymer with a structure that is 75% crystalline. Figure 12-30 will provide appropriate data.

12-120 An age-hardened, Al-Cu bracket is used to hold a heavy electrical-sensing device on the outside of a steel-making furnace. Temperatures may exceed 200°C. Is this a good design? Explain. If it is not, design an appropriate bracket and explain why your choice is acceptable.

12-121 You use an arc-welding process to join a eutectoid steel. Cooling rates may be very high following the joining process. Describe what happens in the heat-affected area of the weld and discuss the problems that might occur. Design a joining process that may minimize these problems.

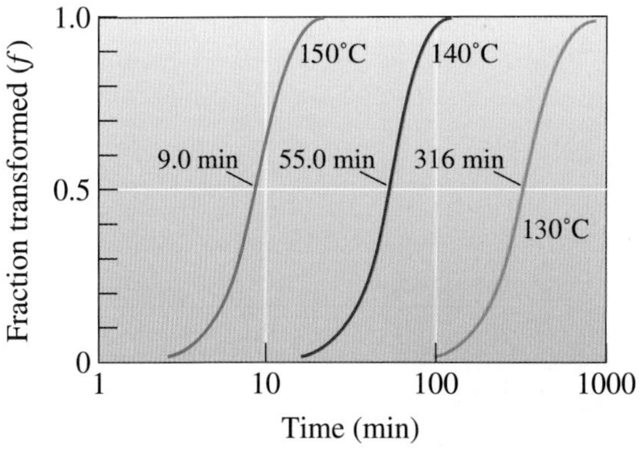

Figure 12-30 The effect of temperature on the crystallization of polypropylene. (Repeated for Problem 12-119.)

Computer Problems

12-122 *Calculation of Phases and Microconstituents in Plain Carbon Steels.* Write a computer program that will calculate the amounts of phases in plain carbon steels in compositions that range from 0 to 1.5% carbon. Assume room temperature for your calculations. The program should ask the user to provide a value for the carbon concentration. The program should make use of this input and provide the amount of phases (ferrite, Fe_3C) in the steel using the lever rule (similar to Example 12-6). The program should also tell the user (as part of the output) whether the steel is eutectoid, hypoeutectoid, or hypereutectoid. Depending on the composition, the program should also make available the amounts of microconstituents present (e.g., how much pearlite). The program should also print an appropriate descriptive message that describes the expected microstructure (assume slow cooling). For example, if the composition chosen is that of a hypoeutectoid steel, the program should output a message stating that the microstructure will primarily consist of α grains and pearlite.

Knovel® Problems

K12-1 What are the effects of thermal cycling on NiTi shape-memory alloys (SMA)?

K12-2 Some precipitation-hardened iron superalloys are used for high-temperature service. Describe heat treatment procedures used for these steels.

Steels constitute the most widely used family of materials for structural, load-bearing applications. Most buildings, bridges, tools, automobiles, and numerous other applications make use of ferrous alloys. With a range of heat treatments that can provide a wide assortment of microstructures and properties, steels are probably the most versatile family of engineering materials.

The Golden Gate Bridge in California is made from steel. The steel is painted to increase its corrosion resistance. (The color is famously known as International Orange.) Each of the main cables that suspend the road over the San Francisco Bay contains 27,572 wires for a total length of approximately 129,000 km. (*Henrik Lehnerer/Shutterstock.com*)

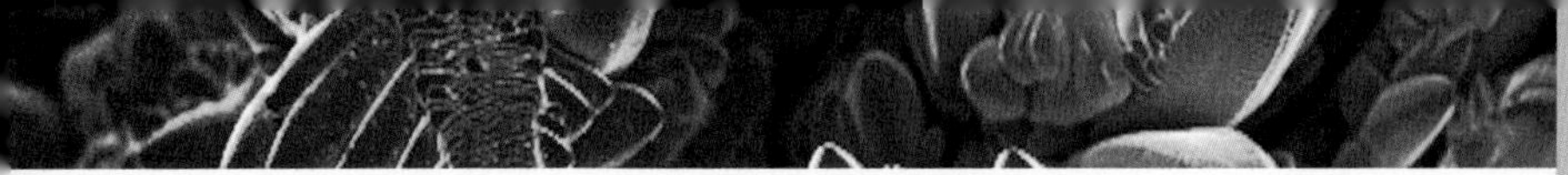

CHAPTER 13

Heat Treatment of Steels and Cast Irons

Have You Ever Wondered?

- *What is the most widely used engineered material?*
- *What makes stainless steels "stainless?"*
- *What is the difference between cast iron and steel?*
- *Are stainless steels ferromagnetic?*

Chapter Learning Objectives

The key objectives of this chapter are to

- Recognize and explain the various designations for different classes of steels.
- Design various heat treatments to produce specific microstructures and properties for steels using time-temperature-transformation and continuous cooling diagrams as needed.
- Differentiate between the classes of steels: plain carbon steels, alloy steels, stainless steels, tool steels, specialty steels, and the subclasses within each of these.
- List common alloying elements of steel and their purposes.
- Explain and recognize the various designations for different cast irons.

Ferrous alloys, which are based on iron-carbon alloys, include plain carbon steels, alloy and tool steels, stainless steels, and cast irons. These are the most widely used materials in the world. In the history of civilization, these materials made their mark by defining the *Iron Age*. Steels typically are produced in two ways: by refining iron ore or by recycling scrap steel.

In producing primary steel, iron ore (processed to contain 50 to 70% iron oxide, Fe_2O_3 or Fe_3O_4) is heated in a blast furnace in the presence of coke (a form of carbon) and oxygen. The coke reduces the iron oxide into a crude molten iron known as **hot metal or pig iron**. At ~1600°C, this material contains about

95% iron; 4% carbon; 0.3 to 0.9% silicon; 0.5% manganese; and 0.025 to 0.05% of sulfur, phosphorus, and titanium. Slag is a byproduct of the blast furnace process. It contains silica, CaO, and other impurities in the form of a silicate melt. Steel has a carbon content up to a maximum of ~2% on a weight basis. Because the liquid pig iron contains a large amount of carbon, oxygen is blown into it in the *basic oxygen furnace* (BOF) to eliminate the excess carbon and produce liquid steel.

In the second method, scrap is often melted in an **electric arc furnace** in which the heat of the arc melts the scrap. Many alloy and specialty steels, such as stainless steels, are produced using electric melting. Molten steels (including stainless steels) often undergo further refining. The goal here is to reduce the levels of impurities such as phosphorus and sulfur and to bring the carbon to the desired level. Molten steel is poured into molds to produce finished steel castings or cast into shapes that are later processed through metal-forming techniques such as rolling or forging.

All of the strengthening mechanisms discussed in the previous chapter apply to at least some of the ferrous alloys. In this chapter, we will discuss how to use the eutectoid reaction to control the structure and properties of steels through heat treatment and alloying. We will also examine two special classes of ferrous alloys: stainless steels and cast irons.

13-1 Designations and Classification of Steels

The dividing point between "steels" and "cast irons" is 2.11% C, where the eutectic reaction becomes possible. For steels, we concentrate on the eutectoid portion of the diagram (Figure 13-1) in which the solubility lines and the eutectoid isotherm are specially identified. The A_3 shows the temperature at which ferrite starts to form on cooling; the A_{cm} shows the temperature at which cementite starts to form; and the A_1 is the eutectoid temperature.

Almost all of the heat treatments of steel are directed toward producing the mixture of ferrite and cementite that gives the proper combination of properties. Figure 13-2 shows the three important microconstituents, or arrangements of ferrite and cementite, that are usually sought. Pearlite is a microconstituent consisting of a lamellar mixture of ferrite and cementite. In bainite, which is obtained by transformation of austenite at a large undercooling, the cementite is more rounded than in pearlite. Tempered martensite, a mixture of very fine and nearly round cementite in ferrite, forms when martensite is reheated following its formation.

Designations

The AISI (American Iron and Steel Institute) and SAE (Society of Automotive Engineers) provide designation systems (Table 13-1) that use a four- or five-digit number to identify each steel alloy. The first two numbers refer to the major alloying elements present, and the last two or three numbers refer to the percentage of carbon. An AISI 1040 steel is a plain carbon steel with 0.40% C. An SAE 10120 steel is a plain carbon steel containing 1.20% C. An AISI 4340 steel is an alloy steel containing 0.40% C. Note that the American Society for Testing of Materials (ASTM) has a different way of classifying steels. The ASTM has a list of specifications that describe steels suitable for different applications.

Classifications

Steels can be classified based on their composition or the way they have been processed. Carbon steels contain up to ~2% carbon. These steels may also contain other elements, such as Si (maximum 0.6%), Cu (up to 0.6%), and

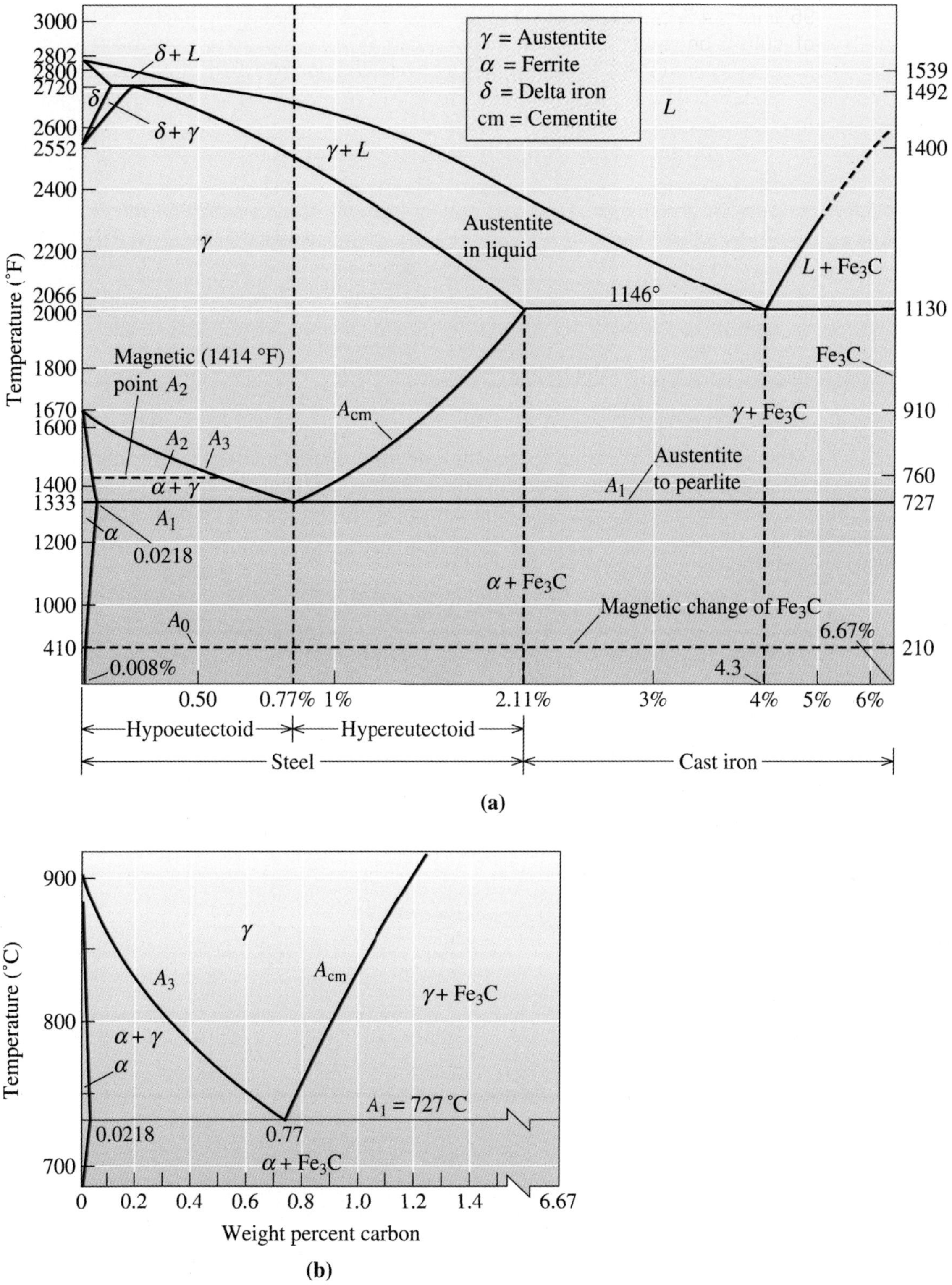

Figure 13-1 (a) The Fe-Fe_3C phase diagram. (b) An expanded version of the eutectoid portion of the Fe-Fe_3C diagram, adapted from several sources.

Mn (up to 1.65%). Decarburized steels contain less than 0.005% C. Ultra-low carbon steels contain a maximum of 0.03% carbon. They also contain very low levels of other elements such as Si and Mn. Low-carbon steels contain 0.04 to 0.15% carbon. These low-carbon steels are used for making car bodies and hundreds of other applications.

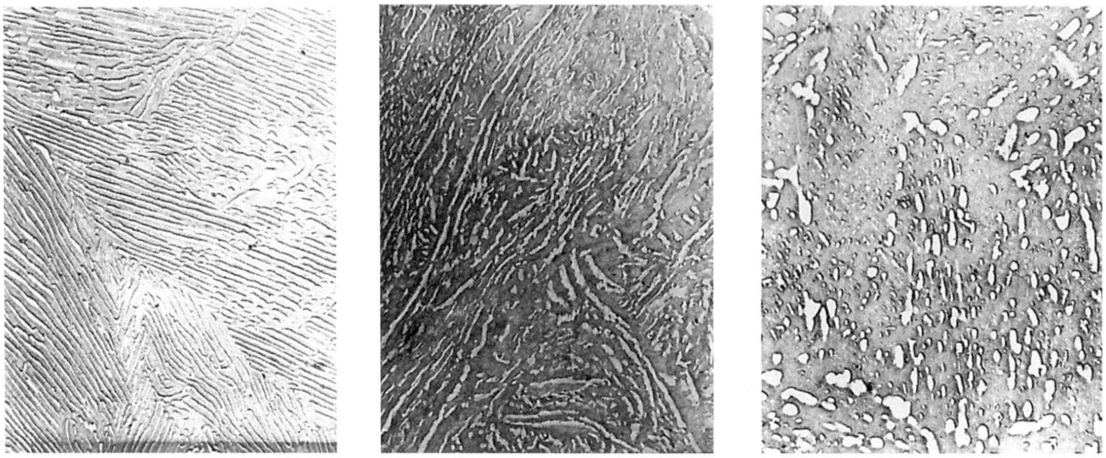

(a) (b) (c)

Figure 13-2 Micrographs of (a) pearlite, (b) bainite, and (c) tempered martensite, illustrating the differences in cementite size and shape among the three microconstituents (× 7500). (*From* The Making, Shaping and Treating of Steel, *10th Ed. Courtesy of the Association for Iron & Steel Technology.*)

Table 13-1 Compositions of selected AISI-SAE steels

AISI-SAE Number	% C	% Mn	% Si	% Ni	% Cr	Others
1020	0.18–0.23	0.30–0.60				
1040	0.37–0.44	0.60–0.90				
1060	0.55–0.65	0.60–0.90				
1080	0.75–0.88	0.60–0.90				
1095	0.90–1.03	0.30–0.50				
1140	0.37–0.44	0.70–1.00				0.08–0.13% S
4140	0.38–0.43	0.75–1.00	0.15–0.30		0.80–1.10	0.15–0.25% Mo
4340	0.38–0.43	0.60–0.80	0.15–0.30	1.65–2.00	0.70–0.90	0.20–0.300% Mo
4620	0.17–0.22	0.45–0.65	0.15–0.30	1.65–2.00		0.20–0.30% Mo
52100	0.98–1.10	0.25–0.45	0.15–0.30		1.30–1.60	
8620	0.18–0.23	0.70–0.90	0.15–0.30	0.40–0.70	0.40–0.60	0.15–0.25% Y
9260	0.56–0.64	0.75–1.00	1.80–2.20			

Example 13-1 *Design of a Method to Determine AISI Number*

An unalloyed steel tool used for machining aluminum automobile wheels has been found to work well, but the purchase records have been lost and you do not know the steel's composition. The microstructure of the steel is tempered martensite. Assume that you cannot estimate the composition of the steel from the structure. Design a treatment that may help determine the steel's carbon content.

SOLUTION

Assume that there is no access to equipment that would permit you to analyze the chemical composition directly. Since the entire structure of the steel is a very fine tempered martensite, we can do a simple heat treatment to produce a structure that can be analyzed more easily. This can be done in two different ways.

The first way is to heat the steel to a temperature just below the A_1 temperature and hold for a long time. The steel overtempers and large Fe_3C spheres form in a ferrite matrix. We then estimate the amount of ferrite and cementite and calculate the carbon content using the lever law. If we measure 16% Fe_3C using this method, the carbon content is

$$\% \, Fe_3C = \left(\frac{x - 0.0218}{6.67 - 0.0218}\right) \times 100 = 16 \quad \text{or} \quad x = 1.086\% \, C$$

A better approach, however, is to heat the steel above the A_{cm} to produce all austenite. If the steel then cools slowly, it transforms to pearlite and a primary microconstituent. If, when we do this, we estimate that the structure contains 95% pearlite and 5% primary Fe_3C, then

$$\% \, \text{Pearlite} = \left(\frac{6.67 - x}{6.67 - 0.77}\right) \times 100 = 95 \quad \text{or} \quad x = 1.065\% \, C$$

The carbon content is on the order of 1.065 to 1.086%, consistent with a 10110 steel. In this procedure, we assumed that the weight and volume percentages of the microconstituents are the same, which is nearly the case in steels.

Mild steel contains 0.15 to 0.3% carbon. This steel is used in piping and structures such as buildings and bridges. Medium-carbon steels contain 0.3 to 0.6% carbon. These are used for example, in making machinery, tractors, and mining equipment. High-carbon steels contain above 0.6% carbon. These are used in making springs, railroad car wheels, and the like. Note that cast irons are Fe-C alloys containing 2 to 4% carbon.

Alloy steels are compositions that contain more significant levels of alloying elements. We will discuss the effect of alloying elements later in this chapter. They improve the hardenability of steels. The AISI defines alloy steels as steels that exceed one or more of the following composition limits: $\geq$1.65% Mn, 0.6% Si, or 0.6% Cu. The total carbon content is up to 1%, and the total alloying element content is below 5%. A material is also an alloy steel if a definite concentration of alloying elements (often Ni, Cr, Mo, or Ti) is specified. These steels are used for making tools (hammers, chisels, etc.) and also in making parts such as axles, shafts, and gears.

Certain specialty steels may contain higher levels of sulfur (>0.1%) or lead (~0.15 to 0.35%) to provide machinability. These, however, cannot be welded easily. In "green steel," lead, an environmental toxin, is replaced with tin and/or antimony. Steels can also be classified based on their processing. For example, the term "concast steels" refers to continuously cast steels. Galvanized steels have a zinc coating for corrosion resistance (Chapter 23). Similarly, tin-plated steel is used to make corrosion-resistant cans and other products. Tin is deposited using electroplating—a process known as "continuous web electrodeposition." "E-steels" are steels that are melted using an electric furnace, while "B-steels" contain a small (0.0005 to 0.003%), yet significant, concentration of boron. "Germ-resistant" coated stainless steels also exist.

13-2 Simple Heat Treatments

Four simple heat treatments—process annealing, annealing, normalizing, and spheroidizing—are commonly used for steels (Figure 13-3). These heat treatments are used to accomplish one of three purposes: (1) eliminating the effects of cold work, (2) controlling dispersion strengthening, or (3) improving machinability.

Process Annealing—Eliminating Cold Work The recrystallization heat treatment used to eliminate the effect of cold working in steels with less than about 0.25% C is called a **process anneal**. The process anneal is done 80°C to 170°C

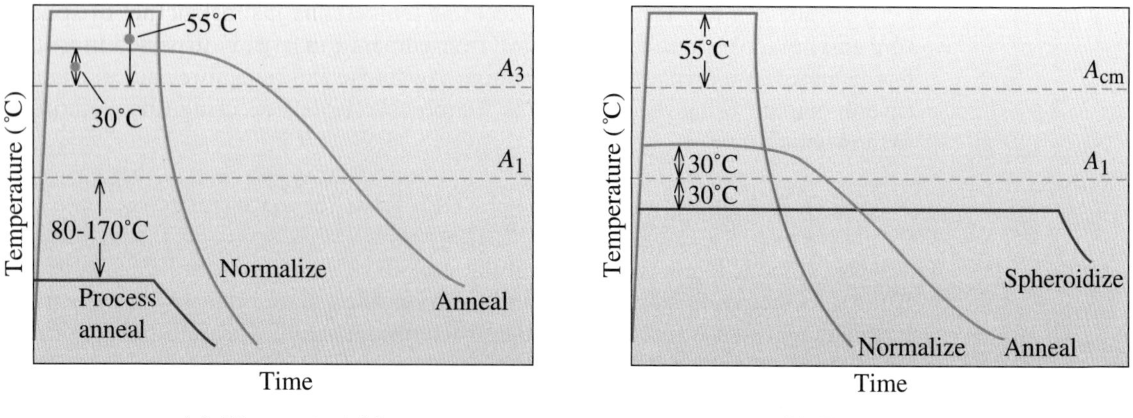

Figure 13-3 Schematic summary of the simple heat treatments for (a) hypoeutectoid steels and (b) hypereutectoid steels.

below the A_1 temperature. The process anneal treatment for steels is similar to the annealing of inorganic glasses in that the intent is to significantly reduce or eliminate residual stresses.

Annealing and Normalizing—Dispersion Strengthening

Steels can be dispersion-strengthened by controlling the fineness of pearlite. The steel is initially heated to produce homogeneous austenite (FCC γ phase), a step called **austenitizing**. **Annealing**, or a full anneal, allows the steel to cool slowly in a furnace, producing coarse pearlite. **Normalizing** allows the steel to cool more rapidly, in air, producing fine pearlite. Figure 13-4 shows the typical properties obtained by annealing and normalizing plain carbon steels.

For annealing, austenitizing of hypoeutectoid steels is conducted about 30°C above the A_3, producing 100% γ; however, austenitizing of a hypereutectoid steel is done at about 30°C above the A_1, producing austenite and Fe_3C. This process prevents the

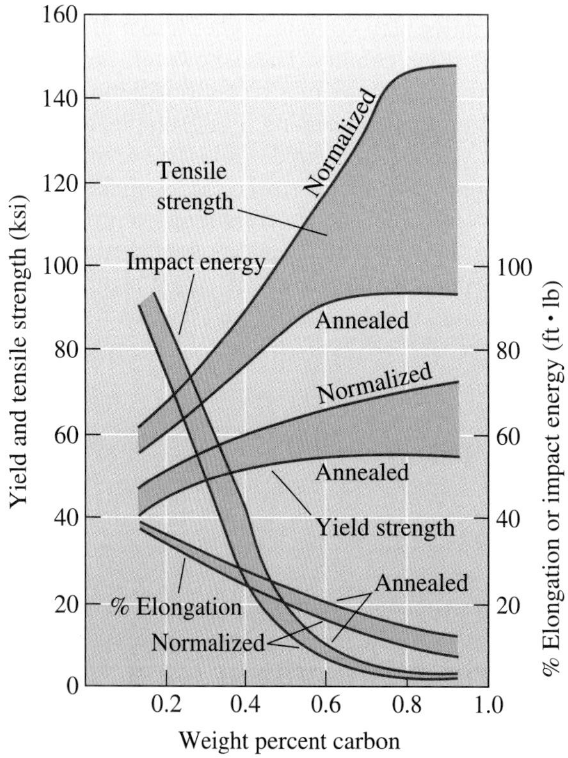

Figure 13-4 The effect of carbon and heat treatment on the properties of plain carbon steels.

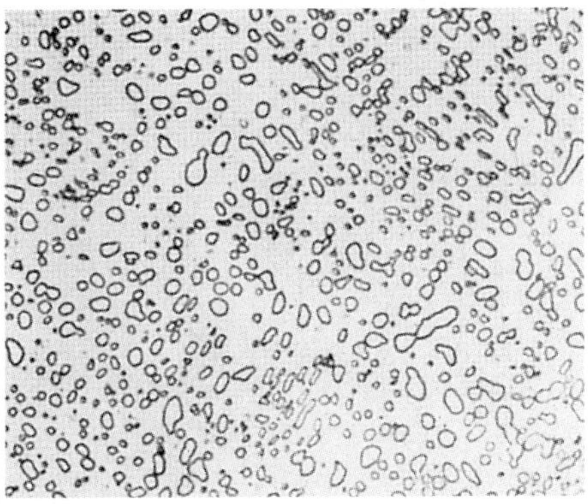

Figure 13-5 The microstructure of spheroidite with Fe_3C particles dispersed in a ferrite matrix (× 850). [*From* ASM Handbook, *Vol. 7, (1972), ASM International, Materials Park, OH 44073-0002. Reprinted with permission of ASM International. All rights reserved. www.asminternational.org.*]

formation of a brittle, continuous film of Fe_3C at the grain boundaries that occurs on slow cooling from the 100% γ region. In both cases, the slow furnace cool and coarse pearlite provide relatively low strength and good ductility.

For normalizing, austenitizing is done at about 55°C above the A_3 or A_{cm}; the steel is then removed from the furnace and cooled in air. The faster cooling gives fine pearlite and provides higher strength.

Spheroidizing—Improving Machinability Steels that contain a large concentration of Fe_3C have poor machining characteristics. It is possible to transform the morphology of Fe_3C using *spheroidizing*. During the spheroidizing treatment, which requires several hours at about 30°C below the A_1, the Fe_3C phase morphology changes into large, spherical particles in order to reduce boundary area. The microstructure, known as **spheroidite**, has a continuous matrix of soft, machinable ferrite (Figure 13-5). After machining, the steel is given a more sophisticated heat treatment to produce the required properties. A similar microstructure occurs when martensite is tempered just below the A_1 for long periods of time. As noted before, alloying elements such as Pb and S are also added to improve machinability of steels, and lead-free "green steels" that have very good machinability have been developed.

The following example shows how different heat treatment conditions can be developed for a given composition of steel.

Example 13-2 *Determination of Heat-Treating Temperatures*

Recommend temperatures for the process annealing, annealing, normalizing, and spheroidizing of 1020, 1077, and 10120 steels.

SOLUTION

From Figure 13-1, we find the critical A_1, A_3, or A_{cm}, temperatures for each steel. We can then specify the heat treatment based on these temperatures.

Steel Type	1020	1077	10120
Critical temperatures	A_1 = 727°C A_3 = 830°C	A_1 = 727°C	A_1 = 727°C A_{cm} = 895°C
Process annealing	727 − (80 to 170) = 557°C to 647°C	Not done	Not done
Annealing	830 + 30 = 860°C	727 + 30 = 757°C	727 + 30 = 757°C
Normalizing	830 + 55 = 885°C	727 + 55 = 782°C	895 + 55 = 950°C
Spheroidizing	Not done	727 − 30 = 697°C	727 − 30 = 697°C

13-3 Isothermal Heat Treatments

The effect of transformation temperature on the properties of a 1080 (eutectoid) steel was discussed in Chapter 12. As the isothermal transformation temperature decreases, pearlite becomes progressively finer before bainite begins to form. At very low temperatures, martensite is obtained.

Austempering and Isothermal Annealing

The isothermal transformation heat treatment used to produce bainite, called **austempering**, simply involves austenitizing the steel, quenching to some temperature below the nose of the TTT curve, and holding at that temperature until all of the austenite transforms to bainite (Figure 13-6).

Annealing and normalizing are usually used to control the fineness of pearlite; however, pearlite formed by an **isothermal anneal** (Figure 13-6) may give more uniform properties, since the cooling rates and microstructure obtained during annealing and normalizing vary across the cross-section of the steel. *Note that the TTT diagrams only describe isothermal heat treatments (i.e., we assume that the sample begins and completes a heat treatment at a given temperature).* Thus, we cannot exactly describe heat treatments by superimposing cooling curves on a TTT diagram such as those shown in Figure 13-6.

Effect of Changes in Carbon Concentration on the TTT Diagram

In either a hypoeutectoid or a hypereutectoid steel, the TTT diagram must reflect the possible formation of a primary phase. The isothermal transformation diagrams for a 1050 and a 10110 steel are shown in Figure 13-7. The most remarkable change is the presence of a "wing" that begins at the nose of the curve and becomes asymptotic to the A_3 or A_{cm} temperature. The wing represents the ferrite start (F_s) time in hypoeutectoid steels or the cementite start (C_s) time in hypereutectoid steels.

When a 1050 steel is austenitized, quenched, and held between the A_3 and the A_1, primary ferrite nucleates and grows. Eventually, equilibrium amounts of ferrite and austenite result. Similarly, primary cementite nucleates and grows to its equilibrium amount in a 10110 steel held between the A_{cm} and A_1 temperatures.

If an austenitized 1050 steel is quenched to a temperature between the A_1 temperature and the nose, primary ferrite again nucleates and grows until reaching

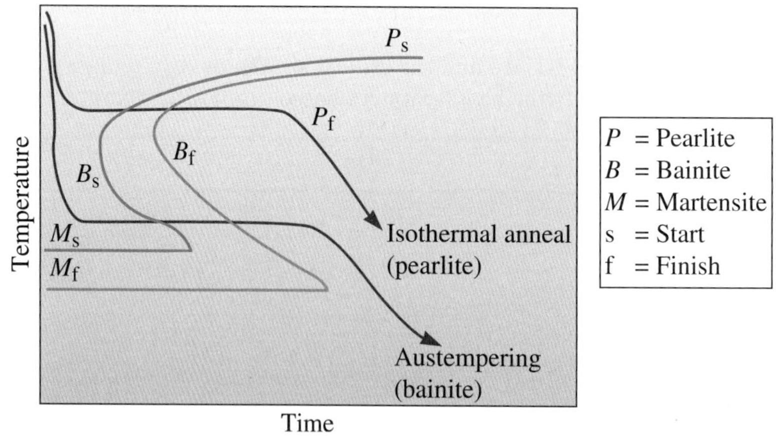

Figure 13-6 The austempering and isothermal anneal heat treatments in a 1080 steel.

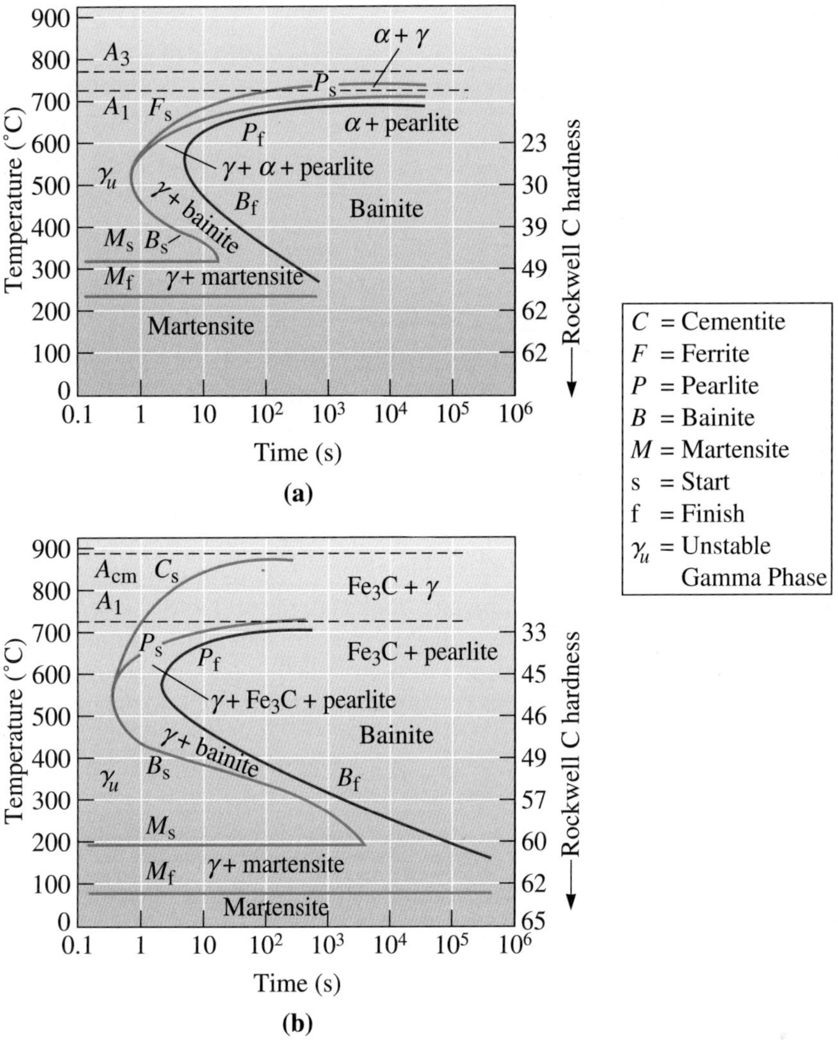

Figure 13-7 The TTT diagrams for (a) a 1050 and (b) a 10110 steel.

the equilibrium amount. The remainder of the austenite then transforms to pearlite. A similar situation, producing primary cementite and pearlite, is found for the hypereutectoid steel.

If we quench the steel below the nose of the curve, only bainite forms, regardless of the carbon content of the steel. If the steels are quenched to temperatures below the M_s, martensite will form. Example 13-3 shows how the phase diagram and TTT diagram can guide development of the heat treatment of steels.

Interrupting the Isothermal Transformation Complicated microstructures are produced by interrupting the isothermal heat treatment. For example, we could austenitize the 1050 steel (Figure 13-8) at 800°C, quench to 650°C and hold for 10 s (permitting some ferrite and pearlite to form), then quench to 350°C and hold for 1 h (3600 s). Whatever unstable austenite remained before quenching to 350°C transforms to bainite. The final structure is ferrite, pearlite, and bainite. We could complicate the treatment further by interrupting the treatment at 350°C after 1 min (60 s) and quenching. Any austenite remaining after 1 min at 350°C forms martensite. The final structure now

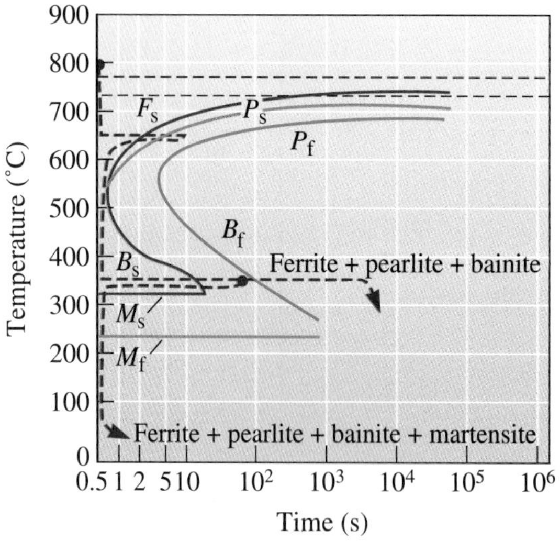

Figure 13-8 Producing complicated structures by interrupting the isothermal heat treatment of a 1050 steel.

contains ferrite, pearlite, bainite, and martensite. Note that each time we change the temperature, we start at zero time! In practice, temperatures cannot be changed instantaneously (i.e., we cannot go instantly from 800 to 650 or 650 to 350°C). This is why it is better to use the continuous cooling transformation (CCT) diagrams.

Example 13-3 *Design of a Heat Treatment for an Axle*

A heat treatment is needed to produce a uniform microstructure and hardness of HRC 23 in a 1050 steel axle.

SOLUTION

We might attempt this task in several ways. We could austenitize the steel, then cool at an appropriate rate by annealing or normalizing to obtain the correct hardness. By doing this, however, we find that the structure and hardness vary from the surface to the center of the axle.

A better approach is to use an isothermal heat treatment. From Figure 13-7, we find that a hardness of HRC 23 is obtained by transforming austenite to a mixture of ferrite and pearlite at 600°C. From Figure 13-1, we find that the A_3 temperature is 770°C. Therefore, our heat treatment is

1. Austenitize the steel at 770 + (30 to 55) = 800°C to 825°C, holding for 1 h and obtaining 100% γ.
2. Quench the steel to 600°C and hold for a minimum of 10 s. Primary ferrite begins to precipitate from the unstable austenite after about 1.0 s. After 1.5 s, pearlite begins to grow, and the austenite is completely transformed to ferrite and pearlite after about 10 s. After this treatment, the microconstituents present are

$$\text{Primary } \alpha = \left(\frac{0.77 - 0.50}{0.77 - 0.0218}\right) \times 100 = 36\%$$

$$\text{Pearlite} = \left(\frac{0.50 - 0.0218}{0.77 - 0.0218}\right) \times 100 = 64\%$$

3. Cool in air to room temperature, preserving the equilibrium amounts of primary ferrite and pearlite. The microstructure and hardness are uniform because of the isothermal anneal.

13-4 Quench and Temper Heat Treatments

Quenching hardens most steels, and tempering increases the toughness. This has been known for perhaps thousands of years. For example, a series of such heat treatments has been used for making Damascus steel and Japanese samurai swords. We can obtain an exceptionally fine dispersion of Fe_3C and ferrite (known as tempered martensite) if we first quench the austenite to produce martensite and then temper. The tempering treatment controls the final properties of the steel (Figure 13-9). Note that this is different from a spheroidizing heat treatment (Figure 13-5). The following example shows how a combination of heat treatments is used to obtain steels with desired properties.

Retained Austenite There is a large volume expansion when martensite forms from austenite. As the martensite plates form during quenching, they surround and isolate small pools of austenite (Figure 13-10), which deform to accommodate the lower density martensite. As the transformation progresses, however, for the remaining pools of austenite to transform, the surrounding martensite must deform. Because the strong martensite resists the transformation, either the existing martensite cracks or the austenite remains trapped in the structure as **retained austenite**. Retained austenite can be a serious problem. Martensite softens and becomes more ductile during tempering. After tempering, the retained austenite cools below the M_s and M_f temperatures and transforms to martensite since the surrounding **tempered martensite** can deform, but now the steel contains more of the hard, brittle martensite! A second

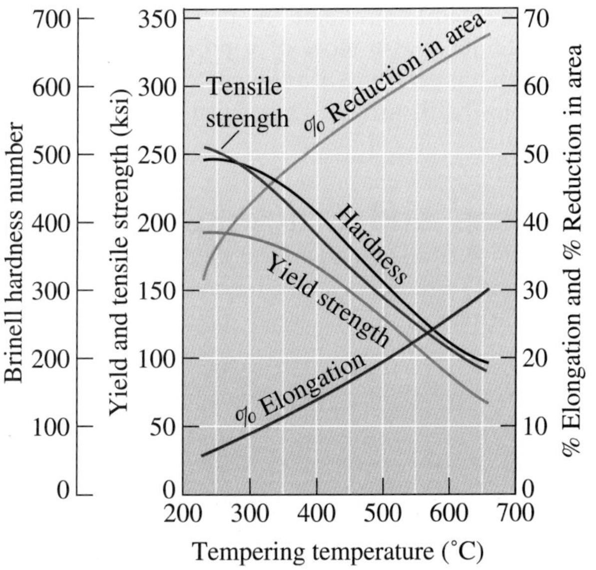

Figure 13-9 The effect of tempering temperature on the mechanical properties of a 1050 steel.

Figure 13-10 Retained austenite (white) trapped between martensite needles (black) (× 1000). [*From* ASM Handbook, *Vol. 8, (1973), ASM International, Materials Park, OH 44073-0002. Reprinted with permission of ASM International. All rights reserved. www.asminternational.org.*]

Example 13-4 *Design of a Quench and Temper Treatment*

A rotating shaft that delivers power from an electric motor is made from a 1050 steel. Its yield strength should be at least 145,000 psi, yet it should also have at least 15% elongation in order to provide toughness. Design a heat treatment to produce this part.

SOLUTION

We are not able to obtain this combination of properties by annealing or normalizing (Figure 13-4); however, a quench and temper heat treatment produces a microstructure that can provide both strength and toughness. Figure 13-9 shows that the yield strength exceeds 145,000 psi if the steel is tempered below 460°C, whereas the elongation exceeds 15% if tempering is done above 425°C. The A_3 temperature for the steel is 770°C. A possible heat treatment is

1. Austenitize above the A_3 temperature of 770°C for 1 h. An appropriate temperature may be 770 + 55 = 825°C.
2. Quench rapidly to room temperature. Since the M_f is about 250°C, martensite will form.
3. Temper by heating the steel to 440°C. Normally, 1 h will be sufficient if the steel is not too thick.
4. Cool to room temperature.

tempering step may be needed to eliminate the martensite formed from the retained austenite. Retained austenite is also more of a problem for high-carbon steels. The martensite start and finish temperatures are reduced when the carbon content increases (Figure 13-11), and thus high-carbon steels must be refrigerated to produce all martensite.

Residual Stresses and Cracking Residual stresses are produced because of cold working. A stress-relief anneal can be used to remove or minimize residual stresses due to cold working. In steels, there is one more mechanism that causes stress. When steels are quenched, the surface of the quenched steel cools rapidly and

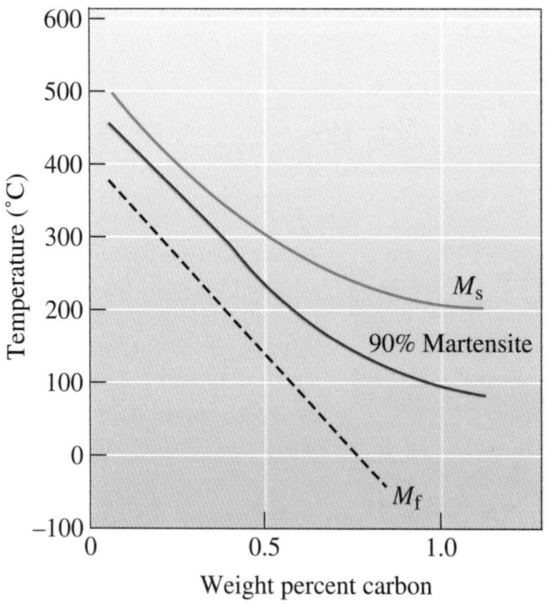

Figure 13-11 Increasing carbon reduces the M_s and M_f temperatures in plain carbon steels.

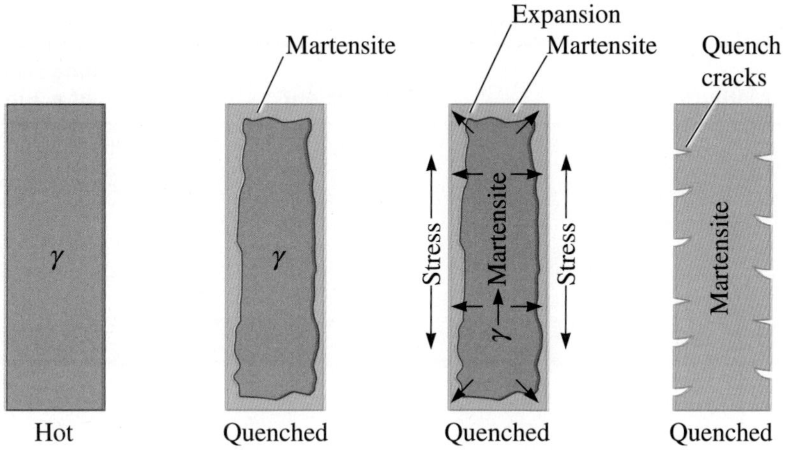

Figure 13-12 Formation of quench cracks caused by residual stresses produced during quenching. The figure illustrates the development of stresses as the austenite transforms to martensite during cooling.

transforms to martensite. When the austenite in the center later transforms, the hard surface is placed in tension, while the center is compressed. If the residual stresses exceed the tensile strength, **quench cracks** form at the surface (Figure 13-12). To avoid this, we can first cool to just above the M_s and hold until the temperature equalizes in the steel; subsequent quenching permits all of the steel to transform to martensite at about the same time. This heat treatment is called **marquenching** or **martempering** (Figure 13-13). Note that as discussed presently, strictly speaking, the CCT diagrams should be used to examine non-isothermal heat treatments.

Quench Rate In using the TTT diagram, we assumed that we could cool from the austenitizing temperature to the transformation temperature instantly. Because this does not occur in practice, undesired microconstituents may form during the quenching process. For example, pearlite may form as the steel cools past the nose of the curve, particularly because the time of the nose is less than one second in plain carbon steels.

The rate at which the steel cools during quenching depends on several factors. First, the surface always cools faster than the center of the part. In addition, as the size of the part increases, the cooling rate at any location is slower. Finally, the cooling rate depends on the temperature and heat transfer characteristics of the quenching medium (Table 13-2).

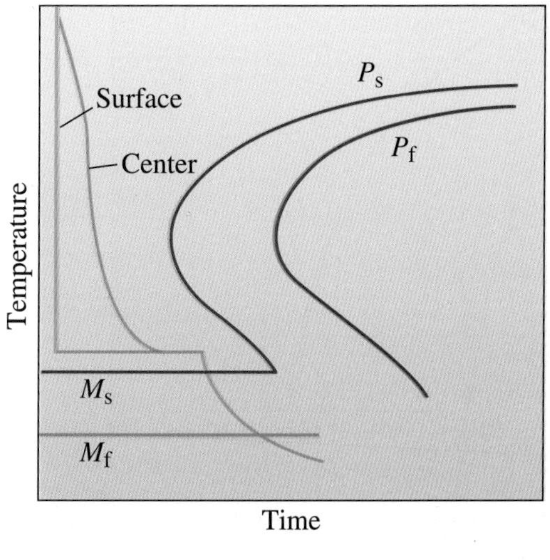

Figure 13-13 The marquenching heat treatment, designed to reduce residual stresses and quench cracking.

Table 13-2 The H coefficient, or severity of the quench, for several quenching media

Medium	H Coefficient	Cooling Rate at the Center of a 1 in. Bar (°C/s)
Oil (no agitation)	0.25	18
Oil (agitation)	1.0	45
H_2O (no agitation)	1.0	45
H_2O (agitation)	4.0	190
Brine (no agitation)	2.0	90
Brine (agitation)	5.0	230

Quenching in oil, for example, produces a lower H coefficient, or slower cooling rate, than quenching in water or brine. The H coefficient is equivalent to the heat transfer coefficient. Agitation helps break the vapor blanket (e.g., when water is the quenching medium) and improves the overall heat transfer rate by bringing cooler liquid into contact with the parts being quenched.

Continuous Cooling Transformation Diagrams We can develop a *continuous cooling transformation* (CCT) diagram by determining the microstructures produced in the steel at various rates of cooling. The CCT curve for a 1080 steel is shown in Figure 13-14.

If we cool a 1080 steel at 5°C/s, the CCT diagram tells us that we obtain coarse pearlite; we have annealed the steel. Cooling at 35°C/s gives fine pearlite and is a normalizing heat treatment. Cooling at 100°C/s permits pearlite to start forming, but the reaction is incomplete and the remaining austenite changes to martensite. We obtain 100% martensite and thus are able to perform a quench and temper heat treatment only if we cool faster than 140°C/s. Other steels, such as the low-carbon steel in Figure 13-15 have more complicated CCT diagrams. You can find a compilation of TTT and CCT diagrams for different grades of steels in various handbooks.

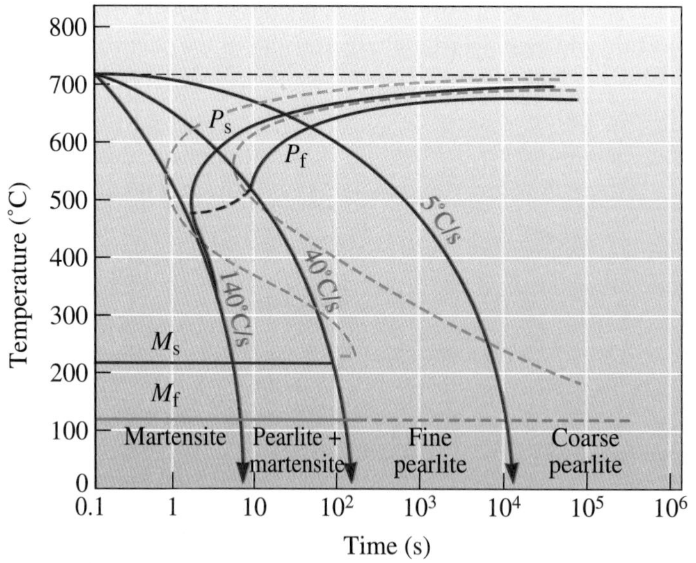

Figure 13-14 The CCT diagram (solid lines) for a 1080 steel compared with the TTT diagram (dashed lines).

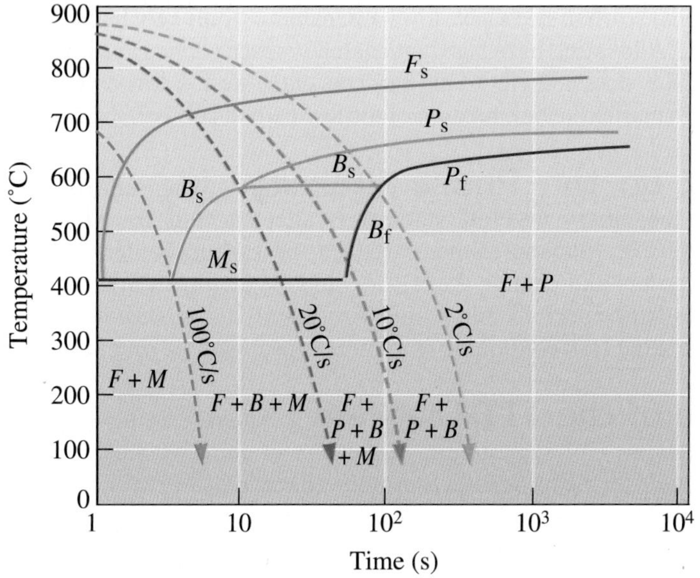

Figure 13-15 The CCT diagram for a low-alloy, 0.2% C steel.

13-5 Effect of Alloying Elements

Alloying elements are added to steels to (a) provide solid-solution strengthening of ferrite, (b) cause the precipitation of alloy carbides rather than that of Fe_3C, (c) improve corrosion resistance and other special characteristics of the steel, and (d) improve **hardenability**. The term hardenability describes the ease with which steels can form martensite in a thick section of steel that is quenched. With a more hardenable steel we can "get away" with a relatively slow cooling rate and still form martensite. Improving hardenability is most important in alloy and tool steels.

Hardenability In plain carbon steels, the nose of the TTT and CCT curves occurs at very short times; hence, very fast cooling rates are required to produce all martensite. In thin sections of steel, the rapid quench produces distortion and cracking. In thick steels, we are unable to produce martensite. All common alloying elements in steel shift the TTT and CCT diagrams to longer times, permitting us to obtain all martensite even in thick sections at slow cooling rates. Figure 13-16 shows the TTT and CCT curves for a 4340 steel.

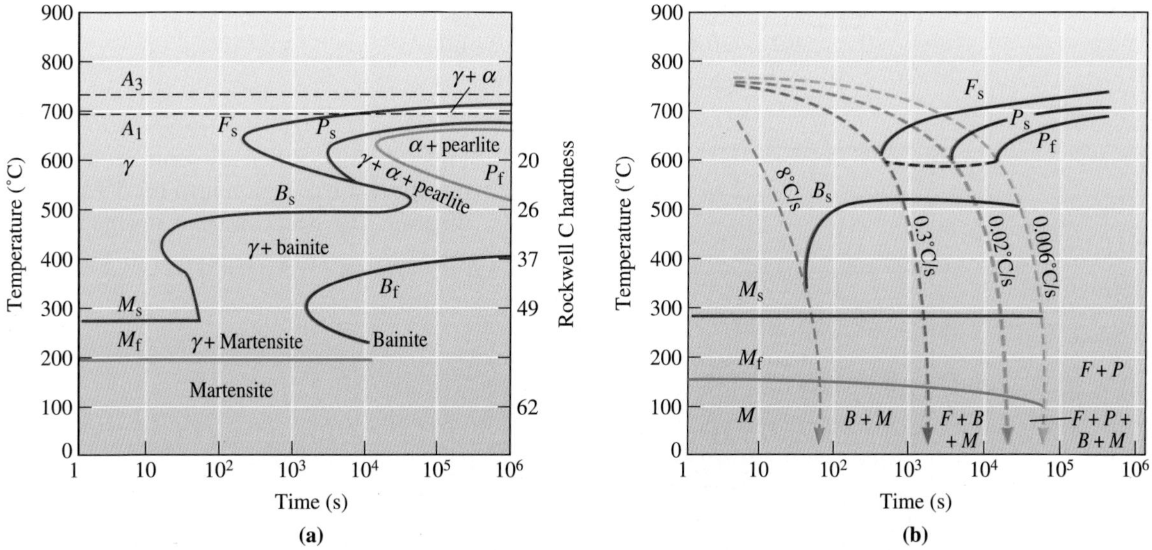

Figure 13-16 (a) TTT and (b) CCT diagrams for a 4340 steel.

Plain carbon steels have low hardenability—only very high cooling rates produce all martensite. Alloy steels have high hardenability—even cooling in air may produce martensite. Hardenability does not refer to the hardness of the steel. A low-carbon, high-alloy steel may easily form martensite but, because of the low-carbon content, the martensite is not hard.

Effect on the Phase Stability When alloying elements are added to steel, the binary Fe-Fe_3C stability is affected, and the phase diagram is altered (Figure 13-17). Alloying elements reduce the carbon content at which the eutectoid reaction occurs and change the A_1, A_3, and A_{cm} temperatures. A steel containing only 0.6% C is hypoeutectoid and would operate at 700°C without forming austenite; the otherwise same steel containing 6% Mn is hypereutectoid and austenite forms at 700°C.

Shape of the TTT Diagram Alloying elements may introduce a "bay" region into the TTT diagram, as in the case of the 4340 steel (Figure 13-16). The bay region is used as the basis for a thermomechanical heat treatment known as **ausforming**. A steel can be austenitized, quenched to the bay region, plastically deformed, and finally quenched to produce martensite (Figure 13-18). Steels subjected to this treatment are known as *ausformed steels.*

Tempering Alloying elements reduce the rate of tempering compared with that of a plain carbon steel (Figure 13-19). This effect may permit the alloy steels to operate more successfully at higher temperatures than plain carbon steels since overaging will not occur during service.

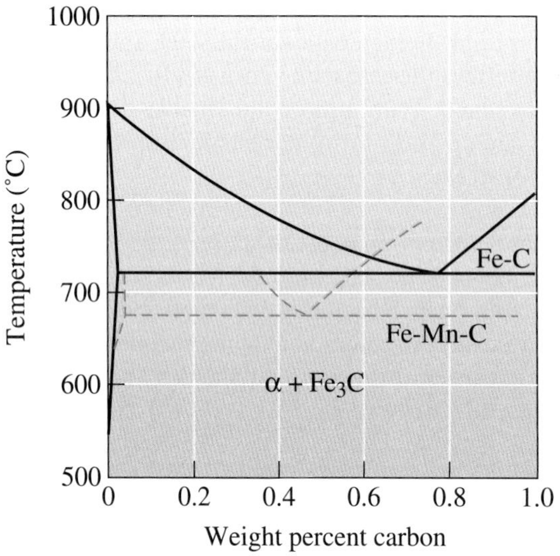

Figure 13-17 The effect of 6% manganese on the stability ranges of the phases in the eutectoid portion of the Fe-Fe_3C phase diagram.

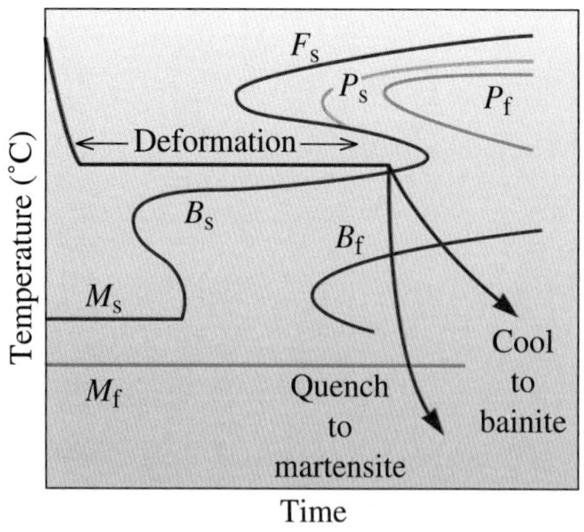

Figure 13-18 When alloying elements introduce a bay region into the TTT diagram, the steel can be ausformed.

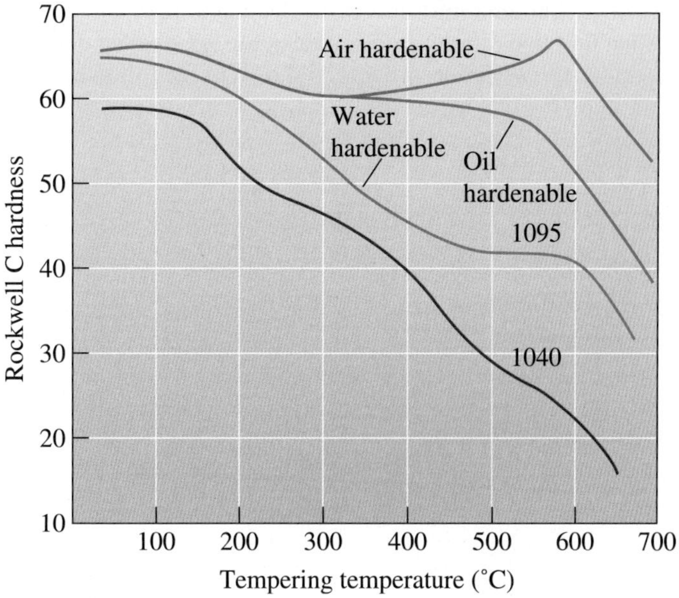

Figure 13-19 The effect of alloying elements on the phases formed during the tempering of steels. The air-hardenable steel shows a secondary hardening peak.

13-6 Application of Hardenability

A **Jominy test** (Figure 13-20) is used to compare hardenabilities of steels. A steel bar 4 in. long and 1 in. in diameter is austenitized, placed into a fixture, and sprayed at one end with water. This procedure produces a range of cooling rates—very fast at the quenched end, almost air cooling at the opposite end. After the test, hardness measurements are made along the test specimen and plotted to produce a **hardenability curve** (Figure 13-21). The distance from the quenched end is the **Jominy distance** and is related to the cooling rate (Table 13-3).

Virtually any steel transforms to martensite at the quenched end. Thus, the hardness at zero Jominy distance is determined solely by the carbon content of the steel. At larger Jominy distances, there is a greater likelihood that bainite or pearlite will form instead of martensite. An alloy steel with a high hardenability (such as 4340) maintains a rather flat hardenability curve; a plain carbon steel (such as 1050) has a curve that drops off quickly. The hardenability is determined primarily by the alloy content of the steel.

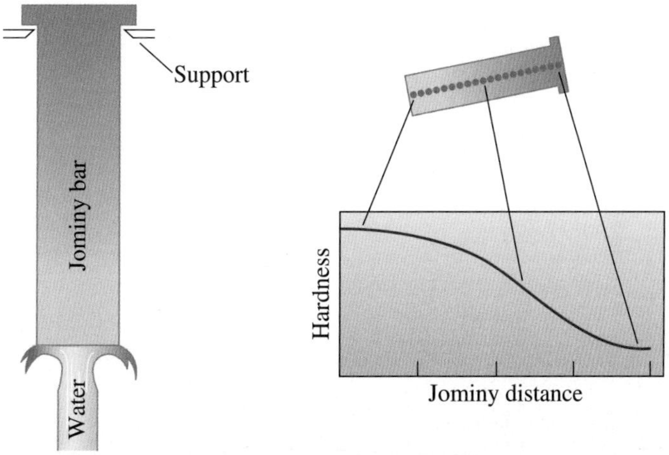

Figure 13-20 The set-up for the Jominy test used for determining the hardenability of a steel.

We can use hardenability curves in selecting or replacing steels in practical applications. The fact that two different steels cool at the same rate if quenched under identical conditions helps in this selection process as shown in the following example.

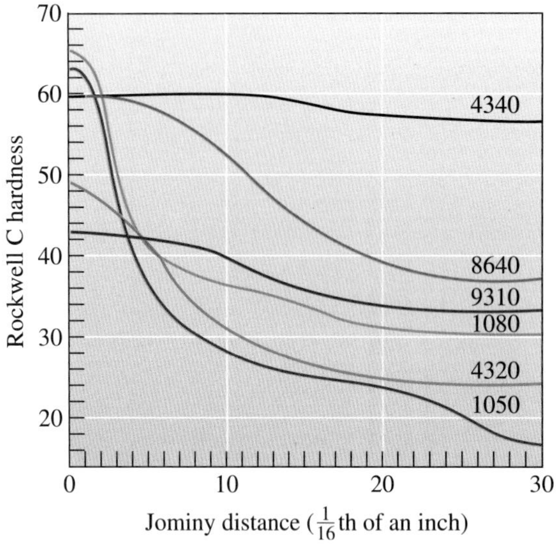

Figure 13-21 The hardenability curves for several steels.

Table 13-3 The relationship between cooling rate and Jominy distance

Jominy Distance (in.)	Cooling Rate (°C/s)
$\frac{1}{16}$	315
$\frac{2}{16}$	110
$\frac{3}{16}$	50
$\frac{4}{16}$	36
$\frac{5}{16}$	28
$\frac{6}{16}$	22
$\frac{7}{16}$	17
$\frac{8}{16}$	15
$\frac{10}{16}$	10
$\frac{12}{16}$	8
$\frac{16}{16}$	5
$\frac{20}{16}$	3
$\frac{24}{16}$	2.8
$\frac{28}{16}$	2.5
$\frac{36}{16}$	2.2

Example 13-5 *Design of a Wear-Resistant Gear*

A gear made from 9310 steel, which has an as-quenched hardness at a critical location of HRC 40, wears at an excessive rate. Tests have shown that an as-quenched hardness of at least HRC 50 is required at that critical location. Design a steel that would be appropriate.

SOLUTION

We know that if different steels of the same size are quenched under identical conditions, their cooling rates or Jominy distances are the same. From Figure 13-21, a hardness of HRC 40 in a 9310 steel corresponds to a Jominy distance of 10/16 in. (10°C/s). If we assume the same Jominy distance, the other steels shown in Figure 13-21 have the following hardnesses at the critical location:

1050	HRC 28
4320	HRC 31
1080	HRC 36
8640	HRC 52
4340	HRC 60

Both the 8640 and 4340 steels are appropriate. The 4320 steel has too low a carbon content to ever reach HRC 50; the 1050 and 1080 have enough carbon, but the hardenability is too low. In Table 13-1, we find that the 86xx steels contain lower quantities of alloying elements than the 43xx steels; thus the 8640 steel is probably less expensive than the 4340 steel and might be our best choice.

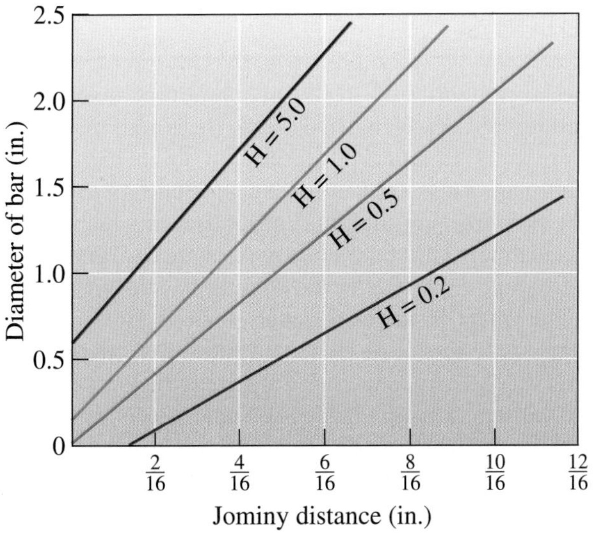

Figure 13-22 The Grossman chart used to determine the hardenability at the center of a steel bar for different quenchants.

In another simple technique, we utilize the severity of the quench and the Grossman chart (Figure 13-22) to determine the hardness at the *center* of a round bar. The bar diameter and H coefficient, or severity of the quench in Table 13-2, give the Jominy distance at the center of the bar. We can then determine the hardness from the hardenability curve of the steel. (See Example 13-6.)

Example 13-6 *Design of a Quenching Process*

Design a quenching process to produce a minimum hardness of HRC 40 at the center of a 1.5-in. diameter 4320 steel bar.

SOLUTION

Several quenching media are listed in Table 13-2. We can find an approximate H coefficient for each of the quenching media, then use Figure 13-22 to estimate the Jominy distance in a 1.5-in. diameter bar for each medium. Finally, we can use the hardenability curve (Figure 13-21) to find the hardness of the 4320 steel. The results are listed below.

	H Coefficient	Jominy Distance	HRC
Oil (no agitation)	0.25	12/16	28
Oil (agitation)	1.0	6/16	39
H_2O (no agitation)	1.0	6/16	39
H_2O (agitation)	4.0	4/16	44
Brine (no agitation)	2.0	5/16	42
Brine (agitation)	5.0	3/16	46

The last three methods, based on brine or agitated water, are satisfactory. Using an unagitated brine quenchant might be least expensive, since no extra equipment is needed to agitate the quenching bath; however, H_2O is less corrosive than the brine quenchant.

13-7 Specialty Steels

There are many special categories of steels, including tool steels, interstitial-free steels, high-strength-low-alloy (HSLA) steels, dual-phase steels, and maraging steels.

Tool steels are usually high-carbon steels that obtain high hardnesses by a quench and temper heat treatment. Their applications include cutting tools in machining operations, dies for die casting, forming dies, and other uses in which a combination of high strength, hardness, toughness, and temperature resistance is needed.

Alloying elements improve the hardenability and high-temperature stability of the tool steels. The water-hardenable steels such as 1095 must be quenched rapidly to produce martensite and also soften rapidly even at relatively low temperatures. Oil-hardenable steels form martensite more easily, temper more slowly, but still soften at high temperatures. The air-hardenable and special tool steels may harden to martensite while cooling in air. In addition, these steels may not soften until near the A_1 temperature. In fact, the highly alloyed tool steels may pass through a **secondary hardening peak** near 500°C as the normal cementite dissolves and hard alloy carbides precipitate (Figure 13-19). The alloy carbides are particularly stable, resist growth or spheroidization, and are important in establishing the high-temperature resistance of these steels.

High-strength-low-alloy (HSLA) steels are low-carbon steels containing small amounts of alloying elements. The HSLA steels are specified on the basis of yield strength with grades up to 80,000 psi; the steels contain the least amount of alloying element that still provides the proper yield strength without heat treatment. In these steels, careful processing permits precipitation of carbides and nitrides of Nb, V, Ti, or Zr, which provide dispersion strengthening and a fine grain size.

Dual-phase steels contain a uniform distribution of ferrite and martensite, with the dispersed martensite providing yield strengths of 60,000 to 145,000 psi. These low-carbon steels do not contain enough alloying elements to have good hardenability using the normal quenching processes, but when the steel is heated into the ferrite-plus-austenite portion of the phase diagram, the austenite phase becomes enriched in carbon, which provides the needed hardenability. During quenching, only the austenite portion transforms to martensite [Figure 13-23(a)].

In a category of steels known as transformation induced plasticity steels, or **TRIP steels**, deformation causes metastable retained austenite to transform to martensite or bainite (i.e., the phase transformation is induced by plasticity). The resulting microstructure is effectively a composite of multiple phases: martensite, bainite, and any remaining retained austenite in a continuous matrix of soft ferrite. TRIP steels are distinguished by their combination of high strength (both tensile and yield) and high ductility due to a high work hardening rate. They are often used in automotive applications. The microstructure of a TRIP steel is shown in Figure 13–23(b).

Maraging steels are low-carbon, highly alloyed steels. The steels are austenitized and quenched to produce a soft martensite that contains less than 0.3% C. When the martensite is aged at about 500°C, intermetallic compounds such as Ni_3Ti, Fe_2Mo, and Ni_3Mo precipitate.

Interstitial-free steels are steels containing Nb and Ti. They react with C and S to form precipitates of carbides and sulfides. Thus, virtually no carbon remains in the ferrite. These steels are very formable and therefore attractive for the automobile industry.

Grain-oriented steels containing silicon are used as soft magnetic materials and are used in transformer cores. Nearly pure iron powder (known as carbonyl iron), obtained by the decomposition of iron pentacarbonyl [$Fe(CO)_5$] and sometimes a reducing heat treatment, is used to make magnetic materials. Pure iron powder is also used as an additive for food supplements in breakfast cereals and other iron-fortified food products under the name reduced iron.

As mentioned before, many steels are also coated, usually to provide good corrosion protection. *Galvanized* steel is coated with a thin layer of zinc (Chapter 23), *terne* steel is coated with lead, and other steels are coated with aluminum or tin.

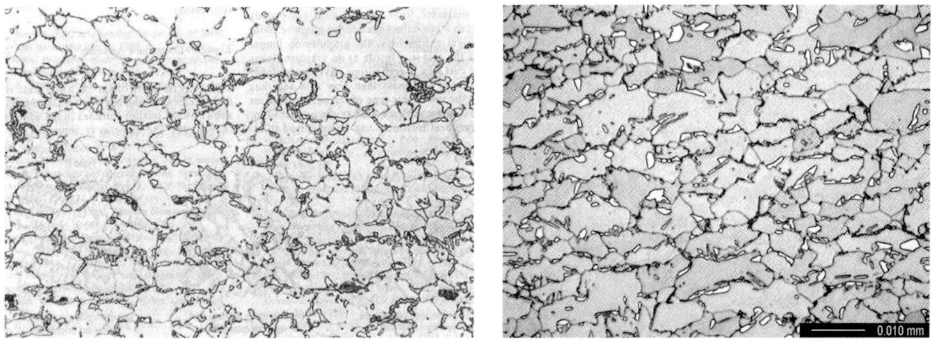

(a) (b)

Figure 13-23 (a) Microstructure of a dual-phase steel, showing islands of white martensite in a light gray ferrite matrix. (b) Microstructure of a TRIP steel, showing ferrite (light gray) + bainite (black along grain boundaries) + retained austenite (white). (*D. P. Hoydick, D. M. Haezebrouck, and E. A. Silva, United States Steel Corporation Research and Technology Center, 2005.*) [Metals Handbook, *Desk Edition (1998), ASM International, Materials Park, OH 44073-0002. Reprinted with permission of ASM International. All rights reserved. www.asminternational.org.*]

13-8 Surface Treatments

We can, by proper heat treatment, produce a structure that is hard and strong at the surface, so that excellent wear and fatigue resistance are obtained, but at the same time gives a soft, ductile, tough core that provides good resistance to impact failure. We have seen principles of carburizing in Chapter 5, when we discussed diffusion. In this section, we examine this and other similar processes.

Selectively Heating the Surface We could begin by rapidly heating the surface of a medium-carbon steel above the A_3 temperature (the center remains below the A_1). After the steel is quenched, the center is still a soft mixture of ferrite and pearlite, while the surface is martensite (Figure 13-24). The depth of the martensite layer is the **case depth**. Tempering produces the desired hardness at the surface. We can provide local heating of the surface by using a gas flame, an induction coil, a laser beam, or an electron beam. We can, if we wish, harden only selected areas of the surface that are most subject to failure by fatigue or wear.

Carburizing and Nitriding These techniques involve controlled diffusion of carbon and nitrogen, respectively (Chapter 5). For best toughness, we start with a low-carbon steel. In **carburizing**, carbon is diffused into the surface of the steel at a temperature above the A_3 (Figure 13-25). A high carbon content is produced at the surface due to rapid diffusion and the high solubility of carbon in austenite. When the steel is then quenched and tempered, the surface becomes a high-carbon tempered martensite, while the ferritic center remains soft and ductile. The thickness of the hardened surface, again called the case depth, is much smaller in carburized steels than in flame- or induction-hardened steels.

Nitrogen provides a hardening effect similar to that of carbon. In **cyaniding**, the steel is immersed in a liquid cyanide bath that permits both carbon and nitrogen to diffuse

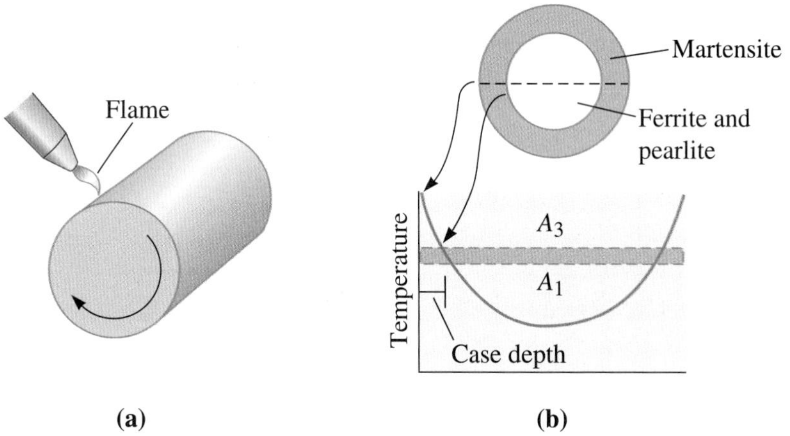

Figure 13-24 (a) Surface hardening by localized heating. (b) Only the surface heats above the A_3 temperature and is quenched to martensite.

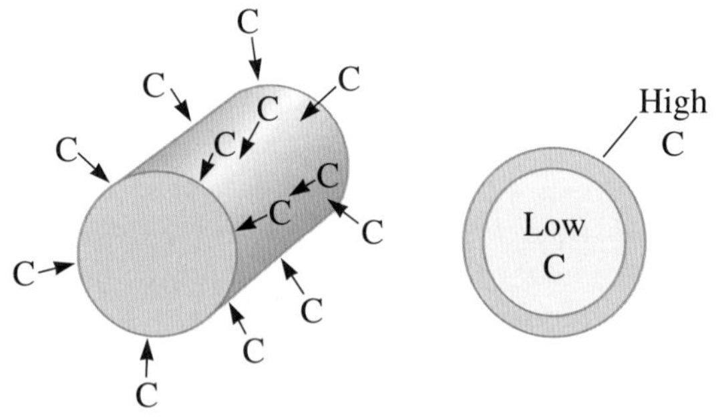

Figure 13-25 Carburizing of a low-carbon steel to produce a high-carbon, wear-resistant surface.

into the steel. In **carbonitriding**, a gas containing carbon monoxide and ammonia is generated, and both carbon and nitrogen diffuse into the steel. Finally, only nitrogen diffuses into the surface from a gas in **nitriding**. Nitriding is carried out below the A_1 temperature.

In each of these processes, compressive residual stresses are introduced at the surface, providing excellent fatigue resistance (Chapter 7) in addition to a good combination of hardness, strength, and toughness. The following example explains considerations for heat treatments such as quenching and tempering and surface hardening.

Example 13-7 *Design of Surface-Hardening Treatments for a Drive Train*

Design the materials and heat treatments for an automobile axle and drive gear (Figure 13-26).

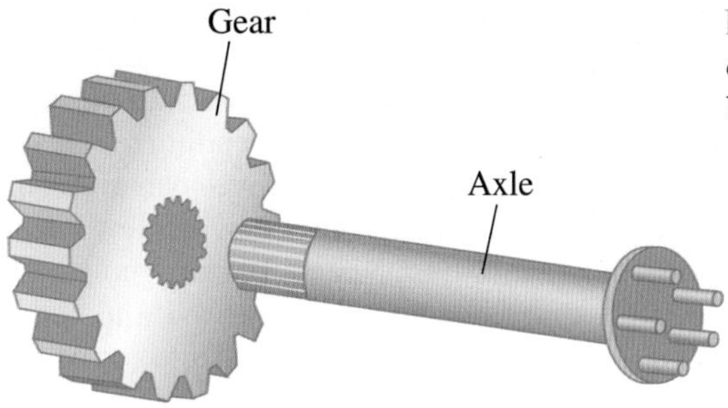

Figure 13-26 Sketch of axle and gear assembly (for Example 13-7).

SOLUTION

Both parts require good fatigue resistance. The gear also should have a good hardness to avoid wear, and the axle should have good overall strength to withstand bending and torsional loads. Both parts should have good toughness. Finally, since millions of these parts will be made, they should be inexpensive.

Quenched and tempered alloy steels might provide the required combination of strength and toughness; however, the alloy steels are expensive. An alternative approach for each part is described below.

The axle might be made from a forged 1050 steel containing a matrix of ferrite and pearlite. The axle could be surface-hardened, perhaps by moving the axle through an induction coil to selectively heat the surface of the steel above the A_3 temperature (about 770°C). After the coil passes any particular location of the axle, the cold interior quenches the surface to martensite. Tempering then softens the martensite to improve ductility. This combination of carbon content and heat treatment meets our requirements. The plain carbon steel is inexpensive; the core of ferrite and pearlite produces good toughness and strength; and the hardened surface provides good fatigue and wear resistance.

The gear is subject to more severe loading conditions, for which the 1050 steel does not provide sufficient toughness, hardness, and wear resistance. Instead, we might carburize a 1010 steel for the gear. The original steel contains mostly ferrite, providing good ductility and toughness. By performing a gas carburizing process above the A_3 temperature (about 860°C), we introduce about 1.0% C in a very thin case at the surface of the gear teeth. This high-carbon case, which transforms to martensite during quenching, is tempered to control the hardness. Now we obtain toughness due to the low-carbon ferrite core, wear resistance due to the high-carbon surface, and fatigue resistance due to the high-strength surface containing compressive residual stresses introduced during carburizing. In addition, the plain carbon 1010 steel is an inexpensive starting material that is easily forged into a near-net shape prior to heat treatment.

13-9 Weldability of Steel

In Chapter 9, we discussed welding and other joining processes. We noted that steels are the most widely used structural materials. In bridges, buildings, and many other applications, steels must be welded. The structural integrity of steel structures not only depends upon the strength of the steel but also the strength of the welded joints. This is why the weldability of steel is always an important consideration.

Many low-carbon steels weld easily. Welding of medium- and high-carbon steels is comparatively more difficult since martensite can form in the heat-affected zone rather easily, thereby causing a weldment with poor toughness. Several strategies such as preheating the material or minimizing incorporation of hydrogen have been developed to counter these problems. The incorporation of hydrogen causes the steel to become brittle. In low-carbon steels, the strength of the welded regions in these materials is higher than the base material. This is due to the finer pearlite microstructure that forms during cooling of the heat-affected zone. Retained austenite along ferrite grain boundaries also limits recrystallization and thus helps retain a fine grain size, which contributes to the strength of the welded region. During welding, the metal nearest the weld heats above the A_1 temperature and austenite forms (Figure 13-27). During cooling, the austenite in this heat-affected zone transforms to a new structure, depending on the cooling rate and the CCT diagram for the steel. Plain low-carbon steels have such a low hardenability that normal cooling rates seldom produce martensite; however, an alloy steel may have to be preheated to slow down the cooling rate or post-heated to temper any martensite that forms.

A steel that is originally quenched and tempered has two problems during welding. First, the portion of the heat-affected zone that heats above the A_1 may form martensite after cooling. Second, a portion of the heat-affected zone below the A_1 may overtemper. Normally, we should not weld a steel in the quenched and tempered condition. The following example shows how the heat-affected zone microstructure can be accounted for using CCT diagrams.

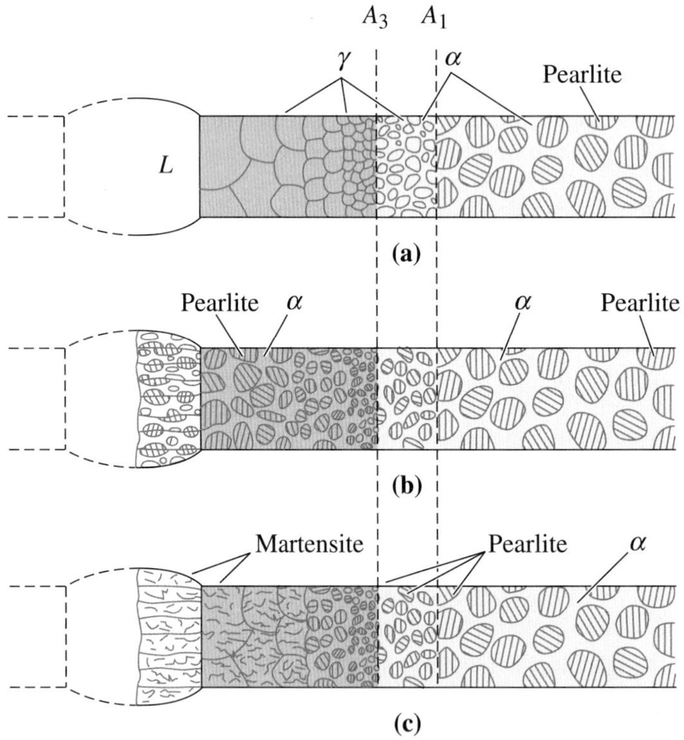

Figure 13-27 The development of the heat-affected zone in a weld: (a) the structure at the maximum temperature, (b) the structure after cooling in a steel of low hardenability, and (c) the structure after cooling in a steel of high hardenability.

Example 13-8 ***Structures of Heat-Affected Zones***

Compare the structures in the heat-affected zones of welds in 1080 and 4340 steels if the cooling rate in the heat-affected zone is 5°C/s.

SOLUTION

From the CCT diagrams, Figures 13-14 and 13-16, the cooling rate in the weld produces the following structures:

1080: 100% pearlite

4340: Bainite and martensite

The high hardenability of the alloy steel reduces the weldability, permitting martensite to form and embrittle the weld.

13-10 Stainless Steels

Stainless steels are selected for their excellent resistance to corrosion. All true stainless steels contain a minimum of about 11% Cr, which permits a thin, protective surface layer of chromium oxide to form when the steel is exposed to oxygen. The chromium is what makes stainless steels stainless. Chromium is also a *ferrite stabilizing element*. Figure 13-28(a) illustrates the effect of chromium on the iron-carbon phase diagram. Chromium causes the austenite region to shrink, while the ferrite region increases in size. For high-chromium, low-carbon compositions, ferrite is present as a single phase up to the solidus temperature.

There are several categories of stainless steels based on crystal structure and strengthening mechanism. Typical properties are included in Table 13-4.

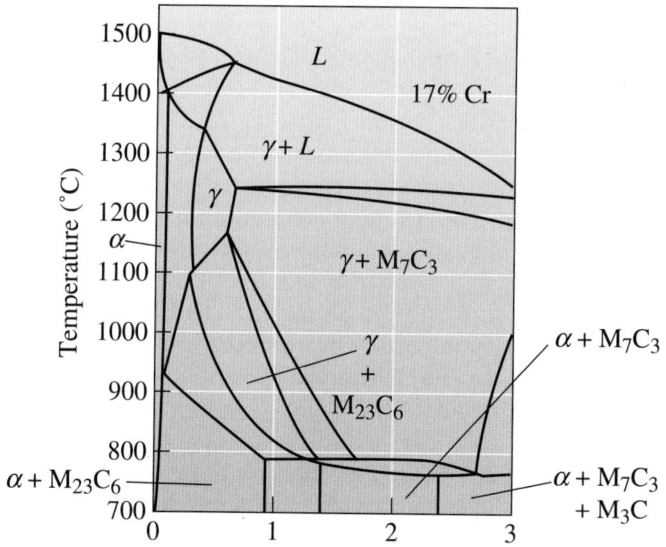

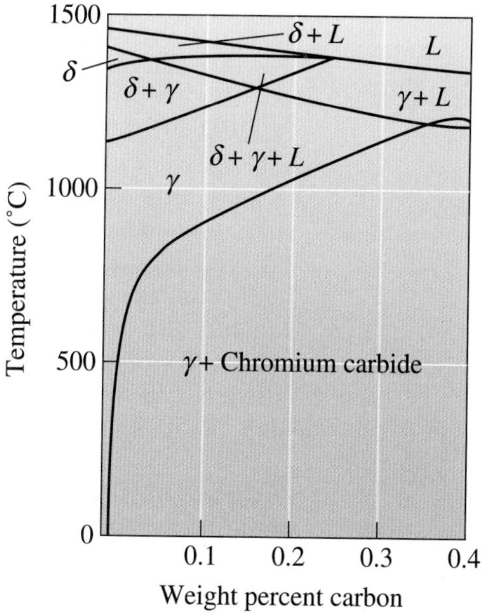

Figure 13-28 (a) The effect of 17% chromium on the iron-carbon phase diagram. At low carbon contents, ferrite is stable at all temperatures. Note that "M" stands for "metal" such as Cr and Fe or other alloying additions. (b) A section of the iron-chromium-nickel-carbon phase diagram at a constant 18% Cr-18% Ni. At low carbon contents, austenite is stable at room temperature.

Table 13-4 Typical compositions and properties of stainless steels

Steel	% C	% Cr	% Ni	Others	Tensile Strength (psi)	Yield Strength (psi)	% Elongation	Condition
Austenitic								
201	0.15	17	5	6.5% Mn	95,000	45,000	40	Annealed
304	0.08	19	10		75,000	30,000	30	Annealed
					185,000	140,000	9	Cold-worked
304L	0.03	19	10		75,000	30,000	30	Annealed
316	0.08	17	12	2.5% Mo	75,000	30,000	30	Annealed
321	0.08	18	10	0.4% Ti	85,000	35,000	55	Annealed
347	0.08	18	11	0.8% Nb	90,000	35,000	50	Annealed
Ferritic								
430	0.12	17			65,000	30,000	22	Annealed
442	0.12	20			75,000	40,000	20	Annealed
Martensitic								
416	0.15	13		0.6% Mo	180,000	140,000	18	Quenched and tempered
431	0.20	16	2		200,000	150,000	16	Quenched and tempered
440C	1.10	17		0.7% Mo	285,000	275,000	2	Quenched and tempered
Precipitation hardening								
17-4	0.07	17	4	0.4% Nb	190,000	170,000	10	Age-hardened
17-7	0.09	17	7	1.0% Al	240,000	230,000	6	Age-hardened

Ferritic Stainless Steels Ferritic stainless steels contain up to 30% Cr and less than 0.12% C. Because of the BCC structure, the ferritic stainless steels have good strengths and moderate ductilities derived from solid-solution strengthening and strain hardening. Ferritic stainless steels are ferromagnetic. They are not heat treatable. They have excellent corrosion resistance, moderate formability, and are relatively inexpensive.

Martensitic Stainless Steels From Figure 13-28(a), we find that a 17% Cr-0.5% C alloy heated to 1200°C forms 100% austenite, which transforms to martensite on quenching in oil. The martensite is then tempered to produce high strengths and hardnesses [Figure 13-29(a)]. The chromium content is usually less than 17% Cr; otherwise, the austenite field becomes so small that very stringent control over both the austenitizing temperature and carbon content is required. Lower chromium contents also permit the carbon content to vary from about 0.1% to 1.0%, allowing martensites of different hardnesses to be produced. The combination of hardness, strength, and corrosion resistance makes the alloys attractive for applications such as high-quality knives, ball bearings, and valves.

Austenitic Stainless Steels Nickel, which is an *austenite stabilizing element*, increases the size of the austenite field, while nearly eliminating ferrite from the iron-chromium-carbon alloys [Figure 13-28(b)]. If the carbon content is below about 0.03%, the carbides do not form and the steel is virtually all austenite at room temperature [Figure 13-29(b)].

The FCC austenitic stainless steels have excellent ductility, formability, and corrosion resistance. Strength is obtained by extensive solid-solution strengthening, and the austenitic stainless steels may be cold worked to higher strengths than the ferritic stainless steels. These are not ferromagnetic, which is an advantage for many applications. For example, cardiovascular stents are often made from 316 stainless steels. The steels have excellent low-temperature impact properties, since they have no transition temperature.

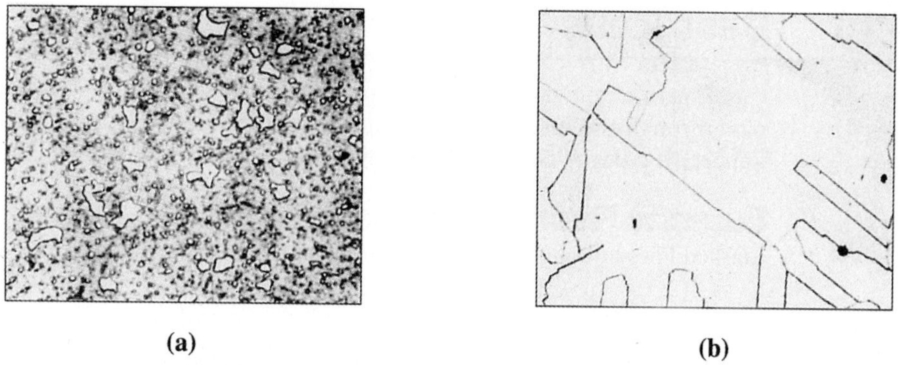

Figure 13-29 (a) Martensitic stainless steel containing large primary carbides and small carbides formed during tempering (× 350). (b) Austenitic stainless steel (× 500). [*From* ASM Handbook, *Vols. 7 and 8, (1972,1973), ASM International, Materials Park, OH 44073-0002. Reprinted with permission of ASM International. All rights reserved. www.asminternational.org.*]

Unfortunately, the high-nickel and chromium contents make the alloys expensive. The 304 alloy containing 18% Cr and 8% nickel (also known as 18-8 stainless) is the most widely used grade of stainless steel. Although stainless, this alloy can undergo **sensitization**. When heated to a temperature of ~480–860°C, chromium carbides precipitate along grain boundaries rather than within grains. This causes chromium depletion in the interior of the grains that leaves the stainless steel sensitive to corrosion.

Precipitation-Hardening (PH) Stainless Steels The precipitation-hardening (or PH) stainless steels contain Al, Nb, or Ta and derive their properties from solid-solution strengthening, strain hardening, age hardening, and the martensitic reaction. The steel is first heated and quenched to permit the austenite to transform to martensite. Reheating permits precipitates such as Ni_3Al to form from the martensite. High strengths are obtained even with low carbon contents.

Duplex Stainless Steels In some cases, mixtures of phases are deliberately introduced into the stainless steel structure. By appropriate control of the composition and heat treatment, a **duplex stainless steel** containing approximately 50% ferrite and 50% austenite can be produced. This combination provides a set of mechanical properties, corrosion resistance, formability, and weldability not obtained in any one of the usual stainless steels.

Most stainless steels are recyclable and the following example shows how differences in properties can be used to separate different types of stainless steels.

Example 13-9 *Design of a Test to Separate Stainless Steels*

In order to efficiently recycle stainless steel scrap, we wish to separate the high-nickel stainless steel from the low-nickel stainless steel. Design a method for doing this.

SOLUTION

Performing a chemical analysis on each piece of scrap is tedious and expensive. Sorting based on hardness might be less expensive; however, because of the different types of treatments—such as annealing, cold working, or quench and tempering—the hardness may not be related to the steel composition.

The high-nickel stainless steels are ordinarily austenitic, whereas the low-nickel alloys are ferritic or martensitic. An ordinary magnet will be attracted to the low-nickel ferritic and martensitic steels, but will not be attracted to the high-nickel austenitic steel. We might specify this simple and inexpensive magnetic test for our separation process.

13-11 Cast Irons

Cast irons are iron-carbon-silicon alloys, typically containing 2–4% C and 0.5–3% Si, that pass through the eutectic reaction during solidification. The microstructures of the five important types of cast irons are shown schematically in Figure 13-30.

Eutectic Reaction in Cast Irons Based on the Fe-Fe_3C phase diagram (dashed lines in Figure 13-31), the eutectic reaction that occurs in Fe-C alloys at 1140°C is

$$L \rightarrow \gamma + Fe_3C \tag{13-1}$$

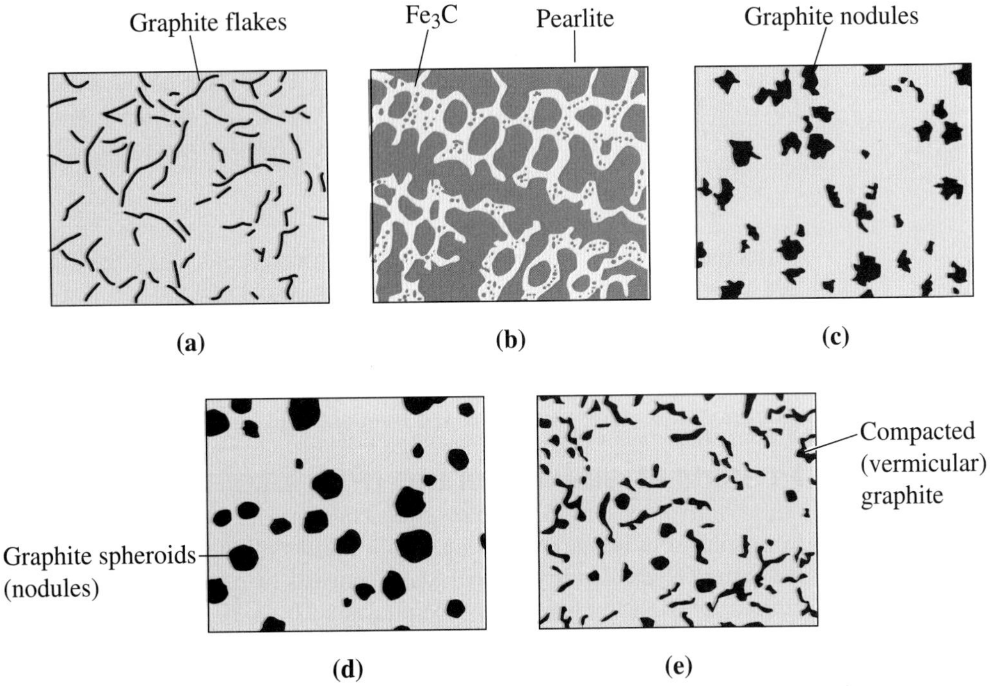

Figure 13-30 Schematic drawings of the five types of cast iron: (a) gray iron, (b) white iron, (c) malleable iron, (d) ductile iron, and (e) compacted graphite iron.

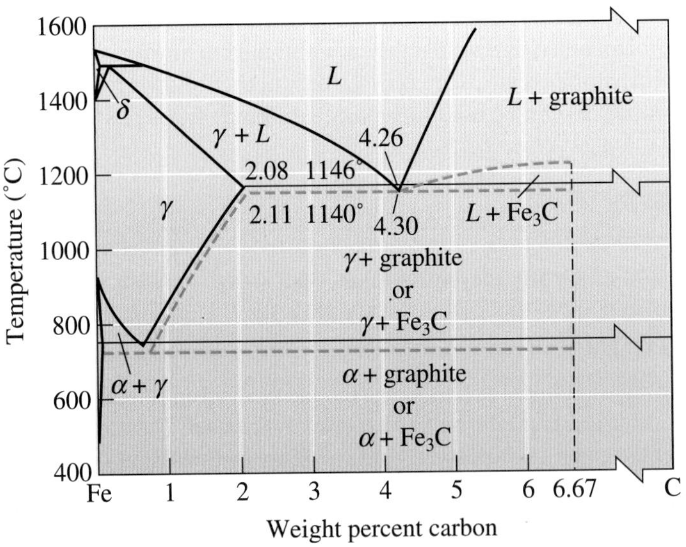

Figure 13-31 The iron-carbon phase diagram showing the relationship between the stable iron-graphite equilibria (solid lines) and the metastable iron-cementite reactions (dashed lines).

This reaction produces **white cast iron** with a microstructure composed of Fe_3C and pearlite. The Fe-Fe_3C system, however, is really a metastable phase diagram. Under truly equilibrium conditions, the eutectic reaction is

$$L \rightarrow \gamma + \text{graphite} \tag{13-2}$$

The Fe-C phase diagram is shown as solid lines in Figure 13-31. When the stable $L \rightarrow \gamma +$ graphite eutectic reaction occurs at 1146°C, gray, ductile, or compacted graphite cast iron forms.

In Fe-C alloys, the liquid easily undercools 6°C (the temperature difference between the stable and metastable eutectic temperatures), and white iron forms. Adding about 2% silicon to the iron increases the temperature difference between the eutectics, permitting larger undercoolings to be tolerated and more time for the stable graphite eutectic to nucleate and grow. Silicon, therefore, is a *graphite stabilizing* element. Elements such as chromium and bismuth have the opposite effect and encourage white cast iron formation. We can also introduce inoculants, such as silicon (as Fe-Si ferrosilicon), to encourage the nucleation of graphite, or we can reduce the cooling rate of the casting to provide more time for the growth of graphite.

Silicon also reduces the amount of carbon contained in the eutectic. We can take this effect into account by defining the **carbon equivalent** (CE):

$$\text{CE} = \%\,\text{C} + \tfrac{1}{3}\%\,\text{Si} \tag{13-3}$$

The eutectic composition is always near 4.3% CE. A high carbon equivalent encourages the growth of the graphite eutectic.

Eutectoid Reaction in Cast Irons The matrix structure and properties of each type of cast iron are determined by how the austenite transforms during the eutectoid reaction. In the Fe-Fe_3C phase diagram used for steels, the austenite transformed to ferrite and cementite, often in the form of pearlite; however, silicon also encourages the *stable* eutectoid reaction:

$$\gamma \rightarrow \alpha + \text{graphite} \tag{13-4}$$

Under equilibrium conditions, carbon atoms diffuse from the austenite to existing graphite particles, leaving behind the low-carbon ferrite. The transformation diagram (Figure 13-32) describes how the austenite might transform during heat treatment. **Annealing** (or furnace cooling) of cast iron gives a soft ferritic matrix (not coarse

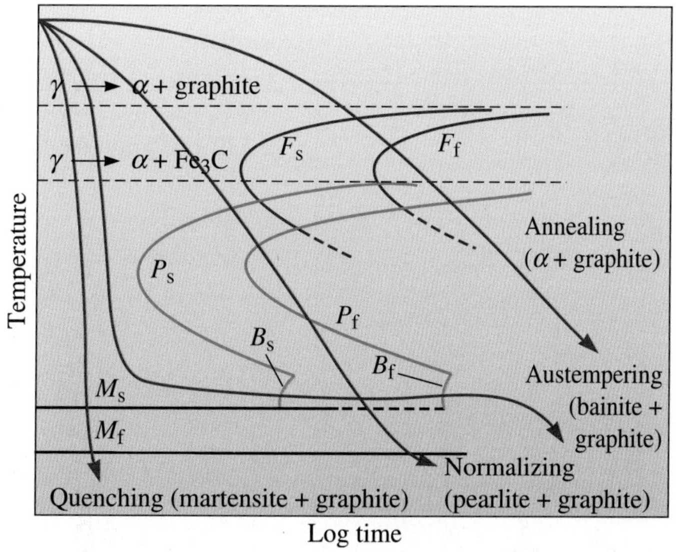

Figure 13-32 The transformation diagram for austenite in a cast iron.

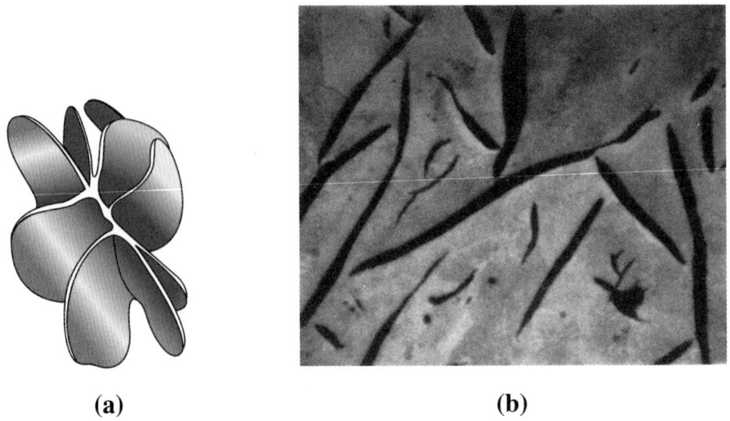

Figure 13-33 (a) Sketch and (b) micrograph of the flake graphite in gray cast iron (× 100). *(Reprinted Courtesy of Don Askeland)*

pearlite as in steels!). Normalizing, or air cooling, gives a pearlitic matrix. The cast irons can also be austempered to produce bainite or can be quenched to martensite and tempered. Austempered ductile iron, with strengths of up to 200,000 psi, is used for high-performance gears.

Gray cast iron is the most widely used cast iron and is named for the dull gray color of the fracture surface. Gray cast iron contains many clusters, or **eutectic cells**, of interconnected graphite flakes (Figure 13-33) that cause low strength and ductility. The point at which the flakes are connected is the original graphite nucleus. Inoculation helps produce smaller eutectic cells, thus improving strength. The gray irons are specified by a class number of 20 to 80. A class 20 gray iron has a nominal tensile strength of 20,000 psi. In thick castings, coarse graphite flakes and a ferrite matrix produce tensile strengths as low as 12,000 psi (Figure 13-34), whereas in thin castings, fine graphite and pearlite form and give tensile strengths near 40,000 psi. Higher strengths are obtained by reducing the carbon equivalent, by alloying, or by heat treatment. Although the graphite flakes concentrate stresses and cause low strength and ductility, gray iron has a number of attractive properties, including high compressive strength, good machinability, good resistance to sliding wear, good resistance to thermal fatigue, good thermal conductivity, and good vibration damping.

White cast iron is a hard, brittle alloy containing massive amounts of Fe_3C. A fracture surface of this material appears white, hence the name. A group of highly alloyed white irons is used for their hardness and resistance to abrasive wear. Elements

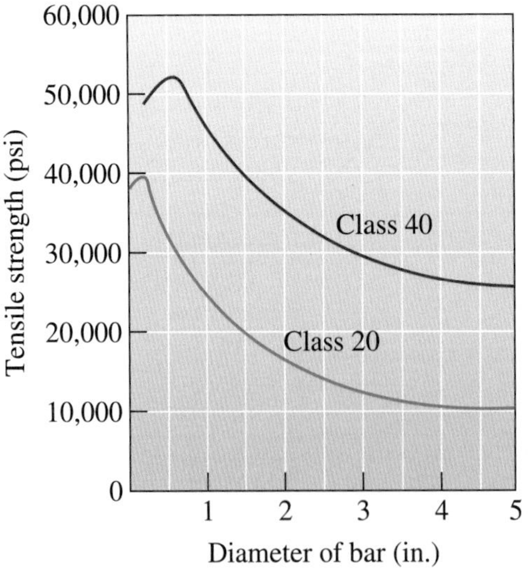

Figure 13-34 The effect of the cooling rate or casting size on the tensile properties of two gray cast irons.

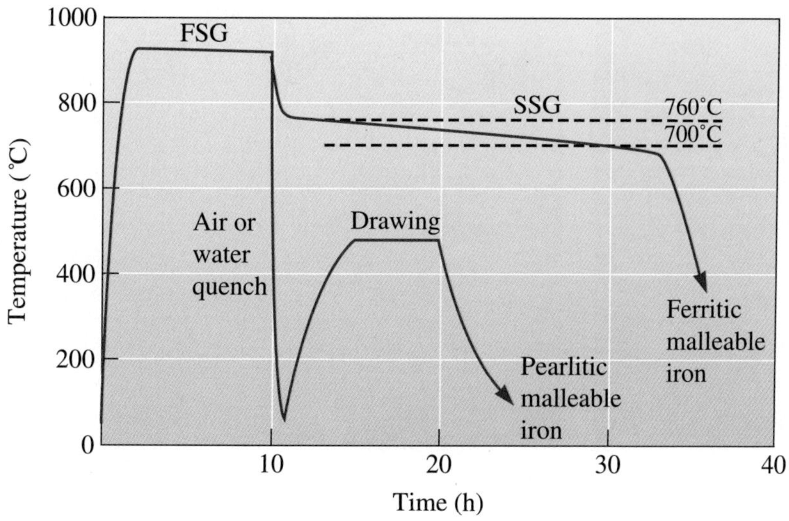

Figure 13-35 The heat treatments for ferritic and pearlitic malleable irons.

such as chromium, nickel, and molybdenum are added so that, in addition to the alloy carbides formed during solidification, martensite is formed during subsequent heat treatment.

Malleable cast iron, formed by the heat treatment of white cast iron, produces rounded clumps of graphite. It exhibits better ductility than gray or white cast irons. It is also very machinable. Malleable iron is produced by heat treating unalloyed 3% carbon equivalent (2.5% C, 1.5% Si) white iron. During the heat treatment, the cementite formed during solidification decomposes and graphite clumps, or nodules, are produced. The nodules, or temper carbon, often resemble popcorn. The rounded graphite shape permits a good combination of strength and ductility. The production of malleable iron requires several steps (Figure 13-35). Graphite nodules nucleate as the white iron is slowly heated. During **first stage graphitization** (FSG), cementite decomposes to the stable austenite and graphite phases as the carbon in Fe_3C diffuses to the graphite nuclei. Following FSG, the austenite transforms during cooling. Figure 13-36 shows the microstructures of the original white iron (a) and the two types of malleable iron that can be produced [(b) and (c)]. To make *ferritic malleable iron*, the casting is cooled slowly through the eutectoid temperature range to cause **second stage graphitization** (SSG). The ferritic malleable iron has good toughness compared with that of other irons because its low-carbon equivalent reduces the transition temperature below room temperature. *Pearlitic malleable iron* is obtained when austenite is cooled in air or oil to form pearlite or martensite. In either case, the matrix is hard and brittle. The iron is then drawn at a temperature below the eutectoid. **Drawing** is a heat treatment that tempers the martensite or spheroidizes the pearlite. A higher drawing temperature decreases strength and increases ductility and toughness.

Ductile or nodular cast iron [Figure 13-36(d), (e), and (f)] contains spheroidal graphite particles. Ductile iron is produced by treating liquid iron with a carbon equivalent of near 4.3% with magnesium, which causes spheroidal graphite (called nodules) to grow during solidification, rather than during a lengthy heat treatment. Several steps are required to produce this iron. These include desulfurization, **nodulizing**, and inoculation. In desulfurization, any sulfur and oxygen in the liquid metal is removed by adding desulfurizing agents such as calcium oxide (CaO). In nodulizing, Mg is added, usually in a dilute form such as a MgFeSi alloy. If pure Mg is added, the nodulizing reaction is very violent, since the boiling point of Mg is much lower than the temperature of the liquid iron, and most of the Mg will be lost. A residual of about

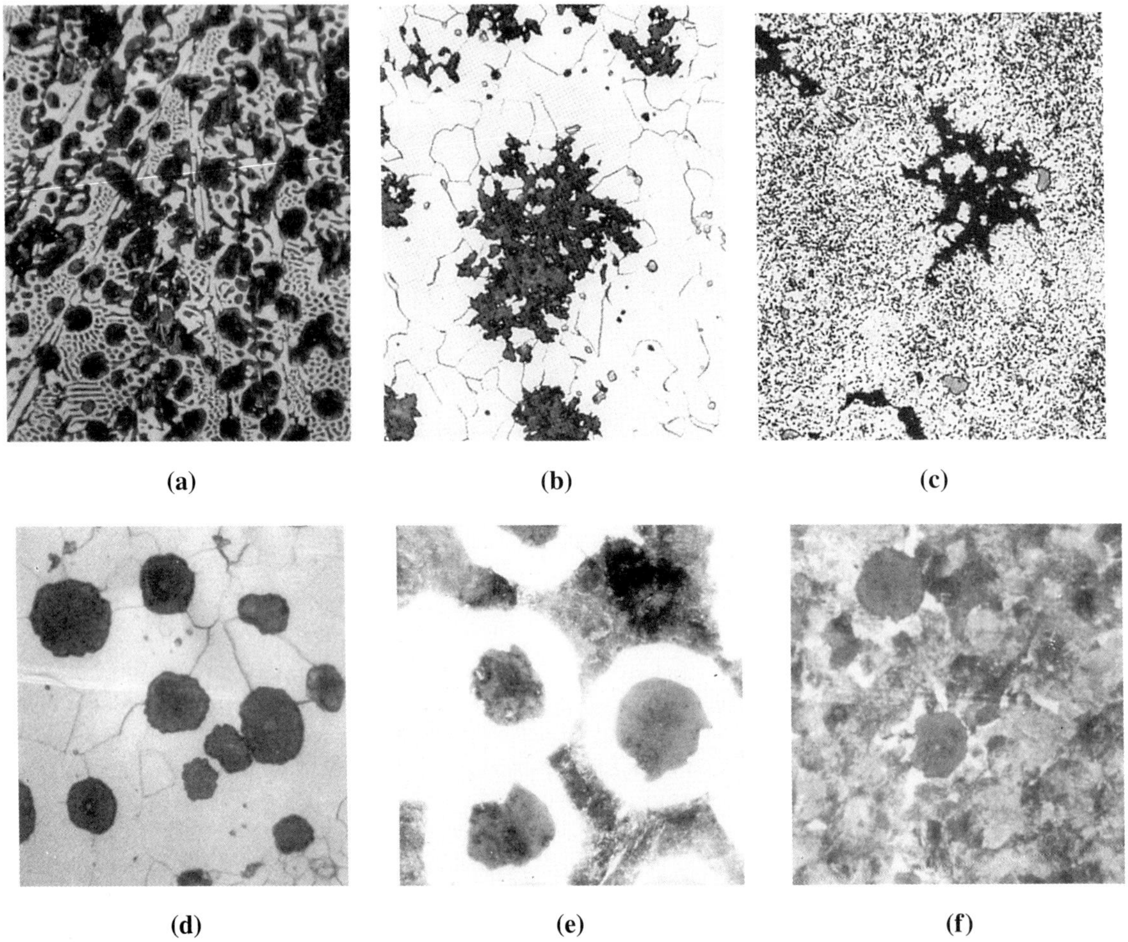

Figure 13-36 (a) White cast iron prior to heat treatment (× 100). (b) Ferritic malleable iron with graphite nodules and small MnS inclusions in a ferrite matrix (× 200). (c) Pearlitic malleable iron drawn to produce a tempered martensite matrix (× 500). [*Images (b) and (c): From* ASM Handbook, *Vols. 7 and 8, (1972,1973), ASM International, Materials Park, OH 44073-0002. Reprinted with permission of ASM International. All rights reserved. www.asminternational.org.*] (d) Annealed ductile iron with a ferrite matrix (× 250). (e) As-cast ductile iron with a matrix of ferrite (white) and pearlite (× 250). (f) Normalized ductile iron with a pearlite matrix (× 250). [*Images (a), (d), (e), and (f) Reprinted Courtesy of Don Askeland*]

0.03% Mg must be in the liquid iron after treatment in order for spheroidal graphite to grow. Finally, **inoculation** with FeSi compounds to cause heterogeneous nucleation of the graphite is essential; if inoculation is not effective, white iron will form instead of ductile iron. The nodulized and inoculated iron must then be poured into molds within a few minutes to avoid fading. **Fading** occurs by the gradual, nonviolent loss of Mg due to vaporization and/or reaction with oxygen, resulting in flake or compacted graphite instead of spheroidal graphite. In addition, the inoculant effect will also fade, resulting in white iron.

Compared with gray iron, ductile cast iron has excellent strength and ductility. Due to the higher silicon content (typically around 2.4%) in ductile irons compared with 1.5% Si in malleable irons, the ductile irons are stronger but not as tough as malleable irons.

Compacted graphite cast iron contains rounded but interconnected graphite also produced during solidification. The graphite shape in compacted graphite cast iron is intermediate between flakes and spheres with numerous rounded rods of graphite that

Table 13-5 Typical properties of cast irons

	Tensile Strength (psi)	Yield Strength (psi)	% Elongation	Notes
Gray irons				
Class 20	12,000–40,000	—	—	
Class 40	28,000–54,000	—	—	
Class 60	44,000–66,000	—	—	
Malleable irons				
32510	50,000	32,500	10	Ferritic
35018	53,000	35,000	18	Ferritic
50005	70,000	50,000	5	Pearlitic
70003	85,000	70,000	3	Pearlitic
90001	105,000	90,000	1	Pearlitic
Ductile irons				
60–40–18	60,000	40,000	18	Annealed
65–45–12	65,000	45,000	12	As-cast ferritic
80–55–06	80,000	55,000	6	As-cast pearlitic
100–70–03	100,000	70,000	3	Normalized
120–90–02	120,000	90,000	2	Quenched and tempered
Compacted graphite irons				
Low strength	40,000	28,000	5	90% Ferritic
High strength	65,000	55,000	1	80% Pearlitic

are interconnected to the nucleus of the eutectic cell. This compacted graphite, sometimes called **vermicular graphite**, also forms when ductile iron fades. The compacted graphite permits strengths and ductilities that exceed those of gray cast iron, but allows the iron to retain good thermal conductivity and vibration damping properties. The treatment for the compacted graphite iron is similar to that for ductile iron; however, only about 0.015% Mg is introduced during nodulizing. A small amount of titanium is added to ensure the formation of the compacted graphite. Typical properties of cast irons are given in Table 13-5.

Summary

- The properties of steels, determined by dispersion strengthening, depend on the amount, size, shape, and distribution of cementite. These factors are controlled by alloying and heat treatment.
- A process anneal recrystallizes cold-worked steels.
- Spheroidizing produces large, spherical Fe_3C and good machinability in high-carbon steels.
- Annealing, involving a slow furnace cool after austenitizing, produces a coarse pearlitic structure.
- Normalizing, involving an air cool after austenitizing, produces a fine pearlitic structure and higher strength compared to annealing.
- In isothermal annealing, pearlite with a uniform interlamellar spacing is obtained by transforming the austenite at a constant temperature.
- Austempering is used to produce bainite, containing rounded Fe_3C, by an isothermal transformation.

- Quench and temper heat treatments require the formation and decomposition of martensite, providing exceptionally fine dispersions of round Fe_3C.
- We can better understand heat treatments by use of TTT diagrams, CCT diagrams, and hardenability curves.
- The TTT diagrams describe how austenite transforms to pearlite and bainite at a constant temperature.
- The CCT diagrams describe how austenite transforms during continuous cooling. These diagrams give the cooling rates needed to obtain martensite during quenching.
- The hardenability curves compare the ease with which different steels transform to martensite.
- Alloying elements increase the times required for transformations in the TTT diagrams, reduce the cooling rates necessary to produce martensite in the CCT diagrams, and improve the hardenability of the steel.
- Specialty steels and heat treatments provide unique properties or combinations of properties. Of particular importance are surface-hardening treatments, such as carburizing, that produce an excellent combination of fatigue and impact resistance. Stainless steels, which contain a minimum of 11% Cr, have excellent corrosion resistance.
- Cast irons, by definition, undergo the eutectic reaction during solidification. Depending on the composition and treatment, either γ and Fe_3C or γ and graphite form during freezing.

Glossary

Annealing (cast iron) A heat treatment used to produce a ferrite matrix in a cast iron by the transformation of austenite via furnace cooling.

Annealing (steel) A heat treatment used to produce a soft, coarse pearlite in steel by austenitizing, then furnace cooling.

Ausforming A thermomechanical heat treatment in which austenite is plastically deformed below the A_1 temperature, then permitted to transform to bainite or martensite.

Austempering The isothermal heat treatment by which austenite transforms to bainite.

Austenitizing Heating a steel or cast iron to a temperature at which homogeneous austenite can form. Austenitizing is the first step in most of the heat treatments for steels and cast irons.

Carbon equivalent The sum of the weight percentage of carbon plus one-third of the weight percentage of silicon in a cast iron.

Carbonitriding Hardening the surface of steel with carbon and nitrogen obtained from a special gas atmosphere.

Carburizing A group of surface-hardening techniques by which carbon diffuses into steel.

Case depth The depth below the surface of a steel to which hardening occurs by surface hardening and carburizing processes.

Cast iron Ferrous alloys containing sufficient carbon so that the eutectic reaction occurs during solidification.

Compacted graphite cast iron A cast iron treated with small amounts of magnesium and titanium to cause graphite to grow during solidification as an interconnected, coral-shaped precipitate, giving properties midway between gray and ductile iron.

Cyaniding Hardening the surface of steel with carbon and nitrogen obtained from a bath of liquid cyanide solution.

Drawing Reheating a malleable iron in order to reduce the amount of carbon combined as cementite by spheroidizing pearlite, tempering martensite, or graphitizing both.

Dual-phase steels Special steels treated to produce martensite dispersed in a ferrite matrix.

Ductile cast iron Cast iron treated with magnesium to cause graphite to precipitate during solidification as spheres, permitting excellent strength and ductility. (Also known as *nodular* cast iron.)

Duplex stainless steel A special class of stainless steels containing a microstructure of ferrite and austenite.

Electric arc furnace A furnace used to melt steel scrap using electricity. Often, specialty steels are made using electric arc furnaces.

Eutectic cell A cluster of graphite flakes produced during solidification of gray iron that are all interconnected to a common nucleus.

Fading The loss of the nodulizing or inoculating effect in cast irons as a function of time, permitting undesirable changes in microstructure and properties.

First stage graphitization The first step in the heat treatment of a malleable iron, during which the carbides formed during solidification are decomposed to graphite and austenite.

Gray cast iron Cast iron which, during solidification, contains graphite flakes, causing low strength and poor ductility. This is the most widely used type of cast iron.

Hardenability The ease with which a steel can be quenched to form martensite. Steels with high hardenability form martensite even on slow cooling.

Hardenability curves Graphs showing the effect of the cooling rate on the hardness of as-quenched steel.

Hot metal The molten iron produced in a blast furnace, also known as pig iron. It contains about 95% iron, 4% carbon, 0.3–0.9% silicon, 0.5% manganese, and 0.025–0.05% each of sulfur, phosphorus, and titanium.

Inoculation The addition of an agent to molten cast iron that provides nucleation sites at which graphite precipitates during solidification.

Interstitial-free steels These are steels containing Nb and Ti. These elements react with C and S to form precipitates of carbides and sulfides, leaving the ferrite nearly free of interstitial elements.

Isothermal annealing Heat treatment of a steel by austenitizing, cooling rapidly to a temperature between the A_1 and the nose of the TTT curve, and holding until the austenite transforms to pearlite.

Jominy distance The distance from the quenched end of a Jominy bar. The Jominy distance is related to the cooling rate.

Jominy test The test used to evaluate hardenability. An austenitized steel bar is quenched at one end only, thus producing a range of cooling rates along the bar.

Malleable cast iron Cast iron obtained by a lengthy heat treatment, during which cementite decomposes to produce rounded clumps of graphite. Good strength, ductility, and toughness are obtained as a result of this structure.

Maraging steels A special class of alloy steels that obtain high strengths by a combination of the martensitic and age-hardening reactions.

Marquenching Quenching austenite to a temperature just above the M_s and holding until the temperature is equalized throughout the steel before further cooling to produce martensite. This process reduces residual stresses and quench cracking. (Also known as *martempering*.)

Nitriding Hardening the surface of steel with nitrogen obtained from a special gas atmosphere.

Nodulizing The addition of magnesium to molten cast iron to cause the graphite to precipitate as spheres rather than as flakes during solidification.

Normalizing A simple heat treatment obtained by austenitizing and air cooling to produce a fine pearlitic structure. This can be done for steels and cast irons.

Pig iron The molten iron produced in a blast furnace also known as hot metal. It contains about 95% iron, 4% carbon, 0.3–0.9% silicon, 0.5% manganese, and 0.025–0.05% each of sulfur, phosphorus, and titanium.

Process anneal A low-temperature heat treatment used to eliminate all or part of the effect of cold working in steels.

Quench cracks Cracks that form at the surface of a steel during quenching due to tensile residual stresses that are produced because of the volume change that accompanies the austenite-to-martensite transformation.

Retained austenite Austenite that is unable to transform into martensite during quenching because of the volume expansion associated with the reaction.

Second stage graphitization The second step in the heat treatment of malleable irons that are to have a ferritic matrix. The iron is cooled slowly from the first stage graphitization temperature so that austenite transforms to ferrite and graphite rather than to pearlite.

Secondary hardening peak Unusually high hardness in a steel tempered at a high temperature caused by the precipitation of alloy carbides.

Sensitization When heated to a temperature of ~480–860°C, chromium carbides precipitate along grain boundaries rather than within grains, causing chromium depletion in the interior. This causes stainless steel to corrode very easily.

Spheroidite A microconstituent containing coarse spherical cementite particles in a matrix of ferrite, permitting excellent machining characteristics in high-carbon steels.

Stainless steels A group of ferrous alloys that contain at least 11% Cr, providing extraordinary corrosion resistance.

Tempered martensite The microconstituent of ferrite and cementite formed when martensite is tempered.

Tool steels A group of high-carbon steels that provide combinations of high hardness, toughness, and resistance to elevated temperatures.

TRIP steels A group of steels with a microstructure that consists of a continuous ferrite matrix, a harder second phase (martensite and/or bainite), and retained austenite. TRIP stands for transformation induced plasticity.

Vermicular graphite The rounded, interconnected graphite that forms during the solidification of cast iron. This is the intended shape in compacted graphite iron, but it is a defective shape in ductile iron.

White cast iron Cast iron that produces cementite rather than graphite during solidification. The white irons are hard and brittle.

Problems

Section 13-1 Designations and Classification of Steels

13-1 What is the difference between cast iron and steels?

13-2 To what do the A_1, A_3, and A_{cm} temperatures refer? Are these temperatures constant?

13-3 A manufacturing plant requires 15 lb of AISI 8620 per final product. The plant makes 700 units per day (three shifts) six days a week, 50 weeks a year. How many tons of yttrium is this plant consuming per year if nothing is wasted?

13-4 Calculate the amounts of ferrite, cementite, primary microconstituent, and pearlite in the following steels:
(a) 1015;
(b) 1035;
(c) 1095; and
(d) 10130.

13-5 Estimate the AISI-SAE number for steels having the following microstructures:
(a) 38% pearlite–62% primary ferrite;
(b) 93% pearlite–7% primary cementite;
(c) 97% ferrite–3% cementite; and
(d) 86% ferrite–14% cementite.

13-6 What do the terms low-, medium-, and high-carbon steel mean?

13-7 Two samples of steel contain 93% pearlite. Estimate the carbon content of each sample if one is known to be hypoeutectoid and the other hypereutectoid.

Section 13-2 Simple Heat Treatments

Section 13-3 Isothermal Heat Treatments

13-8 Explain the following heat treatments: (a) process anneal; (b) austenitizing; (c) annealing; (d) normalizing; and (e) quenching.

13-9 Define austempering and explain its purpose.

13-10 Complete the following table:

	1035 Steel	10115 Steel
A_1 temperature		
A_3 or A_{cm} temperature		
Full annealing temperature		
Normalizing temperature		
Process annealing temperature		
Spheroidizing temperature		

13-11 Explain why, strictly speaking, TTT diagrams can be used for isothermal treatments only.

13-12 Determine the constants c and n in the Avrami relationship (Equation 12-2) for the transformation of austenite to pearlite for a 1050 steel. Assume that the material has been subjected to an isothermal heat treatment at 600°C and make a log–log plot of f versus t given the following information:

$f = 0.2$ at $t = 2$ s;
$f = 0.5$ at $t = 4$ s; and
$f = 0.8$ at $t = 7$ s.

13-13 In a pearlitic 1080 steel, the cementite platelets are 4×10^{-5} cm thick, and the ferrite platelets are 14×10^{-5} cm thick. In a spheroidized 1080 steel, the cementite spheres are 4×10^{-3} cm in diameter. Estimate the total interface area between the ferrite and cementite in a cubic centimeter of each steel. Determine the percent reduction in surface area when the pearlitic steel is spheroidized. The density of ferrite is 7.87 g/cm^3 and that of cementite is 7.66 g/cm^3.

13-14 Describe the microstructure present in a 1050 steel after each step in the following heat treatments:
(a) heat at 820°C, quench to 650°C and hold for 90 s, and quench to 25°C;
(b) heat at 820°C, quench to 450°C and hold for 90 s, and quench to 25°C;
(c) heat at 820°C, and quench to 25°C;
(d) heat at 820°C, quench to 720°C and hold for 100 s, and quench to 25°C;
(e) heat at 820°C, quench to 720°C and hold for 100 s, quench to 400°C and hold for 500 s, and quench to 25°C;
(f) heat at 820°C, quench to 720°C and hold for 100 s, quench to 400°C and hold for 10 s, and quench to 25°C; and
(g) heat at 820°C, quench to 25°C, heat to 500°C and hold for 10^3 s, and air cool to 25°C.

13-15 Describe the microstructure present in a 10110 steel after each step in the following heat treatments:
(a) heat to 900°C, quench to 400°C and hold for 10^3 s, and quench to 25°C;
(b) heat to 900°C, quench to 600°C and hold for 50 s, and quench to 25°C;
(c) heat to 900°C and quench to 25°C;
(d) heat to 900°C, quench to 300°C and hold for 200 s, and quench to 25°C;
(e) heat to 900°C, quench to 675°C and hold for 1 s, and quench to 25°C;
(f) heat to 900°C, quench to 675°C and hold for 1 s, quench to 400°C and hold for 900 s, and slowly cool to 25°C;
(g) heat to 900°C, quench to 675°C and hold for 1 s, quench to 300°C and hold for 10^3 s, and air cool to 25°C; and
(h) heat to 900°C, quench to 300°C and hold for 100 s, quench to 25°C, heat to 450°C for 3600 s, and cool to 25°C.

13-16 Recommend appropriate isothermal heat treatments to obtain the following, including appropriate temperatures and times:
(a) an isothermally annealed 1050 steel with HRC 23;
(b) an isothermally annealed 10110 steel with HRC 40;
(c) an isothermally annealed 1080 steel with HRC 38;
(d) an austempered 1050 steel with HRC 40;
(e) an austempered 10110 steel with HRC 55; and
(f) an austempered 1080 steel with HRC 50.

13-17 Compare the minimum times required to isothermally anneal the following steels at 600°C. Discuss the effect of the carbon content of the steel on the kinetics of nucleation and growth during the heat treatment.
(a) 1050;
(b) 1080; and
(c) 10110.

Section 13-4 Quench and Temper Heat Treatments

13-18 Explain the following terms: (a) quenching; (b) tempering; (c) retained austenite; and (d) marquenching/martempering.

13-19 Typical media used for quenching include air, brine (10% salt in water), water, and various oils.
(a) Rank the four media in order of the cooling rate from fastest to slowest.

(b) Describe a situation when quenching in air would be undesirable.

(c) During quenching in liquid media, typically either the part being cooled or the bath is agitated. Explain why.

13-20 We wish to produce a 1050 steel that has a Brinell hardness of at least 330 and an elongation of at least 15%.

(a) Recommend a heat treatment, including appropriate temperatures, that permits this to be achieved. Determine the yield strength and tensile strength that are obtained by this heat treatment. (See Figure 13-9.)

(b) What yield and tensile strengths would be obtained in a 1080 steel using the same heat treatment? (See Figure 12-27.)

(c) What yield strength, tensile strength and % elongation would be obtained in the 1050 steel if it were normalized? (See Figure 13-4.)

13-21 We wish to produce a 1050 steel that has a tensile strength of at least 175,000 psi and a reduction in area of at least 50%.

(a) Recommend a heat treatment, including appropriate temperatures, that permits this to be achieved. Determine the Brinell hardness number, % elongation, and yield strength that are obtained by this heat treatment. (See Figure 13-9.)

(b) What yield and tensile strengths would be obtained in a 1080 steel using the same heat treatment? (See Figure 12-27.)

(c) What yield strength, tensile strength, and % elongation would be obtained in the 1050 steel if it were annealed? (See Figure 13-4.)

13-22 A 1030 steel is given an improper quench and temper heat treatment, producing a final structure composed of 60% martensite and 40% ferrite. Estimate the carbon content of the martensite and the austenitizing temperature that was used. What austenitizing temperature would you recommend?

13-23 A 1050 steel should be austenitized at 820°C, quenched in oil to 25°C, and tempered at 400°C for an appropriate time.

(a) What yield strength, hardness, and % elongation would you expect to obtain from this heat treatment? (See Figure 13-9.)

(b) Suppose the actual yield strength of the steel is found to be 125,000 psi. What might have gone wrong in the heat treatment to cause this low strength?

(c) Suppose the Brinell hardness is found to be HB 525. What might have gone wrong in the heat treatment to cause this high hardness?

13-24 A part produced from a low-alloy, 0.2% C steel has a microstructure containing ferrite, pearlite, bainite, and martensite after quenching. What microstructure would be obtained if we used a 1080 steel? What microstructure would be obtained if we used a 4340 steel? [See Figures 13-14, 13-15, and 13-16(b).]

13-25 Fine pearlite and a small amount of martensite are found in a quenched 1080 steel. What microstructure would be expected if we used a low-alloy, 0.2% C steel? What microstructure would be expected if we used a 4340 steel? [See Figures 13-14, 13-15, and 13-16(b).]

13-26 Predict the phases formed when a bar of 1080 steel is quenched from slightly above the eutectoid temperature under the following conditions:

(a) oil (without agitation);

(b) oil (with agitation);

(c) water (with agitation); and

(d) brine (no agitation).

Suggest a quenching medium if we wish to obtain coarse pearlite. (See Figure 13-14 and Table 13-2 for this problem.)

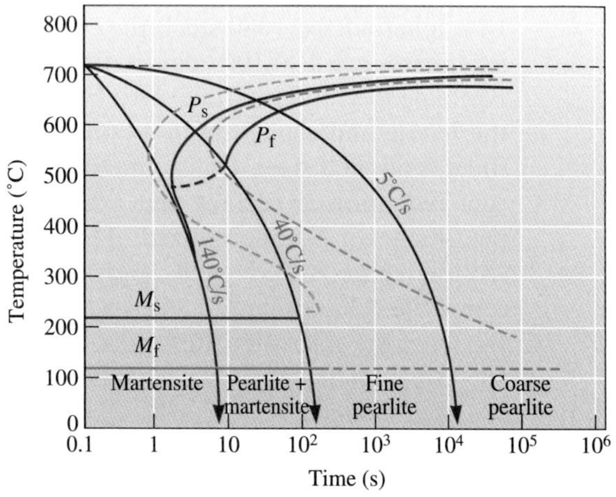

Figure 13-14 The CCT diagram (solid lines) for a 1080 steel compared with the TTT diagram (dashed lines). (Repeated for Problem 13-26.)

Section 13-5 Effect of Alloying Elements

Section 13-6 Application of Hardenability

13-27 Define ausforming and explain its purpose.

13-28 Explain the difference between hardenability and hardness. Explain using a sketch how hardenability of steels is measured.

13-29 We have found that a 1070 steel, when austenitized at 750°C, forms a structure containing pearlite and a small amount of grain-boundary ferrite that gives acceptable strength and ductility. What changes in the microstructure, if any, would be expected if the 1070 steel contained an alloying element such as Mo or Cr? Explain.

13-30 Using the TTT diagrams, compare the hardenabilities of 4340 and 1050 steels by determining the times required for the isothermal transformation of ferrite and pearlite (F_s, P_s, and P_f) to occur at 650°C. [See Figures 13-8 and 13-16(a).]

13-31 Water hardenable 1095 steel is supposed to be tempered at 500°C, but a negligent operator does not check the automatic process control and so it is tempered at 450°C. What is the hardness of the as-produced steel compared to the specified steel?

13-32 Determine the cooling rate for AISI 9310, 4320 and 1080 steels to have the same hardness. (See Figure 13-21.)

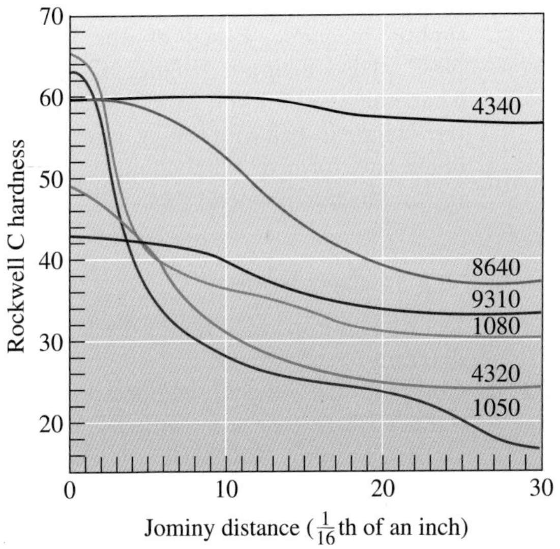

Figure 13-21 The hardenability curves for several steels. (Repeated for Problems 13-32 and 13-59.)

13-33 We would like to obtain a hardness of HRC 38 to 40 in a quenched steel. What range of cooling rates would we have to obtain for the following steels? Are some steels inappropriate for achieving these levels of hardness? (See Figure 13-21 and Table 13-3.)
(a) 4340;
(b) 8640;
(c) 9310;
(d) 4320;
(e) 1050; and
(f) 1080.

13-34 A steel part must have an as-quenched hardness of HRC 35 in order to avoid excessive-wear rates during use. When the part is made from 4320 steel, the hardness is only HRC 32. Determine the hardness if the part were made under identical conditions, but with the following steels. Which, if any, of these steels would be better choices than 4320?
(a) 4340;
(b) 8640;
(c) 9310;
(d) 1050; and
(e) 1080.

13-35 A part produced from a 4320 steel has a hardness of HRC 35 at a critical location after quenching. Determine
(a) the cooling rate at that location, and
(b) the microstructure and hardness that would be obtained if the part were made from a 1080 steel.

13-36 A 1080 steel is cooled at the fastest possible rate that still permits all pearlite to form. What is this cooling rate? What Jominy distance and hardness are expected for this cooling rate?

13-37 Determine the hardness and microstructure at the center of a 1.5-in.-diameter 1080 steel bar produced by quenching in
(a) unagitated oil;
(b) unagitated water; and
(c) agitated brine.

13-38 A 2-in.-diameter bar of 4320 steel is to have a hardness of at least HRC 35. What is the minimum severity of the quench (H coefficient)? What type of quenching medium would you recommend to produce the desired hardness with the least chance of quench cracking?

13-39 A steel bar is to be quenched in agitated water. Determine the maximum diameter of the bar that will produce a minimum hardness of HRC 40 if the bar is
(a) 1050;
(b) 1080;
(c) 4320;
(d) 8640; and
(e) 4340.

13-40 The center of a 1-in.-diameter bar of 4320 steel has a hardness of HRC 40. Determine the hardness and microstructure at the center of a 2-in. bar of 1050 steel quenched in the same medium.

13-41 You encounter a specification that lists a Jominy distance of 1.27 cm and a bar diameter of 3.81 cm. What is the H coefficient for this?

Section 13-7 Specialty Steels

Section 13-8 Surface Treatments

Section 13-9 Weldability of Steel

13-42 What are tool steels?

13-43 What is the principle of the surface hardening of steels using carburizing and nitriding?

13-44 A 1010 steel is to be carburized using a gas atmosphere that produces 1.0% C at the surface of the steel. The case depth is defined as the distance below the surface that contains at least 0.5% C. If carburizing is done at 1000°C, determine the time required to produce a case depth of 0.01 in. (See Chapter 5.)

13-45 A 1015 steel is to be carburized at 1050°C for 2 h using a gas atmosphere that produces 1.2% C at the surface of the steel. Plot the percent carbon versus the distance from the surface of the steel. If the steel is slowly cooled after carburizing, determine the amount of each phase and microconstituent at 0.002-in. intervals from the surface. (See Chapter 5.)

13-46 Why is it that the strength of the heat-affected zone is higher for low-carbon steels? What is the role of retained austenite in this case?

13-47 Why is it easy to weld low-carbon steels, but difficult to weld high-carbon steels?

13-48 A 1050 steel is welded. After cooling, hardnesses in the heat-affected zone are obtained at various locations from the edge of the fusion zone. Determine the hardnesses expected at each point if a 1080 steel were welded under the same conditions. Predict the microstructure at each location in the as-welded 1080 steel.

Distance from Edge 1050 of Fusion Zone	Hardness in Weld
0.05 mm	HRC 50
0.10 mm	HRC 40
0.15 mm	HRC 32
0.20 mm	HRC 28

Section 13-10 Stainless Steels

13-49 What is a stainless steel? Why are stainless steels stainless?

13-50 Surplus stock inventory contains 15 ton 304, 5.2 ton 304L, 1 ton 316, and 4.7 ton 440C stainless steel. A spike in molybdenum prices makes it worthwhile to sell the steels that contain it. How much molybdenum is contained in each of the steels and how much total mass is being sold?

13-51 A food processing plant that is designed to use 5.75 metric tons of 316 stainless steel will actually use 115% of the design specification. How many moles of chromium must be refined for this? Assume the refinery loses 5% of what it refines.

13-52 We wish to produce a martensitic stainless steel containing 17% Cr. Recommend a carbon content and austenitizing temperature that would permit us to obtain 100% martensite during the quench. What microstructure would be produced if the martensite were then tempered until the equilibrium phases formed?

13-53 Occasionally, when an austenitic stainless steel is welded, the weld deposit may be slightly magnetic. Based on the Fe-Cr-Ni-C phase diagram [Figure 13-28(b)], what phase would you expect is causing the magnetic behavior? Why might this phase have formed? What could you do to restore the nonmagnetic behavior?

Section 13-11 Cast Irons

13-54 Define cast iron using the Fe-Fe_3C phase diagram.

13-55 Compare the eutectic temperatures of a Fe-4.3% C cast iron with a Fe-3.6% C-2.1% Si alloy. Which alloy is expected to be more machinable and why?

13-56 What are the different types of cast irons? Explain using a sketch.

13-57 A bar of a class 40 gray iron casting is found to have a tensile strength of 50,000 psi. Why is the tensile strength greater than that given by the class number? What is the diameter of the test bar?

13-58 You would like to produce a gray iron casting that freezes with no primary austenite or graphite. If the carbon content in the iron is 3.5%, what percentage of silicon must you add?

Design Problems

13-59 We would like to produce a 2-in.-thick steel wear plate for a rock-crushing unit. To avoid frequent replacement of the wear plate, the hardness should exceed HRC 38 within 0.25 in. of the steel surface. The center of the plate should have a hardness of no more than HRC 32 to ensure some toughness. We have only a water quench available to us. Design the plate, assuming that we only have the steels given in Figure 13-21 available to us.

13-60 A quenched and tempered 10110 steel is found to have surface cracks that cause the heat-treated part to be rejected by the customer. Why did the cracks form? Design a heat treatment, including appropriate temperatures and times, to minimize these problems.

13-61 Design a corrosion-resistant steel to use for a pump that transports liquid helium at 4 K in a superconducting magnet.

13-62 Design a heat treatment for a hook made from a 1-in.-diameter steel rod having a microstructure containing a mixture of ferrite, bainite, and martensite after quenching. Estimate the mechanical properties of your hook.

13-63 Design an annealing treatment for a 1050 steel. Be sure to include details of temperatures, cooling rates, microstructures, and properties.

13-64 Design a process to produce a 0.5-cm-diameter steel shaft having excellent toughness as well as excellent wear and fatigue resistance. The surface hardness should be at least HRC 60, and the hardness 0.01 cm beneath the surface should be approximately HRC 50. Describe the process, including details of the heat-treating atmosphere, the composition of the steel, temperatures, and times.

Computer Problems

13-65 *Empirical Relationships for Transformations in Steels.* Heat treatment of steels depends upon knowing the A_1, A_3, A_{cm}, M_s, and M_f temperatures. Steel producers often provide their customers with empirical formulas that define these temperatures as a function of alloying element concentrations. For example, one such formula described in the literature is

$$M_s(\text{in °C}) = 561 - 474(\%\text{ C}) - 33(\%\text{ Mn}) - 17(\%\text{ Ni}) - 17(\%\text{ Cr}) - 21(\%\text{ Mo})$$

Write a computer program that will ask the user to provide information for the concentrations of different alloying elements and provide the M_s and M_f temperatures. Assume that the M_f temperature is 215°C below M_s.

Knovel® Problems

K13-1 Find a time-temperature-transformation (TTT) diagram for 1340 steel. Assume that the steel is held at 1330°F for 1 h, quenched to 800°F and held for 1000 s at this temperature, and then cooled in air to room temperature.

(a) What is the microstructure and hardness of the steel after the heat treatment?

(b) What are the designation, hardenability, and applications of this steel?

(c) What are the mechanical properties of this steel delivered as hot rolled, normalized, and annealed when tempered at 800°F and oil-quenched?

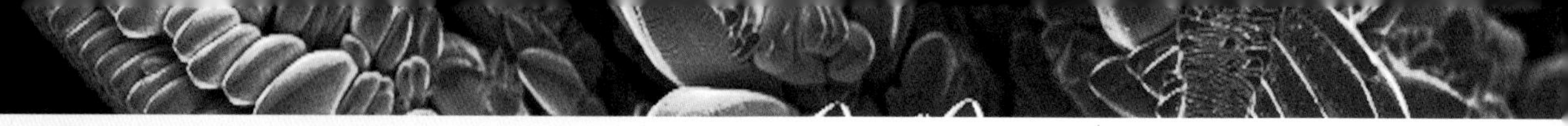

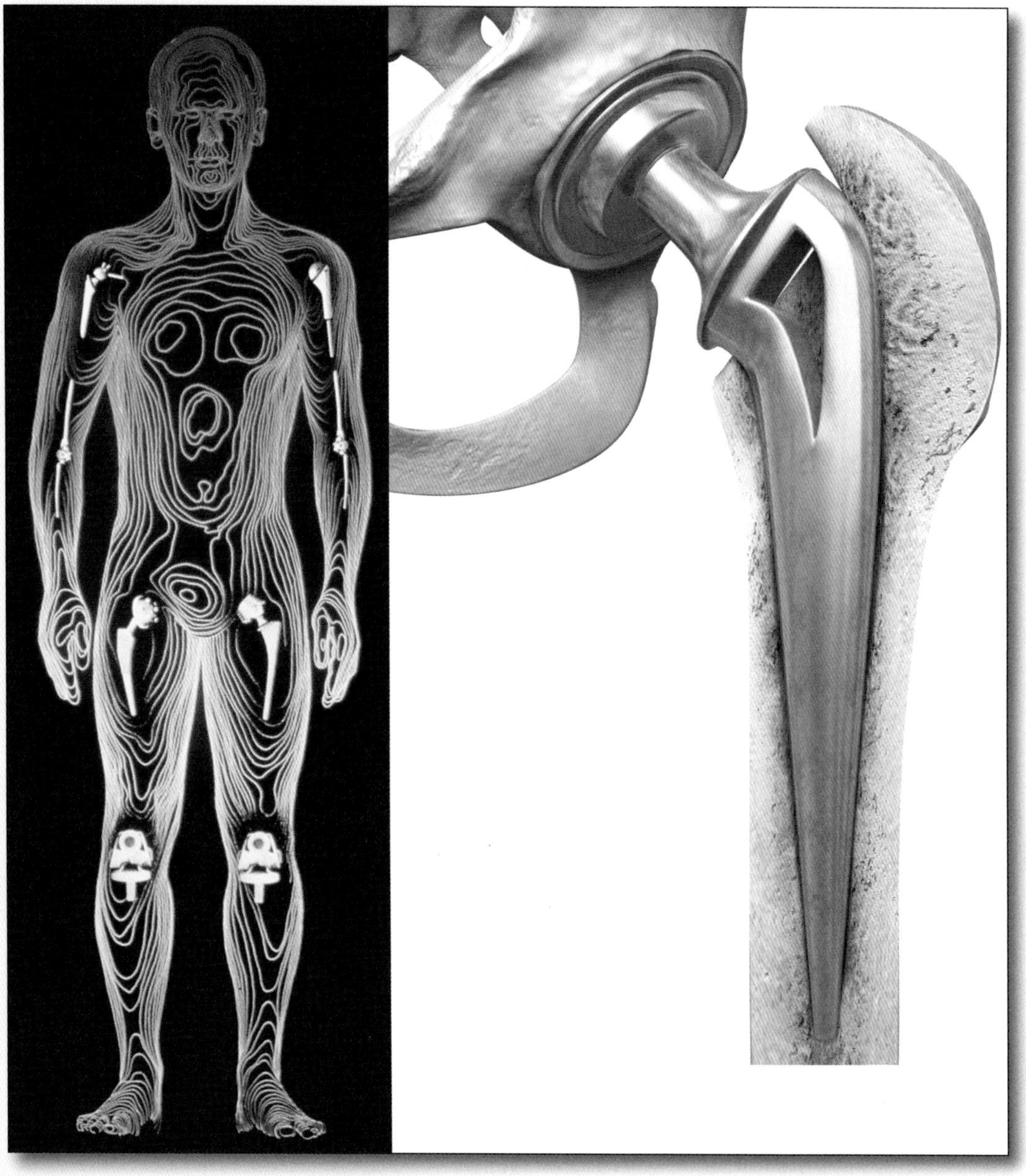

Increasingly, artificial implants are being incorporated into the human body. Some examples of commonly replaced joints are shown in the diagram on the left (*PhotoLink/Photodisc Green/Getty Images*). Every year, approximately 500,000 people receive a hip implant such as that shown in the photo on the right (© *MedicalRF.com / Alamy*). The implant must be made from a biocompatible material that has strong fracture toughness and an excellent fatigue life, and it must be corrosion resistant and possess a stiffness similar to that of bone. Nonferrous alloys based on titanium are often used for this application, as are cobalt-chrome alloys.

CHAPTER

14

Nonferrous Alloys

Have You Ever Wondered?

- *In the history of mankind, which came first: copper or steel?*
- *Why does the Statue of Liberty have a green patina?*
- *What materials are used to manufacture biomedical implants for hip prostheses?*
- *Are some metals toxic?*
- *What materials are used as "catalysts" in the automobile catalytic converter? What do they "convert"?*

Chapter Learning Objectives

The key objectives of this chapter are to

- Recognize and explain the various designations for different alloys of nonferrous metals.
- Identify the uses for common nonferrous metals and alloys.
- Identify the materials properties of common nonferrous metals and alloys.

Nonferrous alloys (i.e., alloys of elements other than iron) include, but are not limited to, alloys based on aluminum, copper, nickel, cobalt, zinc, precious metals (such as Pt, Au, Ag, Pd), and other metals (e.g., Nb, Ta, W). In this chapter, we will briefly explore the properties and applications of Cu, Al, and Ti alloys in load-bearing applications. We will not discuss the electronic, magnetic, and other applications of nonferrous alloys.

In many applications, weight is a critical factor. To relate the strength of the material to its weight, a **specific strength**, or strength-to-weight ratio, is defined as

$$\text{Specific strength} = \frac{\text{strength}}{\text{density}} \tag{14-1}$$

Table 14-1 Specific strength and cost of nonferrous alloys, steels, and polymer composites

Metal	Density (g/cm³)	Density (lb/in.³)	Tensile Strength (psi)	Specific Strength (in.)	Cost per lb ($)[b]
Aluminum	2.70	(0.098)	83,000	8.6×10^5	0.80
Beryllium	1.85	(0.067)	55,000	8.2×10^5	425.00
Copper	8.93	(0.322)	150,000	4.7×10^5	3.20
Lead	11.36	(0.410)	10,000	0.2×10^5	1.00
Magnesium	1.74	(0.063)	55,000	8.7×10^5	2.75
Nickel	8.90	(0.321)	180,000	5.6×10^5	8.00
Titanium	4.51	(0.163)	160,000	9.8×10^5	8.00
Tungsten	19.25	(0.695)	150,000	2.2×10^5	24.00
Zinc	7.13	(0.257)	75,000	2.9×10^5	0.95
Steels	~7.87	(0.284)	200,000	7.0×10^5	0.30
Aramid/epoxy (Kevlar, vol. fraction of fibers 0.6, longitudinal tension)	1.4	(0.05)	200,000	4.0×10^5	—
Aramid/epoxy (Kevlar, vol. fraction of fibers 0.6, transverse tension)[a]	1.4	(0.05)	4300	8.6×10^3	—
Glass/epoxy (Vol. fraction of E-glass fibers 0.6, longitudinal tension)	2.1	(0.075)	150,000	2.0×10^6	—
Glass/epoxy (Vol. fraction of E-glass fibers 0.6, transverse tension)	2.1	(0.075)	7000	9.3×10^4	—

[a]*Data for composites based on Harper, C.A.,* Handbook of Materials Product Design, *3rd ed., McGraw-Hill, 2001. Commodity composites are relatively inexpensive; high-performance components are expensive.*
[b]*Costs were compiled in 2014.*

Table 14-1 compares the specific strengths of steel, some high-strength nonferrous alloys, and polymer-matrix composites. Another factor to consider in designing with nonferrous metals is their cost, which also varies considerably. Table 14-1 gives the approximate price of different materials. One should note, however, that materials prices fluctuate with global economic trends and that the price of the material is only a small portion of the cost of a part. Fabrication and finishing, not to mention marketing and distribution, often contribute much more to the overall cost of a part. Composites based on carbon and other fibers also have significant advantages with respect to their specific strengths. Their properties are typically anisotropic, and the temperature at which they can be used is limited. In practice, to overcome the anisotropy, composites are often made in many layers. The directions of fibers are changed in different layers so as to minimize the anisotropy in properties. Composites will be discussed in Chapter 17.

14-1 Aluminum Alloys

Aluminum is the third most plentiful element on earth (next to oxygen and silicon) but, until the late 1800s, was expensive and difficult to produce. The six pound cap installed on the top of the Washington Monument in 1884 was one of the largest aluminum parts made at that time.

General Properties and Uses of Aluminum Aluminum has a density of 2.70 g/cm^3, or one-third the density of steel, and a modulus of elasticity of 10×10^6 psi (70 GPa). Although aluminum alloys have lower tensile strengths than steels, their specific strengths (or strength-to-weight ratios) are excellent. The Wright brothers used an Al-Cu alloy for their engine for this very reason (Chapter 12). Aluminum can be formed easily, has high thermal and electrical conductivities, and does not show a ductile-to-brittle transition at low temperatures. It is nontoxic and can be recycled with only about 5% of the energy that was needed to make it from alumina. This is why the recycling of aluminum is so successful. Aluminum's beneficial physical properties include its non-ferromagnetic behavior and resistance to oxidation and corrosion. Aluminum does not display a true endurance limit, however, so failure by fatigue may eventually occur, even at low stresses. As suggested by its low melting temperature, aluminum does not perform well at elevated temperatures. Finally, aluminum alloys have low hardnesses, leading to poor wear resistance. Aluminum responds readily to strengthening mechanisms. Table 14-2 compares the strength of pure, annealed aluminum with that of alloys strengthened by various techniques. The alloys may be 30 times stronger than pure aluminum.

About 25% of the aluminum produced today is used in the transportation industry, another 25% in the manufacturing of beverage cans and other packaging, about 15% in construction, 15% in electrical applications, and 20% in other applications. About 200 pounds of aluminum were used in an average car made in the United States in 2010. Aluminum reacts with oxygen, even at room temperature, to produce an extremely thin aluminum-oxide layer that protects the underlying metal from many corrosive environments (Chapter 23). This is counter to the common belief that Al does not "rust." It does, but the rust product is highly adherent and also seals the surface against subsequent corrosion. We should be careful, though, not to generalize this behavior. For example, fine aluminum powder with a large surface area to volume ratio, when in the presence of an oxidizer and a catalyst, can serve as the fuel for solid rocket boosters. Such boosters used ~200,000 lbs of atomized aluminum powder every time the space shuttle took off and generated enough force for the shuttle to reach a speed of ~3000 miles per hour. New developments related to aluminum include the development of aluminum alloys containing higher Mg concentrations for use in automotive applications.

Designation Aluminum alloys can be divided into two major groups—wrought and casting alloys—depending on their method of fabrication. **Wrought alloys**, which are shaped by plastic deformation, have compositions and microstructures significantly different from casting alloys, reflecting the different requirements of the manufacturing process. Within each major group, we can divide the alloys into two subgroups: heat-treatable and nonheat-treatable alloys.

Aluminum alloys are designated by the numbering system shown in Table 14-3. The first number specifies the principal alloying elements, and the remaining numbers refer to the specific composition of the alloy.

Table 14-2 The effect of strengthening mechanisms in aluminum and aluminum alloys

Material	Tensile Strength (psi)	Yield Strength (psi)	% Elongation	Ratio of Alloy-to-Metal Yield Strengths
Pure Al	6,500	2,500	60	1
Commercially pure Al (at least 99% pure)	13,000	5,000	45	2.0
Solid-solution-strengthened Al alloy	16,000	6,000	35	2.4
Cold-worked Al	24,000	22,000	15	8.8
Dispersion-strengthened Al alloy	42,000	22,000	35	8.8
Age-hardened Al alloy	83,000	73,000	11	29.2

Table 14-3 Designation system for aluminum alloys

Wrought alloys:		
1xxx[a]	Commercially pure Al (>99% Al)	Not age hardenable
2xxx	Al-Cu and Al-Cu-Li	Age hardenable
3xxx	Al-Mn	Not age hardenable
4xxx	Al-Si and Al-Mg-Si	Age hardenable if magnesium is present
5xxx	Al-Mg	Not age hardenable
6xxx	Al-Mg-Si	Age hardenable
7xxx	Al-Mg-Zn	Age hardenable
8xxx	Al-Li, Sn, Zr, or B	Age hardenable
9xxx	Not currently used	
Casting alloys:		
1xx.x.[b]	Commercially pure Al	Not age hardenable
2xx.x.	Al-Cu	Age hardenable
3xx.x.	Al-Si-Cu or Al-Mg-Si	Some are age hardenable
4xx.x.	Al-Si	Not age hardenable
5xx.x.	Al-Mg	Not age hardenable
7xx.x.	Al-Mg-Zn	Age hardenable
8xx.x.	Al-Sn	Age hardenable
9xx.x.	Not currently used	

[a] *The first digit shows the main alloying element, the second digit shows modification, and the last two digits shows the decimal % of the Al concentration (e.g., 1060 is a 99.6% Al alloy).*
[b] *Last digit indicates product form, 1 or 2 is ingot (depends upon purity) and 0 is for casting.*

The degree of strengthening is given by the **temper designation** T or H, depending on whether the alloy is heat treated or strain hardened (Table 14-4). Other designations indicate whether the alloy is annealed (O), solution-treated (W), or used in the as-fabricated condition (F). The numbers following the T or H indicate the amount of strain hardening,

Table 14-4 Temper designations for aluminum alloys

F	As-fabricated (hot-worked, forged, cast, etc.)
O	Annealed (in the softest possible condition)
H	Cold-worked
	H1x—cold-worked only. (x refers to the amount of cold work and strengthening.)
	H12—cold work that gives a tensile strength midway between the O and H14 tempers.
	H14—cold work that gives a tensile strength midway between the O and H18 tempers.
	H16—cold work that gives a tensile strength midway between the H14 and H18 tempers.
	H18—cold work that gives about 75% reduction in area.
	H19—cold work that gives a tensile strength 2000 psi greater than that obtained by the H18 temper.
	H2x—cold-worked and partly annealed.
	H3x—cold-worked and stabilized at a low temperature to prevent age hardening of the structure.
W	Solution-treated
T	Age-hardened
	T1—cooled from the fabrication temperature and naturally aged.
	T2—cooled from the fabrication temperature, cold-worked, and naturally aged.
	T3—solution-treated, cold-worked, and naturally aged.
	T4—solution-treated and naturally aged.
	T5—cooled from the fabrication temperature and artificially aged.
	T6—solution-treated and artificially aged.
	T7—solution-treated and stabilized by overaging.
	T8—solution-treated, cold-worked, and artificially aged.
	T9—solution-treated, artificially aged, and cold-worked.
	T10—cooled from the fabrication temperature, cold-worked, and artificially aged.

Table 14-5 Properties of typical aluminum alloys

Alloy		Tensile Strength (psi)	Yield Strength (psi)	% Elongation	Applications
Nonheat-treatable wrought alloys:					
1100-O	>99% Al	13,000	5,000	40	Electrical components,
1100-H18		24,000	22,000	10	foil, food processing,
3004-O	1.2% Mn-1.0% Mg	26,000	10,000	25	beverage can
3004-H18		41,000	36,000	9	bodies, architectural
4043-O	5.2% Si	21,000	10,000	22	uses, filler metal for
4043-H18		41,000	39,000	1	welding, beverage
5182-O	4.5% Mg	42,000	19,000	25	can tops, and
5182-H19		61,000	57,000	4	marine components
Heat-treatable wrought alloys:					
2024-T4	4.4% Cu	68,000	47,000	20	Truck wheels, aircraft
2090-T6	2.4% Li-2.7% Cu	80,000	75,000	6	skins, pistons,
4032-T6	12% Si-1% Mg	55,000	46,000	9	canoes, railroad cars,
6061-T6	1% Mg-0.6% Si	45,000	40,000	15	and aircraft frames
7075-T6	5.6% Zn-2.5% Mg	83,000	73,000	11	
Casting alloys:					
201-T6	4.5% Cu	70,000	63,000	7	Transmission
319-F	6% Si-3.5% Cu	27,000	18,000	2	housings, general
356-T6	7% Si-0.3% Mg	33,000	24,000	3	purpose castings,
380-F	8.5% Si-3.5% Cu	46,000	23,000	3	aircraft fittings,
390-F	17% Si-4.5% Cu	41,000	35,000	1	motor housings,
443-F	5.2% Si (sand		8,000	8	automotive engines,
	cast)	19,000	9,000	10	food-handling
	(permanent mold)	23,000	16,000	9	equipment, and
	(die cast)	33,000			marine fittings

the exact type of heat treatment, or other special aspects of the processing of the alloy. Typical alloys and their properties are included in Table 14-5.

Wrought Alloys The 1xxx, 3xxx, 5xxx, and most of the 4xxx wrought alloys are not age hardenable. The 1xxx and 3xxx alloys are single-phase alloys except for the presence of small amounts of inclusions or intermetallic compounds [Figure 14-1(a)].

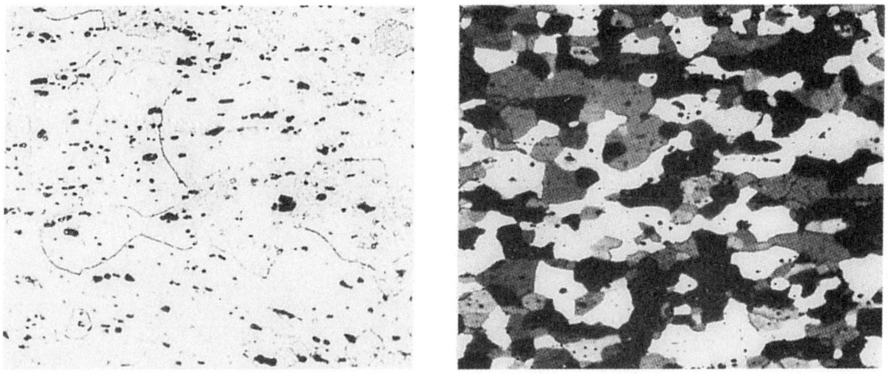

(a) (b)

Figure 14-1 (a) $FeAl_3$ inclusions in annealed 1100 aluminum (× 350). (b) Mg_2Si precipitates in annealed 5457 aluminum alloy (× 75). [*From* ASM Handbook, *Vol. 7, (1972), ASM International, Materials Park, OH 44073-0002. Reprinted with permission of ASM International. All rights reserved. www.asminternational.org.*]

Their properties are controlled by strain hardening, solid-solution strengthening, and grain-size control. Because the solubilities of the alloying elements in aluminum are limited at room temperature, the degree of solid-solution strengthening is limited. The 5xxx alloys contain two phases at room temperature—α, a solid solution of magnesium in aluminum, and Mg_2Al_3, a hard, brittle intermetallic compound (Figure 14-2). The aluminum-magnesium alloys are strengthened by a fine dispersion of Mg_2Al_3, as well as by strain hardening, solid-solution strengthening, and grain-size control. Because Mg_2Al_3 is not coherent, age-hardening treatments are not possible. The 4xxx series alloys also contain two phases, α and nearly pure silicon, β (Chapter 11). Alloys that contain both silicon and magnesium can be age hardened by permitting Mg_2Si to precipitate [Figure 14-1(b)]. The 2xxx, 6xxx, and 7xxx alloys are age-hardenable alloys. Although excellent specific strengths are obtained for these alloys, the amount of precipitate that can form is limited. In addition, they cannot be used at temperatures above approximately 175°C in the aged condition. Alloy 2024 is a widely used aircraft alloy. There is also a renewed interest in the development of precipitation-hardened Al-Li alloys due to their high Young's modulus and low density; however, high processing costs, anisotropic properties, and lower fracture toughness have proved to be limiting factors. Al-Li alloys (2195) were used to make the second-generation space shuttle fuel tanks, and they will be used in NASA's planned Orion Multi-purpose Crew Vehicle for missions to the moon and ultimately to Mars. Al-Sc and Al-Ti alloys are also of interest in high strength applications.

Casting Alloys Many of the common aluminum casting alloys shown in Table 14-5 contain enough silicon to cause the eutectic reaction, giving the alloys low melting points, good fluidity, and good castability. **Fluidity** is the ability of the liquid metal to flow through a mold without prematurely solidifying, and **castability** refers to the ease with which a good casting can be made from the alloy.

The properties of the aluminum-silicon alloys are controlled by solid-solution strengthening of the α aluminum matrix, dispersion strengthening by the β phase, and solidification, which controls the primary grain size and shape as well as the nature of the eutectic microconstituent. Fast cooling obtained in die casting or permanent mold casting (Chapter 9) increases strength by refining grain size and the eutectic microconstituent (Figure 14-3).

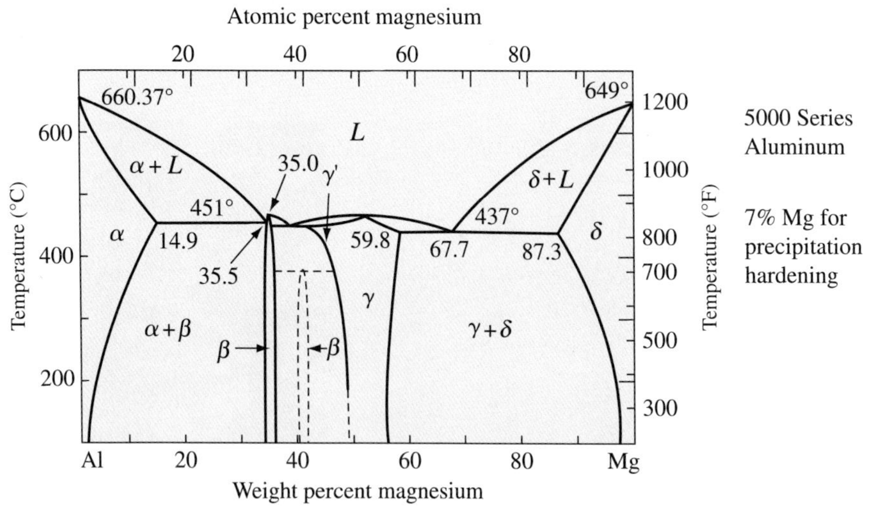

Figure 14-2 Aluminum-magnesium phase diagram. Dotted lines indicate uncertain solubility limits. (*Based on* American Society for Metals Handbook, *8th ed., Vol. 8, Metallography, Structures and Phase Diagrams, ASM International, 1973.*)

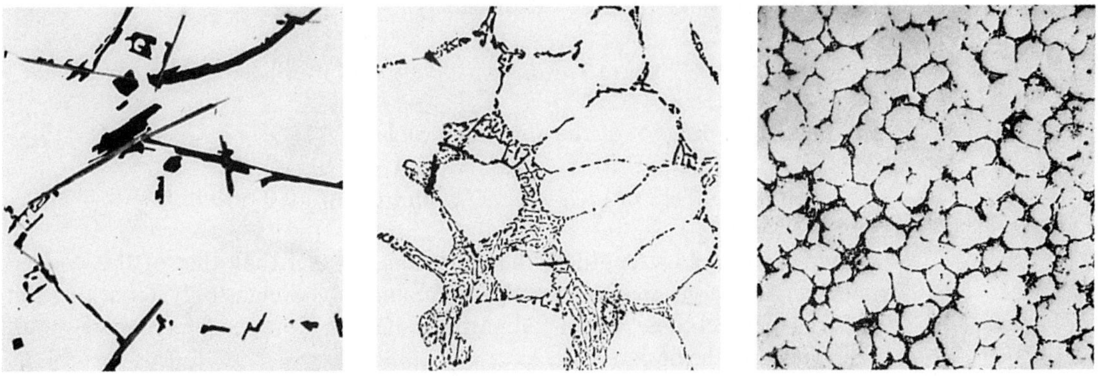

(a) **(b)** **(c)**

Figure 14-3 (a) Sand cast 443 aluminum alloy containing coarse silicon and inclusions (× 350). (b) Permanent mold 443 alloy containing fine dendrite cells and fine silicon due to faster cooling (× 350). (c) Die cast 443 alloy with a still finer microstructure (× 350). [*From* ASM Handbook, *Vol. 7, (1972), ASM International, Materials Park, OH 44073-0002. Reprinted with permission of ASM International. All rights reserved. www.asminternational.org.*]

Grain refinement using boron and titanium additions, modification using sodium or strontium to change the eutectic structure, and hardening with phosphorus to refine the primary silicon (Chapter 9) are all done in certain alloys to improve the microstructure and, thus, the degree of dispersion strengthening. Many alloys also contain copper, magnesium, or zinc, thus permitting age hardening. The following example illustrates applications of aluminum alloys.

Example 14-1 *Strength-to-Weight Ratio in Design*

A steel cable 0.5 in. in diameter has a yield strength of 70,000 psi. The density of steel is about 7.87 g/cm^3. Based on the data in Table 14-5, determine (a) the maximum load that the steel cable can support without yielding, (b) the diameter of a cold-worked aluminum-manganese alloy (3004-H18) required to support the same load as the steel, and (c) the weight per foot of the steel cable versus the aluminum alloy cable.

SOLUTION

a. The maximum load F that the steel cable can support without yielding is given by

$$F_{steel} = (S_y^{steel} A_0^{steel}) = 70{,}000 \text{ psi} \left[\frac{\pi}{4}(0.5 \text{ in.})^2\right] = 13{,}745 \text{ lb}$$

where S_y is the yield strength and A_0 is the initial cross-sectional area.

b. The yield strength of the aluminum alloy is 36,000 psi. Thus, the diameter of the aluminum cable d_{Al} is

$$\frac{\pi}{4} d_{Al}^2 = \frac{F_{Al}}{S_y^{Al}} = \frac{13{,}745 \text{ lb}}{36{,}000 \text{ psi}} = 0.382 \text{ in.}^2$$

$$d_{Al} = 0.697 \text{ in.}$$

c. The densities of steel and aluminum are

$$\rho_{steel} = 7.87 \text{ g/cm}^3 = 0.284 \text{ lb/in.}^3$$

$$\rho_{Al} = 2.70 \text{ g/cm}^3 = 0.098 \text{ lb/in.}^3$$

Therefore, the weight per foot of the steel cable W_{steel} is

$$W_{steel} = A_0^{steel}(12 \text{ in.})\, \rho_{steel} = \frac{\pi}{4}(0.5 \text{ in.})^2(12 \text{ in.})(0.284 \text{ lb/in.}^3) = 0.669 \text{ lb}$$

and the weight per foot of the aluminum cable is

$$W_{Al} = A_0^{Al}(12 \text{ in.})\, \rho_{Al} = \frac{\pi}{4}(0.697 \text{ in.})^2(12 \text{ in.})(0.098 \text{ lb/in.}^3) = 0.449 \text{ lb}$$

Although the yield strength of the aluminum is lower than that of the steel and the cable must be larger in diameter, the aluminum cable weighs only about two-thirds as much as the steel cable. When comparing materials, a proper safety factor should also be included in the design.

14-2 Magnesium and Beryllium Alloys

Magnesium, which is often extracted electrolytically from concentrated magnesium chloride in seawater, is lighter than aluminum with a density of 1.74 g/cm^3, and it melts at a slightly lower temperature than aluminum. In many environments, the corrosion resistance of magnesium approaches that of aluminum; however, exposure to salts, such as that near a marine environment, causes rapid deterioration. Although magnesium alloys are not as strong as aluminum alloys, their specific strengths are comparable. Consequently, magnesium alloys are used in aerospace applications, high-speed machinery, and transportation and materials handling equipment. The drawbacks of Mg include that it has a low modulus of elasticity (6.5×10^6 psi or 45 GPa) and poor resistance to fatigue, creep, and wear. Magnesium also poses a hazard during casting and machining, since it combines easily with oxygen and burns. Magnesium strips can be easily ignited in the lab, and magnesium powder is exceptionally dangerous. Finally, the response of magnesium to strengthening mechanisms is relatively poor.

Structure and Properties Magnesium, which has the HCP structure, is less ductile than aluminum. The Mg alloys do have some ductility, however, because alloying increases the number of active slip planes. Some deformation and strain hardening can be accomplished at room temperature, and the alloys can be readily deformed at elevated temperatures. Strain hardening produces a relatively small effect in pure magnesium because of the low strain-hardening coefficient (Chapter 8).

As in aluminum alloys, the solubility of alloying elements in magnesium at room temperature is limited, causing only a small degree of solid-solution strengthening. The solubility of many alloying elements increases with temperature, however, as typified by the Mg-Al phase diagram (Figure 14-4). Therefore, alloys may be strengthened by either dispersion strengthening or age hardening. Some age-hardened magnesium alloys, such as those containing Zr, Th, Ag, or Ce, have good resistance to overaging at temperatures as high as 300°C. Alloys containing up to 9% Li have exceptionally light weight. Properties of typical magnesium alloys are listed in Table 14-6.

Advanced magnesium alloys include those with very low levels of impurities and those containing large amounts (>5%) of cerium and other rare earth elements. These alloys form a protective MgO film that improves corrosion resistance. Rapid solidification processing permits larger amounts of alloying elements to be dissolved in the magnesium, further improving corrosion resistance. Improvements in strength, particularly at high temperatures, can be obtained by introducing ceramic particles or fibers such as silicon carbide into the metal.

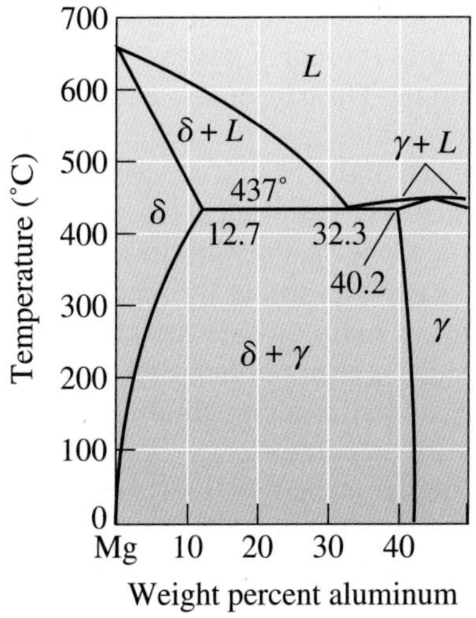

Figure 14-4 A portion of the magnesium-aluminum phase diagram. (*Based on* American Society for Metals Handbook, *8th ed., Vol. 8, Metallography, Structures and Phase Diagrams, ASM International, 1973.*)

Table 14-6 Properties of typical magnesium alloys

Alloy	Composition	Tensile Strength (psi)	Yield Strength (psi)	% Elongation
Pure Mg:				
Annealed		23,000	13,000	3–15
Cold-worked		26,000	17,000	2–10
Casting alloys:				
AM 100-T6	10% Al-0.1% Mn	40,000	22,000	1
AZ81A-T4	7.6% Al-0.7% Zn	40,000	12,000	15
ZK61A-T6	6% Zn-0.7% Zr	45,000	28,000	10
Wrought alloys:				
AZ80A-T5	8.5% Al-0.5% Zn	55,000	40,000	7
ZK40A-T5	4% Zn-0.45% Zr	40,000	37,000	4
HK31A-H24	3% Th-0.6% Zr	38,000	30,000	8

Beryllium is lighter than aluminum, with a density of 1.848 g/cm^3, yet it is stiffer than steel, with a modulus of elasticity of 42×10^6 psi (290 GPa). Beryllium alloys, which have yield strengths of 30,000 to 50,000 psi (200–350 MPa), have high specific strengths and maintain both strength and stiffness to high temperatures. Instrument grade beryllium is used in inertial guidance systems where the elastic deformation must be minimal; structural grades are used in aerospace applications; other applications take advantage of the transparency of beryllium to electromagnetic radiation. X-ray tubes, which must be sealed to the atmosphere, typically include a window of Be through which the x-rays easily are emitted. Unfortunately, beryllium is expensive (Table 14-1), brittle, reactive, and toxic *in powder form*. Its production is quite complicated, and hence, the applications of Be alloys are very limited. Beryllium oxide (BeO), which also is toxic in a powder form, is used to make high-thermal conductivity ceramics.

14-3 Copper Alloys

Copper occurs in nature as elemental copper and was extracted successfully from minerals long before iron was, since the relatively lower temperatures required for the copper extraction could be achieved more easily. Copper ores, as they are found in mines, contain not only copper but also other minerals and elements. One particularly common addition is sulfur. Copper is extracted from the ores through a process of crushing and separation into what is known as a copper concentrate. The copper concentrate is then converted to a liquid in a flash smelter. A flux combines with iron in the melt, thereby creating a slag that rises to the top of the containment vessel, and a flow of oxygen converts sulfur to sulfur dioxide gas that is subsequently exhausted. The liquid left at the bottom of the containment vessel after flash smelting is known as a copper matte. Copper matte still contains copper sulfide and iron sulfide. In a separate reactor, known as a copper converter, oxygen introduced to the matte converts more sulfur to sulfur dioxide gas and the copper sulfide to an impure copper called **blister copper**, which is then purified electrolytically. Other methods for copper extraction include leaching copper from low-sulfur ores with a weak acid, then electrolytically extracting the copper from the solution.

Copper-based alloys have higher densities than steels. Although the yield strengths of some alloys are high, their specific strengths are typically less than that of aluminum or magnesium alloys. These alloys have better resistance to fatigue, creep, and wear than the lightweight aluminum and magnesium alloys. Many of these alloys have excellent ductility, corrosion resistance, and electrical and thermal conductivities, and most can easily be joined or fabricated into useful shapes. Applications for copper-based alloys include electrical components (such as wire), pumps, valves, and plumbing parts, where these properties are advantageous.

Copper alloys also are unusual in that they may be selected to produce an appropriate decorative color. Pure copper is red, zinc additions produce a yellow color, and nickel produces a silver color. Copper can corrode easily, forming a basic copper sulfate [$CuSO_4 \cdot 3Cu(OH)_2$]. This is a green compound that is insoluble in water (but soluble in acids). This green patina provides an attractive finish for many applications. The Statue of Liberty is green because of the green patina of the oxidized copper skin that covers the steel skeleton within.

The wide variety of copper-based alloys takes advantage of all of the strengthening mechanisms that we have discussed. The effects of these strengthening mechanisms on the mechanical properties are summarized in Table 14-7.

Copper containing less than 0.1% impurities is used for electrical and microelectronics applications. Small amounts of cadmium, silver, and Al_2O_3 improve hardness

Table 14-7 Properties of typical copper alloys obtained by different strengthening mechanisms

Material	Tensile Strength (psi)	Yield Strength (psi)	% Elongation	Strengthening Mechanism
Pure Cu, annealed	30,300	4,800	60	None
Commercially pure Cu, annealed to coarse grain size	32,000	10,000	55	Solid Solution
Commercially pure Cu, annealed to fine grain size	34,000	11,000	55	Grain size
Commercially pure Cu, cold-worked 70%	57,000	53,000	4	Strain hardening
Annealed Cu-35% Zn	47,000	15,000	62	Solid solution
Annealed Cu-10% Sn	66,000	28,000	68	Solid solution
Cold-worked Cu-35% Zn	98,000	63,000	3	Solid solution + strain hardening
Age-hardened Cu-2% Be	190,000	175,000	4	Age hardening
Quenched and tempered Cu-Al	110,000	60,000	5	Martensitic reaction
Cast manganese bronze	71,000	28,000	30	Eutectoid reaction

without significantly impairing conductivity. The single-phase copper alloys are strengthened by cold working. Examples of this effect are shown in Table 14-7. Because copper is FCC, it has excellent ductility and a high strain-hardening coefficient.

Solid-Solution Strengthened Alloys A number of copper-based alloys contain large quantities of alloying elements, yet remain single phase. Important binary phase diagrams are shown in Figure 14-5. The copper-zinc, or **brass**, alloys with less than 40% Zn form single-phase solid solutions of zinc in copper. The mechanical properties—even elongation—increase as the zinc content increases. These alloys can

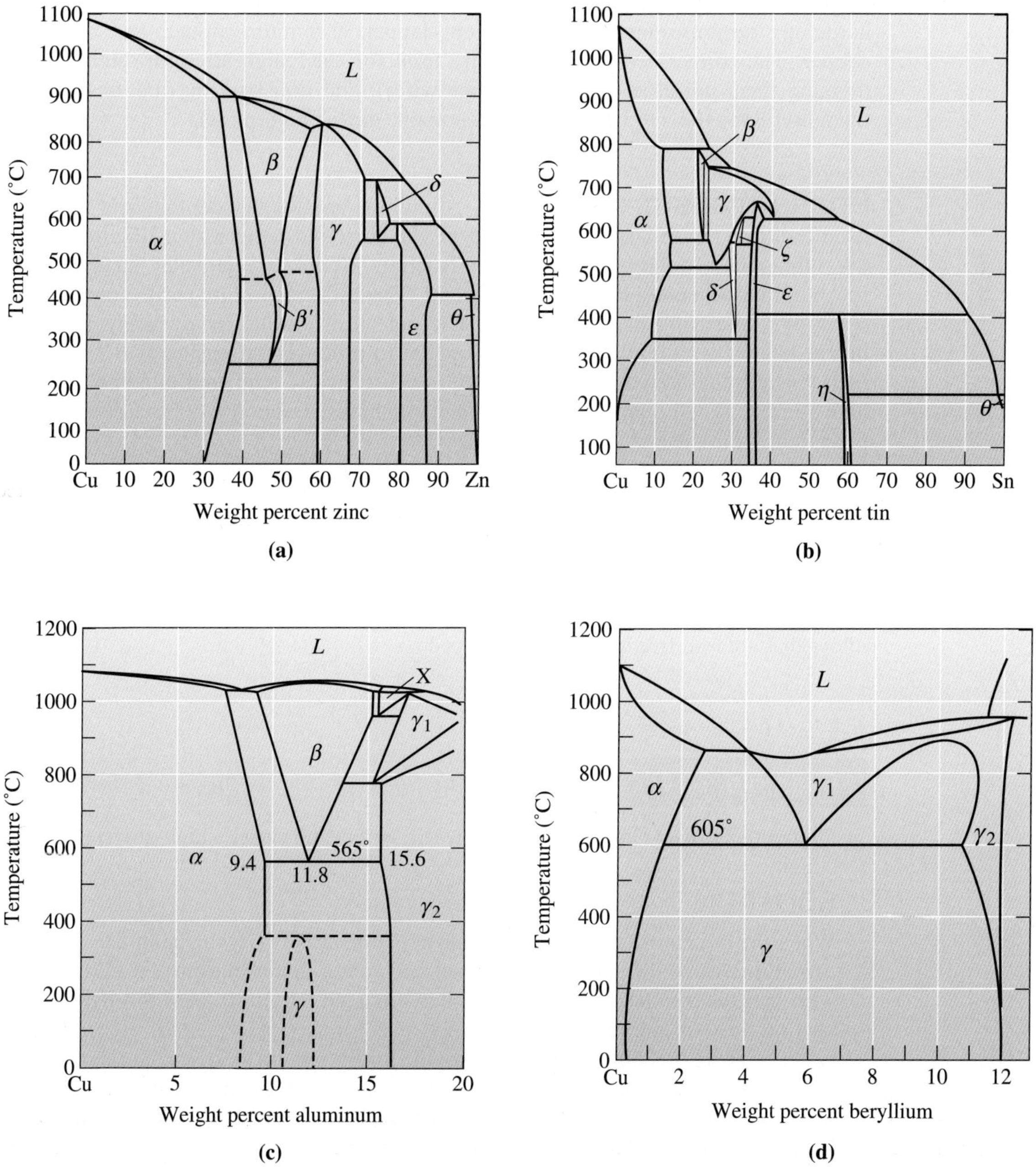

Figure 14-5 Binary phase diagrams for the (a) copper-zinc, (b) copper-tin, (c) copper-aluminum, and (d) copper-beryllium systems.

be cold formed into rather complicated yet corrosion-resistant components. **Bronzes** are generally considered alloys of copper containing tin and can certainly contain other elements. Manganese bronze is a particularly high-strength alloy containing manganese as well as zinc for solid-solution strengthening.

Tin bronzes, often called phosphor bronzes, may contain up to 10% Sn and remain single phase. The phase diagram predicts that the alloy will contain the Cu_3Sn (ε) compound; however, the kinetics of the reaction are so slow that the precipitate may not form.

Alloys containing less than about 9% Al or less than 3% Si are also single phase. These aluminum bronzes and silicon bronzes have good forming characteristics and are often selected for their good strength and excellent toughness.

Age-Hardenable Alloys A number of copper-based alloys display an age-hardening response, including zirconium-copper, chromium-copper, and beryllium-copper. The copper-beryllium alloys are used for their high strength, high stiffness (making them useful as springs and fine wires), and nonsparking qualities (making them useful for tools to be used near flammable gases and liquids).

Phase Transformations Aluminum bronzes that contain over 9% Al can form the β phase on heating above 565°C, the eutectoid temperature [Figure 14-5(c)]. On subsequent cooling, the eutectoid reaction produces a lamellar structure (like pearlite) that contains a brittle γ_2 compound. The low-temperature peritectoid reaction, $\alpha + \gamma_2 \rightarrow \gamma$, normally does not occur. The eutectoid product is relatively weak and brittle, but we can rapidly quench the β phase to produce martensite, or β', which has high strength and low ductility. When β' is subsequently tempered, a combination of high strength, good ductility, and excellent toughness is obtained as fine platelets of α precipitate from the β'.

Lead-Copper Alloys Virtually any of the wrought alloys may contain up to 4.5% Pb. The lead forms a monotectic reaction with copper and produces tiny lead spheres as the last liquid to solidify. The lead improves machining characteristics. Use of lead-copper alloys, however, has a major environmental impact and, consequently, new alloys that are lead free have been developed.

Example 14-2 *Design/Materials Selection for an Electrical Switch*

Design the contacts for a switch or relay that opens and closes a high-current electrical circuit.

SOLUTION

When the switch or relay opens and closes, contact between the conductive surfaces can cause wear and result in poor contact and arcing. A high hardness would minimize wear, but the contact materials must allow the high current to pass through the connection without overheating or arcing.

Therefore, our design must provide for both good electrical conductivity and good wear resistance. A relatively pure copper alloy dispersion strengthened with a hard phase that does not disturb the copper lattice would, perhaps, be ideal. In a $Cu\text{-}Al_2O_3$ alloy, hard ceramic-oxide particles provide wear resistance but do not interfere with the electrical conductivity of the copper matrix.

14-4 Nickel and Cobalt Alloys

Nickel and cobalt alloys are used for corrosion protection and for high-temperature resistance, taking advantage of their high melting points and high strengths. Nickel is FCC and has good formability; cobalt is an allotropic metal, with an FCC structure above 417°C and an HCP structure at lower temperatures. Special cobalt alloys are used for exceptional wear resistance and, because of resistance to human body fluids, for prosthetic devices. Typical alloys and their applications are listed in Table 14-8.

Rapidly solidified powders of nickel- and cobalt-based superalloys are used to make the rings that retain turbine blades, as well as turbine blades for aircraft engines. In Chapter 12, we discussed applications of shape-memory alloys based on Ni-Ti. Iron, nickel, and cobalt are ferromagnetic. Certain Fe-Ni and Fe-Co-based alloys form very good magnetic materials. A Ni-36% Fe alloy (Invar) displays practically no expansion during heating; this effect is exploited in producing bimetallic composite materials. Cobalt is used in WC-Co cutting tools.

Nickel and Monel

Nickel and its alloys have excellent corrosion resistance and forming characteristics. When copper is added to nickel, the maximum strength is obtained near 60% Ni. A number of alloys, called **Monels**, with approximately this composition are used for their strength and corrosion resistance in salt water and at elevated temperatures. Some of the Monels contain small amounts of aluminum and titanium. These alloys show an age-hardening response by the precipitation of γ', a coherent Ni_3Al or Ni_3Ti precipitate that nearly doubles the tensile strengths. The precipitates resist overaging at temperatures up to 425°C (Figure 14-6).

Superalloys

Superalloys are nickel, iron-nickel, and cobalt alloys that contain large amounts of alloying elements intended to produce a combination of high strength at elevated temperatures, resistance to creep at temperatures up to 1000°C, and resistance to corrosion. These excellent high-temperature properties are obtained even though the

Table 14-8 Compositions, properties, and applications for selected nickel and cobalt alloys

Material	Tensile Strength (psi)	Yield Strength (psi)	% Elongation	Strengthening Mechanism	Applications
Pure Ni (99.9% Ni)	50,000	16,000	45	Annealed	Corrosion resistance
	95,000	90,000	4	Cold-worked	Corrosion resistance
Ni-Cu alloys:					
Monel 400 (Ni-31.5% Cu)	78,000	39,000	37	Annealed	Valves, pumps, heat exchangers
Monel K-500 (Ni-29.5% Cu-2.7% Al-0.6% Ti)	150,000	110,000	30	Aged	Shafts, springs, impellers
Ni superalloys:					
Inconel 600 (Ni-15.5% Cr-8% Fe)	90,000	29,000	49	Carbides	Heat-treatment equipment
Hastelloy B-2 (Ni-28% Mo)	130,000	60,000	61	Carbides	Corrosion resistance
DS-Ni (Ni-2% ThO_2)	71,000	48,000	14	Dispersion	Gas turbines
Fe-Ni superalloys:					
Incoloy 800 (Ni-46% Fe-21% Cr)	89,000	41,000	37	Carbides	Heat exchangers
Co superalloys:					
Stellite 6B (60% Co-30% Cr-4.5% W)	177,000	103,000	4	Carbides	Abrasive wear resistance

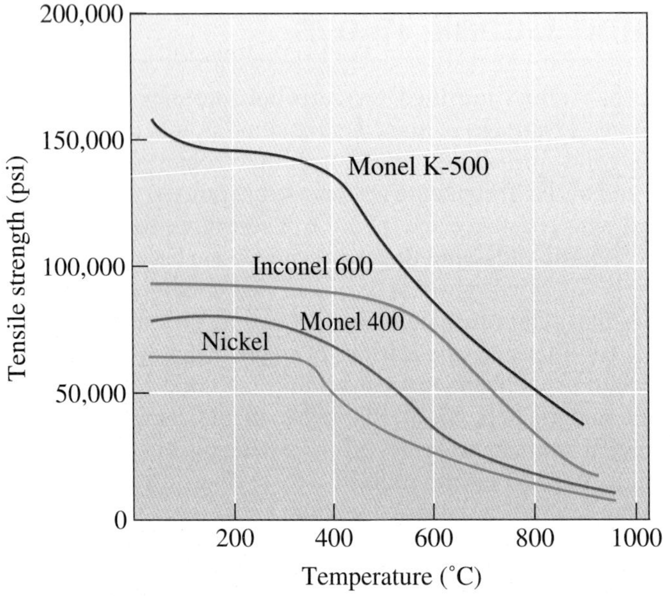

Figure 14-6 The effect of temperature on the tensile strength of several nickel-based alloys.

melting temperatures of the alloys are about the same as that for steels. Typical applications include vanes and blades for turbine and jet engines, heat exchangers, chemical reaction vessel components, and heat-treating equipment.

To obtain high strength and creep resistance, the alloying elements must produce a strong, stable microstructure at high temperatures. Solid-solution strengthening, dispersion strengthening, and precipitation hardening are generally employed as discussed below. Inconel™ 713C, for example, is a typical superalloy used in turbochargers and comprises Ni, Nb, Cr, Mo, Ti, Al, and trace amounts of C, B, and Zr.

Large additions of chromium, molybdenum, and tungsten and smaller additions of tantalum, zirconium, niobium, and boron provide solid-solution strengthening to a superalloy. The effects of solid-solution strengthening are stable and, consequently, make the alloy resistant to creep, particularly when large atoms such as molybdenum and tungsten (which diffuse slowly) are used.

All superalloys contain a small amount of carbon that, by combining with other alloying elements, produces a network of fine, stable carbide particles. The carbide network interferes with dislocation movement and prevents grain boundary sliding. The carbides include TiC, BC, ZrC, TaC, Cr_7C_3, $Cr_{23}C_6$, Mo_6C, and W_6C, although often they are more complex and contain several alloying elements. Stellite 6B, a cobalt-based superalloy, has unusually good wear resistance at high temperatures due to these carbides.

Many of the nickel and nickel-iron superalloys that contain aluminum and titanium form the coherent precipitate γ'(Ni_3Al or Ni_3Ti) during aging. The γ' particles (Figure 14-7) have a crystal structure and lattice parameter similar to that of the nickel matrix; this similarity leads to a low surface energy and minimizes overaging of the alloys, providing good strength and creep resistance even at high temperatures.

By varying the aging temperature, precipitates of various sizes can be produced. Small precipitates, formed at low aging temperatures, can grow between the larger precipitates produced at higher temperatures, therefore increasing the volume percentage of the γ' and further increasing the strength [Figure 14-7(b)].

The high-temperature use of superalloys can be improved when a ceramic coating is used. One method for doing this is to first coat the superalloy with a metallic bond coat composed of a complex NiCoCrAlY alloy and then apply an outer thermal barrier coating of a stabilized ZrO_2-ceramic (Chapter 5). The coating helps reduce oxidation of the superalloy and permits jet engines to operate at higher temperatures and with greater efficiency.

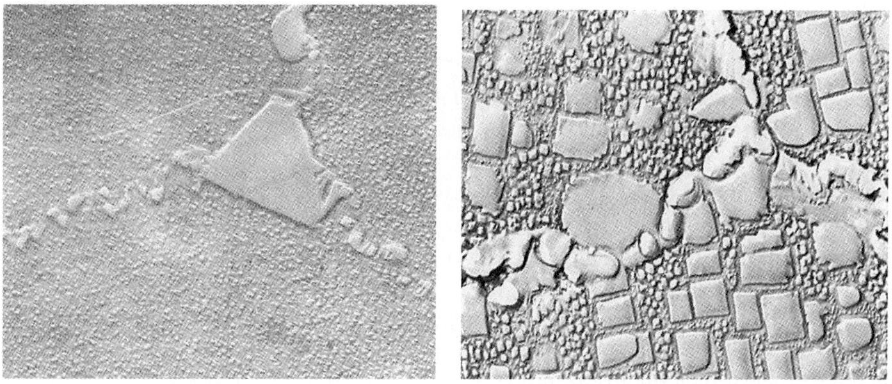

(a) (b)

Figure 14-7 (a) Microstructure of a superalloy, with carbides at the grain boundaries and γ' precipitates in the matrix ($\times$ 15,000). (b) Microstructure of a superalloy aged at two temperatures, producing both large and small cubical γ' precipitates ($\times$ 10,000). [*From* ASM Handbook, *Vol. 9, Metallography and Microstructure (1985), ASM International, Materials Park, OH 44073-0002. Reprinted with permission of ASM International. All rights reserved. www.asminternational.org.*]

Example 14-3 *Design/Materials Selection for a High-Performance Jet Engine Turbine Blade*

Design a nickel-based superalloy for producing turbine blades for a gas turbine aircraft engine that will have a particularly long creep-rupture time at temperatures approaching 1100°C.

SOLUTION

First, we need a very stable microstructure. Addition of aluminum or titanium permits the precipitation of up to 60 vol% of the γ' phase during heat treatment and may permit the alloy to operate at temperatures approaching 0.85 times the absolute melting temperature. Addition of carbon and alloying elements such as tantalum and hafnium permits the precipitation of alloy carbides that prevent grain boundaries from sliding at high temperatures. Other alloying elements, including molybdenum and tungsten, provide solid-solution strengthening.

Second, we might produce a directionally solidified or even single-crystal turbine blade (Chapter 9). In directional solidification, only columnar grains form during freezing, eliminating transverse grain boundaries that might nucleate cracks. In a single crystal, no grain boundaries are present. We might use the investment casting process, being sure to pass the liquid superalloy through a filter to trap any tiny inclusions before the metal enters the ceramic investment mold.

We would then heat treat the casting to ensure that the carbides and γ' precipitate have the correct size and distribution. Multiple aging temperatures might be used to ensure that the largest possible volume percent γ' is formed.

Finally, the blade might contain small cooling channels along its length. Air for combustion in the engine can pass through these channels, providing active cooling to the blade, before reacting with fuel in the combustion chamber.

14-5 Titanium Alloys

Titanium is produced from TiO_2 by the Kroll process. The TiO_2 is converted to $TiCl_4$ (titanium tetrachloride, also informally known as *tickle!*), which is subsequently reduced to titanium metal by sodium or magnesium. The resultant titanium "sponge" is then consolidated, alloyed as necessary, and processed using vacuum arc melting. Titanium provides excellent corrosion resistance, high specific strength, and good high-temperature properties. Strengths up to 200,000 psi (1400 MPa), coupled with a density of 4.505 g/cm^3, provide excellent mechanical properties. An adherent, protective TiO_2 film provides excellent resistance to corrosion and contamination below 535°C. Above 535°C, the oxide film breaks down, and small atoms such as carbon, oxygen, nitrogen, and hydrogen embrittle the titanium.

Titanium's excellent corrosion resistance allows applications in chemical processing equipment, marine components, and biomedical implants such as hip prostheses. Titanium is an important aerospace material, finding applications as airframe and jet engine components. When it is combined with niobium, a superconductive intermetallic compound is formed; when it is combined with nickel, the resulting alloy displays the shape-memory effect; when it is combined with aluminum, a new class of intermetallic alloys is produced. Titanium alloys are used for sports equipment such as the heads of certain golf clubs.

Titanium is allotropic, with the HCP crystal structure (α) at low temperatures and a BCC structure (β) above 882°C. Alloying elements provide solid-solution strengthening and change the allotropic transformation temperature. The alloying elements can be divided into four groups (Figure 14-8). Additions such as tin and zirconium provide solid-solution strengthening without affecting the transformation temperature. Aluminum, oxygen, hydrogen, and other α-stabilizing elements increase the temperature at which α transforms to β. Beta stabilizers such as vanadium, tantalum, molybdenum, and niobium lower the transformation temperature, even causing β to be stable at room temperature. Finally, manganese, chromium, and iron produce a eutectoid reaction, reducing the temperature at which the $\alpha - \beta$ transformation occurs and producing a two-phase structure at room temperature. Several categories of titanium and its alloys are listed in Table 14-9.

Commercially Pure Titanium Unalloyed titanium is used for its superior corrosion resistance. Impurities, such as oxygen, increase the strength of the titanium (Figure 14-9) but reduce corrosion resistance. Applications include heat exchangers, piping, reactors, pumps, and valves for the chemical and petrochemical industries.

Alpha Titanium Alloys The most common of the all-α alloys contains 5% Al and 2.5% Sn, which provide solid-solution strengthening to the HCP α. The α alloys are annealed at high temperatures in the β region. Rapid cooling gives an acicular, or Widmanstätten, α-grain structure (Figure 14-10) that provides good resistance to fatigue. Furnace cooling gives a more platelike α structure that provides better creep resistance.

Beta Titanium Alloys Although large additions of vanadium or molybdenum produce an entirely β structure at room temperature, none of the β alloys actually are alloyed to that extent. Instead, they are rich in β stabilizers, so that rapid cooling produces a metastable structure composed of all β. Strengthening is obtained both from the large amount of solid-solution strengthening alloying elements and by aging the metastable β structure to permit α to precipitate. Applications include high-strength fasteners, beams, and other fittings for aerospace applications.

Alpha-Beta Titanium Alloys With proper balancing of the α and β stabilizers, a mixture of α and β is produced at room temperature. Ti – 6% Al – 4% V, an example of this approach, is by far the most common of all the titanium alloys. Because the alloys contain two phases, heat treatments can be used to control the microstructure and properties.

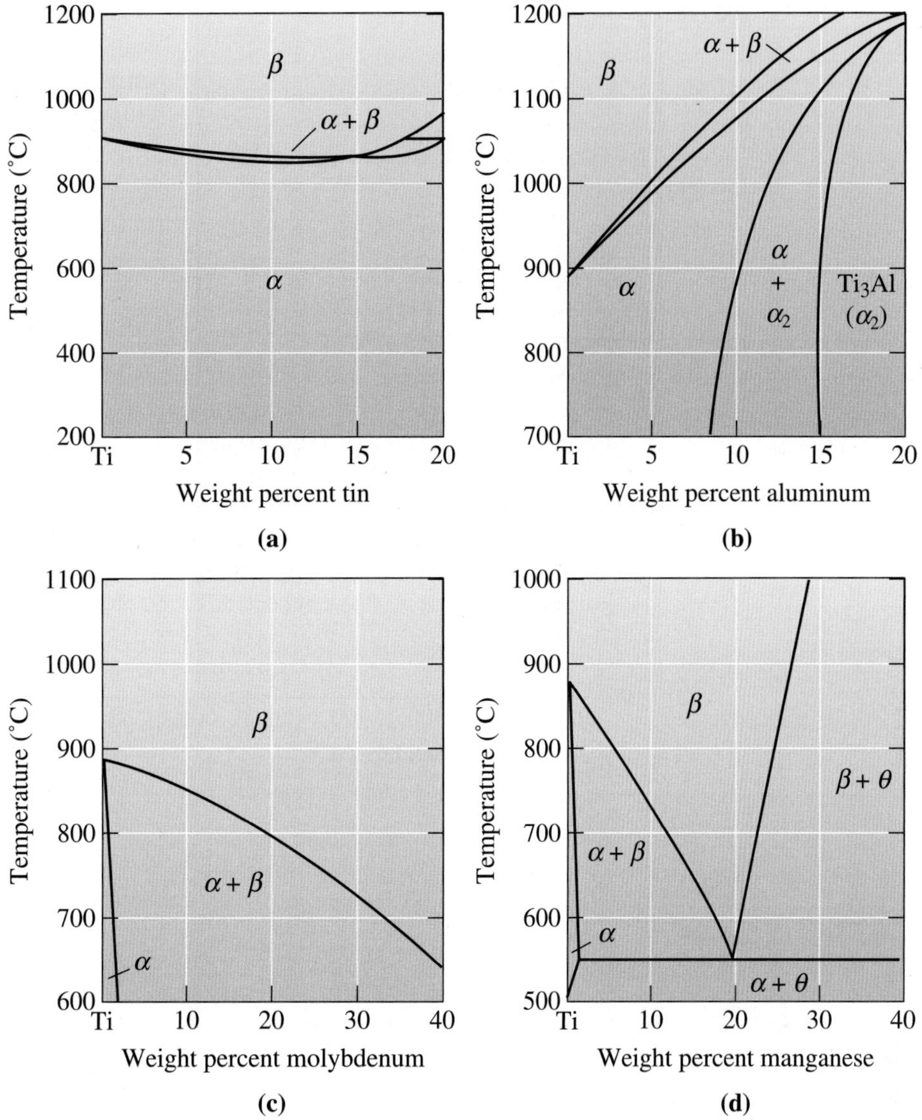

Figure 14-8 Portions of the phase diagrams for (a) titanium-tin, (b) titanium-aluminum, (c) titanium-molybdenum, and (d) titanium-manganese.

Table 14-9 Properties of selected titanium alloys

Material	Tensile Strength (psi)	Yield Strength (psi)	% Elongation
Commercially pure Ti:			
99.5% Ti	35,000	25,000	24
99.0% Ti	80,000	70,000	15
Alpha Ti alloys:			
5% Al-2.5% Sn	125,000	113,000	15
Beta Ti alloys:			
13% V-11% Cr-3% Al	187,000	176,000	5
Alpha-beta Ti alloys:			
6% Al-4% V	150,000	140,000	8

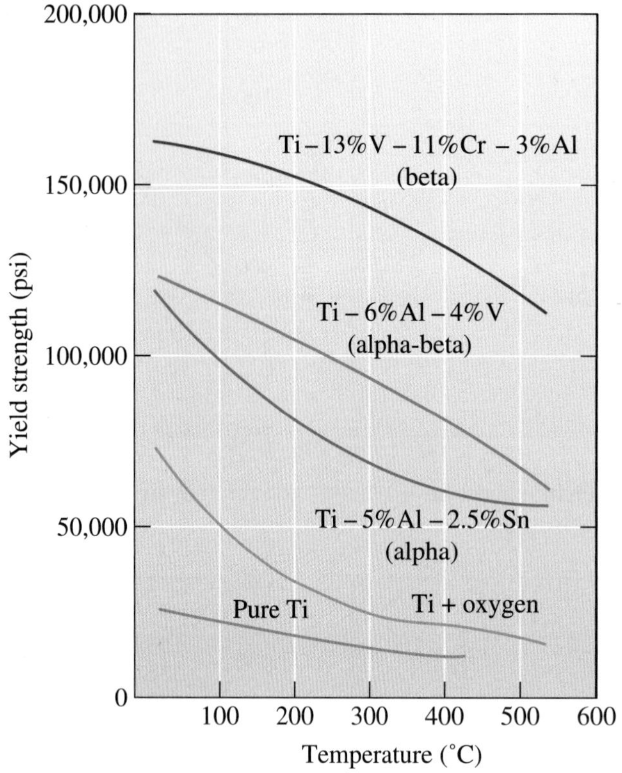

Figure 14-9 The effect of temperature on the yield strength of selected titanium alloys.

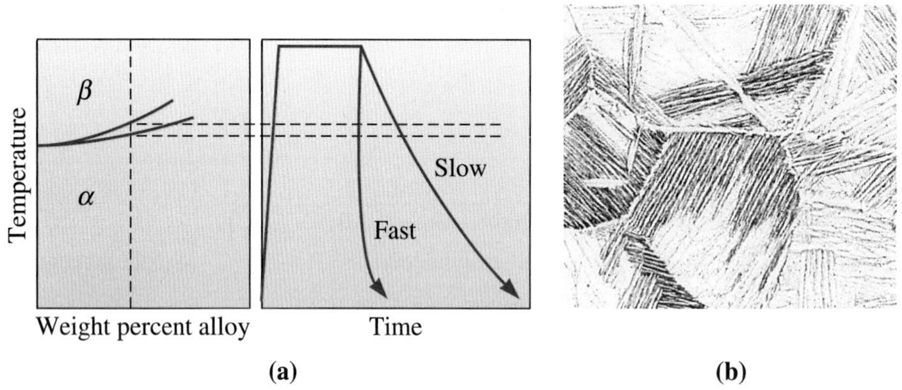

Figure 14-10 (a) Annealing and (b) microstructure of rapidly cooled α titanium (× 100). Both the grain boundary precipitate and the Widmanstätten plates are α. [*From* ASM Handbook, *Vol. 7, (1972), ASM International, Materials Park, OH 44073-0002. Reprinted with permission of ASM International. All rights reserved. www.asminternational.org.*]

Annealing provides a combination of high ductility, uniform properties, and good strength. The alloy is heated just below the β-transus temperature, permitting a small amount of α to remain and prevent grain growth [Figure 14-11(a)]. Slow cooling causes equiaxed α grains to form [Figure 14-11(b)]; the equiaxed structure provides good ductility and formability while making it difficult for fatigue cracks to nucleate. Faster cooling, particularly from above the α-β transus temperature, produces an acicular—or "basketweave"—α phase [Figure 14-11(c)]. Although fatigue cracks may nucleate more easily in this structure, cracks must follow a tortuous path along the boundaries between α and β. This condition results in a low-fatigue crack growth rate, good fracture toughness, and good resistance to creep.

Two possible microstructures can be produced when the β phase is quenched from a high temperature. The phase diagram in Figure 14-12 includes a dashed martensite start line, which provides the basis for a quench and temper treatment.

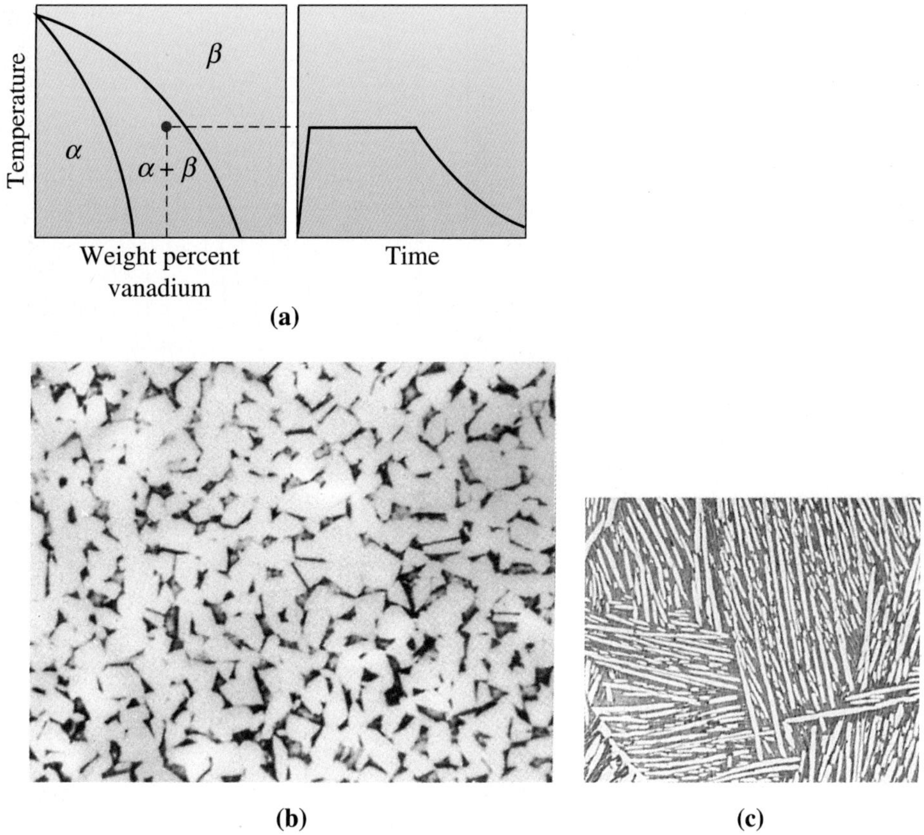

Figure 14-11 Annealing of an α-β titanium alloy. (a) Annealing is done just below the α-β transformation temperature. (b) Slow cooling gives equiaxed α grains ($\times$ 250), and (c) rapid cooling yields acicular α grains ($\times$ 2500). [*From* ASM Handbook, *Vol. 7, (1972), ASM International, Materials Park, OH 44073-0002. Reprinted with permission of ASM International. All rights reserved. www.asminternational.org.*]

The β transforms to titanium martensite (α') in an alloy that crosses the M_s line on cooling. The titanium martensite is a relatively soft, supersaturated phase. When α' is reheated, tempering occurs by the precipitation of β from the supersaturated α':

$$\alpha' \rightarrow \alpha + \beta \text{ precipitates} \tag{14-2}$$

Fine β precipitates initially increase the strength compared with the α', opposite to what is found when a steel martensite is tempered; however, softening occurs when tempering is done at too high a temperature.

More highly alloyed α-β compositions are age hardened. When the β in these alloys is quenched, β_{ss}, which is supersaturated in titanium, remains. When β_{ss} is aged, α precipitates in a Widmanstätten structure, (Figure 14-12):

$$\beta_{ss} \rightarrow \beta + \alpha \text{ precipitates} \tag{14-3}$$

The formation of this structure leads to improved strength and fracture toughness. Components for airframes, rockets, jet engines, and landing gear are typical applications for the heat-treated α-β alloys. Some alloys, including the Ti – 6% Al – 4% V alloy, are superplastic and can be deformed as much as 1000%. This alloy is also used for making implants for human bodies. Titanium alloys are considered **biocompatible** (i.e., they are not rejected by the body). By developing porous coatings of bone-like ceramic compositions known as hydroxyapatite, it is possible to make titanium implants **bioactive** (i.e., the natural bone can grow into the hydroxyapatite coating). The following three examples illustrate applications of titanium alloys.

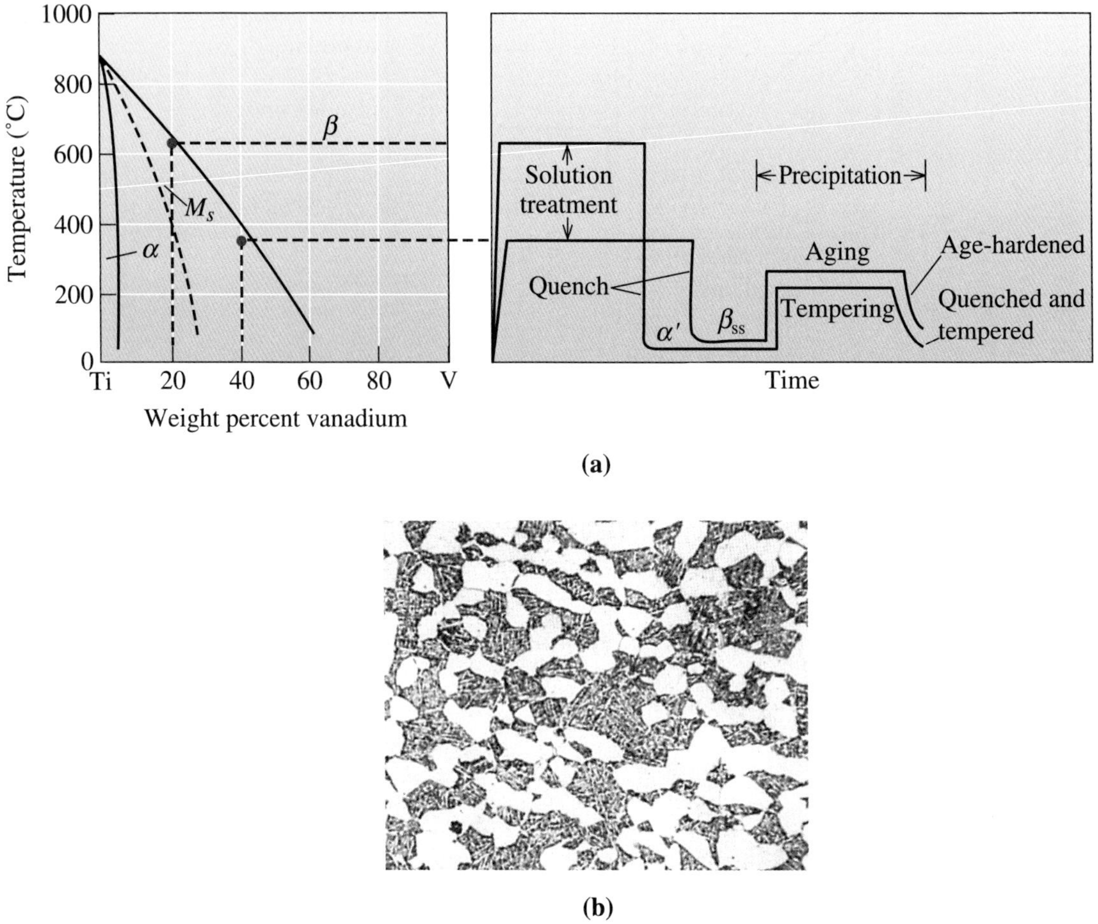

Figure 14-12 (a) Heat treatment and (b) microstructure of the α-β titanium alloys. The structure contains primary α (large white grains) and a dark β matrix with needles of α formed during aging (× 250). [*From* ASM Handbook, *Vol. 7, (1972), ASM International, Materials Park, OH 44073-0002. Reprinted with permission of ASM International. All rights reserved. www.asminternational.org.*]

Example 14-4 *Design of a Heat Exchanger*

Design a 5-ft-diameter, 30-ft-long heat exchanger for the petrochemical industry (Figure 14-13).

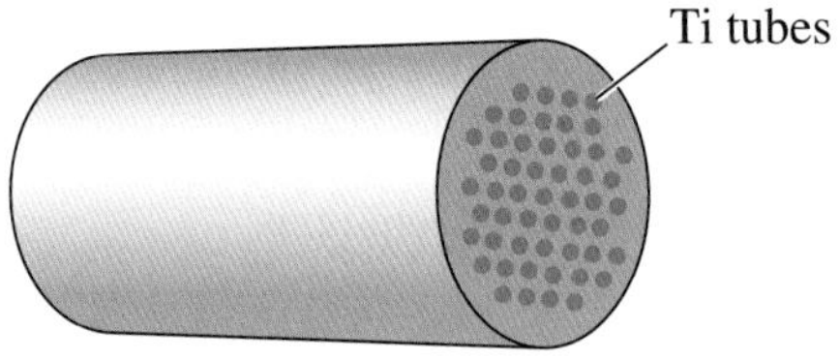

Figure 14-13 Sketch of a heat exchanger using titanium tubes (for Example 14-4).

SOLUTION

The heat exchanger must meet several design criteria. It must have good corrosion resistance to handle aggressive products of the chemical refinery; it must operate at relatively high temperatures; it must be easily formed into the sheet and tubes from which the heat exchanger will be fabricated; and it must have good weldability for joining the tubes to the body of the heat exchanger.

Provided that the maximum operating temperature is below 535°C so that the oxide film is stable, titanium might be a good choice to provide corrosion resistance at elevated temperatures. A commercially pure titanium provides the best corrosion resistance.

Pure titanium also provides superior forming and welding characteristics and would, therefore, be our most logical selection. If pure titanium does not provide sufficient strength, an alternative is an α titanium alloy, still providing good corrosion resistance, forming characteristics, and weldability but also somewhat improved strength.

Example 14-5 *Design of a Connecting Rod*

Design a high-performance connecting rod for the engine of a race car (Figure 14-14).

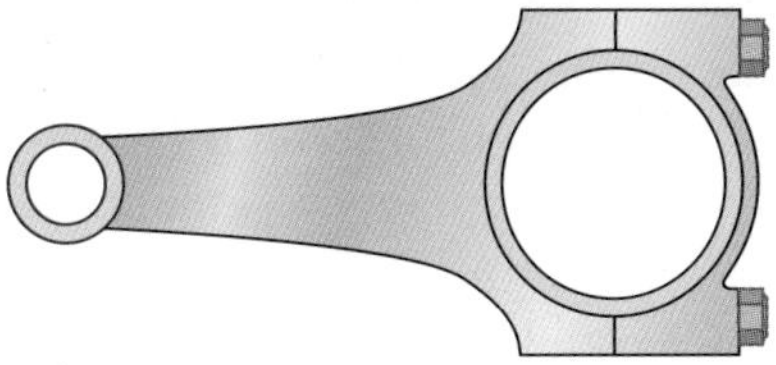

Figure 14-14 Sketch of connecting rod (for Example 14-5).

SOLUTION

A high-performance racing engine requires materials that can operate at high temperatures and stresses while minimizing the weight of the engine. In normal automobiles, the connecting rods are often a forged steel or a malleable cast iron. We might be able to save considerable weight by replacing these parts with titanium.

To achieve high strengths, we might consider an α-β titanium alloy. Because of its availability, the Ti – 6% Al – 4% V alloy is a good choice. The alloy is heated to about 1065°C, which is in the all-β portion of the phase diagram. On quenching, a titanium martensite forms; subsequent tempering produces a microstructure containing β precipitates in an α matrix.

When the heat treatment is performed in the all-β region, the tempered martensite has an acicular structure, which reduces the rate of growth of any fatigue cracks that might develop.

Example 14-6 *Materials for Hip Prosthesis*

What type of a material would you choose for an implant to be used for a total hip replacement implant?

SOLUTION

A hip prosthesis is intended to replace part of the worn out or damaged femur bone. The implant has a metal head and fits down the cavity of the femur. We need to consider the following factors: biocompatibility, corrosion resistance, high fracture toughness, excellent fatigue life (so that implants last for many years since it is difficult to do the surgery as patients get older), and wear resistance. We also need to consider the stiffness. If the alloy chosen is too stiff compared to the bone, most of the load will be carried by the implant. This leads to weakening of the remaining bone and, in turn, can make the implant loose. Thus, we need a material that has a high tensile strength, corrosion resistance, biocompatibility, and fracture toughness. These requirements suggest 316 stainless steel or Ti – 6% Al – 4% V. Neither of these materials are ferromagnetic, and both are opaque to x-rays. This is good for magnetic resonance and x-ray imaging. Titanium alloys are not very hard and can wear out. Stainless steels are harder, but they are much stiffer than bone. Titanium is biocompatible and would be a better choice. Perhaps a composite material in which the stem is made from a Ti – 6% Al – 4% V alloy and the head is made from a wear-resistant, corrosion-resistant, and relatively tough ceramic, such as alumina, may be an answer. The inside of the socket could be made from an ultra-high density (ultra-high molecular weight) polyethylene that has a very low friction coefficient. The surface of the implant could be made porous so as to encourage the bone to grow. Another option is to coat the implant with a material like porous hydroxyapatite to encourage bone growth.

14-6 Refractory and Precious Metals

The **refractory metals**, which include tungsten, molybdenum, tantalum, and niobium (or columbium), have exceptionally high melting temperatures (above 1925°C) and, consequently, have the potential for high-temperature service. Applications include filaments for lightbulbs, rocket nozzles, nuclear power generators, tantalum- and niobium-based electronic capacitors, and chemical processing equipment. These metals, however, have a high density, limiting their specific strengths (Table 14-10).

Oxidation The refractory metals begin to oxidize between 200°C and 425°C and are rapidly contaminated or embrittled. Consequently, special precautions are required during casting, hot working, welding, or powder metallurgy. The metals must also be protected during service at elevated temperatures. For example, the tungsten filament in a lightbulb is protected by a vacuum.

For some applications, the metals may be coated with a silicide or aluminide. The coating must (a) have a high melting temperature, (b) be compatible with the refractory metal, (c) provide a diffusion barrier to prevent contaminants from reaching the underlying metal, and (d) have a coefficient of thermal expansion similar to that of the refractory metal. Coatings are available that protect the metal to about 1650°C. In some applications, such as capacitors for cellular phones, the formation of oxides is useful since we want to make use of the oxide as a nonconducting material.

Forming Characteristics The refractory metals, which have the BCC crystal structure, display a ductile-to-brittle transition temperature. Because the transition temperatures for niobium and tantalum are below room temperature, these two metals can be readily formed. Annealed molybdenum and tungsten, however, normally have a transition temperature above room temperature, causing them to be brittle. Fortunately, if these metals are hot worked, the forming characteristics are improved.

Table 14-10 Properties of some refractory metals

			T = 1000°C		
Metal	Melting Temperature (°C)	Density (g/cm^3)	Tensile Strength (psi)	Yield Strength (psi)	Transition Temperature (°C)
Nb	2468	8.57	17, 000	8,000	−140
Mo	2610	10.22	50,000	30,000	30
Ta	2996	16.6	27,000	24,000	−270
W	3410	19.25	66,000	15,000	300

Alloys Large increases in both room-temperature and high-temperature mechanical properties are obtained by alloying. Tungsten alloyed with hafnium, rhenium, and carbon can operate at temperatures up to 2100°C. These alloys typically are solid-solution strengthened; in fact, tungsten and molybdenum form a complete series of solid solutions, much like copper and nickel. Some alloys, such as W-2% ThO_2, are dispersion strengthened by oxide particles during their manufacturing by powder metallurgy processes. Welding tips are commonly made from W-2% ThO_2. Composite materials, such as niobium reinforced with tungsten fibers, may also improve high-temperature properties.

Precious Metals These include gold, silver, palladium, platinum, and rhodium. As their name suggests, they are expensive; all are used for jewelry. From an engineering viewpoint, these materials resist corrosion and are very good conductors of electricity. As a result, alloys of these materials are often used as electrodes for devices. These electrodes are formed using thin-film deposition (e.g., sputtering or electroplating) or screen printing of metal powder dispersions/pastes. Nanoparticles of Pt/Rh/Pd (loaded onto a ceramic support) are also used as catalysts in automobiles. These metals facilitate the oxidation of CO to CO_2 and NO_x to N_2 and O_2. They are also used as catalysts in petroleum refining.

Summary

- The "light metals" include low-density alloys based on aluminum, magnesium, and beryllium. Aluminum alloys have a high specific strength due to their low density and, as a result, find many aerospace applications. The excellent corrosion resistance and electrical conductivity of aluminum also provide for a vast number of applications. Aluminum and magnesium are limited to use at low temperatures because of the degradation of their mechanical properties as a result of overaging or recrystallization. Copper alloys (brasses and bronzes) are used in many structural and other applications. Titanium alloys have intermediate densities and temperature resistance, along with excellent corrosion resistance, leading to applications in aerospace, chemical processing, and biomedical devices.
- Nickel and cobalt alloys, including superalloys, provide good properties at even higher temperatures. Combined with their good corrosion resistance, these alloys find many applications in aircraft engines and chemical processing equipment.

Glossary

Bioactive A material that is not rejected by the human body and eventually becomes part of the body (e.g., hydroxyapatite).

Biocompatible A material that is not rejected by the human body.

Blister copper An impure form of copper obtained during the copper refining process.

Brass A group of copper-based alloys, normally containing zinc as the major alloying element.

Bronze Generally, copper alloys containing tin, but can contain other elements.

Castability The ease with which a metal can be poured into a mold to make a casting without producing defects or requiring unusual or expensive techniques to prevent casting problems.

Fluidity The ability of liquid metal to fill a mold cavity without prematurely freezing.

Monel The copper-nickel alloy, containing approximately 60% Ni, that gives the maximum strength in the binary alloy system.

Nonferrous alloy An alloy based on some metal other than iron.

Refractory metals Metals having a melting temperature above 1925°C.

Specific strength The ratio of strength to density. Also called the strength-to-weight ratio.

Superalloys A group of nickel, iron-nickel, and cobalt-based alloys that have exceptional heat resistance, creep resistance, and corrosion resistance.

Temper designation A shorthand notation using letters and numbers to describe the processing of an alloy. H tempers refer to cold-worked alloys; T tempers refer to age-hardening treatments.

Wrought alloys Alloys that are shaped by a deformation process.

Problems

14-1 In some cases, we may be more interested in the cost of a material per unit volume than in the cost per unit weight. Rework Table 14-1 to show the cost in terms of $\$/cm^3$. Does this change/alter the relationship between the different materials?

14-2 Determine the specific strengths of the following metals and alloys (use the densities of the major metal component as an approximation of the alloy density where required):

Alloy/Metal	Tensile Strength (MPa)
1100-H18	165
5182-O	290
2024-T4	469
2090-T6	552
201-T6	483
ZK40A-T5	276
Age hardened Cu-2% Be	1310
Alpha Ti alloy	862
W	455
Ta	186

Section 14-1 Aluminum Alloys

14-3 How much would 50 kg of the hypothetical alloy 5074-O cost in raw material prices using the estimates given in Table 14-1?

14-4 Structural steels have traditionally been used in shipbuilding; however, with increasing fuel costs, it is desirable to find alternative lower weight materials. Some of the key properties required for shipbuilding materials are high yield strength, high corrosion resistance, and low cost. Discuss the benefits and disadvantages of aluminum alloys as a replacement for structural steels in ships.

14-5 Assuming that the density remains unchanged, compare the specific strength of the 2090-T6 aluminum alloy to that of a die cast 443-F aluminum alloy. If you considered the actual density, do you think the difference between the specific strengths would increase or decrease? Explain.

14-6 Explain why aluminum alloys containing more than about 15% Mg are not used.

14-7 Would you expect a 2024-T9 aluminum alloy to be stronger or weaker than a 2024-T6 alloy? Explain.

14-8 Estimate the tensile strength expected for the following aluminum alloys:
(a) 1100-H14;
(b) 5182-H12; and
(c) 3004-H16.

14-9 A 1-cm-diameter steel cable with a yield strength of 480 MPa needs to be replaced to reduce the overall weight of the cable. Which of the following aluminum alloys could be a potential replacement?
(a) 3004-H18 (S_y = 248 MPa);
(b) 1100-H18 (S_y = 151 MPa);
(c) 4043-H18 (S_y = 269 MPa); and
(d) 5182-O (S_y = 131 MPa).
The density of the steel used in the cable is 7.87 g/cm^3, and assume that the density of all the aluminum alloys listed above is 2.7 g/cm^3.

14-10 Suppose, by rapid solidification from the liquid state, that a supersaturated Al-7% Li alloy can be produced and subsequently aged. Compare the amount of β that will form in this alloy with that formed in a 2090 alloy. See Figure 11-28.

14-11 Determine the amount of Mg_2Al_3 (β) expected to form in a 5182-O aluminum alloy (see Figure 14-2).

14-12 A 5182-O aluminum alloy part that had been exposed to salt water showed severe corrosion along the grain boundaries. Explain this observation based on the expected phases at room temperature in this alloy. (See Figure 14-2.)

Section 14-2 Magnesium and Beryllium Alloys

14-13 From the data in Table 14-6, estimate the ratio by which the yield strength of magnesium can be increased by alloying and heat treatment and compare with that of aluminum alloys.

14-14 Suppose a 24-in.-long round bar is to support a load of 400 lb without any permanent deformation. Calculate the

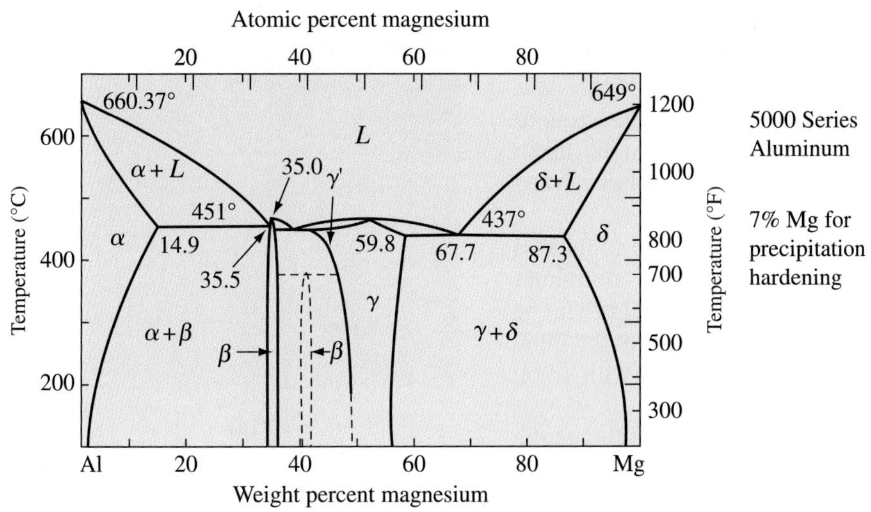

Figure 14-2 The aluminum-magnesium phase diagram. (Repeated for Problems 14-11 and 14-12.)

minimum diameter of the bar if it is made from
(a) AZ80A-T5 magnesium alloy and
(b) 6061-T6 aluminum alloy.
Calculate the weight of the bar and the approximate cost (based on pure Mg and Al) in each case.

14-15 A 10-m rod, 0.5 cm in diameter, must elongate no more than 2 mm under load. Determine the maximum force that can be applied if the rod is made from
(a) aluminium;
(b) magnesium; and
(c) beryllium.

14-16 A new process is proposed to harden an Mg-Al alloy. Part of the patent filing suggests that a 35 wt% Al liquid be solidified at 250°C. Is this realistic or should this application be thrown out as physically impossible?

Section 14-3 Copper Alloys

14-17 (a) Explain how pure copper is made. (b) What are some of the important properties of copper? (c) What is brass? (d) What is bronze? (e) Why is the Statue of Liberty green?

14-18 We say that copper can contain up to 40% Zn or 9% Al and still be single phase. How do we explain this statement in view of the phase diagram for the Cu-Zn system? [See Figure 14-5(a).]

14-19 Compare the percent increase in the yield strength of commercially pure annealed aluminum, magnesium, and copper by strain hardening. Explain the differences observed.

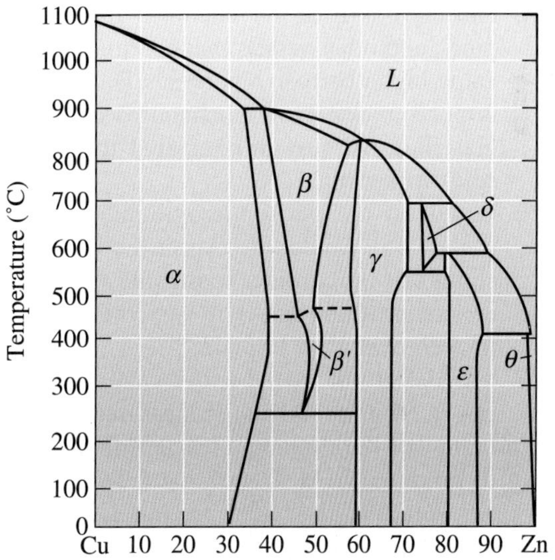

(a)

Figure 14-5 (a) Binary phase diagram for the copper-zinc system. (Repeated for Problem 14-18.)

14-20 What temperature ranges correspond to the phases of Cu-Sn alloys?
(a) η at 60 wt% Sn;
(b) ζ at 34 wt% Sn;
(c) γ at 30 wt% Sn; and
(d) α at 0 wt% Sn.

14-21 We would like to produce a quenched and tempered aluminum bronze containing 13% Al. Recommend a heat treatment, including appropriate temperatures. Calculate the amount of each phase after each step of the treatment.

14-22 A number of casting alloys have very high lead contents; however, the Pb content in wrought alloys is comparatively low. Why isn't more lead added to the wrought alloys? What precautions must be taken when a leaded wrought alloy is hot worked or heat treated?

14-23 Would you expect the fracture toughness of quenched and tempered aluminum bronze to be high or low? Would there be a difference in the resistance of the alloy to crack nucleation compared with crack growth? Explain.

14-24 Based on the phase diagrams, estimate the solubilities of Ni, Zn, Al, Sn, and Be in copper at room temperature. Are these solubilities expected in view of Hume-Rothery's conditions for solid solubility? Explain.

Section 14-4 Nickel and Cobalt Alloys

14-25 Hastelloy is used for chemical reactors, pipes, and other vessels that contain high-value, highly corrosive materials. Hastelloy is mostly nickel. How much more load can Hastelloy B-2 sustain compared to pure, cold-worked nickel?

14-26 Based on the photomicrograph in Figure 14-7(a), would you expect the γ' precipitate or the carbides to provide a greater strengthening effect in superalloys at low temperatures? Explain.

14-27 The density of Ni_3Al is 7.5 g/cm^3. Suppose a Ni-5 wt% Al alloy is heat treated so that all of the aluminum reacts with nickel to produce Ni_3Al. Determine the volume percentage of the Ni_3Al precipitate in the nickel matrix.

Section 14-5 Titanium Alloys

14-28 When steel is joined using arc welding, only the liquid-fusion zone must be protected by a gas or flux. When titanium is welded, both the front and back sides of the welded metal must be protected. Why must these extra precautions be taken when joining titanium?

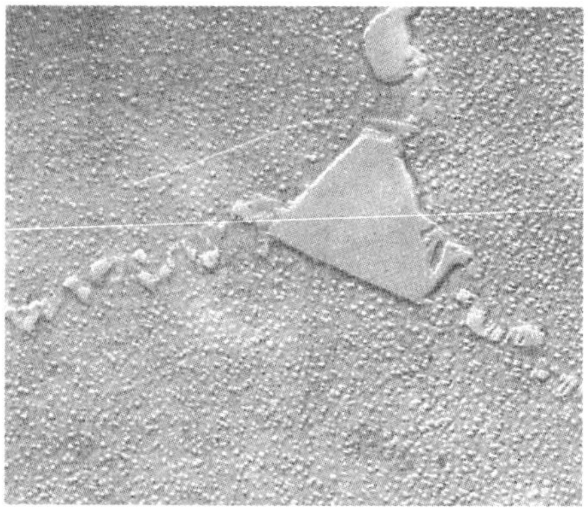

(a)

Figure 14-7 (a) Microstructure of a superalloy with carbides at the grain boundaries and γ' precipitates in the matrix ($\times$ 15,000). [*From* ASM Handbook, *Vol. 9, Metallography and Microstructure (1985), ASM International, Materials Park, OH 44073-0002. Reprinted with permission of ASM International. All rights reserved. www .asminternational.org.*] (Repeated for Problem 14-26.)

14-29 Both a Ti-15% V alloy and a Ti-35% V alloy are heated to the lowest temperature at which all β just forms. They are then quenched and reheated to 300°C. Describe the changes in microstructure during the heat treatment for each alloy, including the amount of each phase. What is the matrix and what is the precipitate in each case? Which is an age-hardening process? Which is a quench and temper process? [See Figure 14-12(a).]

14-30 A Ti-Mn alloy is composed of 20 wt% manganese and is kept at 550°C. What phase(s) are present? See Figure 14-8(d).

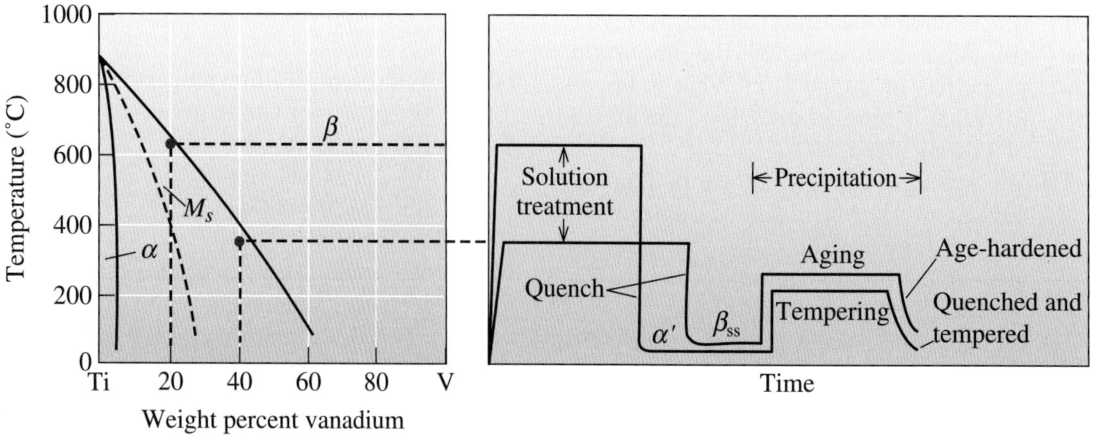

Figure 14-12 Heat treatment of the α-β titanium alloys. (Repeated for Problem 14-29.)

14-31 Determine the specific yield strengths of the strongest Al, Mg, Cu, Ti, and Ni alloys. Use the densities of the pure metals, in lb/in.^{3}, in your calculations. Try to explain their order.

Section 14-6 Refractory and Precious Metals

14-32 What is a refractory metal or refractory alloy? What is a precious metal?

14-33 The temperature of a coated tungsten part is increased. What happens when the protective coating on a tungsten part expands more than the tungsten? What happens when the protective coating on a tungsten part expands less than the tungsten?

14-34 For what applications are Pt, Rh, Pd, and Ag used?

14-35 List the refractory metals by their suitability for cryogenic service.

14-36 Nanoparticles (3 nm in diameter) of platinum with a total weight of 1 milligram are used in automobile catalytic converters to facilitate oxidation reactions. As Pt is an expensive metal, a method to coat iron nanoparticles with Pt is being explored to reduce the cost of the catalyst metal. If 3 nm diameter Fe particles are being coated with 1 nm of Pt to achieve the same effective surface area with fewer particles and less Pt, calculate the difference in cost and weight of the catalyst. Assume that the cost of Pt is \$40/g and that of Fe is \$0.005/g. The density of Fe is 7.87 g/cm^3 and that of Pt is 21.45 g/cm^3.

Design Problems

14-37 A part for an engine mount for a private aircraft must occupy a volume of 60 cm^3 with a minimum thickness of 0.5 cm and a minimum width of 4 cm. The load on the part during service may be as much as 75,000 N. The part is expected to remain below 100°C during service. Design a material and its treatment that will perform satisfactorily in this application.

14-38 You wish to design the rung on a ladder. The ladder should be light in weight so that it can be easily transported and used. The rungs on the ladder should be 0.25 in. × 1 in. and are 12 in. long. Design a material and its processing for the rungs.

14-39 We have determined that we need an alloy having a density of 2.3 ± 0.05 g/cm^3 that must be strong, yet still have some ductility. Design a material and its processing that might meet these requirements.

14-40 We wish to design a mounting device that will position and aim a laser for precision cutting of a composite material. What design requirements might be important? Design a material and its processing that might meet these requirements.

14-41 Design a nickel-titanium alloy that will produce 60 volume percent Ni_3Ti precipitate in a pure nickel matrix.

14-42 An actuating lever in an electrical device must open and close almost instantly and carry a high current when closed. What design requirements would be important for this application? Design a material and its processing to meet these requirements.

14-43 A fan blade in a chemical plant must operate at temperatures as high as 400°C under rather corrosive conditions. Occasionally, solid material is ingested and impacts the fan. What design requirements would be important? Design a material and its processing for this application.

Computer Problems

14-44 *Database Identification System for Alloys.* Write a computer program that will ask the user to input a three- or four-digit code for aluminum alloys (Table 14-3). You will have to ask the user to provide one digit at a time or figure out a way of comparing different digits in a string. This will be followed by a letter and a number (e.g., T and 4).

The program should then provide the user with some more detailed information about the type of alloy. For example, if the user enters 2024 and then T4, the program should provide an output that will specify that (a) the alloy is wrought type, (b) Cu is the major alloying element (since the first digit is 2), and (c) it is naturally aged. Do not make the program too complex. The main idea here is for you to see how databases for alloys are designed.

Knovel® Problems

K14-1 Not all aluminum alloys are easily weldable. Provide a few designations of readily weldable aluminum alloys. List some welding techniques recommended for aluminum alloys.

K14-2 List several wrought aluminum alloys with tensile strengths between 10-15 ksi.

Ceramics comprise a large and diverse class of materials. Their applications are numerous, ranging from fine art such as colorful and intricate glass shapes to state of the art such as the ceramic tiles that were used in the space shuttle. Ceramics play a critical role in a wide array of electronic, magnetic, optical, and energy-related technologies, including for example capacitors, components of memory devices, telescope lenses, and catalyst supports. Traditional ceramics serve as refractories (heat-resistant materials) for metals processing and are also familiar in bath tubs, sinks, and dinnerware. *(Image courtesy of NASA / kaca kaca/Shutterstock.com)*

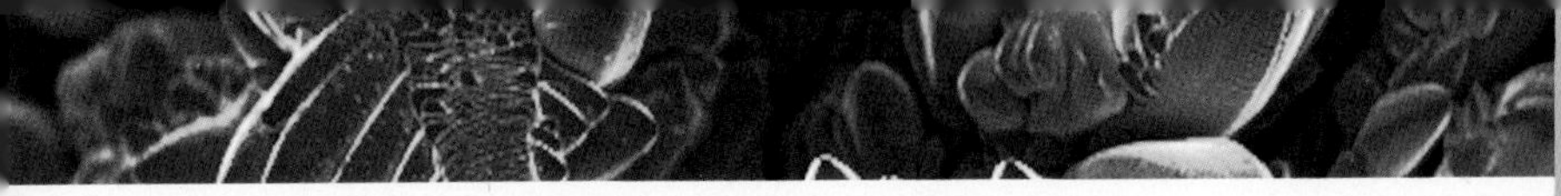

CHAPTER

15

Ceramics

Have You Ever Wondered?

- *What ceramic material is found in bone and teeth?*
- *What makes some diamonds yellow?*
- *What material was used to protect the space shuttle from high temperatures during re-entry?*
- *What is the magnetic strip on a credit card made from?*

Chapter Learning Objectives

The key objectives of this chapter are to

- Define a ceramic and recognize that this materials class encompasses nonmetallic inorganic solids, both crystalline and amorphous.
- Identify the applications for various ceramic materials.
- Identify some of the most common crystal structures for crystalline ceramics.
- Identify some of the most common defects in crystalline ceramics.
- Choose the appropriate process and processing parameters for producing ceramic components.
- Describe the structure of silica and silicate compounds.
- Explain how the properties of silica and silicates are altered by the incorporation of other oxides.

Ceramics, metals, and polymers comprise the primary classes of materials that we discussed in Chapter 1. Metals can be defined by their electronic structure. They typically have one or two valence electrons; this electronic structure gives rise to their unique properties including high strengths and ductilities and high electrical and thermal conductivities. Polymers are formed from long chains of organic molecules as we will see in Chapter 16. Semiconductors, like metals, have a unique electronic structure (Chapter 19) that permits their electrical conductivities to be precisely controlled, thereby making them useful in electronics applications such as transistors (electrical switches). But what is a ceramic? We usually know a ceramic when we see one and are certainly familiar with ceramics in our everyday lives. Examples include toilets, sinks, dishes, windows, the insulating materials on spark plugs, as well as advanced engineering components in microelectronics and optoelectronics. The human body has the amazing ability to make hydroxyapatite, a ceramic found in bones and teeth. Other naturally occurring ceramics are diamonds and gemstones such as sapphire and emerald. Ceramics are usually brittle, poor electrical conductors, good thermal insulators, and may be transparent. They are always inorganic, but may be crystalline, partially crystalline, or amorphous. They may be a combination of a metal and a nonmetal (e.g., oxides, carbides, nitrides, sulfides, fluorides, etc.) and can vary significantly in their electronic properties. Ceramics can be formed as single crystals as in thin films (e.g., gallium nitride for lasers) or nanotubes (e.g., boron nitride or carbon); parts can be fabricated from the sintering of powders as in tungsten carbide cutting tools; and glasses can be blown into intricate shapes using heat and shaping tools. Is there a single definition that encompasses all of these characteristics of ceramics without excluding materials that we typically think of as ceramic? W.D. Kingery, who wrote a seminal text in the field and is the "father of modern ceramics," perhaps defined ceramics best when he wrote that a ceramic is "a nonmetallic inorganic solid." This is a broad, but fully inclusive definition that also includes diamond and graphite since we consider the allotropes of carbon to be inorganic. We recognize that there can be dramatic differences within this range and also that ceramics can be used for different purposes. Ceramics include crystals, glasses, and glass-ceramics. We will discuss each of these three in this chapter. Table 15-1 provides a partial list of functions and applications of some exemplar ceramic materials.

Table 15-1 Functions and applications of some exemplar ceramics.

Function	Application	Examples of Ceramics
Electrical	Capacitor dielectrics	$BaTiO_3$, $SrTiO_3$, Ta_2O_5
	Microwave dielectrics	$Ba(Mg_{1/3}Ta_{2/3})O_3$, $Ba(Zn_{1/3}Ta_{2/3})O_3$ $BaTi_4O_9$, $Ba_2Ti_9O_{20}$, $Zr_xSn_{1-x}TiO_4$, Al_2O_3
	Conductive oxides	In-doped SnO_2 (ITO)
	Superconductors	$YBa_2Cu_3O_{7-x}$ (YBCO)
	Electronic packaging	Al_2O_3
	Insulators	Porcelain
	Solid-oxide fuel cells	ZrO_2, $LaCrO_3$
	Piezoelectric	$Pb(Zr_xTi_{1-x})O_3$ (PZT), $Pb(Mg_{1/3}Nb_{2/3})O_3$
	Electro-optical	La-doped PZT (PLZT), $LiNbO_3$
Magnetic	Recording media	γ-Fe_2O_3
	Ferrofluids, credit cards	Fe_3O_4
	Circulators, isolators	Nickel zinc ferrite
	Inductors, magnets	Manganese zinc ferrite
Optical	Fiber optics	Doped SiO_2
	Glasses	SiO_2 based
	Lasers	Al_2O_3, yttrium aluminum garnate (YAG)
	Lighting	Al_2O_3, glasses
Automotive	Oxygen sensors, fuel cells	ZrO_2
	Catalyst support	Cordierite
	Spark plugs	Al_2O_3
	Tires	SiO_2
	Windshields/windows	SiO_2-based glasses

Table 15-1 Functions and applications of some exemplar ceramics. *(continued)*

Function	Application	Examples of Ceramics
Mechanical/Structural	Cutting tools	WC-Co cermets Silicon-aluminum-oxynitride (Sialon) Al_2O_3
	Composites	SiC, Al_2O_3, silica glass fibers
	Abrasives	SiC, Al_2O_3, diamond, BN, $ZrSiO_4$
Biomedical	Implants	Hydroxyapatite
	Dentistry	Porcelain, Al_2O_3
	Ultrasound imaging	PZT
Construction	Buildings	Concrete Glass Bathroom fixtures
Others	Defense applications Armor materials	PZT, B_4C
	Sensors	SnO_2
	Nuclear	UO_2 Glasses for waste disposal
	Metals processing	Alumina and silica-based refractories, oxygen sensors, casting molds, etc.
Chemical	Catalysis Air, liquid filtration Sensors Paints, rubber	Various oxides (Al_2O_3, ZrO_2, ZnO, TiO_2)
Domestic	Tiles, bathroom fixtures, whiteware, kitchenware, pottery, art, jewelry	Clay, alumina, and silica-based ceramics, glass-ceramics, diamond, ruby, cubic zirconia, and other crystals

In studying ceramics, we will draw on the foundational knowledge we have gained from earlier chapters including bonding (Chapter 2), crystal structures (Chapter 3), defects (Chapter 4), diffusion (Chapter 5), mechanical properties (Chapter 6), and fracture (Chapter 7). We will also examine the synthesis, processing, and applications of ceramic materials. The electrical properties of ceramics and the piezoelectric effect that some ceramics exhibit will be discussed in Chapter 19 and optical properties will be discussed in Chapter 21.

15-1 Bonding in Ceramics

Ceramics are held together by strong primary bonds. These bonds are predominantly ionic or covalent with some mixed bonding including some metallic bonding. As we learned in Chapter 2, a deep potential well gives rise to high melting temperatures (see Figure 2-20). A large amount of thermal energy must be provided to cause atoms or ions to separate to larger interatomic or interionic distances, thereby leading to melting. High melting point materials also tend to have high elastic moduli (Tables 15-2 and 15-3), meaning that the bonds are stiff. Materials that have predominantly ionic or covalent bonds lack free electrons to conduct electricity because there is either a transfer or sharing of electrons to achieve complete s^2p^6 electronic configurations. Because of the lack of free electrons, ceramics also tend to be good thermal insulators although there are some exceptions. One such exception is diamond since diamond has extraordinarily stiff bonds that efficiently transport heat (Sections 2-7 and 22-3). We will see in Section 15-3 that these bonds also typically lead to high strengths and hardnesses as well as brittleness for ceramic materials.

Table 15-2 Properties of commonly encountered polycrystalline ceramics

Material	Melting Point (°C)	Thermal-Expansion Coefficient (× 10^{-6} cm/cm)/°C	Knoop Hardness (HK) (100 g)
Al_2O_3	2000	~6.8	2100
BN	2732	0.57^a, -0.46^b	5000
SiC	2700	~3.7	2500
Diamond	3550	1.02	7000
Mullite	1810	4.5	—
TiO_2	1840	8.8	—
Cubic ZrO_2	2700	10.5	—

[a]*Perpendicular to pressing direction.*
[b]*Parallel to pressing direction.*

Table 15-3 Mechanical properties of selected advanced ceramics

Material	Density (g/cm^3)	Tensile Strength (psi)	Flexural Strength (psi)	Compressive Strength (psi)	Young's Modulus (psi)	Fracture Toughness (psi $\sqrt{\text{in.}}$)
Al_2O_3	3.98	30,000	80,000	400,000	56×10^6	5,000
SiC (sintered)	3.1	25,000	80,000	560,000	60×10^6	4,000
Si_3N_4 (reaction bonded)	2.5	20,000	35,000	150,000	30×10^6	3,000
Si_3N_4 (hot pressed)	3.2	80,000	130,000	500,000	45×10^6	5,000
Sialon	3.24	60,000	140,000	500,000	45×10^6	9,000
ZrO_2 (partially stabilized)	5.8	65,000	100,000	270,000	30×10^6	10,000
ZrO_2 (transformation toughened)	5.8	50,000	115,000	250,000	29×10^6	11,000

Example 15-1 *Bond Stiffness and the Elastic Modulus*

Consider a uniaxial load applied to a simple cubic unit cell and demonstrate that the elastic modulus E of a material is directly proportional to the bond stiffness.

SOLUTION

Assume that the equilibrium distance between atoms is the lattice parameter a. The engineering stress S that develops when a load P is applied to a cross-sectional area a^2 is

$$S = \frac{P}{a^2}$$

For elastic loading, Hooke's law applies such that

$$S = \frac{P}{a^2} = Ee$$

where e is the engineering strain. The engineering strain can also be expressed as

$$e = \frac{\Delta a}{a}$$

where Δa is the extension, and the load P can be written as

$$P = k\Delta a$$

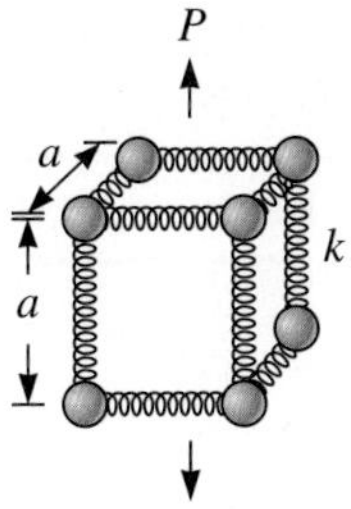

Figure 15-1 The bonds between atoms can be modeled as springs with a stiffness k in this simple approximation. The lattice parameter of the simple cubic unit cell is a.

where k is the stiffness of the bonds, which are modeled as linear springs (see Figure 15-1). Solving for E,

$$E = \frac{P}{a^2 e} = \frac{k\Delta a}{a^2}\frac{a}{\Delta a} = \frac{k}{a}$$

This equation is a rough approximation and should not be considered to be strictly correct, but it does show that the elastic modulus is proportional to the stiffness between atoms or ions in a solid. As the bond stiffness increases, the elastic modulus increases.

15-2 Structures of Crystalline Ceramics

Ceramics may be crystalline, partially crystalline, or amorphous. We will discuss inorganic glasses in Section 15-7 and partially crystalline ceramics in Section 15-8. Crystalline ceramics exhibit hundreds of crystal structures. Many of these crystal structures were covered in Sections 3-7 and 3-8 including the sodium chloride, zincblende, fluorite, corundum, perovskite, and diamond cubic structures. See Table 15-4 for a table of some common crystal structures found in ceramics, their lattice and basis descriptions, and exemplary ceramic materials that adopt these crystal structures.

As discussed in Section 3-7, many of the ionic ceramic structures can be viewed as a packing of anions with smaller cations in interstitial sites. For example, the sodium chloride structure is a face-centered cubic structure of Cl anions with all of the octahedral sites occupied by Na cations. As a reminder, the structures that are adopted by ceramic compounds are a strong function of the relative radii of the atoms or ions in the compound and the need to maintain electric neutrality (Section 3-7). In some cases, the simplest or correct way to describe a ceramic crystal structure is not necessarily as a close-packing of anions with cations in the interstitial sites. It is also important to note that even when the cations occupy interstitial sites, this does not necessarily lead to an atomic packing fraction greater than that of close-packed metals.

Alumina One of the most widely used crystalline ceramics is aluminum oxide, or alumina. It exists as several crystalline polymorphs (Section 3-4), some with quite complicated crystal structures. This is similar to the situation that we will see later in this chapter with silica (silicon dioxide). The form known as α-Al_2O_3, which comprises the commonly found mineral corundum (Section 3-7), exhibits many of the typical characteristics of ceramics. Because it is so hard, for instance, corundum powders are often used as polishing compounds. Variants of alumina are also the basis for many gemstones, including ruby and sapphire. Elements such as titanium, iron, chromium, and copper produce alumina-based gemstones with specific colors. Alumina is used in various electronic devices and for scratch-resistant transparent covers or windows on watches and certain scientific instruments. It is also used as a **refractory** (i.e., a material that retains

Table 15-4 Table of some ceramic crystal structures, lattice and basis descriptions, and exemplary ceramics adopting these forms. This list is by no means exhaustive.

	Structure	Name/Lattice and Basis	Some Examples
1	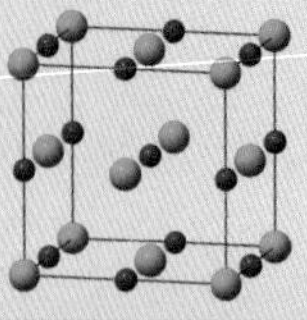	*Sodium chloride (NaCl)* FCC lattice with a basis of Cl (0, 0, 0) and Na (1/2, 0, 0)	BaO, CaO, MgO, SrO, TiC, TiO
2	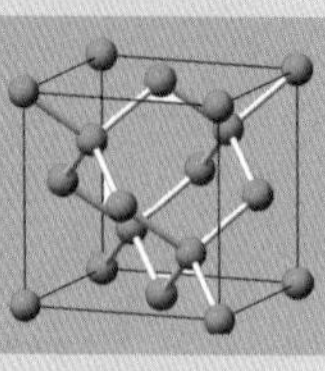	*Diamond cubic* FCC lattice with a basis of C (0, 0, 0) and C (1/4, 1/4, 1/4)	Diamond
3	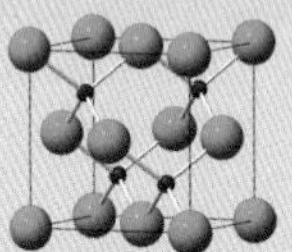	*Cubic zincblende (ZnS) also known as sphalerite* FCC lattice with a basis of S (0, 0, 0) and Zn (1/4, 1/4, 1/4)	β-SiC
4		Fluorite (*CaF_2*) FCC lattice with a basis of Ca (0, 0, 0), F (1/4, 1/4, 1/4), and F (1/4, 1/4, 3/4)	BaF_2, CaF_2, CeO_2, ThO_2, UO_2, ZrO_2
5	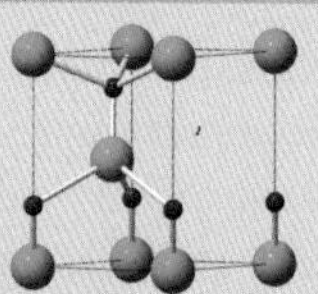	*Hexagonal zincblende (ZnS) also known as wurtzite* Hexagonal lattice with a basis of S (0, 0, 0), S (2/3, 1/3, 1/2), Zn (0, 0, 3/8), Zn (2/3, 1/3, 7/8)	α-SiC, BeO, ZnO,
6	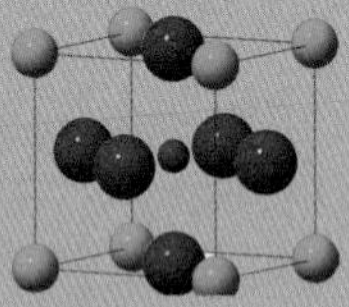	*Perovskite ($CaTiO_3$)* Simple cubic lattice with a basis of Ca (0, 0, 0), Ti (1/2, 1/2, 1/2), and O (0, 1/2, 1/2), O (1/2, 0, 1/2), and O (1/2, 1/2, 0)	$BaTiO_3$, $CaTiO_3$, $PbZrO_3$, $SrTiO_3$
7	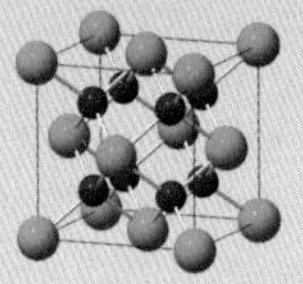	*Antifluorite (Na_2O)* FCC lattice with a basis of O (0, 0, 0), Na (1/4, 1/4, 1/4), and Na (1/4, 1/4, 3/4)	K_2O, Li_2O, Na_2O

its strength at high temperatures) to contain molten metals or in applications where a material must operate at high temperatures with high strength. A classic application of alumina is insulators in spark plugs.

In Section 3-6, we described tetrahedral and octahedral interstitial sites in three basic cubic structures. Similar sites are readily identifiable in hexagonal crystals as well. Crystalline alumina can be thought of as a hexagonal close-packed (HCP) structure of oxygen ions, with aluminum occupying 2/3 of the octahedral sites in a specific and ordered pattern. Since there are equal numbers of lattice sites and octahedral sites in the HCP structure (we saw that the same is true for the FCC structure), the stoichiometry of alumina requires that one-third of the octahedral sites are vacant, while two-thirds are filled with aluminum. The conventional unit cell for alumina contains 30 atoms, or six Al_2O_3 formula units.

Example 15-2 *Silicon Carbide*

Silicon carbide (SiC) provides outstanding oxidation resistance at temperatures even above the melting point of steel. SiC often is used as a coating for metals, carbon-carbon composites, and other ceramics to provide protection at extreme temperatures. SiC is used as an abrasive in grinding wheels and as particulate and fibrous reinforcement in both metal matrix and ceramic matrix composites. It is also used to make heating elements for furnaces. SiC carbide exists as two polymorphs: α–SiC, which has the hexagonal zincblende (wurtzite) crystal structure, and β–SiC, which has the cubic zincblende (sphalerite) structure.

How many atoms of each type are there per unit cell in α–SiC? What type of bonding do you expect dominates the behavior and why?

SOLUTION

The crystal structure of α–SiC can be described as a hexagonal lattice with a basis of Si (0, 0, 0), Si (2/3, 1/3, 1/2), C (0, 0, 3/8), and C (2/3, 1/3, 7/8). There is one lattice point per unit cell of the hexagonal lattice and two Si atoms in the basis so there are two Si atoms per unit cell.

There is one lattice point per unit cell of the hexagonal lattice and two C atoms in the basis so there are two C atoms per unit cell. The unit cell of SiC is shown in Figure 15-2.

Si and C are both Group IV elements. The electronic structure of Si is [Ne] $3s^2 3p^2$, and the electronic structure of C is [He] $2s^2 2p^2$. (In fact, in solid form, each Si and C atom forms four sp^3 hybrid orbitals each containing one electron; these electrons occupying sp^3 orbitals are then shared covalently between the Si and C atoms so that each atom has four complete orbitals.) We can calculate the fraction of covalent bonds according to Equation 2-1 in Section 2-6, which includes a term for the difference between the electronegativities of the two elements. The electronegativity of Si is 1.8, and the electronegativity of C is 2.5.

$$\text{Fraction covalent} = \exp(-0.25\Delta E^2)$$

$$\text{Fraction covalent} = \exp[-0.25(2.5-1.8)^2] = 0.88$$

Thus the majority of the bonds are covalent.

Again, we have not considered the specific interstitial sites of hexagonal structures, but the wurtzite structure can be envisioned as a hexagonal close-packed structure of Si atoms with half of the tetrahedral sites occupied by C atoms. The tetrahedral sites are formed, for example, by three adjoining atoms in the A layer of the HCP structure and a fourth atom in the B layer. Therefore, each carbon atom is surrounded by and bonded to four silicon atoms, and each silicon atom is surrounded by and bonded to four carbon atoms as shown in Figure 15-2.

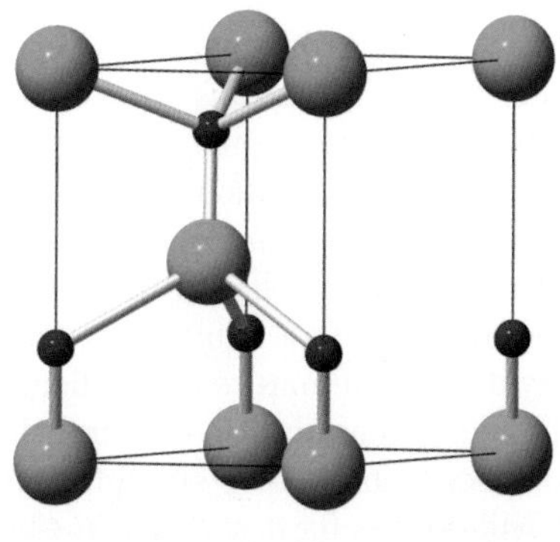

Figure 15-2 The unit cell of SiC. The Si atoms are shown in green, and the C atoms are shown in red.

Example 15-3 *Cubic Zirconia*

Cubic zirconia (ZrO_2) is a less expensive alternative to diamond for jewelry. It is also used to make oxygen gas sensors used in, for example, automotive engines, and to measure dissolved oxygen in molten steels because of its high ionic conductivity. Zirconia is an additive in electronic ceramics, and it is a refractory material. Zirconia exists as several different polymorphs depending on temperature and pressure. Cubic zirconia is unstable at room temperature unless it has been doped with an addition such as Y_2O_3, known as yttria. (Note that the letters *ia* at the end of the terms "yttria" and "zirconia" mean "oxygen-containing," i.e., "yttrium containing oxygen" and "zirconium containing oxygen.") The unit cell of cubic zirconia can be described as a face-centered cubic lattice of zirconium with all tetrahedral sites occupied by oxygen.

What is the crystal structure of cubic zirconia? Explain how the stoichiometry of ZrO_2 is consistent with this crystal structure.

SOLUTION

This is the fluorite crystal structure. (See Table 15-4.) In a single unit cell of cubic zirconia, there are four Zr ions and eight O ions. The ratio of Zr^{4+} ions to O^{2-} ions is 4:8 or 1:2, which is the same as the stoichiometry of ZrO_2. Notice that this is a crystal structure for which we cannot think of the anions as being arranged in a close-packed fashion. In fact, the anions are the larger ions, but in this description of the structure, they occupy the smaller tetrahedral sites. The antifluorite structure (Na_2O) (see Table 15-4) can be envisioned as a close-packing of anions in an FCC fashion with the cations occupying all of the tetrahedral sites.

Example 15-4 *Strontium Oxide*

Strontium oxide (SrO), also known as strontia, has the NaCl crystal structure. Write the lattice and basis description of strontium oxide.

SOLUTION

SrO has an FCC lattice with a basis of O^{2-} (0, 0, 0) and Sr^{2+} (1/2, 0, 0).

Note that SrO has a radius ratio for the ions of 1.12 Å/(1.32 Å) = 0.85. This radius ratio indicates that SrO should have the CsCl crystal structure. This example illustrates that while radius ratios provide a general guideline to indicate a likely structure for a material, the guideline is not strictly followed in nature for every crystal.

15-3 Defects in Crystalline Ceramics

We discussed the nomenclature used to describe defects in crystalline ceramics in Section 4-2 with an emphasis on point defects of various types. One of the guiding principles is that charge balance must be maintained when these defects exist. All of the point defects discussed in Section 4-2 occur in ceramic materials, with Schottky defects being unique to ionically bonded materials. Crystalline ceramics contain dislocations as shown for alumina in Figure 15-3; however, due to the stiffness of the atomic bonds and a typically large Burgers vectors in ceramic materials compared to metals, it is difficult for dislocations to propagate. The highly directional nature of covalent bonds and the alternating charges present in ionically bonded materials also hinder dislocation propagation. These properties lead to high hardnesses, compressive strengths, and elastic moduli as shown in Tables 15-2 and 15-3. We will discuss flaws such as cracks, pores, and voids in ceramic materials in the next section.

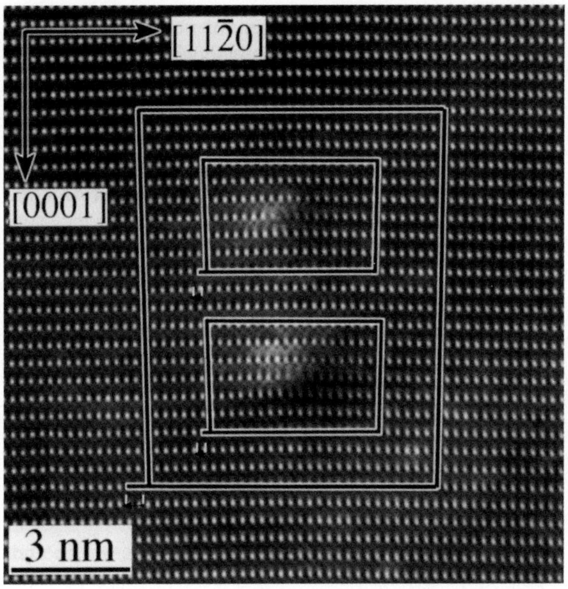

(a)

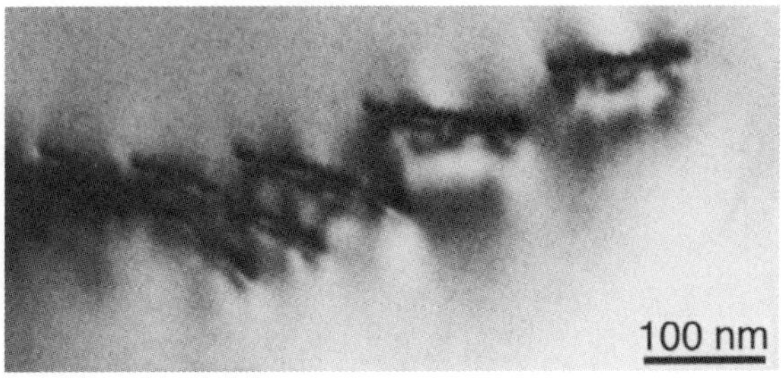

(b)

Figure 15-3 (a) A high resolution transmission electron micrograph of edge dislocations in Al_2O_3. (*Y. Ikuhara*) (b) Diffraction contrast from dislocations in Al_2O_3. (Ceramic materials: science and engineering *by NORTON, M. G. ; CARTER, C. BARRY Reproduced with permission of SPRINGER in the format Book via Copyright Clearance Center.*)

Example 15-5 *Impurities in Diamond*

The canary or fancy yellow diamonds called Type Ia diamonds contain substitutional nitrogen impurities at a concentration of less than 100 ppm. The presence of nitrogen creates absorption levels in the band structure of diamond that make it possible for electrons to absorb light so that the diamond is no longer colorless (see Chapter 19). When nitrogen is present but exists at levels less than 100 parts per million (ppm), a diamond will appear yellow. If the concentration of nitrogen in diamond is 100 ppm, what is the concentration of nitrogen atoms per unit cell and per cm^3? The lattice parameter of diamond is 3.5668×10^{-8} cm.

SOLUTION

The diamond cubic crystal structure contains eight atoms per unit cell. If there are 100 nitrogen atoms per 10^6 carbon atoms, i.e., 100 ppm, then

$$\left(\frac{100 \text{ N atoms}}{10^6 \text{ C atoms}}\right)\left(\frac{8 \text{ C atoms}}{\text{unit cell}}\right) = 8 \times 10^{-4} \text{ N atoms/(unit cell) or 1250 unit cells/(N atom)}$$

Assuming that the presence of nitrogen does not alter the lattice parameter of diamond, then

$$\left(\frac{8 \times 10^{-4} \text{ N atoms}}{\text{unit cell}}\right)\left[\frac{\text{unit cell}}{(3.5668 \times 10^{-8} \text{ cm})^3}\right] = 1.76 \times 10^{19} \text{ N atoms/cm}^3$$

Diamond also has industrial applications. It is the hardest naturally occurring material. Industrial diamonds are used as abrasives for grinding and polishing as well as abrasion-resistant coatings for many different applications (e.g., cutting tools). Synthetic diamond and diamond-like coatings are prepared using chemical vapor deposition processes (Chapter 19) as well as through high temperature-pressure processing.

Example 15-6 *Dislocations in MgO*

A sketch of a dislocation in magnesium oxide (MgO), which has the sodium chloride crystal structure and a lattice parameter of 0.396 nm, is shown in Figure 15-4. Note that the slip system for MgO is $\{\bar{1}10\}$<110>. Determine the length of the Burgers vector.

SOLUTION

In Figure 15-4, we begin a clockwise loop around the dislocation at point x and then move equal atom spacings in the two horizontal directions and equal atom spacings in the two vertical directions to finish at point y. Note that it is necessary that the lengths of the two horizontal segments of the loop be equal and the lengths of the vertical segments be equal, but it is not necessary that the horizontal and vertical segments be equal in length to each other. The chosen loop must close in a perfect crystal. The vector **b** is the Burgers vector. Because **b** is parallel to a [110] direction, it must be perpendicular to (110) planes. The length of **b** is the distance between two adjacent (110) planes. From Equation 3-7,

$$d_{110} = \frac{a_0}{\sqrt{h^2 + k^2 + l^2}} = \frac{0.396 \text{ nm}}{\sqrt{1^2 + 1^2 + 0^2}} = 0.280 \text{ nm}$$

The Burgers vector is a (110) direction that is 0.280 nm in length. Note, however, that two extra half planes of atoms make up the dislocation—one composed of magnesium ions and one of oxygen ions (Figure 15-4). This formula for calculating the magnitude of the Burgers vector becomes more complicated for non-cubic systems. It is better to consider the magnitude of the Burgers vector as equal to the repeat distance in the slip direction.

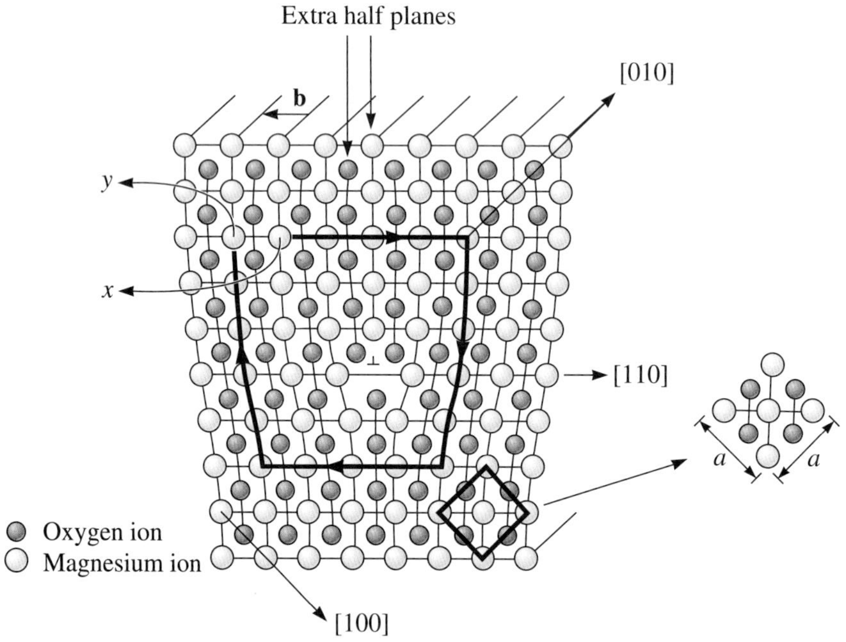

Figure 15-4 An edge dislocation in MgO showing the slip direction and Burgers vector (for Example 15-6). This drawing shows the arrangement of ions on the (001) plane, and therefore, the [001] direction is perpendicular to the plane of the page. The inset shows the outline of a conventional unit cell for MgO. (*Based on W.D. Kingery, H.K. Bowen, and D.R. Uhlmann,* Introduction to Ceramics, *John Wiley, 1976.*)

Example 15-7 Slip in MgO and Mg

Based on the principles of the bonding of MgO and Mg and the Peierls-Nabarro stress, explain why slip in MgO is much more difficult than in Mg. For the purpose of this discussion, consider the {0001}<100> slip system in Mg.

SOLUTION

As determined in Example 15-6, both the spacing between slip planes and **b** are 0.280 nm for MgO. For the {0001}<100> slip system in Mg, the interplanar spacing is half of the height of the HCP unit cell or 0.5209 nm/2 = 0.260 nm, and the Burgers vector is the distance between atoms in the close-packed direction or $a/2$ = 0.32087 nm/2 = 0.160 nm. The Peierls-Nabarro stress is given by Equation 4-2,

$$\tau = c\exp\left(-\frac{kd}{b}\right)$$

where τ is the shear stress required to move a dislocation, d is the interplanar spacing between slip planes, b is the magnitude of the Burgers vectors, and c and k are material-specific constants. Ignoring differences in c and k, we see that the stress required to propagate a dislocation increases as the ratio of d/b decreases. Therefore, we can reasonably expect that MgO with a d/b ratio of 1.0 will require a larger stress to cause dislocations to propagate than Mg (with a d/b ratio of 1.62), thereby making slip more difficult. Additionally, as discussed in Section 4-3, dislocation propagation in ionically bonded materials requires that cations move past other cations and anions move past other anions, creating electronic repulsion that resists dislocation motion.

15-4 Flaws in Ceramics

Since the critical resolved shear stress required to propagate dislocations in ceramics is high, ceramics are susceptible to flaws such as notches, cracks, and pores. Thus ceramics tend to be strong in compression for which crack propagation is suppressed, and weak in tension for which crack propagation is exacerbated. Fracture toughness is the ability of a material containing flaws to withstand an applied load. When the stress intensity factor exceeds the fracture toughness of a material, a crack propagates unstably leading to failure (Section 7-1). Materials that are flaw-sensitive in this way show a distribution of strengths that follow Weibull statistics (Section 7-5). For brittle materials, we can only predict the likelihood of failure at a particular stress, not the actual stress at which failure occurs. Thus, processing and quality control are critical to narrowing the distribution of flaw sizes to achieve mechanical reliability.

Conducting tension tests on a brittle material is often challenging due to the difficulty in gripping the material without causing immediate fracture. Three- and four-point bend tests are a common alternative for determining mechanical properties. One popular method for determining the fracture toughness of brittle ceramics is indentation testing. Indentation testing allows multiple indents to be performed quickly and in large numbers thereby sampling variations in properties and structure throughout the specimen. Indentation causes subsurface radial cracks to form beneath the indenter during the application of the load; these cracks propagate to the surface of the indentation as residual stresses increase when the material is unloaded. The stress intensity factor K_R that exists at the tips of the cracks is given by

$$K_R = \beta_r \frac{P}{c^{3/2}} \qquad (15\text{-}1)$$

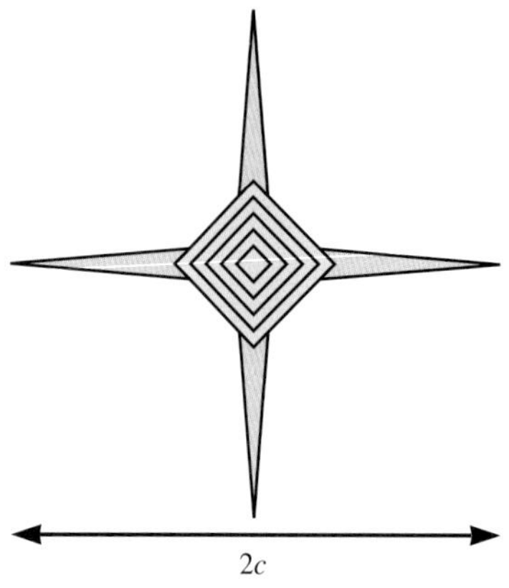

Figure 15-5 A Vickers indent in a brittle material showing the formation of radial cracks with the definition of the crack length $2c$.

where β_r is a constant equal to $0.016\left(\frac{E}{H}\right)^{1/2}$ with elastic modulus E and hardness H, P is the applied load, and $2c$ is the radial crack length (see Figure 15-5). The cracks arrest when K_R decreases to the fracture toughness K_{IC} of the material. See Table 15-3 for the fracture toughness of some advanced ceramics.

Porosity

Pores represent the most important defect in polycrystalline ceramics. The presence of pores is usually detrimental to the mechanical properties of bulk ceramics since pores provide pre-existing locations from which cracks can grow. Elastic modulus and strength decrease as porosity increases. Figure 15-6 shows how strength decreases as porosity increases for cement (cement is the ceramic used to bind concrete and will be discussed in Section 15-11 and Chapter 18). The presence of pores is one of the reasons why ceramics show such brittle behavior under tensile loading. Since there is a distribution of pore sizes and the overall level of porosity changes, the mechanical properties of ceramics vary. This variability is measured using Weibull statistics (Chapter 7). The presence of pores, on the other hand, may be useful for increasing resistance to thermal shock (Chapter 22). In certain applications, such as filters for hot metals or for liquids or gases, the presence of interconnected pores is desirable.

If a material has no pores, then its density is known as its true density ρ. A material with pores has a density less than the true density. Pores in a ceramic may be either interconnected or closed. The **apparent porosity** is the percentage of the total volume of the ceramic that is occupied by interconnected pores. Interconnected pores are those that provide channels through which gases and fluids can flow. It is a measure of the permeability, or the ease with which gases and fluids seep through the ceramic component. It is an important characteristic of a porous ceramic. The apparent porosity is determined by weighing the dry ceramic W_d, then reweighing the ceramic both when it is suspended in water W_s (the ceramic is hung from a balance and immersed in a container of water—the weight of the suspending wire is negligible) and after it is removed from water W_w. Using units of grams:

$$\text{Apparent Porosity} = \frac{W_w - W_d}{W_w - W_s} \times 100 \tag{15-2}$$

Note that the numerator is numerically equal to the volume of the interconnected pores, and the denominator is numerically equal to the exterior volume of the ceramic including pores. This numerical equivalency arises because the density of water is 1 g/cm^3.

Another measure of the porosity is known as the **true porosity**. This value includes the volume of both the interconnected and closed pores (i.e., True Porosity = Apparent Porosity + Closed-Pore Porosity). The closed pores cannot become filled with water. The

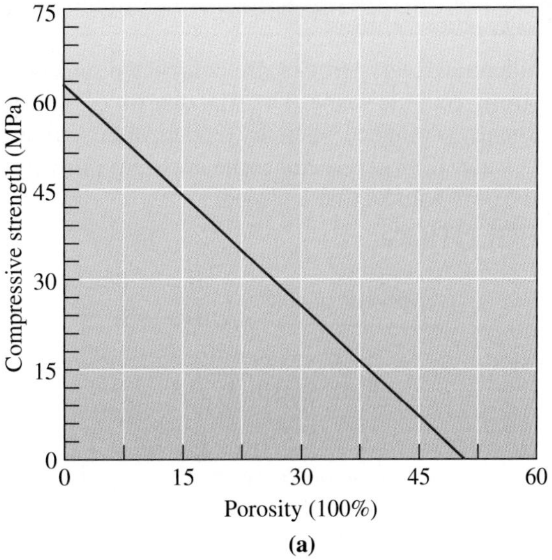

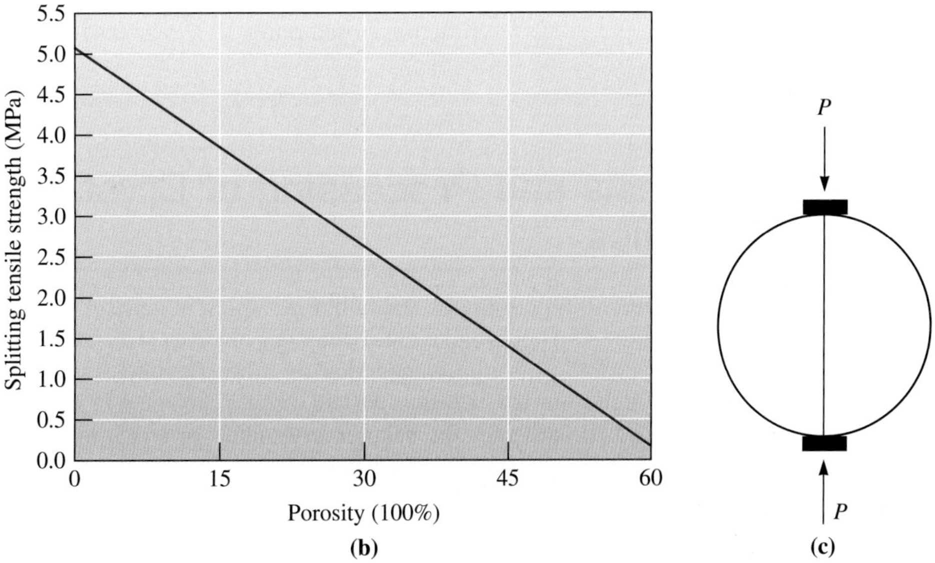

Figure 15-6 (a) The compressive strength of cement as a function of porosity. (b) The splitting tensile strength of cement as a function of porosity as determined using the test method shown in (c), in which a cement cylinder is subjected to a load P on the diameter along the length of the cylinder. The splitting tensile strength is the stress at which the cylinder breaks into two halves.

true porosity, which correlates better with the properties of the ceramic than the apparent porosity, is

$$\text{True Porosity} = \frac{\rho - B}{\rho} \times 100 \tag{15-3}$$

where

$$B = \frac{W_d}{W_w - W_s} \tag{15-4}$$

The **bulk density** B is the ratio of the weight of the dry ceramic to the exterior volume of the ceramic including pores, and again ρ is the true density of the ceramic (with no pores). The following example illustrates how porosity levels in ceramics are determined.

Example 15-8 ***Silicon Carbide Ceramics***

Silicon carbide particles are compacted and fired at a high temperature to produce a strong ceramic shape. The true density of SiC is 3.2 g/cm^3. The ceramic shape subsequently is weighed when dry (360 g), after soaking in water (385 g), and while suspended in water (224 g). Calculate the apparent porosity, the true porosity, and the fraction of the pore volume that is closed.

SOLUTION

$$\text{Apparent porosity} = \frac{W_w - W_d}{W_w - W_s} \times 100 = \frac{385 - 360}{385 - 224} \times 100 = 15.5\%$$

$$\text{Bulk density} = B = \frac{W_d}{W_w - W_s} = \frac{360}{385 - 224} = 2.24 \text{ g/cm}^3$$

$$\text{True porosity} = \frac{\rho - B}{\rho} \times 100 = \frac{3.2 - 2.24}{3.2} \times 100 = 30\%$$

The closed-pore percentage is the true porosity minus the apparent porosity, or $30 - 15.5 = 14.5\%$. Thus,

$$\text{Fraction closed pores} = \frac{14.5}{30} = 0.48$$

15-5 Synthesis and Processing of Crystalline Ceramics

The synthesis and processing of crystalline ceramics, in general, follows the process flow shown in Figure 15-7. In broad strokes, the steps involve (1) *forming* a ceramic powder; (2) *shaping* the ceramic through **compaction** or other methods; and (3) *consolidating* the ceramic through **sintering** or **firing**. We will discuss each step in more depth.

As noted earlier in the chapter, crystalline ceramics melt at high temperatures and exhibit brittle behavior under tension. As a result, the casting and thermomechanical processing used for metals, alloys, and thermoplastics cannot be applied when processing ceramics, and parts made from ceramics are typically formed from ceramic powders. A **powder** is a collection of fine particles. Ceramic powders are synthesized by crushing and grinding ceramic constituents. Impurities are separated, and various powders are blended to produce the desired composition. A **slurry** of powder (a suspension of powder in a liquid) is then sprayed through a nozzle into a chamber in the presence of hot air to form soft agglomerates in a process known as spray drying. The agglomerates are readily shaped through a variety of processes to form a **green ceramic**. The green ceramic has the desired shape, but has to be densified using a high-temperature heat treatment known as sintering or firing. [This is the step used to transform wet clay (a ceramic discussed in Section 15-9) to pottery forms such as vases and bowls.] The ceramic may subsequently be subjected to additional operations such as grinding, polishing, or machining as needed for the final application.

There are a variety of processes used to shape ceramic powders into useful objects. We will discuss several of them here.

Compaction

Compaction is the process by which porosity is eliminated between agglomerates by the application of a load. The different steps of uniaxial compaction, in which the compacting force is applied in one direction, are shown in Figure 15-8. The compaction process can be completed within one minute for smaller parts; thus, uniaxial compaction is well suited for making a large number of small and simple shapes ($<$ 6 inches). Many electronic and magnetic ceramics, **cermet** (e.g., WC-Co) cutting tool bits, and other materials are processed using this technique. Complex or large pieces (up to a few

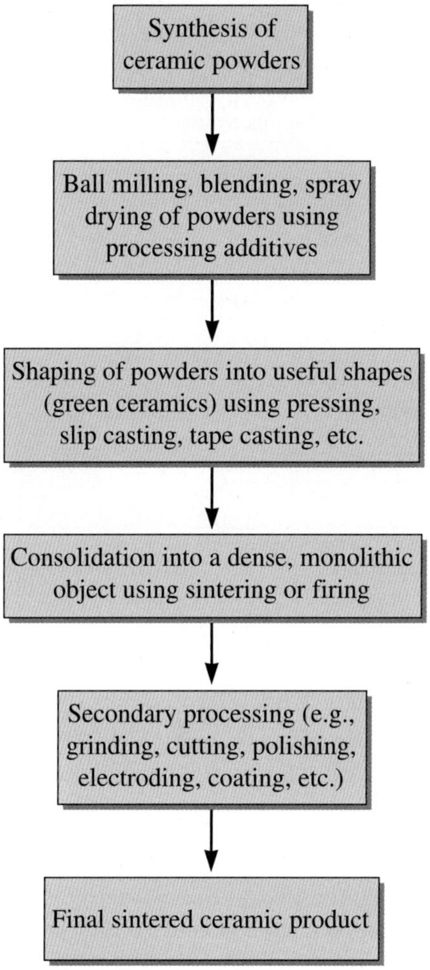

Figure 15-7 Typical steps encountered in the processing of ceramics.

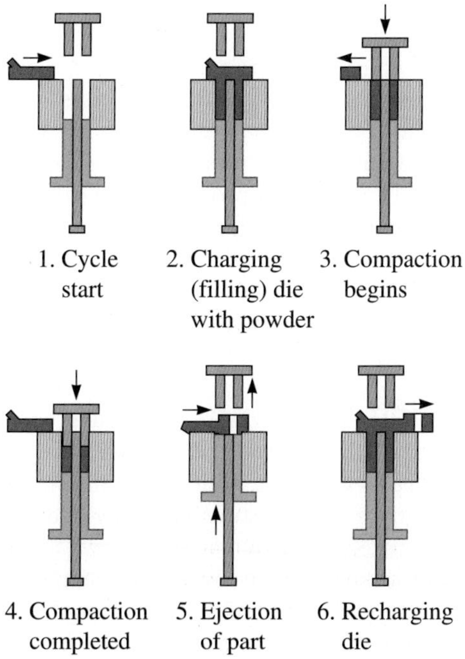

Figure 15-8 Uniaxial powder compaction showing the die-punch assembly during different stages. Typically, for small parts these stages are completed in less than a minute. (*Based on* Materials and Processes in Manufacturing, *Eighth Edition, E.P. DeGarmo, J.T. Black, R.A. Koshe, John Wiley & Sons, Inc., 1997.*)

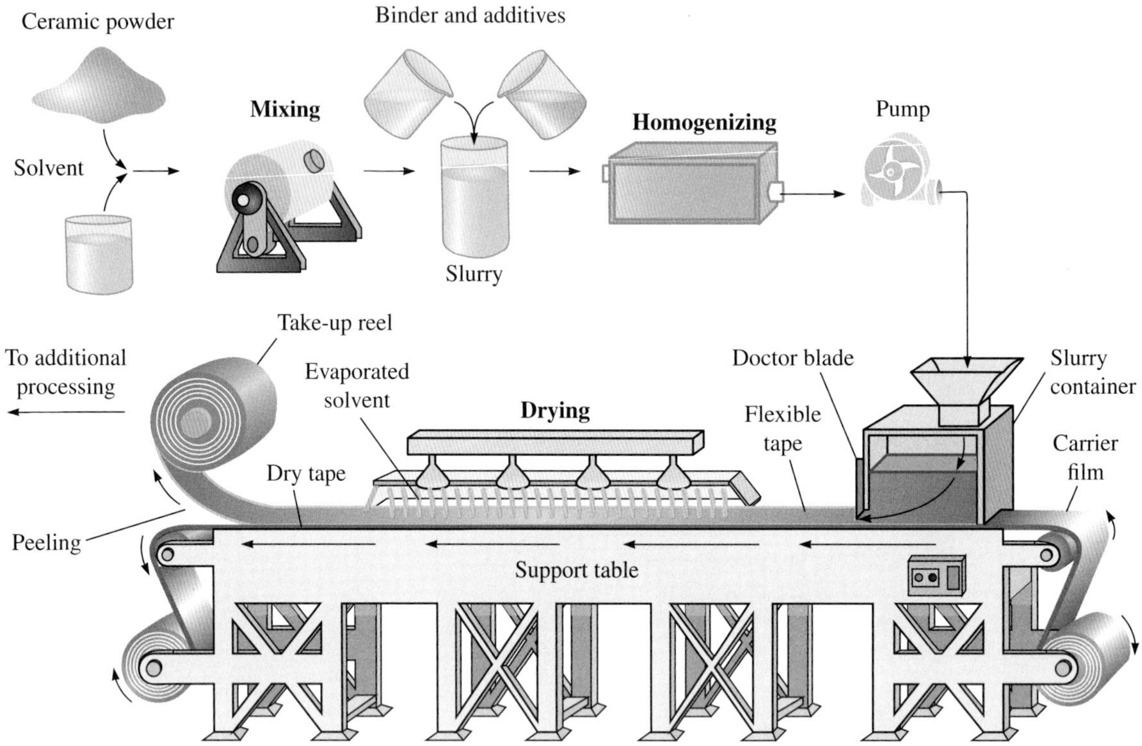

Figure 15-9 Stages of the tape casting process. (*Based on Encyclopædia Britannica, Inc.*)

feet in diameter and six to eight feet long) can be produced when compaction is performed with applied pressure using oil in a process known as **cold isostatic pressing** (CIP).

Tape Casting A technique known as **tape casting** (Figure 15-9) is used for the production of thin ceramic tapes (~3 to 100 μm thick). A slurry is cast with the help of a so-called doctor blade onto a plastic substrate. The resulting green ceramic tape is then subjected to sintering. Many commercially important electronic packages based on alumina substrates and millions of barium titanate capacitors are made using this type of tape casting process.

Slip Casting **Slip casting** (Figure 15-10) typically uses a aqueous slurry of ceramic powder. The slurry, known as the slip, is poured into a plaster of Paris ($CaSO_4 : 2H_2O$) mold. As the water from the slurry begins to move out by capillary action, a thick mass builds along the mold wall. When sufficient product thickness is built up, the rest of the slurry is poured out (this is called drain casting). It is also possible to continue to pour more slurry in to form a solid piece (this is called solid casting). Pressure may also be used to inject the slurry into polymer molds. The green ceramic is then dried and sintered at a high temperature. Slip casting is widely used to make ceramic art (figurines and statues) and bathroom fixtures such as toilets.

Extrusion and Injection Molding **Extrusion** and **injection molding** are popular techniques used for making furnace tubes, bricks, tiles, and insulators. The extrusion process [Figure 15-11(a)] uses a viscous, dough-like mixture of ceramic particles containing a binder and other additives. This mixture has a clay-like consistency, which is then fed to an extruder where it is mixed well in a machine known as a pug mill, sheared, deaerated, and then injected into a die where a continuous shape of green ceramic is produced by the extruder. This material is cut at appropriate lengths and then dried and sintered. Cordierite ceramics (compounds comprising Mg, Fe, Al, Si, and O) used for

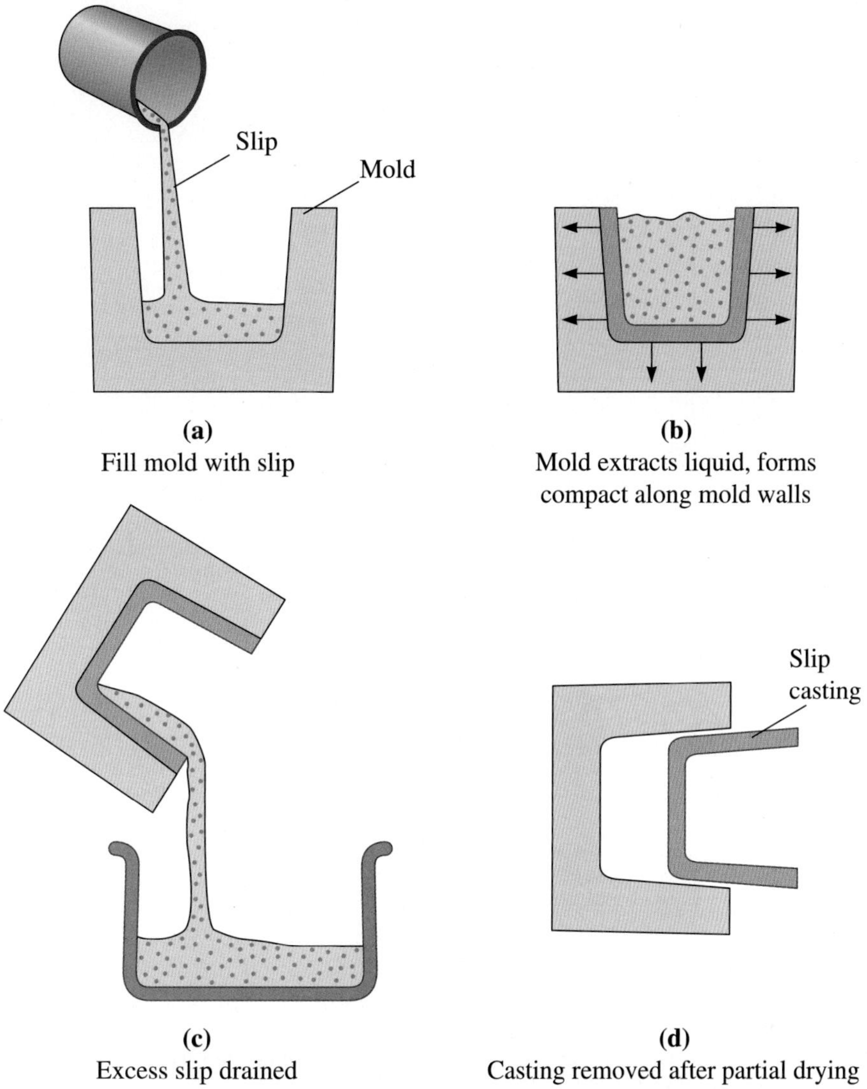

Figure 15-10 Steps in slip casting of ceramics. (Based on Modern Ceramic Engineering: properties, processing, and use in design *by Richerson, David W., Taylor & Francis, 1992.*)

making catalyst honeycomb structures are made using the extrusion process. Injection molding of ceramics [Figure 15-11(b)] is similar to injection molding of polymers (Chapter 16). Extrusion is used for relatively simple shapes of constant cross-section, whereas ceramic injection molding is better suited for complex shapes. In the process of injection molding, ceramic powder is mixed with a thermoplastic plasticizer and other additives. The mixture is then extruded and injected into a die to form a green ceramic. The polymer contained in the injection-molded ceramic is burnt off, and the rest of the ceramic body is sintered at a high temperature.

Sintering After the formation of a green ceramic, the ceramic powder particles are in direct contact with each other, but are not bonded together. The particles bond to each other at the points of contact between them during the high-temperature heat treatment known as sintering. The driving force for this process is the reduction in the surface area of powder since surfaces, with dangling bonds, are high-energy regions of the material. Atoms or ions diffuse along grain boundaries and from the bulk (volume diffusion) to the contact regions (known as the necks) between particles. Doing so causes the size of the pores between the powder particles to shrink, thereby leading to densification of the

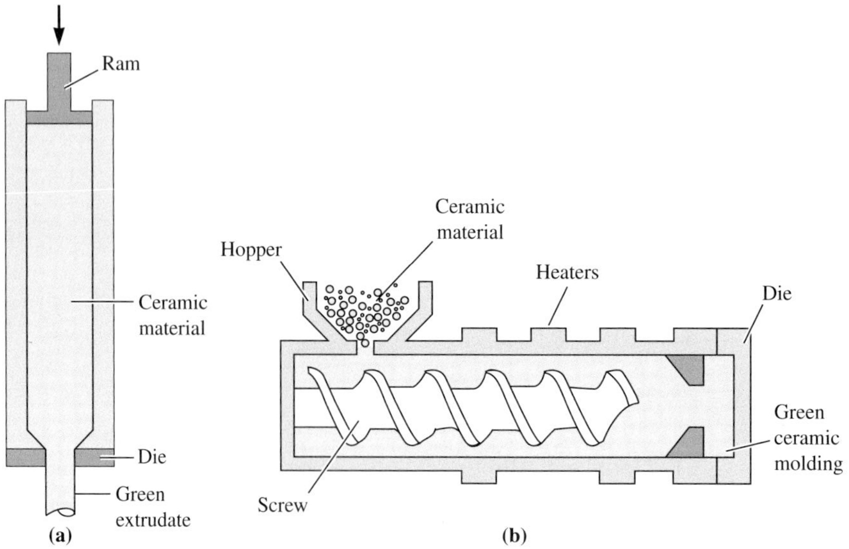

Figure 15-11 (a) Extrusion and (b) injection molding.

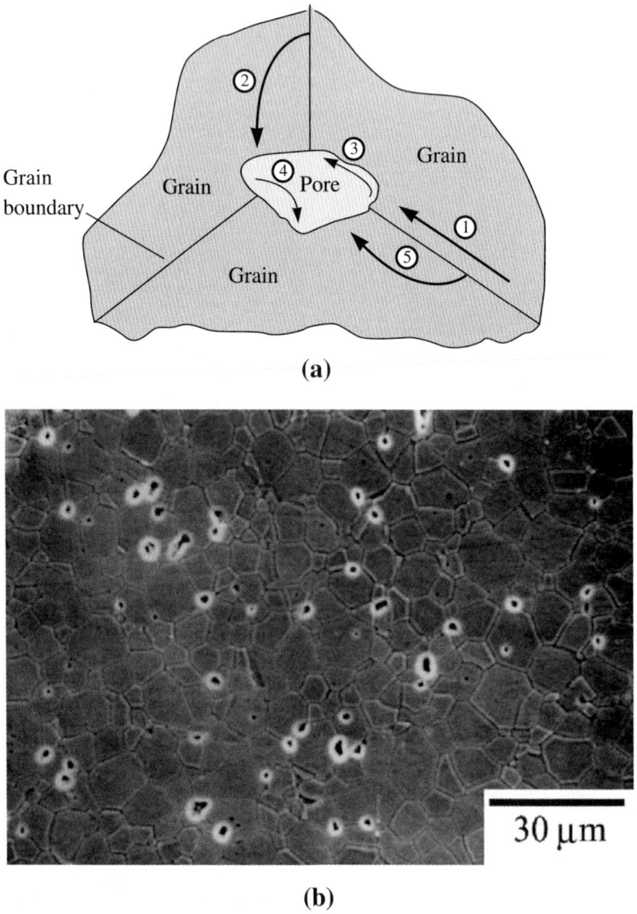

Figure 15-12 (a) Different diffusion mechanisms involved in sintering. The grain boundary and bulk diffusion (1, 2 and 5) to the neck contribute to densification. Evaporation condensation (4) and surface diffusion (3) do not contribute to densification. (*Based on* Physical Ceramics: Principles for Ceramic Science and Engineering, *Y.M. Chiang, D. Birnie, W.D. Kingery, John Wiley & Sons, Inc., 1997.*) (b) A partially sintered powder compact of Al_2O_3. Isolated pores are visible. (*Reprinted from* Sintering: Densification, Grain Growth, and Microstructure, *Suk-Joong L. Kang, Chapter 1: Sintering Processes, Pages 3–8, Copyright (2005), with permission from Elsevier.*)

material. Atoms or ions may also evaporate from the surface of the pore and condense on the same surface, and atoms or ions may diffuse along the pore surfaces. These two latter mechanisms, however, do not contribute to pore shrinkage and densification of the material [Figure 15-12(a)]. Sintering was also discussed in Section 5-9.

Sintering may be performed without the application of pressure, but ceramics are often sintered using applied pressure as well. This technique, known as **hot pressing**, is used for refractory and covalently bonded ceramics that do not show good pressureless sintering behavior. Similarly, large pieces of metals and alloys compacted using CIP can be sintered under hydrostatic pressure in a process known as **hot isostatic pressing** (HIP). In hot pressing or HIP, the applied pressure acts against the internal pore pressure and enhances densification without causing grain growth. Hot pressing and hot isostatic pressing also are used for making ceramics and metallic parts with almost no porosity. Some innovative processes that make use of microwaves (similar to the way food is heated in microwave ovens) have been developed for the drying and sintering of ceramic materials.

The average grain size achieved from sintering of ceramic particles is typically closely related to the primary particle size. An exception to this is if there is grain growth due to long sintering times or exaggerated or **abnormal grain growth**. Typically, ceramics with a small grain size are stronger than coarse-grained ceramics. Finer grain sizes help minimize stresses that develop at grain boundaries due to anisotropic expansion and contraction. Normally, starting with finer ceramic raw materials produces a fine grain size. Magnetic, dielectric, and optical properties of ceramic materials depend upon the average grain size and, in these applications, grain size must be controlled properly. Although we will not discuss this here in detail, in certain applications, it is important to use single crystals of ceramic materials so as to avoid the deleterious grain boundaries that are present in polycrystalline ceramics.

15-6 Silica and Silicate Compounds

Silicon dioxide, or silica, is well known to many people in the form of quartz. This crystalline mineral is abundant in nature, comprising a large fraction of the earth's crust, and it is often the dominant material in sand. Silica also exists as an amorphous material, and as such is the major component in architectural glass. But despite these familiarities and the simple molecular formula of silicon dioxide (SiO_2), silica and silica-based compounds, or "silicates," exist in myriad forms that may exhibit extraordinary structural or behavioral complexities. Some members of this broad family of materials find large-scale commercial applications, while others form the basis for a variety of semi-precious gemstones; still others are quite rare and can be created only under extreme conditions of pressure or temperature, or both.

Silica and silicate compounds exist in numerous structural forms, known as polymorphs (Section 3-4). With some rare exceptions, the common building block of these materials is an SiO_4 molecular unit that has the silicon atom sitting at the center of a tetrahedron of oxygen atoms, as shown in Figure 15-13(a). A single unit would be an anion of -4 charge, but actual silica and silicate materials, both crystalline and amorphous, comprise an extended network of these tetrahedral units that are charge neutral, whether or not they contain cation species.

Pure silica in its simplest forms has the oxygen atoms at every corner of every tetrahedral unit shared by two adjacent tetrahedra, providing the requisite SiO_2 stoichiometry. Even in this corner-sharing arrangement, where every oxygen is "bridging" two tetrahedra, a number of structurally distinct polymorphs are found, including the crystalline minerals tridymite, cristobalite, and quartz, each of which are stable within various temperature ranges. These minerals, in turn, may exhibit a variety of crystallographic structures; tridymite, for example, has been found with at least seven different crystalline variants.

Given the stability of the $(SiO_4)^{4-}$ tetrahedral unit, it is sometimes helpful to think of this molecular unit in the same way that we think of individual atoms when considering the complex crystal structures that silicates often exhibit. We then think of these

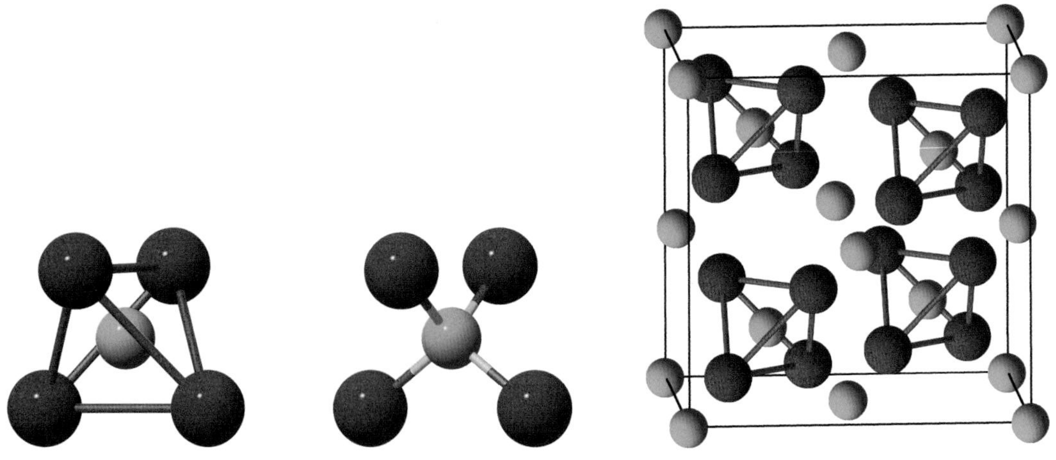

(a) (b)

Figure 15-13 (a) Two views of the SiO_4^{4-} tetrahedral unit; the tetrahedral cage of oxygen atoms (red) surrounding the silicon atom (green) is on the left, while the actual Si—O bonds are illustrated on the right. (b) A unit cell of β-cristobalite with four SiO_4 tetrahedral units shown within.

tetrahedra, rather than single atoms, as stacked or arranged in some repeating fashion and thereby forming a crystal. One form of cristobalite for instance, known as β-cristobalite, can be modeled as $(SiO_4)^{4-}$ tetrahedra occupying the same positions as individual carbon atoms in the diamond cubic structure. For this reason, β-cristobalite can also be thought of as silicon atoms occupying their "normal" positions in crystalline silicon, which exhibits the diamond cubic structure, with a single oxygen atom located half-way between every pair of silicon atoms. Recalling that there are, e.g., silicon atoms at the (0, 0, 0) and (1/4, 1/4, 1/4) basis positions in an FCC lattice, we find oxygen atoms at the (1/8, 1/8, 1/8) and equivalent positions. This structure is illustrated in Figure 15-13(b).

In addition to sharing corners, SiO_4 tetrahedra may share edges or faces, creating new building blocks that in turn are also repeatedly arrayed. Two tetrahedra that share a single edge, for instance, form a unit with SiO_3 stoichiometry which itself may exist in several configurations as the two tetrahedra rotate around their common edge. Furthermore, these building blocks may form long chains, rings, and sheets, that, when combined with additional elements, comprise one of the several thousand silicate materials. Garnet, for example, which is often mistaken for a single material, is a class of silicates with the general formula unit of $A_3B_2(SiO_4)_3$, where the A species are typically divalent (+2) cations of calcium, magnesium, iron, or manganese, and the B species are trivalent (+3) cations of aluminum, iron or chromium.

Exemplary of many sheet structures is another common silicate glass, mica, which demonstrates nicely one of the fundamental principles of materials science, that structure and properties are closely related. Micas, or micaceous silicates, exhibit a structure in which the bonding of the various constituents within each sheet is strong compared to the bonding between the sheets. As a result, many micaceous silicates exhibit highly preferential fracture along a specific plane. This is similar to the easy fracture within pure carbon in the form of graphite, despite the vastly more complex structure of micas. In one family, micas have the general formula unit of $A_3B_4C_8O_{20}(OH, F)_4$ in which the A species may typically be potassium, sodium, or calcium, the B species may be aluminum, magnesium, or iron, and the C species is typically silicon or aluminum. With all of the chemical and configurational possibilities, it becomes easy to understand why the silicate family is so vast.

As mentioned earlier, silica-based materials exist in both crystalline and amorphous forms. The use of the term "glass" or "glassy" to indicate any amorphous material stems from the fact that conventional glasses such as in windows or bottles are all amorphous. Of course it is the transparency of silicate glasses that for millennia have given rise to their architectural and artistic uses. "Soda-lime" glasses are the most common form of structural glasses. "Soda" refers to one of the starting materials – sodium carbonate – that is used to make glass, which incorporates various amounts of silicon, oxygen, sodium, calcium, aluminum, and often other elements in very small quantities.

By varying their composition and processing, glasses with different sets of optical and mechanical properties are achieved. Green bottle glass includes, for instance, iron oxides. **Tempered glasses** are treated thermally or chemically to increase the toughness of conventional "plate" glasses as used in windows and bottles. Pyrex® is a trade brand of boron-containing glasses used in kitchenware, that, because of their low thermal-expansion coefficients, have also been used to manufacture telescope lenses. Silica-based materials are used in thermal insulation, refractories, abrasives, as fibers in fiber-reinforced composites, and laboratory glassware. The space shuttle made use of ~25,000 reusable, lightweight, highly porous ceramic tiles that protected the aluminum frame from the heat generated during re-entry into the earth's atmosphere. These tiles were made from high-purity silica fibers and colloidal silica coated with a borosilicate glass. Powders made using fine particles of silica are used in tires, paints, and many other applications.

Silica in many forms is also at the heart of several classes of high technology products. Optical communication systems are based on thin fibers of silica glass and offer nearly unlimited data transmission rates compared to conventional copper wire-based cables. The faces of many "touch-screen" devices, such as smart phones, use specially toughened glasses that are highly resistant to fracture. And all silicon-based microelectronics (integrated circuits or "chips") are based in part on the highly specific electrical properties of ultrathin layers of amorphous silicon dioxide, some of which are now only a few atomic layers thick.

15-7 Inorganic Glasses

In this section, we will focus more generally on inorganic glasses. While there are many glasses based on sulfides, fluorides, and various alloys, those based on silica, as just described, are the most widely used. A **glass** is a metastable material that has hardened and become rigid without crystallizing. There are many factors that enable crystallization to be bypassed for glasses, but the combination of limited diffusion of the atoms or ions coupled with a sufficiently rapid cooling rate is the most critical. High viscosities reflect the sluggish kinetics of glass forming liquids.

Figure 15-14 shows a schematic diagram for the density of a material that has both crystalline and amorphous forms as a function of temperature. At the melting temperature, a crystalline phase forms from the liquid (in the manner described in Chapter 9),

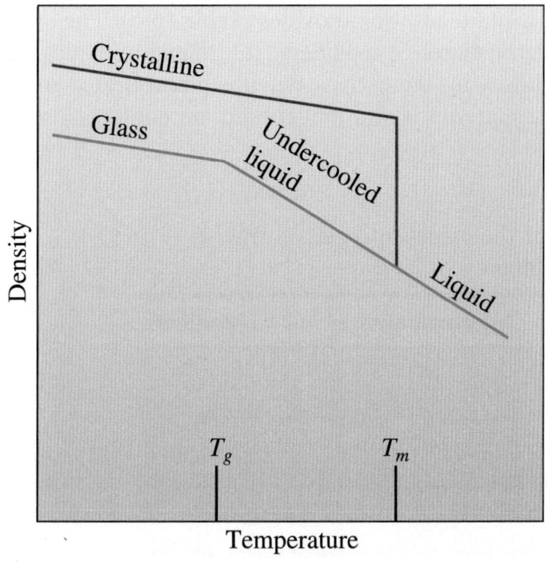

Figure 15-14 When silica crystallizes on cooling, an abrupt change in the density is observed. For glassy silica, however, the change in slope at the glass-transition temperature indicates the formation of a glass from the undercooled liquid. Glass does not have a fixed T_m or T_g. Crystalline materials have a fixed T_m, and they do not have a T_g.

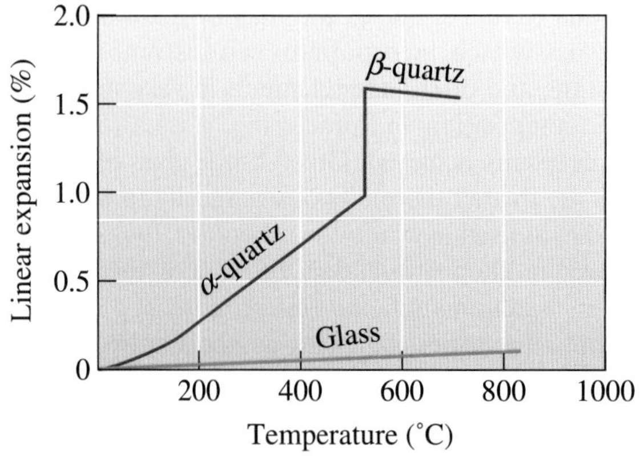

Figure 15-15 The thermal-expansion coefficient of fused silica (amorphous silica) is lower than that of crystallize quartz. The discontinuity in the curve for crystalline quartz occurs due to the α to β phase transition. Note that solidification from the melt to the β-quartz and glassy polymorphs is not shown in this figure.

and an abrupt change in density occurs. A glass that forms from the same melt, however, does not show a change in the slope of the density curve at the melting temperature. A critical temperature in the formation of glass is known as the **glass-transition temperature** T_g. Until the glass transition temperature is reached on cooling, we can think of the material as an undercooled liquid. Upon reaching the glass transition temperature, the material is considered to be a glass rather than an undercooled liquid. A change of slope in the density versus temperature plot is observed, but there is no discontinuity. Note that the glass-transition temperature is not a fixed temperature; for example, it depends on composition and the rate of cooling.

Figure 15-15 is a plot of the linear thermal-expansion coefficient of three polymorphs of pure silica as a function of temperature. Two crystalline forms are shown: β-quartz and α-quartz as well as the glassy form known as fused or vitreous silica. On cooling from high temperatures at slow rates of cooling, the molten liquid first crystallizes as β-quartz. At a temperature of 573°C, the β-quartz transforms to α-quartz. The discontinuity in the value of the thermal-expansion coefficient reflects the different crystal structures of α- and β-quartz. In contrast, when the melt is cooled at a rate greater than the critical cooling rate, a glassy solid forms. There is no abrupt transition in the plot of the thermal-expansion coefficient because there is no abrupt change in structure. The thermal-expansion coefficient of glassy silica is lower than that of the β-quartz and α-quartz because its relatively open structure allows it to accommodate thermal energy by rotation of the tetrahedral units rather than an increase in bond length.

Silica is known as a **glass former**, meaning that it easily produces an amorphous phase(s) during processing. Other glass formers include B_2O_3, GeO_2, and P_2O_5. Generally the silicate glasses contain additional oxides (Table 15-5). The oxides that are added can be classified as **intermediates** or **modifiers**. Intermediate oxides (such as aluminum oxide) do not form glasses by themselves but are incorporated into the structure of glass formers; thus, the material is still amorphous on cooling despite the addition of

Table 15-5 Division of the oxides into glass formers, intermediates, and modifiers

Glass Formers	Intermediates	Modifiers
B_2O_3	TiO_2	Y_2O_3
SiO_2	ZnO	MgO
GeO_2	PbO_2	CaO
P_2O_5	Al_2O_3	PbO
V_2O_3	BeO	Na_2O

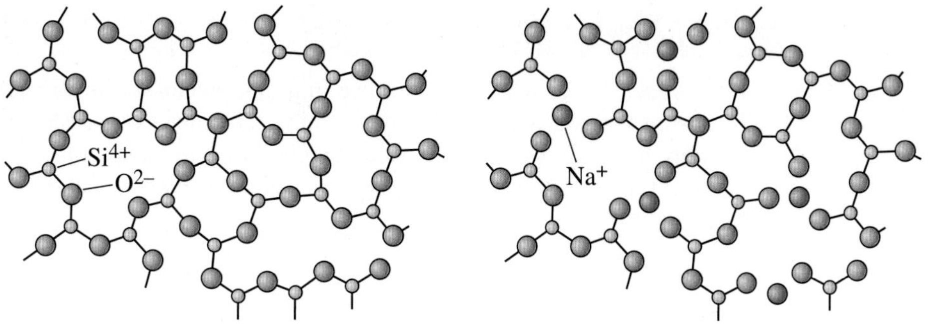

Figure 15-16 The effect of Na_2O on the silica glass network. Sodium oxide is a modifier, disrupting the glassy network and reducing the ability to form a glass.

the intermediates. Intermediates tend to increase the thermal shock resistance of glass by increasing the amount of free volume between atoms (or the empty space in the network structure), making the bonds more readily able to accommodate rapid dimensional change. As the group of oxides known as modifiers are added, they break up the silica structure of the glass formers and eventually cause the glass to **devitrify** (crystallize). In so doing, modifiers reduce the melting point and viscosity of a glass for ease of processing. Modification can be understood by considering the O:Si ratio. For pure silica, this ratio is 2:1. When Na_2O is added to silica, for example as a modifier, the sodium ions enter holes within the network rather than becoming part of the network; however, the oxygen ion that enters with the Na_2O does become part of the network (Figure 15-16). When this happens, there are not enough silicon ions to combine with the extra oxygen ions and keep the network intact. Eventually, a high O:Si ratio causes the remaining silica tetrahedra to form chains, rings, or compounds, and the silica no longer transforms to a glass. When the O:Si ratio is above about 2.5, silica glass is difficult to form; above a ratio of three, a glass forms only when special precautions are taken, such as the use of rapid cooling rates. As you can see in Figure 15-17, the addition of Na_2O produces eutectics with very low melting temperatures. Adding CaO, which reduces the solubility of the glass in water thereby enhancing its chemical stability, further modifies these glasses. The example that follows shows how to design a glass.

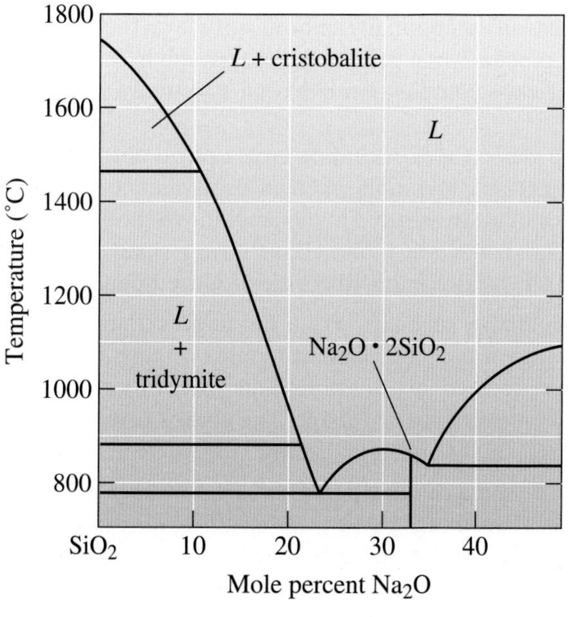

Figure 15-17 The SiO_2-Na_2O phase diagram. Additions of soda (Na_2O) to silica dramatically reduce the melting temperature of silica by forming eutectics.

Example 15-9 *Design of a Glass*

We produce good chemical resistance in a glass when we introduce B_2O_3 into silica. To ensure that we have good glass-forming tendencies, we wish the O:Si ratio to be no more than 2.5, but we also want the glassware to have a low melting temperature to make the glass-forming process easier and more economical. Design such a glass.

SOLUTION

Because B_2O_3 reduces the melting temperature of silica, we would like to add as much as possible. We also, however, want to ensure that the O:Si ratio is no more than 2.5, so the amount of B_2O_3 is limited. As an example, let us determine the amount of B_2O_3 we must add to obtain an O:Si ratio of exactly 2.5. Let f_B be the mole fraction of B_2O_3 added to the glass, and $(1 - f_B)$ be the mole fraction of SiO_2:

$$\frac{O}{Si} = \frac{\left(3\,\frac{\text{O ions}}{\text{B}_2\text{O}_3}\right)(f_B) + \left(2\,\frac{\text{O ions}}{\text{SiO}_2}\right)(1 - f_B)}{\left(1\,\frac{\text{Si ion}}{\text{SiO}_2}\right)(1 - f_B)} = 2.5$$

$$3f_B + 2 - 2f_B = 2.5 - 2.5f_B \quad \text{or} \quad f_B = 0.143$$

Therefore, we must produce a glass containing no more than 14.3 mol% B_2O_3. In weight percent:

$$\text{wt\% B}_2\text{O}_3 = \frac{(f_B)(69.62 \text{ g/mol})}{(f_B)(69.62 \text{ g/mol}) + (1 - f_B)(60.08 \text{ g/mol})} \times 100$$

$$\text{wt\% B}_2\text{O}_3 = \frac{(0.143)(69.62 \text{ g/mol})}{(0.143)(69.62 \text{ g/mol}) + (0.857)(60.08 \text{ g/mol})} \times 100 = 16.2$$

Glasses are manufactured into useful articles at a high temperature by controlling the viscosity so that the glass can be shaped without breaking. Figure 15-18 helps us understand the processing in terms of the viscosity ranges.

1. *Liquid range*. Sheet and plate glass are produced when the glass is in the molten state. Techniques include rolling the molten glass through water-cooled rolls or floating the molten glass over a pool of liquid tin (Figure 15-19). The liquid-tin process produces an exceptionally smooth surface on the glass. The development of the float-glass process was a genuine breakthrough in the area of glass processing. The basic float-glass composition has been essentially unchanged for many years (Table 15-6).

   Some glass shapes, including large optical mirrors, are produced by casting the molten glass into a mold, then ensuring that cooling is as slow as possible to minimize residual stresses and avoid cracking of the glass part. Glass fibers may be produced by drawing the liquid glass through small openings in a platinum die (Figure 15-20). Typically, many fibers are produced simultaneously for a single die.
2. *Working range*. Shapes such as those of containers or lightbulbs can be formed by pressing, drawing, or blowing glass into molds (Figure 15-20). A hot *gob* of liquid glass may be pre-formed into a crude shape (parison), then pressed or blown into a heated die to produce the final shape. The glass is heated to the working range so that the glass is formable, but not "runny."
3. *Annealing range*. Some ceramic parts may be annealed to reduce residual stresses introduced during forming. Large glass castings, for example, are often annealed and slowly cooled to prevent cracking. Some glasses may be heat treated to cause devitrification, or the precipitation of a crystalline phase from the glass.

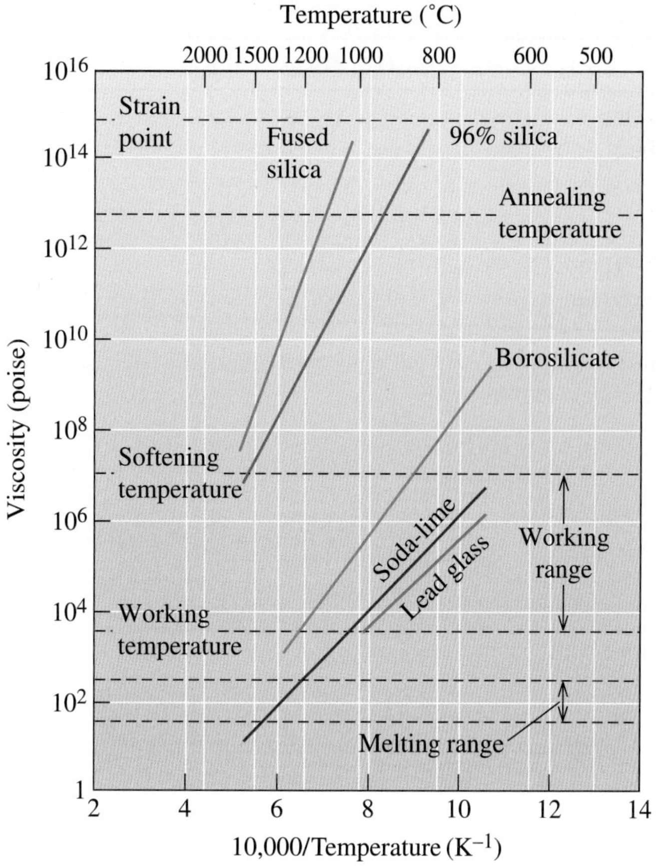

Figure 15-18 The effect of temperature and composition on the viscosity of glass.

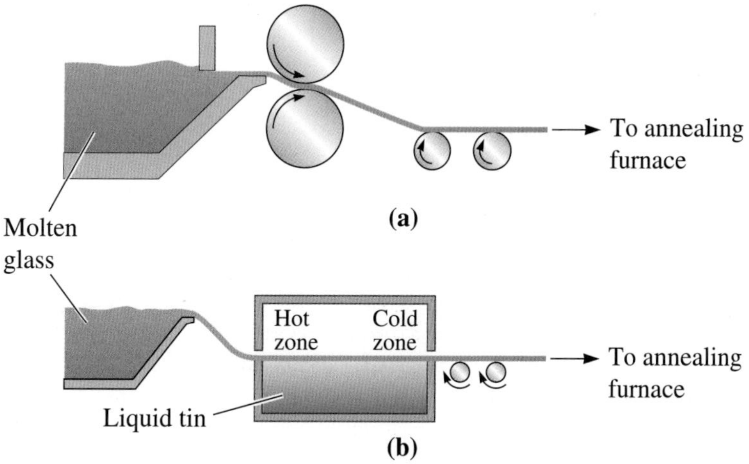

Figure 15-19 Techniques for manufacturing sheet and plate glass: (a) rolling and (b) floating the glass on molten tin.

Tempered glass is produced by quenching the surface of plate glass with air, causing the surface layers to cool and contract. When the center cools, its contraction is restrained by the already rigid surface, which is placed in compression (Figure 15-21). Tempered glass is capable of withstanding much higher tensile stresses and impact blows than untempered glass. Additionally, when tempered glass does fail, it breaks into small, round pieces rather than the sharp shards that form when untempered glass fails. Consequently, tempered glass is used in car and home windows, shelving for refrigerators, ovens, furniture, and many other applications where safety is important. Laminated glass, consisting of two annealed glass pieces with a polymer (such as polyvinyl butyral or PVB) in between, is used to make car windshields.

Table 15-6 Compositions of typical glasses (in weight percent)

Glass	SiO_2	Al_2O_3	CaO	Na_2O	B_2O_3	MgO	PbO	Others
Fused silica	99							
Vycor™	96				4			
Pyrex™	81	2		4	12			
Glass jars	74	1	5	15		4		
Window glass	72	1	10	14		2		
Plate glass/Float glass	73	1	13	13				
Lightbulbs	74	1	5	16		4		
Fibers	54	14	16		10	4		
Thermometer	73	6		10	10			
Lead glass	67			6			17	10% K_2O
Optical flint	50			1			19	13% BaO, 8% K_2O, ZnO
Optical crown	70			8		10		2% BaO, 8% K_2O
E-glass fibers	55	15	20		10			
S-glass fibers	65	25				10		

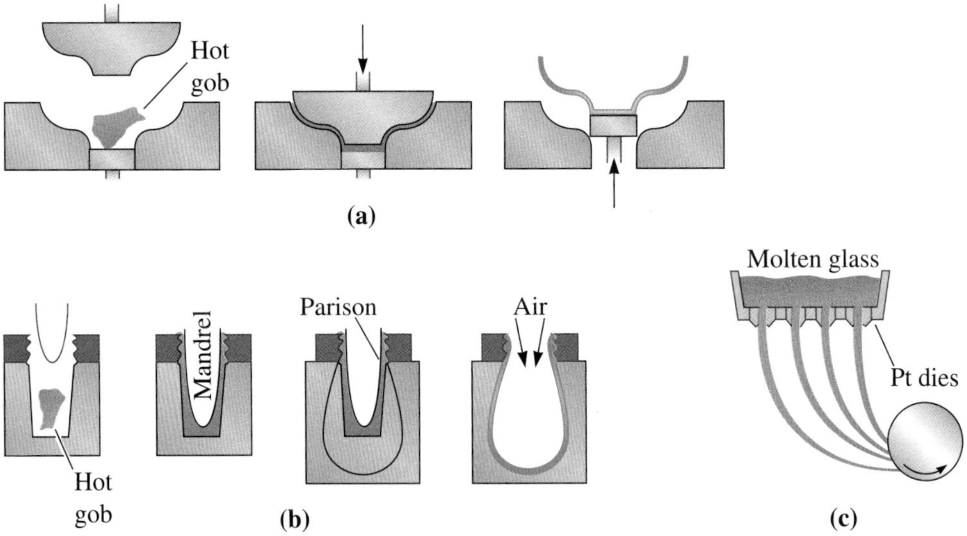

Figure 15-20 Techniques for forming glass products: (a) pressing, (b) press and blow process, and (c) drawing of fibers.

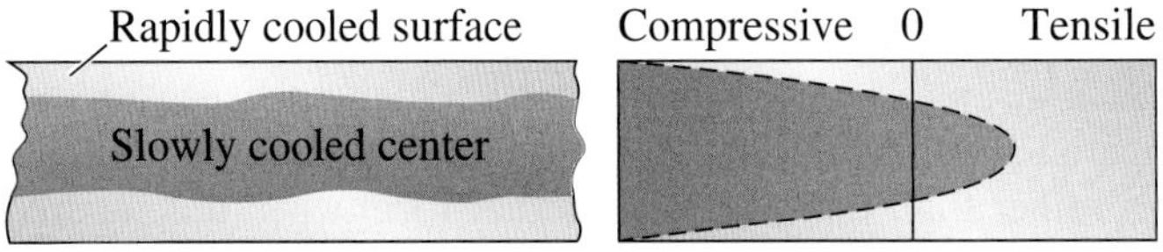

Figure 15-21 Tempered glass is cooled rapidly to produce compressive residual stresses at the surface.

Glass Compositions Pure SiO_2 must be heated to very high temperatures to obtain viscosities that permit economical forming. Most commercial glasses are based on silica modifiers such as soda (Na_2O) to break down the network structure and form eutectics with low melting temperatures, whereas lime (CaO) is added to reduce the solubility of the glasses in water. The most common commercial glass contains approximately 75% SiO_2, 15% Na_2O, and 10% CaO and is known as soda lime glass.

Borosilicate glasses, which contain about 15% B_2O_3, have excellent chemical and dimensional stability. Their uses include laboratory glassware (Pyrex^{TM}), glass-ceramics, and containers for the disposal of radioactive nuclear waste. Calcium aluminoborosilicate glass—or E-glass—is used as a general-purpose fiber for composite materials, such as fiberglass. Aluminosilicate glass, with 20% Al_2O_3 and 12% MgO, and high-silica glasses, with 3% B_2O_3, are excellent for high-temperature resistance and for protection against heat or thermal shock. The S-glass, a magnesium aluminosilicate, is used to produce high-strength fibers for composite materials. Fused silica, or virtually pure SiO_2, has the best resistance to high temperature, thermal shock, and chemical attack, although it is also expensive.

Special optical qualities can also be obtained, including sensitivity to light. Photochromic glass, which is darkened by the ultraviolet portion of sunlight, is used for sunglasses. Photosensitive glass darkens permanently when exposed to ultraviolet light; if only select portions of the glass are exposed and then immersed in hydrofluoric acid, etchings can be produced. Polychromatic glasses are sensitive to all light, not just ultraviolet radiation. Similarly, nanosized crystals of semiconductors such as cadmium sulfide (CdS) are nucleated in silicate glasses in a process known as *striking*. These glasses exhibit lively colors and have useful optical properties.

15-8 Glass-Ceramics

Glass-ceramics are crystalline materials that are derived from amorphous glasses. Usually, glass-ceramics have a substantial level of crystallinity (~70–99%). With glass-ceramics, we can take advantage of the formability and density of glass. Also, a product that contains very low porosity can be obtained by producing a shape with conventional glass-forming techniques, such as pressing or blowing.

The first step in producing a glass-ceramic is to ensure that crystallization does not occur during cooling from the forming temperature. Continuous cooling and isothermal cooling transformation diagrams, much like the CCT and TTT diagrams for steels, can be constructed for silicate-based glasses. Figure 15-22(a) shows a TTT diagram for a glass. If glass cools too slowly, a transformation line is crossed; nucleation and growth of the crystals begin, but in an uncontrolled manner. The addition of modifying oxides to glass, much like the addition of alloying elements to steel, shifts the transformation curve to longer times and prevents devitrification even at slow cooling rates. As noted in previous chapters, strictly speaking, we should make use of CCT (and not TTT) diagrams for this discussion.

Nucleation of the crystalline phase is controlled in two ways. First, the glass contains agents, such as TiO_2, that react with other oxides and form phases that provide the nucleation sites. Second, a heat treatment is designed to provide the appropriate number of nuclei; the temperature should be relatively low in order to maximize the rate of nucleation [Figure 15-22(b)]. The overall rate of crystallization depends on the growth rate of the crystals once nucleation occurs; thus, higher temperatures are then required to maximize the growth rate. Consequently, a heat-treatment schedule similar to that shown in Figure 15-22(c) for the Li_2O-Al_2O_3-SiO_2 glass-ceramics (Pyroceram^{TM}) can be used. The low-temperature step provides nucleation sites, and the high-temperature step speeds the rate of growth of the crystals. As much as 99% of the part may crystallize.

This special structure of glass-ceramics can provide good mechanical strength and toughness, often with a low coefficient of thermal expansion and high-temperature corrosion resistance. Perhaps the most important glass-ceramic is based on the

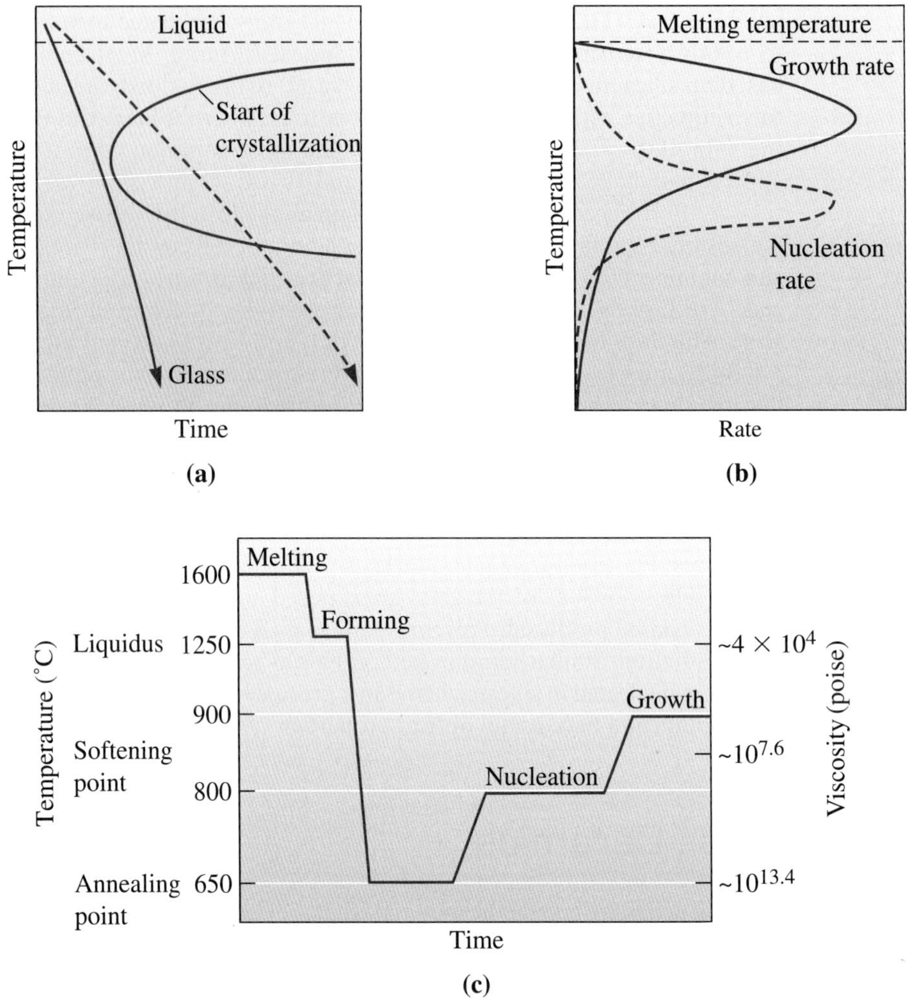

Figure 15-22 Producing a glass-ceramic: (a) Cooling must be rapid to avoid the start of crystallization. (b) The rate of nucleation of precipitates is high at low temperatures, whereas the rate of growth of the precipitates is high at higher temperatures. (c) A typical heat-treatment profile for glass-ceramics fabrication, illustrated here for Li_2O-Al_2O_3-SiO_2 glasses.

Li_2O-Al_2O_3-SiO_2 system. These materials are used for cookware (Corning Ware™) and ceramic tops for stoves. Other glass-ceramics are used in communications, computer, and optical applications.

15-9 Processing and Applications of Clay Products

Clay products are those that use clay and water as a binder for ceramic particles (typically the particles are silica). Clay products have been used since antiquity in the form of pottery and bricks. For this reason, today's clay products are considered in the class of so-called "traditional ceramics." They are still used for pottery and bricks as well as sewer pipes, cookware, dinnerware, and other common products. Clays are hydrous sheet aluminum silicates with a crystalline structure; kaolinite, a mineral with the formula $Al_2Si_2O_5(OH)_4$ is the most common clay material. Other materials, such as feldspar [$(K, Na)_2O \cdot Al_2O_3 \cdot 6SiO_2$], are incorporated to serve as **fluxing** agents during later heat treatment.

Forming Techniques for Clay Products

The powders, clay, flux, and water are mixed and formed into a desired shape. Dry or semi-dry mixtures are mechanically pressed into "green" (unbaked) shapes of sufficient strength to be handled. For more uniform compaction of complex shapes, isostatic pressing may be done; the powders are placed into a rubber mold and subjected to high pressures through a gas or liquid medium. Higher moisture contents permit the powders to be more plastic or formable. **Hydroplastic forming** processes, such as extrusion, jiggering (forming of clay in a rotating mold using a profile tool to form the inner surface), and hand working, can be applied to these pliable, plastic mixes. Ceramic slurries containing large amounts of organic plasticizers (substances that impart properties such as softness), rather than water, can be injected into molds.

Still higher moisture contents permit the formation of a slip, or pourable slurry, containing fine ceramic powder. The slip is poured into a porous mold, and slip casting is performed (Figure 15-10). Slip casting is used in manufacturing washbasins and other commercial products. After forming, the ceramic bodies—or greenware—are still weak, contain water or other lubricants, and are porous, and subsequent drying and firing are required.

Drying and Firing of Clay Products

During drying, excess moisture is removed and large dimensional changes occur. Initially, the water between the clay platelets—the interparticle water—evaporates and provides most of the shrinkage. Relatively little dimensional change occurs as the remaining water in the pores evaporates. The temperature and humidity are controlled to provide uniform drying throughout the part, thus minimizing stresses, distortion, and cracking.

The rigidity and strength of a ceramic part are obtained during **firing**. During heating, the clay dehydrates, eliminating the hydrated water that is part of the kaolinite crystal structure (Figure 15-23). Impurities and the fluxing agent react with the ceramic particles (SiO_2) and clay, producing a low melting-point liquid phase at the grain surfaces. The liquid helps eliminate porosity and, after cooling, changes to a rigid glass that binds the ceramic particles. This glassy phase provides a **ceramic bond**, but it also causes additional shrinkage of the entire ceramic body.

The grain size of the final part is determined primarily by the size of the original powder particles. Furthermore, as the amount of **flux** increases, the melting temperature decreases; more glass forms, and the pores become rounder and smaller. A smaller initial grain size accelerates this process by providing more surface area at which **vitrification** can occur.

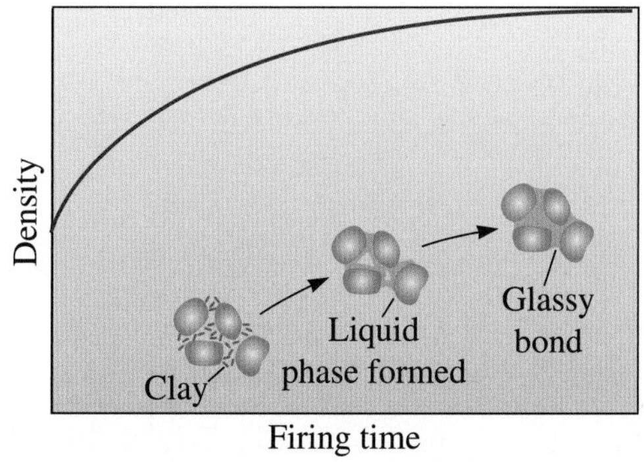

Figure 15-23 During firing, clay and other fluxing materials react with coarser particles to produce a glassy bond and reduce porosity.

Applications of Clay Products Many structural clay products and whitewares are produced using these processes. Brick and tile used for construction are pressed or extruded into shape, dried, and fired to produce the ceramic bond. Higher firing temperatures or finer original particle sizes produce more vitrification, less porosity, and higher density. The higher density improves mechanical properties but reduces the insulating qualities of the brick or tile.

Earthenware are porous clay bodies fired at relatively low temperatures. Little vitrification occurs; the porosity is very high and interconnected, and earthenware ceramics may leak. Consequently, these products must be covered with an impermeable glaze.

Higher firing temperatures, which provide more vitrification and less porosity, produce stoneware. The stoneware, which is used for drainage and sewer pipe, contains only 2% to 4% porosity. Ceramics known as china and porcelain require even higher firing temperatures to cause complete vitrification and virtually no porosity.

15-10 Refractories

Refractory materials are important components of the equipment used in the production, refining, and handling of metals and glasses, for constructing heat-treating furnaces, and for other high-temperature processing equipment. **Refractories** must survive at high temperatures without being corroded or weakened by the surrounding environment. Typical refractories are composed of coarse oxide particles bonded by a finer refractory material. The finer material melts during firing, providing bonding. In some cases, refractory bricks contain about 20% to 25% apparent porosity to provide improved thermal insulation.

Refractories are often divided into three groups—acidic, basic, and neutral—depending on their chemical behavior (Table 15-7). Acidic refractories are acidic by nature (i.e., they form hydrogen ions in aqueous environments). Acidic refractories are used in acidic environments. They cannot be used in basic environments as they will corrode and degrade. Similarly, basic refractories are used in basic environments. Neutral refractories are used in applications for which the refractory is subjected to both acidic and basic environments. We will discuss the materials and applications for each of these groups.

Table 15-7 Compositions of typical refractories (weight percents)

Refractory	SiO_2	Al_2O_3	MgO	Fe_2O_3	Cr_2O_3
Acidic					
Silica	95–97				
Superduty firebrick	51–53	43–44			
High-alumina firebrick	10–45	50–80			
Basic					
Magnesite			83–93	2–7	
Olivine	43		57		
Neutral					
Chromite	3–13	12–30	10–20	12–25	30–50
Chromite-magnesite	2–8	20–24	30–39	9–12	30–50

Based on Ceramic Data Book, Cahners Publishing Co., 1982.

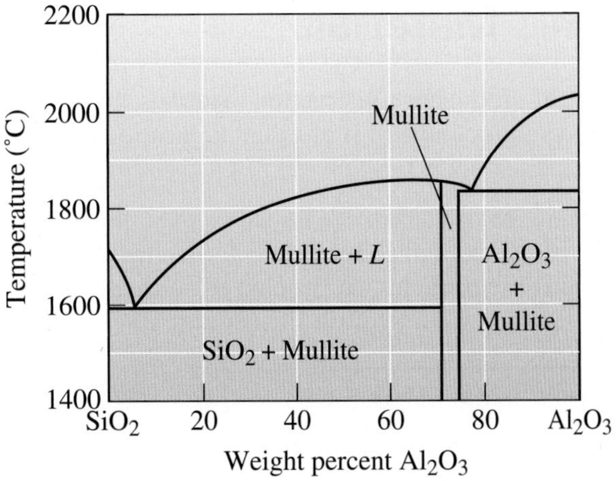

Figure 15-24 A simplified SiO_2-Al_2O_3 phase diagram, the basis for alumina silicate refractories.

Acidic Refractories Common acidic refractories include silica, alumina, and fireclay (a kaolinite with some impurities). Pure silica is sometimes used to contain molten metal. In some applications, the silica may be bonded with small amounts of boron oxide, which melts and produces the ceramic bond. When a small amount of alumina is added to silica, the refractory contains a very low melting-point eutectic microconstituent (Figure 15-24) and is not suited for refractory applications at temperatures above about 1600°C, a temperature often required for steel making. When larger amounts of alumina are added, the microstructure contains increasing amounts of mullite, $3Al_2O_3 \cdot 2SiO_2$, which has a high melting temperature. These refractories are generally relatively weak, but they are inexpensive. Alumina concentrations above about 50% constitute the high-alumina refractories.

Basic Refractories A number of refractories are based on MgO (magnesia, or periclase). Pure MgO has a high melting point, good refractory properties, and good resistance to attack by the basic environments often found in steel making processes. Olivine is a mineral that occurs as a solid solution of Mg_2SiO_4 and Fe_2SiO_4. Pure Mg_2SiO_4 is known as forsterite, and it has an extremely high melting temperature of nearly 1900°C. Olivine is sometimes added to furnaces to form a slag, the by-product of refinement of metal ores. Olivine sand is sometimes used to contain molten metals in sand casting. Other magnesia refractories may include CaO or carbon. Typically, the basic refractories are more expensive than the acidic refractories.

Neutral Refractories These refractories, which include chromite and chromite-magnesite, might be used to separate acidic and basic refractories, preventing them from attacking one another.

Special Refractories Carbon, or graphite, is used in many refractory applications, particularly when oxygen is not present. Other refractory materials include zirconia (ZrO_2), zircon ($ZrO_2 \cdot SiO_2$), and a variety of nitrides, carbides, and borides. Most of the carbides, such as TiC and ZrC, do not resist oxidation well, and their high-temperature applications are best suited to reducing (non-oxidizing) conditions. Silicon carbide is an exception, however; when SiC is oxidized at high temperatures, a thin layer of SiO_2 forms at the surface, protecting the SiC from further oxidation up to about 1500°C. Nitrides and borides also have high melting temperatures and are less susceptible to oxidation. Some of the oxides and nitrides are candidates for use in jet engines.

15-11 Other Ceramic Materials

In addition to their use in producing construction materials, appliances, structural materials, and refractories, ceramics find a host of other applications, including the following.

Cements Ceramic raw materials are joined using a binder that does not require firing or sintering in a process called **cementation**. A chemical reaction converts a liquid resin to a solid that joins the particles. In the case of sodium silicate, the introduction of CO_2 gas acts as a catalyst to dehydrate the sodium silicate solution into a glassy material:

$$x\text{Na}_2\text{O} \cdot y\text{SiO}_2 \cdot \text{H}_2\text{O} + \text{CO}_2 \rightarrow \text{glass (not balanced)}$$

Figure 15-25 shows silica sand grains used to produce molds for metal casting. The liquid sodium silicate coats the sand grains and provides bridges between the sand grains. Introduction of the CO_2 converts the bridges to a solid, joining the sand grains.

Fine alumina powder dispersions catalyzed with phosphoric acid produce an aluminum phosphate cement:

$$\text{Al}_2\text{O}_3 + 2\text{H}_3\text{PO}_4 \rightarrow 2\text{AlPO}_4 + 3\text{H}_2\text{O}$$

When alumina particles are bonded with the aluminum phosphate cement, refractories capable of operating at temperatures as high as 1650°C are produced.

Plaster of Paris, or gypsum, is another material that is hardened by a cementation reaction:

$$\text{CaSO}_4 \cdot \tfrac{1}{2}\text{H}_2\text{O} + \tfrac{3}{2}\text{H}_2\text{O} \rightarrow \text{CaSO}_4 \cdot 2\text{H}_2\text{O}$$

When the liquid slurry reacts, interlocking solid crystals of gypsum ($CaSO_4 \cdot 2H_2O$) grow with very small pores between the crystals. Larger amounts of water in the original slurry provide more porosity and decrease the strength of the final plaster. One of the important uses of this material is for construction of walls in buildings.

The most common and important of the cementation reactions occurs in Portland cement, which is used to produce concrete.

Coatings Ceramics are often used to provide protective coatings to other materials. Common commercial coatings include **glazes** and **enamels**. Glazes are applied to the surface of a ceramic material to seal a permeable clay body, to provide protection and decoration, or for special purposes. Enamels are applied to metal surfaces. The enamels and glazes are clay products that vitrify easily during firing. A common composition is $CaO \cdot Al_2O_3 \cdot 2SiO_2$.

Special colors can be produced in glazes and enamels by the addition of other minerals. Zirconium silicate gives a white glaze, cobalt oxide makes the glaze blue, chromium

Figure 15-25 A micrograph of silica sand grains bonded with sodium silicate through the cementation mechanism (× 60). *(Reprinted Courtesy of Don Askeland.)*

oxide produces green, lead oxide gives a yellow color, and a red glaze may be produced by adding a mixture of selenium and cadmium sulfides.

One of the problems encountered with a glaze or enamel is surface cracking, or crazing, which occurs when the glaze has a coefficient of thermal expansion different than that of the underlying material. This is frequently the most important factor in determining the composition of the coating.

Special coatings are used for advanced ceramics and high-service temperature metals. SiC coatings are applied to carbon-carbon composite materials to improve their oxidation resistance. Zirconia coatings are applied to nickel-based superalloys to provide thermal barriers that protect the metal from melting or adverse reactions.

Thin Films and Single Crystals Thin films of many complex and multi-component ceramics are produced using different techniques such as sputtering, sol-gel, and chemical-vapor deposition (CVD). Usually, the thickness of such films is 0.05 to 10 μm and more likely greater than 2 μm. Many functional electronic ceramic thin films are prepared and integrated onto silicon wafers, glasses, and other substrates. For example, the magnetic strips on credit cards use iron oxide (γ-Fe_2O_3 or Fe_3O_4) thin films for storing data. Indium tin oxide (ITO), a conductive and transparent material, is coated on glass and used in applications such as touch-screen displays (Table 15-1). Many other coatings are used on glass to make the glass energy efficient. Self-cleaning glasses typically have TiO_2 (titania) coatings. Thin films of ceramics, such as lead zirconium titanate (PZT), lanthanum-doped PZT (PLZT), and $BaTiO_3$-$SrTiO_3$ solid solutions, can be prepared and used. Often, the films develop an orientation, or texture, which may be advantageous for a given application. Single crystals of ceramics [e.g., SiO_2 or quartz, lithium niobate ($LiNbO_3$), sapphire, or yttrium aluminum garnate] are used in many electrical and electro-optical applications. These crystals are grown from melts using techniques similar to those described in Chapter 9.

Fibers Fibers are produced from ceramic materials for several uses: as a reinforcement in composite materials, for weaving into fabrics, or for use in fiber-optic systems. Borosilicate glass fibers, the most commonly produced fibers, provide strength and stiffness in fiberglass. Fibers can be produced from a variety of other ceramics, including alumina, silicon carbide, silica, and boron carbide. The sol-gel process is also used to produce commercial fibers for many applications.

A special type of fibrous material is the silica tile that was used to provide the thermal protection system for NASA's space shuttle. Silica fibers were bonded with colloidal silica to produce an exceptionally lightweight tile with densities as low as 0.144 g/cm^3; the tile was coated with special high-emissivity glazes to permit protection up to 1300°C.

Joining and Assembly of Ceramic Components Ceramics are often made as monolithic components rather than assemblies of numerous components. When two ceramic parts are placed in contact under a load, stress concentrations at the brittle surface are created, leading to an increased probability of failure. In addition, methods for joining ceramic parts into a larger assembly are limited. The brittle ceramics cannot be joined by fusion welding or deformation bonding processes. At low temperatures, adhesive bonding using polymer materials may be accomplished; ceramic cements may be used at higher temperatures. **Diffusion bonding** and brazing can be used to join ceramics and to join ceramics to metals. Diffusion bonding is a joining technique in which two surfaces are pressed together at high pressures and temperatures. Diffusion of atoms to the interface fills in voids and produces a strong bond between the surfaces.

Summary

- Ceramics are nonmetallic inorganic materials that have high hardnesses and high melting points. These include single crystal and polycrystalline ceramics, glasses, and glass-ceramics. Typical ceramics are electrical and thermal insulators with good chemical stability and good strength in compression.
- Ceramics play a critical role in a wide array of electronic, magnetic, optical, and energy-related technologies. Many advanced ceramics play an important role in providing thermal insulation and high-temperature properties. Applications of advanced ceramics range from credit cards, housings for silicon chips, optical fibers that enable communication, and safe and energy efficient glasses. Traditional ceramics serve as refractories for metals processing and consumer applications.
- Crystalline ceramics contain dislocations, but most polycrystalline ceramics do not plastically deform. Flaws such as cracks and pores determine the mechanical performance of ceramic materials. Porosity and cracks limit the ability of the material to withstand a tensile load.
- Ceramics processing is commonly conducted using compaction and sintering. For specialized applications, isostatic compaction, hot pressing, and hot isostatic pressing (HIP) are used, especially to achieve higher densification levels.
- Tape casting, slip casting, extrusion, and injection molding are some of the other techniques used to form green ceramics into different shapes. These processes are then followed by a burnout step in which binders and plasticizers are burnt off, and the resultant ceramic is sintered.
- Silicon dioxide, also known as silica, and silicates (silica-based compounds) have important industrial applications in both crystalline and amorphous forms. The structural building block of these materials is an SiO_4 molecular unit that has the silicon atom sitting at the center of a tetrahedron of oxygen atoms.
- Many silicates and other ceramics form glasses rather easily, since the kinetics of crystallization are sluggish. Glasses can be formed as sheets using float-glass or as fibers and other shapes. Silicate glasses are used in a significant number of applications that include window glass, windshields, fiber optics, and fiberglass.
- Glass-ceramics are formed using controlled crystallization of inorganic glasses often to produce a desirable combination of strength and toughness. These materials are used widely for kitchenware and many other applications.
- Clay products use hydrous sheet aluminum silicates with a crystalline structure and water as a binder for ceramic particles. Clay products include pottery, china, and porcelain.
- Refractories are materials that are used in high-temperature applications and must be chosen based on their composition so that they do not react and degrade in the environment.

Glossary

Abnormal grain growth A type of grain growth observed in metals and ceramics. In this mode of grain growth, a bimodal grain size distribution usually emerges as some grains become very large at the expense of smaller grains.

Apparent porosity The percentage of the volume of a ceramic body that is composed of interconnected pores.

Bulk density The mass of a ceramic body per unit volume, including closed and interconnected pores.

Cementation Bonding ceramic raw materials using binders that form a glass or gel without firing at high temperatures.

Ceramic A nonmetallic inorganic solid. Usually hard and brittle.

Ceramic bond Bonding ceramic materials by permitting a glassy product to form at high firing temperatures.

Cermet A ceramic-metal composite (e.g., WC-Co) providing a good combination of hardness with other properties such as toughness.

Compaction Application of a load to a ceramic powder in a die to form the desired shape and eliminate porosity of a green ceramic.

Cold isostatic pressing A powder-shaping technique in which hydrostatic pressure is applied during compaction. This is used for achieving a higher green ceramic density or compaction of more complex shapes.

Devitrification The crystallization of glass.

Diffusion bonding A joining technique in which two surfaces are pressed together at high pressures and temperatures. Diffusion of atoms to the interface fills in voids and produces a strong bond between the surfaces.

Enamel A ceramic coating on metal.

Extrusion A processing technique in which ceramic powder combined with a binder is pushed through a die of constant cross-section.

Firing Heating a ceramic body at a high temperature to cause a ceramic bond to form.

Flux Additions to materials for the purpose of promoting melting.

Glass An amorphous solid. In common parlance, glass refers to silica-based materials such as window glass.

Glass-ceramics Ceramic shapes formed in the glassy state and later allowed to crystallize during heat treatment to achieve improved strength and toughness.

Glass formers Oxides with a high bond strength that easily produce a glass during processing.

Glass-transition temperature The temperature below which an undercooled liquid becomes a glass.

Glaze A ceramic coating applied to glass. The glaze contains glassy and crystalline ceramic phases.

Green ceramic A ceramic that has been shaped into a desired form but has not yet been sintered.

Hot isostatic pressing A processing technique in which large pieces of metals, alloys, and ceramics can be produced using sintering of powders under a hydrostatic pressure generated by a gas.

Hot pressing A processing technique in which sintering is conducted under uniaxial pressure.

Hydroplastic forming A number of processes by which a moist and pliable ceramic clay body is formed into a useful shape.

Injection molding A processing technique in which a thermoplastic mass (loaded with ceramic powder) is mixed in an extruder-like setup and then injected into a die to form complex parts. In the case of ceramics, the thermoplastic is burnt off. Also, widely used for thermoplastics (Chapter 16).

Intermediates Oxides that, when added to a glass, help to extend the glassy network, although the oxides normally do not form a glass themselves.

Laminated glass Annealed glass with a polymer (e.g., polyvinyl butyral, PVB) sandwiched in between, used for car windshields.

Modifiers Oxides that, when added to a glass, break up the network structure and inhibit glass formation.

Parison A crude glassy shape that serves as an intermediate step in a production process. The parison is later formed into a finished product.

Powder A collection of fine particles.

Refractories A group of ceramic materials capable of withstanding high temperatures for prolonged periods of time.

Sintering A process in which a material in particulate form is heated to a high temperature so as to densify it.

Slip A liquid slurry that is poured into a mold. When the slurry begins to harden at the mold surface, the remaining liquid slurry is decanted, leaving behind a hollow ceramic casting.

Slip casting Forming a hollow ceramic part by introducing a pourable slurry into a mold. The water in the slurry is extracted into the porous mold, leaving behind a drier surface. Excess slurry can then be poured out.

Slurry A suspension of particles in a liquid.

Spray drying A slurry of a ceramic powder is sprayed into a large chamber in the presence of hot air. This leads to the formation of soft agglomerates that can flow well into the dies used during powder compaction.

Tape casting A process for making thin sheets of ceramics using a ceramic slurry consisting of binders, plasticizers, etc. The slurry is cast with the help of a blade onto a plastic substrate. The resultant green tape is then dried, cut, and machined and used to make electronic ceramic and other devices.

Tempered glass A high-strength glass that has a surface layer where the stress is compressive, induced thermally during cooling or by the chemical diffusion of ions.

True porosity The percentage of the volume of a ceramic body that is composed of both closed and interconnected pores.

Vitrification Formation of a glass.

Problems

Section 15-1 Bonding in Ceramics

15-1 What are the primary types of atomic bonds in ceramics?

15-2 What are some of the typical characteristics of ceramic materials?

15-3 State any one application of the following ceramics: (a) alumina; (b) silica; (c) barium titanate; (d) zirconia; (e) boron carbide; and (f) diamond.

15-4 Zirconia (cubic) and diamond are tested to determine their expansion when their temperatures increase a set interval. Assuming both materials are pure, how much more should the ZrO_2 expand than the diamond sample? Does this give you any insight about distinguishing between these materials?

15-5 Convert the properties of Al_2O_3 from Table 15-3 into values with SI units.

Section 15-2 Structures of Crystalline Ceramics

15-6 Table 15-4 describes the sodium chloride structure as a face-centered cubic lattice with a basis of Cl (0, 0, 0) and Na (1/2, 0, 0). Using Cl (0, 0, 0) as one ion of the basis, give three equivalent descriptions for the position of the Na ion.

15-7 Table 15-4 describes the sodium chloride structure as a face-centered cubic lattice with a basis of Cl (0, 0, 0) and Na (1/2, 0, 0). Using Na (0, 0, 0) as one ion of the basis, give three equivalent descriptions for the position of the Cl ion.

15-8 Table 15-4 describes the cubic zincblende structure as a face-centered cubic lattice with a basis of S (0, 0, 0) and Zn (1/4, 1/4, 1/4). Using S as one ion of the basis, give three other equivalent descriptions for the position of the Zn ion.

15-9 If a face-centered cubic crystal structure has all of its tetrahedral sites filled with atoms that fit perfectly into each tetrahedral site, what is its atomic packing fraction?

15-10 If a face-centered cubic crystal structure has all of its octahedral sites filled with atoms that fit perfectly into each octahedral site, what is the atomic packing fraction?

15-11 If a hexagonal close-packed crystal structure has all of its tetrahedral sites filled with atoms that fit perfectly into each tetrahedral site, what is the atomic packing fraction? Note that each HCP unit cell has four tetrahedral sites with the same radius ratio as the tetrahedral sites in the FCC structure.

Section 15-3 Defects in Crystalline Ceramics

Section 15-4 Flaws in Ceramics

15-12 Plastic deformation due to dislocation motion is important in metals; however, this is not a very important consideration for the properties of ceramics and glasses. Explain.

15-13 Why is the tensile strength of ceramics much lower than the compressive strength?

15-14 Explain why ceramics tend to show wide scatter in their mechanical properties.

15-15 Define the terms apparent porosity and true porosity. Why are these porosities called "apparent" and "true"?

15-16 The density of Al_2O_3 is 3.96 g/cm^3. A ceramic part is produced by sintering alumina powder. It weighs 80 g when dry, 92 g after it has soaked in water, and 58 g when suspended in water. Calculate the apparent porosity, the true porosity, and the closed porosity.

15-17 Silicon carbide (SiC) has a density of 3.1 g/cm^3. A sintered SiC part is produced, occupying a volume of 500 cm^3 and weighing 1200 g. After soaking in water, the part weighs 1250 g. Calculate the bulk density, the true porosity, and the volume fraction of the total porosity that consists of closed pores.

15-18 A sintered zirconium oxide (ZrO_2) ceramic has a true porosity of 28%, and the closed pore fraction is 0.5. If the weight after soaking in water is 760 g, what is the dry weight of the ceramic? The density of ZrO_2 is 5.68 g/cm^3.

15-19 The quality control team in a ceramics processing facility takes a rather large sample of a ceramic coming off the line and weighs the sample when dry, when immersed in water, and after removing it from the water. The weights are 1.589 lb, 1.575 lb, and 1.633 lb, respectively. Find the apparent porosity and the volume of the interconnected pores.

15-20 The following physical properties are known for a sample: $W_w = 50.0$ g, $\rho = 2.170$ g/cm^3, and true porosity = 39.0%. Find the bulk density.

Section 15-5 Synthesis and Processing of Crystalline Ceramics

15-21 Explain why ceramics typically are processed as powders. How is this similar to or different from the processing of metals?

15-22 What is the driving force for sintering?

15-23 What mechanisms of diffusion play the most important role in the solid-state sintering of ceramics?

15-24 Explain the use of the following processes (use a sketch as needed): (a) uniaxial compaction and sintering, (b) hot pressing, (c) HIP, and (d) tape casting.

15-25 What are some of the important characteristics of sintered ceramics?

Section 15-6 Silica and Silicate Compounds

15-26 What is the main reason why glass formation is easy in silicate systems?

15-27 What material was used to make the tiles that provided thermal protection for NASA's space shuttle?

15-28 How many atoms are there in a unit cell of β-cristobalite?

15-29 Draw the β-cristobalite unit cell when viewed from the [100] and [110] directions.

15-30 How many distinct configurations of just two SiO_4 tetrahedra can you draw?

15-31 Report on the glass known by the trademark "Gorilla Glass." Provide a list of the references or websites that you used.

Section 15-7 Inorganic Glasses

15-32 What does the term "glass transition temperature" mean? Is this a fixed temperature for a given composition of glass?

15-33 A quartz rod 27.500 cm long is heated from room temperature to 600°C in a slow, uniform process. What is the length of the rod after it has been at temperature long enough to equilibrate?

15-34 Define the terms "glass formers," "intermediates," and "modifiers."

15-35 Part of a silica process is outlined in the following steps:

1. Heat 15 mol% soda-85 mol% silica to 1800°C.
2. Cool to 1200°C.
3. At constant temperature, increase soda content to 25 mol%.
4. Cool to 1000°C.
5. At constant temperature, increase soda content to 35 mol%.

What, if any, solids are formed during this process? See Figure 15-17.

15-36 At 1200°C, list the viscosities of the glasses in Figure 15-18 in centipoise (cP).

15-37 How many grams of BaO can be added to 1 kg of SiO_2 before the O:Si ratio exceeds 2.5 and glass-forming tendencies are poor?

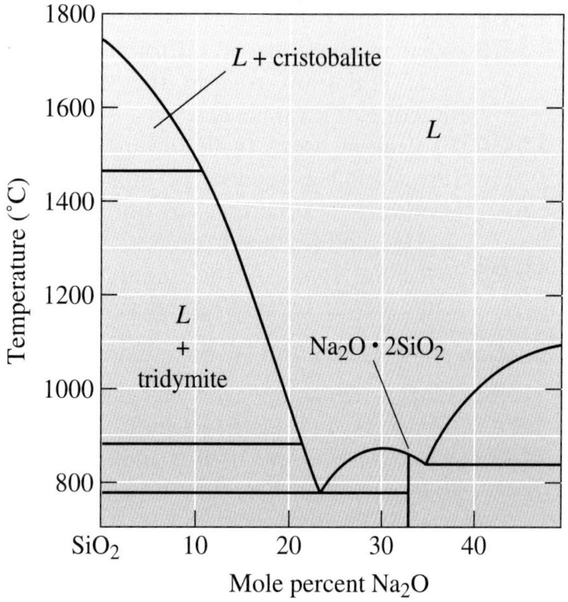

Figure 15-17 The SiO_2-Na_2O phase diagram. Additions of soda (Na_2O) to silica dramatically reduce the melting temperature of silica by forming eutectics. (Repeated for Problem 15-35.)

Compare this with the case when Li_2O is added to SiO_2.

15-38 Calculate the O:Si ratio when 30 wt% Y_2O_3 is added to SiO_2. Will this material have good glass-forming tendencies?

15-39 A silicate glass has 10 mol% Na_2O and 5 mol% CaO. Calculate the O:Si ratio and the composition in wt% for this glass.

15-40 A borosilicate glass (82% SiO_2, 2% Al_2O_3, 4% Na_2O, 12% B_2O_3) has a density of 2.23 g/cm^3, while a fused silica glass (assume 100% SiO_2) has a density of 2.2 g/cm^3. Explain why the density of the borosilicate glass is different from the weighted average of the densities of its components. The densities of Al_2O_3, Na_2O and B_2O_3 are 3.98 g/cm^3, 2.27 g/cm^3 and 2.5 g/cm^3, respectively.

15-41 Rank the following glasses by softening temperature (highest to lowest):
(a) Pyrex^{TM} (81% SiO_2, 2% Al_2O_3, 4% Na_2O, 12% B_2O_3);
(b) Lime glass (72% SiO_2, 1% Al_2O_3, 10% CaO, 14% Na_2O); and
(c) Vycor^{TM} (96% SiO_2, 4% B_2O_3).

15-42 The molten tin method is used to form a sheet of borosilicate glass. What is the solidification temperature of pure tin? Will this reduce the working range of the glass?

Section 15-8 Glass-Ceramics

15-43 How is a glass-ceramic different from a glass and a ceramic?

15-44 What are the advantages of using glass-ceramics as compared to either glasses or ceramics?

15-45 Draw a typical heat-treatment profile encountered in the processing of glass-ceramics.

15-46 What are some of the important applications of glass-ceramics?

Section 15-9 Processing and Applications of Clay Products

Section 15-10 Refractories

Section 15-11 Other Ceramic Materials

15-47 What is the maximum difference between the masses of MgO required to make 10 kg of chromite refractory and 10 kg of chromite-magnesite?

15-48 What is the lowest temperature at which a mixture of SiO_2-Al_3O_3 could still be completely liquid? (See Figure 15-24.)

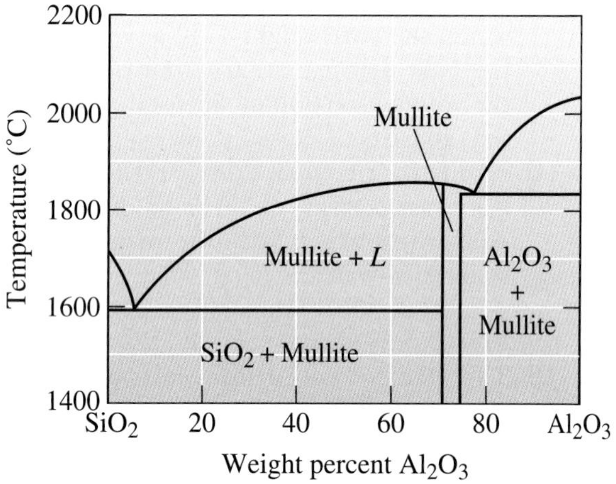

Figure 15-24 A simplified SiO_2-Al_2O_3 phase diagram, the basis for alumina silicate refractories (for Problem 15-48).

15-49 What do the terms "glaze" and "enamel" mean?

Design Problems

15-50 Design a silica soda-lime glass that can be cast at 1300°C and that will have an O:Si ratio of less than 2.3 to ensure good glass-forming tendencies. To ensure adequate

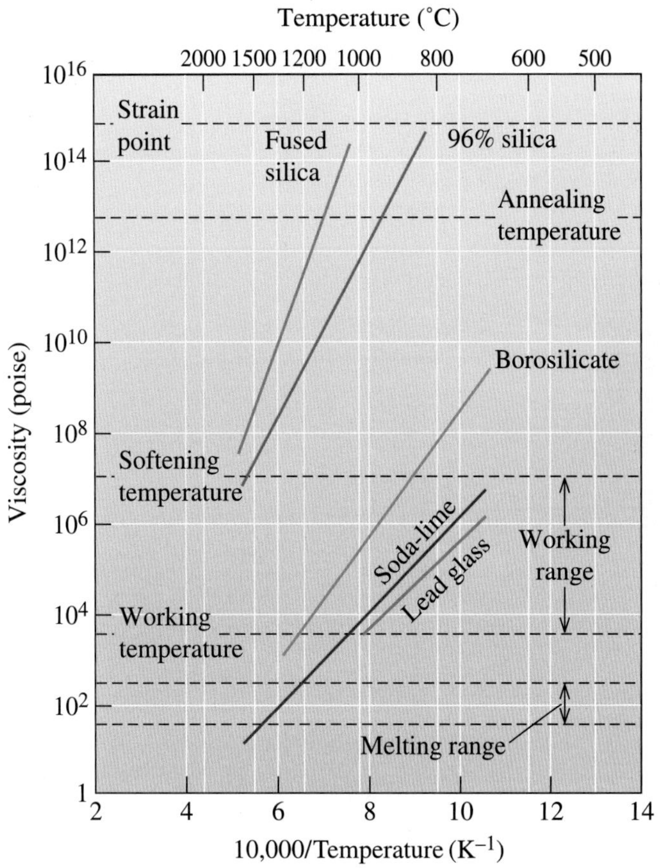

Figure 15-18 The effect of temperature and composition on the viscosity of glass. (Repeated for Problem 15-36.)

viscosity, the casting temperature should be at least 100°C above the liquidus temperature.

Computer Problems

15-51 *Apparent and Bulk Density.* Write a program that will ask the user to provide the information needed to calculate the apparent and bulk density of ceramic parts. The program should also calculate the interconnected and closed porosity.

Knovel® Problems

K15-1 What are typical grain sizes and sintering parameters for nanoceramics?

K15-2 What are the advantages of sol-gel coatings over other ceramic coatings?

K15-3 What are typical values for the percent linear thermal expansion of refractories? Using interactive graphs, determine which refractory experiences a greater linear thermal expansion at 1200°C: zirconia or alumina?

Polymers can be processed into a variety of shapes through numerous manufacturing techniques. This photo shows the extrusion of a molten thermoplastic tube of plastic (at the center of the photo in the background). The tube moves vertically upward, but before the polymer fully solidifies, cold air is blown through it, causing it to expand up to several times in diameter. In this system, the extruded tube is then cut into two sheets (seen in front of the tube) along its length using a razor blade. Finally, the flattened tubes are sealed with heat on two sides to form plastic bags. (*@ Timothy Guth*)

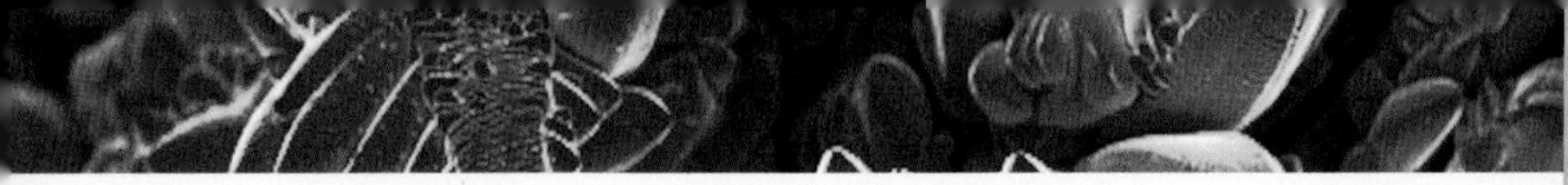

CHAPTER 16

Polymers

Have You Ever Wondered?

- *What are DVDs made from?*
- *What is Silly Putty® made from?*
- *What polymer is used in chewing gum?*
- *Which was the first synthetic fiber ever made?*
- *Why are some plastics "dishwasher safe" and some not?*
- *What are bulletproof vests made from?*
- *What polymer is used for non-stick cookware?*

Chapter Learning Objectives

The key objectives of this chapter are to

- Describe the structure, properties, and processing of the three classes of polymers: thermoplastics, thermosets, and elastomers.
- Describe the processes of condensation polymerization and addition polymerization.
- Explain the process of cross-linking and how cross-linking influences the properties of thermosets.
- Define the glass-transition temperature.
- Calculate the degree of polymerization and molecular weight, and explain their significance in terms of the mechanical behavior of polymers.
- Compare and contrast the physical and mechanical properties of polymers with those of metals.

In this chapter, we will examine the structure, properties, and processing of polymers. The suffix *mer* means "unit." Thus, the term polymer means "many units," and in this context, the term mer refers to a unit group of atoms or molecules that defines a characteristic arrangement for a polymer. **Polymers** consist of chains of molecules. The chains have average molecular weights that range from 10,000 to more than one million g/mol built by joining many mers through chemical bonding to form giant molecules known as macromolecules. Molecular weight is defined as the sum of atomic masses in each molecule. *Polymerization* is the process by which small molecules consisting of one unit (known as a **monomer**) or a few units (known as **oligomers**) are chemically joined to create these giant molecules. Polymerization normally begins with the production of long chains in which the atoms are strongly joined by covalent bonding. Most polymers are organic, meaning that they are carbon-based; however, polymers can be inorganic (e.g., silicones based on a Si-O network).

Plastics are materials that are composed principally of naturally occurring and modified or artificially made polymers often containing additives such as fibers, fillers, pigments, and the like that further enhance their properties. Plastics include thermoplastics, thermosets, and elastomers (natural or synthetic). In this book, we use the terms plastic and polymers interchangeably.

Plastics are used in an amazing number of applications including clothing, toys, home appliances, structural and decorative items, coatings, paints, adhesives, automobile tires, biomedical materials, car bumpers and interiors, foams, and packaging. Polymers are often used in composites, both as fibers and as a matrix. Liquid crystal displays (LCDs) are based on polymers. We also use polymers in photochromic lenses. Plastics are often used to make electronic components because of their insulating ability and low dielectric constant. More recently, significant developments have occurred in the area of flexible electronic devices based on the useful piezoelectric, semiconducting, optical, and electro-optical properties seen in some polymers. Polymers such as polyvinyl acetate (PVA) are water-soluble. Many such polymers can be dissolved in water or organic solvents to be used as binders, surfactants, or plasticizers in processing ceramics and semiconductors and as additives to many consumer products. Polyvinyl butyral (PVB), a polymer, makes up part of the laminated glass used for car windshields (Chapter 15). Polymers are probably used in more technologies than any other class of materials.

Commercial—or standard commodity—polymers are lightweight, corrosion-resistant materials with low strength and stiffness, and they are not suitable for use at high temperatures. These polymers are, however, relatively inexpensive and are readily formed into a variety of shapes, ranging from plastic bags to mechanical gears to bathtubs. *Engineering polymers* are designed to give improved strength or better performance at elevated temperatures. These materials are produced in relatively small quantities and often are expensive. Some of the engineering polymers can perform at temperatures as high as 350°C; others—usually in a fiber form—have strengths that are greater than that of steel.

Polymers also have many useful physical properties. Some polymers, such as acrylics like Plexiglas™ and Lucite™, are transparent and can be substituted for glasses. Although most polymers are electrical insulators, special polymers (such as the acetals) and polymer-based composites possess useful electrical conductivity. Teflon has a low coefficient of friction and is the coating for nonstick cookware. Polymers also resist corrosion and chemical attack.

16-1 Classification of Polymers

Polymers are classified in several ways: by how the molecules are synthesized, by their molecular structure, or by their chemical family. One way to classify polymers is to state if the polymer is a **linear polymer** or a **branched polymer** (Figure 16-1). A linear polymer consists of spaghetti-like molecular chains. In a branched polymer, there are primary polymer chains and secondary offshoots of smaller chains that stem from these main chains. Note that even though we say "linear," the chains are actually not in the form of

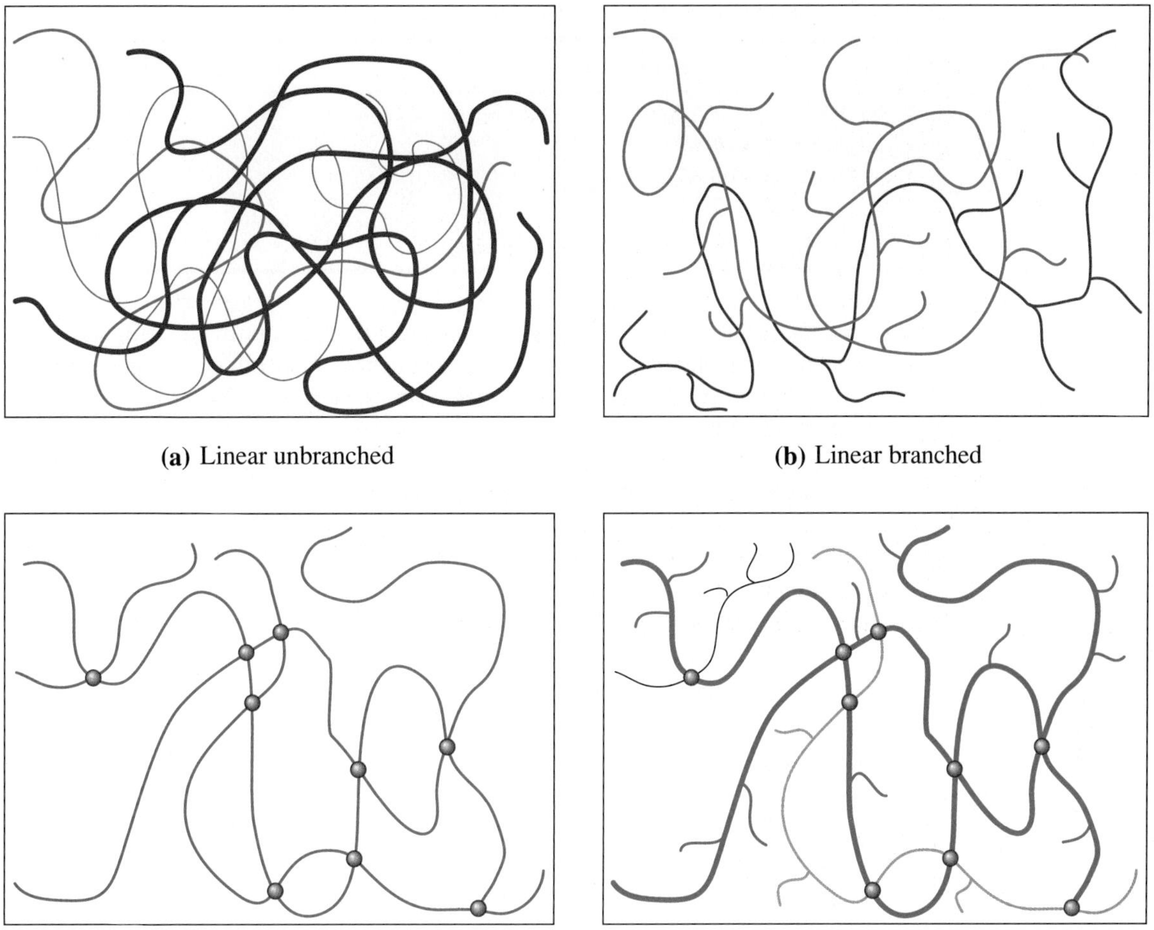

(a) Linear unbranched (b) Linear branched

(c) Thermoset-unbranched (d) Thermoset-branched

Figure 16-1 Schematic showing linear and branched polymers. Note that branching can occur in any type of polymer (thermoplastics, thermosets, and elastomers). (a) Linear unbranched polymer: notice chains are not straight lines and not connected. Different polymer chains are shown using different shades designed to show clearly that each chain is not connected to another. (b) Linear branched polymer: chains are not connected; however, they have branches. (c) Thermoset polymer without branching: chains are connected to one another by covalent bonds, but they do not have branches. Joining points are highlighted with solid circles. (d) Thermoset polymer that has branches and chains that are interconnected via covalent bonds; different chains and branches are shown in different shades for better contrast. Places where chains are actually chemically bonded are shown with solid circles.

straight lines. A better method to describe polymers is in terms of their mechanical and thermal behavior. Table 16-1 compares the three major polymer categories.

Thermoplastics are composed of long chains produced by joining together monomers; they typically behave in a plastic, ductile manner. The chains may or may not have branches. Individual chains are intertwined. There are relatively weak van der Waals bonds between atoms of different chains. This is somewhat similar to a few trees that are tangled up together. The trees may or may not have branches, and they are not physically connected to each other. The chains in thermoplastics can be untangled by application of a tensile stress. Thermoplastics can be amorphous or crystalline. Upon heating, thermoplastics soften and melt. They are processed into shapes by heating to elevated temperatures. Thermoplastics are easily recycled.

Thermosetting polymers are composed of long chains (linear or branched) of molecules that are strongly cross-linked to one another to form three-dimensional network structures. Network or thermosetting polymers are like a bunch of strings that are

Table 16-1 Comparison of the three polymer categories

Behavior	General Structure	Example
Thermoplastic	Flexible linear chains (straight or branched)	Polyethylene
Thermosetting	Rigid three-dimensional network (chains may be linear or branched)	Polyurethanes
Elastomers	Thermoplastics or lightly cross-linked thermosets, consist of spring-like molecules	Natural rubber

knotted to one another in several places and not just tangled up. Each string may have other side strings attached to it. Thermosets are generally stronger, but more brittle, than thermoplastics. Thermosets do not melt upon heating but begin to decompose. They cannot easily be reprocessed after the cross-linking reaction has occurred, and hence, recycling is difficult.

Elastomers sustain elastic deformations greater than 200%. They may be thermoplastics or lightly cross-linked thermosets. The polymer chains consist of coil-like molecules that can reversibly stretch by applying a force. Elastomers are often referred to as rubbers.

Thermoplastic elastomers are a special group of polymers. They have the processing ease of thermoplastics and the elastic behavior of elastomers.

Representative Structures

Figure 16-2 shows three ways we can represent a segment of polyethylene, the simplest of the thermoplastics. The polymer chain consists of a backbone of carbon atoms; two hydrogen atoms are bonded to each carbon atom in the chain. The angle between the two H bonds associated with a given C atom is 109.5°. The chain twists and turns throughout space. As shown in the figure, polyethylene has no branches and hence is a linear thermoplastic. The simple two-dimensional model in Figure 16-2(c) includes the essential elements of the polymer structure and will be used to describe the various polymers. The single lines (—) between the carbon atoms and between the carbon and hydrogen atoms represent a single covalent bond. Two parallel lines

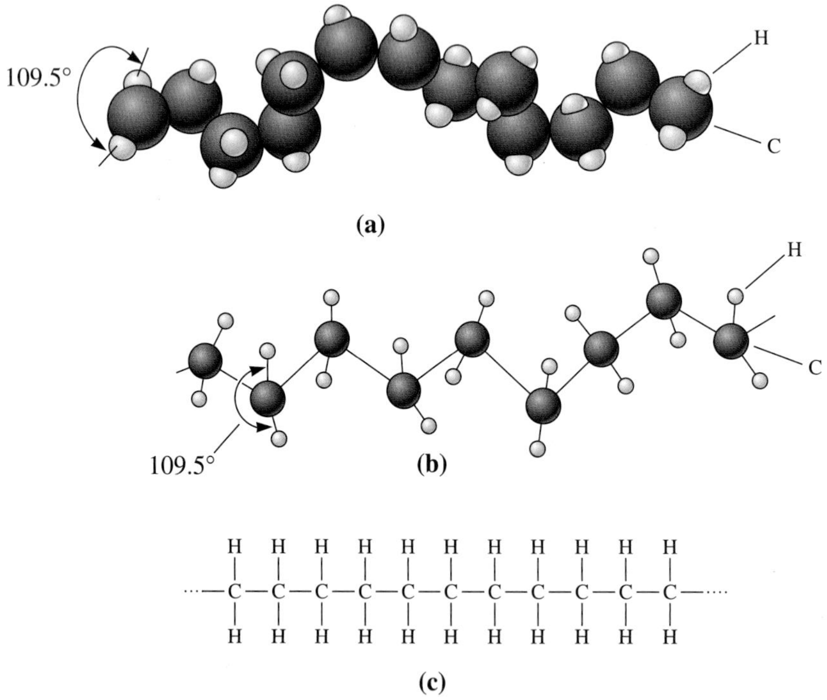

Figure 16-2 Three ways to represent the structure of polyethylene: (a) a solid three-dimensional model, (b) a three-dimensional "space" model, and (c) a simple two-dimensional model.

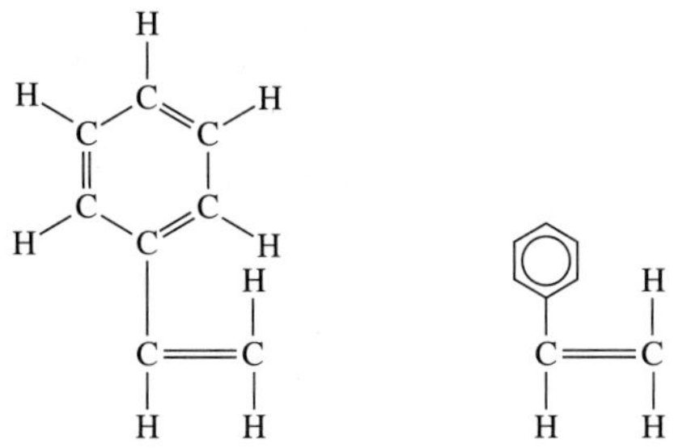

Figure 16-3 Two ways to represent the benzene ring. In this case, the benzene ring is shown attached to a pair of carbon atoms, producing styrene.

(=) represent a double covalent bond between atoms. A number of polymers include ring structures such as the benzene ring found in polystyrene and other polymers (Figure 16-3). Molecules that contain the six-membered benzene ring are known as aromatics.

In the structure shown in Figure 16-2(c) if we replace one of the hydrogen atoms in C_2H_4 with CH_3, a benzene ring, or chlorine, we get the structure of polypropylene, polystyrene, and polyvinyl chloride (PVC), respectively. If we replaced all H atoms in the C_2H_4 groups with fluorine, we get the structure of polytetrafluoroethylene or Teflon. Many polymer structures can thus be derived from the structure of polyethylene. The following example shows how different types of polymers are used.

Example 16-1 *Materials Selection for Polymer Components*

Discuss the type of polymer you might select for the following applications: a surgeon's glove, a beverage container, and a pulley.

SOLUTION

The glove must be capable of stretching a great deal in order to slip onto the surgeon's hand, yet it must conform tightly to the hand to permit the maximum sensation of touch during surgery. A material that undergoes a large amount of elastic strain—particularly with relatively little applied stress—might be appropriate; this requirement describes an elastomer.

The beverage container should be easily and economically produced. It should have some ductility and toughness so that it does not accidentally shatter and leak the contents. If the beverage is carbonated, diffusion of CO_2 is a major concern (Chapter 5). A thermoplastic such as polyethylene terephthalate (PET) will have the necessary formability and strength needed for this application. It is also easily recycled.

The pulley will be subjected to some stress and wear as a belt passes over it. A relatively strong, rigid, hard material is required to prevent wear, so a thermosetting polymer might be most appropriate.

16-2 Addition and Condensation Polymerization

Polymerization by **addition** and **condensation** are the two main ways to create a polymer. The polymers derived from these processes are known as addition and condensation polymers, respectively. The formation of the most common polymer, polyethylene (PE) from ethylene molecules, is an example of addition or chain-growth polymerization. Ethylene, a gas, is the monomer (single unit) and has the formula C_2H_4. The two carbon atoms are joined by a double covalent bond. Each carbon atom shares two of its electrons with the second carbon atom, and two hydrogen atoms are bonded covalently to each of the carbon atoms (Figure 16-4).

In the presence of an appropriate combination of heat, pressure, and catalysts, the double bond between the carbon atoms is broken and replaced with a single covalent

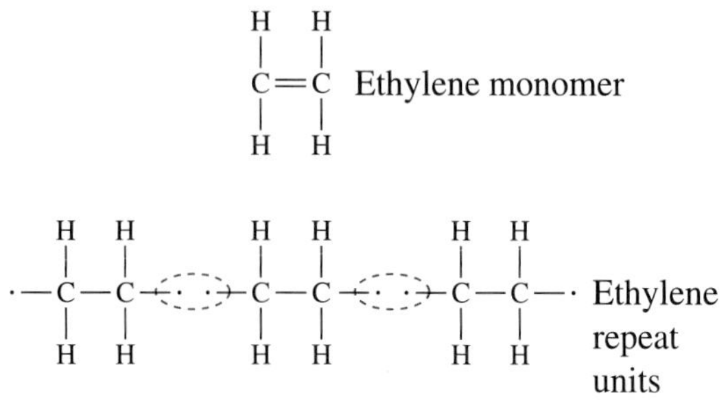

Figure 16-4 The addition reaction for producing polyethylene from ethylene molecules. The unsaturated double bond in the monomer is broken to produce active sites, which then attract additional repeat units to either end to produce a chain.

bond. The ends of the monomer are now *free radicals*; each carbon atom has an unpaired electron that it may share with other free radicals. Addition polymerization occurs because the original monomer contains a double covalent bond between the carbon atoms. The double bond is an **unsaturated bond**. After changing to a single bond, the carbon atoms are still joined, but they become active; other **repeat units** or mers can be added to produce the polymer chain.

Termination of Addition Polymerization We need polymers that have a controlled average molecular weight and molecular weight distribution. Thus, the polymerization reactions must have an "off" switch as well! The chains may be terminated by two mechanisms (Figure 16-5). First, the ends of two growing chains may be joined. This process, called *combination*, creates a single large chain from two smaller chains. Second, the active end of one chain may remove a hydrogen atom from a second chain by a process known as *disproportionation*. This reaction terminates two chains, rather than combining two chains into one larger chain. Sometimes, compounds known as terminators are added to end polymerization reactions. In general, for thermoplastics, a higher average molecular weight leads to a higher melting temperature and higher Young's modulus for the polymer. The following example illustrates an addition polymerization reaction.

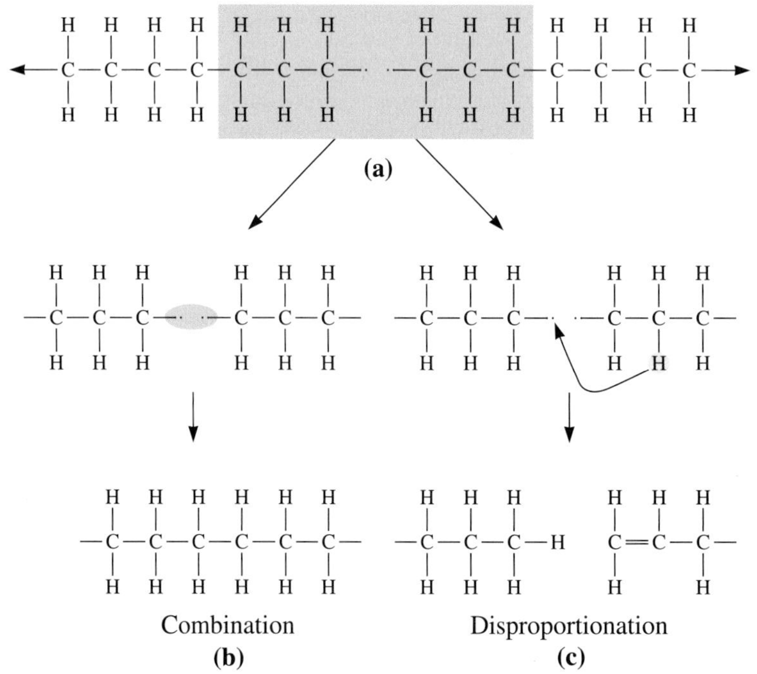

Figure 16-5 Termination of polyethylene chain growth: (a) the active ends of two chains come into close proximity, (b) the two chains undergo combination and become one large chain, and (c) rearrangement of a hydrogen atom and creation of a double covalent bond by disproportionation cause termination of two chains.

Example 16-2 Calculation of Initiator Required

Calculate the amount of benzoyl peroxide [$(C_6H_5CO)_2O_2$] initiator required to produce 1 kg of polyethylene with an average molecular weight of 200,000 g/mol. Each benzoyl peroxide molecule produces two free radicals that are each capable of initiating a polyethylene chain. Assume that 20% of the initiator is effective and that all termination occurs by the combination mechanism.

SOLUTION

One benzoyl peroxide molecule produces two free radicals that initiate two chains that then combine to form one polyethylene chain. Thus, there is a 1:1 ratio between benzoyl peroxide molecules and polyethylene chains. If the initiator were 100% effective, one molecule of benzoyl peroxide would be required per polyethylene chain.

To determine the amount of benzoyl peroxide required, the number of chains with an average molecular weight of 200,000 g/mol in 1 kg of polyethylene must be calculated.

The molecular weight of ethylene = (2 C atoms)(12 g/mol) + (4 H atoms)(1 g/mol) = 28 g/mol. The number of molecules per chain (also known as the degree of polymerization) is given by

$$\frac{200{,}000 \text{ g/mol}}{28 \text{ g/mol}} = 7143 \text{ ethylene molecules/chain}$$

The total number of monomers required to form 1 kg of polyethylene is

$$\frac{(1000 \text{ g})(6.022 \times 10^{23} \text{ monomers/mol})}{28 \text{ g/mol}} = 2.15 \times 10^{25} \text{ monomers}$$

Thus, the number of polyethylene chains in 1 kg is

$$\frac{2.15 \times 10^{25} \text{ ethylene molecules}}{7143 \text{ ethylene molecules/chain}} = 3.0 \times 10^{21} \text{ chains}$$

Since the benzoyl peroxide initiator is only 20% effective, $5(3.0 \times 10^{21}) = 1.5 \times 10^{22}$ benzoyl peroxide molecules are required to initiate 3.0×10^{21} chains. The molecular weight of benzoyl peroxide is (14 C atoms)(12 g/mol) + (10 H atoms)(1 g/mol) + (4 O atoms)(16 g/mol) = 242 g/mol. Therefore, the amount of initiator required is

$$\frac{1.5 \times 10^{22} \text{ molecules } (242 \text{ g/mol})}{6.022 \times 10^{23} \text{ molecules/mol}} = 6.0 \text{ g}$$

Condensation Polymerization Polymer chains can also form by condensation reactions, or *step-growth* polymerization, producing structures and properties that resemble those of addition polymers. In condensation polymerization, a relatively small molecule (such as water, ethanol, methanol, etc.) is formed as a result of the polymerization reaction. This mechanism may often involve different monomers as starting or precursor molecules. The polymerization of dimethyl terephthalate and ethylene glycol (also used as radiator coolant) to produce *polyester* is an important example (Figure 16-6). During polymerization, a hydrogen atom on the end of the ethylene glycol monomer combines with an OCH_3 (methoxy) group from the dimethyl terephthalate. A byproduct, methyl alcohol (CH_3OH), is "condensed" off, and the two monomers combine to produce a larger molecule. Each of the monomers in this example is bifunctional, meaning that both ends of the monomer may react, and the condensation polymerization can continue by the same reaction. Eventually, a long polymer chain—a polyester—is produced. The length of the polymer chain depends on the ease with which the monomers can diffuse to the ends and undergo the condensation reaction. Chain growth ceases when no more monomers reach the end of the chain to continue the reaction. Condensation polymerization reactions also occur in sol-gel processing of ceramic materials.

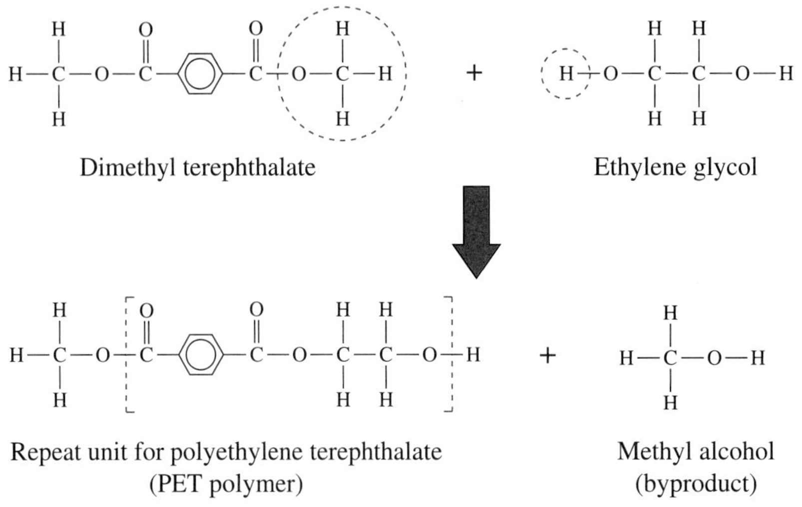

Figure 16-6 The condensation reaction for polyethylene terephthalate (PET), which is a common polyester. The OCH_3 group and a hydrogen atom are removed from the monomers, permitting the two monomers to join and producing methyl alcohol as a byproduct.

The following example describes the discovery of nylon and calculations related to condensation polymerization.

Example 16-3 *Condensation Polymerization of 6,6-Nylon*

Nylon was first reported by Wallace Hume Carothers of Du Pont™ in about 1934. In 1939, Charles Stine, also from Du Pont™, reported the discovery of this first synthetic fiber to a group of 3000 women gathered for the New York World's Fair. The first application was nylon stockings. Today nylon is used in hundreds of applications. Prior to nylon, Carothers had discovered neoprene (an elastomer).

The linear polymer 6,6-nylon is to be produced by combining 1000 g of hexamethylene diamine with adipic acid. A condensation reaction then produces the polymer. The molecular structures of the monomers are shown below. The linear nylon chain is produced when a hydrogen atom from the hexamethylene diamine combines with an OH group from adipic acid to form a water molecule.

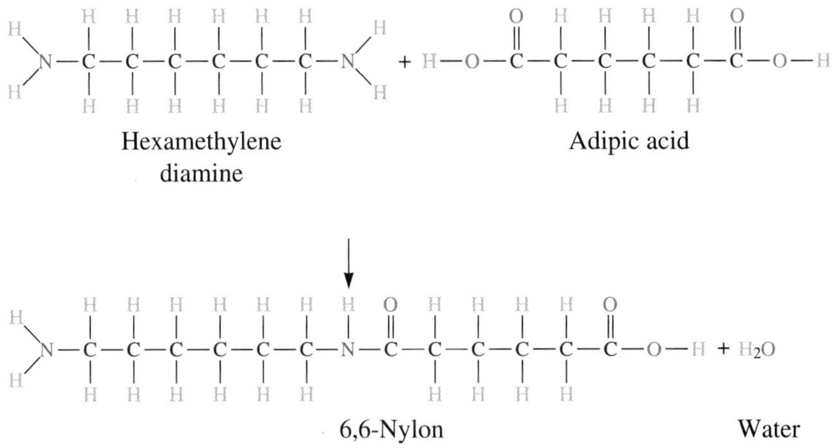

Note that the reaction can continue at both ends of the new molecule; consequently, long chains may form. This polymer is called 6,6-nylon because both monomers contain six carbon atoms.

How many grams of adipic acid are needed, and how much 6,6-nylon is produced, assuming 100% efficiency?

SOLUTION

The molecular weights of hexamethylene diamine, adipic acid, and water are 116, 146, and 18 g/mol, respectively. The number of moles of hexamethylene diamine is equal to the number of moles of adipic acid:

$$\frac{1000 \text{ g}}{116 \text{ g/mol}} = 8.62 \text{ moles} = \frac{x \text{ g}}{146 \text{ g/mol}}$$

$$x = 1259 \text{ g of adipic acid required}$$

One water molecule is lost when hexamethylene diamine reacts with adipic acid. Each time a monomer is added to the chain, one molecule of water is lost. Thus, when a long chain forms, there are (on average) two water molecules released for each repeat unit of the chain (each repeat unit being formed from two monomers). Thus, the number of moles of water lost is 2 (8.62 moles) = 17.24 moles or

$$17.24 \text{ moles } H_2O \text{ (18 g/mol)} = 310 \text{ g } H_2O$$

The total amount of nylon produced is

$$1000 \text{ g} + 1259 \text{ g} - 310 \text{ g} = 1949 \text{ g} = 1.95 \text{ kg}$$

16-3 Degree of Polymerization

Polymers, unlike organic or inorganic compounds, do not have a fixed molecular weight. For example, polyethylene may have a molecular weight that ranges from ~25,000 to 6 million g/mol! The average length of a linear polymer is represented by the **degree of polymerization**, or the number of repeat units in the chain. The degree of polymerization can also be defined as

$$\text{Degree of polymerization} = \frac{\text{average molecular weight of polymer}}{\text{molecular weight of repeat unit}} \quad (16\text{-}1)$$

If the polymer contains only one type of monomer, the molecular weight of the repeat unit is that of the monomer. If the polymer contains more than one type of monomer, the molecular weight of the repeat unit is the sum of the molecular weights of the monomers, less the molecular weight of the byproduct.

The lengths of the chains in a linear polymer vary considerably. Some may be quite short due to early termination; others may be exceptionally long. We can define an average molecular weight in two ways.

The *weight average molecular weight* is obtained by dividing the chains into size ranges and determining the fraction of chains having molecular weights within that range. The weight average molecular weight $\overline{M}_w$ is

$$\overline{M}_w = \sum f_i M_i \quad (16\text{-}2)$$

where M_i is the mean molecular weight of each range and f_i is the weight fraction of the polymer having chains within that range.

The *number average molecular weight* $\overline{M}_n$ is based on the number fraction, rather than the weight fraction, of the chains within each size range. It is always smaller than the weight average molecular weight and is given by

$$\overline{M}_n = \sum x_i M_i \quad (16\text{-}3)$$

where M_i is again the mean molecular weight of each size range, but x_i is the fraction of the total number of chains within each range. Either $\overline{M}_w$ or $\overline{M}_n$ can be used to calculate the degree of polymerization.

The two following examples illustrate these concepts.

Example 16-4 *Degree of Polymerization for 6,6-Nylon*

Calculate the degree of polymerization if 6,6-nylon has a molecular weight of 120,000 g/mol.

SOLUTION

The reaction by which 6,6-nylon is produced was described in Example 16-3. Hexamethylene diamine and adipic acid combine and release a molecule of water. When a long chain forms, there is, on average, one water molecule released for each reacting molecule. The molecular weights are 116 g/mol for hexamethylene diamine, 146 g/mol for adipic acid, and 18 g/mol for water. The repeat unit for 6,6-nylon is

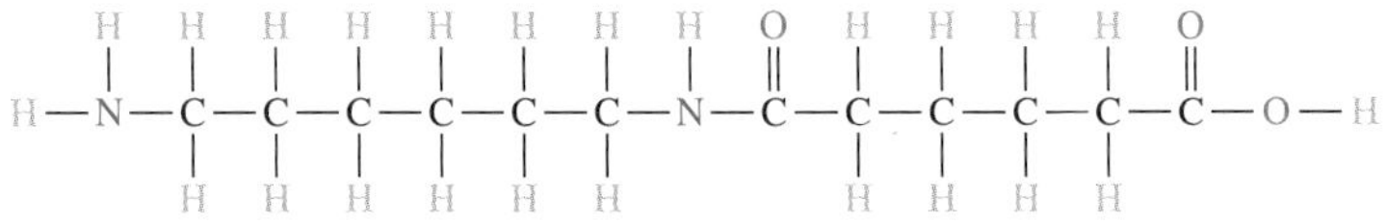

The molecular weight of the repeat unit is the sum of the molecular weights of the two monomers, minus that of the two water molecules that are evolved:

$$M_{\text{repeat unit}} = 116 + 146 - 2(18) = 226 \text{ g/mol}$$

$$\text{Degree of polymerization} = \frac{120{,}000 \text{ g/mol}}{226 \text{ g/mol}} = 531$$

The degree of polymerization refers to the total number of repeat units in the chain.

Example 16-5 *Number and Weight Average Molecular Weights*

We have a polyethylene sample containing 4000 chains with molecular weights between 0 and 5000 g/mol, 8000 chains with molecular weights between 5000 and 10,000 g/mol, 7000 chains with molecular weights between 10,000 and 15,000 g/mol, and 2000 chains with molecular weights between 15,000 and 20,000 g/mol. Determine both the number and weight average molecular weights.

SOLUTION

First we need to determine the number fraction x_i and weight fraction f_i for each of the four ranges. For x_i, we simply divide the number in each range by 21,000, which is the total number of chains. To find f_i, we first multiply the number of chains by the mean molecular weight of the chains in each range, giving the "weight" of each group, then find f_i by dividing by the total weight of 1.925×10^8 g/mol. We can then use Equations 16-2 and 16-3 to find the molecular weights.

Number of Chains	Mean M per Chain (g/mol)	x_i	$x_i M_i$ (g/mol)	Weight (g/mol)	f_i	$f_i M_i$ (g/mol)
4000	2,500	0.191	477.5	1.0×10^7	0.0519	129.75
8000	7,500	0.381	2857.5	6.0×10^7	0.3118	2338.50
7000	12,500	0.333	4162.5	8.75×10^7	0.4545	5681.25
2000	17,500	0.095	1662.5	3.5×10^7	0.1818	3181.50
$\Sigma = 21{,}000$		$\Sigma = 1$	$\Sigma = 9160$	$\Sigma = 1.925 \times 10^8$	$\Sigma = 1$	$\Sigma = 11{,}331$

$$\overline{M}_n = \sum x_i M_i = 9160 \text{ g/mol}$$

$$\overline{M}_w = \sum f_i M_i = 11{,}331 \text{ g/mol}$$

The weight average molecular weight is larger than the number average molecular weight.

16-4 Typical Thermoplastics

Some of the mechanical properties of typical thermoplastics are shown in Table 16-2. Table 16-3 shows the repeat units and applications for several thermoplastics formed by addition polymerization.

Thermoplastics with Complex Structures There is a large number of polymers that are formed from complex monomers, often by the condensation mechanism. For example, oxygen, nitrogen, sulfur, and benzene rings (or aromatic groups) may be incorporated into the chain. These polymers are typically used for special applications and in relatively small quantities. Table 16-4 shows the repeat units and typical applications for a number of these complex polymers. Polyoxymethylene, or acetal, is a simple example

Table 16-2 Properties of selected thermoplastics

	Tensile Strength (psi)	% Elongation	Elastic Modulus (psi)	Density (g/cm^3)	Izod Impact (ft lb/in.)
Polyethylene (PE):					
Low-density	3,000	800	40,000	0.92	9.0
High-density	5,500	130	180,000	0.96	4.0
Ultra-high molecular weight	7,000	350	100,000	0.934	30.0
Polyvinyl chloride (PVC)	9,000	100	600,000	1.40	
Polypropylene (PP)	6,000	700	220,000	0.90	1.0
Polystyrene (PS)	8,000	60	450,000	1.06	0.4
Polyacrylonitrile (PAN)	9,000	4	580,000	1.15	4.8
Polymethyl methacrylate (PMMA) (acrylic, Plexiglas)	12,000	5	450,000	1.22	0.5
Polychlorotrifluoroethylene	6,000	250	300,000	2.15	2.6
Polytetrafluoroethylene (PTFE, Teflon)	7,000	400	80,000	2.17	3.0
Polyoxymethylene (POM) (acetal)	12,000	75	520,000	1.42	2.3
Polyamide (PA) (nylon)	12,000	300	500,000	1.14	2.1
Polyester (PET)	10,500	300	600,000	1.36	0.6
Polycarbonate (PC)	11,000	130	400,000	1.20	16.0
Polyimide (PI)	17,000	10	300,000	1.39	1.5
Polyetheretherketone (PEEK)	10,200	150	550,000	1.31	1.6
Polyphenylene sulfide (PPS)	9,500	2	480,000	1.30	0.5
Polyethersulfone (PES)	12,200	80	350,000	1.37	1.6
Polyamide-imide (PAI)	27,000	15	730,000	1.39	4.0

Table 16-3 Repeat units and applications for selected addition thermoplastics

Polymer	Repeat Unit	Application	Polymer	Repeat Unit	Application
Polyethylene (PE)	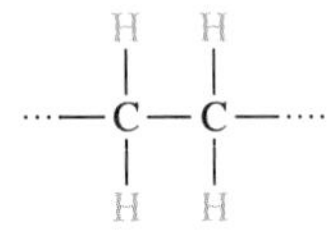	Packaging films, wire insulation, squeeze bottles, tubing, household items	Polyacrylonitrile (PAN)	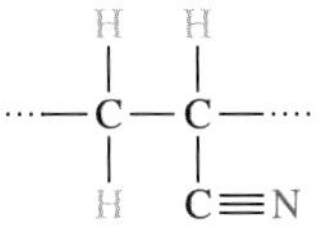	Textile fibers, precursor for carbon fibers, food containers
Polyvinyl chloride (PVC)	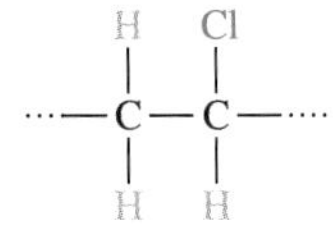	Pipe, valves, fittings, floor tile, wire insulation, vinyl automobile roofs	Polymethyl methacrylate (PMMA) (acrylic-Plexiglas)	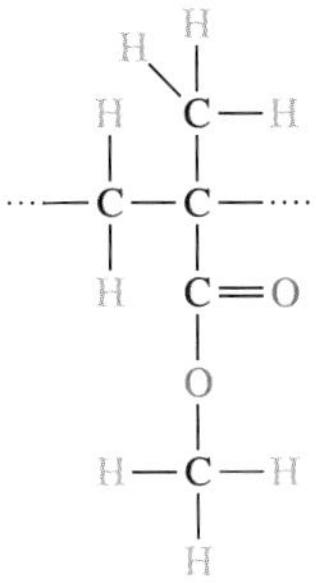	Windows, windshields, coatings, hard contact lenses, lighted signs
Polypropylene (PP)	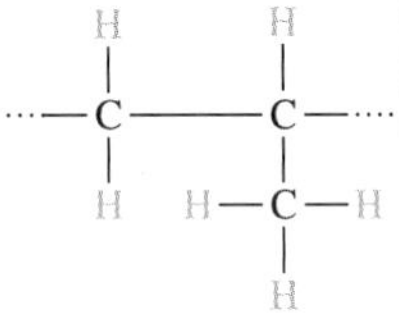	Tanks, carpet fibers, rope, packaging	Polychlorotri-fluoroethylene	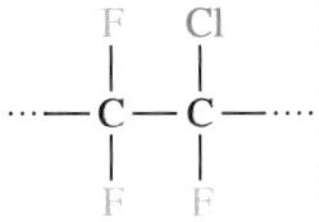	Valve components, gaskets, tubing, electrical insulation
Polystyrene (PS)	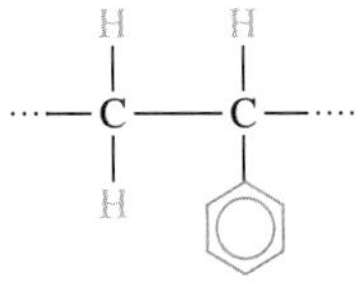	Packaging and insulation foams, lighting panels, appliance components, egg cartons	Polytetrafluoroethylene (Teflon) (PTFE)	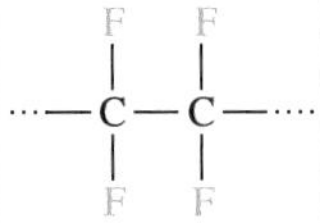	Seals, valves, non-stick coatings

Table 16-4 Repeat units and applications for complex thermoplastics

Polymer	Repeat Unit	Applications
Polyoxymethylene (acetal)(POM)	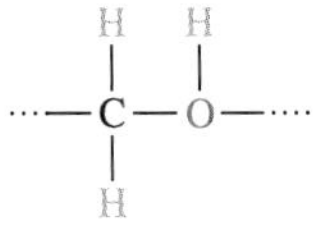	Plumbing fixtures, pens, bearings, gears, fan blades
Polyamide (nylon) (PA)	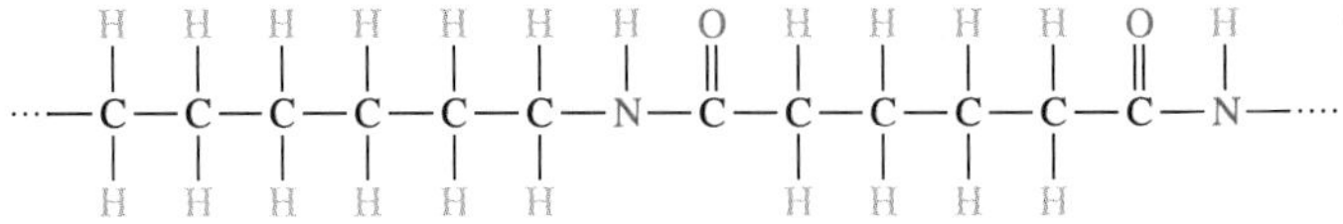	Bearings, gears, fibers, rope, automotive components, electrical components
Polyester (PET)	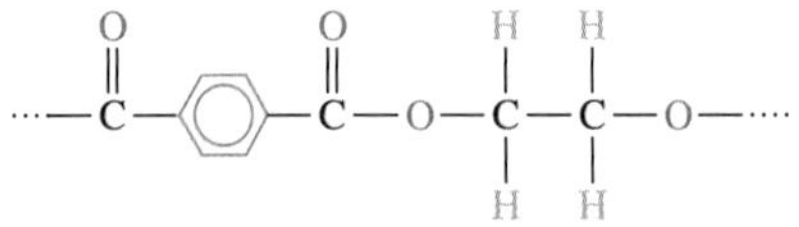	Fibers, photographic film, recording tape, boil-in bag containers, beverage containers

Table 16-4 Repeat units and applications for complex thermoplastics *(Continued)*

Polymer	Repeat Unit	Applications
Polycarbonate (PC)	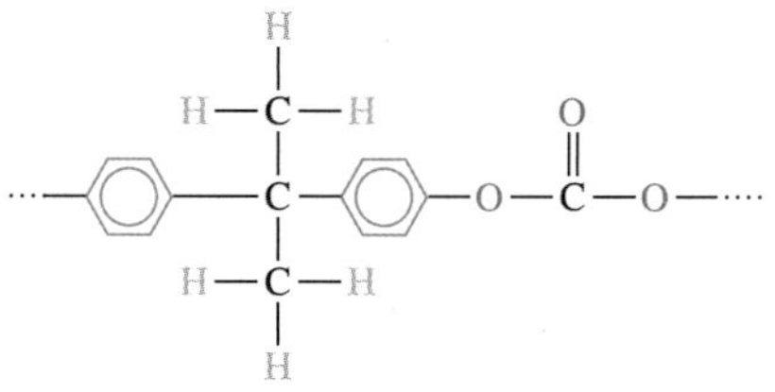	Electrical and appliance housings, automotive components, football helmets, bottles, compact discs (CDs), DVDs
Polyimide (PI)	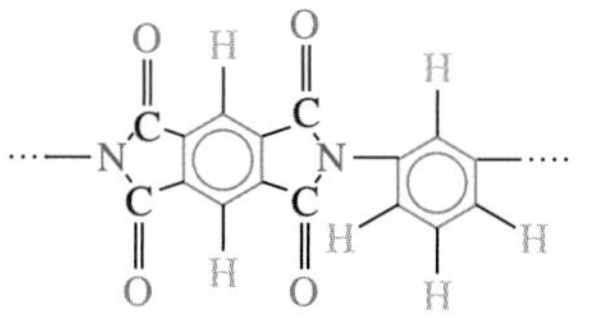	Adhesives, circuit boards, flexible cables
Polyetheretherketone (PEEK)	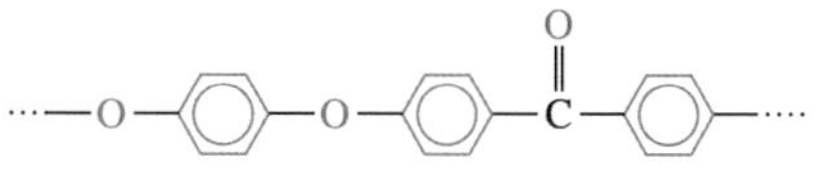	High-temperature electrical insulation and coatings
Polyphenylene sulfide (PPS)	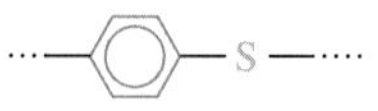	Coatings, fluid-handling components, electronic components, hair dryer components
Polyether sulfone (PES)	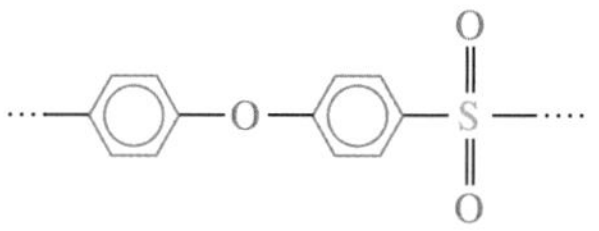	Electrical components, coffeemakers, hair dryers, microwave oven components
Polyamide-imide (PAI)	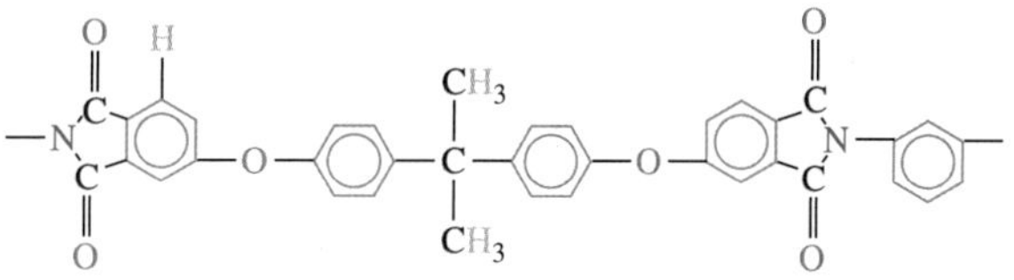	Electronic components, aerospace and automotive applications

in which the backbone of the polymer chain contains alternating carbon and oxygen atoms. A number of these polymers, including polyimides and polyetheretherketone (PEEK), are important aerospace materials. Because bonding within the chains is stronger compared to the simpler addition polymers, rotation and sliding of the chains is more difficult, leading to higher strengths, higher stiffnesses, and higher melting points. In some cases, good impact properties are obtained from these complex chains, with polycarbonates being particularly remarkable. Polycarbonates (Lexan™, Merlon™, and Sparlux™) are used to make bulletproof windows, compact discs for data storage, DVDs, and many other products.

16-5 Structure—Property Relationships in Thermoplastics

Degree of Polymerization In general, for a given type of thermoplastic (e.g., polyethylene), the tensile strength, creep resistance, impact toughness, wear resistance, and melting temperature all increase with increasing average molecular weight or degree of polymerization. The increases in these properties are not linear. As the average molecular weight increases, the melting temperature increases, and this makes the processing more difficult. In fact, we can make use of a bimodal molecular weight distribution in polymer processing. One component has lower molecular weight and helps melting, thereby making processing easier.

Effect of Side Groups In polyethylene, the linear chains easily rotate and slide when stress is applied, and no strong polar bonds are formed between the chains; thus, polyethylene has a low strength.

Vinyl compounds have one of the hydrogen atoms replaced with a different atom or atom group. When R in the side group is chlorine, we produce polyvinyl chloride (PVC); when the side group is CH_3, we produce polypropylene (PP); addition of a benzene ring as a side group gives polystyrene (PS); and a CN group produces polyacrylonitrile (PAN). (See Table 16-3.)

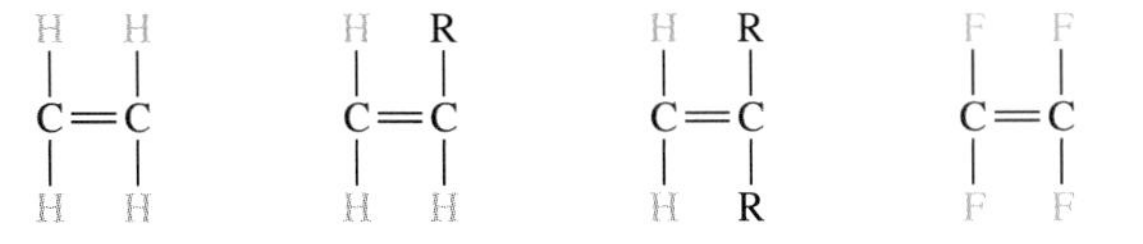

Ethylene | Vinyl compound | Vinylidene compound | Tetrafluoroethylene

When two of the hydrogen atoms are replaced, the monomer is a *vinylidene compound*, important examples of which include polyvinylidene chloride (the basis for Saran Wrap™) and polymethyl methacrylate (acrylics such as Lucite™ and Plexiglas™). Generally, a head-to-tail arrangement of the repeat units in the polymers is obtained (Figure 16-7), but head-to-head arrangements are observed.

The effects of adding other atoms or atom groups to the carbon backbone in place of hydrogen atoms are illustrated by the typical properties given in Table 16-2. Larger atoms such as chlorine or groups of atoms such as methyl (CH_3) and benzene make it more difficult for the chains to rotate, uncoil, disentangle, and deform by viscous flow when a stress is applied or when the temperature is increased. This condition leads to higher strengths, stiffnesses, and melting temperatures than those for polyethylene. The chlorine atom in PVC and the carbon-nitrogen group in PAN are strongly attracted by hydrogen bonding to hydrogen atoms on adjacent chains. This, for example, is the reason why PVC is more rigid than many other polymers. The way to get around the rigidity of PVC is to add low molecular weight compounds such as phthalate esters, known as plasticizers. When PVC contains these compounds, the glass-transition temperature is lowered. This makes PVC more ductile and workable; such PVC is known as vinyl (not to be confused with the vinyl group mentioned here and in other places). PVC is used to make three-ring binders, pipes, tiles, and clear Tygon™ tubing.

In polytetrafluoroethylene (PTFE or Teflon), all four hydrogen atoms in the polyethylene structure are replaced by fluorine. The monomer again is symmetrical, and the strength of the polymer is not much greater than that of polyethylene. The C-F bond permits PTFE to have a high melting point with the added benefit of low friction and nonstick characteristics that make the polymer useful for bearings and cookware.

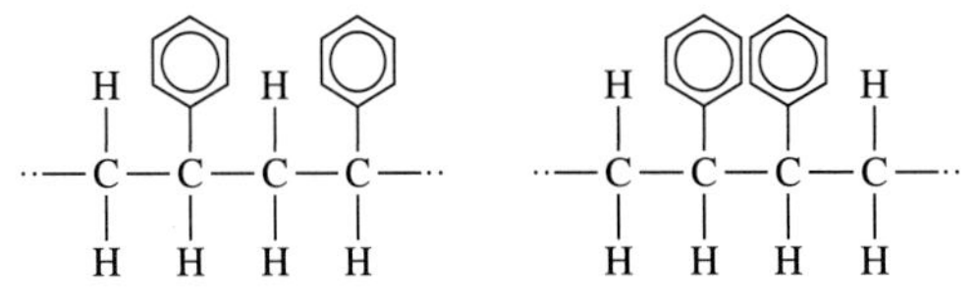

Head-to-tail | Head-to-head

Figure 16-7 Head-to-tail versus head-to-head arrangement of repeat units. The head-to-tail arrangement is most typical.

Teflon was invented by accident by Roy Plunkett, who was working with tetrafluoroethylene gas. He found a tetrafluoroethylene gas cylinder that had no pressure (and, thus, seemed empty) but was heavier than usual. The gas inside had polymerized into solid Teflon!

Branching prevents dense packing of the chains, thereby reducing the density, stiffness, and strength of the polymer. Low-density polyethylene (LPDE), which has many branches, is weaker than high-density polyethylene (HDPE), which has virtually no branching (Table 16-2).

Crystallization and Deformation

Crystallinity is important in polymers since it affects mechanical and optical properties. Crystallinity evolves in the processing of polymers as a result of temperature changes and applied stress (e.g., formation of PET bottles discussed in previous chapters). If crystalline regions become too large, they begin to scatter light and make the plastic translucent. In certain special polymers, localized regions crystallize in response to an applied electric field, and this is the principle by which liquid-crystal displays work. We will discuss this in the following section. Crystallization of the polymer also helps to increase density, resistance to chemical attack, and mechanical properties—even at higher temperatures—because of the stronger bonding between the chains. In addition, the deformation that straightens and aligns the chains, leading to crystallization, also produces a preferred orientation. Deformation of a polymer is often used in producing fibers having mechanical properties in the direction of the fiber that exceed those of many metals and ceramics. In fact, this texture strengthening played a key role in the discovery of nylon fibers. In previous chapters, we have seen how PET bottles develop a biaxial texture and strength along the radial and length directions.

Tacticity

When a polymer is formed from nonsymmetrical repeat units, the structure and properties are determined by the location of the nonsymmetrical atoms or atom groups. This condition is called **tacticity**, or stereoisomerism. In the syndiotactic arrangement, the atoms or atom groups alternately occupy positions on opposite sides of the linear chain. The atoms are all on the same side of the chain in *isotactic* polymers, whereas the arrangement of the atoms is random in *atactic* polymers (Figure 16-8).

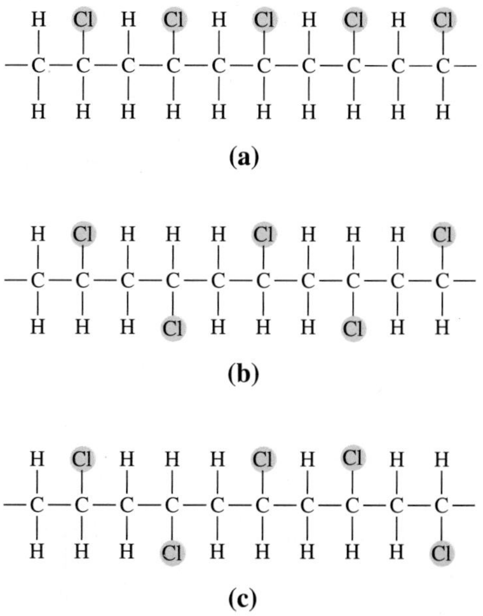

Figure 16-8 Three possible arrangements of nonsymmetrical monomers: (a) isotactic, (b) syndiotactic, and (c) atactic.

The atactic structure, which is the least regular and least predictable, tends to give poor packing, low density, low strength and stiffness, and poor resistance to heat or chemical attack. Atactic polymers are more likely to have an amorphous structure. An example of the importance of tacticity occurs in polypropylene. Atactic polypropylene is an amorphous wax-like polymer with poor mechanical properties, whereas isotactic polypropylene may crystallize and is one of the most widely used commercial polymers.

Copolymers Similar to the concept of solid solutions or the idea of composites, linear-addition chains composed of two or more types of molecules can be arranged to form **copolymers**. This is a powerful way to blend properties of different polymers. The arrangement of the monomers in a copolymer may take several forms (Figure 16-9). These include alternating, random, block, and grafted copolymers. ABS, composed of acrylonitrile, butadiene (a synthetic elastomer), and styrene, is one of the most common polymer materials (Figure 16-10). It contains two copolymers. Styrene and acrylonitrile form a linear copolymer (SAN) that serves as a matrix. Styrene and butadiene also form a linear copolymer, BS rubber, which acts as the filler material. The combination of the two copolymers gives ABS an excellent combination of strength, rigidity, and toughness. Another common copolymer contains repeat units of ethylene and propylene. Whereas polyethylene and polypropylene are both easily crystallized, the copolymer remains amorphous. When this copolymer is cross-linked, it behaves as an elastomer. Dylark^{TM} is based on a copolymer of maleic anhydride and styrene. Styrene provides toughness, while maleic anhydride provides high-temperature properties. Carbon black (for protection from ultraviolet rays and enhancing stiffness), rubber (for toughness), and glass fibers

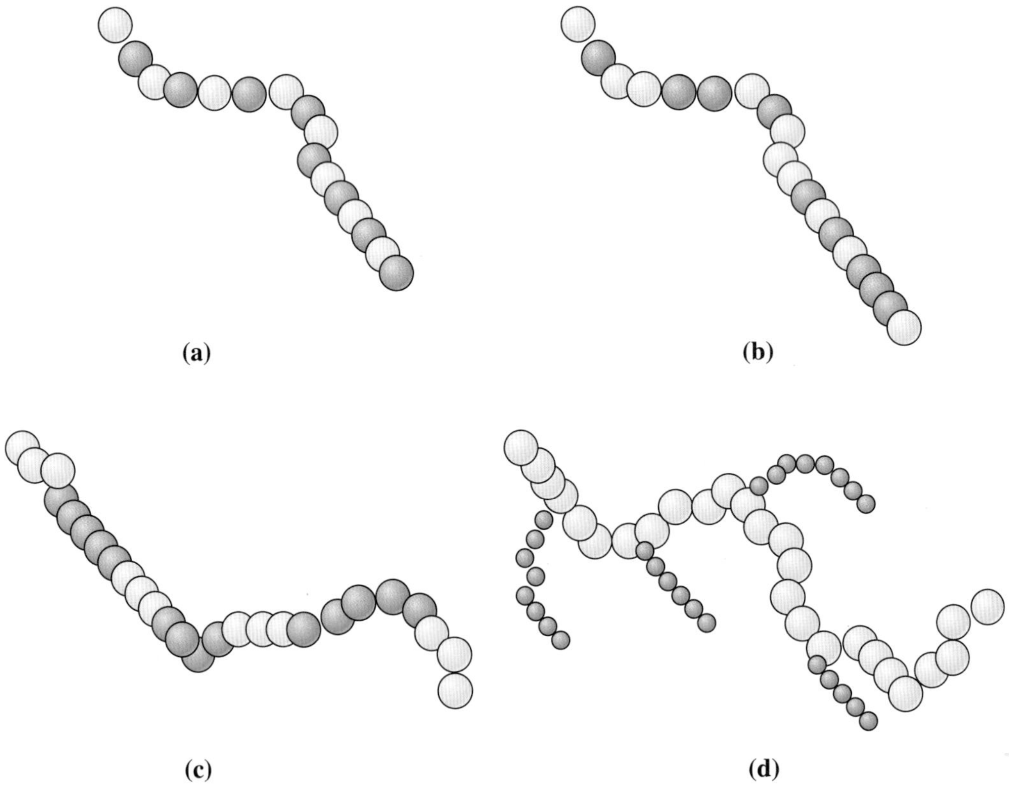

Figure 16-9 Four types of copolymers: (a) alternating monomers, (b) random monomers, (c) block copolymers, and (d) grafted copolymers. Circles of different colors or sizes represent different monomers.

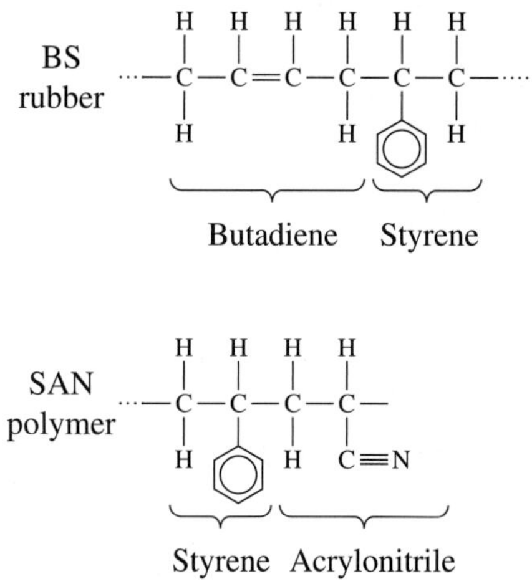

Figure 16-10 Copolymerization produces the polymer ABS, which consists of two copolymers, BS and SAN, grafted together.

(for stiffness) are added to the Dylark copolymer. It has been used to make instrument panels for car dashboards. The Dylark plastic is then coated with vinyl, which provides a smooth and soft finish.

Blending and Alloying We can improve the mechanical properties of many of the thermoplastics by blending or alloying. By mixing an immiscible elastomer with the thermoplastic, we produce a two-phase polymer, as found in ABS. The elastomer does not enter the structure as a copolymer but, instead, helps to absorb energy and improve toughness. Polycarbonates used to produce transparent aircraft canopies are toughened by elastomers in this manner.

Liquid-Crystalline Polymers Some of the complex thermoplastic chains become so stiff that they act as rigid rods, even at high temperatures. These materials are **liquid-crystalline polymers** (LCPs). Some aromatic polyesters and aromatic polyamides (or **aramids**) are examples of liquid-crystalline polymers and are used as high-strength fibers (as will be discussed in Chapter 17). Kevlar^{TM}, an aromatic polyamide, is the most familiar of the LCPs and is used as a reinforcing fiber for aerospace applications and for bulletproof vests. Liquid-crystal polymers are also used to make electronic displays.

16-6 Effect of Temperature on Thermoplastics

Properties of thermoplastics change depending upon temperature. We need to know how these changes occur to (a) better design components and (b) guide the type of processing techniques that need to be used. Several critical temperatures and structures, summarized in Figures 16-11 and 16-12, may be observed.

Thermoplastics can be amorphous or crystalline once they cool below the melting temperature (Figure 16-11). Most often, engineered thermoplastics have both amorphous and crystalline regions. The crystallinity in thermoplastics can be introduced by temperature (slow cooling) or by the application of stress that can untangle chains (**stress-induced crystallization**). Similar to dispersion strengthening of metallic materials, the formation of crystalline regions in an otherwise amorphous matrix helps increase the strength of thermoplastics. In typical thermoplastics, bonding within the chains is covalent, but the long coiled chains are held to one another by weak van der Waals bonds

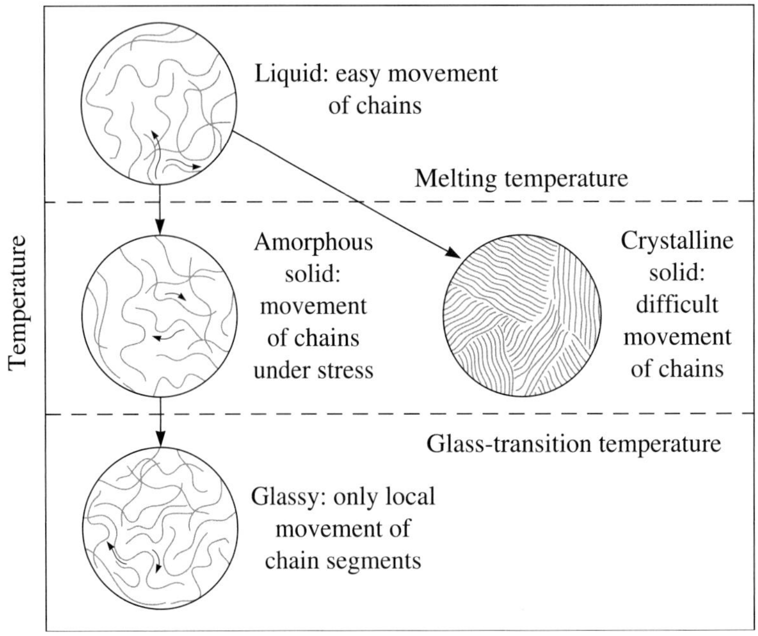

Figure 16-11 The effect of temperature on the structure and behavior of thermoplastics.

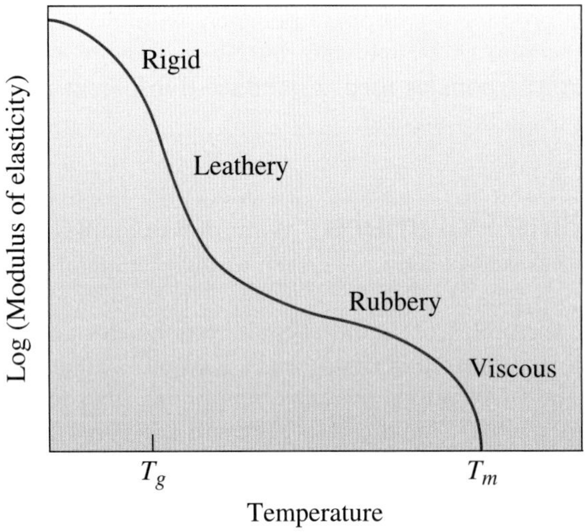

Figure 16-12 The effect of temperature on the modulus of elasticity for an amorphous thermoplastic. Note that T_g and T_m are not fixed.

and by entanglement. When a tensile stress is applied to the thermoplastic, the weak bonding between the chains can be overcome, and the chains can rotate and slide relative to one another. The ease with which the chains slide depends on both temperature and the polymer structure.

Degradation Temperature At very high temperatures, the covalent bonds between the atoms in the linear chain may be destroyed, and the polymer may burn or char. In thermoplastics, decomposition occurs in the liquid state; in thermosets, the decomposition occurs in the solid state. This temperature T_d (not shown in Figure 16-12), is the **degradation** (or decomposition) **temperature**. When plastics burn, they create smoke, which is dangerous. Some materials (such as limestone, talc, alumina, etc.) added to thermoplastics are thermal or heat stabilizers. They absorb heat and protect the polymer matrix. Fire retardant additives include hydrated alumina, antimony compounds, and halogen compounds (e.g., MgBr, PCl_5). Some additives retard fire by excluding oxygen but generate dangerous gases and are not appropriate for certain applications.

Exposure to other forms of chemicals or energy (e.g., oxygen, ultraviolet radiation, and attack by bacteria) also cause a polymer to degrade or **age** slowly, even at low temperatures. Carbon black (up to ~3%) is one of the commonly used additives that helps improve the resistance of plastics to ultraviolet degradation.

Liquid Polymers Thermoplastics usually do not melt at a precise temperature. Instead there is a range of temperatures over which melting occurs. The approximate melting ranges of typical polymers are included in Table 16-5. At or above the melting temperature T_m, bonding between the twisted and intertwined chains is weak. If a force is applied, the chains slide past one another, and the polymer flows with virtually no elastic strain. The strength and modulus of elasticity are nearly zero, and the polymer is suitable for casting and many forming processes. Most thermoplastic melts are shear thinning (i.e., their apparent viscosity decreases with an increase in the steady-state shear rate).

Rubbery and Leathery States Below the melting temperature, the polymer chains are still twisted and intertwined. These polymers have an amorphous structure. Just below the melting temperature, the polymer behaves in a *rubbery* manner. When stress is applied, both elastic and plastic deformation of the polymer occurs. When the stress is removed, the elastic deformation is quickly recovered, but the polymer is

Table 16-5 Melting, glass-transition, and processing temperature ranges (°C) for selected thermoplastics and elastomers

Polymer	Melting Temperature Range	Glass-Transition Temperature Range (T_g)	Processing Temperature Range
Addition polymers			
Low-density (LD) polyethylene	98–115	−90 to −25	149–232
High-density (HD) polyethylene	130–137	−110	177–260
Polyvinyl chloride	175–212	87	
Polypropylene	160–180	−25 to −20	190–288
Polystyrene	240	85–125	
Polyacrylonitrile	320	107	
Polytetrafluoroethylene (Teflon)	327		
Polychlorotrifluoroethylene	220		
Polymethyl methacrylate (acrylic)		90–105	
Acrylonitrile butadiene styrene (ABS)	110–125	100	177–260
Condensation polymers			
Acetal	181	−85	
6,6-nylon	243–260	49	260–327
Cellulose acetate	230		
Polycarbonate	230	149	271–300
Polyester	255	75	
Polyethylene terephthalate (PET)	212–265	66–80	227–349
Elastomers			
Silicone		−123	
Polybutadiene	120	−90	
Polychloroprene	80	−50	
Polyisoprene	30	−73	

permanently deformed due to the movement of the chains. Large permanent elongations can be achieved, permitting the polymer to be formed into useful shapes by molding and extrusion.

At lower temperatures, bonding between the chains is stronger, the polymer becomes stiffer and stronger, and a *leathery* behavior is observed. Many of the commercial polymers, including polyethylene, have a useable strength in this condition.

Glassy State Below the **glass-transition temperature** T_g, the linear amorphous polymer becomes hard, brittle, and glass-like. This is again not a fixed temperature but a range of temperatures. When the polymer cools below the glass-transition temperature, certain properties—such as density or modulus of elasticity—change at a different rate (Figure 16-13). A change in slope in the plot of density as a function of temperature occurs at the glass transition temperature. A change in slope is also observed for a semicrystalline polymer as the amorphous portion transforms to a liquid or vice versa.

Although glassy polymers have poor ductility and formability, they do have good strength, stiffness, and creep resistance. A number of important polymers, including polystyrene and polyvinyl chloride, have glass-transition temperatures above room temperature (Table 16-5).

The glass-transition temperature is typically about 0.5 to 0.75 times the absolute melting temperature T_m. Polymers such as polyethylene, which have no complicated side groups attached to the carbon backbone, have low glass-transition temperatures (even below room temperature) compared with polymers such as polystyrene, which have more complicated side groups.

As pointed out in Chapter 6, many thermoplastics become brittle at lower temperatures. The brittleness of the polymer used for some of the O-rings ultimately caused the 1986 *Challenger* disaster. The lower temperatures that existed during the launch time caused the embrittlement of the rubber O-rings used for the booster rockets.

Observing and Measuring Crystallinity in Polymers Many thermoplastics partially crystallize when cooled below the melting temperature, with the chains becoming closely aligned over appreciable distances. A sharp increase in the density occurs as the coiled and intertwined chains in the liquid are rearranged into a more orderly, close-packed structure (Figure 16-13).

One model describing the arrangement of the chains in a crystalline polymer is shown in Figure 16-14. In this *folded chain* model, the chains loop back on themselves, with

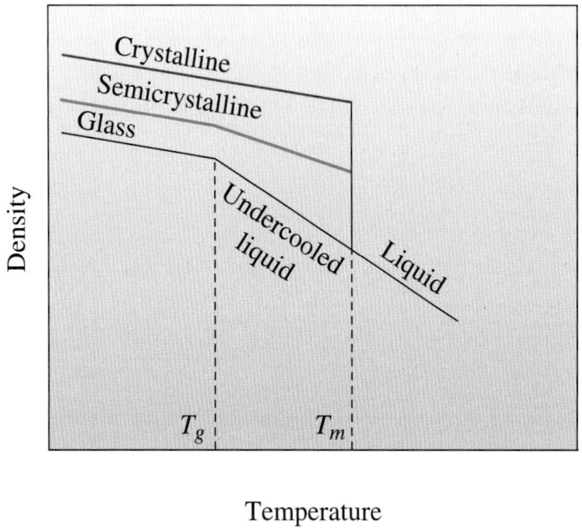

Figure 16-13 The relationship between the density and temperature of a polymer shows the melting and glass-transition temperatures. Note that T_g and T_m are not fixed; rather, they are ranges of temperatures.

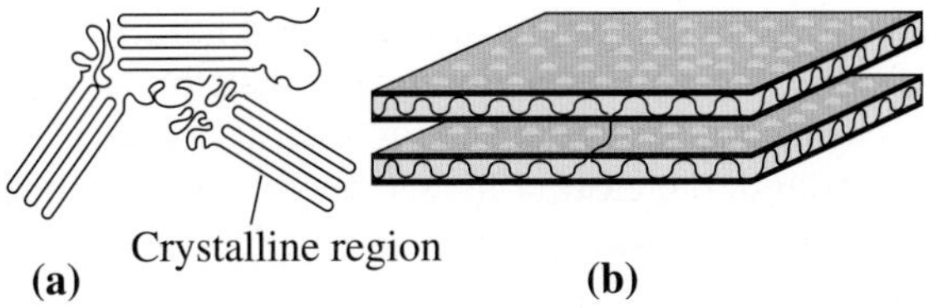

Figure 16-14 The folded chain model for crystallinity in polymers, shown in (a) two dimensions and (b) three dimensions.

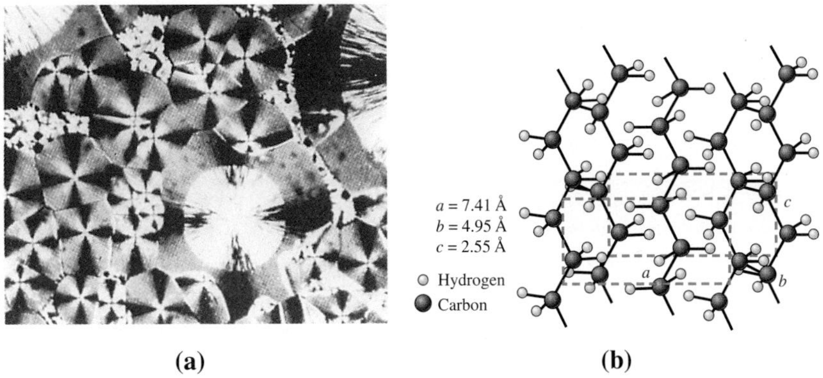

Figure 16-15 (a) Spherulitic crystals in an amorphous matrix of nylon (× 200). (Structure and Properties of Engineering Materials by *BRICK, R. M.; et al. Copyright 1977.*) (b) The unit cell of crystalline polyethylene.

each loop being approximately 100 carbon atoms long. The folded chain extends in three dimensions, producing thin plates or lamellae. The crystals can take various forms, with the spherulitic shape shown in Figure 16-15(a) being particularly common. The crystals have a unit cell that describes the regular packing of the chains. The crystal structure for polyethylene, shown in Figure 16-15(b), describes one such unit cell. Crystal structures for several polymers are listed in Table 16-6. Some polymers are polymorphic, having more than one crystal structure.

Even in crystalline polymers, there are always thin regions between the lamellae, as well as between spherulites, that are amorphous transition zones. The weight percentage of the structure that is crystalline can be calculated from the density of the polymer:

$$\% \text{ Crystalline} = \frac{\rho_c(\rho - \rho_a)}{\rho(\rho_c - \rho_a)} \times 100 \qquad (16\text{-}4)$$

where ρ is the measured density of the polymer, ρ_a is the density of amorphous polymer, and ρ_c is the density of completely crystalline polymer. Similarly, x-ray diffraction can be used to measure the level of crystallinity and determine lattice constants for single crystal polymers.

Table 16-6 Crystal structures of several polymers

Polymer	Crystal Structure	Lattice Parameters (nm)
Polyethylene	Orthorhombic	$a_0 = 0.741$ $b_0 = 0.495$ $c_0 = 0.255$
Polypropylene	Orthorhombic	$a_0 = 1.450$ $b_0 = 0.569$ $c_0 = 0.740$
Polyvinyl chloride	Orthorhombic	$a_0 = 1.040$ $b_0 = 0.530$ $c_0 = 0.510$
Polyisoprene (cis)	Orthorhombic	$a_0 = 1.246$ $b_0 = 0.886$ $c_0 = 0.810$

As the side groups get more complex, it becomes harder to crystallize thermoplastics. For example, polyethylene (H as side group) can be crystallized more easily than polystyrene (benzene ring as side group). High-density polyethylene (HDPE) has a higher level of crystallinity and, therefore, a higher density (0.97 g/cm^3) than low-density polyethylene (LDPE), which has a density of 0.92 g/cm^3. The crystallinity and, hence, the density in LDPE is lower, since the polymer is branched. A completely crystalline polymer would not display a glass-transition temperature; however, the amorphous regions in semicrystalline polymers do transform to a glassy material below the glass-transition temperature (Figure 16-13). Such polymers as acetal, nylon, HDPE, and polypropylene are referred to as crystalline even though the level of crystallinity may be low to moderate. The following examples show how properties of plastics can be accounted for in different applications.

Example 16-6 *Design of a Polymer Insulation Material*

A storage tank for liquid hydrogen will be made from metal, but we wish to coat the metal with a 3-mm-thick polymer as an intermediate layer between the metal and additional insulation layers. The temperature of the intermediate layer may drop to −80°C. Choose a material for this layer.

SOLUTION

We want the material to have reasonable ductility. As the temperature of the tank changes, stresses develop in the coating due to differences in thermal expansion, and we do not want the polymer to fail due to these stresses. A material that has good ductility and/or can undergo large elastic strains is needed. We therefore would prefer either a thermoplastic that has a glass-transition temperature below −80°C or an elastomer, also with a glass-transition temperature below −80°C. Of the polymers listed in Table 16-5, thermoplastics such as polyethylene and acetal are satisfactory. Suitable elastomers include silicone and polybutadiene.

We might prefer one of the elastomers, for they can accommodate thermal stress by elastic, rather than plastic, deformation.

Example 16-7 *Impact Resistant Polyethylene*

A new grade of flexible, impact-resistant polyethylene for use as a thin film requires a density of 0.88 to 0.915 g/cm^3. Design the polyethylene required to produce these properties. The density of amorphous polyethylene is about 0.87 g/cm^3.

SOLUTION

To produce the required properties and density, we must control the percent crystallinity of the polyethylene. We can use Equation 16-4 to determine the crystallinity that corresponds to the required density range. To do so, however, we must know the density of completely crystalline polyethylene. We can use the data in Figure 16-15 and Table 16-6 to calculate this density if we recognize that there are two polyethylene repeat units in each unit cell.

$$\rho_c = \frac{(4\text{ C atoms})(12\text{ g/mol}) + (8\text{ H atoms})(1\text{ g/mol})}{(7.41 \times 10^{-8}\text{ cm})(4.95 \times 10^{-8}\text{ cm})(2.55 \times 10^{-8}\text{ cm})(6.022 \times 10^{23}\text{ atoms/mol})}$$
$$= 0.9942\text{ g/cm}^3$$

We know that $\rho_a = 0.87$ g/cm^3 and that ρ varies from 0.88 to 0.915 g/cm^3. The required crystallinity then varies from

$$\% \text{ crystalline} = \frac{(0.9942)(0.88 - 0.87)}{(0.88)(0.9942 - 0.87)} \times 100 = 9.1$$

$$\% \text{ crystalline} = \frac{(0.9942)(0.915 - 0.87)}{(0.915)(0.9942 - 0.87)} \times 100 = 39.4$$

Therefore, we must be able to process the polyethylene to produce a range of crystallinity between 9.1 and 39.4%.

16-7 Mechanical Properties of Thermoplastics

Most thermoplastics (molten and solid) exhibit non-Newtonian and **viscoelastic behavior**. The behavior is non-Newtonian (i.e., the stress and strain are not linearly related for most parts of the stress strain curve). The viscoelastic behavior means when an external force is applied to a thermoplastic polymer, both elastic and plastic (or viscous) deformation occurs. The mechanical behavior is closely tied to the manner in which the polymer chains move relative to one another under load. The deformation process depends on both time and the rate at which the load is applied. Figure 16-16 shows a stress strain curve for 6,6-nylon.

Elastic Behavior Elastic deformation in thermoplastics is the result of two mechanisms. An applied stress causes the covalent bonds within the chain to stretch and distort, allowing the chains to elongate elastically. When the stress is removed, recovery from this distortion is almost instantaneous. This behavior is similar to that in metals and ceramics, which also deform elastically by the stretching of metallic, ionic, or covalent bonds. In addition, entire segments of the polymer chains may be distorted; when the stress is removed, the segments move back to their original positions over a period of time—often hours or even months. This time-dependent, or viscoelastic, behavior may contribute to some nonlinear elastic behavior.

Plastic Behavior of Amorphous Thermoplastics These polymers deform plastically when the stress exceeds the yield strength. Unlike deformation in the case of metals, however, plastic deformation is not a consequence of dislocation movement. Instead, chains stretch, rotate, slide, and disentangle under load to cause permanent deformation. The drop in the stress beyond the yield point can be explained

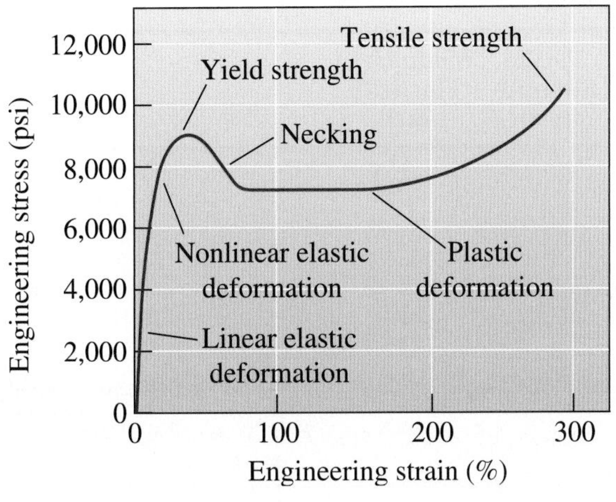

Figure 16-16 The engineering stress strain curve for 6,6-nylon, a typical thermoplastic polymer.

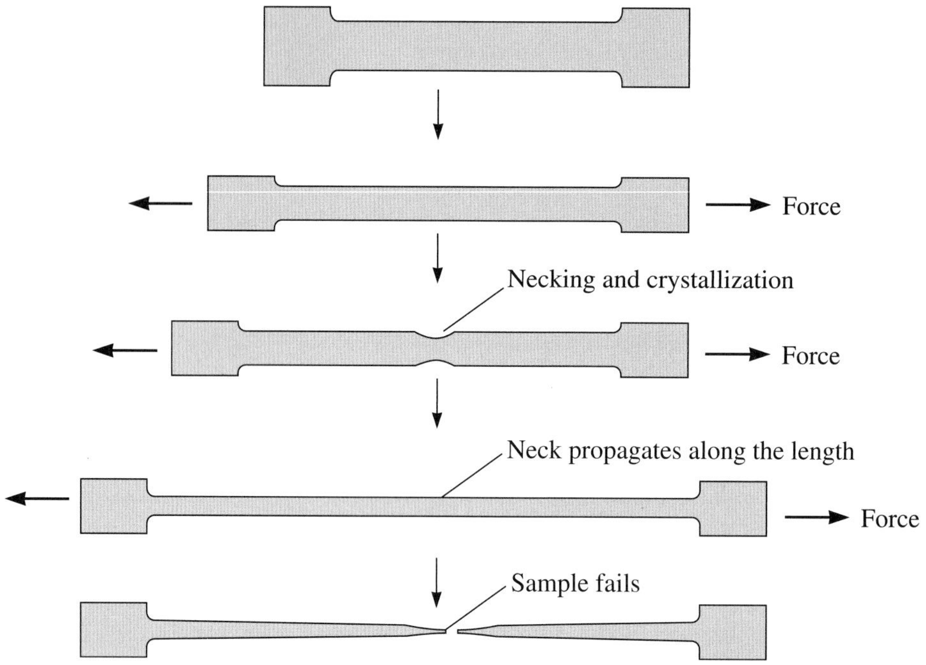

Figure 16-17 Necks can be stable in amorphous polymers because local alignment strengthens the necked region.

by this phenomenon. Initially, the chains may be highly tangled and intertwined. When the stress is sufficiently high, the chains begin to untangle and straighten. Necking also occurs, permitting continued sliding of the chains at a lesser stress. Eventually, however, the chains become almost parallel and close together; stronger van der Waals bonding between the more closely aligned chains requires higher stresses to complete the deformation and fracture process (Figure 16-17).

Example 16-8 *Comparing Mechanical Properties of Thermoplastics*

Compare the mechanical properties of low-density (LD) polyethylene, high-density (HD) polyethylene, polyvinyl chloride, polypropylene, and polystyrene, and explain their differences in terms of their structures.

SOLUTION

Let us look at the maximum tensile strength and modulus of elasticity for each polymer.

Polymer	Tensile Strength (psi)	Modulus of Elasticity (ksi)	Structure
LD polyethylene	3,000	40	Highly branched, amorphous structure with symmetrical monomers
HD polyethylene	5,500	180	Amorphous structure with symmetrical monomers but little branching
Polypropylene	6,000	220	Amorphous structure with small methyl side groups
Polystyrene	8,000	450	Amorphous structure with benzene side groups
Polyvinyl chloride	9,000	600	Amorphous structure with large chlorine atoms as side groups

We can conclude that

1. Branching, which reduces the density and close packing of chains, reduces the strength and elastic modulus of polyethylene.
2. Adding atoms or atom groups other than hydrogen to the chain increases strength and stiffness. The methyl group in polypropylene provides some improvement; the benzene ring in styrene provides higher strength and stiffness; and the chlorine atom in polyvinyl chloride provides a large increase in these properties.

Creep and Stress Relaxation Thermoplastics exhibit creep, a time-dependent permanent deformation with constant stress or load (Figures 16-18 and 16-19).

They also show **stress relaxation** [i.e., under a constant strain, the stress level decreases with time (Chapter 6)]. Stress relaxation, like creep, is a consequence of the viscoelastic behavior of the polymer. Perhaps the most familiar example of this behavior is a rubber band (an elastomer) stretched around a pile of books. Initially, the tension in the rubber band is high when the rubber band is taut. After several weeks, the strain in the rubber band is unchanged (it still completely encircles the books), but the stress will have decreased. Similarly, the nylon strings in tennis rackets are pulled at a higher tension initially since this tension (i.e., stress) decreases with time.

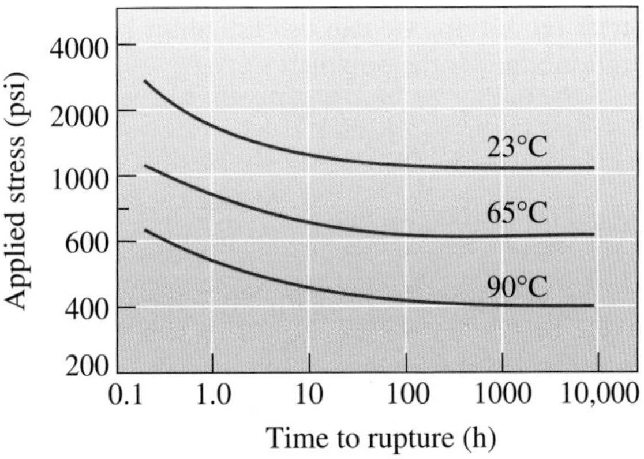

Figure 16-18 The effect of temperature on the stress-rupture behavior of high-density polyethylene.

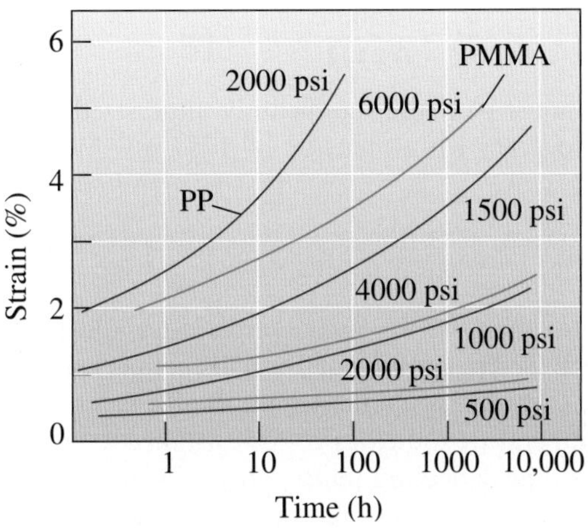

Figure 16-19 Creep curves for acrylic (PMMA) and polypropylene (PP) at 20°C and several applied stresses.

In a simple model, the rate at which stress relaxation occurs is related to the **relaxation time** λ, which is considered a property of the polymer (more complex models consider a distribution of relaxation times). The stress after time t is given by

$$\sigma = \sigma_0 \exp(-t/\lambda) \tag{16-5}$$

where σ_0 is the original stress. The *relaxation time*, in turn, depends on the viscosity and, thus, the temperature:

$$\lambda = \lambda_0 \exp\left(\frac{Q}{RT}\right) \tag{16-6}$$

where λ_0 is a constant and Q is the activation energy related to the ease with which polymer chains slide past each other. Relaxation of the stress occurs more *rapidly* at *higher temperatures* and for polymers with a low viscosity.

The following example shows how stress relaxation can be accounted for while designing with polymers.

Example 16-9 *Determination of Initial Stress in a Polymer*

A band of polyisoprene is to hold together a bundle of steel rods for up to one year. If the stress on the band is less than 1500 psi, the band will not hold the rods tightly. Determine the initial stress that must be applied to a polyisoprene band when it is slipped over the steel. A series of tests showed that an initial stress of 1000 psi decreased to 980 psi after six weeks.

SOLUTION

Although the strain on the elastomer band may be constant, the stress will decrease over time due to stress relaxation. We can use Equation 16-5 and our initial tests to determine the relaxation time for the polymer:

$$\sigma = \sigma_0 \exp\left(-\frac{t}{\lambda}\right)$$

$$980 \text{ psi} = 1000 \text{ psi} \exp\left(-\frac{6 \text{ weeks}}{\lambda}\right)$$

$$-\frac{6}{\lambda} = \ln\left(\frac{980}{1000}\right) = \ln(0.98) = -0.0202$$

$$\lambda = \frac{6}{0.0202} = 297 \text{ weeks}$$

Now that we know the relaxation time, we can determine the stress that must be initially placed onto the band so that it will still be stressed to 1500 psi after 1 year (52 weeks).

$$1500 = \sigma_0 \exp(-52/297) = \sigma_0 \exp(-0.1751) = 0.8394\,\sigma_0$$

$$\sigma_0 = \frac{1500}{0.8394} = 1787 \text{ psi}$$

The polyisoprene band must be made significantly undersized so it can slip over the materials it is holding together with a tension of 1787 psi. After one year, the stress will still be 1500 psi.

One more practical measure for high temperature and creep properties of a polymer is the heat **deflection temperature** or heat **distortion temperature** under load, which is the temperature at which a given deformation of a beam occurs for a standard load. A high deflection temperature indicates good resistance to creep and permits us to compare various polymers. The deflection temperatures for several polymers are shown in Table 16-7, which gives the temperature required to cause a 0.01 in. deflection for a

Table 16-7 Deflection temperatures for selected polymers for a 264 psi load

Polymer	Deflection Temperature (°C)
Polyester	40
Polyethylene (ultra-high density)	40
Polypropylene	60
Phenolic	80
Polyamide (6,6-nylon)	90
Polystyrene	100
Polyoxymethylene (acetal)	130
Polyamide-imide	280
Epoxy	290

264 psi load at the center of a bar resting on supports 4 in. apart. A polymer is "dishwasher safe" if it has a heat distortion temperature greater than ~50°C.

Impact Behavior Viscoelastic behavior also helps us understand the impact properties of polymers. At very high rates of strain, as in an impact test, there is insufficient time for the chains to slide and cause plastic deformation. For these conditions, the thermoplastics behave in a brittle manner and have poor impact values. Polymers may have a transition temperature. At low temperatures, brittle behavior is observed in an impact test, whereas more ductile behavior is observed at high temperatures, where the chains move more easily. These effects of temperature and strain rate are similar to those seen in metals that exhibit a ductile-to-brittle transition temperature; however, the mechanisms are different.

Deformation of Crystalline Polymers A number of polymers are used in the crystalline state. As we discussed earlier, however, the polymers are never completely crystalline. Instead, small regions—between crystalline lamellae and between crystalline spherulites—are amorphous transition regions. Polymer chains in the crystalline region extend into these amorphous regions as tie chains. When a tensile load is applied to the polymer, the crystalline lamellae within the spherulites slide past one another and begin to separate as the tie chains are stretched. The folds in the lamellae tilt and become aligned with the direction of the tensile load. The crystalline lamellae break into smaller units and slide past one another, until eventually the polymer is composed of small aligned crystals joined by tie chains and oriented parallel to the tensile load. The spherulites also change shape and become elongated in the direction of the applied stress. With continued stress, the tie chains disentangle or break, causing the polymer to fail.

Crazing **Crazing** occurs in thermoplastics when localized regions of plastic deformation form in a direction perpendicular to that of the applied stress. In transparent thermoplastics, such as some of the glassy polymers, the craze produces a translucent or opaque region that looks like a crack. The craze can grow until it extends across the entire cross section of the polymer part. The craze is not a crack, and, in fact, it can continue to support an applied stress. The process is similar to that for the plastic deformation of the polymer, but the process can proceed even at a low stress over an extended length of time. Crazing can lead to brittle fracture of the polymer and is often assisted by the presence of a solvent (known as solvent crazing).

Blushing **Blushing** or whitening refers to failure of a plastic because of localized crystallization (due to repeated bending, for example) that ultimately causes voids to form.

16-8 Elastomers [Rubbers]

A number of natural and synthetic polymers called elastomers display a large amount (>200%) of elastic deformation when a force is applied. Rubber bands, automobile tires, O-rings, hoses, and insulation for electrical wires are common uses for these materials. Elastomers are often referred to as "rubbers" because crude natural rubber is an elastomer.

Geometric Isomers Some monomers that have different structures, even though they have the same composition, are called **geometric isomers**. Isoprene, or natural rubber, is an important example (Figure 16-20). The monomer includes two double bonds between carbon atoms; this type of monomer is called a **diene**. Polymerization occurs by breaking both double bonds, creating a new double bond at the center of the molecule and active sites at both ends.

In the *trans* form of isoprene, the hydrogen atom and the methyl group at the center of the repeat unit are located on opposite sides of the newly formed double bond. This arrangement leads to relatively straight chains; the polymer crystallizes and forms a hard rigid polymer called *gutta percha*. This is used to make golf balls and shoe soles.

In the *cis* form, however, the hydrogen atom and the methyl group are located on the same side of the double bond. This different geometry causes the polymer chains to develop a highly coiled structure, preventing close packing and leading to an amorphous, rubbery polymer. If a stress is applied to cis-isoprene, the polymer behaves in a viscoelastic manner. The chains uncoil and bonds stretch, producing elastic deformation, but the chains also slide past one another, producing nonrecoverable plastic deformation. The polymer behaves as a thermoplastic rather than an elastomer.

Cross-Linking We prevent viscous plastic deformation while retaining large elastic deformation by **cross-linking** the chains (Figure 16-21). **Vulcanization**, which uses sulfur atoms, is a common method for cross-linking. Strands of sulfur atoms link the polymer chains as the polymer is processed and shaped at temperatures of about 120 to 180°C (Figure 16-22). The cross-linking steps may include rearranging a hydrogen

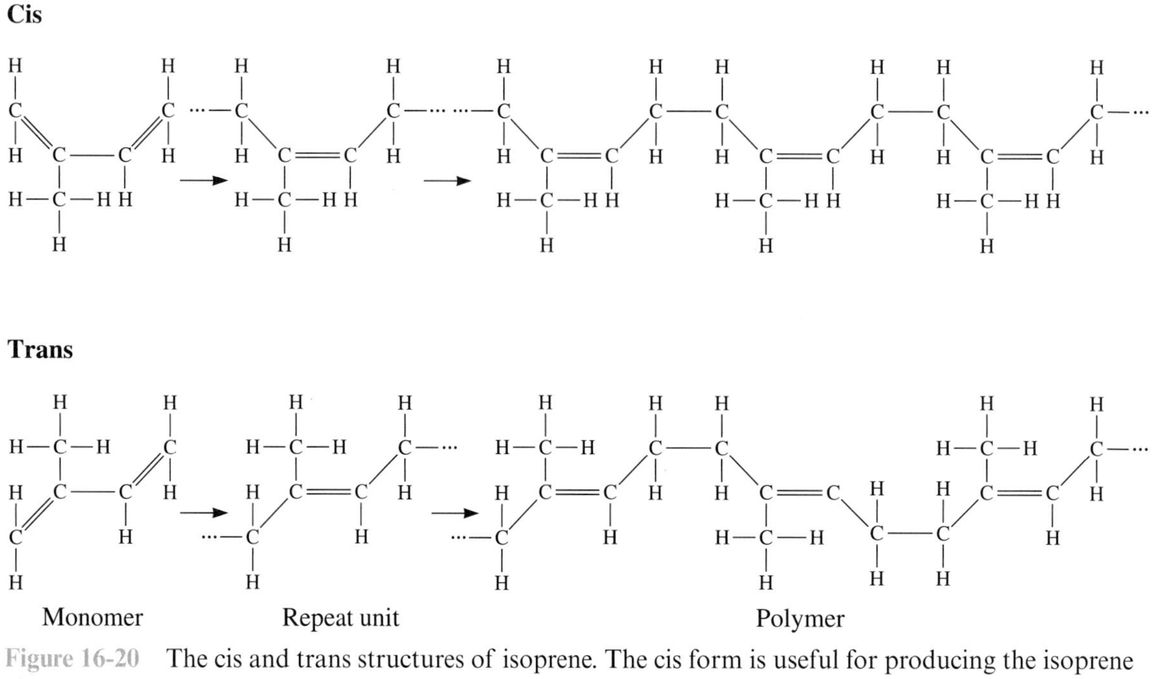

Figure 16-20 The cis and trans structures of isoprene. The cis form is useful for producing the isoprene elastomer.

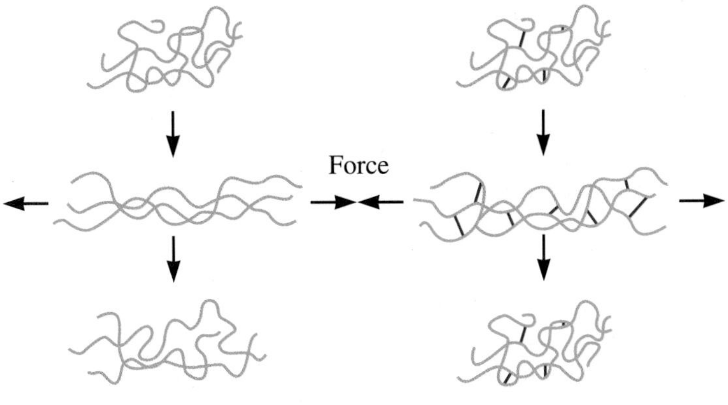

Figure 16-21 (a) When the elastomer contains no cross-links, the application of a force causes both elastic and plastic deformation; after the load is removed, the elastomer is permanently deformed. (b) When cross-linking occurs, the elastomer still may undergo large elastic deformation; however, when the load is removed, the elastomer returns to its original shape.

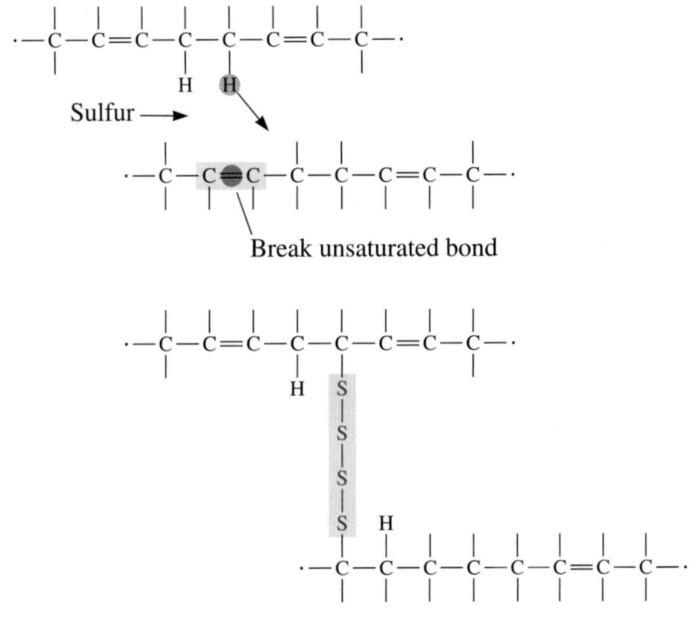

Figure 16-22 Cross-linking of polyisoprene chains may occur by introducing strands of sulfur atoms. Sites for attachment of the sulfur strands occur by rearrangement or loss of a hydrogen atom and the breaking of an unsaturated bond.

atom and replacing one or more of the double bonds with single bonds. The cross-linking process is not reversible; consequently, the elastomer cannot be easily recycled.

The stress strain curve for an elastomer is shown in Figure 16-23. Virtually all of the curve represents elastic deformation; elastomers display a nonlinear elastic behavior. Initially, the modulus of elasticity decreases because of the uncoiling of the chains; however, after the chains have been extended, further elastic deformation occurs by the stretching of the bonds, leading to a higher modulus of elasticity.

The number of cross-links (or the amount of sulfur added to the material) determines the elasticity of the rubber. Low sulfur additions leave the rubber soft and flexible, as in elastic bands or rubber gloves. Increasing the sulfur content restricts the uncoiling of the chains, and the rubber becomes harder, more rigid, and brittle, as in rubber used for motor mounts. Typically, 0.5 to 5% sulfur is added to provide cross-linking in elastomers. Many more *efficient vulcanizing* (EV) systems, which are sulfur free, also have been developed and used in recent years.

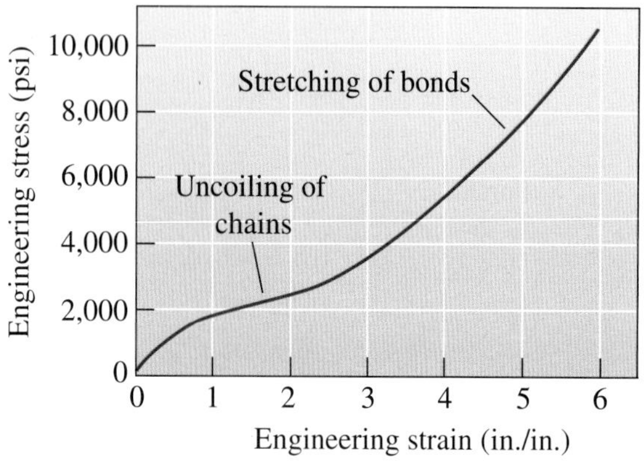

Figure 16-23 The stress strain curve for an elastomer. Virtually all of the deformation is elastic; the modulus of elasticity varies as the strain changes.

Typical Elastomers Elastomers, which are amorphous polymers, do not easily crystallize during processing. They have a low glass-transition temperature, and chains can easily be deformed elastically when a force is applied. The repeat units, applications, and properties for some common elastomers are given in Tables 16-8 and 16-9. Polyisoprene is a natural rubber. Polychloroprene, or Neoprene, is a common material for hoses and electrical insulation. Many of the important synthetic elastomers are copolymers. Polybutadiene (butadiene rubber or Buna-S) is similar to polyisoprene, but the repeat unit has four carbon atoms consisting of one double bond. This is a relatively low-cost rubber, but the resistance to solvents is poor. As a result, it is used as a toughening material to make other elastomers. Butadiene-styrene rubber (BSR or BS), which is also one of the components of ABS (Figure 16-10), is used for automobile tires. Butyl rubber is different from polybutadiene. Butyl rubber, or polyisobutadiene, is used to make inner tubes for tires, vibration mounts, and weather-stripping material. Silicones are another important elastomer based on chains composed of silicon and oxygen atoms. Silly Putty® was invented by James Wright of General Electric. It is made using hydroxyl terminated polydimethyl siloxane, boric oxide, and some other additives. At slow strain rates, you can stretch it significantly, while if you pull it fast, it snaps. The silicone rubbers (also known as polysiloxanes) provide high-temperature resistance, permitting use of the elastomer at temperatures as high as 315°C. Low molecular weight silicones form liquids and are known as silicone oils. Silicones can also be purchased as a two-part system that can be molded and cured. Chewing gum contains a base that is made from natural rubber, styrene butadiene, or polyvinyl acetate (PVA).

Thermoplastic Elastomers (TPEs) This is a special group of polymers that do not rely on cross-linking to produce a large amount of elastic deformation. Figure 16-24 shows the structure of a styrene-butadiene block copolymer engineered so that the styrene repeat units are located only at the ends of the chains. Approximately 25% of the chain is composed of styrene. The styrene ends of several chains form spherical-shaped domains. The styrene has a high glass-transition temperature; consequently, the domains are strong and rigid and tightly hold the chains together. Rubbery areas containing butadiene repeat units are located between the styrene domains; these portions of the polymer have a glass-transition temperature below room temperature and therefore behave in a soft, rubbery manner. Elastic deformation occurs by recoverable movement of the chains; sliding of the chains at normal temperatures is prevented by the styrene domains.

The styrene-butadiene block copolymers differ from the BS rubber discussed earlier in that cross-linking of the butadiene monomers is not necessary and, in fact, is undesirable. When the thermoplastic elastomer is heated, the styrene heats above the glass-transition temperature, the domains are destroyed, and the polymer deforms in a viscous manner—that is, it behaves as any other thermoplastic, making fabrication

Table 16-8 Repeat units and applications for selected elastomers

Polymer	Repeat Unit	Applications
Polyisoprene	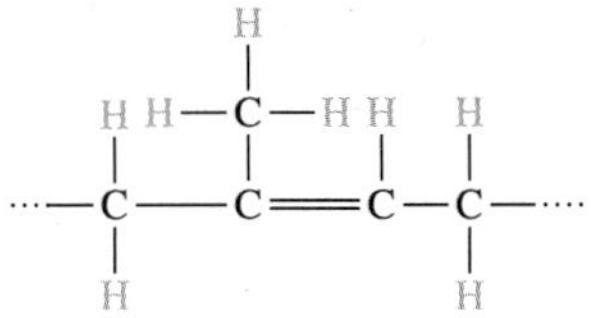	Tires, golf balls, shoe soles
Polybutadiene (or butadiene rubber or Buna-S	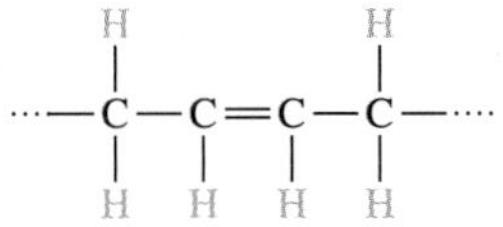	Industrial tires, toughening other elastomers, inner tubes of tires, weather stripping, steam hoses
Polyisobutylene (or butyl rubber)	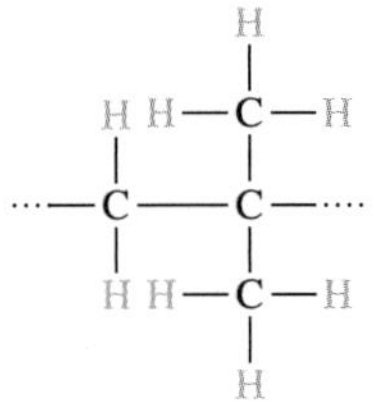	
Polychloroprene (Neoprene)	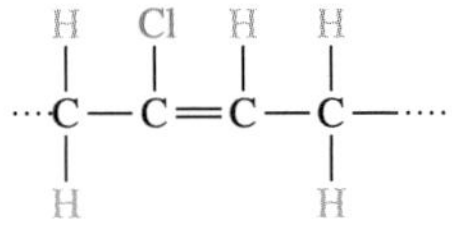	Hoses, cable sheathing
Butadiene-styrene (BS or SBR rubber)	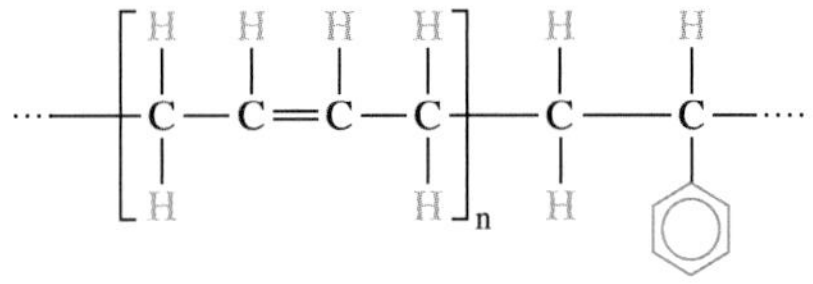	Tires
Butadiene-acrylonitrile (Buna-N)	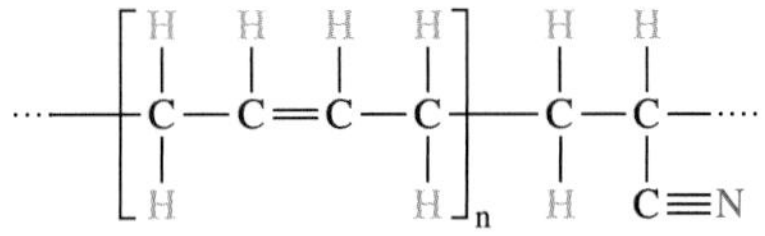	Gaskets, fuel hoses
Silicones	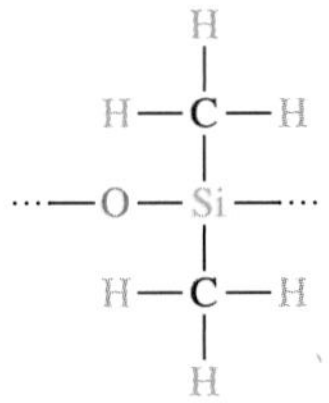	Gaskets, seals

Table 16-9 Properties of selected elastomers

	Tensile Strength (psi)	% Elongation	Density (g/cm^3)
Polyisoprene	3,000	800	0.93
Polybutadiene	3,500		0.94
Polyisobutylene	4,000	350	0.92
Polychloroprene	3,500	800	1.24
Butadiene-styrene	3,000	2000	1.0
Butadiene-acrylonitrile	700	400	1.0
Silicones	1,000	700	1.5
Thermoplastic elastomers	5,000	1300	1.06

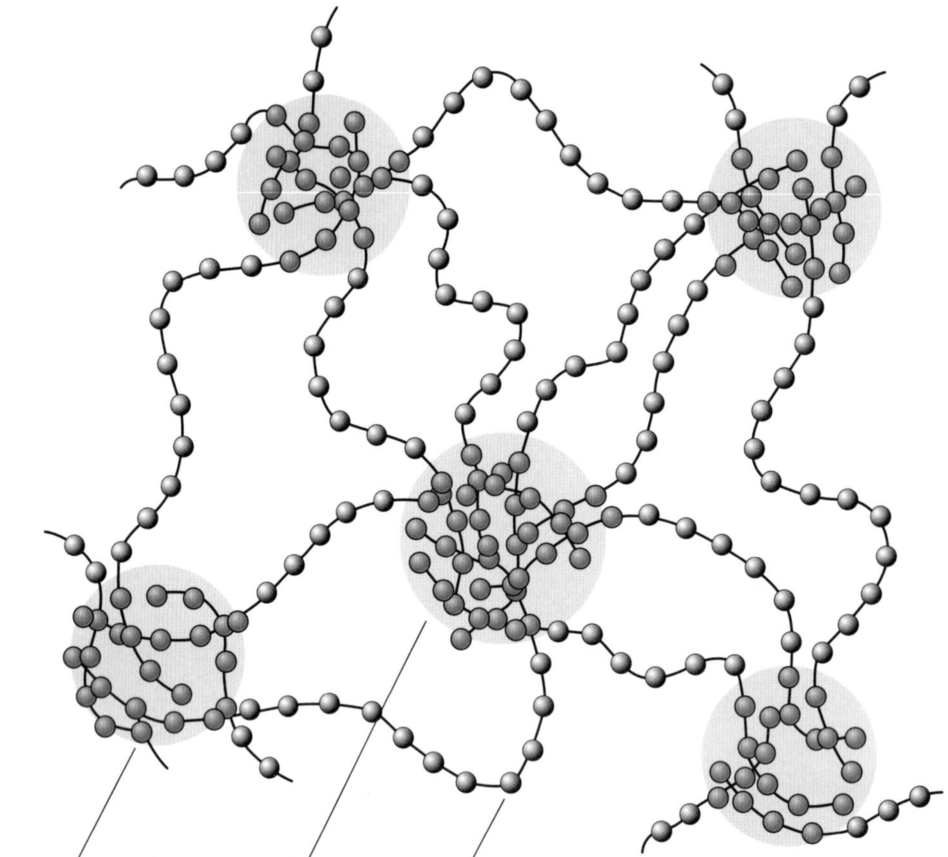

Figure 16-24 The structure of the styrene-butadiene copolymer in a thermoplastic elastomer. The glassy nature of the styrene domains provides elastic behavior without cross-linking of the butadiene.

very easy. When the polymer cools, the domains reform, and the polymer reverts to its elastomeric characteristics. The thermoplastic elastomers consequently behave as ordinary thermoplastics at elevated temperatures and as elastomers at low temperatures. This behavior also permits thermoplastic elastomers to be more easily recycled than conventional elastomers.

16-9 Thermosetting Polymers

Thermosets are highly cross-linked polymer chains that form a three-dimensional network structure. Because the chains cannot rotate or slide, these polymers possess good strength, stiffness, and hardness. Thermosets also have poor ductility and impact properties and a high glass-transition temperature. In a tensile test, thermosetting polymers display the same behavior as a brittle metal or ceramic.

Thermosetting polymers often begin as linear chains. Depending on the type of repeat units and the degree of polymerization, the initial polymer may be either a solid or a liquid resin; in some cases, a two- or three-part liquid resin is used (as in the case of the two tubes of epoxy glue that we often use). Heat, pressure, mixing of the various resins, or other methods initiate the cross-linking process. Cross-linking is not reversible; once formed, the thermosets cannot be reused or conveniently recycled.

Table 16-10 Functional units and applications for selected thermosets

Polymer	Functional Units	Typical Applications
Phenolics	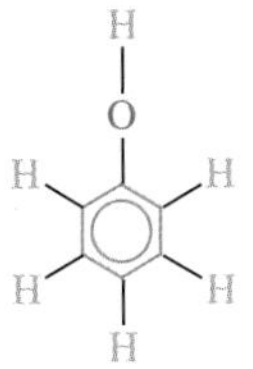	Adhesives, coatings, laminates
Amines	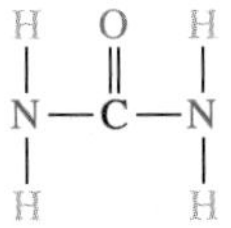	Adhesives, cookware, electrical moldings
Polyesters	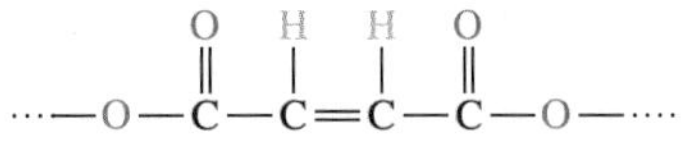	Electrical moldings, decorative laminates, polymer matrix in fiberglass
Epoxies	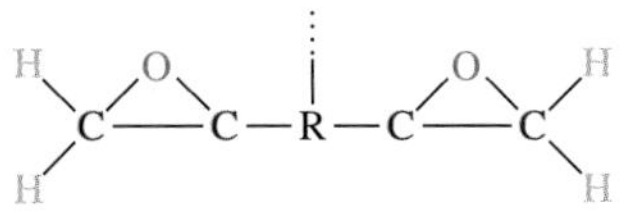	Adhesives, electrical moldings, matrix for composites
Urethanes	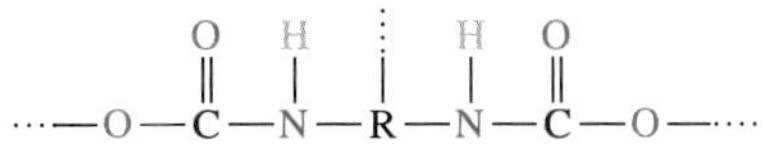	Fibers, coatings, foams, insulation
Silicone	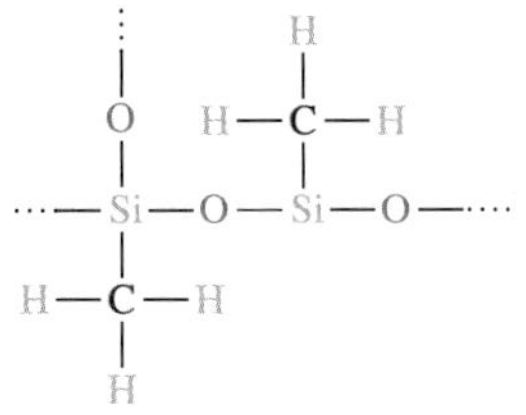	Adhesives, gaskets, sealants

Table 16-11 Properties of typical thermosetting polymers

	Tensile Strength (psi)	% Elongation	Elastic Modulus (ksi)	Density (g/cm^3)
Phenolics	9,000	2	1300	1.27
Amines	10,000	1	1600	1.50
Polyesters	13,000	3	650	1.28
Epoxies	15,000	6	500	1.25
Urethanes	10,000	6		1.30
Silicone	4,000	0	1200	1.55

The functional groups for a number of common thermosetting polymers are summarized in Table 16-10, and representative properties are given in Table 16-11. A functional group is a defined arrangement of atoms with a particular set of properties.

Phenolics Phenolics, the most commonly used thermosets, are often used as adhesives, coatings, laminates, and molded components for electrical or motor applications.

Bakelite™ is one of the common phenolic thermosets. A condensation reaction joining phenol and formaldehyde molecules produces the initial linear phenolic resin. This process continues until a linear phenol-formaldehyde chain is formed; however, the phenol is trifunctional. After the chain has formed, there is a third location on each phenol ring that provides a site for cross-linking with the adjacent chains.

Amines Amino resins, produced by combining urea or melamine monomers with formaldehyde, are similar to the phenolics. The monomers are joined by a formaldehyde link to produce linear chains. Excess formaldehyde provides the cross-linking needed to give strong, rigid polymers suitable for adhesives, laminates, molding materials for cookware, and electrical hardware such as circuit breakers, switches, outlets, and wall plates.

Urethanes Depending on the degree of cross-linking, the urethanes behave as thermosetting polymers, thermoplastics, or elastomers. These polymers find applications as fibers, coatings, and foams for furniture, mattresses, and insulation.

Polyesters Polyesters form chains from acid and alcohol molecules by a condensation reaction, giving water as a byproduct. When these chains contain unsaturated bonds, a styrene molecule may provide cross-linking. Polyesters are used as molding materials for a variety of electrical applications, decorative laminates, boats and other marine equipment, and as a matrix for composites such as fiberglass.

Epoxies Epoxies are thermosetting polymers formed from molecules containing a tight C—O—C ring. During polymerization, the C—O—C rings are opened and the bonds are rearranged to join the molecules. The most common of the commercial epoxies is based on bisphenol A, to which two epoxide units have been added. These molecules are polymerized to produce chains and then co-reacted with curing agents that provide cross-linking. Epoxies are used as adhesives, rigid molded parts for electrical applications, automotive components, circuit boards, sporting goods, and matrices for high-performance fiber-reinforced composite materials for aerospace applications.

Polyimides Polyimides display a ring structure that contains a nitrogen atom. One special group, the bismaleimides (BMI), is important in the aircraft and aerospace industry. They can operate continuously at temperatures of 175°C and do not decompose until reaching 460°C.

Interpenetrating Polymer Networks Some special polymer materials can be produced when linear thermoplastic chains are intertwined through a thermosetting framework, forming **interpenetrating polymer networks**. For example, nylon, acetal, and polypropylene chains can penetrate into a cross-linked silicone thermoset. In more advanced systems, two interpenetrating thermosetting framework structures can be produced.

16-10 Adhesives

Adhesives are polymers used to join other polymers, metals, ceramics, composites, or combinations of these materials. Adhesives are used for a variety of applications. The most critical of these are the "structural adhesives," which find use in the automotive, aerospace, appliance, electronics, construction, and sporting equipment areas.

Chemically Reactive Adhesives These adhesives include polyurethane, epoxy, silicone, phenolics, anaerobics, and polyimides. One-component systems consist of a single polymer resin cured by exposure to moisture, heat, or—in the case of anaerobics—the absence of oxygen. Two-component systems (such as epoxies) cure when two resins are combined.

Evaporation or Diffusion Adhesives The adhesive is dissolved in either an organic solvent or water and is applied to the surfaces to be joined. When the carrier evaporates, the remaining polymer provides the bond. Water-base adhesives are preferred from the standpoint of environmental and safety considerations. The polymer may be completely dissolved in water or may consist of latex, a stable dispersion of polymer in water. A number of elastomers, vinyls, and acrylics are used.

Hot-Melt Adhesives These thermoplastics and thermoplastic elastomers melt when heated. On cooling, the polymer solidifies and joins the materials. Typical melting temperatures of commercial hot-melts are about 80°C to 110°C, which limit the elevated-temperature use of these adhesives. High-performance hot-melts, such as polyamides and polyesters, can be used up to 200°C.

Pressure-Sensitive Adhesives These adhesives are primarily elastomers or elastomer copolymers produced as films or coatings. Pressure is required to cause the polymer to stick to the substrate. They are used to produce electrical and packaging tapes, labels, floor tiles, wall coverings, and wood-grained textured films. Removable pressure-sensitive adhesives are used for medical applications such as bandages and transdermal drug delivery.

Conductive Adhesives A polymer adhesive may contain a filler material such as silver, copper, or aluminum flakes or powders to increase electrical and thermal conductivities. In some cases, thermal conductivity is desired but electrical conductivity is not; alumina, boron nitride, and silica may be used as fillers to provide this combination of properties.

16-11 Polymer Processing and Recycling

There are a number of methods for producing polymer shapes, including molding, extrusion, and manufacturing of films and fibers. The techniques used to form the polymers depend to a large extent on the nature of the polymer—in particular, whether it is thermoplastic or thermosetting. The greatest variety of techniques are used to form the thermoplastics. The polymer is heated to near or above the melting temperature so that it becomes rubbery or liquid. The polymer is then formed in a mold or die to produce the required shape. Thermoplastic elastomers can be formed in the same manner. In these processes, scrap can be easily recycled, and waste is minimized. Fewer forming techniques are used for the thermosetting polymers because, once cross-linking has occurred, the thermosetting polymers are no longer capable of being formed.

The following are some of the techniques mainly used for processing of polymers; most of these, you will note, apply only to thermoplastics (Figure 16-25).

Extrusion This is the most widely used technique for processing thermoplastics. Extrusion can serve two purposes. First, it provides a way to form certain simple shapes continuously [Figures 16-25(a) and 16-26]. Second, extrusion provides an excellent mixer for additives when processing polymers that ultimately may be formed using some other method. A screw mechanism consisting of one or a pair of screws (twin screw) forces heated thermoplastic and additives through a die opening to produce solid shapes, films, sheets, tubes, and pipes (Figure 16-26). An industrial extruder can be up to

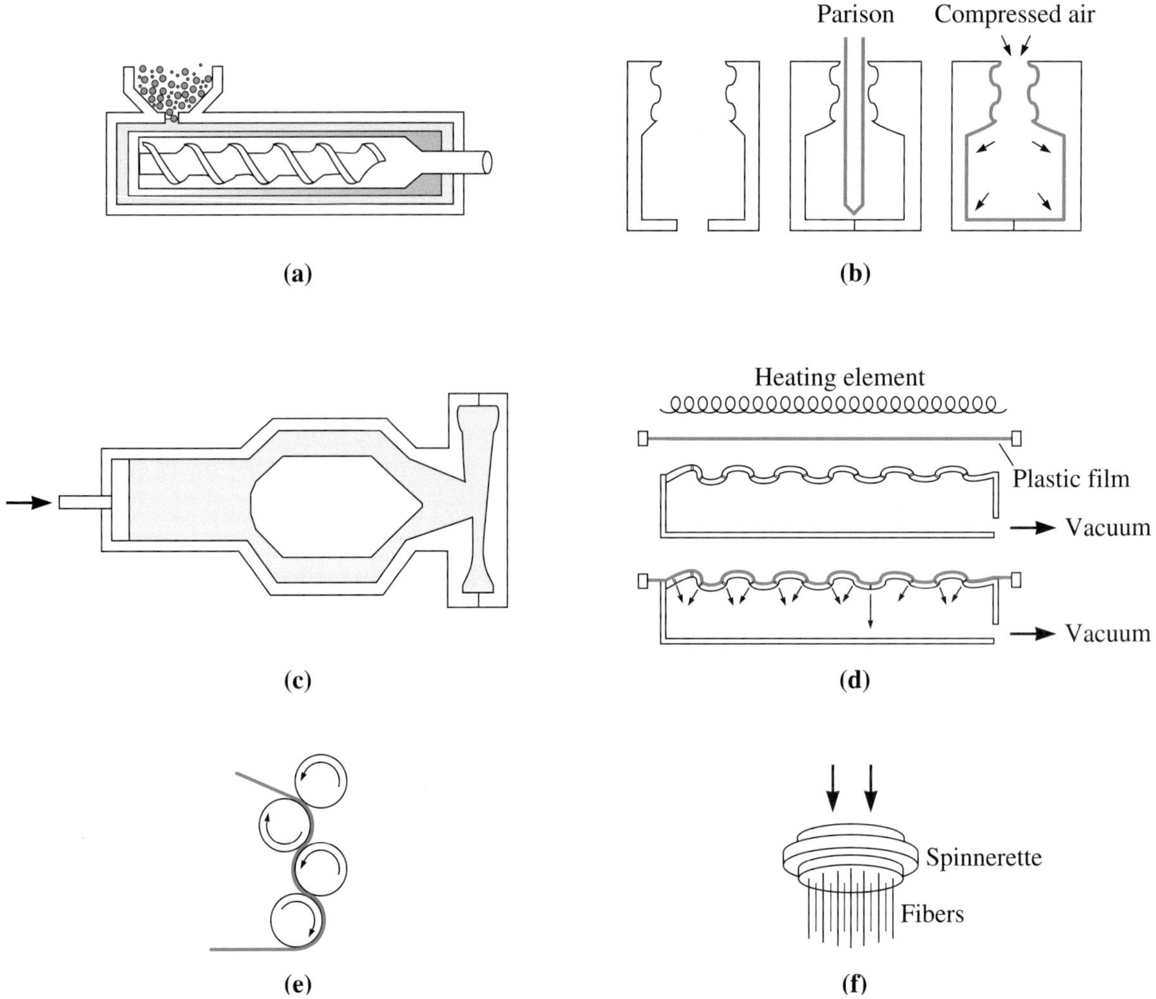

Figure 16-25 Typical forming processes for thermoplastics: (a) extrusion, (b) blow molding, (c) injection molding (the extrusion barrel is not shown), (d) thermoforming, (e) calendaring, and (f) spinning.

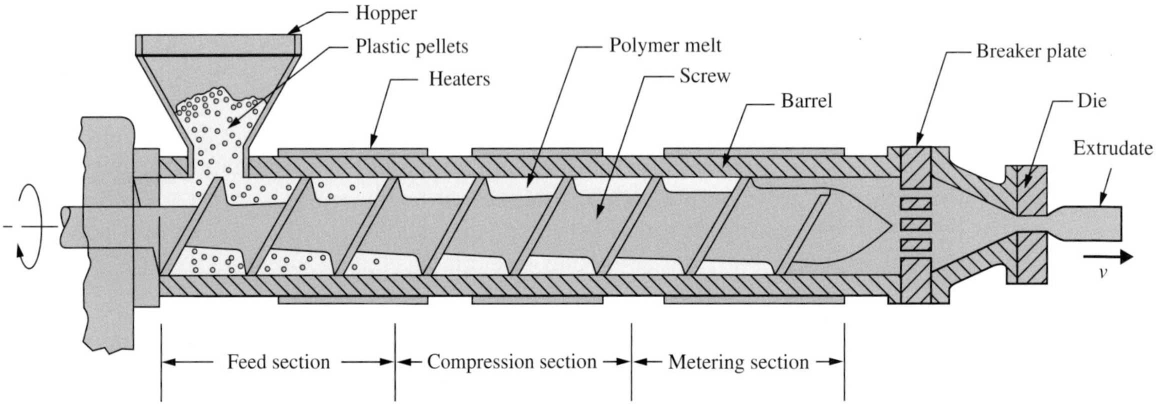

Figure 16-26 Schematic of an extruder used for polymer processing. (*Based on* Fundamentals of Modern Manufacturing: Materials, Processes, and Systems, *Third Edition, Mikell P. Groover, John Wiley & Sons, Inc., 2007.*)

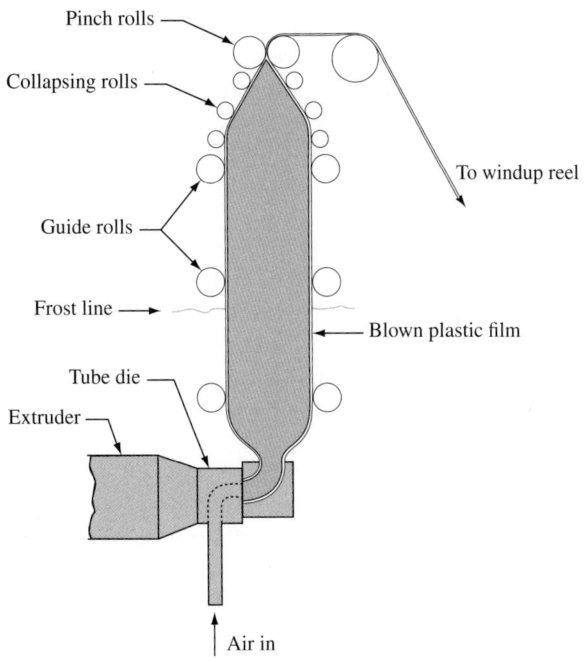

Figure 16-27 Blown-film extrusion process. A thermoplastic polymer is extruded upward in the form of a tube. Air is blown into the tube causing it to expand and cool. Plastic bags are produced in this manner. (*Based on* Fundamentals of Modern Manufacturing: Materials, Processes, and Systems, *Third Edition, Mikell P. Groover, John Wiley & Sons, Inc., 2007.*)

60 to 70 feet long, 2 feet in diameter, and consist of different heating or cooling zones. Since thermoplastics show shear thinning behavior and are viscoelastic, the control of both temperature and viscosity is critical in polymer extrusion. One special extrusion process for producing films is illustrated in Figure 16-27. Extrusion also can be used to coat wires and cables with either thermoplastics or elastomers.

Blow Molding A hollow preform of a thermoplastic called a **parison** is introduced into a die by gas pressure and expanded against the walls of the die [Figure 16-25(b)]. This process is used to produce plastic bottles, containers, automotive fuel tanks, and other hollow shapes.

Injection Molding Thermoplastics heated above the melting temperature using an extruder are forced into a closed die to produce a molding [Figure 16-25(c)]. This process is similar to die casting of molten metals. A plunger or a special screw mechanism applies pressure to force the hot polymer into the die. A wide variety of products, ranging from cups, combs, and gears to garbage cans, can be produced in this manner.

Thermoforming Thermoplastic polymer sheets heated to the plastic region can be formed over a die to produce such diverse products as egg cartons and decorative panels. The forming can be done using matching dies, a vacuum [Figure 16-25(d)], or air pressure.

Calendaring In a calendar, molten plastic is poured onto a set of rolls with a small opening [Figure 16-25(e)]. The rolls, which may be embossed with a pattern, squeeze out a thin sheet of the polymer—often, polyvinyl chloride. Typical products include vinyl floor tile and shower curtains.

Spinning Filaments, fibers, and yarns may be produced by spinning [Figure 16-25(f)]. The molten thermoplastic polymer is forced through a die containing many tiny holes. The die, called a **spinnerette**, can rotate and produce a yarn. For some materials, including nylon, the fiber may subsequently be stretched to align the chains parallel to the axis of the fiber; this process increases the strength of the fibers.

Casting Many polymers can be cast into molds to solidify. The molds may be plate glass for producing individual thick plastic sheets or moving stainless steel belts for continuous casting of thinner sheets. *Rotational molding* is a special casting process in which molten polymer is poured into a mold rotating about two axes. Centrifugal action forces the polymer against the walls of the mold, producing a thin shape such as a camper top.

Compression Molding Thermoset moldings are most often formed by placing the solid material before cross-linking into a heated die. Application of high pressure and temperature causes the polymer to melt, fill the die, and immediately begin to harden. Small electrical housings as well as fenders, hoods, and side panels for automobiles can be produced by this process [Figure 16-28(a)].

Transfer Molding A double chamber is used in the transfer molding of thermosetting polymers. The polymer is heated under pressure in one chamber. After melting, the polymer is injected into the adjoining die cavity. This process permits some of the advantages of injection molding to be used for thermosetting polymers [Figure 16-28(b)].

Reaction Injection Molding (RIM) Thermosetting polymers in the form of liquid resins are first injected into a mixer and then directly into a heated mold to produce a shape. Forming and curing occur simultaneously in the mold. In

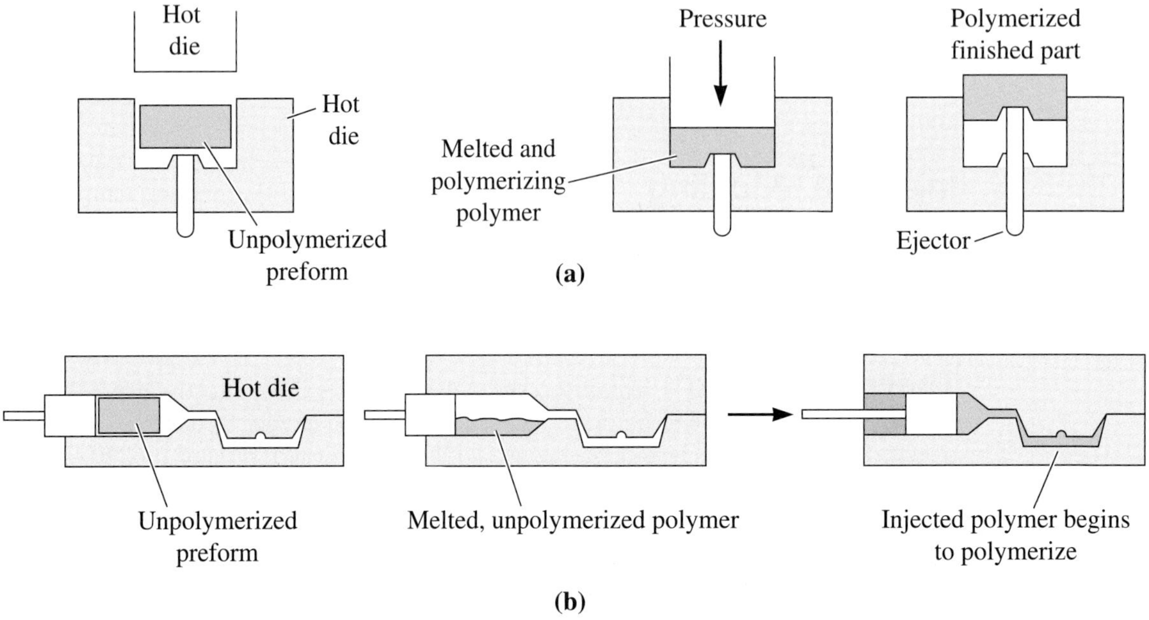

Figure 16-28 Typical forming processes for thermosetting polymers: (a) compression molding and (b) transfer molding.

Example 16-10 *Insulation Boards for Houses*

You want to design a material that can be used for making insulation boards that are approximately 4 ft wide and 8 ft tall. The material must provide good thermal insulation. What material would you choose?

SOLUTION

Glasses tend to be good insulators of heat; however, they will be heavy, more expensive, and prone to fracture. Polymers are lightweight and can be produced inexpensively, and they can be good thermal insulators. We can use foamed polystyrene since the air contained in the beads adds significantly to their effectiveness as thermal insulators. For better mechanical properties, we may want to produce foams that have relatively high density (compared to foams that are used to make coffee cups). Finally, from a safety viewpoint, we want to be sure that some fire and flame retardants are added to the foams. Such panels are made using expanded polystyrene beads containing pentane. A molding process is used to make the foams. The sheets can be cut into the required sizes using a heated metal wire.

reinforced-reaction injection molding (RRIM), a reinforcing material consisting of particles or short fibers is introduced into the mold cavity and is impregnated by the liquid resins to produce a composite material. Automotive bumpers, fenders, and furniture parts are made using this process.

Foams Foamed products can be produced using polystyrene, urethanes, polymethyl methacrylate, and a number of other polymers. The polymer is produced in the form of tiny beads, often containing a blowing agent such as pentane. During the pre-expansion process, the beads increase in diameter by as many as 50 times. The pre-expanded beads are then injected into a die, with the individual beads fusing together, using steam, to form exceptionally lightweight products with densities of perhaps only 0.02 g/cm^3. Expandable polystyrene (EPS) cups, packaging, and insulation are some of the applications for foams. Engine blocks for many automobiles are made using a pattern of expanded polystyrene beads through a process that is similar to investment casting.

Recycling of Plastics Recycling is a very important issue, and a full discussion of the entire process is outside the scope of this book. It is critical to remember, however, that recycling plays an important role in our everyday lives. Material is recycled in many ways. For example, part of the polymer that is scrap from a manufacturing process (known as regrind) is used by recycling plants. The recycling of thermoplastics is relatively easy and practiced widely. Note that many of the everyday plastic products you encounter (bags, soda bottles, yogurt containers, etc.) have numbers stamped on them. For PET products (recycling symbol "PETE" because of trademark issues), the number is 1. For HDPE, vinyl (recycling symbol V), LDPE, PP, and PS, the numbers are 2, 3, 4, 5, and 6, respectively. Other plastics are marked number 7.

Thermosets and elastomers are more difficult to recycle, although they can still be used. For example, tires can be shredded and used to make safer playground surfaces or roads.

Despite enormous recycling efforts, a large portion of the materials in landfills today is plastics (the largest portion is paper). Given the limited amount of petroleum, the threat of global warming, and a need for a cleaner and safer environment, careful use and recycling makes sense for all materials.

Summary

- Polymers are made from large macromolecules produced by the joining of smaller molecules, called monomers, using addition or condensation polymerization reactions. *Plastics* are materials that are based on polymeric compounds, and they often contain other additives that improve their properties. Compared with most metals and ceramics, plastics have low strength, stiffness, and melting temperatures; however, they also have low density and good chemical resistance. Plastics are used in many diverse technologies.
- Thermoplastics have chains that are not chemically bonded to each other, permitting the material to be easily formed into useful shapes, to have good ductility, and to be economically recycled. Thermoplastics can have an amorphous structure, which provides low strength and good ductility when the ambient temperature is above the glass-transition temperature. The polymers are more rigid and brittle when the temperature falls below the glass-transition temperature. Many thermoplastics can also partially crystallize during cooling or by application of a stress. This increases their strength.
- The thermoplastic chains can be made more rigid and stronger by using nonsymmetrical monomers that increase the bonding strength between the chains and make it more difficult for the chains to disentangle when stress is applied. In addition, many monomers containing atoms or groups of atoms other than carbon produce more rigid chains; this structure also produces high-strength thermoplastics.
- Elastomers are thermoplastics or lightly cross-linked thermosets that exhibit greater than 200% elastic deformation. Chains are cross-linked using vulcanization. The cross-linking makes it possible to obtain very large elastic deformations without permanent plastic deformation. Increasing the number of cross-links increases the stiffness and reduces the amount of elastic deformation of the elastomers.
- Thermoplastic elastomers combine features of both thermoplastics and elastomers. At high temperatures, these polymers behave as thermoplastics and are plastically formed into shapes; at low temperatures, they behave as elastomers.
- Thermosetting polymers are highly cross-linked into a three-dimensional network structure. Typically, high glass-transition temperatures, good strength, and brittle behavior result. Once cross-linking occurs, these polymers cannot be easily recycled.
- Manufacturing processes depend on the behavior of the polymers. Processes such as extrusion, injection molding, thermoforming, casting, and spinning are made possible by the viscoelastic behavior of thermoplastics. The non-reversible bonding in thermosetting polymers limits their processing to fewer techniques, such as compression molding, transfer molding, and reaction-injection molding.

Glossary

Addition polymerization Process by which polymer chains are built up by adding monomers together without creating a byproduct.

Aging Slow degradation of polymers as a result of exposure to low levels of heat, oxygen, bacteria, or ultraviolet rays.

Aramids Polyamide polymers containing aromatic groups of atoms in the linear chain.

Blushing Crystallization of small volumes of a thermoplastic material due to repeated bending; this creates voids that ultimately cause the material to fail.

Branched polymer Any polymer comprising chains that consist of a main chain and secondary chains that are joined to the main chain.

Condensation polymerization A polymerization mechanism in which a small molecule (e.g., water, methanol, etc.) forms as a byproduct.

Copolymer An addition polymer produced by joining more than one type of monomer.

Crazing Localized plastic deformation in a polymer.

Cross-linking Attaching chains of polymers together via permanent chemical bonds to produce a three-dimensional network polymer.

Deflection temperature The temperature at which a polymer will deform a given amount under a standard load. (Also called the *distortion* temperature.)

Degradation temperature The temperature above which a polymer burns, chars, or decomposes.

Degree of polymerization The average molecular weight of the polymer divided by the molecular weight of the monomer.

Diene A group of monomers that contain two double-covalent bonds. These monomers are often used in producing elastomers.

Elastomers Polymers (thermoplastics or lightly cross-linked thermosets) that have an elastic deformation $> 200\%$.

Geometric isomer A molecule that has the same composition as, but a structure different from, a second molecule.

Glass-transition temperature The temperature range below which the amorphous polymer assumes a rigid glassy structure.

Interpenetrating polymer networks Polymer structures produced by intertwining two separate polymer structures or networks.

Linear polymer Any polymer in which molecules are in the form of spaghetti-like chains.

Liquid-crystalline polymers Exceptionally stiff polymer chains that act as rigid rods, even at high temperatures.

Mer A unit group of atoms and molecules that defines a characteristic arrangement for a polymer.

Monomer The molecule from which a polymer is produced.

Oligomer Low molecular weight molecules. These may contain two (dimers) or three (trimers) mers.

Parison A hot glob of soft or molten polymer that is blown or formed into a useful shape.

Plastic A predominantly polymeric material containing other additives.

Polymer Materials made from giant (or macromolecular), chain-like molecules having average molecular weights from 10,000 to more than 1,000,000 g/mol built by the joining of many mers or units by chemical bonds. Polymers are usually, but not always, carbon based.

Relaxation time A property of a polymer that is related to the rate at which stress relaxation occurs.

Repeat unit The structural unit from which a polymer is built. Also called a *mer*.

Spinnerette An extrusion die containing many small openings through which hot or molten polymer is forced to produce filaments. Rotation of the spinnerette twists the filaments into a yarn.

Stress-induced crystallization The process of forming crystals by the application of an external stress. Typically, a significant fraction of many amorphous plastics can be crystallized in this fashion, making them stronger.

Stress relaxation A reduction of the stress acting on a material over a period of time at a constant strain due to viscoelastic deformation.

Tacticity Describes the location in the polymer chain of atoms or atom groups in nonsymmetrical monomers.

Thermoplastic elastomers Polymers that behave as thermoplastics at high temperatures, but as elastomers at lower temperatures.

Thermoplastics Linear or branched polymers in which chains of molecules are not interconnected.

Thermosetting polymers Polymers that are heavily cross-linked to produce a strong three-dimensional network structure.

Unsaturated bond The double- or even triple-covalent bond joining two atoms together in an organic molecule. When a single covalent bond replaces the unsaturated bond, polymerization can occur.

Viscoelastic behavior The deformation of a material by elastic deformation and viscous flow when stress is applied.

Vulcanization Cross-linking elastomer chains by introducing sulfur or other chemicals.

Problems

Section 16-1 Classification of Polymers

16-1 What are linear and branched polymers? Can thermoplastics be branched?

16-2 Define the terms (a) thermoplastic; (b) thermosetting plastics; (c) elastomers; and (d) thermoplastic elastomers.

16-3 For what electrical and optical applications are polymers used? Explain using examples.

16-4 What are the major advantages of plastics compared to ceramics, glasses, and metals?

Sections 16-2 Addition and Condensation Polymerization

16-5 What do the terms condensation polymerization, addition polymerization, initiator, and terminator mean?

16-6 Kevlar ($C_{14}H_{10}N_2O_2$) is used in various applications from tires to body armor due to its high strength-to-weight ratio. The polymer is produced from the monomers paraphenylene diamine ($C_6H_8N_2$) and terephthaloyl chloride ($C_8H_4Cl_2O_2$) by the following reaction:

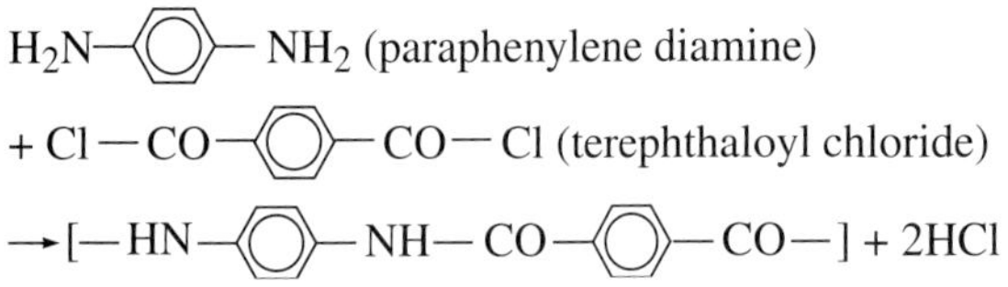

(a) What type of polymerization does the above reaction represent?
(b) Assuming 100% efficiency, calculate the weight of terephthaloyl chloride required to completely combine with 1 kg of paraphenylene diamine.
(c) How much Kevlar is produced?

16-7 Examine Figure 16-6. If 100 kg each of the two reactants were combined and the reaction proceeded to 100% completion, how much methanol would be generated?

16-8 What would be the molecular weight of the polymer shown below where $n = 10^6$?

```
         H
         |
  H — [ U ]n — H
         |
         H
```

Note that uranium can have a +4 oxidation state.

Section 16-3 Degree of Polymerization

Section 16-4 Typical Thermoplastics

16-9 Explain why low-density polyethylene is good for making grocery bags, but super-high molecular weight polyethylene must be used in applications for which strength and high wear resistance are needed.

16-10 Revisit Example 16-5. What are the degrees of polymerization for the weight average and number average molecular weights?

16-11 Calculate the number of chains in a 5-m-long PVC pipe with an inner diameter of 5 cm and a thickness of 0.5 cm if the degree of polymerization is 1000. Assume that the chains are equal in length. The density of PVC is 1.4 g/cm^3. The repeat unit of PVC is shown in Table 16-3.

16-12 The molecular weight of polymethyl methacrylate (see Table 16-3) is 250,000 g/mol. If all of the polymer chains are the same length, calculate
(a) the degree of polymerization, and
(b) the number of chains in 1 g of the polymer.

16-13 The degree of polymerization of polytetrafluoroethylene (see Table 16-3) is 7500. If all of the polymer chains are the same length, calculate
(a) the molecular weight of the chains, and
(b) the total number of chains in 1000 g of the polymer.

16-14 A polyethylene rope weighs 0.25 lb per foot. If each chain contains 7000 repeat units, calculate
(a) the number of polyethylene chains in a 10 ft length of rope, and

(b) the total length of chains in the rope, assuming that carbon atoms in each chain are approximately 0.15 nm apart and the length of one repeat unit is 0.24495 nm.

16-15 Analysis of a sample of polyacrylonitrile (see Table 16-3) shows that there are six lengths of chains, with the following number of chains of each length shown in the table below. Determine
(a) the weight average molecular weight and degree of polymerization and
(b) the number average molecular weight and degree of polymerization.

Number of Chains	Mean Molecular Weight of Chains (g/mol)
10,000	3000
18,000	6000
17,000	9000
15,000	12,000
9000	15,000
4000	18,000

16-16 Calculate the molecular weights of the repeat units in Table 16-3. Show your work for full credit.

16-17 Plot the tensile strength of the polymers in Table 16-2 as a function of the number of aromatic (benzene) rings in the monomer.

Section 16-5 Structure—Property Relationships in Thermoplastics

Section 16-6 Effect of Temperature on Thermoplastics

Section 16-7 Mechanical Properties of Thermoplastics

16-18 Explain what the following terms mean: decomposition temperature, heat distortion temperature, glass-transition temperature, and melting temperature. Why is it that thermoplastics do not have a fixed melting or glass-transition temperature?

16-19 Using Table 16-5, plot the relationship between the glass-transition temperatures and the melting temperatures of the addition thermoplastics. What is the approximate relationship between these two critical temperatures? Do the condensation thermoplastics and the elastomers also follow the same relationship?

16-20 List the addition polymers in Table 16-5 that might be good candidates for making a bracket that holds a side-view mirror onto the outside of an automobile, assuming that temperatures frequently fall below zero degrees Celsius. Explain your choices.

16-21 Based on Table 16-5, which of the elastomers might be suited for use as a gasket in a pump for liquid CO_2 at $-78°C$? Explain.

16-22 How do the glass-transition temperatures of polyethylene, polypropylene, and polymethyl methacrylate compare? Explain their differences, based on the structure of the monomer of each.

16-23 Which of the addition polymers in Table 16-5 are used in their leathery condition at room temperature? How is this condition expected to affect their mechanical properties compared with those of glassy polymers?

16-24 What factors influence the crystallinity of polymers? Explain the development and role of crystallinity in PET and nylon.

16-25 Describe the relative tendencies of the following polymers to crystallize. Explain your answer.
(a) Branched polyethylene versus linear polyethylene;
(b) polyethylene versus polyethylene-polypropylene copolymer;
(c) isotactic polypropylene versus atactic polypropylene; and
(d) polymethyl methacrylate versus acetal (polyoxymethylene).

16-26 The crystalline density of polypropylene is 0.946 g/cm^3, and its amorphous density is 0.855 g/cm^3. What is the weight percent of the structure that is crystalline in a polypropylene that has a density of 0.900 g/cm^3?

16-27 If the strain rate in a polymer can be represented by

$$d\varepsilon/dt = \sigma/\eta + (1/E)d\sigma/dt$$

where ε = strain, σ = stress, η = viscosity, E = modulus of elasticity, and t = time, derive Equation 16-5, assuming constant strain (i.e., $d\varepsilon = 0$). What is the relaxation time (λ) in Equation 16-5 a function of?

16-28 HDPE is loaded to 400 psi and left. What temperature would cause a rupture at about 100 hours of this constant load? See Figure 16-18.

16-29 A polymer component that needs to maintain a stress level above 10 MPa for proper functioning of an assembly is scheduled to be replaced every two years as preventative maintenance. If the initial stress was 18 MPa and dropped to 15 MPa after one year of operation (assuming constant strain), will the part replacement under the specified preventative schedule successfully prevent a failure?

16-30 Explain the meaning of these terms: creep, stress relaxation, crazing, blushing, and aging of polymers.

16-31 A stress of 2500 psi is applied to a polymer serving as a fastener in a complex assembly. At a constant strain, the stress drops to 2400 psi after 100 h. If the stress on the part must remain above 2100 psi in order for the part to function properly, determine the life of the assembly.

16-32 A stress of 1000 psi is applied to a polymer that operates at a constant strain; after six months, the stress drops to 850 psi. For a particular application, a part made from the same polymer must maintain a stress of 900 psi after 12 months. What should be the original stress applied to the polymer for this application?

16-33 Data for the rupture time of polyethylene are shown in Figure 16-18. At an applied stress of 600 psi, the figure indicates that the polymer ruptures in 0.2 h at 90°C but survives 10,000 h at 65°C. Assuming that the rupture time is related to the viscosity, calculate the activation energy for the viscosity of the polyethylene and estimate the rupture time at 23°C.

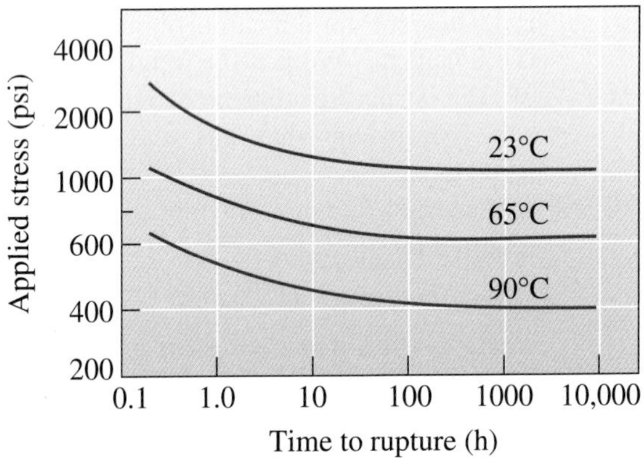

Figure 16-18 The effect of temperature on the stress-rupture behavior of high-density polyethylene. (Repeated for Problems 16-28 and 16-33.)

Section 16-8 Elastomers (Rubbers)

16-34 The polymer ABS can be produced with varying amounts of styrene, butadiene, and acrylonitrile monomers, which are present in the form of two copolymers: BS rubber and SAN.
(a) How would you adjust the composition of ABS if you wanted to obtain good impact properties?
(b) How would you adjust the composition if you wanted to obtain good ductility at room temperature?
(c) How would you adjust the composition if you wanted to obtain good strength at room temperature?

16-35 Figure 16-23 shows the stress strain curve for an elastomer. From the curve, calculate and plot the modulus of elasticity versus strain and explain the results.

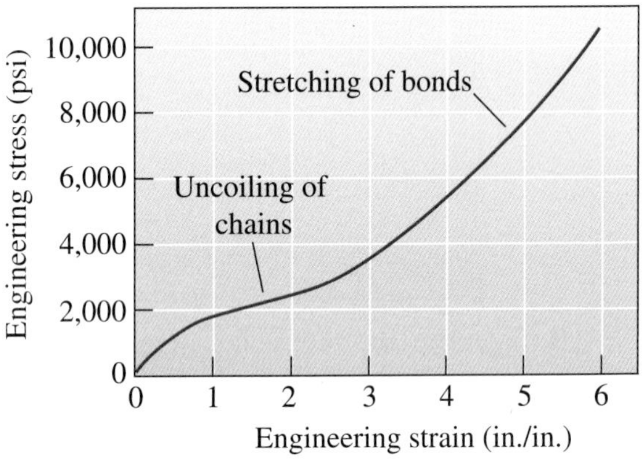

Figure 16-23 The stress strain curve for an elastomer. Virtually all of the deformation is elastic; the modulus of elasticity varies as the strain changes. (Repeated for Problem 16-35.)

Section 16-9 Thermosetting Polymers

16-36 Explain the term thermosetting polymer. A thermosetting polymer cannot be produced using only adipic acid and ethylene glycol. Explain why.

16-37 Explain why the degree of polymerization is not usually used to characterize thermosetting polymers.

16-38 Defend or contradict the choice to use the following materials as hot-melt adhesives for an application in which the assembled part is subjected to impact loading:
(a) polyethylene;
(b) polystyrene;
(c) styrene-butadiene thermoplastic elastomer;
(d) polyacrylonitrile; and
(e) polybutadiene.

16-39 Compare and contrast properties of thermoplastics, thermosetting materials, and elastomers.

Design Problems

16-40 Figure 16-29 shows the behavior of polypropylene, polyethylene, and acetal at two temperatures. We would like to produce a 12-in.-long rod of a polymer that will operate at 40°C for 6 months under a constant load of 500 lb. Design the material and size of the rod such that no more than 5% elongation will occur by creep.

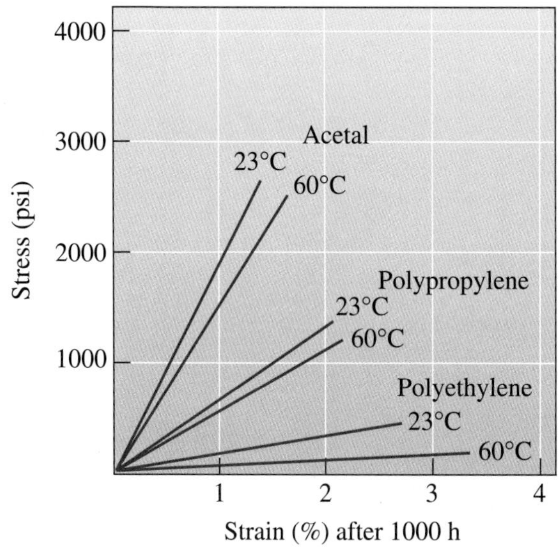

Figure 16-29 The effect of applied stress on the percent creep strain for three polymers (for Problem 16-40).

16-41 Design a polymer material that might be used to produce a 3-in.-diameter gear to be used to transfer energy from a low-power electric motor. What are the design requirements? What class of polymers (thermoplastics, thermosets, elastomers) might be most appropriate? What particular polymer might you first consider? What additional information concerning the application and polymer properties do you need to know to complete your design?

16-42 Design a polymer material and a forming process to produce the case for a desktop computer. What are the design and forming requirements? What class of polymers might be most appropriate? What particular polymer might you first consider? What additional information do you need to know?

16-43 What kind of polymer can be used to line the inside of the head of a hip prosthesis implant? Discuss what requirements would be needed for this type of polymer.

Computer Problems

16-44 *Polymer Molecular Weight Distribution.* The following data were obtained for polyethylene. Determine the average molecular weight and degree of polymerization.

Molecular Weight Range (g/mol)	f_i	x_i
0–3000	0.01	0.03
3000–6000	0.08	0.10
6000–9000	0.19	0.22
9000–12,000	0.27	0.36
12,000–15,000	0.23	0.19
15,000–18,000	0.11	0.07
18,000–21,000	0.06	0.02
21,000–24,000	0.05	0.01

Write a computer program or use a spreadsheet program to solve this problem.

Knovel® Problems

K16-1 Provide an example of a thermoplastic elastomer (TPE) based on an interpenetrating polymer network (IPN). How is it made?

K16-2 How does chain length affect the glass-transition temperature of polymers? Find an equation describing this relationship.

K16-3 Hindered phenolics are common organic antioxidants. For what classes of plastics are they used?

In a drive to increase efficiency and thereby decrease fuel costs, aerospace companies look to incorporate lighter and stronger materials into airplanes. One of the primary innovations of the Boeing 787 is the extensive use of composite materials; composites are formed by incorporating multiple phase components in a material in such a way that the properties of the resultant material are unique and not otherwise attainable. Composite materials comprise half of the Dreamliner's total weight. For example, the fuselage of the Boeing 787 is made from carbon fiber-reinforced plastic. Carbon fiber-reinforced plastic is a composite of carbon fiber in an epoxy matrix. The carbon fibers impart strength and stiffness to the material, and the epoxy matrix binds the fibers together. (*AFP/Getty Images*)

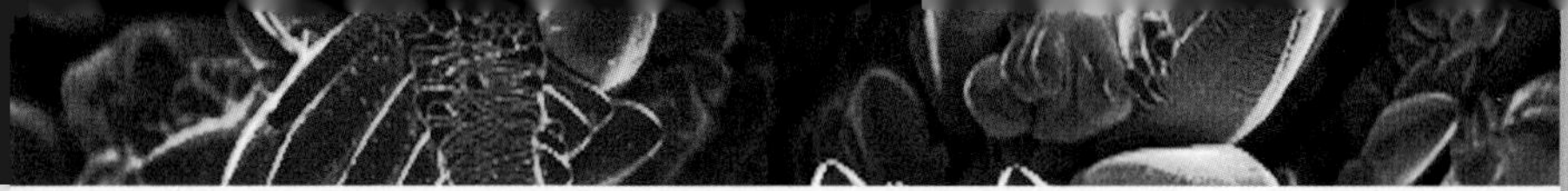

CHAPTER 17

Composites: Teamwork and Synergy in Materials

Have You Ever Wondered?

- *What are some naturally occurring composites?*
- *Why is abalone shell, made primarily from calcium carbonate, so much stronger than chalk, which is also made from calcium carbonate?*
- *What sporting gear applications make use of composites?*
- *Why are composites finding increased use in aircraft and automobiles?*

Chapter Learning Objectives

The key objectives of this chapter are to

- Explain how composites achieve improved strength, fatigue resistance, elastic properties, and strength-to-weight ratios compared to the constituent materials.
- Describe the morphology of particulate, fiber-reinforced, and laminar composites.
- Use the rule of mixtures to calculate various properties for composite materials.
- Calculate the elastic moduli for fiber-reinforced composites parallel and perpendicular to the fiber direction.
- Describe the processes for manufacturing fibers and fiber-reinforced composites.
- Describe the processes for manufacturing laminates and laminar composites.
- Identify the applications for various types of composites.

Composites are produced when two or more materials or phases are used together to give a combination of properties that cannot be attained otherwise. Composite materials may be selected to give unusual combinations of stiffness, strength, density, high-temperature performance, corrosion resistance, hardness, or conductivity. Abalone shell, wood, bone, and teeth are examples of

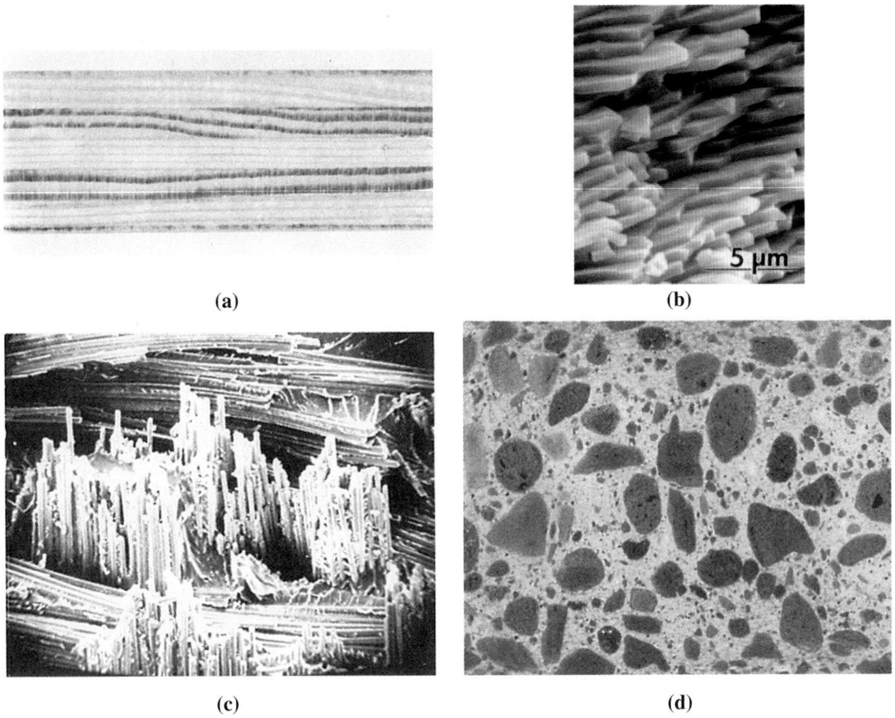

Figure 17-1 Some examples of composite materials: (a) plywood is a laminar composite of layers of wood veneer, (b) abalone shell is a composite of aragonitic ($CaCO_3$-orthorhombic) platelets surrounded by 10 nm thick layers of a proteinaceous organic matrix. (*Courtesy of Professor Mehmet Sarikaya, University of Washington, Seattle.*) (c) Fiberglass is a fiber-reinforced composite containing stiff, strong glass fibers in a softer polymer matrix ($\times$ 175) , and (d) concrete is a particulate composite containing coarse sand or gravel in a cement matrix (reduced 50%). [*Images (a), (c), and (d) Reprinted Courtesy of Don Askeland*]

naturally occurring composites. Microstructures of selected composites are shown in Figure 17-1. An example of a material that is a composite at the macroscale is steel-reinforced concrete. Microscale composites include such materials as carbon or glass fiber-reinforced plastics (CFRP or GFRP). These composites offer significant gains in specific strengths and are finding increasing usage in airplanes, electronic components, automotives, and sporting equipment.

As mentioned in Chapters 12 and 13, dispersion-strengthened and precipitation-hardened alloys are examples of traditional materials that are **nanocomposites**. In a nanocomposite, the **dispersed phase** consists of nanoscale particles and is distributed in a **matrix phase**. Many glass-ceramics are nanoscale composites of different ceramic phases. Many plastics can be considered composites as well. For example, Dylark^{TM} is a composite of maleic anhydride-styrene copolymer. It contains carbon black for stiffness and protection against ultraviolet rays. It also contains glass fibers for increased Young's modulus and rubber for toughness. Epoxies may be filled with silver to increase thermal conductivity. Some dielectric materials are made using multiple phases such that the overall dielectric properties of interest (e.g., the dielectric constant) do not change appreciably with temperature (within a certain range). Some composite structures may consist of different materials arranged in different layers. This leads to what are known as functionally graded materials and structures. For example, a yttria stabilized zirconia (YSZ) coating on a turbine blade will have other layers in between that provide bonding with the turbine blade material. The YSZ coating itself contains levels of porosity that are essential for providing protection against high temperatures (Chapter 5). Similarly, coatings on glass are examples of composite structures. Thus, the *concept* of using composites is a generic one and can be applied at the macro, micro, and nano length scales.

In composites, the properties and volume fractions of individual phases are important. The connectivity of phases is also very important. Usually the matrix phase is the continuous phase, and the other phase is said to be the dispersed phase. Thus, terms such as "metal-matrix" indicate a metallic material used to form the continuous phase.

Connectivity describes how the phases are connected in the composite. Composites are often classified based on the shape or nature of the dispersed phase (e.g., particle-reinforced, whisker-reinforced, or fiber-reinforced composites). **Whiskers** are like fibers, but their length is much smaller. The bonding between the particles, whiskers, or fibers and the matrix is also very important. In structural composites, polymeric molecules known as coupling agents are used. These molecules form bonds with the dispersed phase and become integrated into the continuous matrix phase as well.

In this chapter, we will primarily focus on composites used in structural or mechanical applications. Composites can be placed into three categories—particulate, fiber, and laminar (Figure 17-1). Concrete, a mixture of cement and gravel, is a particulate composite; fiberglass, containing glass fibers embedded in a polymer, is a fiber-reinforced composite; and plywood, having alternating layers of wood veneer, is a laminar composite. If the reinforcing particles are uniformly distributed, particulate composites have isotropic properties; fiber composites may be either isotropic or anisotropic; laminar composites always display anisotropic behavior.

17-1 Dispersion-Strengthened Composites

A special group of dispersion-strengthened nanocomposite materials containing particles 10 to 250 nm in diameter is classified as particulate composites. The **dispersoids,** usually a metallic oxide, are introduced into the matrix by means other than traditional phase transformations (Chapters 12 and 13). Even though the small particles are not coherent with the matrix, they block the movement of dislocations and produce a pronounced strengthening effect.

At room temperature, the dispersion-strengthened composites may be weaker than traditional age-hardened alloys, which contain a coherent precipitate. Because the composites do not catastrophically soften by overaging, overtempering, grain growth, or coarsening of the dispersed phase, the strength of the composite decreases only gradually with increasing temperature (Figure 17-2). Furthermore, their creep resistance is superior to that of metals and alloys.

The dispersoid must have a low solubility in the matrix and must not chemically react with the matrix, but a small amount of solubility may help improve the bonding between the dispersant and the matrix. Copper oxide (Cu_2O) dissolves in copper at high temperatures; thus, the Cu_2O-Cu system would not be effective. Al_2O_3 does not dissolve in aluminum; the Al_2O_3-Al system does give an effective dispersion-strengthened material.

Illustrations of Dispersion-Strengthened Composites

Table 17-1 lists some dispersion-strengthened composites of interest. Perhaps the classic example is the sintered aluminum powder (SAP) composite. SAP has an aluminum matrix strengthened by up to 14% Al_2O_3. The composite is formed by powder metallurgy. In one method, aluminum and alumina powders are blended, compacted at high pressures, and sintered. In a second technique, the aluminum powder is treated to add a continuous oxide film on each particle. When the powder is compacted, the oxide film fractures into tiny flakes that are surrounded by the aluminum metal during sintering.

Another important group of dispersion-strengthened composites includes thoria dispersed (TD) metals such as TD-nickel (Figure 17-3). TD-nickel can be produced by

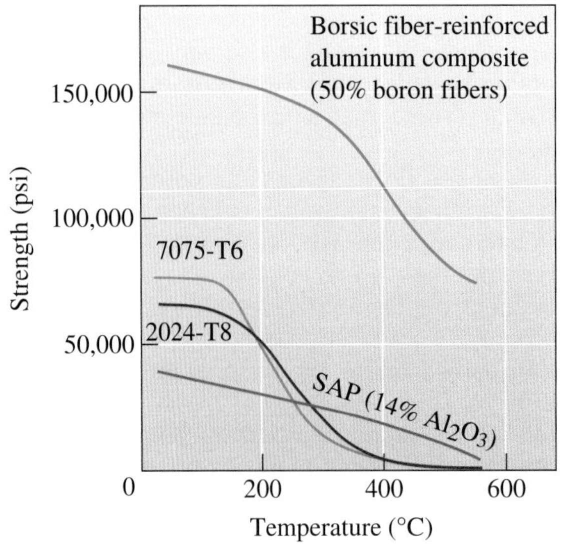

Figure 17-2 Comparison of the yield strength of dispersion-strengthened sintered aluminum powder (SAP) composite with that of two conventional two-phase high-strength aluminum alloys. The composite has benefits above about 300°C. A fiber-reinforced aluminum composite is shown for comparison.

Table 17-1 Applications of selected dispersion-strengthened composites

System	Applications
Ag-CdO	Electrical contact materials
$Al\text{-}Al_2O_3$	Possible use in nuclear reactors
Be-BeO	Aerospace and nuclear reactors
$Co\text{-}ThO_2, Y_2O_3$	Possible creep-resistant magnetic materials
$Ni\text{-}20\%\ Cr\text{-}ThO_2$	Turbine engine components
Pb-PbO	Battery grids
$Pt\text{-}ThO_2$	Filaments, electrical components
$W\text{-}ThO_2, ZrO_2$	Filaments, heaters

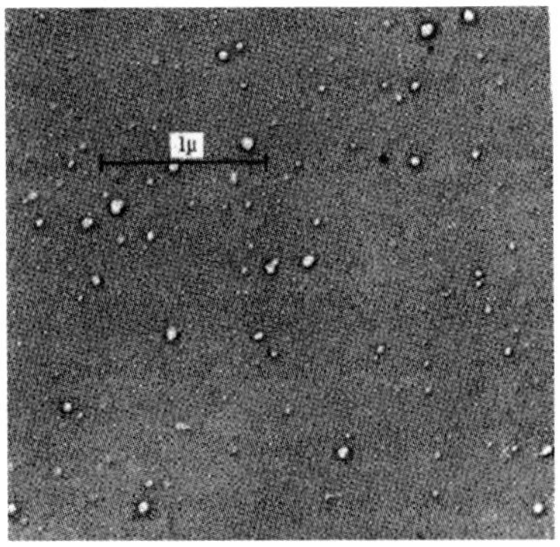

Figure 17-3 Electron micrograph of TD-nickel. (*Cremens, W. S.; Norris, L. F.; Weeton, J. W. "Overlay copy technique to provide high-contrast electron micrographs for automatic metallographic analysis,"* NASA Technical Report, *Glenn Research Center, 1966.*)

internal oxidation. Thorium is present in nickel as an alloying element. After a powder compact is made, oxygen is allowed to diffuse into the metal, react with the thorium, and produce thoria (ThO_2). The following example illustrates calculations related to a dispersion-strengthened composite.

Example 17-1 *TD-Nickel Composite*

Suppose 2 wt% ThO_2 is added to nickel. The density of ThO_2 is 9.69 g/cm³. Each ThO_2 particle has a diameter of 1000 Å. How many particles are present in each cubic centimeter?

SOLUTION

Assume 100 g of material such that 2 g are ThO_2, and 98 g are Ni. The density of nickel is 8.90 g/cm³. The volume fraction of ThO_2 is

$$f_{ThO_2} = \frac{\dfrac{2\text{ g}}{9.69\text{ g/cm}^3}}{\dfrac{2\text{ g}}{9.69\text{ g/cm}^3} + \dfrac{98\text{ g}}{8.90\text{ g/cm}^3}} = 0.0184$$

Therefore, there is 0.0184 cm³ of ThO_2 per cm³ of composite. The volume of each ThO_2 sphere is

$$V_{ThO_2} = \tfrac{4}{3}\pi r^3 = \tfrac{4}{3}\pi(5 \times 10^{-6}\text{ cm})^3 = 5.24 \times 10^{-16}\text{ cm}^3$$

$$\text{Concentration of ThO}_2\text{ particles} = \frac{0.0184}{5.24 \times 10^{-16}} = 3.51 \times 10^{13}\text{ particles/cm}^3$$

17-2 Particulate Composites

The particulate composites are designed to produce unusual combinations of properties rather than to improve strength. The particulate composites contain large amounts of coarse particles that do not block slip effectively.

Rule of Mixtures Certain properties of a particulate composite depend only on the relative amounts and properties of the individual constituents. The **rule of mixtures** can accurately predict these properties. The density of a particulate composite, for example, is

$$\rho_c = \sum(f_i\rho_i) = f_1\rho_1 + f_2\rho_2 + \cdots + f_n\rho_n \tag{17-1}$$

where ρ_c is the density of the composite, $\rho_1, \rho_2, \ldots, \rho_n$ are the densities of each constituent in the composite, and $f_1, f_2, \ldots, f_n$ are the volume fractions of each constituent. Note that the connectivity of different phases (i.e., how the dispersed phase is arranged with respect to the continuous phase) is also very important for many properties.

Cemented Carbides Cemented carbides, or cermets, contain hard ceramic particles dispersed in a metallic matrix. Tungsten carbide inserts used for cutting tools in machining operations are typical of this group. Tungsten carbide (WC) is a hard, stiff, high-melting temperature ceramic. To improve toughness, tungsten carbide particles are combined with cobalt powder and pressed into powder compacts. The compacts are heated above the melting temperature of the cobalt. The liquid cobalt surrounds each of the solid tungsten carbide particles (Figure 17-4). After solidification, the cobalt serves as the binder for tungsten carbide and provides good impact resistance. Other carbides, such as TaC and TiC, may also be included in the cermet. The following example illustrates the calculation of density for cemented carbide.

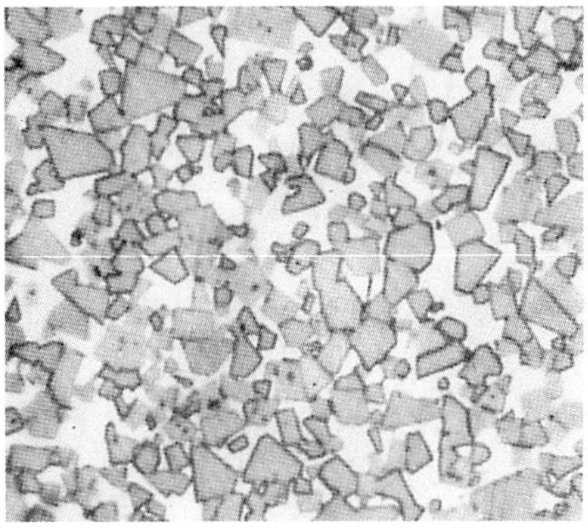

Figure 17-4 Microstructure of tungsten carbide—20% cobalt-cemented carbide (× 1300). [*From* ASM Handbook, *Vol. 7, (1972), ASM International, Materials Park, OH 44073-0002. Reprinted with permission of ASM International. All rights reserved. www.asminternational.org.*]

Example 17-2 *Cemented Carbides*

A cemented carbide cutting tool used for machining contains 75 wt% WC, 15 wt% TiC, 5 wt% TaC, and 5 wt% Co. Estimate the density of the composite.

SOLUTION

First, we must convert the weight percentages to volume fractions. The densities of the components of the composite are

$$\rho_{WC} = 15.77 \text{ g/cm}^3 \qquad \rho_{TiC} = 4.94 \text{ g/cm}^3$$

$$\rho_{TaC} = 14.5 \text{ g/cm}^3 \qquad \rho_{Co} = 8.83 \text{ g/cm}^3$$

$$f_{WC} = \frac{75/15.77}{75/15.77 + 15/4.94 + 5/14.5 + 5/8.83} = \frac{4.76}{8.70} = 0.546$$

$$f_{TiC} = \frac{15/4.94}{8.70} = 0.349$$

$$f_{TaC} = \frac{5/14.5}{8.70} = 0.040$$

$$f_{Co} = \frac{5/8.83}{8.70} = 0.065$$

From the rule of mixtures, the density of the composite is

$$\rho_c = \sum (f_i \rho_i) = (0.546)(15.77) + (0.349)(4.94) + (0.040)(14.5) + (0.065)(8.83)$$
$$= 11.5 \text{ g/cm}^3$$

Abrasives Grinding and cutting wheels are formed from alumina (Al_2O_3), silicon carbide (SiC), and cubic boron nitride (CBN). To provide toughness, the abrasive particles are bonded by a glass or polymer matrix. Diamond abrasives are typically bonded with a metal matrix. As the hard particles wear, they fracture or pull out of the matrix, exposing new cutting surfaces.

Electrical Contacts Materials used for electrical contacts in switches and relays must have a good combination of wear resistance and electrical conductivity. Otherwise, the contacts erode, causing poor contact and arcing. Tungsten-reinforced silver provides this combination of characteristics. A tungsten powder compact is made using conventional powder metallurgy processes (Figure 17-5) to produce highly interconnected porosity. Liquid silver is then vacuum infiltrated to fill the interconnected voids. Both the silver and the tungsten are continuous. Thus, the pure silver efficiently conducts current while the hard tungsten provides wear resistance.

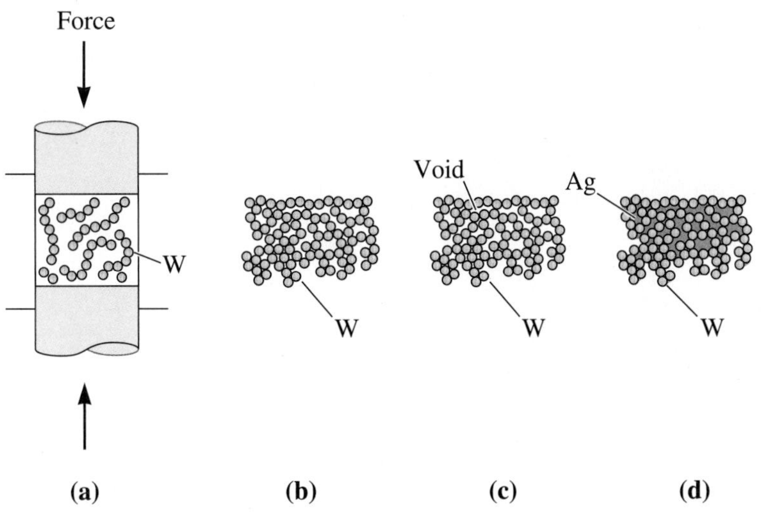

Figure 17-5 The steps in producing a silver-tungsten electrical composite: (a) Tungsten powders are pressed, (b) a low-density compact is produced, (c) sintering joins the tungsten powders, and (d) liquid silver is infiltrated into the pores between the particles.

Example 17-3 *Silver-Tungsten Composite*

A silver-tungsten composite for an electrical contact is produced by first making a porous tungsten powder metallurgy compact, then infiltrating pure silver into the pores. The density of the tungsten compact before infiltration is 14.5 g/cm^3. Calculate the volume fraction of porosity and the final weight percent of silver in the compact after infiltration.

SOLUTION

The densities of pure tungsten f_W and pure silver f_{Ag} are 19.25 g/cm^3 and 10.49 g/cm^3, respectively. We can assume that the density of a pore is zero, so from the rule of mixtures:

$$\rho_c = f_W \rho_W + f_{\text{pore}} \rho_{\text{pore}}$$

$$14.5 \text{ g/cm}^3 = f_W(19.25 \text{ g/cm}^3) + f_{\text{pore}}(0)$$

$$f_W = 0.75$$

$$f_{\text{pore}} = 1 - 0.75 = 0.25$$

After infiltration, the volume fraction of silver equals the volume fraction of pores:

$$f_{Ag} = f_{\text{pore}} = 0.25$$

Assume 1 cm^3 of material:

$$\text{wt\% Ag} = \frac{(0.25 \text{ cm}^3)(10.49 \text{ g/cm}^3)}{(0.25 \text{ cm}^3)(10.49 \text{ g/cm}^3) + (0.75 \text{ cm}^3)(19.25 \text{ g/cm}^3)} \times 100 = 15\%$$

This solution assumes that all of the pores are open, or interconnected.

Polymers Many engineering polymers that contain fillers and extenders are particulate composites. A classic example is carbon black in vulcanized rubber. Carbon black consists of tiny carbon spheroids only 5 to 500 nm in diameter. The carbon black improves the strength, stiffness, hardness, wear resistance, resistance to degradation due to ultraviolet rays, and heat resistance of the rubber. Nanoparticles of silica are added to rubber tires to enhance their stiffness.

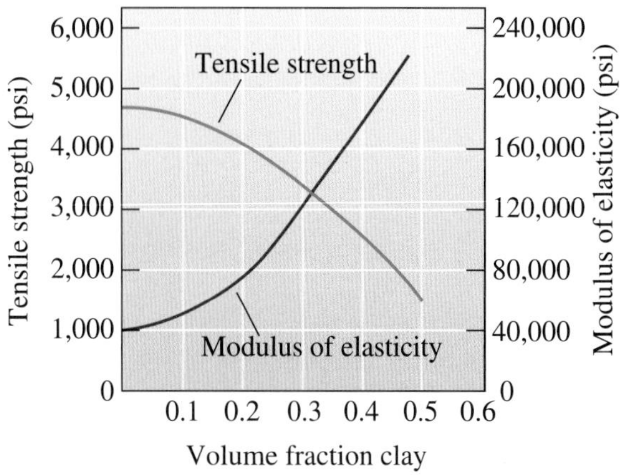

Figure 17-6 The effect of clay on the properties of polyethylene.

Extenders, such as calcium carbonate ($CaCO_3$), solid glass spheres, and various clays, are added so that a smaller amount of the more expensive polymer is required. The extenders may stiffen the polymer, increase the hardness and wear resistance, increase thermal conductivity, or improve resistance to creep; however, strength and ductility normally decrease (Figure 17-6). Introducing hollow glass spheres may impart the same changes in properties while significantly reducing the weight of the composite. Other special properties can be obtained. Elastomer particles are introduced into polymers to improve toughness. Polyethylene may contain metallic powders, such as lead, to improve the absorption of fission products in nuclear applications. The design of a polymer composite is illustrated in the example below.

Example 17-4 *Design of a Particulate Polymer Composite*

Design a clay-filled polyethylene composite suitable for injection molding of inexpensive components. The final part must have a tensile strength of at least 3000 psi and a modulus of elasticity of at least 80,000 psi. Polyethylene costs approximately 50 cents per pound, and clay costs approximately 5 cents per pound. The density of polyethylene is 0.95 g/cm^3 and that of clay is 2.4 g/cm^3.

SOLUTION

From Figure 17-6, a volume fraction of clay below 0.35 is required to maintain a tensile strength greater than 3000 psi, whereas a volume fraction of at least 0.2 is needed for the minimum modulus of elasticity. For lowest cost, we use the maximum allowable clay, or a volume fraction of 0.35.

In 1000 cm^3 of the composite, there are 350 cm^3 of clay and 650 cm^3 of polyethylene, or

$$\frac{(350 \text{ cm}^3)(2.4 \text{ g/cm}^3)}{454 \text{ g/lb}} = 1.85 \text{ lb clay}$$

$$\frac{(650 \text{ cm}^3)(0.95 \text{ g/cm}^3)}{454 \text{ g/lb}} = 1.36 \text{ lb polyethylene}$$

The cost of materials is

$$(1.85 \text{ lb clay})(\$0.05/\text{lb}) = \$0.0925$$
$$(1.36 \text{ lb PE})(\$0.50/\text{lb}) = \$0.68$$
$$\text{total} = \$0.7725 \text{ per } 1000 \text{ cm}^3$$

Suppose that weight is critical. The composite's density is

$$\rho_c = (0.35)(2.4 \text{ g/cm}^3) + (0.65)(0.95 \text{ g/cm}^3) = 1.46 \text{ g/cm}^3$$

We may wish to sacrifice some of the economic savings to obtain lighter weight. If we use only 0.2 volume fraction clay, then (using the same method as above), we find that we need 1.06 lb clay and 1.67 lb polyethylene.

The cost of materials is now

$$(1.06 \text{ lb clay})(\$0.05/\text{lb}) = \$0.053$$
$$(1.67 \text{ lb PE})(\$0.50/\text{lb}) = \$0.835$$
$$\text{total} = \$0.89 \text{ per } 1000 \text{ cm}^3$$

The density of the composite is

$$\rho_c = (0.2)(2.4 \text{ g/cm}^3) + (0.8)(0.95 \text{ g/cm}^3) = 1.24 \text{ g/cm}^3$$

The material costs about 15% more, but there is a weight savings of 15%.

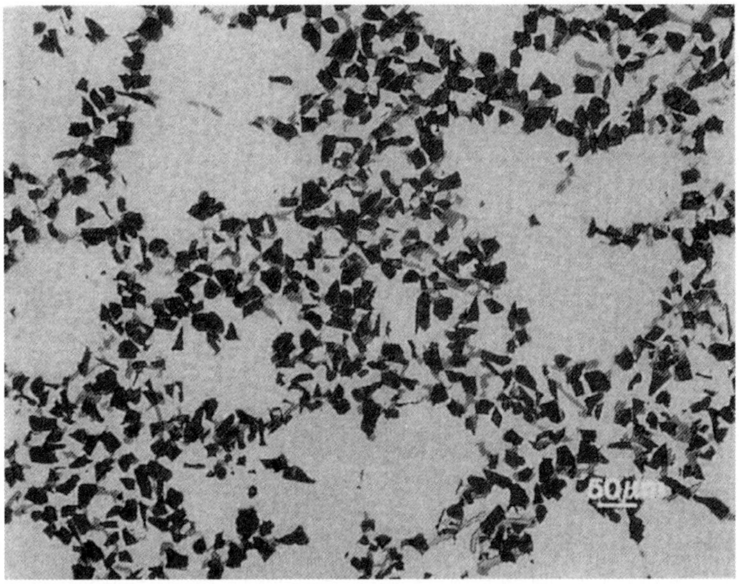

Figure 17-7 An investment cast aluminum alloy reinforced with 20 vol% silicon carbide particles. (High performance composites for the 1990's: *Proceedings of a TMS Northeast Regional Symposium sponsored by TMS-New Jersey Chapter held at Morristown, New Jersey, June 6–8, 1990 by DAS, S. K.; et al. Copyright © 1991 by The Minerals, Metals & Materials Society. Reprinted with permission.*)

Cast Metal Particulate Composites Aluminum castings containing dispersed SiC particles for automotive applications, including pistons and connecting rods, represent an important commercial application for particulate composites (Figure 17-7). With special processing, the SiC particles can be wet by the liquid, helping to keep the ceramic particles from sinking during freezing.

17-3 Fiber-Reinforced Composites

Most fiber-reinforced composites provide improved strength, fatigue resistance, Young's modulus, and strength-to-weight ratio by incorporating strong, stiff, but brittle fibers into a softer, more ductile matrix. The matrix material transmits the force to the fibers, which carry most of the applied force. The matrix also provides protection for the fiber surface and minimizes diffusion of species such as oxygen or moisture that can degrade the mechanical properties of fibers. The strength of the composite may be high at both room temperature and elevated temperatures (Figure 17-2).

Many types of reinforcing materials are employed. Straw has been used to strengthen mud bricks for centuries. Steel-reinforcing bars are introduced into concrete structures. Glass fibers in a polymer matrix produce fiberglass for transportation and aerospace applications. Fibers made from boron, carbon, polymers (e.g., aramids, Chapter 16), and ceramics provide exceptional reinforcement in advanced composites based on matrices of polymers, metals, ceramics, and even intermetallic compounds.

The Rule of Mixtures in Fiber-Reinforced Composites As for particulate composites, the rule of mixtures always predicts the density of fiber-reinforced composites:

$$\rho_c = f_m \rho_m + f_f \rho_f \tag{17-2}$$

where the subscripts m and f refer to the matrix and the fiber. Note that $f_m = 1 - f_f$.

In addition, the rule of mixtures accurately predicts the thermal and electrical conductivities of fiber-reinforced composites along the fiber direction if the fibers are *continuous* and *unidirectional*:

$$k_c = f_m k_m + f_f k_f \tag{17-3}$$

$$\sigma_c = f_m \sigma_m + f_f \sigma_f \tag{17-4}$$

where k is the thermal conductivity and σ is the electrical conductivity. Thermal or electrical energy can be transferred through the composite at a rate that is proportional to the volume fraction of the conductive material. In a composite with a metal matrix and ceramic fibers, the bulk of the energy would be transferred through the matrix; in a composite consisting of a polymer matrix containing metallic fibers, energy would be transferred through the fibers.

When the fibers are not continuous or unidirectional, the simple rule of mixtures may not apply. For example, in a metal fiber-polymer matrix composite, electrical conductivity would be low and would depend on the length of the fibers, the volume fraction of the fibers, and how often the fibers touch one another. This is expressed using the concept of connectivity of phases.

Modulus of Elasticity The rule of mixtures is used to predict the modulus of elasticity when the fibers are continuous and unidirectional. Parallel to the fibers, the modulus of elasticity may be as high as

$$E_{c,\|} = f_m E_m + f_f E_f \tag{17-5}$$

When the applied stress is very large, the matrix begins to deform and the stress strain curve is no longer linear (Figure 17-8). Since the matrix now contributes little to the stiffness of the composite, the modulus can be approximated by

$$E_{c,\|} = f_f E_f \tag{17-6}$$

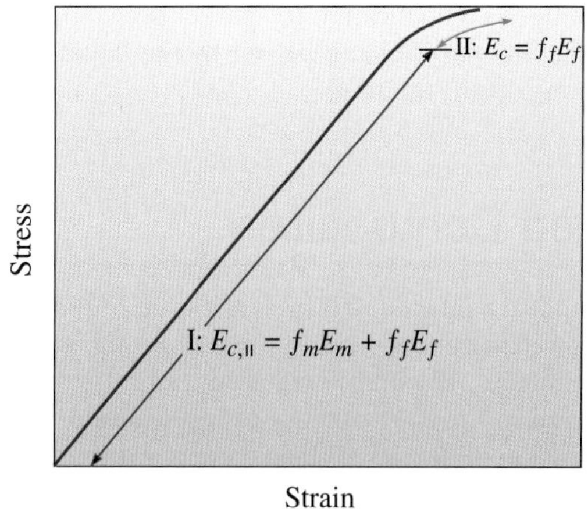

Figure 17-8 The stress strain curve for a fiber-reinforced composite. At low stresses (region I), the modulus of elasticity is given by the rule of mixtures. At higher stresses (region II), the matrix deforms, and the rule of mixtures is no longer obeyed.

When the load is applied perpendicular to the fibers, each component of the composite acts independently of the other. The modulus of the composite is now

$$\frac{1}{E_{c,\perp}} = \frac{f_m}{E_m} + \frac{f_f}{E_f} \quad (17\text{-}7)$$

The following examples further illustrate these concepts.

Example 17-5 Rule of Mixtures for Composites: Stress Parallel to Fibers

Derive the rule of mixtures (Equation 17-5) for the modulus of elasticity of a fiber-reinforced composite when a stress (σ) is applied along the axis of the fibers.

SOLUTION

The total force acting on the composite is the sum of the forces carried by each constituent:

$$F_c = F_m + F_f$$

Since $F = \sigma A$, where A is the cross-sectional area,

$$\sigma_c A_c = \sigma_m A_m + \sigma_f A_f$$

$$\sigma_c = \sigma_m\left(\frac{A_m}{A_c}\right) + \sigma_f\left(\frac{A_f}{A_c}\right)$$

If the fibers have a uniform cross-section, the area fraction equals the volume fraction f:

$$\sigma_c = \sigma_m f_m + \sigma_f f_f$$

From Hooke's law, $\sigma = E\varepsilon$. Therefore,

$$E_{c,\parallel}\,\varepsilon_c = E_m \varepsilon_m f_m + E_f \varepsilon_f f_f$$

If the fibers are rigidly bonded to the matrix, both the fibers and the matrix must stretch equal amounts (iso-strain conditions):

$$\varepsilon_c = \varepsilon_m = \varepsilon_f$$

$$E_{c,\parallel} = f_m E_m + f_f E_f$$

Example 17-6 Modulus of Elasticity for Composites: Stress Perpendicular to Fibers

Derive the equation for the modulus of elasticity of a fiber-reinforced composite when a stress is applied perpendicular to the axis of the fiber (Equation 17-7).

SOLUTION

In this example, the strains are no longer equal; instead, the weighted sum of the strains in each component equals the total strain in the composite, whereas the stresses in each component are equal (iso-stress conditions):

$$\varepsilon_c = f_m \varepsilon_m + f_f \varepsilon_f$$

$$\frac{\sigma_c}{E_{c,\perp}} = f_m\left(\frac{\sigma_m}{E_m}\right) + f_f\left(\frac{\sigma_f}{E_f}\right)$$

Since $\sigma_c = \sigma_m = \sigma_f$,

$$\frac{1}{E_{c,\perp}} = \frac{f_m}{E_m} + \frac{f_f}{E_f}$$

Example 17-7 Boron Aluminum Composites

Boron coated with SiC (or Borsic) reinforced aluminum containing 40 vol% fibers is an important high-temperature, lightweight composite material. Estimate the density, modulus of elasticity, and tensile strength parallel to the fiber axis. Also estimate the modulus of elasticity perpendicular to the fibers.

SOLUTION

The properties of the individual components are shown here.

Material	Density ρ (g/cm^3)	Modulus of Elasticity E (psi)	Tensile Strength TS (psi)
Fibers	2.36	55×10^6	400,000
Aluminum	2.70	10×10^6	5000

From the rule of mixtures:

$$\rho_c = (0.6)(2.70 \text{ g/cm}^3) + (0.4)(2.36 \text{ g/cm}^3) = 2.56 \text{ g/cm}^3$$

$$E_{c,\parallel} = (0.6)(10 \times 10^6 \text{ psi}) + (0.4)(55 \times 10^6 \text{ psi}) = 28 \times 10^6 \text{ psi}$$

$$TS_c = (0.6)(5000 \text{ psi}) + (0.4)(400{,}000 \text{ psi}) = 163{,}000 \text{ psi}$$

Note that the tensile strength calculation is only an approximation.

Perpendicular to the fibers:

$$\frac{1}{E_{c,\perp}} = \frac{0.6}{10 \times 10^6} + \frac{0.4}{55 \times 10^6} = 6.727 \times 10^{-8}$$

$$E_{c,\perp} = 14.9 \times 10^6 \text{ psi}$$

The actual modulus and strength parallel to the fibers are shown in Figure 17-9. The calculated modulus of elasticity (28×10^6 psi) is exactly the same as the measured modulus. The estimated strength (163,000 psi) is substantially higher than the actual strength (about 130,000 psi). We also note that the modulus of elasticity is very anisotropic, with the modulus perpendicular to the fibers being only half the modulus parallel to the fibers.

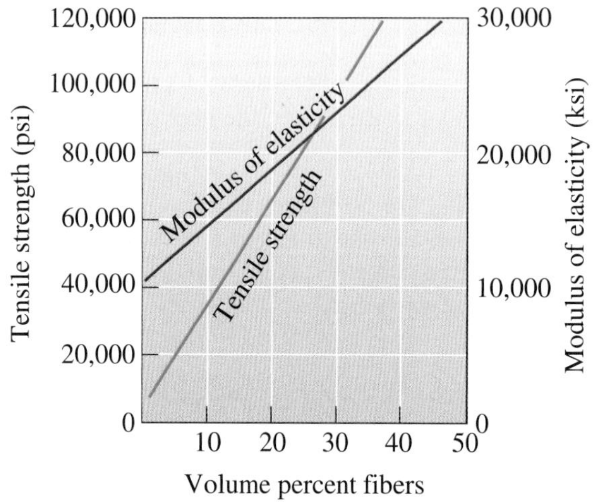

Figure 17-9 The influence of volume percent boron-coated SiC (Borsic) fibers on the properties of Borsic-reinforced aluminum parallel to the fibers (for Example 17-7).

Example 17-8 Nylon-Glass Fiber Composites

Glass fibers in nylon provide reinforcement. If the nylon contains 30 vol% E-glass, what fraction of the force applied parallel to the fiber axis is carried by the glass fibers?

SOLUTION

The modulus of elasticity for each component of the composite is

$$E_{\text{glass}} = 10.5 \times 10^6 \text{ psi} \qquad E_{\text{nylon}} = 0.4 \times 10^6 \text{ psi}$$

Both the nylon and the glass fibers have equal strain if the bonding is good, so

$$\varepsilon_c = \varepsilon_m = \varepsilon_f$$

$$\varepsilon_m = \frac{\sigma_m}{E_m} = \varepsilon_f = \frac{\sigma_f}{E_f}$$

$$\frac{\sigma_f}{\sigma_m} = \frac{E_f}{E_m} = \frac{10.5 \times 10^6}{0.4 \times 10^6} = 26.25$$

Thus the fraction of the force borne by the fibers is given by

$$\text{Fraction} = \frac{F_f}{F_f + F_m} = \frac{\sigma_f A_f}{\sigma_f A_f + \sigma_m A_m} = \frac{\sigma_f(0.3)}{\sigma_f(0.3) + \sigma_m(0.7)}$$

$$= \frac{0.3}{0.3 + 0.7(\sigma_m/\sigma_f)} = \frac{0.3}{0.3 + 0.7(1/26.25)} = 0.92$$

where F_f is the force carried by the fibers and F_m is the force carried by the matrix.

Almost all of the load is carried by the glass fibers.

17-4 Characteristics of Fiber-Reinforced Composites

Many factors must be considered when designing a fiber-reinforced composite, including the length, diameter, orientation, amount, and properties of the fibers; the properties of the matrix; and the bonding between the fibers and the matrix.

Fiber Length and Diameter Fibers can be short, long, or even continuous. Their dimensions are often characterized by the **aspect ratio** l/d, where l is the fiber length and d is the diameter. Typical fibers have diameters varying from 10 μm (10×10^{-4} cm) to 150 μm (150×10^{-4} cm).

The strength of a composite improves when the aspect ratio is large. Fibers often fracture because of surface imperfections. Making the diameter as small as possible gives the fiber less surface area and, consequently, fewer flaws that might propagate during processing or under a load. We also prefer long fibers. The ends of a fiber carry less of the load than the remainder of the fiber; consequently, the fewer the ends, the higher the load-carrying ability of the fibers (Figure 17-10).

In many fiber-reinforced systems, discontinuous fibers with an aspect ratio greater than some critical value are used to provide an acceptable compromise between processing

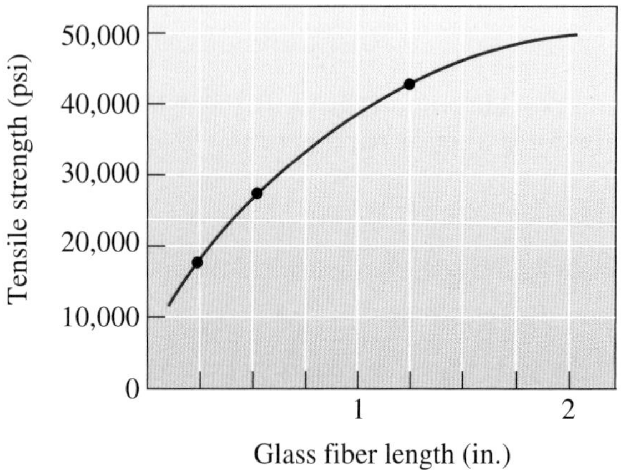

Figure 17-10 Increasing the length of chopped E-glass fibers in an epoxy matrix increases the strength of the composite. In this example, the volume fraction of glass fibers is about 0.5.

ease and properties. A critical fiber length l_c for any given fiber diameter d can be determined according to

$$l_c = \frac{TS_f d}{2\tau_i} \tag{17-8}$$

where TS_f is the tensile strength of the fiber and τ_i is related to the strength of the bond between the fiber and the matrix, or the stress at which the matrix begins to deform. If the fiber length l is smaller than l_c, little reinforcing effect is observed; if l is greater than about $15l_c$, the fiber behaves almost as if it were continuous. The strength of the composite can be estimated from

$$\sigma_c = f_f TS_f\left(1 - \frac{l_c}{2l}\right) + f_m \sigma_m \tag{17-9}$$

where σ_m is the stress on the matrix when the fibers break.

Amount of Fiber A greater volume fraction of fibers increases the strength and stiffness of the composite, as we would expect from the rule of mixtures. The maximum volume fraction is about 80%, beyond which fibers can no longer be completely surrounded by the matrix.

Orientation of Fibers The reinforcing fibers may be introduced into the matrix in a number of orientations. Short, randomly oriented fibers having a small aspect ratio—typical of fiberglass—are easily introduced into the matrix and give relatively isotropic behavior in the composite.

Long, or even continuous, unidirectional arrangements of fibers produce anisotropic properties, with particularly good strength and stiffness parallel to the fibers. These fibers are often designated as 0° plies, indicating that all of the fibers are aligned with the direction of the applied stress. Unidirectional orientations provide poor properties if the load is perpendicular to the fibers (Figure 17-11).

One of the unique characteristics of fiber-reinforced composites is that their properties can be tailored to meet different types of loading conditions. Long, continuous fibers can be introduced in several directions within the matrix (Figure 17-12); in orthogonal arrangements (0°/90° plies), good strength is obtained in two perpendicular directions. More complicated arrangements (such as 0°/±45°/90° plies) provide reinforcement in multiple directions.

Fibers can also be arranged in three-dimensional patterns. In even the simplest of fabric weaves, the fibers in each individual layer of fabric have some small degree of orientation in a third direction. Better three-dimensional reinforcement occurs when the fabric layers are knitted or stitched together. More complicated three-dimensional weaves can also be used (Figure 17-13).

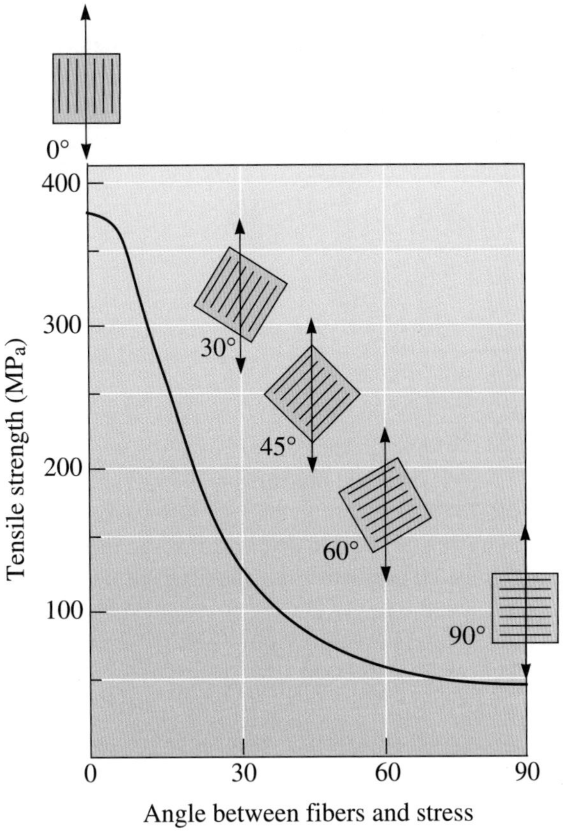

Figure 17-11 Effect of fiber orientation on the tensile strength of E-glass fiber-reinforced epoxy composites.

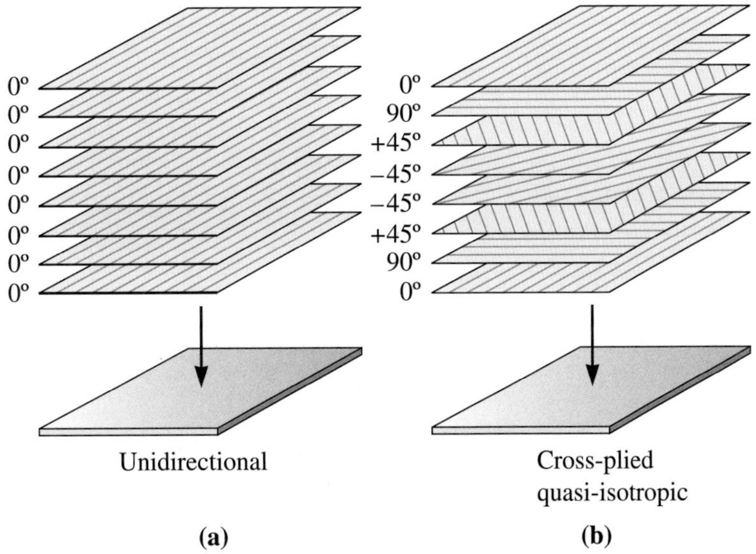

Figure 17-12 (a) Tapes containing aligned fibers can be joined to produce a multi-layered unidirectional composite structure. (b) Tapes containing aligned fibers can be joined with different orientations to produce a quasi-isotropic composite. In this case, a 0°/±45°/90° composite is formed.

Fiber Properties In most fiber-reinforced composites, the fibers are strong, stiff, and lightweight. Thus the **specific strength** and **specific modulus** of the fiber are important characteristics:

$$\text{Specific strength} = \frac{TS}{\rho} \tag{17-10}$$

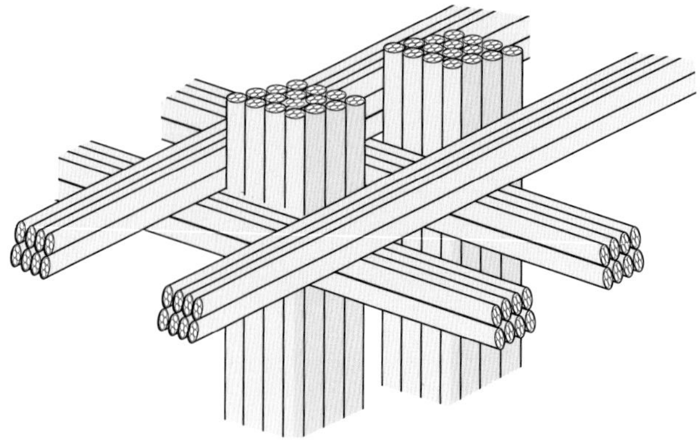

Figure 17-13 A three-dimensional weave for fiber-reinforced composites.

$$\text{Specific modulus} = \frac{E}{\rho} \qquad (17\text{-}11)$$

where TS is the tensile strength, ρ is the density, and E is the modulus of elasticity. If the composite is to be used at elevated temperatures, the fiber should also have a high melting temperature.

Properties of typical fibers are shown in Table 17-2 and Figure 17-14. Note in Table 17-2, the units of density are g/cm^3. Also, note that $1\frac{\text{g}}{\text{cm}^3} = 0.0361\frac{\text{lb}}{\text{in.}^3}$. The highest

Table 17-2 Properties of selected reinforcing materials*

Material	Density ρ (g/cm^3)	Tensile Strength TS (ksi)	Modulus of Elasticity E ($\times 10^6$ psi)	Melting Temperature (°C)	Specific Modulus ($\times 10^7$ in.)	Specific Strength ($\times 10^6$ in.)
Polymers:						
Kevlar™	1.44	650	18.0	500	34.7	12.5
Nylon	1.14	120	0.4	249	1.0	2.9
Polyethylene	0.97	480	25.0	147	71.4	13.7
Metals:						
Be	1.85	185	41.6	1290	65.9	2.8
Boron	2.36	500	55.0	2076	64.6	4.7
W	19.25	580	59.6	3410	8.5	0.8
Glass:						
E-glass	2.55	500	10.5	<1725	11.4	5.6
S-glass	2.50	650	12.6	<1725	14.0	7.2
Carbon:						
HS (high strength)	1.75	820	40.0	3700	63.5	13.0
HM (high modulus)	1.90	270	77.0	3700	112.0	3.9
Ceramics:						
Al_2O_3	3.95	300	56.0	2015	38.8	2.1
B_4C	2.36	330	70.0	2450	82.4	3.9
SiC	3.00	570	60.0	2700	47.3	5.3
ZrO_2	4.84	300	50.0	2677	28.6	1.7
Whiskers:						
Al_2O_3	3.96	3000	62.0	1982	43.4	21.0
Cr	7.20	1290	35.0	1890	13.4	4.9
Graphite	1.66	3000	102.0	3700	170.0	50.2
SiC	3.18	3000	60.0	2700	60.8	26.2
Si_3N_4	3.18	2000	55.0		47.8	17.5

*$1\frac{\text{g}}{\text{cm}^3} = 0.0361\frac{\text{lb}}{\text{in.}^3}$

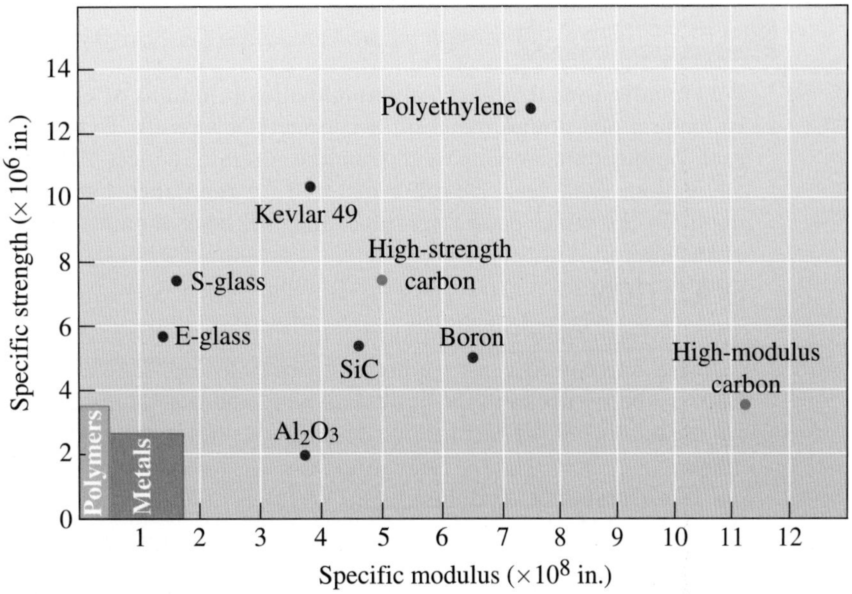

Figure 17-14 Comparison of the specific strength and specific modulus of fibers versus metals and polymers.

specific modulus is usually found in materials having a low atomic number and covalent bonding, such as carbon and boron. These two elements also have high strengths and melting temperatures.

Aramid fibers, of which Kevlar™ is the best known example, are aromatic polyamide polymers strengthened by a backbone containing benzene rings (Figure 17-15) and are examples of liquid-crystalline polymers in that the polymer chains are rod-like and very stiff. Specially prepared polyethylene fibers are also available. Both the aramid and polyethylene fibers have excellent strength and stiffness but are limited to low temperature use. Because of their lower density, polyethylene fibers have superior specific strength and specific modulus.

Ceramic fibers and whiskers, including alumina and silicon carbide, are strong and stiff. Glass fibers, which are the most commonly used, include pure silica, S-glass (25% Al_2O_3, 10% MgO, balance SiO_2), and E-glass (18% CaO, 15% Al_2O_3, balance SiO_2). Although they are considerably denser than polymer fibers, the ceramics can be used at much higher temperatures. Beryllium and tungsten, although metallically bonded, have high moduli that make them attractive fiber materials for certain applications. The following example discusses issues related to designing with composites.

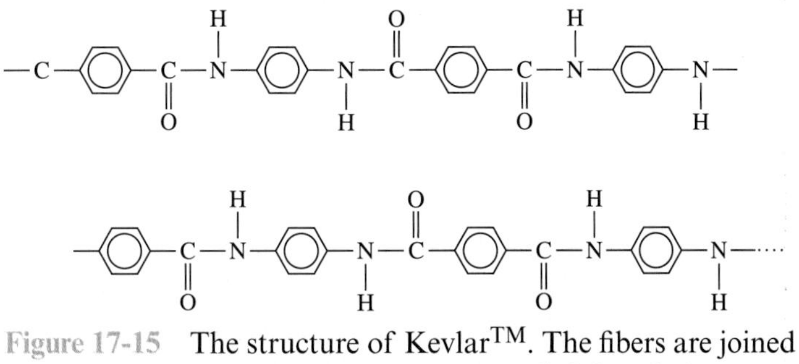

Figure 17-15 The structure of Kevlar™. The fibers are joined by secondary bonds between oxygen and hydrogen atoms on adjoining chains.

Example 17-9 *Design of an Aerospace Composite*

We are now using a 7075-T6 aluminum alloy (modulus of elasticity of 10×10^6 psi) to make a 500-pound panel on a commercial aircraft. Experience has shown that each pound reduction in weight on the aircraft reduces the fuel consumption by 500 gallons each year. Design a material for the panel that will reduce weight, yet maintain the same specific modulus, and will be economical over a 10-year lifetime of the aircraft.

SOLUTION

There are many possible materials that might be used to provide a weight savings. As an example, let's consider using a boron fiber-reinforced Al-Li alloy in the T6 condition. Both the boron fiber and the lithium alloying addition increase the modulus of elasticity; the boron and the Al-Li alloy also have densities less than that of typical aluminum alloys.

The specific modulus of the current 7075-T6 alloy is

$$\text{Specific modulus} = \frac{(10 \times 10^6 \text{ psi})}{\left[\dfrac{\left(2.7 \dfrac{\text{g}}{\text{cm}^3}\right)\left(2.54 \dfrac{\text{cm}}{\text{in.}}\right)^3}{\left(454 \dfrac{\text{g}}{\text{lb}}\right)}\right]}$$

$$= 1.03 \times 10^8 \text{ in.}$$

The density of the boron fibers is approximately 2.36 g/cm^3 (0.085 lb/in.^{3}) and that of a typical Al-Li alloy is approximately 2.5 g/cm^3 (0.090 lb/in.^{3}). The modulus of this alloy is approximately 11×10^6 psi. If we use 0.6 volume fraction boron fibers in the composite, then the density, modulus of elasticity, and specific modulus of the composite are

$$\rho_c = (0.6)(0.085 \text{ lb/in.}^3) + (0.4)(0.090 \text{ lb/in.}^3) = 0.087 \text{ lb/in.}^3$$

$$E_{c,\parallel} = (0.6)(55 \times 10^6 \text{ psi}) + (0.4)(11 \times 10^6 \text{ psi}) = 37 \times 10^6 \text{ psi}$$

$$\text{Specific modulus} = \frac{37 \times 10^6 \text{ psi}}{0.087 \text{ lb/in.}^3} = 4.3 \times 10^8 \text{ in.}$$

If the specific modulus is the only factor influencing the design of the component, the thickness of the part might be reduced by 75%, giving a component weight of 125 pounds rather than 500 pounds. The weight savings would then be 375 pounds, or (500 gal/lb) (375 lb) = 187,500 gal per year. At \$2.50 per gallon, about \$470,000 in fuel savings could be realized each year, or \$4.7 million over the 10-year aircraft lifetime.

This is certainly an optimistic comparison, since strength or fabrication factors may not permit the part to be made as thin as suggested. In addition, the high cost of boron fibers (over \$300/lb) and higher manufacturing costs of the composite compared with those of 7075 aluminum would reduce cost savings.

Matrix Properties The matrix supports the fibers and keeps them in the proper position, transfers the load to the strong fibers, protects the fibers from damage during manufacturing and use of the composite, and prevents cracks in the fiber from propagating throughout the entire composite. The matrix usually provides the major control over electrical properties, chemical behavior, and elevated temperature use of the composite.

Polymer matrices are particularly common. Most polymer materials—both thermoplastics and thermosets—are available in short glass fiber-reinforced grades. These composites are formed into useful shapes by the processes described in Chapter 16. Sheet-molding compounds (SMCs) and bulk-molding compounds (BMCs) are typical of this type of composite. Thermosetting aromatic polyimides are used for somewhat higher temperature applications.

Metal-matrix composites (MMCs) include aluminum, magnesium, copper, nickel, and intermetallic compound alloys reinforced with ceramic and metal fibers. A variety of aerospace and automotive applications utilize MMCs. The metal matrix permits the composite to operate at high temperatures, but producing the composite is often more difficult and expensive than producing the polymer-matrix materials.

The ceramic-matrix composites (CMCs) have good properties at elevated temperatures and are lighter in weight than the high-temperature metal-matrix composites. In a later section, we discuss how to develop toughness in CMCs.

Bonding and Failure Particularly in polymer and metal-matrix composites, good bonding must be obtained between the various constituents. The fibers must be firmly bonded to the matrix material if the load is to be properly transmitted from the matrix to the fibers. In addition, the fibers may pull out of the matrix during loading, reducing the strength and fracture resistance of the composite if bonding is poor. Figure 17-16 illustrates poor bonding of carbon fibers in a silver-copper alloy matrix.

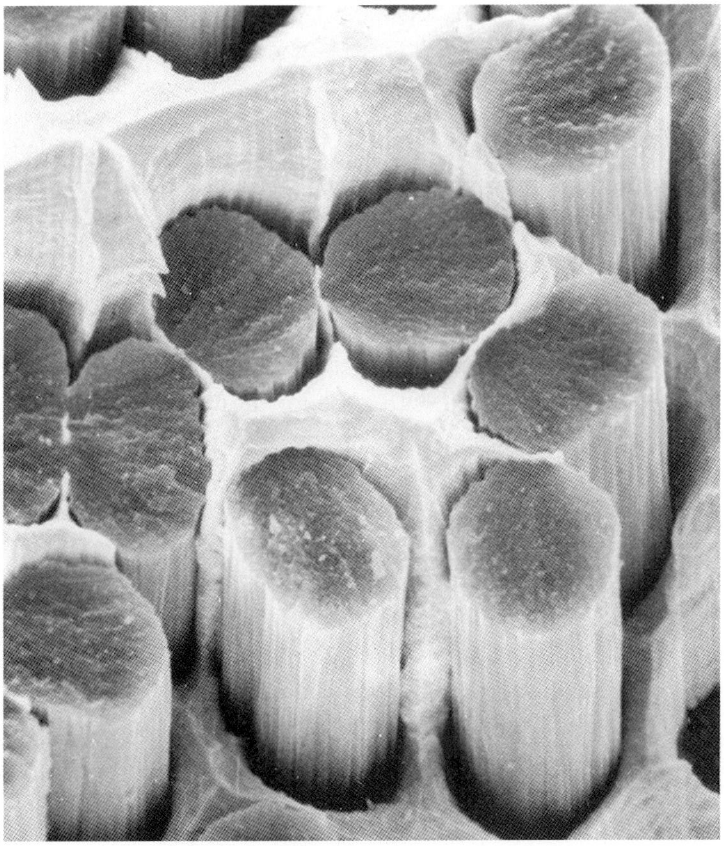

Figure 17-16 Scanning electron micrograph of the fracture surface of a silver-copper alloy reinforced with carbon fibers. Poor bonding causes much of the fracture surface to follow the interface between the metal matrix and the carbon tows (× 3000). [*From* ASM Handbook, *Vol. 9, Metallography and Microstructure (1985), ASM International, Materials Park, OH 44073-0002. Reprinted with permission of ASM International. All rights reserved. www.asminternational.org.*]

In some cases, special coatings may be used to improve bonding. Glass fibers are coated with a silane coupling or "keying" agent (called **sizing**) to improve bonding and moisture resistance in fiberglass composites. Carbon fibers similarly are coated with an organic material to improve bonding. Boron fibers can be coated with silicon carbide or boron nitride to improve bonding with an aluminum matrix; in fact, these fibers are called Borsic fibers to reflect the presence of the silicon carbide (SiC) coating.

Another property that must be considered when combining fibers into a matrix is the similarity between the coefficients of thermal expansion for the two materials. If the fibers expand and contract at a rate much different from that of the matrix, fibers may break or bonding can be disrupted, causing premature failure.

In many composites, individual plies or layers of fabric are joined. Bonding between these layers must also be good or another problem—**delamination**—may occur. Delamination has been suspected as a cause in some accidents involving airplanes using composite-based structures. The layers may tear apart under load and cause failure. Using composites with a three-dimensional weave helps to prevent delamination.

17-5 Manufacturing Fibers and Composites

Producing a fiber-reinforced composite involves several steps, including producing the fibers, arranging the fibers into bundles or fabrics, and introducing the fibers into the matrix.

Making the Fiber

Metal fibers, glass fibers, and many polymer fibers (including nylon, aramid, and polyacrylonitrile) can be formed by drawing processes, as described in Chapter 8 (wire drawing of metal) and Chapter 16 (using the spinnerette for polymer fibers).

Boron, carbon, and ceramics are too brittle and reactive to be worked by conventional drawing processes. Boron fiber is produced by **chemical vapor deposition (CVD)** [Figure 17-17(a)]. A very fine, heated tungsten filament is used as a substrate, passing through a seal into a heated chamber. Vaporized boron compounds such as BCl_3 are introduced into the chamber, decompose, and permit boron to precipitate onto the tungsten wire [Figure 17-18(a)]. SiC fibers are made in a similar manner, with carbon fibers as the substrate for the vapor deposition of silicon carbide [Figure 17-18(b)].

Carbon fibers are made by **carbonizing**, or pyrolizing, an organic filament, which is more easily drawn or spun into thin, continuous lengths [Figure 17-17(b)]. The organic

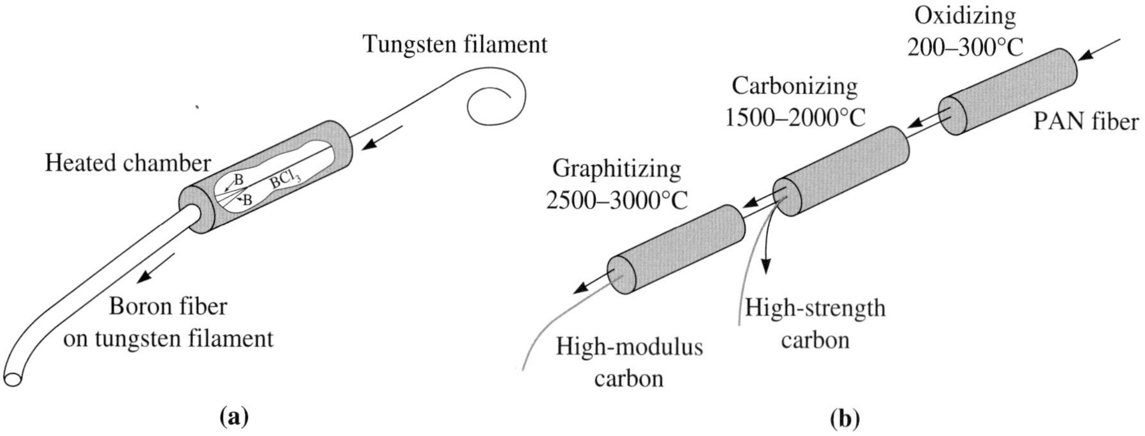

Figure 17-17 Methods for producing (a) boron and (b) carbon fibers.

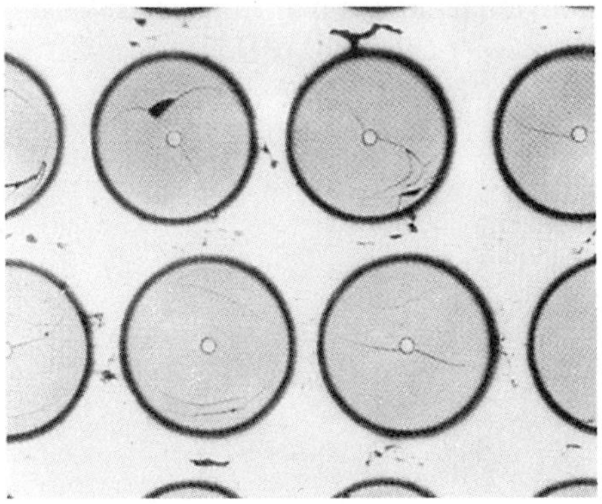

(a)

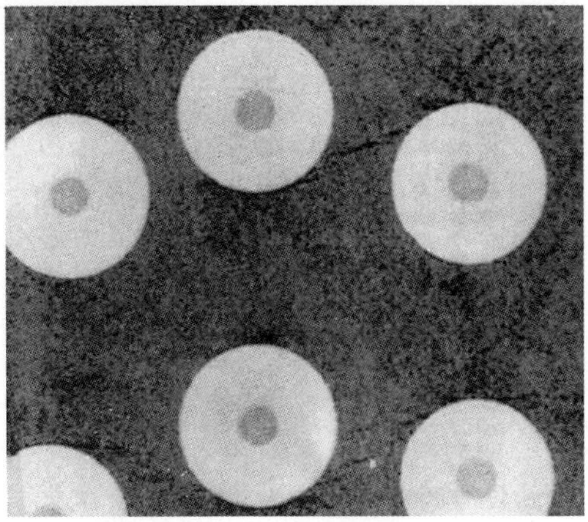

(b)

Figure 17-18 Micrographs of two fiber-reinforced composites: (a) In Borsic fiber-reinforced aluminum, the fibers are composed of a thick layer of boron deposited on a small-diameter tungsten filament (× 1000). [*From* ASM Handbook, *Vol. 9, Metallography and Microstructure (1985), ASM International, Materials Park, OH 44073-0002. Reprinted with permission of ASM International. All rights reserved. www.asminternational.org.*] (b) In this microstructure of a ceramic-fiber-ceramic-matrix composite, silicon carbide fibers are used to reinforce a silicon nitride matrix. The SiC fiber is vapor-deposited on a small carbon precursor-filament (× 125). (*Courtesy of Dr. R. T. Bhatt, NASA Lewis Research Center.*)

filament, known as a **precursor**, is often rayon (a cellulosic polymer), polyacrylonitrile (PAN), or pitch (various aromatic organic compounds). High temperatures decompose the organic polymer, driving off all of the elements but carbon. As the carbonizing temperature increases from 1000°C to 3000°C, the tensile strength decreases while the modulus of elasticity increases (Figure 17-19). Drawing the carbon filaments at critical times during carbonizing may produce desirable preferred orientations in the final carbon filament.

Whiskers are single crystals with aspect ratios of 20 to 1000. Whiskers contain no mobile dislocations; therefore, slip cannot occur, leading to exceptionally high strengths. Because of the complex processing required to produce whiskers, their cost may be quite high.

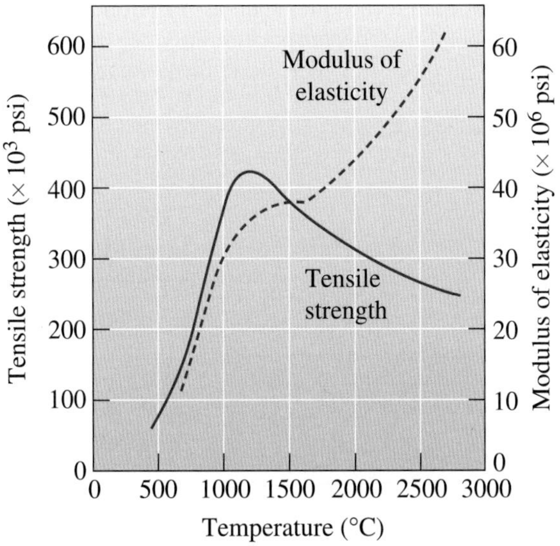

Figure 17-19 The effect of heat-treatment temperature on the strength and modulus of elasticity of carbon fibers.

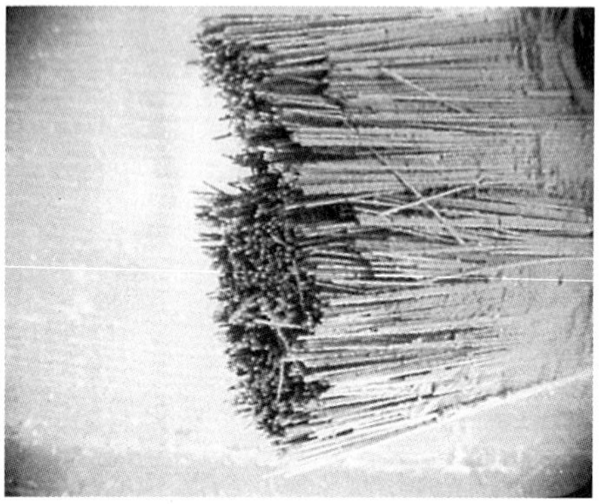

Figure 17-20 A scanning electron micrograph of a carbon tow containing many individual carbon filaments (× 200). (*Reprinted Courtesy of Don Askeland*)

Arranging the Fibers Exceptionally fine filaments are bundled together as rovings, yarns, or tows. In **yarns**, as many as 10,000 filaments are twisted together to produce the fiber. A **tow** contains a few hundred to more than 100,000 untwisted filaments (Figure 17-20). **Rovings** are untwisted bundles of filaments, yarns, or tows.

Often, fibers are chopped into short lengths of 1 cm or less. These fibers, also called **staples**, are easily incorporated into the matrix and are typical of the sheet-molding and bulk-molding compounds for polymer-matrix composites. The fibers often are present in the composite in a random orientation.

Long or continuous fibers for polymer-matrix composites can be processed into mats or fabrics. *Mats* contain non-woven, randomly oriented fibers loosely held together by a polymer resin. The fibers can also be woven, braided, or knitted into two-dimensional or three-dimensional fabrics. The fabrics are then impregnated with a polymer resin. The resins at this point in the processing have not yet been completely polymerized; these mats or fabrics are called **prepregs**.

When unidirectionally aligned fibers are to be introduced into a polymer matrix, **tapes** may be produced. Individual fibers can be unwound from spools onto a mandrel, which determines the spacing of the individual fibers, and prepregged with a polymer resin. These tapes, only one fiber diameter thick, may be up to 48 in. wide. Figure 17-21 illustrates that tapes can also be produced by covering the fibers with upper and lower layers of metal foil that are then joined by diffusion bonding.

Producing the Composite A variety of methods for producing composite parts are used, depending on the application and materials. Short fiber-reinforced composites are normally formed by mixing the fibers with a liquid or plastic matrix, then using relatively conventional techniques such as injection molding for polymer-base composites or casting for metal-matrix composites. Polymer matrix composites can also be produced by a spray-up method, in which short fibers mixed with a resin are sprayed against a form and cured.

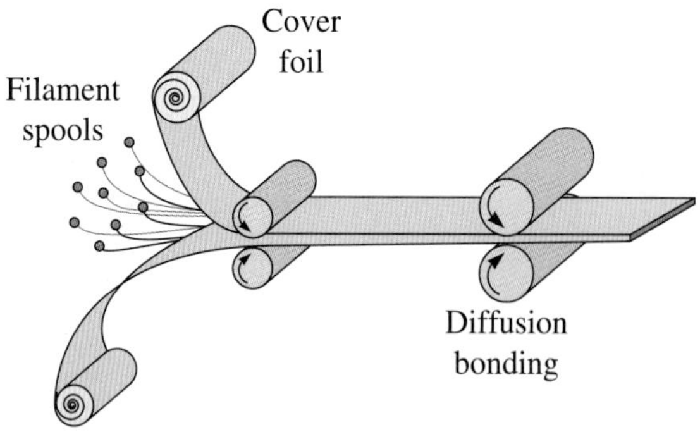

Figure 17-21 Production of fiber tapes by encasing fibers between metal cover sheets by diffusion bonding.

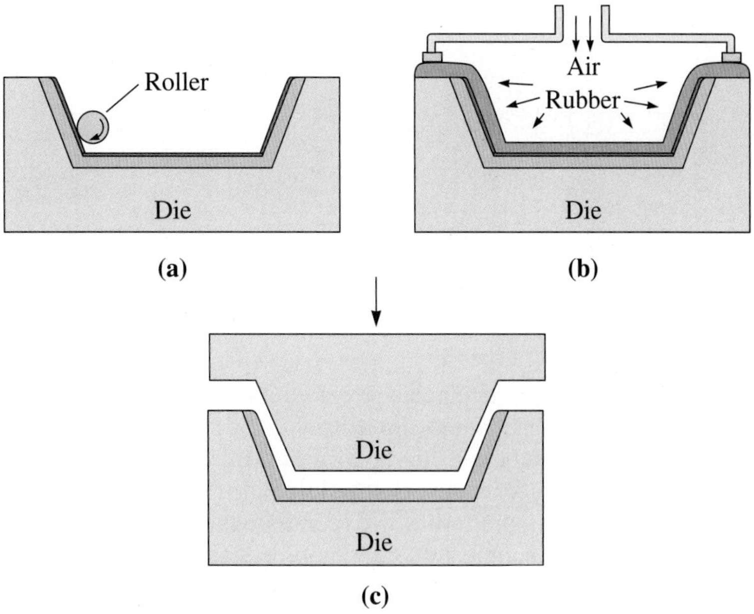

Figure 17-22 Producing composite shapes in dies by (a) hand lay-up, (b) pressure bag molding, and (c) matched die molding.

Special techniques, however, have been devised for producing composites using continuous fibers, either in unidirectionally aligned, mat, or fabric form (Figure 17-22). In hand lay-up techniques, the tapes, mats, or fabrics are placed against a form, saturated with a polymer resin, rolled to ensure good contact and freedom from porosity, and finally cured. Fiberglass car and truck bodies might be made in this manner, which is generally slow and labor intensive.

Tapes and fabrics can also be placed in a die and formed by bag molding. High-pressure gases or a vacuum are introduced to force the individual plies together so that good bonding is achieved during curing. Large polymer matrix components for the skins of military aircraft have been produced by these techniques. In matched die molding, short fibers or mats are placed into a two-part die; when the die is closed, the composite shape is formed.

Filament winding is used to produce products such as pressure tanks and rocket motor casings (Figure 17-23). Fibers are wrapped around a form or mandrel to gradually build up a hollow shape that may be even several feet in thickness. The filament can be dipped in the polymer-matrix resin prior to winding, or the resin can be impregnated around the fiber during or after winding. Curing completes the production of the composite part.

Pultrusion is used to form a simple-shaped product with a constant cross section, such as round, rectangular, pipe, plate, or sheet shapes (Figure 17-24). Fibers or mats are drawn from spools, passed through a polymer resin bath for impregnation, and gathered

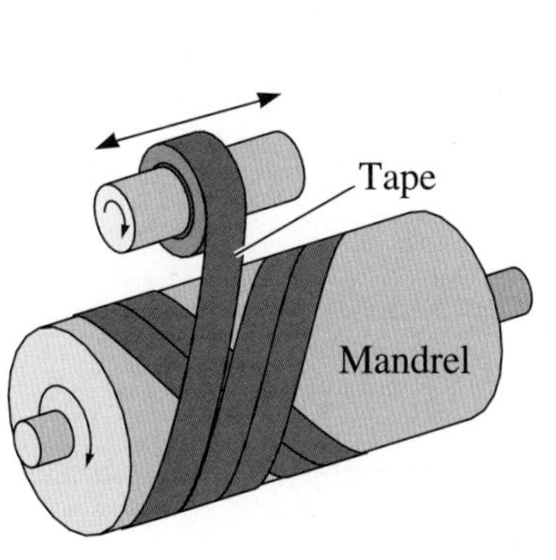

Figure 17-23 Producing composite shapes by filament winding.

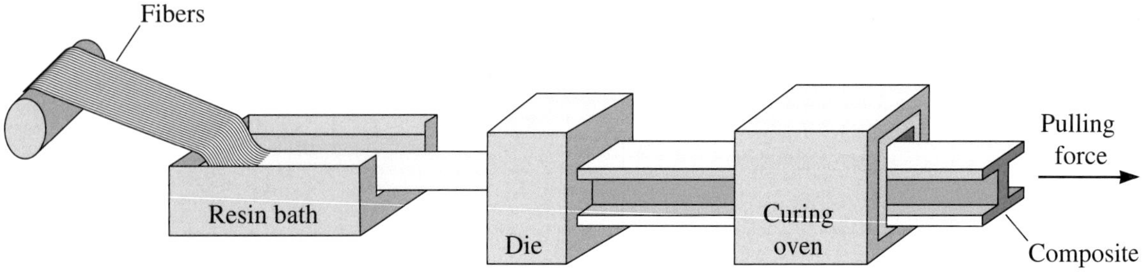

Figure 17-24 Producing composite shapes by pultrusion.

together to produce a particular shape before entering a heated die for curing. Curing of the resin is accomplished almost immediately, so a continuous product is produced. The pultruded stock can subsequently be formed into somewhat more complicated shapes, such as fishing poles, golf club shafts, and ski poles.

Metal-matrix composites with continuous fibers are more difficult to produce than polymer-matrix composites. Casting processes that force liquid around the fibers using capillary rise, pressure casting, vacuum infiltration, or continuous casting are used. Various solid-state compaction processes can also be used.

17-6 Fiber-Reinforced Systems and Applications

Before completing our discussion of fiber-reinforced composites, let's look at the behavior and applications of some of the most common of these materials. Figure 17-25 compares the specific moduli and specific strengths of several composites with those of metals and polymers. Note that the values in this figure are lower than those in Figure 17-14, since we are now looking at the composite, not just the fiber.

Advanced Composites The term advanced composites is often used when the composite is intended to provide service in very critical applications, as in the aerospace industry (Table 17-3). The advanced composites normally are polymer–matrix composites reinforced with high-strength polymer, metal, or ceramic fibers. Carbon fibers

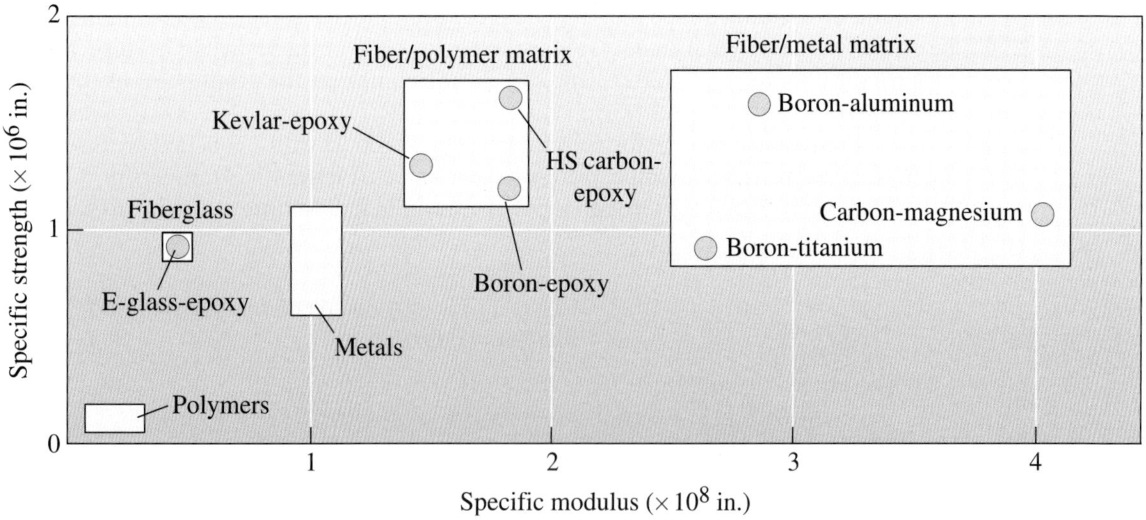

Figure 17-25 A comparison of the specific moduli and specific strengths of several composite materials with those of metals and polymers.

Table 17-3 Examples of fiber-reinforced materials and applications

Material	Applications
Borsic-aluminum	Fan blades in engines, other aircraft and aerospace applications
Kevlar™-epoxy and Kevlar™-polyester	Aircraft, aerospace applications, boat hulls, sporting goods (including tennis rackets, golf club shafts, fishing rods), flak jackets
Graphite-polymer	Aircraft, aerospace, and automotive applications, sporting goods
Glass-polymer	Lightweight automotive applications, water and marine applications, corrosion-resistant applications, sporting goods equipment, aircraft and aerospace components

are used extensively where particularly good stiffness is required; aramid—and, to an even greater extent, polyethylene—fibers are better suited to high-strength applications in which toughness and damage resistance are more important. Unfortunately, the polymer fibers lose their strength at relatively low elevated temperatures, as do all of the polymer matrices (Figure 17-26).

The advanced composites are also frequently used for sporting goods. Tennis rackets, golf clubs, skis, ski poles, and fishing poles often contain carbon or aramid fibers because the higher stiffness provides better performance. In the case of golf clubs, carbon fibers allow less weight in the shaft and therefore more weight in the head. Fabric reinforced with polyethylene fibers is used for lightweight sails for racing yachts.

A unique application for aramid fiber composites is armor. Tough Kevlar™ composites provide better ballistic protection than do other materials, making them suitable for lightweight, flexible bulletproof clothing.

Hybrid composites are composed of two or more types of fibers. For instance, Kevlar™ fibers may be mixed with carbon fibers to improve the toughness of a stiff composite, or Kevlar™ may be mixed with glass fibers to improve stiffness. Particularly

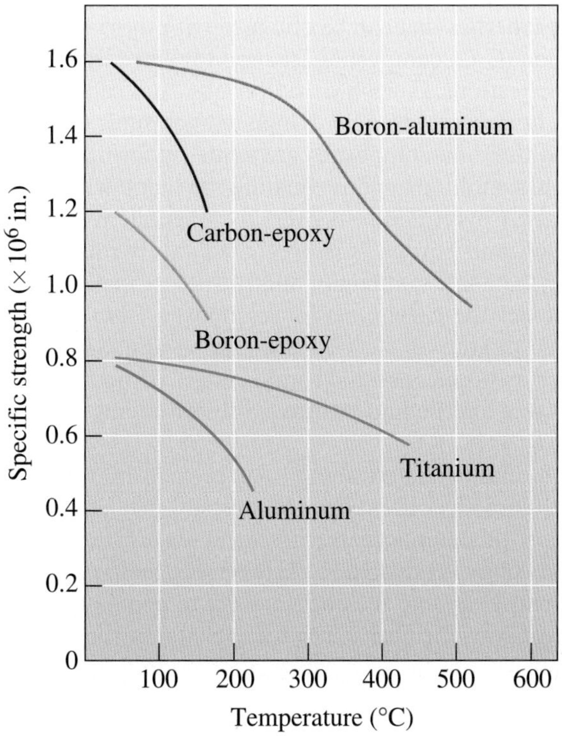

Figure 17-26 Specific strength versus temperature for several composites and metals.

good tailoring of the composite to meet specific applications can be achieved by controlling the amounts and orientations of each fiber.

Tough composites can also be produced if careful attention is paid to the choice of materials and processing techniques. Better fracture toughness in the rather brittle composites can be obtained by using long fibers, amorphous (such as polyetheretherketone known as PEEK or polyphenylene sulfide known as PPS) rather than crystalline or cross-linked matrices, thermoplastic-elastomer matrices, or interpenetrating network polymers.

Metal-Matrix Composites Metal-matrix composites, strengthened by metal or ceramic fibers, provide high-temperature resistance. Aluminum reinforced with borsic fibers has been used extensively in aerospace applications and was used as struts for the space shuttle. Copper-based alloys have been reinforced with SiC fibers, producing high-strength propellers for ships.

Aluminum is commonly used as the matrix in metal-matrix composites. Al_2O_3 fibers reinforce the pistons for some diesel engines; SiC fibers and whiskers are used in aerospace applications, including stiffeners and missile fins; and carbon fibers provide reinforcement for the aluminum antenna mast of the Hubble telescope. Polymer fibers, because of their low melting or degradation temperatures, normally are not used in a metallic matrix. *Polymets*, however, are produced by hot extruding aluminum powder and high melting-temperature liquid-crystalline polymers. A reduction of 1000 to 1 during the extrusion process elongates the polymer into aligned filaments and bonds the aluminum powder particles into a solid matrix.

Metal-matrix composites may find important applications in components for rocket or aircraft engines. Superalloys reinforced with metal fibers (such as tungsten) or ceramic fibers (such as SiC or B_4N) maintain their strength at higher temperatures, permitting jet engines to operate more efficiently. Similarly, titanium and titanium aluminides reinforced with SiC fibers are candidates for turbine blades and discs.

Ceramic-Matrix Composites Composites containing ceramic fibers in a ceramic matrix are also finding applications. Two important uses will be discussed to illustrate the unique properties that can be obtained with these materials.

Carbon-carbon (C-C) composites are used for extraordinary temperature resistance in aerospace applications. Carbon-carbon composites can operate at temperatures of up to 3000°C and, in fact, are stronger at high temperatures than at low temperatures (Figure 17-27). Carbon-carbon composites are made by forming a polyacrylonitrile or carbon fiber fabric into a mold, then impregnating the fabric with an organic resin, such as a phenolic. The part is pyrolyzed to convert the phenolic resin to carbon. The composite, which is still soft and porous, is impregnated and pyrolyzed several more times, continually increasing the density, strength, and stiffness. Finally, the part is coated with silicon carbide to protect the carbon-carbon composite from oxidation. Strengths of 300,000 psi and elastic moduli of 50×10^6 psi can be obtained. Carbon-carbon composites have been used as nose cones and leading edges of high-performance aerospace vehicles such as the space shuttle and as brake discs on racing cars and commercial jet aircraft.

Ceramic-fiber–ceramic-matrix composites provide improved strength and fracture toughness compared with conventional ceramics (Table 17-4). Fiber reinforcements improve the toughness of the ceramic matrix in several ways. First, a crack moving through the matrix encounters a fiber; if the bonding between the matrix and the fiber is poor, the crack is forced to propagate around the fiber in order to continue the fracture process. In addition, poor bonding allows the fiber to begin to pull out of the matrix [Figure 17-28(a)]. Both processes consume energy, thereby increasing fracture toughness. Finally, as a crack

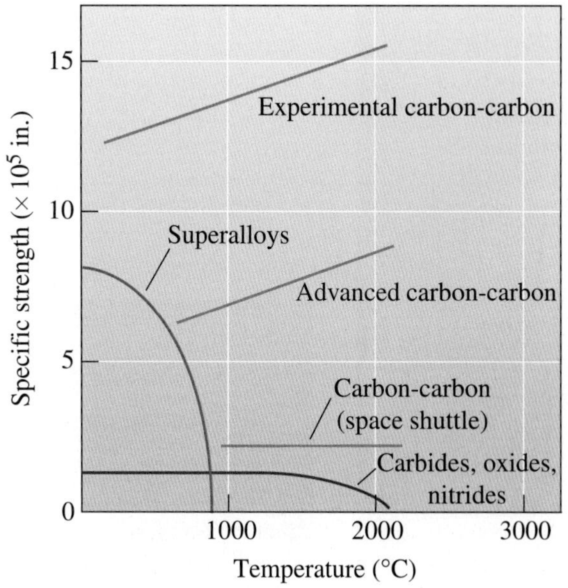

Figure 17-27 A comparison of the specific strengths of various carbon-carbon composites with that of other high-temperature materials relative to temperature.

Table 17-4 Effect of SiC-reinforcement fibers on the properties of selected ceramic materials

Material	Flexural Strength (psi)	Fracture Toughness (psi $\sqrt{\text{in.}}$)
Al_2O_3	80,000	5,000
Al_2O_3/SiC	115,000	8,000
SiC	72,000	4,000
SiC/SiC	110,000	23,000
ZrO_2	30,000	5,000
ZrO_2/SiC	65,000	20,200
Si_3N_4	68,000	4,000
Si_3N_4/SiC	115,000	51,000
Glass	9,000	1,000
Glass/SiC	120,000	17,000
Glass ceramic	30,000	2,000
Glass ceramic/SiC	120,000	16,000

in the matrix propagates, unbroken fibers may bridge the crack and make it more difficult for the crack to open [Figure 17-28(b)].

Unlike polymer and metal matrix composites, poor bonding—rather than good bonding—is required! Consequently, control of the interface structure is crucial. In a glass-ceramic (based on $Al_2O_3 \cdot SiO_2 \cdot Li_2O$) reinforced with SiC fibers and with Nb_2O_5 as an additive, an interface layer containing carbon and NbC is produced that makes debonding of the fiber from the matrix easy. If, however, the composite is heated to a high temperature, the interface is oxidized; the oxide occupies a large volume, exerts a clamping force on the fiber, and prevents easy pull-out. Fracture toughness is then decreased. The following example illustrates some of the cost and property issues that arise while working with composites.

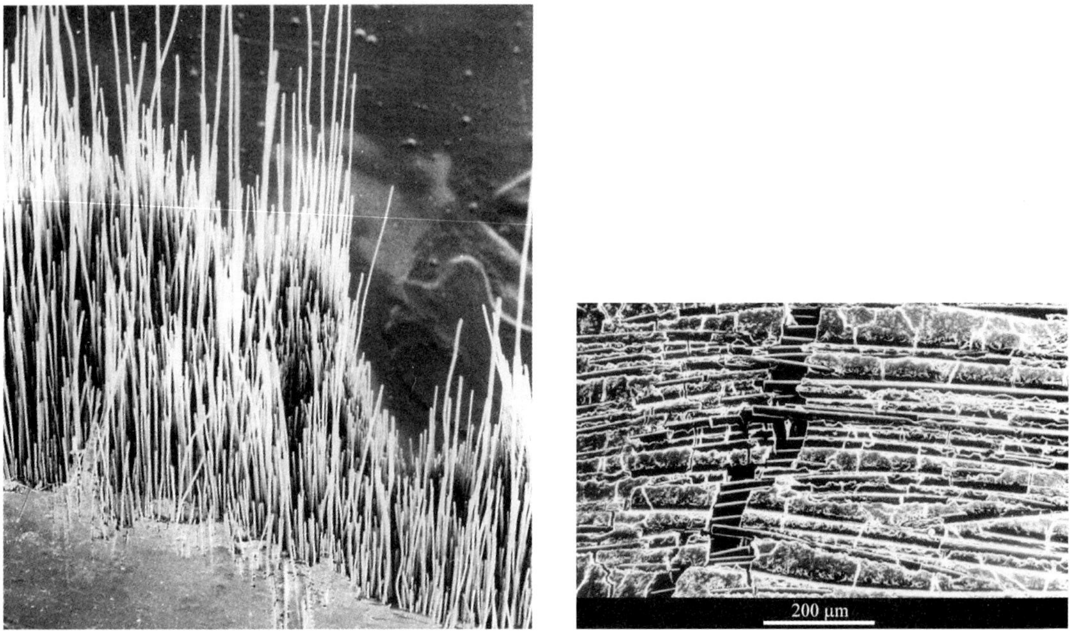

(a) (b)

Figure 17-28 Two failure modes in ceramic-ceramic composites: (a) Extensive pull-out of SiC fibers in a glass matrix provides good composite toughness (× 20). [*From* ASM Handbook, *Vol. 9, Metallography and Microstructure (1985), ASM International, Materials Park, OH 44073-0002. Reprinted with permission of ASM International. All rights reserved. www.asminternational.org.*] (b) Fiber bridging of a cracked matrix in a fiber–reinforced magnesium aluminosilicate ceramic. The toughness of a ceramic matrix composite is enhanced when fibers bridge cracks. (*Courtesy of Dr. Kevin M. Knowles/DoITPoMS Micrograph Library, www.doitpoms.ac.uk*)

Example 17-10 *Design of a Composite Strut*

Design a unidirectional fiber-reinforced epoxy-matrix strut having a round cross-section. The strut is 10 ft long and, when a force of 500 pounds is applied, it should stretch no more than 0.10 in. We want to ensure that the stress acting on the strut is less than the yield strength of the epoxy matrix, 12,000 psi. If the fibers should happen to break, the strut will stretch an extra amount but may not catastrophically fracture. Epoxy costs about \$0.80/lb and has a modulus of elasticity of 500,000 psi and a density of 0.0451 lb/in.^{3}

SOLUTION

Suppose that the strut were made entirely of epoxy (that is, no fibers):

$$\varepsilon_{\text{max}} = \frac{0.10 \text{ in.}}{120 \text{ in.}} = 0.00083 \text{ in./in.}$$

$$\sigma_{\text{max}} = E\varepsilon_{\text{max}} = (500{,}000 \text{ psi})(0.00083) = 415 \text{ psi}$$

$$A_{\text{strut}} = \frac{F}{\sigma_{\text{max}}} = \frac{500 \text{ lb}}{415 \text{ psi}} = 1.2 \text{ in.}^2 \quad \text{or} \quad d = 1.24 \text{ in.}$$

$$\text{Weight}_{\text{strut}} = (0.0451 \text{ lb/in.}^3)(\pi)(1.24 \text{ in./2})^2(120 \text{ in.}) = 6.54 \text{ lb}$$

$$\text{Cost}_{\text{strut}} = (6.54 \text{ lb})(\$0.80/\text{lb}) = \$5.23$$

With no reinforcement, the strut is large and heavy; the materials cost is high due to the large amount of epoxy needed.

In a composite, the maximum strain is still 0.00083 in./in. If we make the strut as small as possible—that is, it operates at 12,000 psi—then the minimum modulus of elasticity E_c of the composite is

$$E_c > \frac{\sigma_{\max}}{\varepsilon_{\max}} = \frac{12{,}000 \text{ psi}}{0.00083} = 14.5 \times 10^6 \text{ psi}$$

Let's look at several possible composite systems. The modulus of glass fibers is less than 14.5×10^6 psi; therefore, glass reinforcement is not a possible choice.

For high modulus carbon fibers, $E = 77 \times 10^6$ psi; the density is 1.9 g/cm^3 = 0.0686 lb/in.^{3}, and the cost is about \$30/lb. The minimum volume fraction of carbon fibers needed to give a composite modulus of 14.5×10^6 psi is

$$E_c = f_c(77 \times 10^6) + (1 - f_c)(0.5 \times 10^6) > 14.5 \times 10^6$$

$$f_c = 0.183$$

The volume fraction of epoxy remaining is 0.817. An area of 0.817 times the total cross-sectional area of the strut must support a 500-lb load with no more than 12,000 psi if all of the fibers should fail:

$$A_{\text{epoxy}} = 0.817A_{\text{total}} = \frac{F}{\sigma_{\max}} = \frac{500 \text{ lb}}{12{,}000 \text{ psi}} = 0.0417 \text{ in.}^2$$

$$A_{\text{total}} = \frac{0.0417 \text{ in.}^2}{0.817} = 0.051 \text{ in.}^2 \quad \text{or} \quad d = 0.255 \text{ in.}$$

$$\text{Volume}_{\text{strut}} = (0.051 \text{ in.}^2)(120 \text{ in.}) = 6.12 \text{ in.}^3$$

$$\begin{aligned}\text{Weight}_{\text{strut}} &= \rho\text{Volume}_{\text{strut}} = [(0.0686 \text{ lb/in.}^3)(0.183) + (0.0451 \text{ lb/in.}^3)(0.817)] \\ &\quad (6.12 \text{ in.}^3) \\ &= 0.302 \text{ lb}\end{aligned}$$

To calculate the weight fraction of carbon, consider 1 in.^{3} of composite. Then the weight fraction of the carbon is given by

$$\begin{aligned}\text{Weight fraction}_{\text{carbon}} &= \frac{(0.183 \text{ in.}^3)(0.0686 \text{ lb/in.}^3)}{(0.183 \text{ in.}^3)(0.0686 \text{ lb/in.}^3) + (0.817 \text{ in.}^3)(0.0451 \text{ lb/in.}^3)} \\ &= 0.254\end{aligned}$$

and the total weight of the carbon in the composite is

$$\text{Weight}_{\text{carbon}} = (0.254)(0.302 \text{ lb}) = 0.077 \text{ lb}$$

Similarly, the weight fraction of epoxy is calculated as

$$\begin{aligned}\text{Weight fraction}_{\text{epoxy}} &= \frac{(0.817 \text{ in.}^3)(0.0451 \text{ lb/in.}^3)}{(0.183 \text{ in.}^3)(0.0686 \text{ lb/in.}^3) + (0.817 \text{ in.}^3)(0.0451 \text{ lb/in.}^3)} \\ &= 1 - 0.254 = 0.746\end{aligned}$$

and the total weight of the epoxy in the composite is

$$\text{Weight}_{\text{epoxy}} = (0.746)(0.302 \text{ lb}) = 0.225 \text{ lb}$$

Therefore,

$$\text{Cost}_{\text{strut}} = (0.077 \text{ lb})(\$30/\text{lb}) + (0.225 \text{ lb})(\$0.80/\text{lb}) = \$2.49$$

The carbon-fiber reinforced strut is less than one-quarter the diameter of an all-epoxy structure, with only 5% of the weight and half of the cost.

We might also repeat these calculations using Kevlar™ fibers, with a modulus of 18×10^6 psi, a density of 1.44 g/cm^3 = 0.052 lb/in.^{3}, and a cost of about \$20/lb. By doing so, we would find that a volume fraction of 0.8 fibers is required. Note that 0.8 volume fraction is at the maximum of fiber volume that can be incorporated into a matrix. We would also find that the required diameter of the strut is 0.515 in. and that the strut weighs 1.266 lb and costs \$20.98. The modulus of the Kevlar™ is not high enough to offset its high cost.

Although the carbon fibers are the most expensive, they permit the lightest weight and the lowest material cost strut. (This calculation does not, however, take into consideration the costs of manufacturing the strut.) Our design, therefore, is to use a 0.255-in.-diameter strut containing 0.183 volume fraction high modulus carbon fiber.

17-7 Laminar Composite Materials

Laminar composites include very thin coatings, thicker protective surfaces, claddings, bimetallics, laminates, and a host of other applications. In addition, the fiber-reinforced composites produced from tapes or fabrics can be considered partly laminar. Many laminar composites are designed to improve corrosion resistance while retaining low cost, high strength, or light weight. Other important characteristics include superior wear or abrasion resistance, improved appearance, and unusual thermal-expansion characteristics.

Rule of Mixtures Some properties of the laminar composite materials parallel to the lamellae are estimated from the rule of mixtures. The density, electrical and thermal conductivities, and modulus of elasticity parallel to the lamellae can be calculated with little error using the following formulas:

$$\text{Density} = \rho_{c,\parallel} = \sum (f_i \rho_i) \qquad (17\text{-}12)$$

$$\text{Electrical conductivity} = \sigma_{c,\parallel} = \sum (f_i \sigma_i)$$

$$\text{Thermal conductivity} = k_{c,\parallel} = \sum (f_i k_i) \qquad (17\text{-}13)$$

$$\text{Modulus of elasticity} = E_{c,\parallel} = \sum (f_i E_i)$$

The laminar composites are very anisotropic. The properties perpendicular to the lamellae are

$$\text{Electrical conductivity} = \frac{1}{\sigma_{c,\perp}} = \sum \left(\frac{f_i}{\sigma_i}\right)$$

$$\text{Thermal conductivity} = \frac{1}{k_{c,\perp}} = \sum \left(\frac{f_i}{k_i}\right) \qquad (17\text{-}14)$$

$$\text{Modulus of elasticity} = \frac{1}{E_{c,\perp}} = \sum \left(\frac{f_i}{E_i}\right)$$

Many other properties, such as corrosion and wear resistance, depend primarily on only one of the components of the composite, so the rule of mixtures is not applicable.

Producing Laminar Composites Several methods are used to produce laminar composites, including a variety of deformation and joining techniques used primarily for metals (Figure 17-29).

Individual plies are often joined by *adhesive bonding*, as is the case in producing plywood. Polymer-matrix composites built up from several layers of fabric or tape prepregs are also joined by adhesive bonding; a film of unpolymerized material is placed between each layer of prepreg. When the layers are pressed at an elevated temperature, polymerization is completed and the prepregged fibers are joined to produce composites that may be dozens of layers thick.

Most of the metallic laminar composites, such as claddings and bimetallics, are produced by *deformation bonding*, such as hot- or cold-roll bonding. The pressure exerted by the rolls breaks up the oxide film at the surface, brings the surfaces into atom-to-atom contact, and permits the two surfaces to be joined. Explosive bonding also can

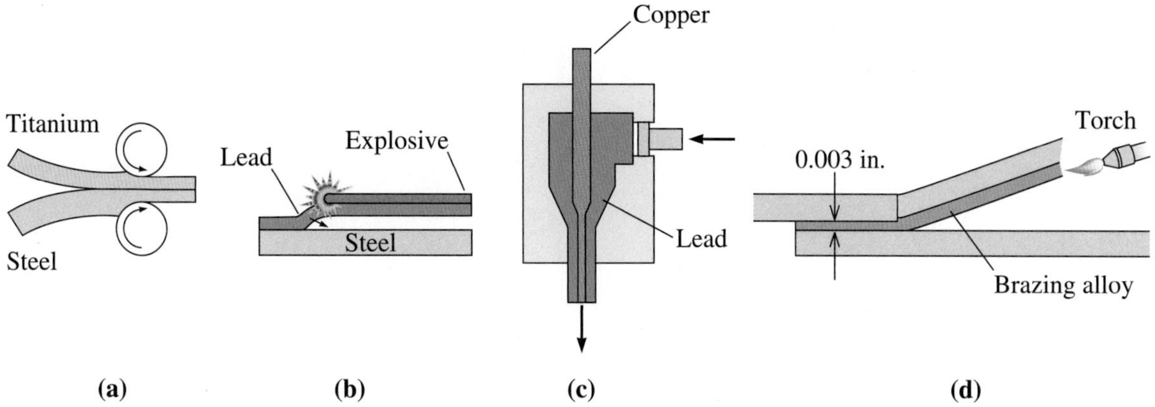

Figure 17-29 Techniques for producing laminar composites: (a) roll bonding, (b) explosive bonding, (c) coextrusion, and (d) brazing.

be used. An explosive charge provides the pressure required to join metals. This process is particularly well suited for joining very large plates that will not fit into a rolling mill. Very simple laminar composites, such as coaxial cable, are produced by coextruding two materials through a die in such a way that the soft material surrounds the harder material. Metal conductor wire can be coated with an insulating thermoplastic polymer in this manner.

Brazing can join composite plates (Chapter 9). The metallic sheets are separated by a very small clearance—preferably about 0.003 in.—and heated above the melting temperature of the brazing alloy. The molten brazing alloy is drawn into the thin joint by capillary action.

17-8 Examples and Applications of Laminar Composites

The types of laminar composites are so varied and their applications are so numerous that we cannot make generalizations concerning their behavior. Instead we will examine the characteristics of a few commonly used examples.

Laminates Laminates are layers of materials joined by an organic adhesive. In laminated safety glass, a plastic adhesive, such as polyvinyl butyral (PVB), joins two pieces of glass; the adhesive prevents fragments of glass from flying about when the glass is broken (Chapter 15). Laminates are used for insulation in motors, for gears, for printed circuit boards, and for decorative items such as Formica® countertops and furniture.

Microlaminates include composites composed of alternating layers of aluminum sheet and fiber-reinforced polymer. *Arall* (aramid-aluminum laminate) and *Glare* (glass-aluminum laminate) are two examples. In Arall, an aramid fiber such as Kevlar™ is prepared as a fabric or unidirectional tape, impregnated with an adhesive, and laminated between layers of aluminum alloy (Figure 17-30). The composite laminate has an unusual combination of strength, stiffness, corrosion resistance, and light weight. Fatigue resistance is improved, since the interface between the layers may block cracks. Glare has similar properties and is used in the fuselage of the Airbus 380. Compared with polymer-matrix composites, the microlaminates have good resistance to lightning-strike damage (which is important in aerospace applications), are formable and machinable, and are easily repaired.

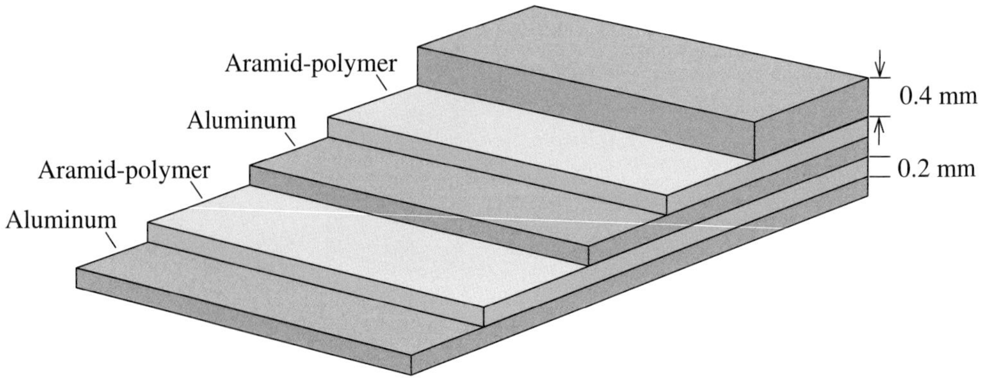

Figure 17-30 Schematic diagram of an aramid-aluminum laminate, Arall.

Clad Metals Clad materials are metal-metal composites. A common example of **cladding** is United States silver coinage. A Cu-80% Ni alloy is bonded to both sides of a Cu-20% Ni alloy. The ratio of thicknesses is about 1/6:2/3:1/6. The high-nickel alloy is a silver color, while the predominantly copper core provides low cost.

Clad materials provide a combination of good corrosion resistance with high strength. *Alclad* is a clad composite in which commercially pure aluminum is bonded to higher strength aluminum alloys. The pure aluminum protects the higher strength alloy from corrosion. The thickness of the pure aluminum layer is about 1% to 15% of the total thickness. Alclad is used in aircraft construction, heat exchangers, building construction, and storage tanks, where combinations of corrosion resistance, strength, and light weight are desired.

Bimetallics Laminar composites made from two metals with different coefficients of thermal expansion are used as temperature indicators and controllers. If two pieces of metal are heated, the metal with the higher coefficient of thermal expansion becomes longer. If the two pieces of metal are rigidly bonded together, the difference in their coefficients causes the strip to bend and produce a curved surface. The amount of movement depends on the temperature. By measuring the curvature or deflection of the strip, we can determine the temperature. Likewise, if the free end of the strip activates a relay, the strip can turn on or off a furnace or air conditioner to regulate temperature. Metals selected for **bimetallics** must have (a) very different coefficients of thermal expansion, (b) expansion characteristics that are reversible and repeatable, and (c) a high modulus of elasticity, so that the bimetallic device can do work. Often the low-expansion strip is made from Invar, an iron-nickel alloy, whereas the high-expansion strip may be brass, Monel, or pure nickel.

Bimetallics can act as circuit breakers as well as thermostats; if a current passing through the strip becomes too high, heating causes the bimetallic to deflect and break the circuit.

17-9 Sandwich Structures

Sandwich materials have thin layers of a facing material joined to a lightweight filler material, such as a polymer foam. Neither the filler nor the facing material is strong or rigid, but the composite possesses both properties. A familiar example is corrugated cardboard. A corrugated core of paper is bonded on either side to flat, thick paper. Neither the corrugated core nor the facing paper is rigid, but the combination is.

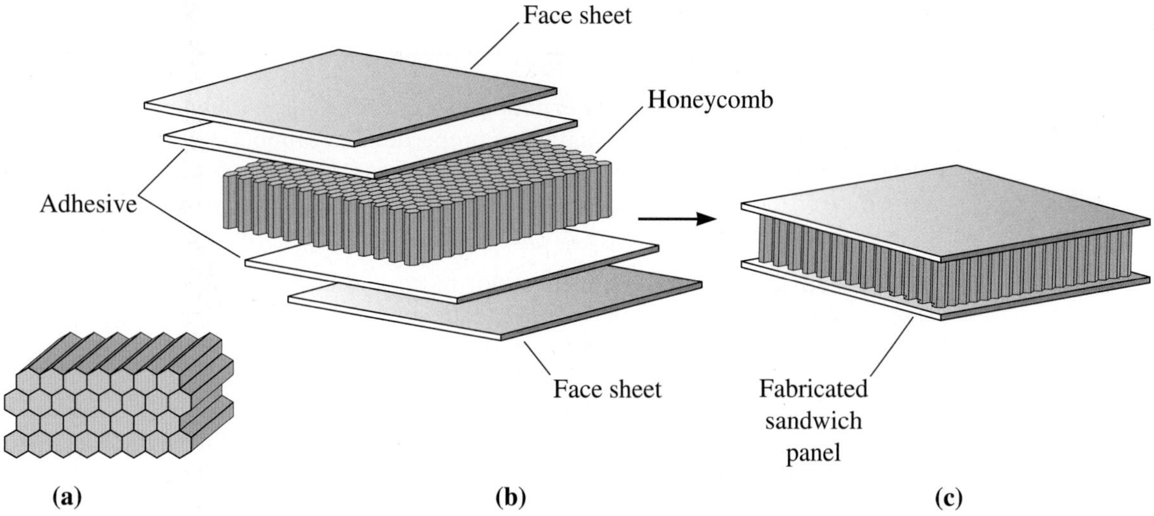

Figure 17-31 (a) A hexagonal cell honeycomb core (b) can be joined to two face sheets by means of adhesive sheets, (c) producing an exceptionally lightweight yet stiff, strong honeycomb sandwich structure.

Another important example is the honeycomb structure used in aircraft applications. A **honeycomb** is produced by gluing thin aluminum strips at selected locations. The honeycomb material is then expanded to produce a very low-density cellular panel that, by itself, is unstable (Figure 17-31). When an aluminum facing sheet is adhesively bonded to either side of the honeycomb, however, a very stiff, strong, and exceptionally lightweight sandwich with a density as low as 0.04 g/cm^3 is obtained.

The honeycomb cells can have a variety of shapes, including hexagonal, square, rectangular, and sinusoidal, and they can be made from aluminum, fiberglass, paper, aramid polymers, and other materials. The honeycomb cells can be filled with foam or fiberglass to provide excellent sound and vibration absorption. Figure 17-32 describes one method by which honeycomb can be fabricated.

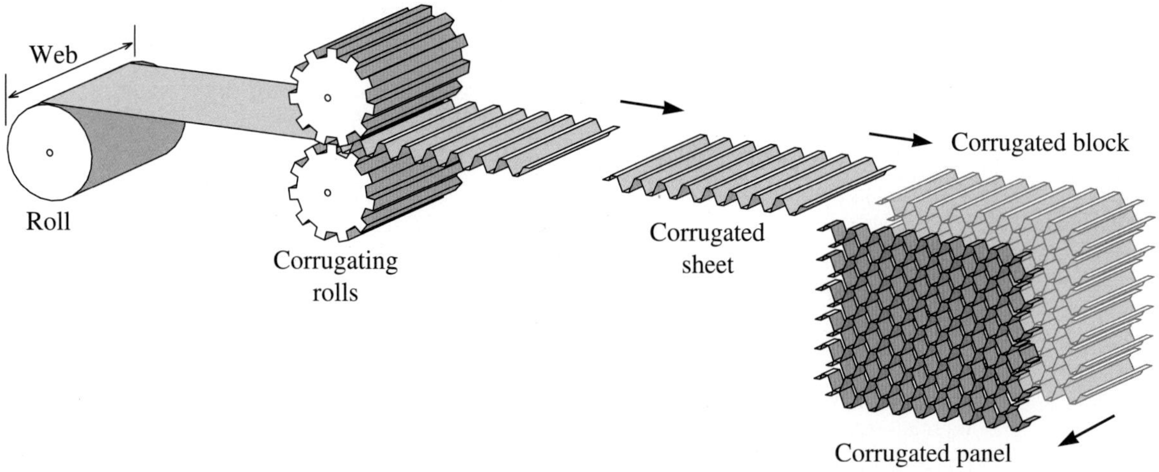

Figure 17-32 In the corrugation method for producing a honeycomb core, the material (such as aluminum) is corrugated between two rolls. The corrugated sheets are joined together with adhesive and then cut to the desired thickness.

Summary

- Composites are made from two or more materials or phases joined or connected in such a way so as to give a combination of properties that cannot be attained otherwise. The volume fraction and the connectivity of the phases or materials in a composite and the nature of the interface between the dispersed phase and matrix are very important in determining the properties.
- Virtually any combination of metals, polymers, and ceramics is possible. In many cases, the rule of mixtures can be used to estimate the properties of the composite.
- Composites have many applications in construction, aerospace, automotive, sports, microelectronics, and other industries.
- Dispersion-strengthened materials contain oxide particles in a metal matrix. The small stable dispersoids interfere with slip, providing good mechanical properties at elevated temperatures.
- Particulate composites contain particles that impart combinations of properties to the composite. Metal-matrix composites contain ceramic or metallic particles that provide improved strength and wear resistance and ensure good electrical conductivity, toughness, or corrosion resistance. Polymer-matrix composites contain particles that enhance stiffness, heat resistance, or electrical conductivity while maintaining light weight, ease of fabrication, or low cost.
- Fiber-reinforced composites provide improvements in strength, stiffness, or high-temperature performance in metals and polymers and impart toughness to ceramics. Fibers typically have low densities, giving high specific strengths and specific moduli, but they often are very brittle. Fibers can be continuous or discontinuous. Discontinuous fibers with a high aspect ratio l/d produce better reinforcement than fibers with a low aspect ratio.
- Fibers are introduced into a matrix in a variety of orientations. Random orientations and isotropic behavior are obtained using discontinuous fibers; unidirectionally aligned fibers produce composites with anisotropic behavior with large improvements in strength and stiffness parallel to the fiber direction. Properties can be tailored to meet the imposed loads by orienting the fibers in multiple directions.
- Laminar composites consist of layers of different materials. These layers may be sheets of different metals, with one metal providing strength and the other providing hardness or corrosion resistance. Layers may also include sheets of fiber-reinforced material bonded to metal or polymer sheets or even to fiber-reinforced sheets having different fiber orientations. The laminar composites are always anisotropic.
- Sandwich materials, including honeycombs, are exceptionally lightweight laminar composites, with solid facings joined to an almost hollow core.

Glossary

Aramid fibers Polymer fibers, such as Kevlar^{TM}, formed from polyamides, which contain the benzene ring in the backbone of the polymer.

Aspect ratio The length of a fiber divided by its diameter.

Bimetallic A laminar composite material produced by joining two strips of metal with different thermal-expansion coefficients, making the material sensitive to temperature changes.

Brazing A process in which a liquid filler metal is introduced by capillary action between two solid materials that are to be joined. Solidification of the brazing alloy provides the bond.

Carbonizing Driving off the non-carbon atoms from a polymer fiber, leaving behind a carbon fiber of high strength. Also known as pyrolizing.

Cemented carbides Particulate composites containing hard ceramic particles bonded with a soft metallic matrix. The composite combines high hardness and cutting ability, yet still has good shock resistance.

Chemical vapor deposition Method for manufacturing materials by condensing the material from a vapor onto a solid substrate.

Cladding A laminar composite produced when a corrosion-resistant or high-hardness layer is formed onto a less expensive or higher strength backing.

Delamination Separation of individual plies of a fiber-reinforced composite.

Dispersed phase The phase or phases that are distributed throughout a continuous matrix of the composite.

Dispersoids Oxide particles formed in a metal matrix that interfere with dislocation movement and provide strengthening, even at elevated temperatures.

Filament winding Process for producing fiber-reinforced composites in which continuous fibers are wrapped around a form or mandrel. The fibers may be prepregged or the filament-wound structure may be impregnated to complete the production of the composite.

Honeycomb A lightweight but stiff assembly of aluminum strip joined and expanded to form the core of a sandwich structure.

Matrix phase The continuous phase in a composite. The composites are named after the continuous phase (e.g., polymer-matrix composites).

Nanocomposite A material in which the dispersed phase has dimensions on the order of nanometers and is distributed in the continuous matrix.

Precursor A starting chemical (e.g., a polymer fiber) that is carbonized to produce carbon fibers.

Prepregs Layers of fibers in unpolymerized resins. After the prepregs are stacked to form a desired structure, polymerization joins the layers together.

Pultrusion A continuous manufacturing process in which fibers or mats are impregnated with a polymer resin and pulled through a die to produce a form with a constant cross-section.

Rovings Untwisted bundles of filaments, yarns, or tows.

Rule of mixtures The statement that some properties of a composite material can be described as the sums of the properties of the individual materials comprising the composite weighted by their volume fractions.

Sandwich A composite material constructed of a lightweight, low-density material surrounded by dense, solid layers. The sandwich combines overall light weight with excellent stiffness.

Sizing Coating glass fibers with an organic material to improve bonding and moisture resistance in fiberglass.

Specific modulus The modulus of elasticity of a material divided by the density.

Specific strength The tensile or yield strength of a material divided by the density.

Staples Fibers chopped into short lengths.

Tapes A strip of prepreg that is only one filament thick. The filaments may be unidirectional or woven. Several layers of tapes can be joined to produce a composite structure.

Tow A bundle of untwisted filaments.

Whiskers Very fine fibers grown in a manner that produces single crystals with no mobile dislocations, thus giving nearly theoretical strengths.

Yarns Continuous fibers produced from a group of twisted filaments.

Problems

Section 17-1 Dispersion-Strengthened Composites

17-1 What is a composite?

17-2 What do the properties of composite materials depend upon?

17-3 Give examples for which composites are used for load-bearing applications.

17-4 Give two examples for which composites are used for non-structural applications.

17-5 What is a dispersion-strengthened composite? How is it different from a particle-reinforced composite?

17-6 What is a nanocomposite? Is it possible for a steel to be described as a nanocomposite? Explain.

17-7 In Figure 17-2, find the largest and smallest difference in strength between Borsic and SAP and also the temperatures at which these differences occur.

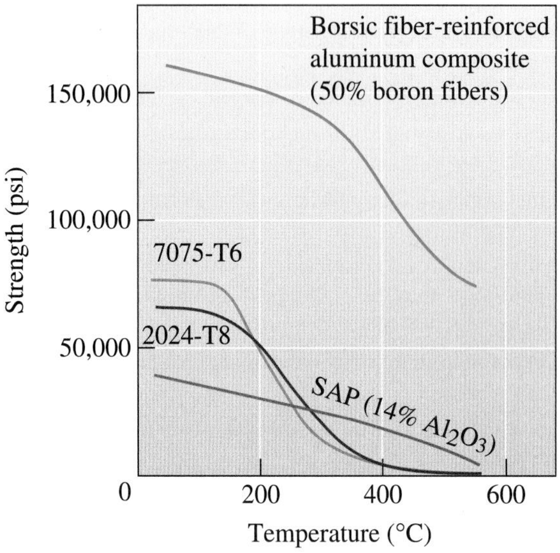

Figure 17-2 Comparison of the yield strength of dispersion-strengthened sintered aluminum powder (SAP) composite with that of two conventional two-phase high-strength aluminum alloys. (Repeated for Problem 17-7.)

17-8 A tungsten matrix with 20% porosity is infiltrated with silver. Assuming that the pores are interconnected, what is the density of the composite before and after infiltration with silver? The density of pure tungsten is 19.25 g/cm^3 and that of pure silver is 10.49 g/cm^3.

17-9 Nickel containing 2 wt% thorium is produced in powder form, consolidated into a part, and sintered in the presence of oxygen, causing all of the thorium to produce ThO_2 spheres 80 nm in diameter. Calculate the number of spheres per cm^3. The density of ThO_2 is 9.69 g/cm^3.

17-10 Spherical aluminum powder (SAP) 0.002 mm in diameter is treated to create a thin oxide layer and is then used to produce an SAP dispersion-strengthened material containing 10 vol% Al_2O_3. Calculate the average thickness of the oxide film prior to compaction and sintering of the powders into the part.

17-11 Yttria (Y_2O_3) particles 750 Å in diameter are introduced into tungsten by internal oxidation. Measurements using an electron microscope show that there are 5×10^{14} oxide particles per cm^3. Calculate the wt% Y originally in the alloy. The density of Y_2O_3 is 5.01 g/cm^3.

17-12 With no special treatment, aluminum is typically found to have an Al_2O_3 layer that is 3 nm thick. If spherical aluminum powder prepared with a total diameter of 0.01 mm is used to produce SAP dispersion-strengthened aluminum, calculate the volume percent Al_2O_3 in the material and the number of oxide particles per cm^3. Assume that the oxide breaks into disk-shaped flakes 3 nm thick and 3×10^{-4} mm in diameter. Compare the number of oxide particles per cm^3 with the number of solid solution atoms per cm^3 when 3 at% of an alloying element is added to aluminum.

Section 17-2 Particulate Composites

17-13 What is a particulate composite?

17-14 What is a cermet? What is the role of WC and Co in a cermet?

17-15 If the volume fractions of parts in a composite are all identical ($f_1 = f_2 = \cdots = f_n$), but the density of each part varies, what does the rule of mixtures become algebraically?

17-16 Calculate the density of a cemented carbide, or cermet, based on a titanium matrix if the composite contains 50 wt% WC, 22 wt% TaC, and 14 wt% TiC. (See Example 17-2 for densities of the carbides.)

17-17 Spherical silica particles (100 nm in diameter) are added to vulcanized rubber in tires to improve stiffness. If the density of the vulcanized rubber matrix is 1.1 g/cm^3, the density of silica is 2.5 g/cm^3, and the tire has a porosity of 4.5%, calculate the number of silica particles lost when a tire wears down 0.4 cm in thickness. The density of the tire is 1.2 g/cm^3; the overall tire diameter is 63 cm; and it is 10 cm wide.

17-18 A typical grinding wheel is 9 in. in diameter, 1 in. thick, and weighs 6 lb. The wheel contains SiC (density of 3.2 g/cm^3) bonded by silica glass (density of 2.5 g/cm^3); 5 vol% of the wheel is porous. The SiC is in the form of 0.04 cm cubes. Calculate

(a) the volume fraction of SiC particles in the wheel and

(b) the number of SiC particles lost from the wheel after it is worn to a diameter of 8 in.

17-19 An electrical contact material is produced by infiltrating copper into a porous tungsten carbide (WC) compact. The density of the final composite is 12.3 g/cm^3. Assuming that all of the pores are filled with copper, calculate
(a) the volume fraction of copper in the composite;
(b) the volume fraction of pores in the WC compact prior to infiltration; and
(c) the original density of the WC compact before infiltration.

17-20 An electrical contact material is produced by first making a porous tungsten compact that weighs 125 g. Liquid silver is introduced into the compact; careful measurement indicates that 105 g of silver is infiltrated. The final density of the composite is 13.8 g/cm^3. Calculate the volume fraction of the original compact that is interconnected porosity and the volume fraction that is closed porosity (no silver infiltration).

17-21 How much clay must be added to 10 kg of polyethylene to produce a low-cost composite having a modulus of elasticity greater than 120,000 psi and a tensile strength greater than 2000 psi? The density of the clay is 2.4 g/cm^3 and that of the polyethylene is 0.92 g/cm^3. (See Figure 17-6.)

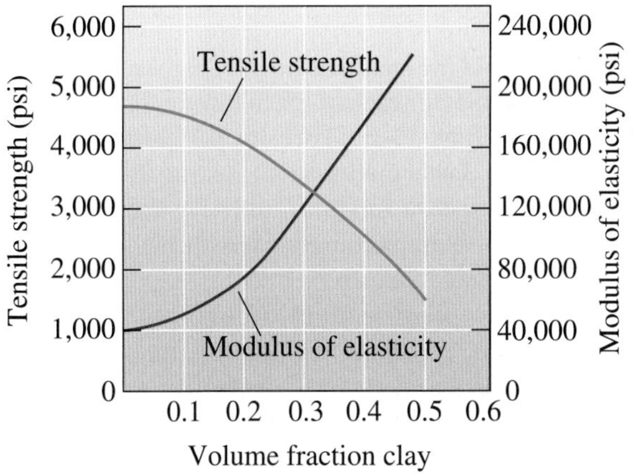

Figure 17-6 The effect of clay on the properties of polyethylene. (Repeated for Problem 17-21.)

17-22 We would like to produce a lightweight epoxy part to provide thermal insulation. We have available hollow glass beads for which the outside diameter is 1/16 in. and the wall thickness is 0.001 in. Determine the weight and number of beads that must be added to the epoxy to produce a one-pound composite with a density of 0.65 g/cm^3. The density of the glass is 2.5 g/cm^3 and that of the epoxy is 1.25 g/cm^3.

Section 17-3 Fiber-Reinforced Composites

Section 17-4 Characteristics of Fiber-Reinforced Composites

17-23 What is a fiber-reinforced composite?

17-24 What fiber-reinforcing materials are commonly used?

17-25 In a fiber-reinforced composite, what is the role of the matrix?

17-26 What do the terms CFRP and GFRP mean?

17-27 Explain briefly how the volume of fiber, fiber orientation, and fiber strength and modulus affect the properties of fiber-reinforced composites.

17-28 Determine the elastic modulus and tensile strength parallel to the fibers of Borsic-reinforced aluminum for 25 volume percent fibers. (See Figure 17-9.)

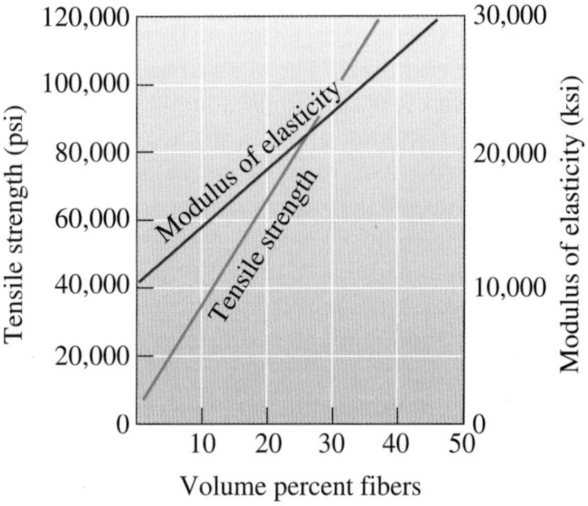

Figure 17-9 The influence of volume percent boron-coated SiC (Borsic) fibers on the properties of Borsic-reinforced aluminum parallel to the fibers. (Repeated for Problem 17-28.)

17-29 Five kilograms of continuous boron fibers are introduced in a unidirectional orientation into 8 kg of an aluminum matrix. Calculate
(a) the density of the composite;
(b) the modulus of elasticity parallel to the fibers; and
(c) the modulus of elasticity perpendicular to the fibers.

17-30 We want to produce 10 pounds of a continuous unidirectional fiber-reinforced composite of HS carbon in a polyimide matrix that has a modulus of elasticity of at least

25×10^6 psi parallel to the fibers. How many pounds of fibers are required? See Chapter 16 for properties of polyimide.

17-31 Carbon nanotubes (CNTs) with low weight (density = 1.3 g/cm^3), high tensile strength (50 GPa), and high modulus of elasticity (1 TPa) in the axial direction have been touted as the strongest material yet discovered. Calculate the specific strength for CNTs. How does this value compare to that of graphite whiskers given in Table 17-2? If a composite was made using an alumina (Al_2O_3) matrix with 1% volume CNT fibers, what fraction of the load would the CNT fibers carry? In practice, ceramic matrix CNT composites do not exhibit the expected improvement in mechanical properties. What could be some possible reasons for this?

17-32 An epoxy matrix is reinforced with 40 vol% E-glass fibers to produce a 2-cm-diameter composite that is to withstand a load of 25,000 N that is applied parallel to the fiber length. Calculate the stress acting on each fiber. The elastic modulus of the epoxy is 0.4×10^6 psi.

17-33 A titanium alloy with a modulus of elasticity of 16×10^6 psi is used to make a 1000-lb part for a manned space vehicle. Determine the weight of a part having the same modulus of elasticity parallel to the fibers, if the part is made from

(a) aluminum reinforced with boron fibers and

(b) polyester (with a modulus of 650,000 psi) reinforced with high-modulus carbon fibers.

(c) Compare the specific modulus for all three materials.

17-34 Short, but aligned, Al_2O_3 fibers with a diameter of 20 μm are introduced into a 6,6-nylon matrix. The strength of the bond between the fibers and the matrix is estimated to be 1000 psi. Calculate the critical fiber length and compare with the case when 1 μm alumina whiskers are used instead of the coarser fibers. What is the minimum aspect ratio in each case?

17-35 We prepare several epoxy-matrix composites using different lengths of 3-μm-diameter ZrO_2 fibers and find that the strength of the composite increases with increasing fiber length up to 5 mm. For longer fibers, the strength is virtually unchanged. Estimate the strength of the bond between the fibers and the matrix.

17-36 Glass fibers with a diameter 50 μm are introduced into a 6,6 nylon matrix. Determine the critical fiber length at which the fibers behave as if they are continuous. The strength of the bond between the fibers and matrix is 10 MPa. The relevant materials properties are as follows:

$E_{glass} = 72$ GPa, $E_{nylon} = 2.75$ GPa,

$TS_{glass} = 3.4$ GPa, and $TS_{nylon} = 827$ MPa.

17-37 A copper-silver bimetallic wire, 1 cm in diameter, is prepared by co-extrusion with copper as the core and silver as the outer layer. The desired properties along the axis parallel to the length of the bimetallic wire are as follows:

(a) Thermal conductivity > 410 W/(m · K);

(b) Electrical conductivity > 60×10^6 Ω^{-1}m^{-1}; and

(c) Weight < 750 g/m.

Determine the allowed range of the diameter of the copper core.

	Copper	Silver
Density (g/cm^3)	8.96	10.49
Electrical conductivity (Ω^{-1}m^{-1})	59×10^6	63×10^6
Thermal conductivity [W/(m · K)]	401	429

17-38 Find the materials with the highest numerical values for each of the following six properties in Table 17-2: (a) density; (b) tensile strength; (c) elastic modulus; (d) melting temperature; (e) specific modulus; and (f) specific strength. For example, tungsten has the highest density at 19.25 g/cm^3.

17-39 If a fiber-reinforced composite contains two separate regions of unidirectional fibers and is tested as shown, what would be the formula for the tensile strength?

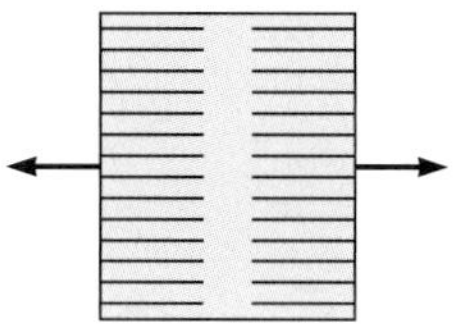

Figure 17-33 The orientation of fibers in a matrix for Problem 17-39.

17-40 Plot seven equally spaced points from the curve in Figure 17-11 and find a non-polynomial equation to fit the line that has an R^2 value of at least 0.900.

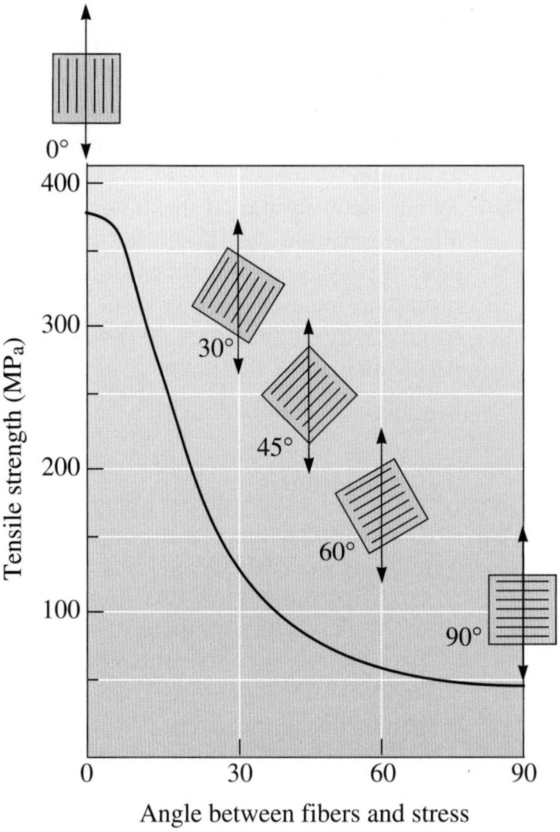

Figure 17-11 Effect of fiber orientation on the tensile strength of E-glass fiber-reinforced epoxy composites. (Repeated for Problem 17-40.)

17-41 What is a coupling agent? What is "sizing" as it relates to the production of glass fibers?

Section 17-5 Manufacturing Fibers and Composites

17-42 Explain briefly how boron and carbon fibers are made.

17-43 Explain briefly how continuous glass fibers are made.

17-44 What is the difference between a fiber and a whisker?

17-45 In one polymer-matrix composite, as produced, discontinuous glass fibers are introduced directly into the matrix; in a second case, the fibers are first "sized." Discuss the effect this difference might have on the critical fiber length and the strength of the composite.

17-46 Explain why bonding between carbon fibers and an epoxy matrix should be excellent, whereas bonding between silicon nitride fibers and a silicon carbide matrix should be poor.

17-47 A polyimide matrix with an elastic modulus of 0.3×10^6 psi is to be reinforced with 70 vol% carbon fibers to give a minimum modulus of elasticity of 40×10^6 psi. Recommend a process for producing the carbon fibers required. Estimate the tensile strength of the fibers that are produced.

17-48 Determine the ratio of the cross-sectional area of fiber to filament in both images in Figure 17-18.

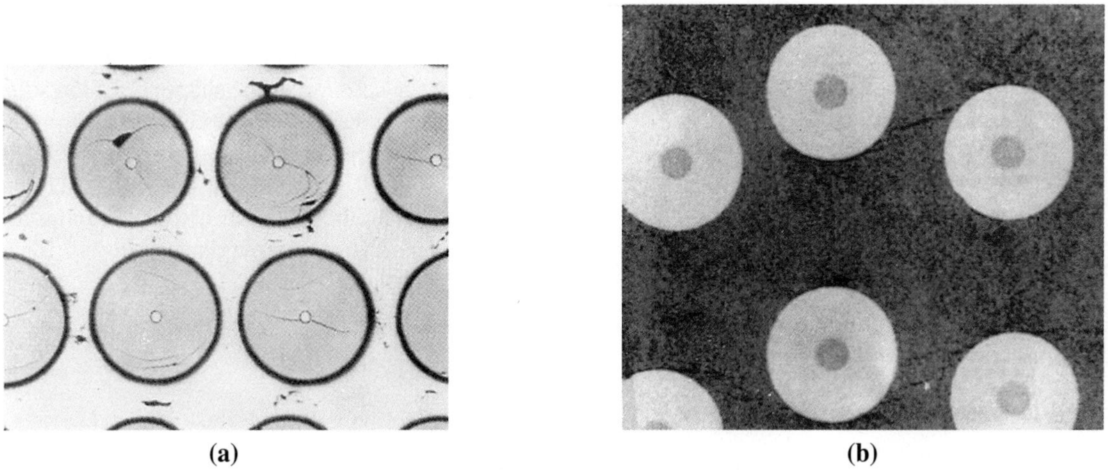

Figure 17-18 Micrographs of two fiber-reinforced composites: (a) In Borsic fiber-reinforced aluminum, the fibers are composed of a thick layer of boron deposited on a small-diameter tungsten filament (× 1000). [*From* ASM Handbook, *Vol. 9, Metallography and Microstructure (1985), ASM International, Materials Park, OH 44073-0002. Reprinted with permission of ASM International. All rights reserved. www.asminternational.org.*] (b) In this microstructure of a ceramic-fiber-ceramic-matrix composite, silicon carbide fibers are used to reinforce a silicon nitride matrix. The SiC fiber is vapor-deposited on a small carbon precursor-filament (× 125). (*Courtesy of Dr. R.T. Bhatt, NASA Lewis Research Center.*) (Repeated for Problem 17-48.)

Section 17-6 Fiber-Reinforced Systems and Applications

17-49 Explain briefly in what sporting equipment composite materials are used. What is the main reason why composites are used in these applications?

17-50 What are the advantages of using ceramic-matrix composites?

17-51 Calculate the steepest and the shallowest apparent (i.e., linear) slopes of the lines in Figure 17-26 and identify the materials to which they belong.

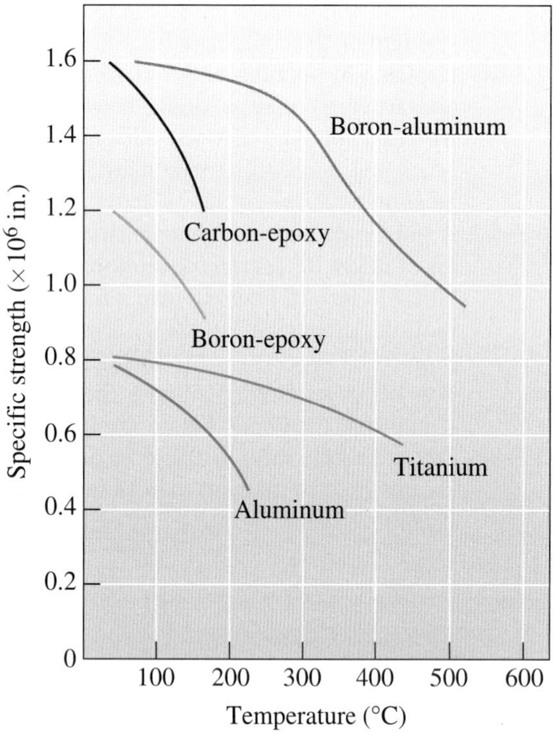

Figure 17-26 Specific strength versus temperature for several composites and metals. (Repeated for Problem 17-51.)

Section 17-7 Laminar Composite Materials

Section 17-8 Examples and Applications of Laminar Composites

Section 17-9 Sandwich Structures

17-52 What is a laminar composite?

17-53 A microlaminate, Arall, is produced using five sheets of 0.4-mm-thick aluminum and four sheets of 0.2-mm-thick epoxy reinforced with unidirectionally aligned Kevlar™ fibers. The volume fraction of Kevlar™ fibers in these intermediate sheets is 55%. The elastic modulus of the epoxy is 0.5×10^6 psi. Calculate the modulus of elasticity of the microlaminate parallel and perpendicular to the unidirectionally aligned Kevlar™ fibers. What are the principle advantages of the Arall material compared with those of unreinforced aluminum?

17-54 A laminate composed of 0.1-mm-thick aluminum sandwiched around a 2-cm-thick layer of polystyrene foam is produced as an insulation material. Calculate the thermal conductivity of the laminate parallel and perpendicular to the layers. The thermal conductivity of aluminum is $\dfrac{0.057 \text{ cal}}{\text{cm} \cdot \text{s} \cdot \text{K}}$ and that of the foam is $\dfrac{7.7 \times 10^{-5} \text{ cal}}{\text{cm} \cdot \text{s} \cdot \text{K}}$.

17-55 A 0.01-cm-thick sheet of a polymer with a modulus of elasticity of 0.7×10^6 psi is sandwiched between two 4-mm-thick sheets of glass with a modulus of elasticity of 12×10^6 psi. Calculate the modulus of elasticity of the composite parallel and perpendicular to the sheets.

17-56 A U.S. quarter is $\frac{15}{16}$ in. in diameter and is about $\frac{1}{16}$ in. thick. Assuming copper costs about \$1.10 per pound and nickel costs about \$4.10 per pound, compare the material cost in a composite quarter versus a quarter made entirely from nickel.

17-57 Calculate the density of a honeycomb structure composed of the following elements: The two 2-mm-thick cover sheets are produced using an epoxy matrix prepreg containing 55 vol% E-glass fibers. The aluminum honeycomb is 2 cm thick; the cells are in the shape of 0.5 cm squares and the walls of the cells are 0.1 mm thick. The density of the epoxy is 1.25 g/cm^3. Compare the weight of a 1 m × 2 m panel of the honeycomb with a solid aluminum panel of the same dimensions.

Design Problems

17-58 Design the materials and processing required to produce a discontinuous, but aligned, fiber-reinforced fiberglass composite that will form the hood of a sports car. The composite should provide a density of less than 1.6 g/cm^3 and a strength of 20,000 psi. Be sure to list all of the assumptions you make in creating your design.

17-59 Design an electrical-contact material and a method for producing the material that will result in a density of no more than 6 g/cm^3, yet at least 50 vol% of the material will be conductive.

17-60 What factors will have to be considered in designing a bicycle frame using an aluminum frame and a frame made using C-C composite?

Computer Problems

17-61 *Properties of a Laminar Composite*. Write a computer program (or use spreadsheet software) to calculate the properties of a laminar composite. For example, if the user provides the value of the thermal conductivity of each phase and the corresponding volume fraction, the program should provide the value of the effective thermal conductivity. Properties parallel and perpendicular to the lamellae should be calculated.

Knovel® Problems

K17-1 Using the rule of mixtures, calculate the density of polypropylene containing 30 vol% talc filler. Compare the calculated density with the typical density for this composite and explain the discrepancy, if any. Assume polypropylene has 40–50% crystallinity.

Construction materials, such as steels, concrete, wood, and glasses, along with composites, such as fiberglass, play a major role in the development and maintenance of a nation's infrastructure. Although many of our current construction materials are well developed and the technologies for producing them are mature, several new directions can be seen in construction materials. The use of smart sensors and actuators that can help control structures or monitor the health of structures is one such area. The use of composites and polymeric adhesives has also increased considerably. Processes such as welding and galvanizing as well as phenomena such as corrosion play important roles in the reliability and durability of structures. (*Martin Puddy/Stone/Getty Images*)

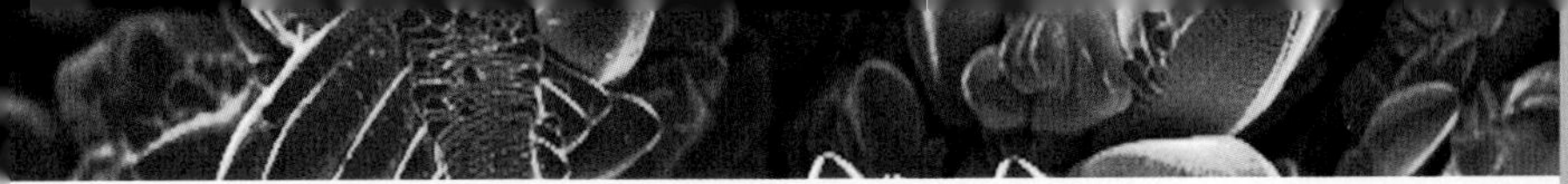

CHAPTER 18

Construction Materials

Have You Ever Wondered?

- *What are the most widely used manufactured construction materials?*
- *What is the difference between cement and concrete?*
- *What construction material is a composite made by nature?*
- *What is reinforced concrete?*

Chapter Learning Objectives

The key objectives of this chapter are to

- Describe the constituents and structure of wood.
- Calculate the density of wood as a function of water content.
- Calculate the expansion and contraction of wood as a function of water content.
- Design the concrete mix necessary to achieve desired properties.
- Distinguish between mortar, cement, concrete, and asphalt.

A number of important materials are used in the construction of buildings, highways, bridges, and much of our country's infrastructure. In this chapter, we examine three of the most important of these materials: wood, concrete, and asphalt. The field of construction materials is indeed very important for engineers, especially civil engineers and highway engineers. The properties and processing of steels used for making reinforced concrete, ceramics (e.g., sand, lime, concrete), plastics (e.g., epoxies and polystyrene foams), glasses, and composites (e.g., fiberglass) play a critical role in the development and use of construction materials. Another area in civil engineering that is becoming increasingly important is related to the use of sensors and actuators in buildings and bridges. Advanced materials developed for microelectronic and optical applications play an important role in this area.

For example, smart bridges and buildings that make use of optical fiber sensors currently are being developed. These sensors can monitor the health of the structures on a continuous basis and thus can provide early warnings of any potential problems. Similarly, researchers in areas of smart structures are also working on many other ideas using sensors that can detect things such as the formation of ice. If ice is detected, the system can start to spray salt water to prevent or delay freezing. Sensors such as this are also installed on steep driveways in some commercial parking garages where activation of the snow/ice sensors initiates heating of that part of the driveway. Similarly, we now have many smart coatings on glasses that can deflect heat and make buildings energy efficient.

There are new coatings that have resulted in self-cleaning glasses. New technologies are also being implemented to develop "green buildings."

There are many other areas of materials that relate to structures. For example, the corrosion of bridges and the limitations it poses on the bridge's life expectancy is a major cost for any nation. Strategies using galvanized steels and the proper paints to protect against corrosion are crucial aspects of bridge design. Similarly, many material joining techniques, such as welding, play a very important role in the construction of bridges and buildings. Long-term environmental impacts of the materials used must be considered (e.g., what are the best materials to use for water pipes, insulation, fire retardancy, etc.?). The goal of this chapter is to present a summary of the properties of wood, concrete, and asphalt.

18-1 The Structure of Wood

Wood, a naturally occuring composite, is one of our most familiar materials. Although it is not a "high-tech" material, we are literally surrounded by it in our homes and value it for its beauty and durability. In addition, wood is a strong, lightweight material that still dominates much of the construction industry.

We can consider wood to be a complex fiber-reinforced composite composed of long, unidirectionally aligned, tubular polymer cells in a polymer matrix. Furthermore, the polymer tubes are composed of bundles of partially crystalline, cellulose fibers aligned at various angles to the axes of the tubes. This arrangement provides excellent tensile properties in the longitudinal direction.

Wood consists of four main constituents. **Cellulose** fibers make up about 40% to 50% of wood. Cellulose is a naturally occurring thermoplastic polymer with a degree of polymerization of about 10,000. The structure of cellulose is shown in Figure 18-1. About 25% to 35% of a tree is **hemicellulose**, a polymer having a degree of polymerization of about 200. Another 20% to 30% of a tree is **lignin**, a low molecular weight, organic cement that bonds the various constituents of the wood. Finally, **extractives** are organic impurities such as oils, which provide color to the wood or act as preservatives against the environment and insects, and inorganic minerals such as silica, which dull saw blades during the cutting of wood. As much as 10% of the wood may be extractives.

There are three important levels in the structure of wood: the fiber structure, the cell structure, and the macrostructure (Figure 18-2).

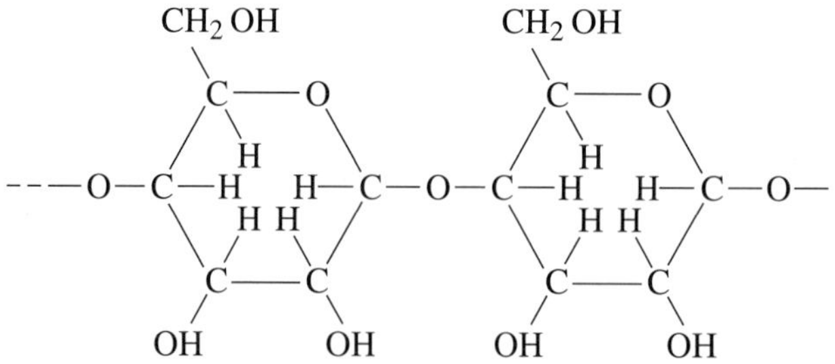

Figure 18-1 The structure of the cellulose filaments in wood.

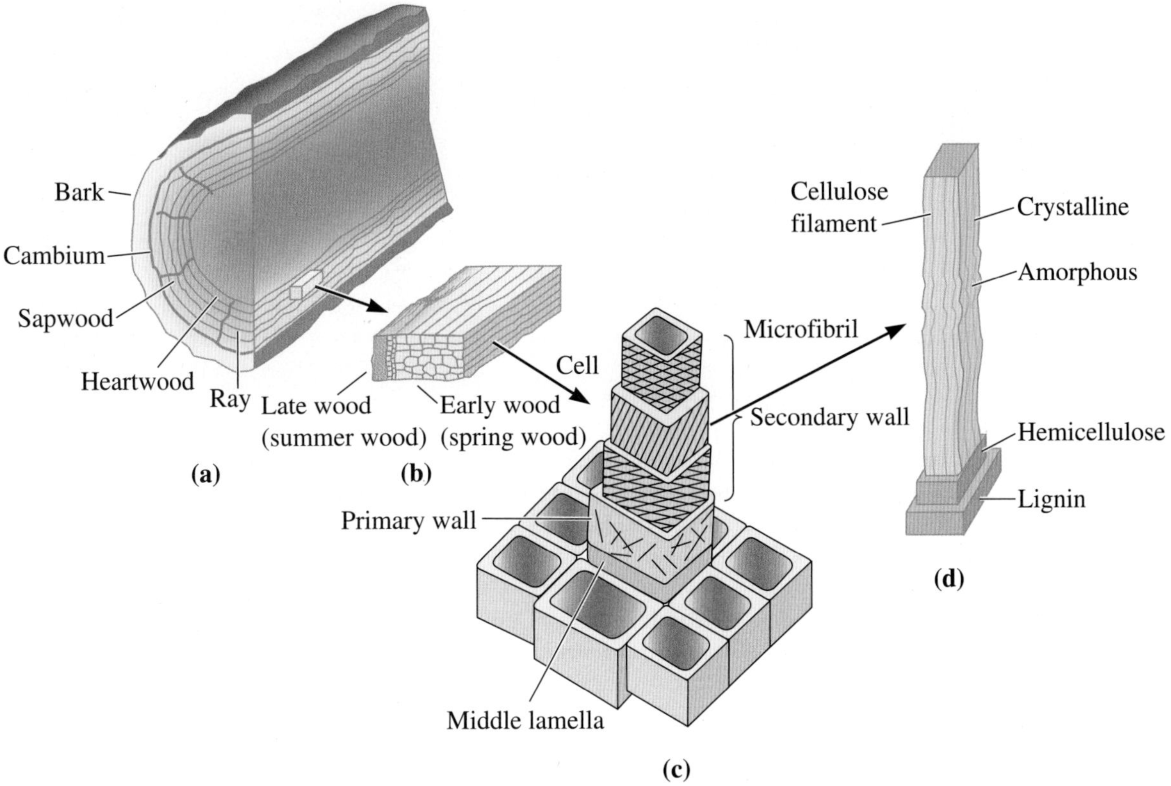

Figure 18-2 The structure of wood: (a) the macrostructure, including a layer structure outlined by the annual growth rings, (b) detail of the cell structure within one annual growth ring, (c) the structure of a cell, including several layers composed of microfibrils of cellulose fibers, hemicellulose fibers, and lignin, and (d) the microfibril's aligned, partly crystalline cellulose chains.

Fiber Structure The basic component of wood is cellulose, $C_6H_{10}O_5$, arranged in polymer chains that form long fibers. Much of the fiber length is crystalline, with the crystalline regions separated by small lengths of amorphous cellulose. A bundle of cellulose chains are encased in a layer of randomly oriented, amorphous hemicellulose chains. Finally, the hemicellulose is covered with lignin [Figure 18-2(d)]. The entire bundle, consisting of cellulose chains, hemicellulose chains, and lignin, is called a **microfibril**; it can have a virtually infinite length.

Cell Structure The tree is composed of elongated cells, often having an aspect ratio of 100 or more, that constitute about 95% of the solid material in wood. The hollow cells are composed of several layers built up from the microfibrils [Figure 18-2(c)]. The first, or primary, wall of the cell contains randomly oriented microfibrils. As the cell walls thicken, three more distinct layers are formed. The outer and inner walls contain microfibrils oriented in two directions that are not parallel to the cell. The middle wall, which is the thickest, contains microfibrils that are unidirectionally aligned, usually at an angle not parallel to the axis of the cell.

Macrostructure A tree is composed of several layers [Figure 18-2(a)]. The outer layer, or *bark*, protects the tree. The **cambium**, just beneath the bark, contains new growing cells. The **sapwood** contains a few hollow living cells that store nutrients and serve as the conduit for water. Finally, the **heartwood**, which contains only dead cells, provides most of the mechanical support for the tree.

The tree grows when new elongated cells develop in the cambium. Early in the growing season, the cells are large; later they have a smaller diameter, thicker

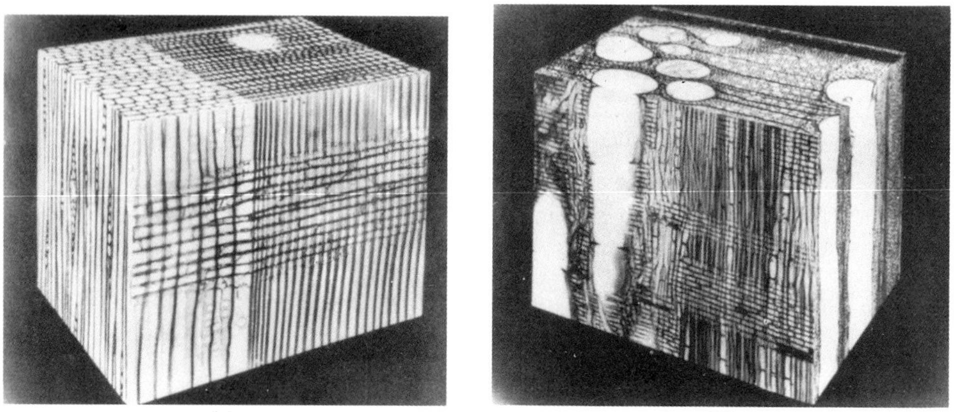

(a) (b)

Figure 18-3 The cellular structure in (a) softwood (*From J.M. Dinwoodie,* Wood: Nature's Cellular Polymeric Fiber-Composite, *The Institute of Metals. By permission from Maney Publishing.*) and (b) hardwood. Softwoods contain larger, longer cells than hardwoods. The hardwoods, however, contain large-diameter vessels. Water is transported through softwoods by the cells and through hardwoods by the vessels. (*From J.M. Dinwoodie,* Wood: Nature's Cellular Polymeric Fiber-Composite, *The Institute of Metals. By permission from Maney Publishing.*)

walls, and a higher density. This difference between the early (or *spring*) wood and the late (or *summer*) wood permits us to observe annual growth rings [Figure 18-2(b)]. In addition, some cells grow in a radial direction; these cells, called *rays*, provide the storage and transport of food.

Hardwood Versus Softwood The hardwoods are deciduous trees such as oak, ash, hickory, elm, beech, birch, walnut, and maple. In these trees, the elongated cells are relatively short, with a diameter of less than 0.1 mm and a length of less than 1 mm. Contained within the wood are longitudinal pores, or vessels, which carry water through the tree (Figure 18-3).

The softwoods are the conifers, evergreens such as pine, spruce, hemlock, fir, spruce, and cedar, and have similar structures. In softwoods, the cells tend to be somewhat longer than in the hardwoods. The hollow center of the cells is responsible for transporting water. In general, the density of softwoods tends to be lower than that of hardwoods because of a greater percentage of void space.

18-2 Moisture Content and Density of Wood

The material making up the individual cells in virtually all woods has essentially the same density—about 1.45 g/cm^3; however, wood contains void space that causes the actual density to be much lower. The density of wood depends primarily on the species of the tree (or the amount of void space particular to that species) and the percentage of water in the wood (which depends on the amount of drying and on the relative humidity to which the wood is exposed during use). Completely dry wood varies in density from about 0.3 to 0.8 g/cm^3, with hardwoods having higher densities than softwoods. The measured density is normally higher due to the water contained in the wood. The percentage water is given by

$$\% \text{ Water} = \frac{\text{weight of water}}{\text{weight of dry wood}} \times 100 \tag{18-1}$$

Table 18-1 Properties of typical woods

Wood	Density (for 12% Water) (g/cm³)	Modulus of Elasticity (Mpsi)
Cedar	0.32	1.1
Pine	0.35	1.2
Fir	0.48	2.0
Maple	0.48	1.5
Birch	0.62	2.0
Oak	0.68	1.8

On the basis of this definition, it is possible to describe a wood as containing more than 100% water. The water is contained both in the hollow cells or vessels, where it is not tightly held, and in the cellulose structure in the cell walls, where it is more tightly bonded to the cellulose fibers. While a large amount of water is stored in a live tree, the amount of water in the wood after the tree is harvested depends, eventually, on the humidity to which the wood is exposed; higher humidity increases the amount of water held in the cell walls. The density of a wood is usually given at a moisture content of 12%, which corresponds to 65% humidity. The density and modulus of elasticity parallel to the grain of several common woods are included in Table 18-1 for this typical water content.

The following example illustrates the calculation for the density of the wood.

Example 18-1 *Density of Dry and Wet Wood*

A green wood has a density of 0.86 g/cm^3 and contains 175% water. Calculate the density of the wood after it has completely dried.

SOLUTION

A 100-cm^3 sample of the wood would have a mass of 86 g. From Equation 18-1, we can calculate the weight (or in this case, mass) of the dry wood to be

$$\% \text{ Water} = \frac{\text{weight of water}}{\text{weight of dry wood}} \times 100 = 175$$

$$= \frac{\text{green weight} - \text{dry weight}}{\text{dry weight}} \times 100 = 175$$

Solving for the dry weight of the wood:

$$\text{Dry weight of wood} = \frac{(100)(\text{green weight})}{275}$$

$$= \frac{(100)(86)}{275} = 31.3\,\text{g}$$

$$\text{Density of dry wood} = \frac{31.3\,\text{g}}{100\,\text{cm}^3} = 0.313\,\text{g}/\text{cm}^3$$

18-3 Mechanical Properties of Wood

The strength of a wood depends on its density, which in turn depends on both the water content and the type of wood. As a wood dries, water is eliminated first from the vessels and later from the cell walls. As water is removed from the vessels, practically no change in the strength or stiffness of the wood is observed (Figure 18-4). On continued drying to less than about 30% water, there is water loss from the actual cellulose fibers. This loss permits the individual fibers to come closer together, increasing the bonding between the fibers and the density of the wood and, thereby, increasing the strength and stiffness of the wood.

The type of wood also affects the density. Because they contain less of the higher-density late wood, softwoods typically are less dense and therefore have lower strengths than hardwoods. In addition, the cells in softwoods are larger, longer, and more open than those in hardwoods, leading to lower density.

The mechanical properties of wood are highly anisotropic. In the longitudinal direction (Figure 18-5), an applied tensile load acts parallel to the microfibrils and cellulose chains in the middle section of the secondary wall. These chains are strong—because they are mostly crystalline—and are able to carry a relatively high load. In the radial and tangential directions, however, the weaker bonds between the microfibrils and cellulose fibers may break, resulting in very low tensile properties. Similar behavior is observed in compression and bending loads. Because of the anisotropic behavior, most lumber is cut in a tangential-longitudinal or radial-longitudinal manner. These cuts maximize the longitudinal behavior of the wood.

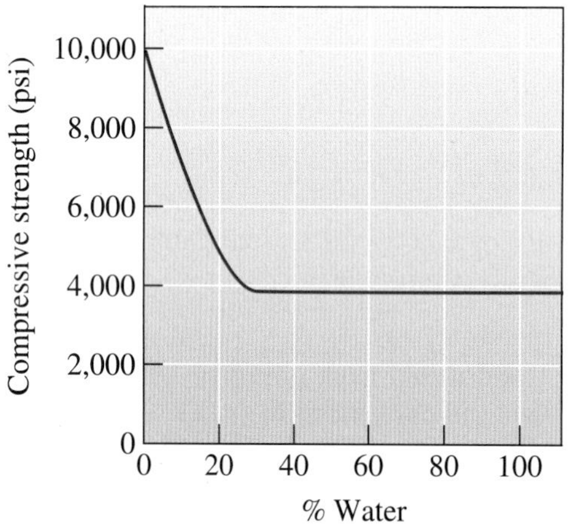

Figure 18-4 The effect of the percentage of water in a typical wood on the compressive strength parallel to the grain.

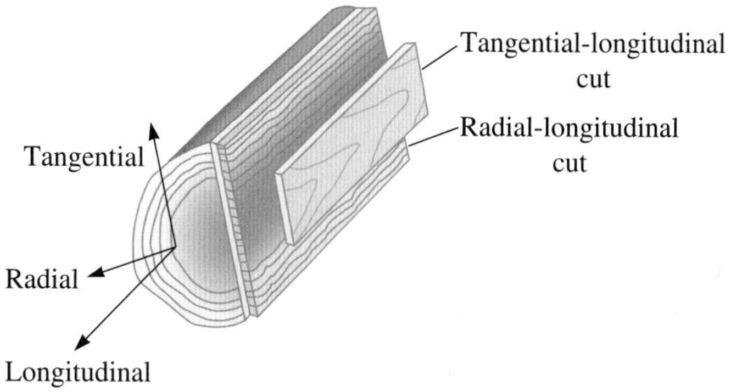

Figure 18-5 The different directions in a log. Because of differences in cell orientation and the grain, wood displays anisotropic behavior.

Wood has poor properties in compression and bending (which produces a combination of compressive and tensile forces). In compression, the fibers in the cells tend to buckle, causing the wood to deform and break at low stresses. Unfortunately, most applications for wood place the component in compression or bending and therefore do not take full advantage of the engineering properties of the material. Similarly, the modulus of elasticity is highly anisotropic; the modulus perpendicular to the grain is about 1/20th that given in Table 18-1 parallel to the grain. Table 18-2 compares the tensile and compressive strengths parallel and perpendicular to the cells for several woods.

Clear wood, free of imperfections such as knots, has a specific strength and specific modulus that compare well with those of other common construction materials (Table 18-3). Wood also has good toughness, largely due to the slight misorientation of the cellulose fibers in the middle layer of the secondary wall. Under load, the fibers straighten, permitting some ductility and energy absorption. Clear wood may have a longitudinal tensile strength of 10,000 to 20,000 psi. Less expensive construction lumber, which usually contains many imperfections, may have a tensile strength below 5000 psi. The knots also disrupt the grain of the wood in the vicinity of the knot, causing the cells to be aligned perpendicular to the tensile load.

Table 18-2 Anisotropic behavior of several woods (at 12% moisture)

	Tensile Strength Longitudinal (psi)	Tensile Strength Radial (psi)	Compressive Strength Longitudinal (psi)	Compressive Strength Radial (psi)
Beech	12,500	1,010	7,300	1,010
Elm	17,500	660	5,520	690
Maple	15,700	1,100	7,830	1,470
Oak	11,300	940	6,200	810
Cedar	6,600	320	6,020	920
Fir	11,300	390	5,460	610
Pine	10,600	310	4,800	440
Spruce	8,600	370	5,610	580

Table 18-3 Comparison of the specific strength and specific modulus of wood with those of other common construction materials

Material	Specific Strength ($\times 10^5$in.)	Specific Modulus ($\times 10^7$in.)
Clear wood	7.0	9.5
Aluminum	5.0	10.5
1020 steel	2.0	10.5
Copper	1.5	5.5
Concrete	0.6	3.5

Based on F. F. Wangaard, "Wood: Its Structure and Properties," J. Educ. Models for Mat. Sci. and Engr., *Vol. 3, No. 3, 1979.*

18-4 Expansion and Contraction of Wood

Like other materials, wood changes dimensions when heated or cooled. Dimensional changes in the longitudinal direction are very small in comparison with those in metals, polymers, and ceramics, but the dimensional changes in the radial and tangential directions are greater than those for most other materials.

In addition to dimensional changes caused by temperature fluctuations, the moisture content of the wood causes significant changes in dimension. Again, the greatest changes occur in the radial and tangential directions, where the moisture content affects the spacing between the cellulose chains in the microfibrils. The change in dimensions Δx in wood in the radial and tangential directions is approximated by

$$\Delta x = x_0[c(M_f - M_i)] \tag{18-2}$$

where x_0 is the initial dimension, M_i is the initial water content, M_f is the final water content, and c is a coefficient that describes the dimensional change and can be measured in either the radial or the tangential direction. Table 18-4 includes the dimensional coefficients for several woods. In the longitudinal direction, no more than 0.1 to 0.2% change is observed.

During the initial drying of wood, the large dimensional changes perpendicular to the cells may cause warping and even cracking. In addition, when the wood is used, its water content may change, depending on the relative humidity in the environment. As the wood gains or loses water during use, shrinkage or swelling continues to occur. If a wood structure does not allow movement caused by moisture fluctuations, warping and cracking can occur—a particularly severe condition in large expanses of wood, such as the floor of a large room. Excessive expansion may cause large bulges in the floor; excessive shrinkage may cause large gaps between individual planks of the flooring.

18-5 Plywood

The anisotropic behavior of wood can be reduced and wood products can be made in larger sizes by producing plywood. Thin layers of wood, called **plies,** are cut from logs—normally, softwoods. The plies are stacked together with the grains between adjacent plies oriented at 90° angles; usually an odd number of plies is used. Ensuring that these angles are as precise as possible is important to ensure that the plywood does not warp or twist when the moisture content in the material changes. The individual plies are generally bonded to one another using a thermosetting phenolic resin. The resin is introduced between the plies, which are then pressed together while hot to cause the resin to polymerize.

Similar wood products are also produced as "laminar" composite materials. The facing (visible) plies may be of a more expensive hardwood with the center plies of a less expensive softwood. Wood particles can be compacted into sheets and laminated between two wood plies, producing a particle board. Wood plies can be used as the facings for honeycomb materials.

Table 18-4 Dimensional coefficient c [in./(in. · % H_2O)] for several woods

Wood	Radial	Tangential
Beech	0.00190	0.00431
Elm	0.00144	0.00338
Maple	0.00165	0.00353
Oak	0.00183	0.00462
Cedar	0.00111	0.00234
Fir	0.00155	0.00278
Pine	0.00141	0.00259
Spruce	0.00148	0.00263

18-6 Concrete Materials

An **aggregate** is a combination of gravel, sand, crushed stones, or slag. Slag is a byproduct from the refining of ore. A **mortar** is made by mixing cement, water, air, and fine aggregate. Concrete contains all of the ingredients of the mortar and coarse aggregates. **Cements** are inorganic materials that set and harden after being mixed into a paste using water. **Concrete** is a particulate composite in which both the particulate and the matrix are ceramic materials. In concrete, sand and a coarse aggregate are bonded in a matrix of **Portland cement**. A cementation reaction between water and the minerals in the cement provides a strong matrix that holds the aggregate in place and provides good compressive strength to the concrete.

Cements

Cements are classified as hydraulic and nonhydraulic. **Hydraulic cements** set and harden under water. **Nonhydraulic cements** (e.g., lime, CaO) cannot harden under water and require air for hardening. Portland cement is the most widely used and manufactured construction material. It was patented by Joseph Aspdin in 1824 and is named as such after the limestone cliffs on the Isle of Portland in England.

Hydraulic cement is made from calcium silicates with an approximate composition of CaO (~60 to 65%), SiO_2 (~20 to 25%), and iron oxide and alumina (~7 to 12%). The cement binder, which is very fine in size, is composed of various ratios of $3CaO \cdot Al_2O_3$, $2CaO \cdot SiO_2$, $3CaO \cdot SiO_2$, $4CaO \cdot Al_2O_3 \cdot Fe_2O_3$, and other minerals. In the cement terminology, CaO, SiO_2, Al_2O_3, and Fe_2O_3 are often indicated as *C*, *S*, *A*, and *F*, respectively. Thus, C_3S means $3CaO\text{-}SiO_2$. When water is added to the cement, a hydration reaction occurs, producing a solid gel that bonds the aggregate particles. Possible reactions include

$$3CaO \cdot Al_2O_3 + 6H_2O \rightarrow Ca_3Al_2(OH)_{12} + \text{heat}$$

$$3CaO + SiO_2 + (x + 1)H_2O \rightarrow Ca_2SiO_4 \cdot xH_2O + Ca(OH)_2 + \text{heat}$$

After hydration, the cement provides the bond for the aggregate particles. Consequently, enough cement must be added to coat all of the aggregate particles. The cement typically constitutes on the order of 15 vol% of the solids in the concrete.

The composition of the cement helps determine the rate of curing and the final properties of the concrete, as shown in Figure 18-6. Nearly complete curing of the concrete is normally expected within 28 days (Figure 18-7), although some additional curing may continue for years.

There are about ten general types of cements used. Some are shown in Table 18-5. In large structures such as dams, curing should be slow in order to avoid excessive heating caused by the hydration reaction. These cements typically contain low percentages of $3CaO \cdot SiO_2$, such as in Types II and IV. Some construction jobs, however, require that

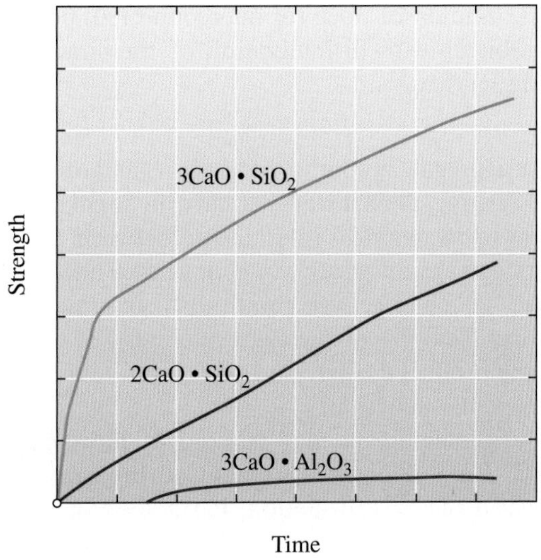

Figure 18-6 The rate of hydration of the minerals in Portland cement. (*Based on Lea,* Chemistry of Cement and Concrete, *Elsevier, 2003.*)

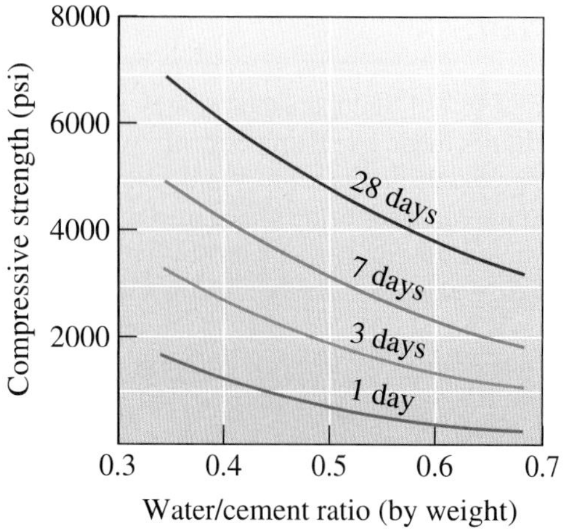

Figure 18-7 The compressive strength of concrete increases with time. After 28 days, the concrete approaches its maximum strength.

Table 18-5 Types of Portland cements

	Approximate Composition				
	3C · S	2C · S	3C · A	4C · A · F	Characteristics
Type I	55	20	12	9	General purpose
Type II	45	30	7	12	Low rate of heat generation, moderate resistance to sulfates
Type III	65	10	12	8	Rapid setting
Type IV	25	50	5	13	Very low rate of heat generation
Type V	40	35	3	14	Good sulfate resistance

concrete forms be removed and reused as quickly as possible; cements for these purposes may contain large amounts of $3CaO \cdot SiO_2$, as in Type III.

The composition of the cement also affects the resistance of the concrete to the environment. For example, sulfates in the soil may attack the concrete; using higher proportions of $4CaO \cdot Al_2O_3 \cdot Fe_2O_3$ and $2CaO \cdot SiO_2$ helps produce concretes more resistant to sulfates, as in Type V.

Sand Chemically, sand is predominantly silica (SiO_2). Sands are composed of fine mineral particles, typically of the order of 0.1 to 1.0 mm in diameter. They often contain at least some adsorbed water, which should be taken into account when preparing a concrete mix. The sand helps fill voids between the coarser aggregate, giving a high packing factor, reducing the amount of open (or interconnected) porosity in the finished concrete, and reducing problems with disintegration of the concrete due to repeated freezing and thawing during service.

Aggregate Coarse aggregate is composed of gravel and rock. Aggregate must be clean, strong, and durable. Aggregate particles that have an angular rather than a round shape provide strength due to mechanical interlocking between particles, but angular particles also provide more surface on which voids or cracks may form. It is normally preferred that the aggregate size be large; this condition also minimizes the surface area at which cracks or voids form. The size of the aggregate must, of course, be matched to the size of the structure being produced; aggregate particles should not be any larger than about 20% of the thickness of the structure.

In some cases, special aggregates may be used. Lightweight concretes can be produced by using mineral slags, which are produced during steel-making operations; these concretes have improved thermal insulation. Particularly heavy concretes can be

Table 18-6 Characteristics of concrete materials

Material	True Density	
Cement	190 lb/ft^3	1 sack = 94 lb
Sand	160 lb/ft^3	
Aggregate	170 lb/ft^3	Normal
	80 lb/ft^3	Lightweight slag
	30 lb/ft^3	Lightweight vermiculite
	280 lb/ft^3	Heavy Fe_3O_4
	390 lb/ft^3	Heavy ferrophosphorus
Water	62.4 lb/ft^3	7.48 gal/ft^3

produced using dense minerals or even metal shot; these heavy concretes can be used in building nuclear reactors to better absorb radiation. The densities of several aggregates are included in Table 18-6.

18-7 Properties of Concrete

Many factors influence the properties of concrete. Some of the most important are the water-cement ratio, the amount of air entrainment, and the type of aggregate.

Water-Cement Ratio The ratio of water to cement affects the behavior of concrete in several ways:

1. A minimum amount of water must be added to the cement to ensure that all of it undergoes the hydration reaction. Too little water causes low strength. Normally, however, other factors such as workability place the lower limit on the water–cement ratio.
2. A high water–cement ratio improves the **workability** of concrete—that is, how easily the concrete slurry can fill all of the space in the form. Air pockets or interconnected porosity caused by poor workability reduce the strength and durability of the concrete structure. Workability can be measured by the *slump test*. For example, a wet concrete shape 12 in. tall is produced (Figure 18-8) and is permitted to stand under its own weight. After some period of time, the shape deforms. The reduction in height of the form is the **slump**. A minimum water–cement ratio of about 0.4 (by weight) is usually required for workability. A larger slump, caused by a higher water–cement ratio, indicates greater workability. Slumps of 1 to 6 in. are typical; high slumps are needed for pouring narrow or complex forms, while low slumps may be satisfactory for large structures such as dams.
3. Increasing the water–cement ratio beyond the minimum required for workability decreases the compressive strength of the concrete. This strength is usually measured by determining the stress required to crush a concrete cylinder 6 in. in diameter and 12 in. tall. Figure 18-9 shows the effect of the water–cement ratio on concrete's strength.
4. High water–cement ratios increase the shrinkage of concrete during curing, creating a danger of cracking.

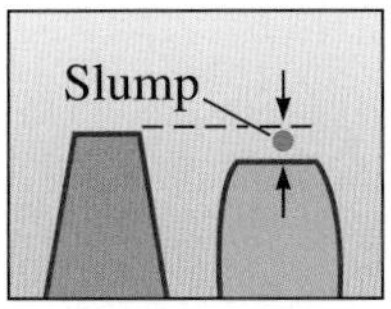

Figure 18-8 The slump test, in which deformation of a concrete shape under its own weight is measured, is used to describe the workability of concrete mix.

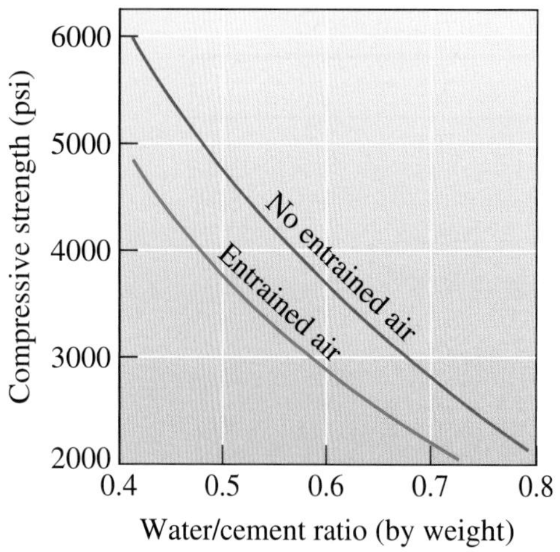

Figure 18-9 The effect of the water–cement ratio and entrained air on the 28-day compressive strength of concrete.

Because of the different effects of the water–cement ratio, a compromise between strength, workability, and shrinkage may be necessary. A weight ratio of 0.45 to 0.55 is typical. To maintain good workability, organic plasticizers may be added to the mix with little effect on strength.

Air-Entrained Concrete Almost always, a small amount of air is entrained into the concrete during pouring. For coarse aggregate, such as 1.5 in. rock, 1% by volume of the concrete may be air. For finer aggregate, such as 0.5 in. gravel, 2.5% air may be trapped.

We sometimes intentionally entrain air into concrete—sometimes as much as 8% for fine gravel. The entrained air improves workability of the concrete and helps minimize problems with shrinkage and freeze–thaw conditions. Air-entrained concrete has a lower strength, however. (See Figure 18-9.)

Type and Amount of Aggregate The size of the aggregate affects the concrete mix. Figure 18-10 shows the amount of water per cubic yard of concrete required to produce the desired slump, or workability; more water is required for smaller aggregates. Figure 18-11 shows the amount of aggregate that should be present in the concrete mix. The volume ratio of aggregate in the concrete is based on the bulk density of the aggregate, which is about 60% of the true density shown in Table 18-6.

The examples that follow show how to calculate the contents for a concrete mixture.

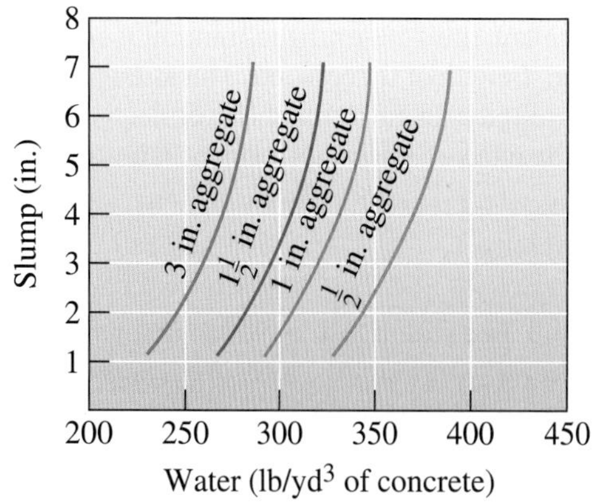

Figure 18-10 The amount of water per cubic yard of concrete required to give the desired workability (or slump) depends on the size of the coarse aggregate.

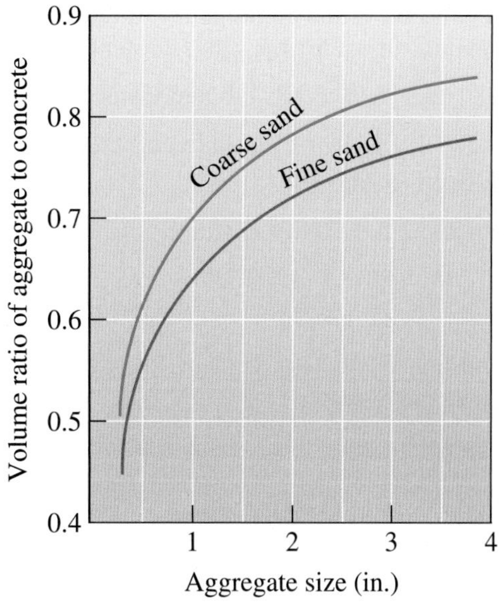

Figure 18-11 The volume ratio of aggregate to concrete depends on the sand and aggregate sizes. Note that the volume ratio uses the bulk density of the aggregate—about 60% of the true density.

Example 18-2 *Composition of Concrete*

Determine the amounts of water, cement, sand, and aggregate in 5 cubic yards of concrete, assuming that we want to obtain a water/cement ratio of 0.4 (by weight) and that the cement/sand/aggregate ratio is 1:2.5:4 (by weight). A "normal" aggregate will be used, containing 1% water, and the sand contains 4% water. Assume that no air is entrained into the concrete.

SOLUTION

One method by which we can calculate the concrete mix is to first determine the volume of each constituent based on one sack (94 lb) of cement. We should remember that after the concrete is poured, there are no void spaces between the various constituents; therefore, we need to consider the true density—not the bulk density—of the constituents in our calculations.

For each sack of cement we use, the volume of materials required is

$$\text{Cement} = \frac{94\,\text{lb/sack}}{190\,\text{lb/ft}^3} = 0.495\ \text{ft}^3$$

$$\text{Sand} = \frac{2.5 \times 94\ \text{lb cement}}{160\,\text{lb/ft}^3} = 1.469\,\text{ft}^3$$

$$\text{Gravel} = \frac{4 \times 94\,\text{lb cement}}{170\,\text{lb/ft}^3} = 2.212\ \text{ft}^3$$

$$\text{Water} = \frac{0.4 \times 94\ \text{lb cement}}{62.4\,\text{lb/ft}^3} = 0.603\,\text{ft}^3$$

$$\text{Total volume of concrete} = 4.78\ \text{ft}^3/\text{sack of cement}$$

Therefore, in 5 cubic yards (or 135 ft^3), we need

$$\text{Cement} = \frac{135\,\text{ft}^3}{4.78\ \text{ft}^3/\text{sack}} = 28\,\text{sacks}$$

$$\text{Sand} = (28\ \text{sacks})(94\ \text{lb/sack})(2.5\ \text{sand/cement}) = 6580\,\text{lb}$$

$$\text{Gravel} = (28\ \text{sacks})(94\ \text{lb/sack})(4\ \text{gravel/cement}) = 10{,}528\,\text{lb}$$

$$\text{Water} = (28\ \text{sacks})(94\ \text{lb/sack})(0.4\ \text{water/cement}) = 1053\,\text{lb}$$

The sand contains 4% water and the gravel contains 1% water. To obtain the weight of the *wet* sand and gravel, we must adjust for the water content of each:

$$\text{Sand} = (6580 \text{ dry})/0.96 = 6854 \text{ lb and water} = 274 \text{ lb}$$

$$\text{Gravel} = (10{,}528 \text{ dry})/0.99 = 10{,}634 \text{ lb and water} = 106 \text{ lb}$$

Therefore, we actually only need to add:

$$\text{Water} = 1053 \text{ lb} - 274 \text{ lb} - 106 \text{ lb} = 673 \text{ lb}$$

$$= \frac{(673 \text{ lb})(7.48 \text{ gal/ft}^3)}{62.4 \text{ lb/ft}^3} = 81 \text{ gal}$$

Accordingly, we recommend that 28 sacks of cement, 6854 lb of sand, and 10,634 lb of gravel be combined with 81 gal of water.

Example 18-3 *Design of a Concrete Mix for a Retaining Wall*

Design a concrete mix that will provide a 28-day compressive strength of 4000 psi in a concrete intended for producing a 5-in.-thick retaining wall 6 ft high. We expect to have about 2% air entrained in the concrete, although we will not intentionally entrain air. The aggregate contains 1% moisture, and we have only coarse sand containing 5% moisture available.

SOLUTION

Some workability of the concrete is needed to ensure that the form will fill properly with the concrete. A slump of 3 in. might be appropriate for such an application. The wall thickness is 5 in. To help minimize cost, we would use a large aggregate. A 1-in. diameter aggregate size would be appropriate (about 1/5 of the wall thickness).

To obtain the desired workability of the concrete using 1 in. aggregate, we should use about 320 lb of water per cubic yard (Figure 18-10).

To obtain the 4000 psi compressive strength after 28 days (assuming no intentionally entrained air), we need a water–cement weight ratio of 0.57 (Figure 18-9).

Consequently, the weight of cement required per cubic yard of concrete is (320 lb water/0.57 water–cement) = 561 lb cement.

Because our aggregate size is 1 in. and we have only coarse sand available, the volume ratio of the aggregate to the concrete is 0.7 (Figure 18-11). Thus, the amount of aggregate required per yard of concrete is 0.7 yd^3; however, this amount is in terms of the bulk density of the aggregate. Because the bulk density is about 60% of the true density, the actual volume occupied by the aggregate in the concrete is $0.7 \text{ yd}^3 \times 0.6 = 0.42 \text{ yd}^3$.

Let's determine the volume of each constituent per cubic yard (27 ft^3) of concrete in order to calculate the amount of sand required:

$$\text{Water} = 320 \text{ lb}/(62.4 \text{ lb/ft}^3) = 5.13 \text{ ft}^3$$

$$\text{Cement} = 561 \text{ lb}/(190 \text{ lb/ft}^3) = 2.95 \text{ ft}^3$$

$$\text{Aggregate} = 0.42 \text{ yd}^3 \times 27 \text{ ft}^3/\text{yd}^3 = 11.34 \text{ ft}^3$$

$$\text{Air} = 0.02 \times 27 \text{ ft}^3 = 0.54 \text{ ft}^3$$

$$\text{Sand} = 27 - 5.13 - 2.95 - 11.34 - 0.54 = 7.04 \text{ ft}^3$$

Or converting to other units, assuming that the aggregate and sand are dry:

$$\text{Water} = 5.13\,\text{ft}^3 \times 7.48\,\text{gal/ft}^3 = 38.4\,\text{gal}$$

$$\text{Cement} = 561\,\text{lb}/(94\,\text{lb/sack}) = 6\,\text{sacks}$$

$$\text{Aggregate} = 11.34\,\text{ft}^3 \times 170\,\text{lb/ft}^3 = 1928\,\text{lb}$$

$$\text{Sand} = 7.04\,\text{ft}^3 \times 160\,\text{lb/ft}^3 = 1126\,\text{lb}$$

The aggregate and the sand are wet. Thus, the actual amounts of aggregate and sand needed are

$$\text{Aggregate} = 1928/0.99 = 1947\,\text{lb}\,(19\,\text{lb water})$$

$$\text{Sand} = 1126/0.95 = 1185\,\text{lb}\,(59\,\text{lb water})$$

The actual amount of water needed is

$$\text{Water} = 38.4\,\text{gal} - \frac{(19 + 59\,\text{lb})(7.48\,\text{gal}/\text{ft}^3)}{62.4\,\text{lb}/\text{ft}^3} = 29.1\,\text{gal}$$

Thus, for each cubic yard of concrete, we will combine six sacks of cement, 1947 lb of aggregate, 1185 lb of sand, and 29.1 gal of water. This should give us a slump of 3 in. (the desired workability) and a compressive strength of 4000 psi after 28 days.

18-8 Reinforced and Prestressed Concrete

Concrete, like other ceramic-based materials, develops good compressive strength. Due to the porosity and interfaces present in the brittle structure, however, it has very poor tensile properties. Several methods are used to improve the load-bearing capability of concrete in tension.

Reinforced Concrete Steel rods (known as rebar), wires, or mesh are frequently introduced into concrete to provide improvement in resisting tensile and bending forces. The tensile stresses are transferred from the concrete to the steel, which has good tensile properties. Polymer fibers, which are less likely to corrode, also can be used as reinforcement. Under flexural stresses, the steel supports the part that is in tension; the part that is under compression is supported by the concrete.

Prestressed Concrete Instead of simply being laid as reinforcing rods in a form, the steel initially can be pulled in tension between an anchor and a jack, thus remaining under tension during the pouring and curing of the concrete. After the concrete sets, the tension on the steel is released. The steel then tries to relax from its stretched condition, but the restraint caused by the surrounding concrete places the concrete in compression. Now higher tensile and bending stresses can be applied to the concrete because of the compressive residual stresses introduced by the pretensioned steel. In order to permit the external tension to be removed in a timely manner, the early-setting Type III cements are often used for these applications.

Poststressed Concrete An alternate method of placing concrete under compression is to place hollow tubes in the concrete before pouring. After the concrete cures, steel rods or cables running through the tubes then can be pulled in tension, acting against the concrete. As the rods are placed in tension, the concrete is placed in compression. The rods or cables then are secured permanently in their stretched condition.

18-9 Asphalt

Asphalt is a composite of aggregate and **bitumen**, which is a thermoplastic polymer most frequently obtained from petroleum. Asphalt is an important material for paving roads. The properties of the asphalt are determined by the characteristics of the aggregate and binder, their relative amounts, and additives.

The aggregate, as in concrete, should be clean and angular and should have a distribution of grain sizes to provide a high packing factor and good mechanical interlocking between the aggregate grains (Figure 18-12). The binder, composed of thermoplastic chains, bonds the aggregate particles. The binder has a relatively narrow useful temperature range, being brittle at sub-zero temperatures and beginning to melt at relatively low temperatures. Additives such as gasoline or kerosene can be used to modify the binder, permitting it to liquefy more easily during mixing and causing the asphalt to cure more rapidly after application.

The ratio of binder to aggregate is important. Just enough binder should be added so that the aggregate particles touch, but voids are minimized. Excess binder permits viscous deformation of the asphalt under load. Approximately 5 to 10% bitumen is present in a typical asphalt. Some void space is also required—usually, about 2 to 5%. When the asphalt is compressed, the binder can squeeze into voids, rather than be squeezed from the surface of the asphalt and lost. Too much void space, however, permits water to enter the structure; this increases the rate of deterioration of the asphalt and may embrittle the binder. Emulsions of asphalt are used for sealing driveways. The aggregate for asphalt is typically sand and fine gravel; however, there is some interest in using recycled glass products as the aggregate. **Glasphalt** provides a useful application for crushed glass. Similarly, there are also applications for materials developed using asphalt and shredded rubber tires.

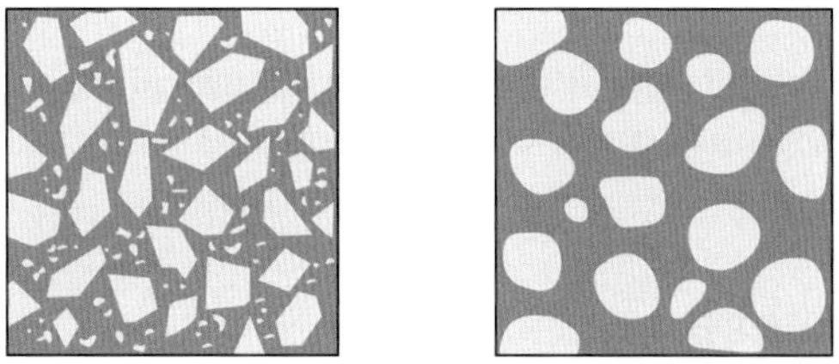

(a) (b)

Figure 18-12 The ideal structure of asphalt (a) compared with the undesirable structure (b) in which round grains, a narrow distribution of grains, and excess binder all reduce the strength of the final material.

Summary

- Construction materials play a vital role in the infrastructure of any nation. Concrete, wood, asphalt, glasses, composites, and steels are some of the most commonly encountered materials.
- Advanced materials used to make sensors and actuators also play an increasingly important role in monitoring the structural health of buildings and bridges. Coatings and thin films on glasses contribute to more energy-efficient buildings.
- Wood is a natural fiber-reinforced polymer composite material. Cellulose fibers constitute aligned cells that provide excellent reinforcement in longitudinal directions in wood, but give poor strength and stiffness in directions perpendicular to the cells and fibers. The properties of wood therefore are highly anisotropic and depend on the species of the tree and the amount of moisture present in the wood. Wood has good tensile strength but poor compressive strength.
- Concrete is a particulate composite. In concrete, ceramic particles such as sand and gravel are used as filler in a ceramic-cement matrix. The water–cement ratio is a particularly important factor governing the behavior of the concrete. This behavior can be modified by entraining air and by varying the composition of the cement and aggregate materials. Concrete has good compressive strength but poor tensile strength.
- Asphalt also is a particulate composite, using the same type of aggregates as in concrete, but with an organic, polymer binder.

Glossary

Aggregate A combination of gravel, sand, crushed stones, or slag.

Bitumen The organic binder, composed of low melting point polymers and oils, for asphalt.

Cambium The layer of growing cells in wood.

Cellulose A naturally occurring polymer fiber that is the major constituent of wood. Cellulose has a high degree of polymerization.

Cements Inorganic materials that set and harden after being mixed into a paste using water.

Concrete A composite material that consists of a binding medium in which particles of aggregate are dispersed.

Extractives Impurities in wood.

Glasphalt Asphalt in which the aggregate includes recycled glass.

Heartwood The center of a tree comprising dead cells, which provides mechanical support to a tree.

Hemicellulose A naturally occurring polymer fiber that is an important constituent of wood. It has a low degree of polymerization.

Hydraulic cement A cement that sets and hardens under water.

Lignin The polymer cement in wood that bonds the cellulose fibers in the wood cells.

Microfibril Bundles of cellulose and other polymer chains that serve as the fiber reinforcement in wood.

Mortar A mortar is made by mixing cement, water, air, and fine aggregate. Concrete contains all of the ingredients of the mortar and coarse aggregate.

Nonhydraulic Cement Cements that cannot harden under water and require air for hardening.

Plies The individual sheet of wood veneer from which plywood is constructed.

Portland cement A hydraulic cement made from calcium silicates; the approximate composition is CaO (~60–65%), SiO_2 (~20–25%), and iron oxide and alumina (~7–12%).

Sapwood Hollow, living cells in wood that store nutrients and conduct water.

Slump The decrease in height of a standard concrete form when the concrete settles under its own weight.

Workability The ease with which a concrete slurry fills all of the space in a form.

Problems

Section 18-1 The Structure of Wood

Section 18-2 Moisture Content and Density of Wood

Section 18-3 Mechanical Properties of Wood

Section 18-4 Expansion and Contraction of Wood

Section 18-5 Plywood

18-1 Table 18-1 lists the densities for typical woods. Calculate the densities of the woods after they are completely dried and at 100% water content.

18-2 Use Table 18-1 to determine if there is a relationship between density and the elastic modulus of the woods listed.

18-3 A sample of wood with the dimensions 3 in. × 4 in. × 12 in. has a dry density of 0.35 g/cm^3.
(a) Calculate the number of gallons of water that must be absorbed by the sample to contain 120% water.
(b) Calculate the density after the wood absorbs this amount of water.

18-4 The density of a sample of oak is 0.90 g/cm^3. Calculate
(a) the density of completely dry oak and
(b) the percent water in the original sample.

18-5 A green wood with a density of 0.82 g/cm^3 contains 150% water. The compressive strength of this wood is 27 MPa. After several days of drying, the compressive strength increases to 41 MPa. What is the water content and density of the dried wood? (See Figure 18-4.)

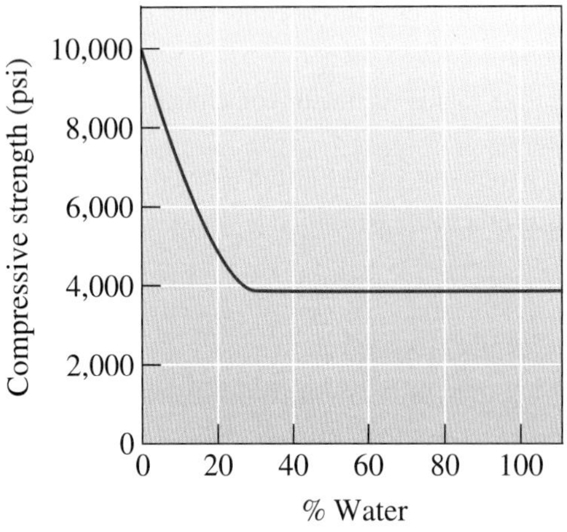

Figure 18-4 The effect of the percentage of water in a typical wood on the compressive strength parallel to the grain. (Repeated for Problem 18-5.)

18-6 Determine the expansion in the radial direction of cedar when it changes from 12% to 27% water content. The initial dimension is 1 ft.

18-7 Boards of oak 0.5 cm thick, 1 m long, and 0.25 m wide are used as flooring for a 10 m × 10 m area. If the floor was laid at a moisture content of 25% and the expected moisture could be as high as 45%, determine the dimensional change in the floor parallel to and perpendicular to the length of the boards. The boards were cut from logs with a tangential-longitudinal cut.

18-8 Boards of maple 1 in. thick, 6 in. wide, and 16 ft long are used as the flooring for a 60 ft × 60 ft hall. The boards were cut from logs with a tangential-longitudinal cut. The floor is laid when the boards have a moisture content of 12%. After some particularly humid days, the moisture content in the boards increases to 45%. Determine the dimensional change in the flooring parallel to the boards and perpendicular to the boards. What will happen to the floor?

18-9 A wall 30 feet long is built using radial-longitudinal cuts of 5-in. wide pine with the boards arranged in a vertical fashion. The wood contains a moisture content of 55% when the wall is built; however, the humidity level in the room is maintained to give 45% moisture in the wood. Determine the dimensional changes in the wood boards, and estimate the size of the gaps that will be produced as a consequence of these changes.

Section 18-6 Concrete Materials

Section 18-7 Properties of Concrete

Section 18-8 Reinforced and Prestressed Concrete

Section 18-9 Asphalt

18-10 What is the maximum water to cement ratio acceptable for a structure that must have a final compressive strength of at least 5 ksi? (See Figure 18-7.)

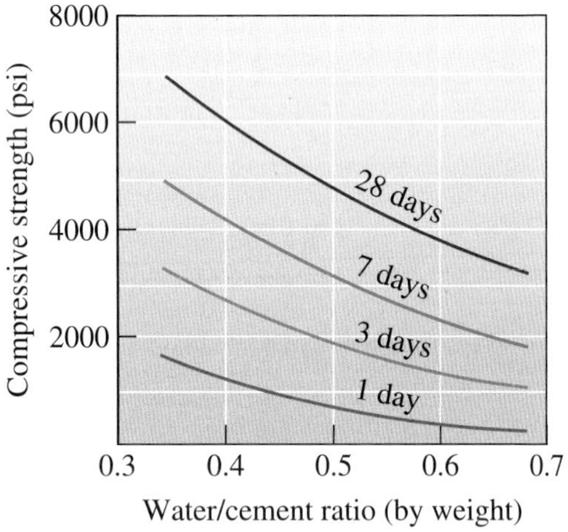

Figure 18-7 The compressive strength of concrete increases with time. After 28 days, the concrete approaches its maximum strength. (Repeated for Problem 18-10.)

18-11 Determine the amounts of water, cement, and sand in 10 m^3 of concrete if the cement–sand–aggregate ratio is 1:2.5:4.5 and the water–cement ratio is 0.4. Assume that no air is entrained into the concrete. The sand used for this mixture contains 4 wt% water, and the aggregate contains 2 wt% water.

18-12 Calculate the amount of cement, sand, aggregate, and water needed to create a concrete mix with a 28-day compressive strength of 34 MPa for a 10 m × 10 m × 0.25 m structure given the following conditions: allowed slump = 10 cm and only 3.8 cm (1.5 in.) aggregate with 2% moisture and coarse sand with 4% moisture are available for this project. Assume no air entrainment in your calculations.

18-13 We have been asked to prepare 100 yd^3 of normal concrete using a volume ratio of cement–sand–coarse aggregate of 1:2:4. The water–cement ratio (by weight) is to be 0.5. The sand contains 6 wt% water, and the coarse aggregate contains 3 wt% water. No entrained air is expected.

(a) Determine the number of sacks of cement that must be ordered, the tons of sand and aggregate required, and the amount of water needed.

(b) Calculate the total weight of the concrete per cubic yard.

(c) What is the weight ratio of cement–sand–coarse aggregate?

18-14 We plan to prepare 10 yd^3 of concrete using a 1:2.5:4.5 weight ratio of cement–sand–coarse aggregate. The water–cement ratio (by weight) is 0.45. The sand contains 3 wt% water; the coarse aggregate contains 2 wt% water; and 5% entrained air is expected. Determine the number of sacks of cement, tons of sand and coarse aggregate, and gallons of water required.

Design Problems

18-15 A wooden structure is functioning in an environment controlled at 65% humidity. Design a wood support column that is to hold a compressive load of 20,000 lb. The distance from the top to the bottom of the column should be 96 ± 0.25 in. when the load is applied.

18-16 Design a wood floor that will be 50 ft by 50 ft and will be in an environment in which humidity changes will cause a fluctuation of plus or minus 5% water in the wood. We want to minimize any buckling or gap formation in the floor.

18-17 We would like to produce a concrete that is suitable for use in building a large structure in a sulfate environment. For these situations, the maximum water-cement ratio should be 0.45 (by weight). The compressive strength of the concrete after 28 days should be at least 4000 psi. We have an available coarse aggregate containing 2% moisture in a variety of sizes, and both fine and coarse sand containing 4% moisture. Design a concrete that will be suitable for this application.

18-18 We would like to produce a concrete sculpture. The sculpture will be as thin as 3 in. in some areas and should be light in weight, but it must have a 28-day compressive strength of at least 2000 psi. Our available aggregate contains 1% moisture, and our sands contain 5% moisture. Design a concrete that will be suitable for this application.

18-19 The binder used in producing asphalt has a density of about 1.3 g/cm^3. Design an asphalt, including the weight and volumes of each constituent, that might be suitable for use as pavement. Assume that the sands and aggregates are the same as those for a normal concrete.

18-20 *Concrete Canoe Design.* Describe what novel materials can be used to make a concrete canoe.

Knovel® Problems

K18-1 What is the difference in lignin content between hardwood and softwood species? What is the chemical structure of lignin?

K18-2 The equilibrium moisture content of wood is the moisture content at which the wood is neither gaining nor losing moisture. What is the moisture content of wood at 70°F with a relative humidity of 50%?

K18-3 What measures are used to control cracking in reinforced concrete?

The first integrated circuit was fabricated in 1958 by Jack Kilby, who was then an engineer at Texas Instruments. It functioned as a phase-shift oscillator and consisted of a transistor, a resistor, and capacitor devices as shown in the upper image. (*Courtesy of Texas Instruments*). For the first time, multiple electronic devices were incorporated on a single substrate. For this innovation, Jack Kilby was awarded the Nobel Prize in Physics in 2000.

Today's microchips may contain hundreds of millions of transistors. Dynamic random access memory (DRAM) chips have passed the billion-transistor milestone. This integration of devices is made possible by sophisticated manufacturing equipment and techniques. The lower image shows the multiple layers of copper interconnects that are needed to make connections between transistors and other circuit elements. The insulating oxide between the wires has been removed. The transistors are not visible in this image. *(Courtesy of International Business Machines Corporation, © International Business Machines Corporation.)*

CHAPTER

19

Electronic Materials

Have You Ever Wondered?

- *Is diamond a good conductor of electricity?*
- *How many devices are there in a single microchip?*
- *How are thin films less than 1 micron thick deposited on a substrate?*
- *How does a microwave oven heat food?*

Chapter Learning Objectives

The key objectives of this chapter are to

- Calculate the electrical conductivity of a metal, semiconductor, or insulator.
- Calculate the number of free carriers in a solid.
- Explain how the Pauli exclusion principle gives rise to the band structure of solids.
- Draw schematic diagrams indicating typical band structures for metals, semiconductors, and insulators and identify and explain the critical features of these diagrams, particularly the band gap.
- Explain how the band structure of solids gives rise to the electrical properties of metals, semiconductors, and insulators.
- Explain how electrical conductivity changes with temperature for metals, semiconductors, and insulators.
- Define the term *hole* in the context of electronic materials.
- Explain the purpose of doping in semiconductors and how doping changes the electrical properties.
- Describe the processes used to manufacture integrated circuits.
- Describe the mechanisms of polarization in materials.

Silicon-based microelectronics are a ubiquitous part of modern life. With microchips in items from laundry machines and microwaves to cell phones, in MP3 players and from personal computers to the world's fastest super computers, silicon was the defining material of the later 20th century and will dominate computer-based and information-related technologies for the foreseeable future.

While silicon is the substrate or base material of choice for most devices, microelectronics include materials of nearly every class, including metals such as copper and gold, other semiconductors such as gallium arsenide, and insulators such as silicon dioxide. Even semiconducting polymers are finding applications in such devices as light-emitting diodes.

In this chapter, we will discuss the principles of electrical conductivity in metals, semiconductors, insulators, and ionic materials. We will see that a defining difference between metals and semiconductors is that as temperature increases, the resistivity of a metal increases, while the resistivity of a semiconductor decreases. This critical difference arises from the **band structures** of these materials. The band structure consists of the array of energy levels that are available to or forbidden for electrons to occupy and determines the electronic behavior of a solid, such as whether it is a conductor, semiconductor, or insulator.

The conductivity of a semiconductor that does not contain impurities generally increases exponentially with temperature. From a reliability standpoint, an exponential dependence of conductivity on temperature is undesirable for electronic devices that generate heat as they operate. Thus, semiconductors are doped (i.e., impurities are intentionally added) in order to control the conductivity of semiconductors with extreme precision. We will learn how **dopants** change the band structure of a semiconductor so that the electrical conductivity can be tailored for particular applications.

Metals, semiconductors, and insulators are all critical components of integrated circuits. Some features of integrated circuits are now approaching atomic-scale dimensions, and the fabrication of integrated circuits is arguably the most sophisticated manufacturing process in existence. It involves simultaneously fabricating hundreds of millions, and even billions, of devices on a single microchip and represents a fundamentally different manufacturing paradigm from that of any other process. We will learn about some of the steps involved in fabricating integrated circuits, including the process of depositing thin films (films on the order of 10 Å to 1 μm in thickness).

We will examine some of the properties of insulating materials. Insulators are used in microelectronic devices to electrically isolate active regions from one another. Insulators are also used in capacitors due to their dielectric properties. Finally, we will consider piezoelectric materials, which change their shape in response to an applied voltage or vice versa. Such materials are used as actuators in a variety of applications.

Superconductors comprise a special class of electronic materials. Superconductors are materials that exhibit zero electrical resistance under certain conditions (which usually includes a very low temperature on the order of 135 K or less) and that completely expel a magnetic field. A discussion of superconductivity is beyond the scope of this text.

19-1 Ohm's Law and Electrical Conductivity

Most of us are familiar with the common form of Ohm's law,

$$V = IR \tag{19-1}$$

where V is the voltage (volts, V), I is the current (amperes or amps, A), and R is the resistance (ohms, Ω) to the current flow. This law is applicable to most but not all materials. The resistance (R) of a resistor is a characteristic of the size, shape, and properties of the material according to

$$R = \rho \frac{l}{A} = \frac{l}{\sigma A} \tag{19-2}$$

where l is the length (cm) of the resistor, A is the cross-sectional area (cm^2) of the resistor, ρ is the electrical resistivity (ohm · cm or $\Omega \cdot$ cm), and σ, which is the reciprocal of ρ, is the electrical conductivity (ohm^{-1} · cm^{-1}). The magnitude of the resistance depends upon the dimensions of the resistor. The resistivity or conductivity does not depend on the dimensions of the material. Thus, resistivity (or conductivity) allows us to compare different materials. For example, silver is a better conductor than copper. Resistivity is a **microstructure-sensitive property**, similar to yield strength. The resistivity of pure copper is much less than that of commercially pure copper, because impurities in commercially pure copper scatter electrons and contribute to increased resistivity. Similarly, the resistivity of annealed, pure copper is slightly lower than that of cold-worked, pure copper because of the scattering effect associated with dislocations.

In components designed to conduct electrical energy, minimizing power losses is important, not only to conserve energy, but also to minimize heating. The electrical power P (in watts, W) lost when a current flows through a resistance is given by

$$P = VI = I^2R \tag{19-3}$$

A high resistance R results in larger power losses. These electrical losses are known as joule heating losses.

A second form of Ohm's law is obtained if we combine Equations 19-1 and 19-2 to give

$$\frac{I}{A} = \sigma \frac{V}{l}$$

If we define I/A as the **current density** J (A/cm^2) and V/l as the **electric field** E (V/cm), then

$$J = \sigma E \tag{19-4}$$

The current density J is also given by

$$J = nq\bar{v}$$

where n is the number of charge carriers (carriers/cm^3), q is the charge on each carrier (1.6×10^{-19} C), and $\bar{v}$ is the average **drift velocity** (cm/s) at which the charge carriers move [Figure 19-1(a)]. Thus,

$$\sigma E = nq\bar{v} \quad \text{or} \quad \sigma = nq\frac{\bar{v}}{E}$$

Diffusion occurs as a result of temperature and concentration gradients, and drift occurs as a result of an applied electric or magnetic field. Conduction may occur as a result of diffusion, drift, or both, but drift is the dominant mechanism in electrical conduction.

The term $\bar{v}/E$ is called the **mobility** $\mu\left(\frac{\text{cm}^2}{\text{V} \cdot \text{s}}\right)$ of the carriers (which in the case of metals is the mobility of electrons):

$$\mu = \frac{\bar{v}}{E}$$

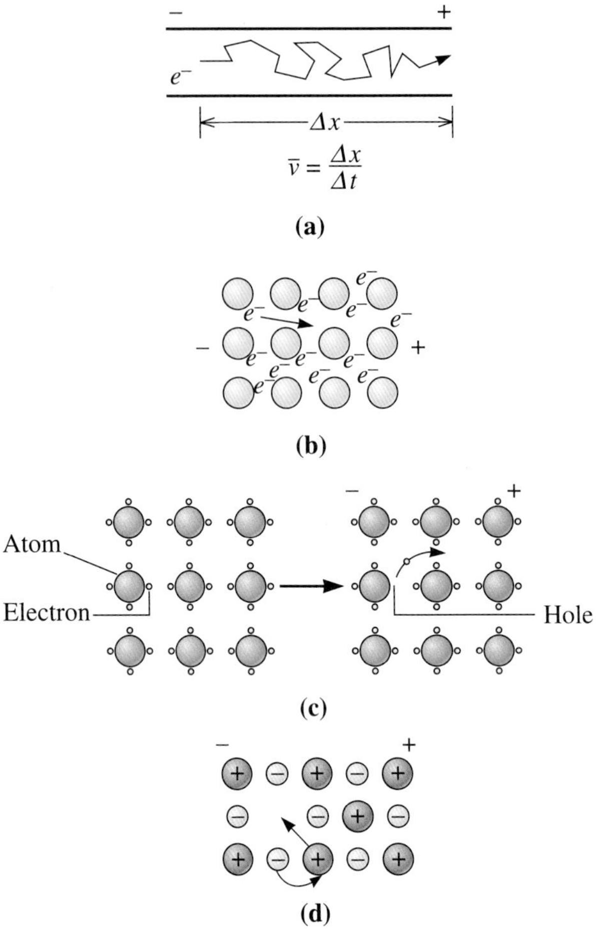

Figure 19-1 (a) Charge carriers, such as electrons, are deflected by atoms or defects and take an irregular path through a conductor. The average rate at which the carriers move is the drift velocity $\bar{v}$. (b) Valence electrons in metals move easily. (c) Covalent bonds must be broken in semiconductors and insulators that do not contain impurities for an electron to be able to move. (d) Entire ions may diffuse via a vacancy mechanism to carry charge in many ionically bonded materials.

Finally,

$$\sigma = nq\mu \tag{19-5a}$$

The charge q is a constant; from inspection of Equation 19-5a, we find that we can control the electrical conductivity of materials by (1) controlling the number of charge carriers in the material or (2) controlling the mobility—or ease of movement—of the charge carriers. The mobility is particularly important in metals, whereas the number of carriers is more important in semiconductors and insulators.

Electrons are the charge carriers in metals [Figure 19-1(b)]. Electrons are, of course, negatively charged. In semiconductors, electrons conduct charge as do positively charged carriers known as **holes** [Figure 19-1(c)]. We will learn more about holes in Section 19-4. In semiconductors, electrons and holes flow in opposite directions in response to an applied electric field, but in so doing, they both contribute to the net current. Thus, Equation 19-5a can be modified as follows for expressing the conductivity of semiconductors:

$$\sigma = nq\mu_n + pq\mu_p \tag{19-5b}$$

In this equation, μ_n and μ_p are the mobilities of electrons and holes, respectively. The terms n and p represent the concentrations of free electrons and holes in the semiconductor.

In ceramics, when conduction does occur, it can be the result of electrons that "hop" from one defect to another or the movement of ions [Figure 19-1(d)]. The mobility depends on atomic bonding, imperfections, microstructure, and, in ionic compounds, diffusion rates.

Because of these effects, the electrical conductivity of materials varies tremendously, as illustrated in Table 19-1. These values are approximate and are for high-purity materials at 300 K (unless noted otherwise). Note that the values of conductivity for metals and semiconductors depend very strongly on temperature. Table 19-2 includes some useful units and relationships.

Table 19-1 Electrical conductivity of selected materials at $T = 300$ K*

Material	Conductivity ($ohm^{-1} \cdot cm^{-1}$)	Material	Conductivity ($ohm^{-1} \cdot cm^{-1}$)
Superconductors		**Semiconductors**	
Hg, Nb_3Sn		Group 4B elements	
$YBa_2Cu_3O_{7-x}$	Infinite (under certain conditions	Si	4×10^{-6}
MgB_2	such as low temperatures)	Ge	0.02
Metals		**Compound semiconductors**	
Alkali metals		GaAs	2.5×10^{-9}
Na	2.13×10^5	AlAs	0.1
K	1.64×10^5	SiC	10^{-10}
Alkali earth metals		**Ionic Conductors**	
Mg	2.25×10^5	Indium tin oxide (ITO)	$\sim 10^4$
Ca	3.16×10^5	Yttria-stabilized zirconia (YSZ)	$<10^{14}$
Group 3B metals		**Insulators, Linear, and Nonlinear Dielectrics**	
Al	3.77×10^5	Polymers	
Ga	0.66×10^5	Polyethylene	10^{-15}
Transition metals		Polytetrafluoroethylene	10^{-18}
Fe	1.00×10^5	Polystyrene	10^{-17} to 10^{-19}
Ni	1.46×10^5	Epoxy	10^{-12} to 10^{-17}
Group 1B metals		Ceramics	
Cu	5.98×10^5	Alumina (Al_2O_3)	10^{-14}
Ag	6.80×10^5	Silicate glasses	10^{-17}
Au	4.26×10^5	Boron nitride (BN)	10^{-13}
		Barium titanate ($BaTiO_3$)	10^{-14}
		C (diamond)	$<10^{-18}$

**Unless specified otherwise, assumes high-purity material.*

Table 19-2 Some useful relationships, constants, and units

Electron volt = 1 eV = 1.6×10^{-19} joule = 1.6×10^{-12} erg
1 amp = 1 coulomb/second
1 volt = 1 amp · ohm
$k_B T$ at room temperature (300 K) = 0.0259 eV
c = speed of light 2.998×10^8 m/s
ε_0 = permittivity of free space = 8.85×10^{-12} F/m
q = charge on electron = 1.6×10^{-19} C
Avogadro constant $N_A = 6.022 \times 10^{23}$
k_B = Boltzmann constant = 8.617×10^{-5} eV/K = 1.38×10^{-23} J/K
h = Planck's constant 6.63×10^{-34} J·s = 4.14×10^{-15} eV·s

Electronic materials can be classified as (a) superconductors, (b) conductors, (c) semiconductors, and (d) dielectrics or insulators, depending upon the magnitude of their electrical conductivity. Materials with conductivity less than 10^{-12} $\Omega^{-1} \cdot cm^{-1}$, or resistivity greater than 10^{12} $\Omega \cdot cm$, are considered insulating or dielectric. Materials with conductivity less than 10^3 $\Omega^{-1} \cdot cm^{-1}$ but greater than 10^{-12} $\Omega^{-1} \cdot cm^{-1}$ are considered semiconductors. Materials with conductivity greater than 10^3 $\Omega^{-1} \cdot cm^{-1}$, or resistivity less than 10^{-3} $\Omega^{-1} \cdot cm^{-1}$, are considered conductors. (These are approximate ranges of values.)

We use the term "dielectric" for materials that are used in applications where the dielectric constant is important. The **dielectric constant (k)** of a material is a microstructure-sensitive property related to the material's ability to store an electrical charge. We use the term "insulator" to describe the ability of a material to stop the flow of DC or AC current, as opposed to its ability to store a charge. A measure of the effectiveness of an insulator is the maximum electric field it can support without an electrical breakdown.

Example 19-1 *Design of a Transmission Line*

Design an electrical transmission line 1500 m long that will carry a current of 50 A with no more than 5×10^5 W loss in power. The electrical conductivity of several materials is included in Table 19-1.

SOLUTION

Electrical power is given by the product of the voltage and current or

$$P = VI = I^2R = (50)^2R = 5 \times 10^5\,\text{W}$$

$$R = 200\,\text{ohms}$$

From Equation 19-2,

$$A = \frac{l}{R \cdot \sigma} = \frac{(1500\,\text{m})(100\,\text{cm/m})}{(200\,\text{ohms})\sigma} = \frac{750}{\sigma}$$

Let's consider three metals—aluminum, copper, and silver—that have excellent electrical conductivity. The table below includes appropriate data and some characteristics of the transmission line for each metal.

	σ ($\text{ohm}^{-1} \cdot \text{cm}^{-1}$)	A (cm^2)	Diameter (cm)
Aluminum	3.77×10^5	0.00199	0.050
Copper	5.98×10^5	0.00125	0.040
Silver	6.80×10^5	0.00110	0.037

Any of the three metals will work, but cost is a factor as well. Aluminum will likely be the most economical choice (Chapter 14), even though the wire has the largest diameter. Other factors, such as whether the wire can support itself between transmission poles, also contribute to the final choice.

Example 19-2 *Drift Velocity of Electrons in Copper*

Assuming that all of the valence electrons contribute to current flow, (a) calculate the mobility of an electron in copper and (b) calculate the average drift velocity for electrons in a 100 cm copper wire when 10 V are applied.

SOLUTION

(a) The valence of copper is one; therefore, the number of valence electrons equals the number of copper atoms in the material. The lattice parameter of copper is 3.6151×10^{-8} cm and, since copper is FCC, there are four atoms/unit cell. From Table 19-1, the conductivity $\sigma = 5.98 \times 10^5\ \Omega^{-1} \cdot \text{cm}^{-1}$

$$n = \frac{(4\ \text{atoms/cell})(1\ \text{electron/atom})}{(3.6151 \times 10^{-8}\,\text{cm})^3/\text{cell}} = 8.466 \times 10^{22}\,\text{electrons/cm}^3$$

$$q = 1.6 \times 10^{-19}\,\text{C}$$

$$\mu = \frac{\sigma}{nq} = \frac{5.98 \times 10^5\ \Omega^{-1} \cdot \text{cm}^{-1}}{(8.466 \times 10^{22}\ \text{electrons/cm}^3)(1.6 \times 10^{-19}\text{C})}$$

$$= 44.1\ \frac{\text{cm}^2}{\Omega \cdot \text{C}} = 44.1\ \frac{\text{cm}^2}{\text{V} \cdot \text{s}}$$

(b) The electric field is

$$E = \frac{V}{l} = \frac{10\ \text{V}}{100\ \text{cm}} = 0.1\ \text{V/cm}$$

The mobility is 44.1 $\text{cm}^2/(\text{V} \cdot \text{s})$; therefore,

$$\bar{v} = \mu E = [44.1\ \text{cm}^2/(\text{V} \cdot \text{s})](0.1\ \text{V/cm}) = 4.41\ \text{cm/s}$$

19-2 Band Structure of Solids

As we saw in Chapter 2, the electrons of atoms in isolation occupy fixed and discrete energy levels. The Pauli exclusion principle is satisfied for each atom because only two electrons, at most, occupy each energy level, or orbital. When N atoms come together to form a solid, the Pauli exclusion principle still requires that no more than two electrons in the solid have the same energy. As two atoms approach each other in order to form a bond, the Pauli exclusion principle would be violated if the energy levels of the electrons did not change. Thus, the energy levels of the electrons "split" in order to form new energy levels.

Figure 19-2 schematically illustrates this concept. Consider two atoms approaching each other to form a bond. The orbitals that contain the valence electrons are located (on average) farther from the nucleus than the orbitals that contain the "core" or innermost electrons. The orbitals that contain the valence electrons of one atom thus interact

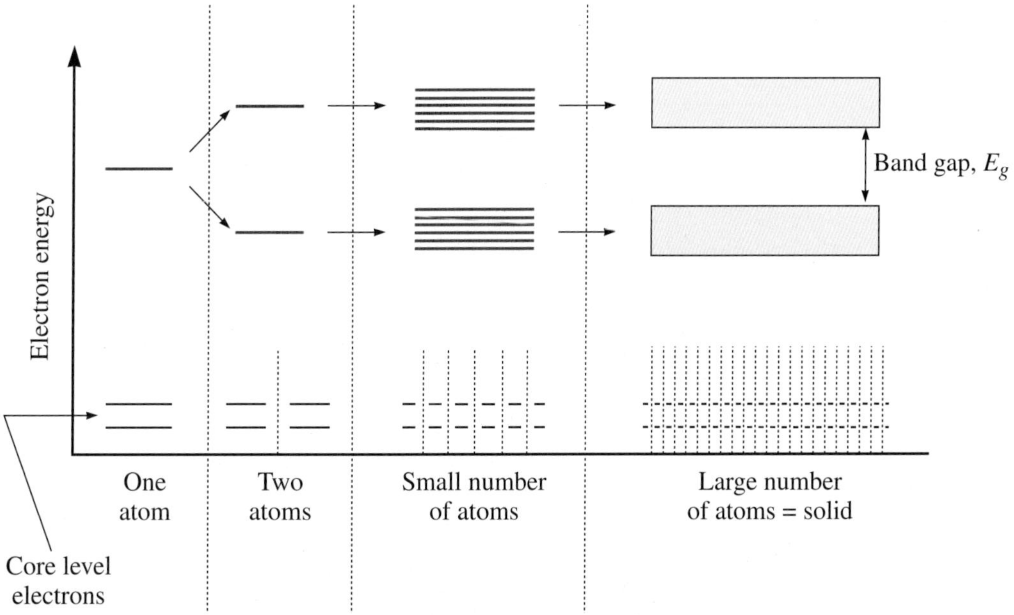

Figure 19-2 The energy levels broaden into bands as the number of electrons grouped together increases. (*Courtesy of John Bravman*)

with the orbitals that contain the valence electrons of the other atom first. Since the orbitals of these electrons have the same energy when the atoms are in isolation, the orbitals shift in energy or "split" so that the Pauli exclusion principle is satisfied. As shown in Figure 19-2, when considering two atoms, each with one orbital of interest, one of the orbitals shifts to a higher energy level while the other orbital shifts to a lower energy level. The electrons of the atoms will occupy these new orbitals by first filling the lowest energy levels. As the number of atoms increases, so does the number of energy levels. A new orbital with its own energy is formed for each orbital of each atom, and as the number of atoms in the solid increases, the separation in energy between orbitals becomes finer, ultimately forming what is called an energy band. For example, when N atoms come together to form a solid, the $2s$ energy band contains N discrete energy levels, one for each atom in the solid since each atom contributes one orbital.

In order for charge carriers to conduct, the carriers must be able to accelerate and increase in energy. The energy of the carriers can increase only if there are available energy states to which the carriers can be promoted. Thus, the particular distribution of energy states in the band structure of a solid has critical implications for its electrical and optical properties.

Depending on the type of material involved (metal, semiconductor, insulator), there may or may not be a sizable energy gap between the energy levels of the orbitals that shifted to a lower energy state and the energy levels of the orbitals that shifted to a higher energy state. This energy gap, if it exists, is known as the bandgap. We will discuss the bandgap in more detail later.

Band Structure of Sodium Sodium is a metal and has the electronic structure $1s^2 2s^2 2p^6 3s^1$. Figure 19-3 shows a schematic diagram of the band structure of sodium as a function of the interatomic separation. (Note that Figure 19-2 shows a general band diagram for a fixed interatomic separation.) The energies within the bands depend on the spacing between the atoms; the vertical line represents the equilibrium interatomic spacing of the atoms in solid sodium.

Sodium and other alkali metals in column 1A of the periodic table have only one electron in the outermost s level. The $3s$ valence band in sodium is half filled and, at absolute zero, only the lowest energy levels are occupied. The **Fermi energy** (E_f) is the energy

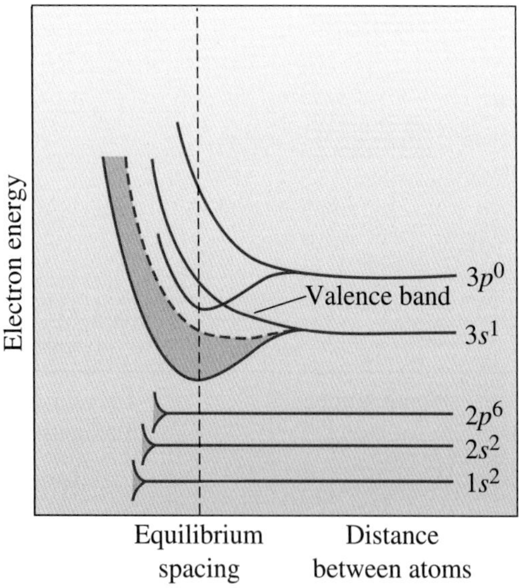

Figure 19-3 The simplified band structure for sodium. The energy levels broaden into bands. The $3s$ band, which is only half filled with electrons, is responsible for conduction in sodium. Note that the energy levels of the orbitals in the $1s$, $2s$, and $2p$ levels do not split at the equilibrium spacing for sodium.

level at which half of the possible energy levels in the band are occupied by electrons. It is the energy level where the probability of finding an electron is 1/2. When electrons gain energy, they are excited into the empty higher energy levels. The promotion of carriers to higher energy levels enables electrical conduction.

Band Structure of Magnesium and Other Metals

Magnesium and other metals in column 2A of the periodic table have two electrons in their outermost *s* band. These metals have a high conductivity because the *p* band overlaps the *s* band at the equilibrium interatomic spacing. This overlap permits electrons to be excited into the large number of unoccupied energy levels in the combined 3*s* and 3*p* band. Overlapping 3*s* and 3*p* bands in aluminum and other metals in column 3B provide a similar effect.

In the transition metals, including scandium through nickel, an unfilled 3*d* band overlaps the 4*s* band. This overlap provides energy levels into which electrons can be excited; however, complex interactions between the bands prevent the conductivity from being as high as in some of the better conductors. In copper, the inner 3*d* band is full, and the atom core tightly holds these electrons. Consequently, there is little interaction between the electrons in the 4*s* and 3*d* bands, and copper has a high conductivity. A similar situation is found for silver and gold.

Band Structure of Semiconductors and Insulators

The elements in Group 4—carbon (diamond), silicon, germanium, and tin—contain two electrons in their outer *p* shell and have a valence of four. Based on our discussion in the previous section, we might expect these elements to have a high conductivity due to the unfilled *p* band, but this behavior is not observed!

These elements are covalently bonded; consequently, the electrons in the outer *s* and *p* bands are rigidly bound to the atoms. The covalent bonding produces a complex change in the band structure. The 2*s* and 2*p* levels of the carbon atoms in diamond can contain up to eight electrons, but there are only four valence electrons available. When carbon atoms are brought together to form solid diamond, the 2*s* and 2*p* levels interact and produce two bands (Figure 19-4). Each hybrid band can contain $4N$ electrons. Since there are only $4N$ electrons available, the lower (or **valence**) band is completely full, whereas the upper (or **conduction**) band is empty at 0 K.

A large **energy gap** or **bandgap** (E_g) separates the valence band from the conduction band in diamond ($E_g \sim 5.5$ eV). Few electrons possess sufficient energy to jump

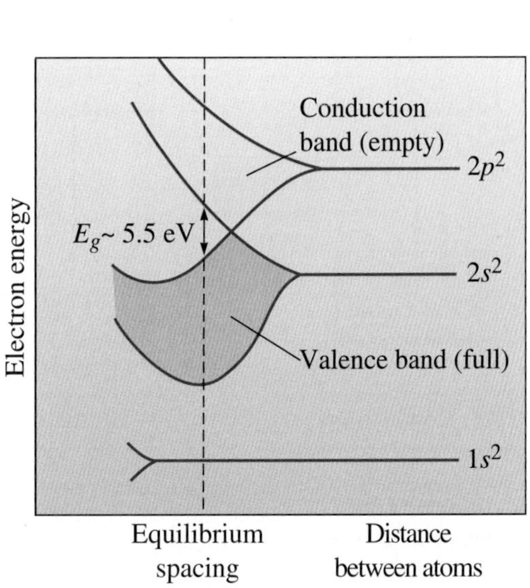

Figure 19-4 The band structure of carbon in the diamond form. The 2*s* and 2*p* levels combine to form two hybrid bands separated by an energy gap, E_g.

the forbidden zone to the conduction band. Consequently, diamond has an electrical conductivity of less than 10^{-18} ohm^{-1} · cm^{-1}. Other covalently and ionically bonded materials have a similar band structure and, like diamond, are poor conductors of electricity. Increasing the temperature supplies the energy required for electrons to overcome the energy gap. For example, the electrical conductivity of boron nitride increases from about 10^{-13} at room temperature to 10^{-4} ohm^{-1} · cm^{-1} at 800°C.

Figure 19-5 shows a schematic of the band structure of typical metals, semiconductors, and insulators. The differing temperature dependencies of the conductivities of metals and semiconductors are a consequence of their different band structures. In a metal, there is a large supply of electrons available (roughly 10^{23} per cubic centimeter, or one per atom) at any temperature. An almost infinitesimally small amount of energy supplied to the electrons will promote them to unoccupied states where they can participate in conduction. In a semiconductor at 0 K, the valence band is completely full and the conduction band is completely empty; the semiconductor does not conduct. As temperature increases, the supply of electrons increases exponentially because the thermal energy frees an increasing number of electrons from the covalent bonds, making the electrons available for conduction, and we find that the resistivity of a semiconductor decreases with increasing temperature. In pure silicon, there are about 10^{10} free electrons per cubic centimeter at room temperature, or about 10^{13} times fewer than in a metal. Thus, an important distinction between metals and semiconductors is that the conductivity of semiconductors *increases* with temperature, as more and more electrons are promoted to the conduction band from the valence band. The conductivity of most metals, on the other hand, *decreases* with increasing temperature. This is because the number of electrons that are already available begin to scatter more (i.e., increasing temperature reduces mobility). We will discuss this more later in this chapter.

Although germanium, silicon, and α-Sn have the same crystal structure and band structure as diamond, their energy gaps are smaller. In fact, the energy gap (E_g) in α-Sn is so small (E_g = 0.1 eV) that α-Sn behaves as a metal. The energy gap is somewhat larger in silicon (E_g = 1.1 eV) and germanium (E_g = 0.67 eV)—these elements behave as semiconductors. Typically, we consider materials with a bandgap greater than 4.0 eV as insulators, dielectrics, or nonconductors; materials with a bandgap less than 4.0 eV are considered semiconductors.

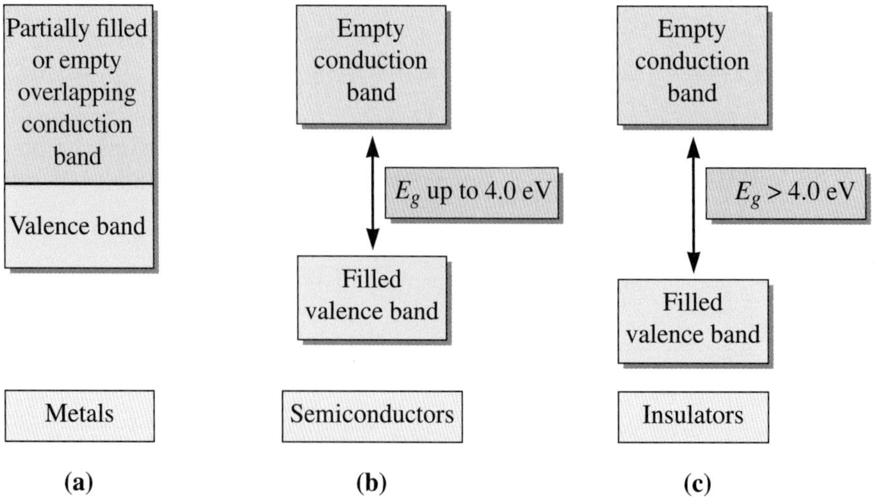

Figure 19-5 Schematic of band structures for (a) metals, (b) semiconductors, and (c) dielectrics or insulators. (Temperature is assumed to be 0 K.)

19-3 Conductivity of Metals and Alloys

The conductivity of a pure, defect-free metal is determined by the electronic structure of the atoms, but we can change the conductivity by influencing the mobility, μ, of the carriers. Recall that the mobility is proportional to the average drift velocity, $\bar{v}$. The average drift velocity is the velocity with which charge carriers move in the direction dictated by the applied field. The paths of electrons are influenced by internal fields due to atoms in the solid and imperfections in the lattice. When these internal fields influence the path of an electron, the drift velocity (and thus the mobility of the charge carriers) decreases. The **mean free path** (λ_e) of electrons is defined as

$$\lambda_e = \tau\bar{v} \tag{19-6}$$

The average time between collisions is τ. The mean free path defines the average distance between collisions; a longer mean free path permits higher mobilities and higher conductivities.

Temperature Effect When the temperature of a metal increases, thermal energy causes the amplitudes of vibration of the atoms to increase (Figure 19-6). This increases the *scattering cross section* of atoms or defects in the lattice. Essentially, the atoms and defects act as larger targets for interactions with electrons, and interactions occur more frequently. Thus, the mean free path decreases, the mobility of electrons

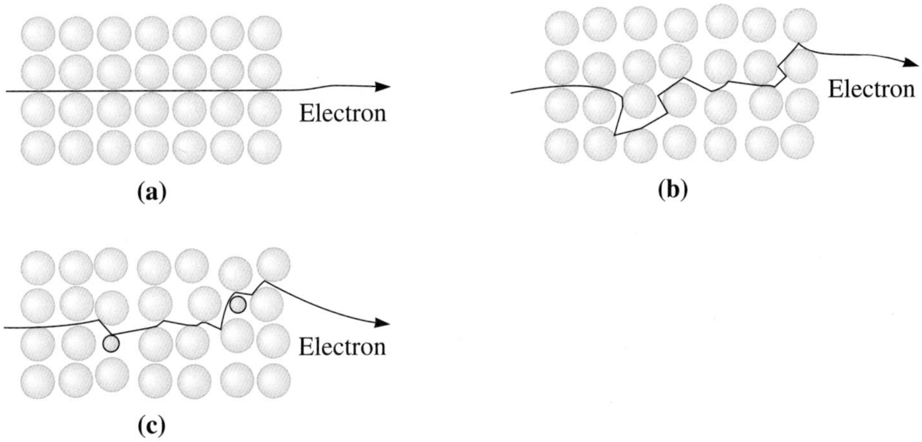

Figure 19-6 Movement of an electron through (a) a perfect crystal, (b) a crystal heated to a high temperature, and (c) a crystal containing atomic level defects. Scattering of the electrons reduces the mobility and conductivity.

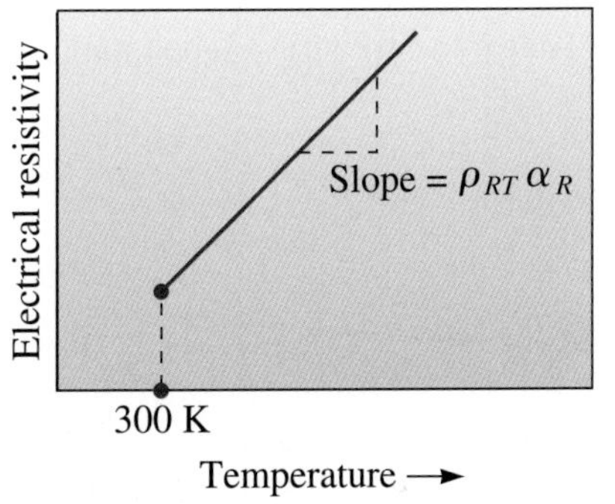

Figure 19-7 The effect of temperature on the electrical resistivity of a metal with a perfect crystal structure.

Table 19-3 The temperature resistivity coefficient α_R for selected metals

Metal	Room Temperature Resistivity (ohm · cm)	Temperature Resistivity Coefficient (α_R) [ohm/(ohm · °C)]
Be	4.0×10^{-6}	0.0250
Mg	4.45×10^{-6}	0.0037
Ca	3.91×10^{-6}	0.0042
Al	2.65×10^{-6}	0.0043
Cr	12.90×10^{-6} (0°C)	0.0030
Fe	9.71×10^{-6}	0.0065
Co	6.24×10^{-6}	0.0053
Ni	6.84×10^{-6}	0.0069
Cu	1.67×10^{-6}	0.0043
Ag	1.59×10^{-6}	0.0041
Au	2.35×10^{-6}	0.0035
Pd	10.8×10^{-6}	0.0037
W	5.3×10^{-6} (27°C)	0.0045
Pt	9.85×10^{-6}	0.0039

(Based on Handbook of Electromagnetic Materials: Monolithic and Composite Versions and Their Applications *by P. S. Neelkanta, Taylor & Francis, 1995.)*

is reduced, and the resistivity increases. The change in resistivity of a pure metal as a function of temperature can be estimated according to

$$\rho = \rho_{RT}(1 + \alpha_R \Delta T) \tag{19-7}$$

where ρ is the resistivity at any temperature T, ρ_{RT} is the resistivity at room temperature (i.e., 25°C), $\Delta T = (T - T_{RT})$ is the difference between the temperature of interest and room temperature, and α_R is the *temperature resistivity coefficient.* The relationship between resistivity and temperature is linear over a wide temperature range (Figure 19-7). Values for the temperature resistivity coefficients for various metals are given in Table 19-3.

The following example illustrates how the resistivity of pure copper can be calculated.

Example 19-3 *Resistivity of Pure Copper*

Calculate the electrical conductivity of pure copper at (a) 400°C and (b) −100°C.

SOLUTION

The resistivity of copper at room temperature is 1.67×10^{-6} ohm · cm, and the temperature resistivity coefficient is 0.0043 ohm/(ohm · °C). (See Table 19-3.)

(a) At 400°C:

$$\rho = \rho_{RT}(1 + \alpha_R \Delta T) = (1.67 \times 10^{-6})[1 + 0.0043(400 - 25)]$$

$$\rho = 4.363 \times 10^{-6} \text{ ohm} \cdot \text{cm}$$

$$\sigma = 1/\rho = 2.29 \times 10^{5} \text{ ohm}^{-1} \cdot \text{cm}^{-1}$$

(b) At −100°C:

$$\rho = (1.67 \times 10^{-6})[1 + 0.0043(-100 - 25)] = 7.724 \times 10^{-7} \text{ ohm} \cdot \text{cm}$$

$$\sigma = 1.29 \times 10^{6} \text{ ohm}^{-1} \cdot \text{cm}^{-1}$$

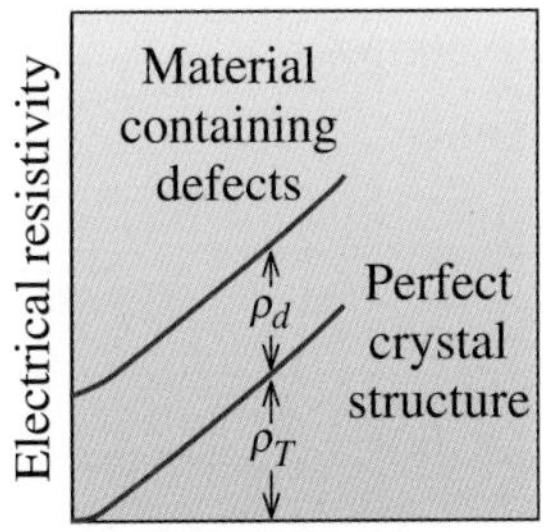

Figure 19-8 The electrical resistivity of a metal is due to a constant defect contribution ρ_d and a variable temperature contribution ρ_T.

Effect of Atomic Level Defects Imperfections in crystal structures scatter electrons, reducing the mobility and conductivity of the metal [Figure 19-6(c)]. For example, the increase in the resistivity due to solid solution atoms for dilute solutions is

$$\rho_d = b(1 - x)x \tag{19-8}$$

where ρ_d is the increase in resistivity due to the defects, x is the atomic fraction of the impurity or solid solution atoms present, and b is the defect resistivity coefficient. In a similar manner, vacancies, dislocations, and grain boundaries reduce the conductivity of the metal. Each defect contributes to an increase in the resistivity of the metal. Thus, the overall resistivity is

$$\rho = \rho_T + \rho_d \tag{19-9}$$

where ρ_d equals the contributions from all of the defects. Equation 19-9 is known as **Matthiessen's rule**. The effect of the defects is *independent* of temperature (Figure 19-8).

Effect of Processing and Strengthening Strengthening mechanisms and metal processing techniques affect the electrical properties of a metal in different ways (Table 19-4). Solid-solution strengthening is *not* a good way to obtain high

Table 19-4 The effect of alloying, strengthening, and processing on the electrical conductivity of copper and its alloys

Alloy	$\frac{\sigma_{alloy}}{\sigma_{Cu}} \times 100$	Remarks
Pure annealed copper	100	Few defects to scatter electrons; the mean free path is long.
Pure copper deformed 80%	98	Many dislocations, but because of the tangled nature of the dislocation networks, the mean free path is still long.
Dispersion-strengthened Cu-0.7% Al_2O_3	85	The dispersed phase is not as closely spaced as solid-solution atoms, nor is it coherent, as in age hardening. Thus, the effect on conductivity is small.
Solution-treated Cu-2% Be	18	The alloy is single phase; however, the small amount of solid-solution strengthening from the supersaturated beryllium greatly decreases conductivity.
Aged Cu-2% Be	23	During aging, the beryllium leaves the copper lattice to produce a coherent precipitate. The precipitate does not interfere with conductivity as much as the solid-solution atoms.
Cu-35% Zn	28	This alloy is solid-solution strengthened by zinc, which has an atomic radius near that of copper. The conductivity is low, but not as low as when beryllium is present.

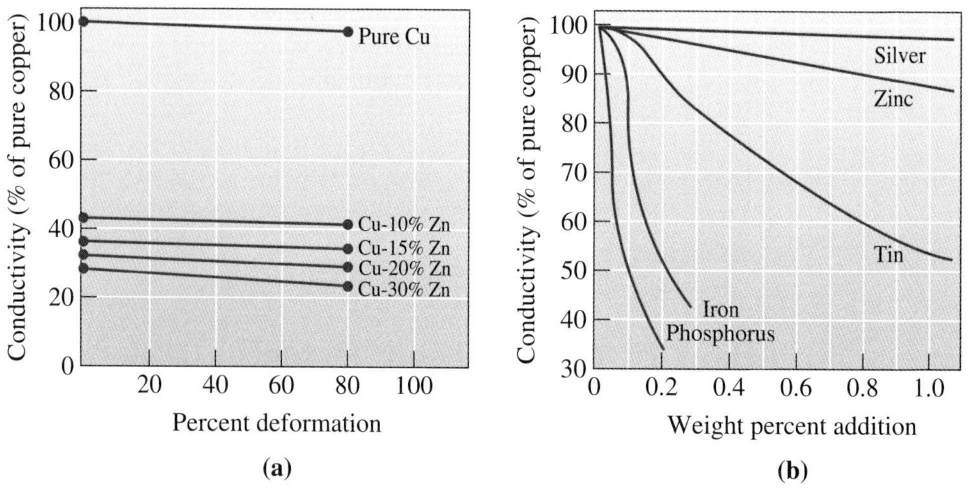

Figure 19-9 (a) The effect of solid-solution strengthening and cold work on the electrical conductivity of copper, and (b) the effect of the addition of selected elements on the electrical conductivity of copper.

strength in metals intended to have high conductivities. The mean free paths are short due to the random distribution of the interstitial or substitutional atoms. Figure 19-9 shows the effect of zinc and other alloying elements on the conductivity of copper; as the amount of alloying element increases, the conductivity decreases substantially.

Age hardening and dispersion strengthening reduce the conductivity to an extent that is less than solid-solution strengthening, since there is a longer mean free path between precipitates, as compared with the path between point defects. Strain hardening and grain-size control have even less effect on conductivity (Figure 19-9 and Table 19-4). Since dislocations and grain boundaries are further apart than solid-solution atoms, there are large volumes of metal that have a long mean free path. Consequently, cold working is an effective way to increase the strength of a metallic conductor without seriously impairing the electrical properties of the material. In addition, the effects of cold working on conductivity can be eliminated by the low-temperature recovery heat treatment in which good conductivity is restored while the strength is retained.

Conductivity of Alloys Alloys typically have higher resistivities than pure metals because of the scattering of electrons due to the alloying additions. For example, the resistivity of pure Cu at room temperature is $\sim 1.67 \times 10^{-6}\ \Omega \cdot \text{cm}$ and that of pure gold is $\sim 2.35 \times 10^{-6}\ \Omega \cdot \text{cm}$. The resistivity of a 35% Au-65% Cu alloy at room temperature is much higher, $\sim 12 \times 10^{-6}\ \Omega \cdot \text{cm}$. Ordering of atoms in alloys by heat treatment can decrease their resistivity. Compared to pure metals, the resistivities of alloys tend to be stable with regard to temperature variation. Relatively high-resistance alloys such as nichrome (~80% Ni-20% Cr) can be used as heating elements. Certain alloys of Bi-Sn-Pb-Cd are used to make electrical fuses due to their low melting temperatures.

19-4 Semiconductors

Elemental semiconductors are found in Group 4B of the periodic table and include germanium and silicon. Compound semiconductors are formed from elements in Groups 2B and 6B of the periodic table (e.g., CdS, CdSe, CdTe, HgCdTe, etc.) and are known as II–VI (two–six) semiconductors. Compound semiconductors also can be formed by combining elements from Groups 3B and 5B of the periodic table (e.g., GaN, GaAs, AlAs, AlP, InP, etc.). These are known as III–V (three–five) semiconductors.

An **intrinsic semiconductor** is one with properties that are not controlled by impurities. An **extrinsic semiconductor** (*n*- or *p*-type) is preferred for devices, since its

Table 19-5 Properties of commonly encountered semiconductors at room temperature

Semiconductor	Bandgap (eV)	Mobility of Electrons (μ_n) $\left(\frac{cm^2}{V\text{-}s}\right)$	Mobility of Holes (μ_p) $\left(\frac{cm^2}{V\text{-}s}\right)$	Dielectric Constant (k)	Resistivity ($\Omega \cdot cm$)	Density $\left(\frac{g}{cm^3}\right)$	Melting Temperature (°C)
Silicon (Si)	1.11	1350	480	11.8	2.5×10^5	2.33	1415
Amorphous Silicon (a:Si:H)	1.70	1	10^{-2}	~11.8	10^{10}	~2.30	—
Germanium (Ge)	0.67	3900	1900	16.0	43	5.32	936
SiC (α)	2.86	500		10.2	10^{10}	3.21	2830
Gallium Arsenide (GaAs)	1.43	8500	400	13.2	4×10^8	5.31	1238
Diamond	~5.50	1800	1500	5.7	$> 10^{18}$	3.52	~3550

properties are stable with temperature and can be controlled using ion implantation or diffusion of impurities known as dopants. Semiconductor materials, including silicon and germanium, provide the building blocks for many electronic devices. These materials have an easily controlled electrical conductivity and, when properly combined, can act as switches, amplifiers, or information storage devices. The properties of some of the commonly encountered semiconductors are included in Table 19-5.

As we learned in Section 19-2, as the atoms of a semiconductor come together to form a solid, two energy bands are formed [Figure 19-5(b)]. At 0 K, the energy levels of the valence band are completely full, as these are the lowest energy states for the electrons. The valence band is separated from the conduction band by a bandgap. At 0 K, the conduction band is empty.

The energy gap E_g between the valence and conduction bands in semiconductors is relatively small (Figure 19-5). As a result, as temperature increases, some electrons possess enough thermal energy to be promoted from the valence band to the conduction band. The excited electrons leave behind unoccupied energy levels, or holes, in the valence band. When an electron moves to fill a hole, another hole is created; consequently, the holes appear to act as positively charged electrons and carry an electrical charge. When a voltage is applied to the material, the electrons in the conduction band accelerate toward the positive terminal, while holes in the valence band move toward the negative terminal (Figure 19-10). Current is, therefore, conducted by the movement of both electrons and holes in semiconductors.

The conductivity is determined by the number of electrons and holes according to

$$\sigma = nq\mu_n + pq\mu_p \tag{19-10}$$

where n is the concentration of electrons in the conduction band, p is the concentration of holes in the valence band, and μ_n and μ_p are the mobilities of electrons and holes, respectively (Table 19-5). This equation is the same as Equation 19-5b.

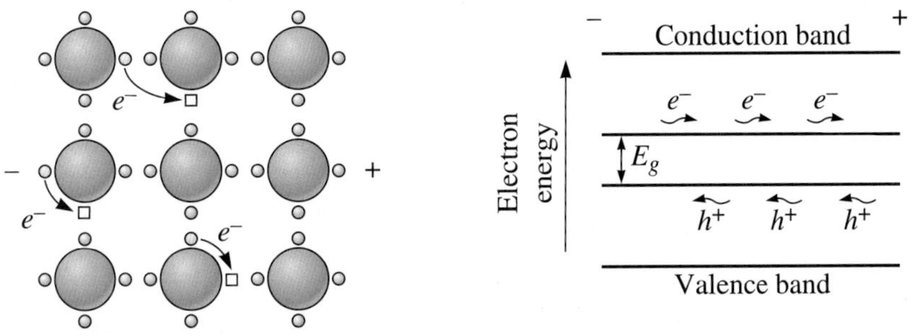

Figure 19-10 When a voltage is applied to a semiconductor, the electrons move through the conduction band, while the holes move through the valence band in the opposite direction.

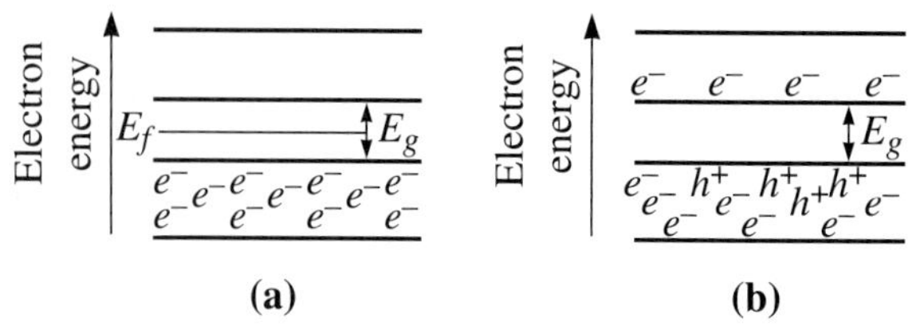

Figure 19-11 The distribution of electrons and holes in the valence and conduction bands (a) at absolute zero and (b) at an elevated temperature.

In intrinsic semiconductors, for every electron promoted to the conduction band, there is a hole left in the valence band, such that

$$n_i = p_i$$

where n_i and p_i are the concentrations of electrons and holes, respectively, in an intrinsic semiconductor. Therefore, the conductivity of an intrinsic semiconductor is

$$\sigma = qn_i(\mu_n + \mu_p) \tag{19-11}$$

In intrinsic semiconductors, we control the number of charge carriers and, hence, the electrical conductivity by controlling the temperature. At absolute zero temperature, all of the electrons are in the valence band, whereas all of the levels in the conduction band are unoccupied [Figure 19-11(a)]. As the temperature increases, there is a greater probability that an energy level in the conduction band is occupied (and an equal probability that a level in the valence band is unoccupied, or that a hole is present) [Figure 19-11(b)]. The number of electrons in the conduction band, which is equal to the number of holes in the valence band, is given by

$$n = n_i = p_i = n_0 \exp\left(\frac{-E_g}{2\mathrm{k_B}T}\right) \tag{19-12a}$$

where n_0 is given by

$$n_0 = 2\left(\frac{2\pi \mathrm{k_B}T}{\mathrm{h}^2}\right)^{3/2}(m_n^* m_p^*)^{3/4} \tag{19-12b}$$

In these equations, $\mathrm{k_B}$ and h are the Boltzmann and Planck's constants and m_n^* and m_p^* are the effective masses of electrons and holes in the semiconductor, respectively. The effective masses account for the effects of the internal forces that alter the acceleration of electrons in a solid relative to electrons in a vacuum. For Ge, Si, and GaAs, the room temperature values of n_i are 2.5×10^{13}, 1.5×10^{10}, and 2×10^6 electrons/cm^3, respectively. The $n_i p_i$ product remains constant at any given temperature for a given semiconductor. This allows us to calculate n_i or p_i values at different temperatures.

Higher temperatures permit more electrons to cross the forbidden zone and, hence, the conductivity increases:

$$\sigma = n_0 q(\mu_n + \mu_p)\exp\left(\frac{-E_g}{2\mathrm{k_B}T}\right) \tag{19-13}$$

Note that both n_i and σ are related to temperature by an Arrhenius equation, rate $= A\exp\left(\frac{-Q}{RT}\right)$. As the temperature increases, the conductivity of a semiconductor also increases because more charge carriers are available for conduction. Note that as for metals, the mobilities of the carriers decrease at high temperatures, but this is a much weaker dependence than the exponential increase in the number of charge carriers. The increase in conductivity with temperature in semiconductors sharply contrasts with the decrease in conductivity of metals with increasing temperature (Figure 19-12). Even at high temperatures, however, the conductivity of a metal is orders of magnitudes higher than the conductivity of a semiconductor. The example that follows shows the calculation for carrier concentration in an intrinsic semiconductor.

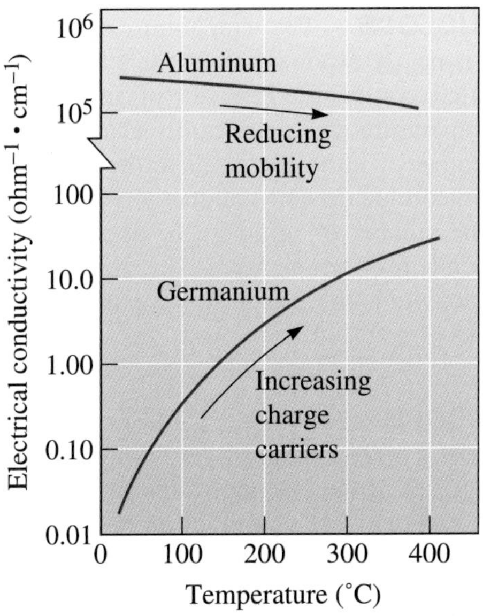

Figure 19-12 The electrical conductivity versus temperature for intrinsic semiconductors compared with metals. Note the break in the vertical axis scale.

Example 19-4 *Carrier Concentrations in Intrinsic Ge*

For germanium at 25°C, estimate (a) the number of charge carriers, (b) the fraction of the total electrons in the valence band that are excited into the conduction band, and (c) the constant n_0 in Equation 19-12a.

SOLUTION

From Table 19-5, $\rho = 43\ \Omega \cdot \text{cm}$, $\therefore \sigma = 0.0233\ \Omega^{-1} \cdot \text{cm}^{-1}$
Also from Table 19-5,

$$E_g = 0.67\,\text{eV},\ \mu_n = 3900\,\frac{\text{cm}^2}{\text{V}\cdot\text{s}},\ \mu_p = 1900\,\frac{\text{cm}^2}{\text{V}\cdot\text{s}}$$

$$2k_BT = 2(8.617 \times 10^{-5}\text{eV/K})(273\text{ K} + 25\text{ K}) = 0.05136\text{eV at } T = 25°\text{C}$$

(a) From Equation 19-10,

$$n = \frac{\sigma}{q(\mu_n + \mu_p)} = \frac{0.0233}{(1.6 \times 10^{-19})(3900 + 1900)} = 2.51 \times 10^{13}\,\frac{\text{electrons}}{\text{cm}^3}$$

There are 2.51×10^{13} electrons/cm^3 and 2.51×10^{13} holes/cm^3 conducting charge in germanium at room temperature.

(b) The lattice parameter of diamond cubic germanium is 5.6575×10^{-8} cm.
The total number of electrons in the valence band of germanium at 0 K is

$$\text{Total electrons} = \frac{(8\,\text{atoms/cell})(4\,\text{electrons/atom})}{(5.6575 \times 10^{-8}\,\text{cm})^3/\text{cell}}$$
$$= 1.77 \times 10^{23}$$

$$\text{Fraction excited} = \frac{2.51 \times 10^{13}}{1.77 \times 10^{23}} = 1.42 \times 10^{-10}$$

(c) From Equation 19-12a,

$$n_0 = \frac{n}{\exp[-E_g/(2k_BT)]} = \frac{2.51 \times 10^{13}\,\text{carriers/cm}^3}{\exp(-0.67/0.05136)}$$
$$= 1.16 \times 10^{19}\,\text{carriers/cm}^3$$

Extrinsic Semiconductors The temperature dependence of conductivity in intrinsic semiconductors is nearly exponential, but this is not useful for practical applications. We cannot accurately control the behavior of an intrinsic semiconductor because slight variations in temperature can significantly change the conductivity. By intentionally adding a small number of impurity atoms to the material (called doping), we can produce an extrinsic semiconductor. The conductivity of the extrinsic semiconductor depends primarily on the number of impurity, or dopant, atoms and in a certain temperature range is independent of temperature. This ability to have a tunable yet temperature-independent conductivity is the reason why we almost always use extrinsic semiconductors to make devices.

***n*-Type Semiconductors** Suppose we add an impurity atom such as antimony (which has a valence of five) to silicon or germanium. Four of the electrons from the antimony atom participate in the covalent-bonding process, while the extra electron enters an energy level just below the conduction band (Figure 19-13). Since the extra electron is not tightly bound to the atoms, only a small increase in energy, E_d, is required for the electron to enter the conduction band. This energy level just below the conduction band is called a donor state. An *n*-type dopant "donates" a free electron for each impurity atom added. The energy gap controlling conductivity is now E_d rather than E_g (Table 19-6). No corresponding holes are created when the donor electrons enter the conduction band. It is still the case that electron-hole pairs are created when thermal energy causes electrons to be promoted to the conduction band from the valence band; however, the number of electron-hole pairs is significant only at high temperatures.

***p*-Type Semiconductors** When we add an impurity such as gallium or boron, which has a valence of three, to Si or Ge, there are not enough electrons to complete the covalent bonding process. A hole is created in the valence band that can be filled by electrons from other locations in the band (Figure 19-14). The holes act as "acceptors" of electrons. These hole sites have a somewhat higher than normal energy and create an acceptor level of possible electron energies just above the valence band (Table 19-6). An electron must gain an energy of only E_a in order to create a hole in the valence band. The hole then carries charge. This is known as a *p*-type semiconductor.

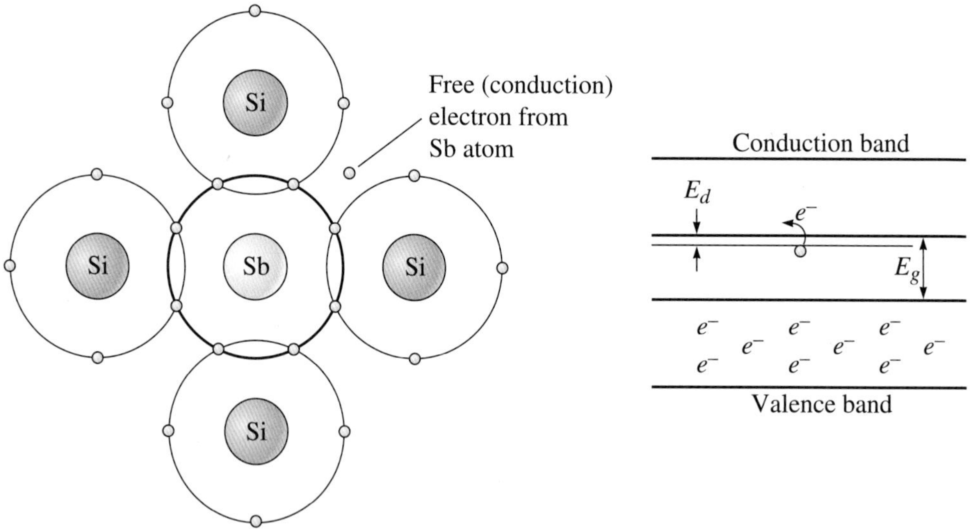

Figure 19-13 When a dopant atom with a valence greater than four is added to silicon, an extra electron is introduced and a donor energy state is created. Now electrons are more easily excited into the conduction band.

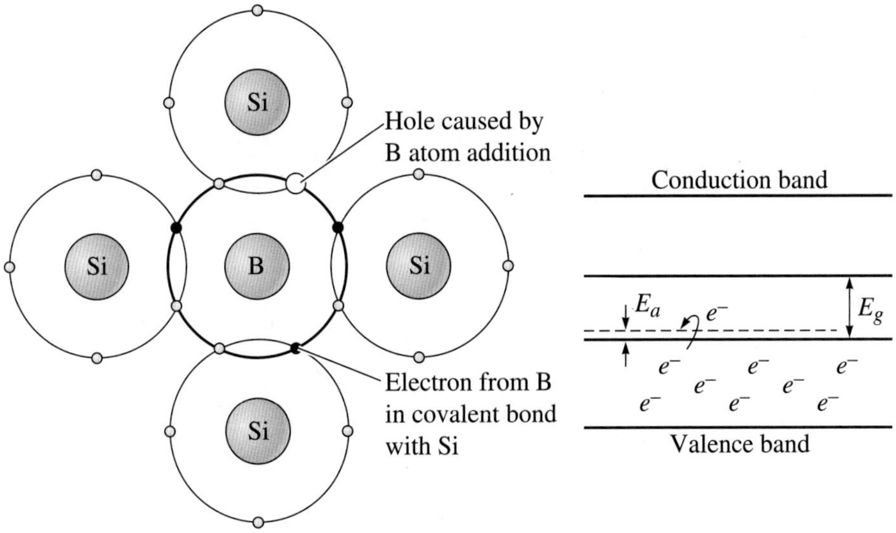

Figure 19-14 When a dopant atom with a valence of less than four is substituted into the silicon structure, a hole is introduced in the structure and an acceptor energy level is created just above the valence band.

Table 19-6 The donor and acceptor energy levels (in electron volts) when silicon and germanium semiconductors are doped

	Silicon		Germanium	
Dopant	E_d	E_a	E_d	E_a
P	0.045		0.0120	
As	0.049		0.0127	
Sb	0.039		0.0096	
B		0.045		0.0104
Al		0.057		0.0102
Ga		0.065		0.0108
In		0.160		0.0112

Charge Neutrality In an extrinsic semiconductor, there has to be overall electrical neutrality. Thus, the sum of the number of donor atoms (N_d) and holes per unit volume (p_{ext}) (both are positively charged) is equal to the number of acceptor atoms (N_a) and electrons per unit volume (n_{ext}) (both are negatively charged):

$$p_{ext} + N_d = n_{ext} + N_a$$

In this equation, n_{ext} and p_{ext} are the concentrations of electrons and holes in an extrinsic semiconductor.

If the extrinsic semiconductor is heavily n-type doped (i.e., $N_d >> n_i$), then $n_{ext} \sim N_d$. Similarly, if there is a heavily acceptor-doped (p-type) semiconductor, then $N_a >> p_i$ and hence $p_{ext} \sim N_a$. This is important, since this says that by adding a considerable amount of dopant, we can dominate the conductivity of a semiconductor by controlling the dopant concentration.

The changes in carrier concentration with temperature are shown in Figure 19-15. From this, the approximate conductivity changes in an extrinsic semiconductor are easy to follow. When the temperature is too low, the donor or acceptor atoms are not ionized and hence the conductivity is very small. As temperature begins to increase, electrons (or holes) contributed by the donors (or acceptors) become available for conduction.

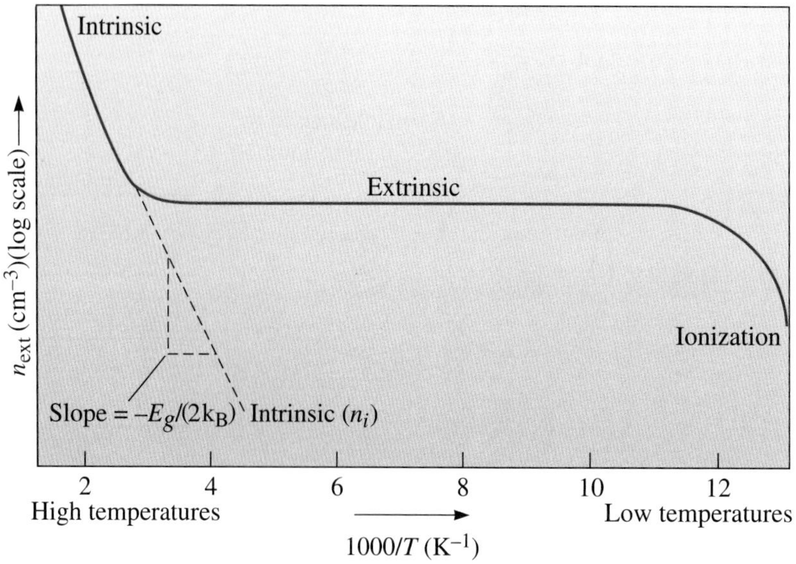

Figure 19-15 The effect of temperature on the carrier concentration of an n-type semiconductor. At low temperatures, the donor atoms are not ionized. As temperature increases, the ionization process is complete, and the carrier concentration increases to a level that is dictated by the level of doping. The conductivity then essentially remains unchanged until the temperature becomes too high and the thermally generated carriers begin to dominate. The effect of dopants is lost at very high temperatures, and the semiconductor essentially shows "intrinsic" behavior.

At sufficiently high temperatures, the conductivity is nearly independent of temperature (region labeled as extrinsic). The value of conductivity at which the plateau occurs depends on the level of doping. When temperatures become too high, the behavior approaches that of an intrinsic semiconductor since the effect of dopants essentially is lost. In this analysis, we have not accounted for the effects of dopant concentration on the mobility of electrons and holes and the temperature dependence of the bandgap. At very high temperatures (not shown in Figure 19-15), the conductivity *decreases* again as scattering of carriers dominates.

Example 19-5 *Design of a Semiconductor*

Design a p-type semiconductor based on silicon, which provides a constant conductivity of 100 ohm$^{-1}\cdot$cm^{-1} over a range of temperatures. Compare the required concentration of acceptor atoms in Si with the concentration of Si atoms.

SOLUTION

In order to obtain the desired conductivity, we must dope the silicon with atoms having a valence of +3, adding enough dopant to provide the required number of charge carriers. If we assume that the number of intrinsic carriers is small compared to the dopant concentration, then

$$\sigma = N_a q \mu_p$$

where $\sigma = 100$ ohm$^{-1}\cdot$cm^{-1} and $\mu_p = 480$ cm^2/(V$\cdot$s). Note that electron and hole mobilities are properties of the host material (i.e., silicon in this case) and not the dopant species. If we remember that a coulomb can be expressed as ampere-seconds and voltage can be expressed as ampere-ohm, the number of charge carriers required is

$$N_a = \frac{\sigma}{q\mu_p} = \frac{100\ \text{ohm}^{-1}\ \text{cm}^{-1}}{(1.6 \times 10^{-19}\ \text{A}\cdot\text{s})[480\ \text{cm}^2/(\text{A}\cdot\text{ohm}\cdot\text{s})]} = 1.30 \times 10^{18}\ \text{acceptor atoms/cm}^3$$

Assume that the lattice constant of Si remains unchanged as a result of doping:

$$N_a = \frac{(1\ \text{hole/dopant atom})(x\ \text{dopant atom/Si atom})(8\ \text{Si atoms/unit cell})}{(5.4307 \times 10^{-8}\ \text{cm})^3/\text{unit cell}}$$

$$x = (1.30 \times 10^{18})(5.4307 \times 10^{-8})^3/8 = 26 \times 10^{-6}\ \text{dopant atom/Si atom}$$

or 26 dopant atoms/10^6 Si atoms

Possible dopants include boron, aluminum, gallium, and indium. High-purity chemicals and clean room conditions are essential for processing since we need 26 dopant atoms per million silicon atoms.

Many other materials that are normally insulating (because the bandgap is too large) can be made semiconducting by doping. Examples of this include $BaTiO_3$, ZnO, TiO_2, and many other oxides. Thus, the concept of *n*- and *p*-type dopants is not limited to Si, Ge, GaAs, etc. We can dope $BaTiO_3$, for example, and make *n*- or *p*-type $BaTiO_3$. Such materials are useful for many sensor applications such as **thermistors**.

Direct and Indirect Bandgap Semiconductors In a direct bandgap semiconductor, an electron can be promoted from the conduction band to the valence band without changing the momentum of the electron. An example of a direct bandgap semiconductor is GaAs. When the excited electron falls back into the valence band, electrons and holes combine to produce light. This is known as **radiative recombination**. Thus, direct bandgap materials such as GaAs and solid solutions of these (e.g., GaAs-AlAs, etc.) are used to make light-emitting diodes (LEDs) of different colors. The bandgap of semiconductors can be tuned using solid solutions. The change in bandgap produces a change in the wavelength [i.e., the frequency of the color (ν) is related to the bandgap E_g as $E_g = h\nu$, where h is Planck's constant]. Since an optical effect is obtained using an electronic material, often the direct bandgap materials are known as optoelectronic materials (Chapter 21). Many lasers and LEDs have been developed using these materials. LEDs that emit light in the infrared range are used in optical-fiber communication systems to convert light waves into electrical pulses. Different colored lasers, such as the blue laser using GaN, have been developed using direct bandgap materials.

In an indirect bandgap semiconductor (e.g., Si, Ge, and GaP), the electrons cannot be promoted to the valence band without a change in momentum. As a result, in materials that have an indirect bandgap (e.g., silicon), we cannot get light emission. Instead, electrons and holes combine to produce heat that is dissipated within the material. This is known as **nonradiative recombination**. Note that both direct and indirect bandgap materials can be doped to form *n*- or *p*-type semiconductors.

19-5 Applications of Semiconductors

We fabricate diodes, transistors, lasers, and LEDs using semiconductors. The ***p-n* junction** is used in many of these devices, such as transistors. Creating an *n*-type region in a *p*-type semiconductor (or vice versa) forms a *p-n* junction [Figure 19-16(a)]. The *n*-type region contains a relatively large number of free electrons, whereas the *p*-type region contains a relatively large number of free holes. This concentration gradient between the *p*-type and *n*-type regions causes diffusion of electrons from the *n*-type material to the *p*-type material and diffusion of holes from the *p*-type material to the *n*-type material. At the junction where the *p*- and *n*-regions meet, free electrons in the *n*-type material recombine with holes in the *p*-type material. This creates a *depleted region* at the junction where the number of available charge carriers is low, and thus, the resistivity is high. Consequently,

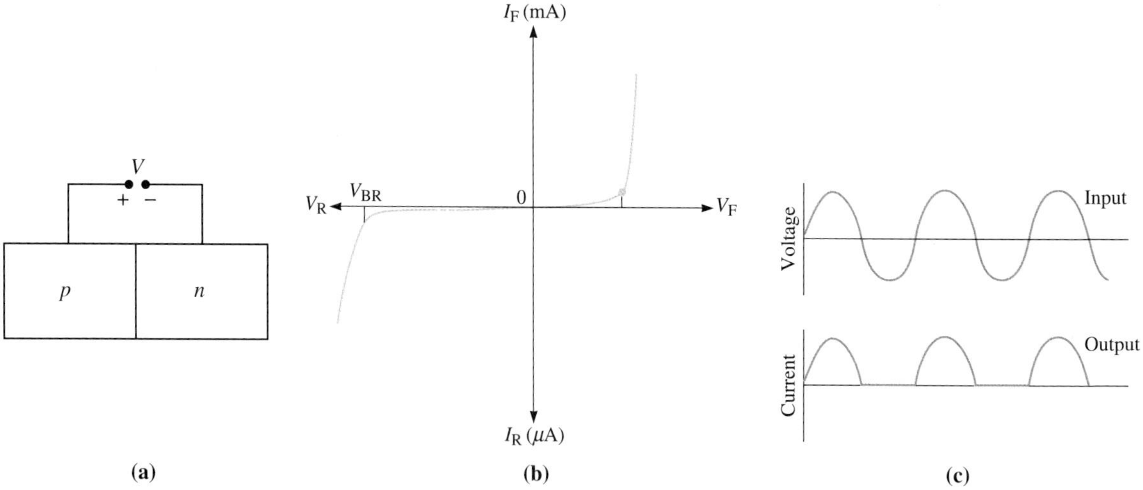

Figure 19-16 (a) A *p-n* junction under forward bias. (b) The current–voltage characteristic for a *p-n* junction. Note the different scales in the first and third quadrants. At sufficiently high reverse bias voltages, "breakdown" occurs, and large currents can flow. Typically this destroys the devices. (c) If an alternating signal is applied, rectification occurs, and only half of the input signal passes the rectifier. (*Based on Thomas L. Floyd,* Electronic Devices *(Conventional Flow Version), 6th, Pearson Education, 2002.*)

an electric field develops due to the distribution of exposed positive ions on the *n*-side of the junction and the exposed negative ions on the *p*-side of the junction. The electric field counteracts further diffusion.

Electrically, the *p-n* junction is conducting when the *p*-side is connected to a positive voltage. This **forward bias** condition is shown in Figure 19-16(a). The applied voltage directly counteracts the electric field at the depleted region, making it possible for electrons from the *n*-side to diffuse across the depleted region to the *p*-side and holes from the *p*-side to diffuse across the depleted region to the *n*-side. When a negative bias is applied to the *p*-side of a *p-n* junction (**reverse bias**), the *p-n* junction does not permit much current to flow. The depleted region simply becomes larger because it is further depleted of carriers. When no bias is applied, there is no current flowing through the *p-n* junction. The forward current can be as large as a few milli-amperes, while the reverse-bias current is a few nano-amperes.

The current–voltage (*I–V*) characteristics of a *p-n* junction are shown in Figure 19-16(b). Because the *p-n* junction permits current to flow in only one direction, it passes only half of an alternating current, therefore converting the alternating current to direct current [Figure 19-16(c)]. These junctions are called **rectifier diodes**.

Bipolar Junction Transistors There are two types of **transistors** based on *p-n* junctions. The term transistor is derived from two words, "transfer" and "resistor." A transistor can be used as a switch or an amplifier. One type of transistor is the *bipolar junction transistor* (BJT). In the era of mainframe computers, bipolar junction transistors often were used in central processing units. A bipolar junction transistor is a sandwich of either *n-p-n* or *p-n-p* semiconductor materials, as shown in Figure 19-17(a). There are three zones in the transistor: the emitter, the base, and the collector. As in the *p-n* junction, electrons are initially concentrated in the *n*-type material, and holes are concentrated in the *p*-type material.

Figure 19-17(b) shows a schematic diagram of an *n-p-n* transistor and its electrical circuit. The electrical signal to be amplified is connected between the base and the emitter, with a small voltage between these two zones. The output from the transistor, or the amplified signal, is connected between the emitter and the collector and operates at a higher voltage. The circuit is connected so that a forward bias is produced between the emitter and the base (the positive voltage is at the *p*-type base), while a

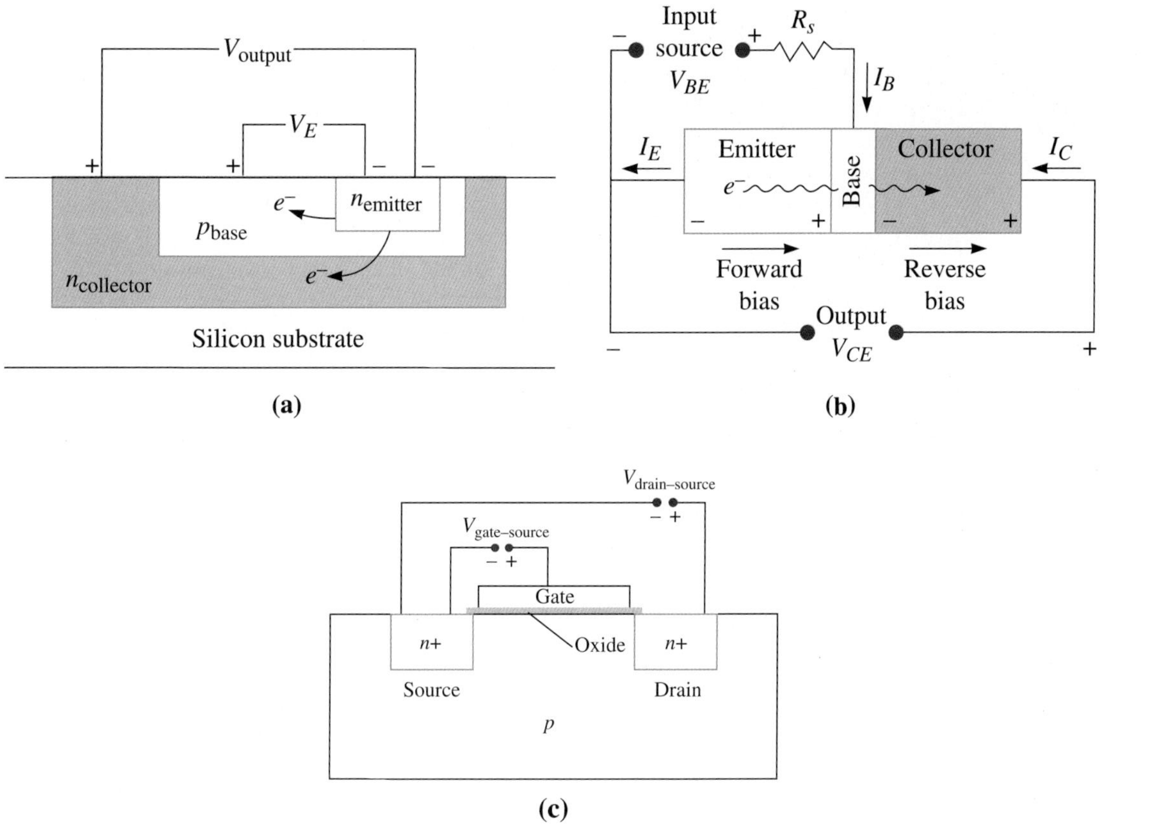

Figure 19-17 (a) Sketch of the cross-section of the transistor. (b) A circuit for an *n-p-n* bipolar junction transistor. The input creates a forward and reverse bias that causes electrons to move from the emitter, through the base, and into the collector, creating an amplified output. (c) Sketch of the cross-section of a metal oxide semiconductor field effect transistor.

reverse bias is produced between the base and the collector (with the positive voltage at the *n*-type collector). The forward bias causes electrons to leave the emitter and enter the base.

Electrons and holes attempt to recombine in the base; however, if the base is exceptionally thin and lightly doped, or if the recombination time τ is long, almost all of the electrons pass through the base and enter the collector. The reverse bias between the base and collector accelerates the electrons through the collector, the circuit is completed, and an output signal is produced. The current through the collector (I_c) is given by

$$I_c = I_0 \exp\left(\frac{V_E}{B}\right) \tag{19-14}$$

where I_0 and B are constants and V_E is the voltage between the emitter and the base. If the input voltage V_E is increased, a very large current I_c is produced.

Field Effect Transistors

A second type of transistor, which is almost universally used today for data storage and processing, is the *field effect transistor* (FET). A metal oxide semiconductor (MOS) field effect transistor (or MOSFET) consists of two highly doped *n*-type regions (*n*+) in a *p*-type substrate or two highly doped *p*-type regions in an *n*-type substrate. (The manufacturing processes by which a device such as this is formed will be discussed in Section 19-6.) Consider a MOSFET that consists of two highly doped *n*-type regions in a *p*-type substrate. One of the *n*-type regions is called the source; the second is called the drain. A potential is applied between the source and the drain with the drain region being positive, but in the absence of a third component of the transistor (a conductor called the gate), electrons cannot flow from the source to the drain under the action of the electric field through the low conductivity p-type region. The gate is separated from the semiconductor by a thin insulating layer of oxide and spans the distance between the two n-type regions. In advanced device structures, the insulator is only several atomic layers thick and comprises materials other than pure silica.

A potential is applied between the gate and the source with the gate being positive. The potential draws electrons to the vicinity of the gate (and repels holes), but the electrons cannot enter the gate because of the silica. The concentration of electrons beneath the gate makes this region (known as the channel) more conductive, so that a large potential between the source and drain permits electrons to flow from the source to the drain, producing an amplified signal ("on" state). By changing the input voltage between the gate and the source, the number of electrons in the conductive path changes, thus also changing the output signal. When no voltage is applied to the gate, no electrons are attracted to the region between the source and the drain, and there is no current flow from the source to the drain ("off" state).

19-6 General Overview of Integrated Circuit Processing

Integrated circuits (ICs, also known as microchips) comprise large numbers of electronic components that have been fabricated on the surface of a substrate material in the form of a thin, circular wafer less than 1 mm thick and as large as 300 mm in diameter. Two particularly important components found on ICs are transistors, which can serve as electrical switches, as discussed in Section 19-5, and **capacitors**, which can store data in a digital format. Each wafer may contain several hundred chips. Intel's well-known Xeon™ microprocessor is an example of an individual chip.

When first developed in the early 1960s, an integrated circuit comprised just a few electrical components, whereas modern ICs may include several billion components, all within the area of a postage stamp. The smallest dimensions of IC components (or "devices") now approach the atomic scale. This increase in complexity and sophistication, achieved simultaneously with a dramatic decrease in the cost per component, has enabled the entire information technology era in which we live. Without these achievements, mobile phones, the Internet, desktop computers, medical imaging devices, and portable music systems—to name just a few icons of contemporary life—could not exist. It has been estimated that humans now produce more transistors per year than grains of rice.

Since the inception of modern integrated circuit manufacturing, a reduction in size of the individual components that comprise ICs has been a goal of researchers and technologists working in this field. A common expression of this trend is known as "Moore's Law," named after Gordon Moore, author of a seminal paper published in 1965. In that paper, Moore, who would go on to cofound Intel Corporation, predicted that the rapid growth in the number of components fabricated on a chip represented a trend that would continue far into the future. He was correct, and the general trends he predicted are still in evidence four decades later.

As a result, for instance, we see that the number of transistors on a microprocessor has grown from a few thousand to several hundred million, while dynamic random access memory (DRAM) chips have passed the billion-transistor milestone. This has led to enormous advances in the capability of electronic systems, especially on a per-dollar cost basis. During this time, various dimensions of every component on a chip have shrunk, often by orders of magnitude, with some dimensions now best measured in nanometers. This scaling, as it is called, drove and continues to drive Moore's Law and plays out in almost every aspect of IC design and fabrication. Maintaining this progress has required the commitment of enormous resources, both financial and human, as IC processing tools (and the physical environments in which they reside) have been serially developed to meet the challenges of reliably producing ever-smaller features.

Fabrication of integrated circuits involves several hundred individual processing steps and may require several weeks to effect. In many instances, the same types of processing step are repeated again and again, with some variations and perhaps with other processing steps interposed, to create the integrated circuit. These so-called "unit processes" include methods to deposit thin layers of materials onto a substrate, means to define and create intricate patterns within a layer of material, and methods to introduce precise quantities of dopants into layers or the surface of the wafer. The length scales involved with some of these processes are approaching atomic dimensions.

The equipment used for these unit processes includes some of the most sophisticated and expensive instruments ever devised, many of which must be maintained in "clean rooms" that are characterized by levels of dust and contamination orders of magnitude lower than that found in a surgical suite. A modern IC fabrication facility may require several billion dollars of capital expenditure to construct and a thousand or more people to operate.

Silicon wafers most often are grown using the Czochralski growth technique [Figure 19-18(a)]. A small seed crystal is used to grow very large silicon single crystals. The seed crystal is slowly rotated, inserted into, and then pulled from a bath of molten silicon. Silicon atoms attach to the seed crystal in the desired orientation as the seed crystal is retracted. Float zone and liquid encapsulated Czochralski techniques also are used. Single crystals are preferred, because the electrical properties of uniformly doped and essentially dislocation-free single crystals are better defined than those of polycrystalline silicon.

Following the production of silicon wafers, which itself requires considerable expense and expertise, there are four major classes of IC fabrication procedures. The first, known as "front end" processing, comprises the steps in which the electrical components (e.g., transistors) are created in the uppermost surface regions of a semiconductor wafer. It is important to note that most of the thickness of the wafer exists merely as a mechanical support; the electrically active components are formed on the surface and typically extend only a few thousandths of a millimeter into the wafer. Front-end processing may include a hundred or more steps. A schematic diagram of some exemplar front-end processing steps to produce a field-effect transistor are shown in Figure 19-18(b).

"Back end" processing entails the formation of a network of "interconnections" on and just above the surface of the wafer. Interconnections are formed in thin films of material deposited on top of the wafer that are patterned into precise networks; these serve as three-dimensional conductive pathways that allow electrical signals to pass between the individual electronic components, as required for the IC to operate and perform mathematical and logical operations or to store and retrieve data. Back-end processing culminates with protective layers of materials applied to the wafers that prevent mechanical and environmental damage. One feature of IC fabrication is that large numbers of wafers—each comprising several hundred to several thousand ICs, which in turn may each include several million to several billion individual components—are often fabricated at the same time.

Once back-end processing has been completed, wafers are subjected to a number of testing procedures to evaluate both the wafer as a whole and the individual chips. As the number of components per chip has increased and the size of the components has diminished, testing procedures have themselves become ever more complex and specialized. Wafers with too small a fraction of properly functioning chips are discarded.

The last steps in producing functioning ICs are collectively known as "packaging," during which the wafers are cut apart to produce individual, physically distinct chips. To protect the chips from damage, corrosion, and the like, and to allow electrical signals to pass into and out of the chips, they are placed in special, hermetically sealed containers, often only slightly larger than the chip itself. A computer powered by a single microprocessor contains many other chips for many other functions.

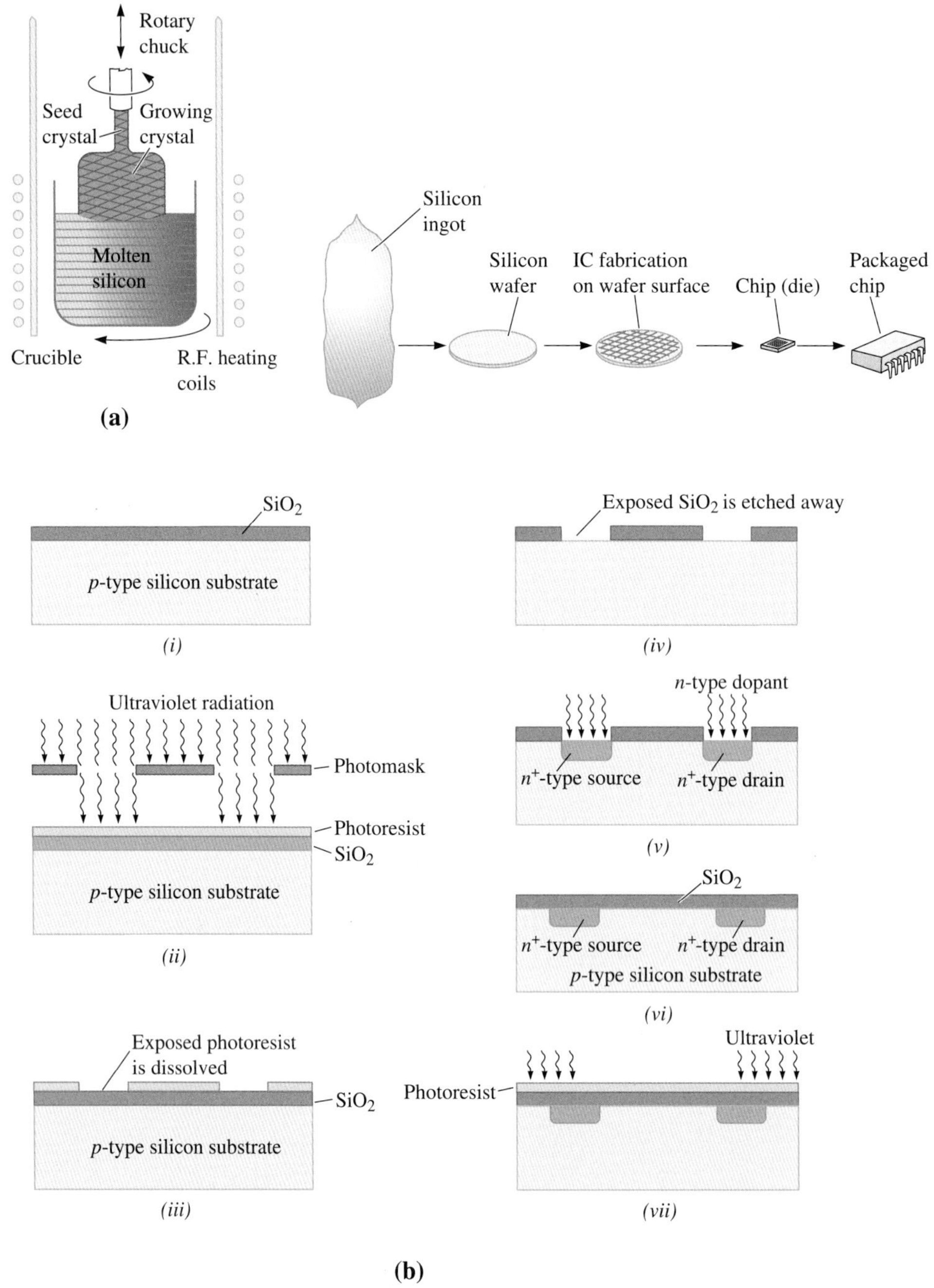

Figure 19-18 (a) Czochralski growth technique for growing single crystals of silicon. (*Based on* Microchip Fabrication, *Third Edition, by P. VanZant, McGraw-Hill, 1997.*) (b) Overall steps encountered in the processing of semiconductors. Production of a FET semiconductor device: (i) A *p*-type silicon substrate is oxidized. (ii) In a process known as photolithography, ultraviolet radiation passes through a photomask (which is much like a stencil), thereby exposing a photosensitive material known as photoresist that was previously deposited on the surface. (iii) The exposed photoresist is dissolved. (iv) The exposed silica is removed by etching. (v) An *n*-type dopant is introduced to produce the source and drain. (vi) The silicon is again oxidized. (vii) Photolithography is repeated to introduce other components, including electrical connections, for the device. (*Based on* Fundamentals of Modern Manufacturing, *by M.P. Groover, John Wiley & Sons, Inc., 1996.*)

19-7 Deposition of Thin Films

As noted in Section 19-6, integrated circuit fabrication depends in part on the deposition of **thin films** of materials onto a substrate. This is similarly true for many technologies that employ films, coatings, or other thin layers of materials, such as wear-resistant coatings on cutting tools, anti-reflective coatings on optical components, and magnetic layers deposited onto aluminum discs for data storage. Thin films may display very different microstructures and physical properties than their bulk counterparts, features that may be exploited in a number of ways. Creating, studying, and using thin films represents a tremendously broad area of materials science and engineering that has enormous impact on modern technology.

Thin films are, as the name implies, very small in one dimension—especially in comparison to their extent in the other two dimensions. There is no well-defined upper bound on what constitutes "thin," but many modern technologies routinely employ thicknesses of several microns down to just a few atomic dimensions. There are myriad ways by which thin films can be deposited, but in general, any technique involves both a *source* of the material to be deposited and a means to *transport* the material from the source to the workpiece surface upon which it is to be deposited. Many deposition techniques require that the source and workpiece be maintained in a vacuum system, while others place the workpiece in a liquid environment.

Physical vapor deposition (PVD) is one very important category of thin-film growth techniques. PVD takes places in a vacuum chamber, and by one means or another creates a low-pressure vapor of the material to be deposited. Some of this vapor will condense on the workpiece and thereby start to deposit as a thin film. Simply melting a material in vacuum, depending on its vapor pressure, may sometimes produce a useful deposit of material.

Sputtering is an example of physical vapor deposition and is the most important PVD method for integrated circuit manufacturing. The interconnections that carry electrical signals from one electronic device to another on an IC chip typically have been made from aluminum alloys that have been sputter deposited. Sputtering can be used to deposit both conducting and insulating materials.

As shown in Figure 19-19, in a sputtering chamber, argon or other atoms in a gas are first ionized and then accelerated by an electric field toward a source of material to be deposited, sometimes called a "target." These ions dislodge and eject atoms from the surface of the source material, some of which drift across a gap toward the workpiece; those that condense on its surface are said to be deposited. Depending on how long the process continues, it is possible to sputter deposit films that are many microns thick.

Chemical vapor deposition (CVD) represents another set of techniques that is widely employed in the IC industry. In CVD, the source of the material to be deposited exists in gaseous form. The source gas and other gases are introduced into a heated vacuum chamber where they undergo a chemical reaction that creates the desired material as a product. This product condenses on the workpiece (as in PVD processes) creating, over time, a layer of the material. In some CVD processes, the chemical reaction may take place preferentially on certain areas of the workpiece, depending on the chemical composition of these areas. Thin films of polycrystalline silicon, tungsten, and titanium nitride are commonly deposited by CVD as part of IC manufacturing. Nanowire growth, which was discussed in Chapter 11, also often proceeds via a CVD process.

Electrodeposition is a third method for creating thin films on a workpiece. Although this is a very old technology, it has recently been adopted for use in IC manufacturing, especially for depositing the copper films that are replacing aluminum films in most advanced integrated circuits. In electrodeposition, the source and workpiece are both immersed in a liquid electrolyte and also are connected by an external electrical circuit. When a voltage is applied between the source and workpiece, ions of the source material dissolve in the electrolyte, drift under the influence of the field towards the workpiece, and chemically bond on its surface. Over time, a thin film is thus deposited. In some circumstances, an external electric field may not be required; this is called electroless deposition. Electrodeposition and electroless deposition are sometimes referred to as "plating," and the deposited film is sometimes said to be "plated out" on the workpiece.

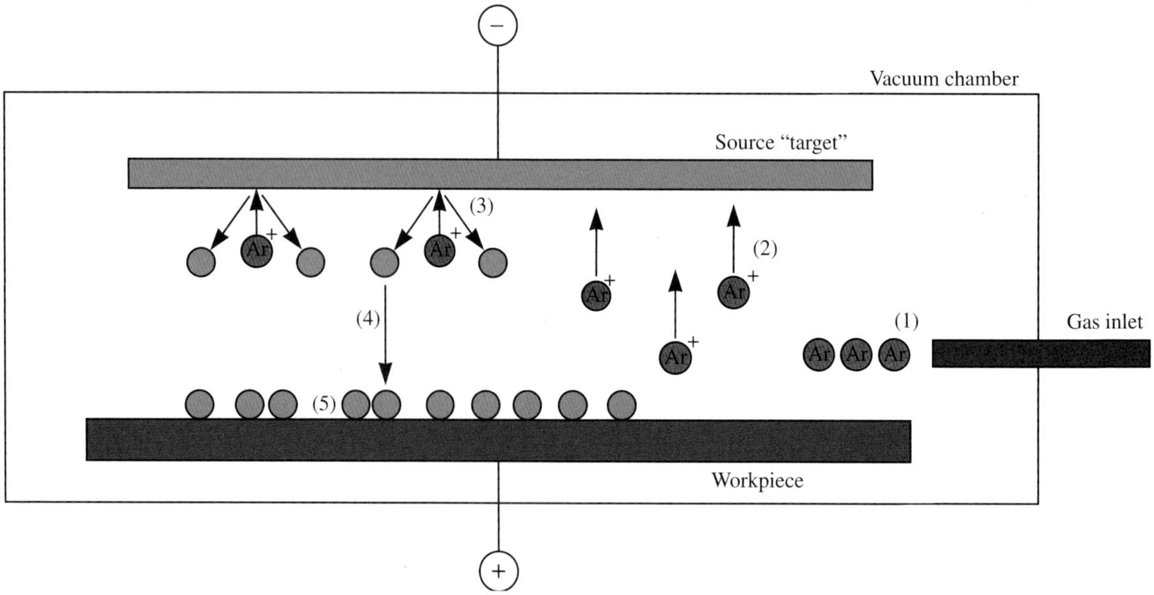

Figure 19-19 Schematic illustration of sputtering. The workpiece and sputter target are placed in a vacuum chamber. (1) An inlet allows a gas such as argon to enter at low pressure. In the presence of the electric field across the target and workpiece, some of the argon atoms are ionized (2) and then accelerated towards the target. By momentum transfer, atoms of the target are ejected (3), drift across the gap towards the workpiece (4), and condense on the workpiece (5), thereby depositing target atoms and eventually forming a film. (*Courtesy of John Bravman.*)

19-8 Conductivity in Other Materials

Electrical conductivity in most ceramics and polymers is low; however, special materials provide limited or even good conduction. In Chapter 4, we saw how the Kröger-Vink notation can be used to explain defect chemistry in ceramic materials. Using dopants, it is possible to convert many ceramics (e.g., $BaTiO_3$, TiO_2, ZrO_2) that are normally insulating into conductive oxides. The conduction in these materials can occur as a result of movement of ions or electrons and holes.

Conduction in Ionic Materials Conduction in ionic materials often occurs by movement of entire ions, since the energy gap is too large for electrons to enter the conduction band. Therefore, most ionic materials behave as insulators.

In ionic materials, the mobility of the charge carriers, or ions, is

$$\mu = \frac{ZqD}{k_B T} \tag{19-15}$$

where D is the diffusion coefficient, k_B is the Boltzmann constant, T is the absolute temperature, q is the electronic charge, and Z is the charge on the ion. The mobility is many orders of magnitude lower than the mobility of electrons; hence, the conductivity is very small:

$$\sigma = nZq\mu \tag{19-16}$$

For ionic materials, n is the concentration of ions contributing to conduction. Impurities and vacancies increase conductivity. Vacancies are necessary for diffusion in substitutional types of crystal structures, and impurities can diffuse and help carry the current. High temperatures increase conductivity because the rate of diffusion increases. The following example illustrates the estimation of mobility and conductivity in MgO.

Example 19-6 *Ionic Conduction in MgO*

Suppose that the electrical conductivity of MgO is determined primarily by the diffusion of the Mg^{2+} ions. Estimate the mobility of the Mg^{2+} ions and calculate the electrical conductivity of MgO at 1800°C. The diffusion coefficient of Mg^{2+} ions in MgO at 1800°C is 10^{-10} cm²/s.

SOLUTION

For MgO, $Z = 2/\text{ion}$, $q = 1.6 \times 10^{-19}$ C, $k_B = 1.38 \times 10^{-23}$ J/K, and $T = 2073$ K:

$$\mu = \frac{ZqD}{k_B T} = \frac{(2)(1.6 \times 10^{-19}\ \text{C})(10^{-10}\ \text{cm}^2/\text{s})}{(1.38 \times 10^{-23}\ \text{J/K})(2073\ \text{K})} = 1.12 \times 10^{-9}\text{C} \cdot \text{cm}^2/(\text{J} \cdot \text{s})$$

Since one coulomb is equivalent to one ampere · second, and one joule is equivalent to one ampere · second · volt:

$$\mu = 1.12 \times 10^{-9}\text{cm}^2/(\text{V} \cdot \text{s})$$

MgO has the NaCl structure with four magnesium ions per unit cell. The lattice parameter is 3.96×10^{-8} cm, so the number of Mg^{2+} ions per cubic centimeter is

$$n = \frac{(4\ \text{Mg}^{2+}\ \text{ions/cell})}{(3.96 \times 10^{-8}\text{cm})^3/\text{cell}} = 6.4 \times 10^{22}\ \text{ions/cm}^3$$

$$\sigma = nZq\mu = (6.4 \times 10^{22})(2)(1.6 \times 10^{-19})(1.12 \times 10^{-9})$$

$$= 23 \times 10^{-6}\text{C} \cdot \text{cm}^2/(\text{cm}^3 \cdot \text{V} \cdot \text{s})$$

Since one coulomb is equivalent to one ampere · second (A · s) and one volt is equivalent to one ampere · ohm (A · Ω),

$$\sigma = 2.3 \times 10^{-5}\text{ohm}^{-1} \cdot \text{cm}^{-1}$$

Applications of Ionically Conductive Oxides The most widely used conductive and transparent oxide is indium tin oxide (ITO), used as a transparent conductive coating on plate glass. Other applications of ITO include touch screen displays for computers and devices such as automated teller machines. Other conductive oxides include ytrria stabilized zirconia (YSZ), which is used as a solid electrolyte in solid oxide fuel cells. Lithium cobalt oxide is used as a solid electrolyte in lithium ion batteries. It is important to remember that, although most ceramic materials behave as electrical insulators, by properly engineering the point defects in ceramics, it is possible to convert many of them into semiconductors.

Conduction in Polymers Because their valence electrons are involved in covalent bonding, polymers have a band structure with a large energy gap, leading to low-electrical conductivity. Polymers are frequently used in applications that require electrical insulation to prevent short circuits, arcing, and safety hazards. Table 19-1 includes the conductivity of four common polymers. In some cases, however, the low conductivity is a hindrance. For example, if lightning strikes the polymer-matrix composite wing of an airplane, severe damage can occur. We can solve these problems by two approaches: (1) introducing an additive to the polymer to improve the conductivity, and (2) creating polymers that inherently have good conductivity.

The introduction of electrically conductive additives can improve conductivity. For example, polymer-matrix composites containing carbon or nickel-plated carbon fibers combine high stiffness with improved conductivity; hybrid composites containing metal fibers, along with normal carbon, glass, or aramid fibers, also produce lightning-safe aircraft skins. Figure 19-20 shows that when enough carbon fibers are introduced to nylon in order to ensure fiber-to-fiber contact, the resistivity is reduced by many orders of

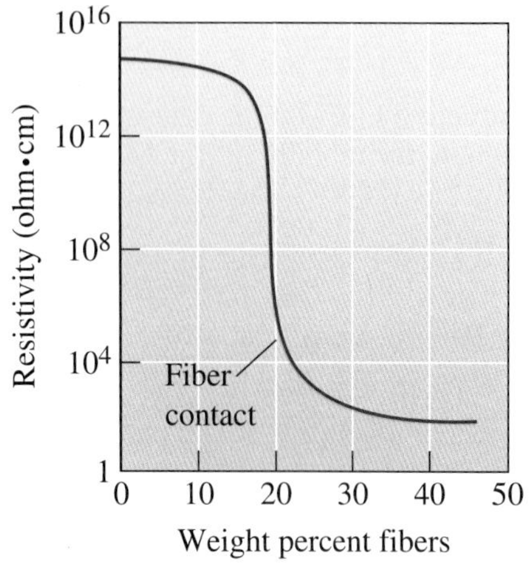

Figure 19-20 Effect of carbon fibers on the electrical resistivity of nylon.

magnitude. Conductive fillers and fibers are also used to produce polymers that shield against electromagnetic radiation.

Some polymers inherently have good conductivity as a result of doping or processing techniques. When acetal polymers are doped with agents such as arsenic pentafluoride, electrons or holes are able to jump freely from one atom to another along the backbone of the chain, increasing the conductivity to near that of metals. Some polymers, such as polyphthalocyanine, can be cross-linked by special curing processes to raise the conductivity to as high as 10^2 ohm^{-1} · cm^{-1}, a process that permits the polymer to behave as a semiconductor. Because of the cross-linking, electrons can move more easily from one chain to another. Organic light-emitting diodes are fabricated from semiconducting polymers including polyanilines.

19-9 Insulators and Dielectric Properties

Materials used to insulate an electric field from its surroundings are required in a large number of electrical and electronic applications. Electrical insulators obviously must have a very low conductivity, or high resistivity, to prevent the flow of current. Insulators must also be able to withstand intense electric fields. Insulators are produced from ceramic and polymeric materials in which there is a large energy gap between the valence and conduction bands; however, the high-electrical resistivity of these materials is not always sufficient. At high voltages, a catastrophic breakdown of the insulator may occur, and current may flow. For example, the electrons may have kinetic energies sufficient to ionize the atoms of the insulator, thereby creating free electrons and generating a current at high voltages. In order to select an insulating material properly, we must understand how the material stores, as well as conducts, electrical charge. Porcelain, alumina, cordierite, mica, and some glasses and plastics are used as insulators. The resistivity of most of these is $> 10^{14}\ \Omega \cdot$cm, and the breakdown electric fields are ~5 to 15 kV/mm.

19-10 Polarization in Dielectrics

When we apply stress to a material, some level of strain develops. Similarly, when we subject materials to an electric field, the atoms, molecules, or ions respond to the applied electric field (E). Thus, the material is said to be polarized. A dipole is a pair of opposite charges separated by a certain distance. If one charge of $+q$ is separated from another charge of $-q$ (q is the electronic charge) and d is the distance between these charges, the dipole moment is defined as the product of q and d. The magnitude of polarization is given by $P = zqd$, where z is the number of charge centers that are displaced per cubic meter.

Any separation of charges (e.g., between the nucleus and electron cloud) or any mechanism that leads to a change in the separation of charges that are already present (e.g., movement or vibration of ions in an ionic material) causes **polarization**. There are four primary mechanisms causing polarization: (1) electronic polarization, (2) ionic polarization, (3) molecular polarization, and (4) space charge (Figure 19-21). Their occurrence depends upon the electrical frequency of the applied field, just like the mechanical behavior of materials depends on the strain rate (Chapters 6 and 8). If we apply a very rapid rate of strain, certain mechanisms of plastic deformation are not activated. Similarly, if we apply a rapidly alternating electric field, some polarization mechanisms may be unable to induce polarization in the material.

Polarization mechanisms play two important roles. First, if we make a **capacitor** from a material, the polarization mechanisms allow charge to be stored, since the dipoles created in the material (as a result of polarization) can bind a certain portion of the charge on the electrodes of the capacitor. Thus, the higher the dielectric polarization, the higher the dielectric constant (k) of the material. The dielectric constant is defined as the ratio of capacitance between a capacitor filled with dielectric material and one with vacuum between its electrodes. This charge storage, in some ways, is similar to the elastic strain in a

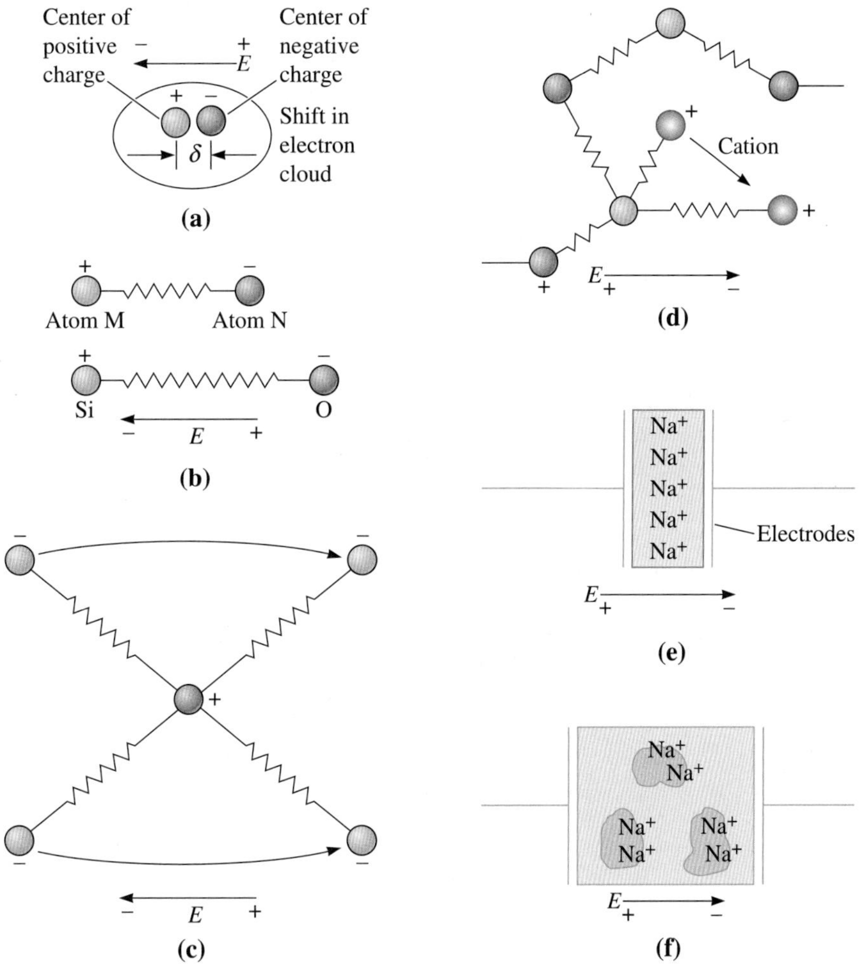

Figure 19-21 Polarization mechanisms in materials: (a) electronic, (b) atomic or ionic, (c) high-frequency dipolar or orientation (present in ferroelectrics), (d) low-frequency dipolar (present in linear dielectrics and glasses), (e) interfacial-space charge at electrodes, and (f) interfacial-space charge at heterogeneities such as grain boundaries. (*Based on* Principles of Electronic Ceramics, *L. L. Hench and J. K. West, Wiley Interscience, John Wiley & Sons, Inc., 1990.*)

material subjected to stress. The second important role played by polarization mechanisms is that when polarization sets in, charges move (ions or electron clouds are displaced). If the electric field oscillates, the charges move back and forth. These displacements are extremely small (typically < 1 Å); however, they cause **dielectric losses**. This energy is lost as heat. The dielectric loss is similar to the viscous deformation of a material. If we want to store a charge, as in a capacitor, dielectric loss is not good; however, if we want to use microwaves to heat up our food, dielectric losses that occur in water contained in the food are great! The dielectric losses are often measured by a parameter known as *tan* δ. When we are interested in extremely low loss materials, such as those used in microwave communications, we refer to a parameter known as the dielectric *quality factor* ($Q_d \sim 1/\tan\delta$). The dielectric constant and dielectric losses depend strongly on electrical frequency and temperature.

Electronic polarization is omnipresent since all materials contain atoms. The electron cloud gets displaced from the nucleus in response to the field seen by the atoms. The separation of charges creates a dipole moment [Figure 19-21(a)]. This mechanism can survive at the highest electrical frequencies ($\sim 10^{15}$ Hz) since an electron cloud can be displaced rapidly, back and forth, as the electrical field switches. Larger atoms and ions have higher electronic polarizability (tendency to undergo polarization), since the electron cloud is farther away from the nucleus and held less tightly. This polarization mechanism is also linked closely to the refractive index of materials, since light is an electromagnetic wave for which the electric field oscillates at a very high frequency ($\sim 10^{14} - 10^{16}$ Hz). The higher the electronic polarizability, the higher the refractive index. We use this mechanism in making "lead crystal," which is really an amorphous glass that contains up to 30% PbO. The large lead ions (Pb^{+2}) are highly polarizable due to the electronic polarization mechanisms and provide a high-refractive index when high enough concentrations of lead oxide are present in the glass. This high-refractive index causes more light to be reflected than by a glass with a lower refractive index, thereby enhancing its appearance for aesthetic purposes.

Example 19-7 *Electronic Polarization in Copper*

Suppose that the average displacement of the electrons relative to the nucleus in a copper atom is 10^{-8} Å when an electric field is imposed on a copper plate. Calculate the electronic polarization.

SOLUTION

The atomic number of copper is 29, so there are 29 electrons in each copper atom. The lattice parameter of copper is 3.6151 Å. Thus,

$$z = \frac{(4\,\text{atoms/cell})(29\ \text{electrons/atom})}{(3.6151 \times 10^{-10}\,\text{m})^3/\text{cell}} = 2.455 \times 10^{30}\,\text{electrons/m}^3$$

$$P = zqd = \left(2.455 \times 10^{30}\,\frac{\text{electrons}}{\text{m}^3}\right)\left(1.6 \times 10^{-19}\,\frac{\text{C}}{\text{electron}}\right)(10^{-8}\text{Å})(10^{-10}\,\text{m/Å})$$

$$= 3.93 \times 10^{-7}\,\text{C/m}^2$$

Frequency and Temperature Dependence of the Dielectric Constant and Dielectric Losses A capacitor is a device that is capable of storing electrical charge. It typically consists of two electrodes with a dielectric material situated between them. The dielectric may or may not be a solid; even an air gap or vacuum can serve as a dielectric. Two parallel, flat-plate electrodes represent the simplest configuration for a capacitor.

Capacitance C is the ability to store charge and is defined as

$$C = \frac{Q}{V} \tag{19-17}$$

where Q is the charge on the electrode plates of a capacitor and V is the applied voltage [Figure 19-22(a)]. Note that a voltage must be applied to create the charge on the electrodes, but that the charge is "stored" in the absence of the voltage until an external circuit allows it to dissipate. In microelectronic devices, this is the basis for digital data storage. If the space between two parallel plates (with surface area A and separated by a distance t) is filled with a material, then the dielectric constant k, also known as the relative permittivity ε_r, is determined according to

$$C = \frac{\mathrm{k}\,\varepsilon_0 A}{t} \tag{19-18}$$

The constant ε_0 is the **permittivity** of a vacuum and is 8.85×10^{-12} F/m. As the material undergoes polarization, it can bind a certain amount of charge on the electrodes, as shown in Figure 19-22(b). The greater the polarization, the higher the dielectric constant, and therefore, the greater the bound charge on the electrodes.

The dielectric constant is the measure of how susceptible the material is to the applied electric field. The dielectric constant depends on the composition, microstructure, electrical frequency, and temperature. The capacitance depends on the dielectric constant, area of the electrodes, and the separation between the electrodes. Capacitors in parallel provide added capacitance (just like resistances add in series). This is the reason why multi-layer capacitors consist of 100 or more layers connected in parallel. These are typically based on $BaTiO_3$ formulations and are prepared using a tape-casting process. Silver-palladium or nickel is used as electrode layers.

For electrical insulation, the **dielectric strength** (i.e., the electric field value that can be supported prior to electrical breakdown) is important. The dielectric properties of some materials are shown in Table 19-7.

Linear and Nonlinear Dielectrics

The dielectric constant, as expected, is related to the polarization that can be achieved in the material. We can show that the dielectric polarization induced in a material depends upon the applied electric field and the dielectric constant according to

$$P = (\mathrm{k} - 1)\varepsilon_0 E \text{ (for linear dielectrics)} \tag{19-19}$$

where E is the strength of the electric field (V/m). For materials that polarize easily, both the dielectric constant and the capacitance are large and, in turn, a large quantity of charge can be stored. In addition, Equation 19-19 suggests that polarization increases, at least until all of the dipoles are aligned, as the voltage (expressed by the strength of the

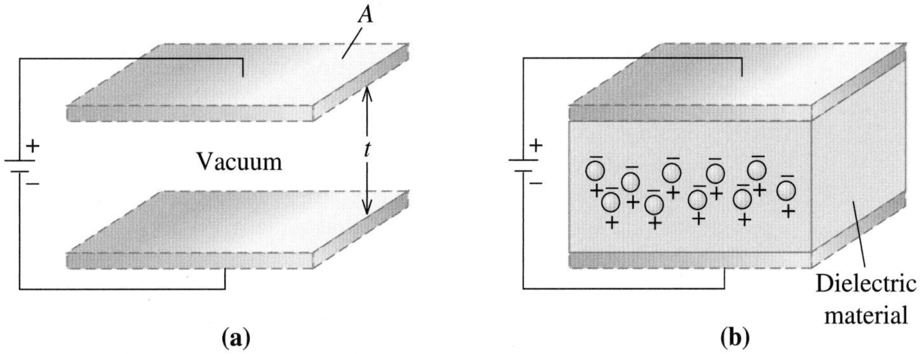

Figure 19-22 (a) A charge can be stored on the conductor plates in a vacuum. (b) When a dielectric is placed between the plates, the dielectric polarizes, and additional charge is stored.

Table 19-7 Properties of selected dielectric materials

Material	Dielectric Constant (at 60 Hz)	Dielectric Constant (at 10^6 Hz)	Dielectric Strength (10^6 V/m)	tan δ (at 10^6 Hz)	Resistivity (ohm · cm)
Polyethylene	2.3	2.3	20	0.00010	$> 10^{16}$
Teflon	2.1	2.1	20	0.00007	10^{18}
Polystyrene	2.5	2.5	20	0.00020	10^{18}
PVC	3.5	3.2	40	0.05000	10^{12}
Nylon	4.0	3.6	20	0.04000	10^{15}
Rubber	4.0	3.2	24		
Phenolic	7.0	4.9	12	0.05000	10^{12}
Epoxy	4.0	3.6	18		10^{15}
Paraffin wax		2.3	10		10^{13}–10^{19}
Fused silica	3.8	3.8	10	0.00004	10^{11}–10^{12}
Soda-lime glass	7.0	7.0	10	0.00900	10^{15}
Al_2O_3	9.0	6.5	6	0.00100	10^{11}–10^{13}
TiO_2		14–110	8	0.00020	10^{13}–10^{18}
Mica		7.0	40		10^{13}
$BaTiO_3$		2000–5000	12	~0.0001	10^{8}–10^{15}
Water		78.3			10^{14}

electric field) increases. The quantity (k − 1) is known as dielectric susceptibility (χ_e). The dielectric constant of vacuum is one, or the dielectric susceptibility is zero. This makes sense since a vacuum does not contain any atoms or molecules.

In **linear dielectrics**, P is linearly related to E and k is constant. This is similar to how stress and strain are linearly related by Hooke's law. In linear dielectrics, k (or χ_e) remains constant with changing E. In materials such as $BaTiO_3$, the dielectric constant changes with E, and hence, Equation 19-19 cannot be used. These materials in which P and E are not related by a straight line are known as **nonlinear dielectrics** or ferroelectrics. These materials are similar to elastomers for which stress and strain are not linearly related and a unique value of the Young's modulus cannot be assigned.

19-11 Electrostriction, Piezoelectricity, and Ferroelectricity

When any material undergoes polarization, its ions and electron clouds are displaced, causing the development of a mechanical strain in the material. This effect is seen in all materials subjected to an electric field and is known as **electrostriction.**

Of the total 32 crystal classes, eleven have a center of symmetry. This means that if we apply a mechanical stress, there is no dipole moment generated since ionic movements are symmetric. Of the 21 that remain, 20 point groups, which lack a center of symmetry, exhibit the development of dielectric polarization when subjected to stress. These materials are known as **piezoelectric.** (The word *piezo* means pressure.) When these materials are stressed, they develop a voltage. This development of a voltage upon the application of stress is known as the *direct* or *motor piezoelectric effect* (Figure 19-23). This effect helps us make devices such as spark igniters, which are often made using lead zirconium titanate (PZT). This effect is also used, for example, in detecting submarines and other objects under water.

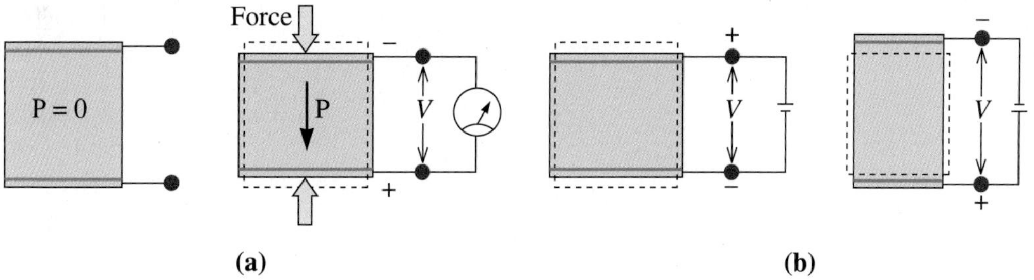

Figure 19-23 The (a) direct and (b) converse piezoelectric effect. In the direct piezoelectric effect, applied stress causes a voltage to appear. In the converse effect (b), an applied voltage leads to the development of strain.

Conversely, when an electrical voltage is applied, a piezoelectric material shows the development of strain. This is known as the *converse* or *generator piezoelectric effect.* This effect is used in making actuators. For example, this movement can be used to generate ultrasonic waves that are used in medical imaging, as well as such applications as ultrasonic cleaners or toothbrushes. Sonic energy can also be created using piezoelectrics to make the high-fidelity "tweeter" found in most speakers. In addition to $Pb(Zr_xTi_{1-x})O_3$ (PZT), other piezoelectrics include SiO_2 (for making quartz crystal oscillators), ZnO, and polyvinylidene fluoride (PVDF). Many naturally occurring materials such as bone and silk are also piezoelectric.

The "d" constant for a piezoelectric is defined as the ratio of strain (ε) to electric field:

$$\varepsilon = d \cdot E \tag{19-20}$$

The "g" constant for a piezoelectric is defined as the ratio of the electric field generated to the stress applied (X):

$$E = g \cdot X \tag{19-21}$$

The d and g piezoelectric coefficients are related by the dielectric constant as:

$$g = \frac{d}{k\varepsilon_0} \tag{19-22}$$

We define **ferroelectrics** as materials that show the development of a spontaneous and reversible dielectric polarization (P_s). Examples include the tetragonal polymorph of barium titanate. Lead zirconium titanate is both ferroelectric and piezoelectric. Ferroelectric materials show a **hysteresis loop** (i.e., the induced polarization is not linearly related to the applied electric field) as seen in Figure 19-24. Ferroelectric materials exhibit ferroelectric domains in which the region (or domain) has uniform polarization (Figure 19-25.) Certain ferroelectrics, such as PZT, exhibit a strong piezoelectric effect, but in order to maximize the piezoelectric effect (e.g., the development of strain or voltage), piezoelectric materials are deliberately "poled" using an electric field to align all domains in one direction. The electric field is applied at high temperature and maintained while the material is cooled.

The dielectric constant of ferroelectrics reaches a maximum near a temperature known as the **Curie temperature**. At this temperature, the crystal structure acquires a center of symmetry and thus is no longer piezoelectric. Even at these high temperatures, however, the dielectric constant of ferroelectrics remains high. $BaTiO_3$ exhibits this behavior, and this is the reason why $BaTiO_3$ is used to make single and multi-layer capacitors. In this state, vibrations and shocks do not generate spurious voltages due to the piezoelectric effect. Since the Curie transition occurs at a high temperature, use of additives in $BaTiO_3$ helps to shift the Curie transition temperature to lower temperatures. Additives also can be used to broaden the Curie transition. Materials such as $Pb(Mg_{1/3}Nb_{2/3})O_3$ or PMN are known as *relaxor ferroelectrics.*

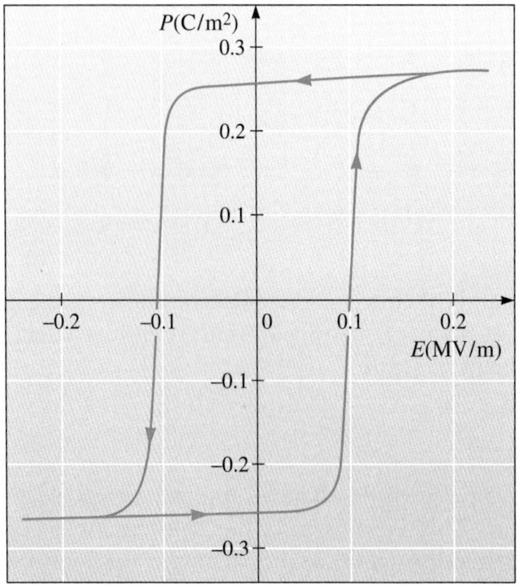

Figure 19-24 The ferroelectric hysteresis loop for a single-domain single crystal of $BaTiO_3$. (*From* Electroceramics: Material, Properties, Applications, *by A. J. Moulson and J. M. Herbert, Chapman and Hall, Kluwer Academic Publishers, 1990.*)

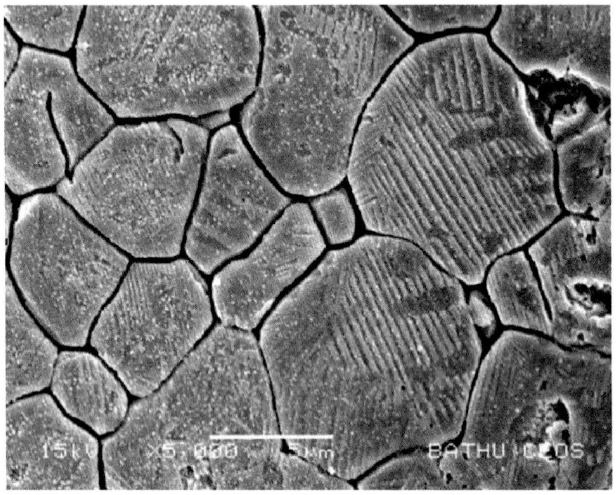

Figure 19-25 Ferroelectric domains can be seen in the microstructure of polycrystalline $BaTiO_3$. (*Courtesy of Professor Chris Bowen*)

These materials show very high dielectric constants (up to 20,000) and good piezoelectric behavior, so they are used to make capacitors and piezoelectric devices.

Example 19-8 *Design of a Multi-Layer Capacitor*

A multi-layer capacitor is to be designed using a $BaTiO_3$ formulation containing $SrTiO_3$. The dielectric constant of the material is 3000. (a) Calculate the capacitance of a multi-layer capacitor consisting of 100 layers connected in parallel using nickel electrodes. The area of each layer is 10 mm × 5 mm, and the thickness of each layer is 10 μm.

SOLUTION

(a) The capacitance of a parallel plate capacitor is given by

$$C = \frac{k\varepsilon_0 A}{t}$$

Thus, the capacitance per layer will be

$$C_{\text{layer}} = \frac{(3000)(8.85 \times 10^{-12}\,\text{F/m})(10 \times 10^{-3}\,\text{m})(5 \times 10^{-3}\,\text{m})}{10 \times 10^{-6}\,\text{m}}$$

$$C_{\text{layer}} = 13.28 \times 10^{-8}\ \text{F}$$

We have 100 layers connected in parallel. Capacitances add in this arrangement. All layers have the same geometric dimensions in this case.

$$C_{\text{total}} = (\text{number of layers}) \cdot (\text{capacitance per layer})$$

$$C_{\text{total}} = (100)(13.28 \times 10^{-8}\,\text{F}) = 13.28\,\mu\text{F}$$

Summary

- Electronic materials include insulators, dielectrics, conductors, semiconductors, and superconductors. These materials can be classified according to their band structures. Electronic materials have enabled many technologies ranging from high-voltage line insulators to solar cells, computer chips, and many sensors and actuators.
- Important properties of conductors include the conductivity and the temperature dependence of conductivity. In pure metals, the resistivity increases with temperature. Resistivity is sensitive to impurities and microstructural defects such as grain boundaries. Resistivity of alloys is typically higher than that of pure metals.
- Semiconductors have conductivities between insulators and conductors and are much poorer conductors than metals. The conductivities of semiconductors can be altered by orders of magnitude by minute quantities of certain dopants. Semiconductors can be classified as elemental (Si, Ge) or compound (GaN, InP, GaAs). Both of these can be intrinsic or extrinsic (*n*- or *p*-type). Some semiconductors have direct bandgaps (e.g., GaAs), while others have indirect bandgaps (e.g., Si).
- Creating an *n*-type region in a *p*-type semiconductor (or vice versa) forms a *p*-*n* junction. The *p*-*n* junction is used to make diodes and transistors.
- Microelectronics fabrication involves hundreds of precision processes that can produce hundreds of millions and even a billion transistors on a single microchip.
- Thin films are integral components of microelectronic devices and also are used for wear-resistant and anti-reflective coatings. Thin films can be deposited using a variety of techniques, including physical vapor deposition, chemical vapor deposition, and electrodeposition.
- Ionic materials conduct electricity via the movement of ions or electrons and holes.
- Dielectrics have large bandgaps and do not conduct electricity. With insulators, the focus is on breakdown voltage or field. With dielectrics, the emphasis is on the dielectric constant, frequency, and temperature dependence. Polarization mechanisms in materials dictate this dependence.
- In piezoelectrics, the application of stress results in the development of a voltage; the application of a voltage causes strain.
- Ferroelectrics are materials that show a reversible and spontaneous polarization. $BaTiO_3$, PZT, and PVDF are examples of ferroelectrics. Ferroelectrics exhibit a large dielectric constant and are often used to make capacitors.

Glossary

Bandgap A range of energy values in which there are no allowed electron energy levels (except those attributed to the presence of impurities). These values fall between the top of the valence band and the bottom of the conduction band.

Band structure The band structure consists of the array of energy levels that are available to or forbidden for electrons to occupy and determines the electronic behavior of a solid, such as whether it is a conductor, semiconductor, or insulator.

Capacitor A device that is capable of storing electrical charge. It typically consists of two electrodes with a dielectric material situated between them, but even an air gap can serve as a dielectric. A capacitor can be a single layer or multi-layer device.

Chemical Vapor Deposition A thin-film growth process in which gases undergo a reaction in a heated vacuum chamber to create the desired product on a substrate.

Conduction band The unfilled energy levels into which electrons are excited in order to conduct.

Curie temperature The temperature above which a ferroelectric is no longer piezoelectric.

Current density Current per unit cross-sectional area.

Dielectric constant The ratio of the permittivity of a material to the permittivity of vacuum, thus describing the relative ability of a material to polarize and store a charge; the same as relative permittivity.

Dielectric loss A measure of how much electrical energy is lost due to motion of charge entities that respond to an electric field via different polarization mechanisms. This energy appears as heat.

Dielectric strength The maximum electric field that can be maintained between two conductor plates without causing a breakdown.

Doping Deliberate addition of controlled amounts of other elements to increase the number of charge carriers in a semiconductor.

Drift velocity The average rate at which electrons or other charge carriers move through a material under the influence of an electric or magnetic field.

Electric field The voltage gradient or volts per unit length.

Electrodeposition A method for depositing materials in which a source and workpiece are connected electrically and immersed in an electrolyte. A voltage is applied between the source and workpiece, and ions from the source dissolve in the electrolyte, drift to the workpiece, and gradually deposit a thin film on its surface.

Electrostriction The dimensional change that occurs in any material when an electric field acts on it.

Energy gap A range of energy values in which there are no allowed electron energy levels (except those attributed to the presence of impurities). These values fall between the top of the valence band and the bottom of the conduction band.

Extrinsic semiconductor A semiconductor prepared by adding dopants, which determine the number and type of charge carriers. Extrinsic behavior can also be seen due to impurities.

Fermi energy The energy level at which the probability of finding an electron is $1/2$.

Ferroelectric A material that shows spontaneous and reversible dielectric polarization.

Forward bias Connecting a *p-n* junction device so that the *p*-side is connected to a positive terminal, thereby enabling current to flow.

Holes Unfilled energy levels in the valence band. Because electrons move to fill these holes, the holes produce a current.

Hysteresis loop The loop traced out by the nonlinear polarization in a ferroelectric material as the electric field is cycled. A similar loop occurs in certain magnetic materials.

Integrated circuit An electronic package that comprises large numbers of electronic devices fabricated on a single chip.

Intrinsic semiconductor A semiconductor in which properties are controlled by the element or compound that is the semiconductor and not by dopants or impurities.

Linear dielectrics Materials in which the dielectric polarization is linearly related to the electric field; the dielectric constant is not dependent on the electric field.

Matthiessen's rule The resistivity of a metal is given by the addition of a base resistivity that accounts for the effect of temperature and a temperature independent term that reflects the effect of atomic level defects, including solutes forming solid solutions.

Mean free path The average distance that electrons move without being scattered by other atoms or lattice defects.

Microstructure-sensitive property Properties that depend on the microstructure of a material (e.g., conductivity, dielectric constant, or yield strength).

Mobility The ease with which a charge carrier moves through a material.

Nonlinear dielectrics Materials in which dielectric polarization is not linearly related to the electric field (e.g., ferroelectric). These have a field-dependent dielectric constant.

Nonradiative recombination The generation of heat when an electron loses energy and falls from the conduction band to the valence band to occupy a hole; this occurs mainly in indirect bandgap materials such as Si.

***p-n* junction** A device made by creating an *n*-type region in a *p*-type material (or vice versa). A *p-n* junction behaves as a diode and multiple *p-n* junctions function as transistors. It is also the basis of LEDs and solar cells.

Permittivity The ability of a material to polarize and store a charge within it.

Physical Vapor Deposition A thin-film growth process in which a low-pressure vapor supplies the material to be deposited on a substrate. Sputtering is one example of PVD.

Piezoelectrics Materials that develop voltage upon the application of a stress and develop strain when an electric field is applied.

Polarization Movement of charged entities (i.e., electron cloud, ions, dipoles, and molecules) in response to an electric field.

Radiative recombination The emission of light when an electron loses energy and falls from the conduction band to the valence band to occupy a hole; this occurs in direct bandgap materials such as GaAs.

Rectifier A *p-n* junction device that permits current to flow in only one direction in a circuit.

Reverse bias Connecting a junction device so that the *p*-side is connected to a negative terminal; very little current flows through a *p-n* junction under reverse bias.

Sputtering A thin-film growth process by which gas atoms are ionized and then accelerated by an electric field towards the source, or "target," of material to be deposited. These ions eject atoms from the target surface, some of which are then deposited on a substrate. Sputtering is one type of a physical vapor deposition process.

Superconductor A material that exhibits zero electrical resistance under certain conditions.

Thermistor A semiconductor device that is particularly sensitive to changes in temperature, permitting it to serve as an accurate measure of temperature.

Thin film A coating or layer that is small or thin in one dimension. Typical thicknesses range from 10 Å to a few microns depending on the application.

Transistor A semiconductor device that amplifies or switches electrical signals.

Valence band The energy levels filled by electrons in their lowest energy states.

Problems

Section 19-1 Ohm's Law and Electrical Conductivity

Section 19-2 Band Structure of Solids

Section 19-3 Conductivity of Metals and Alloys

19-1 Of the metals listed in Table 19-1, which are the most and least conductive?

19-2 Find the resistance of a fiber with a cross-sectional area of 1.34 mm^2 and length of 10 cm that is subjected to a voltage of 225 V for which the current density is 1.25 A/cm^2. Assume the resistor is uniform in material and properties.

19-3 A wire two microns in diameter is made from 10 cm silver, 0.1 cm gallium, and

10 cm silver all joined end to end (i.e., in series). Assume the joints are perfect and have no additional resistance. What is the total resistance of the wire?

19-4 A current of 10 A passes through a 1-mm-diameter wire 1000 m long. Calculate the power loss if the wire is made from
(a) aluminum and
(b) silicon (see Table 19-1).

19-5 A 0.5-mm-diameter fiber, 1 cm in length, made from boron nitride is placed in a 120-V circuit. Using Table 19-1, calculate
(a) the current flowing in the circuit; and
(b) the number of electrons passing through the boron nitride fiber per second.
(c) What would the current and number of electrons be if the fiber were made from magnesium instead of boron nitride?

19-6 The power lost in a 2-mm-diameter copper wire is to be less than 250 W when a 5 A current is flowing in the circuit. What is the maximum length of the wire?

19-7 A current density of 100,000 A/cm^2 flows through a gold wire 50 m in length. The resistance of the wire is found to be 2 ohm. Calculate the diameter of the wire and the voltage applied to the wire.

19-8 We would like to produce a 5000-ohm resistor from boron-carbide fiber having a diameter of 0.1 mm. What is the required length of the fiber?

19-9 Ag has an electrical conductivity of 6.80 $\times$ 10^5 $\Omega^{-1} \cdot$ cm^{-1}. Au has an electrical conductivity of 4.26 $\times$ 10^5 $\Omega^{-1} \cdot$ cm^{-1}. Calculate the number of charge carriers per unit volume and the electron mobility in each in order to account for this difference in electrical conductivity. Comment on your findings.

19-10 If the current density is 1.25 A/cm^2 and the drift velocity is 10^7 cm/s, how many charge carriers are present?

19-11 A current density of 5000 A/cm^2 flows through a magnesium wire. If half of the valence electrons serve as charge carriers, calculate the average drift velocity of the electrons.

19-12 We apply 10 V to an aluminum wire 2 mm in diameter and 20 m long. If 10% of the valence electrons carry the electrical charge, calculate the average drift velocity of the electrons in km/h and miles/h.

19-13 In a welding process, a current of 400 A flows through the arc when the voltage is 35 V. The length of the arc is about 0.1 in., and the average diameter of the arc is about 0.18 in. Calculate the current density in the arc, the electric field across the arc, and the electrical conductivity of the hot gases in the arc during welding.

19-14 Draw schematics of the band structure for a metal and a semiconductor. Explain in detail why the conductivity of pure metals decreases with increasing temperature, while the opposite is true for semiconductors. Separately write the equations for conductivity for metals and semiconductors. Discuss how each term in each equation changes with temperature.

19-15 A typical thickness for a copper conductor (known as an interconnect) in an integrated circuit is 250 nm. The mean free path of electrons in pure, annealed copper is about 40 nm. As the thickness of copper interconnects approaches the mean free path, how do you expect conduction in the interconnect is affected? Explain.

19-16 Calculate the resistivities of cobalt and beryllium at 505 K.

19-17 Calculate the electrical conductivity of platinum at -200°C.

19-18 Calculate the electrical conductivity of nickel at -50°C and at $+500$°C.

19-19 The electrical resistivity of pure chromium is found to be 18 $\times$ 10^{-6} ohm $\cdot$ cm. Estimate the temperature at which the resistivity measurement was made.

19-20 After finding the electrical conductivity of cobalt at 0°C, we decide we would like to double that conductivity. To what temperature must we cool the metal?

19-21 From Figure 19-9(b), estimate the defect resistivity coefficient for tin in copper.

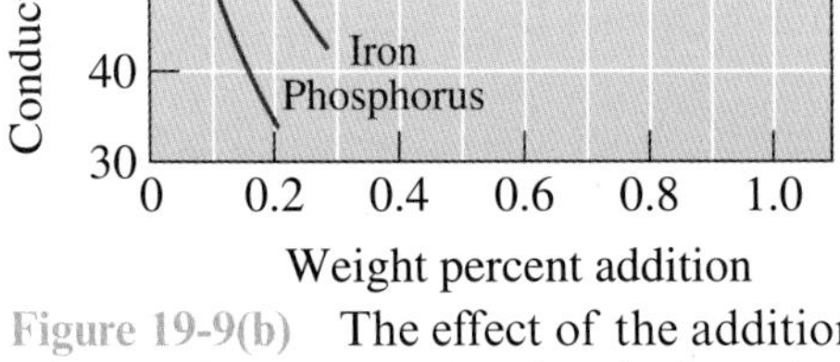

Figure 19-9(b) The effect of the addition of selected elements on the electrical conductivity of copper. (Repeated for Problem 19-21.)

19-22 Find the resistivity of palladium at 443°C if it has a defect contribution of 1.02×10^{-7} ohm·cm.

19-23 (a) Copper and nickel form a complete solid solution. Draw a schematic diagram illustrating the resistivity of a copper and nickel alloy as a function of the atomic percent nickel. Comment on why the curve has the shape that it does.

(b) Copper and gold do not form a complete solid solution. At the compositions of 25 and 50 atomic percent gold, the ordered phases Cu_3Au and CuAu form, respectively. Do you expect that a plot of the resistivity of a copper and gold alloy as a function of the atomic percent gold will have a shape similar to the plot in part (a)? Explain.

19-24 The electrical resistivity of a beryllium alloy containing 5 at% of an alloying element is found to be 50×10^{-6} ohm · cm at 400°C. Determine the contributions to resistivity due to temperature and due to impurities by finding the expected resistivity of pure beryllium at 400°C, the resistivity due to impurities, and the defect resistivity coefficient. What would be the electrical resistivity if the beryllium contained 10 at% of the alloying element at 200°C?

Section 19-4 Semiconductors

Section 19-5 Applications of Semiconductors

Section 19-6 General Overview of Integrated Circuit Processing

Section 19-7 Deposition of Thin Films

19-25 Explain the following terms: semiconductor, intrinsic semiconductor, extrinsic semiconductor, elemental semiconductor, compound semiconductor, direct bandgap semiconductor, and indirect bandgap semiconductor.

19-26 What is radiative and nonradiative recombination? What types of materials are used to make LEDs?

19-27 Report the temperature range (in °C) for which the sample in Figure 19-15 behaves as an extrinsic semiconductor.

19-28 For germanium and silicon, compare, at 25°C, the number of charge carriers per cubic centimeter, the fraction of the total electrons in the valence band that are excited into the conduction band, and the constant n_0.

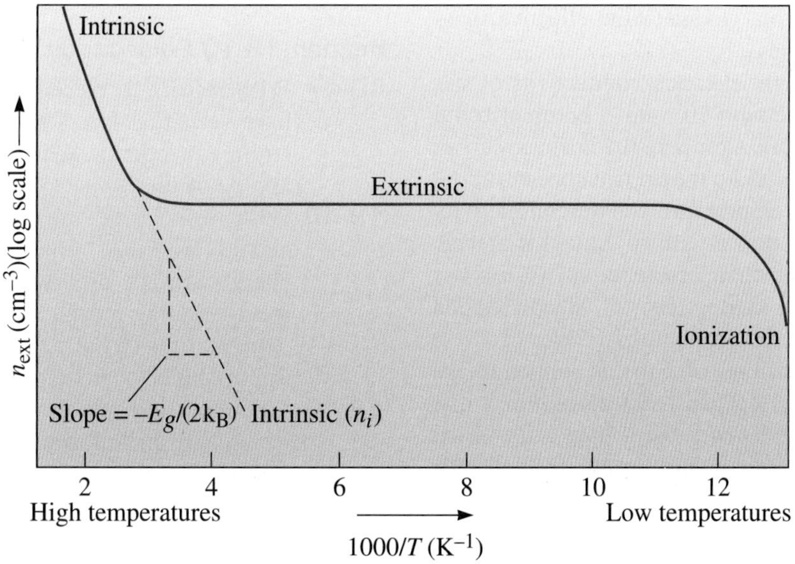

Figure 19-15 The effect of temperature on the carrier concentration of an n-type semiconductor. At low temperatures, the donor or acceptor atoms are not ionized. As temperature increases, the ionization process is complete, and the carrier concentration increases to a level that is dictated by the level of doping. The conductivity then essentially remains unchanged until the temperature becomes too high and the thermally generated carriers begin to dominate. The effect of dopants is lost at very high temperatures, and the semiconductor essentially shows "intrinsic" behavior. (Repeated for Problem 19-27.)

19-29 For germanium and silicon, compare the temperatures required to double the electrical conductivities from the room temperature values.

19-30 Determine the electrical conductivity of silicon when 0.0001 at% antimony is added as a dopant and compare it to the electrical conductivity when 0.0001 at% indium is added.

19-31 We would like to produce an extrinsic germanium semiconductor having an electrical conductivity of 2000 $ohm^{-1} \cdot cm^{-1}$. Determine the amount of phosphorus and the amount of gallium required to make *n*- and *p*-type semiconductors, respectively.

19-32 Estimate the electrical conductivity of silicon doped with 0.0002 at% arsenic at 600°C, which is above the plateau in the conductivity-temperature curve.

19-33 Determine the amount of arsenic that must be combined with 1 kg of gallium to produce a *p*-type semiconductor with an electrical conductivity of 500 $ohm^{-1} \cdot cm^{-1}$ at 25°C. The lattice parameter of GaAs is about 5.65 Å, and GaAs has the zincblende structure.

19-34 Calculate the intrinsic carrier concentration for GaAs at room temperature. Given that the effective mass of electrons in GaAs is $0.067m_e$, where m_e is the mass of the electron, calculate the effective mass of the holes.

19-35 Calculate the electrical conductivity of silicon doped with 10^{18} cm^{-3} boron at room temperature. Compare the intrinsic carrier concentration to the dopant concentration.

19-36 At room temperature, will the conductivity of silicon doped with 10^{17} cm^{-3} of arsenic be greater than, about equal to, or less than the conductivity of silicon doped with 10^{17} cm^{-3} of phosphorus?

19-37 When a voltage of 5 mV is applied to the emitter of a transistor, a current of 2 mA is produced. When the voltage is increased to 8 mV, the current through the collector rises to 6 mA. By what percentage will the collector current increase when the emitter voltage is doubled from 9 mV to 18 mV?

19-38 Design a light-emitting diode that will emit at 870 nm. Is this wavelength in the visible range? What is a potential application for this type of LED?

19-39 How can we make LEDs that emit white light (i.e., light that looks like sunlight)?

19-40 Investigate the scaling relationship known as Moore's Law. Is it expected that this trend will continue to be followed in the future using established microelectronics fabrication techniques? If not, what are some of the alternatives currently being considered?

19-41 Silicon is the material of choice for the substrate for integrated circuits. Explain why silicon is preferred over germanium, even though the electron and hole mobilities are much higher and the bandgap is much smaller for germanium than for silicon.

Section 19-8 Conductivity in Other Materials

19-42 If we want the resistivity of a nylon to be 10^4 ohm · cm, what weight percent carbon fibers should be added?

19-43 Calculate the electrical conductivity of a fiber-reinforced polyethylene part that is reinforced with 20 vol % of continuous, aligned nickel fibers.

19-44 What are ionic conductors? What are their applications?

19-45 How do the touch screen displays on some computers work?

19-46 Can polymers be semiconducting? What would be the advantages in using these instead of silicon?

Section 19-9 Insulators and Dielectric Properties

Section 19-10 Polarization in Dielectrics

19-47 With respect to mechanical behavior, we have seen that stress (a cause) produces strain (an effect). What is the electrical analog of this?

19-48 With respect to mechanical behavior, elastic modulus represents the elastic energy stored, and viscous dissipation represents the mechanical energy lost in deformation. What is the electrical analog of this?

19-49 Calculate the displacement of the electrons or ions for the following conditions:

(a) electronic polarization in nickel of 2×10^{-7} C/m^2;

(b) electronic polarization in aluminum of 2×10^{-8} C/m^2;

(c) ionic polarization in NaCl of 4.3×10^{-8} C/m^2; and

(d) ionic polarization in ZnS of 5×10^{-8} C/m^2.

19-50 A 2-mm-thick alumina dielectric is used in a 60 Hz circuit. Calculate the voltage required to produce a polarization of 5×10^{-7} C/m^2.

19-51 Suppose we are able to produce a polarization of 5×10^{-8} C/m^2 in a cube (5 mm side) of barium titanate. Assume a dielectric constant of 3000. What voltage is produced?

19-52 For the 14 materials with both a dielectric strength and resistivity listed in Table 19-7, plot the dielectric strength as a function of resistivity. Assume log average values for the resistivity ranges.

19-53 What polarization mechanism will be present in (a) alumina; (b) copper; (c) silicon; and (d) barium titanate?

Section 19-11 Electrostriction, Piezoelectricity, and Ferroelectricity

19-54 Define the following terms: electrostriction, piezoelectricity (define both its direct and converse effects), and ferroelectricity.

19-55 Calculate the capacitance of a parallel-plate capacitor containing five layers of mica for which each mica sheet is 1 cm $\times$ 2 cm $\times$ 0.005 cm. The layers are connected in parallel.

19-56 A multi-layer capacitor is to be designed using a relaxor ferroelectric formulation based on lead magnesium niobate (PMN). The apparent dielectric constant of the material is 20,000. Calculate the capacitance of a multi-layer capacitor consisting of ten layers connected in parallel using Ni electrodes. The area of the capacitor is 10 mm $\times$ 10 mm, and the thickness of each layer is 20 μm.

19-57 A force of 20 lb is applied to the face of a 0.5 cm $\times$ 0.5 cm $\times$ 0.1 cm thickness of quartz crystal. Determine the voltage produced by the force. The modulus of elasticity of quartz is 10.4×10^6 psi.

19-58 For the example hysteresis loop in Figure 19-24, what is the polarization (C/m^2) if the electric field starts at 0.2 MV/m and drops to zero?

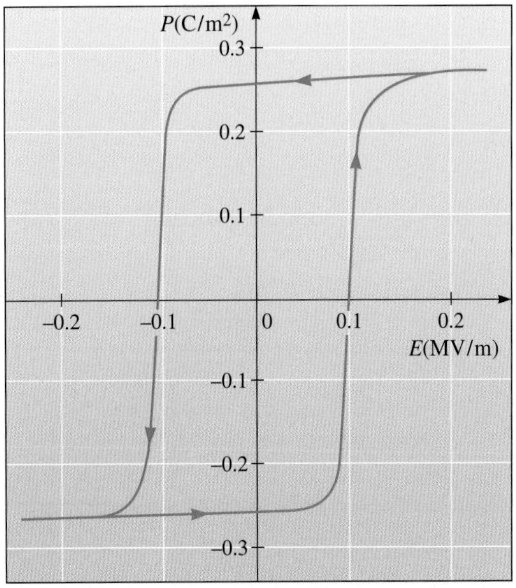

Figure 19-24 The ferroelectric hysteresis loop for a single-domain single crystal of $BaTiO_3$. (*From Electroceramics:* Material, Properties, Applications, *by A. J. Moulson and J. M. Herbert, Chapman and Hall, Kluwer Academic Publishers, 1990.*) (Repeated for Problem 19-58.)

Design Problems

19-59 We would like to produce a 100-ohm resistor using a thin wire of a material. Design such a device.

19-60 Design a capacitor that is capable of storing 1 μF when 100 V is applied.

19-61 Design an epoxy-matrix composite that has a modulus of elasticity of at least 35×10^6 psi and an electrical conductivity of at least 1×10^5 ohm$^{-1} \cdot$ cm^{-1}.

Computer Problems

19-62 *Design of Multi-layer Capacitors.* Write a computer program that can be used to calculate the capacitance of a multi-layer capacitor. The program, for example, should ask the user to provide values of the dielectric constant and the dimensions of the layer. The program should also be flexible in that if the user provides an intended value of capacitance and other dimensions, the program should provide the required dielectric constant.

Knovel® Problems

K19-1 Calculate the resistivity of pure iridium at 673 K using its temperature resistivity coefficient.

K19-2 Electrical conductivity is sometimes given in the units of %IACS. What does IACS stand for? Define the unit using the information found.

K19-3 Can organic materials such as polymers and carbon nanotubes be semiconductors? If they are, what determines their semiconducting properties?

A magnetic hard drive is the heart of many computers and all large-scale data centers, such as those operated by Amazon and Facebook. These disks use magnetic materials that are unique in that information can be easily written to them, but information cannot be easily erased. The hard drive system makes use of nanostructured and nanoscale thin-film magnetic materials for information storage. (*Photodisc Green/Getty Images*)

CHAPTER

20

Magnetic Materials

Have You Ever Wondered?

- *What affects the "lifting strength" of a magnet?*
- *What are "soft" and "hard" magnetic materials?*
- *Are there "nonmagnetic" materials?*
- *Are there materials that develop mechanical strain upon the application of a magnetic field?*

Chapter Learning Objectives

The key objectives of this chapter are to

- Define the terms ferromagnetic, ferrimagnetic, paramagnetic, diamagnetic, anti-ferromagnetic, and superparamagnetic.
- Explain the physical basis for the magnetic behavior of elements based on electron spin and orbital motion of electrons around the nucleus.
- Calculate magnetic field and inductance.
- Define the terms magnetic permeability of vacuum, relative permeability of a material, and magnetic susceptibility.
- Describe the movement of domains in a magnetic field and the accompanying change in magnetization due to an increase in the applied magnetic field.
- Identify saturation magnetization, remanance, and coercive field on ferromagnetic hysteresis loops.

Every material in the world responds to the presence of a magnetic field. Magnetic materials are used to construct such things as electrical motors, generators, and transformers. Much of data storage technology (computer hard disks, computer disks, and the like) is based on magnetic particles. Magnetic materials are also used in loudspeakers, microphones, land-line telephones, and tape-based video recorders. Magnetic materials, such as iron oxide (Fe_3O_4) particles, are used to make exotic compositions of "liquid magnets" or ferrofluids. The same iron oxide particles are also used to bind DNA molecules, cells, and proteins.

In this chapter, we look at the fundamental basis for responses of certain materials to the presence of magnetic fields. We will also examine the properties and applications of different types of magnetic materials.

20-1 Classification of Magnetic Materials

Strictly speaking, there is no such thing as a "nonmagnetic" material. Every material consists of atoms; atoms consist of electrons spinning around them, similar to a current-carrying loop that generates a magnetic field. Thus, every material responds to a magnetic field. The manner in which this response of electrons and atoms in a material is scaled determines whether a material will be strongly or weakly magnetic and also gives rise to many types of magnetic phenomena. Examples of **ferromagnetic** materials are materials such as Fe, Ni, Co, and some of their alloys. Examples of **ferrimagnetic** materials include many ceramic materials such as nickel zinc ferrite and manganese zinc ferrite. The term "nonmagnetic," usually means that the material is neither ferromagnetic nor ferrimagnetic. These "nonmagnetic" materials are further classified as **diamagnetic** (e.g., superconductors) or **paramagnetic**. In some cases, we also encounter materials that are **antiferromagnetic** or **superparamagnetic**. We will discuss these different classes of materials and their applications later in the chapter. Ferromagnetic and ferrimagnetic materials are usually further classified as either soft or hard magnetic materials. High-purity iron or plain carbon steels are examples of a magnetically soft material as they can become magnetized, but when the magnetizing source is removed, these materials lose their magnet-like behavior.

Permanent magnets or **hard magnetic materials** retain their magnetization. These are permanent "magnets." Many ceramic ferrites are used to make inexpensive refrigerator magnets. A hard magnetic material does not lose its magnetic behavior easily.

20-2 Magnetic Dipoles and Magnetic Moments

The magnetic behavior of materials can be traced to the structure of atoms. The orbital motion of the electron around the nucleus and the spin of the electron about its own axis (Figure 20-1) cause separate magnetic moments. These two motions (i.e., spin and orbital) contribute to the magnetic behavior of materials. When the electron spins, there is a magnetic moment associated with that motion. The **magnetic moment** of an electron due to its spin is known as the **Bohr magneton** (μ_B). This is a fundamental constant and is defined as

$$\mu_B = \text{Bohr magneton} = \frac{q\text{h}}{4\pi m_e} = 9.274 \times 10^{-24}\ \text{A}\cdot\text{m}^2 \tag{20-1}$$

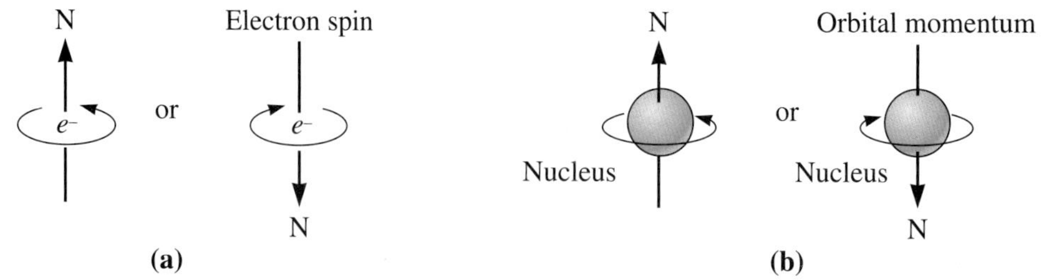

Figure 20-1 Origin of magnetic dipoles: (a) The spin of the electron produces a magnetic field with a direction dependent on the quantum number m_s. (b) Electrons orbiting around the nucleus create a magnetic field around the atom.

where q is the charge on the electron, h is Planck's constant, and m_e is the mass of the electron. This moment is directed along the axis of electron spin.

The nucleus of the atom consists of protons and neutrons. These also have a spin; however, the overall magnetic moment due to their spin is much smaller than that for electrons. We normally do not encounter the effects of a magnetic moment of a nucleus with the exception of such applications as nuclear magnetic resonance (NMR).

We can think of electrons in materials as small elementary magnets. If the magnetic moments due to electrons in materials could line up in the same direction, the world would be a magnetic place! However this, as you know, is not the case. Thus, there must be some mechanism by which the magnetic moments associated with electron spin and their orbital motion get canceled in most materials, leaving behind only a few materials that are "magnetic." There are two effects that, fortunately, make most materials in the world not "magnetic."

First, we must consider the magnetic moment of atoms. According to the Pauli exclusion principle, two electrons within the same orbital must have opposite spins. This means their electron spin derived magnetic moments have opposite signs (one can be considered "spin up ↑" and the other one "spin down ↓") and cancel. The second effect is that the orbital moments of electrons also cancel each other. Thus, in a completely filled shell, all electron spin and orbital moments cancel. This is why atoms of most elements do not have a net magnetic moment. Some elements, such as transition elements (3*d*, 4*d*, 5*d* partially filled), the lanthanides (4*f* partially filled), and actinides (5*f* partially filled), have a net magnetic moment due to an unpaired electron.

Certain elements, such as the transition metals, have an inner energy level that is not completely filled. The elements scandium through copper, the electronic structures of which are shown in Table 20-1, are typical. Except for chromium and copper, the valence electrons in the 4*s* level are paired; the unpaired electrons in chromium and copper are canceled by interactions with other atoms. Copper also has a completely filled 3*d* shell and thus does not display a net magnetic moment.

The electrons in the 3*d* level of the remaining transition elements do not enter the shells in pairs. Instead, as in manganese, the first five electrons have the same spin. Only after half of the 3*d* level is filled do pairs with opposing spins form. Therefore, each atom in a transition metal has a permanent magnetic moment, which is related to the number of unpaired electrons. Each atom behaves as a magnetic dipole.

In many elements, these magnetic moments exist for free individual atoms, however, when the atoms form crystalline materials, these moments are "quenched" or canceled out. Thus, a number of materials made from elements with atoms that have a net magnetic moment do not exhibit magnetic behavior. For example, the Fe^{+2} ion has a net magnetic moment of $4\mu_B$ (four times the magnetic moment of an electron); however, $FeCl_2$ crystals are not magnetic.

Table 20-1 The electron spins in the 3*d* energy level in transition metals with arrows indicating the direction of spin

Metal			3*d*			4*s*
Sc	↑					↓↑
Ti	↑	↑				↓↑
V	↑	↑	↑			↓↑
Cr	↑	↑	↑	↑	↑	↑
Mn	↑	↑	↑	↑	↑	↓↑
Fe	↓↑	↑	↑	↑	↑	↓↑
Co	↓↑	↓↑	↑	↑	↑	↓↑
Ni	↓↑	↓↑	↓↑	↑	↑	↓↑
Cu	↓↑	↓↑	↓↑	↓↑	↓↑	↑

The response of the atom to an applied magnetic field depends on how the magnetic dipoles of each atom react to the field. Most of the transition elements (e.g., Cu, Ti) react in such a way that the sum of the individual atoms' magnetic moments is zero. The atoms in nickel, iron, and cobalt, however, undergo an "exchange interaction," whereby the orientation of the dipole in one atom influences the surrounding atoms to have the same dipole orientation, producing a desirable amplification of the effect of the magnetic field. In the case of Fe, Ni, and Co, the magnetic moments of the atoms line up in the same directions, and these materials are known as ferromagnetic.

In certain materials, such as BCC chromium, the magnetic moments of atoms at the center of the unit cell are opposite in direction to those of the atoms at the corners of the unit cell; thus, the net moment is zero. Materials in which there is a complete cancellation of the magnetic moments of atoms or ions are known as anti-ferromagnetic.

Materials in which magnetic moments of different atoms or ions do not completely cancel out are known as ferrimagnetic materials. We will discuss these materials in a later section.

20-3 Magnetization, Permeability, and the Magnetic Field

Let's examine the relationship between the magnetic field and magnetization. Figure 20-2 depicts a coil having n turns. When an electric current is passed through the coil, a magnetic field H is produced, with the strength of the field given by

$$H = \frac{nI}{l} \tag{20-2}$$

where n is the number of turns, l is the length of the coil (m), and I is the current (A). The units of H are therefore ampere turn/m, or simply A/m. An alternate unit for magnetic field is the oersted, obtained by multiplying A/m by $4\pi \times 10^{-3}$ (see Table 20-2).

When a magnetic field is applied in a vacuum, lines of magnetic flux are induced. The flux density, or *inductance* B, is related to the applied field by

$$B = \mu_0 H \tag{20-3}$$

where B is the inductance, H is the magnetic field, and μ_0 is a constant called the **magnetic permeability of vacuum**. If H is expressed in units of oersted, then B is in gauss and μ_0 is 1 gauss/oersted. In an alternate set of units, H is in A/m, B is in tesla (also called weber/m^2), and μ_0 is $4\pi \times 10^{-7}$ weber/(A · m) (also called henry/m).

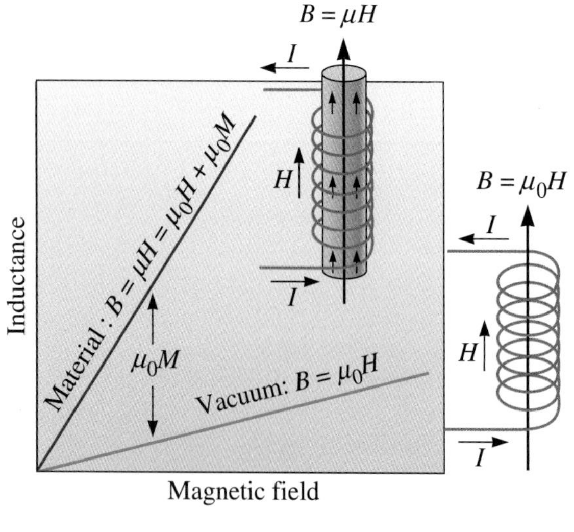

Figure 20-2 A current passing through a coil sets up a magnetic field H with a flux density B. The flux density is higher when a magnetic core is placed within the coil.

Table 20-2 Units, conversions, and values for magnetic materials

	Gaussian and cgs emu (Electromagnetic Units)	SI Units	Conversion
Inductance or magnetic flux density (B)	gauss (G)	tesla [or weber (Wb)/m^2]	1 tesla = 10^4 G, (Wb)/m^2
Magnetic flux (ϕ)	maxwell (Mx), G · cm^2	Wb, volt-second	1 Wb = 10^8 G · cm^2
Magnetic potential difference or magnetic electromotive force (U, F)	gilbert (Gb)	ampere (A)	1 A = $4\pi \times 10^{-1}$ Gb
Magnetic field strength, magnetizing force (H)	oersted (Oe), gilbert (Gb)/cm	A/m	1 A/m = $4\pi \times 10^{-3}$ Oe
(Volume) magnetization (M)	emu/cm^3	A/m	1 A/m = 10^{-3} emu/cm^3
(Volume) magnetization ($4\pi M$)	G	A/m	1 A/m = $4\pi \times 10^{-3}$ G
Magnetic polarization or intensity of magnetization (J or I)	emu/cm^3	T, Wb/m^2	1 tesla = $(1/4\pi) \times 10^4$ emu/cm^3
(Mass) magnetization (σ, M)	emu/g	A · m^2/kg Wb · m/kg	1 A · m^2/kg = 1 emu/g 1 Wb · m/kg = $(1/4\pi) \times 10^7$ emu/g
Magnetic moment (m)	emu, erg/G	A · m^2, joules per tesla (J/T)	1 J/T = 10^3 emu
Magnetic dipole moment (j)	emu, erg/G	Wb · m	1 Wb · m = $(1/4\pi) \times 10^{10}$ emu
Magnetic permeability (μ)	Dimensionless	Wb/(A · m) [henry (H)/m]	1 Wb/(A · m) = $(1/4\pi) \times 10^7$
Magnetic permeability of free space (μ_0)	1 gauss/oersted	$\mu = 4\pi \times 10^{-7}$ H/m	
Relative permeability (μ_r)	Not defined	Dimensionless	
(Volume) energy density, energy product (W)	erg/cm^3	J/m^3	1 J/m^3 = 10 erg/cm^3

When we place a material within the magnetic field, the magnetic-flux density is determined by the manner in which induced and permanent magnetic dipoles interact with the field. The flux density now is

$$B = \mu H \tag{20-4}$$

where μ is the permeability of the material in the field. If the magnetic moments reinforce the applied field, then $\mu > \mu_0$, a greater flux that can accomplish work is created, and the magnetic field is magnified. If the magnetic moments oppose the field, however, $\mu < \mu_0$.

We can describe the influence of the magnetic material by the relative permeability μ_r, where

$$\mu_r = \frac{\mu}{\mu_0} \tag{20-5}$$

A large relative permeability means that the material amplifies the effect of the magnetic field. Thus, the relative permeability has the same importance that conductivity has in dielectrics. A material with higher magnetic permeability (e.g., iron) will carry magnetic flux more readily. We will learn later that the permeability of ferromagnetic or ferrimagnetic materials is not constant and depends on the value of the applied magnetic field (H).

The **magnetization** M represents the increase in the inductance due to the core material, so we can rewrite the equation for inductance as

$$B = \mu_0 H + \mu_0 M \tag{20-6}$$

The first part of this equation is simply the effect of the applied magnetic field. The second part is the effect of the magnetic material that is present. This is similar to our discussion on dielectric polarization and the mechanical behavior of materials. In materials, stress causes strain, electric field (E) induces dielectric polarization (P), and a magnetic field (H) causes magnetization ($\mu_0 M$) that contributes to the total flux density B.

The **magnetic susceptibility** χ_m, which is the ratio between magnetization and the applied field, gives the amplification produced by the material:

$$\chi_m = \frac{M}{H} \tag{20-7}$$

Both μ_r and χ_m refer to the degree to which the material enhances the magnetic field and are therefore related by

$$\mu_r = 1 + \chi_m \tag{20-8}$$

As noted before, the μ_r and, therefore, the χ_m values for ferromagnetic and ferrimagnetic materials depend on the applied field (H). For ferromagnetic and ferrimagnetic materials, the term $\mu_0 M >> \mu_0 H$. Thus, for these materials,

$$B \cong \mu_0 M \tag{20-9}$$

We sometimes interchangeably refer to either inductance or magnetization. Normally, we are interested in producing a high inductance B or magnetization M. This is accomplished by selecting materials that have a high relative permeability or magnetic susceptibility.

The following example shows how these concepts can be applied for comparing actual and theoretical magnetizations in pure iron.

Example 20-1 *Theoretical and Actual Saturation Magnetization in Fe*

Calculate the maximum, or saturation, magnetization that we expect in iron. The lattice parameter of BCC iron is 2.866 Å. Compare this value with 2.1 tesla (a value of saturation flux density experimentally observed for pure Fe).

SOLUTION

Based on the unpaired electronic spins, we expect each iron atom to have four electrons that act as magnetic dipoles. The number of atoms per m^3 in BCC iron is

$$\text{Number of Fe atoms/m}^3 = \frac{2 \text{ atoms/cell}}{(2.866 \times 10^{-10} \text{ m})^3} = 8.496 \times 10^{28}$$

The maximum volume magnetization (M_{sat}) is the total magnetic moment per unit volume:

$$M_{sat} = \left(8.496 \times 10^{28} \frac{\text{atoms}}{\text{m}^3}\right)(9.274 \times 10^{-24} \text{ A} \cdot \text{m}^2)\left(4 \frac{\text{Bohr magnetons}}{\text{atom}}\right)$$

$$M_{sat} = 3.15 \times 10^6 \frac{\text{A}}{\text{m}}$$

To convert the value of saturation magnetization M into saturation flux density B in tesla, we need the value of $\mu_0 M$. In ferromagnetic materials $\mu_0 M >> \mu_0 H$ and therefore, $B \cong \mu_0 M$.

Thus, the saturation induction or saturation flux density in tesla is given by $B_{sat} = \mu_0 M_{sat}$.

$$B_{sat} = \left(4\pi \times 10^{-7} \frac{\text{Wb}}{\text{A} \cdot \text{m}}\right)\left(3.15 \times 10^6 \frac{\text{A}}{\text{m}}\right)$$

$$B_{sat} = 3.96 \frac{\text{Wb}}{\text{m}^2} = 3.96 \text{ tesla}$$

This is almost two times the experimentally observed value of 2.1 tesla. Reversing our calculations, we can show that the each iron atom contributes only about 2.1 Bohr magneton and not 4. This is the difference between behavior of individual atoms and their behavior in a crystalline solid. It can be shown that in the case of iron, the difference is due to the $3d$ electron orbital moment being quenched in the crystal.

20-4 Diamagnetic, Paramagnetic, Ferromagnetic, Ferrimagnetic, and Superparamagnetic Materials

As mentioned before, there is no such thing as a "nonmagnetic" material. All materials respond to magnetic fields. When a magnetic field is applied to a material, several types of behavior are observed (Figure 20-3).

Diamagnetic Behavior A magnetic field acting on any atom induces a magnetic dipole for the entire atom by influencing the magnetic moment caused by the orbiting electrons. These dipoles oppose the magnetic field, causing the magnetization to be less than zero. This behavior, called **diamagnetism**, gives a relative permeability of about 0.99995 (or a negative susceptibility approximately -10^{-6}; note the negative sign). Materials such as copper, silver, silicon, gold, and alumina are diamagnetic at room temperature. Superconductors are perfect diamagnets ($\chi_m = -1$); they lose their superconductivity at higher temperatures or in the presence of a magnetic field. In a diamagnetic material, the magnetization (M) direction is opposite to the direction of applied field (H).

Paramagnetism When materials have unpaired electrons, a net magnetic moment due to electron spin is associated with each atom. When a magnetic field is applied, the dipoles align with the field, causing a positive magnetization. Because the dipoles do not interact, extremely large magnetic fields are required to align all of the dipoles. In addition, the effect is lost as soon as the magnetic field is removed. This effect, called **paramagnetism**, is found in metals such as aluminum, titanium, and alloys of copper. The magnetic susceptibility (χ_m) of paramagnetic materials is positive and

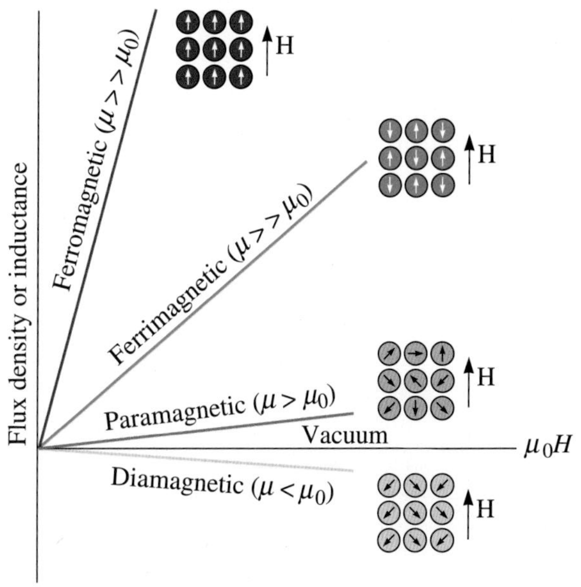

Figure 20-3 The effect of the core material on the flux density. The magnetic moment opposes the field in diamagnetic materials. Progressively stronger moments are present in paramagnetic, ferrimagnetic, and ferromagnetic materials for the same applied field.

lies between 10^{-4} and 10^{-5}. Ferromagnetic and ferrimagnetic materials above the Curie temperature also exhibit paramagnetic behavior.

Ferromagnetism Ferromagnetic behavior is caused by the unfilled energy levels in the 3*d* level of iron, nickel, and cobalt. Similar behavior is found in a few other materials, including gadolinium. In ferromagnetic materials, the permanent unpaired dipoles easily line up with the imposed magnetic field due to the exchange interaction, or mutual reinforcement of the dipoles. Large magnetizations are obtained even for small magnetic fields, giving large susceptibilities approaching 10^6. Similar to ferroelectrics, the susceptibility of ferromagnetic materials depends upon the intensity of the applied magnetic field. This is similar to the mechanical behavior of elastomers with the modulus of elasticity depending upon the level of strain. Above the Curie temperature, ferromagnetic materials behave as paramagnetic materials and their susceptibility is given by the following equation, known as the Curie-Weiss law:

$$\chi_m = \frac{C}{(T - T_c)} \tag{20-10}$$

In this equation, C is a constant that depends upon the material, T_c is the Curie temperature, and T is the temperature above T_c. Essentially, the same equation also describes the change in dielectric permittivity above the Curie temperature of ferroelectrics. Similar to ferroelectrics, ferromagnetic materials show the formation of hystereis loop domains and magnetic domains. These materials will be discussed in the next section.

Antiferromagnetism In materials such as manganese, chromium, MnO, and NiO, the magnetic moments produced in neighboring dipoles line up in opposition to one another in the magnetic field, even though the strength of each dipole is very high. This effect is illustrated for MnO in Figure 20-4. These materials are **antiferromagnetic** and have zero magnetization. The magnetic susceptibility is positive and small. In addition, CoO and $MnCl_2$ are examples of antiferromagnetic materials.

Ferrimagnetism In ceramic materials, different ions have different magnetic moments. In a magnetic field, the dipoles of cation A may line up with the field, while dipoles of cation B oppose the field. Because the strength or number of dipoles is not equal, a net magnetization results. The **ferrimagnetic** materials can provide good amplification of the imposed field. We will look at a group of ceramics called ferrites that display this behavior in a later section. These materials show a large, magnetic-field dependent magnetic susceptibility similar to ferromagnetic materials. They also show Curie-Weiss behavior (similar to ferromagnetic materials) at temperatures above the Curie temperature. Most ferrimagnetic materials are ceramics and are good insulators of electricity. Thus, in these materials, electrical losses (known as eddy current losses) are much smaller compared to those in metallic ferromagnetic materials. Therefore, ferrites are used in many high-frequency applications.

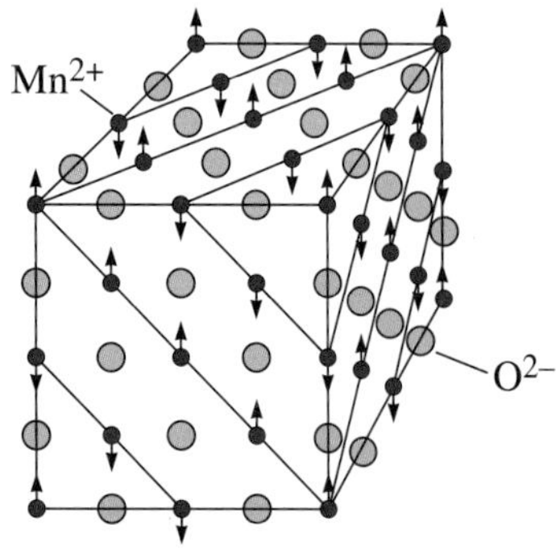

Figure 20-4 The crystal structure of MnO consists of alternating layers of {111} type planes of oxygen and manganese ions. The magnetic moments of the manganese ions in every other (111) plane are oppositely aligned. Consequently, MnO is antiferromagnetic.

Superparamagnetism When the grain size of ferromagnetic and ferrimagnetic materials falls below a certain critical size, these materials behave as if they are paramagnetic. The magnetic dipole energy of each particle becomes comparable to the thermal energy. This small magnetic moment changes its direction randomly (as a result of the thermal energy). Thus, the material behaves as if it has no net magnetic moment. This is known as superparamagnetism. Thus, if we produce iron oxide (Fe_3O_4) particles in a 3 to 5 nm size, they behave as superparamagnetic materials. Such iron-oxide superparamagnetic particles are used to form dispersions in aqueous or organic carrier phases or to form "liquid magnets" or ferrofluids. The particles in the fluid move in response to a gradient in the magnetic field. Since the particles form a stable sol, the entire dispersion moves and, hence, the material behaves as a liquid magnet. Such materials are used as seals in computer hard drives and in loudspeakers as heat transfer (cooling) media. The permanent magnet used in the loudspeaker holds the liquid magnets in place. Superparamagnetic particles of iron oxide (Fe_3O_4) also can be coated with different chemicals and used to separate DNA molecules, proteins, and cells from other molecules.

The following example illustrates how to select a material for a given application.

Example 20-2 *Design/Materials Selection for a Solenoid*

We want to produce a solenoid coil that produces an inductance of at least 2000 gauss when a 10 mA current flows through the conductor. Due to space limitations, the coil should be composed of 10 turns over a 1 cm length. Select a core material for the coil. Refer to Table 20-4.

SOLUTION

First, we can determine the magnetic field H produced by the coil. From Equation 20-2,

$$H = \frac{nI}{l} = \frac{(10)(0.01\text{ A})}{0.01\text{ m}} = 10\text{ A/m}$$

$$H = (10\text{ A/m})[4\pi \times 10^{-3}\text{ oersted/(A/m)}] = 0.12566\text{ oersted}$$

If the inductance B must be at least 2000 gauss, then the permeability of the core material must be

$$\mu = \frac{B}{H} = \frac{2000}{0.12566} = 15{,}916\text{ gauss/oersted}$$

The relative permeability of the core material must be at least

$$\mu_r = \frac{\mu}{\mu_0} = \frac{15{,}916}{1} = 15{,}916$$

If we examine the magnetic materials listed in Table 20-4, we find that 4750 alloy has a maximum relative permeability of 80,000 and might be a good selection for the core material.

20-5 Domain Structure and the Hysteresis Loop

From a phenomenological viewpoint, ferromagnetic materials are similar to ferroelectrics. A single crystal of iron or a polycrystalline piece of low-carbon steel is ferromagnetic; however, these materials ordinarily do not show a net magnetization. Within the single crystal or polycrystalline structure of a ferromagnetic or ferrimagnetic material, a substructure composed of magnetic domains is produced, even in the absence of an external

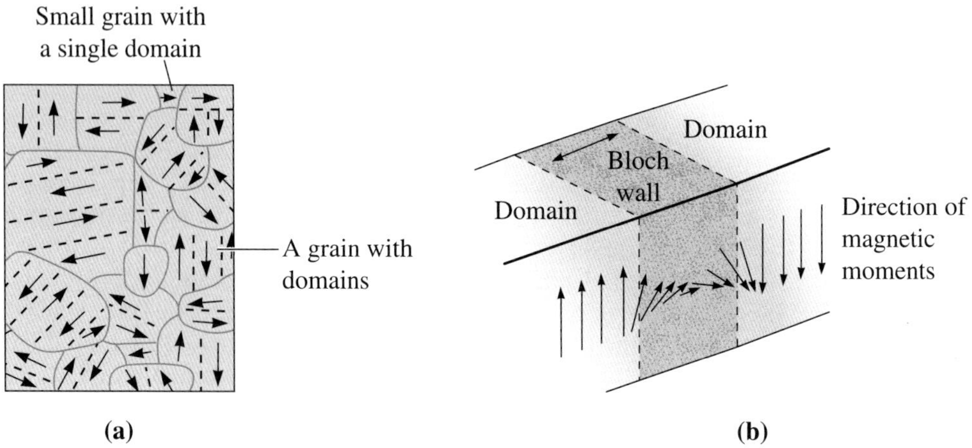

Figure 20-5 (a) A qualitative sketch of magnetic domains in a polycrystalline material. The dashed lines show demarcation between different magnetic domains; the dark curves show the grain boundaries. (b) The magnetic moments change direction continuously across the boundary between domains.

field. This spontaneously happens because the presence of many domains in the material, arranged so that the net magnetization is zero, minimizes what is known as the magnetostatic energy. **Domains** are regions in the material in which all of the dipoles are aligned in a certain direction. In a material that has never been exposed to a magnetic field, the individual domains have a random orientation. Because of this, the net magnetization in the virgin ferromagnetic or ferrimagnetic material as a whole is zero [Figure 20-5(a)]. Similar to ferroelectrics, application of a magnetic field (poling) will coerce many of the magnetic domains to align with the magnetic field direction.

Boundaries, called **Bloch walls**, separate the individual magnetic domains. The Bloch walls are narrow zones in which the direction of the magnetic moment gradually and continuously changes from that of one domain to that of the next [Figure 20-5(b)]. The domains are typically very small, about 0.005 cm or less, while the Bloch walls are about 100 nm thick.

Movement of Domains in a Magnetic Field When a magnetic field is imposed on the material, domains that are nearly lined up with the field grow at the expense of unaligned domains. In order for the domains to grow, the Bloch walls must move; the field provides the force required for this movement. Initially, the domains grow with difficulty, and relatively large increases in the field are required to produce even a little magnetization. This condition is indicated in Figure 20-6 by a shallow slope, which is the initial permeability of the material. As the field increases in strength, favorably oriented domains grow more easily, with permeability increasing as well. A maximum permeability can be defined as shown in the figure. Eventually, the unfavorably oriented domains disappear, and rotation completes the alignment of the domains with the field. The **saturation magnetization**, produced when all of the domains are oriented along with the magnetic field, is the greatest amount of magnetization that the material can obtain. Under these conditions, the permeability of these materials becomes quite small.

Effect of Removing the Field When the field is removed, the resistance offered by the domain walls prevents regrowth of the domains into random orientations. As a result, many of the domains remain oriented near the direction of the original field and a residual magnetization, known as the **remanance** (M_r) is present in the material. The value of B_r (usually in tesla) is known as the retentivity of the magnetic material.

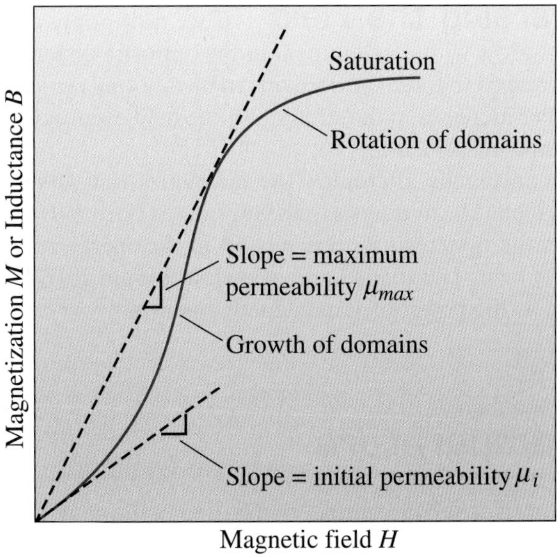

Figure 20-6 When a magnetic field is first applied to a magnetic material, magnetization initially increases slowly, then more rapidly as the domains begin to grow. Later, magnetization slows, as domains must eventually rotate to reach saturation. Notice the permeability values depend upon the magnitude of *H*. By convention, the maximum permeability is taken as the maximum slope of a tangent from the origin to the *B-H* curve.

The material then acts as a permanent magnet. Figure 20-7(a) shows this effect in the magnetization-field curve. Notice that the *M-H* loop shows saturation, but the *B-H* loop does not. The magnetic field needed to bring the induced magnetization back to zero is the **coercivity** of the material. This is a microstructure-sensitive property.

For magnetic recording materials, Fe, γ-Fe_2O_3, Fe_3O_4, and needle-shaped CrO_2 particles are used. The elongated shape of magnetic particles leads to higher coercivity (H_c). The dependence of coercivity on the shape of a particle or grain is known as **magnetic shape anisotropy**. The coercivity of recording materials needs to be smaller than that for permanent magnets since data written onto a magnetic data storage medium should be erasable. On the other hand, the coercivity values should be higher than soft magnetic materials since we want to retain the information stored. Such materials are described as magnetically semi-hard.

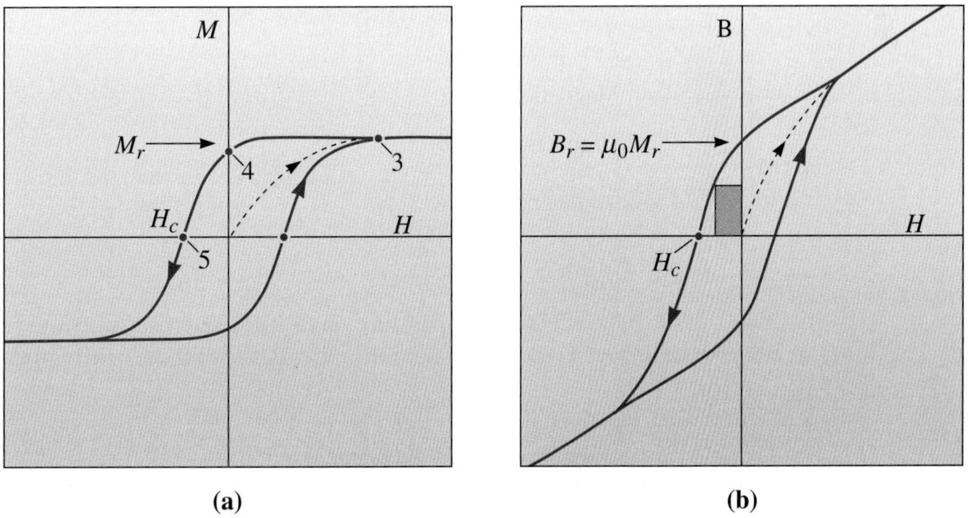

Figure 20-7 (a) The ferromagnetic hysteresis *M-H* loop showing the effect of the magnetic field on inductance or magnetization. The dipole alignment leads to saturation magnetization (point 3), a remanance (point 4), and a coercive field (point 5). (b) The corresponding *B-H* loop. Notice the *B* value does not saturate since $B = \mu_0 H + \mu_0 M$. (*Based on* Permanent Magnetism, *by R. Skomski and J. M. D. Coey, edited by J. M. D. Coey and D. R. Tilley, Institute of Physics Publishing, 1999.*)

Effect of Reversing the Field If we now apply a field in the reverse direction, the domains grow with an alignment in the opposite direction. A coercive field H_c (or coercivity) is required to force the domains to be randomly oriented and cancel one another's effect. Further increases in the strength of the field eventually align the domains to saturation in the opposite direction.

As the field continually alternates, the magnetization versus field relationship traces out a **hysteresis loop**. The hysteresis loop is shown as both *B*-*H* and *M*-*H* plots. The area contained within the hysteresis loop is related to the energy consumed during one cycle of the alternating field. The shaded area shown in Figure 20-7(b) is the largest *B*-*H* product and is known as the power of the magnetic material.

20-6 The Curie Temperature

When the temperature of a ferromagnetic or ferrimagnetic material is increased, the added thermal energy increases the mobility of the domains, making it easier for them to become aligned, but also preventing them from remaining aligned when the field is removed. Consequently, saturation magnetization, remanance, and the coercive field are all reduced at high temperatures (Figure 20-8). If the temperature exceeds the **Curie temperature** (T_c), ferromagnetic or ferrimagnetic behavior is no longer observed. Instead, the material behaves as a paramagnetic material. The Curie temperature (Table 20-3), which depends on the material, can be changed by alloying elements. French scientists Marie and Pierre Curie (the only husband and wife to win a Nobel prize; Marie Curie actually won two Nobel prizes) performed research on magnets, and the Curie temperature refers to their name. The dipoles still can be aligned in a magnetic field above the Curie temperature, but they become randomly aligned when the field is removed.

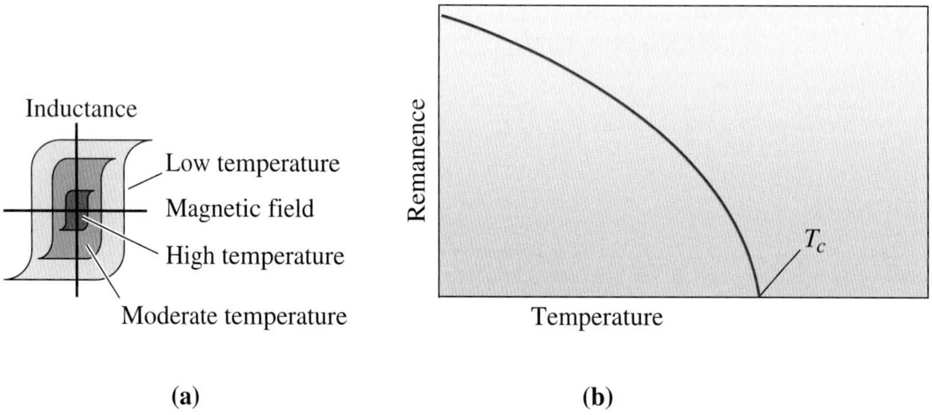

Figure 20-8 The effect of temperature on (a) the hysteresis loop and (b) the remanance. Ferromagnetic behavior disappears above the Curie temperature.

Table 20-3 Curie temperatures for selected materials

Material	Curie Temperature (°C)	Material	Curie Temperature (°C)
Gadolinium	16	Iron	771
$Nd_2Fe_{12}B$	312	Alnico 1	780
Nickel	358	Cunico	855
$BaO \cdot 6Fe_2O_3$	469	Alnico 5	900
Co_5Sm	747	Cobalt	1117

20-7 Applications of Magnetic Materials

Ferromagnetic and ferrimagnetic materials are classified as magnetically soft or magnetically hard depending upon the shape of the hysteresis loop [Figure 20-9(a)]. Generally, if the coercivity value is $> \sim 10^4$ A · m^{-1}, we consider the material as magnetically hard. If the coercivity values are less than 10^3 A · m^{-1}, we consider the material as magnetically soft. Figure 20-9(b) shows the classifications of different commercially important magnetic materials. Note that while the coercivity is a strongly *microstructure-sensitive*

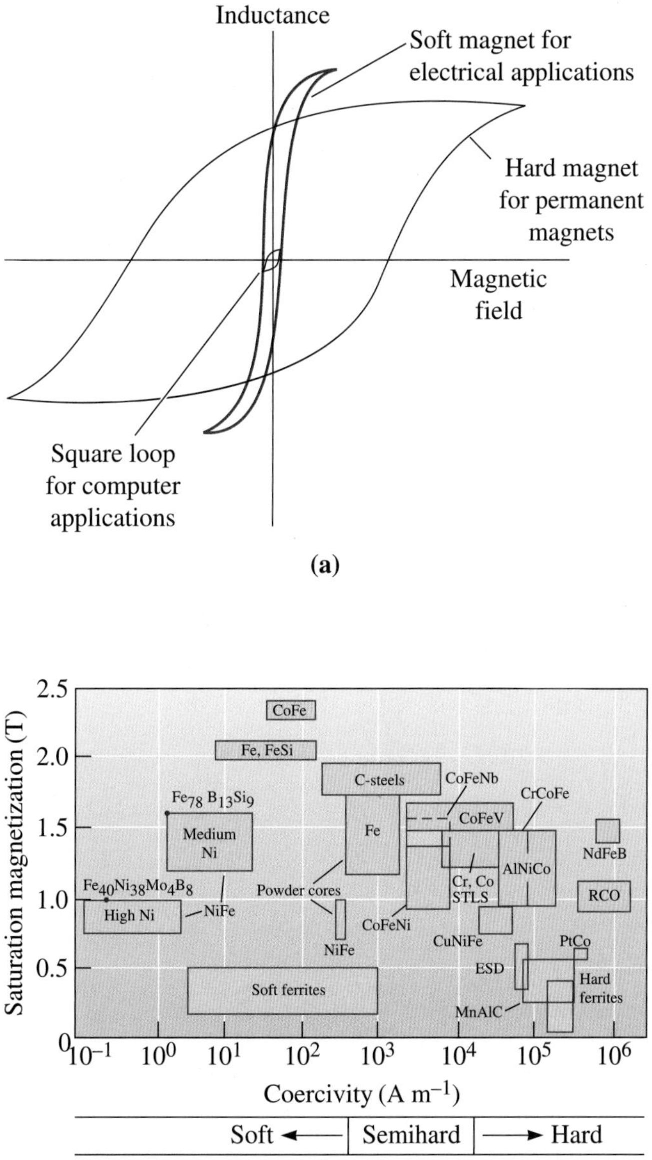

(b)

Figure 20-9 (a) Comparison of the hysteresis loops for three applications of ferromagnetic and ferrimagnetic materials. (*Based on "Magnetic Materials: An Overview, Basic Concepts, Magnetic Measurements, Magnetostrictive Materials," by G. Y. Chin et al. In R. Bloor, M. Flemings, and S. Mahajan (eds.),* Encyclopedia of Advanced Materials, *Vol. 1, Peragamon Press, 1994.*) (b) Saturation magnetization and coercivity values for different magnetic materials.

property, the saturation magnetization is constant (i.e., it is not microstructure dependent) for a material of a given composition. This is similar to the way the yield strength of metallic materials is strongly dependent on the microstructure, while the Young's modulus is not. Many factors, such as the structure of grain boundaries and the presence of pores or surface layers on particles, affect the coercivity values. The coercivity of single crystals depends strongly on crystallographic directions. There are certain directions along which it is easy to align the magnetic domains. There are other directions along which the coercivity is much higher. Coercivity of magnetic particles also depends upon shape of the particles. This is why in magnetic recording media we use acicular and not spherical particles. This effect is also used in Fe-Si steels, which are textured or grain oriented so as to minimize energy losses during the operation of an electrical transformer.

Let's look at some applications for magnetic materials.

Soft Magnetic Materials Ferromagnetic materials are often used to enhance the magnetic flux density (B) produced when an electric current is passed through the material. The magnetic field is then expected to do work. Applications include cores for electromagnets, electric motors, transformers, generators, and other electrical equipment. Because these devices utilize an alternating field, the core material is continually cycled through the hysteresis loop. Table 20-4 shows the properties of selected soft, magnetic materials. *Note that for these materials the value of relative magnetic permeability depends strongly on the strength of the applied field* (Figure 20-6).

These materials often have the following characteristics:

1. High-saturation magnetization.
2. High permeability.
3. Small coercive field.

Table 20-4 Properties of selected soft magnetic materials

		Permeability μ_r					
Name	**Composition**	**Initial**	**Maximum**	**Coercivity H_c (A · m^{-1})**	**Retentivity B_r (T)**	**B_{max} (T)**	**Resistivity ($\mu\Omega \cdot$ m)**
Ingot iron	99.8% Fe	150	5000	80	0.77	2.14	0.10
Low-carbon steel	99.5% Fe	200	4000	100		2.14	1.12
Silicon iron, unoriented	Fe-3% Si	270	8000	60		2.01	0.47
Silicon iron, grain-oriented	Fe-3% Si	1400	50,000	7	1.20	2.01	0.50
4750 alloy	Fe-48% Ni	11,000	80,000	2		1.55	0.48
4-79 permalloy	Fe-4% Mo-79% Ni	40,000	200,000	1		0.80	0.58
Superalloy	Fe-5% Mo-80% Ni	80,000	450,000	0.4		0.78	0.65
2V-Permendur	Fe-2% V-49% Co	800	450,000	0.4		0.78	0.65
Supermendur	Fe-2% V-49% Co		100,000	16	2.00	2.30	0.40
Metglas[a] 2650SC	$Fe_{81}B_{13.5}Si_{3.5}C_2$		300,000	3	1.46	1.61	1.35
Metglas[a] 2650S-2	$Be_{78}B_{13}S_9$		600,000	2	1.35	1.56	1.37
MnZn ferrite	H5C2[b]	10,000		7	0.09	0.40	1.5×10^5
MnZn ferrite	H5E[b]	18,000		3	0.12	0.44	5×10^4
NiZn ferrite	K5[b]	290		80	0.25	0.33	2×10^{12}

[a] *Allied Corporation trademark.*
[b] *TDK ferrite code.*
(Based on "Magnetic Materials: An Overview, Basic Concepts, Magnetic Measurements, Magnetostrictive Materials," by G.Y. Chin et al. In R. Bloor, M. Flemings, and S. Mahajan (eds.), Encyclopedia of Advanced Materials, *Vol. 1, Peragamon Press, 1994.)*

4. Small remanance.
5. Small hysteresis loop.
6. Rapid response to high-frequency magnetic fields.
7. High electrical resistivity.

High saturation magnetization permits a material to do work, while high permeability permits saturation magnetization to be obtained with small imposed magnetic fields. A small coercive field also indicates that domains can be reoriented with small magnetic fields. A small remanance is desired so that almost no magnetization remains when the external field is removed. These characteristics also lead to a small hysteresis loop, therefore minimizing energy losses during operation.

If the frequency of the applied field is so high that the domains cannot be realigned in each cycle, the device may heat due to dipole friction. In addition, higher frequencies naturally produce more heating because the material cycles through the hysteresis loop more often, losing energy during each cycle. For high frequency applications, materials must permit the dipoles to be aligned at exceptionally rapid rates.

Energy can also be lost by heating if eddy currents are produced. During operation, electrical currents can be induced into the magnetic material. These currents produce power losses and joule, or I^2R, heating. Eddy current losses are particularly severe when the material operates at high frequencies. If the electrical resistivity is high, eddy current losses can be held to a minimum. Soft magnets produced from ferrimagnetic ceramic materials have a high resistivity and therefore are less likely to heat than metallic ferromagnetic materials. Recently, a class of smart materials, known as magnetorheological or MR fluids based on soft magnetic carbonyl iron particles, has been introduced in various applications related to vibration control, such as Delphi's MagneRide™ system. These materials are like magnetic paints and can be made to absorb energy from shocks and vibrations by turning on a magnetic field. The stiffening of MR fluids is controllable and reversible. Some of the models of Cadillac and Corvette ofter a suspension based on these smart materials.

Data Storage Materials Magnetic materials are used for data storage. Memory is stored by magnetizing the material in a certain direction. For example, if the "north" pole is up, the bit of information stored is 1. If the "north" pole is down, then a 0 is stored.

For this application, materials with a square hysteresis loop, a low remanance, a low saturation magnetization, and a low coercive field are preferable. Hard ferrites based on Ba, CrO_2, acicular iron particles, and γ-Fe_2O_3 satisfy these requirements. The stripe on credit cards and bank machine cards are made using γ-Fe_2O_3 or Fe_3O_4 particles. The square loop ensures that a bit of information placed in the material by a field remains stored; a steep and abrupt change in magnetization is required to remove the information from storage in the ferromagnet. Furthermore, the magnetization produced by small external fields keeps the coercive field (H_c), saturation magnetization, and remanance (B_r) low.

The B_r and H_c values of some typical magnetic recording materials are shown in Table 20-5.

Many new alloys based on Co-Pt-Ta-Cr have been developed for hard disks. Computer hard disks are made using sputtered thin films of these materials. As discussed in earlier chapters, many different alloys, such as those based on nanostructured Fe-Pt and Fe-Pd, are being developed for data storage applications. More recently, a technology known as *spintronics* (*spin-based electronics*) has evolved. The premise of spintronics is to make use of the spin of electrons as a way of affecting the flow of electrical current (known as spin-polarized current) to make devices such as field effect transistors (FETs). The spin of the electrons (up or down) is also being considered as a way of storing information. A very successful example of a real-world spintronic-based device is a giant magnetoresistance (GMR) sensor that is used for reading information from computer hard disks.

Table 20-5 Properties of typical magnetic recording materials in powder form

	Particle Length (μm)	Aspect Ratio	Magnetization B_r (Wb/m^2)	(emu/cm^3)	Coercivity H_c (kA/m)	(Oe)	Surface Area (m^2/g)	Curie Temp. T_c (°C)
γ-Fe_2O_3	0.20	5:1	0.44	350	22–34	420	15–30	600
Co-γ-Fe_2O_3	0.20	6:1	0.48	380	30–75	940	20–35	700
CrO_2	0.20	10:1	0.50	400	30–75	950	18–55	125
Fe	0.15	10:1	1.40[a]	1100[a]	56–176	2200	20–60	770
Barium ferrite	0.05	0.02 μm thick	0.40	320	56–240	3000	20–25	350

[a]*For overcoated, stable particles use only 50 to 80% of these values due to reduced magnetic particle volume (Based on* The Complete Handbook of Magnetic Recording, Fourth Edition, *by F. Jorgensen, McGraw-Hill, 1996.)*

Permanent Magnets Finally, magnetic materials are used to make strong permanent magnets (Table 20-6). Strong permanent magnets, often called hard magnets, require the following:

1. High remanance (stable domains).
2. High permeability.
3. High coercive field.
4. Large hysteresis loop.
5. High power (or BH product).

The *record* for any energy product is obtained for $Nd_2Fe_{14}B$ magnets with an energy product of ~445 kJ · m^{-3} [~56 mega-gauss-oersted (MGOe)]. These magnets are made in the form of a powder by the rapid solidification of a molten alloy. Powders are either bonded in a polymer matrix or by hot pressing, producing bulk materials. The energy product increases when the sintered magnet is "oriented" or poled. Corrosion resistance, brittleness, and a relatively low Curie temperature of ~312°C are some of the limiting factors of this extraordinary material.

The **power** of the magnet is related to the size of the hysteresis loop, or the maximum product of B and H. The area of the largest rectangle that can be drawn in the second or fourth quadrants of the B-H curve is related to the energy required to demagnetize the magnet [Figure 20-10(a) and 20-10(b)]. For the product to be large, both the remanance and the coercive field should be large.

In many applications, we need to calculate the lifting power of a permanent magnet. The magnetic force obtainable using a permanent magnet is given by

$$F = \frac{\mu_0 M^2 A}{2} \tag{20-11}$$

Table 20-6 Properties of selected hard, or permanent, magnetic materials

Material	Common Name	$\mu_0 M_r$ (T)	$\mu_0 H_c$ (T)	$(BH)_{max}$ (kJ · m^{-3})	T_c (°C)
Fe-Co	Co-steel	1.07	0.02	6	887
Fe-Co-Al-Ni	Alnico-5	1.05	0.06	44	880
$BaFe_{12}O_{19}$	Ferrite	0.42	0.31	34	469
$SmCo_5$	Sm-Co	0.87	0.80	144	723
$Nd_2Fe_{14}B$	Nd-Fe-B	1.23	1.21	290–445	312

(Based on Permanent Magnetism, *by R. Skomski and J. M. D. Coey, p. 23, Table 1.2. Edited by J. M. D. Coey and D. R. Tilley, Institute of Physics Publishing, 1999.)*

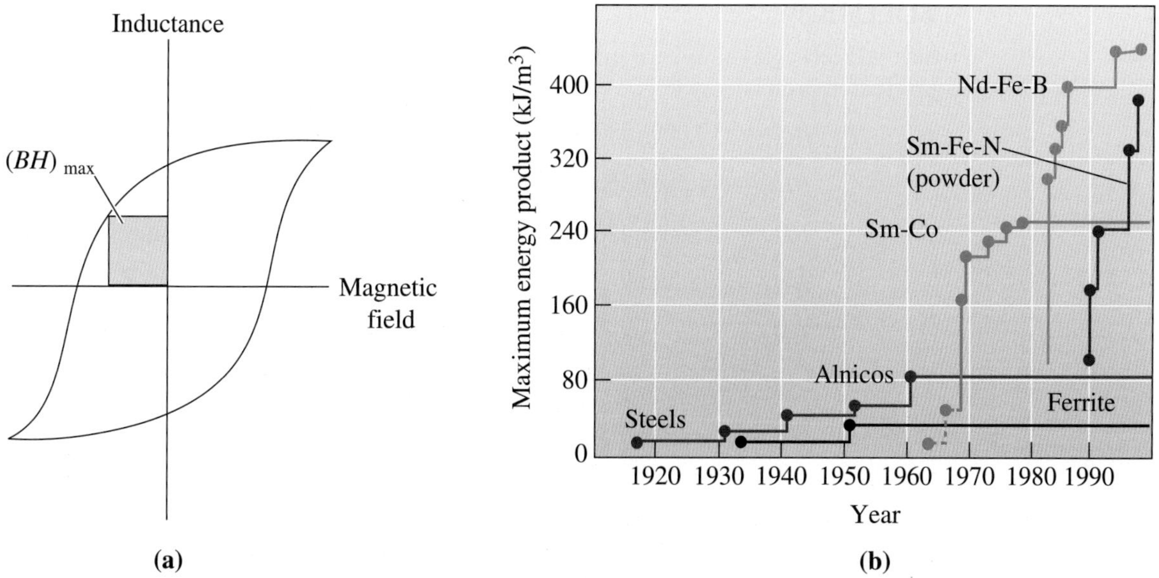

Figure 20-10 (a) The largest rectangle drawn in the second or fourth quadrant of the *B-H* curve gives the maximum *BH* product. $(BH)_{max}$ is related to the power, or energy, required to demagnetize the permanent magnet. (b) Development of permanent magnet materials. The maximum energy product is shown on the vertical axis. *(Based on* Permanent Magnetism, *by R. Skomski and J. M. D. Coey, edited by J. M. D. Coey and D. R. Tilley, Institute of Physics Publishing, 1999.)*

In this equation A is the cross-sectional area of the magnet, M is the magnetization, and μ_0 is the magnetic permeability of free space.

One of the most successful examples of the contributions by materials scientists and engineers in this area is the development of strong rare earth magnets. The progress made in the development of strong permanent magnets is illustrated in Figure 20-10(b). Permanent magnets are used in many applications including loudspeakers, motors, generators, holding magnets, mineral separation, and bearings. Typically, they offer a nonuniform magnetic field; however, it is possible to use geometric arrangements known as Halbach arrays to produce relatively uniform magnetic fields. The following examples illustrate applications of some of these concepts related to permanent magnetic materials.

Example 20-3 *Energy Product for Permanent Magnets*

Determine the power, or *BH* product, for the magnetic material with the properties shown in Figure 20-11.

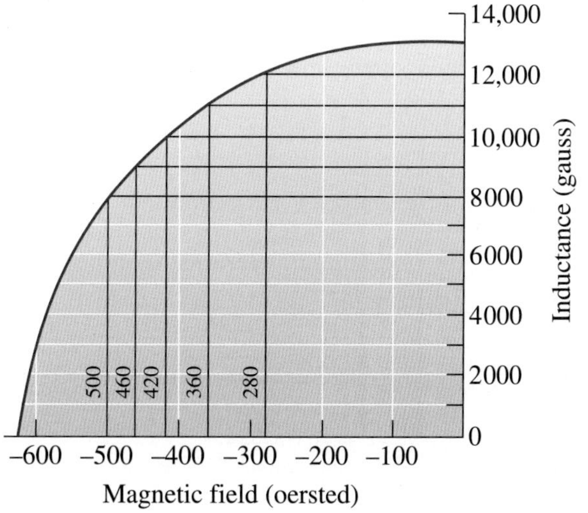

Figure 20-11 The second quadrant of the *B-H* curve for a permanent magnetic material (for Example 20-3).

SOLUTION

Several rectangles have been drawn in the second quadrant of the *B-H* curve. The *BH* product in each is

$$BH_1 = (12{,}000)(280) = 3.4 \times 10^6 \text{ gauss} \cdot \text{oersted}$$

$$BH_2 = (11{,}000)(360) = 4.0 \times 10^6 \text{ gauss} \cdot \text{oersted}$$

$$BH_3 = (10{,}000)(420) = 4.2 \times 10^6 \text{ gauss} \cdot \text{oersted} = \text{maximum}$$

$$BH_4 = (9000)(460) = 4.1 \times 10^6 \text{ gauss} \cdot \text{oersted}$$

$$BH_5 = (8000)(500) = 4.0 \times 10^6 \text{ gauss} \cdot \text{oersted}$$

Thus, the power is about 4.2×10^6 gauss · oersted.

Example 20-4 *Design/Selection of Magnetic Materials*

Select an appropriate magnetic material for the following applications: a high electrical-efficiency motor, a magnetic device to keep cupboard doors closed, a magnet used in an ammeter or voltmeter, and magnetic resonance imaging.

SOLUTION

High electrical-efficiency motor: To minimize hysteresis losses, we might use an oriented silicon iron, taking advantage of its anisotropic behavior and its small hysteresis loop. Since the iron-silicon alloy is electrically conductive, we would produce a laminated structure with thin sheets of the silicon iron sandwiched between a nonconducting dielectric material. Sheets thinner than about 0.5 mm might be recommended.

Magnet for cupboard doors: The magnetic latches used to fasten cupboard doors must be permanent magnets; however, low cost is a more important design feature than high power. An inexpensive ferritic steel or a low-cost ferrite would be recommended.

Magnets for an ammeter or voltmeter: For these applications, alnico alloys (iron alloys that also contain Al, Ni, and Co) are particularly effective. We find that these alloys are among the least sensitive to changes in temperature, ensuring accurate current or voltage readings over a range of temperatures.

Magnetic resonance imaging: One of the applications for MRI is in medical diagnostics. In this case, we want a very powerful magnet. A $Nd_2Fe_{12}B$ magnetic material, which has an exceptionally high *BH* product, might be recommended for this application. We can also make use of very strong electromagnets fabricated from superconductors.

The example that follows shows how the lifting power of a permanent magnet can be calculated.

Example 20-5 *Lifting Power of a Magnet*

Calculate the force in kN for one square meter area of a permanent magnet with a saturation magnetization of 1.61 tesla.

SOLUTION

As noted before, the attractive force from a permanent magnet is given by

$$F = \frac{\mu_0 M^2 A}{2}$$

We have been given the value of $\mu_0 M = 1.61$ tesla. We can rewrite the equation that provides the force due to a permanent magnet as follows:

$$F = \frac{\mu_0 M^2 A}{2} = \frac{(\mu_0 M)^2 A}{2\mu_0}$$

$$\frac{F}{A} = \frac{(1.61\ \mathrm{T})^2}{2\left(4\pi \times 10^{-7}\ \dfrac{\mathrm{H}}{\mathrm{m}}\right)} = 1.03\ \mathrm{MN/m^2}$$

Note that the force in this case will be 1.03 MN since the area (A) has been specified as 1 m^2.

20-8 Metallic and Ceramic Magnetic Materials

Let's look at typical alloys and ceramic materials used in magnetic applications and discuss how their properties and behavior can be enhanced. Some polymeric materials have shown magnetic activity; however, the Curie temperatures of these materials are too low compared to those for metallic and ceramic magnetic materials to be used in engineering applications.

Magnetic Alloys

Iron, nickel, and cobalt, which are ferromagnetic, are not usually used in pure form for electrical applications because they have high electrical conductivities and relatively large hysteresis loops, leading to excessive power loss. They are relatively poor permanent magnets; the domains are easily reoriented and both the remanance and the *BH* product are small compared with those of more complex alloys. Some change in the magnetic properties is obtained by introducing defects into the structure. Dislocations, grain boundaries, boundaries between multiple phases, and point defects help pin the domain boundaries, therefore keeping the domains aligned when the original magnetizing field is removed.

Iron-Nickel Alloys. Some iron-nickel alloys, such as Permalloy, have high permeabilities, making them useful as soft magnets. One example of an application for these magnets is the "head" that stores or reads information on a computer disk (Figure 20-12). As the disk

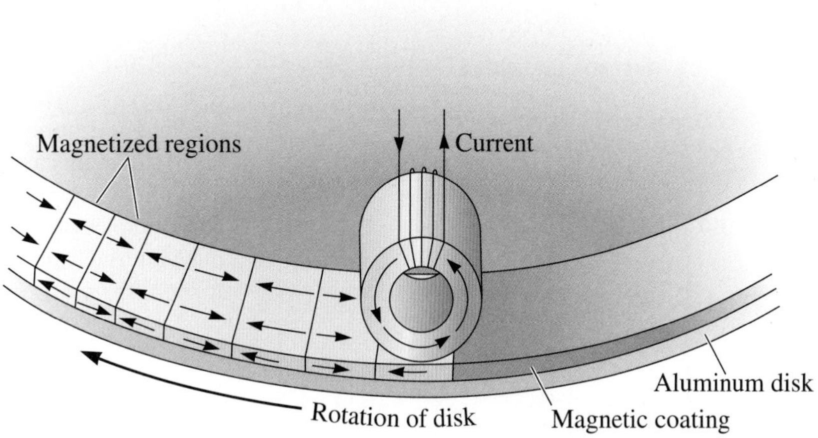

Figure 20-12 Information can be stored or retrieved from a magnetic disk by use of an electromagnetic head. A current in the head magnetizes domains in the disk during storage; the domains in the disk induce a current in the head during retrieval.

rotates beneath the head, a current produces a magnetic field in the head. The magnetic field in the head, in turn, magnetizes a portion of the disk. The direction of the field produced in the head determines the orientation of the magnetic particles embedded in the disk and, consequently, stores information. The information can be retrieved by again spinning the disk beneath the head. The magnetized region in the disk induces a current in the head; the direction of the current depends on the direction of the magnetic field in the disk.

Silicon Iron. Silicon irons are processed into grain-oriented steels. Introduction of 3 to 5% Si into iron produces an alloy that, after proper processing, is useful in electrical applications such as motors and generators. We take advantage of the anisotropic magnetic behavior of silicon iron to obtain the best performance. As a result of rolling and subsequent annealing, a sheet texture is formed in which the ⟨100⟩ directions in each grain are aligned. Because the silicon iron is most easily magnetized in ⟨100⟩ directions, the field required to give saturation magnetization is very small, and both a small hysteresis loop and a small remanance are observed (Figure 20-13). This type of anisotropy is known as **magnetocrystalline anisotropy**.

Composite Magnets. Composite magnets are used to reduce eddy current losses. Thin sheets of silicon iron are laminated with sheets of a dielectric material. The laminated layers are then built up to the desired overall thickness. The laminate increases the resistivity of the composite magnets and makes them successful at low and intermediate frequencies.

At very high frequencies, losses are more significant because the domains do not have time to realign. In this case, a composite material containing domain-sized magnetic particles in a polymer matrix may be used. The particles, or domains, rotate easily, while eddy current losses are minimized because of the high resistivity of the polymer.

Data Storage Materials. Magnetic materials for information storage must have a square loop and a low coercive field, permitting very rapid transmission of information. Hard disks for computer data storage are produced in a similar manner. In a hard disk, magnetic particles are embedded in a polymer film on a flat aluminum substrate. Because of the polymer matrix and the small particles, the domains can rotate quickly in response to a magnetic field. These materials are summarized in Table 20-5.

Complex Metallic Alloys for Permanent Magnets. Improved permanent magnets are produced by making the grain size so small that only one domain is present in each grain. Now the boundaries between domains are grain boundaries rather than Bloch

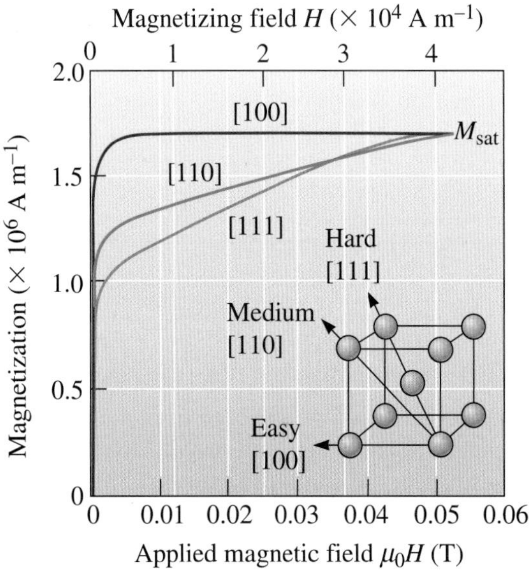

Figure 20-13 The initial magnetization curve for iron is highly anisotropic; magnetization is easiest when the ⟨100⟩ directions are aligned with the field and hardest along [111]. (*Based on* Principles of Electrical Engineering Materials and Devices, *by S. O. Kasap, Irwin, McGraw-Hill, 1997.*)

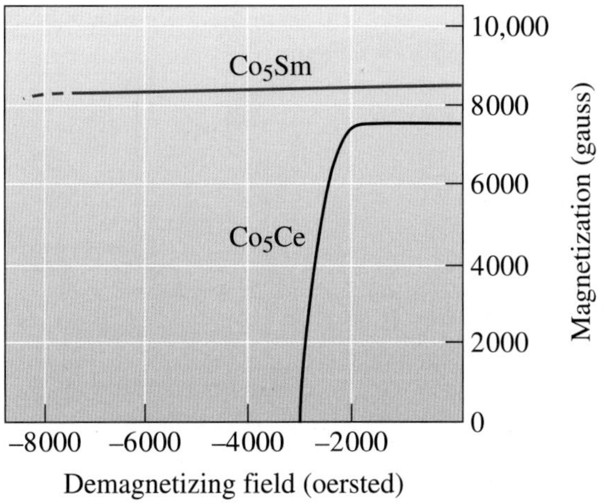

Figure 20-14 Demagnetizing curves for Co_5Sm and Co_5Ce, representing a portion of the hysteresis loop.

walls. The domains can change their orientation only by rotating, which requires greater energy than domain growth. Two techniques are used to produce these magnetic materials: phase transformations and powder metallurgy. Alnico, one of the most common of the complex metallic alloys, has a single-phase BCC structure at high temperatures, but when alnico slowly cools below 800°C, a second BCC phase rich in iron and cobalt precipitates. This second phase is so fine that each precipitate particle is a single domain, producing a very high remanance, coercive field, and power. Often the alloys are permitted to cool and transform while in a magnetic field to align the domains as they form.

A second technique—powder metallurgy—is used for a group of rare earth metal alloys, including samarium-cobalt. A composition giving Co_5Sm, an intermetallic compound, has a high *BH* product (Figure 20-14) due to unpaired magnetic spins in the 4*f* electrons of samarium. The brittle intermetallic is crushed and ground to produce a fine powder in which each particle is a domain. The powder is then compacted while in an imposed magnetic field to align the powder domains. Careful sintering to avoid growth of the particles produces a solid-powder metallurgy magnet. In $Nd_2Fe_{14}B$, a fine-grained intermetallic provides the domains, and a fine HfB_2 precipitate prevents movement of the domain walls.

Ferrimagnetic Ceramic Materials

Common magnetic ceramics are the ferrites, which have a spinel crystal structure (Figure 20-15). These ferrites have nothing to do with the *ferrite phase* we encountered in studying the Fe-C phase diagram (Chapters 12 and 13). Ferrites are used in wireless communications and in microelectronics in such applications as inductors. Ferrite powders are made using ceramic processing techniques.

We can understand the behavior of these ceramic magnets by looking at magnetite, Fe_3O_4. Magnetite contains two different iron ions, Fe^{2+} and Fe^{3+}, so we could rewrite the formula for magnetite as $Fe^{2+}Fe_2^{3+}O_4^{2-}$. The magnetite, or spinel, crystal structure is based on an FCC arrangement of oxygen ions, with iron ions occupying selected interstitial sites. Although the spinel unit cell actually contains eight of the FCC arrangements, we need examine only one of the FCC subcells:

1. Four oxygen ions are in the FCC positions of the subcell.
2. Octahedral sites, which are surrounded by six oxygen ions, are present at each edge and the center of the subcell. One Fe^{2+} and one Fe^{3+} ion occupy octahedral sites.
3. Tetrahedral sites have positions in the subcell such as (1/4, 1/4, 1/4). One Fe^{3+} ion occupies one of the tetrahedral sites.
4. When Fe^{2+} ions form, the two 4*s* electrons of iron are removed, but all of the 3*d* electrons remain. Because there are four unpaired electrons in the 3*d* level of iron, the magnetic strength of the Fe^{2+} dipole is four Bohr magnetons. When Fe^{3+} forms, both 4*s* electrons and one of the 3*d* electrons are removed. The Fe^{3+} ion has five unpaired electrons in the 3*d* level and, thus, has a strength of five Bohr magnetons.

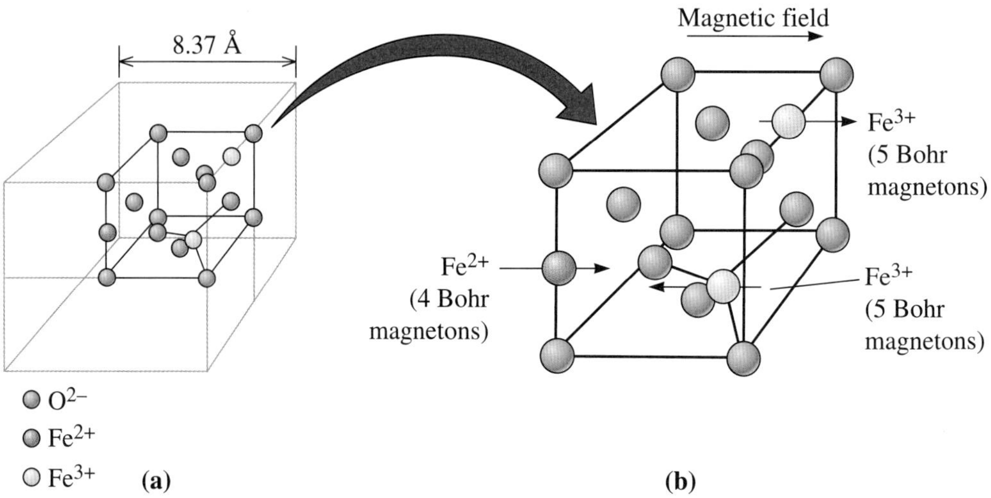

Figure 20-15 (a) The structure of magnetite, Fe_3O_4. (b) The subcell of magnetite. The magnetic moments of ions in the octahedral sites line up with the magnetic field, but the magnetic moments of ions in tetrahedral sites oppose the field. A net magnetic moment is produced by this ionic arrangement.

5. The ions in the tetrahedral sites of the magnetite line up so that their magnetic moments oppose the applied magnetic field, but the ions in the octahedral sites reinforce the field [Figure 20-15(b)]. Consequently, the Fe^{3+} ion in the tetrahedral site neutralizes the Fe^{3+} ion in the octahedral site (the Fe^{3+} ion coupling is antiferromagnetic). The Fe^{2+} ion in the octahedral site is not opposed by any other ion, and it therefore reinforces the magnetic field. The following example shows how we can calculate the magnetization in Fe_3O_4, which is one of the ferrites.

Example 20-6 *Magnetization in Magnetite (Fe_3O_4)*

Calculate the total magnetic moment per cubic centimeter in magnetite. Calculate the value of the saturation flux density (B_{sat}) for this material. The lattice parameter is 8.37×10^{-8} cm.

SOLUTION

In the subcell [Figure 20-15(b)], the total magnetic moment is four Bohr magnetons obtained from the Fe^{2+} ion, since the magnetic moments from the two Fe^{3+} ions located at tetrahedral and octahedral sites are canceled by each other.

In the unit cell overall, there are eight subcells, so the total magnetic moment is 32 Bohr magnetons per cell.

The size of the unit cell, with a lattice parameter of 8.37×10^{-8} cm is

$$V_{cell} = (8.37 \times 10^{-8})^3 = 5.864 \times 10^{-22}\ \text{cm}^3$$

The magnetic moment per cubic centimeter is

$$\text{Total moment} = \frac{32\ \text{Bohr magnetons/cell}}{5.864 \times 10^{-22}\ \text{cm}^3/\text{cell}} = 5.457 \times 10^{22}\ \text{magnetons/cm}^3$$

$$= (5.457 \times 10^{22})(9.274 \times 10^{-24}\ \text{A} \cdot \text{m}^2/\text{magneton})$$

$$= 0.506\ \text{A} \cdot \text{m}^2/\text{cm}^3 = 5.06 \times 10^5\ \text{A} \cdot \text{m}^2/\text{m}^3 = 5.06 \times 10^5\ \text{A/m}$$

This expression represents the magnetization M at saturation (M_{sat}). The value of $B_{sat} \simeq \mu_0 M_{sat}$ will be $[4\pi \times 10^{-7}\ \text{Wb/(A} \cdot \text{m)}](5.06 \times 10^5\ \text{A/m}) = 0.64$ tesla.

Table 20-7 Magnetic moments for ions in the spinel structure

Ion	Bohr Magnetons	Ion	Bohr Magnetons
Fe^{3+}	5	Co^{2+}	3
Mn^{2+}	5	Ni^{2+}	2
Fe^{2+}	4	Cu^{2+}	1
		Zn^{2+}	0

When ions are substituted for Fe^{2+} ions in the spinel structure, the magnetic behavior may be changed. Ions that may not produce ferromagnetism in a pure metal may contribute to ferrimagnetism in the spinels, as shown by the magnetic moments in Table 20-7. Soft magnets are obtained when the Fe^{2+} ion is replaced by various mixtures of manganese, zinc, nickel, and copper. The nickel and manganese ions have magnetic moments that partly cancel the effect of the two iron ions, but a net ferrimagnetic behavior, with a small hysteresis loop, is obtained. The high electrical resistivity of these ceramic compounds helps minimize eddy currents and permits the materials to operate at high frequencies. Ferrites used in computer applications may contain additions of manganese, magnesium, or cobalt to produce a square hysteresis loop behavior.

Another group of soft ceramic magnets is based on garnets, which include yttria iron garnet, $Y_3Fe_5O_{12}$ (YIG). These complex oxides, which may be modified by substituting aluminum or chromium for iron or by replacing yttrium with lanthanum or praseodymium, behave much like the ferrites. Another garnet, based on gadolinium and gallium, can be produced in the form of a thin film. Tiny magnetic domains can be produced in the garnet film; these domains, or *magnetic bubbles*, can then serve as means to store data. Once magnetized, the domains do not lose their memory in case of a sudden power loss.

Hard ceramic magnets used as permanent magnets include another complex oxide family, the hexagonal ferrites. The hexagonal ferrites include $SrFe_{12}O_{19}$ and $BaFe_{12}O_{19}$.

The example that follows highlights materials selection for a ceramic magnet.

Example 20-7 *Design/Materials Selection for a Ceramic Magnet*

Design a cubic ferrite magnet that has a total magnetic moment per cubic meter of 5.50×10^5 A/m.

SOLUTION

We found in Example 20-6 that the magnetic moment per cubic meter for Fe_3O_4 is 5.06×10^5 A/m. To obtain a higher saturation magnetization, we must replace Fe^{2+} ions with ions having more Bohr magnetons per atom. One such possibility (Table 20-7) is Mn^{2+}, which has five Bohr magnetons.

Assuming that the addition of Mn ions does not appreciably affect the size of the unit cell, we find from Example 20-6 that

$$V_{cell} = 5.864 \times 10^{-22}\ \text{cm}^3 = 5.864 \times 10^{-28}\ \text{m}^3$$

Let x be the fraction of Mn^{2+} ions that have replaced the Fe^{2+} ions, which have now been reduced to $1 - x$. Then, the total magnetic moment is

Total moment

$$= \frac{(8\ \text{subcells})[(x)(5\ \text{magnetons}) + (1 - x)(4\ \text{magnetons})](9.274 \times 10^{-24}\ \text{A} \cdot \text{m}^2)}{5.864 \times 10^{-28}\ \text{m}^3}$$

$$= \frac{(8)(5x + 4 - 4x)(9.274 \times 10^{-24})}{5.864 \times 10^{-28}} = 5.50 \times 10^5$$

$x = 0.347$

Therefore we need to replace 34.7 at% of the Fe^{2+} ions with Mn^{2+} ions to obtain the desired magnetization.

Magnetostriction Certain materials can develop strain when their magnetic state is changed. This effect is used in actuators. The magnetostrictive effect can be seen either by changing the magnetic field or by changing the temperature. Iron, nickel, Fe_3O_4, $TbFe_2$, DyFe, and $SmFe_2$ are examples of some materials that show this effect. Terfenol-D, which is named after its constituents terbium, iron, and dysprosium and its developer, the Naval Ordnance Laboratory (NOL), is one of the best known magnetostrictive materials. Its composition is $\sim Tb_xDy_{1-x}Fe_y$ $(0.27 < x < 0.30, 1.9 < y < 2)$. The magnetostriction phenomenon is analogous to electrostriction. Recently, some ferromagnetic alloys that also show magnetostriction have been developed.

Summary

- All materials interact with magnetic fields. The magnetic properties of materials are related to the interaction of magnetic dipoles with a magnetic field. The magnetic dipoles originate with the electronic structure of the atom, causing several types of behavior.
- Magnetic materials have enabled numerous technologies that range from high intensity superconducting magnets for MRI; semi-hard materials used in magnetic data storage; permanent magnets used in loud speakers, motors, and generators; to superparamagnetic materials used to make ferrofluids and for magnetic separation of DNA molecules and cells.
- In diamagnetic materials, the magnetic dipoles oppose the applied magnetic field.
- In paramagnetic materials, the magnetic dipoles weakly reinforce the applied magnetic field, increasing the net magnetization or inductance.
- Ferromagnetic and ferrimagnetic materials are magnetically nonlinear. Their permeability depends strongly on the applied magnetic field. In ferromagnetic materials (such as iron, nickel, and cobalt), the magnetic dipoles strongly reinforce the applied magnetic field, producing large net magnetization or inductance. In ferrimagnetic materials, some magnetic dipoles reinforce the field, whereas others oppose the field. A net increase in magnetization or inductance occurs. Magnetization may remain even after the magnetic field is removed. Increasing the temperature above the Curie temperature destroys the ferromagnetic or ferrimagnetic behavior.
- The structure of ferromagnetic and ferrimagnetic materials includes domains, within which all of the magnetic dipoles are aligned. When a magnetic field is applied, the dipoles become aligned with the field, increasing the magnetization to its maximum, or saturation, value. When the field is removed, some alignment of the domains may remain, giving a remanant magnetization.
- For soft magnetic materials, little remanance exists, only a small coercive field is required to remove any alignment of the domains, and little energy is consumed in reorienting the domains when an alternating magnetic field is applied.

- For hard, or permanent, magnetic materials, the domains remain almost completely aligned when the field is removed, large coercive fields are required to randomize the domains, and a large hysteresis loop is observed. This condition provides the magnet with a high power.
- Magnetostriction is the development of strain in response to an applied magnetic field or a temperature change that induces a magnetic transformation. Terfenol type magnetostrictive materials have been developed for actuator applications.

Glossary

Antiferromagnetism Arrangement of magnetic moments such that the magnetic moments of atoms or ions cancel out causing zero net magnetization.

Bloch walls The boundaries between magnetic domains.

Bohr magneton The strength of a magnetic moment of an electron (μ_B) due to electron spin.

Coercivity The magnetic field needed to force the domains in a direction opposite to the magnetization direction. This is a microstructure-sensitive property.

Curie temperature The temperature above which ferromagnetic or ferrimagnetic materials become paramagnetic.

Diamagnetism The effect caused by the magnetic moment due to the orbiting electrons, which produces a slight opposition to the imposed magnetic field. All materials exhibit this relatively weak effect.

Domains Small regions within a single or polycrystalline material in which all of the magnetization directions are aligned.

Ferrimagnetism Magnetic behavior obtained when ions in a material have their magnetic moments aligned in an antiparallel arrangement such that the moments do not completely cancel out and a net magnetization remains.

Ferromagnetism Alignment of the magnetic moments of atoms in the same direction so that a net magnetization remains after the magnetic field is removed.

Hard magnet Ferromagnetic or ferrimagnetic material that has a coercivity $> 10^4$ A · m^{-1}. This is the same as a permanent magnet.

Hysteresis loop The loop traced out by magnetization in a ferromagnetic or ferrimagnetic material as the magnetic field is cycled.

Magnetic moment The strength of the magnetic field associated with a magnetic dipole.

Magnetic permeability The ratio between inductance or magnetization and magnetic field.

Magnetic susceptibility The ratio between magnetization and the applied field.

Magnetization The total magnetic moment per unit volume.

Magnetocrystalline anisotropy In single crystals, the coercivity depends upon crystallographic direction creating easy and hard axes of magnetization.

Paramagnetism Alignment of the electron spins when a magnetic field is applied.

Permanent magnet A hard magnetic material.

Power The strength of a permanent magnet as expressed by the maximum product of the inductance and magnetic field.

Remanance The polarization or magnetization that remains in a material after it has been removed from a magnetic field. The remanance is due to the permanent alignment of the dipoles.

Saturation magnetization When all of the dipoles have been aligned by the field, producing the maximum magnetization.

Shape anisotropy The dependence of coercivity on the shape of magnetic particles.

Soft magnet Ferromagnetic or ferrimagnetic material that has a coercivity $\leq 10^3$ A · m^{-1}.

Superparamagnetism In the nanoscale regime, materials that are ferromagnetic or ferrimagnetic but behave in a paramagnetic manner (because of their nano-sized grains or particles).

Problems

Section 20-1 Classification of Magnetic Materials

Section 20-2 Magnetic Dipoles and Magnetic Moments

20-1 State any four real-world applications of different magnetic materials.

20-2 Explain the following statement: "Strictly speaking, there is no such thing as a nonmagnetic material."

20-3 Normally, we disregard the magnetic moment of the nucleus. In what application does the nuclear magnetic moment become important?

20-4 What two motions of electrons are important in determining the magnetic properties of materials?

20-5 Explain why only a handful of solids exhibit ferromagnetic or ferrimagnetic behavior.

20-6 Calculate and compare the maximum magnetization we would expect in iron, nickel, cobalt, and gadolinium. There are seven electrons in the 4*f* level of gadolinium. Compare the calculated values with the experimentally observed values.

20-7 Calculate the mass of an electron by using the mass of the Bohr magneton and Planck's constant.

Section 20-3 Magnetization, Permeability, and the Magnetic Field

Section 20-4 Diamagnetic, Paramagnetic, Ferromagnetic, Ferrimagnetic, and Superparamagnetic Materials

Section 20-5 Domain Structure and Hysteresis Loop

20-8 Define the following terms: magnetic induction, magnetic field, magnetic susceptibility, and magnetic permeability.

20-9 Define the following terms: ferromagnetic, ferrimagnetic, diamagnetic, paramagnetic, superparamagnetic, and antiferromagnetic materials.

20-10 What is a ferromagnetic material? What is a ferrimagnetic material? Explain and provide examples of each type of material.

20-11 How does the permeability of ferromagnetic and ferrimagnetic materials change with temperature when the temperature is greater than the Curie temperature?

20-12 Derive the equation $\mu_r = 1 + \chi_m$ using Equations 20-4 through 20-7.

20-13 A 4-79 permalloy solenoid coil needs to produce a minimum inductance of $1.5\ \text{Wb/m}^2$. If the maximum allowed current is 5 mA, how many turns are required in a wire 1 m long?

20-14 An alloy of nickel and cobalt is to be produced to give a magnetization of 2×10^6 A/m. The crystal structure of the alloy is FCC with a lattice parameter of 0.3544 nm. Determine the atomic percent cobalt required, assuming no interaction between the nickel and cobalt.

20-15 Estimate the magnetization that might be produced in an alloy containing nickel and 70 at% copper, assuming that no interaction occurs.

20-16 An Fe-80% Ni alloy has a maximum permeability of 300,000 when an inductance of 3500 gauss is obtained. The alloy is placed in a 20-turn coil that is 2 cm in length. What current must flow through the conductor coil to obtain this field?

20-17 An Fe-49% Ni alloy has a maximum permeability of 64,000 when a magnetic field of 0.125 oersted is applied. What inductance is obtained and what current is needed to obtain this inductance in a 200-turn, 3-cm-long coil?

20-18 Draw a schematic of the *B-H* and *M-H* loops for a typical ferromagnetic material. What is the difference between these two loops?

20-19 Is the magnetic permeability of ferromagnetic or ferrimagnetic materials constant? Explain.

20-20 From a phenomenological viewpoint, what are the similarities between elastomers, ferromagnetic and ferrimagnetic materials, and ferroelectrics?

20-21 What are the major differences between ferromagnetic and ferrimagnetic materials?

20-22 Compare the electrical resistivities of ferromagnetic metals and ferrimagnetic ceramics.

20-23 Why are eddy current losses important design factors in ferromagnetic materials but less important in ferrimagnetic materials?

20-24 Which element has the highest saturation magnetization? What alloys have the highest saturation magnetization of all materials?

20-25 What material has the highest energy product of all magnetic materials?

20-26 Is coercivity of a material a microstructure-sensitive property? Is remanance a microstructure-sensitive property? Explain.

20-27 Is saturation magnetization of a material a microstructure-sensitive property? Explain.

20-28 Can the same material have different hysteresis loops? Explain.

20-29 The following data describe the effect of the magnetic field on the inductance in a silicon steel. Calculate the initial permeability and the maximum permeability for the material.

H (A/m)	*B* (tesla)
0.00	0
20	0.08
40	0.30
60	0.65
80	0.85
100	0.95
150	1.10
250	1.25

20-30 A magnetic material has a coercive field of 167 A/m, a saturation magnetization of 0.616 tesla, and a residual inductance of 0.3 tesla. Sketch the hysteresis loop for the material.

20-31 A magnetic material has a coercive field of 10.74 A/m, a saturation magnetization of 2.158 tesla, and a remanance induction of 1.183 tesla. Sketch the hysteresis loop for the material.

20-32 Using Figure 20-16, determine the following properties of the magnetic material: remanance, saturation magnetization, coercive field, initial permeability, maximum permeability, and power (maximum *BH* product).

20-33 Using Figure 20-17, determine the following properties of the magnetic material: remanance, saturation magnetization, coercive field, initial permeability, maximum permeability, and power (maximum *BH* product).

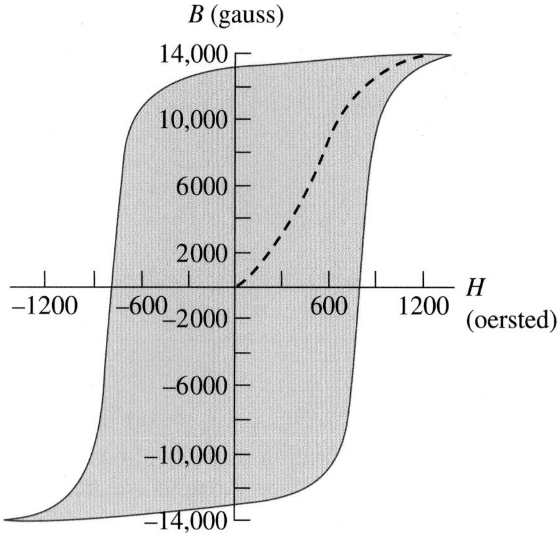

Figure 20-16 Hysteresis curve for a hard magnetic material (for Problem 20-32).

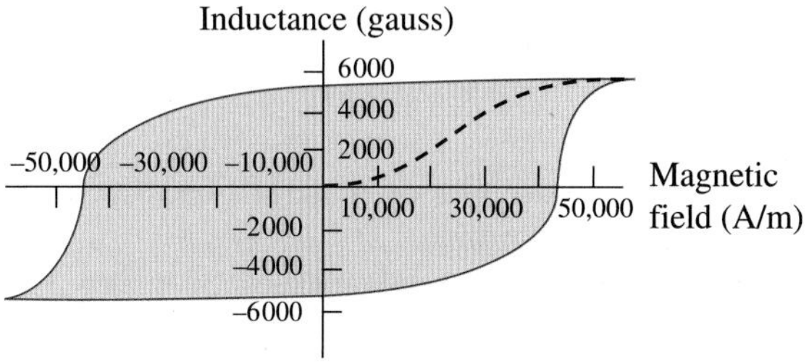

Figure 20-17 Hysteresis curve for a hard magnetic material (for Problem 20-33).

Section 20-6 The Curie Temperature

Section 20-7 Applications of Magnetic Materials

Section 20-8 Metallic and Ceramic Magnetic Materials

20-34 Give the electron configuration of the pure elements in Table 20-3.

20-35 In Figure 20-9(b), what materials have the highest and lowest saturation magnetizations and coercivities?

20-36 Sketch the *M*-*H* loop for Fe at 300 K, 500 K, and 1000 K.

20-37 Define the terms soft and hard magnetic materials. Draw a typical *M*-*H* loop for each material.

20-38 What important characteristics are associated with soft magnetic materials?

20-39 Are materials used for magnetic data storage magnetically hard or soft? Explain.

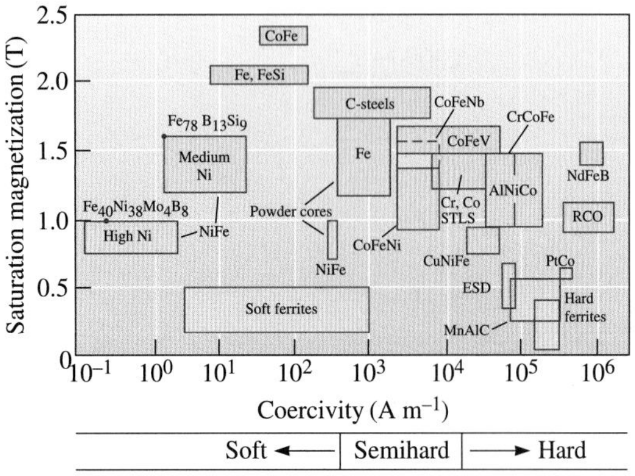

(b)

Figure 20-9(b) Saturation magnetization and coercivity values for different magnetic materials. (*Adapted from* "Magnetic Materials: An Overview, Basic Concepts, Magnetic Measurements, Magnetostrictive Materials," *by G. Y. Chin et al. In D. Bloor, M. Flemings, and S. Mahajan (Eds.), Encyclopedia of Advanced Materials, Vol. 1, 1994, p. 1424, Table 1. Copyright © 1994 Pergamon Press. Reprinted with permission of the editor.*) (Repeated for Problem 20-35.)

20-40 Give examples of materials used in magnetic recording.

20-41 What are the advantages of using Fe-Nd-B magnets? What are some of their disadvantages?

20-42 Use the data in Table 20-4 to correlate maximum permeability with coercivity. Is there a potential trend?

20-43 What is the maximum magnetic power for Co-γ-Fe_2O_3?

20-44 Estimate the power of the Co_5Ce material shown in Figure 20-14.

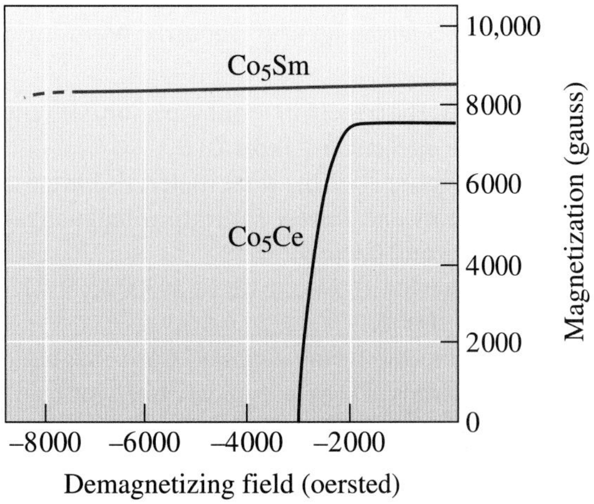

Figure 20-14 Demagnetizing curves for Co_5Sm and Co_5Ce, representing a portion of the hysteresis loop. (Repeated for Problem 20-44.)

20-45 Use the data in Table 20-6 to see if there is a possible relationship between BH_{max} and the Curie temperature.

20-46 Take the maximum power from the problem in Example 20-3 and convert it to SI units.

20-47 Refer to Figure 20-13. What are the coordinates of M_{sat}? Use the units of A/m for both values.

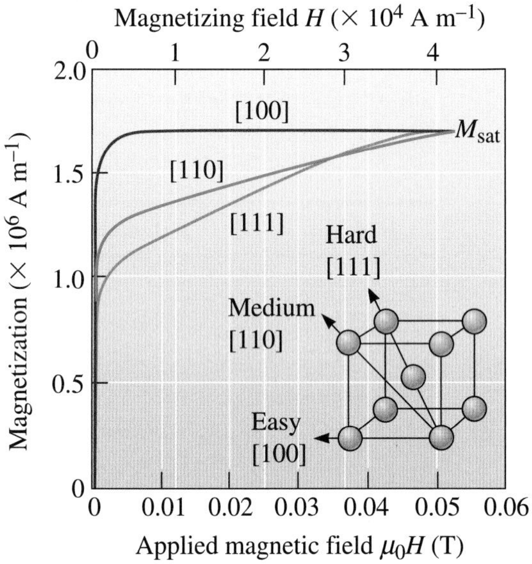

Figure 20-13 The initial magnetization curve for iron is highly anisotropic; magnetization is easiest when the ⟨100⟩ directions are aligned with the field and hardest along [111]. (*Based on* Principles of Electrical Engineering Materials and Devices, *by S. O. Kasap, Irwin, McGraw-Hill, 1997.*) (Repeated for Problem 20-47.)

20-48 What advantages does the Fe-3% Si material have compared with permalloy for use in electric motors?

20-49 The coercive field for pure iron is related to the grain size of the iron by the relationship $H_c = 1.83 + 4.14/\sqrt{A}$, where A is the area of the grain in two dimensions (mm^2) and H_c has units of A/m. If only the grain size influences the 99.95% iron (coercivity 0.9 oersted), estimate the size of the grains in the material. What happens to the coercivity value when the iron is annealed to increase the grain size?

20-50 Calculate the attractive force per square meter from a permanent magnet with a saturation magnetization of 1.0 tesla.

20-51 Suppose we replace 10% of the Fe^{2+} ions in magnetite with Cu^{2+} ions. Determine the total magnetic moment per cubic centimeter.

20-52 Suppose that the total magnetic moment per cubic meter in a spinel structure in which Ni^{2+} ions have replaced a portion of the Fe^{2+} ions is 4.6×10^5 A/m. Calculate the fraction of the Fe^{2+} ions that have been replaced and the wt% Ni present in the spinel.

20-53 What is magnetostriction? How is this similar to electrostriction? How is it different from the piezoelectric effect?

20-54 State examples of materials that show the magnetostriction effect.

20-55 What is spintronics? Give an example of a spintronics-based device.

Design Problems

20-56 Design a solenoid no longer than 1 cm that will produce an inductance of 3000 gauss.

20-57 Design a permanent magnet that will have a remanance of at least 5000 gauss, that will not be demagnetized if exposed to a temperature of 400°C or to a magnetic field of 1000 oersted, and that has good magnetic power.

20-58 Design a spinel-structure ferrite that will produce a total magnetic moment per cubic meter of 5.6×10^5 A/m.

20-59 Design a spinel-structure ferrite that will produce a total magnetic moment per cubic meter of 4.1×10^5 A/m.

20-60 Design a permanent magnet to lift a 1000 kg maximum load under operating temperatures as high as 750°C. Which material(s) listed in Table 20-6 will meet the above requirement?

Computer Problems

20-61 *Converting Magnetic Units.* Write a computer program that will convert magnetic units from the cgs or Gaussian system to the SI system. For example, if the user provides a value of flux density in gauss, the program should provide a value in Wb/m^2 or tesla.

Knovel® Problems

K20-1 Some minerals exhibit strong magnetic properties. One of these minerals is called maghemite. Describe its molecular structure.

K20-2 Some magnetic materials, neodymium-iron in particular, readily oxidize and have to be protected by a coating. What are the coatings and coating techniques used for this purpose?

K20-3 Eddy current losses increase with increasing thickness of laminations in electric motor cores. What happens to the losses with increasing grain size?

Optical materials play a critical role in the infrastructure of our communications and information technology systems. There are millions of miles of optical fiber, such as shown in the image above, installed worldwide. Optical fibers for communications and medical applications, lasers for medical and manufacturing applications, micromachined mirror arrays, light-emitting diodes, and solar cells have enabled a wide range of new technologies. (*PhotoDisc Blue/Getty Images.*)

CHAPTER 21

Photonic Materials

Have You Ever Wondered?

- *Why does the sky appear blue?*
- *How does an optical fiber work?*
- *What factors control the transmission and absorption of light in different materials?*
- *What does the acronym LASER stand for?*
- *What is a ruby laser made from?*
- *Does the operation of a fluorescent tube light involve phosphorescence?*
- *How did the invention of blue lasers enable high definition DVDs?*

Chapter Learning Objectives

The key objectives of this chapter are to

- Define the term *photon* and calculate its energy.
- Calculate the intensity of light that is refracted, reflected, absorbed, and transmitted by a material.
- Explain how the band structure of solids gives rise to the optical properties of metals, semiconductors, and insulators.
- Explain the physical basis for the functioning of optoelectronic devices such as solar cells and light-emitting diodes.

Photonic or optical materials have had a significant impact on the development of the communications infrastructure and information technology. Photonic materials have also played a key role in many other technologies related to medicine, manufacturing, and astronomy, just to name a few. Today, millions of miles of optical fiber have been installed worldwide. The term "optoelectronics" refers to the science and technology that combine optical and electronic materials. Examples include light-emitting diodes (LEDs), solar cells, and semiconductor lasers. Starting with simple mirrors, prisms, and lenses to the latest photonic band gap materials, the field of optical materials and devices has advanced at a very rapid pace. The goal of this chapter is to present a summary of fundamental principles that have guided applications of optical materials.

Optical properties of materials are related to the interaction of a material with electromagnetic radiation in the form of waves or particles of energy called photons. This radiation may have characteristics that fall in the visible light spectrum or may be invisible to the human eye. In this chapter, we explore two avenues by which we can use the optical properties of materials: emission of photons from materials and interaction of photons with materials.

21-1 The Electromagnetic Spectrum

Light is energy, or radiation, in the form of waves or particles called **photons** that can be emitted from a material. The important characteristics of the photons—their energy E, wavelength λ, and frequency ν—are related by the equation

$$E = \mathrm{h}\nu = \frac{\mathrm{h}c}{\lambda} \tag{21-1}$$

where c is the speed of light (in vacuum, the speed c_0 is 3×10^{10} cm/s), and h is Planck's constant (6.626×10^{-34} J · s). Since there are 1.6×10^{-19} J per electron volt (eV), the value of "h" is also given by 4.14×10^{-15} eV · s. This equation permits us to consider the photon either as a particle of energy E or as a wave with a characteristic wavelength and frequency.

The spectrum of electromagnetic radiation is shown in Figure 21-1. Gamma and x-rays have short wavelengths, or high frequencies, and possess high energies; microwaves and radio waves possess low energies; and visible light represents only a very narrow portion of the electromagnetic spectrum. Figure 21-1 also shows the response of the human eye to different colors. Bandgaps (E_g) of semiconductors (in eV) and corresponding wavelengths of light are also shown. As discussed in Chapter 19, these relationships are used to make LEDs of different colors.

21-2 Refraction, Reflection, Absorption, and Transmission

All materials interact in some way with light. Photons cause a number of optical phenomena when they interact with the electronic or crystal structure of a material (Figure 21-2). If incoming photons interact with valence electrons, several things may happen. The photons may give up their energy to the material, in which case *absorption* occurs. Or the photons may give up their energy, but photons of identical energy are immediately emitted by the material; in this case, *reflection* occurs. Finally, the photons may not interact with the electronic structure of the material; in this case, *transmission* occurs. Even in transmission, however, photons are changed in velocity, and *refraction* occurs. A small fraction of the incident light may be scattered with a slightly different frequency (Raman scattering).

As Figure 21-2 illustrates, an incident beam of intensity I_0 may be partly reflected, partly absorbed, and partly transmitted. The intensity of the incident beam I_0 therefore can be expressed as

$$I_0 = I_r + I_a + I_t \tag{21-2}$$

where I_r is the portion of the beam that is reflected, I_a is the portion that is absorbed, and I_t is the portion finally transmitted through the material. Reflection may occur at both the front and back surfaces of the material. Figure 21-2(a) shows reflection only at the front surface. Also, reflection occurs at a certain angle with respect to the normal of

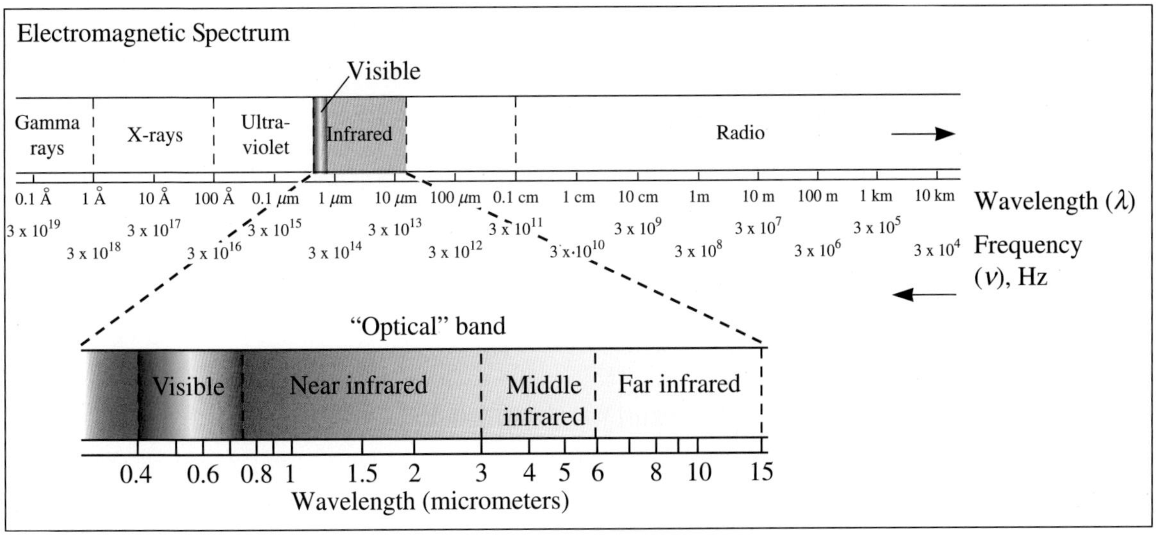

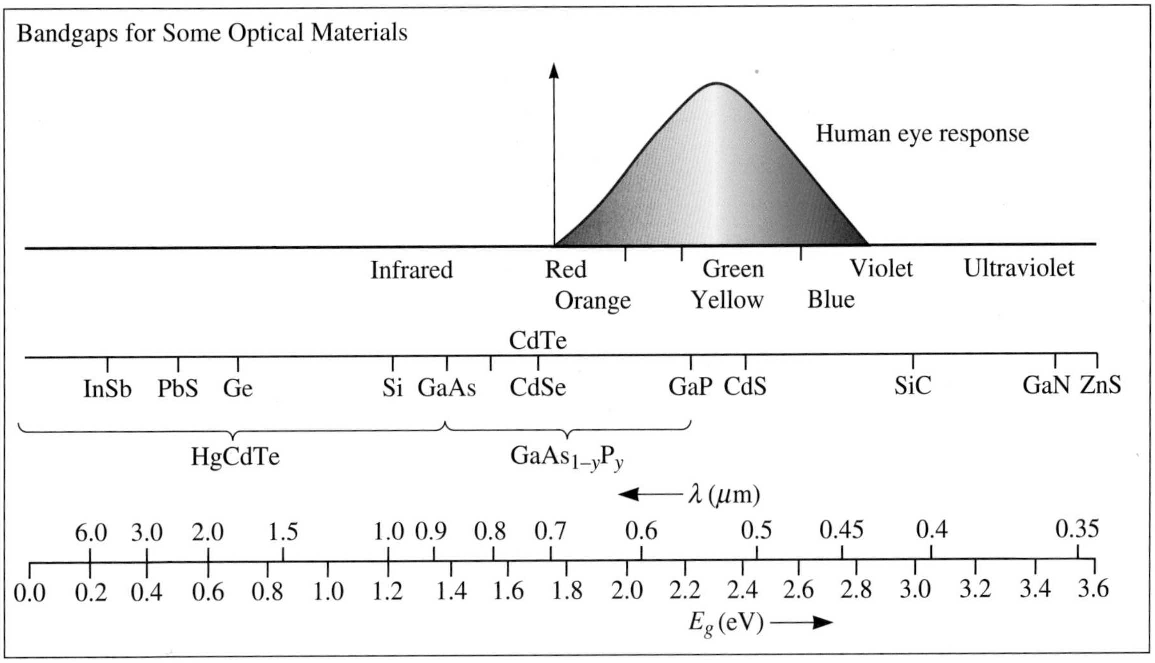

Figure 21-1 The electromagnetic spectrum of radiation; the bandgaps and cutoff frequencies for some optical materials are also shown. (*Based on* Optoelectronics: An Introduction to Materials and Devices, *by J. Singh, McGraw-Hill, 1996.*)

the surface (specular reflection) and also in many other directions [diffuse reflection, not shown in Figure 21-2(a)]. Several factors are important in determining the behavior of the photon, with the energy required to excite an electron to a higher energy state being of particular importance.

Let's examine each of these four phenomena. We begin with refraction, since it is related to reflection and transmission.

Refraction Even when a photon is transmitted, the photon causes polarization of the electrons in the material and, by interacting with the polarized material, loses some of its energy. The speed of light (c) can be related to the ease with which a

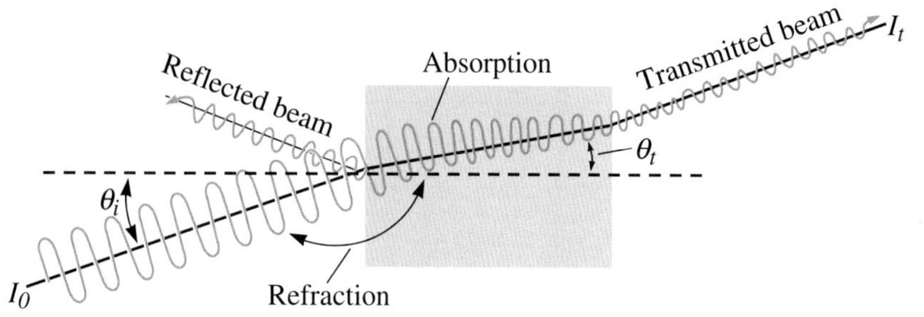

(a)

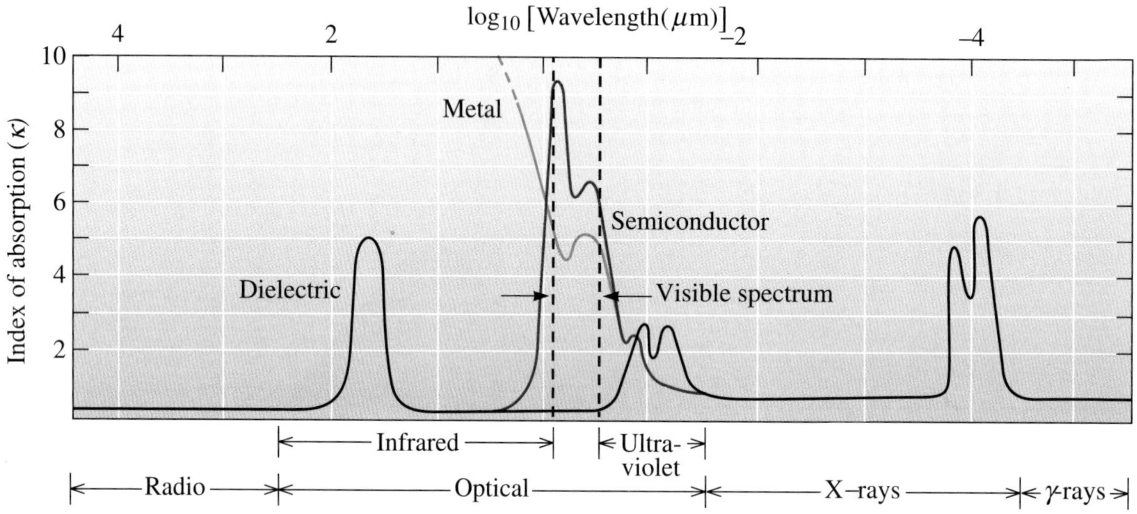

(b)

Figure 21-2 (a) Interaction of photons with a material. In addition to reflection, absorption, and transmission, the beam changes direction, or is refracted. The change in direction is given by the index of refraction n. (b) The absorption index (κ) as a function of wavelength.

material polarizes both electrically (permittivity ε) and magnetically (permeability μ) through

$$c = \frac{1}{\sqrt{\mu\varepsilon}} \tag{21-3}$$

Generally, optical materials are not magnetic, and the permeability can be neglected. Because the speed of the photons decreases, the beam of photons changes direction when it enters the material [Figure 21-2(a)]. Suppose photons traveling in a vacuum impinge on a material. If θ_i and θ_t respectively, are the angles that the incident and refracted beams make with the normal of the surface of the material, then

$$n = \frac{c_0}{c} = \frac{\lambda_{\text{vacuum}}}{\lambda} = \frac{\sin\theta_i}{\sin\theta_t} \tag{21-4}$$

The ratio n is the **index of refraction,** c_0 is the speed of light in a vacuum (3×10^8 m/s), and c is the speed of light in the material. The frequency of light does not change as it is refracted. Typical values of the index of refraction for several materials are listed in Table 21-1.

We can also define a complex refractive index (n^*). This includes κ, a parameter known as the absorption index:

$$n^* = n(1 - i\kappa) \tag{21-5}$$

Table 21-1 Index of refraction of selected materials for photons of wavelength 5890 Å

Material	Index of Refraction (n)	Material	Index of Refraction (n)
Air	1.00	Polystyrene	1.60
Ice	1.309	TiO_2	1.74
Water	1.333	Sapphire (Al_2O_3)	1.8
Teflon™	1.35	Leaded glasses (crystal)	2.50
SiO_2 (glass)	1.46	Rutile (TiO_2)	2.6
Polymethyl methacrylate	1.49	Diamond	2.417
Typical silicate glasses	~1.50	Silicon	3.49
Polyethylene	1.52	Gallium arsenide	3.35
Sodium chloride (NaCl)	1.54	Indium phosphide	3.21
SiO_2 (quartz)	1.55	Germanium	4.0
Epoxy	1.58		

In Equation 21-5, $i = \sqrt{-1}$ is the imaginary number. The absorption index is defined as

$$\kappa = \frac{\alpha\lambda}{4\pi n} \tag{21-6}$$

In Equation 21-6, α is the **linear absorption coefficient** (see Equation 21-12), λ is the wavelength of light, and n is the refractive index. Figure 21-2(b) shows the variation in index of absorption with the frequency of electromagnetic waves. Drawing an analogy, the refractive index is similar to the dielectric constant of materials, and the absorption index is similar to the dielectric loss factor.

If the photons are traveling in Material 1, instead of in a vacuum, and then pass into Material 2, the velocities of the incident and refracted beams depend on the ratio between their indices of refraction, again causing the beam to change direction:

$$\frac{c_1}{c_2} = \frac{n_2}{n_1} = \frac{\sin\theta_i}{\sin\theta_t} \tag{21-7}$$

Equation 21-7 is also known as Snell's law.

When a ray of light enters from a material with refractive index (n_1) into a material of refractive index (n_2), and if $n_1 > n_2$, the ray is bent away from the normal and toward the boundary surface [Figure 21-3(a)]. A beam traveling through Material 1 is reflected rather than transmitted if the angle θ_t becomes 90°.

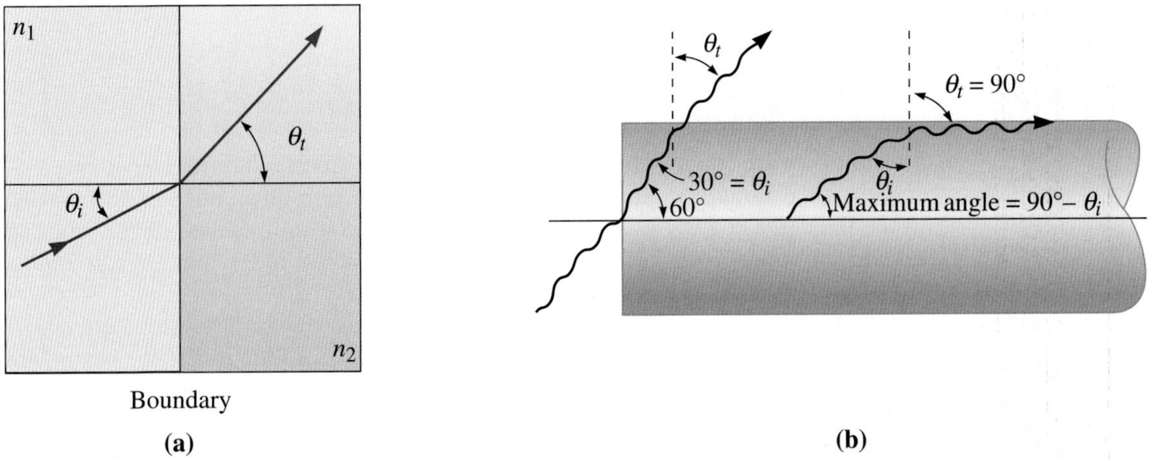

Figure 21-3 (a) When a ray of light enters from Material 1 into Material 2, if the refractive index of Material 1 (n_1) is greater than that of Material 2 (n_2), the ray bends away from the normal and toward the boundary surface. (b) Diagram of a light beam in a glass fiber for Example 21-1.

More interaction of the photons with the electronic structure of the material occurs when the material is easily polarized. We saw different dielectric polarization mechanisms in Chapter 19. Among these, the electronic polarization (i.e., displacement of the electron cloud around the atoms and ions) is the one that controls the refractive index of materials. Consequently, we find a relationship between the index of refraction n and the high-frequency dielectric constant k. of the material. From Equations 21-3 and 21-4 and for nonferromagnetic or nonferrimagnetic materials,

$$n = \frac{c_0}{c} = \sqrt{\frac{\mu\varepsilon}{\mu_0\varepsilon_0}} \cong \sqrt{\frac{\varepsilon}{\varepsilon_0}} = \sqrt{k_\infty} \qquad (21\text{-}8)$$

In Equation 21-8, c_0 and c are the speed of light in a vacuum and in the material, respectively, μ_0 and μ are the magnetic permeabilities of the vacuum and material, respectively, and ε_0 and ε are the dielectric permittivities of the vacuum and material, respectively. As we discussed in Chapter 19, the material known as "lead crystal," which is actually amorphous silicate glass with up to ~30% lead oxide, has a high index of refraction ($n \sim 2.4$), since Pb^{2+} ions have high electronic polarizability. We use a similar strategy to dope silica fibers to enhance the refractive index of the core of optical fibers as compared to the outer cladding region. This helps keep the light (and hence information) in the core of the optical fiber. The difference in the high-frequency dielectric constant and the low-frequency dielectric constant is a measure of the other polarization mechanisms that are contributing to the dielectric constant.

The refractive index n is not a constant for a particular material. The frequency, or wavelength, of the photons affects the index of refraction. **Dispersion** of a material is defined as the variation of the refractive index with wavelength:

$$(\text{Dispersion})_\lambda = \frac{dn}{d\lambda} \qquad (21\text{-}9)$$

This dependence of the refractive index on wavelength is nonlinear. The dispersion within a material means light pulses of different wavelengths, starting at the same time at the end of an optical fiber, will arrive at different times at the other end. Thus, material dispersion plays an important role in fiber optics. This is one of the reasons why we prefer to use a single wavelength source of light for fiber-optic communications. Dispersion also causes chromatic aberration in optical lenses.

Since dielectric polarization (P) is equal to the dipole moment per unit volume, and the high-frequency dielectric constant is related to the refractive index, we expect that (for the same material), a denser form or polymorph will have a higher refractive index (compare the refractive indices of ice and water or glass and quartz).

The following example illustrates how an optical fiber is designed to minimize optical losses during insertion. It is followed by an example calculating the index of refraction.

Example 21-1 *Design of a Fiber Optic System*

Optical fibers are commonly made from high-purity silicate glasses. They consist of a core with a refractive index that is higher than the refractive index of the coating on the fiber (the coating is called the cladding). Even a simple glass fiber in air can serve as an optical fiber because the fiber has a refractive index that is greater than that of air.

Consider a beam of photons that is introduced from a laser into a glass fiber with an index of refraction of 1.5. Choose the angle of introduction of the beam with respect to the fiber axis that will result in a minimum of leakage of the beam from the fiber. Also, consider how this angle might change if the fiber is immersed in water.

SOLUTION

To prevent leakage of the beam, we need total internal reflection, and thus the angle θ_t must be at least 90°. Suppose that the photons enter at a 60° angle to the axis of the fiber. From Figure 21-3(b), we find that $\theta_i = 90 - 60 = 30°$. If we let the glass be Material 1 and if the glass fiber is in air ($n = 1.0$), then from Equation 21-7:

$$\frac{n_2}{n_1} = \frac{\sin\,\theta_i}{\sin\,\theta_t} \quad \text{or} \quad \frac{1}{1.5} = \frac{\sin\,30°}{\sin\,\theta_t}$$

$$\sin\,\theta_t = 1.5\,\sin\,30° = 1.5(0.50) = 0.75 \;\text{ or }\; \theta_t = 48.6°$$

Because θ_t is less than 90°, photons escape from the fiber. To prevent transmission, we must introduce the photons at a shallower angle, giving $\theta_t = 90°$.

$$\frac{1}{1.5} = \frac{\sin\,\theta_i}{\sin\,\theta_t} = \frac{\sin\,\theta_i}{\sin\,90°} = \sin\,\theta_i$$

$$\sin\,\theta_i = 0.6667 \text{ or } \theta_i = 41.8°$$

If the angle between the beam and the axis of the fiber is $90 - 41.8 = 48.2°$ or less, the beam is reflected.

If the fiber were immersed in water ($n = 1.333$), then

$$\frac{1.333}{1.5} = \frac{\sin\,\theta_i}{\sin\,\theta_t} = \frac{\sin\,\theta_i}{\sin 90°} = \sin\theta_i$$

$$\sin\,\theta_i = 0.8887 \text{ or } \theta_i = 62.7°$$

In water, the photons would have to be introduced at an angle of less than $90 - 62.7 = 27.3°$ in order to prevent transmission.

Example 21-2 *Light Transmission in Polyethylene*

Suppose a beam of photons in a vacuum strikes a sheet of polyethylene at an angle of 10° to the normal of the surface of the polymer. Polyethylene has a high-frequency dielectric constant of $k_\infty = 2.3$. Calculate the index of refraction of polyethylene, and find the transmission angle.

SOLUTION

The index of refraction is related to the high-frequency dielectric constant:

$$n = \sqrt{k_\infty} = \sqrt{2.3} = 1.52$$

The angle θ_t is

$$n = \frac{\sin\,\theta_i}{\sin\,\theta_t}$$

$$\sin\,\theta_t = \frac{\sin\,\theta_i}{n} = \frac{\sin\,10°}{1.52} = \frac{0.174}{1.52} = 0.1145$$

$$\theta_t = 6.57°$$

Reflection When a beam of photons strikes a material, the photons interact with the valence electrons and give up their energy. In metals, the radiation of almost any wavelength excites the electrons into higher energy levels. One might expect that, if the photons are totally absorbed, no light would be reflected and the metal would appear black. In aluminum or silver, however, photons of almost identical wavelength are immediately re-emitted as the excited electrons return to their lower energy levels—that is, reflection occurs. Since virtually the entire visible spectrum is reflected, these metals have a white, or silvery color.

The **reflectivity** R gives the fraction of the incident beam that is reflected and is related to the index of refraction. If the material is in a vacuum or in air:

$$R = \left(\frac{n-1}{n+1}\right)^2 \tag{21-10}$$

If the material is in some other medium with an index of refraction of n_i, then

$$R = \left(\frac{n-n_i}{n+n_i}\right)^2 \tag{21-11}$$

These equations apply to reflection from a single surface and assume a normal (perpendicular to the surface) incidence. The value of R depends upon the angle of incidence. Materials with a high index of refraction have a higher reflectivity than materials with a low index. Because the index of refraction varies with the wavelength of the photons, so does the reflectivity.

In metals, the reflectivity is typically on the order of 0.9 to 0.95, whereas the reflectivity of typical glasses is nearer to 0.05. The high reflectivity of metals is one reason that they are *opaque.*

There are many applications for which we want materials to have very good reflectivity. Examples include mirrors and certain types of coatings on glasses. In fact, these coatings must also be designed such that much of a certain part of the electromagnetic spectrum (e.g., infrared, the part that produces heat) must be reflected. Many such coatings have been developed for glasses. There are also many applications for which the reflectivity must be extremely limited. Such coatings are known as antireflective (AR) coatings. These coatings are used for glasses, in automobile rear view mirrors, on windows, or for the glass of picture frames so that you see through the glass without seeing your own reflection.

Absorption That portion of the incident beam that is not reflected by the material is either absorbed or transmitted through the material. The fraction of the beam that is absorbed is related to the thickness of the material and the manner in which the photons interact with the material's structure. The intensity of the beam after passing through the material is given by

$$I = I_0 \exp(-\alpha x) \tag{21-12}$$

where x is the path through which the photons move (usually the thickness of the material), α is the linear absorption coefficient of the material for the photons, I_0 is the intensity of the beam after reflection at the front surface, and I is the intensity of the beam when it reaches the back surface. Equation 21-12 is also known as Bouguer's law, or the Beer–Lambert law. Figure 21-4 shows the linear absorption coefficient as a function of wavelength for several metals.

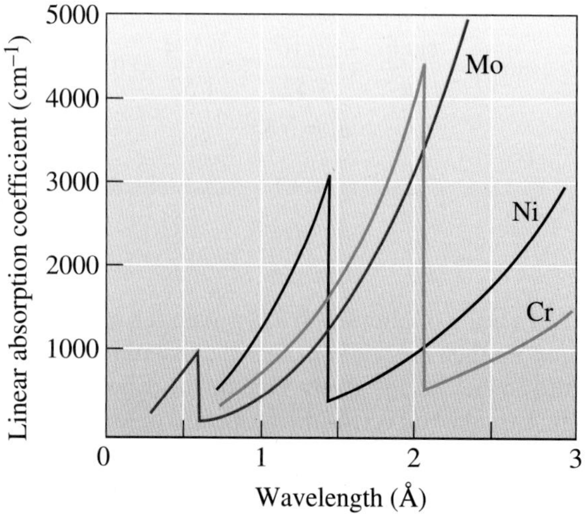

Figure 21-4 The linear absorption coefficient relative to wavelengths for several metals. Note the sudden decrease in the absorption coefficient for wavelengths greater than the absorption edge.

Absorption in materials occurs by several mechanisms. In *Rayleigh scattering*, photons interact with the electrons orbiting an atom and are deflected without any change in photon energy; this is an example of "elastic" scattering. Rayleigh scattering is more significant for higher photon energies and is responsible for the color of the sky. Since the blue (high energy) end of the visible spectrum scatters most efficiently, the sunlight scattered by molecules and atmospheric particles that reaches our eyes is mostly blue. The effect is quite strong as the intensity of scattering is proportional to the inverse wavelength of the photons raised to the fourth power. Rayleigh scattering also depends on the particle size and is most efficient for particles that are much smaller than the wavelength of the light. Scattering from particles much larger than the wavelength of light occurs because of the *Tyndall effect.* This is why clouds, consisting of water droplets, look white. *Compton scattering* also occurs when a photon interacts with an electron, but because the incoming photon loses some of its energy to the electron, the wavelength of the light increases; this is an example of "inelastic" scattering.

The *photoelectric effect* occurs when the energy of a photon is fully absorbed, resulting in the ejection of an electron from an atom. The atom thus becomes ionized. No electrons are ejected when the incoming photons have energies less than the electron binding energy, regardless of the intensity of the light. When the photons have energies greater than the binding energy, electrons are ejected with a kinetic energy equal to the photon energy minus the binding energy. As the energy of the photon increases (or the wavelength decreases, see Figure 21-4), less absorption occurs until the photon has an energy equal to that of the binding energy. At this energy, the absorption coefficient increases abruptly because electrons may now be ejected, thereby providing an efficient means for photons to be absorbed. The energy or wavelength at which this occurs is referred to as the **absorption edge** because of the shape of the peaks in Figure 21-4. The abrupt change in the absorption coefficient corresponds to the energy required to remove an electron from the atom; this absorption edge is important to certain x-ray analytical techniques. Albert Einstein received the Nobel Prize in Physics in 1921 for explaining the photoelectric effect.

In some cases, the effect of scattering can be written as

$$I = I_0 \exp [(-\alpha_i + \alpha_s)x] \tag{21-13}$$

In Equation 21-13, α_i is what we previously termed α, the intrinsic absorption coefficient, and α_s is the scattering coefficient.

Examples of a portion of the characteristic spectra for several elements are included in Table 21-2. The K_α, K_β, and L_α lines correspond to the wavelengths of radiation emitted from transitions of electrons between shells, as discussed later in this chapter.

Table 21-2 Characteristic emission lines and absorption edges for selected elements

Metal	K_α (Å)	K_β (Å)	L_α (Å)	Absorption Edge (Å)
Al	8.337	7.981	—	7.951
Si	7.125	6.768	—	6.745
S	5.372	5.032	—	5.018
Cr	2.291	2.084	—	2.070
Mn	2.104	1.910	—	1.896
Fe	1.937	1.757	—	1.743
Co	1.790	1.621	—	1.608
Ni	1.660	1.500	—	1.488
Cu	1.542	1.392	13.357	1.380
Mo	0.711	0.632	5.724	0.620
W	0.211	0.184	1.476	0.178

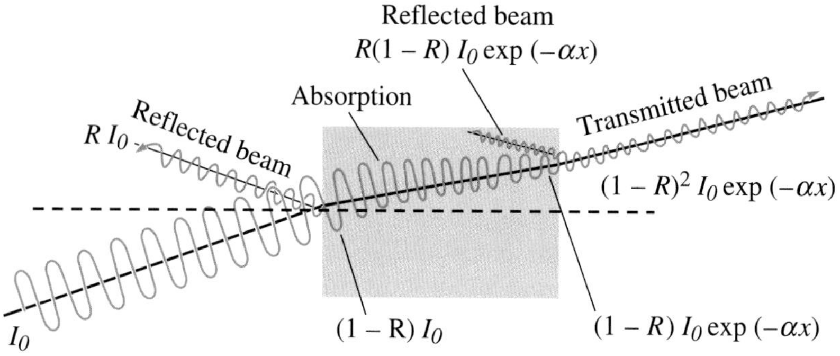

Figure 21-5 Fractions of the original beam that are reflected, absorbed, and transmitted. The frequency of the light does not change as it is refracted.

Transmission The fraction of the beam that is not reflected or absorbed is transmitted through the material. Using the following steps, we can determine the fraction of the beam that is transmitted (see Figure 21-5).

1. If the incident intensity is I_0, then the loss due to reflection at the front face of the material is RI_0. The fraction of the incident beam that actually enters the material is $I_0 - RI_0 = (1 - R)I_0$:

$$I_{\text{reflected at front surface}} = RI_0$$
$$I_{\text{after reflection}} = (1 - R)I_0$$

2. A portion of the beam that enters the material is lost by absorption. The intensity of the beam after passing through a material having a thickness x is

$$I_{\text{after absorption}} = (1 - R)I_0 \exp(-\alpha x)$$

3. Before the partially absorbed beam exits the material, reflection occurs at the back surface. The fraction of the beam that reaches the back surface and is reflected is

$$I_{\text{reflected at back surface}} = R(1 - R)I_0 \exp(-\alpha x)$$

4. Consequently, the fraction of the beam that is completely transmitted through the material is

$$\begin{aligned} I_{\text{transmitted}} &= I_{\text{after absorption}} - I_{\text{reflected at back}} \\ &= (1 - R)I_0 \exp(-\alpha x) - R(1 - R)I_0 \exp(-\alpha x) \\ &= (1 - R)(1 - R)I_0 \exp(-\alpha x) \\ I_t &= (1 - R)^2 I_0 \exp(-\alpha x) \end{aligned} \tag{21-14}$$

The intensity of the transmitted beam may depend on the wavelength of the photons in the beam. In metals, because there is no energy gap, virtually any photon has sufficient energy to excite an electron into a higher energy level, thus absorbing the energy of the excited photon [Figure 21-6(a)]. As a result, even extremely thin samples of metals are opaque. Dielectrics, on the other hand, possess a large energy gap between the valence and conduction bands. If the energy of the incident photons is less than the energy gap, no electrons gain enough energy to escape the valence band and, therefore, absorption does not occur [Figure 21-6(b)]. For intrinsic semiconductors, the energy gap is smaller than that for insulators, and absorption occurs when the photons have energies exceeding the energy gap E_g, whereas transmission occurs for less energetic photons [Figure 21-6(b) also applies to semiconductors]. In other words, only the light below a certain wavelength

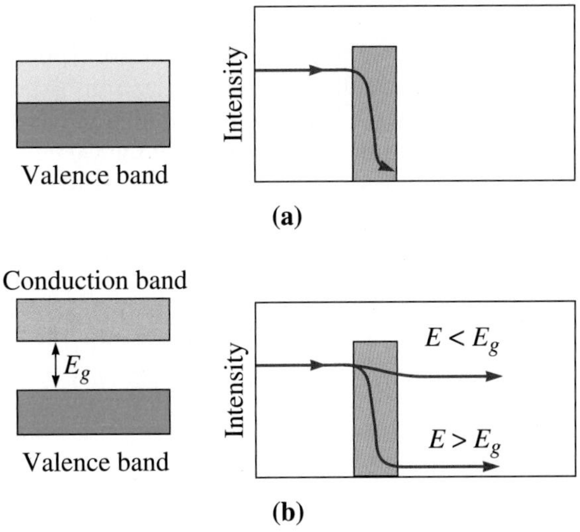

Figure 21-6 Relationships between absorption and the energy gap: (a) metals and (b) dielectrics and intrinsic semiconductors. The diagram on the left represents the band structure of the material under consideration. The diagram on the right represents the intensity of light as it passes from air into the material and back into air.

is absorbed. Extrinsic semiconductors include donor or acceptor energy levels within the bandgap that provide additional energy levels for absorption. Therefore, semiconductors are opaque to short wavelength radiation but transparent to long wavelength photons (see Example 21-3). For example, silicon and germanium appear opaque to visible light, but they are transparent to longer wavelength infrared radiation. Many of the narrow bandgap semiconductors (e.g., HgCdTe) are used for detection of infrared radiation. These detector materials have to be cooled to low temperatures (e.g., using liquid nitrogen) since the thermal energy of electrons at room temperature is otherwise enough to saturate the conduction band.

The intensity of the transmitted beam also depends on microstructural features. Porosity in ceramics scatters photons; even a small amount of porosity (less than 1 volume percent) may make a ceramic opaque. For example, alumina that has relatively low mass density (owing to porosity) is opaque, while high density alumina is optically transparent. High density alumina is often used to manufacture light bulbs. Crystalline precipitates, particularly those that have a much different index of refraction than the matrix material, also cause scattering. These crystalline *opacifiers* cause a glass that normally may have excellent transparency to become translucent or even opaque. Typically, smaller pores or precipitates cause a greater reduction in the transmission of light.

Photoconduction occurs in semiconducting materials if the semiconductor is part of an electrical circuit. If the energy of an incoming photon is sufficient, an electron is excited into the conduction band from the valence band, thereby creating a hole in the valence band. The electron and hole then carry charge through the circuit [Figure 21-7(a)]. The maximum wavelength of the incoming photon that will produce photoconduction is related to the energy gap in the semiconducting material:

$$\lambda_{\text{max}} = \frac{hc}{E_g} \qquad (21\text{-}15)$$

We can use this principle for photodetectors or "electric eyes" that open or close doors or switches when a beam of light focused on a semiconducting material is interrupted.

Solar cells also use the absorption of light to generate voltage [Figure 21-7(b)]. Essentially the electron-hole pairs generated by optical absorption are separated, and this leads to the development of a voltage. This voltage causes a current flow in an external circuit. Solar cells are *p-n* junctions designed so that photons excite electrons into the conduction band. The electrons move to the *n*-side of the junction, while holes move to the *p*-side of the junction. This movement produces a contact voltage due to the charge

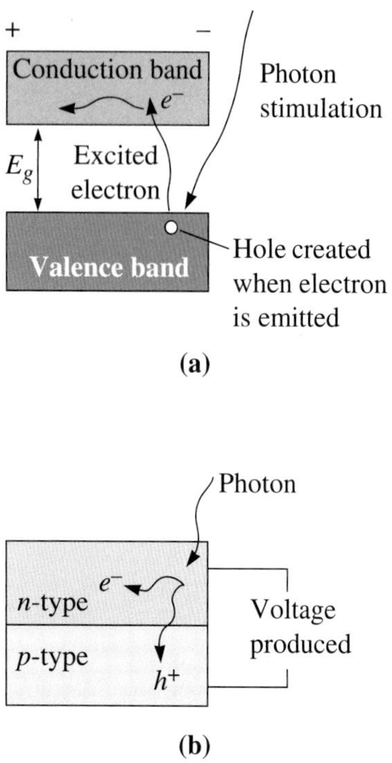

Figure 21-7 (a) Photoconduction in semiconductors involves the absorption of a stimulus by exciting electrons from the valence band to the conduction band. Rather than dropping back to the valence band to cause emission, the excited electrons carry a charge through an electrical circuit. (b) A solar cell takes advantage of this effect.

imbalance. If the *p-n* junction is connected to an electric circuit, the junction acts as a battery to power the circuit. Solar cells make use of antireflective coatings so that maximum key elements of the solar spectrum are captured.

LEDs As discussed in Chapter 19, the light that is absorbed by a direct band gap semiconductor causes electrons to be promoted to the conduction band. When these electrons fall back into the valence band, they combine with the holes and may cause emission of light. Many semiconductor solid solutions can be tailored to have particular bandgaps, producing LEDs of different colors. This phenomenon is also used in semiconductor lasers (Figure 21-8). The design of LEDs is discussed later in this chapter.

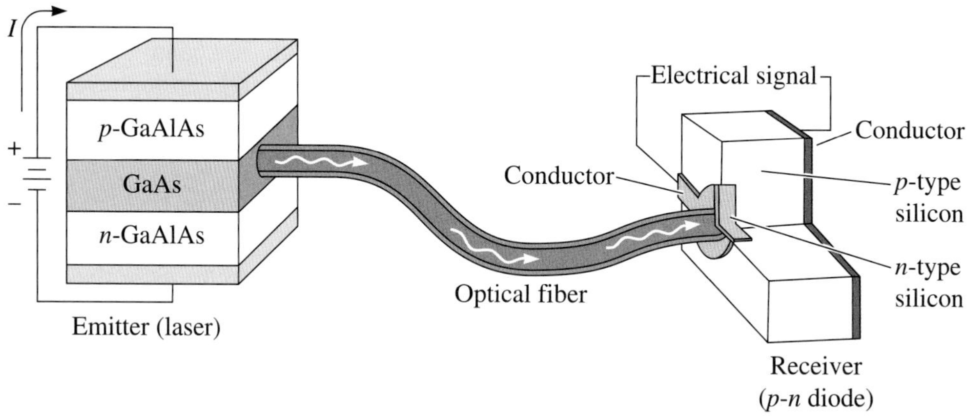

Figure 21-8 Elements of a photonic system for transmitting information involves a laser or LED to generate photons from an electrical signal, optical fibers to transmit the beam of photons efficiently, and an LED receiver to convert the photons back into an electrical signal.

The following examples illustrate applications of many of these concepts related to the absorption and transmission of light.

Example 21-3 *Determining Critical Energy Gaps*

Determine the critical energy gaps for a semiconductor that provide complete transmission and complete absorption of photons in the visible spectrum.

SOLUTION

In order for complete transmission to occur, the bandgap of the semiconductor must be larger than the energies of all photons in the visible spectrum. The visible light spectrum varies from 4×10^{-5} cm to 7×10^{-5} cm. The photons with shorter wavelengths have higher energies. Thus, the minimum band gap energy E_g required to ensure that no photons in the visible spectrum are absorbed (and all photons are transmitted) is

$$E_g = \frac{hc}{\lambda} = \frac{(6.626 \times 10^{-34}\,\text{J} \cdot \text{s})(3 \times 10^{10}\,\text{cm/s})}{(4 \times 10^{-5}\,\text{cm})(1.6 \times 10^{-19}\,\text{J/eV})} = 3.1\,\text{eV}$$

If the semiconductor band gap is 3.1 eV or larger, all photons in the visible spectrum will be transmitted.

In order for a photon to be absorbed, the energy gap of the semiconductor must be less than the photon energy. The photons with longer wavelengths have lower energies. Thus, the maximum bandgap energy that will allow for complete absorption of all wavelengths of the visible spectrum is

$$E_g = \frac{hc}{\lambda} = \frac{(6.626 \times 10^{-34}\,\text{J} \cdot \text{s})(3 \times 10^{10}\,\text{cm/s})}{(7 \times 10^{-5}\,\text{cm})(1.6 \times 10^{-19}\,\text{J/eV})} = 1.8\,\text{eV}$$

If the bandgap is 1.8 eV or smaller, all photons in the visible spectrum will be absorbed. For semiconductors with an E_g between 1.8 eV and 3.1 eV, a portion of the photons in the visible spectrum will be absorbed.

Example 21-4 *Design of a Radiation Shield*

A material has a reflectivity of 0.15 and an absorption coefficient (α) of 100 cm^{-1}. Design a shield that will permit only 1% of the incident radiation to be transmitted through the material.

SOLUTION

From Equation 21-14, the fraction of the incident intensity that will be transmitted is

$$\frac{I_t}{I_0} = (1 - R)^2 \exp(-\alpha x)$$

and the required thickness of the shield can be determined:

$$0.01 = (1 - 0.15)^2 \exp(-100x)$$

$$\frac{0.01}{(0.85)^2} = 0.01384 = \exp(-100x)$$

$$\ln(0.01384) = -4.28 = -100x$$

$$x = 0.0428\text{ cm}$$

The material should have a thickness of 0.0428 cm in order to transmit 1% of the incident radiation.

If we wished, we could determine the amount of radiation lost in each step:

Reflection at the front face: $I_r = RI_0 = 0.15I_0$

Intensity after reflection: $I = I_0 - 0.15I_0 = 0.85I_0$

Intensity after absorption: $I_a = (0.85)I_0 \exp[(-100)(0.0428)] = 0.0118I_0$

Absorbed Intensity: $0.85I_0 - 0.0118I_0 = 0.838I_0$

Reflection at the back face: $I_r = R(1 - R)I_0 \exp(-\alpha x)$

$= (0.15)(1 - 0.15)I_0 \exp[-(100)(0.0428)] = 0.0018I_0$

21-3 Selective Absorption, Transmission, or Reflection

Unusual optical behavior is observed when photons are selectively absorbed, transmitted, or reflected. We have already seen that semiconductors transmit long wavelength photons but absorb short wavelength radiation. There are a variety of other cases in which similar selectivity produces unusual optical properties.

In certain materials, replacement of normal ions by transition or rare earth elements produces a *crystal field*, which creates new energy levels within the structure. This phenomenon occurs when Cr^{3+} ions replace Al^{3+} ions in Al_2O_3. The new energy levels absorb visible light in the violet and green-yellow portions of the spectrum. Red wavelengths are transmitted, giving the reddish color in ruby. In addition, the chromium ion replacement creates an energy level that permits luminescence (discussed later) to occur when the electrons are excited by a stimulus. Lasers made from chromium-doped aluminium oxide produce a characteristic red beam because of this.

Glasses can also be doped with ions that produce selective absorption and transmission (Table 21-3). Similarly, electrons or hole traps called *F-centers*, can be present in crystals. When fluorite (CaF_2) is formed with excess calcium, a fluoride ion vacancy is produced. To maintain electrical neutrality, an electron is trapped in the vacancy, producing energy levels that absorb all visible photons—with the exception of purple.

Polymers—particularly those containing an aromatic ring in the backbone—can have complex covalent bonds that produce an energy level structure which causes selective absorption. For this reason, chlorophyll in plants appears green, and hemoglobin in blood appears red.

Table 21-3 Effect of ions on colors produced in glasses

Ion	Color	Ion	Color
Cr^{2+}	Blue	Mn^{2+}	Orange
Cr^{3+}	Green	Fe^{2+}	Blue-green
Cu^{2+}	Blue-green	U^{6+}	Yellow

21-4 Examples and Use of Emission Phenomena

Let's look at some particular examples of emission phenomena which, by themselves, provide some familiar and important functions.

Gamma Rays—Nuclear Interactions Gamma rays, which are very high-energy photons, are emitted during the radioactive decay of unstable nuclei of certain atoms. Therefore, the energy of the gamma rays depends on the structure of the atom nucleus and varies for different materials. The gamma rays produced from a material have fixed wavelengths. For example, when cobalt 60 decays, gamma rays having energies of 1.17×10^6 and 1.34×10^6 eV (or wavelengths of 1.06×10^{-10} cm and 0.93×10^{-10} cm) are emitted. The gamma rays can be used as a radiation source to detect defects in a material (a nondestructive test).

X-rays—Inner Electron Shell Interactions

X-rays, which have somewhat lower energies than gamma rays, are produced when electrons in the inner shells of an atom are stimulated. The stimulus could be high-energy electrons or other x-rays. When stimulation occurs, x-rays of a wide range of energies are emitted. Both a continuous and a characteristic spectrum of x-rays are produced.

Suppose that a high-energy electron strikes a material. As the electron decelerates, energy is given up and emitted as photons. Each time the electron strikes an atom, more of its energy is given up. Each interaction, however, may be more or less severe, so the electron gives up a different fraction of its energy each time and produces photons of different wavelengths (Figure 21-9). A **continuous spectrum** is produced (the smooth portion of the curves in Figure 21-10). If the electron were to lose all of its energy in one impact, the minimum wavelength of the emitted photons would correspond to the original energy of the stimulus. The minimum wavelength of x-rays produced is called the **short wavelength limit** λ_{swl}. The short wavelength limit decreases, and the number and energy of the emitted photons increase, when the energy of the stimulus increases.

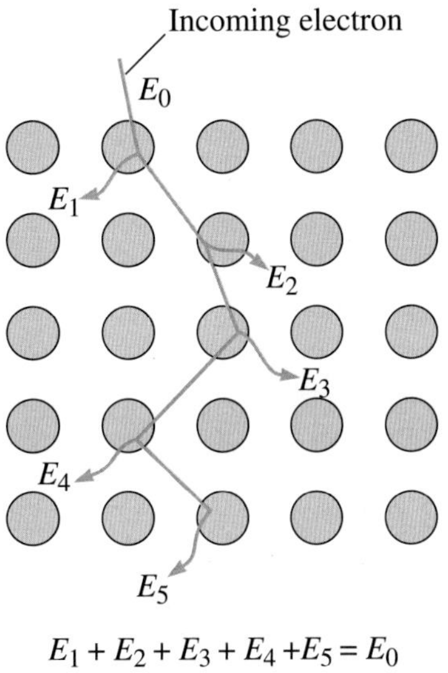

Figure 21-9 When an accelerated electron strikes and interacts with a material, its energy may be reduced in a series of steps. In the process, several photons of different energies E_1 to E_5 are emitted, each with a unique wavelength.

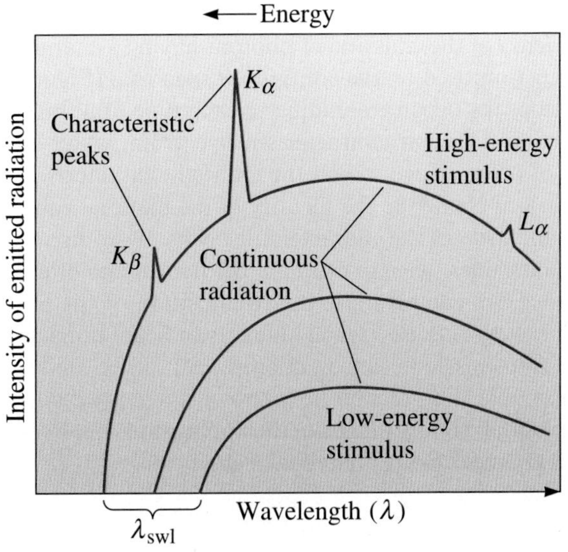

Figure 21-10 The continuous and characteristic spectra of radiation emitted from a material. Low-energy stimuli produce a continuous spectrum of low-energy, long wavelength photons. A more intense, higher energy spectrum is emitted when the stimulus is more energetic until, eventually, characteristic radiation is observed.

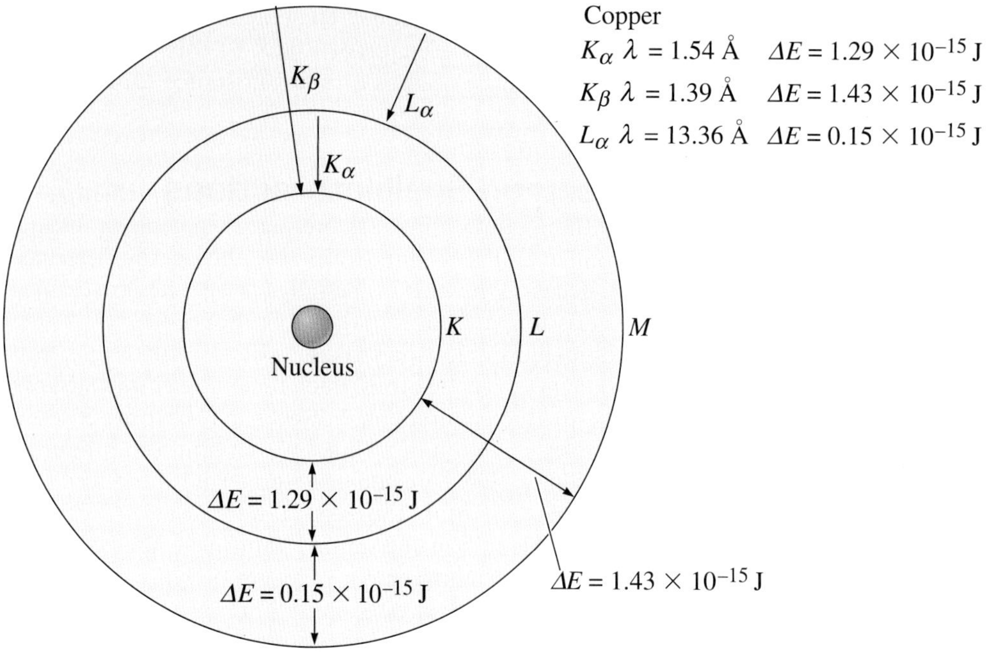

Figure 21-11 Characteristic x-rays are produced when electrons transition from one energy level to a lower energy level, as illustrated here for copper. The energy and wavelength of the x-rays are fixed by the energy differences between the energy levels.

The incoming stimulus may also have sufficient energy to excite an electron from an inner energy level to an outer energy level. The excited electron is not stable and, to restore equilibrium, electrons from a higher level fill the empty inner level. This process leads to the emission of a **characteristic spectrum** of x-rays that is different for each type of atom.

The characteristic spectrum is produced because there are discrete energy differences between any two energy levels. When an electron drops from one level to a second level, a photon having the corresponding energy and wavelength is emitted. This effect is illustrated in Figure 21-11. We typically refer to the energy levels by the *K, L, M, . . .* designation, as described in Chapter 2. If an electron is excited from the *K* shell, electrons may fill that vacancy from an outer shell. Normally, electrons in the closest shells fill the vacancies. Thus, photons with energy $\Delta E = E_K - E_L$ (K_α x-rays) or $\Delta E = E_K - E_M$ (K_β x-rays) are emitted. When an electron from the *M* shell fills the *L* shell, a photon with energy $\Delta E = E_L - E_M$ (L_α x-rays) is emitted; it has a long wavelength, or low energy. Note that we need a more energetic stimulus to produce K_α x-rays than that required for L_α x-rays.

As a consequence of the emission of photons having a characteristic wavelength, a series of peaks is superimposed on the continuous spectrum (Figure 21-10). The wavelengths at which these peaks occur are unique to each type of atom (Table 21-2). Thus, each element produces a different characteristic spectrum, which serves as a "fingerprint" for that type of atom. If we match the emitted characteristic wavelengths with those expected for various elements, the identity of the material can be determined. We can also measure the intensity of the characteristic peaks. By comparing measured intensities with standard intensities, we can estimate the percentage of each type of atom in the material and, hence, we can estimate the composition of the material. The energy (or wavelength) of the x-rays emitted when an electron beam impacts a sample (such as that in a scanning or transmission electron microscope) can be analyzed to get chemical information about a sample. This technique is known as energy dispersive x-ray analysis (EDXA). The examples that follow illustrate the application of x-ray emission as used in x-ray diffraction (XRD) and EDXA analytical techniques.

Example 21-5 *Design/Materials Selection for an X-ray Filter*

Design a filter that preferentially absorbs K_β x-rays from the nickel spectrum but permits K_α x-rays to pass with little absorption. This type of filter is used in x-ray diffraction analysis of materials.

SOLUTION

When determining a crystal structure or identifying unknown materials using various x-ray diffraction techniques, we prefer to use x-rays of a single wavelength. If both K_α and K_β characteristic peaks are present and interact with the material, analysis becomes much more difficult.

To avoid this difficulty, we can use selective absorption to isolate the K_α peak. Table 21-2 includes the information that we need. If a filter material is selected such that the absorption edge lies between the K_α and K_β wavelengths, then the K_β is almost completely absorbed, whereas the K_α is almost completely transmitted. In nickel, $K_\alpha = 1.660$ Å and $K_\beta = 1.500$ Å. A filter with an absorption edge between these characteristic peaks will work. Cobalt, with an absorption edge of 1.608 Å, would be our choice. Figure 21-12 shows how this filtering process occurs.

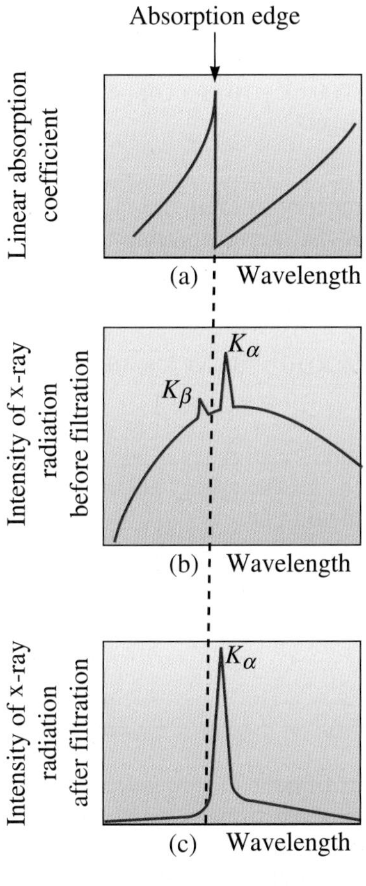

Figure 21-12 Elements have a selective lack of absorption of certain wavelengths. If a filter is selected with an absorption edge between the K_α and K_β peaks of an x-ray spectrum, all x-rays except K_α are absorbed (for Example 21-5). (a) The linear absorption coefficient of a filter material as a function of wavelength; (b) the intensity of the x-ray radiation before filtration; and (c) the intensity of the x-ray radiation after filtration.

Example 21-6 *Design of an X-ray Filter*

Design a filter to transmit at least 95% of the energy of a beam composed of zinc K_α x-rays, using aluminum as the shielding material. (The aluminum has a linear absorption coefficient of 108 cm^{-1}.) Assume no loss to reflection.

SOLUTION

Assuming that no losses are caused by the reflection of x-rays from the aluminum, we simply need to choose the thickness of the aluminum required to transmit 95% of the incident intensity. The final intensity will therefore be $0.95I_0$. Thus, from Equation 21-12,

$$\ln\left(\frac{0.95I_0}{I_0}\right) = -(108)(x)$$

$$\ln(0.95) = -0.051 = -108x$$

$$x = \frac{-0.051}{-108} = 0.00047\,\text{cm}$$

We would like to roll the aluminum to a thickness of 0.00047 cm or less. The filter could be thicker if a material were selected that has a lower linear absorption coefficient for zinc K_α x-rays.

Example 21-7 *Generation of X-rays for X-ray Diffraction*

Suppose an electron accelerated at 5000 eV strikes a copper target. Will K_α, K_β, or L_α x-rays be emitted from the copper target?

SOLUTION

The electron must possess enough energy to excite an electron to a higher level, or its wavelength must be less than that corresponding to the energy difference between the shells:

$$E = (5000\,\text{eV})(1.6 \times 10^{-19}\,\text{J/eV}) = 8 \times 10^{-16}\,\text{J}$$

$$\lambda = \frac{hc}{E} = \frac{(6.626 \times 10^{-34}\,\text{J}\cdot\text{s})(3 \times 10^{10}\,\text{cm/s})}{8 \times 10^{-16}\,\text{J}}$$

$$= 2.48 \times 10^{-8}\,\text{cm} = 2.48\ \text{Å}$$

Note that one electron volt (eV) is the kinetic energy acquired by an electron moving through a potential difference of one volt.

For copper, K_α is 1.542 Å, K_β is 1.392 Å, and L_α is 13.357 Å (Table 21-2). Therefore, the L_α peak may be produced, but K_α and K_β will not.

Example 21-8 *Energy Dispersive X-ray Analysis (EDXA)*

The micrograph in Figure 21-13 was obtained using a scanning electron microscope at a magnification of 1000. The beam of electrons in the SEM was directed at the three different phases, creating x-rays and producing the characteristic peaks. From the energy spectra, determine the probable composition of each phase. Assume each region represents a different phase.

SOLUTION

All three phases have an energy peak of about 1.5 keV = 1500 eV, which corresponds to a wavelength of

$$\lambda = \frac{hc}{E} = \frac{(6.626 \times 10^{-34}\,\text{J}\cdot\text{s})(3 \times 10^{10}\,\text{cm/s})}{(1500\,\text{eV})(1.6 \times 10^{-19}\,\text{J/eV})(10^{-8}\,\text{cm/Å})} = 8.283\text{Å}$$

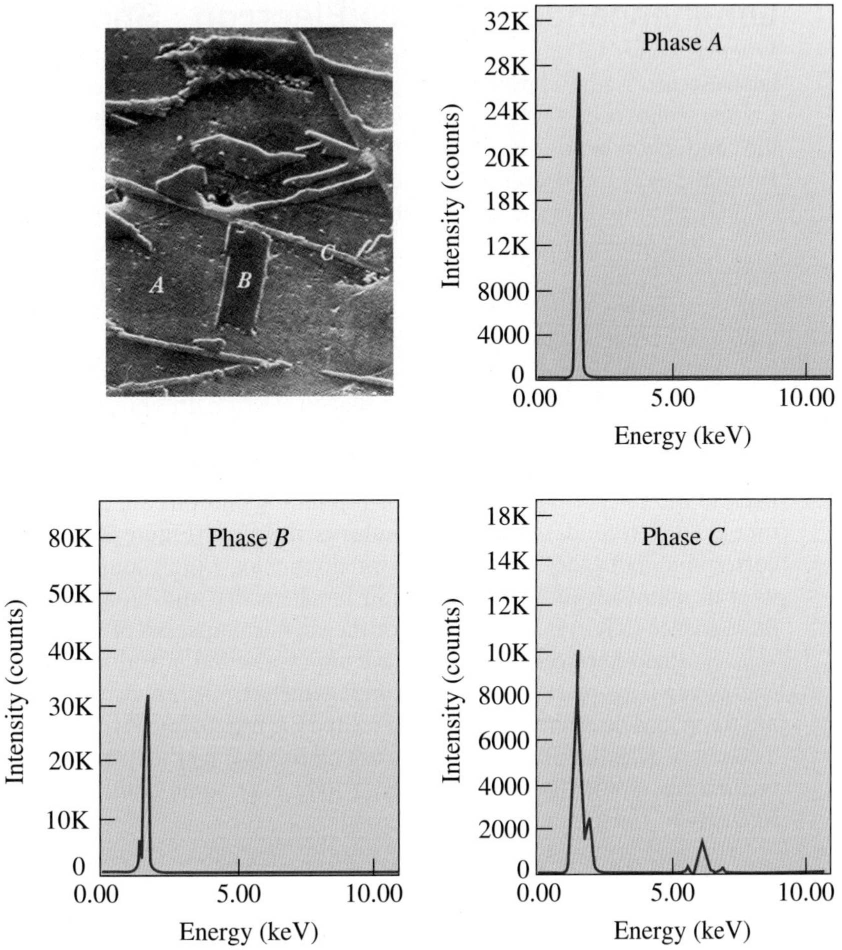

Figure 21-13 Scanning electron micrograph of a multiple phase material. The energy distributions of emitted radiation from the three phases marked *A*, *B*, and *C* are shown. The identity of each phase is determined in Example 21-8. (*Reprinted Courtesy of Don Askeland*)

Phase	Peak Energy	λ	λ (Table 21-2)	Line
A	1.5 keV	8.283 Å	8.337 Å	K_αAl
B	1.5 keV	8.283 Å	8.337 Å	K_αAl
	1.7 keV	7.308 Å	7.125 Å	K_αSi
C	1.5 keV	8.283 Å	8.337 Å	K_αAl
	1.7 keV	7.308 Å	7.125 Å	K_αSi
	5.8 keV	2.142 Å	2.104 Å	K_αMn
	6.4 keV	1.941 Å	1.937 Å	K_αFe
	7.1 keV	1.750 Å	1.757 Å	K_βFe

In a similar manner, energies and wavelengths can be found for the other peaks. These wavelengths are compared with those in Table 21-2, and the identity of the elements in each phase can be found, as summarized in the table.

Thus, Phase *A* appears to be an aluminum matrix, Phase *B* appears to be a silicon needle (perhaps containing some aluminum), and Phase *C* appears to be an Al-Si-Mn-Fe compound. Actually, this is an aluminum-silicon alloy. The stable phases are aluminum and silicon with inclusions forming due to the presence of manganese and iron as impurities.

Luminescence—Outer Electron Shell Interactions

Whereas x-rays are produced by electron transitions in the inner energy levels of an atom, **luminescence** is the conversion of radiation or other forms of energy to visible light. Luminescence occurs when the incident radiation excites electrons from the valence band to the conduction band. The excited electrons remain in the higher energy levels only briefly. When the electrons drop back to the valence band, photons are emitted. If the wavelength of these photons is in the visible light range, luminescence occurs.

Luminescence does not occur in metals. Electrons are merely excited into higher energy levels within the unfilled valence band. When the excited electron returns to the lower energy level, the photon that is produced has a very small energy and a wavelength longer than that of visible light [Figure 21-14(a)].

In certain ceramics and semiconductors, however, the energy gap between the valence and conduction bands is such that an electron dropping through this gap produces a photon in the visible range. Two different effects are observed in these luminescent materials: fluorescence and phosphorescence. In **fluorescence,** all of the excited electrons drop back to the valence band and the corresponding photons are emitted within a very short time ($\sim 10^{-8}$ seconds) after the stimulus is removed [Figure 21-14(b)]. One wavelength, corresponding to the energy gap E_g, predominates. Fluorescent dyes and microscopy are used in many advanced techniques in biochemistry and biomedical engineering. X-ray fluorescence (XRF) is widely used for the chemical analysis of materials.

Phosphorescent materials have impurities that introduce a donor level within the energy gap [Figure 21-14(c)]. The stimulated electrons first drop into the donor level and are trapped. The electrons must then escape the trap before returning to the valence band. There is a delay before the photons are emitted. When the source is removed, electrons in the traps gradually escape and emit light over some additional period of time. The intensity of the luminescence is given by

$$\ln\left(\frac{I}{I_0}\right) = -\frac{t}{\tau} \tag{21-16}$$

where τ is the **relaxation time**, a constant for the material. After time t following removal of the source, the intensity of the luminescence is reduced from I_0 to I. Phosphorescent materials are very important in the operation of certain video monitors. In this case, the relaxation time must not be too long or the images begin to overlap. For color imaging, three types of phosphorescent materials are used; the energy gaps are engineered so that red, green, and blue colors are produced. Oscilloscope and radar screens rely on the same principle. Fluorescent lamps contain mercury vapor. The mercury vapor, in the presence of an electric arc, fluoresces and produces ultraviolet light. The inside of the glass of these lamps is coated with a phosphorescent material. The role of this material is to convert the small wavelength ultraviolet radiation into visible light. The relaxation times range from 5×10^{-9} seconds to about 2 seconds. The following example illustrates the selection of a phosphor for a video screen.

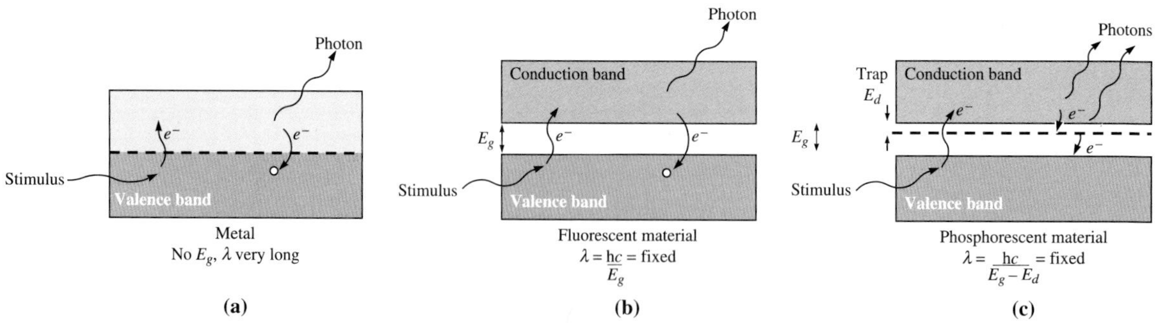

Figure 21-14 Luminescence occurs when photons have a wavelength in the visible spectrum. (a) In metals, there is no energy gap, so luminescence does not occur. (b) Fluorescence may occur if there is an energy gap. (c) Phosphorescence occurs when the photons are emitted over a period of time due to donor traps in the energy gap.

Example 21-9 *Design/Materials Selection for a Television Screen*

Select a phosphor material that will produce a blue image on a television screen.

SOLUTION

Photons having energies that correspond to the color blue have wavelengths of about 4.5×10^{-5} cm (Figure 21-1). The energy of the emitted photons therefore is

$$E = \frac{hc}{\lambda} = \frac{(6.626 \times 10^{-34}\,\text{J} \cdot \text{s})(3 \times 10^{10}\,\text{cm/s})}{(4.5 \times 10^{-5}\,\text{cm})(1.6 \times 10^{-19}\,\text{J/eV})} = 2.76\,\text{eV}$$

Figure 21-1 includes energy gaps for a variety of materials. None of the materials listed has an E_g of 2.76 eV, but ZnS has an E_g of 3.54 eV. If a suitable dopant were introduced to provide a trap 3.54 − 2.76 = 0.78 eV below the conduction band, phosphorescence of photons with energies corresponding to the color blue would occur.

We would also need information concerning the relaxation time to ensure that the phosphorescence would not persist long enough to distort the image. Typical phosphorescent materials for television screens include $CaWO_4$, which produces photons with a wavelength of 4.3×10^{-5} cm (blue). This material has a relaxation time of 4×10^{-6} s. ZnO doped with excess zinc produces photons with a wavelength of 5.1×10^{-5} cm (green), whereas $Zn_3(PO_4)_2$ doped with manganese gives photons with a wavelength of 6.45×10^{-5} cm (red).

Light-Emitting Diodes—Electroluminescence **Light-emitting diodes** (LEDs) are used to provide the display for watches, clocks, calculators, and other electronic devices. The stimulus for these devices is an externally applied voltage, which causes electron transitions and **electroluminescence**. LEDs are *p-n* junction devices engineered so that the E_g is in the visible spectrum (often red). A voltage applied to the diode in the forward-bias direction causes holes and electrons to recombine at the junction and emit photons (Figure 21-15). GaAs, GaP, GaAlAs, and GaAsP are typical materials for LEDs.

Lasers—Amplification of Luminescence The **laser** (*l*ight *a*mplification by *s*timulated *e*mission of *r*adiation) is another example of a special application of luminescence. In certain materials, electrons excited by a stimulus produce photons which, in turn, excite additional photons of identical wavelength. Consequently, a large amplification of the photons emitted in the material occurs. By proper choice of stimulant and material, the wavelength of the photons can be in the visible range. The output of the laser is a beam of photons that are parallel, of the same wavelength, and coherent. In a *coherent* beam, the wavelike nature of the photons is in phase, so that destructive interference does not occur. Lasers are useful in heat treating and melting of metals, welding, surgery, and transmission and processing of information. They are also useful in a variety of other applications including reading of compact discs and DVDs.

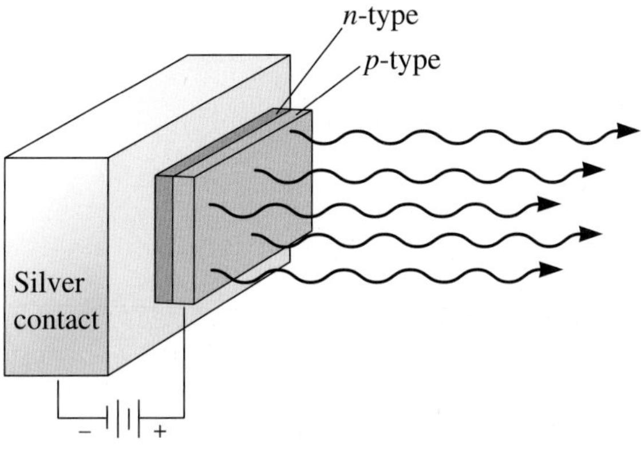

Figure 21-15 Diagram of a light-emitting diode (LED). A forward-bias voltage across the *p-n* junction produces photons.

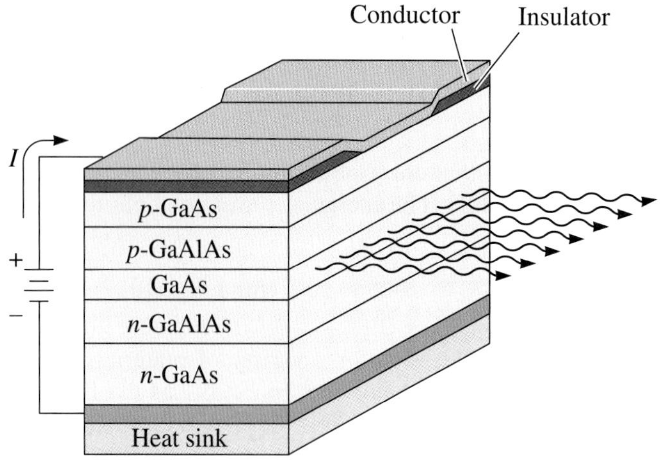

Figure 21-16 Schematic cross-section of a GaAs laser. Because the surrounding *p*- and *n*-type GaAlAs layers have a higher energy gap and a lower index of refraction than GaAs, the photons are trapped in the active GaAs layer.

Blu-ray™ technology has enabled high-resolution DVDs to be commercialized. The shorter wavelength of the blue–violet light from a laser can read finer pits that encode the digital information than longer wavelength light. This has enabled a nearly six-fold increase in data storage capability compared to conventional DVDs, making it possible to store a two hour, high-definition movie on a single disc.

A variety of materials are used to produce lasers. Ruby, which is single crystal Al_2O_3 doped with a small amount of Cr_2O_3 (emits at 6943 Å) and yttrium aluminum garnet ($Y_3Al_5O_{12}$ YAG) doped with neodymium (emits at 1.06 μm) are two common solid-state lasers. Other lasers are based on CO_2 gas.

Semiconductor lasers, such as those based on GaAs solid solutions, which have an energy gap corresponding to a wavelength in the visible range, are also used (Figure 21-16).

Thermal Emission When a material is heated, electrons are thermally excited to higher energy levels, particularly in the outer energy levels where the electrons are less tightly bound to the nucleus. The electrons immediately drop back to their normal levels and release photons, an event known as **thermal emission**.

As the temperature increases, thermal agitation increases, and the maximum energy of the emitted photons increases. A continuous spectrum of radiation is emitted, with a minimum wavelength and an intensity distribution dependent on the temperature. The photons may include wavelengths in the visible spectrum; consequently, the color of the material changes with temperature. At low temperatures, the wavelength of the radiation is too long to be visible. As the temperature increases, emitted photons have shorter wavelengths. At 700°C, we begin to see a reddish tint; at 1500°C, the orange and red wavelengths are emitted (Figure 21-17). Higher temperatures produce all wavelengths

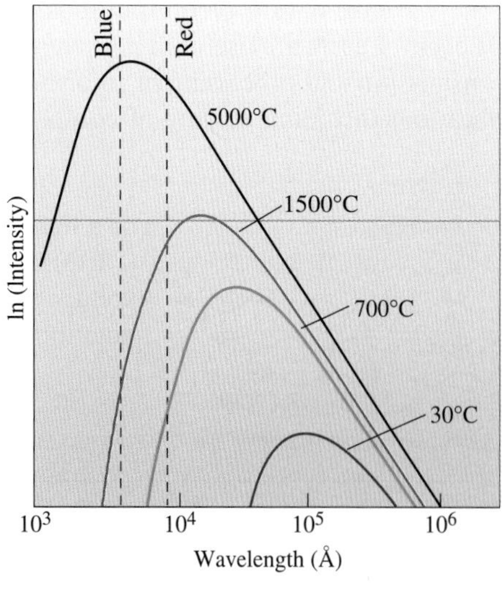

Figure 21-17 Intensity in relation to wavelengths of photons emitted thermally from a material. As the temperature increases, more photons are emitted in the visible spectrum.

in the visible range, and the emitted spectrum is white light. By measuring the intensity of a narrow band of the emitted wavelengths with a pyrometer, we can estimate the temperature of the material.

21-5 Fiber-Optic Communications Systems

In 1880, Alexander Graham Bell invented a light-based communications system known as the photophone, and William Wheeler was granted a patent (U.S. Patent 247,229) in 1881 for a system that used pipes to light distant rooms. These two inventions were the predecessors of the fiber-optic communications systems that exist today. The other key invention that helped commercialize fiber-optics technologies was the invention of the laser in 1960. The laser provided a monochromatic source of light so fiber optics could be used effectively. Another advancement came when high-purity silica glass fibers became available. These fibers provide minimal optical losses and are essential for carrying information over longer distances without the need for equipment to boost the signal. Optical fibers are also free from electromagnetic interference (EMI) since they carry signals as light, not radio waves.

A fiber-optic system transmits a light signal generated from some other source, such as an electrical signal. The fiber-optic system transmits the light to a receiver using an optical fiber, processes the data received, and converts the data to a usable form. Photonic materials are required for this process. Most of the principles and materials presently used in photonic systems have already been introduced in the previous sections.

Summary

- The optical properties of materials include the refractive index, absorption coefficient, and dispersion. These are determined by the interaction of electromagnetic radiation, or photons, with materials. The refractive index of materials depends primarily upon the extent of electronic polarization and is therefore related to the high-frequency dielectric constant of materials.
- As a result of the interaction between light and materials, refraction, reflection, transmission, scattering, and diffraction can occur. These phenomena are used in a wide variety of applications of photonic materials. These applications include fiber optics for communications and lasers for surgery and welding. Devices that use optoelectronic effects include LEDs, solar cells, and photodiodes. Other applications include phosphors for fluorescent lights, televisions, and many analytical techniques.
- Emission of photons occurs by electron transitions or nuclear decay within an atom. Fluorescence, phosphorescence, electroluminescence (used in light-emitting diodes), and lasers are examples of luminescence. Photons are emitted by thermal excitation, with photons in the visible portion of the spectrum produced when the temperature is sufficiently high. X-ray emission from materials is used in EDXA and XRF analysis.

Glossary

Absorption edge The wavelength at which the absorption characteristics of a material abruptly change.

Characteristic spectrum The spectrum of radiation emitted from a material. It shows peaks at fixed wavelengths corresponding to particular electron transitions within an atom. Every element has a unique characteristic spectrum.

Continuous spectrum Radiation emitted from a material having all wavelengths longer than a critical short wavelength limit.

Dispersion Frequency dependence of the refractive index.

Electroluminescence Use of an applied electrical signal to stimulate photons from a material.

Fluorescence Emission of light obtained typically within $\sim 10^{-8}$ seconds of stimulation.

Index of refraction Relates the change in velocity and propagation direction of radiation as it passes through a transparent medium (also known as the refractive index).

Laser A beam of monochromatic coherent radiation produced by the controlled emission of photons. The acronym stands for light amplification by stimulated emission of radiation.

Light-emitting diodes Electronic *p-n* junction devices that convert an electrical signal into visible light.

Linear absorption coefficient Describes the ability of a material to absorb radiation.

Luminescence Conversion of radiation to visible light.

Phosphorescence Emission of radiation from a material after the stimulus is removed.

Photoconduction Production of a voltage due to the stimulation of electrons into the conduction band by radiation.

Photons Energy or radiation produced from atomic, electronic, or nuclear sources that can be treated as particles or waves.

Reflectivity The percentage of incident radiation that is reflected.

Refractive index See Index of refraction.

Relaxation time The time required for $1/e$ of the electrons to drop from the conduction band to the valence band in luminescence.

Short wavelength limit The shortest wavelength or highest energy radiation emitted from a material under particular conditions.

Solar cell A *p-n* junction device that creates a voltage due to excitation by photons.

Thermal emission Emission of photons from a material due to excitation of the material by heat.

X-rays Electromagnetic radiation in the wavelength range ~0.1 to 100 Å.

Problems

Section 21-1 The Electromagnetic Spectrum

Section 21-2 Refraction, Reflection, Absorption, and Transmission

21-1 State the definitions of refractive index and absorption coefficient. Compare these with the definitions of dielectric constant, loss factor, Young's modulus, and viscous deformation.

21-2 What is Snell's law? Illustrate using a diagram.

21-3 Upon what does the index of refraction of a material depend?

21-4 What is "lead crystal"? What makes the refractive index of this material so much higher than that of ordinary silicate glass?

21-5 What polarization mechanism affects the refractive index?

21-6 Why is the refractive index of ice smaller than that of water?

21-7 What is dispersion? What is the importance of dispersion in fiber-optic systems?

21-8 What factors limit the transmission of light through dielectric materials?

21-9 What is the principle by which LEDs and solar cells work?

21-10 Write the formula for wavelength as a function of frequency, and find the frequency of a 560-nm photon.

21-11 What material has a wavelength of about 0.5 μm? 0.88 μm? What is the band gap energy of lead sulfate?

21-12 Find the speed of light through quartz. If the light has a wavelength of 5890 Å in a vacuum, what wavelength does it have in the quartz?

21-13 A beam of photons strikes a material at an angle of 25° to the normal of the surface. Which, if any, of the materials listed in Table 21-1 could cause the beam of photons to continue at an angle of 18 to 20° from the normal of the material's surface?

21-14 A laser beam passing through air strikes a 5-cm-thick polystyrene block at a 20° angle to the normal of the block. By what distance is the beam displaced from its original path when the beam reaches the opposite side of the block?

21-15 A length of 6000 km of fiber-optic cable is laid to connect New York to London. If the core of the cable has a refractive index of 1.48 and the cladding has a refractive index of 1.45, what is the time needed for a beam of photons introduced at 0° in New York to reach London? Assume that dispersion effects can be neglected for this calculation. What is the maximum angle of incidence at which there is no leakage of light from the core?

21-16 A block of glass 10-cm-thick with $n = 1.5$ transmits 90% of light incident on it. Determine the linear absorption coefficient (α) for this material. If this block is placed in water, what fraction of the incident light will be transmitted through it?

21-17 A beam of photons passes through air and strikes a soda-lime glass that is part of an aquarium containing water. What fraction of the beam is reflected by the front face of the glass? What fraction of the remaining beam is reflected by the back face of the glass?

21-18 We find that 20% of the original intensity of a beam of photons is transmitted from air through a 1-cm-thick material having a dielectric constant of 2.3 and back into air. Determine the fraction of the beam that is
(a) reflected at the front surface;
(b) absorbed in the material; and
(c) reflected at the back surface.
Determine the linear absorption coefficient of the material.

21-19 A beam of photons in air strikes a composite material consisting of a 1-cm-thick sheet of polyethylene and a 2-cm-thick sheet of soda-lime glass. The incident beam is 10° from the normal of the composite. Determine the angle of the beam with respect to the normal of the composite as it
(a) passes through the polyethylene;
(b) passes through the glass; and
(c) passes through air on the opposite side of the composite.
By what distance is the beam displaced from its original path when it emerges from the composite?

21-20 A glass fiber ($n = 1.5$) is coated with Teflon. Calculate the maximum angle that a beam of light can deviate from the axis of the fiber without escaping from the inner portion of the fiber.

21-21 A beam of light in water reflects off of a polished germanium sphere. What is its reflectivity?

21-22 If a 10-cm-thick prism reduces the intensity of a light by 25%, what is its linear absorption coefficient?

21-23 A material has a linear absorption coefficient of 591 cm^{-1} for a particular wavelength. Determine the thickness of the material required to absorb 99.9% of the photons.

21-24 Plot K_α as a function of the absorption edge and see if there is any likely trend. (See Table 21-2.) Try and explain the trend(s) if you find any.

Section 21-3 Selective Absorption, Transmission, or Reflection

21-25 What is a photochromic glass?

21-26 How are colored glasses produced?

21-27 Why does ruby have a red color?

Section 21-4 Examples and Uses Of Emission Phenomena

21-28 What is the principle of energy dispersive x-ray analysis (EDXA)?

21-29 What is fluorescence? What is phosphorescence?

21-30 What is XRF?

21-31 How does a fluorescent lamp work?

21-32 What is electroluminescence?

21-33 A scanning electron microscope has three settings for the acceleration voltage (a) 5 keV; (b) 10 keV; and (c) 20 keV. Determine the minimum voltage setting needed to produce K_α peaks for the materials listed in Table 21-2.

21-34 A phosphorescent material has a relaxation time of 75 s. What intensity will it have at exactly 75 seconds after the removal of the energy source?

21-35 The relaxation time of a phosphor used for a video display is 5×10^{-2} seconds. If the refresh frequency is 60 Hz, then what is the reduction in intensity of the luminescence before it is reset to 100% by the refresh?

21-36 Calcium tungstate ($CaWO_4$) has a relaxation time of 4×10^{-6} s. Determine the time required for the intensity of this phosphorescent material to decrease to 1% of the original intensity after the stimulus is removed.

21-37 The intensity of radiation from a phosphorescent material is reduced to 90% of its original intensity after 1.95×10^{-7} s. Determine the time required for the intensity to decrease to 1% of its original intensity.

21-38 A phosphor material with a bandgap of 3.5 eV with appropriate doping will be used to produce blue (475 nm) and green (510 nm) colors. Determine the energy level of the donor traps with respect to the conduction band in each case.

21-39 What is a laser?

21-40 Determine the wavelength of photons produced when electrons excited into the conduction band of indium-doped silicon
(a) drop from the conduction band to the acceptor band; and
(b) then drop from the acceptor band; to the valence band. (See Chapter 19.)

21-41 Which, if any, of the semiconducting compounds listed in Chapters 19 and 21 are capable of producing an infrared laser beam?

21-42 What type of electromagnetic radiation (ultraviolet, infrared, visible) is produced when an electron recombines with a hole in pure germanium and what is its wavelength?

21-43 Which, if any, of the dielectric materials listed in Chapter 19 would reduce the speed of light in air from 3×10^{10} cm/s to less than 0.5×10^{10} cm/s?

21-44 What filter material would you use to isolate the K_α peak of the following x-rays: iron, manganese, or nickel? Explain your answer.

21-45 What voltage must be applied to a tungsten filament to produce a continuous spectrum of x-rays having a minimum wavelength of 0.09 nm?

21-46 A tungsten filament is heated with a 12,400 V power supply. What is
(a) the wavelength; and
(b) the frequency of the highest energy x-rays that are produced?

21-47 What is the minimum accelerating voltage required to produce K_α x-rays in nickel?

21-48 Based on the characteristic x-rays that are emitted, determine the difference in energy between electrons in tungsten for
(a) the K and L shells;
(b) the K and M shells; and
(c) the L and M shells.

21-49 Figure 21-18 shows the results of an x-ray fluorescence analysis in which the intensity of x-rays emitted from a material is plotted relative to the wavelength of the x-rays. Determine
(a) the accelerating voltage used to produce the exciting x-rays; and
(b) the identity of the elements in the sample.

21-50 Figure 21-19 shows the intensity as a function of energy of x-rays produced from an energy-dispersive analysis of radiation emitted from a specimen in a scanning electron microscope. Determine the identity of the elements in the sample.

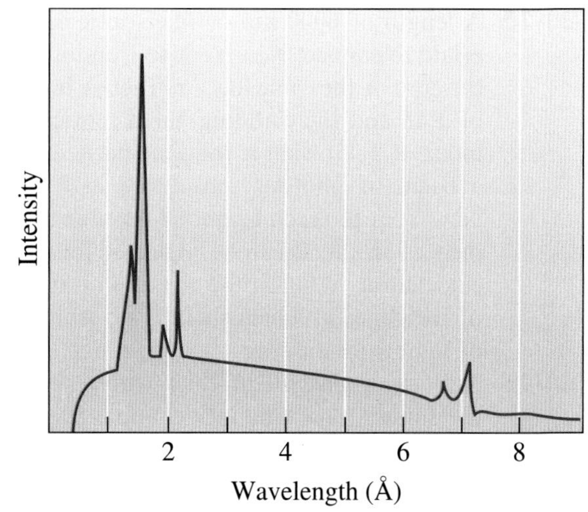

Figure 21-18 Results from an x-ray fluorescence analysis of an unknown metal sample (for Problem 21-49).

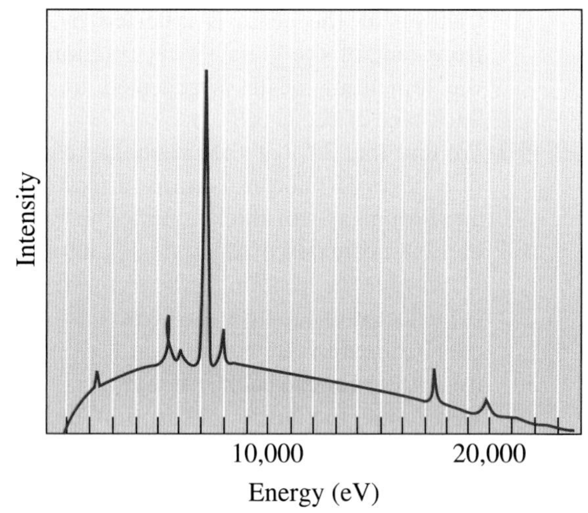

Figure 21-19 X-ray emission spectrum (for Problem 21-50).

21-51 Refer to Figure 21-17. Assume the material is being slowly cooled from 5000°C to absolute zero. What wavelength of photons will be emitted last before the object stops emitting light altogether?

Section 21-5 Fiber-Optic Communication Systems

21-52 How is information transmitted via an optical fiber?

21-53 What material is used to make most optical fibers? From what material is the cladding made? What material is used to enhance the refractive index of the core?

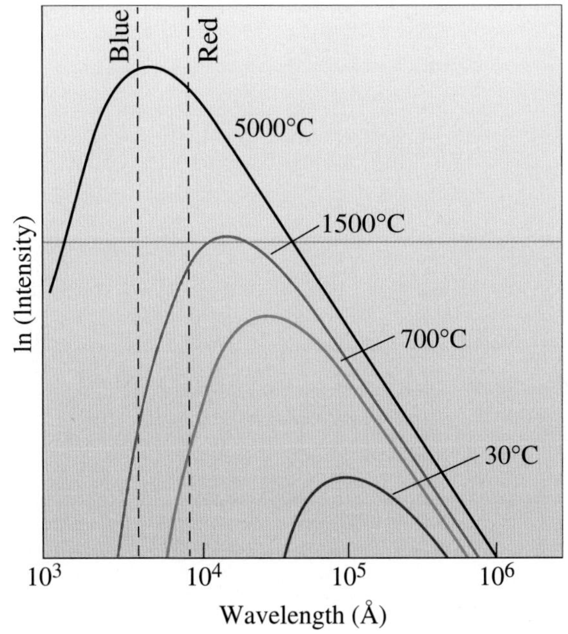

Figure 21-17 Intensity in relation to wavelengths of photons emitted thermally from a material. As the temperature increases, more photons are emitted in the visible spectrum. (Repeated for Problem 21-51.)

Design Problems

21-54 Nickel x-rays are to be generated inside a container, with the x-rays being emitted from the container through only a small slot. Design a container that will ensure that no more than 0.01% of the K_α nickel x-rays escape through the rest of the container walls, yet 95% of the K_α nickel x-rays pass through a thin window covering the slot. The following data give the mass absorption coefficients of several metals for nickel K_α x-rays. The mass absorption coefficient α_m is α/ρ, where α is the linear mass absorption coefficient and ρ is the density of the filter material.

Material	α_m (cm^2/g)
Be	1.8
Al	58.4
Ti	247.0
Fe	354.0
Co	54.4
Cu	65.0
Sn	322.0
Ta	200.0
Pb	294.0

21-55 Design a method by which a photoconductive material might be used to measure the temperature of a material from the material's thermal emission.

21-56 Design a method, based on a material's refractive characteristics, that will cause a beam of photons originally at a 2° angle to the normal of the material to be displaced from its original path by 2 cm at a distance of 50 cm from the material.

Computer Problems

21-57 *Calculating Power in Decibels.* In an optical communications system or electrical power transmission system, the power or signal often is transferred between several components. The decibel (dB) is a convenient unit to measure the relative power levels. If the input power to a device is P_1 and the output power is P_2, then P_2/P_1 is the ratio of power transmitted, thus representing efficiency. This ratio in decibels is written as

$$\text{dB} = 10 \log \frac{P_2}{P_1}$$

The power terms must be expressed in the same units. Write a computer program that will calculate the dB value for the transmission of power between two components of a fiber-optic system (e.g., light source to fiber). Then, extend this calculation to three components (e.g., light transmitted from source to fiber and then fiber to a detector).

Knovel® Problems

K21-1 A beam of light passes through benzene to glass (silicon dioxide). The angle of incidence of the light is 30°. What is the angle of the refracted light?

K21-2 Calculate the reflectivity of mercury in air from the index of refraction n.

Thermal properties of materials play a critical role in numerous technologies. Some applications require materials that minimize heat transfer, while others require efficient transport of heat. Thermal barrier coatings are applied to the surface of turbine blades to minimize the exposure of the turbine blades (which will creep) to the high temperatures of aircraft engines. Refractories are ceramic materials that are used to contain molten metals during casting. As transistors and other electronic devices on microchips become smaller and more numerous to improve computing performance, the power density and therefore heat generation increase. Thermal management of this waste heat is thus increasingly important. The image above shows two types of cooling fins that remove heat from an electronic package allowing the electronics to function more efficiently.

The thermal expansion of materials also plays a key role in many situations; such expansion can lead to the development of stresses that lead to materials failure. Materials scientists and engineers have created many novel materials that have a negative or near-zero thermal-expansion coefficient. This is an important characteristic in optical materials such as those used in telescope mirrors because the focus of the mirrors changes depending on thermal expansion and contraction. (*Lisa J. Goodman/Photographer's Choice/Getty Images*)

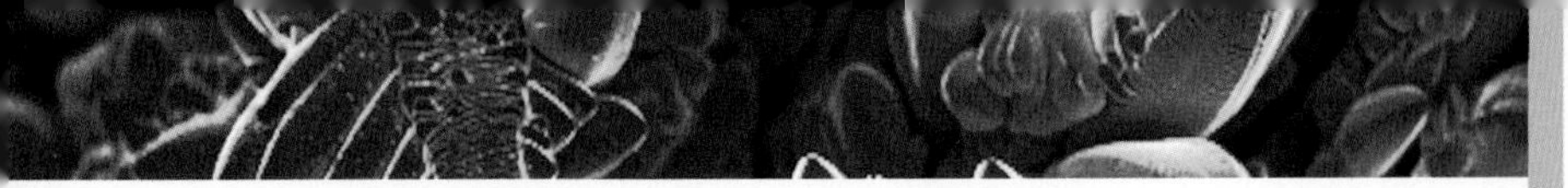

CHAPTER

22

Thermal Properties of Materials

Have You Ever Wondered?

- *What material has the highest thermal conductivity?*
- *Are there materials that have zero or negative thermal-expansion coefficients?*
- *What causes a glass baking dish to break if it is cooled too quickly after being removed from an oven?*

Chapter Learning Objectives

The key objectives of this chapter are to

- Define heat capacity and specific heat.
- Explain the processes by which heat is transported through a material.
- Calculate the changes in dimensions of materials due to heating or cooling according to the thermal-expansion coefficients.
- Calculate the heat flux in a material due to a temperature gradient.
- Understand how microstructure affects the thermal conductivity of materials.
- Quantify the thermal shock resistance of a material.

In previous chapters, we described how a material's properties can change with temperature. In many cases, we found that a material's mechanical and physical properties depend on the service or processing temperature. An appreciation of the thermal properties of materials is helpful to understanding the mechanical failure of materials when the temperature changes; in designing processes in which materials must be heated; or in selecting materials to transfer heat rapidly.

Thermal management has become a very important issue in electronic packaging materials. Some computer memory chips now have more than a billion transistors; the heat that is produced must be dissipated effectively for devices to function properly.

In metallic materials, electrons transfer heat. In ceramic materials, the conduction of heat involves elastic waves. In certain other applications, such as thermal barrier coatings, we want to minimize the heat transfer through the material. Heat transfer is also important in many applications ranging from, for example, polystyrene foam cups used for hot beverages to sophisticated coatings on glasses to make energy efficient buildings. In this chapter, we will discuss heat capacity, thermal-expansion properties, and the thermal conductivity of materials.

22-1 Heat Capacity and Specific Heat

In Chapter 21, we noted that optical behavior depends on how photons are produced and interact with a material. The photon is treated as a particle with a particular energy or as electromagnetic radiation having a particular wavelength or frequency. Some of the thermal properties of materials can also be characterized in the same dual manner; however, these properties are determined by the behavior of **phonons**, rather than photons.

At absolute zero, the atoms in a material have a minimum energy. When heat is supplied, the atoms gain thermal energy and vibrate at a particular amplitude and frequency. The vibration of each atom is transferred to the surrounding atoms and produces an elastic wave called a phonon. The energy of the phonon E can be expressed in terms of the wavelength where h is Planck's constant and c is the speed of light or frequency ν, just as in Equation 21-1:

$$E = \frac{hc}{\lambda} = h\nu \tag{22-1}$$

The energy required to change the temperature of the material one degree is the **heat capacity** or **specific heat**.

The heat capacity is the energy required to raise the temperature of one mole of a material by one degree. The specific heat is defined as the energy needed to increase the temperature of one gram of a material by 1°C. The heat capacity can be expressed either at constant pressure, C_p, or at a constant volume, C_V. At high temperatures, the heat capacity for a given volume of material approaches

$$C_p = 3R \simeq 6\,\frac{\text{cal}}{\text{mol} \cdot \text{K}} \tag{22-2}$$

where R is the gas constant (1.987 cal/mol); however, as shown in Figure 22-1, heat capacity is not a constant. The heat capacity of metals approaches $6\,\frac{\text{cal}}{\text{mol} \cdot \text{K}}$ near room temperature, but this value is not reached in ceramics until near 1000 K.

The relationship between specific heat and heat capacity is

$$\text{Specific heat} = C_p = \frac{\text{heat capacity}}{\text{atomic weight}} \tag{22-3}$$

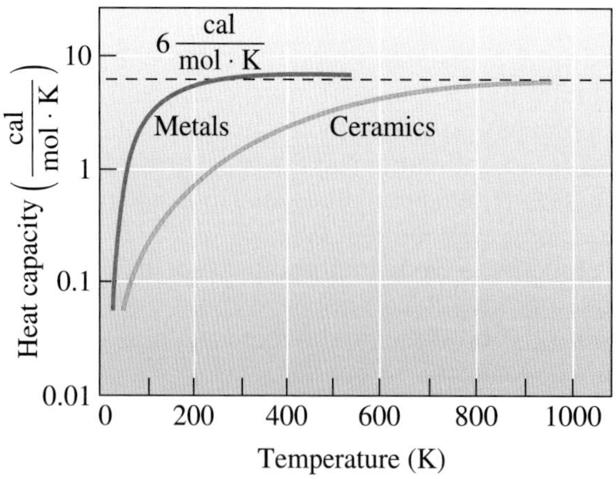

Figure 22-1 Heat capacity as a function of temperature for metals and ceramics.

Table 22-1 The specific heat of selected materials at 300 K

Material	Specific Heat $\left(\frac{cal}{g \cdot K}\right)$	Material	Specific Heat $\left(\frac{cal}{g \cdot K}\right)$
Metals		**Ceramics**	
Al	0.215	Al_2O_3	0.200
Cu	0.092	Diamond	0.124
B	0.245	SiC	0.250
Fe	0.106	Si_3N_4	0.170
Pb	0.038	SiO_2 (silica)	0.265
Mg	0.243	**Polymers**	
Ni	0.106		
Si	0.168	High-density polyethylene	0.440
Ti	0.125		
W	0.032	Low-density polyethylene	0.550
Zn	0.093		
		6,6-nylon	0.400
		Polystyrene	0.280
		Other	
		Water	1.000
		Nitrogen	0.249

Note: $1 \frac{cal}{g \cdot K} = 4184 \frac{J}{kg \cdot K}$

In most engineering calculations, specific heat is used more conveniently than heat capacity. The specific heat of typical materials is given in Table 22-1. Neither the heat capacity nor the specific heat depends significantly on the structure of the material; thus, changes in dislocation density, grain size, or vacancies have little effect.

The most important factor affecting specific heat is the lattice vibrations or phonons; however, other factors affect the heat capacity. One striking example occurs in ferromagnetic materials such as iron (Figure 22-2). An abnormally high heat capacity is observed in iron at the Curie temperature, where the normally aligned magnetic moments of the iron atoms are randomized and the iron becomes paramagnetic. Heat capacity also depends on the crystal structure, as shown in Figure 22-2 for iron.

The following examples illustrate the use of specific heat.

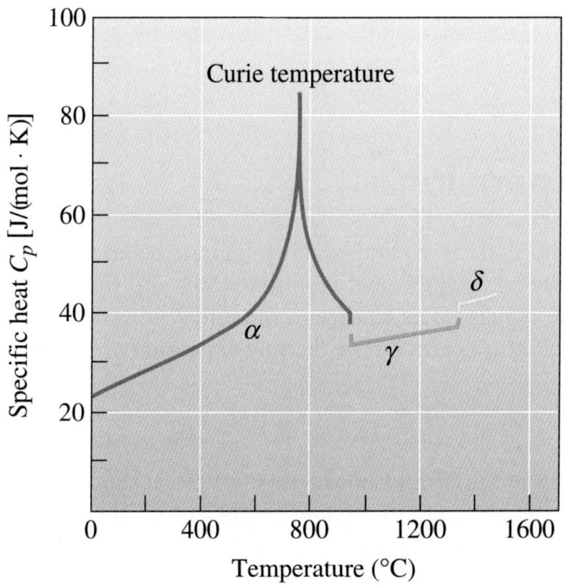

Figure 22-2 The effect of temperature on the specific heat of iron. Both the change in crystal structure and the change from ferromagnetic to paramagnetic behavior are indicated.

Example 22-1 *Specific Heat of Tungsten*

How much heat must be supplied to 250 g of tungsten to raise its temperature from 25°C to 650°C?

SOLUTION

The specific heat of tungsten is $0.032 \frac{\text{cal}}{\text{g} \cdot \text{K}}$. Thus,

$$\begin{aligned}\text{Heat required} &= (\text{specific heat})(\text{mass})(\Delta T)\\ &= [0.032\,\text{cal}/(\text{g}\cdot\text{K})](250\,\text{g})(650-25)\\ &= 5000\,\text{cal}\end{aligned}$$

If no losses occur, 5000 cal (or 20,920 J) must be supplied to the tungsten. A variety of processes might be used to heat the metal. We could use a gas torch, we could place the tungsten in an induction coil to induce eddy currents, we could pass an electrical current through the metal, or we could place the metal into an oven heated by SiC resistors.

Example 22-2 *Specific Heat of Niobium*

Suppose the temperature of 50 g of niobium increases 75°C when heated. Estimate the specific heat and determine the heat in calories required.

SOLUTION

The atomic weight of niobium is 92.91 g/mol. We can use Equation 22-3 to estimate the heat required to raise the temperature of one gram by one °C:

$$C_p \approx \frac{6}{92.91} = 0.0646\,\text{cal}/(\text{g}\cdot{}^\circ\text{C})$$

Thus, the total heat required is

$$\text{Heat} = \left(0.0646 \frac{\text{cal}}{\text{g}\cdot{}^\circ\text{C}}\right)(50\,\text{g})(75^\circ\text{C}) = 242\,\text{cal}$$

Note that a temperature change of 1°C is equivalent to a temperature change of 1 K.

22-2 Thermal Expansion

An atom that gains thermal energy and begins to vibrate behaves as though it has a larger atomic radius. The average distance between the atoms and therefore the overall dimensions of the material increase. The change in the dimensions of the material Δl per unit length is given by the **linear coefficient of thermal expansion** α:

$$\alpha = \frac{l_f - l_0}{l_0(T_f - T_0)} = \frac{\Delta l}{l_0 \Delta T} \tag{22-4}$$

where T_0 and T_f are the initial and final temperatures and l_0 and l_f are the initial and final dimensions of the material. A *volume* coefficient of thermal expansion (α_v) also can be defined to describe the change in volume when the temperature of the material is changed. If the material is isotropic, $\alpha_v = 3\alpha$. An instrument known as a dilatometer

Table 22-2 The linear coefficient of thermal expansion at room temperature for selected materials

Material	Linear Coefficient of Thermal Expansion (α) ($\times 10^{-6}$/°C)	Material	Linear Coefficient of Thermal Expansion (α) ($\times 10^{-6}$/°C)
Al	25.0	6,6-nylon	80.0
Cu	16.6	6,6-nylon—33% glass fiber	20.0
Fe	12.0	Polyethylene	100.0
Ni	13.0	Polyethylene—30% glass fiber	48.0
Pb	29.0	Polystyrene	70.0
Si	3.0	Al_2O_3	6.7
W	4.5	Fused silica	0.55
1020 steel	12.0	Partially stabilized ZrO_2	10.6
3003 aluminum alloy	23.2	SiC	4.3
Gray iron	12.0	Si_3N_4	3.3
Invar (Fe-36% Ni)	1.54	Soda-lime glass	9.0
Stainless steel	17.3		
Yellow brass	18.9		
Epoxy	55.0		

is used to measure the thermal-expansion coefficient. It is also possible to trace thermal expansion using x-ray diffraction (XRD). Coefficients of thermal expansion for several materials are included in Table 22-2.

The coefficient of thermal expansion of a material is related to the atomic bonding. In order for the atoms to vibrate about their equilibrium positions, energy must be supplied to the material. If the potential well is asymmetric, the atoms separate to a larger extent with increased energy compared to a symmetric well, and the material has a high thermal-expansion coefficient (Chapter 2). The thermal-expansion coefficient also tends to be inversely proportional to the depth of the potential well, i.e., materials with high melting temperatures have low coefficients of thermal expansion (Figure 22-3).

Consequently, lead has a much larger coefficient than high melting point metals such as tungsten. Most ceramics, which have strong ionic or covalent bonds, have low coefficients compared with metals. Certain glasses, such as fused silica, also have a poor packing factor, which helps accommodate thermal energy with little dimensional change. Although bonding within the chains of polymers is covalent, the secondary bonds holding the chains together are weak, leading to high coefficients. Polymers that

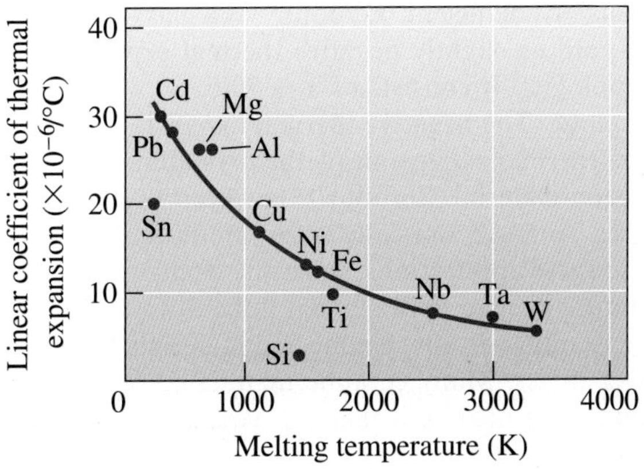

Figure 22-3 The relationship between the linear coefficient of thermal expansion and the melting temperature in metals at 25°C. Higher melting point metals tend to expand to a lesser degree.

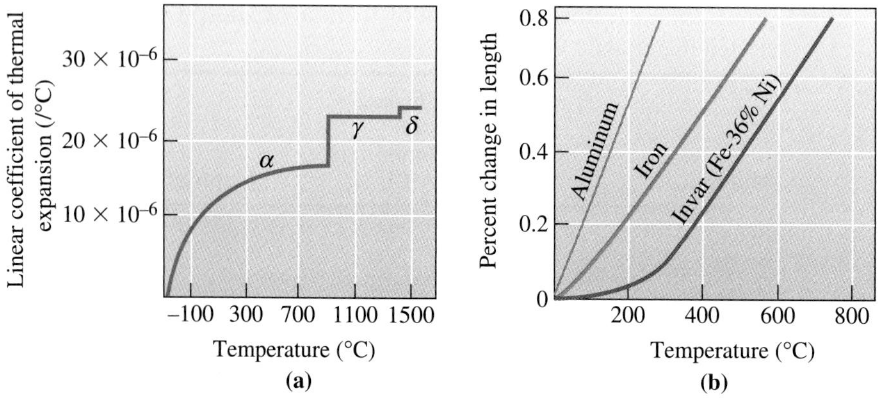

Figure 22-4 (a) The linear coefficient of thermal expansion of iron changes abruptly at temperatures where an allotropic transformation occurs. (b) The expansion of Invar is very low due to the magnetic properties of the material at low temperatures.

contain strong cross-linking typically have lower coefficients than linear polymers such as polyethylene.

Several precautions must be taken when calculating dimensional changes in materials:

1. The expansion characteristics of some materials, particularly single crystals or materials having a preferred orientation, are anisotropic.
2. Allotropic materials have abrupt changes in their dimensions when the phase transformation occurs (Figure 22-4). These abrupt changes contribute to the cracking of refractories on heating or cooling and quench cracks in steels.
3. The linear coefficient of expansion continually changes with temperature. Normally, α either is listed in handbooks as a complicated temperature-dependent function or is given as a constant for only a particular temperature range.
4. Interaction of the material with electric or magnetic fields produced by magnetic domains may prevent normal expansion until temperatures above the Curie temperature are reached. This is the case for Invar, an Fe-36% Ni alloy, which undergoes practically no dimensional changes at temperatures below the Curie temperature (about 200°C). This makes Invar attractive as a material for bimetallics (Figure 22-4).

The thermal expansion of engineered materials can be tailored using multi-phase materials. Upon heating, one phase can show thermal expansion while the other phase can show thermal contraction. Thus, the overall material can show a zero or negative thermal-expansion coefficient. Zerodur™ is a glass-ceramic material that can be controlled to have zero or slightly negative thermal expansion. It was developed by Schott Glass Technologies. It consists of a ~ 70 to 80 wt% crystalline phase. The remainder is a glassy phase. The negative thermal-expansion coefficient of the glassy phase and the positive thermal expansion of the crystalline phase cancel, leading to a zero thermal-expansion material. Zerodur™ has been used as the mirror substrate on a number telescopes. A dense, optically transparent, and zero thermal-expansion material is necessary in these applications, since any changes in dimensions as a result of changes in temperature in space will make it difficult to focus the telescopes properly. Zerodur™ is one example of how engineered materials have helped astronomers learn about far away galaxies. Many ceramic materials based on sodium zirconium phosphate (NZP) that have a near-zero thermal-expansion coefficient also have been developed by materials scientists.

The following examples show the use of linear coefficients of thermal expansion.

Example 22-3 *Bonding and Thermal Expansion*

Explain why, in Figure 22-3, the linear coefficients of thermal expansion for silicon and tin do not fall on the curve. How would you expect germanium to fit into this figure?

SOLUTION

Both silicon and tin are covalently bonded. The strong covalent bonds are more difficult to stretch than the metallic bonds, so these elements have a lower coefficient. Since germanium also is covalently bonded, its thermal expansion should be less than that predicted by Figure 22-3.

Example 22-4 *Design of a Pattern for a Casting Process*

Design the dimensions for a pattern that will be used to produce a rectangular-shaped aluminum casting having dimensions at 25°C of 25 cm × 25 cm × 3 cm.

SOLUTION

To produce a casting having particular final dimensions, the mold cavity into which the liquid aluminum is to be poured must be oversized. After the liquid solidifies, which occurs at 660°C for pure aluminum, the solid casting contracts as it cools to room temperature. If we calculate the amount of contraction that will occur, we can make the original pattern used to produce the mold cavity that much larger.

The linear coefficient of thermal expansion for aluminum is $25 \times 10^{-6}\ °C^{-1}$. The temperature change from the freezing temperature to 25°C is $660 - 25 = 635°C$. The change in any dimension is given by

$$\Delta l = l_0 - l_f = \alpha l_0 \Delta T$$

For the 25 cm dimensions, $l_f = 25\,\text{cm}$. We wish to find l_0:

$$l_0 - 25 = (25 \times 10^{-6})(l_0)(635)$$

$$l_0 - 25 = 0.015875\, l_0$$

$$0.984 l_0 = 25$$

$$l_0 = 25.40\,\text{cm}$$

For the 3 cm dimension, $l_f = 3$ cm.

$$l_0 - 3 = (25 \times 10^{-6})(l_0)(635)$$

$$l_0 - 3 = 0.015875 l_0$$

$$0.984 l_0 = 3$$

$$l_0 = 3.05\,\text{cm}$$

If we design the pattern to the dimensions 25.40 cm × 25.40 cm × 3.05 cm, the casting should contract to the required dimensions.

When an isotropic material is slowly and uniformly heated, the material expands uniformly without creating any residual stress. If, however, the material is constrained, the dimensional changes may not be possible and, instead, stresses develop. These **thermal stresses** are related to the coefficient of thermal expansion, the modulus of elasticity E of the material, and the temperature change ΔT:

$$\sigma_{\text{thermal}} = \alpha E \Delta T \qquad (22\text{-}5)$$

Thermal stresses can arise from a variety of sources. In large rigid structures such as bridges, constraints may develop as a result of the design. Some bridges are designed in sections with steel plates between sections, so that the sections move relative to one another during seasonal temperature changes.

When materials are joined—for example, coating cast iron bathtubs with a ceramic enamel or coating superalloy turbine blades with a yttria stabilized zirconia (YSZ) thermal barrier—changes in temperature cause different amounts of contraction or expansion in the different materials. This disparity leads to thermal stresses that may cause the protective coating to spall off. Careful matching of the thermal properties of the coating to those of the substrate material is necessary to prevent coating cracking (if the coefficient of the coating is less than that of the underlying substrate) or spalling (flaking of the coating due to a high expansion coefficient).

A similar situation may occur in composite materials. Brittle fibers that have a lower coefficient than the matrix may be stretched to the breaking point when the temperature of the composite increases.

Thermal stresses may even develop in a nonrigid, isotropic material if the temperature is not uniform. In producing tempered glass (Chapter 15), the surface is cooled more rapidly than the center, permitting the surface to initially contract. When the center cools later, its contraction is constrained by the rigid surface, placing compressive residual stresses on the surface.

Example 22-5 *Design of a Protective Coating*

A ceramic enamel is to be applied to a 1020 steel plate. The ceramic has a fracture strength of 4000 psi, a modulus of elasticity of 15×10^6 psi, and a coefficient of thermal expansion of $10 \times 10^{-6}\,°C^{-1}$. Determine the maximum temperature change that is allowed without cracking the ceramic.

SOLUTION

Because the enamel is bonded to the 1020 steel, it is essentially constrained. The steel also expands as the temperature increases. Its coefficient of thermal expansion (Table 22-2) is $12 \times 10^{-6}\,°C^{-1}$, and its modulus of elasticity is 30×10^6 psi. Since the steel expands more than the enamel, a stress is introduced into the enamel. The net coefficient of expansion is

$$\Delta\alpha = 12 \times 10^{-6} - 10 \times 10^{-6} = 2 \times 10^{-6}\,°C^{-1}$$

$$\sigma = (2 \times 10^{-6})(15 \times 10^{6})\Delta T = 4000$$

$$\Delta T = 133°C$$

In order to permit greater temperature variations, we might select an enamel that has a higher coefficient of thermal expansion, an enamel that has a lower modulus of elasticity (so that greater strains can be permitted before the stress reaches the fracture stress), or an enamel that has a higher strength.

22-3 Thermal Conductivity

The **thermal conductivity** k is a measure of the rate at which heat is transferred through a material. The treatment of thermal conductivity is similar to that of diffusion (Chapter 5). *Thermal conductivity, similar to the diffusion coefficient, is a microstructure sensitive property*. The conductivity relates the heat Q transferred across a given plane of area A per second when a temperature gradient $\Delta T/\Delta x$ exists (Figure 22-5):

$$\frac{Q}{A} = -k\frac{\Delta T}{\Delta x} \tag{22-6}$$

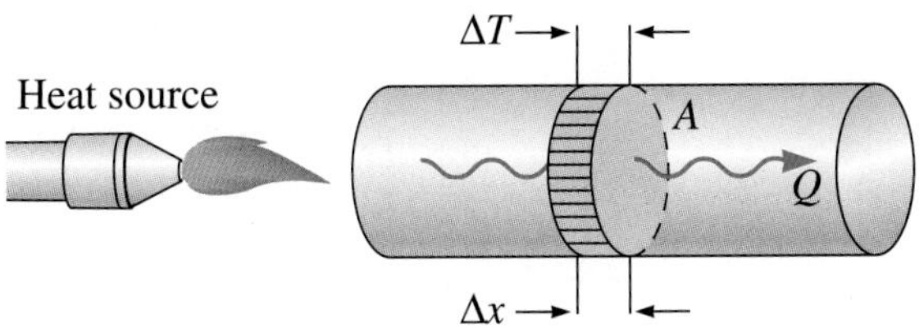

Figure 22-5 When one end of a bar is heated, a heat flux Q/A flows toward the cold end at a rate determined by the temperature gradient produced in the bar.

The negative sign in Equation 22-6 tells us that the heat flux is from higher temperature regions to lower temperature regions, so that if the $\Delta T/\Delta x$ term is negative, Q will be positive. Note that the thermal conductivity k plays the same role in heat transfer that the diffusion coefficient D does in mass transfer. Among all metals, silver has the highest thermal conductivity at room temperature ($430\,W \cdot m^{-1} \cdot K^{-1}$). Copper is next with a thermal conductivity of $400\ W \cdot m^{-1} \cdot K^{-1}$. In general, metals have higher thermal conductivity than ceramics; however, diamond, a ceramic material, has a very high thermal conductivity of $2000\,W \cdot m^{-1} \cdot K^{-1}$. Values for the thermal conductivity of some materials are included in Table 22-3.

Thermal energy is transferred by two important mechanisms: transfer of free electrons and lattice vibrations (or phonons). Valence electrons gain energy, move toward the colder areas of the material, and transfer their energy to other atoms. The amount of energy transferred depends on the number of excited electrons and their mobility; these, in turn, depend on the type of material, lattice imperfections, and temperature. In addition, thermally induced vibrations of the atoms transfer energy through the material.

Metals Because the valence band is not completely filled in metals, electrons require little thermal excitation in order to move and contribute to the transfer of heat.

Table 22-3 Typical values of room temperature thermal conductivity of selected materials

Material	Thermal Conductivity (k) ($W \cdot m^{-1} \cdot K^{-1}$)	Material	Thermal Conductivity (k) ($W \cdot m^{-1} \cdot K^{-1}$)
Pure Metals		**Ceramics**	
Ag	430	Al_2O_3	16–40
Al	238	Carbon (diamond)	2000
Cu	400	Carbon (graphite)	335
Fe	79	Fireclay	0.26
Mg	100	Silicon carbide	up to 270
Ni	90	AlN	up to 270
Pb	35	Si_3N_4	up to 150
Si	150	Soda-lime glass	0.96–1.7
Ti	22	Vitreous silica	1.4
W	171	Vycor™ glass	12.5
Zn	117	ZrO_2	2–3
Zr	23		
		Polymers	
Alloys		6,6-nylon	0.25
		Polyethylene	0.33
1020 steel	100	Polyimide	0.21
3003 aluminum alloy	280	Polystyrene	0.13
304 stainless steel	30	Polystyrene foam	0.029
Cementite	50	Teflon	0.25
Cu-30% Ni	50		
Ferrite	75		
Gray iron	79.5		
Yellow brass	221		

Note: $1 \dfrac{\text{cal}}{\text{s} \cdot \text{cm} \cdot \text{K}} = 418.4 \dfrac{\text{W}}{\text{m} \cdot \text{K}}$

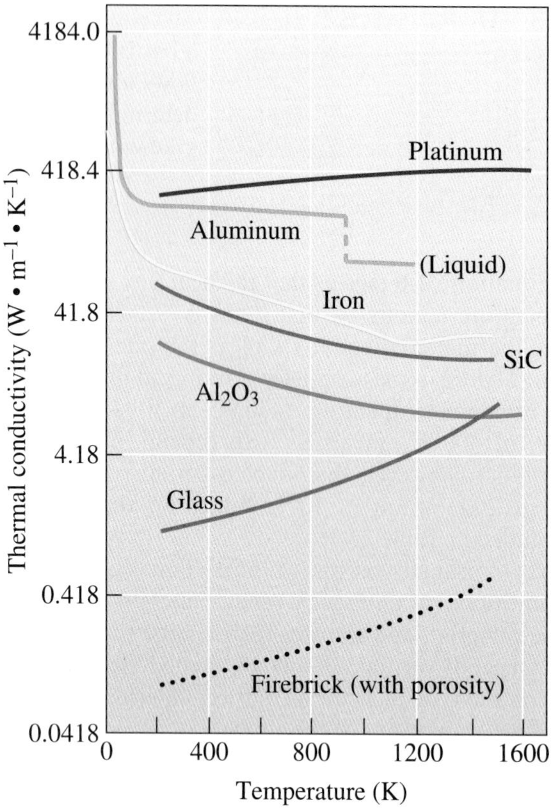

Figure 22-6 The effect of temperature on the thermal conductivity of selected materials. Note the log scale on the vertical axis.

Since the thermal conductivity of metals is due primarily to the electronic contribution, we expect a relationship between thermal and electrical conductivities:

$$\frac{k}{\sigma T} = L = 5.8 \times 10^{-9} \frac{\text{cal} \cdot \text{ohm}}{\text{s} \cdot \text{K}^2} = 2.45 \times 10^{-8} \frac{\text{W} \cdot \text{ohm}}{\text{K}^2} \tag{22-7}$$

where L is the **Lorenz number**. This relationship is followed to a limited extent in many metals.

When the temperature of the material increases, two competing factors affect thermal conductivity. Higher temperatures are expected to increase the energy of the electrons, creating more "carriers" and increasing the contribution from lattice vibrations; these effects increase the thermal conductivity. At the same time, the increased lattice vibrations scatter the electrons, reducing their mobility, and therefore cause the thermal conductivity to decrease. The combined effect of these factors leads to very different behavior for different metals. For iron, the thermal conductivity initially decreases with increasing temperature (due to the lower mobility of the electrons), then increases slightly (due to increased lattice vibrations). The conductivity *decreases* continuously when aluminum is heated but *increases* continuously when platinum is heated (Figure 22-6).

Thermal conductivity in metals also depends on crystal structure defects, microstructure, and processing. Thus, cold-worked metals, solid-solution-strengthened metals, and two-phase alloys might display lower conductivities compared with their defect-free counterparts.

Ceramics The energy gap in ceramics is too large for many electrons to be excited into the conduction band except at very high temperatures. Thus, the transfer of heat in ceramics occurs primarily by lattice vibrations (or phonons). Since the electronic contribution is absent, the thermal conductivity of most ceramics is much lower than that of metals. The main reason why the experimentally observed conductivity of ceramics is low, however, is the level of porosity. Porosity increases scattering. The best insulating brick, for example, contains a large porosity fraction. Effective sintering reduces porosity (and therefore increases thermal conductivity).

Glasses have low thermal conductivity. The amorphous, loosely packed structure minimizes the points at which silicate chains contact one another, making it more difficult for the phonons to be transferred. The thermal conductivity increases as temperature increases; higher temperatures produce more energetic phonons and more rapid transfer of heat. In some applications, such as window glasses, we use double panes of glass and separate them using a gas (e.g., Ar) to provide better thermal insulation. We can also make use of different coatings on glass to make buildings and cars more energy efficient.

The more ordered structures of the crystalline ceramics, as well as glass-ceramics that contain large amounts of crystalline precipitates, cause less scattering of phonons. Compared with glasses, these materials have a higher thermal conductivity. As the temperature increases, however, scattering becomes more pronounced, and the thermal conductivity decreases (as shown for alumina and silicon carbide in Figure 22-6). At still higher temperatures, heat transfer by radiation becomes significant, and the conductivity may increase. The thermal conductivity of polycrystalline ceramic materials is typically lower than that of single crystals.

Some ceramics such as diamond have very high thermal conductivities. Materials with a close-packed structure and high modulus of elasticity produce high energy phonons that encourage high thermal conductivities. It has been shown that many ceramics with a diamond-like crystal structure (e.g., AlN, SiC, BeO, BP, GaN, Si, AlP) have high thermal conductivities. For example, thermal conductivities of $\sim 270\,\text{W}\,\text{m}^{-1}\text{K}^{-1}$ have been reported for SiC and AlN. Although SiC and AlN are good thermal conductors, they are also electrical insulators; therefore, these materials are good candidates for use in electronic packaging substrates where heat dissipation is needed.

Semiconductors Heat is conducted in semiconductors by both phonons and electrons. At low temperatures, phonons are the principal carriers of energy, but at higher temperatures, electrons are excited through the small energy gap into the conduction band, and thermal conductivity increases significantly.

Polymers The thermal conductivity of polymers is very low—even in comparison with silicate glasses. Vibration and movement of the molecular polymer chains transfer energy. Increasing the degree of polymerization, increasing the crystallinity, minimizing branching, and providing extensive cross-linking all produce a more rigid structure and provide for higher thermal conductivity.

The thermal conductivity of many engineered materials depends upon the volume fractions of different phases and their connectivity. Silver-filled epoxies are used in many heat-transfer applications related to microelectronics. Unusually good thermal insulation is obtained using polymer foams, often produced from polystyrene or polyurethane. Styrofoam™ coffee cups are a typical product.

Example 22-6 *Design of a Window Glass*

Design a glass window 4 ft × 4 ft square that separates a room at 25°C from the outside at 40°C and allows no more than 5×10^6 cal of heat to enter the room each day. Assume that the thermal conductivity of glass is $0.96\,\text{W} \cdot \text{m}^{-1} \cdot \text{K}^{-1}$ or $0.0023 \dfrac{\text{cal}}{\text{cm} \cdot \text{s} \cdot \text{K}}$.

SOLUTION

From Equation 22-6,

$$\frac{Q}{A} = -k\frac{\Delta T}{\Delta x}$$

where Q/A is the heat transferred per second through the window.

$$1\text{ day} = (24\,\text{h/day})(3600\,\text{s/h}) = 8.64 \times 10^4\,\text{s}$$

$$A = [(4\text{ ft})(12\text{ in./ft})(2.54\,\text{cm/in.})]^2 = 1.486 \times 10^4\,\text{cm}^2$$

$$Q = \frac{(5 \times 10^6\,\text{cal/day})}{8.64 \times 10^4\,\text{s/day}} = 57.87\,\text{cal/s}$$

$$\frac{Q}{A} = \frac{57.87\,\text{cal/s}}{1.486 \times 10^4\,\text{cm}^2} = 0.00389\,\frac{\text{cal}}{\text{cm}^2 \cdot \text{s}}$$

$$\frac{Q}{A} = 0.00389\,\frac{\text{cal}}{\text{cm}^2 \cdot \text{s}} = -\left(0.0023\,\frac{\text{cal}}{\text{cm} \cdot \text{s} \cdot \text{K}}\right)(25°\text{C} - 40°\text{C})/\Delta x$$

$$\Delta x = 8.9\text{ cm} = \text{thickness}$$

The glass would have to be exceptionally thick to prevent the desired maximum heat flux. We might do several things to reduce the heat flux. Although all of the silicate glasses have similar thermal conductivities, we might use instead a transparent polymer material (such as polymethyl methacrylate). The polymers have thermal conductivities approximately one order of magnitude smaller than the ceramic glasses. We could also use a double-paned glass, with the glass panels separated either by a gas (air or Ar have very low thermal conductivities) or a sheet of transparent polymer.

22-4 Thermal Shock

Stresses leading to the fracture of brittle materials can be introduced thermally as well as mechanically. When a piece of material is cooled quickly, a temperature gradient is produced. This gradient can lead to different amounts of contraction in different areas. If residual tensile stresses become high enough, flaws may propagate and cause failure. This is why it is unsafe to run cold water over a hot glass baking dish after removing it from the oven. If it is cooled too quickly, it will break. Similar behavior can occur if a material is heated rapidly. This failure of a material caused by stresses induced by sudden changes in temperature is known as **thermal shock**. Thermal shock behavior is affected by several factors:

1. *Coefficient of thermal expansion*: A low coefficient minimizes dimensional changes and reduces the ability to withstand thermal shock.
2. *Thermal conductivity*: The magnitude of the temperature gradient is determined partly by the thermal conductivity of the material. A high thermal conductivity helps to transfer heat and reduce temperature differences quickly in the material.
3. *Modulus of elasticity*: A low modulus of elasticity permits large amounts of strain before the stress reaches the critical level required to cause fracture.
4. *Fracture stress*: A high stress required for fracture permits larger strains.
5. *Phase transformations*: Additional dimensional changes can be caused by phase transformations. Transformation of silica from quartz to cristobalite, for example, introduces residual stresses and increases problems with thermal shock. Similarly, we cannot use pure $PbTiO_3$ ceramics in environments subjected to large, sudden temperature changes, since the stresses induced during the cubic to tetragonal transformation will cause the ceramic to fracture.

One method for measuring the resistance to thermal shock is to determine the maximum temperature difference that can be tolerated during a quench without affecting the mechanical properties of the material. Pure (fused) silica glass has a thermal shock resistance of about 3000°C. Figure 22-7 shows the effect of quenching temperature difference on the modulus of rupture in sialon ($Si_3Al_3O_3N_5$) after quenching; no cracks and therefore no change in the properties of the ceramic are evident until the quenching temperature difference approaches 950°C. Other ceramics have poorer resistance. Shock resistance for partially stabilized zirconia (PSZ) and Si_3N_4 is about 500°C; for SiC 350°C; and for Al_2O_3 and ordinary glass, about 200°C.

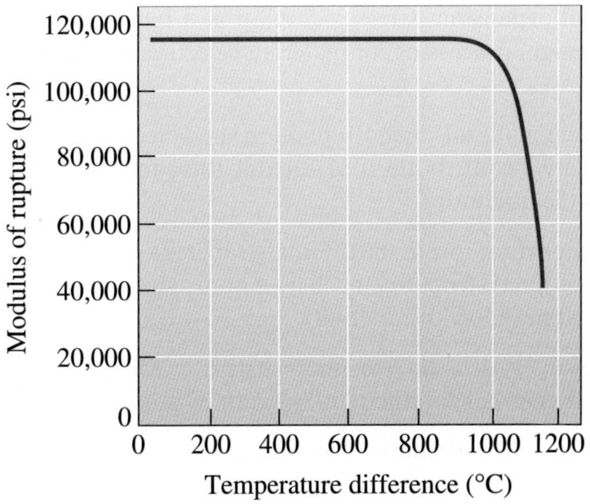

Figure 22-7 The effect of quenching temperature difference on the modulus of rupture of sialon. The thermal shock resistance of the ceramic decreases at about 950°C.

Another way to evaluate the resistance of a material to thermal shock is by the thermal shock parameter (R' or R)

$$R' = \text{Thermal shock parameter} = \frac{\sigma_f k(1 - v)}{E \cdot \alpha} \quad (22\text{-}8a)$$

or

$$R = \frac{\sigma_f(1 - v)}{E \cdot \alpha} \quad (22\text{-}8b)$$

where σ_f is the fracture stress of the material, v is the Poisson's ratio, k is the thermal conductivity, E is the modulus of elasticity, and α is the linear coefficient of thermal expansion. Equation 22-8b is used in situations where the heat transfer rate is essentially infinite. A higher value of the thermal shock parameter means better resistance to thermal shock. The thermal shock parameter represents the maximum temperature change that can occur without fracturing the material.

Thermal shock is usually not a problem in most metals because metals normally have sufficient ductility to permit deformation rather than fracture. As mentioned before, many ceramic compositions developed as zero thermal-expansion ceramics (based on sodium zirconium phosphates) and many glass-ceramics have excellent thermal shock resistance.

Summary

- The thermal properties of materials can be at least partly explained by the movement of electrons and the propagation of phonons.
- Heat capacity and specific heat represent the quantity of energy required to raise the temperature of a given amount of material by one degree, and they are influenced by temperature, crystal structure, and bonding.
- The coefficient of thermal expansion describes the dimensional changes that occur in a material when its temperature changes. Strong bonding leads to a low coefficient of expansion. High melting-point metals and ceramics have low coefficients, whereas low melting-point metals and polymers have high coefficients.
- The thermal expansion of many engineered materials can be made to be near zero by properly controlling the combination of different phases that show positive and negative thermal-expansion coefficients. Zerodur™ glass-ceramics and NZP ceramics are examples of such materials.

- Because of thermal expansion, stresses develop in a material when the temperature changes. Care in design, processing, or materials selection is required to prevent failure due to thermal stresses.
- Heat is transferred in materials by both phonons and electrons. Thermal conductivity depends on the relative contributions of each of these mechanisms, as well as on the microstructure and temperature.
- Phonons play an important role in the thermal properties of ceramics, semiconductors, and polymers. Disordered structures, such as ceramic glasses or amorphous polymers, scatter phonons and have low conductivity.
- Thermal conductivity is sensitive to microstructure. Most commonly encountered ceramics have a low thermal conductivity because of porosity and defects.
- Crystalline ceramics and polymers have higher conductivities than their glassy or amorphous counterparts.
- Electronic contributions to thermal conductivity are most important in metals; consequently, lattice imperfections that scatter electrons reduce conductivity. Increasing temperature not only increases phonon energy but also increases the scattering of both phonons and electrons.

Glossary

Heat capacity The energy required to raise the temperature of one mole of a material by one degree.

Linear coefficient of thermal expansion Describes the amount by which each unit length of a material changes when the temperature of the material changes by one degree.

Lorenz number The constant that relates electrical and thermal conductivity.

Phonon A packet of elastic waves. It is characterized by its energy, wavelength, or frequency, which transfers energy through a material.

Specific heat The energy required to raise the temperature of one gram of a material by one degree.

Thermal conductivity A microstructure-sensitive property that measures the rate at which heat is transferred through a material.

Thermal shock Failure of a material caused by stresses introduced by sudden changes in temperature.

Thermal stresses Stresses introduced into a material due to differences in the amount of expansion or contraction that occur because of a temperature change.

Problems

Section 22-1 Heat Capacity and Specific Heat

22-1 State any two applications for which a high thermal conductivity is desirable.

22-2 State any two applications for which a low thermal conductivity of a material is desirable.

22-3 If the gas constant was exactly 1.987 cal/(mol·K), what is the percent error in assuming $3R$ is 6 cal/(mol · K)?

22-4 In older references in the English language, you may find the specific heat in units of BTU/(lbm · °R), where BTU is the British thermal unit, lbm is pounds, and °R is degrees Rankine. Look up the conversion factors for these units and determine what the specific heat of water is in these units.

22-5 Of the metals listed in Table 22-1, which have the highest and lowest specific heats? Also list their atomic masses.

22-6 Calculate the increase in temperature for 1 kg samples of the following metals when 20,000 joules of heat is supplied to the metal at 25°C:
(a) Al;
(b) Cu;
(c) Fe; and
(d) Ni.

22-7 Calculate the heat (in calories and joules) required to raise the temperature of 1 kg of the following materials by 50°C:
(a) lead;
(b) nickel;
(c) Si_3N_4; and
(d) 6,6-nylon.

22-8 Calculate the temperature of a 100 g sample of the following materials (originally at 25°C) when 3000 calories are introduced:
(a) tungsten;
(b) titanium;
(c) Al_2O_3; and
(d) low-density polyethylene.

22-9 An alumina insulator for an electrical device is also to serve as a heat sink. A 10°C temperature rise in an alumina insulator 1 cm × 1 cm × 0.02 cm is observed during use. Determine the thickness of a high-density polyethylene insulator that would be needed to provide the same performance as a heat sink. The density of alumina is 3.96 g/cm^3.

22-10 A 200 g sample of aluminum is heated to 400°C and then quenched into 2000 cm^3 of water at 20°C. Calculate the temperature of the water after the aluminum and water reach equilibrium. Assume no heat loss from the system.

Section 22-2 Thermal Expansion

22-11 A lead pipe in the basement of an old house travels 15 feet from the water heater outlet elbow to the riser up to a sink. The basement is cool and dank at 20°C, and the water that is suddenly flowing through the pipe is 53°C. How much linear expansion does the pipe undergo?

22-12 A 2-m-long soda-lime glass sheet is produced at 1400°C. Determine its length after it cools to 25°C.

22-13 A copper casting requires the final dimensions to be 2.5 cm × 50 cm × 10 cm. Determine the size of the pattern that must be used to make the mold into which the liquid copper is poured during the manufacturing process.

22-14 A copper casting requires the final dimensions to be 1 in. × 12 in. × 24 in. Determine the size of the pattern that must be used to make the mold into which the liquid copper is poured during the manufacturing process.

22-15 An aluminum casting is made using the permanent mold process. In this process, the liquid aluminum is poured into a gray cast iron mold that is heated to 350°C. We wish to produce an aluminum casting that is 15 in. long at 25°C. Calculate the length of the cavity that must be machined into the gray cast iron mold.

22-16 A Ti-alloy strip and stainless steel strip are roll bonded at 760°C. If the total length of the bimetallic strip is 1 m long when hot rolled, calculate the length of each metal at 25°C. What is the nature of the stress (compressive or tensile) just below the surface on the Ti-alloy side and on the stainless steel side?

22-17 We coat a 100-cm-long, 2-mm-diameter copper wire with a 0.5-mm-thick epoxy insulation coating. Determine the length of the copper and the coating when their temperature increases from 25°C to 250°C. What is likely to happen to the epoxy coating as a result of this heating?

22-18 We produce a 10-in.-long bimetallic composite material composed of a strip of yellow brass bonded to a strip of Invar. Determine the length to which each material would expand when the temperature increases from 20°C to 150°C. Draw a sketch showing what will happen to the shape of the bimetallic strip.

22-19 Give examples of materials that have near-zero thermal-expansion coefficients.

22-20 What is Zerodur™? What are some of the properties and applications of this material?

22-21 A nickel engine part is coated with SiC to provide corrosion resistance at high temperatures. If no residual stresses are present in the part at 20°C, determine the thermal stresses that develop when the part is heated to 1000°C during use. The elastic modulus of SiC is 60×10^6 psi.

22-22 Alumina fibers, 2-cm-long, are incorporated into an aluminum matrix. Assuming good bonding between the ceramic fibers and the aluminum, estimate the thermal stresses acting on the fiber when the temperature of the composite increases

250°C. Are the stresses on the fibers tensile or compressive? The elastic modulus of alumina is 56×10^6 psi.

22-23 A 24-in.-long copper bar with a yield strength of 30,000 psi is heated to 120°C and immediately fastened securely to a rigid framework. Will the copper deform plastically during cooling to 25°C? How much will the bar deform if it is released from the framework after cooling?

22-24 Repeat the previous problem using a silicon carbide rod rather than a copper rod. The elastic modulus of SiC is 60×10^6 psi and the tensile strength of this brittle material is 25 ksi. Assume there is no plastic deformation prior to failure.

Section 22-3 Thermal Conductivity

Section 22-4 Thermal Shock

22-25 Define the terms thermal conductivity and thermal shock.

22-26 The thermal conductivity of all ceramics is low. True or False? Explain.

22-27 Is the thermal conductivity of materials a microstructure-sensitive property? Explain.

22-28 Determine the minimum heat flux (heat flow per unit area) when a 5-mm-thick soda-lime glass is presented with boiling water on one side and freezing water on the other.

22-29 If a 1-m-long silver bar is heated at one end to 300°C and the temperature measured at the other end is 100°C, calculate the heat transferred per unit area.

22-30 A 3-cm plate of silicon carbide separates liquid aluminum (held at 700°C) from a water-cooled steel shell maintained at 20°C. Calculate the heat Q transferred to the steel per cm^2 of silicon carbide each second.

22-31 A sheet of 0.01 in. polyethylene is sandwiched between two 3 ft × 3 ft × 0.125 in. sheets of soda-lime glass to produce a window. Calculate (a) the heat lost through the window each day when the room temperature is 25°C and the outside air is 0°C and (b) the heat entering through the window each day when the room temperature is 25°C and the outside air is 40°C. Use 0.96 W/(m · K) for the thermal conductivity of soda-lime glass.

22-32 We would like to build a heat-deflection plate that permits heat to be transferred rapidly parallel to the sheet but very slowly perpendicular to the sheet. Consequently, we incorporate 1 kg of unidirectional copper wires, each 0.1 cm in diameter, into 5 kg of a polyimide polymer matrix. Estimate the thermal conductivity parallel and perpendicular to the wires.

22-33 An exothermic reaction at a battery electrode releases 80 MW of power per square meter. The electrode consists of a 5-μm-thick graphite layer on a 20-μm-thick copper foil; the reaction occurs at the graphite surface; and heat needs to be transferred out through the graphite to the copper foil. The temperature on the outer side of the copper foil is maintained at 30°C. What is the temperature on the inner surface of the electrode?

22-34 Suppose we just dip a 1-cm-diameter, 10-cm-long rod of aluminum into one liter of water at 20°C. The other end of the rod is in contact with a heat source operating at 400°C. Determine the length of time required to heat the water to 25°C if 75% of the heat is lost by radiation from the bar.

22-35 Determine the Lorenz number using the data for silver found in this book. What is your percent difference from the given value and your explanation for any difference?

22-36 Write down the equations that define the thermal shock resistance (TSR) parameter. Based on this, what can you say about materials that show a near-zero thermal-expansion coefficient?

22-37 Gray cast iron has a higher thermal conductivity than ductile or malleable cast iron. Review Chapter 13 and explain why this difference in conductivity might be expected.

Design Problems

22-38 A chemical-reaction vessel contains liquids at a temperature of 680°C. The wall of the vessel must be constructed so that the outside wall operates at a temperature of 35°C or less. Design the vessel wall and appropriate materials if the maximum heat transfer is 6000 cal/s.

22-39 Design a metal panel coated with glass enamel capable of thermal cycling between 20°C and 150°C. The glasses generally available are expected to have a tensile strength of 5,000 psi and a compressive strength of 50,000 psi.

22-40 What design constraints exist in selecting materials for a turbine blade for a jet engine that is capable of operating at high temperatures?

22-41 Consider the requirements of a low dielectric constant, low dielectric loss, good mechanical strength, and high thermal conductivity for electronic packaging substrates. What materials would you consider?

Computer Problems

22-42 *Thermal Shock Resistance Parameters.* Write a computer program or use spreadsheet software that will provide the values of thermal shock-resistance parameters (TSR) when the values of the fracture stress, thermal conductivity, Young's modulus, Poisson's ratio, and thermal-expansion coefficient are provided.

Knovel® Problems

K22-1 The molar heat capacity is the energy required to increase the temperature of one mole of a material by one degree. How much heat is required to bring 50 moles of ethanol stored at 25°C to a boil? (Note that ethanol is also known as ethyl alcohol.)

K22-2 The thermal-expansion coefficient is the change of length of a material per unit length per degree of temperature change. A gold bar is initially 1 m long. What is the length of the gold bar after it has been heated from 298 to 473 K?

K22-3 Three heat sinks are identical except that they are made from different materials: gold, silver, and copper. Which heat sink will extract heat the fastest?

The process of corrosion affects many important areas of technology, including the construction and manufacturing of bridges, airplanes, ships, and microelectronics; the food industry; space exploration; fiber optic networks; and nuclear and other power utilities, just to name a few. Corrosion can be prevented or contained using different techniques that include the use of coatings and anodic or cathodic protection.

Since steel is the most important structural material and may be highly susceptible to corrosion depending on composition, the corrosion of steel is a significant engineering issue. The photo above shows rusting of steel bar used to reinforce concrete (known as "rebar") when it is exposed to the environment. The rust that is commonly observed on steel is iron oxide that forms as a result of an electrochemical reaction known as galvanic corrosion. In this process, steel is consumed to form the brittle oxide, which may later flake off, thereby degrading the bar (© *Glenn Volkman/Alamy*).

CHAPTER

23

Corrosion and Wear

Have You Ever Wondered?

- *Why does iron rust?*
- *What process makes use of mechanical erosion and chemical corrosion in the processing of semiconductor chips?*
- *Do metals like aluminum and titanium undergo corrosion?*
- *Is the process of corrosion ever useful?*
- *Why do household water-heater tanks contain Mg rods?*
- *What does the acronym "WD-40^{TM}" stand for?*
- *How does wear affect the useful life of different components such as crankshafts?*

Chapter Learning Objectives

The key objectives of this chapter are to

- Summarize the mechanisms and factors that contribute to corrosion in materials.
- Given a pair of metals in electrical contact in an aqueous medium, predict which metal will act as the anode and which will act as the cathode and calculate the potential difference between them.
- Calculate the corrosion rate of an anode using Faraday's equation.
- Calculate the electrode potential of a half-cell in a non-standard solution using the Nernst equation.
- Identify the microstructural features of a metal that promote electrochemical corrosion.
- Explain the operation of a battery and the process of electroplating.
- Choose processing techniques and incorporate design features that limit corrosion and wear for mechanical engineering applications.
- Explain how polymers are susceptible to degradation due to microbes and the environment.

- **Predict the morphology of a metal oxide film.**
- **Explain the physical mechanisms for the processes of wear and erosion.**

The composition and physical integrity of a solid material is altered in a corrosive environment. In chemical corrosion, a corrosive liquid dissolves the material. Metals and certain ceramics react with a gaseous environment, usually at elevated temperatures, and the material may be destroyed by the formation of oxides or other compounds. Polymers degrade when exposed to oxygen at elevated temperatures. Materials may be altered when exposed to radiation or even bacteria. Additionally, a variety of wear and wear-corrosion mechanisms alter the shape of materials.

"Rust" is a word that is commonly associated with corrosion. An abandoned rusty car might come to mind. We will learn in this chapter that the rust that forms, for example on a steel automobile panel, is iron oxide. This particular process of rusting or corrosion occurs because metal atoms are removed from the solid material as a result of an electric circuit that is produced.The loss of atoms degrades the metal, and in some cases, as in the rusting of steel, a solid byproduct is formed. If the same car had been kept in a clean, dry garage, the corrosion would have been far less likely to occur. Thus we can intuit that corrosion is also related to the environment, and in fact, **electrochemical corrosion** such as we are describing here usually occurs in an aqueous medium such as water, soil, or moist air. Corrosion occurs over a period of time and can occur at either high or low temperatures.

Corrosion, in all its various forms, is an expensive phenomenon. Corrosion costs the United States approximately 3% of its gross domestic product (i.e., hundreds of billions of dollars each year). Corrosion affects transportation systems (bridges, pipelines, cars, airplanes, trains, and ships), utilities (electrical, water, telecommunications, and nuclear power plants), production and manufacturing (food industries, microelectronics, and petroleum refining), and materials storage systems to name only a few possibilities. Engineers have to incorporate additional design features or choose more expensive materials to avoid corrosion. When corrosion occurs, repairs have to be made and parts have to be replaced. When failures due to corrosion occur, they are often costly and sometimes fatal.

Although we typically think of corrosion as detrimental, it can be used to deposit metallic coatings on products through a process known as **electroplating** and is the fundamental principle underlying battery technology. The processes of chemical corrosion and erosion are used to make ultra-flat surfaces of silicon wafers for microelectronics processing. Degradation and dissolution of certain biopolymers is useful in some medical applications, such as dissolvable sutures.

The goal of this chapter is to introduce the principles and mechanisms by which corrosion and wear occur under different conditions. This includes electrochemical corrosion, the oxidation of metals, the corrosion of ceramics, and the degradation of polymers. We will see how different factors affect the rate of corrosion. We will predict whether corrosion will occur in a particular materials system by making use of a **galvanic series**, a tabulation of the tendencies of various metals to corrode in a particular environment. We will also discuss how the microstructure of materials influences corrosion and strategies to prevent corrosion.

23-1 Chemical Corrosion

In **chemical corrosion**, or direct dissolution, a material dissolves in a corrosive liquid medium. The material continues to dissolve until either it is consumed or the liquid is saturated. An example is the development of a green patina on the surface of copper-based alloys. This is due to the formation of copper carbonate and copper hydroxides and is why, for example, the Statue of Liberty looks greenish. The chemical corrosion of copper, tantalum, silicon, silicon dioxide, and other materials can be achieved under extremely well-controlled conditions. In the processing of silicon wafers, for example, a process known as chemical mechanical polishing uses a corrosive silica-based slurry to provide mechanical erosion. This process creates extremely flat surfaces that are suitable for the processing of silicon wafers. Chemical corrosion also occurs in nature. For example, the chemical corrosion of rocks by carbonic acid (H_2CO_3) and the mechanical erosion of wind and water play an important role in the formation of canyons and caverns.

Liquid Metal Attack Liquid metals first attack a solid at high-energy locations such as grain boundaries. If these regions continue to be attacked preferentially, cracks eventually grow (Figure 23-1). Often this form of corrosion is complicated by the presence of fluxes that accelerate the attack or by electrochemical corrosion. Aggressive metals such as liquid lithium can also attack ceramics.

Selective Leaching One particular element in an alloy may be selectively dissolved, or leached, from the solid. **Dezincification** occurs in brass containing more than 15% Zn. Both copper and zinc are dissolved by aqueous solutions at elevated temperatures; the zinc ions remain in solution while the copper ions are replated onto the brass (Figure 23-2). Eventually, the brass becomes porous and weak.

Graphitic corrosion of gray cast iron occurs when iron is selectively dissolved in water or soil, leaving behind interconnected graphite flakes and a corrosion product. Localized graphitic corrosion often causes leakage or failure of buried gray iron gas lines, sometimes leading to explosions.

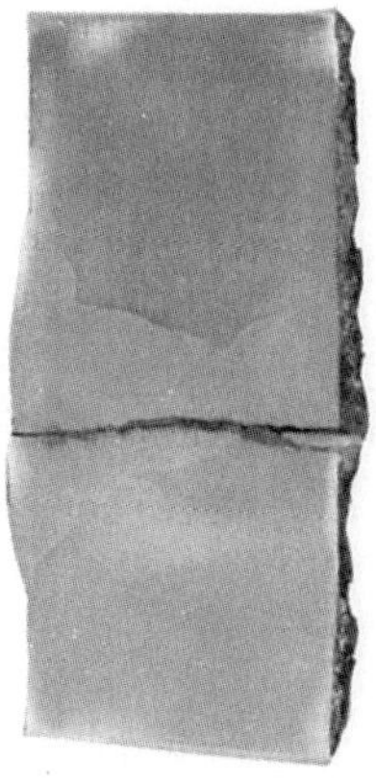

Figure 23-1 Molten lead is held in thick steel pots during refining. In this case, the molten lead has attacked a weld in a steel plate, and cracks have developed. Eventually, the cracks propagate through the steel, and molten lead leaks from the pot. (*Reprinted Courtesy of Don Askeland*)

Figure 23-2 Micrograph of a copper deposit in brass, showing the effect of dezincification (× 50). (*Reprinted Courtesy of Don Askeland*)

Dissolution and Oxidation of Ceramics Ceramic refractories used to contain molten metal during melting or refining may be dissolved by the slags that are produced on the metal surface. For example, an acid (high SiO_2) refractory is rapidly attacked by a basic (high CaO or MgO) slag. A glass produced from SiO_2 and Na_2O is rapidly attacked by water; CaO must be added to the glass to minimize this attack. Nitric acid may selectively leach iron or silica from some ceramics, reducing their strength and density. The strength of silicate glasses depends on flaws that are often created by corrosive interactions with water.

Chemical Attack on Polymers Compared to metals and oxide ceramics, plastics are considered corrosion resistant. Teflon™ and Viton™ are some of the most corrosion-resistant materials and are used in many applications, including the chemical processing industry. These and other polymeric materials can withstand the presence of many acids, bases, and organic liquids. Aggressive solvents do, however, often diffuse into low-molecular-weight thermoplastic polymers. As the solvent is incorporated into the polymer, the smaller solvent molecules force apart the chains, causing swelling. The strength of the bonds between the chains decreases. This leads to softer, lower-strength polymers with low glass-transition temperatures. In extreme cases, the swelling leads to stress cracking.

Thermoplastics may also be dissolved in a solvent. Prolonged exposure causes a loss of material and weakening of the polymer part. This process occurs most easily when the temperature is high and when the polymer has a low molecular weight, is highly branched and amorphous, and is not cross-linked. The structure of the monomer is also important; the CH_3 groups on the polymer chain in polypropylene are more easily removed from the chain than are chloride or fluoride ions in polyvinyl chloride (PVC) or polytetrafluoroethylene (Teflon™). Teflon has exceptional resistance to chemical attack by almost all solvents.

23-2 Electrochemical Corrosion

Electrochemical corrosion, the most common form of attack of metals, occurs when metal atoms lose electrons and become ions. As the metal is gradually consumed by this process, a byproduct of the corrosion process is typically formed. Electrochemical corrosion occurs most frequently in an aqueous medium, in which ions are present in water, soil, or moist air. In this process, an electric circuit is created, and the system is called an **electrochemical cell**. Corrosion of a steel pipe or a steel automobile panel, creating holes in the steel and rust as the byproduct, are examples of this reaction.

Although responsible for corrosion, electrochemical cells may also be useful. By deliberately creating an electric circuit, we can *electroplate* protective or decorative coatings onto materials. In some cases, electrochemical corrosion is even desired. For example, in etching a polished metal surface with an appropriate acid, various features in the microstructure are selectively attacked, permitting them to be observed. In fact, most of the photographs of metal and alloy microstructures in this text were obtained in this way, thus enabling, for example, the observation of pearlite in steel or grain boundaries in copper.

Components of an Electrochemical Cell There are four components of an electrochemical cell (Figure 23-3):

1. The **anode** gives up electrons to the circuit and corrodes.
2. The **cathode** receives electrons from the circuit by means of a chemical, or cathode, reaction. Ions that combine with the electrons produce a byproduct at the cathode.

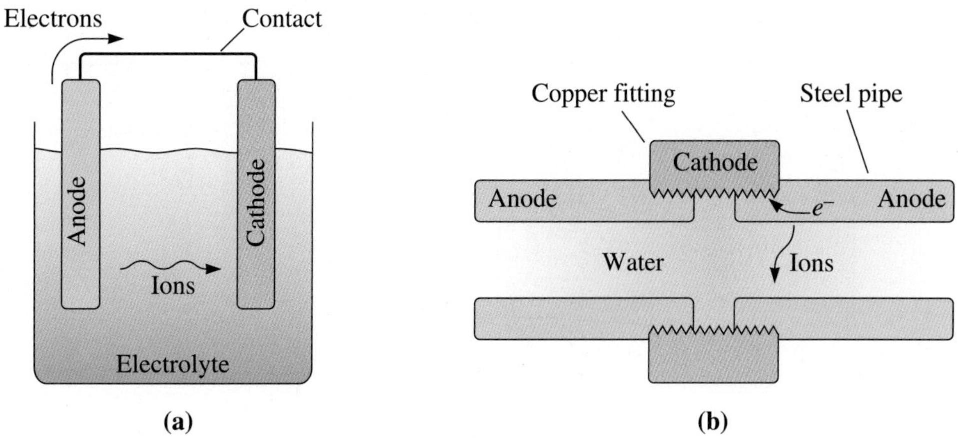

Figure 23-3 The components in an electrochemical cell: (a) a simple electrochemical cell and (b) a corrosion cell between a steel water pipe and a copper fitting.

3. The anode and cathode must be electrically connected, usually by physical contact, to permit the electrons to flow from the anode to the cathode and continue the reaction.
4. A liquid **electrolyte** must be in contact with both the anode and the cathode. The electrolyte is conductive, thus completing the circuit. It provides the means by which metallic ions leave the anode surface and move to the cathode to accept the electrons.

This description of an electrochemical cell defines electrochemical corrosion. If metal ions are deposited onto the cathode, *electroplating* occurs.

Anode Reaction The anode, which is a metal, undergoes an **oxidation reaction** by which metal atoms are ionized. The metal ions enter the electrolytic solution, while the electrons leave the anode through the electrical connection:

$$M \rightarrow M^{n+} + ne^- \tag{23-1}$$

M represents the metal atom or ion, n is the charge on the metal ion, and e^- represents electrons. Because metal ions leave the anode, the anode corrodes, or oxidizes.

Cathode Reaction in Electroplating In electroplating, a cathodic **reduction reaction,** which is the reverse of the anode reaction, occurs at the cathode:

$$M^{n+} + ne^- \rightarrow M \tag{23-2}$$

The metal ions, either intentionally added to the electrolyte or formed by the anode reaction, combine with electrons at the cathode. The metal then plates out and covers the cathode surface. Note, however, that electroplating is a special case and does not always occur.

Cathode Reactions in Corrosion Except in unusual conditions, plating of a metal does not occur during electrochemical corrosion. Instead, the reduction reaction forms a gas, solid, or liquid byproduct at the cathode (Figure 23-4).

1. *The hydrogen electrode*: In oxygen-free liquids, such as hydrochloric acid (HCl) or stagnant water, hydrogen gas may be evolved at the cathode:

$$2H^+ + 2e^- \rightarrow H_2\uparrow(\text{gas}) \tag{23-3}$$

If zinc were placed in such an environment, we would find that the overall reaction is

$$\begin{aligned} Zn &\rightarrow Zn^{2+} + 2e^- \text{ (anode reaction)} \\ 2H^+ + 2e^- &\rightarrow H_2\uparrow\text{(cathode reaction)} \\ Zn + 2H^+ &\rightarrow Zn^{2+} + H_2\uparrow\text{(overall reaction)} \end{aligned} \tag{23-4}$$

The zinc anode gradually dissolves, and hydrogen bubbles continue to evolve at the cathode.

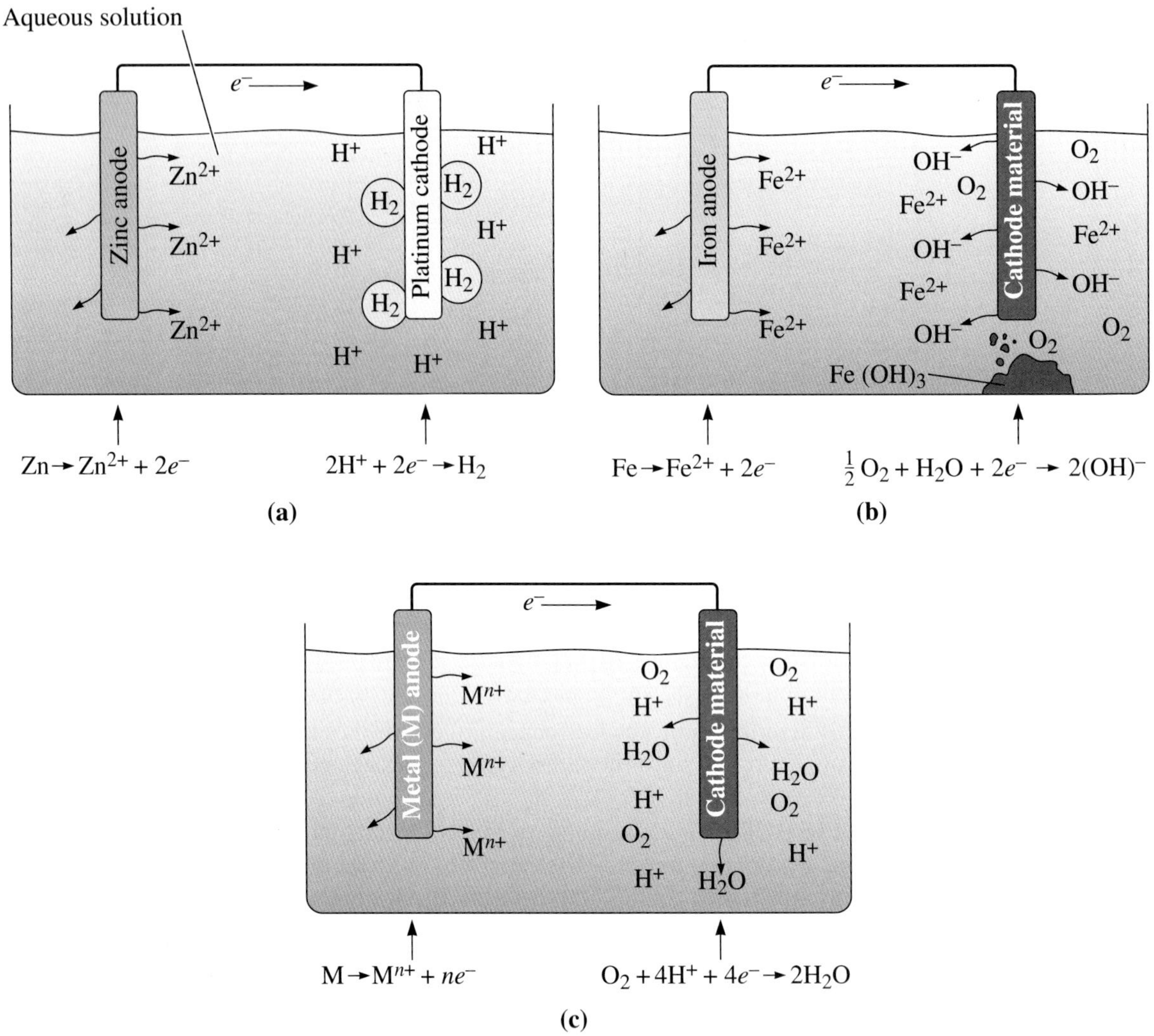

Figure 23-4 The anode and cathode reactions in typical electrolytic corrosion cells: (a) the hydrogen electrode, (b) the oxygen electrode, and (c) the water electrode.

2. *The oxygen electrode*: In aerated water, oxygen is available to the cathode, and hydroxyl, or $(OH)^-$, ions form:

$$\frac{1}{2}O_2 + H_2O + 2e^- \rightarrow 2(OH)^- \tag{23-5}$$

The oxygen electrode enriches the electrolyte in $(OH)^-$ ions. These ions react with positively charged metallic ions and produce a solid product. In the case of rusting of iron:

$$Fe \rightarrow Fe^{2+} + 2e^- \text{ (anode reaction)}$$

$$\left.\begin{array}{l}\frac{1}{2}O_2 + H_2O + 2e^- \rightarrow 2(OH)^- \\ Fe^{2+} + 2(OH)^- \rightarrow Fe(OH)_2\end{array}\right\} \text{(cathode reactions)} \tag{23-6}$$

$$Fe + \frac{1}{2}O_2 + H_2O \rightarrow Fe(OH)_2 \text{ (overall reaction)}$$

The reaction continues as the $Fe(OH)_2$ reacts with more oxygen and water:

$$2Fe(OH)_2 + \frac{1}{2}O_2 + H_2O \rightarrow 2Fe(OH)_3 \tag{23-7}$$

$Fe(OH)_3$ is commonly known as *rust*.

3. *The water electrode*: In oxidizing acids, the cathode reaction produces water as a byproduct:

$$O_2 + 4H^+ + 4e^- \rightarrow 2H_2O \tag{23-8}$$

If a continuous supply of both oxygen and hydrogen is available, the water electrode produces neither a buildup of solid rust nor a high concentration or dilution of ions at the cathode.

23-3 The Electrode Potential in Electrochemical Cells

In corrosion, a potential naturally develops when a material is placed in a solution. Let's see how the potential required to drive the corrosion reaction develops.

Electrode Potential When a pure metal is placed in an electrolyte, an **electrode potential** develops that is related to the tendency of the material to give up its electrons; however, the driving force for the oxidation reaction is offset by an equal but opposite driving force for the reduction reaction. No net corrosion occurs. Consequently, we cannot measure the electrode potential for a single electrode material.

Electromotive Force Series To determine the tendency of a metal to give up its electrons, we measure the potential difference between the metal and a standard electrode using a half-cell (Figure 23-5). The metal electrode to be tested is placed in a 1 molar (M) solution of its ions. A reference electrode is also placed in a 1 M solution of ions. The reference electrode is typically an inert metal such as platinum that conducts electrons but does not react with the electrolyte. The use of a 1 M solution of hydrogen (H^+) ions with the reference electrode is common. The reaction that occurs at this hydrogen electrode is $2H^+ + 2e^- \rightarrow H_2\uparrow$. H_2 gas is supplied to the hydrogen electrode. An electrochemical cell such as this is known as a standard cell when the measurements are made at 25°C and atmospheric pressure with 1 M electrolyte concentrations. The two electrolytes are in electrical contact but are not permitted to mix with one another. Each electrode establishes its own electrode potential. By measuring the voltage between the two electrodes when the circuit is open, we obtain the potential difference. The potential of the hydrogen electrode is arbitrarily set equal to zero volts. If the metal has a greater tendency to give up electrons than the hydrogen electrode, then the potential of the metal is negative—the metal is anodic with respect to the hydrogen electrode. If the potential of the metal is positive, the metal is cathodic with respect to the hydrogen electrode.

The **electromotive force** (or **emf**) **series** shown in Table 23-1 compares the standard electrode potential E^0 for each metal with that of the hydrogen electrode under standard conditions of 25°C and a 1 M solution of ions in the electrolyte. Note that the

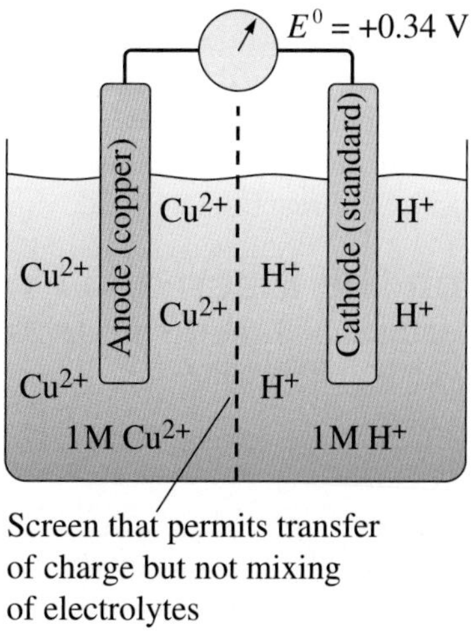

Figure 23-5 The half-cell used to measure the electrode potential of copper under standard conditions. The electrode potential of copper is the potential difference between it and the standard hydrogen electrode in an open circuit. Since E^0 is greater than zero, copper is cathodic compared to the hydrogen electrode.

Table 23-1 The standard reduction potentials for selected elements and reactions

	Metal	Electrode Potential E^0 (Volts)
Anodic ↑	$Li^+ + e^- \rightarrow Li$	−3.05
	$Mg^{2+} + 2e^- \rightarrow Mg$	−2.37
	$Al^{3+} + 3e^- \rightarrow Al$	−1.66
	$Ti^{2+} + 2e^- \rightarrow Ti$	−1.63
	$Mn^{2+} + 2e^- \rightarrow Mn$	−1.63
	$Zn^{2+} + 2e^- \rightarrow Zn$	−0.76
	$Cr^{3+} + 3e^- \rightarrow Cr$	−0.74
	$Fe^{2+} + 2e^- \rightarrow Fe$	−0.44
	$Ni^{2+} + 2e^- \rightarrow Ni$	−0.25
	$Sn^{2+} + 2e^- \rightarrow Sn$	−0.14
	$Pb^{2+} + 2e^- \rightarrow Pb$	−0.13
	$2H^+ + 2e^- \rightarrow H_2$	0.00 — (defined)
	$Cu^{2+} + 2e^- \rightarrow Cu$	+0.34
	$O_2 + 2H_2O + 4e^- \rightarrow 4OH^-$	+0.40
	$Ag^+ + e^- \rightarrow Ag$	+0.80
	$Pt^{4+} + 4e^- \rightarrow Pt$	+1.20
	$O_2 + 4H^+ + 4e^- \rightarrow 2H_2O$	+1.23
Cathodic ↓	$Au^{3+} + 3e^- \rightarrow Au$	+1.50

measurement of the potential is made when the electric circuit is open. The voltage difference begins to change as soon as the circuit is closed.

The more negative the value of potential for the oxidation of metal, the more electropositive is the metal; this means the metal will have a higher tendency to undergo an oxidation reaction. For example, alkali and alkaline earth metals (e.g., Li, K, Ba, Sr, and Mg) are so reactive that they have to be kept under conditions that prevent any contact with oxygen. On the other hand, metals that are toward the bottom of the chart (e.g., Ag, Au, and Pt) will not tend to react with oxygen. This is why we call them "noble metals." Metals such as Fe, Cu, and Ni have intermediate reactivities; however, this is not the only consideration. Aluminum, for example, has a strongly negative standard electrode potential and does react easily with oxygen to form aluminum oxide; it also reacts easily with fluoride to form aluminum fluoride. Both of these compounds form a tenacious and impervious layer that helps stop further corrosion. Titanium also reacts readily with oxygen. The quickly formed titanium oxide creates a barrier that prevents the further diffusion of species and, thus, avoids further oxidation. This is why both aluminum and titanium are highly reactive, but can resist corrosion exceptionally well. Note that while the emf series tells us about the thermodynamic feasibility and driving force; it does *not* tell us about the kinetics of the reaction.

Effect of Concentration on the Electrode Potential The electrode potential depends on the concentration of the electrolyte. At 25°C, the **Nernst equation** gives the electrode potential in nonstandard solutions:

$$E = E^0 + \frac{0.0592}{n} \log (C_{\text{ion}}) \tag{23-9}$$

where E is the electrode potential in a solution containing a concentration C_{ion} of the metal in molar units, n is the charge on the metallic ion, and E^0 is the standard electrode potential in a 1 M solution. Note that when $C_{\text{ion}} = 1$ M, $E = E^0$. The example that follows illustrates the calculation of the electrode potential.

Example 23-1 *Half-Cell Potential for Copper*

Suppose 1 g of copper as Cu^{2+} is dissolved in 1000 g of water to produce an electrolyte. Calculate the electrode potential of the copper half-cell in this electrolyte. Assume $T = 25°C$.

SOLUTION

From chemistry, we know that a standard 1 M solution of Cu^{2+} is obtained when we add 1 mol of Cu^{2+} (an amount equal to the atomic mass of copper) to 1 L (or 1000 g) of water. The atomic mass of copper is 63.55 g/mol. The concentration of the solution when only 1 g of copper is added must be

$$C_{\text{ion}} = \frac{1}{63.55} = 0.0157\,\text{M}$$

From the Nernst equation, with $n = 2$ and $E^0 = +0.34$ V,

$$E = E^0 + \frac{0.0592}{n}\log(C_{\text{ion}}) = 0.34 + \frac{0.0592}{2}\log(0.0157)$$

$$= 0.34 + (0.0296)(-1.80) = 0.29\,\text{V}$$

Rate of Corrosion or Plating

The amount of metal plated on the cathode in electroplating, or removed from the anode by corrosion, can be determined from **Faraday's equation**,

$$w = \frac{ItM}{n\text{F}} \tag{23-10}$$

where w is the mass plated or corroded (g), I is the current (A), M is the atomic mass of the metal (g/mol), n is the charge on the metal ion, t is the time (s), and F is Faraday's constant (96,500 C/mol). The product It (with units of C) is the total charge that is removed from the anode (and transferred to the cathode). The quantity nF/M is the charge per gram of material under consideration. Thus, the ratio of these terms is the mass of deposited (or corroded) material. In other words, this law basically states that one gram equivalent of a metal will be deposited by 96,500 C of charge. Often the current is expressed in terms of current density, $i = I/A$, so Equation 23-10 becomes

$$w = \frac{iAtM}{n\text{F}} \tag{23-11}$$

where the area A is the surface area of the anode or cathode.

The following examples illustrate the use of Faraday's equation to calculate the current density.

Example 23-2 *Design of a Copper Plating Process*

Design a process to electroplate a 0.1-cm-thick layer of copper onto a 1 cm × 1 cm cathode surface.

SOLUTION

In order for us to produce a 0.1-cm-thick layer on a 1 cm^2 surface area, the mass of copper must be

$$\rho_{\text{Cu}} = 8.93\text{ g/cm}^3 \quad A = 1\text{ cm}^2$$

$$\text{Volume of copper} = (1\ \text{cm}^2)(0.1\ \text{cm}) = 0.1\ \text{cm}^3$$
$$\text{Mass of copper} = (8.93\ \text{g/cm}^3)(0.1\ \text{cm}^3) = 0.893\ \text{g}$$

From Faraday's equation, where $M_{Cu} = 63.55$ g/mol and $n = 2$:

$$It = \frac{wn\,F}{M} = \frac{(0.893\ \text{g})(2)(96{,}500\ \text{C/mol})}{(63.55\ \text{g/mol})} = 2712\ \text{A} \cdot \text{s}$$

Therefore, we might use several different combinations of current and time to produce the copper plate:

Current	Time
0.1 A	27,120 s = 7.5 h
1.0 A	2,712 s = 45 min
10.0 A	271.2 s = 4.5 min
100.0 A	27.12 s = 0.45 min

Our choice of the exact combination of current and time might be made on the basis of the rate of production and quality of the plated copper. Low currents require very long plating times, perhaps making the process economically unsound. High currents, however, may reduce plating efficiencies. The plating effectiveness may depend on the composition of the electrolyte containing the copper ions, as well as on any impurities or additives that are present. Currents such as 10 A or 100 A are too high—they can initiate other side reactions that are not desired. The deposit also may not be uniform and smooth. Additional background or experimentation may be needed to obtain the most economical and efficient plating process. A current of ~1 A and a time of ~45 minutes are not uncommon in electroplating operations.

Example 23-3 *Corrosion of Iron*

An iron container 10 cm × 10 cm at its base is filled to a height of 20 cm with a corrosive liquid. A current is produced as a result of an electrolytic cell, and after four weeks, the container has decreased in weight by 70 g. Calculate (a) the current and (b) the current density involved in the corrosion of the iron.

SOLUTION

(a) The total exposure time is

$$t = (4\ \text{wk})(7\ \text{d/wk})(24\ \text{h/d})(3600\ \text{s/h}) = 2.42 \times 10^6\ \text{s}$$

From Faraday's equation, using $n = 2$ and $M = 55.847$ g/mol,

$$I = \frac{wnF}{tM} = \frac{(70\ \text{g})(2)(96{,}500\ \text{C/mol})}{(2.42 \times 10^6\ \text{s})(55.847\ \text{g/mol})}$$
$$= 0.1\ \text{A}$$

(b) The total surface area of iron in contact with the corrosive liquid and the current density are

$$A = (4\ \text{sides})(10\ \text{cm} \times 20\ \text{cm}) + (1\ \text{bottom})(10\ \text{cm} \times 10\ \text{cm}) = 900\ \text{cm}^2$$

$$i = \frac{I}{A} = \frac{0.1\ \text{A}}{900\ \text{cm}^2} = 1.11 \times 10^{-4}\ \text{A/cm}^2$$

Example 23-4 *Copper-Zinc Corrosion Cell*

Suppose that in a corrosion cell composed of copper and zinc, the current density at the copper cathode is 0.05 A/cm^2. The area of both the copper and zinc electrodes is 100 cm^2. Calculate (a) the corrosion current, (b) the current density at the zinc anode, and (c) the zinc lost at the anode in one hour.

SOLUTION

(a) The corrosion current is

$$I = i_{Cu}A_{Cu} = (0.05\ A/cm^2)(100\ cm^2) = 5\ A$$

(b) The current in the cell is the same everywhere. Thus,

$$i_{Zn} = \frac{I}{A_{Zn}} = \frac{5\ A}{100\ cm^2} = 0.05\ A/cm^2$$

(c) The atomic mass of zinc is 65.38 g/mol. There are 3600 s in one hour. From Faraday's equation:

$$w_{Zn} = \frac{ItM}{nF} = \frac{(5\ A)(3600\ s)(65.38\ g/mol)}{(2)(96{,}500\ C/mol)} = 6.1\ g$$

23-4 The Corrosion Current and Polarization

To protect metals from corrosion, we wish to make the current as small as possible. Unfortunately, the corrosion current is very difficult to measure, control, or predict. Part of this difficulty can be attributed to various changes that occur during operation of the corrosion cell. A change in the potential of an anode or cathode, which in turn affects the current in the cell, is called **polarization**. There are three important kinds of polarization: (1) activation, (2) concentration, and (3) resistance polarization.

Activation Polarization This kind of polarization is related to the energy required to cause the anode or cathode reactions to occur. If we can increase the degree of polarization, these reactions occur with greater difficulty, and the rate of corrosion is reduced. Small differences in composition and structure in the anode and cathode materials dramatically change the activation polarization. Segregation effects in the electrodes cause the activation polarization to vary from one location to another. These factors make it difficult to predict the corrosion current.

Concentration Polarization After corrosion begins, the concentration of ions at the anode or cathode surface may change. For example, a higher concentration of metal ions may exist at the anode if the ions are unable to diffuse rapidly into the electrolyte. Hydrogen ions may be depleted at the cathode in a hydrogen electrode, or a high $(OH)^-$ concentration may develop at the cathode in an oxygen electrode. When this situation occurs, either the anode or cathode reaction is stifled because fewer electrons are released at the anode or accepted at the cathode.

In any of these examples, the current density, and thus the rate of corrosion, decreases because of concentration polarization. Normally, the polarization is less pronounced when the electrolyte is highly concentrated, the temperature is increased, or the electrolyte is vigorously agitated. Each of these factors increases the current density and encourages electrochemical corrosion.

Resistance Polarization This type of polarization is caused by the electrical resistivity of the electrolyte. If a greater resistance to the flow of the current is offered, the rate of corrosion is reduced. Again, the degree of resistance polarization may change as the composition of the electrolyte changes during the corrosion process.

23-5 Types of Electrochemical Corrosion

In this section, we will look at some of the more common forms taken by electrochemical corrosion. First, there is *uniform attack*. When a metal is placed in an electrolyte, some regions are anodic to other regions; however, the location of these regions moves and even reverses from time to time. Since the anode and cathode regions continually shift, the metal corrodes uniformly.

Galvanic attack occurs when certain areas always act as anodes, whereas other areas always act as cathodes. These electrochemical cells are called galvanic cells and can be separated into three types: **composition cells, stress cells,** and **concentration cells**.

Composition Cells Composition cells, or *dissimilar metal corrosion*, develop when two metals or alloys, such as copper and iron, form an electrolytic cell. Because of the effect of alloying elements and electrolyte concentrations on polarization, the emf series may not tell us which regions corrode and which are protected. Instead, we use a galvanic series, in which the different alloys are ranked according to their anodic or cathodic tendencies in a particular environment (Table 23-2). We may find a different galvanic series for seawater, freshwater, and industrial atmospheres.

Table 23-2 The galvanic series in seawater

Active, Anodic End	Magnesium and Mg alloys
	Zinc
	Galvanized steel
	5052 aluminum
	3003 aluminum
	1100 aluminum
	Alclad
	Cadmium
	2024 aluminum
	Low-carbon steel
	Cast iron
	50% Pb-50% Sn solder
	316 stainless steel (active)
	Lead
	Tin
	Cu-40% Zn brass
	Nickel-based alloys (active)
	Copper
	Cu-30% Ni alloy
	Nickel-based alloys (passive)
	Stainless steels (passive)
	Silver
	Titanium
	Graphite
	Gold
	Platinum
Noble, Cathodic End	

(*Based on* ASM Metals Handbook, *Vol. 10, Eighth Edition. ASM International, 1975.*)

Example 23-5 *Corrosion of a Soldered Brass Fitting*

A brass fitting used in a marine application is joined by soldering with lead-tin solder. Will the brass or the solder corrode?

SOLUTION

From the galvanic series, we find that all of the copper-based alloys are more cathodic than a 50% Pb-50% Sn solder. Thus, the solder is the anode and corrodes. In a similar manner, the corrosion of solder can contaminate water in freshwater plumbing systems with lead.

Composition cells also develop in two-phase alloys, where one phase is more anodic than the other. Since ferrite is anodic to cementite in steel, small microcells cause steel to galvanically corrode (Figure 23-6). Almost always, a two-phase alloy has less resistance to corrosion than a single-phase alloy of a similar composition.

Intergranular corrosion occurs when the precipitation of a second phase or segregation at grain boundaries produces a galvanic cell. In zinc alloys, for example, impurities such as cadmium, tin, and lead segregate at the grain boundaries during solidification. The grain boundaries are anodic compared to the remainder of the grains, and corrosion of the grain boundary metal occurs (Figure 23-7). In austenitic

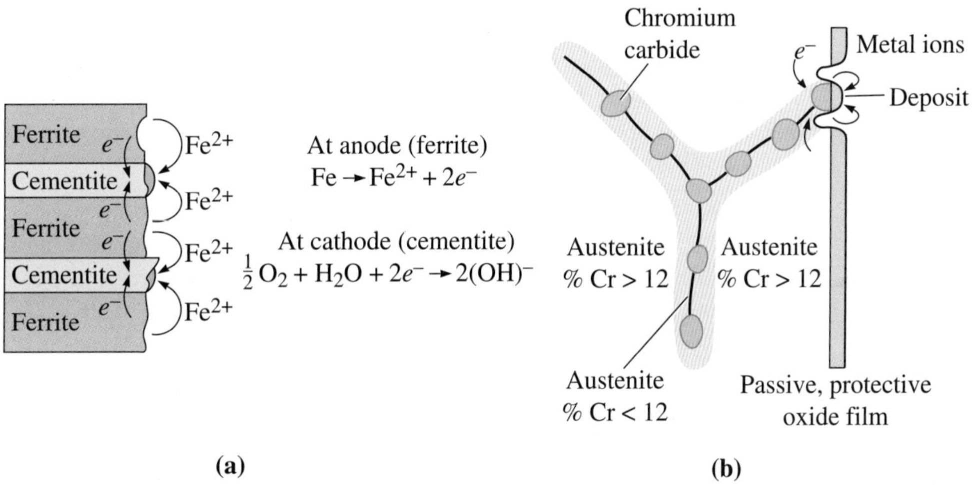

Figure 23-6 Example of microgalvanic cells in two-phase alloys: (a) In steel, ferrite is anodic to cementite. (b) In austenitic stainless steel, precipitation of chromium carbide makes the low Cr austenite in the grain boundaries anodic.

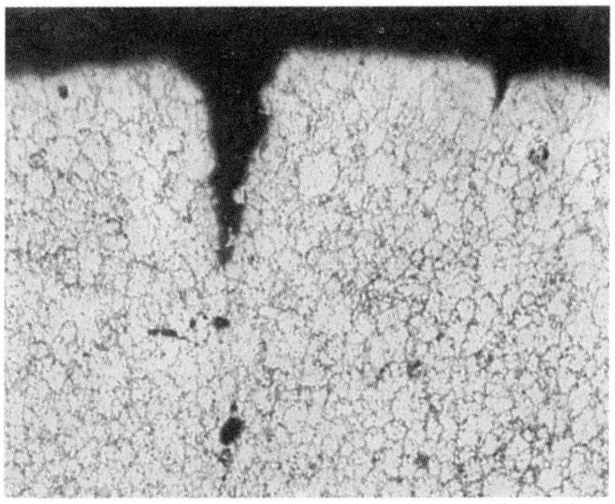

Figure 23-7 Micrograph of intergranular corrosion in a zinc die casting. Segregation of impurities to the grain boundaries produces microgalvanic corrosion cells (× 50). *(Reprinted Courtesy of Don Askeland)*

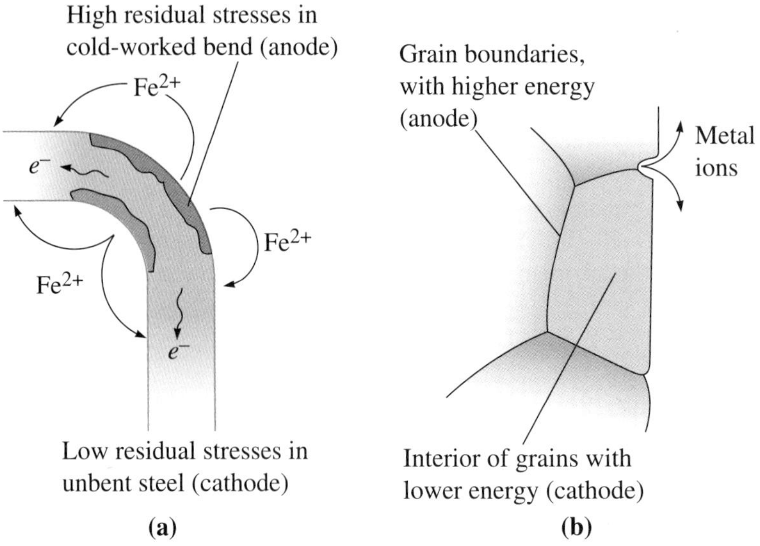

Figure 23-8 Examples of stress cells. (a) Cold work required to bend a steel bar introduces high residual stresses at the bend, which then is anodic and corrodes. (b) Because grain boundaries have high energy, they are anodic and corrode.

stainless steels, chromium carbides can precipitate at grain boundaries [Figure 23-6(b)]. The formation of the carbides removes chromium from the austenite adjacent to the boundaries. The low-chromium ($<12\%$ Cr) austenite at the grain boundaries is anodic to the remainder of the grain and corrosion occurs at the grain boundaries. In certain cold-worked aluminum alloys, grain boundaries corrode rapidly due to the presence of detrimental precipitates. This causes the grains of aluminum to peel back like the pages of a book or leaves. This is known as exfoliation.

Stress Cells Stress cells develop when a metal contains regions with different local stresses. The most highly stressed or high-energy regions act as anodes to the less-stressed cathodic areas (Figure 23-8). Regions with a finer grain size, or a higher density of grain boundaries, are anodic to coarse-grained regions of the same material. Highly cold-worked areas are anodic to less cold-worked areas.

Stress corrosion occurs by galvanic action, but other mechanisms, such as the adsorption of impurities at the tip of an existing crack, may also occur. Failure occurs as a result of corrosion and an applied stress. Higher applied stresses reduce the time required for failure.

Fatigue failures are also initiated or accelerated when corrosion occurs. *Corrosion fatigue* can reduce fatigue properties by initiating cracks (perhaps by producing pits or crevices) and by increasing the rate at which the cracks propagate.

Example 23-6 *Corrosion of Cold-Drawn Steel*

A cold-drawn steel wire is formed into a nail by additional deformation, producing the point at one end and the head at the other. Where will the most severe corrosion of the nail occur?

SOLUTION

Since the head and point have been cold-worked an additional amount compared with the shank of the nail, the head and point serve as anodes and corrode most rapidly.

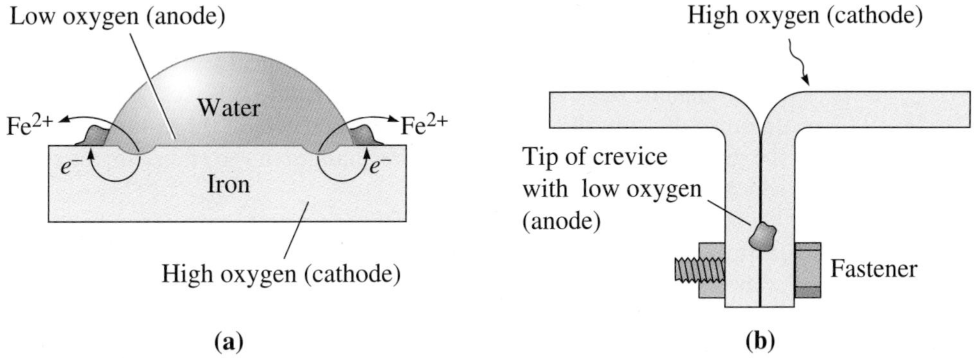

Figure 23-9 Concentration cells: (a) Corrosion occurs beneath a water droplet on a steel plate due to low oxygen concentration in the water. (b) Corrosion occurs at the tip of a crevice because of limited access to oxygen.

Concentration Cells Concentration cells develop due to differences in the concentration of the electrolyte (Figure 23-9). According to the Nernst equation, a difference in metal ion concentration causes a difference in electrode potential. The metal in contact with the most concentrated solution is the cathode; the metal in contact with the dilute solution is the anode.

The oxygen concentration cell (often referred to as **oxygen starvation**) occurs when the cathode reaction is the oxygen electrode, $H_2O + \frac{1}{2}O_2 + 2e^- \rightarrow 2(OH)^-$. Electrons flow from the low oxygen region, which serves as the anode, to the high oxygen region, which serves as the cathode.

Deposits, such as rust or water droplets, shield the underlying metal from oxygen. Consequently, the metal under the deposit is the anode and corrodes. This causes one form of pitting corrosion. Waterline corrosion is similar. Metal above the waterline is exposed to oxygen, while metal beneath the waterline is deprived of oxygen; hence, the metal under water corrodes. Normally, the metal far below the surface corrodes more slowly than metal just below the waterline due to differences in the distance that electrons must travel. Because cracks and crevices have a lower oxygen concentration than the surrounding base metal, the tip of a crack or crevice is the anode, causing **crevice corrosion.**

Pipe buried in soil may corrode because of differences in the composition of the soil. Velocity differences may cause concentration differences. Stagnant water contains low oxygen concentrations, whereas fast-moving, aerated water contains higher oxygen concentrations. Metal near stagnant water is anodic and corrodes. The following is an example of corrosion due to water.

Example 23-7 *Corrosion of Crimped Steel*

Two pieces of steel are joined mechanically by crimping the edges. Why would this be a bad idea if the steel is then exposed to water? If the water contains salt, would corrosion be affected?

SOLUTION

By crimping the steel edges, we produce a crevice. The region in the crevice is exposed to less air and moisture, so it behaves as the anode in a concentration cell. The steel in the crevice corrodes.

Salt in the water increases the conductivity of the water, permitting electrical charge to be transferred at a more rapid rate. This causes a higher current density and, thus, faster corrosion due to less resistance polarization.

Microbial Corrosion Various microbes, such as fungi and bacteria, create conditions that encourage electrochemical corrosion. Particularly in aqueous environments, these organisms grow on metallic surfaces. The organisms typically form colonies that are not continuous. The presence of the colonies and the byproducts of the growth of the organisms produce changes in the environment and, hence, the rate at which corrosion occurs.

Some bacteria reduce sulfates in the environment, producing sulfuric acid, which in turn attacks metal. The bacteria may be either aerobic (which thrive when oxygen is available) or anaerobic (which do not need oxygen to grow). Such bacteria cause attacks on a variety of metals, including steels, stainless steels, aluminum, and copper, as well as some ceramics and concrete. A common example occurs in aluminum fuel tanks for aircraft. When the fuel, typically kerosene, is contaminated with moisture, bacteria grow and excrete acids. The acids attack the aluminum, eventually causing the fuel tank to leak.

The growth of colonies of organisms on a metal surface leads to the development of oxygen concentration cells (Figure 23-10). Areas beneath the colonies are anodic, whereas unaffected areas are cathodic. In addition, the colonies of organisms reduce the rate of diffusion of oxygen to the metal, and—even if the oxygen does diffuse into the colony—the organisms tend to consume the oxygen. The concentration cell produces pitting beneath the regions covered with the organisms. Growth of the organisms, which may include products of the corrosion of the metal, produces accumulations (called **tubercules**) that may plug pipes or clog water-cooling systems in nuclear reactors, submarines, or chemical reactors.

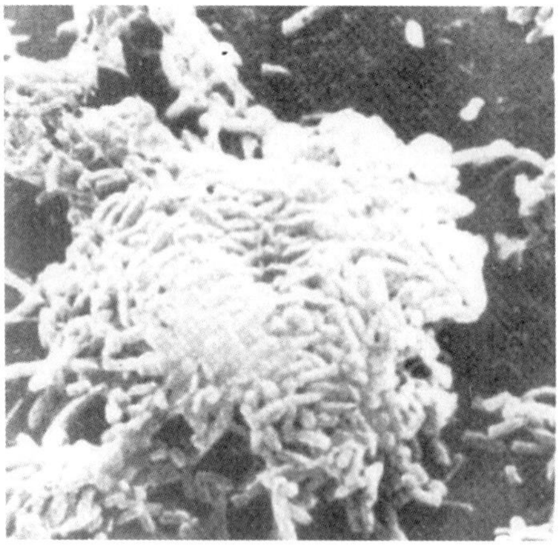

(a)

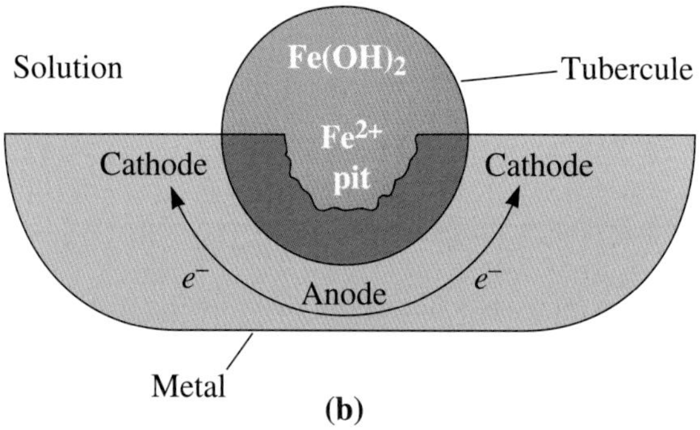

(b)

Figure 23-10 (a) Bacterial cells growing in a colony (× 2700). (b) Formation of a tubercule and a pit under a biological colony. *(Reprinted Courtesy of Don Askeland)*

23-6 Protection Against Electrochemical Corrosion

A number of techniques are used to combat corrosion, including proper design and materials selection and the use of coatings, inhibitors, cathodic protection, and passivation.

Design Proper design of metal structures can slow or even avoid corrosion. Some of the steps that should be taken to combat corrosion are as follows:

1. Prevent the formation of galvanic cells. This can be achieved by using similar metals or alloys. For example, steel pipe is frequently connected to brass plumbing fixtures, producing a galvanic cell that causes the steel to corrode. By using intermediate plastic fittings to electrically insulate the steel and brass, this problem can be minimized.
2. Make the anode area much larger than the cathode area. For example, copper rivets can be used to fasten steel sheet. Because of the small area of the copper rivets, a limited cathode reaction occurs. The copper accepts few electrons, and the steel anode reaction proceeds slowly. If, on the other hand, steel rivets are used for joining copper sheet, the small steel anode area gives up many electrons, which are accepted by the large copper cathode area; corrosion of the steel rivets is then very rapid. This is illustrated in the following example.
3. Design components so that fluid systems are closed, rather than open, and so that stagnant pools of liquid do not collect. Partly filled tanks undergo waterline corrosion. Open systems continuously dissolve gas, providing ions that participate in the cathode reaction, and encourage concentration cells.
4. Avoid crevices between assembled or joined materials (Figure 23-11). Welding may be a better joining technique than brazing, soldering, or mechanical fastening. Galvanic cells develop in brazing or soldering since the filler metals have a different composition from the metal being joined. Mechanical fasteners produce crevices that lead to concentration cells. If the filler metal is closely matched to the base metal, welding may prevent these cells from developing.

Example 23-8 *Effect of Areas on Corrosion Rate for Copper-Zinc Couple*

Consider a copper-zinc corrosion couple. If the current density at the copper cathode is 0.05 A/cm^2, calculate the weight loss of zinc per hour if (1) the copper cathode area is 100 cm^2 and the zinc anode area is 1 cm^2 and (2) the copper cathode area is 1 cm^2 and the zinc anode area is 100 cm^2.

SOLUTION

From Example 23-4 for the small zinc anode area, the amount of zinc lost at the anode in one hour is

$$w_{Zn} = 6.1 \text{ g}$$

For the large zinc anode area,

$$I = i_{Cu}A_{Cu} = (0.05\,\text{A/cm}^2)(1\,\text{cm}^2) = 0.05 \text{ A}$$

$$w_{Zn} = \frac{ItM}{nF} = \frac{(0.05\,\text{A})(3600\,\text{s})(65.38\text{ g/mol})}{(2)(96{,}500\,\text{C/mol})} = 0.061 \text{ g}$$

The rate of corrosion of the zinc is reduced significantly when the zinc anode is much larger than the cathode.

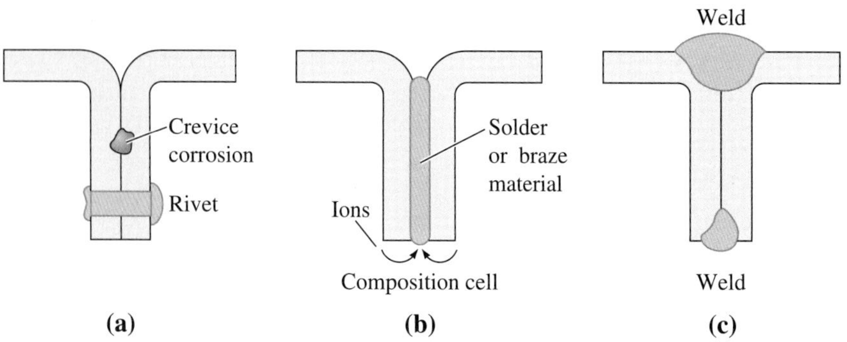

Figure 23-11 Alternative methods for joining two pieces of steel: (a) Fasteners may produce a concentration cell, (b) brazing or soldering may produce a composition cell, and (c) welding with a filler metal that matches the base metal may avoid the formation of galvanic cells.

5. In some cases, the rate of corrosion cannot be reduced to a level that will not interfere with the expected lifetime of the component. In such cases, the assembly should be designed in such a manner that the corroded part can be easily and economically replaced.

Coatings Coatings are used to isolate the anode and cathode regions. Coatings also prevent diffusion of oxygen or water vapor that initiates corrosion or oxidation. Temporary coatings, such as grease or oil, provide some protection but are easily disrupted. Organic coatings, such as paint, or ceramic coatings, such as enamel or glass, provide better protection; however, if the coating is disrupted, a small anodic site is exposed that undergoes rapid, localized corrosion.

Metallic coatings include tin-plated and hot-dip galvanized (zinc-plated) steel (Figure 23-12). A continuous coating of either metal isolates the steel from the electrolyte; however, when the coating is scratched, exposing the underlying steel, the zinc continues to be effective, because zinc is anodic to steel. Since the area of the exposed steel cathode is small, the zinc coating corrodes at a very slow rate and the steel remains protected. In contrast, steel is anodic to tin, so a small steel anode is created when the tin is scratched, and rapid corrosion of the steel subsequently occurs.

Stable, adherent, nonporous, non-conducting oxide layers form on the surface of aluminum, chromium, and stainless steel. These oxides exclude the electrolyte and prevent the formation of galvanic cells. Components such as reaction vessels can also be lined with corrosion-resistant Teflon^{TM} or other plastics.

Inhibitors When added to the electrolyte, some chemicals migrate preferentially to the anode or cathode surface and produce concentration or resistance polarization—that is, they are **inhibitors**. Chromate salts perform this function in automobile radiators.

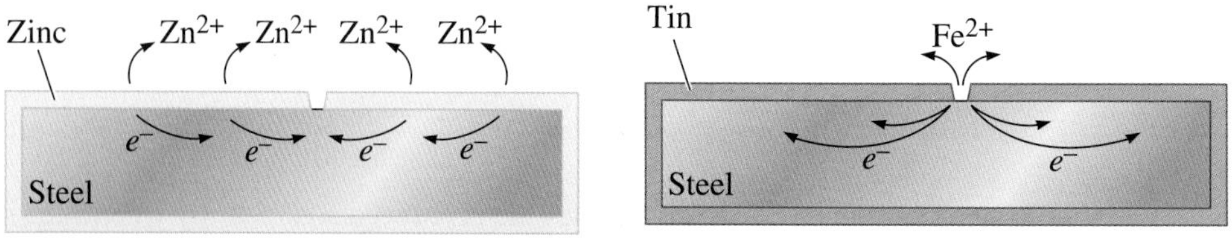

Figure 23-12 Zinc-plated steel and tin-plated steel are protected differently. Zinc protects steel even when the coating is scratched since zinc is anodic to steel. Tin does not protect steel when the coating is disrupted since steel is anodic to tin.

A variety of chromates, phosphates, molybdates, and nitrites produce protective films on anodes or cathodes in power plants and heat exchangers, thus stifling the electrochemical cell. Although the exact contents and the mechanism by which it functions are not clear, the popular lubricant WD-40^{TM} works from the action of inhibitors. The name "WD-40^{TM}" was designated water displacement (WD) that worked on the fortieth experimental attempt!

Cathodic Protection We can protect against corrosion by supplying the metal with electrons and forcing the metal to be a cathode (Figure 23-13). Cathodic protection can use a sacrificial anode or an impressed voltage.

A **sacrificial anode** is attached to the material to be protected, forming an electrochemical circuit. The sacrificial anode corrodes, supplies electrons to the metal, and thereby prevents an anode reaction at the metal. The sacrificial anode, typically zinc or magnesium, is consumed and must eventually be replaced. Applications include preventing the corrosion of buried pipelines, ships, off-shore drilling platforms, and water heaters. A magnesium rod is used in many water heaters. The Mg serves as an anode and undergoes dissolution, thus protecting the steel from corroding.

An **impressed voltage** is obtained from a direct current source connected between an auxiliary anode and the metal to be protected. Essentially, we have connected a battery so that electrons flow to the metal, causing the metal to be the cathode. The auxiliary anode, such as scrap iron, corrodes.

Passivation or Anodic Protection Metals near the anodic end of the galvanic series are active and serve as anodes in most electrolytic cells; however, if these metals are made passive or more cathodic, they corrode at slower rates than normal. **Passivation** is accomplished by producing strong anodic polarization, preventing the normal anode reaction; thus the term anodic protection.

We cause passivation by exposing the metal to highly concentrated oxidizing solutions. If iron is dipped in very concentrated nitric acid, the iron rapidly and uniformly corrodes to form a thin, protective iron hydroxide coating. The coating protects the iron from subsequent corrosion in nitric acid.

We can also cause passivation by increasing the potential on the anode above a critical level. A passive film forms on the metal surface, causing strong anodic polarization, and the current decreases to a very low level. Passivation of aluminum is called **anodizing**, and a thick oxide coating is produced. This oxide layer can be dyed to produce attractive colors. The Ta_2O_5 oxide layer formed on tantalum wires is used to make capacitors.

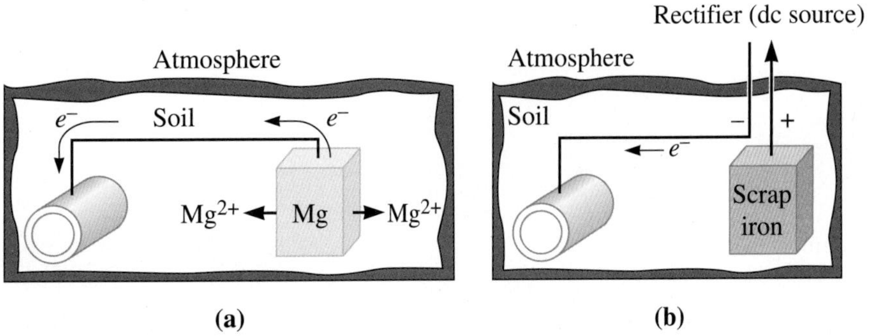

Figure 23-13 Cathodic protection of a buried steel pipeline: (a) A sacrificial magnesium anode ensures that the galvanic cell makes the pipeline the cathode. (b) An impressed voltage between a scrap iron auxiliary anode and the pipeline ensures that the pipeline is the cathode.

Materials Selection and Treatment Corrosion can be prevented or minimized by selecting appropriate materials and heat treatments. In castings, for example, segregation causes tiny, localized galvanic cells that accelerate corrosion. We can improve corrosion resistance with a homogenization heat treatment. When metals are formed into finished shapes by bending, differences in the amount of cold work and residual stresses cause local stress cells. These may be minimized by a stress-relief anneal or a full recrystallization anneal.

The heat treatment is particularly important in austenitic stainless steels (Figure 23-14). When the steel cools slowly from 870 to 425°C, chromium carbides precipitate at the grain boundaries. Consequently the austenite at the grain boundaries may contain less than 12% chromium, which is the minimum required to produce a passive oxide layer. The steel is **sensitized**. Because the grain boundary regions are small and highly anodic, rapid corrosion of the austenite at the grain boundaries occurs. We can minimize the problem by several techniques.

1. If the steel contains less than 0.03% C, chromium carbides do not form.
2. If the percent chromium is very high, the austenite may not be depleted to below 12% Cr, even if chromium carbides form.
3. Addition of titanium or niobium ties up the carbon as TiC or NbC, preventing the formation of chromium carbide. The steel is said to be **stabilized**.
4. The sensitization temperature range—425 to 870°C—should be avoided during manufacturing and service.

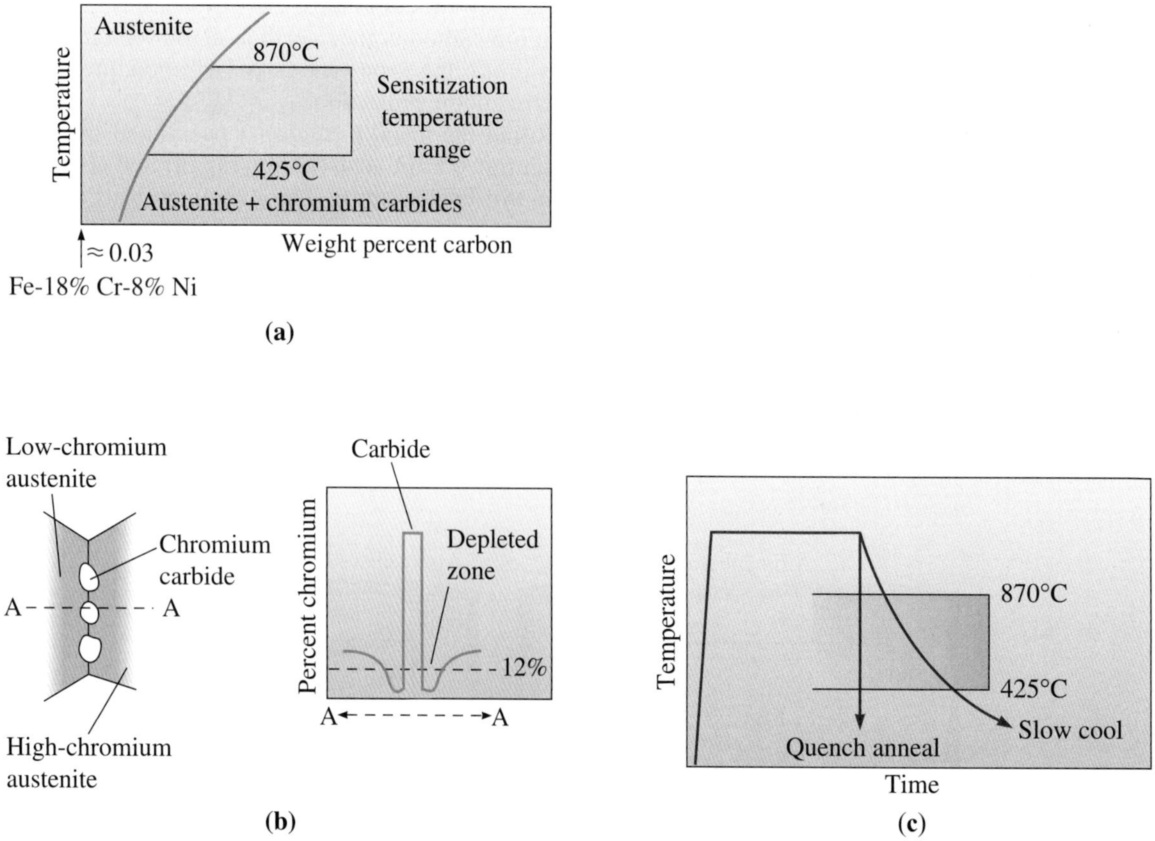

Figure 23-14 (a) Austenitic stainless steels become sensitized when cooled slowly in the temperature range between 870 and 425°C. (b) Slow cooling permits chromium carbides to precipitate at grain boundaries. The local chromium content is depleted. (c) A quench anneal to dissolve the carbides may prevent intergranular corrosion.

5. In a **quench anneal** heat treatment, the stainless steel is heated above 870°C, causing the chromium carbides to dissolve. The structure, now containing 100% austenite, is rapidly quenched to prevent formation of carbides.

Example 23-9 *Design of a Corrosion Protection System*

Steel troughs are located in a field to provide drinking water for a herd of cattle. The troughs frequently rust through and must be replaced. Design a system to prevent or delay this problem.

SOLUTION

The troughs are likely a low-carbon, unalloyed steel containing ferrite and cementite, producing a composition cell. The waterline in the trough, which is partially filled with water, provides a concentration cell. The trough is also exposed to the environment, and the water is contaminated with impurities. Consequently, corrosion of the unprotected steel tank is to be expected.

Several approaches might be used to prevent or delay corrosion. We might, for example, fabricate the trough using stainless steel or aluminum. Either would provide better corrosion resistance than the plain carbon steel, but both are considerably more expensive than the current material.

We might suggest using cathodic protection; a small magnesium anode could be attached to the inside of the trough. The anode corrodes sacrificially and prevents corrosion of the steel. This would require that the farm operator regularly check the tank to be sure that the anode is not completely consumed. We also want to be sure that magnesium ions introduced into the water are not a health hazard.

Another approach would be to protect the steel trough using a suitable coating. Painting the steel (i.e., introducing a protective polymer coating) or using a tin-plated steel provides protection as long as the coating is not disrupted.

The most likely approach is to use a galvanized steel, taking advantage of the protective coating and the sacrificial behavior of the zinc. Corrosion is very slow due to the large anode area, even if the coating is disrupted. Furthermore, galvanized steel is relatively inexpensive, readily available, and does not require frequent inspection.

Example 23-10 *Design of a Stainless Steel Weldment*

A piping system used to transport a corrosive liquid is fabricated from 304 stainless steel. Welding of the pipes is required to assemble the system. Unfortunately, corrosion occurs, and the corrosive liquid leaks from the pipes near the weld. Identify the problem and design a system to prevent corrosion in the future.

SOLUTION

Table 13-4 shows that 304 stainless steel contains 0.08% C, causing the steel to be sensitized if it is improperly heated or cooled during welding. Figure 23-15 shows the maximum temperatures reached in the fusion and heat-affected zones during welding. A portion of the pipe in the HAZ heats into the sensitization temperature range, permitting chromium carbides to precipitate. If the cooling rate of the weld is very slow, the fusion zone and other areas of the heat-affected zone may also be affected. Sensitization of the weld area, therefore, is the likely reason for corrosion of the pipe.

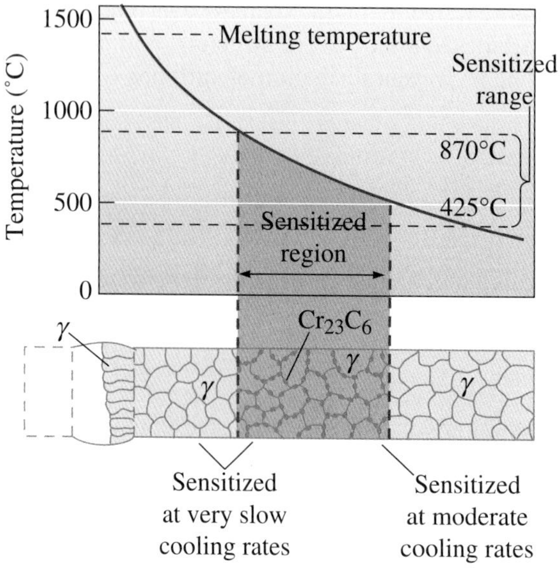

Figure 23-15 The peak temperature surrounding a stainless steel weld and the sensitized structure produced when the weld slowly cools (for Example 23-10).

Several solutions to the problem may be considered. We might use a welding process that provides very rapid rates of heat input, causing the weld to heat and cool very quickly. If the steel is exposed to the sensitization temperature range for only a brief time, chromium carbides may not precipitate. Joining processes such as laser welding or electron-beam welding are high-rate-of-heat-input processes, but they are expensive. In addition, electron beam welding requires the use of a vacuum, and it may not be feasible to assemble the piping system in a vacuum chamber.

We might heat treat the assembly after the weld is made. By performing a quench anneal, any precipitated carbides are re-dissolved during the anneal and do not reform during quenching; however, it may be impossible to perform this treatment on a large assembly.

We might check the original welding procedure to determine if the pipe was preheated before joining in order to minimize the development of stresses due to the welding process. If the pipe were preheated, sensitization would be more likely to occur. We would recommend that any preheat procedure be suspended.

Perhaps our best design is to use a stainless steel that is not subject to sensitization. For example, carbides do not precipitate in a 304L stainless steel, which contains less than 0.03% C. The low-carbon stainless steels are more expensive than the normal 304 steel; however, the extra cost does prevent corrosion and still permits us to use conventional joining techniques.

23-7 Microbial Degradation and Biodegradable Polymers

Attack by a variety of insects and microbes is one form of “corrosion” of polymers. Relatively simple polymers (such as polyethylene, polypropylene, and polystyrene), high-molecular-weight polymers, crystalline polymers, and thermosets are relatively immune to attack.

Some polymers—including polyesters, polyurethanes, cellulosics, and plasticized polyvinyl chloride (which contains additives that reduce the degree of polymerization)—are particularly vulnerable to microbial degradation. These polymers can be broken into low-molecular-weight molecules by radiation or chemical attack until they are small enough to be ingested by the microbes.

We take advantage of microbial attack by producing *biodegradable* polymers, thus helping to remove the material from the waste stream. Biodegradation requires the complete conversion of the polymer to carbon dioxide, water, inorganic salts, and other small byproducts produced by the ingestion of the material by bacteria. Polymers such as cellulosics easily can be broken into molecules with low molecular weights and are therefore biodegradable. In addition, special polymers are produced to degrade rapidly; a copolymer of polyethylene and starch is one example. Bacteria attack the starch portion of the polymer and reduce the molecular weight of the remaining polyethylene. Certain other polymers such as polycaprolactone (PCL), polylactic acid (PLA), and polylactic glycolic acid (PLGA) are useful in a number of biomedical applications such as sutures and scaffolds.

23-8 Oxidation and Other Gas Reactions

Materials of all types may react with oxygen and other gases. These reactions can, like corrosion, alter the composition, properties, or integrity of a material. As mentioned before, metals such as Al and Ti react with oxygen very readily.

Oxidation of Metals

Metals may react with oxygen to produce an oxide at the surface. We are interested in three aspects of this reaction: the ease with which the metal oxidizes, the nature of the oxide film that forms, and the rate at which **oxidation** occurs.

The ease with which oxidation occurs is given by the standard free energy of formation for the oxide. The standard free energy of formation for an oxide can be determined from an Ellingham diagram, such as the one shown in Figure 23-16. Figure 23-16 shows that there is a large driving force for the oxidation of magnesium and aluminum compared to the oxidation of nickel or copper. This is illustrated in the following example.

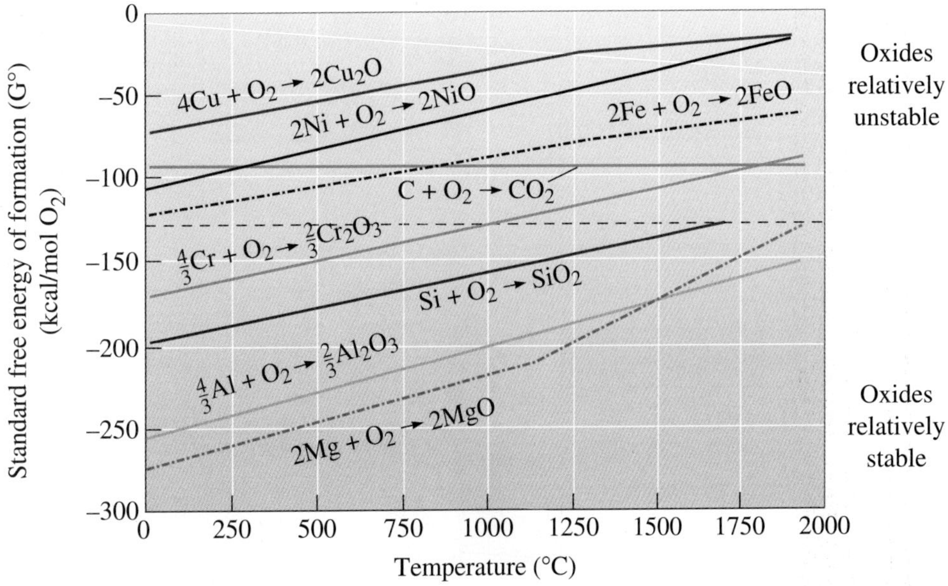

Figure 23-16 The standard free energy of formation for selected oxides as a function of temperature. A large negative free energy indicates a more stable oxide.

Example 23-11 Chromium-Based Steel Alloys

Explain why we should not add alloying elements such as chromium to pig iron before the pig iron is converted to steel in a basic oxygen furnace at 1700°C.

SOLUTION

In a basic oxygen furnace, we lower the carbon content of the metal from about 4% to much less than 1% by blowing pure oxygen through the molten metal. If chromium was already present before the steel making began, chromium would oxidize before the carbon (Figure 23-16), since chromium oxide has a lower free energy of formation (or is more stable) than carbon dioxide (CO_2). Thus, any expensive chromium added would be lost before the carbon was removed from the pig iron.

The type of oxide film influences the rate at which oxidation occurs (Figure 23-17). For the **oxidation reaction**

$$nM + mO_2 \rightarrow M_nO_{2m} \tag{23-12}$$

the **Pilling-Bedworth** (P-B) **ratio** is

$$\text{P-B ratio} = \frac{\text{oxide volume per metal atom}}{\text{metal volume per metal atom}} = \frac{(M_{\text{oxide}})(\rho_{\text{metal}})}{n(M_{\text{metal}})(\rho_{\text{oxide}})} \tag{23-13}$$

where M is the atomic or molecular mass, ρ is the density, and n is the number of metal atoms in the oxide, as defined in Equation 23-12.

If the Pilling-Bedworth ratio is less than one, the oxide occupies a smaller volume than the metal from which it formed; the coating is therefore porous and oxidation continues rapidly—typical of metals such as magnesium. If the ratio is one

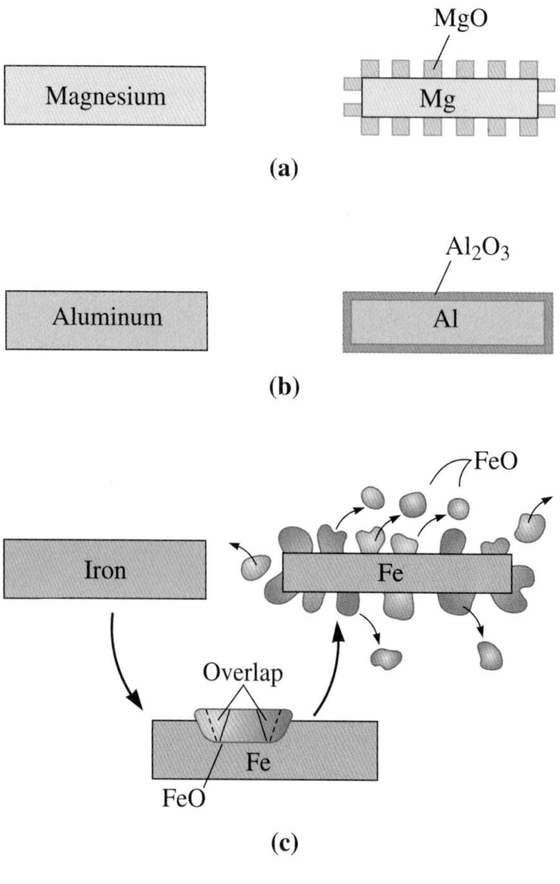

Figure 23-17 Three types of oxides may form, depending on the volume ratio between the metal and the oxide: (a) magnesium produces a porous oxide film, (b) aluminum forms a protective, adherent, nonporous oxide film, and (c) iron forms an oxide film that spalls off the surface and provides poor protection.

to two, the volumes of the oxide and metal are similar and an adherent, non-porous, protective film forms—typical of aluminum and titanium. If the ratio exceeds two, the oxide occupies a large volume and may flake from the surface, exposing fresh metal that continues to oxidize—typical of iron. Although the Pilling-Bedworth equation historically has been used to characterize oxide behavior, many exceptions to this behavior are observed. The use of the P-B ratio equation is illustrated in the following example.

Example 23-12 *Pilling-Bedworth Ratio*

The density of aluminum is 2.7 g/cm^3 and that of Al_2O_3 is about 4 g/cm^3. Describe the characteristics of the aluminum-oxide film. Compare these with the oxide film that forms on tungsten. The density of tungsten is 19.254 g/cm^3 and that of WO_3 is 7.3 g/cm^3.

SOLUTION

For $2Al + 3/2O_2 \rightarrow Al_2O_3$, the molecular weight of Al_2O_3 is 101.96 g/mol and that of aluminum is 26.981 g/mol.

$$\text{P-B} = \frac{M_{Al_2O_3}\rho_{Al}}{nM_{Al}\rho_{Al_2O_3}} = \frac{\left(101.96\,\frac{\text{g}}{\text{mol}}\right)(2.7\text{ g/cm}^3)}{(2)\left(26.981\,\frac{\text{g}}{\text{mol}}\right)(4\text{ g/cm}^3)} = 1.28$$

For tungsten, $W + 3/2O_2 \rightarrow WO_3$. The molecular weight of WO_3 is 231.85 g/mol and that of tungsten is 183.85 g/mol.

$$\text{P-B} = \frac{M_{WO_3}\rho_{W}}{nM_{W}\rho_{WO_3}} = \frac{\left(231.85\,\frac{\text{g}}{\text{mol}}\right)(19.254\text{ g/cm}^3)}{(1)\left(183.85\,\frac{\text{g}}{\text{mol}}\right)(7.3\text{ g/cm}^3)} = 3.33$$

Since P-B $\simeq$ 1 for aluminum, the Al_2O_3 film is nonporous and adherent, providing protection to the underlying aluminum. Since P-B >2 for tungsten, the WO_3 should be nonadherent and nonprotective.

The rate at which oxidation occurs depends on the access of oxygen to the metal atoms. A linear rate of oxidation occurs when the oxide is porous (as in magnesium) and oxygen has continued access to the metal surface:

$$y = \text{k}t \tag{23-14}$$

where y is the thickness of the oxide, t is the time, and k is a constant that depends on the metal and temperature.

A parabolic relationship is observed when diffusion of ions or electrons through a nonporous oxide layer is the controlling factor. This relationship is observed in iron, copper, and nickel:

$$y = \sqrt{\text{k}t} \tag{23-15}$$

Finally, a logarithmic relationship is observed for the growth of thin-oxide films that are particularly protective, as for aluminum and possibly chromium:

$$y = \text{k}\ln(\text{c}t + 1) \tag{23-16}$$

where k and c are constants for a particular temperature, environment, and composition. The example that follows shows the calculation for the time required for a nickel sheet to oxidize completely.

Example 23-13 *Parabolic Oxidation Curve for Nickel*

At 1000°C, pure nickel follows a parabolic oxidation curve given by the constant $k = 3.9 \times 10^{-12}$ cm²/s in an oxygen atmosphere. If this relationship is not affected by the thickness of the oxide film, calculate the time required for a 0.1-cm nickel sheet to oxidize completely.

SOLUTION

Assuming that the sheet oxidizes from both sides:

$$y = \sqrt{kt} = \sqrt{\left(3.9 \times 10^{-12}\frac{\text{cm}^2}{\text{s}}\right)(t)} = \frac{0.1\text{ cm}}{2\text{ sides}} = 0.05\text{ cm}$$

$$t = \frac{(0.05\,\text{cm})^2}{3.9 \times 10^{-12}\,\text{cm}^2/\text{s}} = 6.4 \times 10^8\,\text{s} = 20.3\,\text{years}$$

Temperature also affects the rate of oxidation. In many metals, the rate of oxidation is controlled by the rate of diffusion of oxygen or metal ions through the oxide. If oxygen diffusion is more rapid, oxidation occurs between the oxide and the metal; if the metal ion diffusion is more rapid, oxidation occurs at the oxide-atmosphere interface. Consequently, we expect oxidation rates to follow an Arrhenius relationship, increasing exponentially as the temperature increases.

Oxidation and Thermal Degradation of Polymers Polymers degrade when heated and/or exposed to oxygen. A polymer chain may be ruptured, producing two macroradicals. In rigid thermosets, the macroradicals may instantly recombine (a process called the *cage* effect), resulting in no net change in the polymer. In the more flexible thermoplastics—particularly for amorphous rather than crystalline polymers—recombination does not occur, and the result is a decrease in the molecular weight, viscosity, and mechanical properties of the polymer. Depolymerization continues as the polymer is exposed to high temperatures. Polymer chains can also *unzip*. In this case, individual monomers are removed one after another from the ends of the chain, gradually reducing the molecular weight of the remaining chains. As the degree of polymerization decreases, the remaining chains become more heavily branched or cyclization may occur. In *cyclization*, the two ends of the same chain may be bonded together to form a ring.

Polymers also degrade by the loss of side groups on the chain. Chloride ions [in polyvinyl chloride (PVC)] and benzene rings (in polystyrene) are lost from the chain, forming byproducts. For example, as polyvinyl chloride is degraded, hydrochloric acid (HCl) is produced. Hydrogen atoms are bonded more strongly to the chains; thus, polyethylene does not degrade as easily as PVC or polystyrene. Fluoride ions (in Teflon™) are more difficult to remove than hydrogen atoms, providing Teflon™ with its high temperature resistance. As mentioned before, degradation of polymers, such as PCL, PLGA, and PLA, actually can be useful for certain biomedical applications (e.g., sutures that dissolve) and also in the development of environmentally friendly products.

23-9 Wear and Erosion

Wear and erosion remove material from a component by mechanical attack of solids or liquids. Corrosion and mechanical failure also contribute to this type of attack.

Adhesive Wear **Adhesive wear**—also known as scoring, galling, or seizing—occurs when two solid surfaces slide over one another under pressure. Surface projections, or asperities, are plastically deformed and eventually welded together by the high local

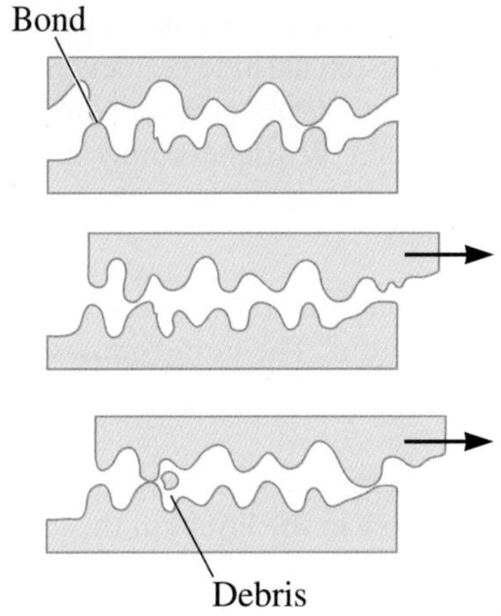

Figure 23-18 The asperities on two rough surfaces may initially be bonded. A sufficient force breaks the bonds, and the surfaces slide. As they slide, asperities may be fractured, wearing away the surfaces and producing debris.

pressures (Figure 23-18). As sliding continues, these bonds are broken, producing cavities on one surface, projections on the second surface, and frequently tiny, abrasive particles—all of which contribute to further wear of the surfaces.

Many factors may be considered in trying to improve the wear resistance of materials. Designing components so that loads are small, surfaces are smooth, and continual lubrication is possible helps prevent adhesions that cause the loss of material.

The properties and microstructure of the material are also important. Normally, if both surfaces have high hardnesses, the wear rate is low. High strength, to help resist the applied loads, and good toughness and ductility, which prevent the tearing of material from the surface, may be beneficial. Ceramic materials, with their exceptional hardness, are expected to provide good adhesive wear resistance.

Wear resistance of polymers can be improved if the coefficient of friction is reduced by the addition of polytetrafluoroethylene (Teflon^{TM}) or if the polymer is strengthened by the introduction of reinforcing fibers such as glass, carbon, or aramid.

Abrasive Wear When material is removed from a surface by contact with hard particles, **abrasive wear** occurs. The particles either may be present at the surface of a second material or may exist as loose particles between two surfaces (Figure 23-19). This type of wear is common in machinery such as plows, scraper blades, crushers, and grinders used to handle abrasive materials and may also occur when hard particles are unintentionally introduced into moving parts of machinery. Abrasive wear is also used for grinding operations to remove material intentionally. In many automotive applications (e.g., dampers, gears, pistons, and cylinders), abrasive wear behavior is a major concern.

Materials with a high hardness, good toughness, and high hot strength are most resistant to abrasive wear. Typical materials used for abrasive-wear applications include

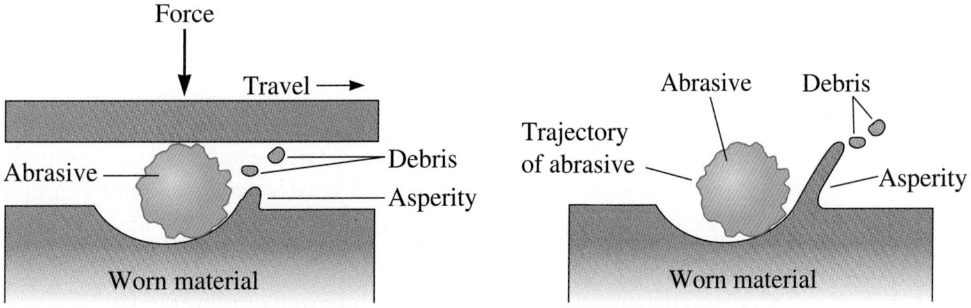

Figure 23-19 Abrasive wear, caused by either trapped or free-flying abrasives, produces troughs in the material, piling up asperities that may fracture into debris.

quenched and tempered steels; carburized or surface-hardened steels; cobalt alloys such as Stellite; composite materials, including tungsten carbide cermets; white cast irons; and hard surfaces produced by welding. Most ceramic materials also resist wear effectively because of their high hardness; however, their brittleness may sometimes limit their usefulness in abrasive wear conditions.

Liquid Erosion The integrity of a material may be destroyed by erosion caused by high pressures associated with a moving liquid. The liquid causes strain hardening of the metal surface, leading to localized deformation, cracking, and loss of material. Two types of liquid erosion deserve mention.

Cavitation occurs when a liquid containing a dissolved gas enters a low pressure region. Gas bubbles, which precipitate and grow in the liquid in the low pressure environment, collapse when the pressure subsequently increases. The high-pressure, local shock wave that is produced by the collapse may exert a pressure of thousands of atmospheres against the surrounding material. Cavitation is frequently encountered in propellers, dams and spillways, and hydraulic pumps.

Liquid impingement occurs when liquid droplets carried in a rapidly moving gas strike a metal surface. High localized pressures develop because of the initial impact and the rapid lateral movement of the droplets from the impact point along the metal surface. Water droplets carried by steam may erode turbine blades in steam generators and nuclear power plants.

Liquid erosion can be minimized by proper materials selection and design. Minimizing the liquid velocity, ensuring that the liquid is deaerated, selection of hard, tough materials to absorb the impact of the droplets, and coating the material with an energy-absorbing elastomer all may help minimize erosion.

Summary

- About 3% of the gross domestic product in the United States is used to combat corrosion. Corrosion causes deterioration of all types of materials. Designers and engineers must know how corrosion occurs in order to consider suitable designs, materials selection, or protective measures.
- In chemical corrosion, a material is dissolved in a solvent, resulting in the loss of material. All materials—metals, ceramics, polymers, and composites—are subject to this form of attack. The choice of appropriate materials having a low solubility in a given solvent or use of inert coatings on materials helps avoid or reduce chemical corrosion.
- Electrochemical corrosion requires that a complete electric circuit develop. The anode corrodes and a byproduct forms at the cathode. Because an electric circuit is required, this form of corrosion is most serious in metals and alloys.
- Composition cells are formed by the presence of two different metals, two different phases within a single alloy, or even segregation within a single phase.
- Stress cells form when the level of residual or applied stresses varies within the metal; the regions subjected to the highest stress are anodic and, consequently, they corrode. Stress corrosion cracking and corrosion fatigue are examples of stress cells.
- Concentration cells form when a metal is exposed to a nonuniform electrolyte; for example, the portion of a metal exposed to the lowest oxygen content corrodes. Pitting corrosion, waterline corrosion, and crevice corrosion are examples of these cells. Microbial corrosion, in which colonies of organisms such as bacteria grow on the metal surface, is another example of a concentration cell.
- Electrochemical corrosion can be minimized or prevented by using electrical insulators to break the electric circuit, designing and manufacturing assemblies without crevices,

designing assemblies so that the anode area is large compared to that of the cathode, using protective and even sacrificial coatings, inhibiting the action of the electrolyte, supplying electrons to the metal by means of an impressed voltage, using heat treatments that reduce residual stresses or segregation, and a host of other actions.

- Oxidation degrades most materials. While an oxide coating provides protection for some metals such as aluminum, most materials are attacked by oxygen. Diffusion of oxygen and metallic atoms is often important; therefore oxidation occurs most rapidly at elevated temperatures.
- Other factors, such as attack by microbes, damage caused by radiation, and wear or erosion of a material, may also cause a material to deteriorate.

Glossary

Abrasive wear Removal of material from surfaces by the cutting action of particles.

Adhesive wear Removal of material from moving surfaces by momentary local bonding and then bond fracture at the surfaces.

Anode The location at which corrosion occurs as electrons and ions are given up in an electrochemical cell.

Anodizing An anodic protection technique in which a thick oxide layer is deliberately produced on a metal surface.

Cathode The location at which electrons are accepted and a byproduct is produced during electrochemical corrosion.

Cavitation Erosion of a material surface by the pressures produced when a gas bubble collapses within a moving liquid.

Chemical corrosion Removal of atoms from a material by virtue of the solubility or chemical reaction between the material and the surrounding liquid.

Composition cells Electrochemical corrosion cells produced between two materials having different compositions. Also known as galvanic cells.

Concentration cells Electrochemical corrosion cells produced between two locations on a material where the composition of the electrolyte is different.

Crevice corrosion A special concentration cell in which corrosion occurs in crevices because of the low concentration of oxygen.

Dezincification A special chemical corrosion process by which both zinc and copper atoms are removed from brass, but the copper is replated back onto the metal.

Electrochemical cell A cell in which electrons and ions can flow by separate paths between two materials, producing a current which, in turn, leads to corrosion or plating.

Electrochemical corrosion Material degradation produced by the development of a current that removes ions from a material.

Electrode potential The voltage produced between the material of interest and a standard. It is related to the tendency of a material to corrode.

Electrolyte The conductive medium through which ions move to carry current in an electrochemical cell.

Electromotive force (emf) series The arrangement of elements according to their electrode potential, or their tendency to corrode.

Faraday's equation The relationship that describes the rate at which corrosion or plating occurs in an electrochemical cell.

Galvanic series The arrangement of alloys according to their tendency to corrode in a particular environment.

Graphitic corrosion A chemical corrosion process by which iron is leached from cast iron, leaving behind a weak, spongy mass of graphite.

Impressed voltage A cathodic protection technique by which a direct current is introduced into the material to be protected, thus preventing the anode reaction.

Inhibitors Additions to the electrolyte that preferentially migrate to the anode or cathode, cause polarization, and reduce the rate of corrosion.

Intergranular corrosion Corrosion at grain boundaries because grain boundary segregation or precipitation produces local galvanic cells.

Liquid impingement Erosion of a material caused by the impact of liquid droplets carried by a gas stream.

Nernst equation The relationship that describes the effect of electrolyte concentration on the electrode potential in an electrochemical cell.

Oxidation Reaction of a metal with oxygen to produce a metallic oxide. This normally occurs most rapidly at high temperatures.

Oxidation reaction The anode reaction by which electrons are given up to the electrochemical cell.

Oxygen starvation In the concentration cell, low-oxygen regions of the electrolyte cause the underlying material to behave as the anode and to corrode.

Passivation Formation of a protective coating on the anode surface that interrupts the electric circuit.

Pilling-Bedworth ratio The ratio of the oxide volume per metal atom to the metal volume per metal atom in a metal oxide film. The value predicts the morphology of the oxide film that forms on a metal surface during oxidation.

Polarization Changing the voltage between the anode and cathode to reduce the rate of corrosion. *Activation* polarization is related to the energy required to cause the anode or cathode reaction; *concentration* polarization is related to changes in the composition of the electrolyte; and *resistance* polarization is related to the electrical resistivity of the electrolyte.

Quench anneal The heat treatment used to dissolve carbides and prevent intergranular corrosion in stainless steels.

Reduction reaction The cathode reaction by which electrons are accepted from the electrochemical cell.

Sacrificial anode Cathodic protection by which a more anodic material is connected electrically to the material to be protected. The anode corrodes to protect the desired material.

Sensitization Precipitation of chromium carbides at the grain boundaries in stainless steels, making the steel sensitive to intergranular corrosion.

Stabilization Addition of titanium or niobium to a stainless steel to prevent intergranular corrosion.

Stress cells Electrochemical corrosion cells produced by differences in imposed or residual stresses at different locations in the material.

Stress corrosion Deterioration of a material in which an applied stress accelerates the rate of corrosion.

Tubercule Accumulations of microbial organisms and corrosion byproducts on the surface of a material.

Problems

Section 23-1 Chemical Corrosion

Section 23-2 Electrochemical Corrosion

Section 23-3 The Electrode Potential in Electrochemical Cells

Section 23-4 The Corrosion Current and Polarization

Section 23-5 Types of Electrochemical Corrosion

23-1 Explain how it is possible to form a simple electrochemical cell using a lemon, a penny, and a galvanized steel nail.

23-2 Silver can be polished by placing it in an aluminum pan containing a solution of baking soda, salt, and water. Speculate as to how the polishing process works.

23-3 A gray cast iron pipe is used in the natural gas distribution system for a city. The pipe fails and leaks, even though no corrosion noticeable to the naked eye has occurred. Offer an explanation for why the pipe failed.

23-4 A brass plumbing fitting produced from a Cu-30% Zn alloy operates in the hot water system of a large office building. After some period of use, cracking and leaking occur. On visual examination, no metal appears to have been corroded. Offer an explanation for why the fitting failed.

23-5 Plot the electronegativity of the elements listed in Table 23-1 as a function of electrode potential. (Refer to Chapter 2 for electronegativity values.) What is the trend that you observe in this data? Using an example, explain how the electronic structure of a metal is consistent with its anodic or cathodic tendency.

23-6 Write the anodic or cathodic reaction (as appropriate) for aluminum, germanium and chromium.

23-7 Suppose 10 g of Sn^{2+} are dissolved in 1000 mL of water to produce an electrolyte. Calculate the electrode potential of the tin half-cell.

23-8 Use the Nernst equation to determine the concentration of chromium in water if the difference from the standard potential is +0.016 volts.

23-9 A half-cell produced by dissolving copper in water produces an electrode potential of +0.32 V. Calculate the amount of copper that must have been added to 1000 mL of water to produce this potential.

23-10 The electrode potential in a platinum half-cell is 1.10 V. Determine the concentration of Pt^{4+} ions in the electrolyte.

23-11 Use Faraday's equation to determine the time for a 5 A current to completely corrode away a block of silver weighing 50 grams.

23-12 A current density of 0.05 A/cm^2 is applied to a 150 cm^2 cathode. What period of time is required to plate out a 1-mm-thick coating of silver onto the cathode?

23-13 We wish to produce 100 g of platinum per hour on a 1000 cm^2 cathode by electroplating. What plating current density is required? Determine the current required.

23-14 A 1-m-square steel plate is coated on both sides with a 0.005-cm-thick layer of zinc. A current density of 0.02 A/cm^2 is applied to the plate in an aqueous solution. Assuming that the zinc corrodes uniformly, determine the length of time required before the steel is exposed.

23-15 A 2-in.-inside-diameter, 12-ft-long copper distribution pipe in a plumbing system is accidentally connected to the power system of a manufacturing plant, causing a current of 0.05 A to flow through the pipe. If the wall thickness of the pipe is 0.125 in., estimate the time required before the pipe begins to leak, assuming a uniform rate of corrosion.

23-16 A steel surface 10 cm × 100 cm is coated with a 0.002-cm-thick layer of chromium. After one year of exposure to an electrolytic cell, the chromium layer is completely removed. Calculate the current density required to accomplish this removal.

23-17 A corrosion cell is composed of a 300 cm^2 copper sheet and a 20 cm^2 iron sheet, with a current density of 0.6 A/cm^2 applied to the copper. Which material is the anode? What is the rate of loss of metal from the anode per hour?

23-18 A corrosion cell is composed of a 20 cm^2 copper sheet and a 400 cm^2 iron sheet, with a current density of 0.7 A/cm^2 applied to the copper. Which material is the anode? What is the rate of loss of metal from the anode per hour?

23-19 List only the pure elements of the galvanic series in seawater from cathodic to anodic in Table 23-2.

23-20 Provide two reasons as to why steel alone in water with no other metal present will corrode.

23-21 Write the anode and cathode half-cell reactions caused by the precipitation of chromium carbide from austenitic stainless steels.

Section 23-6 Protection Against Electrochemical Corrosion

Section 23-7 Microbial Degradation and Biodegradable Polymers

23-22 Explain why titanium is a material that is often used in biomedical implants even though it is one of the most anodic materials listed in Table 23-1.

23-23 Alclad is a laminar composite composed of two sheets of commercially pure aluminum (alloy 1100) sandwiched around a core of 2024 aluminum alloy. Discuss the corrosion resistance of the composite. Suppose that a portion of one of the 1100 layers was machined off, exposing a small patch of the 2024 alloy. How would this affect the corrosion resistance? Explain. Would there be a difference in behavior if the core material were 3003 aluminum? Explain.

23-24 The leaf springs for an automobile are formed from a high-carbon steel. For best corrosion resistance, should the springs be formed by hot working or cold working? Explain. Would corrosion still occur even if you use the most desirable forming process? Explain.

23-25 Several types of metallic coatings are used to protect steel, including zinc, lead, tin, aluminum, and nickel. In which of these cases will the coating provide protection even when the coating is locally disrupted? Explain.

23-26 An austenitic stainless steel corrodes in all of the heat-affected zone (HAZ) surrounding the fusion zone of a weld. Explain why corrosion occurs and discuss the type of welding process or procedure that might have been used. What might you do to prevent corrosion in this region?

23-27 A steel nut is securely tightened onto a bolt in an industrial environment. After several months, the nut is found to contain numerous cracks. Explain why cracking might have occurred.

23-28 The shaft for a propeller on a ship is carefully designed so that the applied stresses are well below the endurance limit for the material. Yet after several months, the shaft cracks and fails. Offer an explanation for why failure might have occurred.

23-29 An aircraft wing composed of carbon fiber-reinforced epoxy is connected to a titanium forging on the fuselage. Will the anode for a corrosion cell be the carbon fiber, the titanium, or the epoxy? Which will most likely be the cathode? Explain.

23-30 The inside surface of a cast iron pipe is covered with tar, which provides a protective coating. Acetone in a chemical laboratory is drained through the pipe on a regular basis. Explain why, after several weeks, the pipe begins to corrode.

23-31 A cold-worked copper tube is soldered, using a lead-tin alloy, into a steel connector. What types of electrochemical cells might develop due to this connection? Which of the materials would you expect to serve as the anode and suffer the most extensive damage due to corrosion? Explain.

23-32 Pure tin is used to provide a solder connection for copper in many electrical uses. Which metal will most likely act as the anode?

23-33 Sheets of annealed nickel, cold-worked nickel, and recrystallized nickel are placed into an electrolyte. Which is the most likely to corrode? Which is the least likely to corrode? Explain.

23-34 A pipeline carrying liquid fertilizer crosses a small creek. A large tree washes down the creek and is wedged against the steel pipe. After some time, a hole is produced in the pipe at the point where the tree touches the pipe, with the diameter of the hole larger on the outside of the pipe than on the inside of the pipe. The pipe then leaks fertilizer into the creek. Offer an explanation for why the pipe corroded.

23-35 Two sheets of a 1040 steel are joined together with an aluminum rivet (Figure 23-20). Discuss the possible corrosion cells that might be created as a result of this joining process. Recommend a joining process that might minimize corrosion for these cells.

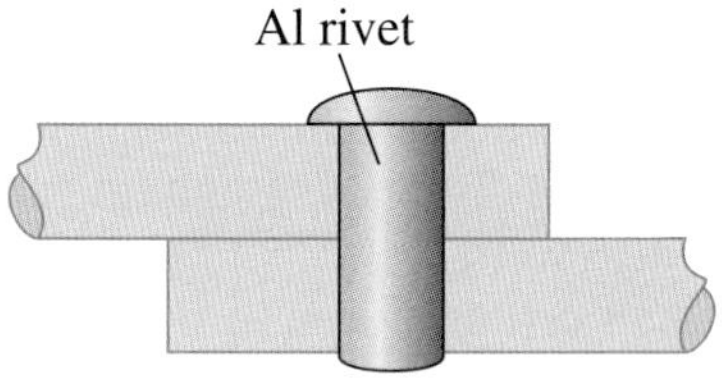

Figure 23-20 Two steel sheets joined by an aluminum rivet (for Problem 23-35).

23-36 Figure 23-21 shows a cross-section through an epoxy-encapsulated integrated circuit, including a microgap between the copper lead frame and the epoxy polymer. Suppose chloride ions from the manufacturing process penetrate the package. What types of corrosion cells might develop? What portions of the integrated circuit are most likely to corrode?

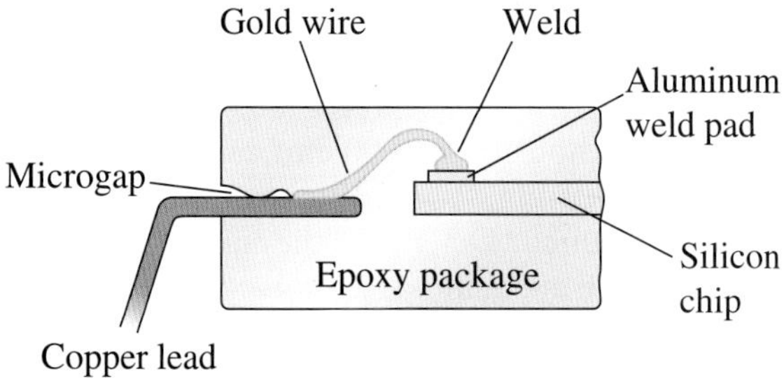

Figure 23-21 Cross-section through an integrated circuit showing the external lead connection to the chip (for Problem 23-36).

23-37 A current density of 0.1 A/cm^2 is applied to the iron in an iron-zinc corrosion cell. Calculate the weight loss of zinc per hour
(a) if the zinc has a surface area of 10 cm^2 and the iron has a surface area of 100 cm^2 and
(b) if the zinc has a surface area of 100 cm^2 and the iron has a surface area of 10 cm^2.

Section 23-8 Oxidation and Other Gas Reactions

Section 23-9 Wear and Erosion

23-38 Determine the Pilling-Bedworth ratio for the following metals and predict the behavior of the oxide that forms on the surface. Is the oxide protective, does it flake off the metal, or is it permeable? (See Appendix A for the metal density.)

	Oxide Density (g/cm^3)
Mg-MgO	3.60
Na-Na_2O	2.27
Ti-TiO_2	5.10
Fe-Fe_2O_3	5.30
Ce-Ce_2O_3	6.86
Nb-Nb_2O_5	4.47
W-WO_3	7.30

23-39 Oxidation of most ceramics is not considered to be a problem. Explain.

23-40 A sheet of copper is exposed to oxygen at 1000°C. After 100 h, 0.246 g of copper are lost per cm^2 of surface area; after 250 h, 0.388 g/cm^2 are lost; and after 500 h, 0.550 g/cm^2 are lost. Determine whether the oxidation is parabolic, linear, or logarithmic, then determine the time required for a 0.75-cm sheet of copper to be completely oxidized. The sheet of copper is oxidized from both sides.

23-41 At 800°C, iron oxidizes at a rate of 0.014 g/cm^2 per hour; at 1000°C, iron oxidizes at a rate of 0.0656 g/cm^2 per hour. Assuming a parabolic oxidation rate, determine the maximum temperature at which iron can be held if the oxidation rate is to be less than 0.005 g/cm^2 per hour.

Design Problems

23-42 A cylindrical steel tank 3 ft in diameter and 8 ft long is filled with water. We find that a current density of 0.015 A/cm^2 acting on the steel is required to prevent corrosion. Design a sacrificial anode system that will protect the tank.

23-43 The drilling platforms for offshore oil rigs are supported on large steel columns resting on the bottom of the ocean. Design an approach to ensure that corrosion of the supporting steel columns does not occur.

23-44 A storage building is to be produced using steel sheet for the siding and roof. Design a corrosion protection system for the steel.

23-45 Design the materials for the scraper blade for a piece of earthmoving equipment.

Computer Problems

23-46 Write a computer program that will calculate the thickness of a coating using different inputs provided by the user (e.g., density and valence of the metal, area or dimensions of the part being plated, desired time of plating, and current).

K Knovel® Problems

K23-1 Describe the mechanism of stray current corrosion.

K23-2 Describe how the corrosion rate of copper changes with increasing solution velocity at different temperatures. Explain the mechanism for this change.

K23-3 Five grams of zirconium are dissolved in 1000 g of water producing an electrolyte of Zr^{4+} ions. Calculate the electrode potential of the copper half-cell in this electrolyte at standard temperature and pressure using the Nernst equation.

APPENDIX A

Selected Physical Properties of Metals

Metal		Atomic Number	Crystal Structure	Lattice Parameters (Å)	Atomic Mass g/mol	Density (g/cm³)	Melting Temperature (°C)
Aluminum	Al	13	FCC	4.04958	26.981	2.699	660.4
Antimony	Sb	51	hex	$a = 4.307$ $c = 11.273$	121.75	6.697	630.7
Arsenic	As	33	hex	$a = 3.760$ $c = 10.548$	74.9216	5.778	816
Barium	Ba	56	BCC	5.025	137.3	3.5	729
Beryllium	Be	4	HCP	$a = 2.2858$ $c = 3.5842$	9.01	1.848	1290
Bismuth	Bi	83	mono	$a = 6.674$ $b = 6.117$ $c = 3.304$ $\beta = 110.3°$	208.98	9.808	271.4
Boron	B	5	rhomb	$a = 10.12$ $\alpha = 65.5°$	10.81	2.36	2076
Cadmium	Cd	48	HCP	$a = 2.9793$ $c = 5.6181$	112.4	8.642	321.1
Calcium	Ca	20	FCC	5.588	40.08	1.55	839
Cerium	Ce	58	HCP	$a = 3.681$ $c = 5.99$	140.12	6.6893	798
Cesium	Cs	55	BCC	6.13	132.91	1.892	28.6
Chromium	Cr	24	BCC	2.8844	51.996	7.19	1907
Cobalt	Co	27	HCP	$a = 2.5071$ $c = 4.0686$	58.93	8.832	1495
Copper	Cu	29	FCC	3.6151	63.55	8.93	1084.9
Gadolinium	Gd	64	HCP	$a = 3.6336$ $c = 5.7810$	157.25	7.901	1313
Gallium	Ga	31	ortho	$a = 4.5258$ $b = 4.5186$ $c = 7.6570$	69.72	5.904	29.8
Germanium	Ge	32	DC	5.6575	72.59	5.324	937.4
Gold	Au	79	FCC	4.0786	196.97	19.302	1064.4
Hafnium	Hf	72	HCP	$a = 3.1883$ $c = 5.0422$	178.49	13.31	2227
Indium	In	49	tetra	$a = 3.2517$ $c = 4.9459$	114.82	7.286	156.6
Iridium	Ir	77	FCC	3.84	192.9	22.65	2447
Iron	Fe	26	BCC FCC BCC	2.866 3.589	55.847 (>912°C) (>1394°C)	7.87	1538

(*Continued*)

Metal		Atomic Number	Crystal Structure	Lattice Parameters (Å)	Atomic Mass g/mol	Density (g/cm³)	Melting Temperature (°C)
Lanthanum	La	57	HCP	$a = 3.774$ $c = 12.17$	138.91	6.146	918
Lead	Pb	82	FCC	4.9489	207.19	11.36	327.4
Lithium	Li	3	BCC	3.5089	6.94	0.534	180.7
Magnesium	Mg	12	HCP	$a = 3.2087$ $c = 5.209$	24.312	1.738	650
Manganese	Mn	25	cubic	8.931	54.938	7.47	1244
Mercury	Hg	80	rhomb		200.59	13.546	−38.9
Molybdenum	Mo	42	BCC	3.1468	95.94	10.22	2623
Nickel	Ni	28	FCC	3.5167	58.71	8.902	1453
Niobium	Nb	41	BCC	3.294	92.91	8.57	2468
Osmium	Os	76	HCP	$a = 2.7341$ $c = 4.3197$	190.2	22.57	3033
Palladium	Pd	46	FCC	3.8902	106.4	12.02	1552
Platinum	Pt	78	FCC	3.9231	195.09	21.45	1769
Potassium	K	19	BCC	5.344	39.09	0.855	63.2
Rhenium	Re	75	HCP	$a = 2.760$ $c = 4.458$	186.21	21.04	3186
Rhodium	Rh	45	FCC	3.796	102.99	12.41	1963
Rubidium	Rb	37	BCC	5.7	85.467	1.532	38.9
Ruthenium	Ru	44	HCP	$a = 2.6987$ $c = 4.2728$	101.07	12.37	2334
Selenium	Se	34	mono	$a = 9.054$ $b = 9.083$ $c = 11.60$ $\beta = 90.8°$	78.96	4.809	217
Silicon	Si	14	DC	5.4307	28.08	2.33	1410
Silver	Ag	47	FCC	4.0862	107.868	10.49	961.9
Sodium	Na	11	BCC	4.2906	22.99	0.967	97.8
Strontium	Sr	38	FCC BCC	6.0849 4.84	87.62	2.6	777 (>557°C)
Tantalum	Ta	73	BCC	3.3026	180.95	16.6	2996
Technetium	Tc	43	HCP	$a = 2.735$ $c = 4.388$	98.9062	11.5	2157
Tellurium	Te	52	hex	$a = 4.4565$ $c = 5.9268$	127.6	6.24	449.5
Thorium	Th	90	FCC	5.086	232	11.72	1775
Tin	Sn	50	tetra DC	$a = 5.832$ $c = 3.182$ 6.4912	118.69	5.765	231.9
Titanium	Ti	22	HCP BCC	$a = 2.9503$ $c = 4.6831$ 3.32	47.9	4.507	1668 (>882°C)
Tungsten	W	74	BCC	3.1652	183.85	19.254	3422
Uranium	U	92	ortho	$a = 2.854$ $b = 5.869$ $c = 4.955$	238.03	19.05	1133

(Continued)

Metal		Atomic Number	Crystal Structure	Lattice Parameters (Å)	Atomic Mass g/mol	Density (g/cm^3)	Melting Temperature (°C)
Vanadium	V	23	BCC	3.0278	50.941	6.1	1910
Yttrium	Y	39	HCP	$a = 3.648$ $c = 5.732$	88.91	4.469	1522
Zinc	Zn	30	HCP	$a = 2.6648$ $c = 4.9470$	65.38	7.133	420
Zirconium	Zr	40	HCP	$a = 3.2312$ $c = 5.1477$	91.22	6.505	1852
			BCC	3.6090			(>862°C)

APPENDIX B

The Atomic and Ionic Radii of Selected Elements

Element	Atomic Radius (Å)	Valence	Ionic Radius (Å)
Aluminum	1.432	+3	0.51
Antimony	1.45	+5	0.62
Arsenic	1.15	+5	2.22
Barium	2.176	+2	1.34
Beryllium	1.143	+2	0.35
Bismuth	1.60	+5	0.74
Boron	0.46	+3	0.23
Bromine	1.19	−1	1.96
Cadmium	1.49	+2	0.97
Calcium	1.976	+2	0.99
Carbon	0.77	+4	0.16
Cerium	1.84	+3	1.034
Cesium	2.65	+1	1.67
Chlorine	0.905	−1	1.81
Chromium	1.249	+3	0.63
Cobalt	1.253	+2	0.72
Copper	1.278	+1	0.96
Fluorine	0.6	−1	1.33
Gallium	1.218	+3	0.62
Germanium	1.225	+4	0.53
Gold	1.442	+1	1.37
Hafnium	1.55	+4	0.78
Hydrogen	0.46	+1	1.54
Indium	1.570	+3	0.81
Iodine	1.35	−1	2.20
Iron	1.241 (BCC) 1.269 (FCC)	+2 +3	0.74 0.64
Lanthanum	1.887	+3	1.15
Lead	1.75	+4	0.84
Lithium	1.519	+1	0.68
Magnesium	1.604	+2	0.66
Manganese	1.12	+2	0.80
		+3	0.66
Mercury	1.55	+2	1.10
Molybdenum	1.363	+4	0.70

(Continued)

Element	Atomic Radius (Å)	Valence	Ionic Radius (Å)
Nickel	1.243	+2	0.69
Niobium	1.426	+4	0.74
Nitrogen	0.71	+5	0.15
Oxygen	0.60	−2	1.32
Palladium	1.375	+4	0.65
Phosphorus	1.10	+5	0.35
Platinum	1.387	+2	0.80
Potassium	2.314	+1	1.33
Rubidium	2.468	+1	1.48
Selenium	1.15	−2	1.91
Silicon	1.176	+4	0.42
Silver	1.445	+1	1.26
Sodium	1.858	+1	0.97
Strontium	2.151	+2	1.12
Sulfur	1.06	−2	1.84
Tantalum	1.43	+5	0.68
Tellurium	1.40	−2	2.11
Thorium	1.798	+4	1.02
Tin	1.405	+4	0.71
Titanium	1.475	+4	0.68
Tungsten	1.371	+4	0.70
Uranium	1.38	+4	0.97
Vanadium	1.311	+3	0.74
Yttrium	1.824	+3	0.89
Zinc	1.332	+2	0.74
Zirconium	1.616	+4	0.79

Note that 1 Å = 10^{-8}cm = 0.1 nanometer (nm)

ANSWERS TO SELECTED PROBLEMS

CHAPTER 2

2-6 (a) 6.7×10^{21} atoms (b) (i) 3.30×10^{22} atoms/cm^3. (ii) 4.63×10^{22} atoms/cm^3.

2-7 (a) 9.78×10^{27} atoms/ton. (b) 4.6 cm^3.

2-8 (a) 5.99×10^{23} atoms. (b) 0.994 mol.

2-41 MgO, MgO has ionic bonds.

2-43 Si, Si has covalent bonds.

CHAPTER 3

3-14 (a) 1.426×10^{-8} cm. (b) 1.4447×10^{-8} cm.

3-17 (a) 5.3349 Å. (b) 2.3101 Å.

3-19 FCC.

3-21 BCT.

3-23 (a) 8 atoms/cell. (b) 0.387.

3-29 (a) 4 Th, 8 O. (b) Yes, 4 Th: 8 O = 1 Th: 2 O. (c) 8. (d) 4. (e) tetrahedral.

3-33 (a) 4. (b) No, the sites are equivalent. (c) 1.3×10^{8} atoms.

3-39 3.53% contraction.

3-49 (c) [602].

3-51 *A*: [00$\bar{1}$]. *B*: [1$\bar{2}$0]. *C*: [$\bar{1}$11]. *D*: [2$\bar{1}\bar{1}$].

3-58 *A*: (1$\bar{1}$01). *B*: (0003). *C*: (1$\bar{1}$00).

3-64 [$\bar{1}$10], [1$\bar{1}$0], [101], [$\bar{1}$0$\bar{1}$], [011], [0$\bar{1}\bar{1}$].

3-66 Tetragonal—4; orthorhombic—2; cubic—12.

3-68 (a) (111). (b) (210). (c) (0$\bar{1}$2). (d) (218).

3-74 [100]: 0.35089 nm, 2.85 nm^{-1}, 0.866. [110]: 0.496 nm, 2.015 nm^{-1}, 0.612. [111]: 0.3039 nm, 3.291 nm^{-1}, 1. The [111] is close packed.

3-76 (100): 1.617×10^{15}/cm^2, packing fraction 0.7854. (110): 1.144×10^{15}/cm^2, packing fraction 0.555. (111): 1.867×10^{15}/cm^2, 0.907. The (111) is close packed.

3-78 4.563×10^{6}.

3-81 (a) 0.2798 Å. (b) 0.6290 Å.

3-83 (a) 4. (b) (1/2, 0, 0), (0, 1/2, 0), (0, 0, 1/2), (1/2, 1/2, 1/2). (c) 0.79.

3-85 (a) 6. (c) 8. (e) 4. (h) 6.

3-88 Fluorite. (a) 5.2885 Å. (b) 12.13 g/cm^3. (c) 0.624.

3-90 Cesium chloride. (a) 4.1916 Å. (b) 4.8 g/cm^3. (c) 0.693.

3-92 (111): 1.473×10^{15}/cm^2, 0.202 (Mg^{+2}). (222): 1.473×10^{15}/cm^2, 0.806 (O^{-2}).

3-98 0.40497 nm.

3-100 (a) BCC. (c) 0.2327 nm.

CHAPTER 4

4-3 4.98×10^{19} vacancies/cm^3.

4-5 (a) 0.00204. (b) 1.39×10^{20} vacancies/cm^3.

4-7 (a) 1.157×10^{20} vacancies/cm^3. (b) 0.532 g/cm^3.

4-11 0.344.

4-13 8.262 g/cm^3.

4-15 (a) 0.0449. (b) one H atom per 22.3 unit cells.

4-17 (a) 0.0522 defects/unit cell. (b) 2.47×10^{20} defects/cm^3.

4-24 (a) $[0\bar{1}1]$, $[01\bar{1}]$, $[\bar{1}10]$, $[1\bar{1}0]$, $[\bar{1}01]$, $[10\bar{1}]$.

(b) $[1\bar{1}1]$, $[\bar{1}1\bar{1}]$, $[\bar{1}\bar{1}1]$, $[11\bar{1}]$.

4-26 $(1\bar{1}0)$, $(\bar{1}10)$, $(0\bar{1}1)$, $(01\bar{1})$, $(10\bar{1})$, $(\bar{1}01)$.

4-28 $(111)[1\bar{1}0]$: b = 2.863 Å, d = 2.338 Å.

$(110)[1\bar{1}1]$: b = 7.014 Å, d = 2.863 Å. Ratio = 0.44.

4-50 (a) $K = 121.8$ psi $\sqrt{\text{in.}}$, $\sigma_0 = 60{,}293$ psi. (b) 103.4 ksi.

4-56 (a) 128 grains/in.^{2}. (b) 1.28×10^6 grains/in.^{2}.

4-57 9.3.

4-58 3.6.

4-62 0.0328°.

4-63 284 Å.

CHAPTER 5

5-10 1.08×10^9 jumps/s.

5-18 $D_H = 1.07 \times 10^{-4}$ cm^2/s versus $D_N = 3.90 \times 10^{-9}$ cm^2/s. Smaller H atoms diffuse more rapidly.

5-20 3.88×10^{-12} cm^2/s.

5-23 (a) 59,390 cal/mol. (b) 0.057 cm^2/s.

5-29 $D_{Fe} = 2.75 \times 10^{-6}$ cm^2/s; Al.

5-32 (a) −0.02495 at% Sb/cm. (b) -1.246×10^{19} Sb/(cm^3 · cm).

5-34 (a) -1.969×10^{11} H atoms/(cm^3 · cm).
(b) 3.3×10^7 H atoms/(cm^2· s).

5-37 1.25×10^{-3} g/h.

5-39 −198°C.

5-49 23.3 years.

5-59 $D_0 = 3.47 \times 10^{-16}$ cm^2/s versus $D_{Al} = 2.48 \times 10^{-13}$ cm^2/s. It is easier for the smaller Al ions to diffuse.

5-62 907°C or 1180 K.

5-63 51 min.

5-64 0.53% C.

5-66 190 s.

5-68 1184 s.

5-70 667°C.

5-78 50,488 cal/mol; yes.

CHAPTER 6

6-14 1100 lb or 4891 N.

6-16 49.7 MPa; 0.003.

6-31 (a) 22.6 GPa. (b) 206 nm.

6-34 (a) l_f = 12.00298 in., d_f = 0.39997 in.

6-35 (b) 54.9 ksi. (d) 37.5% (f) 44.9 ksi. (g) 62.6 ksi.

6-39 (a) 11,600 psi. (b) 13,210 psi. (c) 603 ksi. (d) 4.5%. (e) 3.5%. (f) 11.3 ksi. (g) 76.4 psi.

6-41 (a) 274 MPa. (b) 417 MPa. (c) 172 GPa. (d) 18.55%. (e) 15.8%. (f) 398 MPa. (g) 459 MPa. (h) 0.17 MPa.

6-47 (a) 76.8 ksi. (b) 22.1 Mpsi.

6-49 (a) 41 mm; will not fracture.

6-61 5.6 GPa.

6-71 Not notch-sensitive; poor toughness.

CHAPTER 7

7-3 134 MPa.

7-8 0.99 MPa$\sqrt{m}$.

7-9 No; test will not be sensitive enough.

7-28 280 MPa; −140 MPa.

7-29 15.35 lb.

7-32 d = 1.634 in.

7-34 22 MPa; max = +22 MPa, min = −22 MPa, mean = 0 MPa; a higher frequency will reduce fatigue strength due to heating.

7-36 (a) 2.5 mm. (b) 0.0039 mm.

7-40 $C = 2.061 \times 10^{-3}$; $n = 3.01$.

7-52 101,329 h.

7-54 $n = 6.82$. $m = -5.7$.

7-56 29 days.

7-58 0.52 in.

7-62 2000 lb.

CHAPTER 8

8-6 (a) 1 – the slope is steeper. (b) 1 – it is stronger. (c) 2 – it is more ductile and less strong.

8-9 528 MPa.

8-10 $n = 0.12$; BCC.

8-12 $n = 0.15$.

8-13 0.56.

8-24 0.152 in.

8-28 About 59 ksi.

8-29 25 ksi tensile, 21 ksi yield, 6% elongation.

8-31 First step: 36% CW giving 25 ksi tensile, 21 ksi yield, 6% elongation.
Second step: 64% CW giving 28 ksi tensile, 25 ksi yield, 4% elongation.
Third step: 84% CW giving 30 ksi tensile, 28 ksi yield, 3% elongation.

8-33 0.76 to 0.96 in.

8-37 48% CW, 26 ksi tensile, 23 ksi yield, 4% elongation.

8-53 (a) 1414 lb. (b) Will not break.

8-55 ~3.2.

8-69 (a) 550°C, 750°C, 950°C. (b) 700°C. (c) 900°C. (d) 2285°C.

CHAPTER 9

9-11 35 mJ.

9-19 (a) 6.65 Å. (b) 109 atoms.

9-21 1.136×10^6 atoms.

9-38 (a) 0.0333. (b) 0.333. (c) All.

9-39 1265°C.

9-44 31.15 s.

9-46 $B = 300$ s/cm^2, $n = 1.6$.

9-50 (a) $\sim 4.16 \times 10^{-3}$ cm. (b) 90 s.

9-51 $k = 0.003$ cm/s, $m = 0.35$.

9-53 0.03 s.

9-62 (a) 900°C. (b) 430°C. (c) 470°C. (d) ~250°C/min. (e) 9.7 min.
(f) 7.2 min. (g) 55°C. (h) Zinc. (i) 87.3 min/in.^{2}.

9-72 V/A (riser) = 0.68, V/A (thick) = 1.13, V/A (thin) = 0.89; not effective.

9-75 $D_{Cu} = 1.48$ in. $D_{Fe} = 1.30$ in.

9-77 (a) 46 cm^3. (b) 4.1%.

9-83 23.04 cm.

9-84 −0.78%

9-86 0.046 cm^3/100 g Al.

CHAPTER 10

10-23 3.7 kg Zn.

10-29 (a) Yes. (c) No. (e) No. (g) No.

10-33 Cd should give the smallest decrease in conductivity; none should give unlimited solid solubility.

10-44 (a) 2330°C, 2150°C, 180°C. (c) 2570°C, 2380°C, 190°C.

10-48 (a) 45 wt% Pd in L; 55 wt% Pd in α; 50% L; 50% α.

10-50 (a) 100% L containing 30% MgO. (b) 70.8% L containing 38% MgO, 29.2% S containing 62% MgO. (c) 8.3% L containing 38% MgO, 91.7% S containing 62% MgO. (d) 100% S containing 85% MgO.

10-53 (a) L: 15 mol% MgO or 8.69 wt% MgO. S: 38 mol% MgO or 24.85 wt% MgO.
(b) L: 78.26 mol% or 80.1 wt%; S: 21.74 mol% or 19.9 wt% MgO.
(c) 78.1 vol% L, 21.9 vol% S.

10-55 750 g Ni, Ni/Cu = 1.62.

10-57 331 g MgO.

10-59 64.1 wt% FeO.

10-61 (a) 49 wt% W in L, 70 wt% W in α. (b) Not possible.

10-63 212 lb W; 1200 lb W.

10-65 Ni dissolves; when the liquid reaches 10 wt% Ni, the bath begins to freeze.

10-70 (a) 2900°C, 2690°C, 210°C. (b) 60% L containing 49% W, 40% α containing 70% W.

10-72 (a) 57% W. (b) 17% W.

10-76 (a) 2900°C. (b) 2710°C. (c) 190°C. (d) 2990°C. (e) 90°C. (f) 300 s. (g) 340 s. (h) 60% W.

10-79 (a) 2000°C. (b) 1450°C. (c) 550°C. (d) 40% FeO. (e) 92% FeO.
(f) 65.5% L containing 75% FeO, 34.5% S containing 46% FeO.
(g) 30.3% L containing 88% FeO, 69.7% S containing 55% FeO.

10-80 (a) 3100°C. (b) 2720°C. (c) 380°C. (d) 90% W. (e) 40% W.
(f) 44.4% L containing 70% W, 55.6% α containing 88% W.
(g) 9.1% L containing 50% W, 90.9% α containing 83% W.

CHAPTER 11

11-11 (a) θ, nonstoichiometric. (b) α,β,γ,η. B is allotropic, existing in three different forms at different temperatures. (c) 1100°C: peritectic. 900°C: monotectic. 690°C: eutectic. 600°C: peritectoid. 300°C: eutectoid.

11-15 Cu_3Sn.

11-16 $SiCu_4$.

11-19 (a) 2.5% Mg. (b) 600°C, 470°C, 400°C, 130°C. (c) 74% α containing 7% Mg, 26% L containing 26% Mg. (d) 100% α containing 12% Mg. (e) 67% α containing 1% Mg, 33% β containing 34% Mg.

11-21 (a) Hypereutectic. (b) 98% Sn. (c) 22.8% β containing 97.5% Sn, 77.2% L containing 61.9% Sn. (d) 35% α containing 19% Sn, 65% β containing 97.5% Sn. (e) 22.8% primary β containing 97.5% Sn, 77.2% eutectic containing 61.9% Sn. (f) 30% α containing 2% Sn, 70% β containing 100% Sn.

11-30 (a) Hypoeutectic. (b) 1% Si. (c) 78.5% α containing 1.65% Si, 21.5% L containing 12.6% Si. (d) Phases: 97.6% α containing 1.65% Si, 2.4% β containing 99.83% Si. Microconstituents: 78.5% primary α containing 1.65% Si, 21.5% eutectic containing 12.6% Si. (e) 96% α containing 0% Si, 4% β containing 100% Si.

11-32 56% Sn, hypoeutectic.

11-34 52% Sn.

11-36 16.5% Li, hypereutectic.

11-38 60.5% α, 39.5% β; 27.7% primary α, 72.3% eutectic; 0.54.

11-40 (a) 1150°C. (b) 150°C. (c) 1000°C. (d) 577°C. (e) 423°C.
(f) 10.5 min. (g) 11.5 min. (h) 45% Si.

CHAPTER 12

12-3 $c = 8.9 \times 10^{-6}$, $n = 2.81$.

12-16 1.73.

12-30 (a) For Al–4% Mg: solution treat between 210 and 451°C, quench, and age below 210°C. For Al–6% Mg: solution treat between 280 and 451°C, quench, and age below 280°C. For Al–12% Mg: solution treat between 390 and 451°C, quench, and age below 390°C. (c) The precipitates are not coherent.

12-46 (a) Solution treat between 290 and 400°C, quench, and age below 290°C. (c) Not a good candidate. (e) Not a good candidate.

12-61 27.6%.

12-62 (a) 795°C. (b) Primary ferrite. (c) 56.1% ferrite containing 0.0218% C and 43.9% austenite containing 0.77% C. (d) 95.1% ferrite containing 0.0218% C and 4.9% cementite containing 6.67% C. (e) 56.1% primary ferrite containing 0.0218% C and 43.9% pearlite containing 0.77% C.

12-64 0.53% C, hypoeutectoid.

12-66 0.156% C, hypoeutectoid.

12-68 0.281% C.

12-70 760°C, 0.212% C.

12-74 (a) 900°C; 12% CaO in tetragonal, 3% CaO in monoclinic, 16% CaO in cubic; 31% monoclinic, 69% cubic. (c) 250°C; 47% Zn in β', 36% Zn in α, 59% Zn in γ; 52.2% α, 47.8% γ.

12-86 (a) 615°C. (b) 1.67×10^{-5} cm.

12-88 Bainite with HRC 47.

12-90 Martensite with HRC 66.

12-102 (a) 37.2% martensite with 0.77% C and HRC 65.
(c) 84.8% martensite with 0.35% C and HRC 58.

12-104 (a) 750°C. (b) 0.455% C.

12-106 ~3% expansion.

12-108 Austenitize at 750°C, quench, and temper above 330°C.

CHAPTER 13

13-4 (a) 97.8% ferrite, 2.2% cementite, 82.9% primary ferrite, 17.1% pearlite.
(c) 85.8% ferrite, 14.2% cementite, 3.1% primary cementite, 96.9% pearlite.

13-10 For 1035: A_1 = 727°C; A_3 = 790°C; anneal = 820°C; normalize = 845°C; process anneal = 557–647°C; not usually spheroidized.

13-14 (a) Ferrite and pearlite. (c) Martensite. (e) Ferrite and bainite. (g) Tempered martensite.

13-16 (a) Austenitize at 820°C, hold at 600°C for 10 s, cool.
(c) Austenitize at 780°C, hold at 600°C for 10 s, cool.
(e) Austenitize at 900°C, hold at 320°C for 5000 s, cool.

13-20 (a) Austenitize at 820°C, quench, and temper between 420 and 480°C; 150 to 180 ksi tensile, 140 to 160 ksi yield.
(b) 175 to 180 ksi tensile, 130 to 135 ksi yield. (c) 100 ksi tensile, 65 ksi yield, 20% elongation.

13-22 0.48% C in martensite; austenitized at 770°C; should austenitize at 860°C.

13-24 1080: fine pearlite. 4340: martensite.

13-29 May become hypereutectoid with grain boundary cementite.

13-33 (a) Not applicable. (c) 8 to 10°C/s. (e) 32 to 36°C/s.

13-35 (a) 16°C/s. (b) Pearlite with HRC 38.

13-37 (a) Pearlite with HRC 36. (c) Pearlite and martensite with HRC 46.

13-39 (a) 1.6 in. (c) 1.9 in. (e) Greater than 2.5 in.

13-44 0.30 h.

13-48 0.05 mm: pearlite and martensite with HRC 53. 0.15 mm: pearlite with HRC 38.

13-53 δ-ferrite; nonequilibrium freezing; anneal.

13-58 2.4% Si.

CHAPTER 14

14-6 The β that is present in the eutectic microconstituent embrittles the alloy.

14-10 29.4% β versus 2.3% β.

14-14 (a) 0.113 in., 0.0151 lb, \$0.042.
(b) 0.113 in., 0.0235 lb, \$0.019.

14-19 Al: 340%. Mg: 31%. Cu: 1004%.

14-22 Lead may melt during hot working.

14-26 γ' because it is smaller and more numerous than the carbide.

14-29 Ti-15% V: 100% β transforms to 100% α', which then transforms to 24% β precipitate in an α matrix. This is a quench and temper process. Ti-35% V: 100% β transforms to 100% β_{ss}, which then transforms to 27% α precipitate in a β matrix. This is an age hardening process.

14-31 Al: 7.7×10^5 in. Cu: 5.4×10^5 in. Ni: 3.5×10^5 in.

14-33 Spalls off; cracks.

CHAPTER 15

15-4 1029%.

15-10 0.79.

15-17 $B = 2.4$; true porosity = 22.58%; fraction = 0.557.

15-18 735 g.

15-20 1.324.

15-37 1.276 kg BaO; 0.249 kg Li_2O.

15-39 2.18; 10.32% Na_2O, 4.67% CaO, 85.01% SiO_2.

CHAPTER 16

16-10 327; 405.

16-12 (a) 2500. (b) 2.4×10^{18}.

16-15 (a) 211. (b) 175.

16-21 Polybutadiene and silicone; the T_g must be lower than −78°C.

16-23 Polyethylene and polypropylene.

16-28 91°C.

16-32 1246 psi.

16-33 1.055×10^5 cal/mol.

16-35 At $\varepsilon = 1$, $E = 833$ psi; at $\varepsilon = 4$, $E = 2023$ psi.

CHAPTER 17

17-9 7.8×10^{13} per cm^3.

17-11 2.48%.

17-16 9.41 g/cm^3.

17-19 (a) 0.507. (b) 0.507. (c) 7.77 g/cm^3.

17-21 11.2 to 22.2 kg.

17-29 (a) 2.56 g/cm^3. (b) 29 Mpsi. (c) 15.2 Mpsi.

17-32 188 MPa.

17-34 For $d = 20\ \mu m$, $l_c = 0.30$ cm, $l_c/d = 150$. For $d = 1\ \mu m$, $l_c = 0.15$ cm, $l_c/d = 1500$.

17-45 Sizing improves strength.

17-47 Pyrolize at 2500°C; 250,000 psi.

17-53 $E_{parallel} = 10.0$ Mpsi; $E_{perpendicular} = 2.96$ Mpsi.

17-55 $E_{parallel} = 11.9$ Mpsi; $E_{perpendicular} = 10.0$ Mpsi.

17-57 0.417 g/cm^3; 20.0 kg versus 129.6 kg.

CHAPTER 18

18-3 (a) 0.26 gal. (b) 0.77 g/cm^3.

18-6 0.1998 in.

18-8 Expands 83.9 in. perpendicular to the boards. The floor will buckle.

18-10 0.48.

18-13 (a) 640 sacks, 53.9 tons sand, 111 tons aggregate, 2032 gal. (b) 4000 lb/yd^3. (c) 1 : 1.79 : 3.69.

CHAPTER 19

19-2 13,433 ohm.

19-4 (a) 3377 W. (b) 3.183×10^{14} W.

19-7 $d = 0.0864$ cm; 1174 V.

19-10 7.8×10^{11} cm^{-3}.

19-12 0.0235 km/h; 0.0146 mph.

19-16 2.5×10^{-5} ohm · cm.

19-18 3.03×10^5 $ohm^{-1} \cdot cm^{-1}$ at −50°C; 0.34×10^5 $ohm^{-1} \cdot cm^{-1}$ at 500°C.

19-20 −81.8°C.

19-24 At 400°C, $\rho = 4.15 \times 10^{-5}$ ohm · cm; $\rho_d = 8.5 \times 10^{-6}$ ohm · cm; $b = 1.79 \times 10^{-4}$ ohm · cm. At 200°C, $\rho = 3.76 \times 10^{-5}$ ohm · cm.

19-30 Sb: 10.8 $ohm^{-1} \cdot cm^{-1}$. In: 3.8 $ohm^{-1} \cdot cm^{-1}$.

19-33 0.379 g As.

19-37 2600%.

19-43 Be: 2.92×10^4 $ohm^{-1} \cdot cm^{-1}$.

19-49 (a) 4.85×10^{-19} m. (c) 1.15×10^{-17} m.

19-51 9.4 mV.

19-55 0.001239 μF.

CHAPTER 20

20-6 Fe: 39,604 Oe. Co: 31,573 Oe.

20-13 1194.

20-15 6.07×10^3 Oe.

20-17 8000 G; 1.49 mA.

20-32 (a) 13,000 G. (b) 14,000 G. (c) 800 Oe. (d) 5.8 G/Oe. (e) 15.6 G/Oe. (f) 6.8×10^6 G Oe.

20-44 15×10^6 G · Oe.

20-46 33 kJ/m^3.

20-48 High saturation inductance.

20-51 4.68×10^5 A/m.

CHAPTER 21

21-12 1.94×10^8 m/s.

21-13 Ice, water, Teflon.

21-17 4%; 0.35%.

21-19 (a) 6.56°. (b) 6.65°. (c) 10°. (d) 0.18 cm.

21-20 25.84°.

21-21 0.25.

21-23 0.0117 cm.

21-34 0.37 I_0.

21-36 1.84×10^{-5} s.

21-40 (a) 13.1×10^{-5} cm. (b) 77.6×10^{-5} cm.

21-42 1.85×10^{-4} cm; infrared.

21-45 13,800 V.

21-47 7480 V.

21-49 (a) 24,848 V. (b) Cu, Mn, Si.

CHAPTER 22

22-4 3.73 BTU/(lbm · °R).

22-7 (a) 1900 cal, 7950 J. (c) 8500 cal, 35,564 J.

22-9 0.0375 cm.

22-11 0.172 in.

22-12 1.975 m.

22-15 15.183 in.

22-18 Brass: 10.0246 in., Invar: 10.0020 in.

22-22 256.2 ksi; tensile.

22-24 No; 24.51 ksi; 0.0098 in. decrease in length.

22-28 19.2 kW/m^2.

22-31 (a) 2.45×10^8 J/day. (b) 1.47×10^8 J/day.

22-34 19.6 min.

22-37 Interconnected graphite flakes in gray iron.

CHAPTER 23

23-3 Graphitic corrosion.

23-7 −0.172 V.

23-8 6.5 M.

23-10 0.000034 g/1000 mL.

23-11 2.5 h.

23-13 0.055 A/cm^2; 55 A.

23-15 34 years.

23-17 187.5 g Fe lost/h.

23-23 1100 alloy is anode and continues to protect 2024; 1100 alloy is cathode and the 3003 will corrode.

23-25 Al, Zn.

23-27 Stress corrosion cracking.

23-29 Ti is the anode; carbon is the cathode.

23-33 Cold worked nickel will corrode most rapidly; annealed nickel will corrode most slowly.

23-37 (a) 12.2 g/h. (b) 1.22 g/h.

23-39 Most ceramics are already oxides.

23-41 698°C.

INDEX

A

B

D

E

F

G

H

I

N

O

P

Q

R

S

T

Legend (example cell: 43 (98.91) / [Kr] $4d^5\,5s^2$ / 7 / Tc / Technetium / 2140 11.5 / 5030 * 1.9):
Atomic number — Atomic mass (mean relative); Electron configuration — Oxidation states; Symbol — Name; Melting point [°C] — Density [g/cm^3], for gases [g/l] (0° C, 1013 mbar); Boiling point [°C] — Electronegativity; * Radioactive

Period	Group	Atomic number	Atomic mass	Electron configuration	Oxidation states	Symbol	Name	Melting point [°C]	Density	Boiling point [°C]	Radioactive	Electronegativity
1	1 (1A)	1	1.01	1s	1, −1	H	Hydrogen	−259	0.09	−253		2.1
1	18 (0)	2	4.00	$1s^2$		He	Helium	—	0.18	−269		—
2	1 (1A)	3	6.94	[He] 2s	1	Li	Lithium	181	0.53	1330		1.0
2	2 (2A)	4	9.01	[He] $2s^2$	2	Be	Beryllium	1277	1.85	2970		1.5
2	13 (3B)	5	10.81	[He] $2s^2\,2p$	3	B	Boron	(2030)	2.35	2550		2.0
2	14 (4B)	6	12.01	[He] $2s^2\,2p^2$	4, 2, −4	C	Carbon	(3550)	2.2	4830		2.5
2	15 (5B)	7	14.01	[He] $2s^2\,2p^3$	5, 4, 3, 2, −3	N	Nitrogen	−210	1.25	−196		3.0
2	16 (6B)	8	16.00	[He] $2s^2\,2p^4$	−2, −1	O	Oxygen	−219	1.43	−183		3.5
2	17 (7B)	9	19.00	[He] $2s^2\,2p^5$	−1	F	Fluorine	−220	1.7	−188		4.0
2	18 (0)	10	20.18	[He] $2s^2\,2p^6$		Ne	Neon	−249	0.9	−246		—
3	1 (1A)	11	22.99	[Ne] 3s	1	Na	Sodium	98	0.97	892		0.9
3	2 (2A)	12	24.31	[Ne] $3s^2$	2	Mg	Magnesium	650	1.74	1107		1.2
3	13 (3B)	13	26.98	[Ne] $3s^2\,3p$	3	Al	Aluminum	660	2.70	2450		1.5
3	14 (4B)	14	28.09	[Ne] $3s^2\,3p^2$	4, −4	Si	Silicon	1410	2.33	2680		1.8
3	15 (5B)	15	30.97	[Ne] $3s^2\,3p^3$	5, 3, −3	P	Phosphorus	44	1.82	280		2.1
3	16 (6B)	16	32.06	[Ne] $3s^2\,3p^4$	6, 4, 2, −2	S	Sulfur	119	2.07	445		2.5
3	17 (7B)	17	35.45	[Ne] $3s^2\,3p^5$	7, 5, 3, 1, −1	Cl	Chlorine	−101	3.2	−35		3.0
3	18 (0)	18	39.95	[Ne] $3s^2\,3p^6$		Ar	Argon	−189	1.78	−183		—
4	1 (1A)	19	39.10	[Ar] 4s	1	K	Potassium	64	0.86	760		0.8
4	2 (2A)	20	40.08	[Ar] $4s^2$	2	Ca	Calcium	838	1.55	1440		1.0
4	3 (3A)	21	44.96	[Ar] 3d $4s^2$	3	Sc	Scandium	1539	3.0	2730		1.3
4	4 (4A)	22	47.88	[Ar] $3d^2\,4s^2$	4, 3	Ti	Titanium	1668	4.54	3260		1.5
4	5 (5A)	23	50.94	[Ar] $3d^3\,4s^2$	5, 4, 3, 2, 0	V	Vanadium	1900	6.1	3450		1.5
4	6 (6A)	24	52.00	[Ar] $3d^5$ 4s	6, 3, 2, 0	Cr	Chromium	1875	7.19	2200		1.6
4	7 (7A)	25	54.94	[Ar] $3d^5\,4s^2$	7, 6, 4, 3, 2, 0, −1	Mn	Manganese	1245	7.43	2097		1.5
4	8 (8)	26	55.85	[Ar] $3d^6\,4s^2$	6, 3, 2, 0, −2	Fe	Iron	1536	7.86	3000		1.8
4	9 (8)	27	58.93	[Ar] $3d^7\,4s^2$	3, 2, 0, −1	Co	Cobalt	1495	8.9	2900		1.9
4	10 (8)	28	58.70	[Ar] $3d^8\,4s^2$	3, 2, 0	Ni	Nickel	1453	8.9	2730		1.9
4	11 (1B)	29	63.55	[Ar] $3d^{10}$ 4s	2, 1	Cu	Copper	1453	8.9	2730		1.9
4	12 (2B)	30	65.38	[Ar] $3d^{10}\,4s^2$	2	Zn	Zinc	420	7.13	906		1.6
4	13 (3B)	31	69.72	[Ar] $3d^{10}\,4s^2\,4p$	3	Ga	Gallium	30	5.91	2237		1.6
4	14 (4B)	32	72.59	[Ar] $3d^{10}\,4s^2\,4p^2$	4	Ge	Germanium	937	5.32	2830		1.8
4	15 (5B)	33	74.91	[Ar] $3d^{10}\,4s^2\,4p^3$	5, 3, −3	As	Arsenic	Subl.	5.72	—		2.0
4	16 (6B)	34	74.91	[Ar] $3d^{10}\,4s^2\,4p^4$	6, 4, 2	Se	Selenium	217	4.79	685		2.4
4	17 (7B)	35	79.90	[Ar] $3d^{10}\,4s^2\,4p^5$	7, 5, 3, 1, −1	Br	Bromine	−7	3.12	58		2.8
4	18 (0)	36	83.80	[Ar] $3d^{10}\,4s^2\,4p^6$		Kr	Krypton	−157	3.7	−152		—
5	1 (1A)	37	85.47	[Kr] 5s	1	Rb	Rubidium	39	1.53	688		0.8
5	2 (2A)	38	87.62	[Kr] $5s^2$	2	Sr	Strontium	768	2.6	1380		1.0
5	3 (3A)	39	88.91	[Kr] 4d $5s^2$	3	Y	Yttrium	1509	4.47	2927		1.2
5	4 (4A)	40	91.22	[Kr] $4d^2\,5s^2$	4	Zr	Zirconium	1852	6.49	3580		1.4
5	5 (5A)	41	92.91	[Kr] $4d^4$ 5s	5, 3	Nb	Niobium	2468	8.57	4927		1.6
5	6 (6A)	42	95.94	[Kr] $4d^5$ 5s	6, 5, 4, 3, 2, 0	Mo	Molybdenum	2610	10.2	5560		1.8
5	7 (7A)	43	(98.91)	[Kr] $4d^5\,5s^2$	7	Tc	Technetium	2140	11.5	5030	*	1.9
5	8 (8)	44	101.07	[Kr] $4d^7$ 5s	8, 6, 4, 3, 2, 0, −2	Ru	Ruthenium	2500	12.4	3900		2.2
5	9 (8)	45	102.91	[Kr] $4d^8$ 5s	5, 4, 3, 1, 2, 0	Rh	Rhodium	1966	12.4	3730		2.2
5	10 (8)	46	106.42	[Kr] $4d^{10}$	4, 2, 0	Pd	Palladium	1552	12.0	3140		2.2
5	11 (1B)	47	107.87	[Kr] $4d^{10}$ 5s	2, 1	Ag	Silver	961	10.5	2210		1.9
5	12 (2B)	48	112.41	[Kr] $4d^{10}\,5s^2$	2	Cd	Cadmium	321	8.65	765		1.7
5	13 (3B)	49	114.82	[Kr] $4d^{10}\,5s^2\,5p$	3	In	Indium	156	7.31	2080		1.7
5	14 (4B)	50	118.69	[Kr] $4d^{10}\,5s^2\,5p^2$	4, 2	Sn	Tin	232	7.30	2270		1.8
5	15 (5B)	51	121.75	[Kr] $4d^{10}\,5s^2\,5p^3$	5, 3, −3	Sb	Antimony	631	6.69	1380		1.9
5	16 (6B)	52	127.60	[Kr] $3d^{10}\,5s^2\,5p^4$	6, 4, −2	Te	Tellurium	450	6.24	990		2.1
5	17 (7B)	53	126.90	[Kr] $4d^{10}\,5s^2\,5p^5$	7, 5, 1, −1	I	Iodine	114	4.94	183		2.5
5	18 (0)	54	131.29	[Kr] $4d^{10}\,5s^2\,5p^6$		Xe	Xenon	−112	5.89	−108		—
6	1 (1A)	55	132.91	[Xe] 6s	1	Cs	Cesium	29	1.90	690		0.7
6	2 (2A)	56	137.33	[Xe] $6s^2$	2	Ba	Barium	714	3.76	1640		0.9
6	3 (3A)	71	174.97	[Xe] $4f^{14}$ 5d $6s^2$	3	Lu	Lutetium	1652	9.84	3327		1.2
6	4 (4A)	72	178.49	[Xe] $4f^{14}\,5d^2\,6s^2$	4	Hf	Hafnium	2222	13.31	5400		1.3
6	5 (5A)	73	180.95	[Xe] $4f^{14}\,5d^3\,6s^2$	5	Ta	Tantalum	2996	16.5	5425		1.5
6	6 (6A)	74	183.85	[Xe] $4f^{14}\,5d^4\,6s^2$	6, 5, 4, 3, 2, 0	W	Tungsten	3410	19.3	5930		1.7
6	7 (7A)	75	186.21	[Xe] $4f^{14}\,5d^5\,6s^2$	7, 6, 4, 2, −1	Re	Rhenium	3180	21.0	5900		1.9
6	8 (8)	76	190.20	[Xe] $4f^{14}\,5d^6\,6s^2$	8, 6, 4, 3, 2, 0, −2	Os	Osmium	3050	22.6	5500		2.2
6	9 (8)	77	192.22	[Xe] $4f^{14}\,5d^7\,6s^2$	6, 4, 3, 2, 1, 0, −1	Ir	Iridium	2454	22.7	4500		2.2
6	10 (8)	78	195.08	[Xe] $4f^{14}\,5d^9\,6s^2$	4, 2, 0	Pt	Platinum	1769	21.4	3830		2.2
6	11 (1B)	79	196.97	[Xe] $4f^{14}\,5d^{10}$ 6s	3, 1	Au	Gold	1063	19.3	2970		2.4
6	12 (2B)	80	200.59	[Xe] $4f^{14}\,5d^{10}\,6s^2$	2, 1	Hg	Mercury	−38	13.6	357		1.9
6	13 (3B)	81	204.38	[Xe] $4f^{14}\,5d^{10}\,6s^2\,6p$	3, 1	Tl	Thallium	303	11.85	1457		1.8
6	14 (4B)	82	207.20	[Xe] $4f^{14}\,5d^{10}\,6s^2\,6p^2$	4, 2	Pb	Lead	327	11.4	1725		1.9
6	15 (5B)	83	208.98	[Xe] $4f^{14}\,5d^{10}\,6s^2\,6p^3$	5, 3	Bi	Bismuth	271	9.8	1560		1.9
6	16 (6B)	84	(209)	[Xe] $4f^{14}\,5d^{10}\,6s^2\,6p^4$	6, 4, 2	Po	Polonium	254	9.3	962		2.0
6	17 (7B)	85	(210)	[Xe] $4f^{14}\,5d^{10}\,6s^2\,6p^5$	7, 5, 3, 1, −1	At	Astatine	(302)	—	337		2.2
6	18 (0)	86	(222)	[Xe] $4f^{14}\,5d^{10}\,6s^2\,6p^6$		Rn	Radon	(−71)	9.73	−62		—
7	1 (1A)	87	(223.0)	[Rn] 7s	1	Fr	Francium	(27)	—	677	*	0.7
7	2 (2A)	88	226.03	[Rn] $7s^2$	1	Ra	Radium	700	5.0	1140	*	0.9
7	3 (3A)	103	(262.1)	[Rn] $5f^{14}$ 6d $7s^2$	3	Lr	Lawrencium	—	—	—	*	—
7	4 (4A)	104	(261.1)	[Rn] $5f^{14}\,6d^2\,7s^2$		Rf	Rutherfordium	—	—	—	*	—
7	5 (5A)	105	(262.1)	[Rn] $5f^{14}\,6d^3\,7s^2$		Db	Dubnium	—	—	—	*	—
7	6 (6A)	106	(263.1)	[Rn] $5f^{14}\,6d^4\,7s^2$		Sg	Seaborgium	—	—	—	*	—
7	7 (7A)	107	(264.1)	[Rn] $5f^{14}\,6d^5\,7s^2$		Bh	Bohrium	—	—	—	*	—
7	8 (8)	108	(265.1)	[Rn] $5f^{14}\,6d^6\,7s^2$		Hs	Hassium	—	—	—	*	—
7	9 (8)	109	(268)	[Rn] $5f^{14}\,6d^7\,7s^2$		Mt	Meitnerium	—	—	—	*	—
7	10 (8)	110	(269)	[Rn] $5f^{14}\,6d^8\,7s^2$		Uun		—	—	—	*	—
7	11 (1B)	111	(272)	[Rn] $5f^{14}\,6d^{10}\,7s^2$		Uuu		—	—	—	*	—
7	12 (2B)	112	(277)			Uub		—	—	—	*	—
7	14 (4B)	114	(289)			Uuq		—	—	—	*	—
7	16 (6B)	116	(289)			Uuh		—	—	—	*	—
7	18 (0)	118	(293)			Uuo		—	—	—	*	—

6 Lanthanoids

Atomic number	Atomic mass	Electron configuration	Oxidation states	Symbol	Name	Melting point [°C]	Density	Boiling point [°C]	Radioactive	Electronegativity
57	138.91	[Xe] 5d $6s^2$	3	La	Lanthanum	920	6.17	3470		1.1
58	140.12	[Xe] 4f 5d $6s^2$	4, 3	Ce	Cerium	795	6.67	3468		1.1
59	140.91	[Xe] $4f^3\,6s^2$	4, 3	Pr	Praseodymium	935	6.77	3127		1.1
60	144.24	[Xe] $4f^4\,6s^2$	3	Nd	Neodymium	1024	7.00	3027		1.2
61	(145)	[Xe] $4f^5\,6s^2$	3	Pm	Prometheum	(1027)	7.22	2460	*	—
62	150.36	[Xe] $4f^6\,6s^2$	3, 2	Sm	Samarium	1072	7.54	1790		1.2
63	151.96	[Xe] $4f^7\,6s^2$	3, 2	Eu	Europium	828	5.26	1439		—
64	157.25	[Xe] $4f^7$ 5d $6s^2$	3	Gd	Gadolinium	1312	7.89	3000		1.1
65	158.93	[Xe] $4f^9\,6s^2$	4, 3	Tb	Terbium	1356	8.27	2800		1.2
66	162.50	[Xe] $4f^{10}\,6s^2$	3	Dy	Dysprosium	1407	8.54	2600		—
67	164.93	[Xe] $4f^{11}\,6s^2$	3	Ho	Holmium	1461	8.80	2600		1.2
68	167.26	[Xe] $4f^{12}\,6s^2$	3	Er	Erbium	1497	9.05	2900		1.2
69	168.93	[Xe] $4f^{13}\,6s^2$	3, 2	Tm	Thulium	1545	9.33	1727		1.2
70	173.04	[Xe] $4f^{14}\,6s^2$	3, 2	Yb	Ytterbium	824	6.98	1196		1.1

7 Actinoids

Atomic number	Atomic mass	Electron configuration	Oxidation states	Symbol	Name	Melting point [°C]	Density	Boiling point [°C]	Radioactive	Electronegativity
89	(227)	[Rn] 6d $7s^2$	3	Ac	Actinium	1050	10.1	—	*	1.1
90	232.04	[Rn] $6d^2\,7s^2$	4	Th	Thorium	1750	11.7	3850	*	1.3
91	231.04	[Rn] $5f^2$ 6d $7s^2$	5, 4	Pa	Protactinium	(1230)	15.4	—	*	1.4
92	238.03	[Rn] $5f^3$ 6d $7s^2$	6, 5, 4, 3	U	Uranium	1132	19.07	3818	*	1.4
93	(237.05)	[Rn] $5f^4$ 6d $7s^2$	6, 5, 4, 3	Np	Neptunium	637	19.5	3900	*	1.3
94	(244)	[Rn] $5f^6\,7s^2$	6, 5, 4, 3	Pu	Plutonium	640	19.81	3235	*	1.3
95	(243)	[Rn] $5f^7\,7s^2$	6, 5, 4, 3	Am	Americium	994	13.7	—	*	1.3
96	(247)	[Rn] $5f^7$ 6d $7s^2$	4, 3	Cm	Curium	(1340)	13.51	3100	*	—
97	(247)	[Rn] $5f^9\,7s^2$	4, 3	Bk	Berkelium	—	—	—	*	—
98	(251)	[Rn] $5f^{10}\,7s^2$	4, 3	Cf	Californium	—	—	—	*	—
99	(254)	[Rn] $5f^{11}\,7s^2$	3	Es	Einsteinium	—	—	—	*	—
100	(257)	[Rn] $5f^{12}\,7s^2$	3	Fm	Fermium	—	—	—	*	—
101	(258)	[Rn] $5f^{13}\,7s^2$	3	Md	Mendelevium	—	—	—	*	—
102	(259)	[Rn] $5f^{14}\,7s^2$	3, 2	No	Nobelium	—	—	—	*	—